Antiangiogenic Cancer Therapy

Edited by

Darren W. Davis
Roy S. Herbst
James L. Abbruzzese

CRC Press is an imprint of the
Taylor & Francis Group, an **informa** business

CRC Press
Taylor & Francis Group
6000 Broken Sound Parkway NW, Suite 300
Boca Raton, FL 33487-2742

First issued in paperback 2019

ISBN-13: 978-0-8493-2799-5 (hbk)
ISBN-13: 978-0-367-38881-2 (pbk)

Library of Congress Cataloging-in-Publication Data

Antiangiogenic cancer therapy / edited by Darren W. Davis and Roy S. Herbst, James L. Abbruzzese
p. ; cm.
Includes bibliographical references and index.
ISBN-13: 978-0-8493-2799-5 (hardcover)
ISBN-10: 0-8493-2799-7 (hardcover)
1. Neovascularization inhibitors. I. Davis, Darren W., 1971- II. Herbst, Roy. III. Abbruzzese, James L. IV. Title.
[DNLM: 1. Neoplasms--drug therapy. 2. Angiogenesis Inhibitors--therapeutic use. QZ 267 A6283 2007]

RC271.N46.A585 2007
616.99'4061--dc22 2007000598

Visit the Taylor & Francis Web site at
http://www.taylorandfrancis.com

and the CRC Press Web site at
http://www.crcpress.com

Antiangiogenic Cancer Therapy

Preface

In recent years, tremendous progress has been made in our understanding of molecular mechanisms and cellular regulation of angiogenesis in cancer. Despite this progress, clinical development of angiogenesis inhibitors for the treatment of cancer remains challenging. Given that solid tumors account for more than 85% of cancer mortality, and tumor growth and metastasis are dependent on blood vessels, targeting tumor angiogenesis is one of the most widely pursued therapeutic strategies today. Approaches to target angiogenesis in cancer include destroying the existing vasculature (antivascular) and inhibiting neovascularization (antiangiogenic). We hope that *Antiangiogenic Cancer Therapy* will stimulate the rapid translation and dissemination of basic science discoveries into novel clinical strategies that will provide more effective antiangiogenic therapies for cancer.

Antiangiogenic Cancer Therapy was made possible as a result of a key scientific observation made more than 40 years ago, when Drs Folkman and Becker observed that tumor growth in isolated perfused organs was limited in the absence of tumor vascularization. However, it was arguably Folkman's hypothesis that tumor growth is angiogenesis-dependent in 1971 that led to the notion that angiogenesis could be a relevant target for tumor therapy. Twenty years later, the successful treatment of an angiogenesis-dependent pulmonary hemangioma (a benign tumor) with interferon α-2a enabled physicians and scientists to recognize the potential therapeutic benefit of targeting angiogenesis for cancer therapy. Indeed, in 1999 the development of antiangiogenic therapies for cancer became a top priority of the National Cancer Institute. The first angiogenesis inhibitor, bevacizumab, was approved by the Food and Drug Administration in 2004 for the treatment of metastatic carcinoma of the colon or rectum. Subsequently in 2006, bevacizumab was approved for first-line treatment of patients with advanced nonsquamous nonsmall cell lung cancer.

Although information in the field of angiogenesis is rapidly expanding, our capacity to efficiently process and implement this knowledge has not kept pace. For example, no randomized Phase III trial has demonstrated a survival benefit with currently available antiangiogenic agents when used as a monotherapy. However, the combination of bevacizumab with cytotoxic regimens has led to survival benefit in previously untreated colorectal, lung, and breast cancer, and in previously treated colorectal cancer patients. These results raise important questions about the complexity and use of angiogenesis inhibitors in clinical practice. The thesis of *Antiangiogenic Cancer Therapy* is that by understanding the molecular and cellular regulation of angiogenesis itself, we will be able to understand and implement the most optimal therapeutic strategies. This challenge creates an overwhelming task for clinicians, scientists, teachers, and authors. We have carefully considered what facts and concepts are essential elements to include in this book. An aim of this book is to integrate the fundamental concepts of angiogenesis with therapeutic strategies specific to various cancer types. Thus, although each chapter may stand alone, the scientific details within each chapter provide strength to the overall conceptual framework of the book.

We are deeply grateful to the many people who have helped us compose this book. The experts who contributed to each chapter are the most authoritative in their respective fields. However, their contributions would not be possible without many years of laborious experimental failures and successes by many investigators throughout the world. Therefore, we are also indebted to the many scientists whose contributions have led to remarkable scientific advances, which are cited within each chapter. Finally, we are thankful to the outstanding staff at Taylor & Francis who oversaw the final production of this book.

Since the initial discovery that tumors are angiogenesis-dependent was made four decades ago, this edition is a celebration of the remarkable scientific progress made during that time, and we hope an even better indication of the future to come.

ABSTRACT

Antiangiogenic Cancer Therapy brings together basic scientists and oncologists to provide the most authoritative, up-to-date, and encyclopedic volume currently available on this subject. Part I of this book introduces a series of concepts and topics regarding the role of angiogenesis in cancer. These topics include strategies to prolong the nonangiogenic dormant state of human tumors, molecular mechanisms and cellular regulation of angiogenesis in solid tumors and hematologic malignancies, and the regulation of angiogenesis by the tumor microenvironment. Part II of the book covers specific molecular targets for inhibiting angiogenesis in cancer therapy. Part III discusses clinical trial design and translational research approaches essential for identifying and developing effective angiogenesis inhibitors. These discussions include noninvasive imaging methods and direct analysis of tissue biopsies. Part IV of the book covers antiangiogenic treatment for specific cancer types. These chapters are introduced by state-of-the-art discussions outlining the current understanding of the molecular biology of each cancer type followed by discussions that examine strategies for targeting angiogenesis. Organizing the chapters in this format will allow the reader to easily find the information necessary to understand the fundamental concepts of angiogenesis and the complexities associated with targeting angiogenesis for specific types of cancer. This book will serve to provide information useful to scientists and physicians engaged in the study and development of antiangiogenic agents, as well as medical professionals, medical and graduate students, and allied health professionals interested in learning more about the biology and clinical use of angiogenesis inhibitors.

Editors

Dr Darren W. Davis is president and chief executive officer of ApoCell, Inc., an innovative molecular diagnostic company located near the world famous Texas Medical Center, Houston, Texas. Dr Davis has a BS in biochemical and biophysical sciences and earned his PhD in cancer biology and toxicology at the University of Texas Graduate School of Biomedical Sciences and the University of Texas M.D. Anderson Cancer Center in Houston, Texas. Dr Davis continued his postgraduate training at M.D. Anderson Cancer Center where he developed several methods to analyze the effects of molecular-targeted therapies, including angiogenesis inhibitors, to support clinical drug development. He was shortly promoted to junior faculty and served as one of six investigators of the Goodwin Molecular Monitoring Laboratory for clinical biomarker development, Department of Translational Research, before founding ApoCell in 2004, an M.D. Anderson Cancer Center spin-off company. Dr Davis serves as the principal investigator for numerous biological correlative studies to support clinical trials. His research interests center on molecular mechanisms, apoptosis, and signal transduction of molecular-targeted therapies. Dr Davis has evaluated the pharmacodynamic effects of both conventional and drug-targeted therapies in a wide variety of animal and clinical specimens. Dr Davis has frequently been invited to speak at both national and international conferences and scientific advisory meetings. Dr Davis serves as a consultant for both basic scientists and clinicians and helps identify and select critical end points for clinical trials with leading pharmaceutical companies. Dr Davis is author or coauthor of more than 50 publications, including peer-reviewed journal articles, abstracts, book chapters, and has served as an editor. He has contributed his work to many prominent journals, such as, *Journal of Experimental Medicine*, *Cancer Research*, *Journal of Clinical Oncology*, *Clinical Cancer Research*, *Lung Cancer*, *Cancer*, and *Seminars in Oncology*. His abstracts have been presented at the annual meetings of the American Society of Clinical Oncology, the American Association for Cancer Research, and the European Organization for Research and Treatment of Cancer. Dr Davis is the inventor of four pending patents.

Dr Roy S. Herbst is professor and chief of the Section of Thoracic Medical Oncology in the Department of Thoracic/Head and Neck Medical Oncology, at the University of Texas M.D. Anderson Cancer Center in Houston, Texas. He also serves as professor in the Department of Cancer Biology and codirector of the Phase I working group. Dr Herbst earned his MD at Cornell University Medical College and his PhD in molecular cell biology at the Rockefeller University in New York City, New York. His postgraduate training included an internship and residency in medicine at Brigham and Women's Hospital in Boston, and a chief residency at West Roxbury Veterans Administration Hospital in Dedham, Massachusetts. His clinical fellowships in medicine and hematology were completed at the Dana-Farber Cancer Institute and Brigham and Women's Hospital, respectively. Subsequently, Dr Herbst completed the MS degree in clinical translational research at Harvard University in Cambridge, Massachusetts. Dr Herbst serves as the principal investigator for numerous trials and has conducted research primarily in the treatment of lung cancer, head and neck cancer, and Phase I studies. His Laboratory and Clinical work has focused on the clinical development of molecular-targeted therapies. Dr Herbst has frequently been invited to speak at both national and international conferences. Dr Herbst is author or coauthor of more than 200 publications, including peer-reviewed journal articles, abstracts, and book chapters. He has contributed his work to many prominent journals, such as *Journal of Clinical Oncology*, *Clinical Cancer Research*, *Clinical Lung Cancer*, *Lung Cancer*,

Cancer, *Annals of Oncology*, and *Seminars in Oncology*. His abstracts have been presented at the annual meetings of the American Society of Clinical Oncology, the American Association for Cancer Research, the World Conference on Lung Cancer, the Society of Nuclear Medicine Conference, and the European Organization for Research and Treatment of Cancer. Dr Herbst is an active member of the American Society of Clinical Oncology, the American Association for Cancer Research, the International Association for the Study of Lung Cancer, the Radiation Therapy Oncology Group, and the Southwest Oncology Group Lung Committee. He served as chairman of the American Society of Clinical Oncology—Lung Cancer Program Subcommittee (2001–2002), vice chairman of the Radiation Therapy Oncology Group—Lung Committee, vice chairman of the Southwest Oncology Group—Lung Committee, guest planner of the Annual Meeting Education Program (2003), chairman of the International Association for the Study of Lung Cancer—Targeted Therapy Division of the Translational Research and Targeted Therapy Subcommittee (2003 and 2005), and chairman of the American Society of Clinical Oncology—Cancer Communication Committee (2005–2006). Notably Dr Herbst is the recipient of the American Society of Clinical Oncology Young Investigator Award, the American Society of Clinical Oncology Career Development Award (1999, 2000), and the M.D. Anderson Cancer Center Physician Scientist Program Award (1999–2002).

Dr James L. Abbruzzese is the M.G. and Lillie A. Johnson chair for cancer treatment and research and chairman of the Department of Gastrointestinal Medical Oncology at the University of Texas M.D. Anderson Cancer Center in Houston, Texas. Dr Abbruzzese is a member of numerous scientific advisory boards including the external scientific advisory board for the University of Massachusetts, the Arizona Cancer Center, the Lustgarten Foundation for Pancreatic Cancer Research, and the Pancreatic Cancer Action Network. Born in Hartford, Connecticut, Dr Abbruzzese graduated medical school with honors from the University of Chicago, Pritzker School of Medicine, Chicago, Illinois. He completed residency in internal medicine at the Johns Hopkins Hospital in Baltimore, Maryland, and fellowship in medical oncology at the Dana-Farber Cancer Center, Harvard Medical School in Boston, Massachusetts. He is married and has one child. Dr Abbruzzese has published over 200 peer-reviewed articles, numerous chapters, and reviews. In 2004, he coedited a book entitled *Gastrointestinal Oncology* published by Oxford University Press. His research group was recently awarded a SPORE in pancreatic cancer and U54 grant on angiogenesis. In 2001, Dr Abbruzzese served as a cochair of the American Association for Cancer Research Program Committee. He is a member of the American Association for Cancer Research Fellowships Committee, the American Society of Clinical Oncology Grant Awards and Nominating Committees, and has many other board memberships. Dr Abbruzzese is a deputy editor of *Clinical Cancer Research* and member of several other editorial boards in the past including the *Journal of Clinical Oncology*. His clinical interests center on pancreatic cancer, new drug development, and noninvasive assessment of anticancer drug effects.

Contributors

Abebe Akalu
Departments of Radiation Oncology and Cell Biology
New York University School of Medicine Cancer Institute
New York, New York

Kenneth C. Anderson
Jerome Lipper Multiple Myeloma Center
Department of Medical Oncology
Dana-Farber Cancer Institute and Harvard Medical School
Boston, Massachusetts

Khalid Bajou
Division of Hematology–Oncology
Departments of Pediatrics and Biochemistry and Molecular Biology
University of Southern California Keck School of Medicine and Saban Research Institute of Children's Hospital
Los Angeles, California

Cheryl H. Baker
Department of Biomedical Sciences
University of Central Florida
Orlando, Florida

and

Cancer Research Institute
M.D. Anderson Cancer Center–Orlando
Orlando, Florida

Pablo M. Bedano
Division of Hematology–Oncology
Indiana University School of Medicine
Indianapolis, Indiana

Peter C. Brooks
Departments of Radiation Oncology and Cell Biology
New York University School of Medicine Cancer Institute
New York, New York

Thomas R. Burkard
Institute for Genomics and Bioinformatics and Christian Doppler Laboratory for Genomics and Bioinformatics
Graz University of Technology
Graz, Austria

and

Research Institute of Molecular Pathology
Vienna, Austria

David J. Chaplin
Oxigene, Inc.
Waltham, Massachusetts

Dharminder Chauhan
Jerome Lipper Multiple Myeloma Center
Department of Medical Oncology
Dana-Farber Cancer Institute and Harvard Medical School
Boston, Massachusetts

Ramzi N. Dagher
Center for Drug Evaluation and Research
U.S. Food and Drug Administration
Silver Spring, Maryland

Angus G. Dalgleish
Division of Oncology
Cell and Molecular Sciences
St. Georges University of London
London, United Kingdom

Darren W. Davis
ApoCell, Inc.
Houston, Texas

S. Davis
Regeneron Pharmaceuticals, Inc.
Tarrytown, New York

Yves A. DeClerck
Departments of Pediatrics, Biochemistry, and Molecular Biology
Keck School of Medicine
University of Southern California

and

Saban Research Institute
Childrens Hospital Los Angeles
Los Angeles, California

Bruce J. Dezube
Division of Hematology/Oncology
Beth Israel Deaconess Medical Center
Harvard Medical School
Boston, Massachusetts

Graeme J. Dougherty
Department of Radiation Oncology
University of Arizona
Tucson, Arizona

Keith Dredge
Progen Industries Ltd.
Brisbane, Australia

Dan G. Duda
Department of Radiation Oncology
Massachusetts General Hospital and Harvard Medical School
Boston, Massachusetts

Frank Eisenhaber
Research Institute of Molecular Pathology
Vienna, Austria

Heinrich Elinzano
Neuro-Oncology Branch
National Cancer Institute and National Institutes of Health
Bethesda, Maryland

Napoleone Ferrara
Genetech, Inc.
San Francisco, California

Isaiah J. Fidler
Department of Cancer Biology
University of Texas M.D. Anderson Cancer Center
Houston, Texas

Howard A. Fine
Neuro-Oncology Branch
National Cancer Institute and National Institutes of Health
Bethesda, Maryland

Judah Folkman
Karp Family Research Laboratories
Boston, Massachusetts

Nicholas W. Gale
Regeneron Pharmaceuticals, Inc.
Tarrytown, New York

Francis J. Giles
Department of Leukemia
University of Texas M.D. Anderson Cancer Center
Houston, Texas

Ramaswamy Govindan
Department of Medicine
Washington University School of Medicine
St. Louis, Missouri

Hubert Hackl
Institute for Genomics and Bioinformatics and Christian Doppler Laboratory for Genomics and Bioinformatics
Graz University of Technology
Graz, Austria

Christian Hafner
Department of Dermatology
University of Regensburg
Regensburg, Germany

Kristin Hennenfent
Division of Pharmacy Practice
St. Louis College of Pharmacy
St. Louis, Missouri

John V. Heymach
Departments of Cancer Biology and Thoracic/Head and Neck Oncology
University of M.D. Anderson Cancer Center
Houston, Texas

Daniel J. Hicklin
ImClone Systems, Inc.
New York, New York

Paulo M. Hoff
Department of Gastrointestinal Medical Oncology
University of Texas M.D. Anderson Cancer Center
Houston, Texas

Sakina Hoosen
Clinical R&D
Pfizer, Inc.
New York, New York

Mark A. Horsfield
Cardiovascular Sciences
University of Leicester
Leicester, United Kingdom

Rakesh K. Jain
Department of Radiation Oncology
Massachusetts General Hospital and Harvard Medical School
Boston, Massachusetts

Henry B. Koon
Division of Hematology/Oncology
Beth Israel Deaconess Medical Center
Harvard Medical School
Boston, Massachusetts

Hans-Georg Kopp
Department of Hematology-Oncology
Eberhard-Karls University Tubingen
Tubingen, Germany

Shaji Kumar
Department of Internal Medicine
Mayo Clinic and Foundation
Rochester, Minnesota

Mijung Kwon
Tumor Angiogenesis Section
Surgery Branch
National Cancer Institute
Bethesda, Maryland

Janessa J. Laskin
Division of Medical Oncology
University of British Columbia
Vancouver, British Columbia

Walter E. Laug
Departments of Pediatrics, Biochemistry, and Molecular Biology
Keck School of Medicine
University of Southern California

and

Saban Research Institute
Childrens Hospital Los Angeles
Los Angeles, California

Steven K. Libutti
Tumor Angiogenesis Section
Surgery Branch
National Cancer Institute
Bethesda, Maryland

Glenn Liu
The University of Wisconsin Carbone Comprehensive Cancer Center
Madison, Wisconsin

Kathy D. Miller
Division of Hematology–Oncology
Indiana University School of Medicine
Indianapolis, Indiana

Bruno Morgan
Departments of Cancer Studies and Molecular Medicine
Radiology Department
University of Leicester
Leicester, United Kingdom

Daniel Morgensztern
Washington University School of Medicine
St. Louis, Missouri

Robert J. Motzer
Memorial Sloan-Kettering Cancer Center
New York, New York

George N. Naumov
Department of Surgery
Harvard Medical School and
Vascular Biology Program
Children's Hospital Boston
Boston, Massachusetts

Maria Novatchkova
Research Institute of Molecular Pathology
Vienna, Austria

Liron Pantanowitz
Department of Pathology
Baystate Medical Center
Tufts University School of Medicine
Springfield, Massachusetts

Nicholas Papadopoulos
Regeneron Pharmaceuticals, Inc.
Tarrytown, New York

Richard Pazdur
Center for Drug Evaluation and Research
U.S. Food and Drug Administration
Silver Spring, Maryland

Klaus Podar
Jerome Lipper Multiple Myeloma Center
Department of Medical Oncology
Dana-Farber Cancer Institute and Harvard
Medical School
Boston, Massachusetts

Marco Presta
Department of Biomedical Sciences
and Biotechnology
University of Brescia
Brescia, Italy

Shahin Rafii
Division of Vascular Hematology–Oncology
Department of Genetic Medicine
Cornell University Medical College
New York, New York

Carlos Almeida Ramos
Department of Stem Cell Transplantation
and Cellular Therapy
University of Texas M.D. Anderson
Cancer Center
Houston, Texas

Albrecht Reichle
Department of Hematology and Oncology
University Hospital of Regensburg
Regensburg, Germany

John S. Rudge
Regeneron Pharmaceuticals, Inc.
Tarrytown, New York

Marco Rusnati
Department of Biomedical Sciences
and Biotechnology
University of Brescia
Brescia, Italy

Everardo D. Saad
Multidisciplinary Oncology Group
Federal University of Sao Paulo
Sao Paulo, Brazil

Alan B. Sandler
Division of Hematology and Oncology
Thoracic Oncology Vanderbilt-Ingram
Cancer Center
Nashville, Tennessee

Brian P. Schneider
Division of Hematology–Oncology
Indiana University School of Medicine
Indianapolis, Indiana

Dietmar W. Siemann
Department of Radiation Oncology
University of Florida Shands Cancer Center
Gainesville, Florida

George W. Sledge, Jr.
Division of Hematology–Oncology
Indiana University School of Medicine
Indianapolis, Indiana

David J. Stewart
Department of Thoracic/Head & Neck
Medical Oncology
University of Texas M.D. Anderson
Cancer Center
Houston, Texas

Anita Tandle
Tumor Angiogenesis Section
Surgery Branch
National Cancer Institute
Bethesda, Maryland

Gavin Thurston
Regeneron Pharmaceuticals, Inc.
Tarrytown, New York

Zlatko Trajanoski
Institute for Genomics and Bioinformatics and Christian Doppler Laboratory for Genomics and Bioinformatics
Graz University of Technology
Graz, Austria

Thomas Vogt
Department of Dermatology
University of Regensburg
Regensburg, Germany

Leslie K. Walker
The University of Wisconsin Carbone Comprehensive Cancer Center
Madison, Wisconsin

Stanley J. Wiegand
Regeneron Pharmaceuticals, Inc.
Tarrytown, New York

George Wilding
The University of Wisconsin Carbone Comprehensive Cancer Center
Madison, Wisconsin

Christopher G. Willett
Department of Radiation Oncology
Duke University Medical Center
Durham, North Carolina

Hua-Kang Wu
Department of Cancer Biology
University of Texas M.D. Anderson Cancer Center
Houston, Texas

George D. Yancopoulos
Regeneron Pharmaceuticals, Inc.
Tarrytown, New York

Karen W.L. Yee
Department of Medical Oncology and Hematology
University Health Network—Princess Margaret Hospital
Toronto, Ontario

Zhenping Zhu
ImClone Systems, Inc.
New York, New York

Amado J. Zurita
Department of Genitourinary Medical Oncology
University of Texas M.D. Anderson Cancer Center
Houston, Texas

Table of Contents

Part I

Angiogenesis in Cancer

1 Strategies to Prolong the Nonangiogenic Dormant State of Human Cancer

George N. Naumov and Judah Folkman

CONTENTS

1.1 CLINICAL "LATENCY" IN CANCER RECURRENCE FOLLOWING A PRIMARY TUMOR TREATMENT

Cancer recurrence after treatment of the primary tumor is a major cause of mortality among cancer patients. It may take years to decades before local or distant (i.e., metastatic) recurrence becomes clinically detectable as cancer. This "disease-free" period is a time of uncertainty for patients who appear "cured." For example, Demicheli et al.[1] have demonstrated two hazardous peaks of breast cancer recurrence in patients undergoing mastectomy alone without adjuvant therapy. In a group of 1173 patients, the first peak of cancer recurrence occurred at ~18 months after surgery. A second peak in cancer recurrence developed at ~5 years after surgery and was associated with a plateau-like tail extending up to 15 years.[1] Patients experiencing cancer recurrences within 5 years following surgery have a shorter overall survival than those with recurrences occurring at a later time point.[2]

Similar "latency periods" in cancer recurrence have been documented since the beginning of the twentieth century. Rupert A. Willis has summarized the "time elapsing between the excision of a human malignant tumor and the appearance of a clinically recognizable recurrence" for a variety of human cancers. For example, the latency period in breast cancer patients can be from 6 to 20 years; cutaneous and ocular melanoma, 14 to 32 years; kidney carcinoma,

6 to 8 years; and stomach and colon carcinoma, 5 to 6 years.[3] Moreover, Willis was the first to realize that these latency periods do not correspond with the natural progression of cancer, and he introduced the concept of a "dormant cancer cell" as a possible explanation.

Over the past few years, several hypotheses have been proposed in an attempt to explain the phenomenon of human tumor dormancy. Initially, it was proposed that tumor cells enter a prolonged state of mitotic arrest.[4,5] Others hypothesized that tumor size is controlled by the immune system[6–11] or hormonal deprivation in hormone-dependent tumors.[11–13] In 1972, Folkman and Gimbrone demonstrated that dormancy in human tumors could be due to blocked angiogenesis. In the following years, Folkman and colleagues have presented evidence supporting the concept that most human tumors arise without angiogenic activity and exist in a microscopic dormant state for months to years without neovascularization.[14] Such protection may be attributed in part to host-derived factors, which prevent microscopic tumors from switching to the angiogenic phenotype.

1.2 ANGIOGENESIS DEPENDENCE OF TUMOR GROWTH

Cancer progression is a multistep process (Figure 1.1). With each step, the genetic and epigenetic events in the process become increasingly complex and may be more difficult to

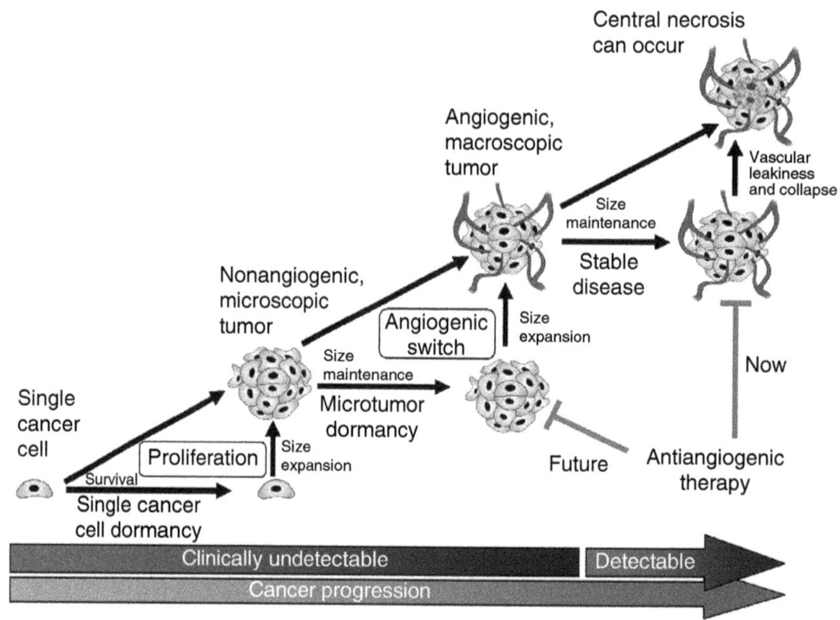

FIGURE 1.1 (See color insert following page 558.) Rate-limiting steps in the tumor progression. Solitary nonproliferating, dormant cancer cells can persist for long periods of time, until they come out of G_0 arrest and start to proliferate. Tumor mass can expand only to a microscopic mass without the recruitment of new blood vessels. Human cancers can remain nonangiogenic and dormant for long periods of time, delaying the tumor progression process. During this microscopic dormant state, nonangiogenic tumors are actively proliferating and undergoing apoptosis. Nonangiogenic tumors can expand in mass after undergoing the angiogenic switch and recruitment of new blood vessels. Angiogenic macroscopic tumors that do not expand in mass are known as "stable disease," although angiogenic tumors can remain at a constant size for prolonged periods of time. Current antiangiogenic therapy targets angiogenic microscopic and macroscopic tumors. However, future antiangiogenic therapy will target nonangiogenic microscopic tumors with the aim of keeping them in a dormant state by preventing the angiogenic switch.

treat. As the cancer progresses from a single neoplastic cell to a large, lethal tumor, it acquires a series of mutations, becoming: (1) self-sufficient in growth signaling, by oncogene activation and loss of tumor suppressor genes, (2) insensitive to antigrowth signaling, (3) unresponsive to apoptotic signaling, (4) capable of limitless cell replications, and (5) tumorigenic and metastatic.[15] Current experimental and clinical evidence indicates that these neoplastic properties may be necessary, but not sufficient, for a cancer cell to progress into a population of tumor cells, which becomes clinically detectable, metastatic, and lethal. For a tumor to develop a highly malignant and deadly phenotype, it must first recruit and sustain its own blood supply, a process known as tumor angiogenesis.[16,17]

For more than a century, it has been observed that surgically removed tumors are hyperemic compared to normal tissues.[18,19] Generally, this phenomenon was explained as simple dilation of existing blood vessels induced by tumor factors. However, Ide et al.[20] demonstrated that tumor-associated hyperemia could be related to new blood vessel growth, and vasodilation may not be the sole explanation for this phenomenon. They showed that when a wound induced in a transparent rabbit ear chamber completely regressed, the implantation of a tumor in the chamber resulted in the growth of new capillary blood vessels.[20] These initial observations were later confirmed by Algire et al.,[21,22] demonstrating that new vessels in the periphery of a tumor implant arose from preexisting host vessels, and not from the tumor implant itself. At the time, this novel concept of tumor-induced neovascularization was generally attributed to an inflammatory reaction, thought to be a side effect of tumor growth, and it was not perceived as a requirement for tumor growth.[23]

In the early 1960s, Folkman and Becker observed that tumor growth in isolated perfused organs was severely restricted in the absence of tumor vascularization.[24–28] In 1971, Folkman proposed the hypothesis that tumor growth is angiogenesis-dependent.[16] This hypothesis suggested that tumor cells and vascular endothelial cells within a neoplasm may constitute a highly integrated, two-compartment system, which dictates tumor growth. This concept indicated that endothelial cells may switch from a resting state to a rapid growth phase because of "diffusible" signals secreted from the tumor cells. Moreover, Folkman proposed that angiogenesis could be a relevant target for tumor therapy (i.e., antiangiogenic therapy).

We now know that angiogenesis plays an important role in numerous physiologic and pathologic processes. The hallmark of pathologic angiogenesis is the persistent growth of blood vessels. Sustained neovascularization can continue for months or years during the progression of many neoplastic and nonneoplastic diseases.[29,30] However, tumor angiogenesis is rarely, if ever, downregulated spontaneously. The fundamental objective of antiangiogenic therapy is to inhibit the progression of pathologic angiogenesis. In contrast, the goal of antivascular therapy is to rapidly occlude new blood vessels so that the blood flow stops. Both therapeutic approaches target the ability of tumors to progress from the nonangiogenic to the angiogenic phenotype, a process termed the "angiogenic switch."[31,40]

Cancer usually becomes clinically detectable only after tumors have become angiogenic and have expanded in mass. Failure of a tumor to recruit new vasculature or to reorganize the existing surrounding vasculature results in a nonangiogenic tumor, which is microscopic in size and unable to increase in mass (Figure 1.1). Without new blood supply, microscopic tumors are usually restricted to a size of <1–2 mm in diameter and are highly dependent on surrounding blood vessels for oxygen and nutrient supply. At sea level, the diffusion limit of oxygen is ~100 μm.[32] Therefore, all mammalian cells, including neoplastic cells, are required to be within 100–200 μm of a blood vessel. As nonangiogenic tumors attempt to expand in mass, attributed to uncontrolled cancer cell proliferation, some tumor cells fall outside the oxygen diffusion limit and become hypoxic. It is well known that hypoxic conditions induce a set of compensatory responses within cancer cells, such as increased transcription of the hypoxia-inducible factor (HIF). Subsequently, this hypoxic signaling leads to upregulation of proangiogenic proteins, such as vascular endothelial growth factor (VEGF), platelet-derived growth factor (PDGF), and nitric oxide synthase (NOS).[33] The angiogenic switch in tumors is presumed to be closely regulated by the presence of pro- and antiangiogenic proteins in the tumor microenvironment. An increase in the local concentration of proangiogenic proteins allows angiogenesis to occur and ultimately permits a tumor to expand in mass. Antiangiogenic therapy offers a fourth anti-cancer modality, in addition to conventional therapeutic approaches, which target well-established and genetically unstable tumors.

1.3 EXPERIMENTAL MODELS OF HUMAN TUMOR DORMANCY

As early as the 1940s, experimental systems involving the transplantation of tumor pieces in isolated perfused organs and in the anterior chamber of the eyes of various species of animals have demonstrated the effects of neovascularization on tumor growth. Greene et al.,[34] observed that H-31 rabbit carcinoma tumor implanted into the eyes of guinea pigs did not vascularize and failed to grow for 16–26 months. During this period, the transplants measured ~2.5 mm in diameter. However, when the same tumors were reimplanted into their original host (i.e., rabbit eyes), they vascularized and grew to fill the anterior chamber within 50 days. Similarly, Folkman et al.[25] showed that in isolated perfused thyroid and intestinal segment tumors, implants grew and arrested at a small size (2–3 mm diameter). This inability of neoplasms to evoke a new blood supply was later attributed to endothelial cell degeneration in the perfused organs that are perfused with platelet-free hemoglobin solution.[35] In 1972, Gimbrone et al.[36] provided in vivo evidence that the progressive growth of a homologous solid tumor can be deliberately arrested at a microscopic size when neovascularization is prevented. In these experiments, two comparable tumor pieces were implanted in each eye of the same animal: one directly on the iris (i.e., angiogenic milieu) and the other suspended in the anterior chamber (i.e., avascular milieu) in the opposing eye. The vascularized tumor implanted on the iris grew to a size 15,000 times the initial volume and filled the anterior chamber of the rabbit eye within 14 days. In contrast, the tumor implant in the avascular anterior chamber remained avascular, and by day 14 after implantation, had only increased by 4 times its initial volume. These "dormant" tumors remained at a size of ~1 mm in diameter for up to 44 days. During this period, the tumors developed a central necrotic core surrounded by a layer of viable tumor cells, in which mitotic figures were observed. Overall, these microscopic tumors remained avascular, as demonstrated through microscopic and histological analyses and fluorescein tests. The malignant growth potential of these microscopic tumors was demonstrated in vivo by reimplanting the tumors directly on the irises of fresh animals. In the irises, the dormant tumors became vascularized and grew rapidly until the anterior chambers of the eyes were filled with tumor, in a manner similar to the control iris implants. These fundamental observations established the relationship between tumor growth and angiogenesis. Moreover, they provided an in vivo experimental model for further investigations of tumor dormancy.

One of the most pressing questions at that time was whether tumor-induced angiogenesis could be inhibited, preventing dormant tumors from progressing to the angiogenic phenotype. Using a V2 rabbit carcinoma in the corneal implant animal model, Brem et al.,[37] demonstrated that tumor angiogenesis could be blocked by diffusible proteins from the cartilage of newborn rabbits. The coimplants of tumor and cartilage pieces completely prevented vascularization in 28% of tumors and significantly delayed the vascularization in the remaining tumors, which eventually became vascularized. In addition, cartilage pieces inhibited vessel formation around a tumor implant in the chorioallantoic membrane (CAM) of chick embryos. The inhibitory proteins found in cartilage were later identified, isolated, and characterized as tissue inhibitors of metalloproteinases (TIMPs).[37a] Subsequently, other angiogenesis inhibitors were identified. Endostatin and angiostatin were discovered as internal peptide fragments of plasminogen and 20 kDa C-terminal fragments of collagen XVIII, respectively.[38,39]

A spontaneous tumor dormancy model in transgenic mice was described by Hanahan and Folkman,[40] in which autochthonous tumors arise in the pancreatic islets as a result of simian virus 40 T antigen (Tag) oncogene expression. In this experimental model, only 4% of tumors become angiogenic after 13 weeks. In contrast, the remaining 96% of pancreatic islet tumors remain microscopic and nonangiogenic.[40,41] The spontaneous progression of non-angiogenic lesions to the angiogenic phenotype in these transgenic tumor-bearing mice led to the development of the "angiogenic switch" concept.[40]

More recently, Achilles et al.[42] reported that human cancers contain subpopulations that differ in their angiogenic potential. These findings suggested that the angiogenic phenotype of a human tumor cell may be controlled by genetic and epigenetic mechanisms. Therefore, human tumors can contain both angiogenic and nonangiogenic tumor cell populations, characterized by their in vivo ability to recruit new blood vessels to a tumor. However, the factors involved in the proportional regulation of these two tumor cell populations are still unknown. The observed heterogeneity of angiogenic activity among human tumor cells allowed for the isolation of these two populations of cancer cells and the development of new and fruitful human tumor dormancy experimental models.

Single-cell cloning of a human tumor cell line was employed as a strategy for the isolation of angiogenic and nonangiogenic tumor cell populations.[42] Achilles et al. established and selected subclones from a human liposarcoma cell line (SW-872) based on high, intermediate, or low proliferation rates in vitro. These clones were expanded in vitro into a population of tumor cells and were then inoculated into immunodeficient (SCID) mice. Three different growth patterns were observed: (1) highly angiogenic and rapidly growing tumors, (2) weakly angiogenic and slowly growing tumors, and (3) nonangiogenic and dormant tumors. In a subsequent experiment, Almog et al.[43] demonstrated that the nonangiogenic tumors spontaneously switch to the angiogenic phenotype and initiate exponential growth ~130 days after subcutaneous inoculation.[43] During the 130 day dormancy period, microscopic (~1–2 mm in diameter) tumors remain avascular and are virtually undetectable by palpation. Because this animal dormancy model was based on the in vitro tumor cell proliferation differences between the angiogenic and nonangiogenic liposarcoma clones, it raised two fundamental questions: (1) Is there a correlation between tumor cell proliferation and angiogenic potential? (2) Can the observed differences in tumor growth be recapitulated using populations of human tumor cells that have not been cloned?

To address these questions, human tumor cell lines were obtained from the American Type Culture Collection (ATCC, Manassas, VA) based on their "no take" phenotype in immunodeficient mice. These cell lines were assessed for in vivo tumor growth over extended time periods. Following a subcutaneous inoculation of a tumor cell suspension, mice were monitored for palpable tumors at the site of inoculation for more than a year (i.e., about half the normal life span of a mouse) and, sometimes, for the life of the animal. Some of the mice inoculated with the "no take" tumor cells spontaneously formed palpable tumors after a dormancy period, which varied from months to more than a year, depending on cancer type (Figure 1.2). With time, tumors became angiogenic, and palpable, expanded in mass exponentially, and within ~50 days of first detection, killed the host animal. Stable cell lines were established from representative angiogenic tumors. When reinoculated into SCID mice, these angiogenic tumor cells formed large (>1 cm in diameter) tumors within a month following inoculation (i.e., without a dormancy period), in 100% of the inoculated mice. It was found that each cancer type had a characteristic and predictable dormancy period and generated a consistent proportion of tumors that switched to the angiogenic phenotype. However, once nonangiogenic tumors switched to the angiogenic phenotype, they escaped

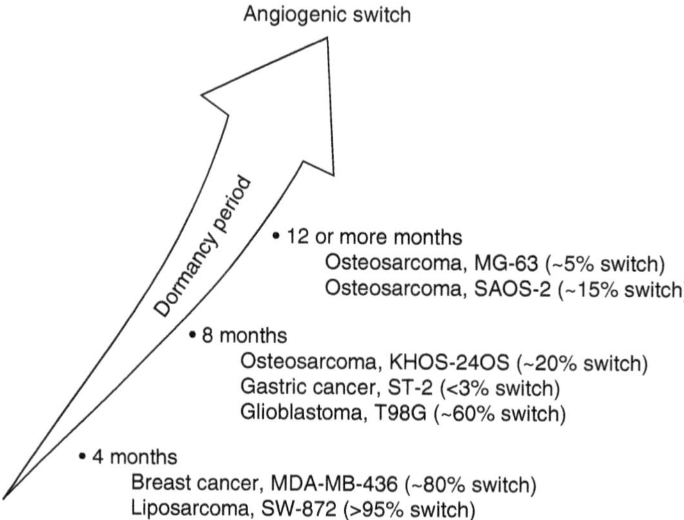

FIGURE 1.2 Human tumor cell lines that spontaneously switch to the angiogenic phenotype after a prolonged dormancy period in immunodeficient mice. The percent of mice that switch to the angiogenic phenotype is shown in brackets and varies between human tumor types.

from the dormancy state and formed lethal tumors in 100% of the mice, regardless of cancer type. At this time, tumor cell population–based animal dormancy models have been developed and characterized for breast cancer, osteosarcoma, and glioblastoma (Figure 1.2).[44] In contrast to the single-cell-derived human liposarcoma animal model, the angiogenic and the nonangiogenic tumor cell populations of the rest of the animal models were derived by in vivo selection for the angiogenic and nonangiogenic phenotypes. These in vivo models of nonangiogenic human tumors permit analysis of the switch to the angiogenic phenotype, and make it possible to address the question of whether the switch can be bi-directional.

1.4 DORMANT TUMORS HAVE BALANCED PROLIFERATION AND APOPTOSIS

After inoculation in animals, human tumor cells can remain dormant for more than a year. However, this does not mean that the tumor cells are in G_0 arrest. Although some tumor cells might be in mitotic arrest, as demonstrated in some tumor dormancy models,[45–47] we reported that the majority of tumor cells are proliferating or undergoing apoptosis. The tumor cell proliferation index in nonangiogenic tumors can be as high as that of large, vascularized tumors. In a human breast cancer (MDA-MB-436) animal model, more than 50% of tumor cells were proliferating and more than 10% were undergoing apoptosis in all microscopic tumors analyzed at various time points during dormancy, as well as in all macroscopic angiogenic tumors.[44] In a different human osteosarcoma (MG-63 and SAOS-2) and gastric (ST-2) cancer dormancy models, microscopic tumors were unable to grow beyond a threshold size of ~1–2 mm in diameter. Within these nonangiogenic tumors, there appears to be a balance between proliferating cells and cells undergoing apoptosis.[48] Tumor cell proliferation in these tumors was ~12%, and tumor cell apoptosis ranged from 4% to 7.5%.

1.5 DEFINITION OF A HUMAN DORMANT TUMOR BASED ON EXPERIMENTAL ANIMAL MODELS

Based on xenograft models of various human tumors inoculated or surgically implanted into immunodeficient animals, a "dormant" tumor can be defined by its microscopic size and nonexpanding mass. In contrast, a "stable" tumor is macroscopic and expanding in mass. In more detail, we define human nonangiogenic tumors as:

1. Unable to induce angiogenic activity, by repulsion of existing blood vessels in the local stroma and/or relative absence of intratumoral microvessels.
2. Remain harmless to the host until they switch to the angiogenic phenotype (i.e., may remain harmless for 1 year or more, which is half the life span of a mouse).
3. Express equal or more antiangiogenic (i.e., thrombospondin-1) than angiogenic (i.e., VEGF, bFGF) proteins.
4. Grow in vivo to ~1 mm in diameter or less, at which time further expansion ceases.
5. Only visible with a hand lens or a dissecting microscope (5–10 × magnification).
6. White or transparent by gross examination.
7. Unable to spontaneously metastasize from the microscopic dormant state.
8. Show active tumor cell proliferation and apoptosis in mice and remain metabolically active during the dormancy period.
9. Can be cloned from a human angiogenic tumor, because human tumors are heterogeneous and contain a mixture of nonangiogenic and angiogenic tumor cells.

In contrast, angiogenic human tumors (as observed in our animal models) are defined as:

1. Able to induce angiogenic activity, by recruiting blood vessels form the surrounding stroma and/or forming new blood vessels within the tumor tissue.
2. Lethal to the host in only few weeks.
3. Express significantly more angiogenic than antiangiogenic proteins.
4. Grow along an exponential curve until they kill the host.
5. Visible and easily detectable based on their macroscopic size.
6. Red by gross examination.
7. Can spontaneously metastasize to various organs.
8. Can be cloned from a human angiogenic tumor, because human tumors are heterogeneous and contain a mixture of nonangiogenic and angiogenic tumor cells.

1.6 IN VIVO IMAGING OF HUMAN DORMANT TUMORS

Traditionally, various in vivo imaging techniques have been used for the detection and quantification of tumors implanted orthotopically or ectopically (i.e., outside their orthotopic site). However, some of these techniques can be employed for the in vivo detection of microscopic dormant tumors. By definition, nonangiogenic, dormant tumors are microscopic in size. Therefore, they are usually undetectable by palpation (limited to tumor sizes 50 mm^3 and smaller) when located in the subcutaneous space or mammary fat pad. It is an even greater challenge to detect microscopic tumors located in internal organs. In the originally published dormancy model of osteosarcoma (MG-63), the presence of dormant tumors in a fraction of the inoculated mice was revealed through careful examination of the hair growth overlying the original tumor inoculation site.[48] The inner side of the skin in the area associated with hair growth contained a microscopic white lesion, from which a histology section showed a viable tumor. Although this detection method clearly reveals this interesting phenomenon, it is terminal (i.e., the animal has to be euthanized) and does not provide longitudinal quantitative information about the tumor size.

Stable infection of tumor cells with fluorescent proteins (such as green fluorescent protein [GFP] and red fluorescent protein [RFP]) or luciferase allows for in vivo longitudinal detection of tumors even at a microscopic size. GFP-expressing tumor cells can be visualized noninvasively from the skin surface by directed blue light (488 nm) epi-illumination. Submillimeter tumors can be localized using this method. The utility of fluorescence visualization of dormant tumors has been reported by Udagawa et al.,[48] using osteosarcoma (MG-63 and SAOS-2) and gastric cancer (ST-2) dormancy models. Tumor-associated blood vessels appear dark against the background of a fluorescent tumor tissue, allowing for morphological (e.g., vessel diameters, tortuousity, branching) and even functional (e.g., red blood cell velocity) quantification of angiogenesis.[49] More recently, this labeling technique was used to determine the minimum number of human tumor cells necessary to form a nonangiogenic, dormant microscopic tumor in mice (Naumov et al., unpublished). However, detecting fluorescently labeled tumors has its limitations. Microscopic tumors in internal organs can only be visualized ex vivo. Certain procedures, such as in vivo videomicroscopy, can be used for visualization of liver and lung metastases.[50,51] However, in the brain, excitation or emission of fluorescently labeled tumor cells is not only limited by tissue depth, but also by light penetration through the skull.

Infection of tumor cells with the luciferase reporter gene allows for the reliable detection in mice of a signal from tumors that are <1 mm in diameter (as verified by histology) in all internal organs, including the brain. Almog et al.[43] has used the luciferase method for the detection of dormant human liposarcoma tumors in the renal fat pad of mice. The method can also be used to monitor the growth of microscopic human glioblastomas stereotactically after tumor cells are inoculated in the brains of mice (Naumov et al., unpublished work). Following intravenous injection of the luciferine substrate, the enzymatic activity of luciferase is rapid and transient. Only viable and metabolically active tumor cells can be detected by luminescence. The transient effect of the enzymatic reaction allows for real-time detection of tumor cells and for monitoring their viability during the dormancy period, as well as at times throughout the angiogenic switch. The persistent luciferase signal during the dormancy period of microscopic human tumors confirms the previous conclusion (based on histology) that dormancy does not result from tumor cell cycle arrest or eradication. Although the intensity of the luciferase signal directly correlates with the size of tumor, this imaging modality does not provide a clear tumor boundary or an anatomical outline of the tumor. However, small animal magnetic resonance imaging (MRI) provides a clear anatomical definition of a microscopic tumor, and it can be effectively used in combination with luciferase imaging (Naumov et al., unpublished work). Recent reports have demonstrated that single cancer cells can be detected in a mouse brain using MRI.[52] Individual tumor cells trapped within the brain microcirculation were detected using MRI and validated using high-resolution confocal microscopy. Graham et al.[53] demonstrated that three-dimensional, high-frequency ultrasound can quantitatively monitor the growth of liver micrometastases as small as 0.5 mm in diameter.

Collectively, these recent advances in animal imaging modalities enable, in most cases, noninvasive, real-time, longitudinal observations of single cancer cell trafficking and detection of nonangiogenic microscopic tumors in vivo during the dormancy period and as they switch to the angiogenic phenotype. Quantitative imaging of tumors throughout their progression to the angiogenic phenotype can be used for evaluating the efficacy of antiangiogenic therapy in primary and metastatic tumors.

1.7 MOLECULAR MECHANISMS OF THE HUMAN ANGIOGENIC SWITCH

Transfection of human osteosarcoma (MG-63 and SAOS-2) and gastric cancer (ST-2) cells with activated c-*Ha-ras* oncogene induces loss of dormancy in otherwise nonangiogenic human cell lines.[48] When inoculated in immunosuppressed mice, wild-type (or control vector-transfected) tumor cells did not form palpable tumors for more than 8 months. White tumor foci, which were avascular or contained sparse vessels, were found throughout the dormancy period at the site of inoculation. However, *ras*-transfected human osteosarcoma (MG-63 and SAOS-2) and gastric cancer (ST-2) cells formed vascularized large tumors within 1 month. The in vivo growth of *ras*-transfected tumor cells was associated with significantly increased angiogenic response, increased proliferation, and decreased apoptosis when compared with wild-type tumor cells. Loss of the dormant phenotype induced by activated *ras* correlated with increased levels (1.5 to 2.5-fold) of $VEGF_{165}$, as assessed in the conditioned media relative to the control tumor cells.[48] Overexpression of $VEGF_{165}$ in the tumor cells also resulted in a loss of dormancy and induced a robust angiogenic response in 30% of animals inoculated with gastric cancer and 40% of animals inoculated with osteosarcoma. In contrast to *ras*-transfected tumor cells, loss of dormancy in $VEGF_{165}$-transfected tumor cells was not associated with an increase in tumor cell proliferation, but was associated with reduced apoptosis. Therefore, the angiogenic response induced by $VEGF_{165}$ was found to be sufficient for the induction of a loss of dormancy by reducing apoptosis. Activation of *ras* can directly induce tumor cell proliferation and confer resistance to apoptosis.[54,55] In addition, *ras* activation can indirectly stimulate an angiogenic response in tumors by inducing proangiogenic proteins, such as VEGF,[56] and by downregulating angiogenesis inhibitors, such as thrombospondin.[57–60] Similar to *ras*, other oncogenes and tumor suppressor genes can indirectly affect tumor growth via an angiogenic mechanism. For example, *p53*, *PTEN*, and *Smad 4* have been shown to increase thrombospondin-1 expression by upregulation of *Tsp-1* gene or by increased mRNA expression.[12,61–63] Thrombospondin-1 expression can be decreased by *Myc*, *Ras*, *Id1*, *WT1*, c-*jun*, and v-*stc* via transcriptional repression, myc phosphorylation, or regulation of mRNA turnover and stability.[59,60,64–69] The inherently low toxicity of natural angiogenesis inhibitors, in addition to their selective effect on pathological neovascularization without harm to normal vasculature, makes them attractive therapeutic agents.

In addition to thrombospondin-1 regulation, the *p53* tumor suppressor gene regulates other currently unidentified inhibitors of angiogenesis.[70] Teodoro et al.[71] recently reported that wild-type *p53* mobilizes endostatin through a specific α(II) collagen prolyl-4-hydroxylase (α(II)PH gene product), which binds to *p53*. *p53* is inactivated in over 50% of all human tumors. Reintroduction of wild-type *p53* into mouse fibrosarcoma (T241) cells correlates with increased thrombospondin-1 expression and induces angiogenesis-restricted dormancy.[72] Inoculation of parental T241 fibrosarcoma cells into a mouse ear resulted in vascularized, visible tumors within 2 weeks. In contrast, when wild-type *p53* was introduced in the same cells, only 12% of the tumors became angiogenic 2 months after inoculation. Therefore, expression of wild-type *p53* resulted in the loss of an angiogenic phenotype. Loss of the angiogenic phenotype was also correlated with the upregulation of the mRNA-encoding thrombospondin-1. This experimental model demonstrated that *p53* can act as a tumor suppressor, independent of its direct effects on cell proliferation and survival. Moreover, *p53* had an indirect antitumor effect by inhibiting angiogenesis and increasing the rate of apoptosis.

In a recent report, Naumov et al.[44] compared the tumor cell secretion and intracellular levels of thrombospondin-1 in nonangiogenic and angiogenic tumor cell populations isolated from a human breast cancer cell line (MDA-MB-436). Angiogenic cells contain 2.5-fold higher levels of c-Myc and p-Myc than their nonangiogenic counterparts, as assessed by Western blot. In contrast, angiogenic tumor cells contain significantly lower levels of thrombospondin-1 than nonangiogenic tumor cells. Moreover, nonangiogenic human breast cancer cells secrete at least 20-fold higher levels of thrombospondin-1 than angiogenic cells. Similar findings were reported using a different human breast cancer cell line (MDA-MB-435).[60] Watnick et al.[60] reported that phosphoinositide 3-kinase (PI3K) can induce a signal transduction cascade leading to the phosphorylation of c-Myc and the subsequent repression of thrombospondin-1. Treatment with LY294002 (a PI3K inhibitor) caused thrombospondin-1 levels within angiogenic cells to increase but had no effect on levels in nonangiogenic cells.[44] Therefore, the PI3K signaling pathway is responsible for the repression of thrombospondin-1, and it is regulated differently in angiogenic and nonangiogenic human tumor cells.

1.8 INDUCTION OF TUMOR DORMANCY USING ANTIANGIOGENIC THERAPY

Tumor progression is highly dependent on the surrounding stroma, including the endothelial cells, fibroblasts, local basement membrane factors, macrophages, platelets, T cells, and other cellular compartments. Antiangiogenic therapy can target the endothelial cell compartment in atleast two distinct ways: directly or indirectly (Figure 1.3).[73]

Direct angiogenesis inhibitors block vascular endothelial cells from proliferating, migrating, or increasing their survival. For example, SU 11248 directly blocks VEGF receptors (among other receptors involved with angiogenic signaling) on endothelial cells (Figure 1.3). Direct angiogenesis inhibitors include: (1) synthetic inhibitors or peptides designed to interfere with specific steps in the angiogenic process (e.g., inhibitors of metalloproteinases, antagonists of the $\alpha_V\beta_3$ or $\alpha_5\beta_1$ integrins), (2) low molecular weight molecules (e.g., TNP-470, caplostatin, thalidomide, 2-methoxyestradiol), and (3) endogenous (i.e., natural) angiogenesis inhibitors (e.g., TSP-1, platelet factor 4, interferon-α, IL-12, angiostatin, endostatin, arrestin, canstatin, tumstatin).[5,39,74–92]

Bouck et al.[93] were the first to demonstrate that a tumor can generate angiogenesis inhibitor (i.e., thrombospondin-1). They subsequently suggested that the angiogenic phenotype was a result of a net imbalance of endogenous angiogenesis stimulators and inhibitors.

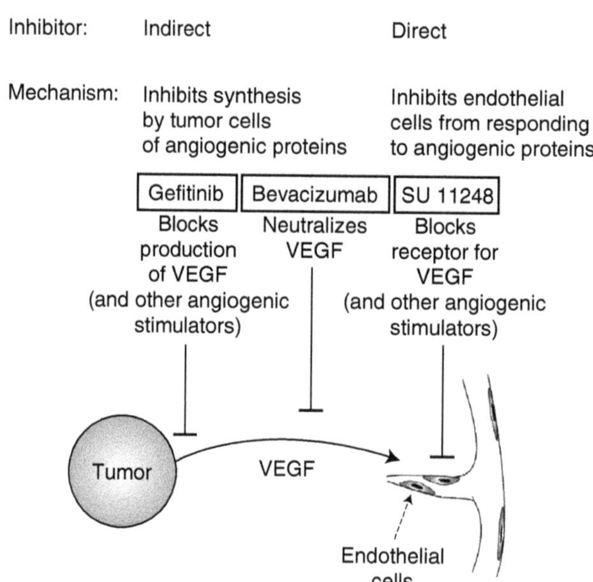

FIGURE 1.3 Examples of direct and indirect angiogenesis inhibitors that can block production of a tumor cell angiogenic protein (Gefitinib), or neutralize a systemic proangiogenic protein (Bevacizumab), or block a receptor for a tumor cell produced angiogenic protein (SU 11248). (Adapted from Folkman, J. et al., *Cancer Medicine*, 7th edn., 2006. With permission.)

In a series of experiments, Folkman and colleagues reported that the surgical removal of a primary Lewis lung carcinoma tumor in mice results in the exponential growth of lung metastases.[94,95] In these experimental animal models, the presence of a primary tumor generated increased circulating angiostatin levels. Angiostatin is a potent antiangiogenic plasminogen fragment,[96] which inhibits the in vivo growth of Lewis lung metastases by preventing neovascularization.[77] However, gene transfer of a cDNA coding for mouse angiostatin into murine T241 fibrosarcoma cells successfully suppressed lung metastatic tumor growth after the removal of the primary tumor.[97] Cao et al.[97] demonstrated that pulmonary micrometastases, expressing angiostatin, remain in a dormant and avascular state for 2–5 months after removal of primary tumors. These dormant micrometastases were characterized as having a high rate of apoptosis counterbalanced by a high proliferation rate.

Holmgren et al.[95] investigated whether treatment with an exogenous angiogenesis inhibitor could replace the endogenous angiogenesis suppressive ability of a primary tumor. In animals with surgically removed primary Lewis lung carcinoma, or T241 mouse sarcomas, treatment with TNP-470 resulted in the suppression of metastases comparable to that observed in the presence of the primary tumor. Therefore, exogenous treatment can be used to replace the endogenous angiogenesis inhibition of a primary tumor. Moreover, it can be used for the systemic suppression of angiogenesis-maintained micrometastases of both Lewis lung cancer and T241 fibrosarcoma in a dormant state. These dormant micrometastases were characterized as having high tumor cell proliferation counterbalanced by high cell death rate (i.e., apoptosis), indicating that inhibition of angiogenesis limits tumor growth by elevating tumor cell apoptosis. Exogenous angiogenesis inhibitors, such as TNP-470, mimic the primary tumor suppression by maintaining high apoptosis in the lung micrometastases, but without having an effect on tumor cell proliferation. A similar mechanism of sustained micrometastatic dormant state has been demonstrated using angiostatin, which maintains a high apoptotic index in lung metastases after the removal of a primary tumor without affecting tumor cell proliferation.[38]

Indirect angiogenesis inhibitors target tumor cell proteins created by oncogenes that drive the angiogenic switch. In general, their mechanism of action is by decreasing or blocking the expression of other tumor cell products, neutralizing the tumor cell product itself, or by blocking receptors on endothelial cells. The impact of oncogenes on tumor angiogenesis has been reviewed by Rak and Kerbel.[59,98,99] For example, gefitinib (Iressa) blocks VEGF production from tumor cells. However, even systemically available VEGF can be neutralized by bevacizumab (Avastin) before it binds to VEGF receptors on endothelial cells (Figure 1.3). There is an emerging group (e.g., thyrosine kinase inhibitors) of anticancer drugs originally developed to target oncogenes, which also have "indirect" antiangiogenic activity. For example, the *ras* farnesyl transferase inhibitors block oncogene signaling pathways, which upregulate tumor cell production of VEGF and downregulate production of Tsp-1.[100] Trastuzumab, an antibody that blocks *HER2/neu* receptor tyrosine kinase signaling, suppresses tumor cell production of angiogenic proteins, such as TGF-α, angiopoietin 1, plasminogen activator inhibitor-1 (PAI-1), and VEGF.[101,102] At the same time, trastuzumab has been shown to upregulate the expression of Tsp-1 (endogenous angiogenesis inhibitor), which may be an important mechanism of its antiangiogenic activity.[102] Upregulation of endogenous antiangiogenic proteins, using direct or indirect angiogenesis inhibitors, can be a useful approach for preventing the angiogenic switch and keeping human tumors in a microscopic dormant state (Figure 1.4).

Acquired drug resistance is a major obstacle in the treatment of cancer. Genetic instability, heterogeneity, and high mutational rates of tumor cells are the major causes of drug resistance.[103] In contrast, antiangiogenic therapy targets endothelial cells, which are genetically stable and have a low mutational rate. In an experimental animal model,

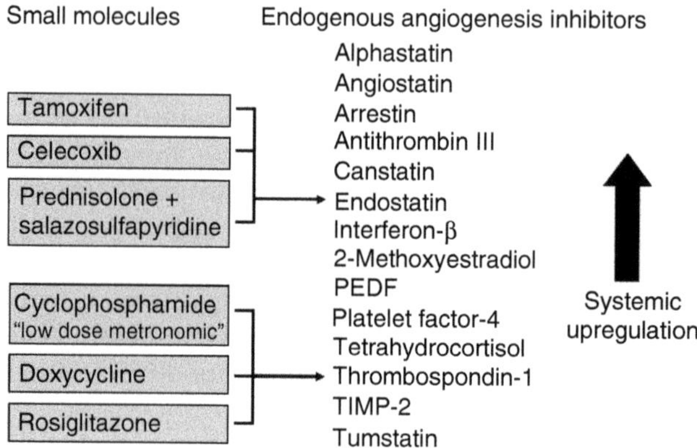

FIGURE 1.4 Summary of a few small molecules and the corresponding increase in endogenous angiogenesis inhibitors. Examples of small molecules that are orally available and may induce increased systemic levels of endogenous angiogenesis inhibitors. Chronic systemic increase of antiangiogenic proteins can prevent the angiogenic switch and delay the progression of cancer. (Adapted from Folkman, J., *Exp. Cell Res.*, 312(5), 594, 2006. With permission.)

Boehm et al.[104] reported that antiangiogenic therapy targeted against tumor-associated endothelial cells does not result in drug resistance. In this study, Lewis lung cancer, T241 fibrosarcoma, and B16F10 melanoma were repeatedly treated with endostatin. When tumors reached the size of ~350–400 mm^3, endostatin treatment was initiated until the tumor became undetectable. Endostatin therapy was then stopped, and the tumor was allowed to regrow. Endostatin therapy was resumed when tumors reached a mean volume of 350–400 mm^3. After the second (for melanoma), fourth (for fibrosarcoma), and sixth (for Lewis lung carcinoma) cycles of endostatin treatment, all tumors remained as barely visible subcutaneous nodules (size 5–50 mm^3) for up to 360 days. In contrast, endostatin resistance developed rapidly when Lewis lung carcinomas were treated with conventional cytotoxic chemotherapy. These studies demonstrated that repeated cycles of endostatin therapy induced tumor dormancy, which persisted indefinitely after therapy.

1.9 METASTATIC DORMANCY

Metastasis, the spread of cancer from a primary tumor to secondary organs, is the major cause of cancer-related deaths. Hematogenous or lymphatic spread of only a few cancer cells from a primary tumor can successfully form a macroscopic tumor at a secondary site.[50,105,106] However, for a macrometastasis to become lethal, it must successfully complete a number of steps (Figure 1.1).[107] Two different types of tumor dormancy have been identified in the metastatic process: (1) solitary dormant cancer cells, which are in G_0 cell cycle arrest and (2) dormant nonangiogenic tumors, in which tumor cells are actively proliferating and dying, but the tumor fails to recruit blood vessels. Both of these steps can contribute to a latency period associated with metastatic growth of human cancer (Figure 1.1).[108]

Previous studies by Holmgren and colleagues[95,109] have identified nonangiogenic micrometastases as a potential contributor to metastatic dormancy. These studies showed that dormant micrometastases did not grow in size beyond 200 μm, but they remained metabolically active. This size limitation was associated with a steady-state balance between the rates of tumor cell proliferation and apoptosis, with no net growth of the metastases. Changes in the intrinsic properties of these dormant micrometastases, or their microenvironment at a later time, triggered metastatic growth associated with a disturbance of the proliferation and apoptosis balance. Progressive growth in such micrometastases was restricted due to suppression of tumor angiogenesis.

Naumov et al.[110] have identified another possible source of metastatic dormancy: viable, solitary dormant tumor cells that are neither proliferating nor undergoing apoptosis after arriving at a metastatic site. These studies showed that more than 50%–80% of breast cancer cells, distributed to mouse liver via the circulation, can remain in the tissue for extended periods of time (up to 77 days) as solitary nonproliferating dormant cells. This surprising phenomenon was observed in populations of breast cancer cells of high and low metastatic ability. In the case of the highly metastatic cell line (D2A1 cells), lethal macrometastases grew from a very small subset of cells (~0.006%), with the majority (~80% cell loss) of injected cells undergoing apoptosis or destroyed by leukocytes. However, ~20% of the injected cells persisted as nonproliferating dormant cells. In contrast, ~80% of poorly metastatic breast cancer cells remained as nonproliferating dormant solitary cells in the mouse liver. A subset of these cells could be recovered and grown under in vitro culture conditions 11 weeks after injection into mice. The recovered tumor cells retained their ability to form primary tumors in the mammary fat pad of mice. These solitary dormant tumor cells may be a potential source of an occasional nonangiogenic metastasis, and of an even rarer, but lethal, angiogenic metastasis.

Taken together, these studies demonstrated that metastasis is a dynamic process, where solitary, nonproliferating dormant cancer cells, nonangiogenic micrometastases, and angiogenic macrometastases can coexist at each stage of the metastatic process. While nonangiogenic micrometastases could be vulnerable to antiangiogenic and cytotoxic chemotherapeutic agents (administered in a metronomic, low-dose regimen, as described by Browder et al.[111] and Kerbel and colleagues[112,113]), solitary dormant cells could remain unaffected because of their inability to proliferate.

Naumov et al.[114] showed that nonproliferating solitary dormant breast cancer cells remained unaffected by doxorubicin treatment. However, the same treatment successfully inhibited actively growing macrometastases in the same mice. Therefore, doxorubicin chemotherapy, which successfully reduced the metastatic burden, failed to affect the number of solitary dormant cells. These findings have important clinical implications for patients undergoing adjuvant chemotherapy. It is possible that dormant nonproliferating tumor cells can remain unaffected by standard chemotherapy and may retain their potential to initiate growth at a later date. Both solitary cancer cells and nonangiogenic metastases can remain dormant and undetectable for months or years, leading to an uncertainty in the prognosis for patients who have already been treated for the primary cancer.

1.10 ANGIOGENIC SWITCH-RELATED BIOMARKERS FOR DETECTION OF DORMANT TUMORS

Even with recent advances in the clinical detection of human cancer, a tumor that is microscopic in size (~1 mm in diameter) remains undetectable. A panel of angiogenic switch-related biomarkers is under development using the human tumor dormancy models. These biomarkers include circulating endothelial progenitor cells (CEPs) and platelets in the blood, as well as matrix metalloproteinases (MMPs) in the urine. The detection of a single microscopic human tumor in existing animal models can be achieved using each one of these biomarkers alone or in combination.[115–117]

We compared the in vivo ability of angiogenic and nonangiogenic human breast tumors (MDA-MB-436 cells) to mobilize mature circulating endothelial cells (CECs) (CD45−, Flk+, CD31+, CD117−) and CEPs (CD45−, Flk+, CD31+, CD117+).[115] The number of blood-borne CECs and CEPs was quantified using a flow cytometer. There was little difference in the percent of mature CECs in the blood of mice inoculated with angiogenic and nonangiogenic cells. However, mice inoculated with nonangiogenic cells had approximately fourfold decrease in CEPs when compared with control mice. Mice inoculated with angiogenic cells had levels of CEPs comparable to those in the control mice. Previous reports[44] have shown that these nonangiogenic breast cancer cells (MDA-MB-436 cells) secrete at least 20-fold higher levels of thrombospondin-1 (Tsp-1) than their angiogenic counterparts. Other studies have suggested that endogenous inhibitors of angiogenesis, such as Tsp-1 and endostatin, may inhibit the mobilization of CEPs.[118] These observations suggest that microscopic dormant (nonangiogenic) tumors may suppress the mobilization of CEPs from the bone marrow via systemic thrombospondin-1.

Klement et al.[117] recently reported that blood platelets can sequester both pro- and antiangiogenic proteins. It is estimated that at least 100 billion platelets are produced per day by megakaryocytes in the bone marrow of an average 70 kg person.[119,120] With a life span of 7–8 days in humans, it is estimated that there are approximately a trillion platelets in constant circulation.[119,121] Folkman and colleagues proposed that the platelet compartment of the blood stream can potentially accumulate angiogenesis-related proteins and possibly release them at a later time.[117] Using a novel "platelet angiogenic proteome," as quantitatively assessed by SELDI-ToF technology (Ciphergen, Freemont, CA), the presence of microscopic human tumors in mice can be detected.[117] Using this technology, the accumulation and reduction in angiogenesis-related proteins sequestered in platelets can be quantitatively followed throughout the angiogenic switch. The identification of proteins that are associated with the angiogenic switch and that may be used as angiogenic switch-related biomarkers is currently under investigation.

In summary, the tumor dormancy animal models presented here permit further clarification of the role of CEC/CEPs, platelets, and MMPs as participants in the "angiogenic switch." Moreover, this angiogenic switch-related biomarker panel may prove to be a useful diagnostic method for the presence of microscopic cancers at primary and metastatic sites long before detection by conventional methods. It may be feasible to develop a panel of angiogenesis-associated biomarkers that can identify the presence of a microscopic human tumor, predict its switch to the angiogenic phenotype, and possibly serve as a guide for antiangiogenic therapy. In the future, it may be possible for a patient who is at risk for cancer recurrence to take an oral drug that can elevate endogenous platelet-associated antiangiogenic proteins and delay, if not prevent, the formation of recurrent tumors.

1.11 CONCLUSION

We speculate that the development of more specific and sensitive biomarkers may permit the very early detection of recurrent cancer, possibly years before symptoms or anatomical location. If this concept can be validated, then relatively nontoxic angiogenesis inhibitors may be used to "treat the biomarkers" without ever seeing the recurrent tumor (i.e., cancer without disease).[14]

ACKNOWLEDGMENTS

The authors thank Ms. Jenny Grillo for critical reading and editing of this chapter. We also thank Kristin Johnson for help with graphics. This work was supported by the Breast Cancer Research Foundation, NIH Program Project (grant #P01CA45548), and an Innovator Award from the Department of Defense.

REFERENCES

1. Demicheli R, Abbattista A, Miceli R, Valagussa P, Bonadonna G. Time distribution of the recurrence risk for breast cancer patients undergoing mastectomy: further support about the concept of tumor dormancy. *Breast Cancer Res Treat* 1996;41(2):177–185.

2. Karrison TG, Ferguson DJ, Meier P. Dormancy of mammary carcinoma after mastectomy. *J Natl Cancer Inst* 1999;91(1):80–85.
3. Willis RA. *Pathology of tumours*. London: Butterworth; 1948.
4. Hadfield G. The dormant cancer cell. *Br Med J* 1954;4888:607–610.
5. Rastinejad F, Polverini PJ, Bouck NP. Regulation of the activity of a new inhibitor of angiogenesis by a cancer suppressor gene. *Cell* 1989;56(3):345–355.
6. Wheelock EF, Weinhold KJ, Levich J. The tumor dormant state. *Adv Cancer Res* 1981;34:107–140.
7. Saudemont A, Jouy N, Hetuin D, Quesnel B. NK cells that are activated by CXCL10 can kill dormant tumor cells that resist CTL-mediated lysis and can express B7-H1 that stimulates T cells. *Blood* 2005;105(6):2428–2435.
8. Stewart TH. Immune mechanisms and tumor dormancy. *Medicina (B Aires)* 1996;56(Suppl 1): 74–82.
9. Saudemont A, Quesnel B. In a model of tumor dormancy, long-term persistent leukemic cells have increased B7-H1 and B7.1 expression and resist CTL-mediated lysis. *Blood* 2004;104(7):2124–2133.
10. Marches R, Scheuermann R, Uhr J. Cancer dormancy: from mice to man. *Cell Cycle* 2006;5(16): 1772–1778.
11. Jain RK, Safabakhsh N, Sckell A et al. Endothelial cell death, angiogenesis, and microvascular function after castration in an androgen-dependent tumor: role of vascular endothelial growth factor. *Proc Natl Acad Sci USA* 1998;95(18):10820–10825.
12. Dameron KM, Volpert OV, Tainsky MA, Bouck N. Control of angiogenesis in fibroblasts by p53 regulation of thrombospondin-1. *Science* 1994;265(5178):1582–1584.
13. O'Reilly MS, Holmgren L, Shing Y et al. Angiostatin: a circulating endothelial cell inhibitor that suppresses angiogenesis and tumor growth. *Cold Spring Harb Symp Quant Biol* 1994;59:471–482.
14. Folkman J, Kalluri R. Cancer without disease. *Nature* 2004;427(6977):787.
15. Folkman J, Heymach J, Kalluri R. *Tumor angiogenesis*. 7th edn. Hamilton, Ontario: B.C. Decker; 2006.
16. Folkman J. Tumor angiogenesis: therapeutic implications. *N Engl J Med* 1971;285(21):1182–1186.
17. Folkman J. What is the evidence that tumors are angiogenesis dependent? *J Natl Cancer Inst* 1990;82(1):4–6.
18. Coman D, Sheldon WF. The significance of hyperemia around tumor implants. *Am J Pathol* 1946;22:821–831.
19. Warren B. The vascular morphology of tumors. In: Peterson H-I, ed. *Tumor blood circulation: angiogenesis, vascular morphology and blood flow of experimental human tumors*. Florida: CRC Press; 1979:1–47.
20. Ide A, Baker NH, Warren SL. Vascularization of the Brown–Pearce rabbit epithelioma transplant as seen in the transparent ear chamber. *Am J Roentgenol* 1939;42:891–899.
21. Algire G, Legallais, FY. Growth rate of transplanted tumors in relation to latent period and host vascular reaction. *Cancer Res* 1947;7:724.
22. Algire G, Chalkely HW, Legallais FY, Park H. Vascular reactions of normal and malignant tumors in vivo: I. Vascular reactions of mice to wounds and to normal and neoplastic transplants. *J Natl Cancer Inst* 1945;6:73–85.
23. Folkman J. Toward an understanding of angiogenesis: search and discovery. *Perspect Biol Med* 1985;29(1):10–36.
24. Folkman J, Long DM, Jr., Becker FF. Growth and metastasis of tumor in organ culture. *Cancer* 1963;16:453–467.
25. Folkman J, Cole P, Zimmerman S. Tumor behavior in isolated perfused organs: in vitro growth and metastases of biopsy material in rabbit thyroid and canine intestinal segment. *Ann Surg* 1966;164(3):491–502.
26. Folkman J. Anti-angiogenesis: new concept for therapy of solid tumors. *Ann Surg* 1972;175(3):409–416.
27. Folkman J. The vascularization of tumors. *Sci Am* 1976;234(5):58–64, 70–73.
28. Folkman J. The intestine as an organ culture. In: Burdette W, ed. *Carcinoma of the colon and antecedent epithelium*. Springfield (IL): CC Thomas; 1970:113–127.
29. Folkman J, Brem H. Angiogenesis and inflammation. In: Gallin JI, Goldstein IM, Snyderman R, eds. *Inflammation: basic principles and clinical correlates*. 2nd edn. New York: Raven Press; 1992:821–839.

30. Folkman J. Angiogenesis in arthritis. In: Smolen J, Lipsky P, eds. *Targeted therapies in rheumatology*. London: Martin Dunitz; 2003:111–131.
31. Folkman J, Watson K, Ingber D, Hanahan D. Induction of angiogenesis during the transition from hyperplasia to neoplasia. *Nature* 1989;339(6219):58–61.
32. Torres Filho IP, Leunig M, Yuan F, Intaglietta M, Jain RK. Noninvasive measurement of microvascular and interstitial oxygen profiles in a human tumor in SCID mice. *Proc Natl Acad Sci USA* 1994;91(6):2081–2085.
33. North S, Moenner M, Bikfalvi A. Recent developments in the regulation of the angiogenic switch by cellular stress factors in tumors. *Cancer Lett* 2005;218(1):1–14.
34. Greene HSN. Heterologous transplantation of mammalian tumors. *J Exp Med* 1941;73:461–486.
35. Gimbrone MA, Jr., Aster RH, Cotran RS, Corkery J, Jandl JH, Folkman J. Preservation of vascular integrity in organs perfused in vitro with a platelet-rich medium. *Nature* 1969;222(188):33–36.
36. Gimbrone MA, Jr., Leapman SB, Cotran RS, Folkman J. Tumor dormancy in vivo by prevention of neovascularization. *J Exp Med* 1972;136(2):261–276.
37. Brem H, Folkman J. Inhibition of tumor angiogenesis mediated by cartilage. *J Exp Med* 1975;141(2):427–439.
37a. Moses MA, Sudhalter J, Langer R. Identification of an inhibitor of neovascularization from the cartilage. *Science* 1990;248:1408–1410.
38. O'Reilly MS, Holmgren L, Chen C, Folkman J. Angiostatin induces and sustains dormancy of human primary tumors in mice. *Nat Med* 1996;2(6):689–692.
39. O'Reilly MS, Boehm T, Shing Y et al. Endostatin: an endogenous inhibitor of angiogenesis and tumor growth. *Cell* 1997;88(2):277–285.
40. Hanahan D, Folkman J. Patterns and emerging mechanisms of the angiogenic switch during tumorigenesis. *Cell* 1996;86(3):353–364.
41. Hanahan D, Christofori G, Naik P, Arbeit J. Transgenic mouse models of tumour angiogenesis: the angiogenic switch, its molecular controls, and prospects for preclinical therapeutic models. *Eur J Cancer* 1996;32A(14):2386–2393.
42. Achilles EG, Fernandez A, Allred EN et al. Heterogeneity of angiogenic activity in a human liposarcoma: a proposed mechanism for "no take" of human tumors in mice. *J Natl Cancer Inst* 2001;93(14):1075–1081.
43. Almog N, Henke V, Flores L et al. Prolonged dormancy of human liposarcoma is associated with impaired tumor angiogenesis. *FASEB J* 2006;20(7):947–949.
44. Naumov GN, Bender E, Zurakowski D et al. A model of human tumor dormancy: an angiogenic switch from the nonangiogenic phenotype. *J Natl Cancer Inst* 2006;98(5):316–325.
45. Aguirre Ghiso JA, Kovalski K, Ossowski L. Tumor dormancy induced by downregulation of urokinase receptor in human carcinoma involves integrin and MAPK signaling. *J Cell Biol* 1999;147(1):89–104.
46. Aguirre-Ghiso JA, Liu D, Mignatti A, Kovalski K, Ossowski L. Urokinase receptor and fibronectin regulate the ERK(MAPK) to p38(MAPK) activity ratios that determine carcinoma cell proliferation or dormancy in vivo. *Mol Biol Cell* 2001;12(4):863–879.
47. Aguirre-Ghiso JA, Estrada Y, Liu D, Ossowski L. ERK(MAPK) activity as a determinant of tumor growth and dormancy; regulation by p38(SAPK). *Cancer Res* 2003;63(7):1684–1695.
48. Udagawa T, Fernandez A, Achilles EG, Folkman J, D'Amato RJ. Persistence of microscopic human cancers in mice: alterations in the angiogenic balance accompanies loss of tumor dormancy. *FASEB J* 2002;16(11):1361–1370.
49. Naumov GN, Wilson SM, MacDonald IC et al. Cellular expression of green fluorescent protein, coupled with high-resolution in vivo videomicroscopy, to monitor steps in tumor metastasis. *J Cell Sci* 1999;112(Pt 12):1835–1842.
50. Chambers AF, Groom AC, MacDonald IC. Dissemination and growth of cancer cells in metastatic sites. *Nat Rev Cancer* 2002;2(8):563–572.
51. Naumov GN, MacDonald IC, Chambers AF, Groom AC. Solitary cancer cells as a possible source of tumour dormancy? *Semin Cancer Biol* 2001;11(4):271–276.
52. Heyn C, Ronald JA, Mackenzie LT et al. In vivo magnetic resonance imaging of single cells in mouse brain with optical validation. *Magn Reson Med* 2006;55(1):23–29.

53. Graham KC, Wirtzfeld LA, MacKenzie LT et al. Three-dimensional high-frequency ultrasound imaging for longitudinal evaluation of liver metastases in preclinical models. *Cancer Res* 2005;65(12):5231–5237.
54. Goustin AS, Leof EB, Shipley GD, Moses HL. Growth factors and cancer. *Cancer Res* 1986;46(3):1015–1029.
55. Bonni A, Brunet A, West AE, Datta SR, Takasu MA, Greenberg ME. Cell survival promoted by the Ras-MAPK signaling pathway by transcription-dependent and -independent mechanisms. *Science* 1999;286(5443):1358–1362.
56. Rak J, Mitsuhashi Y, Bayko L et al. Mutant *ras* oncogenes upregulate VEGF/VPF expression: implications for induction and inhibition of tumor angiogenesis. *Cancer Res* 1995; 55(20):4575–4580.
57. Good DJ, Polverini PJ, Rastinejad F et al. A tumor suppressor-dependent inhibitor of angiogenesis is immunologically and functionally indistinguishable from a fragment of thrombospondin. *Proc Natl Acad Sci USA* 1990;87(17):6624–6628.
58. Sheibani N, Frazier WA. Repression of thrombospondin-1 expression, a natural inhibitor of angiogenesis, in polyoma middle T transformed NIH3T3 cells. *Cancer Lett* 1996;107(1):45–52.
59. Rak J, Yu JL, Klement G, Kerbel RS. Oncogenes and angiogenesis: signaling three-dimensional tumor growth. *J Investig Dermatol Symp Proc* 2000;5(1):24–33.
60. Watnick RS, Cheng YN, Rangarajan A, Ince TA, Weinberg RA. Ras modulates Myc activity to repress thrombospondin-1 expression and increase tumor angiogenesis. *Cancer Cell* 2003;3(3):219–231.
61. Volpert OV, Dameron KM, Bouck N. Sequential development of an angiogenic phenotype by human fibroblasts progressing to tumorigenicity. *Oncogene* 1997;14(12):1495–1502.
62. Chandrasekaran L, He CZ, Al-Barazi H, Krutzsch HC, Iruela-Arispe ML, Roberts DD. Cell contact-dependent activation of alpha3beta1 integrin modulates endothelial cell responses to thrombospondin-1. *Mol Biol Cell* 2000;11(9):2885–2900.
63. Lawler J, Sunday M, Thibert V et al. Thrombospondin-1 is required for normal murine pulmonary homeostasis and its absence causes pneumonia. *J Clin Invest* 1998;101(5):982–992.
64. Volpert OV. Modulation of endothelial cell survival by an inhibitor of angiogenesis thrombospondin-1: a dynamic balance. *Cancer Metastasis Rev* 2000;19(1–2):87–92.
65. Mettouchi A, Cabon F, Montreau N et al. SPARC and thrombospondin genes are repressed by the c-*jun* oncogene in rat embryo fibroblasts. *EMBO J* 1994;13(23):5668–5678.
66. Dejong V, Degeorges A, Filleur S et al. The Wilms' tumor gene product represses the transcription of thrombospondin 1 in response to overexpression of c-*Jun*. *Oncogene* 1999;18(20):3143–3151.
67. Slack JL, Bornstein P. Transformation by v-*src* causes transient induction followed by repression of mouse thrombospondin-1. *Cell Growth Differ* 1994;5(12):1373–1380.
68. Tikhonenko AT, Black DJ, Linial ML. Viral Myc oncoproteins in infected fibroblasts down-modulate thrombospondin-1, a possible tumor suppressor gene. *J Biol Chem* 1996; 271(48):30741–30747.
69. Janz A, Sevignani C, Kenyon K, Ngo CV, Thomas-Tikhonenko A. Activation of the myc oncoprotein leads to increased turnover of thrombospondin-1 mRNA. *Nucleic Acids Res* 2000;28(11):2268–2275.
70. Van Meir EG, Kikuchi T, Tada M et al. Analysis of the *p53* gene and its expression in human glioblastoma cells. *Cancer Res* 1994;54(3):649–652.
71. Teodoro JG, Parker AE, Zhu X, Green MR. p53-mediated inhibition of angiogenesis through up-regulation of a collagen prolyl hydroxylase. *Science* 2006;313(5789):968–971.
72. Holmgren L, Jackson G, Arbiser J. p53 induces angiogenesis-restricted dormancy in a mouse fibrosarcoma. *Oncogene* 1998;17(7):819–824.
73. Kerbel R, Folkman J. Clinical translation of angiogenesis inhibitors. *Nat Rev Cancer* 2002;2(10):727–739.
74. Taylor S, Folkman J. Protamine is an inhibitor of angiogenesis. *Nature* 1982;297(5864):307–312.
75. Ingber D, Fujita T, Kishimoto S et al. Synthetic analogues of fumagillin that inhibit angiogenesis and suppress tumour growth. *Nature* 1990;348(6301):555–557.
76. Brooks PC, Montgomery AM, Rosenfeld M et al. Integrin alpha v beta 3 antagonists promote tumor regression by inducing apoptosis of angiogenic blood vessels. *Cell* 1994;79(7):1157–1164.

77. O'Reilly MS, Holmgren L, Shing Y et al. Angiostatin: a novel angiogenesis inhibitor that mediates the suppression of metastases by a Lewis lung carcinoma. *Cell* 1994;79(2):315–328.
78. O'Reilly MS, Pirie-Shepherd S, Lane WS, Folkman J. Antiangiogenic activity of the cleaved conformation of the serpin antithrombin. *Science* 1999;285(5435):1926–1928.
79. Stetler-Stevenson WG, Krutzsch HC, Liotta LA. Tissue inhibitor of metalloproteinase (TIMP-2). A new member of the metalloproteinase inhibitor family. *J Biol Chem* 1989;264(29):17374–17378.
80. Moses MA, Sudhalter J, Langer R. Identification of an inhibitor of neovascularization from cartilage. *Science* 1990;248(4961):1408–1410.
81. Kruger EA, Figg WD. TNP-470: an angiogenesis inhibitor in clinical development for cancer. *Expert Opin Investig Drugs* 2000;9(6):1383–1396.
82. D'Amato RJ, Loughnan MS, Flynn E, Folkman J. Thalidomide is an inhibitor of angiogenesis. *Proc Natl Acad Sci USA* 1994;91(9):4082–4085.
83. Folkman J. Angiogenesis-dependent diseases. *Semin Oncol* 2001;28(6):536–542.
84. Crum R, Szabo S, Folkman J. A new class of steroids inhibits angiogenesis in the presence of heparin or a heparin fragment. *Science* 1985;230(4732):1375–1378.
85. D'Amato RJ, Lin CM, Flynn E, Folkman J, Hamel E. 2-Methoxyestradiol, an endogenous mammalian metabolite, inhibits tubulin polymerization by interacting at the colchicine site. *Proc Natl Acad Sci USA* 1994;91(9):3964–3968.
86. Fotsis T, Zhang Y, Pepper MS et al. The endogenous oestrogen metabolite 2-methoxyoestradiol inhibits angiogenesis and suppresses tumour growth. *Nature* 1994;368(6468):237–239.
87. Voest EE, Kenyon BM, O'Reilly MS, Truitt G, D'Amato RJ, Folkman J. Inhibition of angiogenesis in vivo by interleukin 12. *J Natl Cancer Inst* 1995;87(8):581–586.
88. Maione TE, Gray GS, Petro J et al. Inhibition of angiogenesis by recombinant human platelet factor-4 and related peptides. *Science* 1990;247(4938):77–79.
89. Colorado PC, Torre A, Kamphaus G et al. Anti-angiogenic cues from vascular basement membrane collagen. *Cancer Res* 2000;60(9):2520–2526.
90. Kamphaus GD, Colorado PC, Panka DJ et al. Canstatin, a novel matrix-derived inhibitor of angiogenesis and tumor growth. *J Biol Chem* 2000;275(2):1209–1215.
91. Maeshima Y, Colorado PC, Torre A et al. Distinct antitumor properties of a type IV collagen domain derived from basement membrane. *J Biol Chem* 2000;275(28):21340–21348.
92. Dawson DW, Volpert OV, Gillis P et al. Pigment epithelium-derived factor: a potent inhibitor of angiogenesis. *Science* 1999;285(5425):245–248.
93. Bouck N. Tumor angiogenesis: the role of oncogenes and tumor suppressor genes. *Cancer Cells* 1990;2(6):179–185.
94. O'Reilly M, Rosenthal R, Sage HE, Smith S, Holmgren L, Moses M, Shing Y, Folkman J. The suppression of tumor metastases by a primary tumor. *Surg Forum* 1993;44:474–476.
95. Holmgren L, O'Reilly MS, Folkman J. Dormancy of micrometastases: balanced proliferation and apoptosis in the presence of angiogenesis suppression. *Nat Med* 1995;1(2):149–153.
96. Lay AJ, Jiang XM, Kisker O et al. Phosphoglycerate kinase acts in tumour angiogenesis as a disulphide reductase. *Nature* 2000;408(6814):869–873.
97. Cao Y, O'Reilly MS, Marshall B, Flynn E, Ji RW, Folkman J. Expression of angiostatin cDNA in a murine fibrosarcoma suppresses primary tumor growth and produces long-term dormancy of metastases. *J Clin Invest* 1998;101(5):1055–1063.

97a. Camphausen K, Moses MA, Beecken WD, Khan MK, Folkman J, O'Reilly MS. Radiation therapy to a primary tumor accelerates metastatic growth in mice. 2001;61(5):2207–2211.

98. Kerbel RS, Viloria-Petit A, Okada F, Rak J. Establishing a link between oncogenes and tumor angiogenesis. *Mol Med* 1998;4(5):286–295.
99. Rak J, Yu JL, Kerbel RS, Coomber BL. What do oncogenic mutations have to do with angiogenesis/vascular dependence of tumors? *Cancer Res* 2002;62(7):1931–1934.
100. Okada F, Rak JW, Croix BS et al. Impact of oncogenes in tumor angiogenesis: mutant K-*ras* up-regulation of vascular endothelial growth factor/vascular permeability factor is necessary, but not sufficient for tumorigenicity of human colorectal carcinoma cells. *Proc Natl Acad Sci USA* 1998;95(7):3609–3614.

101. Petit AM, Rak J, Hung MC et al. Neutralizing antibodies against epidermal growth factor and ErbB-2/neu receptor tyrosine kinases down-regulate vascular endothelial growth factor production by tumor cells in vitro and in vivo: angiogenic implications for signal transduction therapy of solid tumors. *Am J Pathol* 1997;151(6):1523–1530.
102. Izumi Y, Xu L, di Tomaso E, Fukumura D, Jain RK. Tumour biology: herceptin acts as an anti-angiogenic cocktail. *Nature* 2002;416(6878):279–280.
103. Moscow J, Schneider E, Sikic BI, Morrow CS, Cowan KH. Drug resistance and its clinical circumvention. In: Kufe DW, Bast RC, Jr., Hite WH, Hong WK, Pollock RE, Weichselbaum RR, Holland JF, Frei E, III, eds. *Cancer medicine*. Hamilton: B.C. Decker Inc.; 2006:630–647.
104. Boehm T, Folkman J, Browder T, O'Reilly MS. Antiangiogenic therapy of experimental cancer does not induce acquired drug resistance. *Nature* 1997;390(6658):404–407.
105. Chambers AF, Naumov GN, Vantyghem SA, Tuck AB. Molecular biology of breast cancer metastasis. Clinical implications of experimental studies on metastatic inefficiency. *Breast Cancer Res* 2000;2(6):400–407.
106. Weiss L. Metastatic inefficiency. *Adv Cancer Res* 1990;54:159–211.
107. Fidler IJ. Antivascular therapy of cancer metastasis. *J Surg Oncol* 2006;94(3):178–180.
108. Naumov GN, Akslen LA, Folkman J. Role of angiogenesis in human tumor dormancy: animal models of the angiogenic switch. *Cell Cycle* 2006;5(16).
109. Murray C. Tumour dormancy: not so sleepy after all. *Nat Med* 1995;1(2):117–118.
110. Naumov GN, MacDonald IC, Weinmeister PM et al. Persistence of solitary mammary carcinoma cells in a secondary site: a possible contributor to dormancy. *Cancer Res* 2002;62(7):2162–2168.
111. Browder T, Butterfield CE, Kraling BM et al. Antiangiogenic scheduling of chemotherapy improves efficacy against experimental drug-resistant cancer. *Cancer Res* 2000;60(7):1878–1886.
112. Kerbel RS, Viloria-Petit A, Klement G, Rak J. 'Accidental' anti-angiogenic drugs. Anti-oncogene directed signal transduction inhibitors and conventional chemotherapeutic agents as examples. *Eur J Cancer* 2000;36(10):1248–1257.
113. Klement G, Baruchel S, Rak J et al. Continuous low-dose therapy with vinblastine and VEGF receptor-2 antibody induces sustained tumor regression without overt toxicity. *J Clin Invest* 2000;105(8):R15–R24.
114. Naumov GN, Townson JL, MacDonald IC et al. Ineffectiveness of doxorubicin treatment on solitary dormant mammary carcinoma cells or late-developing metastases. *Breast Cancer Res Treat* 2003;82(3):199–206.
115. Naumov GN, Beaudry P, Bender ER, Zurakowski D, Watnick R, Almog N, Heymach J, Folkman J. Suppression of circulating endothelial progenitor cells by human dormant breast cancer cells. *Clin Exp Met* 2004;21(7):636.
116. Harper J, Naumov GN, Exarhopoulos A, Bender E, Louis G, Folkman J, Moses MA. Predicting the switch to the angiogenic phenotype in a human tumor model. In: Proceedings of the American Association for *Cancer Res* 2006:837.
117. Klement G, Kikuchi L, Kieran M, Almog N, Yip T, Folkman J. Early tumor detection using platelet uptake of angiogenesis regulators. Blood 2004;104:239a, abstract 839.
118. Schuch G, Heymach JV, Nomi M et al. Endostatin inhibits the vascular endothelial growth factor-induced mobilization of endothelial progenitor cells. *Cancer Res* 2003;63(23):8345–8350.
119. Harker LA, Finch CA. Thrombokinetics in man. *J Clin Invest* 1969;48(6):963–974.
120. Italiano JE, Hartwig JH. Megakaryocyte development and platelet formation. In: Michelson AD, ed. *Platelets*. Boston: Academic Press; 2002:21–36.
121. Kaushansky K. Lineage-specific hematopoietic growth factors. *N Engl J Med* 2006; 354(19):2034–2045.

2 Vascular Endothelial Growth Factor: Basic Biology and Clinical Implications

Napoleone Ferrara

CONTENTS

The observation that tumor growth can be accompanied by increased vascularity was reported more than one century ago [for review, see (1)]. In 1939, Ide et al. postulated for the first time the existence of a tumor-derived blood vessel growth-stimulating factor (2). In 1945, Algire et al. advanced this concept, hypothesizing that rapid tumor growth is crucially dependent on the development of a neovascular supply (3). In 1971, Folkman (4) proposed that antiangiogenesis may be a valid strategy to treat human cancer and a search for regulators of angiogenesis that may also represent therapeutic targets began.

Neovascularization is essential also for physiological processes such as embryogenesis, tissue repair, and reproductive functions (5). The development of the vascular tree initially occurs by "vasculogenesis," the in situ differentiation of endothelial cell precursors, the angioblasts, from the hemangioblasts (6). The juvenile vascular system then evolves from the primary capillary plexus by subsequent pruning and reorganization of endothelial cells in a process called "angiogenesis" (7). Recent studies suggest that incorporation of bone marrow-derived endothelial progenitor cells (EPC) in the growing vessels complements the sprouting of resident endothelial cells (8–12). Additionally, a subset of perivascular monocytes seems to be particularly important for new vessel growth (13).

Many potential angiogenic factors have been described over the last two decades (14,15). Much evidence indicates that vascular endothelial growth factor (VEGF) is a particularly important regulator of angiogenesis (1). While new vessel growth and maturation are highly complex and coordinated processes, requiring the sequential activation of a series of receptors (e.g., Tie1, Tie2, and platelet-derived growth factor receptor (PDGFR-β)) by numerous ligands in endothelial and mural cells [for recent reviews, see (16,17)], VEGF action often represents a rate-limiting step in angiogenesis. VEGF (referred to also as VEGF-A) belongs to a gene family that includes placenta growth factor (PlGF) (18), VEGF-B (19), VEGF-C (20), and VEGF-D (21,22). VEGF-C and VEGF-D regulate lymphangiogenesis (23).

2.1 IDENTIFICATION OF VEGF

Independent lines of research contributed to the discovery of VEGF, emphasizing the biological complexity of this molecule (1).

In 1983, Senger et al. (24) described the identification in the conditioned medium of a guinea pig tumor cell line of a protein able to induce vascular leakage in the skin, which was named "tumor vascular permeability factor" (VPF). VPF was proposed to be a mediator of the high permeability of tumor blood vessels. However, these efforts did not yield the full purification of the VPF protein. Due to the lack of amino acid sequence information, VPF remained molecularly unknown and thus more definitive studies were not possible at that time.

In 1989, we reported the isolation of an endothelial cell mitogen from medium conditioned by bovine pituitary follicular cells, which we named "vascular endothelial growth factor" (VEGF) (25). NH_2-terminal amino acid sequencing proved that VEGF was distinct from the known endothelial cell mitogens and indeed did not match any known protein in available databases (25). Subsequently, Connolly et al. (26), following up on the work by Senger et al., independently reported the isolation and sequencing of VPF. cDNA cloning of VEGF (27) and VPF (28) revealed that VEGF and VPF were the same molecule. This was surprising, considering that other known endothelial cell mitogens (e.g., bFGF) do not increase vascular permeability.

2.2 BIOLOGICAL ACTIVITIES OF VEGF-A

VEGF-A stimulates the growth of vascular endothelial cells derived from arteries, veins, and lymphatics [for reviews, see (29,30)]. VEGF also induces angiogenesis in three-dimensional in vitro models (31). VEGF-A also induces angiogenesis in a variety of in vivo model systems (30).

VEGF-A is also an important survival factor for endothelial cells (32–35). VEGF prevents endothelial apoptosis induced by serum starvation. Such activity is mediated by the phosphatidylinositol (PI) 3′ kinase/Akt pathway (34,36). In addition, VEGF induces expression of the antiapoptotic proteins Bcl-2, A1 (33), XIAP (37), and survivin (38) in endothelial cells. In vivo, VEGF prosurvival effects are developmentally regulated. VEGF inhibition results in apoptotic changes in the vasculature of neonatal, but not adult mice (39). VEGF dependence has been demonstrated in endothelial cells of newly formed but not of established vessels within tumors (35,40).

Endothelial cells are the primary targets of VEGF-A, but several studies have reported mitogenic and nonmitogenic effects of VEGF-A also on certain nonendothelial cell types, including retinal pigment epithelial cells (41), pancreatic duct cells (42), and Schwann cells (43).

The earliest evidence that VEGF-A can affect blood cells was a report describing its ability to promote monocyte chemotaxis (44). Subsequently, VEGF-A was reported to have hematopoietic effects, inducing colony formation by mature subsets of granulocyte–macrophage progenitor cells (45). VEGF-deficient hematopoietic stem cells (HSCs) and bone marrow mononuclear cells fail to repopulate lethally irradiated hosts, despite coadministration of a large excess of wild-type cells (46).

As previously noted, VEGF is also known as VPF based on its ability to induce vascular leakage (24,47). Such permeability-enhancing activity underlies important roles of this molecule in inflammation and several pathological circumstances, including intraocular neovascular syndromes [reviewed in (48,49)].

2.3 VEGF ISOFORMS

Alternative exon splicing results in the generation of four different VEGF isoforms, that have respectively, 121, 165, 189, and 206 amino acids following signal sequence cleavage ($VEGF_{121}$, $VEGF_{165}$, $VEGF_{189}$, $VEGF_{206}$) (50,51). $VEGF_{165}$, the predominant isoform, lacks the residues encoded by exon 6, whereas $VEGF_{121}$ lacks the residues encoded by exons 6 and 7. Less frequent splice variants have also been reported, including $VEGF_{145}$ (52), $VEGF_{183}$ (53), $VEGF_{162}$ (54), and $VEGF_{165b}$ (55).

Native VEGF is a heparin-binding homodimeric glycoprotein of 45 kDa (25). Such properties closely correspond to those of $VEGF_{165}$, which is now recognized as the major VEGF isoform (56).

$VEGF_{121}$ is an acidic polypeptide, which fails to bind to heparin (56). $VEGF_{189}$ and $VEGF_{206}$ are highly basic and bind to heparin with high affinity (56). $VEGF_{121}$ is a freely diffusible protein. In contrast, $VEGF_{189}$ and $VEGF_{206}$ are almost completely sequestered in the extracellular matrix (ECM). $VEGF_{165}$ has intermediate properties, since it is secreted but a significant fraction remains bound to the cell surface and ECM (57). The ECM-bound isoforms may be released in a diffusible form by heparin or heparinase, which displaces them from their binding to heparin-like moieties, or by plasmin cleavage at the –COOH terminus, which generates a bioactive fragment consisting of the first 110 NH_2-terminal amino acids (56). Given the important role of plasminogen activation during physiological and pathological angiogeneisis processes (58), this proteolytic mechanism can be particularly important in regulating locally the activity and bioavailability of VEGF. More recent studies have shown that matrix metalloproteinase (MMP)-3 can also cleave $VEGF_{165}$ to generate diffusible, nonheparin binding, bioactive proteolytic fragments (59). In addition, Plouet et al. (60) have proposed a role for urokinase in the generation of bioactive VEGF.

2.4 REGULATION OF *VEGF* GENE EXPRESSION

2.4.1 Oxygen Tension

Oxygen tension plays a key role in regulating the expression of a variety of genes (61). VEGF mRNA expression is induced by exposure to low pO_2 in a variety of pathophysiological circumstances (62,63). A 28-base sequence has been identified in the 5′ promoter of the rat and human *VEGF* gene, which mediates hypoxia-induced transcription (64,65). Such a sequence represents a binding site for hypoxia-inducible factor 1 (HIF-1) (66). HIF-1 is a basic, heterodimeric, helix-loop-helix protein consisting of two subunits, HIF-1α and aryl hydrocarbon receptor nuclear translocator (ARNT), also known as HIF-1β (67). Recent studies have demonstrated the critical role of the product of the von Hippel-Lindau (*VHL*) tumor suppressor gene in HIF-1-dependent hypoxic responses [for review, see (68)]. The *VHL* gene is inactivated in patients with VHL disease, an autosomal dominant neoplasia syndrome characterized by capillary hemangioblastomas in retina and cerebellum, and in most sporadic clear cell renal carcinomas (69). The VHL protein is known to interact with a series of proteins including elongins B and C and CUL2, a member of the Cullin family (70). More recent studies demonstrated that indeed one of the functions of VHL is to be part of a ubiquitin ligase complex, which targets HIF subunits for proteasomal degradation (71,72). Oxygen promotes the hydroxylation of HIF at a proline residue (71,72). Recently, a family of prolyl hydroxylases related to *Egl-9 Caenorhabditis elegans* gene product was identified as HIF prolyl hydroxylases (61,73,74).

2.4.2 Growth Factors, Hormones, and Oncogenes

Several growth factors, including EGF, TGF-α, TGF-β, KGF, IGF-1, FGF, and PDGF, upregulate VEGF mRNA expression (75–77), suggesting that paracrine or autocrine release of such factors cooperates with local hypoxia in regulating VEGF release in the microenvironment. In addition, inflammatory cytokines such as IL-1-α and IL-6 induce expression of VEGF in several cell types, including synovial fibroblasts (78,79).

Hormones are also regulators of *VEGF* gene expression. Thyroid-stimulating hormone has been shown to induce *VEGF* expression in several thyroid carcinoma cell lines (80). Shifren et al. (81) have also shown that ACTH is able to induce *VEGF* expression in cultured human fetal adrenal cortical cells, suggesting that *VEGF* may be a local regulator of adrenal cortical angiogenesis and a mediator of the tropic action of ACTH.

A variety of transforming events also result in induction of *VEGF* gene expression. Oncogenic mutations or amplification of ras lead to *VEGF* upregulation (82,83). Mutations in the wnt-signaling pathway, which are frequently associated with premalignant colonic adenomas, result in upregulation of *VEGF* (84). Interestingly, *VEGF* is upregulated in polyps of Apc knockout [Apc(Delta716)] mice, a model for human familial adenomatous polyposis (85).

2.5 VEGF RECEPTORS

VEGF binds two highly related receptor tyrosine kinases (RTKs), VEGF receptor-1 (VEGFR-1) and VEGFR-2. Both VEGFR-1 and VEGFR-2 have seven immunoglobulin (Ig)-like domains in the extracellular domain, a single transmembrane region, and a consensus tyrosine kinase sequence, which is interrupted by a kinase-insert domain (86–88).

A member of the same family of RTKs is VEGFR-3 (Flt-4) (89) which, however, is not a receptor for VEGF-A, but instead binds the lymphangiogenic factors VEGF-C and VEGF-D (23). In addition to these RTKs, VEGF interacts with a family of coreceptors, the neuropilins.

2.5.1 VEGFR-1 (Flt-1)

Although Flt-1 (fms-like tyrosine kinase) was the first RTK to be identified as a VEGF receptor (92), the precise function of this molecule is still a subject of debate. VEGFR-1 binds not only VEGF-A but also PlGF (90) and VEGF-B (91), which in turn fails to bind VEGFR-2 Flt-1, reveals a weak tyrosine autophosphorylation in response to VEGF (92,93). Park et al. (90) initially proposed that VEGFR-1 may not primarily be a receptor transmitting a mitogenic signal, but rather a "decoy" receptor, able to regulate in a negative fashion the activity of VEGF on the vascular endothelium, by sequestering and rendering this factor less available to VEGFR-2. Thus, the observed potentiation of the action of VEGF by PlGF could be explained, at least in part, by displacement of VEGF from VEGFR-1 binding (90). Recent studies have shown that a synergism exists between VEGF and PlGF in vivo, especially during pathological situations, as evidenced by impaired tumorigenesis and vascular leakage in $PlGF^{-/-}$ mice (94).

Gene-targeting studies have demonstrated the essential role of VEGFR-1 during embryogenesis. VEGFR-$1^{-/-}$ mice die in utero between day 8.5 and 9.5 (95,96). Endothelial cells develop but fail to organize in vascular channels. Excessive proliferation of angioblasts has been reported to be responsible for such disorganization and lethality (96), indicating that, at least during early development, VEGFR-1 is a negative regulator of VEGF action. Recently, VEGFR-1 signaling has also been linked to the induction of MMP-9 in lung endothelial cells and to the facilitation of lung metastases (97).

Other studies have emphasized the effects of VEGFR-1 in hematopoiesis and recruitment of bone marrow–derived angiogenic cells (for more details, see Chapter 15). VEGFR-1 activation by PlGF reconstitutes hematopoiesis by recruiting VEGFR-1+HSC (98). In addition, VEGFR-1 activation by enforced expression of PlGF rescues survival and ability to repopulate in $VEGF^{-/-}$ HSC (46). LeCouter et al. (99) recently provided evidence for a novel function of VEGFR-1 in liver sinusoidal endothelial cells (LSEC). VEGFR-1 activation achieved with a receptor-selective VEGF mutant or PlGF resulted in the paracrine release of HGF, IL-6, and other hepatotrophic molecules by LSEC (99).

Furthermore, in some cases, VEGFR-1 is expressed by tumor cells and may mediate a chemotactic signal, thus potentially extending the role of this receptor (100). In this context, it is noteworthy that Bates et al. (101) reported that the epithelial–mesenchymal transformation of colonic organoids results in the increased expression of both VEGF and VEGFR-1, and that the survival of these cells depends on a VEGF/VEGFR-1 autocrine pathway.

2.5.2 VEGFR-2 (KDR, Human; Flk-1, Mouse)

VEGFR-2 binds VEGF-A with lower affinity than VEGFR-1 (K_d 75–250 pM versus 25 pM) (102–104). The key role of this receptor in developmental angiogenesis and hematopoiesis is evidenced by lack of vasculogenesis and failure to develop blood islands and organized blood vessels in Flk-1 null mice, resulting in death in utero between day 8.5 and 9.5 (105). There is now a general agreement that VEGFR-2 is the major mediator of the mitogenic, angiogenic, and permeability-enhancing effects of VEGF.

VEGFR-2 undergoes dimerization and strong ligand-dependent tyrosine phosphorylation in intact cells and results in a mitogenic, chemotactic, and prosurvival signal. Several tyrosine residues have been shown to be phosphorylated [for review, see (106)]. VEGFR-2 activation induces endothelial cell growth by activating the Raf–Mek–Erk pathway. VEGF mutants that bind selectively to VEGFR-2 are fully active endothelial cell mitogens, chemoattractants, and permeability-enhancing agents, whereas mutants specific for VEGFR-1 are devoid of all three activities (107). In addition, VEGF-E, a homolog of VEGF identified in the genome of the parapoxvirus Orf virus (108), which shows VEGF-like mitogenic and permeability-enhancing effects, binds and activates VEGFR-2 but fails to bind VEGFR-1 (109,110).

2.5.3 Neuropilin (NRP)1 and NRP2

Certain tumor and endothelial cells express cell-surface VEGF-binding sites distinct from the two known VEGF RTKs (111). $VEGF_{121}$ fails to bind these sites, indicating that exon 7 encoded basic sequences are required for binding to this putative receptor (111). Soker et al. (112) identified such isoform-specific VEGF receptor as NRP1, a molecule that had been previously shown to bind the collapsin/semaphorin family and was implicated in neuronal guidance [for review, see (113)]. When coexpressed in cells with VEGFR-2, NRP1 enhanced the binding of $VEGF_{165}$ to VEGFR-2 and $VEGF_{165}$-mediated chemotaxis (112). NRP1 appears to present $VEGF_{165}$ to the VEGFR-2 in a manner that enhances the effectiveness of VEGFR-2-mediated signal transduction (113).

2.6 ROLE OF VEGF IN PHYSIOLOGICAL ANGIOGENESIS

Inactivation of a single *VEGF* allele results in embryonic lethality between day 11 and 12, indicating that there is a critical *VEGF-A* gene-dosage requirement at least during development. In contrast, inactivation of *PlGF* (94) or *VEGF-B* (114) genes did not result in any major development abnormalities. Among the other members of the *VEGF* gene family, only *VEGF-C* plays an essential role in development, since its inactivation results in embryonic lethality following defective lymphatic development and fluid accumulation in tissues (115). *VEGF-A* plays an important role in early postnatal life. Administration of a soluble VEGFR-1 chimeric protein (39) or anti-VEGF-A monoclonal antibodies (116) results in growth arrest when the treatment is initiated at day 1 or day 8 postnatally. Such treatment is also accompanied by lethality, primarily due to inhibition of glomerular development and kidney failure (39). However, VEGF neutralization in fully developed normal mice (39) or rats (117) had no marked effects on glomerular function.

Endochondral bone formation is a fundamental mechanism for longitudinal bone growth. Cartilage, an avascular tissue, is replaced by bone in a process named endochondral ossification (118). VEGF-A mRNA is expressed by hypertrophic chondrocytes in the epiphyseal growth plate, suggesting that a VEGF gradient is needed for directional growth and cartilage invasion by metaphyseal blood vessels (119,120). Following VEGF blockade, blood vessel invasion is almost completely suppressed, concomitant with impaired trabecular bone formation, in developing mice and primates (119,121).

Angiogenesis is a key aspect of normal cyclical ovarian function. Follicular growth and the development of the corpus luteum (CL) are dependent on the proliferation of new capillary vessels (122).

Previous studies have shown that the VEGF-A mRNA expression is temporally and spatially related to the proliferation of blood vessels in the ovary (123,124). Administration of VEGF inhibitors delays follicular development (125) and suppresses luteal angiogenesis in rodents (126,127) as well as in primates (121,128). These studies have established that VEGF is the principle regulator of ovarian angiogenesis.

2.7 ROLE OF VEGF IN PATHOLOGIC CONDITIONS

2.7.1 Tumor Angiogenesis

2.7.1.1 Preclinical Studies

Many tumor cell lines secrete VEGF-A in vitro (129). In situ hybridization studies have demonstrated that the VEGF mRNA is expressed in the majority of human tumors (30).

In 1993, we reported that a monoclonal antibody targeting VEGF-A inhibits the growth of several tumor cell lines in nude mice, whereas the antibody had no effect on the proliferation of tumor cells in vitro (130). For a recent review of the preclinical efficacy of this antibody in various tumor models, see (131). Tumor growth inhibition was also demonstrated with other anti-VEGF treatments, including a retrovirus-delivered dominant negative VEGFR-2 mutant (132), small molecule inhibitors of VEGFR-2 signaling (133–135), antisense oligonucleotides (136,137), anti-VEGFR-2 antibodies (138), and soluble VEGF receptors (139–143).

While tumor cells usually represent the major source of VEGF, tumor-associated stroma is also an important site of VEGF production (141,144,145). Recent studies have shown that tumor-derived PDGF-A may be especially important for the recruitment of an angiogenic stroma (146). The growth of a variety of human tumor cells lines transplanted in nude mice is substantially reduced but not completely suppressed by antihuman VEGF-A monoclonal antibodies (130). Administration of mFlt(1–3)-IgG, a chimeric receptor containing the first three Ig-like domains of VEGFR-1, which binds both human and mouse VEGF (141), or recently described cross-reactive anti-VEGF-A monoclonal antibodies, results in a nearly complete suppression of growth in several tumor cell lines in immunodeficient mice (147). Similar results were obtained using a chimeric-soluble receptor consisting of domain 2 of VEGFR-1 fused with domain 3 of VEGFR-2, referred to as "VEGF-trap" (143). Therefore, the use of VEGF inhibitors that only target human VEGF in human xenograft models frequently results in underestimating the contribution of VEGF to the process of tumor angiogenesis.

Several studies have shown that combining anti-VEGF treatment with chemotherapy (148) or radiation therapy (149,150) results in a greater antitumor effects than either treatment alone. An issue that is debated is the mechanism of such potentiation and various hypotheses have been proposed. Klement et al. (148) proposed that chemotherapy, especially when delivered at close regular intervals using relatively low doses, with no prolonged drug-free break periods (metronomic therapy), preferentially damages endothelial cells in tumor blood vessels and that the simultaneous blockade of VEGF-A blunts a key survival signal for endothelial cells, thus selectively amplifying the endothelial cell–targeting effects of chemotherapy. A similar process, in principle, may take place when combining more conventional maximum tolerated dose chemotherapy regimens.

An alternative hypothesis has been proposed by Jain (151). Antiangiogenic agents would "normalize" the abnormal vasculature that is characteristic of many vessels in tumors, resulting in pruning of excessive endothelial and perivascular cells, in a drop in the normally high interstitial pressures detected in solid tumors, and temporarily improved oxygenation and delivery of chemotherapy to tumor cells (151). However, according to recent studies, the tumor vasculature can be "normalized" only transiently and eliciting synergistic effects through this mechanism requires administration of chemotherapy or radiation therapy, over a defined time window, after the angiogenesis inhibitor (152). Considering also that in most clinical protocols no such sequential administration is performed, it is unclear whether such a mechanism may account for the long-term beneficial effects of combination treatments observed in some clinical trials.

2.7.1.2 Clinical Trials in Cancer Patients with VEGF Inhibitors

Several VEGF inhibitors are undergoing clinical development as anticancer agents. These inhibitors include a humanized variant of and anti-VEGF monoclonal antibody that was used in early proof-of-concept studies (bevacizumab; Avastin) (153), an anti-VEGFR-2 antibody (138) and various small molecules inhibiting VEGFR-2 signal transduction (134,135), a VEGF receptor chimeric protein (143) [for recent reviews, see (154,155)].

The pivotal trial with bevacizumab that led to FDA approval was a large randomized placebo-controlled Phase III where bevacizumab was tested in combination with chemotherapy as a first-line therapy for previously untreated metastatic colorectal cancer (CRC) (156). Patients were randomized to receive weekly bolus irinotecan, 5-fluorouracil, and leucovorin (IFL) plus bevacizumab (5 mg/kg every two weeks), or IFL plus bevacizumab placebo. Survival was significantly increased in the IFL/Avastin arm compared to the IFL/placebo arm. Hypertension was more common in the IFL/bevacizumab-treated group, but was readily managed in all cases with oral antihypersensive agents (156). Bevacizumab was approved by the Food and Drug Administration (FDA) on February 26, 2004 as a first-line treatment for metastatic CRC in combination with 5-fluorouracil-based chemotherapy regimens (157). The role of bevacizumab in other tumor types and settings is currently under investigation and Phase III clinical trials in nonsmall cell lung cancer, renal cell cancer, and metastatic breast cancer are ongoing. Recently, an interim analysis of a Phase III study of women with previously untreated metastatic breast cancer treated with bevacizumab in combination with weekly paclitaxel chemotherapy showed that the study met its primary efficacy end point of improving progression-free survival, compared to paclitaxel alone (158).

In addition, administration of bevacizumab in combination with paclitaxel and carboplatin to patients with nonsmall cell lung carcinoma (NSCLC) resulted in increased response rate and time to progression relative to chemotherapy alone in a randomized Phase II trial (159). The most significant adverse event was serious hemoptysis. This was primarily associated with centrally located tumors with squamous histology, cavitation and central necrosis, and proximity of disease to large vessels (159).

Besides bevacizumab, several other VEGF inhibitors are clinically pursued [reviewed in (154,155)]. Among these, a variety of small molecule RTK inhibitors targeting the VEGF receptors are at different stages of clinical development. The most advanced are SU11248 and Bay 43-9006. SU11248 inhibits VEGFRs, PDGFR, c-kit, and Flt-3 (160) and has been reported to have considerable efficacy in imatinib-resistant gastrointestinal stromal tumor (161). An interim analysis of Phase III data indicates that Bay 43-9006 monotherapy results in a significant increase in progression-free survival in patients with advanced renal cell carcinoma (162), which has a similar spectrum of kinase inhibition as SU11248, has also shown promise in metastatic renal cell carcinoma in a Phase II monotherapy study (163).

PTK787 is also a VEGF RTK inhibitor in late-stage clinical trials (134). This molecule is in Phase III in CRC patients. Recently, interim findings of this trial have been presented (164). According to investigator-based assessment, there was a statistically significant increase in progression-free survival in patients treated with PTK787 in combination with FOLFOX4 chemotherapy compared to chemotherapy alone. However, a central review failed to document any significant difference.

2.8 INTRAOCULAR NEOVASCULAR SYNDROMES

Diabetes mellitus, occlusion of central retinal vein, or prematurity can all be associated with retinal ischemia and intraocular neovascularization, which may result in vitreous hemorrhages, retinal detachment, neovascular glaucoma, and blindness (165,166). Expression of VEGF mRNA spatially and temporally correlates with neovascularization in several animal models of retinal ischemia (32,167,168). Elevations of VEGF levels in the aqueous and vitreous of human eyes with proliferative retinopathy secondary to diabetes and other conditions have been previously described (169,170). Similar to the animal models, these studies demonstrated a temporal correlation between VEGF elevations and active proliferative retinopathy (169). Subsequently, animal studies using various VEGF inhibitors, including soluble VEGF receptor chimeric proteins (171), monoclonal antibodies (172), antisense oligonucleotides (173), and small molecule VEGFR-2 kinase inhibitors (174), have directly demonstrated the role of VEGF in ischemia-induced intraocular neovascularization.

Neovascularization and vascular leakage are a major cause of visual loss also in the wet form of age-related macular degeneration (AMD) (166). Anti-VEGF strategies are explored in clinical trials in AMD patients. The most clinically advanced VEGF inhibitors are an aptamer that selectively binds $VEGF_{165}$ (pegaptanib, Macugen) (175) and a recombinant humanized anti-VEGF Fab that neutralizes all VEGF-A isoforms and proteolytic fragments, which is derived from bevacizumab (ranibizumab, Lucentis) (176). Pegaptanib was approved by the FDA in December 2004 for the treatment of AMD, following Phase III studies showing that intraocular administrations of the drug reduced visual loss relative to placebo (177). Very recently, a controlled Phase III study shows that administration of ranibizumab not only maintains but also improves vision in patients with wet AMD (178). In June 2006, ranibizumab was approved by the FDA for the treatment of neovascular AMD.

2.9 PERSPECTIVES

Research conducted over the last 15 years has clearly established that VEGF plays an essential role in the regulation of embryonic and postnatal physiologic angiogenesis processes, such as normal growth processes (39,119) and cyclical ovarian function (126). Furthermore, VEGF inhibition has been shown to suppress pathological angiogenesis in a wide variety of preclinical models, including genetic models of cancer, leading to the clinical development of a variety of VEGF inhibitors. Definitive clinical studies have provided proof that VEGF inhibition, using bevacizumab in combination with chemotherapy, may provide a significant clinical benefit, including increased survival, in patients with previously untreated metastatic CRC (156). Ongoing clinical studies are testing the hypothesis that bevacizumab may have efficacy in other tumor types as well.

It would be of great importance to have reliable markers to monitor the activity of antiangiogenic drugs. So far, the absence of such biomarkers may have impaired clinical development of various antiangiogenic drugs. Circulating endothelial cells and their progenitor subset, MRI dynamic measurement of vascular permeability/flow in response to angiogenesis inhibitors are potential candidates, although their long-term predictive value remains to be established (179–181). Emphasizing such difficulties in identifying predictive markers, a recent study found that VEGF and thrombospondin expression or microvessel density in tumor sections do not correlate with clinical response to bevacizumab in patients with metastatic CRC and patients showed a survival benefit from the treatment irrespective of these parameters (182).

REFERENCES

1. Ferrara, N. VEGF and the quest for tumour angiogenesis factors. *Nat Rev Cancer*, *2*: 795–803, 2002.
2. Ide, A.G., Baker, N.H., and Warren, S.L. Vascularization of the Brown Pearce rabbit epithelioma transplant as seen in the transparent ear chamber. *Am J Roentgenol*, *42*: 891–899, 1939.
3. Algire, G.H., Chalkley, H.W., Legallais, F.Y., and Park, H.D. Vascular reactions of normal and malignant tissues in vivo. I. Vascular reactions of mice to wounds and to normal and neoplastic transplants. *J Natl Cancer Inst*, *6*: 73–85, 1945.
4. Folkman, J. Tumor angiogenesis: therapeutic implications. *N Engl J Med*, *285*: 1182–1186, 1971.
5. Folkman, J. Angiogenesis in cancer, vascular, rheumatoid and other disease. *Nat Med*, *1*: 27–31, 1995.
6. Risau, W. and Flamme, I. Vasculogenesis. *Ann Rev Cell Dev Biol*, *11*: 73–91, 1995.
7. Risau, W. Mechanisms of angiogenesis. *Nature*, *386*: 671–674, 1997.
8. Asahara, T., Murohara, T., Sullivan, A., Silver, M., van der Zee, R., Li, T., Witzenbichler, B., Schatteman, G., and Isner, J.M. Isolation of putative progenitor endothelial cells for angiogenesis. *Science*, *275*: 964–967, 1997.
9. Rafii, S., Meeus, S., Dias, S., Hattori, K., Heissig, B., Shmelkov, S., Rafii, D., and Lyden, D. Contribution of marrow-derived progenitors to vascular and cardiac regeneration. *Semin Cell Dev Biol*, *13*: 61–67, 2002.
10. De Palma, M., Venneri, M.A., Roca, C., and Naldini, L. Targeting exogenous genes to tumor angiogenesis by transplantation of genetically modified hematopoietic stem cells. *Nat Med*, *9*: 789–795, 2003.

11. Lyden, D., Hattori, K., Dias, S., Costa, C., Blaikie, P., Butros, L., Chadburn, A., Heissig, B., Marks, W., Witte, L., Wu, Y., Hicklin, D., Zhu, Z., Hackett, N.R., Crystal, R.G., Moore, M.A., Hajjar, K.A., Manova, K., Benezra, R., and Rafii, S. Impaired recruitment of bone-marrow-derived endothelial and hematopoietic precursor cells blocks tumor angiogenesis and growth. *Nat Med*, *7*: 1194–1201, 2001.
12. Ruzinova, M.B., Schoer, R.A., Gerald, W., Egan, J.E., Pandolfi, P.P., Rafii, S., Manova, K., Mittal, V., and Benezra, R. Effect of angiogenesis inhibition by Id loss and the contribution of bone-marrow-derived endothelial cells in spontaneous murine tumors. *Cancer Cell*, *4*: 277–289, 2003.
13. De Palma, M., Venneri, M.A., Galli, R., Sergi, L.S., Politi, L.S., Sampaolesi, M., and Naldini, L. Tie2 identifies a hematopoietic lineage of proangiogenic monocytes required for tumor vessel formation and a mesenchymal population of pericyte progenitors. *Cancer Cell*, *8*: 211–226, 2005.
14. Klagsbrun, M. and D'Amore, P.A. Regulators of angiogenesis. *Annu Rev Physiol*, *53*: 217–239, 1991.
15. Yancopoulos, G.D., Davis, S., Gale, N.W., Rudge, J.S., Wiegand, S.J., and Holash, J. Vascular-specific growth factors and blood vessel formation. *Nature*, *407*: 242–248, 2000.
16. Carmeliet, P. Angiogenesis in health and disease. *Nat Med*, *9*: 653–660, 2003.
17. Jain, R.K. Molecular regulation of vessel maturation. *Nat Med*, *9*: 685–693, 2003.
18. Maglione, D., Guerriero, V., Viglietto, G., Delli-Bovi, P., and Persico, M.G. Isolation of a human placenta cDNA coding for a protein related to the vascular permeability factor. *Proc Natl Acad Sci USA*, *88*: 9267–9271, 1991.
19. Olofsson, B., Pajusola, K., Kaipainen, A., von Euler, G., Joukov, V., Saksela, O., Orpana, A., Pettersson, R.F., Alitalo, K., and Eriksson, U. Vascular endothelial growth factor B, a novel growth factor for endothelial cells. *Proc Natl Acad Sci USA*, *93*: 2576–2581, 1996.
20. Joukov, V., Pajusola, K., Kaipainen, A., Chilov, D., Lahtinen, I., Kukk, E., Saksela, O., Kalkkinen, N., and Alitalo, K. A novel vascular endothelial growth factor, VEGF-C, is a ligand for the Flt4 (VEGFR-3) and KDR (VEGFR-2) receptor tyrosine kinases. *EMBO J*, *15*: 1751, 1996.
21. Orlandini, M., Marconcini, L., Ferruzzi, R., and Oliviero, S. Identification of a c-*fos*-induced gene that is related to the platelet-derived growth factor/vascular endothelial growth factor family [corrected; erratum to be published]. *Proc Natl Acad Sci USA*, *93*: 11675–11680, 1996.
22. Achen, M.G., Jeltsch, M., Kukk, E., Makinen, T., Vitali, A., Wilks, A.F., Alitalo, K., and Stacker, S.A. Vascular endothelial growth factor D (VEGF-D) is a ligand for the tyrosine kinases VEGF receptor 2 (Flk1) and VEGF receptor 3 (Flt4). *Proc Natl Acad Sci USA*, *95*: 548–553, 1998.
23. Karkkainen, M.J., Makinen, T., and Alitalo, K. Lymphatic endothelium: a new frontier of metastasis research. *Nat Cell Biol*, *4*: E2–E5, 2002.
24. Senger, D.R., Galli, S.J., Dvorak, A.M., Perruzzi, C.A., Harvey, V.S., and Dvorak, H.F. Tumor cells secrete a vascular permeability factor that promotes accumulation of ascites fluid. *Science*, *219*: 983–985, 1983.
25. Ferrara, N. and Henzel, W.J. Pituitary follicular cells secrete a novel heparin-binding growth factor specific for vascular endothelial cells. *Biochem Biophys Res Commun*, *161*: 851–858, 1989.
26. Connolly, D.T., Olander, J.V., Heuvelman, D., Nelson, R., Monsell, R., Siegel, N., Haymore, B.L., Leimgruber, R., and Feder, J. Human vascular permeability factor. Isolation from U937 cells. *J Biol Chem*, *264*: 20017–20024, 1989.
27. Leung, D.W., Cachianes, G., Kuang, W.J., Goeddel, D.V., and Ferrara, N. Vascular endothelial growth factor is a secreted angiogenic mitogen. *Science*, *246*: 1306–1309, 1989.
28. Keck, P.J., Hauser, S.D., Krivi, G., Sanzo, K., Warren, T., Feder, J., and Connolly, D.T. Vascular permeability factor, an endothelial cell mitogen related to PDGF. *Science*, *246*: 1309–1312, 1989.
29. Ferrara, N., Gerber, H.P., and LeCouter, J. The biology of VEGF and its receptors. *Nat Med*, *9*: 669–676, 2003.
30. Ferrara, N. Vascular endothelial growth factor: basic science and clinical progress. *Endocr Rev*, *25*: 581–611, 2004.
31. Pepper, M.S., Ferrara, N., Orci, L., and Montesano, R. Potent synergism between vascular endothelial growth factor and basic fibroblast growth factor in the induction of angiogenesis in vitro. *Biochem Biophys Res Commun*, *189*: 824–831, 1992.
32. Alon, T., Hemo, I., Itin, A., Pe'er, J., Stone, J., and Keshet, E. Vascular endothelial growth factor acts as a survival factor for newly formed retinal vessels and has implications for retinopathy of prematurity. *Nat Med*, *1*: 1024–1028, 1995.

33. Gerber, H.P., Dixit, V., and Ferrara, N. Vascular endothelial growth factor induces expression of the antiapoptotic proteins Bcl-2 and A1 in vascular endothelial cells. *J Biol Chem, 273*: 13313–13316, 1998.
34. Gerber, H.P., McMurtrey, A., Kowalski, J., Yan, M., Keyt, B.A., Dixit, V., and Ferrara, N. VEGF regulates endothelial cell survival by the PI3-kinase/Akt signal transduction pathway. Requirement for Flk-1/KDR activation. *J Biol Chem, 273*: 30336–30343, 1998.
35. Benjamin, L.E., Golijanin, D., Itin, A., Pode, D., and Keshet, E. Selective ablation of immature blood vessels in established human tumors follows vascular endothelial growth factor withdrawal [see comments]. *J Clin Invest, 103*: 159–165, 1999.
36. Fujio, Y. and Walsh, K. Akt mediates cytoprotection of endothelial cells by vascular endothelial growth factor in an anchorage-dependent manner. *J Biol Chem, 274*: 16349–16354, 1999.
37. Tran, J., Rak, J., Sheehan, C., Saibil, S.D., LaCasse, E., Korneluk, R.G., and Kerbel, R.S. Marked induction of the IAP family antiapoptotic proteins survivin and XIAP by VEGF in vascular endothelial cells. *Biochem Biophys Res Commun, 264*: 781–788, 1999.
38. Tran, J., Master, Z., Yu, J.L., Rak, J., Dumont, D.J., and Kerbel, R.S. A role for surviving in chemoresistance of endothelial cells mediated by VEGF. *Proc Natl Acad Sci USA, 99*: 4349–4354, 2002.
39. Gerber, H.P., Hillan, K.J., Ryan, A.M., Kowalski, J., Keller, G.-A., Rangell, L., Wright, B.D., Radtke, F., Aguet, M., and Ferrara, N. VEGF is required for growth and survival in neonatal mice. *Development, 126*: 1149–1159, 1999.
40. Yuan, F., Chen, Y., Dellian, M., Safabakhsh, N., Ferrara, N., and Jain, R.K. Time-dependent vascular regression and permeability changes in established human tumor xenografts induced by an anti-vascular endothelial growth factor/vascular permeability factor antibody. *Proc Natl Acad Sci USA, 93*: 14765–14770, 1996.
41. Guerrin, M., Moukadiri, H., Chollet, P., Moro, F., Dutt, K., Malecaze, F., and Plouet, J. Vasculotropin/vascular endothelial growth factor is an autocrine growth factor for human retinal pigment epithelial cells cultured in vitro. *J Cell Physiol, 164*: 385–394, 1995.
42. Oberg-Welsh, C., Sandler, S., Andersson, A., and Welsh, M. Effects of vascular endothelial growth factor on pancreatic duct cell replication and the insulin production of fetal islet-like cell clusters in vitro. *Mol Cell Endocrinol, 126*: 125–132, 1997.
43. Sondell, M., Lundborg, G., and Kanje, M. Vascular endothelial growth factor has neurotrophic activity and stimulates axonal outgrowth, enhancing cell survival and Schwann cell proliferation in the peripheral nervous system. *J Neurosci, 19*: 5731–5740, 1999.
44. Clauss, M., Gerlach, M., Gerlach, H., Brett, J., Wang, F., Familletti, P.C., Pan, Y.C., Olander, J.V., Connolly, D.T., and Stern, D. Vascular permeability factor: a tumor-derived polypeptide that induces endothelial cell and monocyte procoagulant activity, and promotes monocyte migration. *J Exp Med, 172*: 1535–1545, 1990.
45. Broxmeyer, H.E., Cooper, S., Li, Z.H., Lu, L., Song, H.Y., Kwon, B.S., Warren, R.E., and Donner, D.B. Myeloid progenitor cell regulatory effects of vascular endothelial cell growth factor. *Int J Hematol, 62*: 203–215, 1995.
46. Gerber, H.-P., Malik, A., Solar, G.P., Sherman, D., Liang, X.-H., Meng, G., Hong, K., Marsters, J., and Ferrara, N. Vascular endothelial growth factor regulates hematopoietic stem cell survival by an internal autocrine loop mechanism. *Nature, 417*: 954–958, 2002.
47. Dvorak, H.F., Brown, L.F., Detmar, M., and Dvorak, A.M. Vascular permeability factor/vascular endothelial growth factor, microvascular hyperpermeability, and angiogenesis. *Am J Pathol, 146*: 1029–1039, 1995.
48. Dvorak, H.F. Vascular permeability factor/vascular endothelial growth factor: a critical cytokine in tumor angiogenesis and a potential target for diagnosis and therapy. *J Clin Oncol, 20*: 4368–4380, 2002.
49. Weis, S.M. and Cheresh, D.A. Pathophysiological consequences of VEGF-induced vascular permeability. *Nature, 437*: 497–504, 2005.
50. Houck, K.A., Ferrara, N., Winer, J., Cachianes, G., Li, B., and Leung, D.W. The vascular endothelial growth factor family: identification of a fourth molecular species and characterization of alternative splicing of RNA. *Mol Endocrinol, 5*: 1806–1814, 1991.
51. Tischer, E., Mitchell, R., Hartman, T., Silva, M., Gospodarowicz, D., Fiddes, J.C., and Abraham, J.A. The human gene for vascular endothelial growth factor. Multiple protein forms are encoded through alternative exon splicing. *J Biol Chem, 266*: 11947–11954, 1991.

52. Poltorak, Z., Cohen, T., Sivan, R., Kandelis, Y., Spira, G., Vlodavsky, I., Keshet, E., and Neufeld, G. VEGF145, a secreted vascular endothelial growth factor isoform that binds to extracellular matrix. *J Biol Chem*, *272*: 7151–7158, 1997.
53. Jingjing, L., Xue, Y., Agarwal, N., and Roque, R.S. Human Muller cells express VEGF183, a novel spliced variant of vascular endothelial growth factor. *Invest Ophthalmol Vis Sci*, *40*: 752–759, 1999.
54. Lange, T., Guttmann-Raviv, N., Baruch, L., Machluf, M., and Neufeld, G. VEGF162, a new heparin-binding vascular endothelial growth factor splice form that is expressed in transformed human cells. *J Biol Chem*, *278*: 17164–17169, 2003.
55. Bates, D.O., Cui, T.G., Doughty, J.M., Winkler, M., Sugiono, M., Shields, J.D., Peat, D., Gillatt, D., and Harper, S.J. VEGF(165)b, an inhibitory splice variant of vascular endothelial growth factor, is down-regulated in renal cell carcinoma. *Cancer Res*, *62*: 4123–4131, 2002.
56. Houck, K.A., Leung, D.W., Rowland, A.M., Winer, J., and Ferrara, N. Dual regulation of vascular endothelial growth factor bioavailability by genetic and proteolytic mechanisms. *J Biol Chem*, *267*: 26031–26037, 1992.
57. Park, J.E., Keller, G.-A., and Ferrara, N. The vascular endothelial growth factor isoforms (VEGF): differential deposition into the subepithelial extracellular matrix and bioactivity of extracellular matrix-bound VEGF. *Mol Biol Cell*, *4*: 1317–1326, 1993.
58. Pepper, M.S. Extracellular proteolysis and angiogenesis. *Thromb Haemost*, *86*: 346–355, 2001.
59. Lee, S., Jilani, S.M., Nikolova, G.V., Carpizo, D., and Iruela-Arispe, M.L. Processing of VEGF-A by matrix metalloproteinases regulates bioavailability and vascular patterning in tumors. *J Cell Biol*, *169*: 681–691, 2005.
60. Plouet, J., Moro, F., Bertagnolli, S., Coldeboeuf, N., Mazarguil, H., Clamens, S., and Bayard, F. Extracellular cleavage of the vascular endothelial growth factor 189-amino acid form by urokinase is required for its mitogenic effect. *J Biol Chem*, *272*: 13390–13396, 1997.
61. Safran, M. and Kaelin, W.J., Jr. HIF hydroxylation and the mammalian oxygen-sensing pathway. *J Clin Invest*, *111*: 779–783, 2003.
62. Dor, Y., Porat, R., and Keshet, E. Vascular endothelial growth factor and vascular adjustments to perturbations in oxygen homeostasis. *Am J Physiol*, *280*: C1367–C1374, 2001.
63. Semenza, G.L. Angiogenesis in ischemic and neoplastic disorders. *Annu Rev Med*, *54*: 17–28, 2003.
64. Levy, A.P., Levy, N.S., Wegner, S., and Goldberg, M.A. Transcriptional regulation of the rat vascular endothelial growth factor gene by hypoxia. *J Biol Chem*, *270*: 13333–13340, 1995.
65. Liu, Y., Cox, S.R., Morita, T., and Kourembanas, S. Hypoxia regulates vascular endothelial growth factor gene expression in endothelial cells. Identification of a 5′ enhancer. *Circ Res*, *77*: 638–643, 1995.
66. Madan, A. and Curtin, P.T. A 24-base-pair sequence 3′ to the human erythropoietin gene contains a hypoxia-responsive transcriptional enhancer. *Proc Natl Acad Sci USA*, *90*: 3928–3932, 1993.
67. Wang, G.L. and Semenza, G.L. Purification and characterization of hypoxia-inducible factor 1. *J Biol Chem*, *270*: 1230–1237, 1995.
68. Mole, D.R., Maxwell, P.H., Pugh, C.W., and Ratcliffe, P.J. Regulation of H*IF* by the von Hippel-Lindau tumour suppressor: implications for cellular oxygen sensing. *IUBMB Life*, *52*: 43–47, 2001.
69. Lonser, R.R., Glenn, G.M., Walther, M., Chew, E.Y., Libutti, S.K., Linehan, W.M., and Oldfield, E.H. von Hippel-Lindau disease. *Lancet*, *361*: 2059–2067, 2003.
70. Lonergan, K.M., Iliopoulos, O., Ohh, M., Kamura, T., Conaway, R.C., Conaway, J.W., and Kaelin, W.G., Jr. Regulation of hypoxia-inducible mRNAs by the von Hippel-Lindau tumor suppressor protein requires binding to complexes containing elongins B/C and Cul2. *Mol Cell Biol*, *18*: 732–741, 1998.
71. Jaakkola, P., Mole, D.R., Tian, Y.M., Wilson, M.I., Gielbert, J., Gaskell, S.J., Kriegsheim, A., Hebestreit, H.F., Mukherji, M., Schofield, C.J., Maxwell, P.H., Pugh, C.W., and Ratcliffe, P.J. Targeting of HIF-alpha to the von Hippel-Lindau ubiquitylation complex by O_2-regulated prolyl hydroxylation. *Science*, *292*: 468–472, 2001.
72. Ivan, M., Kondo, K., Yang, H., Kim, W., Valiando, J., Ohh, M., Salic, A., Asara, J.M., Lane, W.S., and Kaelin, W.G., Jr. HIF-alpha targeted for VHL-mediated destruction by proline hydroxylation: implications for O_2 sensing. *Science*, *292*: 464–468, 2001.
73. Maxwell, P.H. and Ratcliffe, P.J. Oxygen sensors and angiogenesis. *Semin Cell Dev Biol*, *13*: 29–37, 2002.

74. Pugh, C.W. and Ratcliffe, P.J. Regulation of angiogenesis by hypoxia: role of the HIF system. *Nat Med*, *9*: 677–684, 2003.
75. Frank, S., Hubner, G., Breier, G., Longaker, M.T., Greenhalg, D.G., and Werner, S. Regulation of VEGF expression in cultured keratinocytes. Implications for normal and impaired wound healing. *J Biol Chem*, *270*: 12607–12613, 1995.
76. Pertovaara, L., Kaipainen, A., Mustonen, T., Orpana, A., Ferrara, N., Saksela, O., and Alitalo, K. Vascular endothelial growth factor is induced in response to transforming growth factor-beta in fibroblastic and epithelial cells. *J Biol Chem*, *269*: 6271–6274, 1994.
77. Warren, R.S., Yuan, H., Matli, M.R., Ferrara, N., and Donner, D.B. Induction of vascular endothelial growth factor by insulin-like growth factor 1 in colorectal carcinoma. *J Biol Chem*, *271*: 29483–29488, 1996.
78. Ben-Av, P., Crofford, L.J., Wilder, R.L., and Hla, T. Induction of vascular endothelial growth factor expression in synovial fibroblasts. *FEBS Lett*, *372*: 83–87, 1995.
79. Cohen, T., Nahari, D., Cerem, L.W., Neufeld, G., and Levi, B.Z. Interleukin 6 induces the expression of vascular endothelial growth factor. *J Biol Chem*, *271*: 736–741, 1996.
80. Soh, E.Y., Sobhi, S.A., Wong, M.G., Meng, Y.G., Siperstein, A.E., Clark, O.H., and Duh, Q.Y. Thyroid-stimulating hormone promotes the secretion of vascular endothelial growth factor in thyroid cancer cell lines. *Surgery*, *120*: 944–947, 1996.
81. Shifren, J.L., Mesiano, S., Taylor, R.N., Ferrara, N., and Jaffe, R.B. Corticotropin regulates vascular endothelial growth factor expression in human fetal adrenal cortical cells. *J Clin Endocrinol Metab*, *83*: 1342–1347, 1998.
82. Grugel, S., Finkenzeller, G., Weindel, K., Barleon, B., and Marme, D. Both v-Ha-Ras and v-Raf stimulate expression of the vascular endothelial growth factor in NIH 3T3 cells. *J Biol Chem*, *270*: 25915–25919, 1995.
83. Okada, F., Rak, J.W., Croix, B.S., Lieubeau, B., Kaya, M., Roncari, L., Shirasawa, S., Sasazuki, T., and Kerbel, R.S. Impact of oncogenes in tumor angiogenesis: mutant K-ras up-regulation of vascular endothelial growth factor/vascular permeability factor is necessary, but not sufficient for tumorigenicity of human colorectal carcinoma cells. *Proc Natl Acad Sci USA*, *95*: 3609–3614, 1998.
84. Zhang, X., Gaspard, J.P., and Chung, D.C. Regulation of vascular endothelial growth factor by the Wnt and K-ras pathways in colonic neoplasia. *Cancer Res*, *61*: 6050–6054, 2001.
85. Seno, H., Oshima, M., Ishikawa, T.O., Oshima, H., Takaku, K., Chiba, T., Narumiya, S., and Taketo, M.M. Cyclooxygenase 2- and prostaglandin E(2) receptor EP(2)-dependent angiogenesis in Apc(Delta716) mouse intestinal polyps. *Cancer Res*, *62*: 506–511, 2002.
86. Shibuya, M., Yamaguchi, S., Yamane, A., Ikeda, T., Tojo, A., Matsushime, H., and Sato, M. Nucleotide sequence and expression of a novel human receptor-type tyrosine kinase (flt) closely related to the fms family. *Oncogene*, *8*: 519–527, 1990.
87. Matthews, W., Jordan, C.T., Gavin, M., Jenkins, N.A., Copeland, N.G., and Lemischka, I.R. A receptor tyrosine kinase cDNA isolated from a population of enriched primitive hematopoietic cells and exhibiting close genetic linkage to c-kit. *Proc Natl Acad Sci USA*, *88*: 9026–9030, 1991.
88. Terman, B.I., Carrion, M.E., Kovacs, E., Rasmussen, B.A., Eddy, R.L., and Shows, T.B. Identification of a new endothelial cell growth factor receptor tyrosine kinase. *Oncogene*, *6*: 1677–1683, 1991.
89. Pajusola, K., Aprelikova, O., Korhonen, J., Kaipainen, A., Pertovaara, L., Alitalo, R., and Alitalo, K. FLT4 receptor tyrosine kinase contains seven immunoglobulin-like loops and is expressed in multiple human tissues and cell lines [published erratum appears in Cancer Res 1993; 53(16): 3845]. *Cancer Res*, *52*: 5738–5743, 1992.
90. Park, J.E., Chen, H.H., Winer, J., Houck, K.A., and Ferrara, N. Placenta growth factor. Potentiation of vascular endothelial growth factor bioactivity, in vitro and in vivo, and high affinity binding to Flt-1 but not to Flk-1/KDR. *J Biol Chem*, *269*: 25646–25654, 1994.
91. Olofsson, B., Korpelainen, E., Pepper, M.S., Mandriota, S.J., Aase, K., Kumar, V., Gunji, Y., Jeltsch, M.M., Shibuya, M., Alitalo, K., and Eriksson, U. Vascular endothelial growth factor B (VEGF-B) binds to VEGF receptor-1 and regulates plasminogen activator activity in endothelial cells. *Proc Natl Acad Sci USA*, *95*: 11709–11714, 1998.
92. de Vries, C., Escobedo, J.A., Ueno, H., Houck, K., Ferrara, N., and Williams, L.T. The fms-like tyrosine kinase, a receptor for vascular endothelial growth factor. *Science*, *255*: 989–991, 1992.

93. Waltenberger, J., Claesson Welsh, L., Siegbahn, A., Shibuya, M., and Heldin, C.H. Different signal transduction properties of KDR and Flt1, two receptors for vascular endothelial growth factor. *J Biol Chem*, *269*: 26988–26995, 1994.
94. Carmeliet, P., Moons, L., Luttun, A., Vincenti, V., Compernolle, V., De Mol, M., Wu, Y., Bono, F., Devy, L., Beck, H., Scholz, D., Acker, T., DiPalma, T., Dewerchin, M., Noel, A., Stalmans, I., Barra, A., Blacher, S., Vandendriessche, T., Ponten, A., Eriksson, U., Plate, K.H., Foidart, J.M., Schaper, W., Charnock-Jones, D.S., Hicklin, D.J., Herbert, J.M., Collen, D., and Persico, M.G. Synergism between vascular endothelial growth factor and placental growth factor contributes to angiogenesis and plasma extravasation in pathological conditions. *Nat Med*, *7*: 575–583, 2001.
95. Fong, G.H., Rossant, J., Gertsenstein, M., and Breitman, M.L. Role of the Flt-1 receptor tyrosine kinase in regulating the assembly of vascular endothelium. *Nature*, *376*: 66–70, 1995.
96. Fong, G.H., Zhang, L., Bryce, D.M., and Peng, J. Increased hemangioblast commitment, not vascular disorganization, is the primary defect in flt-1 knock-out mice. *Development*, *126*: 3015–3025, 1999.
97. Hiratsuka, S., Nakamura, K., Iwai, S., Murakami, M., Itoh, T., Kijima, H., Shipley, J.M., Senior, R.M., and Shibuya, M. MMP9 induction by vascular endothelial growth factor receptor-1 is involved in lung-specific metastasis. *Cancer Cell*, *2*: 289–300, 2002.
98. Hattori, K., Heissig, B., Wu, Y., Dias, S., Tejada, R., Ferris, B., Hicklin, D.J., Zhu, Z., Bohlen, P., Witte, L., Hendrikx, J., Hackett, N.R., Crystal, R.G., Moore, M.A., Werb, Z., Lyden, D., and Rafii, S. Placental growth factor reconstitutes hematopoiesis by recruiting VEGFR1(+) stem cells from bone-marrow microenvironment. *Nat Med*, *8*: 841–849, 2002.
99. LeCouter, J., Moritz, D.R., Li, B., Phillips, G.L., Liang, X.H., Gerber, H.P., Hillan, K.J., and Ferrara, N. Angiogenesis-independent endothelial protection of liver: role of VEGFR-1. *Science*, *299*: 890–893, 2003.
100. Wey, J.S., Fan, F., Gray, M.J., Bauer, T.W., McCarty, M.F., Somcio, R., Liu, W., Evans, D.B., Wu, Y., Hicklin, D.J., and Ellis, L.M. Vascular endothelial growth factor receptor-1 promotes migration and invasion in pancreatic carcinoma cell lines. *Cancer*, *104*: 427–438, 2005.
101. Bates, R.C., Goldsmith, J.D., Bachelder, R.E., Brown, C., Shibuya, M., Oettgen, P., and Mercurio, A.M. Flt-1-dependent survival characterizes the epithelial-mesenchymal transition of colonic organoids. *Curr Biol*, *13*: 1721–1727, 2003.
102. Terman, B.I., Dougher Vermazen, M., Carrion, M.E., Dimitrov, D., Armellino, D.C., Gospodarowicz, D., and Bohlen, P. Identification of the KDR tyrosine kinase as a receptor for vascular endothelial cell growth factor. *Biochem Biophys Res Commun*, *187*: 1579–1586, 1992.
103. Quinn, T.P., Peters, K.G., De Vries, C., Ferrara, N., and Williams, L.T. Fetal liver kinase 1 is a receptor for vascular endothelial growth factor and is selectively expressed in vascular endothelium. *Proc Natl Acad Sci USA*, *90*: 7533–7537, 1993.
104. Millauer, B., Wizigmann Voos, S., Schnurch, H., Martinez, R., Moller, N.P., Risau, W., and Ullrich, A. High affinity VEGF binding and developmental expression suggest Flk-1 as a major regulator of vasculogenesis and angiogenesis. *Cell*, *72*: 835–846, 1993.
105. Shalaby, F., Rossant, J., Yamaguchi, T.P., Gertsenstein, M., Wu, X.F., Breitman, M.L., and Schuh, A.C. Failure of blood-island formation and vasculogenesis in Flk-1-deficient mice. *Nature*, *376*: 62–66, 1995.
106. Matsumoto, T. and Claesson-Welsh, L. VEGF receptor signal transduction. *Science*'s *STKE*, *112 RE21*: 1–17, 2001.
107. Gille, H., Kowalski, J., Li, B., LeCouter, J., Moffat, B., Zioncheck, T., Pelletier, N., and Ferrara, N. Analysis of biological effects and signaling properties of Flt-1 (VEGFR-1) and KDR (VEGFR-2). A reassessment using novel receptor-specific VEGF mutants. *J Biol Chem*, *276*: 3222–3230, 2001.
108. Lyttle, D.J., Fraser, K.M., Flemings, S.B., Mercer, A.A., and Robinson, A.J. Homologs of vascular endothelial growth factor are encoded by the poxvirus orf virus. *J Virol*, *68*: 84–92, 1994.
109. Ogawa, S., Oku, A., Sawano, A., Yamaguchi, S., Yazaki, Y., and Shibuya, M. A novel type of vascular endothelial growth factor, VEGF-E (NZ-7 VEGF), preferentially utilizes KDR/Flk-1 receptor and carries a potent mitotic activity without heparin-binding domain. *J Biol Chem*, *273*: 31273–31282, 1998.

110. Wise, L.M., Veikkola, T., Mercer, A.A., Savory, L.J., Fleming, S.B., Caesar, C., Vitali, A., Makinen, T., Alitalo, K., and Stacker, S.A. Vascular endothelial growth factor (VEGF)-like protein from orf virus NZ2 binds to VEGFR2 and neuropilin-1. *Proc Natl Acad Sci USA*, *96*: 3071–3076, 1999.
111. Soker, S., Fidder, H., Neufeld, G., and Klagsbrun, M. Characterization of novel vascular endothelial growth factor (VEGF) receptors on tumor cells that bind VEGF165 via its exon 7-encoded domain. *J Biol Chem*, *271*: 5761–5767, 1996.
112. Soker, S., Takashima, S., Miao, H.Q., Neufeld, G., and Klagsbrun, M. Neuropilin-1 is expressed by endothelial and tumor cells as an isoform-specific receptor for vascular endothelial growth factor. *Cell*, *92*: 735–745, 1998.
113. Klagsbrun, M. and Eichmann, A. A role for axon guidance receptors and ligands in blood vessel development and tumor angiogenesis. *Cytokine Growth Factor Rev*, *16*: 535–548, 2005.
114. Bellomo, D., Headrick, J.P., Silins, G.U., Paterson, C.A., Thomas, P.S., Gartside, M., Mould, A., Cahill, M.M., Tonks, I.D., Grimmond, S.M., Townson, S., Wells, C., Little, M., Cummings, M.C., Hayward, N.K., and Kay, G.F. Mice lacking the vascular endothelial growth factor-B gene (VEGFB) have smaller hearts, dysfunctional coronary vasculature, and impaired recovery from cardiac ischemia. *Circ Res (Online)*, *86*: E29–E35, 2000.
115. Karkkainen, M.J., Haiko, P., Sainio, K., Partanen, J., Taipale, J., Petrova, T.V., Jeltsch, M., Jackson, D.G., Talikka, M., Rauvala, H., Betsholtz, C., and Alitalo, K. Vascular endothelial growth factor C is required for sprouting of the first lymphatic vessels from embryonic veins. *Nat Immunol*, *5*: 74–80, 2004.
116. Malik, A.K., Baldwin, M.E., Peale, F., Fuh, G., Liang, W.C., Lowman, H., Meng, G., Ferrara, N., and Gerber, H.P. Redundant roles of VEGF-B and PlGF during selective VEGF-A blockade in mice. *Blood*, *107*: 550–557, 2006.
117. Ostendorf, T., Kunter, U., Eitner, F., Loos, A., Regele, H., Kerjaschki, D., Henninger, D.D., Janjic, N., and Floege, J. VEGF(165) mediates glomerular endothelial repair. *J Clin Invest*, *104*: 913–923, 1999.
118. Poole, A.R. (ed.) Cartilage: Molecular Aspects, pp. 179–211. Boca Raton, FL: CRC Press, 1991.
119. Gerber, H.P., Vu, T.H., Ryan, A.M., Kowalski, J., Werb, Z., and Ferrara, N. VEGF couples hypertrophic cartilage remodeling, ossification and angiogenesis during endochondral bone formation. *Nat Med*, *5*: 623–628, 1999.
120. Carlevaro, M.F., Cermelli, S., Cancedda, R., and Descalzi Cancedda, F. Vascular endothelial growth factor (VEGF) in cartilage neovascularization and chondrocyte differentiation: auto-paracrine role during endochondral bone formation. *J Cell Sci*, *113*: 59–69, 2000.
121. Ryan, A.M., Eppler, D.B., Hagler, K.E., Bruner, R.H., Thomford, P.J., Hall, R.L., Shopp, G.M., and O'Neill, C.A. Preclinical safety evaluation of rhuMAbVEGF, an antiangiogenic humanized monoclonal antibody. *Toxicol Pathol*, *27*: 78–86, 1999.
122. Bassett, D.L. The changes in the vascular pattern of the ovary of the albino rat during the estrous cycle. *Am J Anat*, *73*: 251–278, 1943.
123. Phillips, H.S., Hains, J., Leung, D.W., and Ferrara, N. Vascular endothelial growth factor is expressed in rat corpus luteum. *Endocrinology*, *127*: 965–967, 1990.
124. Ravindranath, N., Little-Ihrig, L., Phillips, H.S., Ferrara, N., and Zeleznik, A.J. Vascular endothelial growth factor messenger ribonucleic acid expression in the primate ovary. *Endocrinology*, *131*: 254–260, 1992.
125. Zimmermann, R.C., Xiao, E., Husami, N., Sauer, M.V., Lobo, R., Kitajewski, J., and Ferin, M. Short-term administration of antivascular endothelial growth factor antibody in the late follicular phase delays follicular development in the rhesus monkey. *J Clin Endocrinol Metab*, *86*: 768–772, 2001.
126. Ferrara, N., Chen, H., Davis-Smyth, T., Gerber, H.-P., Nguyen, T.-N., Peers, D., Chisholm, V., Hillan, K.J., and Schwall, R.H. Vascular endothelial growth factor is essential for corpus luteum angiogenesis. *Nat Med*, *4*: 336–340, 1998.
127. Zimmermann, R.C., Hartman, T., Bohlen, P., Sauer, M.V., and Kitajewski, J. Preovulatory treatment of mice with anti-VEGF receptor 2 antibody inhibits angiogenesis in corpora lutea. *Microvasc Res*, *62*: 15–25, 2001.
128. Fraser, H.M., Dickson, S.E., Lunn, S.F., Wulff, C., Morris, K.D., Carroll, V.A., and Bicknell, R. Suppression of luteal angiogenesis in the primate after neutralization of vascular endothelial growth factor. *Endocrinology*, *141*: 995–1000, 2000.

129. Ferrara, N., Houck, K., Jakeman, L., and Leung, D.W. Molecular and biological properties of the vascular endothelial growth family of proteins. *Endocr Rev*, 18–32, 1992.
130. Kim, K.J., Li, B., Winer, J., Armanini, M., Gillett, N., Phillips, H.S., and Ferrara, N. Inhibition of vascular endothelial growth factor-induced angiogenesis suppresses tumor growth in vivo. *Nature*, *362*: 841–844, 1993.
131. Gerber, H.P. and Ferrara, N. Pharmacology and pharmacodynamics of bevacizumab as monotherapy or in combination with cytotoxic therapy in preclinical studies. *Cancer Res*, *65*: 671–680, 2005.
132. Millauer, B., Shawver, L.K., Plate, K.H., Risau, W., and Ullrich, A. Glioblastoma growth inhibited in vivo by a dominant-negative Flk-1 mutant. *Nature*, *367*: 576–579, 1994.
133. Strawn, L.M., McMahon, G., App, H., Schreck, R., Kuchler, W.R., Longhi, M.P., Hui, T.H., Tang, C., Levitzki, A., Gazit, A., Chen, I., Keri, G., Orfi, L., Risau, W., Flamme, I., Ullrich, A., Hirth, K.P., and Shawver, L.K. Flk-1 as a target for tumor growth inhibition. *Cancer Res*, *56*: 3540–3545, 1996.
134. Wood, J.M., Bold, G., Buchdunger, E., Cozens, R., Ferrari, S., Frei, J., Hofmann, F., Mestan, J., Mett, H., O'Reilly, T., Persohn, E., Rosel, J., Schnell, C., Stover, D., Theuer, A., Towbin, H., Wenger, F., Woods-Cook, K., Menrad, A., Siemeister, G., Schirner, M., Thierauch, K.H., Schneider, M.R., Drevs, J., Martiny-Baron, G., and Totzke, F. PTK787/ZK 222584, a novel and potent inhibitor of vascular endothelial growth factor receptor tyrosine kinases, impairs vascular endothelial growth factor-induced responses and tumor growth after oral administration. *Cancer Res*, *60*: 2178–2189, 2000.
135. Wedge, S.R., Ogilvie, D.J., Dukes, M., Kendrew, J., Curwen, J.O., Hennequin, L.F., Thomas, A.P., Stokes, E.S., Curry, B., Richmond, G.H., and Wadsworth, P.F. ZD4190: an orally active inhibitor of vascular endothelial growth factor signaling with broad-spectrum antitumor efficacy. *Cancer Res*, *60*: 970–975, 2000.
136. Saleh, M., Stacker, S.A., and Wilks, A.F. Inhibition of growth of C6 glioma cells in vivo by expression of antisense vascular endothelial growth factor sequence. *Cancer Res*, *56*: 393–401, 1996.
137. Oku, T., Tjuvajev, J.G., Miyagawa, T., Sasajima, T., Joshi, A., Joshi, R., Finn, R., Claffey, K.P., and Blasberg, R.G. Tumor growth modulation by sense and antisense vascular endothelial growth factor gene expression: effects on angiogenesis, vascular permeability, blood volume, blood flow, fluorodeoxyglucose uptake, and proliferation of human melanoma intracerebral xenografts. *Cancer Res*, *58*: 4185–4192, 1998.
138. Prewett, M., Huber, J., Li, Y., Santiago, A., O'Connor, W., King, K., Overholser, J., Hooper, A., Pytowski, B., Witte, L., Bohlen, P., and Hicklin, D.J. Antivascular endothelial growth factor receptor (fetal liver kinase 1) monoclonal antibody inhibits tumor angiogenesis. *Cancer Res*, *59*: 5209–5218, 1999.
139. Kong, H.L., Hecht, D., Song, W., Kovesdi, I., Hackett, N.R., Yayon, A., and Crystal, R.G. Regional suppression of tumor growth by in vivo transfer of a cDNA encoding a secreted form of the extracellular domain of the flt-1 vascular endothelial growth factor receptor. *Hum Gene Ther*, *9*: 823–833, 1998.
140. Goldman, C.K., Kendall, R.L., Cabrera, G., Soroceanu, L., Heike, Y., Gillespie, G.Y., Siegal, G.P., Mao, X., Bett, A.J., Huckle, W.R., Thomas, K.A., and Curiel, D.T. Paracrine expression of a native soluble vascular endothelial growth factor receptor inhibits tumor growth, metastasis, and mortality rate. *Proc Natl Acad Sci USA*, *95*: 8795–8800, 1998.
141. Gerber, H.P., Kowalski, J., Sherman, D., Eberhard, D.A., and Ferrara, N. Complete inhibition of rhabdomyosarcoma xenograft growth and neovascularization requires blockade of both tumor and host vascular endothelial growth factor. *Cancer Res*, *60*: 6253–6258, 2000.
142. Kuo, C.J., Farnebo, F., Yu, E.Y., Christofferson, R., Swearigen, R.A., Carter, R., von Recum, H.A., Yuan, J., Kumihara, J., Flynn, E., D'Amato, R., Folkman, J., and Mulligan, R.C. Comparative evaluation of the antitumor activity of antiangiogenic proteins delivered by gene transfer. *Proc Natl Acad Sci USA*, *98*: 4605–4610, 2001.
143. Holash, J., Davis, S., Papadopoulos, N., Croll, S.D., Ho, L., Russell, M., Boland, P., Leidich, R., Hylton, D., Burova, E., Ioffe, E., Huang, T., Radziejewski, C., Bailey, K., Fandl, J.P., Daly, T., Wiegand, S.J., Yancopoulos, G.D., and Rudge, J.S. VEGF-Trap: a VEGF blocker with potent antitumor effects. *Proc Natl Acad Sci USA*, *99*: 11393–11398, 2002.

144. Tsuzuki, Y., Fukumura, D., Oosthuyse, B., Koike, C., Carmeliet, P., and Jain, R.K. Vascular endothelial growth factor (VEGF) modulation by targeting hypoxia-inducible factor-1alpha→hypoxia response element→VEGF cascade differentially regulates vascular response and growth rate in tumors. *Cancer Res*, *60*: 6248–6252, 2000.
145. Kishimoto, J., Ehama, R., Ge, Y., Kobayashi, T., Nishiyama, T., Detmar, M., and Burgeson, R.E. In vivo detection of human vascular endothelial growth factor promoter activity in transgenic mouse skin. *Am J Pathol*, *157*: 103–110, 2000.
146. Dong, J., Grunstein, J., Tejada, M., Peale, F., Frantz, G., Liang, W.C., Bai, W., Yu, L., Kowalski, J., Liang, X., Fuh, G., Gerber, H.P., and Ferrara, N. VEGF-null cells require PDGFR alpha signaling-mediated stromal fibroblast recruitment for tumorigenesis. *EMBO J*, *23*: 2800–2810, 2004.
147. Liang, W.C., Wu, X., Peale, F.V., Lee, C.V., Meng, Y.G., Gutierrez, J., Fu, L., Malik, A.K., Gerber, H.P., Ferrara, N., and Fuh, G. Cross-species VEGF-blocking antibodies completely inhibit the growth of human tumor xenografts and measure the contribution of stromal VEGF. *J Biol Chem*, *281*: 951–961, 2006.
148. Klement, G., Baruchel, S., Rak, J., Man, S., Clark, K., Hicklin, D.J., Bohlen, P., and Kerbel, R.S. Continuous low-dose therapy with vinblastine and VEGF receptor-2 antibody induces sustained tumor regression without overt toxicity [see comments]. *J Clin Invest*, *105*: R15–R24, 2000.
149. Lee, C.G., Heijn, M., di Tomaso, E., Griffon-Etienne, G., Ancukiewicz, M., Koike, C., Park, K.R., Ferrara, N., Jain, R.K., Suit, H.D., and Boucher, Y. Anti-vascular endothelial growth factor treatment augments tumor radiation response under normoxic or hypoxic conditions. *Cancer Res*, *60*: 5565–5570, 2000.
150. Kozin, S.V., Boucher, Y., Hicklin, D.J., Bohlen, P., Jain, R.K., and Suit, H.D. Vascular endothelial growth factor receptor-2-blocking antibody potentiates radiation-induced long-term control of human tumor xenografts. *Cancer Res*, *61*: 39–44, 2001.
151. Jain, R.K. Normalization of tumor vasculature: an emerging concept in antiangiogenic therapy. *Science*, *307*: 58–62, 2005.
152. Winkler, F., Kozin, S.V., Tong, R.T., Chae, S.S., Booth, M.F., Garkavtsev, I., Xu, L., Hicklin, D.J., Fukumura, D., di Tomaso, E., Munn, L.L., and Jain, R.K. Kinetics of vascular normalization by VEGFR2 blockade governs brain tumor response to radiation: role of oxygenation, angiopoietin-1, and matrix metalloproteinases. *Cancer Cell*, *6*: 553–563, 2004.
153. Presta, L.G., Chen, H., O'Connor, S.J., Chisholm, V., Meng, Y.G., Krummen, L., Winkler, M., and Ferrara, N. Humanization of an anti-VEGF monoclonal antibody for the therapy of solid tumors and other disorders. *Cancer Res*, *57*: 4593–4599, 1997.
154. Gasparini, G., Longo, R., Toi, M., and Ferrara, N. Angiogenic inhibitors: a new therapeutic strategy in oncology. *Nat Clin Pract Oncol*, *2*: 562–577, 2005.
155. Ferrara, N. and Kerbel, R.S. Angiogenesis as a therapeutic target. *Nature*, *438*: 967–974, 2005.
156. Hurwitz, H., Fehrenbacher, L., Novotny, W., Cartwright, T., Hainsworth, H., Helm, W., Berlin, J., Baron, A., Griffing, S., Holmgren, E., Ferrara, N., Fyfe, G., Rogers, B., Ross, R., and Kabbinavar, F. Bevacizumab plus irinotecan, fluorouracil, and leucovorin for metastatic colorectal cancer. *N Engl J Med*, *350*: 2335–2342, 2004.
157. Ellis, L.M. Bevacizumab. *Nat Rev Drug Discov*, *Suppl*: S8–S9, 2005.
158. Miller, K. *Proc Am Soc Clin Oncol*, May 2005.
159. Johnson, D.H., Fehrenbacher, L., Novotny, W.F., Herbst, R.S., Nemunaitis, J.J., Jablons, D.M., Langer, C.J., DeVore, R.F., 3rd, Gaudreault, J., Damico, L.A., Holmgren, E., and Kabbinavar, F. Randomized phase II trial comparing bevacizumab plus carboplatin and paclitaxel with carboplatin and paclitaxel alone in previously untreated locally advanced or metastatic non-small-cell lung cancer. *J Clin Oncol*, *22*: 2184–2191, 2004.
160. Smith, J.K., Mamoon, N.M., and Duhe, R.J. Emerging roles of targeted small molecule protein-tyrosine kinase inhibitors in cancer therapy. *Oncol Res*, *14*: 175–225, 2004.
161. Maki, R.G., Fletcher, J.A., Heinrich, M.C., Morgan, J.A., George, S., Desai, J., Scheu, K., Fletcher, C.D., Baum, C., and Demetri, G.D. Results from a continuation trial of SU11248 in patients (pts) with imatinib (IM)-resistant gastrointestinal stromal tumor (GIST). ASCO Annual Meeting Proceedings, *Abstract* 9011, 2005.

162. Escudier, B., Szczylik, C., Eisen, T., Stadler, W.M., Schwartz, B., Shan, M., and Bukowski, R.M. Randomized Phase III trial of the Raf kinase and VEGFR inhibitor sorafenib (BAY 43-9006) in patients with advanced renal cell carcinoma (RCC). ASCO Annual Meeting Proceedings, *Abstract* 4510, 2005.
163. Rini, B., Rixe, O., Bukowski, R., Michaelson, M.D., Wilding, G., Hudes, G., Bolte, O., Steinfeldt, H., Reich, S.D., and Motzer, R. AG-013736, a multitarget tyrosine kinase receptor inhibitor, demonstrates anti-tumor activity in a Phase 2 study of cytokine-refractory, metastatic renal cell cancer (RCC). ASCO Annual Meeting Proceedings, *Abstract* 4509, 2005.
164. Hecht, J.R., Trarbach, T., Jaeger, E., Hainsworth, J., Wolff, R., Lloyd, K., Bodoky, G., Borner, M., Laurent, D., and Jacques, C. A randomized, double blind, placebo-controlled, phase III study in patients (Pts) with metastatic adenocarcinomas of the colon or rectum receiving first-line chemotherapy with oxaliplatin/5-fluorouracil/leucovorin and PTK787/ZK222584 or placebo (CONFIRM-1). ASCO Annual Meeting Proceedings, *Plen. Sess. I Abstract 3*, 2005.
165. Patz, A. Studies on retinal neovascularization. *Invest Ophthal Vis Sci*, *19*: 1133–1138, 1980.
166. Garner, A. Vascular diseases. In: A. Garner and G.K. Klintworth (eds.), *Pathobiology of Ocular Disease*, 2nd ed., pp. 1625–1710. New York: Marcel Dekker, 1994.
167. Miller, J.W., Adamis, A.P., Shima, D.T., D'Amore, P.A., Moulton, R.S., O'Reilly, M.S., Folkman, J., Dvorak, H.F., Brown, L.F., Berse, B., Yeo, T.K., and Yeo, K.T. Vascular endothelial growth factor/vascular permeability factor is temporally and spatially correlated with ocular angiogenesis in a primate model. *Am J Pathol*, *145*: 574–584, 1994.
168. Pierce, E.A., Foley, E.D., and Smith, L.E. Regulation of vascular endothelial growth factor by oxygen in a model of retinopathy of prematurity [see comments]. *Arch Ophthalmol*, *114*: 1219–1228, 1996.
169. Aiello, L.P., Avery, R.L., Arrigg, P.G., Keyt, B.A., Jampel, H.D., Shah, S.T., Pasquale, L.R., Thieme, H., Iwamoto, M.A., Park, J.E., Nguyen, H., Aiello, L.M., Ferrara, N., and King, G.L. Vascular endothelial growth factor in ocular fluid of patients with diabetic retinopathy and other retinal disorders [see comments]. *N Engl J Med*, *331*: 1480–1487, 1994.
170. Malecaze, F., Clemens, S., Simorer-Pinotel, V., Mathis, A., Chollet, P., Favard, P., Bayard, F., and Plouet, J. Detection of vascular endothelial growth factor mRNA and vascular endothelial growth factor-like activity in proliferative diabetic retinopathy. *Arch Ophthalmol*, *112*: 1476–1482, 1994.
171. Aiello, L.P., Pierce, E.A., Foley, E.D., Takagi, H., Chen, H., Riddle, L., Ferrara, N., King, G.L., and Smith, L.E. Suppression of retinal neovascularization in vivo by inhibition of vascular endothelial growth factor (VEGF) using soluble VEGF-receptor chimeric proteins. *Proc Natl Acad Sci USA*, *92*: 10457–10461, 1995.
172. Adamis, A.P., Shima, D.T., Tolentino, M.J., Gragoudas, E.S., Ferrara, N., Folkman, J., D'Amore, P.A., and Miller, J.W. Inhibition of vascular endothelial growth factor prevents retinal ischemia-associated iris neovascularization in a nonhuman primate. *Arch Ophthalmol*, *114*: 66–71, 1996.
173. Robinson, G.S., Pierce, E.A., Rook, S.L., Foley, E., Webb, R., and Smith, L.E. Oligodeoxynucleotides inhibit retinal neovascularization in a murine model of proliferative retinopathy. *Proc Natl Acad Sci USA*, *93*: 4851–4856, 1996.
174. Ozaki, H., Seo, M.S., Ozaki, K., Yamada, H., Yamada, E., Okamoto, N., Hofmann, F., Wood, J.M., and Campochiaro, P.A. Blockade of vascular endothelial cell growth factor receptor signaling is sufficient to completely prevent retinal neovascularization. *Am J Pathol*, *156*: 697–707, 2000.
175. Ruckman, J., Green, L.S., Beeson, J., Waugh, S., Gillette, W.L., Henninger, D.D., Claesson-Welsh, L., and Janjic, N. 2′-Fluoropyrimidine RNA-based aptamers to the 165-amino acid form of vascular endothelial growth factor (VEGF165). Inhibition of receptor binding and VEGF-induced vascular permeability through interactions requiring the exon 7-encoded domain. *J Biol Chem*, *273*: 20556–20567, 1998.
176. Chen, Y., Wiesmann, C., Fuh, G., Li, B., Christinger, H.W., McKay, P., de Vos, A.M., and Lowman, H.B. Selection and analysis of an optimized anti-VEGF antibody: crystal structure of an affinity-matured Fab in complex with antigen. *J Mol Biol*, *293*: 865–881, 1999.
177. Gragoudas, E.S., Adamis, A.P., Cunningham, E.T., Jr., Feinsod, M., and Guyer, D.R. Pegaptanib for neovascular age-related macular degeneration. *N Engl J Med*, *351*: 2805–2816, 2004.

178. Rosenfeld, P.J., Brown, D.M., Heier, J.S., Boyer, D.S., Kaiser, P.K., Chung, C.Y., and Kim, R.Y. Ranibizumab for neovascular age-related macular degeneration: 1-year results of the MARINA study. *N Engl J Med 355*: 1432–1444, 2006.
179. Morgan, B., Thomas, A.L., Drevs, J., Hennig, J., Buchert, M., Jivan, A., Horsfield, M.A., Mross, K., Ball, H.A., Lee, L., Mietlowski, W., Fuxuis, S., Unger, C., O'Byrne, K., Henry, A., Cherryman, G.R., Laurent, D., Dugan, M., Marme, D., and Steward, W.P. Dynamic contrast-enhanced magnetic resonance imaging as a biomarker for the pharmacological response of PTK787/ZK 222584, an inhibitor of the vascular endothelial growth factor receptor tyrosine kinases, in patients with advanced colorectal cancer and liver metastases: results from two phase I studies. *J Clin Oncol, 21*: 3955–3964, 2003.
180. Willett, C.G., Boucher, Y., di Tomaso, E., Duda, D.G., Munn, L.L., Tong, R.T., Chung, D.C., Sahani, D.V., Kalva, S.P., Kozin, S.V., Mino, M., Cohen, K.S., Scadden, D.T., Hartford, A.C., Fischman, A.J., Clark, J.W., Ryan, D.P., Zhu, A.X., Blaszkowsky, L.S., Chen, H.X., Shellito, P.C., Lauwers, G.Y., and Jain, R.K. Direct evidence that the VEGF-specific antibody bevacizumab has antivascular effects in human rectal cancer. *Nat Med, 10*: 145–147, 2004.
181. Shaked, Y., Bertolini, F., Man, S., Rogers, M.S., Cervi, D., Foutz, T., Rawn, K., Voskas, D., Dumont, D.J., Ben-David, Y., Lawler, J., Henkin, J., Huber, J., Hicklin, D.J., D'Amato, R.J., and Kerbel, R.S. Genetic heterogeneity of the vasculogenic phenotype parallels angiogenesis; implications for cellular surrogate marker analysis of antiangiogenesis. *Cancer Cell, 7*: 101–111, 2005.
182. Jubb, A.M., Hurwitz, H.I., Bai, W., Holmgren, E.B., Tobin, P., Guerrero, A.S., Kabbinavar, F., Holden, S.N., Novotny, W.F., Frantz, G.D., Hillan, K.J., and Koeppen, H. Impact of vascular endothelial growth factor-A expression, thrombospondin-2 expression, and microvessel density on the treatment effect of bevacizumab in metastatic colorectal cancer. *J Clin Oncol, 24*: 217–227, 2006.

3 Angiogenesis in Solid Tumors

Rakesh K. Jain and Dan G. Duda

CONTENTS

3.1 VASCULAR PATHOPHYSIOLOGY OF SOLID TUMORS

Angiogenesis is the process by which new blood vessels form from preexisting vessels. Vasculogenesis is the de novo formation of blood vessels from endothelial precursor cells. Under physiological conditions, the formation of new blood vessels during embryogenesis occurs predominantly by vasculogenesis, whereas in adults, angiogenesis is predominant and critical to wound healing, the menstrual cycle, and pregnancy [1].

Normal angiogenesis is initiated when angiogenic factors like vascular endothelial growth factor A (VEGF-A or VEGF) produced by growing, injured, or hypoxic tissues bind to their receptors on nearby vascular endothelial cells (ECs), resulting in endothelial cell activation [2,3]. Activated endothelial cells then secrete proteases that locally digest their basement membrane, opening the vessel wall architecture to active modeling. Endothelial cells proliferate and migrate through the new openings in the basement membrane, forming a sprout. As the sprout extends, it forms a lumen and connects with other developing vessels. As new vessels mature, they lay down basement membrane. They simultaneously signal for neighboring cells—pericytes and smooth muscle cells—to migrate in to stabilize their structure [1,4].

In contrast, pathological angiogenesis that accompanies tumor growth is a largely abnormal process [5–7]. Tumors begin their life cycle as avascular masses of proliferating neoplastic cells. These growing masses derive oxygen and nutrients from surrounding tissues and capillaries. At 1–3 mm in diameter, with its scavenged resources no longer sufficient, cellular mitosis and cell death (by apoptosis) start to occur at approximately similar rates, and a period of dormancy ensues. As cancer cells acquire additional mutations, aggressive clones develop and the tumor begins to overexpress various vascular growth signaling proteins in large amounts, mainly VEGF, which tip the balance of pro- and antiangiogenic factors toward angiogenesis (process referred to as the angiogenic switch) [8]. Since these growth factors are also chemoattractants for hematopoietic cells, a large number of inflammatory cells are recruited to the tumor site. The normal function of these cells is to recognize and destroy nonself and to subsequently promote wound healing. But since tumors are similar to "wounds that never heal" [9], the accumulation of growth factors and proteases released by inflammatory cells may actually promote tumor angiogenesis and growth [10–12]. Finally, fibroblasts are an important component of tumor angiogenesis. These cells migrate into tumor tissues, where they produce growth factors such as VEGF, and some become perivascular cells [13,14]. Other fibroblasts differentiate into myofibroblasts and release increased amounts of proteases [15]. The growth factors produced in tumors target receptors on adjacent vascular endothelial cells and initiate angiogenesis, attracting the ingrowth of vessels toward the hypoxic center of the expanding cancer. This process is facilitated by the locally increased proteolytic activity, which allows the dissociation and rearrangement of basement membrane and extracellular matrices. Once the new vessels are established, the tumor resumes growth.

3.1.1 Structural Abnormalities of Tumor Vasculature

Despite the critical role of blood vessels in tumor growth and metastasis, the structure and function of tumor vasculature is abnormal (Table 3.1 and Figure 3.1). The organized structure of the vascular network is lost. The system lacks defined arterioles, venules,

TABLE 3.1
Differences between Normal Vasculature and Tumor Vasculature

Normal Vasculature	Tumor Vasculature
Organized	Disorganized
Evenly distributed	Unevenly distributed
Uniformly shaped	Twisted
Nonpermeable	Leaky
Vascular pressure is greater than interstitial pressure	Vascular pressure is similar to tumor interstitial pressure
Properly matured	Immature
Supporting cells present (e.g., pericytes)	Supporting cells absent
Appropriate membrane protein expression	Inappropriate membrane protein expression
Independent of cell survival factors	Dependent on cell survival factors (e.g., VEGF)

Source: Reproduced from Jain, R.K., *Oncology* (*Williston Park*), 19(Suppl), 7, 2005. With permission.

and capillaries, and connections among vessels are sometimes incomplete. The vessels themselves are irregularly shaped with areas of dilation and constriction. Due to excessive endothelial cell proliferation during pathological angiogenesis, new vessels grow in tortuous, irregular shapes, with saccular ends [1]. Endothelial cell arrangement is abnormal with the cells separated by wide gaps at one location or stacked on one another nearby. Endothelial cells can lose their reactivity to common endothelial markers [16]. These activated endothelial cells are often "stretched out," and large gaps develop between tumor endothelial cells [17–21,129]. Frequently, tumor vessels present defective endothelium where the ECs and basement membrane are completely absent. Similarly, the patterns and functions of mural cells are also abnormal [4]. Insufficient numbers of pericytes migrate to support the new vessels, which remain immature. Tumor-associated pericytes demonstrate abnormal protein expression and morphology. Significantly, abnormal pericytes have a loose association with endothelial cells, contributing to the high vascular permeability. Increasing evidence has been offered to support the idea that blood circulating hematopoietic and mesenchymal precursor cells home to tumors to assume a perivascular cell phenotype [22–24], but their

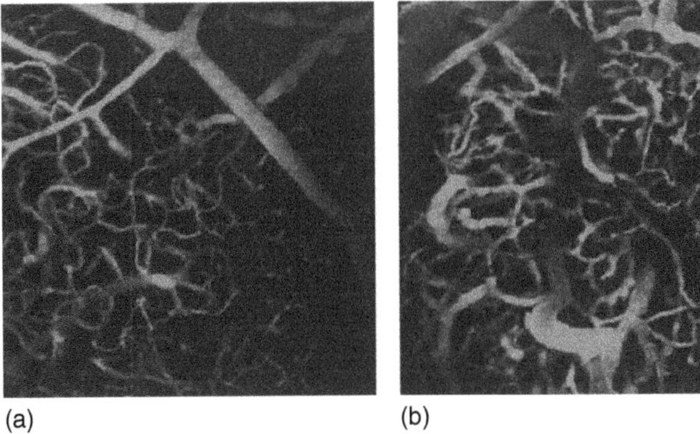

FIGURE 3.1 Normal and tumor vasculature. (a) Photomicrograph of normal vasculature in mouse brain. The vasculature is organized, appropriately connected, and optimally shaped to provide nutrients to all parenchymal cells. (b) Photomicrograph of mouse glioblastoma vasculature. This vasculature is disorganized, poorly connected, and tortuous, resulting in regions of hypoxia. (Reproduced from Jain, R.K., *Oncology* (*Williston Park*), 19(Suppl), 7, 2005. With permission.)

function as pericytes may be abnormal. Finally, the structure and the production of basement membrane, which is normally deposited between the ECs and perivascular cells, are also profoundly affected by the deregulated production of proteases in tumors. In some tumors, certain vessels completely lack a basement membrane [25], whereas in others the vessels have abnormally increased deposition of basement components such as collagen IV [26]. The differences between normal and abnormal vasculature are summarized in Table 3.1. All these structural irregularities translate into an abnormal function of the vasculature.

3.1.2 Functional Abnormalities of Tumor Vasculature

The structural abnormalities result in uneven tumor perfusion and high-tumoral interstitial fluid pressure (IFP) [27–29]. High-tumoral IFP is caused in part by tumor vessel hyperpermeability. In normal tissues, the vessel is able to maintain a gradient of fluid pressure from inside the vessel to the outside. In tumors, this gradient disappears and the pressure outside the blood vessels (IFP) tends to become equal to that inside, that is, microvascular pressure (MVP). Similarly, in normal tissues, the colloid osmotic pressure (osmotic pressure exerted by large proteins) inside blood vessels is much higher compared with that of outside. In tumors, these two become approximately equal because of vessel leakiness. The loss of these pressure gradients between the vessels and the tumor impedes the delivery of large molecular weight therapeutics to the tumor. Uneven tumor perfusion impedes the delivery of all blood-borne molecules, including oxygen and nutrients as well as chemotherapeutics. Without proper perivascular cell coverage, new vessels can collapse under the pressure created by proliferating cancer cells [30], and develop a haphazard pattern of interconnection, further compromising efficient blood flow. Tumor vessel hyperpermeability also contributes to sluggish blood flow in tumors, which results in regions of hypoxia and acidosis. Hypoxia contributes to resistance to some drugs and radiotherapy by decreasing the availability of reactive oxygen species. In addition, it can induce genetic instability and upregulate angiogenesis and metastasis genes [31]. Furthermore, both hypoxia and acidosis can impede the cytotoxic effects of immune cells infiltrating the tumor. Thus pathologic tumor vasculature results in conditions that protect the tumor from cytotoxic therapy and from host immune cells [28].

Thus, pathological angiogenesis impedes, protects, and promotes the cancer simultaneously.

3.1.3 Role of Vascularization in Tumor Progression and Relapse

As described earlier, progression of a tumor beyond a diameter of ~1–2 mm requires new blood vessels. As the inhibition of this process may prevent the growth of in situ or dormant tumors in the future [7], the current anticancer therapies call for strategies that would prevent or delay the local or distant invasion of the primary neoplasm. Angiogenesis is a valid target in these advanced tumors because continued tumor growth as well as metastatic dissemination and growth depend on the formation of new vessels as much as initial tumors do [5,32]. But with tumor progression, the genetic heterogeneity increases in parallel to the focal morphologic and functional changes. This is the result of acquired mutations during uncontrolled proliferation of the neoplastic cells [33]. In turn, this genetic instability results into an overexpression of multiple other proangiogenic factors, such as fibroblast growth factors (FGF), interleukin 8 (IL-8), stromal-derived factor 1 (SDF-1), and others. This directly implies that in locally advanced and metastatic tumors, an efficient blockade of angiogenesis is more likely to be achieved by blocking multiple angiogenic pathways, and not just by anti-VEGF monotherapy.

3.1.4 Conclusion

Like any organ, tumors require new vessel formation for growth [1]. In most solid tumors, the newly formed vessels are plagued by structural and functional abnormalities because of the continued and excessive exposure to angiogenic factors produced by the growing tumor [4]. Although abnormal, these new vessels allow tumor expansion at early stages of carcinogenesis, and progression from in situ lesions to locally invasive, and eventually to metastatic tumors. The hypothesis that tumor progression is dependent on neovascularization [5] was experimentally confirmed by a large and growing body of evidence reported in the literature of the last three decades. The known mechanisms of tumor neovascularization are discussed in the following section.

3.2 MECHANISMS OF NEOVASCULARIZATION IN TUMORS

New blood vessel formation in physiological and pathological conditions is a very complex process that involves the proliferation and migration of several types of cells and requires the expression of multiple growth factors. This active process may involve local, preexisting vasculature either by sprouting (a process termed angiogenesis) or by remodeling (referred to as intussusception). Alternatively, tumor neovascularization may rely on endothelial precursor cells from blood circulation (known as adult vasculogenesis).

3.2.1 Angiogenesis

Angiogenesis is considered the predominant mode of neovascularization in established tumors. In situ tumors become vascularized when the so-called angiogenic switch is turned on by tumor-derived factors. More specifically, angiogenesis occurs when the local balance between endogenous proangiogenic and antiangiogenic factors (Box 3.1 and Figure 3.2) is tipped in favor of the former. A multitude of factors affects tumor angiogenesis positively or negatively (Box 3.1 and Table 3.2, discussed in Section 3.4.11.2). A key event appears to be the local expression of MMP-9 (gelatinase B). In addition to its role in the breakdown of basement membrane (a critical step in EC sprouting), MMP-9 can release the VEGF which exists in the matrix and is bound to heparan sulfate [1,34]. Proteases may be secreted by ECs, cancer cells, mesenchymal cells, as well as immune cells [35–38]. Interestingly, fragments of MMP-9-digested collagen IV such as tumstatin are potent antiangiogenic factors, suggesting a dual role for this protease at different stages of tumor growth [39].

3.2.2 Intussusception

In addition to generating and modifying capillaries or terminal vessels during new vessel formation, the supplying vascular system is subjected to remodeling as well (Figure 3.1). Intussusception, that is, nonsprouting, transluminal pillar formation, is one essential mechanism for growth, arborization, bifurcation, remodeling, and pruning. Complex and efficient vascular beds can thus be generated by local interactions between vascular cells and hemodynamic conditions. Cytokines, which are critical for angiogenesis, also appear to be involved in intussusception: VEGF, angiopoietin-1 (Ang-1), platelet-derived growth factor B (PDGF-B), transforming growth factor-β (TGF-β), Notch 4, and others [40].

3.2.3 Vasculogenesis

In 1997, Isner and coworkers reported the existence of putative endothelial progenitor cells—angioblasts—in adults, and proposed the bone marrow as a source of these cells [41]. Initially, these ideas stirred intense controversy and generated great skepticism. Six years

BOX 3.1
Cellular Mechanisms of Tumor Angiogenesis

Tumor vessels can grow by sprouting, intussusception, or by incorporation of bone marrow–derived endothelial precursors. In addition, tumor cells can co-opt existing vessels. Several molecules have been implicated in these processes. During sprouting angiogenesis, vessels initially dilate and become leaky in response to VEGF. Ang-1 and the junctional molecules VE-cadherin and platelet–endothelial cell-adhesion molecule (PECAM) tighten vessels and their action needs to be overcome during angiogenesis. Ang-2 and proteinases mediate dissolution of the existing basement membrane and the interstitial matrix. Numerous molecules stimulate endothelial proliferation, migration, and assembly, including VEGF, Ang-1, and bFGF. Cell-matrix receptors such as the $\alpha_v\beta_3$ and $\alpha_5\beta_1$ integrins mediate cell spreading and migration. Maturation of nascent vessels involves formation of a new basement membrane and investment of new vessels with pericytes and smooth muscle cells. PDGF-B recruits smooth muscle cells, whereas signaling by TGF-β1 and Ang-1/Tie-2 stabilizes the interaction between endothelial and smooth muscle cells. Proteinase inhibitors (e.g., PAI-1) prevent degradation of the provisional extracellular matrix around nascent vessels. Maintenance of new vessels depends on the survival of endothelial cells. In a normal adult, quiescent endothelial cells can survive for several years. VEGF (through an interaction with VE-cadherin) and Ang-1 are vital survival factors. In contrast, most angiogenesis inhibitors cause endothelial apoptosis. By binding VEGF, soluble VEGFRs (e.g., VEGFR-1, neuropilin-1) reduce the angiogenic activity of VEGF. Molecules that initially induce angiogenesis are subsequently (proteolytically) processed to angiogenesis inhibitors, thereby providing a negative feedback. Most angiogenesis inhibitors suppress tumor angiogenesis, but can also affect normal adult vasculature [41b].

VEGF, bFGF, SDF-1, granulocyte macrophage–colony stimulating factor (GM-CSF), IGF-1, and angiopoietins have been implicated in the mobilization of endothelial precursors, whereas angiopoietins are important in vessel co-option. Multiple molecules are involved in tumor angiogenesis (among them, $\alpha_v\beta_3$, PAI-1, NO, cyclooxygenase-2 (COX-2), thrombospondin-2 (TSP-2), and a large list of angiogenesis inhibitors. The mechanism of action of some of these regulators is poorly understood. For instance, although proteinases might be expected to stimulate tumor angiogenesis by "clearing the path" for migrating endothelial cells, the proteinase inhibitor PAI-1 is a poor prognostic factor. Indeed, PAI-1 is required to prevent uncontrolled plasmin proteolysis, as this causes widespread matrix dissolution and prevents endothelial assembly. (Reproduced from Carmeliet, P. and Jain, R.K., *Nature*, 407, 249, 2000.)

later, the existence of endothelial progenitor cells and their involvement in vessel formation became widely accepted, but the controversy shifted to their role in solid tumors [42]. Emerging initially as an unexpected contribution of the bone marrow to tumor endothelium [43], the focus shifted to the molecular characterization of bone marrow–derived cells (BMDCs). Rafii and coworkers defined endothelial progenitor cells as VEGF receptor-2 (VEGFR-2)-positive BMDCs. However, these authors reported that VEGFR-1-positive BMDCs also home to tumors, contributing indirectly to neovascularization. They provided genetic evidence for the critical role of bone marrow in tumor neovascularization by rescuing tumor growth and angiogenesis in Id mutant mice—which have defective angiogenesis—by transplanting bone marrow from wild-type/nonmutant mice [44]. Simultaneously, Carmeliet and coworkers reported rescuing pathological angiogenesis in placental growth factor (PlGF) null mice by transplanting bone marrow from nonmutant mice [45]. More recent data

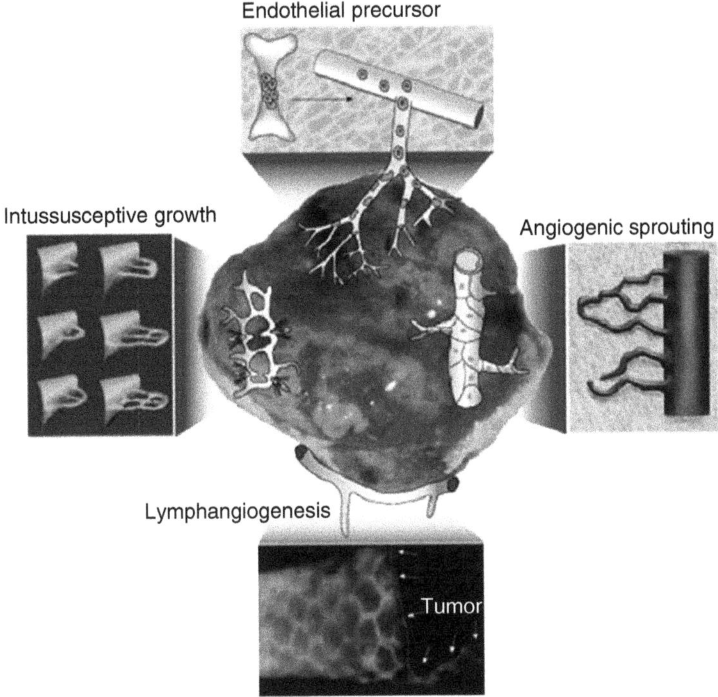

FIGURE 3.2 (See color insert following page 558.) Cellular mechanisms of tumor angiogenesis. Tumor vessels grow by various mechanisms: (1) the host vascular network expands by budding of endothelial sprouts or formation of bridges (angiogenesis); (2) tumor vessels remodel and expand by the insertion of interstitial tissue columns into the lumen of preexisting vessels (intussusception); and (3) endothelial cell precursors (angioblasts) home from the bone marrow or peripheral blood into tumors and contribute to the endothelial lining of tumor vessels (vasculogenesis). Lymphatic vessels around tumors drain the interstitial fluid and provide a gateway for metastasizing tumor cells. (Reproduced from Carmeliet, P. and Jain, R.K., *Nature*, 407, 249, 2000. With permission.)

suggested that hematopoietic cells such as myeloid precursors participate both in vasculogenesis formation of endothelial tubes [46–48] and angiogenesis (by production of growth factors and functioning as perivascular "support" cells [22,24,49]) (Figure 3.3). Vasculogenesis by endothelial BMDCs occurs in tumors, and varies with tumor type, site, and stage; however, this contribution is generally low (<10% [50]).

3.2.4 Conclusion

Neovascularization in tumors occurs via different mechanisms, which may include sprouting of new vessels from preexisting ones (angiogenesis), remodeling of the neighboring/existing vasculature (intussusception), or de novo formation of blood vessels by endothelial precursor cells (vasculogenesis). The relative contribution of each of these processes to neovascularization in tumors is largely uncharacterized, and may vary among tumors and during progression. Nevertheless, the molecular mechanisms driving these three modes of neovascularization are very similar, and will be discussed in the following section.

3.3 MOLECULAR REGULATION OF NEOVASCULARIZATION IN CANCER

New vessel formation in tumors is a complex phenomenon controlled by genetic and epigenetic events that take place in neoplastic and stromal cells. Neoplastic transformation

TABLE 3.2
Ligand and Receptor Signaling Pathways

Ligand/Receptor (Cell Type)	Putative Roles
VEGF/VEGFR-1, VEGFR-2 (EC)	Upregulates proteases for matrix organization Generates provisional matrix by increasing permeability Upregulates PDGF-B to recruit mural cells to stabilize vessels Suppresses apoptosis to stabilize vessels Induces EC specialization (such as vesiculo-vacuolar organelles (VVOs) and fenestration)
VEGF-154/VEGFR-2 and NRP1 (EC)	Promotes arterial growth
VEGFC/VEGFR-3 and NRP2	Guides lymphatic development
EG0EVGF/PKR1, PKR2 (EC)	Induces EC specialization in endocrine organs (such as fenestration)
Notch pathway (EC, mural cell)	Determines fate of the common progenitor cell (EC vs. mural cell?) Establishes vessel fate [artery vs. vein, upstream of ephrin signaling (in zebra fish)] Involved in tip cell formation
Ephrin B2/EphB4 (EC)	Determines arterial and venous EC specialization Guides vessel branching
PDGF/PDGFR (EC, mural cell)	Promotes proliferation, migration, and recruitment of mural cells
S1P1/EDG1 (EC, mural cell)	Promotes recruitment of mural cells (downstream of PDGF-B signaling)
Ang-1/Tie-2 (EC)	Stabilizes vessels by facilitating interaction (EC-mural cell and EC-matrix) Suppresses apoptosis of ECs Induces hierarchical arrangement of vascular branching in the absence of mural cells
Ang-2/Tie-2 (EC)	Induces apoptosis of ECs in the absence of VEGF Determines lymphatic patterning
Ang-1/Tie-1, Tie-2 (EC)	Coordinates vascular polarity
TGF-β1/TGF-βRII (EC, mural cell)	Promotes the production of ECM and proteases Promotes differentiation of fibroblast to myofibroblast to mural cell (through serum response factor)
TGF-β1/ALK1 (EC)	Regulates EC proliferation and migration (activation phase)
TGF-β1/ALK5 (EC)	Regulates vessel maturation (resolution phase)
TGF-β1/ALK1 and endoglin (EC)	Promotes arteriovenous specialization (through Notch/ephrin signaling)
Syk/SLP76 pathway	Separates lymphatic vessels from blood vessels
BDNF/TrkB (EC, hematopoietic precursor cells)	EC survival Promotes angiogenesis
Semaphorin4D/plexinD1 (EC)	Vascular patterning
ING4/NF-kB pathway	Inhibits angiogenesis
Slit2/Robo1 pathway (EC)	EC migration and tube formation

Molecules (Cell Type)	Putative Roles
Molecules governing cell–cell interactions	
VE-cadherin (EC)	Forms EC–EC junctions
N-cadherin (EC, mural cell)	Facilitates EC junctions–mural cell communication
Connexins (EC, mural cell)	Facilitates EC junctions–EC junctions and EC junctions–mural cell communications
Occludins, claudins, zona occludins (ZO1, 2, 3) (EC)	Form tight junctions in brain and retinal capillaries
CD148 (EC)	Regulates EC–mural cell interaction
Molecules governing cell–matrix interactions	
Integrins $\alpha_5\beta_1$, $\alpha_1\beta_1$, $\alpha_2\beta_1$, $\alpha_v\beta_3$, $\alpha_v\beta_5$	Suppress EC apoptosis Modulate EC migration and proliferation

TABLE 3.2 (continued)
Ligand and Receptor Signaling Pathways

Ligand/Receptor (Cell Type)	Putative Roles
Matrix metalloproteinases	Control degradation of basement membrane and extracellular matrix, increase vascular hyperpermeability, cause release of other angiogenic factors Provide cues for vascular patterning by releasing growth factors
Proteases (EC, mural cell)	Cleave matrix molecules (such as collagen XVIII to endostatin), plasma proteins (such as plasminogen to angiostatin), and protease molecules (such as MMP2 to PEX); cleaved products cause EC apoptosis
Protease inhibitors	Stabilize vessels by preventing dissolution of matrix
Thrombospondin-1, thrombospondin-2	Inhibit EC migration, growth, adhesion, survival
Netrin1-UNC-5B pathway	Causes endothelial filopodial retraction: role in vascular morphogenesis

Source: Adapted from Jain, R.K., *Nat. Med.*, 9, 685, 2003. With permission.

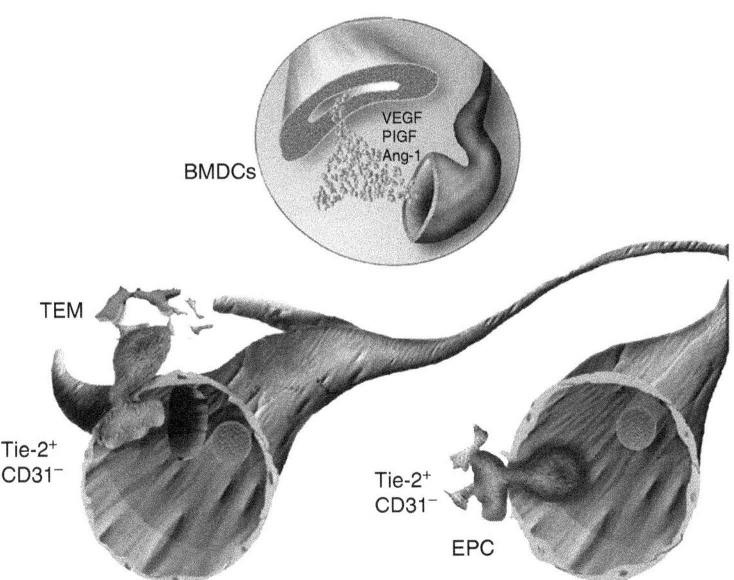

FIGURE 3.3 (See color insert following page 558.) Participation of bone marrow–derived cells in tumor neovascularization. Tumors overexpress angiogenic growth factors such as VEGF, PlGF, and Ang-1. These growth factors augment mobilization of bone marrow cells, which in turn facilitate the tumor neovascularization. These cell populations include VEGFR-2^+ endothelial progenitor cells (EPCs), which purportedly form tumor endothelium, and Tie-2^+ mononuclear cells (TEMs), which control indirectly the angiogenesis during tumor growth and liver regeneration. Two recent reports offer novel insights into the role of BMDCs in tumor angiogenesis. However, the data on BMDC incorporation in angiogenic vessels are conflicting, even for the same tumor cell lines. It is also not clear if these TEMs are the VEGFR-1^+ hematopoietic precursors previously described by Hattori et al. (2002) [174] as the stem cells responsible for hematopoietic reconstitution. Whereas the molecular definition of these cells, the extent and kinetics of their incorporation into vessel wall, and the mechanisms involved are largely unknown, these findings have multiple therapeutic implications and warrant urgent and careful mechanistic and phenotypic characterization. (Reproduced from Jain, R.K. and Duda, D.G., *Cancer Cell*, 3, 515, 2003. With permission.)

consists of inactivating mutations or deletion of tumor suppressor genes (such as *P53*, *PTEN*, *VHL*, *Smad4*) and hyperactivation of oncogenes (such as *Ras*, *Akt*, *mTOR*) [33]. These genetic alterations result in survival advantage and increased proliferation of the neoplastic cells. All these pathways are also involved in generating the increased production of proangiogenic molecules and the downregulation of endogenous inhibitors of angiogenesis. Tumor stromal cells such as fibroblasts, hematopoietic cells, pericytes, and endothelial cells themselves can produce proangiogenic factors, which may be equally critical during tumor growth and relapse after treatment [13,51,52].

3.3.1 Key Angiogenic Growth Factors Secreted in Tumors

Of all the known angiogenic molecules, VEGF (also referred to as VEGF-A) appears to be the most critical [1,6,53]. VEGF promotes the survival and proliferation of endothelial cells, increases the display of adhesion molecules on these cells, and increases vascular permeability. During mouse embryonic development, the exquisite regulation of VEGF expression sets in motion a chain of events that lead to the development of a mature vasculature from primordial cells [4]. Deletion of a single allele of VEGF results in embryonic lethality. So, too, does overexpression of VEGF. In adults, ectopic overexpression of VEGF results in a highly abnormal vasculature [54]. Collectively, these results indicate that the normal vasculature requires precise spatial and temporal control of VEGF levels.

VEGF is overexpressed in the majority of solid tumors, and plays critical roles in all forms of neovascularization in tumors. This highly pleiotropic angiogenic growth factor is a ligand for VEGFRs (VEGFR-1 and VEGFR-2) and neuropilin 1 (NP-1) [2]. VEGF expression is upregulated by tumor-derived signals, such as hypoxia, acidosis, inactivation of tumor suppressor genes, oncogene activation, and hormonal activity. VEGF can allow a tumor to progress out of dormancy and in sufficient quantities, acts as a survival factor for newly formed vessels. VEGF also promotes dysfunctional angiogenesis and poor blood flow, which is most pronounced in the tumor center [27].

Other VEGF family members include VEGF-B and the PlGFs (all of which specifically bind to VEGFR-1), and VEGF-C and VEGF-D (which are ligands for both VEGFR-2 and VEGFR-3). The latter two growth factors are characterized for their role in lymphangiogenesis by signaling via VEGFR-3 on lymphatic endothelial cells [55]. Nevertheless, they may play a role in tumor angiogenesis, as the processed form of these two cytokines binds to VEGFR-2 on endothelial cells [56]. Moreover, tumor endothelial cells have been reported to express VEGFR-3 [57].

Angiopoietins—Ang-1 and Ang-2—have been described as important players in angiogenesis and vascular remodeling in both embryos and adults [58,59]. Ang-1 and Ang-2 compete for binding to the Tie-2/Tek receptor on endothelial cells and appear to antagonize each other. The functions of two other family members, Ang-3 and Ang-4, are less well characterized.

Other families of factors play important roles in tumor angiogenesis depending on tumor type, stage, or treatment. They include IL-8, FGFs, PDGFs, SDF-1 or CXCL12, Eph/Ephrin receptors, Notch, netrin-Unc5b, certain integrins (such as $\alpha_v\beta_3$ and $\alpha_5\beta_1$), and brain-derived neurotrophic factor (BDNF), among others [1,60–66]. So far, clinical translation of antiangiogenic therapies has been successful primarily for anti-VEGF approaches (see Section 3.4). Nevertheless, blocking other pathways simultaneously may be critical in optimizing the effect of antiangiogenic therapy. It may also prevent tumor escape when antiangiogenic therapy is directed against one single factor. Therefore, a better understanding of VEGF-independent angiogenic pathways, as well as developing optimal approaches to target them in tumors, is warranted.

3.3.2 Growth Factor Receptor–Mediated Signaling Pathways in Tumor Endothelial Cells

Signaling through all the receptors for VEGFs may be active in endothelial cells in tumors. VEGF pathways are critical for developmental neovascularization, and is the only known factor to produce embryonic lethality after heterozygous deletion [67,68]. In fact, even mutations that result in subtle changes in VEGF expression, regardless of direction, result in embryonic lethality [69]. Similarly, embryos null for VEGFRs are embryonically lethal [70,71]. Generally, VEGFR-1 and VEGFR-2 are expressed on vascular endothelium, whereas VEGFR-3 is expressed on lymphatic endothelial cells, but this specificity does not apply to tumor endothelial cells. All these three VEGFRs are tyrosine kinase receptors, in contrast to NP-1 and NP-2, which lack an intracellular tyrosine kinase domain. VEGF homo- or heterodimers bind to VEGFR-1 or VEGFR-2, resulting in receptor dimerization and activation. NP-1 can associate with the VEGFRs and modulate their signaling [72]. This stimulates cellular signaling cascades that direct the migration and proliferation of vascular endothelial cells into organized vascular structures. Signaling via VEGFR-2 in endothelial cells is recognized as the main angiogenic pathway. Activation of this pathway leads to increased vascular permeability, endothelial cell survival and, to a lesser extent, proliferation. Signaling via VEGFR-1 is distinct from VEGFR-2 and is thought to be involved primarily in cell migration and production of proteases such as matrix metalloproteinase 9 (MMP-9 or gelatinase B [36]). Nevertheless, an interaction between VEGFR-1 and VEGFR-2 signaling has also been reported [73]. Several downstream pathways have been described for VEGFRs, some of which are common to VEGF–VEGFR and other pathways such as Ang-1–Tie-2, FGF–FGFR, and others. They include MAPK, Raf, PI3K, Src, and eNOS, among others, and their activation is responsible for the increase in vascular permeability, and the proliferation, survival, and migration of endothelial cells [21].

Tie-2 is a receptor ubiquitously expressed by endothelial cells, including endothelial cells in tumors. Ang-1 activation of Tie-2 conveys a potent survival signal, which mediated via PI3K. Both the Ang-1 and Tie-2 deficiencies cause embryonic lethality in mice [74,75]. Ang-2 is a hypoxia-responsive product, and it is believed to antagonize with the effect of Ang-1 (i.e., including induction of apoptosis in endothelial cells [76]). Nevertheless, it has been shown also to activate Tie-2, in a context-dependent manner [77,78].

Other proangiogenic factors affect, in addition to endothelial cells, a variety of other cells. Signaling by FGFs (which comprise over 20 family members) via the FGFRs potently induces endothelial cell proliferation. The chemokine IL-8 (also referred to as CXCL8) has been shown to be expressed in response to nuclear factor kappa B (NFκB) activation and to mediate angiogenesis in some tumors such as gliomas by signaling through IL-8R on tumor endothelial cells [79]. In the absence of hypoxia-inducible factor-1α (HIF-1α) IL-8 may be mediated by hypoxia and be an alternative angiogenic pathway to VEGF [60]. Another hypoxia-responsive factor is SDF-1, which may signal through its receptor CXCR4, and stimulates angiogenesis in a synergistic manner with VEGF [62]. The bidirectional Eph–Ephrin signaling in endothelial cells has also been shown to play a role in tumor angiogenesis, although the mechanisms are poorly understood [80]. Recently, Notch has been shown to be activated in endothelial cells via MAPK, and to be involved in tumor angiogenesis in multiple cancer types [65, summarized in 80b]. PDGF is another complex pathway that can be active in endothelial cells. PDGF receptor-α (PGDFR-α) is typically present on endothelial cells, but endothelial cells in some tumors may also express PGDFR-β [80c]. These two pathways may play important roles in tumor angiogenesis [61,81]. BDNF and neurotrophin-4 (NT-4) can act on their cognate receptor tyrosine kinase TrkB in endothelial cells and induce angiogenesis [66,82]. In fact, in addition to BDNF and Ephrins, a series of molecules common to neurogenesis and angiogenesis have been recently described for their role in vascular branching, and they include semaphorins, netrins, and slits [83]. Finally, certain integrins may play important roles in tumor angiogenesis. Integrins are heterodimeric receptors, which connect cells to extracellular matrices and transduce intracellular signals important for cell survival, adhesion, and migration. In tumors, expression of $\alpha_v\beta_3$, $\alpha_v\beta_5$, $\alpha_5\beta_1$, and $\alpha_6\beta_4$ has been described for both endothelial and tumor cells [84,85]. Surprisingly, integrin knockout mice showed enhanced pathological angiogenesis, whereas antiangiogenic results were obtained with the inhibitors [86]. Nevertheless, disruption of individual integrins using antibodies or small-molecule inhibitors produced encouraging results in animal models, and some of these drugs are now in clinical trials [87]. The results in clinical trials of the $\alpha_v\beta_5$ and $\alpha_v\beta_3$ antagonists will be the key in establishing the appropriate targeting approach for integrins.

3.3.3 Molecular Players in Mural Cell Biology in Tumors

Maturation of the wall involves recruitment of mural cells, development of the surrounding matrix and elastic laminae, and organ-specific specialization of endothelial cells, mural cells, and matrix (such as interendothelial junctions, fenestrations, apical–basal polarity, surface receptors, and foot processes; Figure 3.4). Maturation of the network involves optimal patterning of the network by branching, expanding, and pruning to meet local demands (Figure 3.5). Certain antiangiogenic therapy approaches target these pericytes, and thus interrupt the survival signals and structural support provided by these cells to endothelial cells. Nascent vessels are stabilized by recruiting mural cells and by generating an extracellular matrix. At least four molecular pathways are involved in regulating this process and may constitute valid targets: PDGF-B–PDGFR-β; sphingosine-1-phosphate-1 (S1P1)–endothelial differentiation sphingolipid G-protein-coupled receptor-1 (EDG1); Ang-1–Tie-2; and the TGF-β pathway. PDGF-B is secreted by endothelial cells, presumably in response to VEGF, and facilitates recruitment of mural cells. Although PDGF-B is expressed by a number of cells, including endothelial cells and mural cells, signaling through PDGFR-β, which is expressed on mural cells, is responsible for their proliferation and migration during vascular maturation. Compelling support for this hypothesis comes from studies of *Pdgfb* knockout mice, which undergo embryonic lethality, lack pericytes in certain vessels, and exhibit microvascular aneurysm.

The similarity between the phenotypes of *Pdgfb–Pdgfrb* and *Edg1* knockout mice (failure of mural cells to migrate to blood vessels) indicates that signaling through the EDG1 receptor, which is expressed by mural cells, is another key pathway for mural cell recruitment [4]. EDG1 receptor signaling may occur downstream of PDGF signaling, although this hypothesis has recently been questioned. Alternatively, the lack of EDG1 may alter the endothelial cell matrix production or endothelial–mural cell interaction, and interfere with vessel maturation.

In addition, critical for vessel formation and stabilization are the Tie-2 receptor and its two ligands, Ang-1 and Ang-2. Main sources of Ang-1 and Ang-2 are the mural cells and endothelial cells, respectively. Ang-1 is known to stabilize nascent vessels and make them leak-resistant, presumably by facilitating communication between endothelial cells and mural cells. Notably, in the absence of mural cells, recombinant Ang-1 restored a hierarchical order of the larger vessels, and rescued edema and hemorrhage, in the growing retinal vasculature of mouse neonates [88]. Thus, the mechanism of vessel maturation by Ang-1 is far from clear. The role of Ang-2 appears to be contextual. In the absence of VEGF, Ang-2 acts as an antagonist of Ang-1 and destabilizes vessels, ultimately leading to vessel regression. In the presence of VEGF, Ang-2 facilitates vascular sprouting.

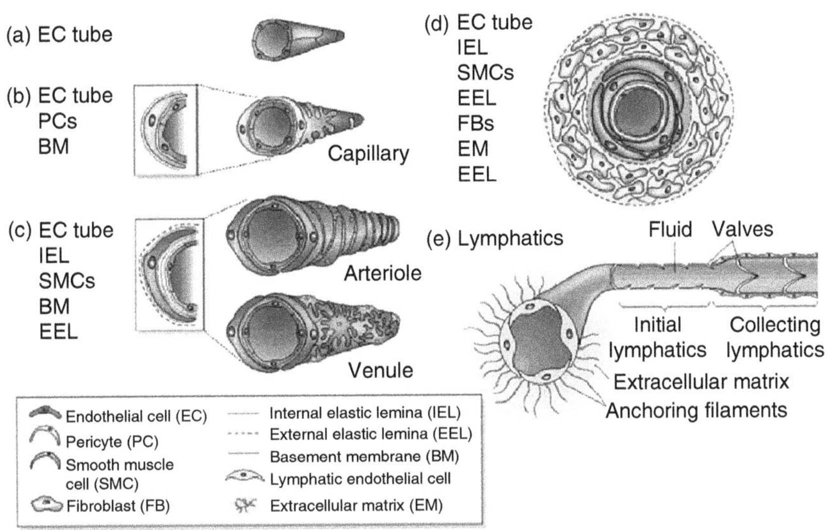

FIGURE 3.4 (See color insert following page 558.) Wall composition of nascent vs. mature vessels. (a) Nascent vessels consist of a tube of ECs. These mature into the specialized structures of capillaries, arteries, and veins. (b) Capillaries, the most abundant vessels in our body, consist of ECs surrounded by basement membrane and a sparse layer of pericytes embedded within the EC basement membrane. Due to their wall structure and large surface-area-to-volume ratio, these vessels form the main site of exchange of nutrients between blood and tissue. Depending on the organ or tissue, the capillary endothelial layer is continuous (as in muscle), fenestrated (as in kidney or endocrine glands), or discontinuous (as in liver sinusoids). The endothelia of the blood–brain barrier or blood–retina barrier are further specialized to include tight junctions, and are thus impermeable to various molecules. (c) Arterioles and venules have an increased coverage of mural cells compared with capillaries. Precapillary arterioles are completely invested with vascular SMCs, which form their own basement membrane and are circumferentially arranged, closely packed, and tightly associated with the endothelium. Extravasation of macromolecules and cells from the blood stream typically occurs from postcapillary venules. (d) The walls of larger vessels consist of three specialized layers: an intima composed of endothelial cells, a media of SMCs, and an adventitia of fibroblasts, together with matrix and elastic laminae. The advential layer has its own blood supply—known as *vasa vasorum*—which extends in part into the media. SMCs and elastic laminae contribute to the vessel tone and mediate the control of vessel diameter and blood flow. Additional control of blood flow is provided by arteriovenous shunts, which can divert blood away from a capillary bed when necessary. (e) Lymphatic capillaries lack pericytes. Larger (collecting) lymphatic vessels are invested in a basement membrane and contain valves that permit lymph flow only in one (proximal) direction; the lymphatic capillaries (initial lymphatics) contain microvalves in their walls. The lymphatic endothelial cells are connected to the surrounding connective tissue through anchoring filaments. (Reproduced from Jain, R.K., *Nat. Med.*, 9, 685, 2003. With permission.)

TGF-β, a multifunctional cytokine, promotes vessel maturation by stimulating extracellular matrix production and by inducing differentiation of mesenchymal cells to mural cells [89]. It is expressed in a number of cell types, including endothelial cells and mural cells and, depending on the context and concentration, is both pro- and antiangiogenic [90].

In tumors, the data on the mural cell coverage of vessels are somewhat controversial, with some studies showing a paucity of mural cells and others showing their abundance [19]. Whether this is due to the different tumor types examined or the different molecular markers (such as smooth muscle actin and desmin) used to identify these cells is unknown: also unknown is the origin of the mural cells. One possibility is that fibroblasts at the tumor–host interface are triggered by components of the tumor microenvironment (such as TGF-β) to differentiate first into myofibroblasts, and then into pericyte-like cells. Indeed, these

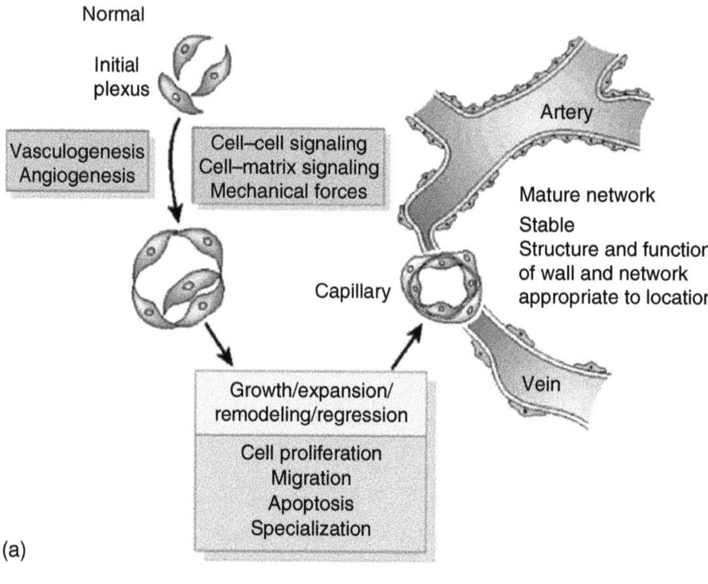

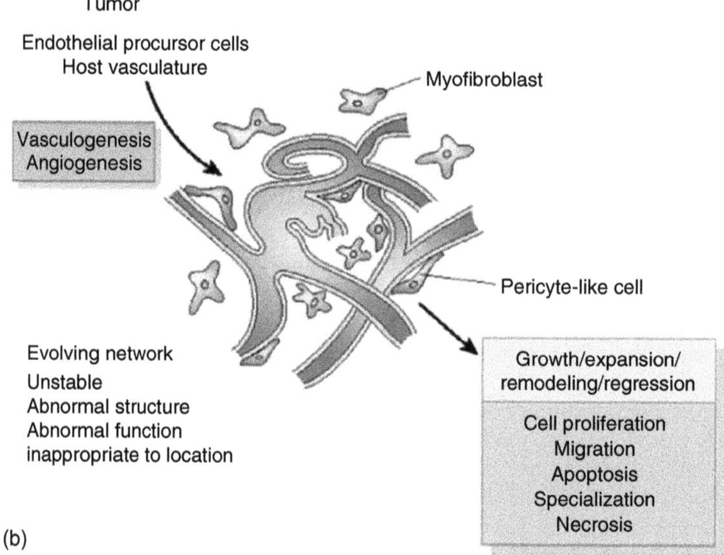

FIGURE 3.5 (See color insert following page 558.) Steps in network formation and maturation during embryonic (physiological) angiogenesis (a) and tumor (pathological) angiogenesis (b). The nascent vascular network forms from an initial cell plexus by processes of vasculogenesis or angiogenesis. This is regulated by cell–cell and cell–matrix signaling molecules and mechanical forces, as its further growth and expansion of the network (by proliferating and migrating cells) alongside its normal remodeling by cell death (apoptosis). Ordered patterns of growth, organization, and specialization (including the investment of vascular channels by mural cells) produce mature networks of arteries, capillaries, and veins—networks that are structurally and functionally stable and appropriate to organ and location. (b) ECs and mural cells derived from circulating precursor cells or from the host vasculature form networks, which are structurally and functionally abnormal. Continual remodeling by inappropriate patterns of growth and regression (cell apoptosis and necrosis) contributes to the instability of these networks. (Reproduced from Jain, R.K., *Nat. Med.*, 9, 685, 2003. With permission.)

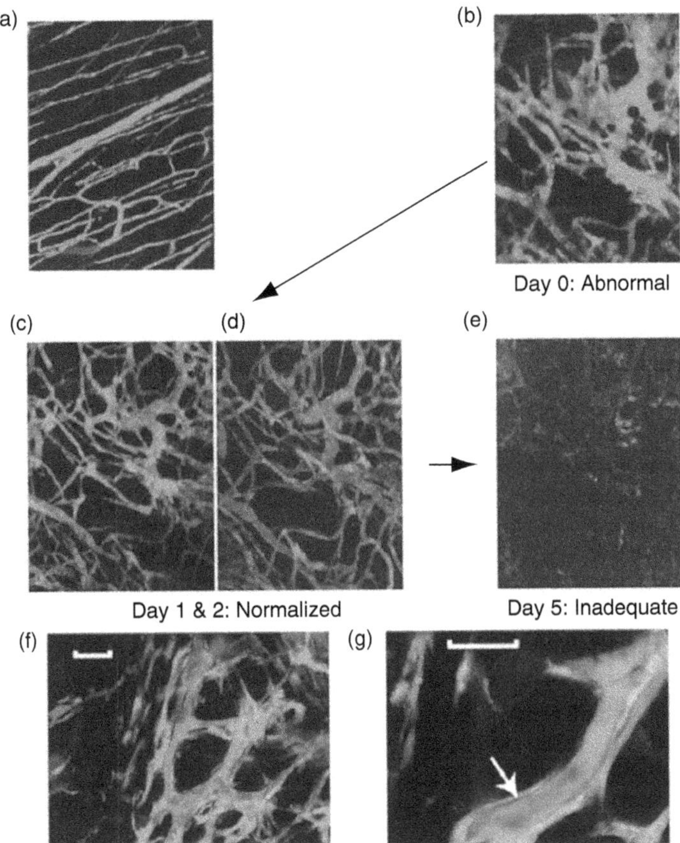

FIGURE 3.6 (See color insert following page 558.) Vessel normalization and EC–mural cell interactions in tumors growing in dorsal windows in mice. (a) Normal capillary bed (dorsal skin and striated muscle). (b) Tumor vasculature (human tumor xenograft). (c, d) Anti-VEGFR-2 therapy prunes immature vessels, leading to a progressively "normalized" vasculature by day 1 and day 2. (e) Further treatment leads to a vasculature that is inadequate to sustain tumor growth by day 5. (f) Perivascular cells expressing GFP (under the control of the VEGF promoter) envelope some vessels in the tumor interior. (g) A perivascular cell, presumably a fibroblast, leading the endothelial sprout (arrow). Scale bar (f, g) = 50 m. Images were obtained using a two-photon microscope. (Reproduced from Jain, R.K., *Nat. Med.*, 7, 987, 2001; Jain, R.K., *Nat. Med.*, 9, 685, 2003. With permission.)

pericyte-like cells have been proposed to guide the endothelial sprouts in tumors [14,19]. Both intravital and immunostaining studies show that the tumor-associated pericytes have abnormal morphology and form tenuous contacts with the endothelial cells and the matrix (Figure 3.6). Intravital studies also show that these perivascular cells produce VEGF, which can serve as a survival factor for the endothelial cells and make these vessels leaky [13,14]. These contrasting roles of the tumor-associated pericyte—stabilizing vs. rendering tumor vessels leaky—are enigmatic and need further investigation.

3.3.4 Molecular Regulation in Mesenchymal, Stromal, and Hematopoietic Cells during Tumor Neovascularization

An important functional component in tumor neovascularization is carried out by stromal and hematopoietic cells. These cells migrate and populate the tumor tissue where they can promote angiogenesis by multiple pathways. One mechanism by which fibroblasts and many hematopoietic cells control new vessel formation is expression of a multitude of proangiogenic factors (VEGF, MMPs, IL-8, Ang-2, among others). Another mechanism is direct incorporation in the vessel wall as part of the endothelial lining [46–48] or perivascular layer [14,22,24].

Fibroblasts may influence carcinogenesis directly. Normal fibroblasts can prevent [91,92] the progression of transformed epithelial cells in a TGF-β-dependent manner [93]. In contradistinction, tumor-activated fibroblasts can enhance malignant epithelial transformation [94,95]. Fibroblasts in tumors express VEGF [13], SDF-1 [95,96], and constitute promising cellular targets for cancer therapy. Some of the same molecular players that control endothelial and mural cell in tumors also affect the fibroblast and hematopoietic cell function. Fibroblasts express PDGFR-α and migrate to tumors in response to factors such as PDGF-A or PDGF-C [97]. TGF-β has been shown to control differentiation of these cells into myofibroblasts in tumors, which results in an increased production of MMPs.

Growth factors produced by the tumors act as potent chemoattractants for hematopoietic cells: VEGF, SDF-1, Ang-1, PlGF, and BDNF [66,98]. The recruitment of hematopoietic BMDCs and the infiltration of these cells into diseased tissue are often massive in scale, particularly in tumors. The predominant populations are myeloid cells (macrophages, neutrophils, etc.), and they also produce chemokines, angiogenic growth factors, and MMPs, thus promoting tumor angiogenesis and progression [10,11]. However, T and B cells may also play important roles at certain stages of carcinogenesis [99,100]. But if the presence of these cells is indicative of a response of the immune system directed against the cancer cells or of a nonspecific inflammatory response that favors tumor growth remains to be determined. Alternatively, if the immune cell infiltrate in tumors is the result of the exponential expansion of precursors (i.e., stem or progenitor cells), the function of BMDCs in the tumor might be significantly different. Increasing evidence suggests that the contribution of hematopoietic stem cells and myeloid progenitor cells is, in fact, critical for new vessel formation in tumors [22]. However, the mechanism of action and the degree of plasticity of these cells are not sufficiently understood. Once they have homed to the tumor, multipotent BMDCs have the potential to modulate neovascularization and to become part of all three nonmalignant tumor compartments: hematopoietic, endothelial, and mesenchymal. Moreover, a recent study showed that gastric cancer induced by *Helicobacter pylori* originated from BMDCs [101]. Both the relative contribution of different BMDCs to solid tumors' stroma and the timing of their incorporation remain unclear [82].

The difficulty of assessing the relative contribution of each lineage of BMDCs to tumor neovascularization has led to debate and hampered the identification of new clinical targets, inhibiting clinical translation of the progress made in preclinical models. Nevertheless, these data suggested that a generic approach of inhibiting chronic inflammation might prevent carcinogenesis. This approach is supported by epidemiological data that have demonstrated that nonsteroidal anti-inflammatory drugs (NSAIDs) reduce their risk of developing colorectal cancer [102]. In addition, two promising therapeutic approaches have been tested in preclinical or clinical models that targeted hematopoietic cell contribution to angiogenesis in tumors. One was targeting these cells using cytokines that also participate in angiogenesis (by inhibition of VEGF, SDF-1, TGF-β, IL-8, COX-2, and so on, or alternatively, promotion of IL-12, IFN-α, or IFN-γ). Another strategy was targeting molecules produced by hematopoietic cells in tumors, such as urokinase plasminogen activator receptor (uPAR [103]), MMP-9 [35,104], cathepsins [105] among others.

3.3.5 Conclusion

New vessel formation in tumors is controlled spatially and temporally by a constellation of molecules. Among them, VEGF is involved in angiogenesis in most tumors and is an attractive therapeutic target. However, beyond the role of VEGF, identifying other key molecular and cellular players and finding ways to block their function in tumor angiogenesis will be crucial for optimizing antiangiogenic therapies for cancer patients.

3.4 TRANSLATION OF THERAPY APPROACHES TARGETING TUMOR VASCULATURE

While basic researchers continue to explore a multitude of antiangiogenic pathways as therapeutic targets, so far only the approaches blocking VEGF have proven their clinical utility. The initial trials for antiangiogenic agents (AAs) yielded disappointing results. This was in contrast to the successes of antiangiogenic drugs in preclinical models of cancer [106]. But after three decades of basic research and clinical development, two antiangiogenic approaches have yielded survival benefit in patients with metastatic cancer in randomized Phase III trials. In one approach, the addition of bevacizumab (a VEGF-specific antibody) to standard chemotherapy improved overall and/or progression-free survival in colorectal, lung, and breast cancer patients. In the second approach, multitargeted agents that block growth factor pathways in both endothelial and cancer cells demonstrated clinical benefit in gastrointestinal stromal tumor (GIST) and renal cell carcinoma patients. In contrast, bevacizumab failed to increase survival with chemotherapy in patients with previously treated and refractory metastatic breast cancer. Furthermore, the addition of vatalanib, a VEGFR-specific tyrosine kinase inhibitor, to conventional cytotoxic therapy did not show a similar benefit in metastatic colorectal cancer patients. These contrasting responses raise critical questions about how these agents work in patients and how to combine them optimally [107]. To this end, we summarize in this section the current understanding of the mechanisms of action of AAs: normalization of tumor vasculature and microenvironment for improved delivery and efficacy of therapeutics, antiangiogenic effects of cytotoxic therapeutics, and cytostatic and cytotoxic effects of AAs on cancer and other stromal cells. In addition, we discuss the progress on the identification of potential biomarkers for efficacy of AAs in humans: molecular and cellular parameters obtained from tissue biopsies, IFP, circulating endothelial cells (CECs), protein levels in bodily fluids, and physiological parameters measured with various imaging techniques. Further research in these key areas will provide invaluable insight into the optimal use of AAs.

3.4.1 ANTIANGIOGENIC THERAPY APPROACH IN CANCER PATIENTS

Solid tumors account for more than 85% of cancer mortality. Tumor angiogenesis is a rational target for therapy given the dependence of solid tumor growth and metastasis on blood vessels [5,27,53,108,109]. Strategies have been pursued to inhibit neovascularization, or destroy existing tumor vessels. These include direct targeting of endothelial cells, supporting perivascular cells, or indirect targeting by inhibition of proangiogenic growth factors release by cancer or stromal cells. Unlike preclinical studies in mice, no randomized Phase III trial has demonstrated a survival benefit with currently available targeted (e.g., anti-VGEF) AAs used as monotherapy. However, the addition of a VEGF-specific antibody, bevacizumab (Avastin, Genentech), to current cytotoxic regimens led to improved outcomes in previously untreated colorectal, breast, and lung cancer patients and in previously treated colorectal cancer patients [110,111].

In contrast, adding bevacizumab to cytotoxic therapy did not enhance survival in previously treated breast cancer patients [112]. Moreover, replacing bevacizumab with vatalanib, a potent VEGFR tyrosine kinase inhibitor, in the combined regimen did not show similar efficacy in chemonaïve or previously treated colorectal cancer patients [113]. Nevertheless, monotherapy using agents with a broader spectrum of inhibitory effect on growth factor pathways (i.e., targeting both endothelial and cancer cells) has resulted in significant antitumor activity against renal cell cancer and increased survival in GIST patients [114,115]. Finally, several agents that indirectly inhibit angiogenesis by targeting oncogenic signaling pathways (such as the EGFR/HER-specific antibodies) have yielded increased survival in clinical trials and are FDA approved.

These contrasting results raise important questions about the use of AAs in clinical practice. Why has anti-VEGF monotherapy not been shown to produce increased survival in randomized trials? How can tumor vessel destruction by combined anti-VEGF treatment—instead of compromising the delivery and efficacy of cytotoxic treatment—prolong survival in previously treated colorectal cancer and chemotherapy-naïve colorectal, lung, and breast cancer patients? Why do anti-VEGF agents not prolong survival in certain previously treated cancer patients? Why the agents that target both endothelial and cancer cells are effective as monotherapy? What biomarkers for efficacy of antiangiogenic treatment can be used to guide the optimal use of AAs in cancer patients? In this section, we summarize the results of recent Phase III clinical trials for AAs, address questions raised above, and suggest new avenues for further investigation.

3.4.2 Clinical Development of Anti-VEGF Therapy

As mentioned in the previous sections, VEGF-A or VEGF is a potent proangiogenic growth factor expressed by most cancer cell types and tumor stromal cells [6,13,53]. VEGF stimulates endothelial cell proliferation, migration, survival, and expression of adhesion molecules, and is a potent inducer of vascular permeability [1]. VEGF may also affect new vessel formation in tumors by acting as a chemoattractant for bone marrow–derived progenitor/stem cells [98]. VEGF expression can be triggered at early stages of neoplastic transformation by environmental stimuli (e.g., hypoxia, low pH) or by genetic mutations (e.g., K-RAS, P53), and persists during progression. Proteolytic enzymes produced by cancer cells or stromal cells can also release existing VEGF bound to the extracellular matrix [1]. Treatment (e.g., radiation, bevacizumab, anti-VEGFR-2 antibody [116–118]) may increase VEGF production or accumulation and inhibiting its activity may be important in maintaining antitumor efficacy. Thus, blocking this pathway appears as a promising antiangiogenic approach to treat multiple types of solid tumors. Such inhibition can be achieved by direct or indirect targeting of the ligand (VEGF) at the mRNA or protein level, its receptors (VEGFR-1, VEGFR-2, and NP-1), and upstream or downstream signaling pathways. Each of these strategies is currently investigated in clinical trials (Table 3.3 and Table 3.4).

Direct approaches include antibodies, soluble receptor, low molecular weight tyrosine kinase inhibitors (TKIs), antisense oligonucleotides, aptamers, and RNA interference (Table 3.3). RNAi-based approaches (e.g., ICS-283, Intradigm) to target VEGF are currently under preclinical development. Of interest, aptamer (pegaptanib sodium, Macugen, Pfizer), as well as antibody fragment (ranibizumab, Lucentis, Genentech)–targeting VEGF are FDA approved for age-related macular degeneration patients. Approaches targeting VEGFR-1 and its specific ligand PlGF using monoclonal antibodies (ImClone, BioInvent International and ThromboGenics) are also under development. Finally, as the VEGF pathway is downstream of most oncogenes, treatment with several agents that target these oncogenic pathways can indirectly inhibit VEGF expression or signaling (Table 3.4).

To date, anti-VEGF monotherapy has not been shown to provide overall survival benefit in patients (indeed, in several trials bevacizumab monotherapy was discontinued for lack of efficacy). Encouraging data from a Phase II trial for renal cell carcinoma [119]—a highly VEGF-dependent tumor—have yet to be validated in a Phase III trial.

TABLE 3.3
Antiangiogenic Agents with Direct Mechanism of Action

Class	Phase of Development	Drug	Targets	Description
Anti-VEGF	Marketed/ Phases III–IV	Bevacizumab (Avastin)	VEGF	Monoclonal antibody
	Phase I	VEGF-trap	VEGF, PlGF, VEGF-B, VEGF-C, VEGF-D	Soluble receptor
		VEGF-AS (Veglin)	VEGF, VEGF-C, VEGF-D	Antisense oligonucleotide
		Aplidin	VEGF	Peptide
Anti-VEGFR	Phase III	Vatalanib (PTK787/ ZK22258)	VEGFR-1, VEGFR-2, VEGFR-3, (PDGFR-β), and c-Kit	Small-molecule tyrosine kinase receptor inhibitor
		Sunitinib (SU11248, Sutent)	VEGFR-2, PDGFR-β	Small-molecule tyrosine kinase receptor inhibitor
		Sorafenib (BAY 43-9006, Nexavar)	VEGFR-2, PDGFR-β, Flt-3, c-Kit	Small-molecule raf kinase and tyrosine kinase inhibitor
		AE-941 (Neovastat)	VEGF–VEGFR binding plus MMP-2, MMP-9	Shark-cartilage component
	Phases I–II	AG-013736	VEGFR, PDGFR-β, c-Kit	Small-molecule tyrosine kinase receptor inhibitor
		AMG 706	VEGFR-1, VEGFR-2, PDGFR-β, c-Kit	Small-molecule tyrosine kinase receptor inhibitor
		Cediranib (AZD2171, Recentin)	VEGFR-1, VEGFR-2, VEGFR-3	Small-molecule tyrosine kinase receptor inhibitor
		CEP-7055	VEGFR-1, VEGFR-2, VEGFR-3	Small-molecule tyrosine kinase receptor inhibitor
		CHIR258	VEGFR-1, VEGFR-2, FGFR1/3	Small-molecule tyrosine kinase receptor inhibitor
		CP-547632	VEGFR-2	Small-molecule tyrosine kinase receptor inhibitor
		GW786034	VEGFR-2	Small-molecule tyrosine kinase receptor inhibitor
		IMC-C1121b	VEGFR-2	Monoclonal antibody
		KRN-951	VEGFR-1, VEGFR-2, VEGFR-3, PDGFR-β, c-Kit	Small-molecule tyrosine kinase receptor inhibitor
		OSI-930	VEGFR, c-Kit	Small-molecule tyrosine kinase receptor inhibitor
		XL999	FGFR, VEGFRs, PDGFR, and Flt-3	Small-molecule tyrosine kinase receptor inhibitor

(*continued*)

TABLE 3.3 (continued)
Antiangiogenic Agents with Direct Mechanism of Action

Class	Phase of Development	Drug	Targets	Description
		ZK-CDK	VEGFRs, PDGFR, CDKs	Small-molecule tyrosine kinase receptor inhibitor
		Vandetanib (ZD6474, Zactima)	VEGFR-2, EGFR	Small-molecule tyrosine kinase receptor inhibitor

Source: Adapted from Jain, R.K. et al., *Nat. Clin. Pract. Oncol.*, 3, 24, 2006. With permission.

Nevertheless, VEGF blockade by bevacizumab has yielded improved survival in cancer patients in four Phase III trials when added to standard chemotherapy (Table 3.5). These results emphasize the need for targeting both stromal and neoplastic cells to enhance survival in most cancer types. In the clinic, this has been achieved by combining an endothelial-specific AA with contemporary cytotoxic agents or by using multitargeted AAs that block growth factor pathways in both cell types (Figure 3.7).

3.4.2.1 Concurrent Targeting of Cancer Cells and Endothelial and Other Stromal Cells by Combination Therapies

Bevacizumab is a humanized VEGF-specific antibody with a reported half-life of 17–21 days [53]. Phase I and II trials documented objective responses (including a few complete responses [118,120]) and the relative safety of bevacizumab in combination with traditional cytotoxic

TABLE 3.4
Antiangiogenic Agents with Indirect Mechanism of Action

Class	Phase	Agents
EGFR/HER inhibitors	Marketed/Phase III	Cetuximab Gefitinib Erlotinib Panitumumab Pertuzumab Trastuzumab
mTOR inhibitors	Phase III	CCI-779
COX-2 inhibitors	Marketed	Celecoxib
PPAR-γ agonists	Marketed	Rosiglitazone
Proteosome inhibitors	Marketed	Bortezomib
Bisphosphonates	Marketed	Zoledronic acid
Cytokines	Marketed	Interferon-α
Thalidomide and analogs	Marketed/Phase III	Thalidomide Lenalidomide
Hormone derivatives	Phase II	2-Methoxyestradiol

Source: Adapted from Jain, R.K. et al., *Nat. Clin. Pract. Oncol.*, 3, 24, 2006. With permission.

TABLE 3.5
Completed Phase III Trials for Bevacizumab (BV) with Standard Contemporary Chemotherapy

Tumor Type	Selection Criteria: Stage	Selection Criteria: Previous Treatment	Regimen	Pts.	Outcome of Trial	Common Toxicity of BV
Breast cancer	Metastatic	Anthracycline/taxane/herceptin	BV (15 mg/kg) + capecitabine (3 week cycles)	462	Increased RR (19.8% vs. 9.1%; $P = 0.001$), comparable PFS (4.86 months vs. 4.17 months; hazard ratio 0.98) and OS (15.1 months vs. 14.5 months)	Hypertension, proteinuria, thromboembolic events, bleeding, pulmonary embolism
Colorectal cancer	Metastatic	No	BV (5 mg/kg) + IFL (2 week cycles)	813	Increased RR (44.8% vs. 34.8%, $P = 0.004$), prolonged PFS (10.6 months vs. 6.2 months, $P < 0.001$) and OS (20.3 months vs. 15.6 months, $P < 0.001$)	Hypertension, thromboembolic events, diarrhea, leukopenia, deep thrombophlebitis, bleeding, gastrointestinal perforation
Lung cancer	Nonsquamous nonsmall cell	No	BV (15 mg/kg) + paclitaxel/carboplatin (3 week cycles)	878	Increased RR (27.2% vs. 10.0%), prolonged PFS (6.4 months vs. 4.5 months, $P < 0.0001$) and OS (12.5 months vs. 10.2 months, $P < 0.007$)	Leukopenia, hypertension, thrombosis, life-threatening or lethal (eight patients), bleeding
Colorectal cancer	Metastatic	5-fluorouracil/irinotecan	BV (10 mg/kg) + FOLFOX4 (2 week cycles)	829	Increased RR (21.8% vs. 9.2%, $P = 0.0001$), prolonged PFS (7.2 months vs. 4.8 months, $P < 0.0001$) and OS (12.9 months vs. 10.8 months, $P < 0.001$)	Neuropathy, hypertension, bleeding, bowel perforation
Breast cancer	Recurrent/metastatic	No	BV (10 mg/m^2) + paclitaxel (2 week cycles)	722	Increased RR (28 months vs. 14 months), prolonged PFS (10.97 months vs. 6.11 months) OS—N/A	Neuropathy, hypertension, proteinuria, bleeding

Source: Adapted from Jain, R.K. et al., *Nat. Clin. Pract. Oncol.*, 3, 24, 2006. With permission.

BC, breast carcinoma; CRC, colorectal carcinoma; LC, lung carcinoma; IFL, irinotecan–5-fluorouracil–leucovorin; FOLFOX4, oxaliplatin, 5-fluorouracil, and leucovorin; RR, response rate; PFS, progression-free survival; OS, overall survival; N/A, not available.

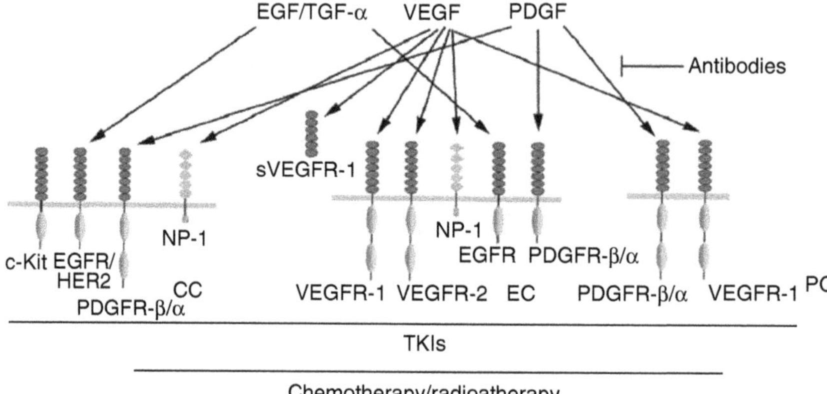

FIGURE 3.7 Combined and direct targeting of cancer cells (CC) and endothelial (EC) and perivascular cells (PC) has yielded increased survival in Phase III trials of anti-VEGF agents. This has been achieved by two approaches: one combines traditional cytotoxic agents (which may kill any proliferating cell) with the VEGF-specific antibody bevacizumab. VEGF blockade inhibits its signaling pathway in endothelial cells responsible for cell survival, migration, proliferation, and vascular permeability; VEGF blockade may also affect cancer cells, when their survival depends on VEGF (e.g., via NP-1). The other approach uses small-molecule receptor TKIs with broad inhibitory spectra (active against VEGFRs, EGFRs, PDGFRs, c-Kit receptors, and downstream soluble kinases such as Raf), which may be present in all three cell populations. Alternatively, combinations of antibodies which block the ligands (VEGF and EGFR/HER or PDGF) may be effective in targeting both cancer and stromal cells. (Reproduced from Jain, R.K. et al., *Nat. Clin. Pract. Oncol.*, 3, 24, 2006. With permission.)

regimens [121]. However, the first randomized placebo-controlled Phase III trial of bevacizumab failed to show increased survival when bevacizumab was combined with chemotherapy in previously treated metastatic breast cancer patients [112]. In this trial, an increased response rate, but no survival benefit, was seen in patients receiving bevacizumab with capecitabine vs. patients receiving capecitabine alone.

The clinical breakthrough for antiangiogenic therapy was demonstrated with the use of bevacizumab and standard chemotherapy in previously untreated, metastatic colorectal cancer patients [111]. A survival increase was seen for patients receiving bevacizumab with irinotecan–5-fluorouracil–leucovorin (IFL) regimen vs. patients receiving IFL only. In February 2004, based on these data, bevacizumab became the first AA to be approved by the FDA for cancer patients.

Three other randomized Phase III trials have shown positive results. One trial investigated the efficacy of bevacizumab with standard chemotherapy (paclitaxel) in patients with previously untreated recurrent or metastatic breast cancer and showed that the primary end point of increased progression-free survival was achieved. In another trial, previously treated patients with advanced colorectal cancer who received bevacizumab in combination with the proven second-line therapy (an oxaliplatin regimen FOLFOX4) had a survival benefit compared to patients who received FOLFOX4 alone. Whether combining bevacizumab with FOLFOX4 or other chemotherapy regimens is the best option for first-line therapy for colorectal cancer under investigation (reviewed and discussed in Ref. [122]). Finally, a trial for patients with previously untreated advanced nonsquamous and nonsmall cell lung cancer showed a survival benefit when bevacizumab was added to standard chemotherapy (paclitaxel and carboplatin) [123]. This was the first randomized trial to show a median survival of >1 year in one of the arms, and the first trial with a targeted agent which has shown survival advantage in combination with chemotherapy in patients with untreated metastatic lung cancer. The utility and feasibility of including bevacizumab in the treatment of previously treated lung cancer patients is unknown.

A Phase III trial of bevacizumab and interferon-α vs. interferon-α alone for renal cell cancer has been completed, but results are not yet available. Lastly, Phase III trials combining bevacizumab with chemotherapy in pancreatic and prostate patients are underway.

3.4.2.2 Concurrent Targeting of Cancer Cells and Endothelial and Other Stromal Cells by Multitargeted Agents

The development of small molecules that inhibit signaling pathways for VEGF and other angiogenic growth factors by blocking receptor tyrosine kinase activity has generated great excitement and promise. Some of these AAs have reached Phase III clinical trials (Table 3.6). One of the first such agents with broad spectrum targets to advance to Phase III trials was vatalanib (PTK787/ZK 222584, Novartis). One trial (CONFIRM-1) compared the efficacy of oral vatalanib in combination with FOLFOX4 vs. FOLFOX4 alone for the first-line therapy of metastatic colorectal cancer. Mature survival data are not available until 2006, but interim analyses showed a reduction in risk of disease progression which was significant with the investigator analysis, but not with an independent central review. A significant effect

TABLE 3.6
Completed Phase III Trials for Agents with VEGF Receptor 2 as One of Their Targets

Tumor Type	Selection Criteria	Regimen Used	Target	Pts.	Outcome of Trial	Grades 3 and 4 Toxicity
Colorectal cancer	Metastatic CRC	Vatalanib or placebo with FOLFOX4	VEGFR-1, VEGFR-2, VEGFR-3, PDGFR-β, and c-Kit	1168	12% reduction in risk of disease progression ($P = 0.118$)	Hypertension, neutropenia, diarrhea Nausea Neuropathy Vomiting, venous thrombosis, dizziness, thrombocytopenia Neutropenia, pulmonary embolism
Renal cancer	Advanced RCC	Sorafenib	RAF kinase, VEGFR-2, PDGFR-α and β, Flt-3 and c-Kit	~900	Improved progression-free survival (24 weeks vs. 12 weeks, $P < 0.000001$)	Rash Diarrhea Hand and foot syndrome Nausea Fatigue
Stomach cancer	Previously treated GIST	SU11248	VEGFR-2, PDGFR-β, Flt-3, and c-Kit	312	Prolonged time to progression (6.3 months vs. 1.5 months, $P < 0.00001$)	Hypertension, fatigue, diarrhea

Source: Adapted from Jain, R.K. et al., *Nat. Clin. Pract. Oncol.*, 3, 24, 2006. With permission.

CRC, colorectal carcinoma; RCC, renal cell carcinoma; sorafenib (formerly BAY 43-9006); GIST, gastrointestinal stromal tumor.

(33% reduction in risk of progression) for vatalanib was seen in patients with high levels of lactate dehydrogenase (>1.5 times the upper limit of normal), a poor prognosis group. Interim results from a Phase III trial of vatalanib in combination with FOLFOX4 chemotherapy as second-line treatment for metastatic colorectal cancer (CONFIRM-2) also showed no significant effect in the overall population, and a significant benefit in patients with high LDH.

Monotherapy with other broad spectrum TKIs has shown efficacy in randomized Phase III trials for tumors with limited treatment options. Sorafenib (BAY 43-9006, Bayer and Onyx) efficiently inhibited both tumor cell proliferation and angiogenesis in preclinical models. Interim data from a Phase III trial showed that renal cell carcinoma patients taking sorafenib had a clinically significant improvement in progression-free survival [114,115]; thus the drug was offered to all study participants. In July 2005, a new drug application with the FDA was filed for sorafenib for use in patients with advanced renal cell carcinoma. Another recent success was obtained with Sutent (SU11248, Pfizer) in patients with GIST. These tumors often express the activated form of the c-Kit receptor, and thus may be good candidates for treatment with TKIs inhibiting c-Kit activity such as imatinib mesylate (STI571, Gleevec, Novartis), which also targets PDGFR-β and bcr-abl. In January 2005, Pfizer's Phase III study assessing Sutent in the treatment of imatinib-resistant GIST successfully met its predetermined efficacy and safety end points [124]. A planned interim analysis of the Phase III study data led to the recommendation that the study be "unblinded" to provide all enrolled patients access to Sutent. Both agents received FDA approval for these indications in 2006.

Finally, the use of approved targeted agents that are indirect inhibitors of angiogenesis with chemotherapy has also shown increased survival in breast cancer patients. For example, we have shown that trastuzumab (Herceptin, Genentech, approved for the treatment of advanced breast cancer since 1998) acts as an antiangiogenic cocktail in a HER2 + breast tumor [125]. Recently, two Phase III trials showed that HER2-overexpressing breast cancer patients who received trastuzumab in combination with doxorubicin, cyclophosphamide, and paclitaxel had a 52% decrease in disease recurrence compared to patients treated with chemotherapy alone [110]. This difference was highly statistically significant. Most of the over 3300 patients enrolled in these studies had aggressive, lymph node–positive disease.

3.4.3 Safety of Antiangiogenic Agents

Most AAs have been well tolerated by patients either as single agents or in combination with standard chemotherapy in clinical trials. As expected, dose escalations and broadening the spectra of targets also increased some expected side effects: hypertension, diarrhea, leukopenia, bleeding, and proteinuria (Table 3.5 and Table 3.6). Serious and unusual toxicities of bevacizumab consisted of neuropathy, asthenia, thromboembolic events, gastrointestinal perforations, and life threatening or fatal hemorrhage. Vatalanib was associated with a higher incidence of dizziness. Addition of trastuzumab to chemotherapy increased the risk of congestive heart failure.

3.4.4 Mechanisms of Actions of Antiangiogenic Agents in Cancer Patients

The widely held view is that these antiangiogenic therapies should destroy the tumor vasculature, thereby depriving the tumor of oxygen and nutrients. The failure of bevacizumab to increase survival in heavily treated breast cancer patients was initially explained by (1) the highly refractory and advanced nature of the tumors in the patients enrolled and (2) the increased expression of other angiogenic factors during breast cancer progression through chemotherapy, which rendered VEGF less critical for continued tumor growth. These hypotheses were partially supported by the success of a subsequent trial of bevacizumab with chemotherapy in treatment-naïve advanced breast cancer patients. Nevertheless, they seem to be contradicted by the efficiency of bevacizumab with chemotherapy in heavily treated colorectal cancer. In contrast, vatalanib does not seem to confer the same survival advantage as bevacizumab in colorectal cancer patients when combined with chemotherapy. These contrasting results raise questions about the mechanisms of action of these agents alone and in combination.

There are at least three mechanisms of action, which can reconcile the differing outcomes in clinical trials.

3.4.4.1 Normalization of Tumor Vasculature and Microenvironment

More than a decade ago, Teicher [126] proposed that combining antiangiogenic therapy with cytotoxic treatments has synergistic effects because it allows targeting of both the malignant cell compartment and the vascular stroma (Figure 3.8). However, the destruction of tumor

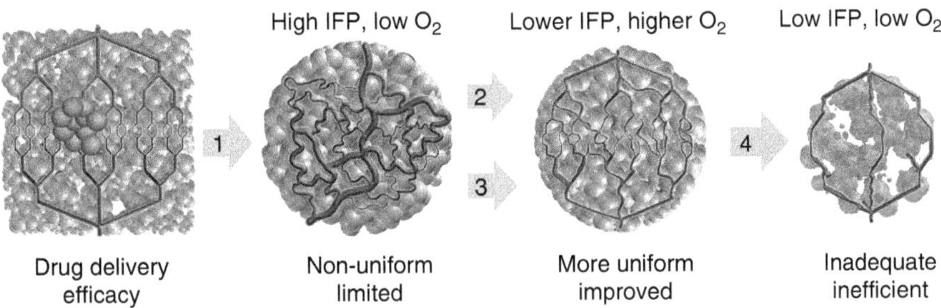

FIGURE 3.8 Proposed mechanisms of action of bevacizumab on tumor vasculature. Owing to high levels of proangiogenic molecules produced locally, such as VEGF, tumors make the transition from in situ carcinoma to frank carcinoma (1). At this stage, tumors become hypervascular, but the vessels are leaky and the blood flow is spatially and temporally heterogeneous. This leads to increased interstitial fluid pressure (IFP) and focal hypoxia, creating barriers for delivery of therapeutics. The proposed mechanism of action of the VEGF-specific antibody bevacizumab is twofold: inhibition of new vessel formation and killing of immature tumor vessels (2); and transient normalization of the remaining vasculature by decrease in macromolecular permeability (and thus the IFP) and hypoxia, and improvement of blood perfusion (3). Another effect of bevacizumab may be the direct killing of cancer cells in subsets of tumors in which the cells express VEGF receptors. Regardless of the mechanisms involved, monotherapy with bevacizumab (or other current targeted AAs) is not curative because it cannot kill all cancer cells, and in the longer term leads to a vasculature which is inefficient for drug delivery (4), and to tumor relapse using alternative pathways for neovascularization. Therefore, combinations of bevacizumab with chemotherapeutics have been pursued in Phase III trials and have led to a survival benefit in patients with chemosensitive tumors, showing the synergistic effect of the two treatment modalities. Synergy may have been achieved because of increased cell killing following tumor vascular normalization: the lowered IFP leads to improved delivery of chemo- and molecularly targeted agents; the improved oxygenation sensitizes cancer cells to cytotoxic therapeutics and reduces the selection of more malignant phenotype; and finally, increased cellular proliferation in normalized tissues increases the cytotoxicity of chemotherapy. Normalization of the vasculature may also benefit the direct killing of cancer cells by bevacizumab, in synergy with the chemotherapeutics. Of interest, cytotoxic therapeutics may directly target the endothelial cells, when administered at certain doses and schedules. Metronomic chemotherapy may also normalize the tumor vasculature and thus improve drug delivery to tumors. (Adapted from Jain, R.K. et al., *Nat. Clin. Pract. Oncol.*, 3, 24, 2006. With permission.)

vasculature by AAs should antagonize chemo- and radiotherapy by impeding the delivery of therapeutics and oxygen, respectively. Indeed, a number of preclinical studies have demonstrated such antagonism [27,127]. At the same time, such combinations have been successful in a number of preclinical and clinical studies. To resolve this paradox, we proposed in 2001 that AAs can transiently "normalize" tumor vasculature [128]. Normalization of tumor vasculature can be defined as the structural and functional changes that allow more efficient delivery of drugs and oxygen, ultimately leading to improved outcomes [128]. A better understanding of the molecular and cellular underpinnings of vascular normalization may ultimately lead to more effective therapies not only for cancer but also for diseases with abnormal vasculature, as well as regenerative medicine, in which the goal is to create and maintain a functionally normal vasculature.

3.4.5 Why Normalize the Tumor Vasculature?

As described in Section 3.1, tumor vasculature is structurally and functionally abnormal. Blood vessels are leaky, tortuous, dilated, and saccular and have a haphazard pattern of interconnection (Figure 3.9). The endothelial cells lining these vessels have aberrant morphology; pericytes (cells that provide support for the endothelial cells) are loosely attached or absent; and the basement membrane is often abnormal—unusually thick at times, entirely absent at others. These structural abnormalities contribute to spatial and

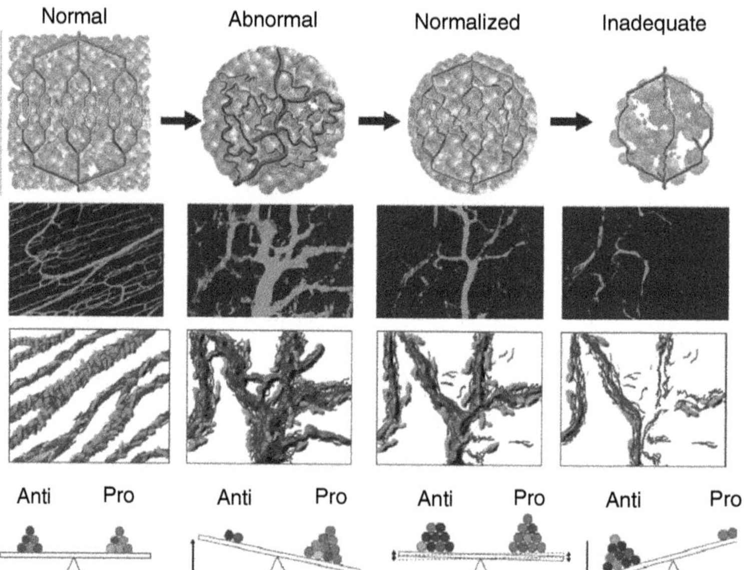

FIGURE 3.9 (See color insert following page 558.) Proposed role of vessel normalization in the response of tumors to antiangiogenic therapy. (a) Tumor vasculature is structurally and functionally abnormal. It is proposed that antiangiogenic therapies initially improve both the structure and the function of tumor vessels. However, sustained or aggressive antiangiogenic regimens may eventually prune away these vessels, resulting in a vasculature that is both resistant to further treatment and inadequate for the delivery of drugs or oxygen. (b) Dynamics of vascular normalization induced by VEGFR-2 blockade. On the *left* is a two-photon image showing normal blood vessels in skeletal muscle; subsequent images show human colon carcinoma vasculature in mice at day 0, day 3, and day 5 after administration of VEGR2-specific antibody. (c) Diagram depicting the concomitant changes in pericyte and basement membrane coverage during vascular normalization. (d) These phenotypic changes in the vasculature may reflect changes in the balance of pro- and antiangiogenic factors in the tissue. (Reproduced from Jain, R.K., *Science*, 307, 58, 2005. With permission.)

temporal heterogeneity in tumor blood flow. In addition, solid pressure generated by proliferating cancer cells compresses intratumor blood and lymphatic vessels, which further impairs not only the blood flow but also the lymphatic flow [29]. Collectively, these vascular abnormalities lead to an abnormal tumor microenvironment characterized by interstitial hypertension (elevated hydrostatic pressure outside the blood vessels), hypoxia, and acidosis. Impaired blood supply and interstitial hypertension interfere with the delivery of therapeutics to solid tumors. Hypoxia renders tumor cells resistant to both radiation and several cytotoxic drugs. Independent of these effects, hypoxia also induces genetic instability and selects for more malignant cells with increased metastatic potential [30]. Hypoxia and low pH also compromise the cytotoxic functions of immune cells that infiltrate a tumor. Unfortunately, cancer cells are able to survive in this abnormal microenvironment. In essence, the abnormal vasculature of tumors and the resulting abnormal microenvironment together pose a formidable barrier to the delivery and efficacy of cancer therapy. This suggests that if we knew how to correct the structure and function of tumor vessels, we would have a chance to normalize the tumor microenvironment and ultimately to improve cancer treatment. The fortified tumor vasculature may also inhibit the shedding of cancer cells into the circulation—a prerequisite for metastasis. In the past, higher doses of drugs and hyperbaric oxygenation have been used to increase the tumor concentrations of drugs and oxygen, respectively. These strategies have not shown much success in the clinic, however. One reason for this failure is that tumor vessels have large holes in their walls [129]. As stated earlier, this leakiness leads to interstitial hypertension as well as spatially and temporally nonuniform blood flow. If the delivery system is flawed, it does not matter how much material is pumped into it. The drugs and oxygen become concentrated in regions which already have enough and still not reach the inaccessible regions [130]. However, if we fix the delivery system, more cells are likely to encounter an effective concentration of drugs and oxygen. This is the rationale for developing therapies that normalize the tumor vasculature. These therapies do not merely increase the total uptake of drugs and oxygen but also distribute these molecules to a larger fraction of the tumor cells by fixing the delivery system.

3.4.6 How Should We Normalize the Tumor Vasculature?

In normal tissues, the collective action of angiogenic stimulators (e.g., VEGF) is counterbalanced by the collective action of angiogenic inhibitors such as thrombospondin-1 (Figure 3.9). This balance tips in favor of the stimulators in both pathological and physiological angiogenesis [27]. However, in pathological angiogenesis, the imbalance persists. Therefore, restoring the balance may render the tumor vasculature close to normal. On the other hand, tipping this balance in favor of inhibitors may lead to vascular regression and, ultimately, to tumor regression. If we had AAs that completely destroyed tumor vessels without harming normal vessels, we would not need to add cytotoxic therapy. Unfortunately, such agents are not currently available. It is conceivable that increased doses of currently available AAs could produce complete tumor regression, but such doses are likely to adversely affect the vasculature of normal tissues, including the cardiovascular, endocrine, and nervous systems (Table 3.5 and Table 3.6). Indeed, antiangiogenic therapy with agents such as bevacizumab is associated with an increased risk of arterial thromboembolic events, and such adverse effects could be more pronounced with increased doses. Furthermore, excessive vascular regression may be counterproductive because it compromises the delivery of drugs and oxygen (Figure 3.9). Indeed, suboptimal doses or scheduling of AAs might lower tumor oxygenation and drug delivery and, thus, antagonize rather than augment the response to radiotherapy or chemotherapy [131–133]. This need for a delicate balance between normalization and excessive vascular regression emphasizes the requirement for careful selection of the dose and administration schedule for AAs.

3.4.7 Can Blocking VEGF Signaling Normalize Tumor Vessels?

VEGF is overexpressed in the majority of solid tumors. Thus, if one were to judiciously downregulate VEGF signaling in tumors, then the vasculature might revert back to a more "normal" state. Indeed, blockade of VEGF signaling passively prunes the immature and leaky vessels of transplanted tumors in mice and actively remodels the remaining vasculature so that it more closely resembles the normal vasculature (Figure 3.9). This "normalized" vasculature is characterized by less leaky, less dilated, and less tortuous vessels with a more normal basement membrane and greater coverage by pericytes (Figure 3.9). These morphological changes are accompanied by functional changes—decreased IFP, increased tumor oxygenation, and improved penetration of drugs in these tumors (Table 3.7) [27].

3.4.8 What about Human Tumors?

Thousands of patients worldwide have received anti-VEGF therapy. The effect of VEGF blockade on human tumors was recently studied in rectal carcinoma patients receiving an antibody to VEGF, bevacizumab, together with radiation and chemotherapy [118,134]. The results in patients mirrored those seen in transplanted tumors in mice: 2 weeks after a single injection of bevacizumab alone, the global (mean) blood flow of tumors, as measured by contrast-enhanced computed tomography (CT), decreased by 30% to 50% in six consecutive patients. Tumor microvascular density, vascular volume, and IFP were also found to be reduced. Surprisingly, however, there was no concurrent decrease in the uptake of radioactive tracers in tumors, which suggests that vessels in the residual "normalized" tumor vasculature were more efficient in delivering these agents to tumor parenchyma than they were before bevacizumab treatment. Similar reductions in blood flow, as measured by magnetic resonance imaging (MRI), had been noted previously in patients treated daily with small-molecule inhibitors of VEGFR tyrosine kinase activity (PTK787 and SU6668 [135,136]). Interestingly, however, positron emission tomography (PET) analysis of patients

TABLE 3.7
Effect of VEGF Blockade in Tumor-Bearing Mice and Cancer Patients

	Preclinical Data[a]			Clinical Data[b]		
Properties	**Control**	**Treatment**	**Change**	**Control**	**Treatment**	**Change**
Blood volume	19.3 ± 2.2	5.4 ± 1.0	−72%	6.8 ± 2.1	5.0 ± 0.9	−26%
Vascular density	52.1 ± 4.6	41.9 ± 3.0	−19%	13.0 ± 3.2	6.9 ± 1.8	−47%
Permeability (BSA)	7.3 ± 0.8	2.8 ± 0.8	−62%			
PS product (small molecules)				14 ± 2	12.9 ± 3.1	−7.9%
Interstitial fluid pressure	6.1 ± 1.0	3.1 ± 0.5	−49%	14.0 ± 1.2	4.0 ± 1.5	−71%
Perivascular cell coverage	0.67 ± 0.04	0.81 ± 0.04	+21%	9.9 ± 3.8	17.8 ± 1.5	+80%
Ang-2 level	10.4 ± 1.3	2.2 ± 0.5	−79%	0.046 ± 0.002	0.020 ± 0.001	−57%
Tumor apoptosis	0.86 ± 0.24	2.50 ± 0.31	+190%	1.7 ± 0.2	3.6 ± 0.7	+112%
Plasma VEGF level	Nondetectable	182.5 ± 135.8		22.5 ± 8.3	272 ± 22.5	+1109%

Source: Adapted from Jain, R.K., *Science*, 307, 58, 2005. With permission.

[a] Tong, R.T. et al., *Cancer Res.*, 64, 3731, 2004; Winkler, F. et al., *Cancer Cell*, 6, 553, 2004.
[b] Willett, C.G. et al., *Nat. Med.*, 10, 145, 2004; Willett, C.G. et al., *J. Clin. Oncol.*, 23, 8136, 2005.

treated with endostatin, an endogenous inhibitor of angiogenesis, revealed a biphasic response—that is, an increase in tumor blood flow at lower doses and a decrease at intermediate doses [137]. As we interpret these human data obtained from MRI, PET, and CT, two key limitations of these imaging methods must be kept in mind. First, as most of these methods yield a parameter that depends on both blood flow and permeability, the blood flow and permeability cannot be calculated unambiguously. Second, tumor blood flow is highly heterogeneous. It is not the total blood flow, but the distribution of blood flow, which determines the distribution of a drug or oxygen in tumors. Therefore, the global (total) blood flow, as estimated by the currently available resolution of MRI, CT, or PET, does not inform us about the degree of spatial heterogeneity in vascular normalization or drug distribution [27,130]. Thus, improved imaging techniques, which can measure the spatial and temporal changes in blood flow and other physiological parameters with higher resolution, are needed to definitively establish the effects of antiangiogenic treatment on vascular function in human tumors growing at different sites.

3.4.9 Is There an Optimal Time or Drug Dose for Normalization?

Optimal scheduling of antiangiogenic therapy with chemotherapy or radiation therapy requires knowledge of the time window during which the vessels initially become normalized, as well as an understanding of how long they remain in that state. Recent studies, in which human tumors growing in mice were treated with an antibody to VEGFR-2, have identified such a "normalization window," that is, a period during which the addition of radiation therapy yields the best therapeutic outcome (Figure 3.10) [26]. This window was short-lived (~6 days) and was characterized by an increase in tumor oxygenation, which enhances

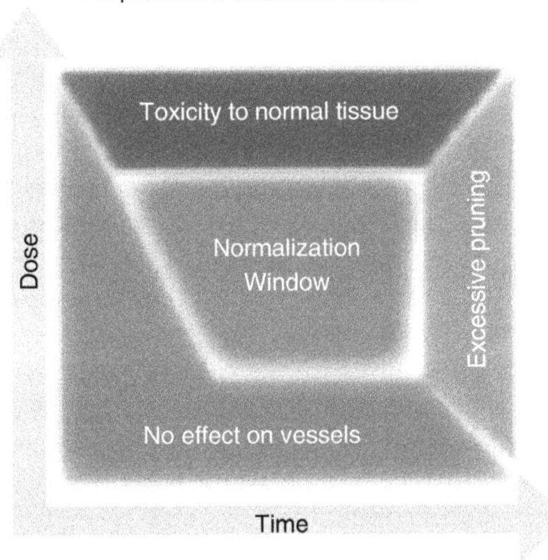

FIGURE 3.10 Proposed effect of drug dose and schedule on tumor vascular normalization. The efficacy of cancer therapies which combine antiangiogenic and cytotoxic drugs depends on the dose and delivery schedule of each drug. The vascular normalization model posits that a well-designed strategy should passively prune away immature, dysfunctional vessels and actively fortify those remaining, while incurring minimal damage to normal tissue vasculature. During this "normalization" window (green), cancer cells may be more vulnerable to traditional cytotoxic therapies and to novel targeted therapies. The degree of normalization is spatially and temporally dependent in a tumor. Vascular normalization occurs only in regions of the tumor where the imbalance of pro- and antiangiogenic molecules has been corrected. (Reproduced from Jain, R.K., *Science*, 307, 58, 2005. With permission.)

radiation therapy by increasing the concentration of reactive oxygen species created by the radiation. During the normalization window, but not before or after it, VEGFR-2 blockade was found to increase pericyte coverage of vessels in a human brain tumor grown in mice. Vessel normalization was accompanied by upregulation of angiopoietin-1 and activation of MMPs. The prevailing hypothesis is that VEGF blockade passively prunes nascent vessels that are not covered with pericytes. In contrast, this study found that pericyte coverage increased before vascular pruning. Improved understanding of the molecular mechanisms of vessel normalization may suggest new strategies for extending the normalization window to provide ample time for cytotoxic therapy. The dose of AAs also determines the efficacy of combination therapy. Although it is tempting to increase the dose of AAs or to use a more potent angiogenic blocker, as one would for chemotherapeutic agents, doing so might lead to normal tissue toxicity and compromise the tumor vessels to the point that drug delivery is impaired. Indeed, renal cell carcinoma patients on a high dose of bevacizumab (10 mg/kg of body weight every 2 weeks) were more likely to develop hypertension and proteinuria than those on a low dose, although the sample size was too small for comparison of the rates of serious adverse events [119]. Even the low dose of bevacizumab (5 mg/kg) given in combination with chemotherapy has contributed to an increased risk of cardiovascular problems, including death, in some cancer patients [138]. Although no dose comparison has yet been made in large clinical trials, it is conceivable that such serious adverse events may increase with higher doses. In studies of mice, more potent blockers of VEGF signaling have induced regression of normal tracheal and thyroid vessels [139].

3.4.10 Is Tumor Growth Accelerated during Vascular Normalization?

One would expect that the improved delivery of oxygen and nutrients during vascular normalization would enhance tumor growth. However, both preclinical and clinical studies to date show that, despite normalization, tumor growth is not accelerated during antiangiogenic monotherapy. There are several possible explanations for this apparent paradox. (1) It is important to remember that vascular normalization occurs in the context of antiangiogenic treatment and that the main effect of this treatment is a reduction in the number of blood vessels (vessel density), which should lead to tumor regression. Moreover, tumors are highly heterogeneous; not all regions are equally vascularized; some tumor vessels are more mature than others; and the balance of pro- and antiangiogenic molecules differs from region to region and from moment to moment. Hence, the effects of vessel normalization in some regions of the tumor are swamped by simultaneous vessel regression in other regions. In addition, the inability of tumors to grow new vessels during antiangiogenic therapy limits the ability of this transient increase in vascular efficiency to expand the tumor mass. If it were easy to achieve complete tumor regression with antiangiogenic monotherapy, vascular normalization would be of marginal importance, because it is expected to affect only a subset of cells and to do so only temporarily. Unfortunately, some tumor cells are able to survive antiangiogenic monotherapy, and these cells must be targeted with combined therapy. (2) The transient normalization of tumor vessels produces a temporary increase in oxygen and nutrient delivery to the cancer cells which surround these "normalized" vessels. This might be expected to enhance the proliferation of these cells and hence to accelerate tumor growth. However, this intuitive notion is not supported by published data. For example, Gullino found no correlation between tumor growth rate in vivo and blood flow rate, vascular volume, or use of oxygen or glucose. Even if the proliferation rate of cancer cells around normalized vessels was increased, this may well enhance therapeutic efficacy, because rapidly proliferating cells are more sensitive to radiation and to many cytotoxic drugs. (3) It is widely assumed that hypoxia leads to the death of cells. Therefore, alleviation of hypoxia during transient normalization of tumor vasculature should accelerate tumor growth. However, a growing body of evidence indicates that hypoxia may in fact promote cancer progression [140]. These two competing effects of antiangiogenic therapy may cancel out each other. (4) Finally, in some tumors, cancer cells depend on the same angiogenic growth factors (e.g., VEGF) for their survival as do the endothelial cells. In these tumors, AAs may kill both cancer cells and endothelial cells and likely induce tumor regression—similar to hormone withdrawal from a hormone-dependent tumor [141]—despite vessel normalization. For all these reasons, any acceleration in tumor growth during transient normalization is presumably masked by indirect and direct killing of cancer cells by AAs. Thus, it is not surprising that tumor regression is slow or modest after antiangiogenic monotherapy despite a significant decrease in microvascular density [134].

3.4.11 Clinical Impact of Vascular Normalization

Emerging preclinical and clinical data from our laboratory and others support the concept of tumor vascular normalization by antiangiogenic therapy [25,26,118,134,142–147]. Vascular normalization could explain why bevacizumab was efficacious in combination with chemotherapy, despite its limited efficacy as monotherapy [27,127]. It may also explain why bevacizumab increased survival in chemotherapy-naïve breast cancer patients (by increasing the delivery of chemotherapeutics to chemoresponsive tumors), and not in chemorefractory breast cancer patients (in which an increase in delivery of chemotherapeutics may have less of an effect). Vascular stabilization during VEGF blockade may decrease the shedding of metastatic cancer cells from the primary tumors. Alleviation of hypoxia by bevacizumab may make the tumors more chemosensitive and less aggressive. Finally, increased cancer cell proliferation, if any, during vascular normalization may render them more sensitive to cytotoxic agents [27,118,127].

Then why did VEGF blockade by vatalanib treatment not show a clear benefit with FOLFOX4 regimen in metastatic colorectal cancer patients? An explanation could be the off-target effects. For example, vatalanib may target PDGFR-β on perivascular cells. This was suggested to be beneficial for vascular targeting since the PDGF-B–PDGFR-β axis is known to control vascular stabilization/maturation by recruitment of supporting perivascular cells. But blocking PDGFR-β may interfere with vascular normalization and thus prevent the synergistic effect of combined therapy [4]. Another explanation may be the timing or extent of VEGF blockade: vatalanib has a considerably shorter half-life (~6 h) compared to that of bevacizumab (~20 days) and the Phase III trials for vatalanib used a single daily dose of the drug. However, pharmacokinetic data suggest that an active dose of vatalanib is maintained in the blood circulation for 24 h [135]. On the other hand, daily dosing of vatalanib may shorten or preclude the normalization of the vasculature.

Despite the failure of vatalanib to improve overall survival with FOLFOX4, other TKIs, with broader inhibitory spectra, have proved efficacious in other tumor types. Multitargeted TKIs may better mimic an effective combination of antiangiogenic and cytotoxic therapy depending on their relative potency against specific tyrosine kinases, a concept supported by the progression-free survival gain produced by sorafenib in renal cell carcinoma, and increase in overall survival produced by Sutent in imatinib-resistant GIST patients. Although renal cell carcinoma is a highly VEGF-dependent malignancy, increase in overall survival by bevacizumab monotherapy has yet to be demonstrated.

These considerations imply that if we are to optimally use single- or multitargeted AAs, the treatment schemes may need to be tailored for each agent. For example, it is not yet clear whether the addition of existing TKIs (which target a number of kinases) to chemotherapy will impact outcome (e.g., vatalanib with FOLFOX4 for metastatic colorectal cancer) to an extent comparable with the responses seen for the combination of bevacizumab (which specifically targets VEGF) with chemotherapy. In choosing a multitargeted agent, the ability to define its spectrum and to match it at the molecular level with the disease will be critical, and can be tested in appropriate preclinical models.

How the AAs with indirect mechanisms of action should be combined remains unclear. These agents target cancer cells directly and block angiogenesis without directly targeting endothelial cells. In a preclinical model of HER2 overexpressing human breast cancer, trastuzumab (a HER2-specific antibody) decreased expression of several angiogenic factors and induced features consistent with vascular normalization [125]. In this tumor model, trastuzumab increased VEGF expression in stromal cells. Thus, a regimen combining the tumor cell targeting of an anti-EGFR agent with a direct AAs such as bevacizumab, or alternatively, a TKI with multiple targets including the EGFR and VEGFR pathways, may achieve efficacious, simultaneous targeting of endothelial and cancer cells. Trials combining trastuzumab, cetuximab, or erlotinib with bevacizumab as well as trials of ZD6474 (Zactima, AstraZeneca, a VEGFR, EFGR, and RET-selective TKI) have reached Phase II or III (in breast, colorectal, lung, pancreatic, and head and neck cancer patients) and promising interim data have already emerged [148].

3.4.11.1 Antiangiogenic Effect of Cytotoxic Agents

Another mechanism for the beneficial effects seen for combined cytotoxic and antiangiogenic therapy may be the direct killing of endothelial cells by chemotherapeutics. In particular, continuous, low-dose (metronomic) chemotherapy may augment the antiangiogenic therapy by killing tumor blood vessels. Preclinical data support this hypothesis [149,150]. Metronomic chemotherapy may also normalize the tumor vasculature and thus improve drug delivery to tumors [27]. Use of metronomic chemotherapy regimens has reached Phase III trials. Importantly, in preclinical models "metronomic" chemotherapy inhibited both local tumor angiogenesis and bone marrow–derived endothelial precursor cell mobilization [151]. In contrast, in the same models, a maximal tolerated dose (MTD) chemotherapy regimen had limited effect on angiogenesis and actively mobilized bone marrow–derived endothelial precursor cells [152]. These cells may promote tumor regrowth, and thus be responsible for the relapses seen with MTD chemotherapy in preclinical models and clinical trials.

To date, no Phase III trial has reported the efficacy or safety of bevacizumab with radiotherapy, although more than half of cancer patients ultimately receive radiation treatment. Direct targeting of tumor endothelial cells at certain radiation doses has been documented in mice. Fractionation and dose scheduling has been suggested to determine the relative toxicity of radiation to endothelial vs. cancer cells [153]. Given the role of tissue oxygenation in tumor response to radiation, as well as the protective role of VEGF for endothelial cell response to radiation, a greater understanding of these issues in conjunction with vascular normalization and anti-VEGF therapies will be critical.

3.4.11.2 Cytotoxicity of Antiangiogenic Agents

Treatment with AAs was designed to target tumor endothelial cells and was expected to have cytostatic effects. However, VEGF inhibition may also have cytotoxic effects on cancer cells that express VEGFRs [154–156]. Many cancer cells express the VEGFR NP-1, and its blockade has been reported to mediate apoptosis [157]. The blockade of the ligand (VEGF) prevents its binding to NP-1, while TKIs have no direct effect on this receptor because NP-1 does not have kinase activity. We have found a significant increase in rectal carcinoma cell apoptosis after bevacizumab treatment [118]. The increase in apoptosis is due to direct killing of cancer cells by bevacizumab or due to indirect killing by destruction of blood vessels is not known. This issue is even more critical for agents that target multiple growth factor pathways in endothelial and cancer cells.

3.4.12 Potential Biomarkers for Antiangiogenic Therapy

Biomarkers of efficacy of antiangiogenic therapy need to be established to validate these mechanistic hypotheses, to identify responsive patients and optimal doses, and to predict efficacy of regimens that include AAs. Optimizing the dose and schedule of bevacizumab for colorectal, breast, and lung cancer, and extrapolating the existing efficacy data to other tumor types are major undertakings, particularly since the bevacizumab doses and schedules varied from one trial to the next (Table 3.8). Intriguingly, efficacy and toxicity of bevacizumab with

TABLE 3.8
Surrogate Markers under Testing for the Evaluation of the Efficacy of Antiangiogenic Agents

Type	Marker	Parameter Evaluated	Comments/Limitation
Invasive	Tissue biopsy	Immunohistochemistry: Protein expression as a marker Microvascular density Perivascular cell coverage of tumor vessels Cell proliferation/apoptosis, genomic analyses	Not easily available in some tumors
	Interstitial fluid pressure measurement	Tumor interstitial fluid pressure	Difficult accessibility in some tumors
	Measurements of tissue oxygenation	Tumor interstitial oxygen tension	Lack of accessibility in some tumors
	Skin wound healing	Wound healing time	
Minimally invasive	Blood CECs	Concentration of viable CECs	Unclear origin, viability and surface phenotype of CECs
	Blood CPCs	Concentration of CPCs	Low concentration of CPCs in humans, heterogeneous population
	Protein level in plasma	Plasma VEGF, PlGF, TSP1	Inability to detect active vs. protein-bound bevacizumab, vatalanib, etc. (for VEGF, PlGF and VEGFRs)
	Protein level in ascites	Ascites	Limited to certain tumors, may be altered by changes in permeability induced by an AA
	Protein level in pleural effusions	Pleural fluid	Limited to certain tumors, may be altered by changes in permeability induced by an AA
Noninvasive	CT imaging	Blood flow and volume, permeability-surface area product mean transit time	Resolution, measurement of composite parameters
	PET imaging	Tracer uptake	Resolution, measurement of composite parameters
	MR imaging	Blood flow, permeability	Resolution, measurement of composite parameters
	Protein level in urine	Urine MMPs, VEGF, etc.	Limited to excreted proteins, depends on factors that may be altered by treatment such as renal function (e.g., proteinuria)

Source: Adapted from Jain, R.K. et al., *Nat. Clin. Pract. Oncol.*, 3, 24, 2006. With permission.

chemotherapy were not always dose-dependent [158], and retrospective subset analyses on primary colorectal tumors and metastases suggested that the survival benefit from the addition of bevacizumab to IFL was independent of K-RAS, B-RAF, or P53 mutation status or P53 expression [159]. Another key issue will be detecting the polymorphism of the target protein in individual patients, or changes resulting from further mutations with disease progression or in response to treatment (e.g., c-Kit in response to imatinib). Finally, biomarkers may prove crucial for identifying alternative AAs when cancer progresses through a given antiangiogenic treatment.

Considering tumor size as a marker of response to AAs has been tempted. Our own experience has demonstrated that major functional, structural, cellular, and molecular changes can occur in tumors in response to VEGF blockade without a significant reduction of tumor volume [25,26,118,134]. Thus, in the absence of an overt cytotoxic effect of the AA, other surrogate markers for efficacy must be identified. In the clinic, surrogate markers for antiangiogenic therapy have yet to be validated. Some of the candidate markers include classic diagnostic or prognostic biomarkers, as well as newly developed, target- and mechanism-based biomarkers (Table 3.6). Significant advances have been made in identifying candidate markers [160]; however, none has been shown to be predictive in clinical settings.

Biopsy of tumor tissue is a routine but highly invasive diagnostic and prognostic method, which has great potential to identify valuable markers for therapeutic efficacy. Immunostaining for the protein/cell of interest and gene profiling using laser capture–microdissection of the cellular components of interest may yield invaluable information on the effect of a given AA. For example, we have evaluated microvascular density, α-SMA and Ang-2 expression, as well as tumor cell apoptosis and proliferation during bevacizumab treatment in rectal cancer patients [118,134]. Measurement of IFP and tissue oxygenation are parameters that reflect vascular function, and by extension, delivery of therapeutics. Changes in tumor IFP [118,134] or tissue oxygen level during treatment may be valuable surrogate markers of efficacy and vascular normalization during antiangiogenic therapy, and can be sampled in several areas of the tumor.

Less invasive methods include measuring changes in protein concentration (e.g., growth factors) in bodily fluids as surrogate markers for therapy. For example, blood plasma levels of VEGF were significantly increased by VEGFR-2 blockade in mice and proposed as a surrogate marker for VEGFR-2 blockade [117]. In a trial for bevacizumab alone followed by bevacizumab and radiochemotherapy in rectal cancer patients, VEGF blockade induced a significant increase in the plasma levels of VEGF and PlGF in all patients analyzed at day 12 after bevacizumab initiation (which is close to the half-life of bevacizumab in circulation) [118]. In the case of antibody therapies, evaluation of the amount of free and active VEGF protein (in the tumor or systemic) is critical. The level of VEGF stored in platelets [161] may also serve as a surrogate marker. Despite the increase in VEGF and PlGF (which act as survival, mitogenic, and chemoattractant factors for endothelial and progenitor/stem cells), we found that bevacizumab decreased the concentration of viable CECs and progenitor cells in rectal cancer patients [118]. Notably, we found that the total number of CECs (which includes nonviable/apoptotic CECs) did not decrease, but rather increased with bevacizumab treatment, consistent with recent preclinical data [162]. This may reflect the shedding of nonviable tumor endothelial cells following antiangiogenic treatment, and may become an independent surrogate marker for antiangiogenic treatment and vascular normalization. In preclinical studies, progenitor cells have been extensively studied both as surrogate markers and for their role in cancer growth and treatment [98,151,152,163], but their concentration in whole blood in humans is very low (i.e., two orders of magnitude lower than CECs [118]). Elucidation of the biology of different subsets of progenitor cells and development of improved techniques to reliably quantify cells in this concentration range may allow the future use of progenitor cells as surrogate markers in the clinic.

Noninvasive techniques have the potential to measure functional parameters and offer surrogate markers for therapy regardless of tumor type or location [164]. They include dynamic MR, CT, and PET imaging techniques and are aggressively pursued in clinical trials by our own team and others [118,134,135,165–167]. One main limitation of these imaging methods is that they measure composite parameters, which depend on both blood flow and permeability. Another has to do with the high heterogeneity of blood flow. The spatial distribution of blood flow determines the distribution of chemotherapeutics and oxygen in tumor tissue, and cannot yet be evaluated with high spatial resolution by available imaging techniques. With improvement of these techniques to measure the spatial and temporal changes in tumor blood flow and vascular permeability with higher resolution, we may be able to assess more precisely the efficacy of AAs. Nevertheless, the cost of such investigations may be prohibitive for larger trials.

Finally, protein measurements in urine have become increasingly feasible [168–170] and exploring them during therapy may offer independent surrogate makers for the effect of AAs. Combining and comparing multiple parameters obtained using different assays in Phase I–II clinical trials hold great promise for defining the effects of individual AAs and identifying simple and meaningful surrogate markers of efficacy [118,134,165,171]. The best candidates should eventually be validated and used in larger Phase III trials and integrated into routine clinical practice [164,172,173].

3.4.13 Future Directions for the Anticancer Therapy Approach Using Antiangiogenic Agents

The approval of the first AA for clinical use in patients with colorectal carcinoma has taught us many lessons, and the most important is that these agents must be used in combination with agents that target cancer cells to have an appreciable impact on patient survival (Box 3.1). Increasing the dose of AA may harm normal tissues and destroy too much of the tumor vasculature, leading to hypoxia and poor drug delivery in the tumor and to toxicity in normal tissues. However, optimal doses and schedules of these reagents tailored to the angiogenic profile of tumors can normalize tumor vasculature and microenvironment without harming normal tissue. At least three major challenges must be met before therapies based on this vascular normalization model can be successfully translated to the clinic. The first challenge is to determine which other direct or indirect antiangiogenic therapies lead to vascular normalization. In principle, any therapy that restores the balance between pro- and antiangiogenic molecules should induce normalization [173]. Indeed, withdrawing hormones from a hormone-dependent tumor lowers VEGF levels and leads to vascular normalization [141]. Recently, metronomic therapy—a drug delivery method in which low doses of chemotherapeutic agents are given at frequent intervals—has also been shown to increase the expression of thrombospondin-1, which is a potent endogenous angiogenesis inhibitor [150]. Conceivably, this therapy might also induce normalization and improve oxygenation and drug penetration into tumors. Whether various synthetic kinase inhibitors other than AZD2171 [173] (e.g., Novartis PTK787, sorafenib, sunitinib, etc.), endogenous inhibitors other than (e.g., angiostatin, endostatin, and tumstatin), antivasocrine agents (i.e., razoxane [49]), conventional chemotherapeutic agents (e.g., taxol [50]), and vascular targeting agents [51–54] do the same remains to be described. Some of these agents may be effective because they target both stromal and cancer cells. To date, most clinical trials are designed primarily to measure changes in the size of the tumor and may therefore not shed light on changes in the vascular biology of tumors. Clinical studies, such as the rectal carcinoma study described earlier and other ongoing translational clinical trials should help bridge the gaps in this aspect of our knowledge. The second challenge is to identify suitable surrogate markers of changes in the structure and function of the tumor vasculature and to develop imaging technology, which helps to identify the timing of the normalization window during antiangiogenesis therapy. Measurement of blood vessel density requires tissue biopsy and provides little information on vessel function. Although imaging techniques are expensive and far from optimal, they can provide serial measures of vascular permeability, vascular volume, blood perfusion, and uptake of some drugs and can therefore be used to monitor the window of normalization in patients. The number of circulating mature endothelial cells and their less differentiated progenitors does decrease after VEGF blockade, both in animals and in patients, but whether this decline coincides with the normalization window is not known. During the course of therapy, serial blood measurements of molecules involved in vessel maturation have the potential to identify surrogate markers. PET with 18-fluoromisonidzole and MRI can provide some indication of tumor oxygenation and might be useful for tracking the normalization window. Finally, the measurement of the IFP is minimally invasive, inexpensive, and easy to implement for anatomically accessible tumors. Hence, this parameter could be used in the interim as a useful indicator of vessel function until novel noninvasive methods are developed. The third challenge is to fill gaps in our understanding of the molecular and cellular mechanisms of the vascular normalization process. With rapid advances in genomic and proteomic technology and access to tumor tissues during the course of therapy, we can begin to monitor tumor response to antiangiogenic therapies at the molecular level.

3.4.14 Conclusion

The recent successes of the AAs have raised great hope and have taught us important lessons about the significance of the target, timing, and dosage of each agent (Box 3.2). AAs are now expected to make a difference in cancer patients with a wide variety of tumor types. With the advent of specific and potent new agents—approved or in the process of approval—oncologists have a variety of direct and indirect AAs to choose from when designing therapy protocols. But whether the regimens used in the successful trials are optimal, and whether AAs will work in patients outside the rigorous inclusion criteria used for those trials remain to

BOX 3.2
Lessons from Current Clinical Trials of Antiangiogenic Agents

1. *Clinical application of antiangiogenic therapy is more complex than initially thought*

   The goal of this therapeutic strategy was to control solid tumors by targeting their blood vessels. Anti-VEGF agents can destroy tumor vessels in humans, but no current antiangiogenic agent (AA) has been shown to achieve this in a Phase III trial. Nevertheless, normalization of tumor vasculature and microenvironment by antiangiogenic treatment may improve delivery and efficacy of cytotoxic therapeutics. Direct cancer cell killing by the AAs may augment this effect.

2. *To date, antiangiogenic monotherapy has not been shown to increase survival in cancer patients*

   Data from current Phase III trials suggest little or no efficacy for AA monotherapy. Monotherapy is active, but it may lead to tumor escape by alternative proangiogenic pathways. One approach, yet to be proven successful, is targeting multiple angiogenesis pathways. Another approach, supported by the results of bevacizumab with chemotherapy (see later), is concomitant killing of cancer cells and blood vessels, either by combination approaches, or by using multitargeted small-molecule TKIs.

BOX 3.2 (continued)

3. *An anti-VEGF antibody increases survival rates when added to chemotherapy in first-line treatment of metastatic colorectal, breast, and lung tumor patients*

   Bevacizumab with chemotherapy enhanced survival in previously untreated metastatic colorectal cancer and lung cancer patients, and progression-free survival in chemotherapy-naïve metastatic or recurrent breast cancer patients over standard chemotherapy alone. These outcomes support the concept that bevacizumab enhances the efficacy of cytotoxic agents by normalizing tumor vasculature and microenvironment, and by enhancing the killing of cancer cells that express VEGFRs.

4. *In previously treated cancer patients, bevacizumab with chemotherapy increased survival in metastatic colorectal, but not breast cancer patients*

   These contrasting results may be due to the fact that the breast tumors were chemorefractory, while FOLFOX4, used in the colorectal cancer trial, is a proven second-line treatment.

5. *VEGF-blockade by vatalanib (a TKI specific to VEGFRs) did not enhance survival conferred by FOLFOX4 in metastatic colorectal cancer patients*

   The discrepancy between the benefits of anti-VEGF therapy with bevacizumab or vatalanib is yet to be explained. Vatalanib affects PDGFR-β—present on tumor stromal cells and involved in perivascular cell recruitment—and thus may interfere with vascular normalization. Vatalanib also blocks VEGFR-3 and c-Kit. On the other hand, vatalanib does not affect the VEGFR NP-1, present on endothelial and cancer cells.

6. *TKIs or antibodies that target multiple pathways, critical for both cancer cell and EC growth, increase survival in monotherapy*

   Multitargeted TKIs such as Sutent (which targets VEGFR-2, PDGFR-β, Flt-3, and c-Kit) and sorafenib (an inhibitor of RAF kinase, VEGFR-2, PDGFR-α and PDGFR-β, Flt-3, and c-Kit) affect pathways involved in both EC and cancer cell growth. They are the first to show increased progression-free survival in monotherapy.

7. *Several molecularly targeted therapies—approved or under trial—target directly cancer cells and indirectly tumor angiogenesis*

   FDA-approved EGFR/HER-specific antibodies (e.g., trastuzumab, cetuximab, or erlotinib) block the EGF/TGF-α signaling in cancer cells and decrease the expression of multiple angiogenic factors. Adding a VEGF-specific blocker such as bevacizumab may benefit patients by targeting VEGF and this hypothesis is currently tested. Alternatively, TKIs under clinical development may be used to block both EGFR and VEGFR.

8. *AA combination therapies have rare but serious toxicities associated with them*

   Side effects of AAs are moderate compared to other therapies, but the etiology is poorly understood. Major safety concerns have been raised by a number of

(*continued*)

BOX 3.2 (continued)

treatment-related deaths due to bowel perforation and hemorrhage (particularly in lung cancer patients). VEGF inhibition may affect the function of organs with fenestrated endothelium (e.g., kidney, thyroid, etc.). In the clinic, over 50% of cancer patients receive radiotherapy. Blockade of VEGF in this context may increase the toxicity of radiotherapy.

9. *There is an urgent need for biomarkers to guide AA monotherapy and combination therapy*

   Robust biomarkers to guide patient selection and protocol design are yet to be clinically validated. Ongoing efforts may identify such surrogate markers (e.g., proteins in tumor tissue or bodily fluids, circulating endothelial and progenitor cells, tumor physiological parameters including the measurements of tumor interstitial fluid pressure and evaluations by imaging techniques).

10. *Optimized treatment strategies are needed to prevent tumor escape from/resistance to anti-VEGF therapy and extend survival beyond 2–5 months*

    Several strategies may significantly improve treatment outcome. Patients who may draw benefit from anti-VEGF treatment need to be identified. For these patients, novel doses and schedules of combination therapies based on tumor biology and validated biomarkers should improve survival more than currently shown with combination AA and cytotoxic therapies. Moreover, the molecular mechanisms underlying tumor escape from anti-VEGF therapy and relapse should be identified and tumor angiogenesis targeted with alternative available or novel AAs. (Reproduced from Jain, R.K. et al., *Nat. Clin. Pract. Oncol.*, 3, 24, 2006.)

be determined. Establishing the most advantageous combinations to prevent tumor escape from therapy (Box 3.3) will require a better understanding of the mechanisms of action of each AA and the sensitivity of each tumor type, as well as development of robust biomarkers

BOX 3.3
Potential Mechanisms of Tumor Escape after Antiangiogenic Therapy

Beyond achieving proof-of-the-principle confirmation in clinical trials, there is a need to optimize antiangiogenic therapy to delay tumor relapse for more than a few months. The mechanisms of tumor escape from the antiangiogenic agent (AA) may be different—yet not mutually exclusive—during therapy. At advanced stages of progression, tumors secrete a variety of proangiogenic factors and become hypervascular, but the vessels are leaky and the blood flow is spatially and temporally heterogeneous. To achieve an antivascular effect, an AA should target critical angiogenic pathways in a given tumor; otherwise tumors escape early after the onset of therapy. When efficacious, the effect of antiangiogenic therapy is twofold: inhibition of new vessel formation and killing of immature tumor vessels and transient normalization of the remaining vasculature by decrease in macromolecular permeability (and thus the IFP) and hypoxia, and improvement of blood perfusion. If the cytotoxic effect of AA monotherapy is insufficient for direct killing of all cancer cells,

BOX 3.3 (continued)

or the AA is combined with a cytotoxic agent to which the tumors are refractory, the tumors relapse. Thus, efficacious cytotoxic agents should be added to efficient AAs. Even if a synergistic effect is achieved by combining the AA and the cytotoxic agents, relapse may occur after combination therapy by at least three mechanisms. First, long-term or high-dose AA therapy may lead to a vasculature which is inefficient for drug delivery (in addition to increasing the rate and grade of its side effects). Optimizing the dose and schedule of the AA may prevent these outcomes. Second, tumors may relapse using alternative pathways for neovascularization. These pathways could be targeted with available or experimental AA. Third, cancer cell clones may acquire resistance to the chemotherapeutics used. These clones (so-called "tumor stem cells") should be specifically targeted with available or novel therapeutics.

and imaging techniques to guide patient selection and protocol design. A deeper understanding of the mechanisms of antitumor activity of AAs, how they can best be combined with other treatment approaches such as chemotherapy and radiation therapy, and how these effects can best be monitored clinically, should contribute to significantly improving cancer treatment and extending survival of cancer patients in the near future and enhance the prospects of developing curative treatment for different cancers in the more distant future.

ACKNOWLEDGMENTS

This chapter contains updated information from several published reviews (References 1, 4, 27, 42, 107, 127, and 128, with permission).

REFERENCES

1. Carmeliet, P. and Jain, R.K., Angiogenesis in cancer and other diseases, *Nature*, 407, 249, 2000.
2. Ferrara, N., Gerber, H.P., and LeCouter, J., The biology of VEGF and its receptors, *Nat Med*, 9, 669, 2003.
3. Jain, R.K. and Carmeliet, P.F., Vessels of death or life, *Sci Am*, 285, 38, 2001.
4. Jain, R.K., Molecular regulation of vessel maturation, *Nat Med*, 9, 685, 2003.
5. Folkman, J., Tumor angiogenesis: therapeutic implications, *N Engl J Med*, 285, 1182, 1971.
6. Dvorak, H.F., Vascular permeability factor/vascular endothelial growth factor: a critical cytokine in tumor angiogenesis and a potential target for diagnosis and therapy, *J Clin Oncol*, 20, 4368, 2002.
7. Folkman, J. and Kalluri, R., Cancer without disease, *Nature*, 427, 787, 2004.
8. Folkman, J., Role of angiogenesis in tumor growth and metastasis, *Semin Oncol*, 29, 15, 2002.
9. Dvorak, H.F., Tumors: wounds that do not heal. Similarities between tumor stroma generation and wound healing, *N Engl J Med*, 315, 1650, 1986.
10. Pollard, J.W., Tumour-educated macrophages promote tumour progression and metastasis, *Nat Rev Cancer*, 4, 71, 2004.
11. Coussens, L.M. and Werb, Z., Inflammation and cancer, *Nature*, 420, 860, 2002.
12. Balkwill, F. and Coussens, L.M., Cancer: an inflammatory link, *Nature*, 431, 405, 2004.
13. Fukumura, D., Xavier, R., Sugiura, T., Chen, Y., Park, E.C., Lu, N., Selig, M., Nielsen, G., Taksir, T., Jain, R.K., and Seed, B., Tumor induction of VEGF promoter activity in stromal cells, *Cell*, 94, 715, 1998.

14. Brown, E.B., Campbell, R.B., Tsuzuki, Y., Xu, L., Carmeliet, P., Fukumura, D., and Jain, R.K., In vivo measurement of gene expression, angiogenesis, and physiological function in tumors using multiphoton laser scanning microscopy, *Nat Med*, 7, 864, 2001.
15. Egeblad, M. and Werb, Z., New functions for the matrix metalloproteinases in cancer progression, *Nat Rev Cancer*, 2, 161, 2002.
16. di Tomaso, E., Capen, D., Haskell, A., Hart, J., Logie, J.J., Jain, R.K., McDonald, D.M., Jones, R., and Munn, L.L., Mosaic tumor vessels: cellular basis and ultrastructure of focal regions lacking endothelial cell markers, *Cancer Res*, 65, 5740, 2005.
17. Chang, Y.S., di Tomaso, E., McDonald, D.M., Jones, R., Jain, R.K., and Munn, L.L., Mosaic blood vessels in tumors: frequency of cancer cells in contact with flowing blood, *Proc Natl Acad Sci USA*, 97, 14608, 2000.
18. Baluk, P., Morikawa, S., Haskell, A., Mancuso, M., and McDonald, D.M., Abnormalities of basement membrane on blood vessels and endothelial sprouts in tumors, *Am J Pathol*, 163, 1801, 2003.
19. Morikawa, S., Baluk, P., Kaidoh, T., Haskell, A., Jain, R.K., and McDonald, D.M., Abnormalities in pericytes on blood vessels and endothelial sprouts in tumors, *Am J Pathol*, 160, 985, 2002.
20. Hashizume, H., Baluk, P., Morikawa, S., McLean, J.W., Thurston, G., Roberge, S., Jain, R.K., and McDonald, D.M., Openings between defective endothelial cells contribute to tumor vessel leakiness, *Am J Pathol*, 156, 1363, 2000.
21. Weis, S.M. and Cheresh, D.A., Pathophysiological consequences of VEGF-induced vascular permeability, *Nature*, 437, 497, 2005.
22. De Palma, M., Venneri, M.A., Roca, C., and Naldini, L., Targeting exogenous genes to tumor angiogenesis by transplantation of genetically modified hematopoietic stem cells, *Nat Med*, 9, 789, 2003.
23. Direkze, N.C., Hodivala-Dilke, K., Jeffery, R., Hunt, T., Poulsom, R., Oukrif, D., Alison, M.R., and Wright, N.A., Bone marrow contribution to tumor associated myofibroblasts and fibroblasts, *Cancer Res*, 64, 8485, 2004.
24. Rajantie, I., Ilmonen, M., Alminaite, A., Ozerdem, U., Alitalo, K., and Salven, P., Adult bone marrow–derived cells recruited during angiogenesis comprise precursors for periendothelial vascular mural cells, *Blood*, 104, 2084, 2004.
25. Tong, R.T., Boucher, Y., Kozin, S.V., Winkler, F., Hicklin, D.J., and Jain, R.K., Vascular normalization by vascular endothelial growth factor receptor 2 blockade induces a pressure gradient across the vasculature and improves drug penetration in tumors, *Cancer Res*, 64, 3731, 2004.
26. Winkler, F., Kozin, S.V., Tong, R., Chae, S., Booth, M.F., Garkavtsev, I., Xu, L., Hicklin, D.J., Fukumura, D., di Tomaso, E., Munn, L.L., and Jain, R.K., Kinetics of vascular normalization by VEGFR2 blockade governs brain tumor response to radiation: role of oxygenation, angiopoietin-1 and matrix metalloproteinases, *Cancer Cell*, 6, 553, 2004.
27. Jain, R.K., Normalization of tumor vasculature: an emerging concept in antiangiogenic therapy, *Science*, 307, 58, 2005.
28. Jain, R.K., Barriers to drug delivery in solid tumors, *Sci Am*, 271, 58, 1994.
29. Jain, R.K., Tong, R.T., and Munn, L.L., Effect of vascular normalization by anti-angiogenic therapy on interstitial hypertension, peri-tumor edema and lymphatic metastasis: insights from a mathematical model, *Cancer Res*, 67, 2729, 2007.
30. Padera, T.P., Stoll, B.R., Tooredman, J.B., Capen, D., di Tomaso, E., and Jain, R.K., Pathology: cancer cells compress intratumour vessels, *Nature*, 427, 695, 2004.
31. Semenza, G.L., Targeting HIF-1 for cancer therapy, *Nat Rev Cancer*, 3, 721, 2003.
32. Fidler, I.J., The pathogenesis of cancer metastasis: the 'seed and soil' hypothesis revisited, *Nat Rev Cancer*, 3, 453, 2003.
33. Vogelstein, B. and Kinzler, K.W., Cancer genes and the pathways they control, *Nat Med*, 10, 789, 2004.
34. Bergers, G., Brekken, R., McMahon, G., Vu, T.H., Itoh, T., Tamaki, K., Tanzawa, K., Thorpe, P., Itohara, S., Werb, Z., and Hanahan, D., Matrix metalloproteinase-9 triggers the angiogenic switch during carcinogenesis, *Nat Cell Biol*, 2, 737, 2000.
35. Coussens, L.M., Tinkle, C.L., Hanahan, D., and Werb, Z., MMP-9 supplied by bone marrow–derived cells contributes to skin carcinogenesis, *Cell*, 103, 481, 2000.

36. Hiratsuka, S., Nakamura, K., Iwai, S., Murakami, M., Itoh, T., Kijima, H., Shipley, J.M., Senior, R.M., and Shibuya, M., MMP9 induction by vascular endothelial growth factor receptor-1 is involved in lung-specific metastasis, *Cancer Cell*, 2, 289, 2002.
37. Huang, S., Van Arsdall, M., Tedjarati, S., McCarty, M., Wu, W., Langley, R., and Fidler, I.J., Contributions of stromal metalloproteinase-9 to angiogenesis and growth of human ovarian carcinoma in mice, *J Natl Cancer Inst*, 94, 1134, 2002.
38. Ben-Yosef, Y., Lahat, N., Shapiro, S., Bitterman, H., and Miller, A., Regulation of endothelial matrix metalloproteinase-2 by hypoxia/reoxygenation, *Circ Res*, 90, 784, 2002.
39. Hamano, Y., Zeisberg, M., Sugimoto, H., Lively, J.C., Maeshima, Y., Yang, C., Hynes, R.O., Werb, Z., Sudhakar, A., and Kalluri, R., Physiological levels of tumstatin, a fragment of collagen IV alpha3 chain, are generated by MMP-9 proteolysis and suppress angiogenesis via alphaV beta3 integrin, *Cancer Cell*, 3, 589, 2003.
40. Kurz, H., Burri, P.H., and Djonov, V.G., Angiogenesis and vascular remodeling by intussusception: from form to function, *News Physiol Sci*, 18, 65, 2003.
41. Asahara, T., Murohara, T., Sullivan, A., Silver, M., van der Zee, R., Li, T., Witzenbichler, B., Schatteman, G., and Isner, J.M., Isolation of putative progenitor endothelial cells for angiogenesis, *Science*, 275, 964, 1997.
41b. Baffert, F., Le T., Scnnino, B., Thurston, G., Kuo, C.J., Hu-Lowe, D., and McDonald, D.M., Cellular changes in normal blood capillaries undergoing regression after inhibition of VEGF signaling. *Am J Physiol Heart Circ Physiol*, 290, H547, 2006.
42. Jain, R.K. and Duda, D.G., Role of bone marrow–derived cells in tumor angiogenesis and treatment, *Cancer Cell*, 3, 515, 2003.
43. Takahashi, T., Kalka, C., Masuda, H., Chen, D., Silver, M., Kearney, M., Magner, M., Isner, J.M., and Asahara, T., Ischemia- and cytokine-induced mobilization of bone marrow–derived endothelial progenitor cells for neovascularization, *Nat Med*, 5, 434, 1999.
44. Lyden, D., Hattori, K., Dias, S., Costa, C., Blaikie, P., Butros, L., Chadburn, A., Heissig, B., Marks, W., Witte, L., Wu, Y., Hicklin, D., Zhu, Z., Hackett, N.R., Crystal, R.G., Moore, M.A., Hajjar, K.A., Manova, K., Benezra, R., and Rafii, S., Impaired recruitment of bone-marrow–derived endothelial and hematopoietic precursor cells blocks tumor angiogenesis and growth, *Nat Med*, 7, 1194, 2001.
45. Carmeliet, P., Moons, L., Luttun, A., Vincenti, V., Compernolle, V., De Mol, M., Wu, Y., Bono, F., Devy, L., Beck, H., Scholz, D., Acker, T., DiPalma, T., Dewerchin, M., Noel, A., Stalmans, I., Barra, A., Blacher, S., Vandendriessche, T., Ponten, A., Eriksson, U., Plate, K.H., Foidart, J.M., Schaper, W., Charnock-Jones, D.S., Hicklin, D.J., Herbert, J.M., Collen, D., and Persico, M.G., Synergism between vascular endothelial growth factor and placental growth factor contributes to angiogenesis and plasma extravasation in pathological conditions, *Nat Med*, 7, 575, 2001.
46. Conejo-Garcia, J.R., Benencia, F., Courreges, M.C., Kang, E., Mohamed-Hadley, A., Buckanovich, R.J., Holtz, D.O., Jenkins, A., Na, H., Zhang, L., Wagner, D.S., Katsaros, D., Caroll, R., and Coukos, G., Tumor-infiltrating dendritic cell precursors recruited by a beta-defensin contribute to vasculogenesis under the influence of VEGF-A, *Nat Med*, 10, 950, 2004.
47. Conejo-Garcia, J.R., Buckanovich, R.J., Benencia, F., Courreges, M.C., Rubin, S.C., Carroll, R.G., and Coukos, G., Vascular leukocytes contribute to tumor vascularization, *Blood*, 105, 679, 2005.
48. Yang, L., DeBusk, L.M., Fukuda, K., Fingleton, B., Green-Jarvis, B., Shyr, Y., Matrisian, L.M., Carbone, D.P., and Lin, P.C., Expansion of myeloid immune suppressor Gr + CD11b + cells in tumor-bearing host directly promotes tumor angiogenesis, *Cancer Cell*, 6, 409, 2004.
49. De Palma, M., Venneri, M.A., Galli, R., Sergi, L.S., Politi, L.S., Sampaolesi, M., and Naldini, L., Tie2 identifies a hematopoietic lineage of proangiogenic monocytes required for tumor vessel formation and a mesenchymal population of pericyte progenitors, *Cancer Cell*, 8, 211, 2005.
50. Duda, D.G., Cohen, K.S., Kozin, S.V., Perentes, J.Y., Fukumura, D., Scadden, D.T., and Jain, R.K., Evidence for the endothelial phenotype of bone marrow–derived cells in perfused blood vessels in tumors and adipose tissue, *Blood*, 107, 2774, 2006.
51. Kuperwasser, C., Chavarria, T., Wu, M., Magrane, G., Gray, J.W., Carey, L., Richardson, A., and Weinberg, R.A., Reconstruction of functionally normal and malignant human breast tissues in mice, *Proc Natl Acad Sci USA*, 101, 4966, 2004.

52. Joyce, J.A., Therapeutic targeting of the tumor microenvironment, *Cancer Cell*, 7, 513, 2005.
53. Ferrara, N., Hillan, K.J., Gerber, H.P., and Novotny, W., Discovery and development of bevacizumab, an anti-VEGF antibody for treating cancer, *Nat Rev Drug Discov*, 3, 391, 2004.
54. Nagy, J.A., Vasile, E., Feng, D., Sundberg, C., Brown, L.F., Detmar, M.J., Lawitts, J.A., Benjamin, L., Tan, X., Manseau, E.J., Dvorak, A.M., and Dvorak, H.F., Vascular permeability factor/vascular endothelial growth factor induces lymphangiogenesis as well as angiogenesis, *J Exp Med*, 196, 1497, 2002.
55. Paavonen, K., Puolakkainen, P., Jussila, L., Jahkola, T., and Alitalo, K., Vascular endothelial growth factor receptor-3 in lymphangiogenesis in wound healing, *Am J Pathol*, 156, 1499, 2000.
56. Kadambi, A., Mouta Carreira, C., Yun, C.O., Padera, T.P., Dolmans, D.E., Carmeliet, P., Fukumura, D., and Jain, R.K., Vascular endothelial growth factor (VEGF)-C differentially affects tumor vascular function and leukocyte recruitment: role of VEGF-receptor 2 and host VEGF-A, *Cancer Res*, 61, 2404, 2001.
57. Bussolati, B., Deambrosis, I., Russo, S., Deregibus, M.C., and Camussi, G., Altered angiogenesis and survival in human tumor–derived endothelial cells, *FASEB J*, 17, 1159, 2003.
58. Papapetropoulos, A., Garcia-Cardena, G., Dengler, T.J., Maisonpierre, P.C., Yancopoulos, G.D., and Sessa, W.C., Direct actions of angiopoietin-1 on human endothelium: evidence for network stabilization, cell survival, and interaction with other angiogenic growth factors, *Lab Invest*, 79, 213, 1999.
59. Zagzag, D., Hooper, A., Friedlander, D.R., Chan, W., Holash, J., Wiegand, S.J., Yancopoulos, G.D., and Grumet, M., In situ expression of angiopoietins in astrocytomas identifies angiopoietin-2 as an early marker of tumor angiogenesis, *Exp Neurol*, 159, 391, 1999.
60. Mizukami, Y., Jo, W.S., Duerr, E.M., Gala, M., Li, J., Zhang, X., Zimmer, M.A., Iliopoulos, O., Zukerberg, L.R., Kohgo, Y., Lynch, M.P., Rueda, B.R., and Chung, D.C., Induction of interleukin-8 preserves the angiogenic response in HIF-1alpha-deficient colon cancer cells, *Nat Med*, 11, 992, 2005.
61. Kim, S.J., Uehara, H., Yazici, S., Langley, R.R., He, J., Tsan, R., Fan, D., Killion, J.J., and Fidler, I.J., Simultaneous blockade of platelet-derived growth factor-receptor and epidermal growth factor-receptor signaling and systemic administration of paclitaxel as therapy for human prostate cancer metastasis in bone of nude mice, *Cancer Res*, 64, 4201, 2004.
62. Kryczek, I., Lange, A., Mottram, P., Alvarez, X., Cheng, P., Hogan, M., Moons, L., Wei, S., Zou, L., Machelon, V., Emilie, D., Terrassa, M., Lackner, A., Curiel, T.J., Carmeliet, P., and Zou, W., CXCL12 and vascular endothelial growth factor synergistically induce neoangiogenesis in human ovarian cancers, *Cancer Res*, 65, 465, 2005.
63. Relf, M., LeJeune, S., Scott, P.A., Fox, S., Smith, K., Leek, R., Moghaddam, A., Whitehouse, R., Bicknell, R., and Harris, A.L., Expression of the angiogenic factors vascular endothelial cell growth factor, acidic and basic fibroblast growth factor, tumor growth factor beta-1, platelet-derived endothelial cell growth factor, placenta growth factor, and pleiotrophin in human primary breast cancer and its relation to angiogenesis, *Cancer Res*, 57, 963, 1997.
64. Shin, D., Garcia-Cardena, G., Hayashi, S., Gerety, S., Asahara, T., Stavrakis, G., Isner, J., Folkman, J., Gimbrone, M.A., Jr., and Anderson, D.J., Expression of ephrinB2 identifies a stable genetic difference between arterial and venous vascular smooth muscle as well as endothelial cells, and marks subsets of microvessels at sites of adult neovascularization, *Dev Biol*, 230, 139, 2001.
65. Zeng, Q., Li, S., Chepeha, D.B., Giordano, T.J., Li, J., Zhang, H., Polverini, P.J., Nor, J., Kitajewski, J., and Wang, C.Y., Crosstalk between tumor and endothelial cells promotes tumor angiogenesis by MAPK activation of Notch signaling, *Cancer Cell*, 8, 13, 2005.
66. Kermani, P., Rafii, D., Jin, D.K., Whitlock, P., Schaffer, W., Chiang, A., Vincent, L., Friedrich, M., Shido, K., Hackett, N.R., Crystal, R.G., Rafii, S., and Hempstead, B.L., Neurotrophins promote revascularization by local recruitment of TrkB + endothelial cells and systemic mobilization of hematopoietic progenitors, *J Clin Invest*, 115, 653, 2005.
67. Carmeliet, P., Ferreira, V., Breier, G., Pollefeyt, S., Kieckens, L., Gertsenstein, M., Fahrig, M., Vandenhoeck, A., Harpal, K., Eberhardt, C., Declercq, C., Pawling, J., Moons, L., Collen, D., Risau, W., and Nagy, A., Abnormal blood vessel development and lethality in embryos lacking a single VEGF allele, *Nature*, 380, 435, 1996.
68. Ferrara, N., Carver-Moore, K., Chen, H., Dowd, M., Lu, L., O'Shea, K.S., Powell-Braxton, L., Hillan, K.J., and Moore, M.W., Heterozygous embryonic lethality induced by targeted inactivation of the VEGF gene, *Nature*, 380, 439, 1996.

69. Damert, A., Miquerol, L., Gertsenstein, M., Risau, W., and Nagy, A., Insufficient VEGFA activity in yolk sac endoderm compromises haematopoietic and endothelial differentiation, *Development*, 129, 1881, 2002.
70. Fong, G.H., Zhang, L., Bryce, D.M., and Peng, J., Increased hemangioblast commitment, not vascular disorganization, is the primary defect in flt-1 knock-out mice, *Development*, 126, 3015, 1999.
71. Schuh, A.C., Faloon, P., Hu, Q.L., Bhimani, M., and Choi, K., In vitro hematopoietic and endothelial potential of flk-1(−/−) embryonic stem cells and embryos, *Proc Natl Acad Sci USA*, 96, 2159, 1999.
72. Klagsbrun, M., Takashima, S., and Mamluk, R., The role of neuropilin in vascular and tumor biology, *Adv Exp Med Biol*, 515, 33, 2002.
73. Autiero, M., Waltenberger, J., Communi, D., Kranz, A., Moons, L., Lambrechts, D., Kroll, J., Plaisance, S., De Mol, M., Bono, F., Kliche, S., Fellbrich, G., Ballmer-Hofer, K., Maglione, D., Mayr-Beyrle, U., Dewerchin, M., Dombrowski, S., Stanimirovic, D., Van Hummelen, P., Dehio, C., Hicklin, D.J., Persico, G., Herbert, J.M., Communi, D., Shibuya, M., Collen, D., Conway, E.M., and Carmeliet, P., Role of PlGF in the intra- and intermolecular cross talk between the VEGF receptors Flt1 and Flk1, *Nat Med*, 9, 936, 2003.
74. Patan, S., TIE1 and TIE2 receptor tyrosine kinases inversely regulate embryonic angiogenesis by the mechanism of intussusceptive microvascular growth, *Microvasc Res*, 56, 1, 1998.
75. Suri, C., Jones, P.F., Patan, S., Bartunkova, S., Maisonpierre, P.C., Davis, S., Sato, T.N., and Yancopoulos, G.D., Requisite role of angiopoietin-1, a ligand for the TIE2 receptor, during embryonic angiogenesis, *Cell*, 87, 1171, 1996.
76. Yu, Q. and Stamenkovic, I., Angiopoietin-2 is implicated in the regulation of tumor angiogenesis, *Am J Pathol*, 158, 563, 2001.
77. Oliner, J., Min, H., Leal, J., Yu, D., Rao, S., You, E., Tang, X., Kim, H., Meyer, S., Han, S.J., Hawkins, N., Rosenfeld, R., Davy, E., Graham, K., Jacobsen, F., Stevenson, S., Ho, J., Chen, Q., Hartmann, T., Michaels, M., Kelley, M., Li, L., Sitney, K., Martin, F., Sun, J.R., Zhang, N., Lu, J., Estrada, J., Kumar, R., Coxon, A., Kaufman, S., Pretorius, J., Scully, S., Cattley, R., Payton, M., Coats, S., Nguyen, L., Desilva, B., Ndifor, A., Hayward, I., Radinsky, R., Boone, T., and Kendall, R., Suppression of angiogenesis and tumor growth by selective inhibition of angiopoietin-2, *Cancer Cell*, 6, 507, 2004.
78. Kim, I., Kim, J.H., Moon, S.O., Kwak, H.J., Kim, N.G., and Koh, G.Y., Angiopoietin-2 at high concentration can enhance endothelial cell survival through the phosphatidylinositol 3′-kinase/Akt signal transduction pathway, *Oncogene*, 19, 4549, 2000.
79. Garkavtsev, I., Kozin, S.V., Chernova, O., Xu, L., Winkler, F., Brown, E., Barnett, G.H., and Jain, R.K., The candidate tumour suppressor protein ING4 regulates brain tumour growth and angiogenesis, *Nature*, 428, 328, 2004.
80. Noren, N.K., Lu, M., Freeman, A.L., Koolpe, M., and Pasquale, E.B., Interplay between EphB4 on tumor cells and vascular ephrin-B2 regulates tumor growth, *Proc Natl Acad Sci USA*, 101, 5583, 2004.
80b. Gridley T., Vascular biology: vessel-guidance, *Nature*, 445, 722, 2007.
80c. Hermansson, M., Nister, M., Betsholtz, C., Heldin, C.H., Westermark, B., and Funa, K., Endothelial cell hyperplasia in human glioblastoma: coexpression of mRNA for platelet-derived growth factor (PDGF) B-chain and PDGF receptor suggests autocrine growth stimulation. *Proc Natl Acad Sci USA*, 85, 7748, 1988.
81. Li, X., Tjwa, M., Moons, L., Fons, P., Noel, A., Ny, A., Zhou, J.M., Lennartsson, J., Li, H., Luttun, A., Ponten, A., Devy, L., Bouche, A., Oh, H., Manderveld, A., Blacher, S., Communi, D., Savi, P., Bono, F., Dewerchin, M., Foidart, J.M., Autiero, M., Herbert, J.M., Collen, D., Heldin, C.H., Eriksson, U., and Carmeliet, P., Revascularization of ischemic tissues by PDGF-CC via effects on endothelial cells and their progenitors, *J Clin Invest*, 115, 118, 2005.
82. Duda, D.G. and Jain, R.K., Pleiotropy of tissue-specific growth factors: from neurons to vessels via the bone marrow, *J Clin Invest*, 115, 596, 2005.
83. Carmeliet, P. and Tessier-Lavigne, M., Common mechanisms of nerve and blood vessel wiring, *Nature*, 436, 193, 2005.
84. Guo, W. and Giancotti, F.G., Integrin signalling during tumour progression, *Nat Rev Mol Cell Biol*, 5, 816, 2004.
85. Hood, J.D., Bednarski, M., Frausto, R., Guccione, S., Reisfeld, R.A., Xiang, R., and Cheresh, D.A., Tumor regression by targeted gene delivery to the neovasculature, *Science*, 296, 2404, 2002.

86. Hynes, R.O., A reevaluation of integrins as regulators of angiogenesis, *Nat Med*, 8, 918, 2002.
87. Jin, H. and Varner, J., Integrins: roles in cancer development and as treatment targets, *Br J Cancer*, 90, 561, 2004.
88. Uemura, A., Ogawa, M., Hirashima, M., Fujiwara, T., Koyama, S., Takagi, H., Honda, Y., Wiegand, S.J., Yancopoulos, G.D., and Nishikawa, S., Recombinant angiopoietin-1 restores higher-order architecture of growing blood vessels in mice in the absence of mural cells, *J Clin Invest*, 110, 1619, 2002.
89. Chambers, R.C., Leoni, P., Kaminski, N., Laurent, G.J., and Heller, R.A., Global expression profiling of fibroblast responses to transforming growth factor-beta1 reveals the induction of inhibitor of differentiation-1 and provides evidence of smooth muscle cell phenotypic switching, *Am J Pathol*, 162, 533, 2003.
90. Gohongi, T., Fukumura, D., Boucher, Y., Yun, C.O., Soff, G.A., Compton, C., Todoroki, T., and Jain, R.K., Tumor–host interactions in the gallbladder suppress distal angiogenesis and tumor growth: involvement of transforming growth factor beta1, *Nat Med*, 5, 1203, 1999.
91. Hayashi, N. and Cunha, G.R., Mesenchyme-induced changes in the neoplastic characteristics of the Dunning prostatic adenocarcinoma, *Cancer Res*, 51, 4924, 1991.
92. Olumi, A.F., Grossfeld, G.D., Hayward, S.W., Carroll, P.R., Tlsty, T.D., and Cunha, G.R., Carcinoma-associated fibroblasts direct tumor progression of initiated human prostatic epithelium, *Cancer Res*, 59, 5002, 1999.
93. Bhowmick, N.A., Chytil, A., Plieth, D., Gorska, A.E., Dumont, N., Shappell, S., Washington, M.K., Neilson, E.G., and Moses, H.L., TGF-beta signaling in fibroblasts modulates the oncogenic potential of adjacent epithelia, *Science*, 303, 848, 2004.
94. Bhowmick, N.A., Neilson, E.G., and Moses, H.L., Stromal fibroblasts in cancer initiation and progression, *Nature*, 432, 332, 2004.
95. Orimo, A., Gupta, P.B., Sgroi, D.C., Arenzana-Seisdedos, F., Delaunay, T., Naeem, R., Carey, V.J., Richardson, A.L., and Weinberg, R.A., Stromal fibroblasts present in invasive human breast carcinomas promote tumor growth and angiogenesis through elevated SDF-1/CXCL12 secretion, *Cell*, 121, 335, 2005.
96. Allinen, M., Beroukhim, R., Cai, L., Brennan, C., Lahti-Domenici, J., Huang, H., Porter, D., Hu, M., Chin, L., Richardson, A., Schnitt, S., Sellers, W.R., and Polyak, K., Molecular characterization of the tumor microenvironment in breast cancer, *Cancer Cell*, 6, 17, 2004.
97. Dong, J., Grunstein, J., Tejada, M., Peale, F., Frantz, G., Liang, W.C., Bai, W., Yu, L., Kowalski, J., Liang, X., Fuh, G., Gerber, H.P., and Ferrara, N., VEGF-null cells require PDGFR alpha signaling-mediated stromal fibroblast recruitment for tumorigenesis, *EMBO J*, 23, 2800, 2004.
98. Rafii, S., Lyden, D., Benezra, R., Hattori, K., and Heissig, B., Vascular and haematopoietic stem cells: novel targets for anti-angiogenesis therapy? *Nat Rev Cancer*, 2, 826, 2002.
99. de Visser, K.E., Korets, L.V., and Coussens, L.M., De novo carcinogenesis promoted by chronic inflammation is B lymphocyte dependent, *Cancer Cell*, 7, 411, 2005.
100. Hanahan, D., Lanzavecchia, A., and Mihich, E., Fourteenth Annual Pezcoller Symposium: the novel dichotomy of immune interactions with tumors, *Cancer Res*, 63, 3005, 2003.
101. Houghton, J., Stoicov, C., Nomura, S., Rogers, A.B., Carlson, J., Li, H., Cai, X., Fox, J.G., Goldenring, J.R., and Wang, T.C., Gastric cancer originating from bone marrow–derived cells, *Science*, 306, 1568, 2004.
102. Thun, M.J., Namboodiri, M.M., and Heath, C.W., Jr., Aspirin use and reduced risk of fatal colon cancer, *N Engl J Med*, 325, 1593, 1991.
103. Hildenbrand, R., Wolf, G., Bohme, B., Bleyl, U., and Steinborn, A., Urokinase plasminogen activator receptor (CD87) expression of tumor-associated macrophages in ductal carcinoma in situ, breast cancer, and resident macrophages of normal breast tissue, *J Leukoc Biol*, 66, 40, 1999.
104. Giraudo, E., Inoue, M., and Hanahan, D., An amino-bisphosphonate targets MMP-9-expressing macrophages and angiogenesis to impair cervical carcinogenesis, *J Clin Invest*, 114, 623, 2004.
105. Joyce, J.A., Baruch, A., Chehade, K., Meyer-Morse, N., Giraudo, E., Tsai, F.Y., Greenbaum, D.C., Hager, J.H., Bogyo, M., and Hanahan, D., Cathepsin cysteine proteases are effectors of invasive growth and angiogenesis during multistage tumorigenesis, *Cancer Cell*, 5, 443, 2004.
106. Garber, K., Angiogenesis inhibitors suffer new setback, *Nat Biotechnol*, 20, 1067, 2002.

107. Jain, R.K., Duda, D.G., Clark, J.W., and Loeffler, L.S., Lessons from Phase III clinical trials for anti-VEGF therapy for cancer, *Nat Clin Pract Oncol*, 3, 24, 2006.
108. Kerbel, R. and Folkman, J., Clinical translation of angiogenesis inhibitors, *Nat Rev Cancer*, 2, 727, 2002.
109. Hicklin, D.J. and Ellis, L.M., Role of the vascular endothelial growth factor pathway in tumor growth and angiogenesis, *J Clin Oncol*, 23, 1011, 2005.
110. http://www.nci.nih.gov/newscenter, Accessed in August 2005.
111. Hurwitz, H., Fehrenbacher, L., Novotny, W., Cartwright, T., Hainsworth, J., Heim, W., Berlin, J., Baron, A., Griffing, S., Holmgren, E., Ferrara, N., Fyfe, G., Rogers, B., Ross, R., and Kabbinavar, F., Bevacizumab plus irinotecan, fluorouracil, and leucovorin for metastatic colorectal cancer, *N Engl J Med*, 350, 2335, 2004.
112. Miller, K.D., Chap, L.I., Holmes, F.A., Cobleigh, M.A., Marcom, P.K., Fehrenbacher, L., Dickler, M., Overmoyer, B.A., Reimann, J.D., Sing, A.P., Langmuir, V., and Rugo, H.S., Randomized phase III trial of capecitabine compared with bevacizumab plus capecitabine in patients with previously treated metastatic breast cancer, *J Clin Oncol*, 23, 792, 2005.
113. News in Brief, *Nat Rev Drug Discov*, 4, 448, 2005.
114. Rini, B.I., Sosman, J.A., and Motzer, R.J., Therapy targeted at vascular endothelial growth factor in metastatic renal cell carcinoma: biology, clinical results and future development, *BJU Int*, 96, 286, 2005.
115. Branca, M.A., Multi-kinase inhibitors create buzz at ASCO, *Nat Biotechnol*, 23, 639, 2005.
116. Gorski, D.H., Beckett, M.A., Jaskowiak, N.T., Calvin, D.P., Mauceri, H.J., Salloum, R.M., Seetharam, S., Koons, A., Hari, D.M., Kufe, D.W., and Weichselbaum, R.R., Blockage of the vascular endothelial growth factor stress response increases the antitumor effects of ionizing radiation, *Cancer Res*, 59, 3374, 1999.
117. Bocci, G., Man, S., Green, S.K., Francia, G., Ebos, J.M., du Manoir, J.M., Weinerman, A., Emmenegger, U., Ma, L., Thorpe, P., Davidoff, A., Huber, J., Hicklin, D.J., and Kerbel, R.S., Increased plasma vascular endothelial growth factor (VEGF) as a surrogate marker for optimal therapeutic dosing of VEGF receptor-2 monoclonal antibodies, *Cancer Res*, 64, 6616, 2004.
118. Willett, C.G., Boucher, Y., Duda, D.G., di Tomaso, E., Munn, L.L., Tong, R., Petit, L., Chung, D.C., Sahani, D.V., Kalva, S.P., Kozin, S.V., Cohen, K.S., Scadden, D.T., Fischman, A.J., Clark, J.W., Ryan, D.P., Zhu, A.X., Blaszkowsky, L.S., Mino-Knudson, M., Shellito, P.C., Lauwers, G.Y., and Jain, R.K., Surrogate markers for antiangiogenic therapy and dose-limiting toxicities for bevacizumab with radio-chemotherapy: continued experience of a Phase I trial in rectal cancer patients, *J Clin Oncol*, 23, 8136, 2005.
119. Yang, J.C., Haworth, L., Sherry, R.M., Hwu, P., Schwartzentruber, D.J., Topalian, S.L., Steinberg, S.M., Chen, H.X., and Rosenberg, S.A., A randomized trial of bevacizumab, an anti-vascular endothelial growth factor antibody, for metastatic renal cancer, *N Engl J Med*, 349, 427, 2003.
120. Tuma, R.S., Success of bevacizumab trials raises questions for future studies, *J Natl Cancer Inst*, 97, 950, 2005.
121. Chen, H.X., Expanding the clinical development of bevacizumab, *Oncologist*, 9(Suppl 1), 27, 2004.
122. Rosen, L.S., VEGF-targeted therapy: therapeutic potential and recent advances, *Oncologist*, 10, 382, 2005.
123. Sandler, A.B., Clinical trials comparing carboplatin/paclitaxel with or without bevacizumab in patients with metastatic NSCLC, *Lung Cancer Update*, 2, 6, 2005.
124. Marx, J., Cancer. Encouraging results for second-generation antiangiogenesis drugs, *Science*, 308, 1248, 2005.
125. Izumi, Y., Xu, L., di Tomaso, E., Fukumura, D., and Jain, R.K., Herceptin acts as an anti-angiogenic cocktail, *Nature*, 416, 279, 2002.
126. Teicher, B.A., A systems approach to cancer therapy (antioncogenics + standard cytotoxics→mechanism(s) of interaction), *Cancer Metastasis Rev*, 15, 247, 1996.
127. Jain, R.K., Antiangiogenic therapy for cancer: current and emerging concepts, *Oncology* (*Williston Park*), 19(Suppl), 7, 2005.
128. Jain, R.K., Normalizing tumor vasculature with anti-angiogenic therapy: a new paradigm for combination therapy, *Nat Med*, 7, 987, 2001.

129. Hobbs, S.K., Monsky, W.L., Yuan, F., Roberts, W.G., Griffith, L., Torchilin, V.P., and Jain, R.K., Regulation of transport pathways in tumor vessels: role of tumor type and microenvironment, *Proc Natl Acad Sci USA*, 95, 4607, 1998.
130. Jain, R.K., Understanding barriers to drug delivery: high resolution in vivo imaging is key, *Clin Cancer Res*, 5, 1605, 1999.
131. Fenton, B.M., Paoni, S.F., and Ding, I., Effect of VEGF receptor-2 antibody on vascular function and oxygenation in spontaneous and transplanted tumors, *Radiother Oncol*, 72, 221, 2004.
132. Ma, J., Pulfer, S., Li, S., Chu, J., Reed, K., and Gallo, J.M., Pharmacodynamic-mediated reduction of temozolomide tumor concentrations by the angiogenesis inhibitor TNP-470, *Cancer Res*, 61, 5491, 2001.
133. Murata, R., Nishimura, Y., and Hiraoka, M., An antiangiogenic agent (TNP-470) inhibited reoxygenation during fractionated radiotherapy of murine mammary carcinoma, *Int J Radiat Oncol Biol Phys*, 37, 1107, 1997.
134. Willett, C.G., Boucher, Y., di Tomaso, E., Duda, D.G., Munn, L.L., Tong, R.T., Chung, D.C., Sahani, D.V., Kalva, S.P., Kozin, S.V., Mino, M., Cohen, K.S., Scadden, D.T., Hartford, A.C., Fischman, A.J., Clark, J.W., Ryan, D.P., Zhu, A.X., Blaszkowsky, L.S., Chen, H.X., Shellito, P.C., Lauwers, G.Y., and Jain, R.K., Direct evidence that the VEGF-specific antibody bevacizumab has antivascular effects in human rectal cancer, *Nat Med*, 10, 145, 2004.
135. Morgan, B., Thomas, A.L., Drevs, J., Hennig, J., Buchert, M., Jivan, A., Horsfield, M.A., Mross, K., Ball, H.A., Lee, L., Mietlowski, W., Fuxuis, S., Unger, C., O'Byrne, K., Henry, A., Cherryman, G.R., Laurent, D., Dugan, M., Marme, D., and Steward, W.P., Dynamic contrast-enhanced magnetic resonance imaging as a biomarker for the pharmacological response of PTK787/ZK 222584, an inhibitor of the vascular endothelial growth factor receptor tyrosine kinases, in patients with advanced colorectal cancer and liver metastases: results from two phase I studies, *J Clin Oncol*, 21, 3955, 2003.
136. Xiong, H.Q., Herbst, R., Faria, S.C., Scholz, C., Davis, D., Jackson, E.F., Madden, T., McConkey, D., Hicks, M., Hess, K., Charnsangavej, C.A., and Abbruzzese, J.L., A phase I surrogate end point study of SU6668 in patients with solid tumors, *Invest New Drugs*, 22, 459, 2004.
137. Herbst, R.S., Mullani, N.A., Davis, D.W., Hess, K.R., McConkey, D.J., Charnsangavej, C., O'Reilly, M.S., Kim, H.W., Baker, C., Roach, J., Ellis, L.M., Rashid, A., Pluda, J., Bucana, C., Madden, T.L., Tran, H.T., and Abbruzzese, J.L., Development of biologic markers of response and assessment of antiangiogenic activity in a clinical trial of human recombinant endostatin, *J Clin Oncol*, 20, 3804, 2002.
138. Ratner, M., Genentech discloses safety concerns over Avastin, *Nat Biotechnol*, 22, 1198, 2004.
139. Baffert, F., Thurston, G., Rochon-Duck, M., Le, T., Brekken, R., and McDonald, D.M., Age-related changes in vascular endothelial growth factor dependency and angiopoietin-1-induced plasticity of adult blood vessels, *Circ Res*, 94, 984, 2004.
140. Nelson, D.A., Tan, T.T., Rabson, A.B., Anderson, D., Degenhardt, K., and White, E., Hypoxia and defective apoptosis drive genomic instability and tumorigenesis, *Genes Dev*, 18, 2095, 2004.
141. Jain, R.K., Safabakhsh, N., Sckell, A., Chen, Y., Jiang, P., Benjamin, L., Yuan, F., and Keshet, E., Endothelial cell death, angiogenesis, and microvascular function after castration in an androgen-dependent tumor: role of vascular endothelial growth factor, *Proc Natl Acad Sci USA*, 95, 10820, 1998.
142. Ansiaux, R., Baudelet, C., Jordan, B.F., Beghein, N., Sonveaux, P., De Wever, J., Martinive, P., Gregoire, V., Feron, O., and Gallez, B., Thalidomide radiosensitizes tumors through early changes in the tumor microenvironment, *Clin Cancer Res*, 11, 743, 2005.
143. Huber, P.E., Bischof, M., Jenne, J., Heiland, S., Peschke, P., Saffrich, R., Grone, H.J., Debus, J., Lipson, K.E., and Abdollahi, A., Trimodal cancer treatment: beneficial effects of combined antiangiogenesis, radiation, and chemotherapy, *Cancer Res*, 65, 3643, 2005.
144. Inai, T., Mancuso, M., Hashizume, H., Baffert, F., Haskell, A., Baluk, P., Hu-Lowe, D.D., Shalinsky, D.R., Thurston, G., Yancopoulos, G.D., and McDonald, D.M., Inhibition of vascular endothelial growth factor (VEGF) signaling in cancer causes loss of endothelial fenestrations, regression of tumor vessels, and appearance of basement membrane ghosts, *Am J Pathol*, 165, 35, 2004.
145. Salnikov, A.V., Roswall, P., Sundberg, C., Gardner, H., Heldin, N.E., and Rubin, K., Inhibition of TGF-beta modulates macrophages and vessel maturation in parallel to a lowering of interstitial fluid pressure in experimental carcinoma, *Lab Invest*, 85, 512, 2005.

146. Vosseler, S., Mirancea, N., Bohlen, P., Mueller, M.M., and Fusenig, N.E., Angiogenesis inhibition by vascular endothelial growth factor receptor-2 blockade reduces stromal matrix metalloproteinase expression, normalizes stromal tissue, and reverts epithelial tumor phenotype in surface heterotransplants, *Cancer Res*, 65, 1294, 2005.
147. Wildiers, H., Guetens, G., De Boeck, G., Verbeken, E., Landuyt, B., Landuyt, W., de Bruijn, E.A., and van Oosterom, A.T., Effect of antivascular endothelial growth factor treatment on the intratumoral uptake of CPT-11, *Br J Cancer*, 88, 1979, 2003.
148. Herbst, R.S., Johnson, D.H., Mininberg, E., Carbone, D.P., Henderson, T., Kim, E.S., Blumenschein, G., Jr., Lee, J.J., Liu, D.D., Truong, M.T., Hong, W.K., Tran, H., Tsao, A., Xie, D., Ramies, D.A., Mass, R., Seshagiri, S., Eberhard, D.A., Kelley, S.K., and Sandler, A., Phase I/II trial evaluating the anti-vascular endothelial growth factor monoclonal antibody bevacizumab in combination with the HER-1/epidermal growth factor receptor tyrosine kinase inhibitor erlotinib for patients with recurrent non-small-cell lung cancer, *J Clin Oncol*, 23, 2544, 2005.
149. Browder, T., Butterfield, C.E., Kraling, B.M., Shi, B., Marshall, B., O'Reilly, M.S., and Folkman, J., Antiangiogenic scheduling of chemotherapy improves efficacy against experimental drug-resistant cancer, *Cancer Res*, 60, 1878, 2000.
150. Kerbel, R.S. and Kamen, B.A., The anti-angiogenic basis of metronomic chemotherapy, *Nat Rev Cancer*, 4, 423, 2004.
151. Shaked, Y., Emmenegger, U., Man, S., Cervi, D., Bertolini, F., Ben-David, Y., and Kerbel, R.S., The optimal biological dose of metronomic chemotherapy regimens is associated with maximum antiangiogenic activity, *Blood*, 106, 3058, 2005.
152. Bertolini, F., Paul, S., Mancuso, P., Monestiroli, S., Gobbi, A., Shaked, Y., and Kerbel, R.S., Maximum tolerable dose and low-dose metronomic chemotherapy have opposite effects on the mobilization and viability of circulating endothelial progenitor cells, *Cancer Res*, 63, 4342, 2003.
153. Ch'ang, H.J., Maj, J.G., Paris, F., Xing, H.R., Zhang, J., Truman, J.P., Cardon-Cardo, C., Haimovitz-Friedman, A., Kolesnick, R., and Fuks, Z., ATM regulates target switching to escalating doses of radiation in the intestines, *Nat Med*, 11, 484, 2005.
154. Wey, J.S., Fan, F., Gray, M.J., Bauer, T.W., McCarty, M.F., Somcio, R., Liu, W., Evans, D.B., Wu, Y., Hicklin, D.J., and Ellis, L.M., Vascular endothelial growth factor receptor-1 promotes migration and invasion in pancreatic carcinoma cell lines, *Cancer*, 104, 427, 2005.
155. Fan, F., Wey, J.S., McCarty, M.F., Belcheva, A., Liu, W., Bauer, T.W., Somcio, R.J., Wu, Y., Hooper, A., Hicklin, D.J., and Ellis, L.M., Expression and function of vascular endothelial growth factor receptor-1 on human colorectal cancer cells, *Oncogene*, 24, 2647, 2005.
156. Akagi, M., Kawaguchi, M., Liu, W., McCarty, M.F., Takeda, A., Fan, F., Stoeltzing, O., Parikh, A.A., Jung, Y.D., Bucana, C.D., Mansfield, P.F., Hicklin, D.J., and Ellis, L.M., Induction of neuropilin-1 and vascular endothelial growth factor by epidermal growth factor in human gastric cancer cells, *Br J Cancer*, 88, 796, 2003.
157. Bachelder, R.E., Crago, A., Chung, J., Wendt, M.A., Shaw, L.M., Robinson, G., and Mercurio, A.M., Vascular endothelial growth factor is an autocrine survival factor for neuropilin-expressing breast carcinoma cells, *Cancer Res*, 61, 5736, 2001.
158. Kabbinavar, F., Hurwitz, H.I., Fehrenbacher, L., Meropol, N.J., Novotny, W.F., Lieberman, G., Griffing, S., and Bergsland, E., Phase II, randomized trial comparing bevacizumab plus fluorouracil (FU)/leucovorin (LV) with FU/LV alone in patients with metastatic colorectal cancer, *J Clin Oncol*, 21, 60, 2003.
159. Ince, W.L., Jubb, A.M., Holden, S.N., Holmgren, E.B., Tobin, P., Sridhar, M., Hurwitz, H.I., Kabbinavar, F., Novotny, W.F., Hillan, K.J., and Koeppen, H., Association of k-ras, b-raf, and p53 status with the treatment effect of bevacizumab, *J Natl Cancer Inst*, 97, 981, 2005.
160. Park, J.W., Kerbel, R.S., Kelloff, G.J., Barrett, J.C., Chabner, B.A., Parkinson, D.R., Peck, J., Ruddon, R.W., Sigman, C.C., and Slamon, D.J., Rationale for biomarkers and surrogate end points in mechanism-driven oncology drug development, *Clin Cancer Res*, 10, 3885, 2004.
161. Verheul, H.M. and Pinedo, H.M., Tumor growth: a putative role for platelets? *Oncologist*, 3, II, 1998.
162. Beaudry, P., Force, J., Naumov, G.N., Wang, A., Baker, C.H., Ryan, A., Soker, S., Johnson, B.E., Folkman, J., and Heymach, J.V., Differential effects of vascular endothelial growth factor receptor-2 inhibitor ZD6474 on circulating endothelial progenitors and mature circulating endothelial cells: implications for use as a surrogate marker of antiangiogenic activity, *Clin Cancer Res*, 11, 3514, 2005.

163. Shaked, Y., Bertolini, F., Man, S., Rogers, M.S., Cervi, D., Foutz, T., Rawn, K., Voskas, D., Dumont, D.J., Ben-David, Y., Lawler, J., Henkin, J., Huber, J., Hicklin, D.J., D'Amato, R.J., and Kerbel, R.S., Genetic heterogeneity of the vasculogenic phenotype parallels angiogenesis; implications for cellular surrogate marker analysis of antiangiogenesis, *Cancer Cell*, 7, 101, 2005.
164. Collins, J.M., Imaging and other biomarkers in early clinical studies: one step at a time or re-engineering drug development? *J Clin Oncol*, 23, 5417, 2005.
165. Liu, G., Rugo, H.S., Wilding, G., McShane, T.M., Evelhoch, J.L., Ng, C., Jackson, E., Kelcz, F., Yeh, B.M., Lee, F.T., Jr., Charnsangavej, C., Park, J.W., Ashton, E.A., Steinfeldt, H.M., Pithavala, Y.K., Reich, S.D., and Herbst, R.S., Dynamic contrast-enhanced magnetic resonance imaging as a pharmacodynamic measure of response after acute dosing of AG-013736, an oral angiogenesis inhibitor, in patients with advanced solid tumors: results from a Phase I study, *J Clin Oncol*, 23, 5464, 2005.
166. Miller, J.C., Pien, H.H., Sahani, D., Sorensen, A.G., and Thrall, J.H., Imaging angiogenesis: applications and potential for drug development, *J Natl Cancer Inst*, 97, 172, 2005.
167. Jennens, R.R., Rosenthal, M.A., Lindeman, G.J., and Michael, M., Complete radiological and metabolic response of metastatic renal cell carcinoma to SU5416 (semaxanib) in a patient with probable von Hippel-Lindau syndrome, *Urol Oncol*, 22, 193, 2004.
168. Moses, M.A., Wiederschain, D., Loughlin, K.R., Zurakowski, D., Lamb, C.C., and Freeman, M.R., Increased incidence of matrix metalloproteinases in urine of cancer patients, *Cancer Res*, 58, 1395, 1998.
169. Chan, L.W., Moses, M.A., Goley, E., Sproull, M., Muanza, T., Coleman, C.N., Figg, W.D., Albert, P.S., Menard, C., and Camphausen, K., Urinary VEGF and MMP levels as predictive markers of 1-year progression-free survival in cancer patients treated with radiation therapy: a longitudinal study of protein kinetics throughout tumor progression and therapy, *J Clin Oncol*, 22, 499, 2004.
170. Roy, R., Wewer, U.M., Zurakowski, D., Pories, S.E., and Moses, M.A., ADAM 12 cleaves extracellular matrix proteins and correlates with cancer status and stage, *J Biol Chem*, 279, 51323, 2004.
171. Rugo, H.S., Herbst, R.S., Liu, G., Park, J.W., Kies, M.S., Steinfeldt, H.M., Pithavala, Y.K., Reich, S.D., Freddo, J.L., and Wilding, G., Phase I trial of the oral antiangiogenesis agent AG-013736 in patients with advanced solid tumors: pharmacokinetic and clinical results, *J Clin Oncol*, 23, 5474, 2005.
172. McShane, L.M., Altman, D.G., and Sauerbrei, W., Identification of clinically useful cancer prognostic factors: what are we missing? *J Natl Cancer Inst*, 97, 1023, 2005.
173. Batchelor, T.T., Sorensen, A.G., di Tomaso, E., Zhang, W.T., Duda, D.G., Cohen, K.S., Kozak, K.R., Cahill, D.P., Chen, P.J., Zhu, M., Ancukiewicz, M., Mrugala, M.M., Plotkin, S., Drappatz, J., Louis, D.N., Ivy, P., Scadden, D.T., Benner, T., Loeffler, J.S., Wen, P.Y., and Jain, R.K. AZD2171, a pan-VEGF receptor tyrosine kinase inhibitor, normalizes tumor vasculature and alleviates edema in glioblastoma patients, *Cancer Cell*, 11, 83, 2007.
174. Hattori, K., Heissig, B., Wu, Y., Dias, S., Tejada, R., Ferris, B., Hicklin, D.J., Zhu, Z., Bohlen, P., Witte, L., Hendrikx, J., Hackett, N.R., Crystal, R.G., Moore, M.A., Werb, Z., Lyden, D., and Rafii, S. Placental growth factor reconstitutes hematopoiesis by recruiting VEGFR1 (+) stem cells from bone-marrow microenvironment. *Nat Med*, 8, 841, 2002.

4 Pathophysiologic Role of VEGF in Hematologic Malignancies

Klaus Podar, Shaji Kumar, Dharminder Chauhan, and Kenneth C. Anderson

CONTENTS

4.1 VEGF ISOFORMS AND EXPRESSION

Identified in the 1980s, the vascular endothelial growth factor (VEGF) was first described as vascular permeability factor (VPF) and is now denoted as VEGF-A.[1–5] The pivotal importance of this protein is reflected by its heterozygous embryonic lethality, which is induced by targeted inactivation of the *VEGF* gene,[6,7] as well as the high homology of VEGF across species.[8,9] Specifically, the "VEGF family" of structurally related dimeric glycoproteins of the platelet-derived growth factor (PDGF) superfamily of growth factors includes VEGF-A, VEGF-B,[10] VEGF-C,[11] VEGF-D,[12] VEGF-E, and placenta growth factor (PlGF).[13,14] Analysis of the crystal structure of VEGF-A shows an antiparallel homodimer, which is covalently linked by two disulfide bridges.[15] In addition to homodimers, active forms of VEGF are also synthesized and secreted as heterodimers with other VEGF family members, such as PlGF, which only binds VEGF receptor-1 (VEGFR-1). Importantly, in contrast to pure PlGF homodimers, VEGF–PlGF heterodimers demonstrate potent mitogenic and chemotactic effects on endothelial cells, which are mediated via VEGFR-2.[16,17] VEGF-A is located at chromosome 6p21.3.[18,19] The coding region spans ~14 kb and contains eight exons separated by seven introns. The VEGF promoter contains binding sites for Spl, AP-1, and AP-2, through which protein kinase C (PKC) and protein kinase A (PKA) can mediate VEGF expression.[20] Most importantly, the hypoxia response element (HRE) upstream of the *VEGF* gene binds hypoxia-induced factor-1 (HIF-1), thereby increasing VEGF expression.

To date, six VEGF isoforms generated by alternative mRNA splicing have been identified. The major isoforms of VEGF-A are[20,21]: $VEGF_{121}$, the predominant isoform, $VEGF_{165}$, $VEGF_{189}$, and $VEGF_{206}$.[4] These forms differ primarily in their bioavailability, which is conferred by heparin and heparan sulfate–binding domains encoded by exon 6 and exon 7. $VEGF_{189}$ and $VEGF_{206}$ contain additional stretches of basic residues, resulting in their nearly complete retention in the extracellular matrix (ECM) and to a lesser extent at the cell surface. $VEGF_{206}$ binds most strongly to heparin. $VEGF_{165}$ lacks exon 6 and is secreted, but also remains bound by 50%–70% to the cell surface and the ECM via its heparin-binding sites. $VEGF_{121}$, which lacks both exon 6 and exon 7, fails to bind heparin and is therefore a freely diffusible protein. The $VEGF_{121}$, $VEGF_{165}$, and $VEGF_{189}$ forms are abundant and usually produced simultaneously. $VEGF_{121}$ and $VEGF_{165}$ isoforms induce mitogenic and permeability-enhancing activity on endothelial cells, whereas the other longer isoforms trigger only permeability-enhancing activity.[21–23] In addition, uPA can cleave $VEGF_{189}$ within exon 6, thereby generating a truncated factor uPA-$VEGF_{189}$ with mitogenic potential equivalent to $VEGF_{165}$.[24] Moreover, plasmin-induced cleavage of both $VEGF_{165}$ and $VEGF_{189}$ releases a bioactive carboxy-terminal domain (111–165) of VEGF.[25]

$VEGF_{165}$, $VEGF_{189}$, and $VEGF_{206}$ levels are increased in several human malignancies including breast,[26] lung,[27] brain,[28] pancreatic,[29] ovarian,[30] kidney, and bladder carcinomas.[31] Less frequently expressed are the VEGF-A spliced forms $VEGF_{145}$, $VEGF_{183}$, $VEGF_{162}$, and $VEGF_{165b}$.[32] $VEGF_{145}$, for example, shows similar heparin affinity as $VEGF_{165}$, and is expressed by both human multiple myeloma (MM) cells[33] and Kaposi sarcoma–associated herpesvirus or human herpesvirus-8 (KSHV or HHV-8)–associated primary effusion lymphomas (PELs).[34] Importantly, ECM-bound $VEGF_{145}$ remains an active endothelial cell growth factor.[35]

VEGF is secreted by a variety of cells, including human hematologic tumor and MM cells.[36] Hypoxia is a key regulator of VEGF expression via the HIF-1/von Hippel-Lindau tumor suppressor gene (VHL) pathway[37,38]; therefore, VEGF is the most highly expressed adjacent to necrotic areas.[39,40] Besides hypoxia, growth factors and cytokines including PDGF, epidermal growth factor (EGF), fibroblast growth factor (FGF), tumor necrosis factor-α (TNF-α), tumor growth factor-α and β1 (TGF-α, TGF-β1), keratinocyte growth factor, insulin-like growth factor-1 (IGF-1), interleukin-1β (IL-1β), and IL-6[41] upregulate transcription of VEGF mRNA.[42,43] VEGF expression is also induced by UV-B and H_2O_2[44]; mutant *p53* (via PKC)[45] and mutant *Ras* oncogenes[46,47]; proangiogenic oncogenes including *Src*, c-*Myc*, *Fos*, and *Bcl-2*[48]; and acidosis.[49] Recent studies in zebra fish demonstrate that bone morphogenetic protein (BMP)–activated Smad-binding elements including Smad1 and Smad5 as well as hedgehog signaling regulate VEGF expression and thereby contribute to hemangiogenic cell proliferation and adult stem cell formation.[50,51] In addition, extracellular matrix metalloproteinase (EMMPRIN) contributes to tumor angiogenesis and growth by stimulating VEGF and matrix metalloproteinases (MMP) expression (Figure 4.1a).

4.2 VEGF RECEPTORS AND SIGNAL TRANSDUCTION IN ENDOTHELIAL CELLS

VEGF-A, VEGF-B, VEGF-C, VEGF-D, and PlGF bind with different affinities to three related receptor tyrosine kinases: VEGFR-1 (fms-like tyrosine kinase-1; Flt-1), VEGFR-2 (kinase domain region, KDR; homolog to murine fetal liver kinase-1, Flk-1), and VEGFR-3. VEGF-A mediates its activity mainly via two receptor tyrosine kinases (RTKs): the high affinity receptor VEGFR-1 (K_D 10–20 pM) located at chromosome 13q12–13[52] and VEGFR-2 (K_D 75–125 pM) located at chromosome 4q11–12.[53–58] Both receptors have a crucial role in the development of the vascular system, evidenced by embryonic lethality on their disruption.[59–61] These VEGF–RTKs are single-pass transmembrane receptors with seven immunoglobulin (Ig)-like loops in the extracellular domain and a cytoplasmic tyrosine kinase domain, separated by an intervening, noncatalytic, 70-amino acid residue sequence. The second and third Ig-like domains of VEGFR-1 and VEGFR-2 mediate high-affinity VEGF binding; the fourth Ig-like domain mediates dimerization of VEGFR-1; the fifth and sixth Ig-like domains of VEGFR-2 are required for VEGF retention after binding.[62–65] Both VEGFR-1 and VEGFR-2 are modified posttranslationally by N-linked glycosylation and phosphorylation on serine, threonine, and tyrosine; however, only glycosylation of VEGFR-2 is required for ligand binding and subsequent autophosphorylation.[66] On SDS-PAGE electrophoresis, VEGFR-1 migrates at a molecular mass of 180 kDa, and VEGFR-2 at 230 kDa. Hypoxia directly upregulates *VEGFR-1* gene expression via a HIF-1α-binding site within the VEGFR-1, but not VEGFR-2 promoter.[67] Importantly, the expression of both VEGFR-1 and VEGFR-2 is also upregulated by platelet endothelial cell adhesion molecule-1 (PECAM-1).[68] The VEGF-A amino acid residues Arg82, Lys84, and His85 on exon 4 are required for binding to VEGFR-2,

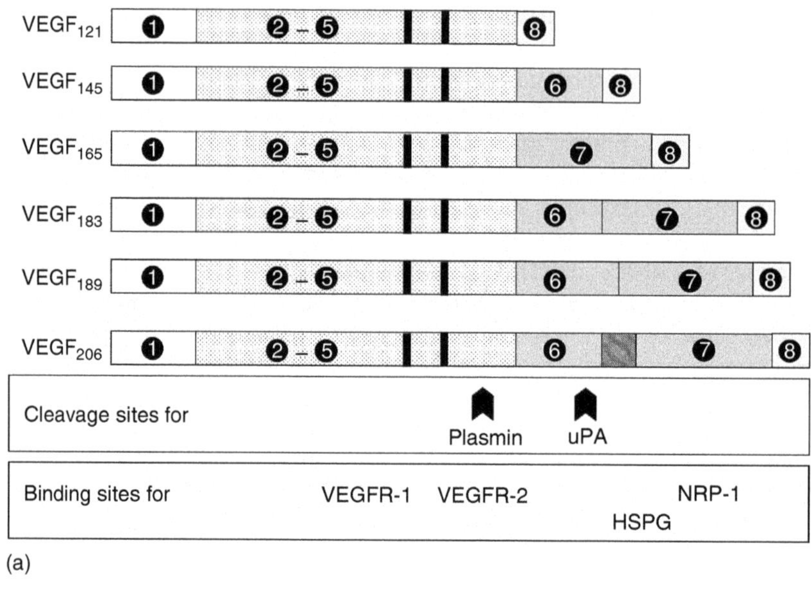

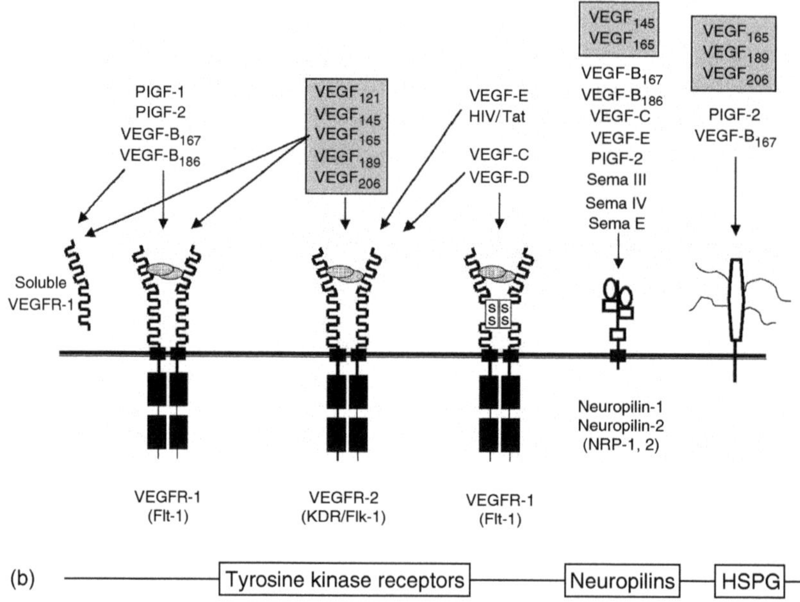

FIGURE 4.1 VEGF isoforms and their interaction with receptors of the VEGF receptor family: VEGF-A is located at chromosome 6p21.3, containing eight exons separated by seven introns. To date, six VEGF isoforms generated by alternative mRNA splicing have been identified. These isoforms differ primarily in their bioavailability, which is conferred by heparin and HSPG–binding domains. $VEGF_{189}$ and $VEGF_{206}$ contain both exon 6 and exon 7 and are nearly entirely retained in the extracellular matrix (ECM); $VEGF_{165}$ lacking exon 6 is secreted, but also remains bound to the cell surface and the ECM; $VEGF_{121}$ lacking exon 6 and exon 7 fails to bind heparin and is therefore freely diffusible. $VEGF_{121}$, $VEGF_{165}$, and $VEGF_{189}$ are abundant and usually produced simultaneously. $VEGF_{145}$ shows similar heparin affinity as $VEGF_{164}$. In addition, urokinase type of plasminogen activator (uPA) and plasmin can cleave $VEGF_{189}$ or $VEGF_{189}$ and $VEGF_{165}$, respectively, thereby generating further bioactive VEGF forms. VEGF-binding sites to VEGFR-1 and VEGFR-2 are located on exon 3 and exon 4, respectively. VEGF-induced signaling is further modified by the association of VEGFR with neuropilins (NRPs). Numbers represent exons.

whereas the VEGF-A amino acid residues Asp63, Glu64, and Glu67 on exon 3 are required for binding to VEGFR-1.[69]

VEGFR-1 and VEGFR-2 are expressed in all adult vascular endothelial cells except in the brain.[70] Besides endothelial cells, VEGFR-1 is expressed on hematopoietic stem cells (HSCs)[42] and monocytes[71,72]; human trophoblast, choriocarcinoma cells[73]; renal mesangial cells[74]; vascular smooth muscle cells[75]; as well as MM and leukemic cells.[36,76] VEGFR-2 is also expressed on circulating endothelial progenitor cells (CEPs),[77–79] pancreatic duct cells,[80] retinal progenitor cells,[81] and megakaryocytes.[82] Coexpression of VEGFR-1 and VEGFR-2 is found on normal human testicular tissue[83] and in the myometrium. Significantly increased levels of both VEGFR-1 and VEGFR-2 are also present on cancers of kidney, bladder,[84] ovary,[30] and brain.[28]

The role of VEGFR-2 in endothelial cells has been extensively studied: VEGFR-2 mediates developmental angiogenesis and hematopoiesis by triggering vascular endothelial cell proliferation, migration, differentiation, and survival, as well as by inducing vascular permeability and blood island formation.[61] VEGF induces receptor dimerization, thereby triggering kinase activation of both the receptor itself and several cytoplasmic signal transduction molecules including PLC-γ,[85–88] VEGFR-associated protein (VRAP),[89] Ras GTPase–activating protein (Ras GAP),[90] FAK,[87,91,92] Sck,[93,94] Src family of tyrosine kinases,[95,96] Grb2,[86,97] PI3-kinase/Akt,[87,98–100] PKC,[101–104] Raf-1,[103,105] MEK/ERK,[87,103,106–108] p38MAPK,[92,109] Nck,[90,98] Crk,[86] Shc,[97] and STAT.[110] VEGFR kinase activity is additionally regulated by (1) naturally occurring soluble forms of VEGFR-1 and VEGFR-2; (2) coreceptors including cadherins, integrin $\alpha_v\beta_3$, neuropilin-1 and neuropilin-2[111–114]; (3) polysaccharides heparin and heparan sulfate[22,115]; (4) phosphatases SHP-1 and SHP-2[97,98] and HCPTPA[116]; and (5) small guanosine triphosphate (GTPase) RhoA.[117]

The functional role of VEGFR-1 is complex and dependent on both the developmental stage and cell type.[60,118,119] Although the role of VEGFR-1 signaling cascades in endothelial cells is not fully delineated, VEGFR-1 signaling is required in hematopoiesis,[120,121] monocyte migration,[71,72,122] and paracrine release of growth factors.[123,124] In addition, a recent report links VEGFR-1 signaling with production of MMP-9 in lung endothelial cells, thereby facilitating lung metastasis.[125] Blockade of VEGFR-1 attenuates blood vessel formation in the setting of cancer, ischemic retinopathy, and rheumatoid arthritis; importantly, blocking VEGFR-1 abrogates inflammatory processes such as atherosclerosis and rheumatoid arthritis, whereas inhibiting VEGFR-2 does not.[123] In addition to the membrane-bound VEGFR-1, a soluble form of VEGFR-1 (sFLT-1) has been reported. sFlt-1, although containing only the first six Ig-like domains, binds VEGF as strongly as full-length VEGFR-1 thereby inhibiting VEGF activity and additionally modulating VEGF effects.[126,127]

Signaling pathways which are initiated by binding of VEGF to VEGFR-1 and VEGFR-2 are further modified by their association with neuropilin-1 and neuropilin-2.[114,128,129] Specifically, neuropilin-1 binds $VEGF_{165}$, VEGF-E, VEGF-B, and PlGF[114,130–132]; and neuropilin-2 binds $VEGF_{165}$, $VEGF_{145}$, and PlGF.[128] For example, the formation of a VEGFR-2/neuroplilin-1 complex is responsible for the differential signaling potency of neuropilin-1-binding $VEGF_{165}$ versus non-neuropilin-1-binding $VEGF_{121}$[133] (Figure 4.1b).

4.3 VEGF FUNCTIONS

VEGF has a multitude of different functions both on endothelial cells and on nonendothelial cells, dependent both on developmental stages as well as tissue and cell specificity. Due to close association between hematopoiesis and vasculogenesis, a progenitor cell for both blood and endothelial cells the hemangioblast was postulated. Only recently the mechanisms that control and integrate VEGF functions have been described, including VEGFR or VEGF

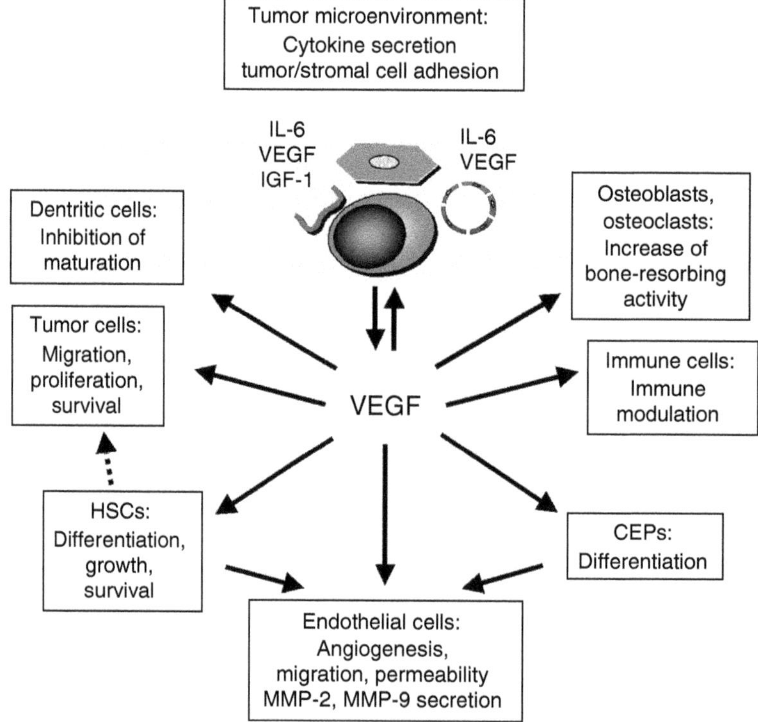

FIGURE 4.2 VEGF functions. VEGF increases endothelial cell growth, migration, and permeability; as well as triggers secretion of MMP-2 and MMP-9, which dissolve the basement membrane and breakdown the ECM to release endothelial cells from anchorage, thereby allowing them to migrate and proliferate. Furthermore, VEGF inhibits maturation of dendritic cells; increases bone-resorbing activity; modulates immune response; triggers circulatory endothelial progenitor (CEP) differentiation; and triggers MM cell growth, migration, and survival. In addition, VEGF stimulates cytokine and growth factor release within the MM BM microenvironment.

homo- or heterooligomerization; modification of VEGFR activity by associated receptor molecules like NRP-1, 2 or heparan sulfate proteoglycans (HSPGs) and heparin; and expression of different VEGF splice forms triggering the formation of distinct VEGFR-associated signaling complexes (Figure 4.2).

4.3.1 Angiogenesis and Tumor Progression

Tumor growth and progression are dependent on blood vessel formation, providing vital oxygen and nutrients within the diffusion limit for oxygen (100 to 200 μm). VEGF is the most potent vascular growth factor known to date and is most closely correlated with spatial and temporal events of blood vessel growth. Besides promoting endothelial cell growth, VEGF also stimulates endothelial cell migration because of its ability to form chemotactic gradients and stimulate secretion of proteases. Specifically, the formation of primary embryonic vasculature begins with the differentiation of endothelial precursors into endothelial cells (vasculogenesis). This is followed by angiogenesis, characterized by the expansion of this primitive network by four divergent morphological pathways: (1) muscular artery/vein formation; (2) vascular bridging; (3) intussusceptive microvascular growth; and (4) sprouting angiogensis vessels (insertion of interstitial tissue columns into the lumen of preexisting vessels). The focal dissolution of endothelial basement membrane and the breakdown of the

ECM are required to release endothelial cells from anchorage, thereby allowing them to migrate into surrounding tissues and proliferate into new blood vessels. Enzymes that catalyze these events include proteolytic enzymes, secreted by activated endothelial cells and tumor cells such as plasminogen activators (e.g., the urokinase-type and tissue-type plasminogen activators, uPA and tPA)[134,135] and MMPs (predominantly the family members MMP-2 (gelatinase A) and MMP-9 (gelatinase B)).[136] Strings of new endothelial cells then organize into vascular tubes, dependent on the interaction between cell-associated surface proteins (hybrid oligosaccharides, galectin-2, PECAM-1, and VE-cadherin) and the ECM.[137–139] Finally, newly formed vessels are stabilized through the recruitment of smooth muscle cells and pericytes mediated via binding of angiopoietin-1 (Ang-1) to the Tie-2 receptor.

In addition to sprouting and cooption[140] of neighboring preexisting vessels, tumor-derived angiogenic factors like VEGF promote formation of the endothelial lining of tumor vessels (vasculogenesis) by recruitment of highly proliferative circulating endothelial precursors (CEPs, angioblasts) from the bone marrow, HSCs, progenitor cells, monocytes, and macrophages.[141] Moreover, tumor cells (i.e., melanoma cells) can act as endothelial cells and form functional avascular blood conduits or mosaic blood vessels, which are lined partially by tumor cells and vessel walls.[142–146]

CEPs, but not circulating endothelial cells (CECs) sloughed from the vessel wall, are highly proliferative and contribute to tumor neoangiogenesis.[141,147] HSCs and CEPs likely originate from a common precursor, the hemangioblast. Typically, CEPs express VEGFR-2, c-KIT, CD133, and CD146,[77–79] whereas HSCs express VEGFR-1, Sca-1, and c-KIT; the lineage-specific differentiation of these HSCs into erythroid, myeloid, megakaryocytic, and lymphoid cells is dependent on the availability of specific cytokines including IL-3, G-CSF, GM-CSF, and TPO. Subsequently, hematopoietic progenitors and terminally differentiated precursor cells produce and secrete factors including VEGF, FGFs, brain-derived nerve growth factor (BDNF), and angiopoietin; together with ECM proteins like fibronectin and collagen, these factors promote differentiation of CEPs, thereby contributing to new vessel formation. In addition, direct cellular contact with stromal cells also regulates the expansion of undifferentiated CEPs.[141] Importantly, VEGFR-1 expressing hematopoietic cells and CEPs colocalize to cooperate in the formation of functional tumor vessels.[148,149]

Angiogenesis is tightly regulated by proangiogenic and antiangiogenic molecules. In tumorigenesis this balance is derailed,[150] thereby triggering tumor growth, invasion, and metastasis.[151] Specifically, a rapid phase of tumor growth occurs when the tumor switches to its angiogenic phenotype. This process has best been studied in the transgenic mouse model of multistage pancreatic islet cell carcinogenesis (Rip1-Tag2), where clonal expansion of a subset of hyperplastic cells results in tumor progression.[152,153] Pro- and antiangiogenic molecules arise from cancer cells, stromal cells, endothelial cells, the ECM, and blood.[154] Importantly, the relative contribution of these molecules is dependent on the tumor type and site, and their expression changes with tumor growth, regression, and relapse. The "angiogenic switch"[155] is triggered by oncogene-mediated tumor expression of angiogenic proteins including VEGF, FGF, PDGF, EGF, lysophosphatic acid (LPA), and angiopoietin; as well as by metabolic stress, mechanical stress, genetic mutations, and the immune response.[150,151,156–160] In contrast, inhibitors of angiogenesis including angiostatin, endostatin, interferons, platelet factor-4, thrombospondin, tissue inhibitors of metalloproteinases-1 through metalloproteinases-3, pigment epithelium-derived factor, 2-methoxy-estradiol, vasostatin, and canstatin decrease this angiogenic stimulus.[161–164] The imbalance of angiogenic regulators (i.e., VEGF) accounts for an abnormal structure of tumor vessels, which in turn results in chaotic, variable blood flow and vessel leakiness, thereby lowering drug delivery and selecting for more malignant tumor cells.[165–168]

In the normal adult, only 0.01% of endothelial cells undergo division,[150] whereas up to 25% of endothelial cells divide in tumor vessels.[169] In addition, recent studies indicate that the gene expression pattern of normal endothelial cells differs from endothelial cells in tumors[170]: that is, 79 differentially expressed gene products were found in endothelial cells from colon cancer compared with normal endothelial cells.[171] In addition, a wider array of angiogenic molecules can be produced as tumor cells grow; therefore, if one proapoptotic molecule is blocked, tumors may utilize another molecule. A cocktail of antiangiogenic therapies may therefore be required to effectively prevent angiogenesis.[150]

4.3.2 Additional Biologic Functions of VEGF

Besides its role as an essential regulator of physiological endothelial cell growth, permeability, and migration in vitro and in vivo,[1–5,172,173] VEGF is a pivotal factor in hematopoiesis which affects the differentiation of multiple hematopoietic lineages.[6,7,61,121] VEGF triggers the differentiation of hematopoietic and endothelial lineages from a common potential precursor cell within the blood islands, the hemangioblasts.[61,174] Specifically, several recent reports show the importance of VEGF in erythroid differentiation.[175] Moreover, VEGF mediates HSC survival and repopulation via an autocrine loop,[121] whereas angiogenesis is regulated via a paracrine VEGF loop. Interestingly, the *Drosophila* PDGF/VEGFR (PVR) directly controls survival of a hemocyte cell line, thereby demonstrating a striking homology with mammalian hematopoiesis. This finding suggests *Drosophila* to be a useful model system for the study of hematopoietic cell survival in development and disease.[176] In addition, VEGF inhibits maturation of dendritic cells through inhibition of nuclear factor-κB activation[177]; increases both osteoclastic bone-resorbing activity[178] as well as osteoclast chemotaxis[179]; induces migration, parathyroid hormone (PTH)–dependent cAMP accumulation, and alkaline phosphatase in osteoblasts[180]; acts as an autocrine factor for osteoblast differentiation[181]; enhances natural killer (NK) cell adhesion to tumor endothelium[182]; recruits monocyte[71] and endothelial cell progenitors[148] to the vasculature; stimulates surfactant production by alveolar type II cells[183]; and mediates a direct neuroprotective effect on motor neurons in vitro.[184] In the context of cancer, VEGF is an important growth, migration, and survival factor in Kaposi's sarcoma (KS)[185–190]; as well as leukemia, and MM[76,191] (Figure 4.2).

4.4 PATHOPHYSIOLOGIC ROLE OF VEGF IN MULTIPLE MYELOMA AND OTHER HEMATOLOGIC MALIGNANCIES

4.4.1 Angiogenesis and Microvessel Density in Multiple Myeloma and Other Hematologic Malignancies: Prognostic Value and Therapeutic Response

The concept of increased, abnormal tumor vascularity was first put forth and studied in the context of solid tumors.[192–198] However, during the last decade increased angiogenesis in the bone marrow has been identified as an important component of hematological malignancies as well. The phenomenon was first described in MM.[199] Subsequent studies have demonstrated that angiogenesis is an integral part of the disease biology in almost all hematological malignancies, including acute leukemias, chronic leukemias, lymphomas, myelodysplastic syndrome (MDS), and chronic myeloproliferative disorders (Table 4.1). Specifically, increased vascularity was observed in the lymph nodes of B-cell non-Hodgkin's lymphoma (B-NHL)[200] and B-cell chronic lymphocytic leukemia (B-CLL)[36,201,202]; as well as in bone marrow (BM) specimens from patients with childhood acute lymphoid leukemia (ALL),[203] acute myeloid leukemia (AML),[204] chronic myelocytic leukemia (CML),[205] MDS,[206] and idiopathic myelofibrosis.[207] The initial clinical and correlative studies confirmed the prognostic value of angiogenesis in these disorders. For example, in B-NHL BM angiogenesis is highly correlated to tumor grade and the percentage of VEGF-positive cells.[208] Furthermore, enhanced BM angiogenesis also has prognostic value in early B-CLL.[209]

TABLE 4.1
Angiogenic Markers in Hematological Malignancies

Malignancy	Increased MVD	Increased VEGF	Increased bFGF
MM	Vacca[199,357]	Yes	Yes
	Rajkumar[210,214]	Yes	Yes
	Kumar[266,358]	—	—
	Munshi[213]	Yes	—
	Di Raimondo[359]; Sezer[211]	—	Yes
ALL	Perez-Atayde[203]	Yes	Yes
	Aguayo[201,360]	—	Yes
CLL	Aguayo[201]	Yes	Yes
	Kini[361]	—	Yes
AML	Aguayo[201,360]	Yes	Yes
	de Bont[362]	Yes	—
MDS	Pruneri[206]	Yes	—
	Aguayo[201,360]	Yes	Yes
	Korkolopoulou[365]	—	—
CML	Aguayo[201,360]	Yes	Yes
	Panteli[364]	—	—
CMML	Aguayo[201]	Yes	Yes
HD	Korkolopoulou[365]	—	—
HCL	Korkolopoulou[366]	—	—

HD, Hodgkin's disease; HCL, hairy cell leukemia; CMML, chronic myelomonocytic leukemia.

Although much has been learned about the biological basis of increased angiogenesis in hematological malignancies, many questions remain unanswered. Angiogenesis is felt to be a valid therapeutic target in MM leading to several clinical trials evaluating antiangiogenic agents.

4.4.2 Increased Angiogenesis in Myeloma

The presence of increased angiogenesis in the BM of patients with MM was initially reported by Vacca et al. a decade ago.[199] Using microvessel area as an estimate of angiogenesis, the authors were able to correlate increased angiogenesis with disease activity; presence of increased angiogenesis discriminated those with active disease from those with monoclonal gammopathy of undetermined significance (MGUS). Vacca et al. also observed a correlation between the degree of angiogenesis and the proliferative status of the tumor. Subsequent studies by us and others have confirmed the presence of increased BM angiogenesis in all stages of MM[210–214] (Figure 4.3). Though different methodologies have been used to estimate the degree of BM angiogenesis in MM, results from different studies have been consistent. BM angiogenesis in MM increases with disease progression from MGUS to active disease. In a study of 400 patients including those with MGUS, smoldering myeloma (SMM), newly diagnosed and relapsed MM, Rajkumar et al. demonstrated progressively increasing microvessel density (MVD) as patients progressed from MGUS to relapsed MM.[214] The median MVD was 1.3 in normal controls, 1.7 in primary amyloidosis, 3 in MGUS, 4 in SMM, 11 in MM, and 20 in RMM, $P < 0.001$. Over 40% of those with relapsed disease had high-grade angiogenesis, compared with only 1%–3% of patients with MGUS or SMM. In addition, the study demonstrated a correlation between MVD and the plasma cell-labeling index (PCLI) (a measure of plasma cell proliferative rate), as well as the BM plasma cell percentage. These results have been confirmed by other studies as well (Table 4.1).[199,215]

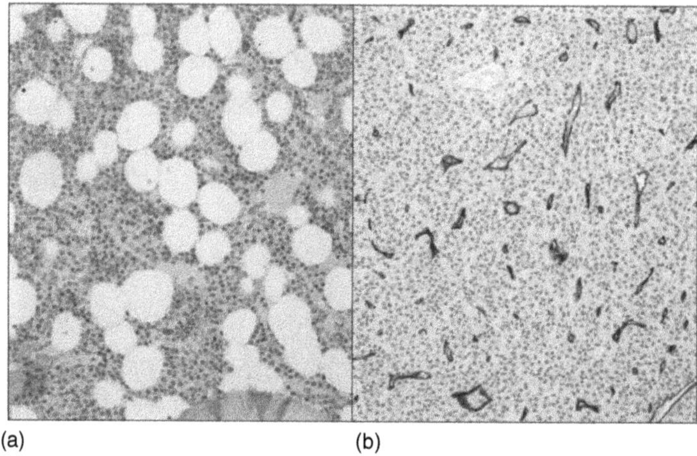

FIGURE 4.3 Immunohistochemistry using antibodies against CD34 demonstrates increased microvessel density (×400 magnification) in bone marrow from patient with MM (b) compared with normal BM (a).

4.4.3 Prognostic Value of Angiogenesis in Myeloma

Extent of BM angiogenesis is also a strong prognostic factor in MM.[210,211,213] In a study of 75 patients with newly diagnosed MM from Mayo Clinic, a significant difference in overall survival was seen among those with high-, intermediate-, and low-grade angiogenesis. Patients with high MVD (>50/×400 field) had a median survival of 2.6 years compared with 5.1 years for those with low MVD.[210] In another study of 88 newly diagnosed patients treated uniformly and undergoing early stem cell transplant, a similar effect of BM MVD was observed.[216] Furthermore, a similar poor prognostic effect of increased BM angiogenesis was noted in a study of newly diagnosed patients undergoing remission induction followed by tandem autotransplants and interferon maintenance (total therapy protocol) at the University of Arkansas.[213] When compared with other known prognostic factors including chromosome 13 deletion, β2-microglobulin, and C-reactive protein, BM MVD emerged as a powerful prognostic determinant. In addition to demonstrating the powerful prognostic effect of BM angiogenesis in MM, these studies also highlight the inability of high-dose therapy (single or tandem transplant) to negate the adverse effect of this phenomenon. These findings are also consistent with the observation that marrow MVD does not decrease significantly following conventional[217] or even high-dose therapy[218] in patients with MM.

A strong correlation has been observed between the BM MVD in MM and other important prognostic factors. BM plasma cell labeling index (PCLI), the percentage of plasma cells with bromodeoxyuridine uptake by immunofluorescence microscopy, is a powerful prognostic factor in all stages of MM.[210,219] There is a strong positive correlation between BM MVD and the PCLI. Increased BM MVD is also associated with increased numbers of circulating plasma cells in MM.[220] MM is typically characterized by accumulation of plasma cells in the BM with plasma cells seen in the circulation only in late stages of the disease or in plasma cell leukemia (PCL). Increased BM MVD has also been reported in association with deletion of 13q in the setting of MM (Table 4.1).[221]

4.4.4 Epiphenomenon versus Causal Relationship

The earlier data do not establish a cause and effect relationship between induction of angiogenesis and MM. Angiogenesis may be an epiphenomenon related to release of angiogenic cytokines by proliferating MM cells. However, given that the lytic lesions in

MM are in effect multiple, discrete, localized, solid tumors in bone, there is no reason why angiogenesis would play any less of a role in this disease compared with solid tumors. In fact, other than the concept of MM as a hematologic malignancy that is not expected to need a source of oxygenated blood, data supporting the role of angiogenesis in MM pathogenesis are similar to that in solid tumors.

The effect of angiogenesis induction has been studied in solitary bone plasmacytoma, a localized plasma cell malignancy considered to be a solid tumor equivalent. Solitary bone plasmacytoma is cured in ~50% of patients with localized radiation therapy; in the remaining patients, it progresses to MM after months to years.

Angiogenesis in baseline plasmacytoma biopsy samples and BM biopsies was studied in 25 patients with solitary bone plasmacytoma.[222] High-grade angiogenesis was present in 64% of patients in the plasmacytoma biopsy specimen; in contrast, BM angiogenesis was low in all patients, confirming lack of disease spread at the time of initial diagnosis. Patients with high-grade angiogenesis in the plasmacytoma sample had a significantly higher chance to progress to MM than patients with low-grade angiogenesis (median PFS for the high MVD group was 17.5 months compared with 57 months for those with low MVD), suggesting a pathogenetic role for angiogenesis in MM.

4.4.5 Alternative Indicators of Therapeutic Response

Although increased MVD in BM is associated with initial tumor burden and relapse in MM, it may not be a good indicator of therapeutic response[218,223] when tumor cytotoxicity exceeds the disappearance of capillaries. Therefore, a decrease in MVD during treatment indicates the effectiveness of the agent; however, the absence of such a decrease in MVD does not necessarily indicate lack of response.[224] Alternative methods to estimate the degree of tumor angiogenesis include measurement of the interstitial fluid pressure and determination of the tumor size. Moreover, imaging technologies to show the degree of disorganized and highly permeable tumor vasculature are in preclinical development. Furthermore, Shaked et al. recently provided evidence that levels of CECs and CEPs correlate with the degree of tumor angiogenesis or the response to antiangiogenic therapy in a murine model. CECs and CEPs may therefore be used as surrogate markers.[225] These findings are very likely to have an important impact in clinical development and application of antiangiogenic therapies, and have already been translated in MM. Specifically, in MM patients CECs are sixfold higher compared with controls and correlate positively with serum M protein and β2-microglobulin. Moreover, elevated levels of CECs were downregulated in response to thalidomide or CC5013, Revlimid, further demonstrating a pivotal role of angiogenesis in MM pathogenesis.[226]

4.4.6 Tumor-Specific Endothelial Cells

The existence of tumor-specific endothelial cells has been suggested: in MM, endothelial cells of the BM (MMECs) represent a heterogeneous cell population forming tortuous, uneven vessels with profuse branching and shunts.[199] When compared with healthy quiescent human umbilical vein endothelial cells (HUVECs), MMECs demonstrate: (1) enhanced expression of specific antigens including Tie-2/Tek, CD105/endoglin, VEGFR-2, bFGFR-2, CD133 (AC133), aquaporin 1, and CD61 (β_3-integrin); (2) enhanced capillarogenic activity; and (3) secretion of growth and invasive factors for plasma cells including bFGF, VEGF, MMP-2, and MMP-9. Moreover, increased levels of expression and secretion are also found in IL-8, I-TAC, SDF-1α, and MCP-1 compared with endothelial cells derived from HUVECs.[227] These findings therefore indicate that MMECs facilitate tumor cell growth, invasion, and dissemination; and conversely, raise the possibility of MM-induced endothelial cell growth in the BM.[228]

Moreover, a recent study of B-cell lymphomas found identical primary and secondary genetic aberrations simultaneously in endothelial cells and tumor cells, suggesting a close relationship between the genetic events in these two cell types. The authors suggest four different mechanisms that could explain this finding: (1) both cell types are derived from a common malignant precursor cell; (2) the endothelial cell carrying the genetic alteration of the lymphoma cell has arisen from a cell that was already committed to the lymphoid lineage; (3) lymphoma cell–endothelial cell fusion has occurred; and (4) apoptotic bodies from tumor cells have been taken up by endothelial cells.[229] Ongoing studies are evaluating whether this is a general phenomenon in hematologic malignancies including leukemia and MM; as well as its functional relevance in tumor pathogenesis.

4.4.7 VEGF and VEGF Receptors in Multiple Myeloma and Other Hematologic Malignancies

Until the early 1990s, the role of VEGF and other angiogenic molecules was not defined in leukemias and other hematological malignancies. First, VEGF was isolated from the HL-60 myeloid leukemia cell line,[4] and bFGF was shown to be expressed by BM stromal and peripheral blood cells.[230] Moreover, VEGF and VEGFR expression was reported in acute and chronic leukemias and MDS,[201,231–235] as well as in cell lines derived from hematopoietic malignancies including MM.[36,201,231–236]

Specifically, in MM most studies found VEGFR-1 to be more commonly expressed than VEGFR-2 on both MM cell lines and patient MM cells. The role for paracrine and juxtacrine VEGF-mediated tumor cell growth and survival is well established, and was recently further supported by a novel in vitro coculture system using primary MM cells and fetal BM stromal cells as a feeder layer.[33,36,237] In addition, coexpression of both VEGF and VEGFR in leukemia, lymphoma, and MM, coupled with direct effects of VEGF on tumor cell survival, migration, and proliferation, confirms the pivotal role for autocrine VEGF loops in the pathogenesis of these malignancies. Only recently, signaling pathways mediating these effects in hematological malignancies have been delineated.[33,233,238–241]

Importantly VEGFR-1, but not VEGFR-2, is associated with inhibition of hematopoietic stem cell cycling, differentiation, and hematopoietic recovery in adults.[121] The potential role of VEGF and VEGFR-1 in development of the MM cell clone in particular, and in leukemogenesis in general, remains to be investigated.

4.4.8 Functional Role of VEGF in the Pathogenesis of Hematologic Malignancies

4.4.8.1 VEGF and the Tumor Microenvironment

Stroma and tumor-associated endothelium nourish and support the growing cell mass of solid tumors.[242] Similarly the stroma of hematopoietic organs support processes of normal and malignant hematopoiesis, that is, "liquid" tumors or leukemias and MM. For example, Friend Murine leukemia virus (F-MuLV)-infected splenocytes secrete elevated levels of IL-6, VEGF, MCP-5, sTNFR1, IL-12p70, TNF-α, and IL-2 compared with normal splenocytes, thereby sustaining proliferation of primary erythroleukemic cells in vitro. Importantly, in vivo administration of a neutralizing VEGF antibody extends survival times of erythroleukemic mice in comparison to controls.[243] Moreover, CLL, marginal zone NHL, hairy cell leukemia, and chronic myelogenous leukemia (CML) are associated with splenomegaly, suggesting that growth response elements are contributed by the spleen.[244–246] Conversely, the BM microenvironment is a heterogeneous population of cells including HSCs; endothelial cells; stromal cells including fibroblasts, macrophages, T lymphocytes; as well as cells involved in bone homeostasis such as chondroclasts, osteoclasts, and osteoblasts (Figure 4.2). Differentiation, maintenance, and expansion of MM cells within the BM microenvironment is a highly coordinated process involving: (1) multiple growth factors, cytokines, and chemokines secreted by tumor cells (autocrine loop), stromal cells (paracrine loop), as well as nonhematopoietic organs (e.g., kidney and liver); as well as (2) direct MM cell–stromal cell contact (juxtacrine loop).

The function of VEGF and its receptors is one component of regulatory processes contributing to pathogenesis of MM in particular, and hematologic malignancies in general. In MM, VEGF is present in the patient BM microenvironment and associated with neovascularization at sites of MM cell infiltration.[247] The secreted isoforms $VEGF_{121}$, $VEGF_{145}$, and $VEGF_{165}$, as well as ECM- and surface-bound $VEGF_{189}$ and $VEGF_{206}$, are produced by MM cell lines, as well as by patient MM and PCL cells.[33] Importantly, the ECM serves as a reservoir of ECM-bound VEGF isoforms. In addition to their effects on basement membrane and the ECM, uPA, plasmin, heparin, heparan sulfate, and heparinases release this ECM-sequestered VEGF, thereby further regulating the bioavailability of VEGF.[22,24]

In hematologic malignancies including leukemia and MM, there is a "public" (or external) autocrine loop: for example, VEGF secreted by tumor cells activates receptors on tumor cells, as well as other cells. "Private" (or internal) autocrine loops, where the factor activates autonomous cell growth via an intracellular receptor without secretion,[248] have been described in AML cells and in murine experimental models. In contrast to the "public" autocrine loop, cell proliferation mediated by a "private" VEGF autocrine loop is not cell density dependent, and neutralizing antibodies do not prevent continued cell growth or differentiation.[248–254] Importantly, this independence of factor secretion contrasts the regulatory role of VEGF during hematopoiesis versus angiogenesis. The role of "private" autocrine loops in MM pathogenesis, specifically in the development of the tumor cell clone as well as in tumor cell proliferation under nutrient-deprived conditions, remains to be determined. Autocrine VEGF and $\alpha4\beta1$-integrin are involved in chemokine-dependent motility of patient CLL, but not normal, B cells on and through endothelium.[255]

In MM, IL-6 secreted by BMSCs enhances the production and secretion of VEGF by MM cells, thereby augmenting MM cell growth and survival[36,256,257]; conversely, binding of MM cells to BMSCs enhances IL-6, IGF-1, as well as VEGF secretion in BMSCs.[256,258,259] Specifically, IGF-1 induces HIF-1α, which triggers VEGF expression[260]; consequently, inhibition of IGFR-1 activity markedly decreases VEGF secretion in MM/BMSC cocultures.[258] Other mechanisms regulating VEGF expression in MM cells include c-maf and CD40: (1) c-maf-driven expression of integrin-$\beta7$ enhances adhesion to BM stroma and thereby increases VEGF production[261] and (2) CD40 activation induces p53-dependent VEGF secretion.[262]

What is the functional role of VEGF in MM pathogenesis? In addition to stimulating angiogenesis, our studies show that VEGF directly stimulates MM cell migration on fibronectin, proliferation, and survival via autocrine and paracrine loops. Importantly, the range of VEGF target cells within the BM compartment of MM may be even broader since VEGF: (1) dramatically affects the differentiation of multiple hematopoietic lineages in vivo; (2) increases the production of B cells and generation of myeloid cells[177,263]; (3) regulates HSC survival by an internal autocrine loop mechanism[121]; (4) increases both osteoclastic bone-resorbing activity[178] and osteoclast chemotaxis[180]; and (5) inhibits maturation of dendritic cells.[177] For example, we have shown that addition of anti-VEGF antibodies to MM patients' BM sera abrogates its inhibitory effect on dendritic cell maturation.[264] Therefore besides angiogenesis, dysregulation of VEGF plays an important role in MM pathogenesis and clinical manifestations, including lytic lesions of the bone and immune deficiency. Inhibition of VEGF in MM may therefore target not only endothelial cells and MM cells, but also a broad array of cells contributing to MM pathogenesis.

4.4.8.2 VEGF Signal Transduction

In MM, tumor cell growth, survival, and migration are necessary for homing of MM cells to the BM, their expansion within the BM microenvironment, and their egress into the peripheral blood. Binding of exogenous $VEGF_{165}$ to MM cells triggers Flt-1 tyrosine phosphorylation. Subsequently, several downstream signaling pathways are activated: (1) a PI3-kinase/PKC-α-dependent cascade mediating MM cell migration on fibronectin, evidenced by using the PKC inhibitor bisindolylmaleimide I and LY294002; (2) a MEK extracellular signal-regulated protein kinase (ERK) pathway mediating MM cell proliferation, evidenced by use of anti-VEGF antibody and PD098059[33]; and (3) a pathway mediating MM cell survival via upregulation of Mcl-1 and survivin in a dose- and time-dependent manner (Figure 4.4).[265] In addition, STAT3 phosphorylation was observed in the JJN3 MM cell line, but not in the MM 1S cell line.[33,266]

As in MM, autocrine VEGF stimulation of VEGFR-2 triggers leukemic cell proliferation and migration, thereby inducing a more invasive tumor phenotype.[233] Moreover, VEGF induces the expression of heat shock protein 90 (Hsp90) and its binding to Bcl-2 and Apaf-1, thereby increasing leukemic cell resistance to serum deprivation-induced apoptosis.[241]

In addition to the known paracrine/external autocrine/juxtacrine loops, an internal autocrine loop of VEGF similar to AML may contribute to growth factor and cell density-independent proliferation of MM cells. Internalization and functional intracellular trafficking of VEGFR-1 may be mediated by lipid rafts in general, and by caveolae in particular. Functionally caveolae, vesicular flask-shaped invaginations of the plasma membrane composed of caveolins, cholesterol, and sphingolipids, have been implicated in transmembrane

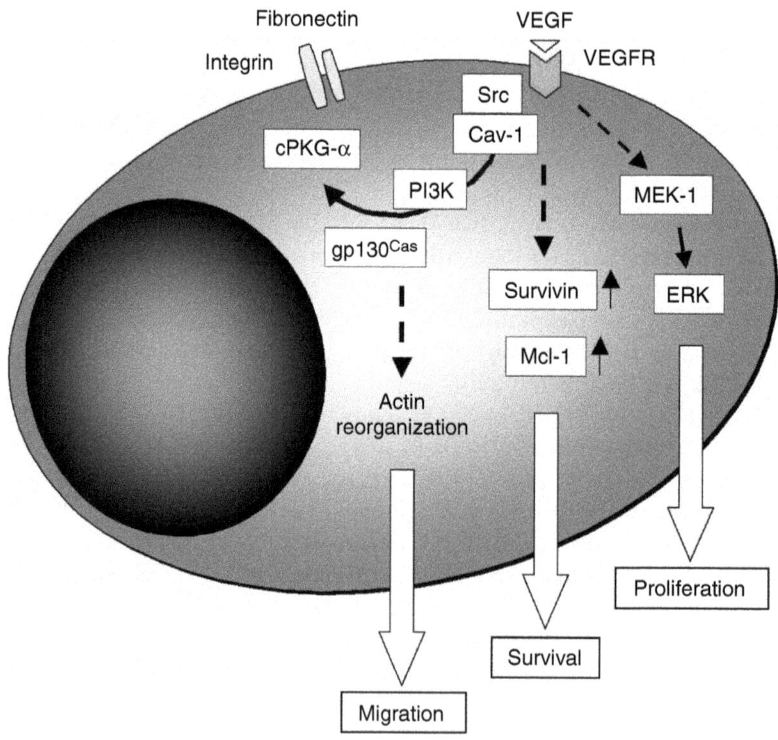

FIGURE 4.4 VEGF signal transduction in MM cells. VEGF mediates MM cell proliferation via MEK-1/ERK signaling and survival via upregulation of Mcl-1 and survivin. VEGF-induced MM cell migration on fibronectin is dependent on the localization of VEGFR-1 within caveolae, as well as by Src tyrosine kinase (STK) family-dependent phosphorylation of caveolin-1, PI3-kinase, and PKC-α.

transport and signal transduction.[267,268] Specifically, we have previously demonstrated that caveolae and caveolin-1 are present in MM cells and required for IL-6- and IGF-1-triggered Akt-1-mediated survival of MM cells.[269] Moreover, our recent data show that caveolin-1 is also associated with VEGF-triggered Src-dependent MM cell migration.[239] Ongoing studies are exploring whether caveolae mediate VEGFR-1 trafficking in MM cells, thereby mediating both "private" and "public" autocrine MM cell proliferation and survival; as well as their role in development of the tumor cell clone (Figure 4.4).

4.5 OVERVIEW OF NOVEL THERAPEUTIC APPROACHES TARGETING VEGF AND THE VEGF RECEPTORS IN MULTIPLE MYELOMA AND OTHER HEMATOLOGIC MALIGNANCIES

4.5.1 Stem Cell Transplantation and Tumor Angiogenesis

Stem cell transplantation, especially allogeneic stem cell transplantation, offers a potential for cure for many hematological malignancies. In the setting of MM autologous stem cell transplantation has prolonged survival in randomized studies.[270] As previously mentioned, BM MVD remains a powerful prognostic factor in patients with MM undergoing stem cell transplantation. Although most of the initial studies in angiogenesis were done on solid tumors, there are data suggesting the importance of angiogenesis in hematologic malignancies. BM MVD was studied in 13 patients with MM, before autologous stem cell transplantation and at the time of response (seven complete and six partial responders).[218] It was significantly higher in patients with MM compared with normal BM. Following autologous stem cell transplantation, MVD remained high in MM samples compared with normal samples and there was no difference in MVD at the time of complete or partial response compared with pretransplantation. These results are similar to those reported after conventional therapies for MM and differ from results seen with antiangiogenic agents like thalidomide.[271] It is conceivable that persistent microvessels following therapy in the BM of patients with MM may have a role in their eventual relapse; however, this has not been proven. In one study of BM MVD in 21 patients with MM undergoing high-dose chemotherapy and autologous SCT,[272] there was a significant decrease in MVD associated with response. Sixteen of twenty-one patients (76.2%) had decreased MVD after SCT, and five patients were found to have a >50% decrease in MVD after SCT. However, there was no difference in overall survival between the patient groups with and without decreased MVD after SCT. The BM MVD remained higher than normal even in responders.

The effect of allogeneic stem cell transplantation on BM angiogenesis in hematological malignancies has also been studied. BM MVD was studied on samples obtained just prior to and at 3–5 months after autologous stem cell transplant (ASCT) in 24 patients with CML.[273] The median MVD pretransplant was 14 (4–37), with 11 patients having high-grade angiogenesis and 13 having low grade. At a median follow up of 4 months from transplant, the median MVD was 20 (range 5–36), with 12 patients having high-grade angiogenesis and 12 low grade. The microvessels in the posttransplant BM appeared morphologically different, with striking dilatation and sinusoidal appearance compared with the pretransplant marrow. However, there was no significant change in MVD following transplant. Abnormal BM angiogenesis also appears to persist in the following ASCT for CML, at least in the short term, and its prognostic value remains unanswered.

4.5.2 Therapeutic Approaches to Target VEGF

From the clinical studies and our understanding of the disease biology, it is clear that increased tumor-associated angiogenesis is important in MM. It is only rational that new

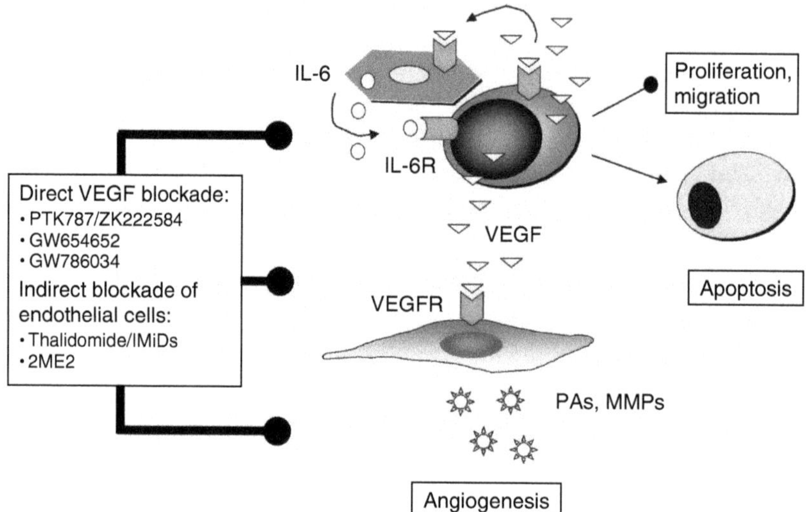

FIGURE 4.5 Direct and indirect inhibition of VEGF.

therapeutic approaches against this incurable disease be targeted toward disruption of this vascular network. The pathological vascular network in the marrow provides support for the MM cells through multiple mechanisms including secretion of important cytokines and presents an ideal target. Several novel therapeutic agents are currently undergoing testing in MM and other malignancies: drugs having a direct effect on the microvessels or the endothelial cells; and those that inhibit mediators stimulating angiogenesis (http://www.cancer.gov/clinicaltrials/developments/anti-angio-table). Approaches to disrupt the VEGF/VEGFR signaling pathways range from small molecule ATP-competitive VEGFR inhibitors to biological agents like soluble receptors, anti-VEGF and anti-VEGFR antibodies, small molecule inhibitors, and VEGF transcription inhibitors. The earliest study showing that VEGF blockade by a monoclonal antibody suppressed angiogenesis and tumor growth in vivo was in a glioblastoma cell line.[274] The antitumor effect of VEGF inactivation in vivo was more recently shown in a murine insulinoma model (Figure 4.5).[275]

4.5.3 Anti-VEGF Antibody (Bevacizumab; Avastin)

The most successful approach to date to therapeutically target VEGF is the use of a humanized monoclonal antibody against VEGF, bevacizumab (Avastin),[276] which was U.S. Food and Drug Administration (FDA) approved for use as a first-line therapy for metastatic colorectal cancer in February 2004; specifically, bevacizumab in combination with intravenous 5-FU-based chemotherapy is a new treatment option.[277] Furthermore, bevacizumab also significantly prolonged time to progression of disease in patients with metastatic renal cell cancer in a clinical phase II trial.[278,279] Effects of bevacizumab were also seen in combination approaches: with chemotherapy in NSCLC,[280,281] pancreatic[282] and breast carcinoma[283]; with interferon-α in both melanoma and metastatic renal cell carcinoma[284]; with radiotherapy in rectal cancer; and with thalidomide in metastatic renal cell carcinoma. Interestingly, bevacicumab reduces the number of CEPs and increases the fraction of tumor endothelial cells with pericyte coverage, thereby reflecting the dropout of immature endothelial cells and potentially providing a novel biomarker.[285] Ongoing studies in hematological malignancies are evaluating the efficacy of bevacizumab in patients with: (1) relapsed or refractory MM (with or without thalidomide); (2) blastic phase CML (CML-BP); (3) MDS; (4) relapsed aggressive NHL.

4.5.4 VEGF-Trap

Another approach is to target VEGFR: Flt (1–3) IgG, a Fc fusion with the first three Ig-like domains of VEGFR-1, inhibits tumor growth in a murine rhabdomyosarcoma xenograft model.[286] A hybrid Fc construct in which domain 2 of VEGFR-1 is joined to domain 3 of the VEGFR-2 (VEGF-trap)[287] causes regression of coopted vessels in a model of neuroblastoma.[288] Importantly, ongoing clinical studies are evaluating the efficacy of the VEGF-trap in patients with incurable relapsed or refractory solid tumors or NHL.

4.5.4.1 PTK787/ZK222584

This orally available tyrosine kinase inhibitor (Novartis Pharmaceuticals) binds to the ATP-binding sites of VEGFRs.[289,290] In MM, we have reported that PTK787/ZK222584: (1) acts directly on MM cells to inhibit VEGF-induced MM cell growth and migration and (2) inhibits paracrine IL-6-mediated MM cell growth in the BM milieu.[291] PTK787/ZK222584 is now under evaluation in: (1) a phase III trial for colorectal cancer; (2) a phase I trial together with imatinib mesylate (Gleevec), for AML, AMM, CML-BP; (3) a phase II trail for primary or secondary MDS; and (4) a phase I trial for MM.

4.5.4.2 Pan-VEGFR Inhibitors

The indazolylpyrimidine GW654652 is a small molecule tyrosine kinase inhibitor, which inhibits all three VEGFRs with similar potency. Preclinical data demonstrate that GW654652 is a potent inhibitor of both VEGF- and bFGF-mediated angiogenesis, as well as VEGF-induced vascular permeability in vivo. In addition, daily oral dosing with GW654652 inhibits the growth of human head and neck, colon, melanoma, and prostate cancer xenografts in vivo.[292] Moreover, GW654652 has a favorable pharmacokinetic profile in murine and canine studies, suggesting its potential clinical application.[293] In MM, the indazolylpyrimidine class of pan-VEGFR inhibitors acts both directly on tumor cells and in the BM microenvironment to overcome drug resistance. Specifically, GW654652 inhibits VEGF-triggered migrational activity and proliferation of MM cell lines, including those sensitive and resistant to conventional therapy, in a dose-dependent fashion. Furthermore, GW654652 blocks both VEGF-induced Flt-1 phosphorylation and downstream activation of AKT-1 and MAPK-signaling pathways. Importantly, GW654652 also acts in the BM microenvironment, since it blocks HUVEC proliferation and inhibits both IL-6 and VEGF secretions, as well as proliferation of MM cells induced by MM cell binding to BM stromal cells. Importantly, inhibition of MM cell growth, survival, and migration is not reversed by removal of GW654652 after treatment; in contrast, its effects on HUVEC survival and proliferation are partially reversible after drug removal. The higher sensitivity of MM cells may be due to lower VEGFR expression levels in MM cells than in HUVECs. Alternatively, these data may indicate that long-term exposure of MM cells to GW654652 leads to stable destruction of Flt-1 on MM cells. These results therefore suggest potential utility of this drug class to target both tumor cells and their microenvironment.[294] A phase I clinical trial in MM is planned at our institution.

4.5.4.3 Other VEGF Inhibitors

Significant efficacy of RTK inhibitors including molecule SU5416 (Sugen),[295–297] SU11248 (Sugen),[298] AG013676 and CP-547,632 (Pfizer),[299] ZD6474 (AstraZeneca),[300] and BAY 43-9006 (Bayer/Onyx)[301,302] has been demonstrated in both preclinical models and clinical trials of solid tumors and hematological malignancies including AML, MDS, and MM.

4.5.5 Therapeutic Approaches Directly Targeting Endothelial Cells

4.5.5.1 Thalidomide/Immunomodulatory Drugs

Thalidomide (Thal) was the first "antiangiogenic agent" proven to have activity in MM. Thal was initially introduced as a sedative hypnotic in the 1960s and was withdrawn due to its severe teratogenic effects. Studies done by D'Amato in the early 1990s were the first to recognize the antiangiogenic properties of Thal,[303] evidenced by inhibition of angiogenesis in rabbit cornea micropocket assays. Subsequently, Thal was tested to treat advanced MM and found to have significant benefit. In their landmark study, Singhal et al. demonstrated an impressive response rate among patients with relapsed refractory MM treated with thalidomide.[223] Multiple studies have since validated these results in different stages of MM; Thal, with or without dexamethasone, has become a valuable therapy for patients with MM.

Subsequently, a series of immunomodulatory drugs (IMiDs) have been developed including CC5013 (lenalidomide; Revimid now Revlimid) and CC4047 (Actimid).[304] Importantly, both Thal and the more potent IMiDs can overcome the growth and survival advantage conferred by the BM milieu, including downregulating VEGF.[305,306] Furthermore, Thal inhibits the proliferation of endothelial cells isolated from MM BM and decreases capillary formation by these cells in matrigel assays.[228] A significant decrease in BM MVD has also been seen in patients obtaining a response to treatment, suggesting that an antiangiogenic mechanism contributes to its therapeutic effect.[271] Interestingly, a recent report using the zebra fish model demonstrated that Thal-induced antiangiogenic action is mediated by ceramide through depletion of VEGFRs, and can be antagonized by sphingosine-1-phosphate.[307] Moreover, the IMiDs costimulate T cells, enhance antitumor immunity mediated by IFN-γ and IL-2, and augment NK cell cytotoxicity.[264,308,309] Finally, the use of CC5013 in SCID mouse human MM models suppressed growth of tumor, as well as decreased tumor MVD compared with untreated animals.[310] These studies provided the basis for the use of IMiD CC5013 in a phase 1 dose-escalation trial in patients with relapsed and refractory MM, which demonstrated either response or stabilization of disease in 79% cases.[311] Two clinical phase II trials have confirmed these data, achieving complete responses with favorable side effect profiles; and two clinical phase III trials comparing Revlimid to dexamethasone/Revlimid treatment of relapsed MM were unblinded because of higher response rates and prolonged time to progression in the patients treated with dexamethasone/Revlimid.

A phase I/II study in patients with advanced MM using IMiD CC4047 (Actimid) (*Celgene*) showed anti-MM activity and an acceptable safety profile.[312] In addition to MM, clinical studies are ongoing in MDS, large B-cell lymphoma, CLL, and small lymphocytic lymphomas.[304]

4.5.5.2 2-Methoxyestradiol

2-Methoxyestradiol (2ME2), a natural metabolite of estradiol, is a potent antitumor and antiangiogenic agent in leukemic cells.[313–318] However, the mechanisms mediating its biological effects remain unclear. Our recent studies show that 2ME2 also inhibits growth and induces apoptosis in MM cell lines and patient cells. Importantly, VEGF secretion induced by adhesion of MM cells to BMSCs is inhibited by 2ME2. Conversely, 2ME2 inhibits MM cell growth, prolongs survival, and decreases angiogenesis in a murine model.[319] Clinical phase I studies are underway in both solid tumors and in MM.

4.5.5.3 LY317615 (PKC-β Inhibitor)

Preclinical data show that the PKC-β inhibitor LY317615 decreases plasma VEGF levels in human tumor xenograft-bearing mice.[320–322] A clinical phase II trail is now active which tests the efficacy of LY317615 in patients with relapsed or refractory lymphoma.

4.5.5.4 Targeting Caveolin-1

Caveolin-1 mediates both VEGF-triggered angiogenesis/permeability and MM cell migration. Importantly, recent data demonstrate that a peptide derived from the eNOS inhibitory CSD sequence of caveolin-1 (Cavtratin) selectively inhibits tumor microvascular permeability and vasculature of prooncogenic macromolecules, thereby blocking tumor progression.[323] Therefore, targeting caveolin-1 in MM potentially decreases both VEGF-triggered MM cell migration and eNOS-enhanced permeability.

4.5.6 Therapeutic Approach Inhibiting Endothelial-Specific Integrin/Survival Signaling

The cyclic pentapeptide EMD121974 (cilengitide) mediates its antiangiogenic activity via selective inhibition of integrins $\alpha_v\beta_3$ and $\alpha_v\beta_5$.[324–327] Clinical trails are ongoing in AML and lymphomas.

4.5.7 Additional Approaches Directly or Indirectly Targeting VEGF

4.5.7.1 Lysophosphatidic Acid Acyltransferase-β Inhibitors

LPA, an intracellular lipid mediator with growth factor–like activities, stimulates a specific G-protein-coupled receptor present in numerous cell types, thereby triggering a multitude of biological responses.[328] LPA stimulates VEGF expression and secretion in ovarian cancer cells, thereby increasing tumor angiogenesis and subsequent tumor growth and metastasis.[329] Conversely, we have recently demonstrated that LPA acyltransferase (LPAAT) inhibitors (Cell Therapeutics, Inc., Seattle, WA) have antitumor activity in MM, at least in part due to their inhibitory effect on VEGF expression.[330]

4.5.7.2 Bortezomib

Bortezomib (previously denoted PS341) is a novel proteasome inhibitor approved by the FDA for therapy of patients with progressive MM after previous treatment.[331] It induces apoptosis in drug-resistant MM cells, and inhibits both binding of MM cells in the BM microenvironment, as well as production and secretion of cytokines which mediate MM cell growth and survival. IL-6 triggered phosphorylation of ERK, but not of STAT3, is blocked by bortezomib.[332] Moreover, bortezomib mediates anti-MM activity by triggering phosphorylation of both p53 protein and JNK, cleavage of DNA-PKCs and ATM,[333] and caspase-dependent downregulation of gp130.[334] The antiangiogenic effect of bortezomib[335,336] is another potential mechanism of its anti-MM activity.[337] Moreover, our recent studies show that bortezomib: (1) downregulates caveolin-1 expression and inhibits caveolin-1 tyrosine phosphorylation, which are required for VEGF-mediated MM cell migration on fibronectin and (2) blocks VEGF-induced tyrosine phosphorylation of caveolin-1 in HUVECs, thereby inhibiting ERK-dependent endothelial cell proliferation.

4.5.7.3 CD40 Antibody

CD40 activation induces p53-dependent VEGF secretion[262]; conversely, a humanized anti-CD40 antibody induces cytotoxicity in human MM cells.[338] A clinical trial is ongoing at our institution.

4.5.7.4 Combination Therapy/Metronomic Chemotherapy

The production of pro- and antiapoptotic molecules changes during the course of conventional therapy for cancer. Specifically, prior studies show that surgery and chemotherapy, in contrast to irradiation, may even enhance tumor angiogenesis by stimulating production of VEGF and other endothelial cell survival and growth factors in tumor cells.[339] High local VEGF concentrations in the BM microenvironment of MM patients suppress the antiproliferative effects of several chemotherapeutics, thereby promoting multidrug resistance.[340,341] Therefore, combining chemotherapies and irradiation with drugs that block VEGF signaling may enhance antitumor efficacy, for example, by "normalizing" tumor vasculature and thereby improving oxygenation and delivery of chemotherapies to tumor cells.[342] Enhanced antitumor activity of conventional chemotherapy[343–345] and irradiation[346,347] regimens has been achieved when combined with antiangiogenic drugs. Novel therapeutic strategies of metronomic chemotherapy[348,349] use frequent uninterrupted administration of conventional chemotherapeutics in doses significantly below the maximum tolerated dose (MTD) for prolonged periods,[350] thereby both reducing toxic side effects and improving antitumor effects.[351–353] Another advantage of metronomic chemotherapy is the possibility that it may be combined with antiangiogenic drugs like bevacizumab (Avastin).[277]

4.6 CONCLUSION AND FUTURE THERAPEUTIC PERSPECTIVES

The complexity of VEGF actions is determined by a multitude of target cells, as well as molecular mechanisms including VEGFR or VEGF homo- or heterooligomerization; modification of VEGFR activity by associated receptor molecules like NRP-1, 2 or HSPGs and heparin; and the expression of different VEGF splice forms, which trigger the formation of distinct VEGFR-associated signaling complexes.

Besides its role as an essential regulator of physiological and pathological angiogenesis, VEGF is now known: to trigger growth, survival, and migration of leukemia and MM cells via paracrine and autocrine pathways; to inhibit maturation of dendritic cells; and to increase bone-resorbing activity. Moreover, VEGF is also associated with HSC differentiation and hematopoietic recovery in adults. VEGF levels and increased vascularity are correlated with clinical outcome in hematological malignancies including leukemias, lymphomas, and MM. In addition, effects of VEGF on target cells other than endothelial cells may contribute to the clinical manifestations of leukemias, lymphomas, and MM. Direct and indirect targeting of VEGF and its receptors are therefore promising novel therapeutic approaches to improve patient outcome. When tumor cells grow, a wide array of angiogenic molecules may be produced. This, at least in part, may be the reason why single-agent therapy inhibiting VEGF is only effective initially, and that a cocktail of antiangiogenic therapies is required to effectively prevent tumor angiogenesis. Enhancement of antitumor efficacy can also be achieved by combining drugs that block VEGF signaling with chemotherapies or irradiation, thereby "normalizing" and sensitizing tumor vasculature and improving oxygenation and delivery of chemotherapies to tumor cells and endothelial cells. In addition, VEGF actions and the array of its target cells may vary, depending on the stage of the malignancy and the tumor type. Ongoing studies are addressing these dynamic interactions and translating them into therapeutic strategies. Furthermore, since VEGF plays a pivotal role in HSC differentiation, the potential role of VEGF and VEGFR-1 in clonal leukemogenic development is also under investigation. Finally comorbidities, drug regimens, environmental and dietary variables are likely to influence angiogenic responses. This may, at least in part, explain why rapid and dramatic antitumor responses in mice triggered by antiangiogenic agents can generally not be recapitulated in clinical phase I and II trials in human cancer patients. The development of an optimal animal model for studying antiangiogenic agents in solid and hematologic malignancies is therefore a major goal. Moreover, differences in patient responses are also influenced by immunologic, BM, and monocyte characteristics.[354–356]

A close collaboration between basic researchers and clinicians will be required both to enhance our understanding of the pathophysiological role of VEGF in hematological malignancies, and to derive related targeted clinical trials, for example, combination therapy of antiangiogenic agents with chemotherapy agents, to improve patient outcome.

ACKNOWLEDGMENT

This work was supported by National Institutes of Health Grants RO 50947 and PO-1 78378, and the Doris Duke Distinguished Clinical Research Scientist Award (to K.C.A.).

REFERENCES

1. Senger DR, Galli SJ, Dvorak AM, Perruzzi CA, Harvey VS, Dvorak HF. Tumor cells secrete a vascular permeability factor that promotes accumulation of ascites fluid. *Science* 1983;219:983–985.
2. Ferrara N, Henzel WJ. Pituitary follicular cells secrete a novel heparin-binding growth factor specific for vascular endothelial cells. *Biochem Biophys Res Commun* 1989;161:851–858.
3. Connolly DT, Olander JV, Heuvelman D, Nelson R, Monsell R, Siegel N, Haymore BL, Leimgruber R, Feder J. Human vascular permeability factor. Isolation from U937 cells. *J Biol Chem* 1989;264:20017–20024.
4. Leung DW, Cachianes G, Kuang WJ, Goeddel DV, Ferrara N. Vascular endothelial growth factor is a secreted angiogenic mitogen. *Science* 1989;246:1306–1309.
5. Keck PJ, Hauser SD, Krivi G, Sanzo K, Warren T, Feder J, Connolly DT. Vascular permeability factor, an endothelial cell mitogen related to PDGF. *Science* 1989;246:1309–1312.
6. Ferrara N, Carver-Moore K, Chen H, Dowd M, Lu L, O'Shea KS, Powell-Braxton L, Hillan KJ, Moore MW. Heterozygous embryonic lethality induced by targeted inactivation of the *VEGF* gene. *Nature* 1996;380:439–442.
7. Carmeliet P, Ferreira V, Breier G, Pollefeyt S, Kieckens L, Gertsenstein M, Fahrig M, Vandenhoeck A, Harpal K, Eberhardt C, Declercq C, Pawling J, Moons L, Collen D, Risau W, Nagy A. Abnormal blood vessel development and lethality in embryos lacking a single VEGF allele. *Nature* 1996;380:435–439.
8. Conn G, Bayne ML, Soderman DD, Kwok PW, Sullivan KA, Palisi TM, Hope DA, Thomas KA. Amino acid and cDNA sequences of a vascular endothelial cell mitogen that is homologous to platelet-derived growth factor. *Proc Natl Acad Sci USA* 1990;87:2628–2632.
9. Senger DR, Connolly DT, Van de Water L, Feder J, Dvorak HF. Purification and NH_2-terminal amino acid sequence of guinea pig tumor-secreted vascular permeability factor. *Cancer Res* 1990;50:1774–1778.
10. Olofsson B, Pajusola K, Kaipainen A, von Euler G, Joukov V, Saksela O, Orpana A, Pettersson RF, Alitalo K, Eriksson U. Vascular endothelial growth factor B, a novel growth factor for endothelial cells. *Proc Natl Acad Sci USA* 1996;93:2576–2581.
11. Joukov V, Pajusola K, Kaipainen A, Chilov D, Lahtinen I, Kukk E, Saksela O, Kalkkinen N, Alitalo K. A novel vascular endothelial growth factor, VEGF-C, is a ligand for the Flt4 (VEGFR-3) and KDR (VEGFR-2) receptor tyrosine kinases. *EMBO J* 1996;15:1751.
12. Achen MG, Jeltsch M, Kukk E, Makinen T, Vitali A, Wilks AF, Alitalo K, Stacker SA. Vascular endothelial growth factor D (VEGF-D) is a ligand for the tyrosine kinases VEGF receptor 2 (Flk1) and VEGF receptor 3 (Flt4). *Proc Natl Acad Sci USA* 1998;95:548–553.
13. Maglione D, Guerriero V, Viglietto G, Delli-Bovi P, Persico MG. Isolation of a human placenta cDNA coding for a protein related to the vascular permeability factor. *Proc Natl Acad Sci USA* 1991;88:9267–9271.
14. Maglione D, Guerriero V, Viglietto G, Ferraro MG, Aprelikova O, Alitalo K, Del Vecchio S, Lei KJ, Chou JY, Persico MG. Two alternative mRNAs coding for the angiogenic factor, placenta growth factor (PlGF), are transcribed from a single gene of chromosome 14. *Oncogene* 1993;8:925–931.

15. Muller YA, Li B, Christinger HW, Wells JA, Cunningham BC, de Vos AM. Vascular endothelial growth factor: crystal structure and functional mapping of the kinase domain receptor binding site. *Proc Natl Acad Sci USA* 1997;94:7192–7197.
16. DiSalvo J, Bayne ML, Conn G, Kwok PW, Trivedi PG, Soderman DD, Palisi TM, Sullivan KA, Thomas KA. Purification and characterization of a naturally occurring vascular endothelial growth factor placenta growth factor heterodimer. *J Biol Chem* 1995;270:7717–7723.
17. Cao Y, Chen H, Zhou L, Chiang MK, Anand-Apte B, Weatherbee JA, Wang Y, Fang F, Flanagan JG, Tsang ML. Heterodimers of placenta growth factor/vascular endothelial growth factor. Endothelial activity, tumor cell expression, and high affinity binding to Flk-1/KDR. *J Biol Chem* 1996;271:3154–3162.
18. Potgens AJ, Lubsen NH, van Altena MC, Vermeulen R, Bakker A, Schoenmakers JG, Ruiter DJ, de Waal RM. Covalent dimerization of vascular permeability factor/vascular endothelial growth factor is essential for its biological activity. Evidence from Cys to Ser mutations. *J Biol Chem* 1994;269:32879–32885.
19. Vincenti V, Cassano C, Rocchi M, Persico G. Assignment of the vascular endothelial growth factor gene to human chromosome 6p21.3. *Circulation* 1996;93:1493–1495.
20. Tischer E, Mitchell R, Hartman T, Silva M, Gospodarowicz D, Fiddes JC, Abraham JA. The human gene for vascular endothelial growth factor. Multiple protein forms are encoded through alternative exon splicing. *J Biol Chem* 1991;266:11947–11954.
21. Houck KA, Ferrara N, Winer J, Cachianes G, Li B, Leung DW. The vascular endothelial growth factor family: identification of a fourth molecular species and characterization of alternative splicing of RNA. *Mol Endocrinol* 1991;5:1806–1814.
22. Houck KA, Leung DW, Rowland AM, Winer J, Ferrara N. Dual regulation of vascular endothelial growth factor bioavailability by genetic and proteolytic mechanisms. *J Biol Chem* 1992;267:26031–26037.
23. Park JE, Keller GA, Ferrara N. The vascular endothelial growth factor (VEGF) isoforms: differential deposition into the subepithelial extracellular matrix and bioactivity of extracellular matrix-bound VEGF. *Mol Biol Cell* 1993;4:1317–1326.
24. Plouet J, Moro F, Bertagnolli S, Coldeboeuf N, Mazarguil H, Clamens S, Bayard F. Extracellular cleavage of the vascular endothelial growth factor 189-amino acid form by urokinase is required for its mitogenic effect. *J Biol Chem* 1997;272:13390–13396.
25. Keyt BA, Berleau LT, Nguyen HV, Chen H, Heinsohn H, Vandlen R, Ferrara N. The carboxyl-terminal domain (111–165) of vascular endothelial growth factor is critical for its mitogenic potency. *J Biol Chem* 1996;271:7788–7795.
26. Yoshiji H, Gomez DE, Shibuya M, Thorgeirsson UP. Expression of vascular endothelial growth factor, its receptor, and other angiogenic factors in human breast cancer. *Cancer Res* 1996;56:2013–2016.
27. Volm M, Koomagi R, Mattern J. Prognostic value of vascular endothelial growth factor and its receptor Flt-1 in squamous cell lung cancer. *Int J Cancer* 1997;74:64–68.
28. Hatva E, Kaipainen A, Mentula P, Jaaskelainen J, Paetau A, Haltia M, Alitalo K. Expression of endothelial cell-specific receptor tyrosine kinases and growth factors in human brain tumors. *Am J Pathol* 1995;146:368–378.
29. Ellis LM, Takahashi Y, Fenoglio CJ, Cleary KR, Bucana CD, Evans DB. Vessel counts and vascular endothelial growth factor expression in pancreatic adenocarcinoma. *Eur J Cancer* 1998;34:337–340.
30. Boocock CA, Charnock-Jones DS, Sharkey AM, McLaren J, Barker PJ, Wright KA, Twentyman PR, Smith SK. Expression of vascular endothelial growth factor and its receptors flt and KDR in ovarian carcinoma. *J Natl Cancer Inst* 1995;87:506–516.
31. Brown LF, Berse B, Jackman RW, Tognazzi K, Manseau EJ, Dvorak HF, Senger DR. Increased expression of vascular permeability factor (vascular endothelial growth factor) and its receptors in kidney and bladder carcinomas. *Am J Pathol* 1993;143:1255–1262.
32. Ferrara N, Gerber HP, LeCouter J. The biology of VEGF and its receptors. *Nat Med* 2003;9:669–676.
33. Podar K, Tai YT, Davies FE, Lentzsch S, Sattler M, Hideshima T, Lin BK, Gupta D, Shima Y, Chauhan D, Mitsiades C, Raje N, Richardson P, Anderson KC. Vascular endothelial growth

factor triggers signaling cascades mediating multiple myeloma cell growth and migration. *Blood* 2001;98:428–435.

34. Aoki Y, Tosato G. Vascular endothelial growth factor/vascular permeability factor in the pathogenesis of primary effusion lymphomas. *Leuk Lymphoma* 2001;41:229–237.
35. Poltorak Z, Cohen T, Sivan R, Kandelis Y, Spira G, Vlodavsky I, Keshet E, Neufeld G. $VEGF_{145}$, a secreted vascular endothelial growth factor isoform that binds to extracellular matrix. *J Biol Chem* 1997;272:7151–7158.
36. Bellamy WT, Richter L, Frutiger Y, Grogan TM. Expression of vascular endothelial growth factor and its receptors in hematopoietic malignancies. *Cancer Res* 1999;59:728–733.
37. Wang GL, Semenza GL. Purification and characterization of hypoxia-inducible factor 1. *J Biol Chem* 1995;270:1230–1237.
38. Iliopoulos O, Levy AP, Jiang C, Kaelin WG, Jr., Goldberg MA. Negative regulation of hypoxia-inducible genes by the von Hippel-Lindau protein. *Proc Natl Acad Sci USA* 1996;93:10595–10599.
39. Shweiki D, Itin A, Soffer D, Keshet E. Vascular endothelial growth factor induced by hypoxia may mediate hypoxia-initiated angiogenesis. *Nature* 1992;359:843–845.
40. Plate KH, Breier G, Weich HA, Risau W. Vascular endothelial growth factor is a potential tumour angiogenesis factor in human gliomas in vivo. *Nature* 1992;359:845–848.
41. Cohen T, Nahari D, Cerem LW, Neufeld G, Levi BZ. Interleukin 6 induces the expression of vascular endothelial growth factor. *J Biol Chem* 1996;271:736–741.
42. Ferrara N, Davis-Smyth T. The biology of vascular endothelial growth factor. *Endocr Rev* 1997;18:4–25.
43. Neufeld G, Cohen T, Gengrinovitch S, Poltorak Z. Vascular endothelial growth factor (VEGF) and its receptors. *FASEB J* 1999;13:9–22.
44. Brauchle M, Funk JO, Kind P, Werner S. Ultraviolet B and H_2O_2 are potent inducers of vascular endothelial growth factor expression in cultured keratinocytes. *J Biol Chem* 1996;271:21793–21797.
45. Kieser A, Weich HA, Brandner G, Marme D, Kolch W. Mutant p53 potentiates protein kinase C induction of vascular endothelial growth factor expression. *Oncogene* 1994;9:963–969.
46. Rak J, Mitsuhashi Y, Bayko L, Filmus J, Shirasawa S, Sasazuki T, Kerbel RS. Mutant ras oncogenes upregulate VEGF/VPF expression: implications for induction and inhibition of tumor angiogenesis. *Cancer Res* 1995;55:4575–4580.
47. Grugel S, Finkenzeller G, Weindel K, Barleon B, Marme D. Both v-Ha-Ras and v-Raf stimulate expression of the vascular endothelial growth factor in NIH 3T3 cells. *J Biol Chem* 1995;270:25915–25919.
48. Kerbel R, Folkman J. Clinical translation of angiogenesis inhibitors. *Nat Rev Cancer* 2002;2:727–739.
49. D'Arcangelo D, Facchiano F, Barlucchi LM, Melillo G, Illi B, Testolin L, Gaetano C, Capogrossi MC. Acidosis inhibits endothelial cell apoptosis and function and induces basic fibroblast growth factor and vascular endothelial growth factor expression. *Circ Res* 2000;86:312–318.
50. He C, Chen X. Transcription regulation of the *vegf* gene by the BMP/Smad pathway in the angioblast of zebrafish embryos. *Biochem Biophys Res Commun* 2005;329:324–330.
51. Gering M, Patient R. Hedgehog signaling is required for adult blood stem cell formation in zebrafish embryos. *Dev Cell* 2005;8:389–400.
52. Satoh H, Yoshida MC, Matsushime H, Shibuya M, Sasaki M. Regional localization of the human c-ros-1 on 6q22 and flt on 13q12. *Jpn J Cancer Res* 1987;78:772–775.
53. de Vries C, Escobedo JA, Ueno H, Houck K, Ferrara N, Williams LT. The fms-like tyrosine kinase, a receptor for vascular endothelial growth factor. *Science* 1992;255:989–991.
54. Shibuya M, Yamaguchi S, Yamane A, Ikeda T, Tojo A, Matsushime H, Sato M. Nucleotide sequence and expression of a novel human receptor-type tyrosine kinase gene (*flt*) closely related to the fms family. *Oncogene* 1990;5:519–524.
55. Terman BI, Jani-Sait S, Carrion ME, Shows TB. The *KDR* gene maps to human chromosome 4q31.2–q32, a locus which is distinct from locations for other type III growth factor receptor tyrosine kinases. *Cytogenet Cell Genet* 1992;60:214–215.
56. Terman BI, Dougher-Vermazen M, Carrion ME, Dimitrov D, Armellino DC, Gospodarowicz D, Bohlen P. Identification of the KDR tyrosine kinase as a receptor for vascular endothelial cell growth factor. *Biochem Biophys Res Commun* 1992;187:1579–1586.

57. Quinn TP, Peters KG, De Vries C, Ferrara N, Williams LT. Fetal liver kinase 1 is a receptor for vascular endothelial growth factor and is selectively expressed in vascular endothelium. *Proc Natl Acad Sci USA* 1993;90:7533–7537.
58. Sait SN, Dougher-Vermazen M, Shows TB, Terman BI. The kinase insert domain receptor gene (*KDR*) has been relocated to chromosome 4q11→q12. *Cytogenet Cell Genet* 1995;70:145–146.
59. Fong GH, Rossant J, Gertsenstein M, Breitman ML. Role of the Flt-1 receptor tyrosine kinase in regulating the assembly of vascular endothelium. *Nature* 1995;376:66–70.
60. Fong GH, Zhang L, Bryce DM, Peng J. Increased hemangioblast commitment, not vascular disorganization, is the primary defect in flt-1 knock-out mice. *Development* 1999;126:3015–3025.
61. Shalaby F, Rossant J, Yamaguchi TP, Gertsenstein M, Wu XF, Breitman ML, Schuh AC. Failure of blood-island formation and vasculogenesis in Flk-1-deficient mice. *Nature* 1995;376:62–66.
62. Davis-Smyth T, Chen H, Park J, Presta LG, Ferrara N. The second immunoglobulin-like domain of the VEGF tyrosine kinase receptor Flt-1 determines ligand binding and may initiate a signal transduction cascade. *EMBO J* 1996;15:4919–4927.
63. Barleon B, Totzke F, Herzog C, Blanke S, Kremmer E, Siemeister G, Marme D, Martiny-Baron G. Mapping of the sites for ligand binding and receptor dimerization at the extracellular domain of the vascular endothelial growth factor receptor FLT-1. *J Biol Chem* 1997;272:10382–10388.
64. Wiesmann C, Fuh G, Christinger HW, Eigenbrot C, Wells JA, de Vos AM. Crystal structure at 1.7 A resolution of VEGF in complex with domain 2 of the Flt-1 receptor. *Cell* 1997;91:695–704.
65. Shinkai A, Ito M, Anazawa H, Yamaguchi S, Shitara K, Shibuya M. Mapping of the sites involved in ligand association and dissociation at the extracellular domain of the kinase insert domain-containing receptor for vascular endothelial growth factor. *J Biol Chem* 1998;273:31283–31288.
66. Takahashi T, Shibuya M. The 230 kDa mature form of KDR/Flk-1 (VEGF receptor-2) activates the PLC-gamma pathway and partially induces mitotic signals in NIH3T3 fibroblasts. *Oncogene* 1997;14:2079–2089.
67. Gerber HP, Condorelli F, Park J, Ferrara N. Differential transcriptional regulation of the two vascular endothelial growth factor receptor genes. Flt-1, but not Flk-1/KDR, is up-regulated by hypoxia. *J Biol Chem* 1997;272:23659–23667.
68. Sheibani N, Frazier WA. Down-regulation of platelet endothelial cell adhesion molecule-1 results in thrombospondin-1 expression and concerted regulation of endothelial cell phenotype. *Mol Biol Cell* 1998;9:701–713.
69. Keyt BA, Nguyen HV, Berleau LT, Duarte CM, Park J, Chen H, Ferrara N. Identification of vascular endothelial growth factor determinants for binding KDR and FLT-1 receptors. Generation of receptor-selective VEGF variants by site-directed mutagenesis. *J Biol Chem* 1996;271:5638–5646.
70. Millauer B, Wizigmann-Voos S, Schnurch H, Martinez R, Moller NP, Risau W, Ullrich A. High affinity VEGF binding and developmental expression suggest Flk-1 as a major regulator of vasculogenesis and angiogenesis. *Cell* 1993;72:835–846.
71. Barleon B, Sozzani S, Zhou D, Weich HA, Mantovani A, Marme D. Migration of human monocytes in response to vascular endothelial growth factor (VEGF) is mediated via the VEGF receptor flt-1. *Blood* 1996;87:3336–3343.
72. Clauss M, Weich H, Breier G, Knies U, Rockl W, Waltenberger J, Risau W. The vascular endothelial growth factor receptor Flt-1 mediates biological activities. Implications for a functional role of placenta growth factor in monocyte activation and chemotaxis. *J Biol Chem* 1996;271:17629–17634.
73. Charnock-Jones DS, Sharkey AM, Boocock CA, Ahmed A, Plevin R, Ferrara N, Smith SK. Vascular endothelial growth factor receptor localization and activation in human trophoblast and choriocarcinoma cells. *Biol Reprod* 1994;51:524–530.
74. Takahashi T, Shirasawa T, Miyake K, Yahagi Y, Maruyama N, Kasahara N, Kawamura T, Matsumura O, Mitarai T, Sakai O. Protein tyrosine kinases expressed in glomeruli and cultured glomerular cells: Flt-1 and VEGF expression in renal mesangial cells. *Biochem Biophys Res Commun* 1995;209:218–226.
75. Grosskreutz CL, Anand-Apte B, Duplaa C, Quinn TP, Terman BI, Zetter B, D'Amore PA. Vascular endothelial growth factor-induced migration of vascular smooth muscle cells in vitro. *Microvasc Res* 1999;58:128–136.

76. Ria R, Roccaro AM, Merchionne F, Vacca A, Dammacco F, Ribatti D. Vascular endothelial growth factor and its receptors in multiple myeloma. *Leukemia* 2003;17:1961–1966.
77. Solovey A, Lin Y, Browne P, Choong S, Wayner E, Hebbel RP. Circulating activated endothelial cells in sickle cell anemia. *N Engl J Med* 1997;337:1584–1590.
78. Peichev M, Naiyer AJ, Pereira D, Zhu Z, Lane WJ, Williams M, Oz MC, Hicklin DJ, Witte L, Moore MA, Rafii S. Expression of VEGFR-2 and AC133 by circulating human CD34(+) cells identifies a population of functional endothelial precursors. *Blood* 2000;95:952–958.
79. Gill M, Dias S, Hattori K, Rivera ML, Hicklin D, Witte L, Girardi L, Yurt R, Himel H, Rafii S. Vascular trauma induces rapid but transient mobilization of VEGFR2(+)AC133(+) endothelial precursor cells. *Circ Res* 2001;88:167–174.
80. Oberg C, Waltenberger J, Claesson-Welsh L, Welsh M. Expression of protein tyrosine kinases in islet cells: possible role of the Flk-1 receptor for beta-cell maturation from duct cells. *Growth Factors* 1994;10:115–126.
81. Yang K, Cepko CL. Flk-1, a receptor for vascular endothelial growth factor (VEGF), is expressed by retinal progenitor cells. *J Neurosci* 1996;16:6089–6099.
82. Katoh O, Tauchi H, Kawaishi K, Kimura A, Satow Y. Expression of the vascular endothelial growth factor (VEGF) receptor gene, *KDR*, in hematopoietic cells and inhibitory effect of VEGF on apoptotic cell death caused by ionizing radiation. *Cancer Res* 1995;55:5687–5692.
83. Ergun S, Kilic N, Fiedler W, Mukhopadhyay AK. Vascular endothelial growth factor and its receptors in normal human testicular tissue. *Mol Cell Endocrinol* 1997;131:9–20.
84. Brown LF, Detmar M, Tognazzi K, Abu-Jawdeh G, Iruela-Arispe ML. Uterine smooth muscle cells express functional receptors (flt-1 and KDR) for vascular permeability factor/vascular endothelial growth factor. *Lab Invest* 1997;76:245–255.
85. Cunningham SA, Arrate MP, Brock TA, Waxham MN. Interactions of FLT-1 and KDR with phospholipase C gamma: identification of the phosphotyrosine binding sites. *Biochem Biophys Res Commun* 1997;240:635–639.
86. Ito N, Wernstedt C, Engstrom U, Claesson-Welsh L. Identification of vascular endothelial growth factor receptor-1 tyrosine phosphorylation sites and binding of SH_2 domain-containing molecules. *J Biol Chem* 1998;273:23410–23418.
87. Wu LW, Mayo LD, Dunbar JD, Kessler KM, Baerwald MR, Jaffe EA, Wang D, Warren RS, Donner DB. Utilization of distinct signaling pathways by receptors for vascular endothelial cell growth factor and other mitogens in the induction of endothelial cell proliferation. *J Biol Chem* 2000;275:5096–5103.
88. Takahashi T, Yamaguchi S, Chida K, Shibuya M. A single autophosphorylation site on KDR/Flk-1 is essential for VEGF-A-dependent activation of PLC-gamma and DNA synthesis in vascular endothelial cells. *EMBO J* 2001;20:2768–2778.
89. Wu LW, Mayo LD, Dunbar JD, Kessler KM, Ozes ON, Warren RS, Donner DB. VRAP is an adaptor protein that binds KDR, a receptor for vascular endothelial cell growth factor. *J Biol Chem* 2000;275:6059–6062.
90. Guo D, Jia Q, Song HY, Warren RS, Donner DB. Vascular endothelial cell growth factor promotes tyrosine phosphorylation of mediators of signal transduction that contain SH_2 domains. Association with endothelial cell proliferation. *J Biol Chem* 1995;270:6729–6733.
91. Abedi H, Zachary I. Vascular endothelial growth factor stimulates tyrosine phosphorylation and recruitment to new focal adhesions of focal adhesion kinase and paxillin in endothelial cells. *J Biol Chem* 1997;272:15442–15451.
92. Rousseau S, Houle F, Kotanides H, Witte L, Waltenberger J, Landry J, Huot J. Vascular endothelial growth factor (VEGF)-driven actin-based motility is mediated by VEGFR2 and requires concerted activation of stress-activated protein kinase 2 (SAPK2/p38) and geldanamycin-sensitive phosphorylation of focal adhesion kinase. *J Biol Chem* 2000;275:10661–10672.
93. Igarashi K, Shigeta K, Isohara T, Yamano T, Uno I. Sck interacts with KDR and Flt-1 via its SH_2 domain. *Biochem Biophys Res Commun* 1998;251:77–82.
94. Warner AJ, Lopez-Dee J, Knight EL, Feramisco JR, Prigent SA. The Shc-related adaptor protein, Sck, forms a complex with the vascular-endothelial-growth-factor receptor KDR in transfected cells. *Biochem J* 2000;347:501–509.

95. He H, Venema VJ, Gu X, Venema RC, Marrero MB, Caldwell RB. Vascular endothelial growth factor signals endothelial cell production of nitric oxide and prostacyclin through flk-1/KDR activation of c-Src. *J Biol Chem* 1999;274:25130–25135.
96. Eliceiri BP, Paul R, Schwartzberg PL, Hood JD, Leng J, Cheresh DA. Selective requirement for Src kinases during VEGF-induced angiogenesis and vascular permeability. *Mol Cell* 1999;4:915–924.
97. Kroll J, Waltenberger J. The vascular endothelial growth factor receptor KDR activates multiple signal transduction pathways in porcine aortic endothelial cells. *J Biol Chem* 1997;272:32521–32527.
98. Igarashi K, Isohara T, Kato T, Shigeta K, Yamano T, Uno I. Tyrosine 1213 of Flt-1 is a major binding site of Nck and SHP-2. *Biochem Biophys Res Commun* 1998;246:95–99.
99. Gerber HP, McMurtrey A, Kowalski J, Yan M, Keyt BA, Dixit V, Ferrara N. Vascular endothelial growth factor regulates endothelial cell survival through the phosphatidylinositol 3′-kinase/Akt signal transduction pathway. Requirement for Flk-1/KDR activation. *J Biol Chem* 1998;273:30336–30343.
100. Dimmeler S, Dernbach E, Zeiher AM. Phosphorylation of the endothelial nitric oxide synthase at Ser-1177 is required for VEGF-induced endothelial cell migration. *FEBS Lett* 2000;477:258–262.
101. Xia P, Aiello LP, Ishii H, Jiang ZY, Park DJ, Robinson GS, Takagi H, Newsome WP, Jirousek MR, King GL. Characterization of vascular endothelial growth factor's effect on the activation of protein kinase C, its isoforms, and endothelial cell growth. *J Clin Invest* 1996;98:2018–2026.
102. Wu HM, Yuan Y, Zawieja DC, Tinsley J, Granger HJ. Role of phospholipase C, protein kinase C, and calcium in VEGF-induced venular hyperpermeability. *Am J Physiol* 1999;276:H535–H542.
103. Takahashi T, Ueno H, Shibuya M. VEGF activates protein kinase C-dependent, but Ras-independent Raf-MEK-MAP kinase pathway for DNA synthesis in primary endothelial cells. *Oncogene* 1999;18:2221–2230.
104. Gliki G, Abu-Ghazaleh R, Jezequel S, Wheeler-Jones C, Zachary I. Vascular endothelial growth factor-induced prostacyclin production is mediated by a protein kinase C (PKC)-dependent activation of extracellular signal-regulated protein kinases 1 and 2 involving PKC-delta and by mobilization of intracellular Ca^{2+}. *Biochem J* 2001;353:503–512.
105. Hood J, Granger HJ. Protein kinase G mediates vascular endothelial growth factor-induced Raf-1 activation and proliferation in human endothelial cells. *J Biol Chem* 1998;273:23504–23508.
106. Parenti A, Morbidelli L, Cui XL, Douglas JG, Hood JD, Granger HJ, Ledda F, Ziche M. Nitric oxide is an upstream signal of vascular endothelial growth factor-induced extracellular signal-regulated kinase 1/2 activation in postcapillary endothelium. *J Biol Chem* 1998;273:4220–4226.
107. Doanes AM, Hegland DD, Sethi R, Kovesdi I, Bruder JT, Finkel T. VEGF stimulates MAPK through a pathway that is unique for receptor tyrosine kinases. *Biochem Biophys Res Commun* 1999;255:545–548.
108. Thakker GD, Hajjar DP, Muller WA, Rosengart TK. The role of phosphatidylinositol 3-kinase in vascular endothelial growth factor signaling. *J Biol Chem* 1999;274:10002–10007.
109. Rousseau S, Houle F, Landry J, Huot J. p38 MAP kinase activation by vascular endothelial growth factor mediates actin reorganization and cell migration in human endothelial cells. *Oncogene* 1997;15:2169–2177.
110. Huser M, Luckett J, Chiloeches A, Mercer K, Iwobi M, Giblett S, Sun XM, Brown J, Marais R, Pritchard C. MEK kinase activity is not necessary for Raf-1 function. *EMBO J* 2001;20:1940–1951.
111. Rahimi N, Kazlauskas A. A role for cadherin-5 in regulation of vascular endothelial growth factor receptor 2 activity in endothelial cells. *Mol Biol Cell* 1999;10:3401–3407.
112. Caveda L, Martin-Padura I, Navarro P, Breviario F, Corada M, Gulino D, Lampugnani MG, Dejana E. Inhibition of cultured cell growth by vascular endothelial cadherin (cadherin-5/VE-cadherin). *J Clin Invest* 1996;98:886–893.
113. Borges E, Jan Y, Ruoslahti E. Platelet-derived growth factor receptor beta and vascular endothelial growth factor receptor 2 bind to the beta 3 integrin through its extracellular domain. *J Biol Chem* 2000;275:39867–39873.
114. Soker S, Takashima S, Miao HQ, Neufeld G, Klagsbrun M. Neuropilin-1 is expressed by endothelial and tumor cells as an isoform-specific receptor for vascular endothelial growth factor. *Cell* 1998;92:735–745.
115. Dougher AM, Wasserstrom H, Torley L, Shridaran L, Westdock P, Hileman RE, Fromm JR, Anderberg R, Lyman S, Linhardt RJ, Kaplan J, Terman BI. Identification of a heparin binding

peptide on the extracellular domain of the KDR VEGF receptor. *Growth Factors* 1997;14:257–268.

116. Huang L, Sankar S, Lin C, Kontos CD, Schroff AD, Cha EH, Feng SM, Li SF, Yu Z, Van Etten RL, Blanar MA, Peters KG. HCPTPA, a protein tyrosine phosphatase that regulates vascular endothelial growth factor receptor-mediated signal transduction and biological activity. *J Biol Chem* 1999;274:38183–38188.
117. Gingras D, Lamy S, Beliveau R. Tyrosine phosphorylation of the vascular endothelial-growth-factor receptor-2 (VEGFR-2) is modulated by Rho proteins. *Biochem J* 2000;348(Pt 2):273–280.
118. Matsumoto T, Claesson-Welsh L. VEGF receptor signal transduction. *Sci STKE* 2001;2001:RE21.
119. Hiratsuka S, Minowa O, Kuno J, Noda T, Shibuya M. Flt-1 lacking the tyrosine kinase domain is sufficient for normal development and angiogenesis in mice. *Proc Natl Acad Sci USA* 1998;95:9349–9354.
120. Hattori K, Heissig B, Wu Y, Dias S, Tejada R, Ferris B, Hicklin DJ, Zhu Z, Bohlen P, Witte L, Hendrikx J, Hackett NR, Crystal RG, Moore MA, Werb Z, Lyden D, Rafii S. Placental growth factor reconstitutes hematopoiesis by recruiting VEGFR1(+) stem cells from bone-marrow microenvironment. *Nat Med* 2002;8:841–849.
121. Gerber HP, Malik AK, Solar GP, Sherman D, Liang XH, Meng G, Hong K, Marsters JC, Ferrara N. VEGF regulates haematopoietic stem cell survival by an internal autocrine loop mechanism. *Nature* 2002;417:954–958.
122. Sawano A, Iwai S, Sakurai Y, Ito M, Shitara K, Nakahata T, Shibuya M. Flt-1, vascular endothelial growth factor receptor 1, is a novel cell surface marker for the lineage of monocyte–macrophages in humans. *Blood* 2001;97:785–791.
123. Luttun A, Tjwa M, Carmeliet P. Placental growth factor (PlGF) and its receptor Flt-1 (VEGFR-1): novel therapeutic targets for angiogenic disorders. *Ann NY Acad Sci* 2002;979:80–93.
124. LeCouter J, Moritz DR, Li B, Phillips GL, Liang XH, Gerber HP, Hillan KJ, Ferrara N. Angiogenesis-independent endothelial protection of liver: role of VEGFR-1. *Science* 2003;299:890–893.
125. Hiratsuka S, Nakamura K, Iwai S, Murakami M, Itoh T, Kijima H, Shipley JM, Senior RM, Shibuya M. MMP9 induction by vascular endothelial growth factor receptor-1 is involved in lung-specific metastasis. *Cancer Cell* 2002;2:289–300.
126. Kendall RL, Wang G, Thomas KA. Identification of a natural soluble form of the vascular endothelial growth factor receptor, FLT-1, and its heterodimerization with KDR. *Biochem Biophys Res Commun* 1996;226:324–328.
127. He Y, Smith SK, Day KA, Clark DE, Licence DR, Charnock-Jones DS. Alternative splicing of vascular endothelial growth factor (VEGF)-R1 (FLT-1) pre-mRNA is important for the regulation of VEGF activity. *Mol Endocrinol* 1999;13:537–545.
128. Gluzman-Poltorak Z, Cohen T, Herzog Y, Neufeld G. Neuropilin-2 is a receptor for the vascular endothelial growth factor (VEGF) forms VEGF-145 and VEGF-165. *J Biol Chem* 2000;275:29922.
129. Gluzman-Poltorak Z, Cohen T, Shibuya M, Neufeld G. Vascular endothelial growth factor receptor-1 and neuropilin-2 form complexes. *J Biol Chem* 2001;276:18688–18694.
130. Tordjman R, Ortega N, Coulombel L, Plouet J, Romeo PH, Lemarchandel V. Neuropilin-1 is expressed on bone marrow stromal cells: a novel interaction with hematopoietic cells? *Blood* 1999;94:2301–2309.
131. Migdal M, Huppertz B, Tessler S, Comforti A, Shibuya M, Reich R, Baumann H, Neufeld G. Neuropilin-1 is a placenta growth factor-2 receptor. *J Biol Chem* 1998;273:22272–22278.
132. Makinen T, Olofsson B, Karpanen T, Hellman U, Soker S, Klagsbrun M, Eriksson U, Alitalo K. Differential binding of vascular endothelial growth factor B splice and proteolytic isoforms to neuropilin-1. *J Biol Chem* 1999;274:21217–21222.
133. Whitaker GB, Limberg BJ, Rosenbaum JS. Vascular endothelial growth factor receptor-2 and neuropilin-1 form a receptor complex that is responsible for the differential signaling potency of VEGF(165) and VEGF(121). *J Biol Chem* 2001;276:25520–25531.
134. Pepper MS, Montesano R, Orci L, Vassalli JD. Plasminogen activator inhibitor-1 is induced in microvascular endothelial cells by a chondrocyte-derived transforming growth factor-beta. *Biochem Biophys Res Commun* 1991;176:633–638.

135. Unemori EN, Ferrara N, Bauer EA, Amento EP. Vascular endothelial growth factor induces interstitial collagenase expression in human endothelial cells. *J Cell Physiol* 1992;153:557–562.
136. Nelson AR, Fingleton B, Rothenberg ML, Matrisian LM. Matrix metalloproteinases: biologic activity and clinical implications. *J Clin Oncol* 2000;18:1135–1149.
137. Nangia-Makker P, Honjo Y, Sarvis R, Akahani S, Hogan V, Pienta KJ, Raz A. Galectin-3 induces endothelial cell morphogenesis and angiogenesis. *Am J Pathol* 2000;156:899–909.
138. Gamble J, Meyer G, Noack L, Furze J, Matthias L, Kovach N, Harlant J, Vadas M. B1 integrin activation inhibits in vitro tube formation: effects on cell migration, vacuole coalescence and lumen formation. *Endothelium* 1999;7:23–34.
139. Yang S, Graham J, Kahn JW, Schwartz EA, Gerritsen ME. Functional roles for PECAM-1 (CD31) and VE-cadherin (CD144) in tube assembly and lumen formation in three-dimensional collagen gels. *Am J Pathol* 1999;155:887–895.
140. Holash J, Maisonpierre PC, Compton D, Boland P, Alexander CR, Zagzag D, Yancopoulos GD, Wiegand SJ. Vessel cooption, regression, and growth in tumors mediated by angiopoietins and VEGF. *Science* 1999;284:1994–1998.
141. Rafii S, Lyden D, Benezra R, Hattori K, Heissig B. Vascular and haematopoietic stem cells: novel targets for anti-angiogenesis therapy? *Nat Rev Cancer* 2002;2:826–835.
142. Folberg R, Hendrix MJ, Maniotis AJ. Vasculogenic mimicry and tumor angiogenesis. *Am J Pathol* 2000;156:361–381.
143. Chang YS, di Tomaso E, McDonald DM, Jones R, Jain RK, Munn LL. Mosaic blood vessels in tumors: frequency of cancer cells in contact with flowing blood. *Proc Natl Acad Sci USA* 2000;97:14608–14613.
144. Yancopoulos GD, Davis S, Gale NW, Rudge JS, Wiegand SJ, Holash J. Vascular-specific growth factors and blood vessel formation. *Nature* 2000;407:242–248.
145. Hendrix MJ, Seftor EA, Meltzer PS, Gardner LM, Hess AR, Kirschmann DA, Schatteman GC, Seftor RE. Expression and functional significance of VE-cadherin in aggressive human melanoma cells: role in vasculogenic mimicry. *Proc Natl Acad Sci USA* 2001;98:8018–8023.
146. Folkman J. Can mosaic tumor vessels facilitate molecular diagnosis of cancer? *Proc Natl Acad Sci USA* 2001;98:398–400.
147. Rafii S. Circulating endothelial precursors: mystery, reality, and promise. *J Clin Invest* 2000;105:17–19.
148. Lyden D, Hattori K, Dias S, Costa C, Blaikie P, Butros L, Chadburn A, Heissig B, Marks W, Witte L, Wu Y, Hicklin D, Zhu Z, Hackett NR, Crystal RG, Moore MA, Hajjar KA, Manova K, Benezra R, Rafii S. Impaired recruitment of bone-marrow-derived endothelial and hematopoietic precursor cells blocks tumor angiogenesis and growth. *Nat Med* 2001;7:1194–1201.
149. Luttun A, Tjwa M, Moons L, Wu Y, Angelillo-Scherrer A, Liao F, Nagy JA, Hooper A, Priller J, De Klerck B, Compernolle V, Daci E, Bohlen P, Dewerchin M, Herbert JM, Fava R, Matthys P, Carmeliet G, Collen D, Dvorak HF, Hicklin DJ, Carmeliet P. Revascularization of ischemic tissues by PlGF treatment, and inhibition of tumor angiogenesis, arthritis and atherosclerosis by anti-Flt1. *Nat Med* 2002;8:831–840.
150. Carmeliet P, Jain RK. Angiogenesis in cancer and other diseases. *Nature* 2000;407:249–257.
151. Hanahan D, Weinberg RA. The hallmarks of cancer. *Cell* 2000;100:57–70.
152. Bergers G, Javaherian K, Lo KM, Folkman J, Hanahan D. Effects of angiogenesis inhibitors on multistage carcinogenesis in mice. *Science* 1999;284:808–812.
153. Folkman J, Watson K, Ingber D, Hanahan D. Induction of angiogenesis during the transition from hyperplasia to neoplasia. *Nature* 1989;339:58–61.
154. Fukumura D, Xavier R, Sugiura T, Chen Y, Park EC, Lu N, Selig M, Nielsen G, Taksir T, Jain RK, Seed B. Tumor induction of VEGF promoter activity in stromal cells. *Cell* 1998;94:715–725.
155. Bergers G, Benjamin LE. Tumorigenesis and the angiogenic switch. *Nat Rev Cancer* 2003;3:401–410.
156. Folkman J. Angiogenesis and angiogenesis inhibition: an overview. *Experienta* 1997;79:1–8.
157. Carmeliet P. Developmental biology. Controlling the cellular brakes. *Nature* 1999;401:657–658.
158. Kerbel RS. Tumor angiogenesis: past, present and the near future. *Carcinogenesis* 2000;21:505–515.
159. Dameron KM, Volpert OV, Tainsky MA, Bouck N. Control of angiogenesis in fibroblasts by p53 regulation of thrombospondin-1. *Science* 1994;265:1582–1584.
160. Bouck N, Stellmach V, Hsu SC. How tumors become angiogenic. *Adv Cancer Res* 1996;69:135–174.

161. Holmgren L. Antiangiogenis restricted tumor dormancy. *Cancer Metastasis Rev* 1996;15:241–245.
162. O'Reilly MS, Holmgren L, Chen C, Folkman J. Angiostatin induces and sustains dormancy of human primary tumors in mice. *Nat Med* 1996;2:689–692.
163. Pike SE, Yao L, Jones KD, Cherney B, Appella E, Sakaguchi K, Nakhasi H, Teruya-Feldstein J, Wirth P, Gupta G, Tosato G. Vasostatin, a calreticulin fragment, inhibits angiogenesis and suppresses tumor growth. *J Exp Med* 1998;188:2349–2356.
164. Kamphaus GD, Colorado PC, Panka DJ, Hopfer H, Ramchandran R, Torre A, Maeshima Y, Mier JW, Sukhatme VP, Kalluri R. Canstatin, a novel matrix-derived inhibitor of angiogenesis and tumor growth. *J Biol Chem* 2000;275:1209–1215.
165. Dvorak HF, Nagy JA, Dvorak JT, Dvorak AM. Identification and characterization of the blood vessels of solid tumors that are leaky to circulating macromolecules. *Am J Pathol* 1988;133:95–109.
166. Morikawa S, Baluk P, Kaidoh T, Haskell A, Jain RK, McDonald DM. Abnormalities in pericytes on blood vessels and endothelial sprouts in tumors. *Am J Pathol* 2002;160:985–1000.
167. Baish JW, Jain RK. Fractals and cancer. *Cancer Res* 2000;60:3683–3688.
168. Helmlinger G, Yuan F, Dellian M, Jain RK. Interstitial pH and pO_2 gradients in solid tumors in vivo: high-resolution measurements reveal a lack of correlation. *Nat Med* 1997;3:177–182.
169. Brien SE, Zagzag D, Brem S. Rapid in situ cellular kinetics of intracerebral tumor angiogenesis using a monoclonal antibody to bromodeoxyuridine. *Neurosurgery* 1989;25:715–719.
170. Ruoslahti E. Specialization of tumour vasculature. *Nat Rev Cancer* 2002;2:83–90.
171. St Croix B, Rago C, Velculescu V, Traverso G, Romans KE, Montgomery E, Lal A, Riggins GJ, Lengauer C, Vogelstein B, Kinzler KW. Genes expressed in human tumor endothelium. *Science* 2000;289:1197–1202.
172. Breier G, Albrecht U, Sterrer S, Risau W. Expression of vascular endothelial growth factor during embryonic angiogenesis and endothelial cell differentiation. *Development* 1992;114:521–532.
173. Jakeman LB, Winer J, Bennett GL, Altar CA, Ferrara N. Binding sites for vascular endothelial growth factor are localized on endothelial cells in adult rat tissues. *J Clin Invest* 1992;89:244–253.
174. Choi K, Kennedy M, Kazarov A, Papadimitriou JC, Keller G. A common precursor for hematopoietic and endothelial cells. *Development* 1998;125:725–732.
175. Cerdan C, Rouleau A, Bhatia M. VEGF-A165 augments erythropoietic development from human embryonic stem cells. *Blood* 2004;103:2504–2512.
176. Bruckner K, Kockel L, Duchek P, Luque CM, Rorth P, Perrimon N. The PDGF/VEGF receptor controls blood cell survival in *Drosophila*. *Dev Cell* 2004;7:73–84.
177. Gabrilovich DI, Chen HL, Girgis KR, Cunningham HT, Meny GM, Nadaf S, Kavanaugh D, Carbone DP. Production of vascular endothelial growth factor by human tumors inhibits the functional maturation of dendritic cells. *Nat Med* 1996;2:1096–1103.
178. Nakagawa M, Kaneda T, Arakawa T, Morita S, Sato T, Yomada T, Hanada K, Kumegawa M, Hakeda Y. Vascular endothelial growth factor (VEGF) directly enhances osteoclastic bone resorption and survival of mature osteoclasts. *FEBS Lett* 2000;473:161–164.
179. Henriksen K, Karsdal M, Delaisse JM, Engsig MT. RANKL and vascular endothelial growth factor (VEGF) induce osteoclast chemotaxis through an ERK1/2-dependent mechanism. *J Biol Chem* 2003;278:48745–48753.
180. Midy V, Plouet J. Vasculotropin/vascular endothelial growth factor induces differentiation in cultured osteoblasts. *Biochem Biophys Res Commun* 1994;199:380–386.
181. Mayer H, Bertram H, Lindenmaier W, Korff T, Weber H, Weich H. Vascular endothelial growth factor (VEGF-A) expression in human mesenchymal stem cells: autocrine and paracrine role on osteoblastic and endothelial differentiation. *J Cell Biochem* 2005;95(4):827–839.
182. Melder RJ, Koenig GC, Witwer BP, Safabakhsh N, Munn LL, Jain RK. During angiogenesis, vascular endothelial growth factor and basic fibroblast growth factor regulate natural killer cell adhesion to tumor endothelium. *Nat Med* 1996;2:992–997.
183. Compernolle V, Brusselmans K, Acker T, Hoet P, Tjwa M, Beck H, Plaisance S, Dor Y, Keshet E, Lupu F, Nemery B, Dewerchin M, Van Veldhoven P, Plate K, Moons L, Collen D, Carmeliet P. Loss of HIF-2alpha and inhibition of VEGF impair fetal lung maturation, whereas treatment with VEGF prevents fatal respiratory distress in premature mice. *Nat Med* 2002;8:702–710.
184. Oosthuyse B, Moons L, Storkebaum E, Beck H, Nuyens D, Brusselmans K, Van Dorpe J, Hellings P, Gorselink M, Heymans S, Theilmeier G, Dewerchin M, Laudenbach V, Vermylen P, Raat H,

Acker T, Vleminckx V, Van Den Bosch L, Cashman N, Fujisawa H, Drost MR, Sciot R, Bruyninckx F, Hicklin DJ, Ince C, Gressens P, Lupu F, Plate KH, Robberecht W, Herbert JM, Collen D, Carmeliet P. Deletion of the hypoxia-response element in the vascular endothelial growth factor promoter causes motor neuron degeneration. *Nat Genet* 2001;28:131–138.

185. Arora N, Masood R, Zheng T, Cai J, Smith DL, Gill PS. Vascular endothelial growth factor chimeric toxin is highly active against endothelial cells. *Cancer Res* 1999;59:183–188.
186. Cornali E, Zietz C, Benelli R, Weninger W, Masiello L, Breier G, Tschachler E, Albini A, Sturzl M. Vascular endothelial growth factor regulates angiogenesis and vascular permeability in Kaposi's sarcoma. *Am J Pathol* 1996;149:1851–1869.
187. Nakamura S, Murakami-Mori K, Rao N, Weich HA, Rajeev B. Vascular endothelial growth factor is a potent angiogenic factor in AIDS-associated Kaposi's sarcoma-derived spindle cells. *J Immunol* 1997;158:4992–5001.
188. Sakurada S, Kato T, Mashiba K, Mori S, Okamoto T. Involvement of vascular endothelial growth factor in Kaposi's sarcoma associated with acquired immunodeficiency syndrome. *Jpn J Cancer Res* 1996;87:1143–1152.
189. Samaniego F, Markham PD, Gendelman R, Watanabe Y, Kao V, Kowalski K, Sonnabend JA, Pintus A, Gallo RC, Ensoli B. Vascular endothelial growth factor and basic fibroblast growth factor present in Kaposi's sarcoma (KS) are induced by inflammatory cytokines and synergize to promote vascular permeability and KS lesion development. *Am J Pathol* 1998;152:1433–1443.
190. Weindel K, Marme D, Weich HA. AIDS-associated Kaposi's sarcoma cells in culture express vascular endothelial growth factor. *Biochem Biophys Res Commun* 1992;183:1167–1174.
191. List AF. Vascular endothelial growth factor signaling pathway as an emerging target in hematologic malignancies. *Oncologist* 2001;6(Suppl 5):24–31.
192. Li VW, Folkerth RD, Watanabe H, Yu C, Rupnick M, Barnes P, Scott RM, Black PM, Sallan SE, Folkman J. Microvessel count and cerebrospinal fluid basic fibroblast growth factor in children with brain tumours. *Lancet* 1994;344:82–86.
193. Maeda K, Chung YS, Ogawa Y, Takatsuka S, Kang SM, Ogawa M, Sawada T, Sowa M. Prognostic value of vascular endothelial growth factor expression in gastric carcinoma. *Cancer* 1996;77:858–863.
194. Vermeulen PB, Verhoeven D, Fierens H, Hubens G, Goovaerts G, Van Marck E, De Bruijn EA, Van Oosterom AT, Dirix LY. Microvessel quantification in primary colorectal carcinoma: an immunohistochemical study. *Br J Cancer* 1995;71:340–343.
195. Weidner N, Folkman J, Pozza F, Bevilacqua P, Allred EN, Moore DH, Meli S, Gasparini G. Tumor angiogenesis: a new significant and independent prognostic indicator in early-stage breast carcinoma. *J Natl Cancer Inst* 1992;84:1875–1887.
196. Dickinson AJ, Fox SB, Persad RA, Hollyer J, Sibley GN, Harris AL. Quantification of angiogenesis as an independent predictor of prognosis in invasive bladder carcinomas. *Br J Urol* 1994;74:762–766.
197. Gasparini G, Barbareschi M, Boracchi P, Verderio P, Caffo O, Meli S, Palma PD, Marubini E, Bevilacqua P. Tumor angiogenesis predicts clinical outcome of node-positive breast cancer patients treated with adjuvant hormone therapy or chemotherapy. *Cancer J Sci Am* 1995;1:131.
198. Gasparini G, Bonoldi E, Viale G, Verderio P, Boracchi P, Panizzoni GA, Radaelli U, Di Bacco A, Guglielmi RB, Bevilacqua P. Prognostic and predictive value of tumour angiogenesis in ovarian carcinomas. *Int J Cancer* 1996;69:205–211.
199. Vacca A, Ribatti D, Roncali L, Ranieri G, Serio G, Silvestris F, Dammacco F. Bone marrow angiogenesis and progression in multiple myeloma. *Br J Haematol* 1994;87:503–508.
200. Vacca A, Ribatti D, Roncali L, Dammacco F. Angiogenesis in B cell lymphoproliferative diseases. Biological and clinical studies. *Leuk Lymphoma* 1995;20:27–38.
201. Aguayo A, Kantarjian H, Manshouri T, Gidel C, Estey E, Thomas D, Koller C, Estrov Z, O'Brien S, Keating M, Freireich E, Albitar M. Angiogenesis in acute and chronic leukemias and myelodysplastic syndromes. *Blood* 2000;96:2240–2245.
202. Ridell B, Norrby K. Intratumoral microvascular density in malignant lymphomas of B-cell origin. *Apmis* 2001;109:66–72.
203. Perez-Atayde AR, Sallan SE, Tedrow U, Connors S, Allred E, Folkman J. Spectrum of tumor angiogenesis in the bone marrow of children with acute lymphoblastic leukemia. *Am J Pathol* 1997;150:815–821.

204. Hussong JW, Rodgers GM, Shami PJ. Evidence of increased angiogenesis in patients with acute myeloid leukemia. *Blood* 2000;95:309–313.
205. Lundberg LG, Lerner R, Sundelin P, Rogers R, Folkman J, Palmblad J. Bone marrow in polycythemia vera, chronic myelocytic leukemia, and myelofibrosis has an increased vascularity. *Am J Pathol* 2000;157:15–19.
206. Pruneri G, Bertolini F, Soligo D, Carboni N, Cortelezzi A, Ferrucci PF, Buffa R, Lambertenghi-Deliliers G, Pezzella F. Angiogenesis in myelodysplastic syndromes. *Br J Cancer* 1999;81:1398–1401.
207. Mesa RA, Hanson CA, Rajkumar SV, Schroeder G, Tefferi A. Evaluation and clinical correlations of bone marrow angiogenesis in myelofibrosis with myeloid metaplasia. *Blood* 2000;96:3374–3380.
208. Ribatti D, Vacca A, Nico B, Fanelli M, Roncali L, Dammacco F. Angiogenesis spectrum in the stroma of B-cell non-Hodgkin's lymphomas. An immunohistochemical and ultrastructural study. *Eur J Haematol* 1996;56:45–53.
209. Molica S, Vacca A, Ribatti D, Cuneo A, Cavazzini F, Levato D, Vitelli G, Tucci L, Roccaro AM, Dammacco F. Prognostic value of enhanced bone marrow angiogenesis in early B-cell chronic lymphocytic leukemia. *Blood* 2002;100:3344–3351.
210. Rajkumar SV, Leong T, Roche PC, Fonseca R, Dispenzieri A, Lacy MQ, Lust JA, Witzig TE, Kyle RA, Gertz MA, Greipp PR. Prognostic value of bone marrow angiogenesis in multiple myeloma. *Clin Cancer Res* 2000;6:3111–3116.
211. Sezer O, Niemoller K, Eucker J, Jakob C, Kaufmann O, Zavrski I, Dietel M, Possinger K. Bone marrow microvessel density is a prognostic factor for survival in patients with multiple myeloma. *Ann Hematol* 2000;79:574–577.
212. Ahn MJ, Park CK, Choi JH, Lee WM, Lee YY, Choi IY, Kim IS, Lee WS, Ki M. Clinical significance of microvessel density in multiple myeloma patients. *J Korean Med Sci* 2001;16:45–50.
213. Munshi NC, Wilson C. Increased bone marrow microvessel density in newly diagnosed multiple myeloma carries a poor prognosis. *Semin Oncol* 2001;28:565–569.
214. Rajkumar SV, Mesa RA, Fonseca R, Schroeder G, Plevak MF, Dispenzieri A, Lacy MQ, Lust JA, Witzig TE, Gertz MA, Kyle RA, Russell SJ, Greipp PR. Bone marrow angiogenesis in 400 patients with monoclonal gammopathy of undetermined significance, multiple myeloma, and primary amyloidosis. *Clin Cancer Res* 2002;8:2210–2216.
215. Ribatti D, Vacca A, Nico B, Quondamatteo F, Ria R, Minischetti M, Marzullo A, Herken R, Roncali L, Dammacco F. Bone marrow angiogenesis and mast cell density increase simultaneously with progression of human multiple myeloma. *Br J Cancer* 1999;79:451–455.
216. Kumar S, Gertz MA, Dispenzieri A, Lacy MQ, Wellik LA, Fonseca R, Lust JA, Witzig TE, Kyle RA, Greipp PR, Rajkumar SV. Prognostic value of bone marrow angiogenesis in patients with multiple myeloma undergoing high-dose therapy. *Bone Marrow Transplant* 2004;34:235–239.
217. Kumar S, Fonseca R, Dispenzieri A, Lacy MQ, Lust JA, Witzig TE, Gertz MA, Kyle RA, Greipp PR, Rajkumar SV. Bone marrow angiogenesis in multiple myeloma: effect of therapy. *Br J Haematol* 2002;119:665–671.
218. Rajkumar SV, Fonseca R, Witzig TE, Gertz MA, Greipp PR. Bone marrow angiogenesis in patients achieving complete response after stem cell transplantation for multiple myeloma. *Leukemia* 1999;13:469–472.
219. Greipp PR, Lust JA, O'Fallon WM, Katzmann JA, Witzig TE, Kyle RA. Plasma cell labeling index and beta 2-microglobulin predict survival independent of thymidine kinase and C-reactive protein in multiple myeloma. *Blood* 1993;81:3382–3387.
220. Kumar S, Witzig TE, Greipp PR, Rajkumar SV. Bone marrow angiogenesis and circulating plasma cells in multiple myeloma. *Br J Haematol* 2003;122:272–274.
221. Schreiber S, Ackermann J, Obermair A, Kaufmann H, Urbauer E, Aletaha K, Gisslinger H, Chott A, Huber H, Drach J. Multiple myeloma with deletion of chromosome 13q is characterized by increased bone marrow neovascularization. *Br J Haematol* 2000;110:605–609.
222. Kumar S, Fonseca R, Dispenzieri A, Lacy MQ, Lust JA, Wellik L, Witzig TE, Gertz MA, Kyle RA, Greipp PR, Rajkumar SV. Prognostic value of angiogenesis in solitary bone plasmacytoma. *Blood* 2003;101:1715–1717.
223. Singhal S, Mehta J, Desikan R, Ayers D, Roberson P, Eddlemon P, Munshi N, Anaissie E, Wilson C, Dhodapkar M, Zeddis J, Barlogie B. Antitumor activity of thalidomide in refractory multiple myeloma. *N Engl J Med* 1999;341:1565–1571.

224. Hlatky L, Hahnfeldt P, Folkman J. Clinical application of antiangiogenic therapy: microvessel density, what it does and doesn't tell us. *J Natl Cancer Inst* 2002;94:883–893.
225. Shaked Y, Bertolini F, Man S, Rogers MS, Cervi D, Foutz T, Rawn K, Voskas D, Dumont DJ, Ben-David Y, Lawler J, Henkin J, Huber J, Hicklin DJ, D'Amato RJ, Kerbel RS. Genetic heterogeneity of the vasculogenic phenotype parallels angiogenesis; implications for cellular surrogate marker analysis of antiangiogenesis. *Cancer Cell* 2005;7:101–111.
226. Zhang H, Vakil V, Braunstein M, Smith EL, Maroney J, Chen L, Dai K, Berenson JR, Hussain MM, Klueppelberg U, Norin AJ, Akman HO, Ozcelik T, Batuman OA. Circulating endothelial progenitor cells in multiple myeloma: implications and significance. *Blood* 2005;105:3286–3294.
227. Pellegrino A, Ria R, Pietro GD, Cirulli T, Surico G, Pennisi A, Morabito F, Ribatti D, Vacca A. Bone marrow endothelial cells in multiple myeloma secrete CXC-chemokines that mediate interactions with plasma cells. *Br J Haematol* 2005;129:248–256.
228. Vacca A, Ria R, Semeraro F, Merchionne F, Coluccia M, Boccarelli A, Scavelli C, Nico B, Gernone A, Battelli F, Tabilio A, Guidolin D, Petrucci MT, Ribatti D, Dammacco F. Endothelial cells in the bone marrow of patients with multiple myeloma. *Blood* 2003;102:3340–3348.
229. Streubel B, Chott A, Huber D, Exner M, Jager U, Wagner O, Schwarzinger I. Lymphoma-specific genetic aberrations in microvascular endothelial cells in B-cell lymphomas. *N Engl J Med* 2004;351:250–259.
230. Brunner G, Nguyen H, Gabrilove J, Rifkin DB, Wilson EL. Basic fibroblast growth factor expression in human bone marrow and peripheral blood cells. *Blood* 1993;81:631–638.
231. Fiedler W, Graeven U, Ergun S, Verago S, Kilic N, Stockschlader M, Hossfeld DK. Vascular endothelial growth factor, a possible paracrine growth factor in human acute myeloid leukemia. *Blood* 1997;89:1870–1875.
232. Chen H, Treweeke AT, West DC, Till KJ, Cawley JC, Zuzel M, Toh CH. In vitro and in vivo production of vascular endothelial growth factor by chronic lymphocytic leukemia cells. *Blood* 2000;96:3181–3187.
233. Dias S, Hattori K, Zhu Z, Heissig B, Choy M, Lane W, Wu Y, Chadburn A, Hyjek E, Gill M, Hicklin DJ, Witte L, Moore MA, Rafii S. Autocrine stimulation of VEGFR-2 activates human leukemic cell growth and migration. *J Clin Invest* 2000;106:511–521.
234. Krauth MT, Simonitsch I, Aichberger KJ, Mayerhofer M, Sperr WR, Sillaber C, Schneeweiss B, Mann G, Gadner H, Valent P. Immunohistochemical detection of VEGF in the bone marrow of patients with chronic myeloid leukemia and correlation with the phase of disease. *Am J Clin Pathol* 2004;121:473–481.
235. Ghannadan M, Wimazal F, Simonitsch I, Sperr WR, Mayerhofer M, Sillaber C, Hauswirth AW, Gadner H, Chott A, Horny HP, Lechner K, Valent P. Immunohistochemical detection of VEGF in the bone marrow of patients with acute myeloid leukemia. Correlation between VEGF expression and the FAB category. *Am J Clin Pathol* 2003;119:663–671.
236. Bellamy WT. Expression of vascular endothelial growth factor and its receptors in multiple myeloma and other hematopoietic malignancies. *Semin Oncol* 2001;28:551–559.
237. Vincent L, Jin DK, Karajannis MA, Shido K, Hooper AT, Rashbaum WK, Pytowski B, Wu Y, Hicklin DJ, Zhu Z, Bohlen P, Niesvizky R, Rafii S. Fetal stromal-dependent paracrine and intracrine vascular endothelial growth factor-a/vascular endothelial growth factor receptor-1 signaling promotes proliferation and motility of human primary myeloma cells. *Cancer Res* 2005;65:3185–3192.
238. Podar K, Tai YT, Lin BK, Narsimhan RP, Sattler M, Kijima T, Salgia R, Gupta D, Chauhan D, Anderson KC. Vascular endothelial growth factor-induced migration of multiple myeloma cells is associated with beta 1 integrin- and phosphatidylinositol 3-kinase-dependent PKC alpha activation. *J Biol Chem* 2002;277:7875–7881.
239. Podar K, Shringarpure R, Tai YT, Simoncini M, Sattler M, Ishitsuka K, Richardson PG, Hideshima T, Chauhan D, Anderson KC. Caveolin-1 is required for vascular endothelial growth factor-triggered multiple myeloma cell migration and is targeted by bortezomib. *Cancer Res* 2004;64:7500–7506.
240. Dias S, Hattori K, Heissig B, Zhu Z, Wu Y, Witte L, Hicklin DJ, Tateno M, Bohlen P, Moore MA, Rafii S. Inhibition of both paracrine and autocrine VEGF/VEGFR-2 signaling pathways is essential to induce long-term remission of xenotransplanted human leukemias. *Proc Natl Acad Sci USA* 2001;98:10857–10862.

241. Dias S, Shmelkov SV, Lam G, Rafii S. VEGF(165) promotes survival of leukemic cells by Hsp90-mediated induction of Bcl-2 expression and apoptosis inhibition. *Blood* 2002;99:2532–2540.
242. Liotta LA, Kohn EC. The microenvironment of the tumour–host interface. *Nature* 2001;411:375–379.
243. Shaked Y, Cervi D, Neuman M, Chen L, Klement G, Michaud CR, Haeri M, Pak BJ, Kerbel RS, Ben-David Y. The splenic microenvironment is a source of pro-angiogenesis/inflammatory mediators accelerating the expansion of murine erythroleukemic cells. *Blood* 2005;105:4500–4507.
244. Goldman JM, Nolasco I. The spleen in myeloproliferative disorders. *Clin Haematol* 1983;12:505–516.
245. Golomb HM. Hairy cell leukemia: an unusual lymphoproliferative disease: a study of 24 patients. *Cancer* 1978;42:946–956.
246. Coad JE, Matutes E, Catovsky D. Splenectomy in lymphoproliferative disorders: a report on 70 cases and review of the literature. *Leuk Lymphoma* 1993;10:245–264.
247. Yaccoby S, Barlogie B, Epstein J. Primary myeloma cells growing in SCID-hu mice: a model for studying the biology and treatment of myeloma and its manifestations. *Blood* 1998;92:2908–2913.
248. Browder TM, Dunbar CE, Nienhuis AW. Private and public autocrine loops in neoplastic cells. *Cancer Cells* 1989;1:9–17.
249. Huang SS, Huang JS. Rapid turnover of the platelet-derived growth factor receptor in sis-transformed cells and reversal by suramin. Implications for the mechanism of autocrine transformation. *J Biol Chem* 1988;263:12608–12618.
250. Lang RA, Metcalf D, Gough NM, Dunn AR, Gonda TJ. Expression of a hemopoietic growth factor cDNA in a factor-dependent cell line results in autonomous growth and tumorigenicity. *Cell* 1985;43:531–542.
251. Kitani A, Hara M, Hirose T, Harigai M, Suzuki K, Kawakami M, Kawaguchi Y, Hidaka T, Kawagoe M, Nakamura H. Autostimulatory effects of IL-6 on excessive B cell differentiation in patients with systemic lupus erythematosus: analysis of IL-6 production and IL-6R expression. *Clin Exp Immunol* 1992;88:75–83.
252. Lu C, Kerbel RS. Interleukin-6 undergoes transition from paracrine growth inhibitor to autocrine stimulator during human melanoma progression. *J Cell Biol* 1993;120:1281–1288.
253. Rogers SY, Bradbury D, Kozlowski R, Russell NH. Evidence for internal autocrine regulation of growth in acute myeloblastic leukemia cells. *Exp Hematol* 1994;22:593–598.
254. Santos SC, Dias S. Internal and external autocrine VEGF/KDR loops regulate survival of subsets of acute leukemia through distinct signaling pathways. *Blood* 2004;103:3883–3889.
255. Till KJ, Spiller DG, Harris RJ, Chen H, Zuzel M, Cawley JC. CLL, but not normal, B cells are dependent on autocrine VEGF and {alpha}4{beta}1 integrin for chemokine-induced motility on and through endothelium. *Blood* 2005;105:4813–4819.
256. Dankbar B, Padro T, Leo R, Feldmann B, Kropff M, Mesters RM, Serve H, Berdel WE, Kienast J. Vascular endothelial growth factor and interleukin-6 in paracrine tumor–stromal cell interactions in multiple myeloma. *Blood* 2000;95:2630–2636.
257. Gupta D, Treon SP, Shima Y, Hideshima T, Podar K, Tai YT, Lin B, Lentzsch S, Davies FE, Chauhan D, Schlossman RL, Richardson P, Ralph P, Wu L, Payvandi F, Muller G, Stirling DI, Anderson KC. Adherence of multiple myeloma cells to bone marrow stromal cells upregulates vascular endothelial growth factor secretion: therapeutic applications. *Leukemia* 2001;15:1950–1961.
258. Mitsiades CS, Mitsiades NS, McMullan CJ, Poulaki V, Shringarpure R, Akiyama M, Hideshima T, Chauhan D, Joseph M, Libermann TA, Garcia-Echeverria C, Pearson MA, Hofmann F, Anderson KC, Kung AL. Inhibition of the insulin-like growth factor receptor-1 tyrosine kinase activity as a therapeutic strategy for multiple myeloma, other hematologic malignancies, and solid tumors. *Cancer Cell* 2004;5:221–230.
259. Menu E, Kooijman R, Van Valckenborgh E, Asosingh K, Bakkus M, Van Camp B, Vanderkerken K. Specific roles for the PI3K and the MEK–ERK pathway in IGF-1-stimulated chemotaxis, VEGF secretion and proliferation of multiple myeloma cells: study in the 5T33MM model. *Br J Cancer* 2004;90:1076–1083.
260. Fukuda R, Hirota K, Fan F, Jung YD, Ellis LM, Semenza GL. Insulin-like growth factor 1 induces hypoxia-inducible factor 1-mediated vascular endothelial growth factor expression, which is dependent on MAP kinase and phosphatidylinositol 3-kinase signaling in colon cancer cells. *J Biol Chem* 2002;277:38205–38211.

261. Hurt EM, Wiestner A, Rosenwald A, Shaffer AL, Campo E, Grogan T, Bergsagel PL, Kuehl WM, Staudt LM. Overexpression of c-maf is a frequent oncogenic event in multiple myeloma that promotes proliferation and pathological interactions with bone marrow stroma. *Cancer Cell* 2004;5:191–199.
262. Tai YT, Podar K, Gupta D, Lin B, Young G, Akiyama M, Anderson KC. CD40 activation induces p53-dependent vascular endothelial growth factor secretion in human multiple myeloma cells. *Blood* 2002;99:1419–1427.
263. Hattori K, Dias S, Heissig B, Hackett NR, Lyden D, Tateno M, Hicklin DJ, Zhu Z, Witte L, Crystal RG, Moore MA, Rafii S. Vascular endothelial growth factor and angiopoietin-1 stimulate postnatal hematopoiesis by recruitment of vasculogenic and hematopoietic stem cells. *J Exp Med* 2001;193:1005–1014.
264. Hayashi T, Hideshima T, Akiyama M, Raje N, Richardson P, Chauhan D, Anderson KC. Ex vivo induction of multiple myeloma-specific cytotoxic T lymphocytes. *Blood* 2003;102:1435–1442.
265. Le Gouill S, Podar K, Amiot M, Hideshima T, Chauhan D, Itshitsuka K, Kumar S, Raje N, Richardson PG, Harousseau JL, Anderson KC. VEGF induces MCL-1 upregulation and protects multiple myeloma cells against apoptosis. *Blood* 2004;104:2886–2892.
266. Kumar S, Witzig TE, Timm M, Haug J, Wellik L, Fonseca R, Greipp PR, Rajkumar SV. Expression of VEGF and its receptors by myeloma cells. *Leukemia* 2003;17:2025–2031.
267. Smart EJ, Graf GA, McNiven MA, Sessa WC, Engelman JA, Scherer PE, Okamoto T, Lisanti MP. Caveolins, liquid-ordered domains, and signal transduction. *Mol Cell Biol* 1999;19:7289–7304.
268. Simons K, Toomre D. Lipid rafts and signal transduction. *Nat Rev Mol Cell Biol* 2000;1:31–39.
269. Podar K, Tai YT, Cole CE, Hideshima T, Sattler M, Hamblin A, Mitsiades N, Schlossman RL, Davies FE, Morgan GJ, Munshi NC, Chauhan D, Anderson KC. Essential role of caveolae in interleukin-6- and insulin-like growth factor I-triggered Akt-1-mediated survival of multiple myeloma cells. *J Biol Chem* 2003;278:5794–5801.
270. Attal M, Harousseau JL, Stoppa AM, Sotto JJ, Fuzibet JG, Rossi JF, Casassus P, Maisonneuve H, Facon T, Ifrah N, Payen C, Bataille R. A prospective, randomized trial of autologous bone marrow transplantation and chemotherapy in multiple myeloma. Intergroupe Francais du Myelome. *N Engl J Med* 1996;335:91–97.
271. Kumar S, Witzig TE, Dispenzieri A, Lacy MQ, Wellik LE, Fonseca R, Lust JA, Gertz MA, Kyle RA, Greipp PR, Rajkumar SV. Effect of thalidomide therapy on bone marrow angiogenesis in multiple myeloma. *Leukemia* 2004;18:624–627.
272. Oh HS, Choi JH, Park CK, Jung CW, Lee SI, Park Q, Suh C, Kim SB, Chi HS, Lee JH, Cho EK, Bang SM, Ahn MJ. Comparison of microvessel density before and after peripheral blood stem cell transplantation in multiple myeloma patients and its clinical implications: multicenter trial. *Int J Hematol* 2002;76:465–470.
273. Kumar S, Litzow MR, Rajkumar SV. Effect of allogeneic stem cell transplantation on bone marrow angiogenesis in chronic myelogenous leukemia. *Bone Marrow Transplant* 2003;32:1065–1069.
274. Kim KJ, Li B, Winer J, Armanini M, Gillett N, Phillips HS, Ferrara N. Inhibition of vascular endothelial growth factor-induced angiogenesis suppresses tumour growth in vivo. *Nature* 1993;362:841–844.
275. Inoue M, Hager JH, Ferrara N, Gerber HP, Hanahan D. VEGF-A has a critical, nonredundant role in angiogenic switching and pancreatic beta cell carcinogenesis. *Cancer Cell* 2002;1:193–202.
276. Ferrara N, Hillan KJ, Gerber HP, Novotny W. Discovery and development of bevacizumab, an anti-VEGF antibody for treating cancer. *Nat Rev Drug Discov* 2004;3:391–400.
277. Hurwitz H, Fehrenbacher L, Novotny W, Cartwright T, Hainsworth J, Heim W, Berlin J, Baron A, Griffing S, Holmgren E, Ferrara N, Fyfe G, Rogers B, Ross R, Kabbinavar F. Bevacizumab plus irinotecan, fluorouracil, and leucovorin for metastatic colorectal cancer. *N Engl J Med* 2004;350:2335–2342.
278. Yang JC, Haworth L, Sherry RM, Hwu P, Schwartzentruber DJ, Topalian SL, Steinberg SM, Chen HX, Rosenberg SA. A randomized trial of bevacizumab, an anti-vascular endothelial growth factor antibody, for metastatic renal cancer. *N Engl J Med* 2003;349:427–434.
279. Yang JC, Sherry RM, Steinberg SM, Topalian SL, Schwartzentruber DJ, Hwu P, Seipp CA, Rogers-Freezer L, Morton KE, White DE, Liewehr DJ, Merino MJ, Rosenberg SA. Randomized study of high-dose and low-dose interleukin-2 in patients with metastatic renal cancer. *J Clin Oncol* 2003;21:3127–3132.

280. Johnson DH, Fehrenbacher L, Novotny WF, Herbst RS, Nemunaitis JJ, Jablons DM, Langer CJ, DeVore RF, 3rd, Gaudreault J, Damico LA, Holmgren E, Kabbinavar F. Randomized phase II trial comparing bevacizumab plus carboplatin and paclitaxel with carboplatin and paclitaxel alone in previously untreated locally advanced or metastatic non-small-cell lung cancer. *J Clin Oncol* 2004;22:2184–2191.
281. Sandler AB, Johnson DH, Herbst RS. Anti-vascular endothelial growth factor monoclonals in non-small cell lung cancer. *Clin Cancer Res* 2004;10:4258s–4262s.
282. Diaz-Rubio E. New chemotherapeutic advances in pancreatic, colorectal, and gastric cancers. *Oncologist* 2004;9:282–294.
283. Rugo HS. Bevacizumab in the treatment of breast cancer: rationale and current data. *Oncologist* 2004;9(Suppl 1):43–49.
284. Rini BI, Halabi S, Taylor J, Small EJ, Schilsky RL. Cancer and Leukemia Group B 90206: a randomized phase III trial of interferon-alpha or interferon-alpha plus anti-vascular endothelial growth factor antibody (bevacizumab) in metastatic renal cell carcinoma. *Clin Cancer Res* 2004;10:2584–2586.
285. Willett CG, Boucher Y, di Tomaso E, Duda DG, Munn LL, Tong RT, Chung DC, Sahani DV, Kalva SP, Kozin SV, Mino M, Cohen KS, Scadden DT, Hartford AC, Fischman AJ, Clark JW, Ryan DP, Zhu AX, Blaszkowsky LS, Chen HX, Shellito PC, Lauwers GY, Jain RK. Direct evidence that the VEGF-specific antibody bevacizumab has antivascular effects in human rectal cancer. *Nat Med* 2004;10:145–147.
286. Gerber HP, Kowalski J, Sherman D, Eberhard DA, Ferrara N. Complete inhibition of rhabdomyosarcoma xenograft growth and neovascularization requires blockade of both tumor and host vascular endothelial growth factor. *Cancer Res* 2000;60:6253–6258.
287. Holash J, Davis S, Papadopoulos N, Croll SD, Ho L, Russell M, Boland P, Leidich R, Hylton D, Burova E, Ioffe E, Huang T, Radziejewski C, Bailey K, Fandl JP, Daly T, Wiegand SJ, Yancopoulos GD, Rudge JS. VEGF-trap: a VEGF blocker with potent antitumor effects. *Proc Natl Acad Sci USA* 2002;99:11393–11398.
288. Kim ES, Serur A, Huang J, Manley CA, McCrudden KW, Frischer JS, Soffer SZ, Ring L, New T, Zabski S, Rudge JS, Holash J, Yancopoulos GD, Kandel JJ, Yamashiro DJ. Potent VEGF blockade causes regression of coopted vessels in a model of neuroblastoma. *Proc Natl Acad Sci USA* 2002;99:11399–11404.
289. Wood JM, Bold G, Buchdunger E, Cozens R, Ferrari S, Frei J, Hofmann F, Mestan J, Mett H, O'Reilly T, Persohn E, Rosel J, Schnell C, Stover D, Theuer A, Towbin H, Wenger F, Woods-Cook K, Menrad A, Siemeister G, Schirner M, Thierauch KH, Schneider MR, Drevs J, Martiny-Baron G, Totzke F. PTK787/ZK222584, a novel and potent inhibitor of vascular endothelial growth factor receptor tyrosine kinases, impairs vascular endothelial growth factor-induced responses and tumor growth after oral administration. *Cancer Res* 2000;60:2178–2189.
290. Thomas AL, Morgan B, Drevs J, Unger C, Wiedenmann B, Vanhoefer U, Laurent D, Dugan M, Steward WP. Vascular endothelial growth factor receptor tyrosine kinase inhibitors: PTK787/ZK 222584. *Semin Oncol* 2003;30:32–38.
291. Lin B, Podar K, Gupta D, Tai YT, Li S, Weller E, Hideshima T, Lentzsch S, Davies F, Li C, Weisberg E, Schlossman RL, Richardson PG, Griffin JD, Wood J, Munshi NC, Anderson KC. The vascular endothelial growth factor receptor tyrosine kinase inhibitor PTK787/ZK222584 inhibits growth and migration of multiple myeloma cells in the bone marrow microenvironment. *Cancer Res* 2002;62:5019–5026.
292. Kumar R, Hopper TM, Miller CG, Johnson JH, Crosby RM, Onori JA, Mullin RJ, Truesdale AT, Epperly AH, Hinkle KW, Cheung M, Stafford JA, Luttrell DK. Discovery and biological evaluation of GW654652: a pan inhibitor of VEGF receptors. *Proc Am Assoc Cancer Res* 2003;44:9 (abstract #39).
293. Cheung M, Boloor A, Hinkle KW, Davis-Ward RG, Harris PA, Mook RA, Veal JM, Truesdale AT, Johnson JH, Crosby RM, Rudolph SK, Knick VB, Epperly AH, Kumar R, Luttrell DK, Stafford JA. Discovery of indazolylpyrimidines as potent inhibitors of VEGFR2 tyrosine kinase. *Proc Am Assoc Cancer Res* 2003;44:9 (abstract #40).
294. Podar K, Catley LP, Tai YT, Shringarpure R, Carvalho P, Hayashi T, Burger R, Schlossman RL, Richardson PG, Pandite LN, Kumar R, Hideshima T, Chauhan D, Anderson KC. GW654652, the

pan-inhibitor of VEGF receptors, blocks the growth and migration of multiple myeloma cells in the bone marrow microenvironment. *Blood* 2004;103:3474–3479.

295. O'Farrell AM, Yuen HA, Smolich B, Hannah AL, Louie SG, Hong W, Stopeck AT, Silverman LR, Lancet JE, Karp JE, Albitar M, Cherrington JM, Giles FJ. Effects of SU5416, a small molecule tyrosine kinase receptor inhibitor, on FLT3 expression and phosphorylation in patients with refractory acute myeloid leukemia. *Leuk Res* 2004;28:679–689.
296. Giles FJ, Stopeck AT, Silverman LR, Lancet JE, Cooper MA, Hannah AL, Cherrington JM, O'Farrell AM, Yuen HA, Louie SG, Hong W, Cortes JE, Verstovsek S, Albitar M, O'Brien SM, Kantarjian HM, Karp JE. SU5416, a small molecule tyrosine kinase receptor inhibitor, has biologic activity in patients with refractory acute myeloid leukemia or myelodysplastic syndromes. *Blood* 2003;102:795–801.
297. Fiedler W, Mesters R, Tinnefeld H, Loges S, Staib P, Duhrsen U, Flasshove M, Ottmann OG, Jung W, Cavalli F, Kuse R, Thomalla J, Serve H, O'Farrell AM, Jacobs M, Brega NM, Scigalla P, Hossfeld DK, Berdel WE. A phase 2 clinical study of SU5416 in patients with refractory acute myeloid leukemia. *Blood* 2003;102:2763–2767.
298. O'Farrell AM, Abrams TJ, Yuen HA, Ngai TJ, Louie SG, Yee KW, Wong LM, Hong W, Lee LB, Town A, Smolich BD, Manning WC, Murray LJ, Heinrich MC, Cherrington JM. SU11248 is a novel FLT3 tyrosine kinase inhibitor with potent activity in vitro and in vivo. *Blood* 2003;101:3597–3605.
299. Beebe JS, Jani JP, Knauth E, Goodwin P, Higdon C, Rossi AM, Emerson E, Finkelstein M, Floyd E, Harriman S, Atherton J, Hillerman S, Soderstrom C, Kou K, Gant T, Noe MC, Foster B, Rastinejad F, Marx MA, Schaeffer T, Whalen PM, Roberts WG. Pharmacological characterization of CP-547,632, a novel vascular endothelial growth factor receptor-2 tyrosine kinase inhibitor for cancer therapy. *Cancer Res* 2003;63:7301–7309.
300. Bates D. ZD-6474. AstraZeneca. *Curr Opin Investig Drugs* 2003;4:1468–1472.
301. Lowinger TB, Riedl B, Dumas J, Smith RA. Design and discovery of small molecules targeting raf-1 kinase. *Curr Pharm Des* 2002;8:2269–2278.
302. Richly H, Kupsch P, Passage K, Grubert M, Hilger RA, Kredtke S, Voliotis D, Scheulen ME, Seeber S, Strumberg D. A phase I clinical and pharmacokinetic study of the Raf kinase inhibitor (RKI) BAY 43-9006 administered in combination with doxorubicin in patients with solid tumors. *Int J Clin Pharmacol Ther* 2003;41:620–621.
303. D'Amato RJ, Loughnan MS, Flynn E, Folkman J. Thalidomide is an inhibitor of angiogenesis. *Proc Natl Acad Sci USA* 1994;91:4082–4085.
304. Bartlett JB, Dredge K, Dalgleish AG. The evolution of thalidomide and its IMiD derivatives as anticancer agents. *Nat Rev Cancer* 2004;4:314–322.
305. Hideshima T, Chauhan D, Shima Y, Raje N, Davies FE, Tai YT, Treon SP, Lin B, Schlossman RL, Richardson P, Muller G, Stirling DI, Anderson KC. Thalidomide and its analogs overcome drug resistance of human multiple myeloma cells to conventional therapy. *Blood* 2000;96:2943–2950.
306. D'Amato RJ, Lentzsch S, Anderson KC, Rogers MS. Mechanism of action of thalidomide and 3-aminothalidomide in multiple myeloma. *Semin Oncol* 2001;28:597–601.
307. Yabu T, Tomimoto H, Taguchi Y, Yamaoka S, Igarashi Y, Okazaki T. Thalidomide-induced antiangiogenic action is mediated by ceramide through depletion of VEGF receptors, and antagonized by sphingosine-1-phosphate. *Blood* 2005;106:125–134.
308. LeBlanc R, Hideshima T, Catley LP, Shringarpure R, Burger R, Mitsiades N, Mitsiades C, Cheema P, Chauhan D, Richardson PG, Anderson KC, Munshi NC. Immunomodulatory drug costimulates T cells via the B7-CD28 pathway. *Blood* 2004;103:1787–1790.
309. Davies FE, Raje N, Hideshima T, Lentzsch S, Young G, Tai YT, Lin B, Podar K, Gupta D, Chauhan D, Treon SP, Richardson PG, Schlossman RL, Morgan GJ, Muller GW, Stirling DI, Anderson KC. Thalidomide and immunomodulatory derivatives augment natural killer cell cytotoxicity in multiple myeloma. *Blood* 2001;98:210–216.
310. Lentzsch S, LeBlanc R, Podar K, Davies F, Lin B, Hideshima T, Catley L, Stirling DI, Anderson KC. Immunomodulatory analogs of thalidomide inhibit growth of Hs Sultan cells and angiogenesis in vivo. *Leukemia* 2003;17:41–44.
311. Richardson PG, Schlossman RL, Weller E, Hideshima T, Mitsiades C, Davies F, LeBlanc R, Catley LP, Doss D, Kelly K, McKenney M, Mechlowicz J, Freeman A, Deocampo R, Rich R,

Ryoo JJ, Chauhan D, Balinski K, Zeldis J, Anderson KC. Immunomodulatory drug CC-5013 overcomes drug resistance and is well tolerated in patients with relapsed multiple myeloma. *Blood* 2002;100:3063–3067.

312. Schey SA, Fields P, Bartlett JB, Clarke IA, Ashan G, Knight RD, Streetly M, Dalgleish AG. Phase I study of an immunomodulatory thalidomide analog, CC-4047, in relapsed or refractory multiple myeloma. *J Clin Oncol* 2004;22:3269–3276.
313. Cushman M, He HM, Katzenellenbogen JA, Lin CM, Hamel E. Synthesis, antitubulin and antimitotic activity, and cytotoxicity of analogs of 2-methoxyestradiol, an endogenous mammalian metabolite of estradiol that inhibits tubulin polymerization by binding to the colchicine binding site. *J Med Chem* 1995;38:2041–2049.
314. Klauber N, Parangi S, Flynn E, Hamel E, D'Amato RJ. Inhibition of angiogenesis and breast cancer in mice by the microtubule inhibitors 2-methoxyestradiol and taxol. *Cancer Res* 1997;57:81–86.
315. Lottering ML, Haag M, Seegers JC. Effects of 17 beta-estradiol metabolites on cell cycle events in MCF-7 cells. *Cancer Res* 1992;52:5926–5932.
316. Mukhopadhyay T, Roth JA. Induction of apoptosis in human lung cancer cells after wild-type p53 activation by methoxyestradiol. *Oncogene* 1997;14:379–384.
317. Schumacher G, Neuhaus P. The physiological estrogen metabolite 2-methoxyestradiol reduces tumor growth and induces apoptosis in human solid tumors. *J Cancer Res Clin Oncol* 2001;127:405–410.
318. Kumar AP, Garcia GE, Slaga TJ. 2-methoxyestradiol blocks cell-cycle progression at G(2)/M phase and inhibits growth of human prostate cancer cells. *Mol Carcinog* 2001;31:111–124.
319. Chauhan D, Catley L, Hideshima T, Li G, Leblanc R, Gupta D, Sattler M, Richardson P, Schlossman RL, Podar K, Weller E, Munshi N, Anderson KC. 2-Methoxyestradiol overcomes drug resistance in multiple myeloma cells. *Blood* 2002;100:2187–2194.
320. Keyes K, Cox K, Treadway P, Mann L, Shih C, Faul MM, Teicher BA. An in vitro tumor model: analysis of angiogenic factor expression after chemotherapy. *Cancer Res* 2002;62:5597–5602.
321. Keyes KA, Mann L, Sherman M, Galbreath E, Schirtzinger L, Ballard D, Chen YF, Iversen P, Teicher BA. LY317615 decreases plasma VEGF levels in human tumor xenograft-bearing mice. *Cancer Chemother Pharmacol* 2004;53:133–140.
322. Herbst RS. Targeted therapy using novel agents in the treatment of non-small-cell lung cancer. *Clin Lung Cancer* 2002;3(Suppl 1):S30–S38.
323. Gratton JP, Lin MI, Yu J, Weiss ED, Jiang ZL, Fairchild TA, Iwakiri Y, Groszmann R, Claffey KP, Cheng YC, Sessa WC. Selective inhibition of tumor microvascular permeability by cavtratin blocks tumor progression in mice. *Cancer Cell* 2003;4:31–39.
324. MacDonald TJ, Taga T, Shimada H, Tabrizi P, Zlokovic BV, Cheresh DA, Laug WE. Preferential susceptibility of brain tumors to the antiangiogenic effects of an alpha(v) integrin antagonist. *Neurosurgery* 2001;48:151–157.
325. Burke PA, DeNardo SJ, Miers LA, Lamborn KR, Matzku S, DeNardo GL. Cilengitide targeting of alpha(v)beta(3) integrin receptor synergizes with radioimmunotherapy to increase efficacy and apoptosis in breast cancer xenografts. *Cancer Res* 2002;62:4263–4272.
326. Raguse JD, Gath HJ, Bier J, Riess H, Oettle H. Cilengitide (EMD 121974) arrests the growth of a heavily pretreated highly vascularised head and neck tumour. *Oral Oncol* 2004;40:228–230.
327. Nisato RE, Tille JC, Jonczyk A, Goodman SL, Pepper MS. Alphav beta 3 and alphav beta 5 integrin antagonists inhibit angiogenesis in vitro. *Angiogenesis* 2003;6:105–119.
328. Moolenaar WH, Kranenburg O, Postma FR, Zondag GC. Lysophosphatidic acid: G-protein signalling and cellular responses. *Curr Opin Cell Biol* 1997;9:168–173.
329. Hu YL, Tee MK, Goetzl EJ, Auersperg N, Mills GB, Ferrara N, Jaffe RB. Lysophosphatidic acid induction of vascular endothelial growth factor expression in human ovarian cancer cells. *J Natl Cancer Inst* 2001;93:762–768.
330. Hideshima T, Chauhan D, Hayashi T, Podar K, Akiyama M, Mitsiades C, MItsiades N, Gong B, Bonham L, de Vries P, Munshi N, Richardson PG, Singer JW, Anderson KC. Antitumor activity of lysophosphatidic acid acyltransferase-beta inhibitors, a novel class of agents, in multiple myeloma. *Cancer Res* 2003;63:8428–8436.
331. Kane RC, Bross PF, Farrell AT, Pazdur R. Velcade: U.S. FDA approval for the treatment of multiple myeloma progressing on prior therapy. *Oncologist* 2003;8:508–513.

332. Hideshima T, Richardson P, Chauhan D, Palombella VJ, Elliott PJ, Adams J, Anderson KC. The proteasome inhibitor PS-341 inhibits growth, induces apoptosis, and overcomes drug resistance in human multiple myeloma cells. *Cancer Res* 2001;61:3071–3076.
333. Hideshima T, Mitsiades C, Akiyama M, Hayashi T, Chauhan D, Richardson P, Schlossman R, Podar K, Munshi NC, Mitsiades N, Anderson KC. Molecular mechanisms mediating antimyeloma activity of proteasome inhibitor PS-341. *Blood* 2003;101:1530–1534.
334. Hideshima T, Chauhan D, Hayashi T, Akiyama M, Mitsiades N, Mitsiades C, Podar K, Munshi NC, Richardson PG, Anderson KC. Proteasome inhibitor PS-341 abrogates IL-6 triggered signaling cascades via caspase-dependent downregulation of gp130 in multiple myeloma. *Oncogene* 2003;22:8386–8393.
335. Nawrocki ST, Bruns CJ, Harbison MT, Bold RJ, Gotsch BS, Abbruzzese JL, Elliott P, Adams J, McConkey DJ. Effects of the proteasome inhibitor PS-341 on apoptosis and angiogenesis in orthotopic human pancreatic tumor xenografts. *Mol Cancer Ther* 2002;1:1243–1253.
336. Oikawa T, Sasaki T, Nakamura M, Shimamura M, Tanahashi N, Omura S, Tanaka K. The proteasome is involved in angiogenesis. *Biochem Biophys Res Commun* 1998;246:243–248.
337. LeBlanc R, Catley LP, Hideshima T, Lentzsch S, Mitsiades CS, Mitsiades N, Neuberg D, Goloubeva O, Pien CS, Adams J, Gupta D, Richardson PG, Munshi NC, Anderson KC. Proteasome inhibitor PS-341 inhibits human myeloma cell growth in vivo and prolongs survival in a murine model. *Cancer Res* 2002;62:4996–5000.
338. Tai YT, Catley LP, Mitsiades CS, Burger R, Podar K, Shringpaure R, Hideshima T, Chauhan D, Hamasaki M, Ishitsuka K, Richardson P, Treon SP, Munshi NC, Anderson KC. Mechanisms by which SGN-40, a humanized anti-CD40 antibody, induces cytotoxicity in human multiple myeloma cells: clinical implications. *Cancer Res* 2004;64:2846–2852.
339. Gorski DH, Beckett MA, Jaskowiak NT, Calvin DP, Mauceri HJ, Salloum RM, Seetharam S, Koons A, Hari DM, Kufe DW, Weichselbaum RR. Blockage of the vascular endothelial growth factor stress response increases the antitumor effects of ionizing radiation. *Cancer Res* 1999;59:3374–3378.
340. Sweeney CJ, Miller KD, Sissons SE, Nozaki S, Heilman DK, Shen J, Sledge GW, Jr. The antiangiogenic property of docetaxel is synergistic with a recombinant humanized monoclonal antibody against vascular endothelial growth factor or 2-methoxyestradiol but antagonized by endothelial growth factors. *Cancer Res* 2001;61:3369–3372.
341. Tran J, Master Z, Yu JL, Rak J, Dumont DJ, Kerbel RS. A role for survivin in chemoresistance of endothelial cells mediated by VEGF. *Proc Natl Acad Sci USA* 2002;99:4349–4354.
342. Jain RK. Normalizing tumor vasculature with anti-angiogenic therapy: a new paradigm for combination therapy. *Nat Med* 2001;7:987–989.
343. Teicher BA, Sotomayor EA, Huang ZD. Antiangiogenic agents potentiate cytotoxic cancer therapies against primary and metastatic disease. *Cancer Res* 1992;52:6702–6704.
344. Kakeji Y, Teicher BA. Preclinical studies of the combination of angiogenic inhibitors with cytotoxic agents. *Invest New Drugs* 1997;15:39–48.
345. Klement G, Baruchel S, Rak J, Man S, Clark K, Hicklin DJ, Bohlen P, Kerbel RS. Continuous low-dose therapy with vinblastine and VEGF receptor-2 antibody induces sustained tumor regression without overt toxicity. *J Clin Invest* 2000;105:R15–R24.
346. Lee CG, Heijn M, di Tomaso E, Griffon-Etienne G, Ancukiewicz M, Koike C, Park KR, Ferrara N, Jain RK, Suit HD, Boucher Y. Anti-vascular endothelial growth factor treatment augments tumor radiation response under normoxic or hypoxic conditions. *Cancer Res* 2000;60:5565–5570.
347. Kozin SV, Boucher Y, Hicklin DJ, Bohlen P, Jain RK, Suit HD. Vascular endothelial growth factor receptor-2-blocking antibody potentiates radiation-induced long-term control of human tumor xenografts. *Cancer Res* 2001;61:39–44.
348. Hanahan D, Bergers G, Bergsland E. Less is more, regularly: metronomic dosing of cytotoxic drugs can target tumor angiogenesis in mice. *J Clin Invest* 2000;105:1045–1047.
349. Kerbel RS, Kamen BA. The anti-angiogenic basis of metronomic chemotherapy. *Nat Rev Cancer* 2004;4:423–436.
350. Browder T, Butterfield CE, Kraling BM, Shi B, Marshall B, O'Reilly MS, Folkman J. Antiangiogenic scheduling of chemotherapy improves efficacy against experimental drug-resistant cancer. *Cancer Res* 2000;60:1878–1886.

351. Gasparini G. Metronomic scheduling: the future of chemotherapy? *Lancet Oncol* 2001;2:733–740.
352. Kamen BA, Rubin E, Aisner J, Glatstein E. High-time chemotherapy or high time for low dose. *J Clin Oncol* 2000;18:2935–2937.
353. Kerbel RS, Klement G, Pritchard KI, Kamen B. Continuous low-dose anti-angiogenic/metronomic chemotherapy: from the research laboratory into the oncology clinic. *Ann Oncol* 2002;13:12–15.
354. Locigno R, Antoine N, Bours V, Daukandt M, Heinen E, Castronovo V. TNP-470, a potent angiogenesis inhibitor, amplifies human T lymphocyte activation through an induction of nuclear factor-kappaB, nuclear factor-AT, and activation protein-1 transcription factors. *Lab Invest* 2000;80:13–21.
355. Asahara T, Masuda H, Takahashi T, Kalka C, Pastore C, Silver M, Kearne M, Magner M, Isner JM. Bone marrow origin of endothelial progenitor cells responsible for postnatal vasculogenesis in physiological and pathological neovascularization. *Circ Res* 1999;85:221–228.
356. Schultz A, Lavie L, Hochberg I, Beyar R, Stone T, Skorecki K, Lavie P, Roguin A, Levy AP. Interindividual heterogeneity in the hypoxic regulation of VEGF: significance for the development of the coronary artery collateral circulation. *Circulation* 1999;100:547–552.
357. Vacca A, Ribatti D, Roccaro AM, Frigeri A, Dammacco F. Bone marrow angiogenesis in patients with active multiple myeloma. *Semin Oncol* 2001;28:543–550.
358. Kumar S, Witzig TE, Timm M, Haug J, Wellik L, Kimlinger TK, Greipp PR, Rajkumar SV. Bone marrow angiogenic ability and expression of angiogenic cytokines in myeloma: evidence favoring loss of marrow angiogenesis inhibitory activity with disease progression. *Blood* 2004;104:1159–1165.
359. Di Raimondo F, Azzaro MP, Palumbo G, Bagnato S, Giustolisi G, Floridia P, Sortino G, Giustolisi R. Angiogenic factors in multiple myeloma: higher levels in bone marrow than in peripheral blood. *Haematologica* 2000;85:800–805.
360. Aguayo A, Giles F, Albitar M. Vascularity, angiogenesis and angiogenic factors in leukemias and myelodysplastic syndromes. *Leuk Lymphoma* 2003;44:213–222.
361. Kini AR, Kay NE, Peterson LC. Increased bone marrow angiogenesis in B cell chronic lymphocytic leukemia. *Leukemia* 2000;14:1414–1418.
362. de Bont ES, Rosati S, Jacobs S, Kamps WA, Vellenga E. Increased bone marrow vascularization in patients with acute myeloid leukaemia: a possible role for vascular endothelial growth factor. *Br J Haematol* 2001;113:296–304.
363. Korkolopoulou P, Apostolidou E, Pavlopoulos PM, Kavantzas N, Vyniou N, Thymara I, Terpos E, Patsouris E, Yataganas X, Davaris P. Prognostic evaluation of the microvascular network in myelodysplastic syndromes. *Leukemia* 2001;15:1369–1376.
364. Panteli K, Zagorianakou N, Bai M, Katsaraki A, Agnantis NJ, Bourantas K. Angiogenesis in chronic myeloproliferative diseases detected by CD34 expression. *Eur J Haematol* 2004;72:410–415.
365. Korkolopoulou P, Thymara I, Kavantzas N, Vassilakopoulos TP, Angelopoulou MK, Kokoris SI, Dimitriadou EM, Siakantaris MP, Anargyrou K, Panayiotidis P, Tsenga A, Androulaki A, Doussis-Anagnostopoulou IA, Patsouris E, Pangalis GA. Angiogenesis in Hodgkin's lymphoma: a morphometric approach in 286 patients with prognostic implications. *Leukemia* 2005;19:894–900.
366. Korkolopoulou P, Gribabis DA, Kavantzas N, Angelopoulou MK, Siakantaris MP, Patsouris E, Androulaki A, Thymara I, Kokoris SI, Kyrtsonis MC, Kittas C, Pangalis GA. A morphometric study of bone marrow angiogenesis in hairy cell leukaemia with clinicopathological correlations. *Br J Haematol* 2003;122:900–910.

5 Tumor Microenvironment and Angiogenesis

Cheryl H. Baker and Isaiah J. Fidler

CONTENTS

5.1 CANCER METASTASIS

The major clinical challenge of systemic cancer therapy is not eradication of the primary tumor, which can be treated with radiation or surgery, but eradication of metastases, which are usually present at the time of initial diagnosis and are likely to be resistant to conventional chemotherapy (1–3). A principal barrier to the destruction of disseminated cancer is the heterogeneous nature of cancer. This heterogeneity is exhibited in a wide range of biologic entities such as cell-surface receptors, enzymes, and karyotypes, and in cellular features, such as morphologic characteristics, growth properties, sensitivity to chemotherapeutic drugs, and the ability to invade and metastasize (3,4–6). Neoplastic transformation involves genetic alterations, such as the activation or dysregulation of oncogenes (7), and cells that are able to circumvent normal growth-control mechanisms may undergo continuous selection pressures. Unfortunately, this continuous evolution of genetically unstable neoplasms eventually favors the emergence of subpopulations of cells with metastatic potential.

The sites of growth and metastasis of tumors are not random. Clinical observations of cancer patients and studies with experimental rodent tumors have revealed that certain tumors metastasize to specific organs independently of vascular anatomy, the rate of blood flow, and the number of tumor cells delivered to each organ (2,3,8,9). For example, a study of the distribution and fate of radiolabeled melanoma cells in an experimental rodent system demonstrated that tumor cells reach the microvasculature of many organs but only a few cells can extravasate into a particular organ's parenchyma and grow in only some organs (10).

In 1889, Stephen Paget (11) proposed that the process of metastasis did not occur by chance but that some tissues provided a better environment than others did for the growth of certain tumors. Only when the tumor cells (the "seeds") were compatible with a particular organ's tissue (the "soil") did metastasis occur. The current interpretation of the seed-and-soil hypothesis consists of two principles. The first, which deals with the seeds, is that tumor cells are genetically unstable, so their division yields variant cells with different biological properties; as a result, all neoplasms are biologically heterogeneous (3). Conventional therapy is therefore likely to produce selection and the emergence of resistant cells (12–17). In fact, the emergence of primary and metastatic tumors that are resistant to chemotherapy and that respond differently to therapeutic agents has been well documented (4).

The second principle, which deals with the soil, is that the outcome of the growth and spread of cancer depends on multiple interactions, or cross-talk, between tumor cells and their host cells within the organ's microenvironment. These interactions influence the diversification of tumor cells, thereby adding complexity to the development and use of appropriate therapies. Thus, when designing treatments for cancer and metastases, one must consider both tumor heterogeneity (18) and organ-specific microenvironments (3,18,19).

5.2 TUMOR ANGIOGENESIS

Angiogenesis occurs because a tumor mass that exceeds 0.25 mm in diameter can no longer receive oxygen and nutrients by diffusion; thus, for a tumor to grow and for metastases to develop, new vasculature must be generated (10,20,21). This process of angiogenesis consists of a series of linked, sequential steps, which lead to the establishment of a new vascular bed. For new capillaries to be generated for neovascularization, endothelial cells must sprout, proliferate, migrate, degrade the basement membrane, and form a structure, that is, organize into a new lumen organization (22). Then, both tumor cells and host cells must secrete a variety of factors to stimulate angiogenesis (Table 5.1).

At present, more than a dozen proangiogenic molecules have been reported, including basic fibroblast growth factor (bFGF), vascular endothelial growth factor, also known as vascular permeability factor (VEGF/VPF) (5,23–25), interleukin-8 (IL-8), angiogenin, angiotropin, platelet-derived endothelial cell growth factor (PD-ECGF), platelet-derived growth factor (PDGF), transforming growth factor-α (TGF-α), TGF-β, epidermal growth factor (EGF), and tumor necrosis factor-α (TNF-α) (5,8,20,23). Many tissues and tumors, however,

TABLE 5.1
Tumor and Host-Derived Stimulating Factors of Angiogenesis

Basic fibroblast growth factor
Vascular endothelial growth factor/vascular permeability factor
Platelet-derived endothelial cell growth factor
Platelet-derived growth factor
Transforming growth factor-α
Transforming growth factor-β
Epidermal growth factor
Tumor necrosis factor-α
Angiogenin
Angiopoietin-1
Angiotropin
Interleukins-1,3,4,8

also generate factors that inhibit angiogenesis, such as angiostatin, endostatin, thrombospondins, interferon-α (IFN-α), and IFN-β (20,24–30). The angiogenic phenotype of a tissue or a tumor is therefore determined by the net balance between positive and negative regulators of neovascularization (31).

Both pro- and antiangiogenic molecules are expressed at different levels in various tissues, enabling an organ's microenvironment to greatly influence the process of angiogenesis (20,21). For example, angiogenesis is favored in rapidly dividing tissues because of the shift in balance toward proangiogenic molecules; in most normal tissues, however, the shift is toward antiangiogenic factors (32). This complexity of the angiogenic process suggests the existence of multiple and redundant system controls, which can be temporarily turned on and off (24).

Since angiogenesis is essential for the growth of neoplasms, antiangiogenic strategies are a potentially unifying concept in the development of effective cancer therapies. Such strategies have several advantages over other types of therapy: (i) they have a broader spectrum of activity, (ii) they target a more genetically stable cell population (endothelial versus tumor cells), thereby reducing the chance of drug resistance, (iii) they are less toxic, and (iv) they may have synergism with other anticancer therapies (33).

5.3 REGULATION OF ANGIOGENESIS BY THE TUMOR MICROENVIRONMENT

Cancer metastasis consists of a series of sequential steps, which only a small fraction of cells within a primary tumor can complete (34, Figure 5.1). Metastasis begins with the invasion of the cells into the host stroma surrounding the primary neoplasm and ends with the destruction of the basement membrane, enhanced motility of some tumor cells, and eventual penetration of blood or lymphatic vessels (35). During this process, tumor cells must survive transport in the circulatory system and interaction with host immune cells and must eventually express adhesion molecules, which favor the cells' arrest in the capillary bed. These steps do not always take place, however. For example, in one study, it was shown that 24 h after the entry of radiolabeled melanoma cells into the circulation, <1% of the cells were viable and <0.1% eventually formed metastases (9). Therefore, the mere presence of circulating tumor cells does not guarantee metastasis. In another study, tumors grew better and produced metastases better subsequent to the implantation of the tumor cells into their orthotopic organs than when they were implanted into ectopic organs (36). Differences in angiogenesis may account for this fact.

The production of bFGF, VEGF, and IL-8 by tumor or host cells or the release of proangiogenic molecules from the extracellular matrix is known to induce the growth of endothelial cells and the formation of blood vessels. Further, the organ microenvironment can directly contribute to the induction and maintenance of the proangiogenic factors bFGF (30,37) and IL-8 (38).

For example, in two studies, patients with renal cell carcinoma had levels of bFGF in either the serum or the urine that inversely correlated with their survival time (39,40). In another study, human renal cancer cells implanted into different organs of nude mice had different metastatic potentials. Those implanted into the kidney were highly metastatic to the lung, whereas those implanted subcutaneously were not metastatic to the lung (30). The subcutaneous or intramuscular tumors had a lower level of mRNA transcripts for bFGF than did continuously cultured cells, whereas the renal tumors had 20 times the bFGF mRNA and protein levels than the cultured cells. A histopathologic examination of all the tumors revealed that subcutaneous tumors had few blood vessels, whereas renal tumors had many blood vessels (30,37).

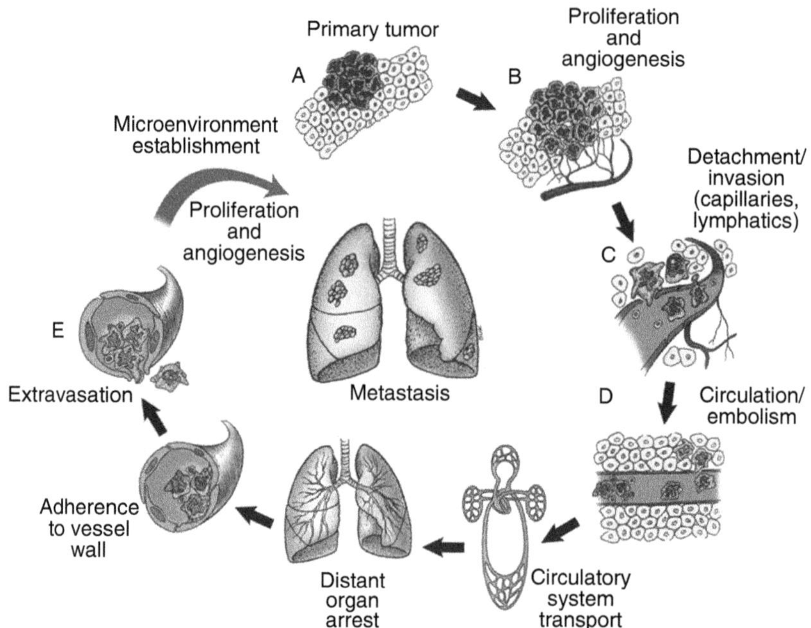

FIGURE 5.1 (See color insert following page 558.) Primary steps in the formation of cancer metastasis. (A) For cellular transformation and tumor growth, the growth of neoplastic cells must be progressive, with an adequate supply to the expanding tumor cells. (B) Extensive vascularization must occur for a tumor mass to exceed 1–2 mm in diameter. The synthesis and secretion of angiogenic factors establish a new capillary network from the surrounding host tissue. (C–E) Local invasion of the host stroma by some tumor cells occurs through several parallel mechanisms. (C) The thin-walled lymphatic channels offer very little resistance to penetration by tumor cells and provide the most common route for tumor cell entry into the circulatory system. (D) Detachment and embolization of single tumor cells or aggregates occur next, with most circulating tumor cells rapidly destroyed. After some tumor cells have survived the circulation, they become trapped in the capillary beds of distant organs by adhering to the endothelial cells or exposed basement membranes. (E) Extravasation occurs next by mechanisms similar to those that operate during intravasation.

In still another study, in situ hybridization and northern blot analysis revealed that the tumors of patients diagnosed with Duke's C or D colon carcinoma had markedly higher levels of bFGF than did those in patients diagnosed with Duke's B. The in situ hybridization also revealed that bFGF was overexpressed at the tumor periphery, where the cells were rapidly dividing. Northern blot analysis detected no mRNA transcripts for bFGF (41). In a follow-up study of patients with colon cancer, bFGF expression was found to be highest in the primary tumors of patients who presented with metastatic disease. That study, therefore, was able to identify a cohort of patients who appeared to be free of metastatic disease at the time of surgery (low bFGF expression) and another cohort of patients who developed metastatic disease later (high bFGF) (42). As previously mentioned, bFGF expression was found to be higher in xenografts of human renal cancers growing in the kidney (orthotopic) than in xenografts growing in subcutaneous tissues (ectopic). In sharp contrast, however, a different study showed that IFN-β expression was high in the epithelial cells and the fibroblasts surrounding subcutaneous tumors, whereas, no IFN-β was found in or around tumors growing in the kidney (37). That study also found that the inhibitory effect of IFN-α and IFN-β on bFGF expression depended on cell density and was independent of the antiproliferative effects of the IFNs (37). Indeed, IFN-α and IFN-β but not IFN-γ have been shown to downregulate the expression of bFGF mRNA and protein in human renal cell carcinoma and in human bladder, prostate, colon, and breast carcinoma (30).

In yet another study, immunohistochemical and in situ hybridization analyses found that secreted IFN-β was expressed in different types of murine epithelial cells, which line portals of entry into the body (43). Specifically, epithelial cells lining the skin, digestive tract, urinary tract, reproductive tract, and upper respiratory tract constitutively expressed IFN-β. The epithelial cells at risk of environmental exposure expressed IFN-β (protein and mRNA), whereas those of internal organs, which are not directly exposed to external pathogens, did not express IFN-β (43).

Constitutive expression of IL-8 is known to directly correlate with the metastatic potential of human melanoma cells (38). IL-8 contributes to angiogenesis by inducing proliferation, migration, and invasion of endothelial cells (44). Several organ-derived cytokines (produced by inflammatory cells) can upregulate IL-8 expression in normal and tumorigenic cells (45). Human melanoma cells, A375P (low-metastatic parental) and A375SM (highly metastatic cloned), were implanted into the subcutis, spleen (producing liver metastases), and tail vein (producing lung metastases) of nude mice. The subcutaneous tumors (melanomas) expressed the most IL-8, followed by lung and then liver lesions. This differential IL-8 expression was not due to selection of a subpopulation of cells but rather due to the fact that the IL-8 mRNA level was always higher in the skin and lower in the liver regardless of the origin of the melanoma cells (45). Furthermore, IL-8 expression was upregulated in melanoma cells cocultured with keratinocytes, whereas it was inhibited in cells cocultured with hepatocytes (45). Similar results obtained with conditioned media from keratinocyte and hepatocyte cultures suggested that organ-derived factors (e.g., IL-1 and TGF-β) could modulate the expression of IL-8 in human melanoma cells (44).

The influence of the microenvironment on the expression of VEGF/VPF and on angiogenesis, tumor cell proliferation, and metastasis was also investigated using human gastric cancer cells implanted orthotopically and ectopically into nude mice. The tumors growing in the stomach wall were highly vascularized and expressed higher levels of VEGF/VPF than did the subcutaneous tumors (46). Moreover, only the stomach tumors metastasized, suggesting that the expression of VEGF/VPF-derived vascularization and the metastasis of human gastric cancer cells are regulated by an organ's microenvironment (46).

5.4 INFLUENCE OF LYMPHOID CELLS ON ANGIOGENESIS

The regulation of angiogenesis by lymphoid cells, including T lymphocytes, macrophages, and mast cells, is well established. It occurs in the tumor microenvironment (47–54). For example, invasive cutaneous melanoma is often associated with a local inflammatory reaction involving T lymphocytes and macrophages, which is often associated with an increased risk of metastasis. This phenomenon suggests that angiogenesis induced by inflammation contributes to tumor progression and metastasis (55,56).

The relatively slow growth of tumors in senescent mice has been closely linked with decreased vascularization (57), which may be due to a diminished immunological response associated with aging (58–60). Previous studies investigated the role of tumor vascularization and its effect on tumor growth in immunosuppressed mice and found that the growth of the immunogenic B16 melanoma was indeed a growth delay in the myelosuppressed mice than in the control mice (61).

In another study, tumor growth in mice pretreated with doxorubicin (DXR) that had been injected with normal splenocytes 1 day before tumor challenge was comparable to that in the control mice, implicating myelosuppression as a cause of retardation of tumor growth and vascularization (61). Similar results were obtained with athymic nude mice, suggesting that the tumor vascularization observed in the DXR-treated mice reconstituted with normal splenocytes was not mediated solely by T lymphocytes. Instead, the enhanced vascularization suggested that myelosuppressive chemotherapeutic drugs such as DXR can inhibit host-mediated vascularization and thus growing tumors can use homeostatic mechanisms to their advantage (61).

In a study of human colon cancer, the role of infiltrating cells in the angiogenesis of the cancer contributed to high expression of PD-ECGF in macrophages and lymphocytes. Low expression, on the other hand, was observed in the colon cancer epithelium. The strong correlation between PD-ECGF and vessel count suggested the involvement of these infiltrating cells in human colon cancer angiogenesis (62).

Many other studies have recognized the importance of macrophages in angiogenesis (53,54,63,64). One reason was that macrophages produce >20 molecules that influence endothelial cell proliferation, migration, and differentiation in vitro (53) and are potentially angiogenic in vivo. Another observation was that macrophages may modify the extracellular matrix, thereby modulating angiogenesis; this modulation happens either through the direct production of extracellular matrix components or through the production of proteases, which effectively alter the structure and composition of the extracellular matrix (54). A third reason is that macrophages have been shown to produce substances that suppress angiogenesis, such as thrombospondin-1 (65–68). Finally, macrophage-derived metalloelastase has been shown to be responsible for the generation of angiostatin in Lewis lung carcinoma (69), and the addition of plasminogen to 3LL Lewis lung carcinoma cells cultured in vitro did not result in the generation of angiostatin. However, its addition to cocultured macrophages and carcinoma cells resulted in angiostatin generation (69). Those researchers concluded that the elastase activity in macrophages was significantly enhanced by the cytokine granulocyte/macrophage-colony stimulating factor (GM-CSF), which was secreted by the tumor cells (70) and that this activity led to the generation of plasminogen.

5.5 TARGETING THE VASCULATURE

Because systemic antitumor therapy fails primarily because of the genetic instability and biologic heterogeneity of neoplasms, therapeutic agents that target a tumor's vasculature (a genetically stable and essential component of tumors) have been explored as an alternative to conventional therapy. The structure and architecture of tumor vasculature can differ dramatically from those of normal organs (71–73). Modern techniques, such as phage display targeting, have defined "vascular addresses" that may be distinct for different organs and for tumors in those organs (74). In normal human tissues, for example, endothelial cells are long lived and recycle every 3–5 years, whereas tumor-associated endothelial cells recycle every 30–40 days. The rate of endothelial cell turnover is <0.1% in normal human tissues but 2%–9.9% in malignant human tumor cells (71).

Endothelial cells, which line blood vessels in different organs, are also phenotypically distinct. The expression of various cytokines and growth factors and their relevant receptors on both tumor cells and organ-specific endothelial cells differs in different organ microenvironments. The interaction of these growth factor ligands (i.e., VEGF, TGF-α, EGF, and PDGF) with their relevant receptors is essential for endothelial cell survival and growth (75–78). For example, tumor cells that produce growth factor ligands are likely to upregulate growth factor receptors and the activation of these receptors on tumor cells (autocrine effect) and on tumor-associated endothelial cells (paracrine effect). These effects are likely to lead to downstream signaling, which can activate the mitogenic and survival pathways, such as the mitogen-activated protein kinases and phosphatidylinositol-3 kinase pathways and lead to downregulation of the apoptotic protein caspases 3 and 8 (75,76,79). Therefore, activation of the growth factor receptors on tumor cells and more so on the tumor-associated endothelial cells is likely to increase the expression of antiapoptotic proteins, causing endothelial cell resistance to chemotherapy, even though these cells divide every 30–40 days. Inhibition of this growth factor receptor activation by receptor antagonists (i.e., monoclonal antibodies and tyrosine kinase inhibitors) in tumor-associated endothelial cells (and tumor cells) can lead to BAX induction, activation of caspase-8, and downregulation of BCL-2 and NF-κβ (80,81). It can also increase a tumor's susceptibility to chemotherapy. Indeed, if these endothelial cells continue to cycle, the use of receptor antagonists (77,78,82–87) combined with anticycling drugs can cause the destruction of the vasculature within neoplasms, leading to the apoptosis of adjacent tumor cells.

5.6 TARGETING ENDOTHELIAL CELLS

As previously stated, human tumors of any given histologic type are extremely heterogeneous, and in most types of cancer, only a subset of patients respond to a particular antitumor agent. Modern techniques such as phage display targeting have defined vascular addresses, which may be distinct for different organs as well as tumors in those organs (74). Furthermore, under the selective pressure of a cytotoxic therapy, the genetic diversity within human tumors leads to a rapid outgrowth of drug-resistant cells. A vast array of drug-resistance mechanisms, such as mutations or amplifications of target genes, upregulation of antiapoptotic pathways, and overexpression of drug efflux transporters, can abrogate a drug's potential therapeutic effects. For all of these reasons, understanding the biological mechanisms of human tumors is clearly essential in the design of effective anticancer therapies.

In recent years, the search for novel anticancer agents has moved from cytotoxic agents to biologically targeted agents to avoid the usual toxic effects of therapy and to develop more precisely targeted therapies. In fact, preclinical animal studies of molecular interactions between tumor cells and tumor-associated endothelial cells have identified unique targets on both dividing tumor cells and tumor-associated endothelial cells. One such promising biological target is growth factor receptors. Targeted (direct) inhibition of growth factor receptors has been accomplished through the use of both antibodies designed to inhibit growth factor receptor signaling and the use of protein tyrosine kinase inhibitors designed to inhibit intracellular activation (autophosphorylation) of growth factor receptors.

5.6.1 Inhibition of Epidermal Growth Factor Receptor Signaling

Angiogenesis and EGF receptor (EGFR) signaling have been independently evaluated as a target for cancer therapy (88,89). EGFR and its ligands TGF-α and EGF are commonly expressed in many human carcinomas, including pancreatic, prostate, renal, and bladder, and their expression is associated with progressive disease (90). The coexpression of both TGF-α and EGF and the EGFR on tumor cells has led to the discovery that the EGFR-dependent autocrine loop is essential for tumor progression and malignancy. However, the paracrine effect of these ligands is not understood well.

Protein tyrosine kinase inhibitors of the EGFR have emerged as promising anticancer agents. A number of preclinical studies have shown these inhibitors' antitumor and antiangiogenic activity (91–96). For example, decreased production of proangiogenic molecules and inhibition of tumor-associated angiogenesis have been demonstrated for small molecule EGFR inhibitors, such as Gefitinib (Iressa) (97,98), PKI116 (77,78,82,83,85,86,99), and others (100). Recent work in our laboratory has shown that oral administration of PKI166 (99) combined with gemcitabine inhibited progressive growth and metastasis of L3.6pl human pancreatic cancer cells implanted orthotopically in nude mice (82,83,85). These cancer cells highly express TGF-α and EGFR, and the combination of PKI166 and gemcitabine produced significant therapeutic effects mediated in part by the induction of apoptosis in tumor-associated endothelial cells (82,83,85).

Another study showed that human pancreatic carcinoma growing orthotopically in nude mice regressed after therapy with the IM-C225 antibody and gemcitabine (84). Although both IM-C225 and gemcitabine were active in this model as single agents, superior results were obtained when they were combined. The combination-treated mice had a lower tumor burden and a lower incidence of liver metastases than did the mice treated with either agent alone. Of particular interest was the finding that apoptosis of tumor-associated endothelial cells was induced by the downregulation of VEGF and IL-8 production by the tumor cells by binding of the IM-C225. The striking observation that receptor inactivation and signaling could lead to endothelial cell apoptosis was further confirmed when IM-C225 was used with paclitaxel in an orthotopic bladder tumor model (101).

Most recently, studies have shown that the expression and activation of EGFR by tumor-associated endothelial cells are influenced by interaction with specific growth factors in the microenvironment (78). For example, we found that endothelial cells in tumors that produced high levels of TGF-α or EGF expressed EGFR and activated EGFR, whereas endothelial cells within tumors that do not produce these ligands did not express EGFR and activated EGFR. In the presence of chemotherapeutic agents, however, those endothelial cells were highly susceptible to apoptosis induced by the specific inhibitor of EGFR protein tyrosine kinase (PKI166). The efficacy of PKI166 was then tested either as a single agent or in combination with chemotherapy in TGF-α/EGF-positive 263J-BV bladder cancer and TGF-α/EGF-negative SN12-PM6 renal cell carcinoma cells growing in the bladder and kidney of nude mice, respectively. Treatment with orally administered PKI166 alone, intraperitoneally administered paclitaxel alone (253J-BV), intraperitoneally administered gemcitabine alone (SN12-PM6), or combined PKI166 and paclitaxel (253J-BV) or gemcitabine (SN12-PM6) produced 60%, 32%, and 81% reductions in the volume of 253J-BV bladder tumors, respectively, and 26%, 23%, and 51% reductions in the tumor volume of SN12-PM6 kidney tumors, respectively. Immunohistochemical analysis using specific anti-EGFR antibodies and antibodies against activated EGFR showed that tumors from all treatment groups expressed similar levels of EGFR, whereas tumors from the untreated mice and the mice treated with chemotherapy alone stained positive for activated EGFR, indicating that tumor cells expressing activated EGFR served as a target for treatment. Finally, the CD31/TUNEL fluorescent double-labeling technique revealed an increase in apoptosis of tumor-associated endothelial cells only in the TGF-α/EGF-positive bladder tumors. No apoptotic endothelial cells were detected in the TGF-α/EGF-negative kidney tumors. These data indicated that the expression of EGFR (total and activated) on tumor-associated endothelial cells indeed depends on TGF-α/EGF expression by the tumors and provides another major target for therapy (78).

The selective nature of EGFR expression within tumors was also shown in an orthotopic model of prostate cancer metastasized to bone (77). After the prostate cancer cells were injected into the tibia of nude mice, prostate cancer cells growing adjacent to the bone expressed high levels of EGF, EGFR, and activated EGFR, whereas prostate cancer cells growing in the adjacent muscle did not (77). Even more striking was the finding that tumor-associated endothelial cells within the bone lesions expressed high levels of activated EGFR, whereas endothelial cells 1–2 mm away from the lesions did not. Oral administration of PKI166 alone or in combination with paclitaxel reduced the incidence and size of the bone tumors and the destruction of the bone. Treatment with PKI166 and paclitaxel also significantly inhibited activation of the EGFR on tumor and tumor-associated endothelial cells and resulted in a significant increase in tumor and tumor-associated endothelial cell apoptosis (77).

The findings described here emphasize the important contribution of the tumor microenvironment to angiogenesis and are examples that support Paget's seed-and-soil hypothesis. The fact that endothelial cells exposed to a particular ligand produced by tumor cells express an activated receptor offers an exciting new approach to developing specific targeted therapies for human cancers.

5.6.2 Blockade of Platelet-Derived Growth Factor Receptor Signaling

PDGF and activation of its receptor, PDGFR, a tyrosine kinase, are associated with the growth and metastasis of many human carcinomas, including those of the stomach, pancreas, lung, and prostate (102). PDGFR is encoded by two genes, *PDGFR-α* and *PDGFR-β* (103) and is itself a potent mitogen in both normal and tumor cells (104). The binding of PDGF to the PDGFR can stimulate cell division (105–107), cell migration (108), and angiogenesis (109). The correlation between the level of expression of PDGF and a tumor's grade provides evidence that PDGF is involved in tumor progression. Compounds that inhibit PDGFR kinases, such as imatinib mesylate (STI571 or Gleevec [Novartis Pharmaceuticals, East Hanover, NJ]), have recently been developed. A recent study examined whether imatinib administered orally as a single agent or combined with intraperitoneally injected paclitaxel can inhibit the growth of androgen-independent human prostate cancer cells growing progressively in the tibia of nude mice (87). The data showed conclusively that these cells express high levels of PDGF, PDGFR, and activated PDGFR. In sharp contrast, prostate cancer cells growing in the musculature of the leg (distant from the bone environment) expressed low levels of PDGF and PDGFR. Furthermore, PDGFR was expressed on endothelial cells within tumor lesions in the bone but not in the muscle, and activated PDGFR also expressed on tumor-associated endothelial cells within the bone lesions but not on the tumor-associated endothelial cells in muscle located only 2–3 mm away from the bone lesions. This differential expression was likely caused by the production of PDGF by the tumor cell, or seed, adjacent to the bone, or soil, but not by the tumor cells growing at a distance from the bone. This provides more evidence that an organ's environment can influence the phenotype of tumor cells (11,110). Administration of imatinib and intraperitoneal paclitaxel resulted in decreased PDGFR activation, which correlated with increased apoptosis in endothelial cells and tumor cells that expressed PDGFR. No changes in PDGFR activation were observed in the tumor cells of untreated mice. Treatment with imatinib and paclitaxel significantly reduced bone lysis and the incidence of both tumors and lymph node metastasis (87).

One recent study examined the response of cultures of murine bone microvascular endothelial cells (75,76) and examined their response to stimulation with the PDGF BB ligand and to the blockade of PDGFR signaling with imatinib (76). The addition of imatinib to the cultures of those cells blocked PDGF BB-induced phosphorylation in a dose-dependent manner and completely abrogated the activation of downstream targets Akt and Erk1/2. Together, imatinib and paclitaxel also induced activation of procaspase-3 and downregulation of Bcl-2 and Bcl-xl. These data suggested that activation of PDGFR stimulated survival pathways in bone endothelial cells and that by selectively inhibiting PDGFR signaling with imatinib, the cells were rendered sensitive to the paclitaxel. This therapeutic combination may prove to be a powerful tool for targeting tumor-associated endothelial cells in the skeleton. Consequently, considerable effort is made to identify those components in the bone microenvironment that support tumor growth. Eventually, continued investigations into the signaling networks used by other types of tumors in the bone microenvironment should lead to the generation of effective therapeutic interventions.

Human pancreatic cancer cells growing in culture have been found to express high levels of PDGF AA and BB (111). Western blot analysis showed that the cancer cells also expressed low levels of PDGFR-α and PDGFR-β. However, when these cells were injected into the pancreas of nude mice, protein lysates from the resulting pancreatic tumors expressed high levels of PDGFR-α and PDGFR-β. Immunohistochemical analyses showed that for both PDGF AA and BB ligands and the PDGFR-α and PDGFR-β receptors were highly expressed, and that the receptors were activated. Resected specimens of human pancreatic adenocarcinomas revealed that all samples were positive for the PDGF ligands and receptors and that >90% of the pancreatic cancers analyzed expressed activated PDGFR. In contrast, only 10% of adjacent nonmalignant pancreatic samples were positive for activated PDGFR. The fact that the PDGFR-α and PDGFR-β levels were higher in the pancreatic tumor lysates than in the cultured cancer cells suggested that the tumor microenvironment and in vivo mechanisms are responsible for an autocrine loop in pancreatic cancers (111). The presence of PDGFR on pericytes and fibroblasts in the stromal tissue surrounding the pancreatic cancer may contribute to a paracrine loop, resulting in increased PDGF expression and growth stimulation of the tumor cells (104,112). Finally, double immunofluorescence staining has identified activated PDGFR on endothelial cells in clinical specimens of human pancreatic cancer and in liver metastases from primary pancreatic adenocarcinomas (112). In this same study, human pancreatic cancer cells growing in the pancreas of nude mice were treated with imatinib administered orally and gemcitabine injected intraperitoneally. After both agents, these orthotopic tumors were 70% smaller than the tumors in untreated mice and 36% smaller than those in mice treated with gemcitabine alone. The antitumor effect of imatinib was enhanced when combined with gemcitabine and was associated with increased apoptosis and decreased proliferation. Finally, we observed a decrease in microvessel density, suggesting that imatinib alone or in combination with gemcitabine-induced apoptosis in both the tumor cells and the dividing tumor-associated endothelial cells.

5.6.3 Blockade of Vascular Endothelial Growth Factor Receptor Signaling

Some tumor cells may constitutively overexpress angiogenic factors or they may respond to external stimuli. The most potent external stimulus of angiogenic factor expression is hypoxia (113,114), which is typically a consequence of poor perfusion. Increased angiogenic factor expression represents a survival response by cells. Hypoxia increases angiogenic factor expression by inducing signaling cascade pathways, which eventually lead to an increase in the transcription of VEGF/VPF and stabilization of the mRNA transcript.

Hypoxia induction of VEGF is probably mediated through Src kinase activity, which then leads to the downstream induction of signaling cascade pathways and eventually to increased activity of hypoxia-inducible factor Iα (HIF-Iα) (115). HIF-Iα then increases the transcription of the *VEGF/VPF* gene, which in turn, leads to the induction of angiogenesis (116–119). A mouse-specific monoclonal antibody (DC101) against the VEGF receptor (VEGFR) inhibited tumor growth, decreased metastasis, and increased the induction of apoptosis in both the tumor cells and the tumor-associated endothelial cells in human pancreatic cancer cells growing orthotopically in nude mice (120). All these data suggested that either growth factor deprivation or receptor signaling blockade leads to dual targeting within the tumor microenvironment.

One subsequent study showed inhibition of tumor growth and hepatic metastasis in human pancreatic cancer cells growing orthotopically in nude mice after therapy with gemcitabine and PTK787, a tyrosine kinase inhibitor of the VEGFR (121). Again, apoptosis of tumor cells and tumor-associated endothelial cells was observed. Combining PTK787 and PKI166 tyrosine kinase inhibitor with gemcitabine produced a 97% reduction in the pancreatic tumor volume in the orthotopic murine model (85). The efficacy of this therapy corresponded to a decrease in the circulating proangiogenic molecules VEGF and IL-8; a decrease in staining for dividing tumor cells as measured by proliferating nuclear antigen; and an increase in the apoptosis of tumor cells and endothelial cells (85).

5.6.4 Antiangiogenic Properties of Interferon-α

The IFN family consists of three major glycoproteins, which exhibit species specificity: leukocyte-derived IFN-α, fibroblast-derived IFN-β, and immune-cell-produced IFN-γ. IFNs were originally captured as antiviral agents but are now thought to regulate multiple biological activities, such as cell growth (122,123), cell differentiation (124), oncogene expression (125,126), host immunity (127–129), and tumorigenicity (130). A growing body of evidence has implicated IFNs in inhibiting a number of steps in angiogenesis. The antiangiogenic properties of IFN-α have been defined in studies using syngeneic murine tumor models and human tumors implanted into nude mice (131). As was the case with chronic administration of IFN-α in hemangiomas (33), frequent, low doses of free-form IFN-α administered to nude mice implanted with human bladder cancer cells resulted in the downregulation of angiogenesis-related genes *bFGF* and *MMP-9* and a reduction in the tumor burden (131). Both in vitro and in vivo, the downregulation of *bFGF* required a long exposure of the cells to low concentrations of IFNs; when the IFN was withdrawn, the cells resumed production of *bFGF*. High doses of IFN-α were not effective, indicating that the optimal biological dose was not the maximum tolerated dose (131).

This study was subsequently expanded to include therapy of pancreatic cancer growing orthotopically in nude mice. Therapy with low-dose IFN-α in combination with gemcitabine reduced pancreatic tumor volume by 87%, a reduction directly correlated with decreased expression of *bFGF* and *MMP-9* and with increased induction of apoptosis of the tumor cells and the tumor-associated endothelial cells (132). Similar findings were reported using low-dose pegylated IFN-α and paclitaxel in human ovarian cancer growing in the peritoneal cavity of nude mice (133) and pegylated IFN-α and docetaxel in prostate cancer growing orthotopically in the prostate of nude mice (134).

5.7 CONCLUSION

Tumor growth and metastasis depend on adequate blood supply. The extent of angiogenesis within and around tumors is regulated by the balance between proangiogenic molecules (e.g., bFGF, VEGF, VPF, IL-8) and antiangiogenic molecules (e.g., IFN, angiostatin, thrombospondin). The cross-talk between tumor cells, normal cells, leukocytes, and stromal cells occurs through growth factors, growth factor receptors, and cytokine signaling, all of which are released and act on cells within the tumor microenvironment. These interactions enhance the establishment of new blood supplies (neovascularization) and therefore regulate the proliferation and survival of metastatic cells. Interruption of one or more of these interactions can lead to the inhibition or regression of cancer metastasis. Therefore, understanding that angiogenesis is essential for tumor growth and metastasis formation has led to a huge effort to discover effective antiangiogenic compounds. Because primary tumor growth is often controlled with surgery or irradiation, antiangiogenic agents may be most beneficial to prevent widespread and metastatic disease. To succeed, several principles must be considered. First, physiologic angiogenesis is important in reproduction and in wound healing, as well as in physiological responses to cardiac ischemia or peripheral vascular disease. Thus, a balance must be maintained between the therapeutic and toxic effects, limiting angiogenesis in the tumor and preventing physiological angiogenesis crucial to homeostasis. Second, antiangiogenic therapy is complicated by its chronic nature, because this type of therapy is not cytotoxic, but rather, only prevents progressive growth of a tumor. Because antiangiogenic therapy is designed to inhibit the development of new blood vessels, the end points for success or failure must be redefined. For example, the early end point may require the measurement of surrogate markers and tumor stabilization, not shrinkage. The best clinical results may be obtained with a combination therapy, as observed in the treatment of many carcinomas described earlier. Nevertheless, strategies that target tumor cells and modulate the host microenvironment may yield promising new treatments for cancer and metastasis.

REFERENCES

1. Sugarbaker, E.V. Cancer metastasis: a product of tumor–host interactions. *Curr. Probl. Cancer*, 3, 1, 1979.
2. Weiss, L. *Principles of Metastasis*. Academic Press, Orlando, FL, 1985.
3. Fidler, I.J. Critical factors in the biology of human cancer metastasis: twenty-eighth G.H.A. Clowes memorial award lecture. *Cancer Res.*, 50, 6130, 1990.
4. Fidler, I.J. and Balch, C.M. The biology of cancer metastasis and implications for therapy. *Curr. Probl. Surg.*, 24, 131, 1987.
5. Folkman, J. and Klagsbrun, M. Angiogenic factors. *Science*, 235, 444, 1987.
6. Aukerman, S.L. and Fidler, I.J. Heterogeneous nature of metastatic neoplasms: relevance to biotherapy, in *Principles of Cancer Biotherapy*, Oldham, R.K., Ed., Marcel Dekker, New York, 23, 1991
7. Bishop, M. The molecular genetics of cancer. *Science*, 235, 305, 1987.
8. Liotta, L.A., Steeg, P.S., and Stetler-Stevenson, W.G. Cancer metastasis and angiogenesis: an imbalance of positive and negative regulation. *Cell*, 64, 327, 1991.
9. Fidler, I.J. Metastasis: quantitative analysis of distribution and fate of tumor emboli labeled with ^{125}I-5-iodo-2′-deoxyuridine. *J. Nat. Cancer Inst.*, 45, 773, 1970.
10. Fujimaki, T. et al. Selective growth of human melanoma cells in the brain parenchyma of nude mice. *Melanoma Res.*, 6, 363, 1996.
11. Paget, S. The distribution of secondary growths in cancer of the breast. *Lancet*, 1, 571, 1889.
12. Yu, J.L., Coomber, B.L., and Kerbel, R.S. A paradigm for therapy-induced microenvironmental changes in solid tumors leading to drug resistance. *Differentiation*, 70, 599, 2002.
13. Kerbel, R.S. Molecular and physiologic mechanisms of drug resistance in cancer: an overview. *Cancer Metastasis Rev.*, 20, 1, 2001.
14. Pak, B.J. et al. Lineage-specific mechanism of drug and radiation resistance in melanoma mediated by tyrosine-related protein 2. *Cancer Metastasis Rev.*, 20, 27, 2001.
15. St. Croix, B., Man, S., and Kerbel, R.S. Reversal of intrinsic and acquired forms of drug resistance by hyaluronidase treatment of solid tumors. *Cancer Lett.*, 1, 35, 1998.
16. Kerbel, R.S. Impact of multicellular resistance on the survival of solid tumors, including micrometastases. *Invasion Metastasis*, 14, 50, 1994.
17. Kerbel, R.S. et al. Multicellular resistance: a new paradigm to explain aspects of acquired drug resistance of solid tumors. *Cold Spring Harb. Symp. Quant. Biol.*, 59, 661, 1994.
18. Fidler, I.J. and Kripke, M.L. Metastasis results from preexisting variant cells within a malignant tumor. *Science*, 197, 893, 1977.
19. Fidler, I.J. and Talmadge, J.E. Evidence that intravenously derived murine pulmonary melanoma metastases can originate from the expansion of a single tumor cell. *Cancer Res.*, 46, 5167, 1986.
20. Fidler, I.J. and Ellis, L.M. The implication of angiogenesis to the biology and therapy of cancer metastasis. *Cell*, 47, 185, 1994.
21. Folkman, J. Angiogenesis in cancer, vascular, rheumatoid, and other disease. *Nat. Med.*, 1, 27, 1995.
22. Auerbach, W. and Auerbach, R. Angiogenesis inhibition: a review. *Pharmac. Ther.*, 63, 265, 1994.
23. Bouck, N., Stellmach, V., and Hsu, T.C. How tumors become angiogenic. *Adv. Cancer Res.*, 69, 135, 1996.
24. Folkman, J. Seminars in medicine of the Beth Israel Hospital, Boston. Clinical applications of research on angiogenesis. *N. Engl. J. Med.*, 333, 1757, 1995.
25. Cockerill, G.W., Gamble, J.R., and Vadas, M.A. Angiogenesis: models and modulators. *Int. Rev. Cytol.*, 159, 113, 1995.
26. O'Reilly, M.S. et al. Angiostatin: a novel angiogenesis inhibitor that mediates the suppression of metastases by a Lewis lung carcinoma. *Cell*, 79, 185, 1994.

27. O'Reilly, M.S. et al. Endostatin: an endogenous inhibitor of angiogenesis and tumor growth. *Cell*, 88, 277, 1997.
28. Miyata, Y. et al. Expression of thrombospondin-derived 4N1K peptide-containing proteins in renal cell carcinoma tissues is associated with a decrease in tumor growth and angiogenesis. *Clin. Cancer Res.*, 9, 1734, 2003.
29. Folkman, J. How is blood vessel growth regulated in normal and neoplastic tissue? G.H.A. Clowes memorial award lecture. *Cancer Res.*, 46, 467, 1986.
30. Singh, R.K. et al. Interferons alpha and beta downregulate the expression of basic fibroblast growth factor in human carcinomas. *Proc. Natl. Acad. Sci. USA* 92, 4562, 1995.
31. Hanahan, D. and Folkman, J. Patterns and emerging mechanisms of the angiogenic switch during tumorigenesis. *Cell*, 86, 353, 1996.
32. Fidler, I.J. Regulation of neoplastic angiogenesis monogram. *J. Natl. Cancer Inst.*, 28, 10, 2000.
33. Kerbel, R.S. Inhibition of tumor angiogenesis as a strategy to circumvent acquired resistance to anticancer therapeutic agents. *Bioassays*, 13, 31, 1991.
34. Liotta, L.A. and Stetler-Stevenson, W.G. Metalloproteinases and cancer invasion. *Semin. Cancer Biol.*, 1, 99, 1990.
35. Poste, G. and Fider, I.J. The pathogenesis of cancer metastasis. *Nature*, 283, 139, 1980.
36. Fidler, I.J. Experimental orthotopic models of organ-specific metastasis by human neoplasms. *Adv. Mol. Cell Biol.*, 9, 191, 1994.
37. Singh, R.K. et al. Organ site-dependent expression of basic fibroblast growth factor in human renal cell carcinoma cells. *Am. J. Pathol.*, 145, 365, 1994.
38. Singh, R.K. et al. Expression of interleukin 8 correlates with the metastatic potential of human melanoma cells in nude mice. *Cancer Res.*, 54, 3242, 1994.
39. Nanus, D.M. et al. Expression of basic fibroblast growth factor in primary human renal tumors: correlation with poor survival. *J. Natl. Cancer Inst.*, 85, 1587, 1994.
40. Nguyen, M. et al. Elevated levels of an angiogenic peptide, basic fibroblast growth factor, in the urine of patients with a wide spectrum of cancers. *J. Natl. Cancer Inst.*, 86, 356, 1994.
41. Kitadia, Y. et al. Multiparametric in situ mRNA hybridization analysis to detect metastasis-related genes in surgical specimens of human colon carcinoma. *Clin. Cancer Res.*, 1, 1095, 1995.
42. Kitadai, Y. et al. Multiparametric in situ mRNA hybridization analysis to predict disease recurrence in patients with colon carcinoma. *Am. J. Pathol.*, 149, 1541, 1996.
43. Bielenberg, D.R., Fidler, I.J., and Bucana, C.D. Constitutive expression of interferon-beta in differentiated epithelial cells exposed to environmental stimuli. *Cancer Biother. Radiopharmaceut.*, 13, 375, 1998.
44. Leek, R.D., Harris, A.L., and Lewis, C.E. Cytokine networks in solid human tumors: regulation of angiogenesis. *J. Leukoc. Biol.*, 56, 423, 1994.
45. Gutman, M. et al. Regulation of IL-8 expression in human melanoma cells by the organ environment. *Cancer Res.*, 55, 2470, 1995.
46. Takahashi, Y. et al. Site-dependent expression of vascular endothelial growth factor, angiogenesis, and proliferation in human gastric carcinoma. *Int. J. Oncol.*, 8, 701, 1996.
47. Sidky, Y.A. and Auerbach, R. Lymphocyte-induced angiogenesis in tumor-bearing mice. *Science*, 192, 1237, 1976.
48. Menininger, I.J. and Zetter, B.R. Mast cells and angiogenesis. *Semin. Cancer Biol.*, 3, 73, 1992.
49. Fidler, I.J. Lymphocytes are not only immunocytes. *Biomedicine*, 32, 1, 1980.
50. Fidler, I.J., Gersten, D.M., and Kripke, M.L. Influence of the immune status on the metastasis of three murine fibrosarcomas of different immunogenicities. *Cancer Res.*, 39, 3816, 1979.
51. Miguez, M., Davel, L., and deLustig, E.S. Lymphocyte-induced angiogenesis: correlation with the metastatic incidence of two murine mammary adenocarcinomas. *Invasion Metastasis*, 6, 313, 1986.
52. Freeman, M.R. et al. Peripheral blood T-lymphocytes and lymphocytes infiltrating human cancers express vascular endothelial growth factor: a potential role for T cells in angiogenesis. *Cancer Res.*, 55, 4140, 1995.
53. Polverini, P. et al. Activated macrophages induce vascular proliferation. *Nature*, 269, 804, 1977.
54. Sunderkotter, C. et al. Macrophages and angiogenesis. *J. Leukoc. Biol.*, 55, 410, 1994.
55. Ruiter, D.J. et al. Major histocompatibility antigens and the mononuclear inflammatory infiltrate in benign nevomelanocytic proliferation and malignant melanoma. *J. Immunol.*, 129, 2808, 1982.

56. Brocker, E.G. et al. Macrophages in melanocytic naevi. *Arch. Dermatol. Res.*, 284, 127, 1992.
57. Kreisle, R.A., Stebler, B.A., and Ershler, W.B. Effect of host age on tumor-associated angiogenesis in mice. *J. Natl. Cancer Inst.*, 82, 44, 1990.
58. Ding, A., Hwang, S., and Schwab, R. Effect of aging on murine macrophages: diminished response to IFN-g for enhanced oxidative metabolism. *J. Immunol.*, 153, 2146, 1994.
59. Goidl, E., Ed. *Aging and the Immune Response: Cellular and Humoral Aspects.* Marcel Dekker, New York, 1987.
60. Weskler, M.E. and Schwab, R. The immunogenetics of immune senescence. *Exp. Clin. Immunogenet.*, 199, 182, 1992.
61. Gutman, M. et al. Leukocyte-induced angiogenesis and subcutaneous growth of B16 melanoma. *Cancer Biother.*, 9, 163, 1994.
62. Takahashi, Y. et al. Platelet derived endothelial cell growth factor in human colon cancer angiogenesis: role of infiltrating cells. *J. Natl. Cancer Inst.*, 88, 1146, 1996.
63. Polverini, J.J. and Leibovich, S.J. Induction of neovascularization in vivo and endothelial cell proliferation in vitro by tumor-associated macrophages. *Lab. Invest.*, 51, 635, 1984.
64. Sunderkotter, C. et al. Macrophage-derived angiogenic factors. *Pharmacol. Ther.*, 51, 195, 1991.
65. DiPietro, L.A. and Polverini, P.J. Angiogenic macrophages produce the angiogenic factor thrombospondin 1. *Am. J. Pathol.*, 143, 678, 1993.
66. Polverini, P.J. How the extracellular matrix and macrophages contribute to angiogenesis-dependent diseases. *Eur. J. Cancer*, 32A, 2430, 1996.
67. Lingen, M.W., Polverini, P.J., and Bouck, N. Inhibition of squamous cell carcinoma angiogenesis by direct interaction of retinoic acid with endothelial cells. *Lab. Invest.*, 74, 476, 1996.
68. Lingen, M.W., Polverini, P.J., and Bouck, N. Retinoic acid induces cells cultured from oral squamous cell carcinomas to become antiangiogenic. *Am. J. Pathol.*, 149, 247, 1996.
69. Dong, Z. et al. Macrophage-derived metalloelastase is responsible for the generation of angiostatin in Lewis lung carcinoma. *Cell*, 88, 801, 1997.
70. Kumar, R., Dong, Z., and Fidler, I.J. Differential regulation of metalloelastase activity in murine peritoneal macrophages by GM-CSF and M-CSF. *J. Immunol.*, 157, 5104, 1996.
71. Eberhard, A. et al. Heterogeneity of angiogenesis and blood vessel maturation in human tumors: implications for antiangiogenic tumor therapies. *Cancer Res.*, 60, 1388, 2000.
72. Nels, V., Denzer, K., and Drenchahn, D. Pericyte involvement in capillary sprouting during angiogenesis in situ. *Cell Tissue Res.*, 270, 469, 1992.
73. Nor, J.E. and Polverni, P.J. Role of endothelial cell survival and death signals in angiogenesis. *Angiogenesis*, 3, 101, 1993.
74. Pasqualini, R., Arap, W., and McDonald, D.M. Probing the structural and molecular diversity of tumor vasculature. *Trends Mol. Med.*, 8, 563, 2002.
75. Langley, R.R. et al. Tissue-specific microvascular endothelial cell lines from $H\text{-}2K^b\text{-}ts$A58 mice for studies of angiogenesis and metastasis. *Cancer Res.*, 63, 2971, 2003.
76. Langley, R.R. et al. Activation of the platelet-derived growth factor-receptor enhances survival of murine bone endothelial cells. *Cancer Res.*, 11, 3727, 2004.
77. Kim, S.J. et al. Blockade of epidermal growth factor receptor signaling in tumor cells and tumor-associated endothelial cells for therapy of androgen-independent human prostate cancer growing in the bone of nude mice. *Clin. Cancer Res.*, 3, 1200, 2003.
78. Baker, C.H. et al. Blockade of epidermal growth factor receptor signaling on tumor cells and tumor-associated endothelial cells for therapy of human carcinomas. *Am. J. Pathol.*, 161, 929, 2002.
79. Bancroft, C.C. et al. Effects of pharmacologic antagonists of epidermal growth factor receptor, PI3K and MEK signal kinases on NF-kappaB and AP-1 activation and IL-8 and VEGF expression in human head and neck squamous cell carcinoma lines. *Int. J. Cancer*, 99, 538, 2002.
80. Busse, D. et al. Tyrosine kinase inhibitors: rationale, mechanisms of action, and implications for drug resistance. *Semin. Oncol.*, 28, 47, 2001.
81. Arteaga, C.L. The epidermal growth factor receptor: from mutant oncogene in nonhuman cancers to therapeutic target in human neoplasia. *J. Int. Oncol.*, 19, 32s, 2001.
82. Bruns, C.J. et al. Blockade of epidermal growth factor receptor signaling by a novel tyrosine kinase inhibitor leads to apoptosis of endothelial cells and therapy of human pancreatic carcinoma. *Cancer Res.*, 60, 2926, 2000.

83. Solorzano, C.C. et al. Optimization for the blockade of epidermal growth factor receptor signaling for therapy of human pancreatic carcinoma. *Clin. Cancer Res.* 7, 2563, 2001.
84. Bruns, C.J. et al. Epidermal growth factor receptor blockade with C225 plus gemcitabine results in regression of human pancreatic carcinoma growing orthotopically in nude mice by antiangiogenic mechanisms. *Clin. Cancer Res.*, 6, 1936, 2000.
85. Baker, C.H., Solorzano, C.C., and Fidler, I.J. Blockade of the vascular endothelial growth factor receptor and epidermal growth factor receptor signaling for therapy of metastatic human pancreatic cancer. *Cancer Res.*, 62, 1996, 2002.
86. Kim, S.J. et al. Simultaneous blockade of platelet-derived growth factor-receptor and epidermal growth factor-receptor signaling and systemic administration of paclitaxel as therapy for human prostate cancer metastasis in bone of nude mice. *Cancer Res.*, 12, 4201, 2004.
87. Uehara, H. et al. Effects of blocking platelet-derived growth factor-receptor signaling in a mouse model of experimental prostate cancer bone metastases. *J. Natl. Cancer Inst.*, 95, 58, 2003.
88. Kullenberg, B. et al. Transforming growth factor alpha increases cell number in a human pancreatic cancer cell line but not in normal mouse pancreas. *Int. J. Pancreatol.*, 3, 199, 2000.
89. Xiong, H.Q. and Abbruzzese, J.L. Epidermal growth factor receptor-targeted therapy for pancreatic cancer. *Semin. Oncol.*, 29, 31, 2002.
90. Yamanaka, Y. et al. Coexpression of epidermal growth factor receptor and ligands in human pancreatic cancer is associated with enhanced tumor aggressiveness. *Anticancer Res.*, 13, 565, 1993.
91. Honegger, A.M. et al. A mutant epidermal growth factor receptor with defective protein tyrosine kinase is unable to stimulate proto-oncogene expression and DNA synthesis. *Mol. Cell. Biol.*, 7, 4568, 1987.
92. Honegger, A.M. et al. Point mutation at the ATP binding site of EGF receptor abolishes protein-tyrosine kinase activity and alters cellular routing. *Cell*, 51, 199, 1987.
93. Redemann, N. et al. Anti-oncogenic activity of signaling-defective epidermal growth factor receptor mutants. *Mol. Cell. Biol.*, 12, 491, 1992.
94. Fry, D.W. et al. A specific inhibitor of the epidermal growth factor receptor tyrosine kinase. *Science*, 265, 1093, 1994.
95. Osherov, N. and Levitzki, A. Epidermal growth factor-dependent activation of the src-family kinases. *Eur. J. Biochem.*, 225, 1047, 1994.
96. Wakeling, A.E. et al. Specific inhibition of epidermal growth factor receptor tyrosine kinase by 4-anilinoquinazolines. *Breast Cancer Res. Treat.*, 38, 67, 1996.
97. Wakeling, A.E. et al. ZD1839 (Iressa): an orally active inhibitor of epidermal growth factor signaling with potential for cancer therapy. *Cancer Res.*, 62, 749, 2002.
98. Ciardiello, F. et al. Inhibition of growth factor production and angiogenesis in human cancer cells by ZD1839 (Iressa), a selective epidermal growth factor receptor tyrosine kinase inhibitor. *Clin. Cancer Res.*, 7, 1459, 2001.
99. Traxler, P. et al. Preclinical profile of PKI166—a novel and potent EGFR tyrosine kinase inhibitor for clinical development. *Clin. Cancer Res.*, 5, 3750s, 1999.
100. Petit, A.M. et al. Neutralizing antibodies against epidermal growth factor and ErbB-2/neu receptor tyrosine kinases downregulate vascular endothelial growth factor production by tumor cells in vitro and in vivo: angiogenic implications for signal transduction therapy of solid tumors. *Am. J. Pathol.*, 151, 1523, 1997.
101. Inoue, K. et al. Paclitaxel enhances the effects of the anti-epidermal growth factor receptor monoclonal antibody ImClone C225 in mice with metastatic human bladder transitional cell carcinoma. *Clin. Cancer Res.*, 6, 4874, 2000.
102. Schiffer, C.A. Signal transduction inhibition: changing paradigms in cancer care. *Semin. Oncol.*, 28, 34, 1986.
103. Ross, P., Raines, E.W., and Bowen-Pope, D.F. The biology of platelet-derived growth factor. *Cell*, 46, 155, 1986.
104. Heldin, C.H. and Westermark, B. Mechanism of action and in vivo role of platelet-derived growth factor. *Physiol. Rev.*, 79, 1283, 1999.
105. Xie, J. et al. A role of PDGFR-beta in basal cell carcinoma proliferation. *Proc. Natl. Acad. Sci. USA*, 98, 9255, 2001.

106. Fuma, K. et al. Expression of platelet-derived growth factor alpha receptors on stromal tissue cells in human carcinoid tumors. *Cancer Res.*, 50, 747, 1990.
107. Liu, Y.C. et al. Platelet-derived growth factor is an autocrine stimulator for the growth and survival of human esophageal carcinoma cell lines. *Exp. Cell Res.*, 228, 206, 1996.
108. Bornfeldt, K.E. et al. Insulin-like growth factor-I and platelet-derived growth factor-BB induce directed migration of human arterial smooth muscle cells via signaling pathways that are distinct from those of proliferation. *J. Clin. Invest.*, 93, 1266, 1994.
109. Plate, K.H. et al. Platelet-derived growth factor receptor-beta is induced during tumor development and upregulated during tumor progression in endothelial cells in human gliomas. *Lab. Invest.*, 67, 529, 1994.
110. Fidler, I.J. et al. The seed and soil hypothesis: vascularization and brain metastasis. *Lancet Oncol.*, 3, 57, 2002.
111. Ebert, M. et al. Induction of platelet-derived growth factor A and B chains and over-expression of their receptors in human pancreatic cancer. *Int. J. Cancer*, 62, 529, 1995.
112. Hwang, R.F. et al. Inhibition of platelet-derived growth factor phosphorylation by STI571 (Gleevec) reduces growth and metastasis of human pancreatic carcinoma in an orthotopic nude mouse model. *Clin. Cancer Res.*, 9, 6534, 2003.
113. Shweiki, D. et al. Vascular endothelial growth factor induced by hypoxia may mediate hypoxia-initiated angiogenesis. *Nature*, 359, 843, 1992.
114. Levy, A.P. et al. Transcriptional regulation of the rat vascular endothelial growth factor gene by hypoxia. *J. Biol. Chem.*, 270, 13333, 1995.
115. Ellis, L.M. et al. Downregulation of vascular endothelial growth factor in human colon carcinoma cell line transfected with an antisense expression vector specific for c-src. *J. Biol. Chem.*, 273, 1052, 1998.
116. Marti, H.J. et al. Hypoxia-induced vascular endothelial growth factor expression precedes neovascularization after cerebral ischemia. *Am. J. Pathol.*, 156, 965, 2000.
117. Minet, E. et al. Role of HIF-1 as a transcription factor involved in embryonic development, cancer progression and apoptosis. *Int. J. Molec. Med.*, 5, 252, 2000.
118. Kimura, H. et al. Hypoxia response element of the human vascular endothelial growth factor gene mediated transcriptional regulation by nitric oxide: control of hypoxia-inducible factor-1 activity by nitric oxide. *Blood*, 95, 189, 2000.
119. Ryan, H.E., Lo, J., and Johnson, R.S. HIF-1α is required for solid tumor formation and embryonic vascularization. *EMBO J.*, 17, 3005, 1998.
120. Bruns, C.J. et al. Effect of the vascular endothelial growth factor receptor-2-antibody DC101 plus gemcitabine on growth, metastasis, and angiogenesis of human pancreatic cancer growing orthotopically in nude mice. *Int. J. Cancer*, 102, 101, 2002.
121. Solorzano, C.C. et al. Inhibition of growth and metastasis of human pancreatic cancer growing in nude mice by PTK787/ZK222584, an inhibitor of the vascular endothelial growth factor receptor tyrosine kinases. *Cancer Biother. Radiopharm.*, 16, 359, 2002.
122. Yaar, M. et al. Effects of alpha and beta interferons on cultured human keritinocytes. *J. Invest. Dermatol.*, 85, 70, 1985.
123. Tamm, I., Lin, S.L., Pfeffer, L.M., and Sehgal, P.B. Interferons α and β as cellular regulatory molecules, in *Interferon 9*, Gresser, I., Ed., Academic Press, London, 13, 1987.
124. Rossi, G. Interferons and cell differentiation, in *Interferon 6*, Gresser, I., Ed., Academic Press, London, 31, 1985.
125. Chatterjee, D. and Savarese, T.M. Posttranscriptional regulation of c-myc proto-oncogene expression and growth inhibition by recombinant human interferon-β ser17 in a human colon carcinoma cell line. *Cancer Chemother. Pharmacol.*, 30, 12, 1992.
126. Reznitzky, D., Yarden, A., Zipori, D., and Kimichi, A. Autocrine β-related interferon controls c-myc suppression and growth arrest during hematopoetic cell differentiation. *Cell*, 46, 31, 1986.
127. deMaeyer-Guignard, J. and deMaeyer, E. Immunomodulation by interferons: recent developments, in *Interferon 6*, Gresser, I., Ed., Academic Press, London, 1985, 69.
128. Strander, H. Interferon treatment of human neoplasia: effects on the immune system. *Adv. Cancer Res.*, 46, 36, 1986.

129. Gresser, I. et al. Host humoral and cellular immune mechanisms in the continued suppression of Friend erythroleukemia metastasis after interferon α/β treatment in mice. *J. Exp. Med.*, 173, 1193, 1991.
130. Fleischmann, W.R. and Fleischmann, C.M. Mechanisms of interferons antitumor actions, in *Interferon: Principles and Medical Applications*, Baron, S., Coppenhaver, D.H., Dianzani, F., Fleischmann, W.R., Jr., Hughs, T.K., Jr., Klimpel, G.R. et al., Eds., University of Texas Press, UTMB-Galveston, TX, 299, 1992.
131. Slaton, J.W. et al. Interferon-α-mediated downregulation of angiogenesis-related genes and therapy of bladder cancer are dependent on optimization of biological dose and schedule. *Clin. Cancer Res.*, 5, 2726, 1999.
132. Solorzano, C.C. et al. Administration of optimal biological dose and schedule of interferon alpha combined with gemcitabine induces apoptosis in tumor-associated endothelial cells and reduces growth of human pancreatic carcinoma implanted orthotopically in nude mice. *Clin. Cancer Res.*, 9, 1858, 2003.
133. Tedjarati, S. et al. Synergistic therapy of human ovarian carcinoma implanted orthotopically in nude mice by optimal biological dose of pegylated interferon alpha combined with paclitaxel. *Clin. Cancer Res.*, 8, 2413, 2002.
134. Huang, S.F. et al. Inhibition of growth and metastasis of orthotopic human prostate cancer in athymic nude mice by combination therapy with pegylated interferon-alpha-2b and docetaxel. *Cancer Res.*, 62, 5720, 2002.

Part II

Targeting Angiogenesis for Cancer Therapy

6 Tyrosine Kinase Inhibitors of Angiogenesis

Janessa J. Laskin and Alan B. Sandler

CONTENTS

The premise behind targeted therapy is to identify factors that make a cancer cell unique and to take advantage of those properties. The upregulation of angiogenesis is one general example of this idea, and the blockage of the tyrosine kinase component of angiogenesis receptors is a very specific example of this theory in practice.

6.1 BIOLOGY OF TYROSINE KINASE INHIBITION

Broadly speaking, tyrosine kinases (TKs) can be divided into two main categories based on whether or not they are directly associated with a transmembranous receptor. Receptor TKs are transmembrane proteins with an extracellular binding domain for ligands and an intracellular kinase domain, which, when activated, stimulates an intracellular signaling cascade. Nonreceptor TKs are not directly associated with proteins on the cell membrane but are primarily found in the cytoplasm, nucleus, and on the inner surface of the plasma membrane.[1]

In noncancerous and nonproliferative cells, both kinds of TKs are under strict regulation by an intricate system of negative feedback loops, and rigid stereotactic conformations. Given the complexity of TK regulation, there are multiple ways in which TKs can become dysregulated in cancer cells. For example, a subset of patients with nonsmall cell lung cancer was noted to be exquisitely sensitive to TK inhibition because of activating mutations found in the TK domain of epidermal growth factor receptor (EGFR) gene.[2–4] These small mutations caused a slight conformational change in the TK domain and this led to an increased cellular reliance on EGFR and thus an increased sensitivity to EGFR TK inhibition. Other mechanisms of deregulation include the fusion of the TK with a stimulatory protein such that the TK is always "on," for example, the BCR–ABL complex in chronic myelogenous leukemia; a mutation that results in the loss of TK autoregulation so that the TK can be activated without ligand binding (e.g., Fms-like TK 3 receptor in acute myeloid leukemia [AML]); and the overexpression of the TK receptor itself as in HER-2/*neu* in breast cancer.[1]

The loss of control of TK activity can affect many aspects of cellular function and, particularly in cancer cells, can lead to decreased apoptosis, dysregulated proliferation, invasiveness, and increased angiogenesis. This suggests that TKs are a rationale and important target for cancer therapy. This strategy is particularly attractive because many of these agents are orally bioavailable small molecule inhibitors of TKs, which are often relatively easy to manufacture and there are obvious practical advantages to oral chemotherapies. Thus, there has been an explosion in the production of TK inhibitors and it is currently impossible to predict which strategies will successfully evolve out of the abundance of preclinical work done. At this point in time there are two main routes that are at the forefront of clinical trials, the vascular endothelial growth factor (VEGF) pathway and the platelet-derived growth factor (PDGF) pathway. This chapter emphasizes these two interrelated domains and the associated receptor TK inhibitors, which are in clinical development since they encompass the majority of our current clinical experience.

6.2 VASCULAR ENDOTHELIAL GROWTH FACTOR

The interaction between VEGF and its TK receptor is one of the most important processes regulating angiogenesis. The VEGF produced by tumor and associated stromal cells acts on endothelial cells promoting proliferation, migration, and invasion leading to angiogenesis.[5] VEGF itself is a homodimeric protein with at least five isoforms, VEGF-A through VEGF-E and placental growth factor; though in reality there are likely numerous VEGF variants (see Chapter 5).[6,7] Similarly, three distinct isoforms of the VEGF receptor (VEGFR) have been described; each with their own individual role to play in angiogenesis. Unfortunately, the nomenclature of these ligands and receptors is somewhat complicated. VEGFR-1, also known as Fms-like TK-1 (Flt-1), has the highest binding affinity for VEGF-A. VEGFR-2, also known as kinase domain region (KDR) or fetal liver kinase-1 (Flk-1), is most commonly associated with endothelial cell proliferation. Finally, VEGFR-3, also known as Flt-4, primarily regulates lymphangiogenesis.[6]

The development of VEGF TK inhibitors has been somewhat problematic. A number of VEGFR TK inhibitors have shown promise in preclinical models and early trials but have ultimately failed to make a significant clinical impact. For example, one of the earliest TK inhibitors of VEGFR-2 (KDR) was SU5416 (semaxanib). In the preclinical setting, SU5416 inhibited the growth and metastasis of a wide variety of solid tumors.[8–10] Phase I and II trials were particularly interesting in patients with Kaposi's sarcoma and colorectal cancer leading to further testing.[11–16] Unfortunately, semaxanib failed to improve survival when added to combination chemotherapy in a large randomized phase III study in patients with metastatic colorectal cancer.[17] Because of the logistics and toxicity of drug administration, further development of semaxanib was abandoned. This experience exemplifies the frustration investigators have had with biologically targeted therapies; but fortunately, there are many more in production.

Since the VEGF pathway involves several ligands and receptors, some of the newer agents are designed to target multiple VEGF Rs. One of these drugs, vatalanib (PTK787/ZK 222584), is a TK inhibitor of all of the currently known VEGFR TKs, primarily VEGFR-2, but also VEGFR-1 and VEGFR-3, in addition to the PDGF receptor (PDGFR) TK and the c-kit protein TK. This agent has good oral bioavailability and a series of phase I studies have been conducted in a number of clinical settings.[18,21] For example, Thomas et al. conducted a phase I single agent trial in 43 patients with advanced solid tumors.[18] Twenty-six patients were entered into the dose escalation portion of the trial and a further 17 were treated in the expansion phase at the maximally tolerated dose. The majority of patients (24) had colorectal cancer. Although not designed to look at efficacy, of the 36 patients evaluable for response,

24 patients had at least stable disease as their best response and some very durable responses were noted. This oral agent was generally well tolerated and the dose-limiting toxicity was lightheadedness. Additional side effects common to other phase I studies included mild nausea, fatigue, dizziness, headache, and ataxia. Phase I and II studies of vatalanib in combination with a variety of chemotherapy agents have also been conducted and no significant unexpected toxicities or detrimental effects of the combination were noted.[21–23] These studies have led to ongoing phase III trials in the first- and second-line treatment of metastatic colorectal cancer, for example, the colorectal oral novel therapy for inhibition of angiogenesis and retarding of metastasis CONFIRM-1 and -2 (CONFIRM-1 in first and CONFIRM-2 in second line). CONFIRM-1 has already completed enrollment in June 2004 with 1168 patients randomized to treatment with oxaliplatin/5-fluorouracil/leucovorin (FOLFOX-4) plus vatalanib or placebo. Preliminary results suggest that the combination was tolerable with no increase in bleeding or bowel perforation in the vatalanib arm compared to the placebo. There was a trend for improvement in progression-free survival associated with the experimental arm and the final results are eagerly anticipated.[24]

Vatalanib has been particularly exciting in hematologic malignancies, in particular in patients with relapsed or refractory AML it had impressive single agent activity and in combination with chemotherapy where it produced complete remissions in patients with secondary AML or myelodysplastic syndrome.[25] Further clinical trials in hematological malignancies are also ongoing.

ZD6474 is a novel, orally available, potent TK inhibitor of the VEGFR; principally it acts on VEGFR-2. Though considered primarily a VEGF inhibitor, interestingly, at higher concentrations it appears to have in vitro inhibitory activity against the EGFR as well.[26,27] Phase I studies demonstrated the safety and tolerability of this agent with toxicities such as diarrhea and thrombocytopenia at higher dose levels and rash and asymptomatic QTc prolongation (on electrocardiograms) noted at the recommended phase II dose level. There have been a number of phase II trials of ZD6474 alone and in combination with traditional chemotherapy in a variety of tumor types.[28,29] One example is an ongoing trial of ZD6474 as maintenance chemotherapy in small cell lung cancer (SCLC). Despite advances in radiation and chemotherapy the majority of patients with SCLC still die of their disease, which tends to recur within the first 1–2 years after initial treatment providing the rationale for a trial of a well-tolerated oral maintenance therapy. Elevated serum VEGF levels are thought to be a bad prognostic indicator in SCLC and this is one surrogate marker that is evaluated in this study.[30]

A novel therapeutic strategy explored with ZD6474 is the delivery of sequential targeted agents, for example, the phase II study of the oral EGFR TK inhibitor gefitinib in patients with previously treated advanced NSCLC. In this study, 168 patients were randomized to receive either gefitinib ($N = 85$) or ZD6474 ($N = 83$) and at the time of disease progression patients had the opportunity to switch to the alternate agent.[31] In the preliminary reports of this study, the median time to progression was 12 weeks and 8 weeks ($p < 0.011$) for the initial treatment with ZD6474 and gefitinib, respectively, and the final analysis is pending.

AZD2171 is a highly potent and orally available inhibitor of VEGFR TK activity, which has shown antitumor activity in a wide range of tumor xenograft models.[32] Increased activity was noted in colorectal and nonsmall cell lung cancers leading to ongoing phase II studies of AZD6171 alone and in combination with chemotherapy as well as with other targeted agents such as gefitinib.[33]

In summary, a number of receptor TK inhibitors of the VEGF family have been developed and are at various stages of clinical testing. Fatigue, diarrhea, and hypertension appear to be common toxic effects of the VEGFR TK inhibitors observed in phase I and II clinical trials. Which cancers benefit most from this approach remains to be defined.

6.3 PLATELET-DERIVED GROWTH FACTOR

The PDGF family consists of four polypeptides, A through D, which form dimeric proteins that signal through two TK receptors PDGFR-α and PDGFR-β. The ligands and receptors can form homo- or heterodimers depending on the availability of either the cell type or the expression of the receptor. Like the VEGF pathway, the interactions of PDGFR and its ligands are complex and occur through autocrine and paracrine stimulation; this pathway is critical for angiogenesis.[34] For example, PDGF-β is expressed on pericytes, endothelial, and the smooth muscle cells that provide support for the development of functional vasculature.

Some new agents in development are more focused on targeting small but key components of the PDGF pathway such as CP-673,451, which selectively inhibits PDGF-BB-stimulated angiogenesis in multiple human xenograft models.[35] These models may help tease out the relative impact of the various components of angiogenesis. Such agents are still early in clinical development but should be watched with interest.

6.4 MULTITARGETED AGENTS

Intuitively, the inhibition of one receptor or even one aspect of angiogenesis does not routinely overcome the complexity of these dysregulated processes. Thus the next generations of TK inhibitors are designed to be multitargeted. Indeed, one might contest that there are so many TKs that it is impossible to isolate only one or two even with a highly selective targeted therapy and since cancers are so heterogeneous it is perhaps a prudent strategy to target this system in a more comprehensive manner.

SU6668 was one of the earliest multitargeted TK inhibitors in clinical testing, with activity, at least in vitro, against VEGFR-2, PDGFR, fibroblast growth factor receptor-1, and stem cell factor receptor (c-kit).[36] Although it was well tolerated in phase I trials and there was some anticancer activity noted, the pharmacokinetic analysis revealed a short plasma half-life, with plasma concentrations exceeding those that inhibited TK activity in preclinical models in the first day, but not following prolonged administration. Due to these disappointing results, the compound was withdrawn from further development.[37,38]

SU011248 (sunitinib) is an orally administered small molecule designed to target a wide variety of receptor TKs including PDGFR, VEGFR, KIT, and FLT-3.[39] This agent is widely tested in a number of solid tumors and so far in phase I and II trials has been given to ~400 patients. Fatigue is the primary toxicity, though as with many of these agents some nausea, diarrhea, hypertension, and asthenia have been noted.[40,41] In early reports, sunitinib appears to have great potential in metastatic renal cancer. An abstract presented by Motzer et al. described the preliminary results for 2 sequential phase II trials in 169 patients with previously treated metastatic renal cancer.[42] In addition to the previously observed toxicities of fatigue, nausea, and diarrhea, a modest degree of neutropenia and anemia was also observed, perhaps related to patients' prior therapies. The response rate for these two trials was ~40% (with one complete response and the remainder achieving partial responses). In addition, ~25% of patients had stable disease lasting for longer than 3 months. Because of these intriguing results, sunitinib is now investigated as first-line therapy in metastatic renal cell cancers. Sunitinib is also actively tested in combination with other targeted cancer treatments; for example, there is an ongoing phase II trial of sunitinib plus gefitinib (the EGFR inhibitor) in metastatic renal cancer, and a phase II trial of maintenance sunitinib in patients with advanced nonsmall cell lung cancer following standard carboplatin/taxol chemotherapy.

There are a plethora of similar multitargeted novel agents in various stages of clinical development (Table 6.1). Although dosing schedules and toxicities are different, currently it is difficult to discriminate on activity since they have yet to be tested in a randomized phase III setting.

TABLE 6.1
Receptor Tyrosine Kinase Inhibitors of Angiogenesis and Their Primary Targets

Agent	VEGFR	PDGFR	Other
AG103736 (Pfizer)	VEGFR-1, VEGFR-2	Yes	Kit
AG13925 (Pfizer)	VEGFR-1, VEGFR-2		
AEE788 (Novartis)	VEGFR-2		EGFR
AMG706 (Amgen)	VEGFR-1, VEGFR-2, VEGFR-3	Yes	Kit, Ret
AZD2171 (Astra-Zeneca)	VEGFR-1, VEGFR-2	Yes	Kit; +/−EGFR
BAY93-1106 (Bayer)	VEGFR-1, VEGFR-2	Yes	Raf
CEP-7055 (Cephalon)	VEGFR-1, VEGFR-2, VEGFR-3		
CP-547,632 (Pfizer)	VEGFR-1, VEGFR-2		
GW786024 (Glaxo-Smithkline)	VEGFR-1, VEGFR-2, VEGFR-3		
PTK 787 (Novartis)	VEGFR-1, VEGFR-2	Yes	c-Kit
SU11248 (Pfizer)	VEGFR-1, VEGFR-2	Yes	
ZD6474 (Astra-Zeneca)	VEGFR-1, VEGFR-2, VEGFR-3		EGFR

6.5 PERSPECTIVES

With rare exception, combination therapy has always been the most successful strategy in cancer treatment. The TK inhibitors of angiogenesis are generally well tolerated and as oral medications they are relatively easy to deliver. It is likely that they will need to be given in combination with chemotherapy or together with other biologically targeted agents or perhaps as maintenance therapy to keep cancers at bay. With the majority of clinical trials in the early stages and abstract form only it remains to be seen which specific agents will eventually win out in phase III testing; however, it is clear that this antiangiogenic approach is a vital component of our cancer treatment armamentarium.

REFERENCES

1. Krause DS, Van Etten RA. Tyrosine kinases as targets for cancer therapy. *N Engl J Med* 2005;353:172–187.
2. Paez JG, Janne PA, Lee JC et al. EGFR mutations in lung cancer: Correlation with clinical response to gefitinib therapy. *Science* 2004;304:1497–1500.
3. Pao W, Miller V, Zakowski M et al. EGF receptor gene mutations are common in lung cancers from "never smokers" and are associated with sensitivity of tumors to gefitinib and erlotinib. *Proc Natl Acad Sci USA* 2004;101:13306–13311.
4. Lynch TJ, Bell DW, Sordella R et al. Activating mutations in the epidermal growth factor receptor underlying responsiveness of non-small-cell lung cancer to gefitinib. *N Engl J Med* 2004;350:2129–2139.
5. Ferrara N. Molecular and biological properties of vascular endothelial growth factor. *J Mol Med* 1999;77:527–543.
6. Herbst RS, Hidalgo M, Pierson AS, Holden SN, Bergen M, Eckhardt SG. Angiogenesis inhibitors in clinical development for lung cancer. *Semin Oncol* 2002;29:66–77.
7. Hicklin DJ, Ellis LM. Role of the vascular endothelial growth factor pathway in tumor growth and angiogenesis. *J Clin Oncol* 2005;23:1011–1027.
8. Fong TA, Shawver LK, Sun L et al. SU5416 is a potent and selective inhibitor of the vascular endothelial growth factor receptor (Flk-1/KDR) that inhibits tyrosine kinase catalysis, tumor vascularization, and growth of multiple tumor types. *Cancer Res* 1999;59:99–106.
9. Angelov L, Salhia B, Roncari L, McMahon G, Guha A. Inhibition of angiogenesis by blocking activation of the vascular endothelial growth factor receptor 2 leads to decreased growth of neurogenic sarcomas. *Cancer Res* 1999;59:5536–5541.

10. Shaheen RM, Davis DW, Liu W et al. Antiangiogenic therapy targeting the tyrosine kinase receptor for vascular endothelial growth factor receptor inhibits the growth of colon cancer liver metastasis and induces tumor and endothelial cell apoptosis. *Cancer Res* 1999;59:5412–5416.
11. Zangari M, Anaissie E, Stopeck A et al. Phase II study of SU5416, a small molecule vascular endothelial growth factor tyrosine kinase receptor inhibitor, in patients with refractory multiple myeloma. *Clin Cancer Res* 2004;10:88–95.
12. Stopeck A, Sheldon M, Vahedian M, Cropp G, Gosalia R, Hannah A. Results of a Phase I dose-escalating study of the antiangiogenic agent, SU5416, in patients with advanced malignancies. *Clin Cancer Res* 2002;8:2798–2805.
13. Kuenen BC, Tabernero J, Baselga J et al. Efficacy and toxicity of the angiogenesis inhibitor SU5416 as a single agent in patients with advanced renal cell carcinoma, melanoma, and soft tissue sarcoma. *Clin Cancer Res* 2003;9(5):1648–1655.
14. Rosen LM, Mulay M, Mayers A, Kabbinavar F, Rosen P, Cropp G, Hannah A. Phase I dose-escalating trial of SU5416, a novel angiogenesis inhibitor in patients with advanced malignancies. *Proc Am Soc Clin Oncol* (meeting abstract) 1999;18:161a (abstract 618).
15. Cropp GR, Rosen L, Mulay M, Langecker P, Hannah A. Pharmacokinetics and pharmacodynamics of SU5416 in a Phase I, dose escalating trial in patients with advanced malignancies. *Proc Am Soc Clin Oncol* (meeting abstract) 1999;18:161a (abstract 619).
16. O'Donnell A, Trigo J, Banerji U, Raynaud F, Padhani A, Hannah A, Hardcastle A, Aherne W, Workman P, Judson IR. A Phase I trial of the VEGF inhibitor SU5416, incorporating dynamic contrast MRI assessment of vascular permeability. *Proc Am Soc Clin Oncol* 2000;19:177 (abstract 685).
17. Pharmacia cancer drug halted [press release]. *Forbes* 2002.
18. Thomas AL, Morgan B, Horsfield MA et al. Phase I study of the safety, tolerability, pharmacokinetics, and pharmacodynamics of PTK787/ZK 222584 administered twice daily in patients with advanced cancer. *J Clin Oncol* 2005;23:4162–4171.
19. Drevs J, Mross K, Medinger M, Muller M, Laurent D, Reitsma D, Henry A, Xia J, Marme D, Unger C. Phase I dose-escalation and pharmacokinetic (PK) study of the VEGF inhibitor PTK787/ZK 222584 (PTK/ZK) in patients with liver metastases. *Proc Am Soc Clin Oncol* 2003;22 (abstract 1142).
20. George D, Michaelson D, Oh WK, Reitsma D, Laurent D, Mietlowski W, Wang Y, Dugan M, Kaelin WG, Kantoff P. Phase I study of PTK787/ZK 222584 (PTK/ZK) in metastatic renal cell carcinoma. *Proc Am Soc Clin Oncol* 2003;22:385 (abstract 1548).
21. Trarbach T, Schleucher N, Tewes M et al. Phase I/II study of PTK787/ZK 222584 (PTK/ZK), a novel, oral angiogenesis inhibitor in combination with FOLFIRI as first-line treatment for patients with metastatic colorectal cancer (CRC). *J Clin Oncol* (meeting abstract) 2005;23:3605.
22. Steward WP, Thomas A, Morgan B et al. Expanded phase I/II study of PTK787/ZK 222584 (PTK/ZK), a novel, oral angiogenesis inhibitor, in combination with FOLFOX-4 as first-line treatment for patients with metastatic colorectal cancer (meeting abstract). *J Clin Oncol* 2004;22:3556.
23. Reardon D, Friedman H, Yung WK, Brada M, Conrad C, Provenzale J, Jackson EF, Serajuddin H, Laurent D, Reitsma D. A phase I trial of PTK787/ZK 222584 (PTK/ZK), an oral VEGF tyrosine kinase inhibitor, in combination with either temozolomide or lomustine for patients with recurrent glioblastoma multiforme (GBM). *Proc Am Soc Clin Oncol* 2003;22 (abstract 412).
24. Hecht JR, Trarbach T, Jaeger E et al. A randomized, double-blind, placebo-controlled, phase III study in patients (Pts) with metastatic adenocarcinoma of the colon or rectum receiving first-line chemotherapy with oxaliplatin/5-fluorouracil/leucovorin and PTK787/ZK 222584 or placebo (CONFIRM-1). *J Clin Oncol* (meeting abstract) 2005;23:LBA3.
25. Roboz GJ, Giles FJ, List AF, Apostolopoulou E, Rae PE, Dugan M, Oasman SJ, Schuster MW, Laurent D, Feldman EJ. Phase I trial PTK787/ZK 222584 (PTK/ZK), an inhibitor of vascular endothelial growth factor receptor tyrosine kinases, in acute myelogenous leukemia (AML) and myelodysplastic syndrome (MDS). *Proc Am Soc Clin Oncol* 2003;22:568 (abstract 2284).
26. Hurwitz H, Holden SN, Eckhardt SG, Rosenthal M, deBoer R, Rischin D, Green M, Basser R. Clinical evaluation of ZD6474, an orally active inhibitor of VEGF signaling, in patients with solid tumors. *Proc Am Soc Clin Oncol* 2002;21:82a (abstract 325).

27. Minami H, Ebi H, Tahara M, Sasaki Y, Yamamoto N, Yamada Y, Tamura T, Saijo N. A phase I study of an oral VEGF receptor tyrosine kinase inhibitor ZD6474, in Japanese patients with solid tumors. *Proc Am Soc Clin Oncol* 2003;22:194 (abstract 778).
28. Heymach JV, Johnson BE, Rowbottom JA et al. A randomized, placebo-controlled phase II trial of ZD6474 plus docetaxel, in patients with NSCLC. *J Clin Oncol* (meeting abstract) 2005;23:3023.
29. Herbst R, Johnson B, Rowbottom JA, Fidias P, Lu C, Prager D, Roubec J, Csada E, Dimery I, Heymach JV. ZD6474 plus docetaxel in patients with previously treated NSCLC: Results of a randomized, placebo-controlled Phase II trial. *Lung Cancer* 2005;49(2):S35.
30. Salven P, Ruotsalainen T, Mattson K, Joensuu H. High pre-treatment serum level of vascular endothelial growth factor (VEGF) is associated with poor outcome in small-cell lung cancer. *Int J Cancer* 1998;79:144–146.
31. Natale R, Bodkin D, Govindan R, Sleckman B, Rizvi N, Capo A, Germonpré P, Dimery I, Webster A, Ranson M. A comparison of the antitumour efficacy of ZD6474 and gefitinib (Iressa™) in patients with NSCLC: Results of a randomized, double-blind Phase II study. *Lung Cancer* 2005;49(2):S36.
32. Drevs J, Medinger M, Mross K et al. Phase I clinical evaluation of AZD2171, a highly potent VEGF receptor tyrosine kinase inhibitor, in patients with advanced tumors. *J Clin Oncol* (meeting abstract) 2005;23:3002.
33. van Cruijsen H, Voest EE, van Herpen CML et al. Phase I clinical evaluation of AZD2171 in combination with gefitinib, in patients with advanced tumors. *J Clin Oncol* (meeting abstract) 2005;23:3030.
34. Ostman A, Heldin CH. Involvement of platelet-derived growth factor in disease: Development of specific antagonists. *Adv Cancer Res* 2001;80:1–38.
35. Roberts WG, Whalen PM, Soderstrom E et al. Antiangiogenic and antitumor activity of a selective PDGFR tyrosine kinase inhibitor, CP-673,451. *Cancer Res* 2005;65:957–966.
36. Laird AD, Vajkoczy P, Shawver LK et al. SU6668 is a potent antiangiogenic and antitumor agent that induces regression of established tumors. *Cancer Res* 2000;60:4152–4160.
37. Britten CD, Rosen LS, Kabbinavar F, Rosen P, Mulay M, Hernandez L, Brown J, Bello C, Kelsey SM, Scigalla P. Phase I trial of SU6668, a small molecule receptor tyrosine kinase inhibitor, given twice daily in patients with advanced cancers. *Proc Am Soc Clin Oncol* 2002;21 (abstract 1922).
38. Kuenen BC, Giaccone G, Ruijter R et al. Dose-finding study of the multitargeted tyrosine kinase inhibitor SU6668 in patients with advanced malignancies. *Clin Cancer Res* 2005;11(17):6240–6246.
39. Mendel DB, Laird AD, Xin X et al. In vivo antitumor activity of SU11248, a novel tyrosine kinase inhibitor targeting vascular endothelial growth factor and platelet-derived growth factor receptors: Determination of a pharmacokinetic/pharmacodynamic relationship. *Clin Cancer Res* 2003;9:327–337.
40. Rosen L, Mulay M, Long J, Wittner J, Brown J, Martino A-M, Bello CL, Walter S, Scigalla P, Zhu J. Phase I trial of SU011248, a novel tyrosine kinase inhibitor in advanced solid tumors. *Proc Am Soc Clin Oncol* 2003;22:191 (abstract 765).
41. Raymond E, Faivre S, Vera K, Delbaldo C, Robert C, Spatz A, Bello C, Brega N, Scigalla P, Armand JP. Final results of a phase I and pharmacokinetic study of SU11248, a novel multi-target tyrosine kinase inhibitor, in patients with advanced cancers. *Proc Am Soc Clin Oncol* 2003;22:192 (abstract 769).
42. Motzer RJ, Rini BI, Michaelson MD et al. Phase 2 trials of SU11248 show antitumor activity in second-line therapy for patients with metastatic renal cell carcinoma (RCC). *J Clin Oncol* (meeting abstract) 2005;23:4508.

7 Development of Antiangiogenic Monoclonal Antibodies for Cancer Therapy

Zhenping Zhu and Daniel J. Hicklin

CONTENTS

Angiogenesis is essential for the development and progression of malignant disease. Insight into the molecular basis of angiogenesis and the discovery of agents that disrupt this process are a key area of research and oncology drug development. Monoclonal antibodies (mAbs) have rapidly emerged as a validated therapeutic class and offer a promising approach for targeting tumor angiogenesis. Antibodies provide important therapeutic advantages compared with other classes of agents, such as high specificity for a particular target and more predictable toxicity profiles. In this chapter, we discuss the fundamental aspects of antibody therapeutics with emphasis on the development of antibodies that target molecular mechanisms related to tumor angiogenesis.

New blood vessel growth (angiogenesis) plays a key role in the development and progression of human cancer. Recruitment of sprouting vessels from existing blood vessels is required to supply rapidly growing tumor cells with sufficient concentrations of oxygen and nutrients, and for dissemination of tumor cells to distal sites. Angiogenesis is a dynamic process regulated by pro- and antiangiogenic factors. These factors regulate endothelial cell proliferation, migration and invasion, organization of endothelial cells into functional tubular structures, maturation of vessels, and vessel regression. Supporting stromal cells such as fibroblasts, pericytes, smooth muscles cells, and various hematopoietic cells also contribute to the angiogenesis process by release of various angiogenic factors. Research over the last decade has led to a better understanding of the molecular players that regulate tumor angiogenesis and to the development of agents that target these mechanisms as potential cancer therapies.

Antibodies have emerged as an important drug class in pharmaceutical development and clinical practice. At the beginning of the twentieth century, Paul Ehrlich originally proposed the use of antibodies as a "magic bullet" to specifically target malignant cells [1]. The inherent specificity of antibodies was postulated to enhance antitumor activity while reducing the nonspecific side effects that are inevitably associated with conventional chemo- and radiotherapy. However, early progress was hindered by the inability to produce specific antitumor antibodies in sufficient quality and quantity for clinical application. The use of polyclonal antibodies was limited by their heterogeneity with respect to size, antigen specificity and affinity, low immunoreactive fraction, and the contamination of unwanted immunoglobulins (Ig). Moreover, the production of highly specific polyclonal antitumor antisera/antibodies was difficult and unreliable. In 1975, Kohler and Milstein, by employing a method of somatic hybridization, successfully generated "hybridoma" cell lines producing mAbs of defined specificities [2]. The principal advantages of mAb over the conventional polyclonal antibodies are obvious, including their defined specificity, homogeneity, and availability of mAb in practically unlimited quantities. These properties of mAb render them as one of the most attractive classes of therapeutic agents to malignant tumors. Unfortunately, early enthusiasm was short-lived as one clinical trial after another failed to show efficacy, mainly due to several intrinsic properties associated with rodent-derived mAbs, such as immunogenicity of the antibodies, inefficient means of production, and a poor understanding of the appropriate mechanisms of action of antibodies as pharmacologic agents [3,4]. During the past decade, mAbs have made a strong comeback, led in part by newly developed molecular engineering methods to produce chimeric, humanized, and fully human mAbs which are less immunogenic [3,4]. Antibody engineering technologies have matured to the point where mAbs therapeutics can be readily customized to obtain the desired size, binding specificity, affinity, and effector function. In addition, immunoconjugate technology has made remarkable advances over the last decade to the point where this is now a feasible and proven antibody therapy approach. Lastly, technological advance in mammalian expression of mAb has enabled large-scale production with reduced cost [4].

Today, mAb therapeutics offers several advantages over other therapeutic classes (Table 7.1). The higher specificity and generally lower toxicity profile observed for antibody therapeutics

TABLE 7.1
Advantages and Disadvantages of Antibody Therapeutics

Advantages	Disadvantages
Target specificity	Limited to soluble and cell surface targets
Predictable pharmacology	Intravenous administration
Reduced toxicity and drug interactions	Potential for immunogenicity/hypersensitivity reactions
Exceptional pharmacokinetics (long half-life)	Costly manufacturing and quality control

TABLE 7.2
FDA Approved Monoclonal Antibody Therapeutics in Oncology

Antibody	Target	Antibody Type	Indication	Year Approved	Company
Rituxan (rituximab)	CD20	Chimeric IgG1	NHL	1997	Biogen Idec/ Genentech/Roche
Herceptin (trastuzumab)	HER2	Humanized IgG1	Breast cancer	1998	Genentech
Mylotarg (gemtuzumab ozogamicin)	CD33	Calicheamicin-labeled humanized IgG4	AML	2000	Wyeth
Campath-1H (alemtuzumab)	CD52	Humanized IgG1	B cell CLL	2001	Millenium/Ilex/Berlex
Zevalin (ibritumomab tituxetan)	CD20	^{90}Y-labeled mouse IgG1	NHL	2002	Biogen Idec
Bexxar (tositumomab)	CD20	^{131}I-labeled mouse IgG2a	NHL	2003	Corixa/ GlaxoSmithKline
Erbitux (cetuximab)	EGFR	Chimeric IgG1	Colorectal & head and neck cancer	2004	ImClone Systems/ Bristol Myers Squibb
Avastin (bevacizumab)	VEGF	Humanized IgG1	Colorectal & non-small cell lung cancer	2004	Genentech/Roche
Vectibix	EGFR	Human IgG2	Colorectal cancer	2006	Amgen

often contributes to increased utility in combination with other anticancer therapies and more rapid development timelines. Since 1994, the Food and Drug Administration (FDA) has approved 20 therapeutic mAb for clinical use in the United States, including a 9 for oncology indications (Table 7.2), with over 400 mAb clinical trials currently ongoing worldwide for oncology indications. In this chapter, we discuss a number of basic aspects regarding the development of antibody-based therapeutics, including general antibody structure, discovery, genetic engineering, and large-scale production for clinical applications. We also discuss the mechanisms of action of antibody therapeutics, with focus on several products which have demonstrated clinical or preclinical efficacy in cancer treatment mainly through an antiangiogenesis mechanism.

7.1 ANTIBODY STRUCTURE

The majority of antibodies that are used for targeted therapy fall into the IgG class of immunoglobulins. IgG is a tetrameric glycoprotein, consisting of four polypeptide chains: two identical light chains and two identical heavy chains. Each IgG contains two antigen-binding fragments (Fab) and an Fc domain joined together by a hinge region (Figure 7.1). The Fab fragment is responsible for specific antigen binding whereas the Fc domain binds to the Fc receptors on effector cells, fixes complement, and elicits other in vivo biological responses. The variable light (VL) region and the variable heavy (VH) region within the Fab fragment directly contact antigen and are responsible for the specificity and diversity of antibodies. Within each of VL and VH, there exist three hypervariable regions, also called complementarity-determining regions (CDRs), which form the binding surface that contacts the antigen. The Fv fragment is a heterodimer of VL and VH domains; the fragment is usually unstable in solution since the two domains are noncovalently linked (Figure 7.1). The instability of Fv fragments was overcome by the invention of single chain Fv (scFv) in which the VL and the VH domains are connected via a peptide linker [5]. The scFv fragment

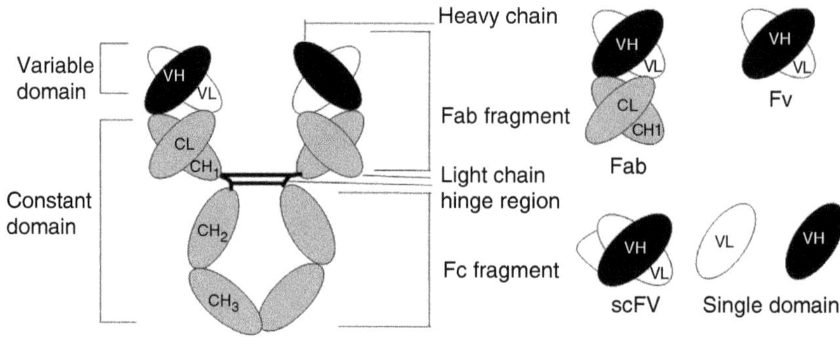

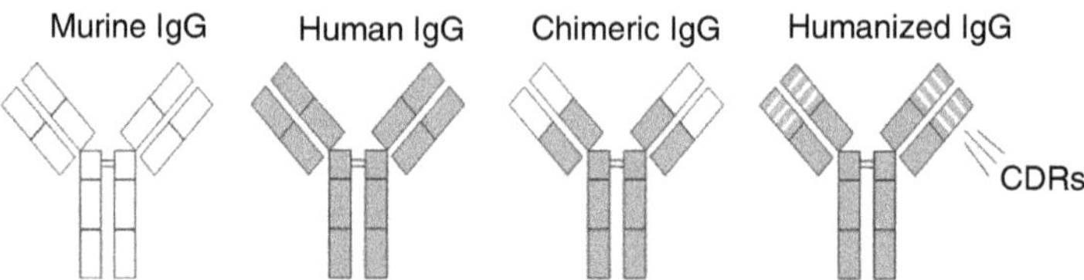

FIGURE 7.1 Schematic representation of an IgG antibody and its fragments. *Top panels*: Each IgG molecules consist of four polypeptide chains: two identical light and two identical heavy chains paired together by interchain disulfide bonds. The Y-shaped IgG contains two Fab (antigen-binding) fragments and an Fc (crystalline) fragment linked together via the hinge region. The smallest module of an antibody required for specific binding is Fv fragment comprising only the variable domains of light (VL) and heavy (VH) chains. Each VL and VH domains contain three hypervariable regions called complementarity-determining regions (CDRs) that form the antigen-binding surface and are responsible for antibody specificity and diversity. Introduction of a polypeptide linker between the VL and the VH domains (in either orientations) results in the formation of an scFv or diabody (or triabody or tetrabody) depending on the length of the linker. Alternatively, a disulfide bond can be introduced into the interface between the VL and the VH domain to form a disulfide bond-stabilized Fv (dsFv). Single domain antibodies, in which a VL or a VH alone comprises the binding unit, are also being exploited in the development of therapeutic and diagnostic agents. *Bottom panels*: mAbs produced from the traditional hybridoma technique are generally of murine origin. Chimeric antibody is generated by joining the VL and the VH domains of a murine mAb to human constant domains: mouse VL to human CL and mouse VH to human CH1–hinge–CH2–CH3, respectively. In antibody humanization, only the CDRs of the murine mAb, along with one to several other mouse residues determined to be critical in maintaining the antibody affinity, are grafted into a human framework. Fully human antibodies can be routinely obtained nowadays with the availability of human antibody phage display library and human Ig transgenic mouse. *Note*: All drawings are not to scale.

in solution exists mostly as monomer, when the linker is ≥12 to 15 amino acids, or as a dimer (as so-called diabody) when the linker is between 5 and 12 amino acids [6,7]. Interestingly, several groups reported recently that the scFv mostly form trimers (triabody) or tetramers (tetrabody) when the VL and VH domains are fused together with a linker of 0 to 2 amino acids [8]. An alternative approach for increasing the stability of an Fv fragment is to introduce an interchain disulfide bond between the interface of the VL and the VH domains [9]. The residues in the interface to be mutated into cysteine, ideally locate outside the CDR regions, were identified with the aid of computer molecular modeling [10,11]. Several disulfide bond-stabilized Fv fragments have been produced with increased stability and antigen-binding affinity [12,13]. Recently, it has been observed that some camel antibodies are composed of only heavy chains [14]. This observation has led to the development of "single domain" antibody fragments, in which a VL or a VH alone comprises the binding unit (Figure 7.1).

These single domain binders are much smaller (~11 to 13 kDa) than the conventional antibodies and fragments, and may themselves represent good candidates for the development of therapeutic and diagnostic agents. Further, they may serve as excellent "building blocks" for the more sophisticated antibody engineering process to create "ideal" molecules with tailor-designed properties, for example, bi- or multispecific antibody molecules [15].

7.2 ANTIBODY DISCOVERY AND ENGINEERING

To date, the majority of hybridoma-derived antitumor mAbs are of murine origin. These antibodies are immunogenic and can elicit a human antimouse antibody (HAMA) response in humans. The HAMA can form immunocomplexes with subsequent administrated therapeutic antibody, leading to increased clearance accompanied with decreased tumor localization of the antibody, and in some case, serious side effects such as an allergic anaphylactic reaction. Smaller antibody fragments, such as Fab, Fv, and scFv, are usually less immunogenic than the intact IgG in humans due to the lack of the Fc domain. Other approaches attempting to reduce the immunogenicity of rodent-derived antibodies include chemical modification of the antibodies such as conjugation of the antibodies to polyethylene glycol or oxidized dextran, and coadministration of immunosuppressive agents such as cyclosporin A, cyclophosphamide, and steroids to patients. Recent advancement in antibody engineering technologies has not only enabled the ability to tailor-make antibody molecules with predefined characteristics such as size, valency, and multispecificities to suit the intended applications, but also led to the production of chimeric and humanized antibodies with greatly reduced immunogenicity (Figure 7.1). Chimeric antibodies were the first generation of this generic engineering approach where the variable domains (both VL and VH) of a murine mAb were cloned and fused to the constant domains of a human IgG to create a new hybrid IgG molecule, which retains the original antigen-binding affinity and specificity, but only contains, in theory, one-third of murine amino acid sequences [16]. Humanization of antibodies takes this approach one step further by genetically grafting only the CDRs of the murine antibody, along with a few murine residues outside the CDRs believed to be important for antigen-binding affinity, into a human IgG framework. These residues are identified with the help of computer-based molecular modeling and site-directed mutagenesis because of their critical roles in maintaining the correct conformation of the antigen-binding surface and direct contacting the antigen [17]. Approximately 90% to 95% of the amino acid sequence of a humanized antibody is of human origin. Both chimeric and humanized antibodies have proven to be much less immunogenic in clinical studies (for reviews, see [18–20]).

Human antibody phage display libraries [21–23] and human Ig transgenic mice [24–26] became available in the mid-1990s enabling the generation of fully human antibodies with desired specificities. Phage display technique, by linking the phenotype (antibody fragments) and the genotype (DNA encoding the corresponding antibody fragments) together in a phage particle, mimics the in vivo antibody response and selection process in vitro. In constructing the antibody library, polymerase chain reaction (PCR) (or other similar/alternative cloning methods) is utilized to amplify (or capture) human B-cell antibody repertoires from pooled lymphocytes isolated from a collection of either healthy or biased (e.g., diseased or immunized) donors. The amplified antibody variable domain genes, that is, both VL and VH, are assembled randomly into scFv or Fab formats, followed by fusion to a phage coat protein (the mostly used is gene III protein in filamentous phage M13) for surface displaying. Additional diversity can be further introduced into the antibody repertoire by randomization of one or more CDR regions in both VL and VH, error-prone or mismatch PCR, or other means of randomized oligonucleotide cloning. In the past several years, with significant improvement in the efficiencies in gene cloning, cell transformation, and antibody expression, libraries of size of 10^9 to 10^{11} members are routinely constructed in both scFv and

Fab formats. These libraries are used by a variety of research and industrial laboratories for the identification of antibody leads. Since the entire selection process is performed in vitro, that is, no immunizations are required, antibody candidates are generally obtained within several weeks compared with the several months for the traditional hybridoma technique. Further, the libraries can be used, in theory, to generate antibodies to any kind of targets, including those of extremely toxic (too toxic to be used as an immunogen in animals), labile (e.g., RNA and enzymes), nonimmunogenic antigens such as those of very small size (e.g., small molecular compounds) and cross-species highly conservative molecules, and even self-antigens. Similar to phage display, yeast and bacteria are also employed as the vehicle for antibody displaying [27–30]. Selection of binders in the latter cases can be simplified since the yeast and bacteria particle are large enough to be sorted out via flow cytometry provided the antigens are labeled with a fluorescent dye [31,32]. Finally, ribosome display technologies are also exploited [33,34]. In this case, mRNA is transcribed from a human antibody cDNA library, followed by translation in vitro in test tubes to produce mRNA/ribosome/antibody complex. After antigen binding, mRNA in the selected complex is amplified by RT-PCR to recover the cDNA sequence encoding the antibody binders. Since there is no need to transform cells in order to generate and select libraries, ribosome display allows much higher library diversity than phage/yeast/bacteria display methods. Further, ribosome display is also suitable for generating toxic, proteolytically sensitive, and unstable proteins (antigens) for antibody selection, and allows the incorporation of modified amino acids at defined positions since all the procedures are carried out in a testing tube. An apparent drawback associated with all the displaying technologies is the necessity of recloning of the library-derived antibody fragments into expression vectors containing the appropriate antibody constant domains for the production of full-length IgG molecules.

Fully human antibodies are also generated in transgenic mice expressing human Ig repertoire [24–26]. In these "humanized" mice, the endogenous mouse Ig genes are deleted and replaced with their human counterparts. On antigen exposure (by immunization), mouse B lymphocytes producing human antibodies can be isolated and high-affinity mAbs are generated using the traditional hybridoma technique. The advantages of this approach include cognate VL/VH pairing, natural in vivo antibody maturation process for high-affinity binders, and full length IgG antibodies are produced by the hybridoma cells without the need for further subcloning. Large-scale antibody production can be readily managed using the existing hybridoma cell lines, or alternately, the genes encoding the antibody sequences can be isolated from the hybridoma, subcloned into appropriate expression vectors, and transfected into other hosts for antibody manufacturing (see later for detailed discussion). The disadvantages are usually associated with those of the hybridoma technique, for example, long period of immunization procedure, and the technical limitations in cell fusion and hybridoma selection/cloning processes.

7.3 ANTIBODY PRODUCTION AND MANUFACTURING

One of the major obstacles in the development of mAb-based therapeutics has been the difficulty in producing the antibodies in sufficient quality and quantity for broad clinical applications. Significant advance has been made in the past years in the production of antibody fragments at high-expressing level (>1 g/L) in *Escherichia coli* via high-density fermentation process [35]. These fragments are smaller than full-length IgG, so they have better solid tumor penetration rates, but their small size and lack of an intact Fc result not only in rapid clearing from circulation, leading to a short in vivo half-life, but also in lacking of effector functions including antibody-dependent cellular cytotoxicity (ADCC) and complement-mediated cytotoxicity (CMC). On the other hand, although full-length IgG was recently expressed successfully in an engineered bacteria host [36], only mammalian cell–derived materials possess proper glycosylation in the Fc region, hence the capability of supporting effector functions. Currently the most commonly used mammalian cell lines consist mainly those of rodent origin, including Chinese hamster ovary (CHO) cells and mouse myeloma (SP0/2 and NS0) cells. The large-scale manufacture of full-length mAb in mammalian cells has been costly and challenging in both scientific and facility engineering fields. Extensive research in past decade in biotech industry has led to a number of technological progresses in mammalian cell expression systems, for example, engineering and optimization of expression vectors and cell lines, specific gene targeting plus choice of good selection markers, efficient high-throughput screening, and optimization of culture media and conditions, and so on [37]. Taken together these achievements have significantly enhanced the production of mAb in mammalian cells.

Recently, production of full-length IgG in transgenic animals and plants has been exploited with reasonable success. The ability to produce mAb in transgenic animals and plants may provide several advantages including the capability of efficient large-scale production (in hundreds of kilograms to multitons) and the potential of significantly lower cost than mammalian cell culture system—as the transgenic systems usually do not require the establishment of costly and sophisticated manufacturing infrastructure, yet the downstream purification and processing are similar to those needed for mammalian cell culture–derived materials [38,39]. A number of therapeutic proteins, such as antithrombin III [40], CD4–IgG2 fusion protein [41], and α_1-antitrypsin [42], have been produced in different transgenic animals including goat, sheep, and chicken. Some of these proteins are currently in various phases of clinical development [40–42]. On the other hand, engineered corn, tobacco, and soybean plants have also been tested for their ability to produce therapeutic mAb [43–46]. One of the major concerns associated with the plant-derived mAb is that the difference in glycosylation between plant and mammalian cells may affect both the pharmacodynamic and the pharmacokinetic properties of the product [44,47,48]. For example, the immune effector functions, such as ADCC and CMC, which may be important to the activity of a therapeutic mAb, may be altered (or even abolished) in the plant-derived mAb due to the difference in glycosylation profile. Further, the plant glycosylation may present significant immunogenicity issue in human therapy. A number of options are explored to enhance the therapeutic potential of the plant-derived mAb, for example, by making an aglycosylated mAb or by genetic engineering to produce transgenic plants with human glycosylation machinery, in order to reduce the immunogenicity and retain the immune effector functions of the antibody. To date, it is unclear whether these alternative means of antibody production in transgenic animals or plants will be adopted by the biopharmaceutical industry and regulatory agencies.

7.4 MECHANISMS OF ACTION OF mAb-BASED THERAPEUTICS

mAbs were originally developed to confer passive immunity against tumor cells via targeting of tumor-associated antigens. However, it has been recognized over the last two decades that mAb can function as potent and specific molecular antagonists. Currently, mAbs are developed as cancer therapeutics to block molecular function, elicit immune effector function or as immunoconjugates for tumor-specific drug delivery [49] (Figure 7.2). The effectiveness of an antibody depends on its capability to induce one or more of these key mechanisms:

1. Molecular inhibition of key regulatory pathways involved in tumor growth, such as blocking growth factor/receptor interaction and downregulating expression of oncogenic proteins (e.g., growth factor receptors), is a major mechanism of antibody-based therapies. The antibody may bind a soluble antigen (ligand) and prevents it

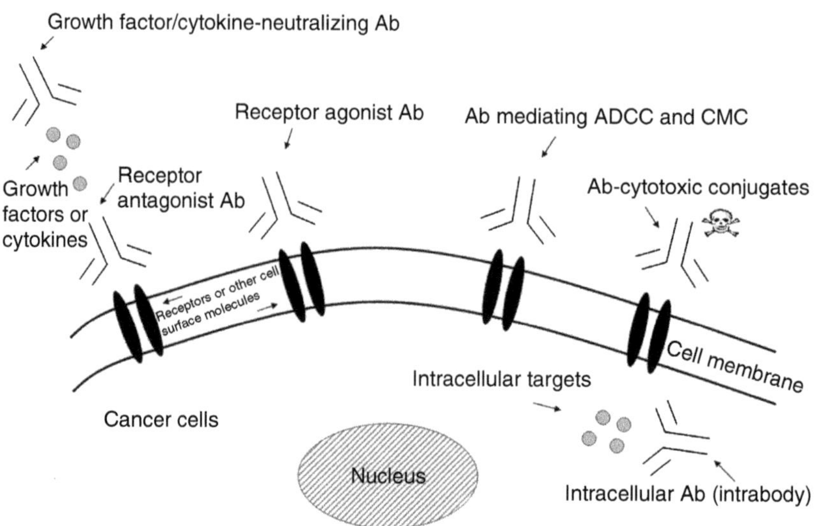

FIGURE 7.2 Mechanisms of action of antibody-based therapies in cancer. A number of antibodies block the molecular function of important signaling pathways in tumor or stromal cells by targeting growth factors or the cognate receptors. Examples of this mechanism are the antibodies Erbitux and Avastin. Antibodies that bind to growth factor receptors may also cause internalization of the receptor and thereby further downregulate the receptor pathway. Antibodies may also act as agonists of receptor function; however, there are currently no antibodies of this mechanism of action approved for cancer therapies. Antibodies may bind to tumor targets and elicit an immune effector function such as ADCC or CMC. This mechanism has been proposed for the activity of the antibody Rituxan. Lastly, antibodies may deliver cytotoxic agents such as chemotherapeutic drugs, toxins, or radionucleotides to tumor cells. Examples of this approach are the immunoconjugates Mylotarg and Zevalin. An experimental antibody mechanism under study is the use of intrabodies to target intracellular molecules.

from interacting with other molecules, for example, its receptors. Avastin, an antiangiogenic therapy for colorectal cancer (CRC), binds vascular endothelial growth factor (VEGF) and blocks its interaction with the receptors [50]. Alternatively, the antibody binds a cell surface receptor and inactivates it by blocking the binding site of an activating ligand. Erbitux, an antibody for the treatment of refractory CRC, binds epidermal growth factor (EGF) receptor at the EGF-binding site and blocks activation by both EGF and transforming growth factor-α [51]. In other cases, the antibody may not directly block ligand/receptor interaction but rather exerts its effects though preventing receptor dimerization/multimerization, which is required for activation. Omnitarg (2C4), an anti-HER2 antibody currently under clinical development, is believed to inhibit tumor cell growth by blocking HER2 from dimerizing with EGFR and HER3 (there are no ligands identified so far for HER2) [52]. Finally, the antibody may bind to and cross-link multiple membrane-bound receptors, mimicking the function of a natural ligand, and activating the receptor. For example, a number of antibodies mimic the function of Apo2L (ligand) and trigger apoptosis by activation of death receptor 5 [53]. By interfering with important growth factor/receptor signaling pathways, these antibodies can not only influence the growth and survival of tumor cells, but may also potentiate the cytotoxic effects of chemotherapeutic drugs and radiation. For example, a number of studies have demonstrated that Erbitux, Rituxan, and Herceptin could significantly enhance the therapeutic efficacy of various cytotoxic agents in both laboratory and clinical settings [54–56].

2. Antibodies may recruit effector mechanisms of the immune system, such as ADCC and CMC. The role of this mechanism in the clinical efficacy of mAb therapies in cancer is still controversial. However, there is evidence to suggest that immune effector mechanisms play a role in the clinical antitumor efficacy of the two FDA-approved antibody products, the anti-lymphoma antibody Rituxan and the anti-breast cancer antibody Herceptin. For example, both Rituxan and Herceptin have been shown to mediate significant levels of ADCC and CMC effects on a number of malignant cell lines [57,58]. Further, the antitumor effect of these two mAb in vivo was severely diminished in mice with deficiency of Fc receptor on its effector cell surface or when mutants of the mAb with reduced Fc-binding efficiency were used [59].
3. Antibodies may be utilized as a carrier molecule to deliver an attached chemotherapeutic agent, toxin, or radioisotope to malignant cells displaying a specific antigen. A number of antibody conjugates have been approved by the FDA for oncology indications, including Mylotarg, an anti-CD33 antibody–calichimicin conjugate for the treatment of CD33-positive acute myeloid leukemia [60], Zevalin, a ^{90}Y-labeled anti-CD20 antibody [61], and Bexxar, a ^{131}I-labeled anti-CD20 antibody for non-Hodgkin's lymphoma (NHL) [62]. Alternatively, the antibodies may act as a means to redirect immune effector cells to target sites, for example, tumors, in the form of a bispecific antibody (BsAb) [63]. In this setting, one arm of the BsAb binds an antigen on tumor cell, and the other binds a determinant expressed on effector cells, such as CD3, CD16, or CD64, which are expressed on T lymphocytes, natural killer cells, or other mononuclear cells. By cross-linking tumor and effector cells, the BsAb not only brings the effector cells within the proximity of the tumor cells but also simultaneously triggers their activation, leading to effective tumor cell killing [64]. Preliminary but promising clinical benefits have been observed in a number of early stage trials [65]. Similarly, BsAb has also been used to redirect and enrich the tumor/normal tissue localization ratio of cytotoxic agents by replacing the effector cell-binding arm with specificity to a chemotherapeutic drug, radioisotope, or toxin [63].
4. Other potential mechanisms for antibody therapeutics include those that stimulate the antiidiotype network to generate antitumor anti-antiidiotypic antibody response [66]; catalytic antibodies that function as catalysts to induce hydrolysis of cell membrane or proteins, or to convert molecular oxygen into hydrogen peroxide to achieve cell killing [67]; or antibodies that enhance patient's immune response to tumors by stimulating cytotoxic T lymphocytes via CD40 [68] or by antagonizing endogenous immune inhibitory factors such as CTLA-4 [69]. Recently, functional antibodies have been expressed intracellularly as "intrabodies." These intrabodies exert their biological effects through interfering with the function of the targeted molecules via a variety of mechanisms, including altering their intracellular trafficking and localization, blocking their interaction with other molecules, or directly neutralizing their enzymatic activity (for kinase targets) [70,71].

7.5 ANTIANGIOGENIC ANTIBODY THERAPEUTICS

A significant preclinical and clinical research effort has been carried out over the last 10 years to identify and develop effective angiogenesis inhibitors for therapeutic purposes [72–76]. Whereas many of the known angiogenesis inhibitors have been found serendipitously, their underlying mechanisms of action are often poorly understood. Our understanding of the complex series of events that are collectively referred to as angiogenesis has improved dramatically over the past decade [77,78]. Stages along the angiogenesis process include the migration of endothelial stem cells, migration and invasion of endothelial cells, proliferation

TABLE 7.3
Monoclonal Antibodies Targeting Angiogenesis-Related Molecules

Antibody	Antibody Type	Target	Stage of Development	Company
Bevacizumab (Avastin)	Humanized IgG1	VEGF-A	FDA approved	Genentech
MEDI-522 (Vitaxin)	Humanized IgG1	$\alpha_v\beta_3$	Phase II	MedImmune
M200 (volociximab)	Chimeric IgG4	$\alpha_5\beta_1$	Phase II	PDL BioPharma and UCB
CDP-791	Pegylated humanized DiFab	VEGFR-2	Phase II	
IMC-1121b	Fully human IgG1	VEGFR-2	Phase I	ImClone Systems
CNTO 95	Fully human IgG1	$\alpha\upsilon$ integrin	Phase I	Centocor
IMC-18F1	Fully human IgG1	VEGFR-1	Phase I	ImClone Systems Incorporated
2C5	Fully human IgG1	VEGFR-3	Preclinical	ImClone Systems Incorporated
E4G10	Rat IgG1	VE-cadherin	Preclinical	ImClone Systems Incorporated
2C3	Mouse IgG1	VEGF-A	Preclinical	Peregrine

of endothelial cells, organization into tubular structures, formation of circulatory systems, maturation of vessels, and vessel regression. New information regarding how each of these steps participate along the angiogenesis pathway, and the molecules responsible for these events, has led to a variety of novel and increasingly mechanism-based approaches for the development of angiogenesis inhibitors. However, the number of validated angiogenesis targets applicable to antibody therapy remains a challenge. At present, there is one antiangiogenesis mAb therapy approved for clinical use and 6 mAbs in clinical testing (Table 7.3). There are additional, but surprisingly few, mAbs at earlier stages of preclinical testing. Thus far, the majority of antiangiogenic mAb therapies have focused on blocking the molecular function of mediators of endothelial cell growth and adhesion. Attempts to elicit immune effector function as well as vascular targeting approaches with antibodies specific for angiogenesis targets are also exploited. Later, we review several major efforts in the development of mAb-based therapeutics that inhibit tumor-associated angiogenesis, with focus on those strategies that interfere with endothelial cell growth and adhesion as their primary mechanism of action.

7.5.1 VEGF/VEGFR Pathway

A large number of growth factors, cytokines and other regulatory proteins that stimulate endothelial cell growth either directly or indirectly by various mechanisms have been identified. The VEGF pathway is well established as one of the key regulators of this process. Consequently, considerable effort has been invested in generating and testing various approaches to inhibit VEGF or its receptors (VEGFR) including mAb therapeutics (for reviews, see [72,75,79]). VEGF is a key regulator of vasculogenesis during embryonic development and angiogenic processes during adult life such as wound healing, diabetic retinopathy, rheumatoid arthritis, psoriasis, inflammatory disorders, tumor growth, and metastasis [80–85]. VEGF is a strong inducer of vascular permeability, stimulator of endothelial cell migration and proliferation, and is an important survival factor for newly formed blood vessels. VEGF binds to and mediates its activity mainly through two tyrosine kinase receptors [86], VEGFR-1 (fms-like tyrosine kinase 1 or Flt-1) [87,88] and VEGFR-2 (kinase insert domain-containing receptor, or KDR in humans, and fetal liver kinase or flk1 in mice) [89,90]. Targeted deletion of genes encoding VEGF, VEGFR-1, or VEGFR-2 in mice is lethal to the embryo, demonstrating the physiological importance of the VEGF pathway in blood vessel formation. Mice lacking even a single VEGF allele die before birth from vascular abnormalities [91,92]. VEGFR-1 null embryos fail to develop normal vasculature due to defective formation of vascular tubes [93]. Interestingly, inactivation of VEGFR-1 by truncation of the tyrosine kinase domain does not impair embryonic angiogenesis, suggesting that signaling through the VEGFR-1 receptor is not important for development of the vasculature in the embryo [94]. VEGFR-2-deficient mice have impaired blood island formation and lack mature endothelial cells [95,96].

Several factors related to VEGF have been identified: VEGF-B, VEGF-C, VEGF-D, VEGF-E, and placenta growth factor (PlGF) [97]. VEGF-B and PlGF bind exclusively to VEGFR-1, VEGF-E is specific for VEGFR-2, whereas VEGF-C and VEGF-D can bind to VEGFR-2 and another receptor, VEGFR-3 (fms-like tyrosine kinase 4 or Flt-4), which is expressed predominantly on lymphatic endothelium [98]. Additional distinct biological functions may be attributed to the complexity in binding specificity between various VEGF and receptor combinations. Furthermore, some VEGF ligands (i.e., VEGF and PlGF) and receptors (i.e., VEGFR-1 and VEGFR-2) may form homo- or heterodimers that bind differentially to various VEGF/VEGFR family members and signal through different pathways. The various angiogenic processes that these VEGF/VEGFR combinations control are yet to be determined.

Numerous studies have shown that overexpression of VEGF and VEGFR-2 is strongly associated with invasion and metastasis in human malignant diseases [99–101]. VEGF is expressed at high levels in various types of human and mouse tumors and is strongly upregulated under hypoxic conditions such as those associated with rapidly growing tumors. The importance of VEGF and VEGFR-2 in tumor angiogenesis has been directly demonstrated in studies where VEGF or VEGFR-2 is inhibited through neutralizing antibody, antisense, soluble receptor, or kinase inhibitors [79,83,99–101]. Recently, VEGFR-2 and VEGFR-1 have also been shown to be expressed on the surface of a variety of malignant cells, including those of both solid and hematological origin [102–104]. Neutralizing antibodies specific to human receptors significantly inhibited the growth of human VEGFR-1^+ or VEGFR-2^+ xenografts in mouse models irrespective of their effects in host/tumor angiogenesis, indicating a direct antitumor activity [102–105]. Taken together these data suggest that blockade of VEGF/VEGFR pathway by mAb therapy would be a useful therapeutic strategy for inhibiting angiogenesis and tumor growth. To this end, antibodies that neutralize VEGF, VEGFR-2 and, more recently VEGFR-1 and VEGFR-3, have been developed and have shown potent antiangiogenic and antitumor activities in both laboratory and clinic settings.

7.5.2 Anti-VEGF Antibody, Bevacizumab

In 1993, Kim et al. [106] reported that an antihuman VEGF antibody, A4.6.1, inhibited the growth of human xenograft tumors in nude mice models. This finding provided the first evidence that inhibition of an endogenous angiogenic factor may result in suppression of tumor growth. Subsequently, tumor inhibition was observed in a variety of human xenograft tumors treated with the anti-VEGF antibody including carcinomas of colorectal, prostate, and ovarian origin (for review, see [79]). Intravital videomicroscopy has provided further evidence that anti-VEGF antibody treatment not only resulted in reduction of tumor vascular permeability [107], but also caused nearly complete suppression of tumor angiogenesis [108]. A humanized version of A.4.6.1, rhuMab-VEGF (IgG1), also known as bevacizumab or Avastin (Genentech Inc., South San Francisco, CA, www.gene.com), recognizes all isoforms of VEGF with high affinity (K_d, 0.8 nM), and inhibits VEGF-induced proliferation of endothelial cells in vitro and tumor growth in vivo with potency and efficacy very similar to those of the parent murine antibody [109]. Toxicological studies in primates have shown that rhuMab-VEGF is safe even after prolonged treatment, and its effects are limited to inhibition of angiogenesis in the female reproductive tract and induction of growth plate dysplasia in animals that have not completed statural growth [110].

7.5.2.1 Bevacizumab in Colorectal Cancer

In an open-label phase II trial [111], 104 previously untreated patients with metastatic CRC were randomized into three arms receiving either 5-fluorouracil (5-FU)/leucovorin (LV) alone or the same doses of 5-FU/leucovorin in combination with bevacizumab at 5 or 10 mg/kg. The response rates were 40% (14/35 patients) in the low dose and 24% (8/33 patients) in the high-dose antibody and chemotherapy combination arms, compared with a response rate of 17% (6/36 patients) in the chemotherapy alone arm. Time to disease progression and medial survival was 9.0 and 21.5 months in the low dose, 7.2 and 16.1 months in the high-dose antibody arms, respectively, compared with those of 5.2 and 13.8 months with chemotherapy alone. Thrombosis was the most significant adverse event and was fatal in one patient. Hypertension, proteinuria, and epistaxis were other potential safety concerns. In a pivotal phase III trial [112], over 800 previously untreated metastatic CRC patients were given bolus IFL (irinotecan, 5-FU, and leucovorin) plus placebo or bevacizumab. The overall response rates were 44.8% in the IFL plus antibody group and 34.8% in the IFL plus placebo group ($p = 0.0036$), with duration of response of 10.4 and 7.1 months ($p = 0.0014$), respectively. The median progression-free survival was 10.55 months for patients received the antibody group and 6.24 months for those received IFL only ($p < 0.00001$). Further, patients received the antibody survived significantly longer (20.34 months) than those received IFL only (15.61 months, $p = 0.00003$). The main toxicities associated with bevacizumab were grade 3 hypertension (11% in the antibody group versus 2.3% in the IFL only group), proteinuria, and arterial thromboembolic event (4.4% in the antibody group versus 1.9% in the IFL only group). These promising outcomes led to the FDA approval of the bevacizumab as first-line therapy in combination with IFL in metastatic CRC patients in February 2004. Recently, preliminary results from an interim analysis of a phase III study [113] showed that bevacizumab plus the FOLFOX4 chemotherapy regimen also extended survival of metastatic CRC patients in second-line settings, compared with FOLFOX4 alone—patients who received bevacizumab plus FOLFOX4 had a 26% reduction in risk of death, compared with patients who received FOLFOX4 alone. Median survival for patients received the combination was 12.5 months, compared with that of 10.7 months for those received FOLFOX4 alone. Adverse events observed in the study were consistent with other clinical trials where bevacizumab was combined with chemotherapy. In June 2006, bevacizumab was approved by the FDA in combination with intravenous 5-FU-based chemotherapy for second-line metastatic CRC patients.

In a mechanistic study [114], six patients with primary and locally advanced CRC were enrolled in a preoperative treatment protocol of bevacizumab administration alone (5 mg/kg intravenously), followed after 2 weeks, the approximate half-life of bevacizumab in circulation, by concurrent administration of bevacizumab with 5-FU and external beam radiation therapy to the pelvis and surgery, 7 weeks after treatment completion. Twelve days after bevacizumab infusion, flexible sigmoidoscopy revealed that bevacizumab induced tumor regression of >30% in one patient, and no change in tumor size in the other five patients. Functional computed tomography scans indicated significant decreases in tumor blood perfusion and blood volume, accompanied by a significant decrease in tumor microvessel density, reduction of interstitial fluid pressure, and decrease in number of viable, circulating endothelial and progenitor cells in the patients. Taken together, these data indicate that VEGF blockade has a direct and rapid antivascular effect in human CRC tumors.

In another phase II trail in CRC [115], 209 patients not eligible for first-line irinotecan were randomized into two groups and treated with 5-FU/LV ± bevacizumab. The study failed to meet the primary end point: duration of survival—the median survival in the antibody combination group was 16.6 months compared with that of 12.9 months in the chemotherapy alone group ($p = 0.159$). The overall response rates (26% versus 15.2%) and the duration of response (9.2 months versus 6.8 months) were also not statistic significant between the two groups ($p = 0.055$ and 0.088, respectively). The progression-free survival of the patients received the antibody was, however, significantly longer than those received chemotherapy alone (9.2 months versus 5.5 months, $p = 0.0002$). Grade 3 hypertension was more common with bevacizumab treatment (16% versus 3%) but was controlled with oral medication and did not cause study drug discontinuation. Finally, bevacizumab plus FU/LV also failed to demonstrate significant clinical benefits in CRC patients who have exhausted standard therapeutic options including both irinotecan- and oxaliplatin-based regimens [116]. In a single arm, multicenter trial, 350 refractory patients were treated with bevacizumab plus bolus or infusional FU/LV. Of the first 100 patients evaluated, objective response was confirmed in only one patient. This result led the investigators to conclude that there did not appear to be sufficient evidence of efficacy for bevacizumab in CRC patients in the third-line setting, and the antibody should, therefore, only be used in additional clinical trials with other agents in this patient population.

7.5.2.2 Bevacizumab in Lung Cancer

In a phase II trial [117], 99 patients with advanced or recurrent nonsmall cell lung cancer (NSCLC) were randomly assigned to bevacizumab 7.5 ($n = 32$) or 15 mg/kg ($n = 35$) plus carboplatin and paclitaxel every 3 weeks or carboplatin and paclitaxel alone ($n = 32$). Compared with the control arm, treatment with carboplatin and paclitaxel plus bevacizumab (15 mg/kg) resulted in a higher response rate (31.5% versus 18.8%), longer median time to progression (7.4 versus 4.2 months) and a modest increase in survival (17.7 months versus 14.9 months). In this trial, six patients treated with bevacizumab experienced sudden, serious pulmonary hemorrhage with fatal outcome in four patients. All patients who died had squamous cell carcinoma, which tends to occur in the larger airways. Other adverse events seen more often in the combination arms included nosebleeds and hypertension. Most recently, the interim results from a phase III multicenter study was announced [118]. In this study, 878 NSCLC patients were randomized to receive paclitaxel and carboplatin with or without bevacizumab. Patients with squamous cell-type carcinoma were excluded from this trial due to their higher risk of experiencing life threatening of fatal pulmonary bleeding as demonstrated in earlier trials. Patients who received bevacizumab with chemotherapy had a median overall survival and median progression-free survival of 12.5 and 6.4 months, respectively, compared with that of 10.2 and 4.5 months in patients treated with chemotherapy alone. The response rate in patients with measurable disease was 27% (97/357) in the group received bevacizumab plus chemotherapy, compared with 10% (35/350) in the group received chemotherapy alone. A preliminary assessment of adverse events by the investigators showed that grade 3/4/5 bleeding occurred in 4.5% of patients in the bevacizumab plus chemotherapy arm, compared with 1% of patients in the chemotherapy alone arm, with fatal hemoptysis occurred at a rate of 1% (5/420) in the combination arm. Treatment-related deaths occurred at a rate of 2% (8/420) in the bevacizumab plus chemotherapy arm, compared with <1% (2/427) in the chemotherapy alone arm. The most common adverse events were neutropenia, hypertension, and thrombotic events. Grade 3/4 neutropenia, hypertension, venous thrombosis, and arterial thrombosis occurred in 24%, 6%, 4%, and 2%, respectively, of patients treated with bevacizumab plus chemotherapy, compared with that of 16%, 1%, 3%, and 1% of patients who received chemotherapy alone [118]. Bevacizumab was approved by the FDA in October 2006 to be used in combination with carboplatin and paclitaxel for the first-line treatment of patients with unvesectable, locally advanced, recurrent or metastatic non-squamos non-small cell lung cancer.

7.5.2.3 Bevacizumab in Renal Cell Carcinoma

Bevacizumab has also demonstrated clinical benefit in renal cell carcinoma (RCC) patients. In a randomized, double-blinded phase II trial [119], patients with metastatic clear cell RCC were randomized into three groups receiving placebo, bevacizumab at 3 or 10 mg/kg. A total of 150 patients were planned to detect a twofold hazard ratio between placebo and either dose of antibody. At the time of interim analysis, there were 40 patients enrolled in the placebo group, 39 in the high-dose group and 37 in the low-dose group. Four partial responses were achieved in the high-dose group with response duration of 6, 9, 15, and 44+ months. Time to disease progression was significantly longer in both antibody groups than in the placebo group. The probability of remaining progression-free for patients given high-dose antibody, low-dose antibody, and placebo was 64%, 39%, and 20%, respectively at 4 months and 30%, 14%, and 5% at 8 months. The trial was terminated prematurely due to the overwhelming evidence of antibody activity. At the last analysis, there were, however, no significant differences in overall survival between groups ($p > 0.20$ for all comparisons). A CLAGB phase III trial has been completed recently to compare IFN-α to IFN-α plus bevacizumab as frontline therapy in metastatic unresectable clear cell RCC patients. Preliminary data analysis indicated that adding bevacizumab to IFN-α significantly improved patient's progression free survival.

7.5.2.4 Bevacizumab in Breast Cancer

In an open-label phase I/II trial [120], 75 breast cancer patients, who had relapsed following at least one chemotherapy treatment for metastatic disease, were given bevacizumab at 3, 10, or 20 mg/kg i.v. every other week. Tumor response was assessed before the 6th (70 days) and 12th (154 days) doses. Eighteen patients were treated at 3 mg/kg, 41 at 10 mg/kg, and 16 at 20 mg/kg. Four patients discontinued study treatment because of an adverse event. Hypertension was reported as an adverse event in 17 patients (22%). The overall response rate was 9.3% (confirmed response rate, 6.7%). The median duration of confirmed response was 5.5 months (range 2.3–13.7 months). At the final tumor assessment on day 154, 12 of 75 patients (16%) had stable disease (SD) or an ongoing response. In a pivotal phase III trial [121], 462 women with metastatic breast cancer who had previously received treatment with both anthracycline- and taxane-based chemotherapy regimens were randomized to receive bevacizumab (15 mg/kg) plus capecitabine or capecitabine alone. No difference was observed in both the progression-free time (4.86 months versus 4.17 months) and the 12 month overall survival (the primary end point) (15.1 months versus 14.5 months) between the two groups, although the antibody combination group yielded higher overresponse rate (19.8% versus 9.1%, $p = 0.001$). Grade 3 or 4 hypertension requiring treatment was more frequent in patients receiving bevacizumab (17.9% versus 0.5%).

Most recently, the interim results from a randomized phase III multicenter study were announced [122]. This study enrolled 722 women with previously untreated metastatic breast cancer, who were randomized to receive treatment with paclitaxel with or without bevacizumab. Results from an interim analysis of this study showed that patients received bevacizumab plus paclitaxel had a median progression-free survival of 11 months, compared with that of 6 months for patients treated with chemotherapy alone. At interim analysis, a 49% improvement in the secondary end point of overall survival was observed. In patients with measurable disease, the overall response rate was 28% (93/330) in the bevacizumab plus chemotherapy arm, and 14% (45/316) in the chemotherapy alone arm. A preliminary assessment of safety showed that grade 3/4 adverse events, which occurred more often in the bevacizumab arm, included neuropathy, hypertension, and proteinuria. Neuropathy, hypertension, and proteinuria occurred in 21%, 13%, and 2%, respectively, of patients in the bevacizumab plus chemotherapy arm, compared with that of 14%, 0%, and 0% of patients in the chemotherapy alone arm. One patient in the bevacizumab plus chemotherapy arm developed symptomatic congestive heart failure. Serious bleeding and blood clots were rare in this study. An sBLA for bevacizumab plus chemotherapy in first-line metastatic breast cancer was filed with the FDA in May 2006. In September 2006, the FDA requested a substantial safety and efficacy update from the phase III trial.

7.5.3 Anti-VEGFR-2 Antibodies

A panel of anti-VEGFR-2 mAb has been generated against the extracellular domain of the receptor [76,99,100]. These mAbs function as potent antagonists for VEGF binding, VEGFR-2 signaling, and VEGF-induced endothelial cell growth in vitro. A rat antimouse VEGFR-2 (Flk1) monoclonal antibody (DC101) was developed by ImClone Systems (New York, NY, www.imclone.com) using conventional hybridoma technique [123] to conduct proof-of-concept studies both in vitro and in animal models. In vitro studies demonstrated that DC101 binds with high affinity and specificity to Flk1, functions as a potent antagonist to VEGF binding, Flk1 signaling, and endothelial cell proliferation [124]. DC101 has been studied extensively in mouse models of angiogenesis, mouse tumors, and human tumor xenografts, demonstrating potent antiangiogenic and antitumor activity in these models ([124], for review, see [76,83,99,100]). In addition, DC101 treatment inhibited the dissemination and growth of metastases in mouse and human tumor metastasis models. Histological examination of DC101-treated tumors showed evidence of decreased microvessel density, reduced tumor cell proliferation along with increased tumor cell apoptosis and extensive tumor necrosis. Further, DC101 showed synergistic or additive antitumor activities when combined with chemotherapeutic drugs or radiation. In one study, DC101 significantly decreased the dose of radiation required to control the growth of xenografted human tumors in nude mice [125]. In other studies, DC101 treatment has been shown to enhance the antitumor activity of chemotherapeutic agents such as paclitaxel, cyclophosphamide, and gemcitabine [126,127], and, in some cases, led to significant regression of implanted tumors. Combination therapy with DC101 and chemotherapy has also been studied in the context of "metronomic" dosing of cytotoxic agents [128]. Metronomic therapy refers to frequent, low-dose administration of cytotoxic drugs with the aim of affecting new blood vessel formation. Since VEGF has been shown to act as a survival factor for endothelial cells in response to chemotherapy or radiation, the addition of anti-Flk1 antibody may enhance the antiangiogenic effect of metronomic therapy on proliferating tumor vasculature. This hypothesis was tested in number of experiments where DC101 was combined with chronic, low-dose of various chemotherapeutic agents, including vinblastine, cyclophosphamide, and doxorubicin [129–131], in the treatment of different human xenografts in nude mice. Remarkably, these treatment regimens resulted in significant enhanced tumor inhibition, and, in some cases, complete regression of large, established tumors, which were sustained for a long period without significant toxicities. These data support the notion that anti-Flk1 treatment potentiates the antivascular effects of low-dose chemotherapy on proliferating tumor endothelium. No overt toxicity has been observed in long-term DC101 treatment experiments of tumor-bearing or nontumor-bearing mice. These findings are important, since low levels of Flk1 expression are present on the endothelium of some normal tissues and required for normal angiogenic processes. Indeed, DC101 treatment did have an impact on normal angiogenesis associated with reproduction [132,133] and bone formation (our unpublished data). The lack of toxicity observed during DC101 therapy may be due to the limited dependence of resting endothelium for Flk1 stimulation. In contrast, tumor angiogenesis is expected to be more dependent on upregulation and function of Flk1 on tumor vasculature and thus more susceptible to anti-Flk1 blockade. The apparent lack of toxicity associated with anti-Flk1 antibody treatment can also be attributed to the high specificity of an antibody antagonist.

As DC101 does not cross-react with human VEGFR-2 (KDR) a panel of new antibodies directed against VEGFR-2 was generated, using both the traditional hybridoma method and the antibody phage display technique [76,99,100]. This effort gave rise to a lead candidate, IMC-IC11 [134], a mouse/human chimeric IgG1 derived from a single-chain Fv isolated from a phage display library [135,136]. The antibody binds both soluble and cell surface–expressed VEGFR-2 with high affinity (K_d, ~300 pM), and competes efficiently with radiolabeled VEGF for binding to VEGFR-2-expressing human endothelial cells. Furthermore, it strongly blocks VEGF-induced phosphorylation of both VEGFR-2 and MAPK, and inhibits VEGF-stimulated mitogenesis of human endothelial cells [135,136]. The binding epitopes for IMC-1C11 are located within the first three N-terminal extracellular immunoglobulin-like domains of the receptor, the same domains that encompass the binding site for VEGF [137]. Cross-species examination revealed that IMC-1C11 cross-reacts with VEGFR-2 expressed on endothelial cells of monkeys and dogs, but not with those on rat and mouse. In a canine retinopathy model, IMC-1C11 significantly inhibited retinal neovascularization in newborn dogs induced by high concentration of oxygen [138]. Furthermore, administration of IMC-1C11 to primate rhesus monkey demonstrated a significant impact on the ovary follicle development during the menstrual cycles, an angiogenesis-related event [139]. Recently, it was shown that certain human leukemia cells, including both primary and cultured cell lines, also express functional VEGFR-2 on cell surface [105,140]. IMC-1C11 strongly inhibited VEGF-stimulated leukemia cell proliferation and migration, and significantly prolonged the survival of NOD-SCID mice inoculated with these cells [105,106,141]. Since IMC-1C11 does not cross-react with mouse Flk1, the in vivo antileukemia effect of the antibody is likely to be due to a direct inhibition of cell growth via blockade of the VEGF/VEGFR-2 autocrine loop in human leukemia cells. ImClone Systems initiated a dose-escalating phase I clinical trial in May 2000 in patients with liver metastatic CRC. When IMC-1C11 was infused at 0.2, 0.6, 2.0, and 4.0 mg/kg for 4 weeks, no serious toxicities were observed. Five out of total 14 enrolled patients had SD by week 4 and continued on therapy, with one patient maintaining SD for 6 months [142].

ImClone Systems is currently developing a fully human anti-VEGFR-2 antibody for the treatment of solid tumors and certain leukemias [141,143,144]. This fully human anti-VEGFR-2 IgG1 antibody, IMC-1121B, was generated from a Fab fragment originally isolated from a large antibody phage display library [143,144]. The antibody specifically binds VEGFR-2 with high affinity of 50 pM and block VEGF/VEGFR-2 interaction with an IC_{50} value of ~1 nM. It strongly inhibited VEGF-induced migration of human leukemia cells in vitro, and, when administered in vivo, significantly prolonged survival of NOD-SCID mice inoculated with VEGFR-2^+ human leukemia cells [143,144]. Phase I clinical trials of IMC-1121B were initiated in January 2005 in patients with advanced malignancies.

Another anti-VEGFR-2 antibody currently in clinical development is CDP-791. CDP-791 is a pegylated antibody product comprising a humanized anti-VEGFR-2 F(ab′)2 fragment conjugated to a polyethylene glycol molecule. In August 2003, UCB (Brussels, Belgium, www.ucb-group.com) initiated phase I clinical trials of CDP-791 in patients with a variety of advanced solid tumors to assess the safety of ascending doses of the antibody and its pharmacological activity. Phase II studies are currently being performed in patients with non-small cell lung cancer.

7.5.4 Anti-VEGFR-1 Antibody

The role of VEGFR-1 in tumor angiogenesis is considerably less clear than that of VEGFR-2. A recent study demonstrated that ribozyme-specific targeting of VEGFR-1 in mouse models resulted in inhibition of angiogenesis, tumor growth, and metastasis [145]. Neutralizing antibodies to VEGFR-1 have been produced and were shown to inhibit receptor signaling and cell migration in response to VEGF stimulation [102,146,147]. A neutralizing antimouse VEGFR-1 mAb, MF1, significantly inhibited neovascularization in a Matrigel plug assay and the mouse corneal pocket assay [147]. In a human tumor A431 xenograft model, treatment with MF1 resulted in a significant inhibition of tumor angiogenesis and tumor growth [147], albeit at a level that was not as potent as that observed with DC101. Histological examination of MF1-treated tumors showed decreased microvessel density and extensive tumor necrosis, similar to that found in studies with DC101 antibody. These studies demonstrate a role for VEGFR-1 in tumor angiogenesis. In addition, several recent reports have also demonstrated that VEGFR-1 may play an important role in promoting hematopoietic stem cell mobilization [148] and angiogenesis in both ischemic tissues [149] and tumors [150]. To this end, MF1 has been shown to inhibit tumor-associated angiogenesis and tumor growth through blocking mobilization of VEGFR-1$^+$ hematopoietic stem cells from bone marrow compartment and integration of these cells into newly formed tumor vessels [148,151].

Accumulating evidence suggests that VEGFR-1 is also expressed on surface of a variety of human tumor cells, including those of breast [152] and colorectal carcinoma [102] and lymphoma [104]. Treatment of human colorectal carcinoma cell lines (HT-29 and SW480) with VEGF-B (a ligand specific for VEGFR-1) led to activation of MAPK and significant induction of cell motility and invasiveness (but nor cell proliferation) [102]. VEGFR-1 was observed in >50% of breast carcinoma cell lines and clinical specimen examined. PlGF stimulation of these cells lines resulted in increased level of phosphorylation of VEGFR-1, MAPK, and Akt, cell migration, and efficiency in colony formation [103,152]. Treatment with an antihuman VEGFR-1 antibody (clone 6.12) significantly inhibited the growth of several human breast cancer xenografts in mouse models [103,152]. The same antibody also blocked VEGF-induced lymphoma cell growth in vitro and prolonged the survival of NOD-SCID mice engrafted with a human diffuse large B-cell lymphoma [104]. These results suggest that the antibody exerts its antitumor effect via targeting directly the tumor cells, since it does not cross-react with the mouse receptor. It is noteworthy that combination of the antimouse receptor (MF1, for antiangiogenic activity) and the antihuman receptor (clone 6.12, for direct antitumor activity) antibodies led to significantly enhanced tumor inhibitory activity, compared with each individual antibody therapy [103]. Taken together, these observations strongly indicate that a neutralizing antihuman VEGFR-1 antibody may demonstrate its activity via both the antiangiogenesis and the direct antitumor mechanisms in human therapy. A fully human anti-VEGFR-1 antibody, IMC-18F1 has been recently generated [153], entered phase I clinical trials in 2006 (ImClone Systems Incorporated, www.imclone.com).

7.5.5 Anti-VEGFR-3 Antibody

VEGFR-3, a specific receptor for VEGF-C and VEGF-D, has been identified as a receptor that is critical for the development of the embryonic vascular system but to be postnatally restricted to the endothelial cells of lymphatic vessels and specialized fenestrated capillaries [154,155]. Transgenic expression of VEGFR-3-specific mutant of VEGF-C in mouse skin resulted in increased growth of dermal lymphatic but not vascular endothelium [156]. Similarly, missense mutations that inactivate VEGFR-3 primarily disrupt lymphatic but not blood vessels [157]. Recent reports suggest that, however, VEGFR-3 may also be expressed and functioning on adult vascular endothelial cells, at least under pathological conditions. For example, overexpression of VEGF-C and VEGF-D has been associated with abnormal growth and enlargement of both vascular and lymphatic vessels certain human tumor specimen [158–161]. In addition, enhanced expression of VEGFR-3 has been observed in neoplastic colonic mucosa [162], and the level of VEGFR-3 expression was correlated with prognosis in patients of breast cancer and cutaneous melanoma [163,164]. Finally, a rat antagonist antibody to VEGFR-3, AFL-4, has been shown to suppress growth of tumor xenografts via disruption of microvasculature [165], and to reduce induction of lymphangiogenesis induced by fibroblast growth factor 2 in a mouse corneal model [166]. Taken together, these observations indicate that VEGFR-3, like VEGFR-2, may also represent a good therapeutic target for inhibiting tumor growth and metastasis [167–169].

Neutralizing mAb specific for both mouse and human VEGFR-3 has been generated to further study the role of the receptor in angiogenesis/lymphangiogenesis (ImClone Systems Incorporated, www.imclone.com). A rat mAb to mouse VEGFR-3, mF4-31C1, was tested in a mouse tail skin model of lymphatic regeneration. Whereas normal mice regenerated complete and functional lymphatic vessels within 60 days of surgery, nude mice implanted with VEGF-C-overexpressing human breast carcinoma cells fully recovered within 25 days with hyperplastic vessels. Treatment with mF-31C1 completely inhibited lymphatic regeneration under both conditions without affecting angiogenesis and preexisting lymphatic vessels [170]. The antibody also significantly inhibited tumor-associated lymphangiogenesis and strongly suppressed regional and distant metastasis in a xenografted metastasis mouse model [171]. A fully human antagonist antibody to the human VEGFR-3, hF4-3C5, was recently generated. The antibody strongly inhibits the binding of VEGFR-3 to VEGF-C, abolishes VEGF-C-mediated mitogenic response, cell migration, and tube formation [172].

7.6 TARGETING ENDOTHELIAL CELL ADHESION

Endothelial cells are highly dependent on appropriate interactions with their environment. During migration and invasion, new vessel-forming endothelial cells must be able to adhere to certain extracellular matrix components for anchorage and survival. Endothelial cells must also be able to recognize each other, for the purpose of assembling and maintaining vascular tubular structures. Finally, endothelial cells need to interact with other cells, such as pericytes and smooth muscle cells to form mature blood vessels, or certain inflammatory cells that can activate endothelial cells and initiate the angiogenic process. A considerable number of cell surface proteins have been identified on endothelial cells, which facilitate the endothelial cell–cell adhesion (cadherins) and contact with other cells, including certain integrins (e.g., $\alpha_V\beta_3$, $\alpha_V\beta_5$), which are involved in endothelial cell interaction with extracellular matrix components such as vitronectin and fibronectin [173,174]. Many adhesion molecules are upregulated on activated endothelium and are, therefore, representing potential antiangiogenesis targets. However, the rationale and effects of adhesion inhibitors on angiogenesis are generally less well understood than those of endothelial cell growth inhibitors. A number of mAbs have been generated to validate some of these targets, as outlined later.

7.6.1 ANTI-$\alpha_V\beta_3$ INTEGRIN ANTIBODY

Of the various integrins reported to be relevant for angiogenesis, $\alpha_V\beta_3$ integrin has been studied in greatest detail. Integrin $\alpha_V\beta_3$, rather specifically expressed on endothelial cells, is upregulated on activated endothelium [175]. $\alpha_V\beta_3$ serves as an adhesion receptor for extracellular matrix components (vitronectin, fibrinogen), providing a means for attachment and, thus, survival of endothelial cells in provisional matrices of tissues undergoing neovascularization. Ligand binding triggers intracellular signaling through $\alpha_V\beta_3$ to survival mechanisms [176] and has also been shown to play a role in the activation of VEGFR-2 [177]. It is interesting to note that blood vessels in mice with the integrin α_V gene deleted develop normally [178], although one would expect from antibody studies (see later) that interference with the function of α_V integrin would have detrimental effects.

A mAb against human $\alpha_V\beta_3$ integrin, LM609, has been shown to interfere with angiogenesis in the quail embryo by preventing proper maturation of newly formed blood capillaries [179]. LM609 also inhibits angiogenesis and invasiveness of human vessels in the SCID mouse/human skin transplant tumor model and growth of human breast cancer in the transplanted human skin [180]. The antibody induces apoptosis of endothelial cells in newly forming blood vessels but not in preexisting vessels of the chick chorioallantoic membrane, resulting in regression of human tumors transplanted into the membrane [181]. A humanized form of LM609, Vitaxin, is currently undergoing clinical testing (MedImmune, Inc., www.medimmune.com). In a phase I study [182], cohorts of three patients with metastatic cancer who failed standard therapy received intravenous doses of 10, 50, or 200 mg on days 0 and 21 of a treatment cycle. There was no significant toxicity noted in these three dose levels. Three patients received two cycles of therapy and had SD at day 85 when taken off study. Pharmacokinetic data suggest a half-life of ~7 days with saturable clearance at the highest dose, and sustainable circulating antibody levels of 5 μg/mL with a dose of 200 mg/patient given every 3 weeks. In a phase II study [183], 15 patients with leiomyosarcomas were treated with 0.25 mg/kg Vitaxin once weekly for up to 6 months. As a previous phase I study of patients with various cancers had suggested a response from a leiomyosarcoma patient, this study failed to demonstrate any objective evidence of efficacy. A similar antibody to Vitaxin, MEDI-522, is also developed. MEDI-522 is a humanized antibody that binds to $\alpha_v\beta_3$ with high affinity (K_d, 2.2 to 2.5 nM). In preclinical studies, the antibody inhibited endothelial migration, FGF-induced corneal angiogenesis, and cancer cell attachment to bone and bone resorption by osteoclasts. It also inhibited tumor growth in mice without affecting wound-healing process and clotting time. In a phase I clinical trial [184], MEDI-522 was given to refractory solid tumor patients at 1, 2, 4, or 6 mg/kg per week for up to 1 year. Sixteen patients were treated for a total of 309 doses (mean, 19.3 doses per patient). The half-life of the antibody was 49, 117, 176, and 185 h in the dose group of 1, 2, 4, and 6 mg/kg, respectively. Disease stabilization was achieved in 7 out of 13 evaluated patients (for 2.5 to 9.9 months). The treatment was generally well tolerated, with few grade 3 to 4 toxicities, including 3/16 patients of hyponatremia, and one patient each for asthenia, hypophosphatemia, and infusion reaction, respectively. No antibody response to MEDI-522 was observed in the treated patients. Other current clinical trials include two randomized open-label phase II trials in which MEDI-522 is used in combination with various chemotherapeutic agents in unresectable stage 4 malignant melanoma patients and in prostate cancer patients with bone metastasis.

7.6.2 Anti-$\alpha_5\beta_1$ Integrin Antibody

$\alpha_5\beta_1$ integrin binds to fibronectin and is required by activated, but not quiescent endothelial cells for migration and survival. M200 (volociximab), an anti-$\alpha_5\beta_1$ integrin mouse–human chimeric IgG4 antibody, is developed by PDL BioPharma (www.pdl.com). M200 blocks $\alpha_5\beta_1$ binding to ligand, inhibits tube formation in an in vitro HUVEC cell angiogenesis model, induces endothelial cell apoptosis through activation of caspase-3, and inhibits the migration of endothelial cells and cancer cells [185]. No effect on quiescent endothelial cells was seen. In a monkey model, M200, given via intravitreal or intravenous injection, inhibited chorionic neovascularization stimulated by laser burns at the back of the monkey eyes [186]. Finally, M200/IIA1 (the parent murine antibody of M200) showed direct antitumor activity in an MB-231 breast cancer xenograft model. A phase I study of M200 has been completed [187]. In this trial, a total of 16 patients with mixed solid tumors were given five infusions of volociximab on days 1, 15, 22, 29, and 36 at dose of 0.5, 1, 2.5, 5, or 10 mg/kg. Of 15 patients, 10 achieved SD, including 5 of 6 receiving 10 mg/kg volociximab. Four patients with SD continued volociximab treatment, three of which maintained SD for >16 weeks. Possible drug-related adverse events were mild-to-moderate nausea, fever, vomiting, headache, anorexia, and asthenia and there were no dose-limiting toxicities. A 10 mg/kg dose was well tolerated, achieved monocyte saturation and a mean trough level of 82 μg/mL 2 weeks after the first dose. M200 entered a phase II trial in patients with RCC in January 2005.

7.6.3 Anti-α_v Integrin Antibody

CNTO 95 (Centocor, Inc., www.centocor.com) is a fully human antibody directed against α_v integrins with 1–20 nM binding affinity to both tumor and endothelial cells. CNTO 95 blocks both $\alpha_v\beta_3$ and $\alpha_v\beta_5$, and inhibits cell adhesion, migration, invasion, proliferation, and microvessel sprouting [188]. The antibody is also capable of inducing apoptosis of both tumor and endothelial cells, and mediating ADCC activity against tumor cells [189]. CNTO 95 inhibited the growth of a number of human xenografted tumors in animal models [189]. Further CNTO 95, when combined with paclitaxel, was more effective than either agent alone in reducing proliferation and inducing apoptosis of human melanoma cells, and in inhibiting in vivo tumor growth [190]. In clinical phase I trial, cohorts of 3 to 6 cancer patients were administered the antibody at doses of 0.1, 0.3, 1, 3, or 10 mg/kg every week. The antibody was well tolerated. No objective responses were achieved, but SD was observed in 7 out of 21 patients treated at different antibody doses.

7.6.4 VE-Cadherin

VE-cadherin (VEC) is an endothelial cell–specific cadherin, which mediates adherens junction formation between endothelial cells [191]. VEC is crucial for the proper assembly of vascular structures during angiogenesis and in maintaining vascular integrity in established vessels. VEC null mouse embryos exhibit severely impaired assembly of vascular structures which result in embryonic lethality at day E9.5, demonstrating VEC as an important mediator in developmental angiogenesis. VECs highly restricted distribution and its unique role in vascular function distinguish VEC as a potential target for function-blocking antiangiogenesis or vascular targeting approaches.

Several anti-VEC mAbs have been developed and were shown to inhibit VEC reorganization, increase paracellular permeability, induce endothelial cell apoptosis, and block angiogenesis in vitro and in vivo [192]. The first antitumor proof-of-principle studies targeting VEC with a monoclonal antibody (BV13) were carried out by Liao et al. [193]. These studies demonstrated that BV13 inhibits angiogenesis, tumor growth, and metastasis in several mouse models and validated VEC as a potential target for antiangiogenic therapy. However, systemic treatment with BV13 also resulted in increased vascular permeability and edema in the lung due to disassembly of VEC junctions in normal vasculature [193,194]. The permeability effect of BV13 on normal tissues is not entirely unexpected, since BV13 does not preferentially distribute to tumor blood vessels but rather, binds to vessels in several tissues [194]. Therefore, antiangiogenic therapy with a VEC mAb such as BV13 would not be feasible due to its disrupting activity on existing adherens junctions. Additional studies by Liao et al. [195] described the generation and characterization of another VEC mAb (E4G10), which selectively binds to tumor vasculature, and was shown in preclinical models to inhibit tumor angiogenesis in preclinical models. The E4G10 mAb does not bind to normal vascular, and, importantly, does not induce vascular permeability in normal tissues. Recent work by May et al. [196] has provided insight into the structural basis for E4G10s unique binding specificity and activity. The E4G10 epitope was mapped to the first 10 amino acids of the mature form of VEC. This region of VEC is thought to mediate transadhesion and the E4G10 epitope is absent when VEC molecules are engaged. Thus, E4G10 can inhibit junction formation and angiogenesis but is unable to target normal vasculature because its epitope is masked. In contrast, mAb BV13 targets a different epitope that is not involved in transadhesion and is accessible in assembled adherens junctions. This target region recognized by the E4G10 mAb may represent a distinctive angiogenic epitope and could potentially provide the basis for a novel approach to angiogenic-specific targeting.

7.7 ANTIBODIES DIRECTED AGAINST OTHER ENDOTHELIAL TARGETS

Anionic phospholipids are usually absent from the surface of normal cells, but were exposed during cell activation, apoptosis, necrosis, and malignant transformation [197,198]. Hypoxia, acidity, thrombin, and inflammatory cytokines strongly induce exposure on endothelial cell surface of anionic phospholipids, especially phosphatidylserine (PS), without causing cell death. From 4% to 40% of tumor-associated vessels were observed to be PS-positive depending on tumor types [197,198]. 3G4 is a mouse IgG3, which binds to PS and recognizes tumor blood vessels but not normal tissues [199]. 3G4 localizes to tumor vessels and necrotic tumor cells in mouse models and retards tumor growth in all systems examined, including L540cy tumors and Meth A tumors, without causing severe toxicities [199]. The antibody was an efficient agent in mediating phagocytosis by microphages, ADCC, and CMC—properties that are most likely the basis of their antiangiogenic and antitumor activity [199]. A mouse/human chimeric antibody, bavituximab, was developed from 3G4. Radiolabeled bavituximab localized to prostate tumors in rats and showed inhibition of breast tumor xenograft growth in mice. A toxicological study of bavituximab has been conducted in cynomolgus monkeys administered up to 100 mg/kg i.v. as a single bolus (10–100 times the predicted therapeutic dose). A transient two to fourfold increase in APTT and a 1.4-fold increase in PT were observed at 9 mg/kg dose levels and higher. No other changes in blood cell counts were seen. Phase I trials of bavituximab in advanced cancer patients were initiated by the end of 2005 (Peregrine Pharmaceuticals Inc., www.peregrineinc.com).

Another interesting antibody-based antiangiogenic targeting approach focuses on the extra domain B (ED-B) of the oncofetal fibronectin B, an alternative splice form of fibronectin, which is present in the stroma of tumor tissues and is expressed in neoplastic but not normal blood vessels [200,201]. A high-affinity antibody to ED-B, scFv L19, was generated via screening a phage display library [202]. This antibody strongly stains neovascular structures (but not healthy tissues and mature blood vessels) in aggressive tumors, physiological angiogenesis and in a number of angiogenesis-related disorders [203]. When labeled with a radioisotope or a photosensitive dye, L19 scFv accumulated in tumors and other angiogenic tissues with high selectivity [204,205]. The antibody has been used extensively as a targeting device in fusion protein forms to specifically deliver to tumor sites other biological molecules, including tissue factor [206], tumor necrosis factor [207], interleukin-2 [208], interleukin-12 [209], interferon-γ [210], and so on. In each of the case, the fusion protein selectively accumulated in tumor sites, leading to local vessel occlusion/tumor infarction (for tissue factor fusion protein), infiltration of immune cells as well as increased interferon-γ concentration in tumor sites (for cytokine fusion proteins), thus resulting in significant tumor inhibition and regression.

Other angiogenesis-related targets of interest for mAb therapy in cancer include the angiopoietin-1 and angiopoietin-2 [211], the Tie-1 [212] and Tie-2 [211] receptors, endoglin (CD105) [213], adrenomedullin [214], NG2 [215], TES-23 (CD44) [216], and TEM-1, TEM-5, TEM-7, or TEM-8 [217]. However, at this time there is only scanty information regarding the generating of antibodies to these targets and validation of their targeting in preclinical models.

7.8 CONCLUSION

The progress achieved in development of therapeutic antibodies over the last decade is remarkable. Antibody-based cancer therapeutics is now a mainstay of oncology drug development due to the clinical and commercial success of such agents as Rituxan, Herceptin, and Erbitux. Traditional obstacles in antibody therapy, such as immunogenicity of rodent-derived antibodies and difficulty in producing antibodies in sufficient quantity and quality for commercial application, are rapidly superseded by the advancement in antibody engineering and expression technologies. Antibody chimerization and humanization have greatly reduced the immunogenicity of murine antibodies, thus making repeated dosing schedule in the clinic a reality. With the availability of human antibody transgenic mice and human antibody phage display libraries, fully human antibodies with desired specificities can be readily isolated. Further, the development in new recombinant techniques to genetically manipulate antibodies has enabled us to tailor-make antibody molecules with predefined characteristics such as size, valency, and multispecificities to suit the intended applications. Finally, achievement in technologies related to antibody production, such as high-level mammalian expression systems and antibody production in transgenic plants and animals, will make large-scale antibody production more feasible and economical than ever. It can be expected that development and use of antibody therapeutics in cancer therapy will continue to expand over the next several years.

The recent approval of the anti-VEGF mAb Avastin has validated antiangiogenesis approaches in cancer therapy and further demonstrated the potential of antibody therapeutics. Nevertheless, development of antibody therapeutics against angiogenesis targets and their application in cancer therapy is still at an early stage. To date, there are only a relatively small number of antiangiogenic mAbs in clinical testing. The future direction of mAb therapies against angiogenesis targets will require a return to challenges of the past. Namely, a better understanding of the key regulators of angiogenesis and identification of robustly validated targets is crucial for future development of antiangiogenic antibody therapies. In addition, the feasibility to deliver cytotoxic therapies specifically to tumor vasculature via antibody targeting remains an open question and area for continued research. These questions and challenges are met with considerable interest and investigation both in academia and in industry. Hence, it can be anticipated that a large number of antibody-based therapeutics against angiogenesis targets will enter development in the near future.

REFERENCES

1. Ehrlich, P. On immunity with specific reference to cell life. *Proc R Soc London*, 66, 429, 1900.
2. Kohler, G. and Milstein, C. Continuous cultures of fused cells secreting antibody of predefined specificity. *Nature*, 256, 495, 1975.
3. Glennie, M.J. and Johnson, W.M. Clinical trials of antibody therapy. *Immunol Today*, 21, 403, 2000.
4. Little, M. et al. Of mice and men: hybridoma and recombinant antibodies. *Immunol Today*, 21, 364, 2000.
5. Huston, J.S. et al. Protein engineering of antibody binding sites: recovery of specific activity in an anti-digoxin single-chain Fv analogue produced in *Escherichia coli. Proc Natl Acad Sci USA*, 85, 5879, 1988.
6. Holliger, P., Prospero, T., and Winter, G. "Diabodies": small bivalent and bispecific antibody fragments. *Proc Nat Acad Sci USA*, 90, 6444, 1993.
7. Zhu, Z. et al. High level secretion of a humanized bispecific diabody from *Escherichia coli. Biotechnology* (*NY*), 14, 192, 1996.
8. Todorovska, A. et al. Design and application of diabodies, triabodies, and tetrabodies for cancer targeting. *J Immunol Methods*, 248, 47, 2001.
9. Reiter, Y. et al. Engineering antibody Fv fragments for cancer detection and therapy: disulfide-stabilized Fv fragments. *Nat Biotechnol*, 14, 1239, 1996.
10. Reiter, Y. et al. Engineering interchain disulfide bonds into conserved framework regions of Fv fragments: improved biochemical characteristics of recombinant immunotoxins containing disulfide-stabilized Fv. *Protein Eng*, 7, 697, 1994.
11. Reiter, Y. et al. Disulfide stabilization of antibody Fv: computer predictions and experimental evaluation. *Protein Eng*, 8, 1323, 1995.
12. Reiter, Y. et al. Improved binding and antitumor activity of a recombinant anti-erbB2 immunotoxin by disulfide stabilization of the Fv fragment. *J Biol Chem*, 269, 18327, 1994.

13. Reiter, Y. and Pastan, I. Antibody engineering of recombinant Fv immunotoxins for improved targeting of cancer: disulfide-stabilized Fv immunotoxins. *Clin Cancer Res*, 2, 245, 1996.
14. Muyldermans, S. et al. Sequence and structure of VH domain from naturally occurring camel heavy chain immunoglobulins lacking light chains. *Protein Eng*, 7, 1129, 1994.
15. Els Conrath, K. et al. Camel single-domain antibodies as modular building units in bispecific and bivalent antibody constructs. *J Biol Chem*, 276, 7346, 2001.
16. Morrison, S.L. et al. Chimeric human antibody molecules: mouse antigen-binding domains with human constant region domains. *Proc Natl Acad Sci USA*, 81, 6851, 1984.
17. Jones, P.T. et al. Replacing the complementarity-determining regions in a human antibody with those from a mouse. *Nature*, 321, 522, 1986.
18. Pendley, C., Schantz, A., and Wagner, C. Immunogenicity of therapeutic monoclonal antibodies. *Curr Opin Mol Ther*, 5, 172, 2003.
19. Schellekens, H. Immunogenicity of therapeutic proteins: clinical implications and future prospects. *Clin Ther*, 24, 1720, 2002.
20. Gonzales, N.R. et al. Minimizing the immunogenicity of antibodies for clinical application. *Tumour Biol*, 26, 31, 2005.
21. McCafferty, J. et al. Phage antibodies: filamentous phage displaying antibody variable domains. *Nature*, 348, 552, 1990.
22. Clackson, T. et al. Making antibody fragments using phage display libraries. *Nature*, 352, 624, 1991.
23. Winter, G. et al. Making antibodies by phage display technology. *Annu Rev Immunol*, 12, 433, 1994.
24. Green, L.L. et al. Antigen-specific human monoclonal antibodies from mice engineered with human Ig heavy and light chain YACs. *Nat Genet*, 7, 13, 1994.
25. Lonberg, N. et al. Antigen-specific human antibodies from mice comprising four distinct genetic modifications. *Nature*, 368, 856, 1994.
26. Lonberg, N. and Huszar, D. Human antibodies from transgenic mice. *Int Rev Immunol*, 13, 65, 1995.
27. Boder, E.T. and Wittrup, K.D. Yeast surface display for directed evolution of protein expression, affinity, and stability. *Methods Enzymol*, 328, 430, 2000.
28. Feldhaus, M.J. and Siegel, R.W. Yeast display of antibody fragments: a discovery and characterization platform. *J Immunol Methods*, 290, 69, 2004.
29. Daugherty, P.S. et al. Antibody affinity maturation using bacterial surface display. *Protein Eng*, 11, 825, 1998.
30. Harvey, B.R. et al. Anchored periplasmic expression, a versatile technology for the isolation of high-affinity antibodies from *Escherichia coli*-expressed libraries. *Proc Natl Acad Sci USA*, 101, 9193, 2004.
31. Feldhaus, M.J. et al. Flow-cytometric isolation of human antibodies from a nonimmune *Saccharomyces cerevisiae* surface display library. *Nat Biotechnol*, 21, 163, 2003.
32. Chen, G. et al. Isolation of high-affinity ligand-binding proteins by periplasmic expression with cytometric screening (PECS). *Nat Biotechnol*, 19, 537, 2001.
33. Irving, R.A. et al. Ribosome display and affinity maturation: from antibodies to single V-domains and steps towards cancer therapeutics. *J Immunol Methods*, 248, 31, 2001.
34. Lipovsek, D. and Pluckthun, A. In-vitro protein evolution by ribosome display and mRNA display. *J Immunol Methods*, 290, 51, 2004.
35. Humphreys, D.P. Production of antibodies and antibody fragments in *Escherichia coli* and a comparison of their functions, uses and modification. *Curr Opin Drug Discov Devel*, 6, 188, 2003.
36. Simmons, L.C. et al. Expression of full-length immunoglobulins in *Escherichia coli*: rapid and efficient production of aglycosylated antibodies. *J Immunol Methods*, 263, 133, 2002.
37. Chadd, H.E. and Chamow, S.M. Therapeutic antibody expression technology. *Curr Opin Biotechnol*, 12, 188, 2001.
38. Giddings, G. et al. Transgenic plants as factories for biopharmaceuticals. *Nat Biotechnol*, 18, 1151, 2000.
39. Hood, E.E., Woodard, S.L., and Horn, M.E. Monoclonal antibody manufacturing in transgenic plants–myths and realities. *Curr Opin Biotechnol*, 13, 630, 2002.

40. Yeung, P.K. Technology evaluation: transgenic antithrombin III (rhAT-III), Genzyme transgenics. *Curr Opin Mol Ther*, 2, 336, 2000.
41. Mukhtar, M., Parveen, Z., and Pomerantz, R.J. Technology evaluation: PRO-542, Progenics Pharmaceuticals Inc. *Curr Opin Mol Ther*, 2, 697, 2000.
42. Carver, A.S. et al. Transgenic livestock as bioreactors: stable expression of human alpha-1-antitrypsin by a flock of sheep. *Biotechnology (NY)*, 11, 1263, 1993.
43. Twyman, R.M., Schillberg, S., and Fischer, R. Transgenic plants in the biopharmaceutical market. *Expert Opin Emerg Drugs*, 10, 185, 2005.
44. Russell, D.A. Feasibility of antibody production in plants for human therapeutic use. *Curr Top Microbiol Immunol*, 240, 119, 1999.
45. Zeitlin, L. et al. A humanized monoclonal antibody produced in transgenic plants for immunoprotection of the vagina against genital herpes. *Nat Biotechnol*, 16, 1361, 1998.
46. Ludwig, D.L. et al. Conservation of receptor antagonist anti-tumor activity by epidermal growth factor receptor antibody expressed in transgenic corn seed. *Hum Antibodies*, 13, 81, 2004.
47. Bakker, H. et al. Galactose-extended glycans of antibodies produced by transgenic plants. *Proc Natl Acad Sci USA*, 98, 2899, 2001.
48. Cabanes-Macheteau, M. et al. N-Glycosylation of a mouse IgG expressed in transgenic tobacco plants. *Glycobiology*, 9, 365, 1999.
49. Carter, P. Improving the efficacy of antibody-based cancer therapies. *Nat Rev Cancer*, 1, 118, 2001.
50. Ferrara, N. et al. Discovery and development of bevacizumab, an anti-VEGF antibody for treating cancer. *Nat Rev Drug Discov*, 3, 391, 2004.
51. Waksal, H. Role of an anti-epidermal growth factor receptor in treating cancer. *Cancer Metastasis Rev*, 18, 427, 1999.
52. Franklin, M.C. et al. Insights into ErbB signaling from the structure of the ErbB2–pertuzumab complex. *Cancer Cell*, 5, 317, 2004.
53. Yagita, H. et al. TRAIL and its receptors as targets for cancer therapy. *Cancer Sci*, 95, 777, 2004.
54. Coiffier, B. Rituximab in combination with CHOP improves survival in elderly patients with aggressive non-Hodgkin's lymphoma. *Semin Oncol*, 29, 18, 2002.
55. Ligibel, J.A. and Winer, E.P. Trastuzumab/chemotherapy combinations in metastatic breast cancer. *Semin Oncol*, 29, 38, 2002.
56. Saltz, L.B. et al. Phase II trial of cetuximab in patients with refractory colorectal cancer that expresses the epidermal growth factor receptor. *J Clin Oncol*, 22, 1201, 2004.
57. Lewis, G.D. et al. Differential responses of human tumor cell lines to anti-p185HER2 monoclonal antibodies. *Cancer Immunol Immunother*, 37, 255, 1993.
58. Reff, M.E. et al. Depletion of B cells in vivo by a chimeric mouse human monoclonal antibody to CD20. *Blood*, 83, 435, 1994.
59. Clynes, R.A. et al. Inhibitory Fc receptors modulate in vivo cytotoxicity against tumor targets. *Nat Med*, 6, 443, 2000.
60. Sievers, E.L. and Linenberger, M. Mylotarg: antibody-targeted chemotherapy comes of age. *Curr Opin Oncol*, 13, 522, 2001.
61. Krasner, C. and Joyce, R.M. Zevalin: 90yttrium labeled anti-CD20 (ibritumomab tiuxetan), a new treatment for non-Hodgkin's lymphoma. *Curr Pharm Biotechnol*, 2, 341, 2001.
62. Cheson, B. Bexxar (Corixa/GlaxoSmithKline). *Curr Opin Investig Drugs*, 3, 165, 2002.
63. Cao, Y. and Lam, L. Bispecific antibody conjugates in therapeutics. *Adv Drug Deliv Rev*, 55, 171, 2003.
64. Kontermann, R.E. Recombinant bispecific antibodies for cancer therapy. *Acta Pharmacol Sin*, 26, 1, 2005.
65. James, N.D. et al. A phase II study of the bispecific antibody MDX-H210 (anti-HER2 × CD64) with GM-CSF in HER2+ advanced prostate cancer. *Br J Cancer*, 85, 152, 2001.
66. Bhattacharya-Chatterjee, M., Chatterjee, S.K., and Foon, K.A. Anti-idiotype antibody vaccine therapy for cancer. *Expert Opin Biol Ther*, 2, 869, 2002.
67. Wentworth, A.D. et al. Antibodies have the intrinsic capacity to destroy antigens. *Proc Natl Acad Sci USA*, 97, 10930, 2000.
68. French, R.R. et al. CD40 antibody evokes a cytotoxic T-cell response that eradicates lymphoma and bypasses T-cell help. *Nat Med*, 5, 548, 1999.

69. Hurwitz, A.A. et al. Combination immunotherapy of primary prostate cancer in a transgenic mouse model using CTLA-4 blockade. *Cancer Res*, 60, 2444, 2000.
70. Wheeler, Y.Y., Chen, S.Y., and Sane, D.C. Intrabody and intrakine strategies for molecular therapy. *Mol Ther*, 8, 355, 2003.
71. Lobato, M.N. and Rabbitts, T.H. Intracellular antibodies as specific reagents for functional ablation: future therapeutic molecules. *Curr Mol Med*, 4, 519, 2004.
72. Hicklin, D.J. et al. Monoclonal antibody strategies to block angiogenesis. *Drug Discov Today*, 6, 517, 2001.
73. Zhu, Z., Bohlen, P., and Witte, L. Clinical development of angiogenesis inhibitors to vascular endothelial growth factor and its receptors as cancer therapeutics. *Curr Cancer Drug Targets*, 2, 135, 2002.
74. Kerbel, R. and Folkman, J. Clinical translation of angiogenesis inhibitors. *Nat Rev Cancer*, 2, 727, 2002.
75. Ferrara, N. et al. Discovery and development of bevacizumab, an anti-VEGF antibody for treating cancer. *Nat Rev Drug Discov*, 3, 391, 2004.
76. Manley, P.W. et al. Advances in the structural biology, design and clinical development of VEGF-R kinase inhibitors for the treatment of angiogenesis. *Biochim Biophys Acta*, 1697, 17, 2004.
77. Auguste, P. et al. Molecular mechanisms of tumor vascularization. *Crit Rev Oncol Hematol*, 54, 53, 2005.
78. Ferrara, N. The role of VEGF in the regulation of physiological and pathological angiogenesis. *EXS* (94), 209, 2005.
79. Paz, K. and Zhu, Z. Development of angiogenesis inhibitors to vascular endothelial growth factor receptor 2. Current status and future perspective. *Front Biosci*, 10, 1415, 2005.
80. Folkman, J. Angiogenesis-dependent diseases. *Semin Oncol*, 28, 536, 2001.
81. Folkman, J. Role of angiogenesis in tumor growth and metastasis. *Semin Oncol*, 29, 15, 2002.
82. Bergers, G. and Benjamin, L.E. Tumorigenesis and the angiogenic switch. *Nat Rev Cancer*, 3, 401, 2003.
83. Blagosklonny, M.V. Antiangiogenic therapy and tumor progression. *Cancer Cell*, 5, 13, 2004.
84. Leung, D.W. et al. Vascular endothelial growth factor is a secreted angiogenic mitogen. *Science*, 246, 1306, 1989.
85. Ferrara, N. Vascular endothelial growth factor: basic science and clinical progress. *Endocr Rev*, 25, 581, 2004.
86. Ferrara, N., Gerber, H.P., and LeCouter, J. The biology of VEGF and its receptors. *Nat Med*, 9, 669, 2003.
87. Shibuya, M. et al. Nucleotide sequence and expression of a novel human receptor-type tyrosine kinase gene (*flt-1*) closely related to the fms family. *Oncogene*, 5, 519, 1990.
88. De Vries, C. et al. The *fms*-like tyrosine kinase, a receptor for vascular endothelial growth factor. *Science*, 255, 989, 1992.
89. Terman, B.I. et al. Identification of the KDR tyrosine kinase receptor as a receptor for vascular endothelial growth factor. *Biochem Biophys Res Commun*, 187, 1579, 1992.
90. Millauer, B. et al. High affinity VEGF binding and developmental expression suggest Flk-1 as a major regulator of vasculogenesis and angiogenesis. *Cell*, 72, 835, 1993.
91. Ferrara, N. et al. Heterozygous embryonic lethality induced by targeted inactivation of the *VEGF* gene. *Nature*, 380, 439, 1996.
92. Carmeliet, P. et al. Abnormal blood vessel development and lethality in embryos lacking a single VEGF allele. *Nature*, 380, 435, 1996.
93. Fong, G.H. et al. Role of the Flt-1 receptor tyrosine kinase in regulating the assembly of vascular endothelium. *Nature*, 376, 66, 1995.
94. Hiratsuka, S. et al. Flt-1 lacking the tyrosine kinase domain is sufficient for normal development and angiogenesis in mice. *Proc Natl Acad Sci USA*, 95, 9349, 1998.
95. Shalaby, F. et al. Failure of blood-island formation and vasculogenesis in Flk-1-deficient mice. *Nature*, 376, 62, 1995.
96. Shalaby, F. et al. A requirement for Flk1 in primitive and definitive hematopoiesis and vasculogenesis. *Cell*, 89, 981, 1997.
97. Veikkola, T. and Alitalo, K. VEGFs, receptors and angiogenesis. *Semin Cancer Biol*, 3, 211, 1999.
98. Taipale, J. Vascular endothelial growth factor receptor-3. *Curr Top Microbiol Immunol*, 237, 85, 1999.

99. Witte, L. et al. Monoclonal antibodies targeting the VEGF receptor-2 (Flk1/KDR) as an anti-angiogenic therapeutic strategy. *Cancer Metastasis Rev*, 17, 155, 1998.
100. Zhu, Z. and Witte, L. Inhibition of tumor growth and metastasis by targeting tumor-associated angiogenesis with antagonists to the receptors of vascular endothelial growth factor. *Invest New Drugs*, 17, 195, 1999.
101. Hicklin, D.J. and Ellis, L.M. Role of the vascular endothelial growth factor pathway in tumor growth and angiogenesis. *J Clin Oncol*, 23, 1011, 2005.
102. Fan, F. et al. Expression and function of vascular endothelial growth factor receptor-1 on human colorectal cancer cells. *Oncogene*, 24, 2647, 2005.
103. Wang, E.S. et al. Targeting autocrine and paracrine VEGF receptor pathways inhibits human lymphoma xenografts in vivo. *Blood*, 104, 2893, 2004.
104. Dias, S. et al. Autocrine stimulation of VEGFR-2 activates human leukemic cell growth and migration. *J Clin Invest*, 106, 511, 2000.
105. Dias, S. et al. Inhibition of both paracrine and autocrine VEGF/VEGFR-2 signaling pathways is essential to induce long-term remission of xenotransplanted human leukemias. *Proc Natl Acad Sci USA*, 98, 10857, 2001.
106. Kim, K.J. et al. Inhibition of vascular endothelial growth factor-induced angiogenesis suppresses tumour growth in vivo. *Nature*, 362, 841, 1993.
107. Yuan, F. et al. Time-dependent vascular regression and permeability changes in established human tumor xenografts induced by an anti-vascular endothelial growth factor/vascular permeability factor antibody. *Proc Natl Acad Sci USA*, 93, 14765, 1996.
108. Borgstrom, P. et al. Complete inhibition of angiogenesis and growth of microtumours by anti-vascular endothelial growth factor neutralizing antibody: novel concepts of angiostatic therapy from intravital videomicroscopy. *Cancer Res*, 56, 4032, 1996.
109. Presta, L.G. et al. Humanization of an anti-vascular endothelial growth factor mAb for the therapy of solid tumors and other disorders. *Cancer Res*, 57, 4593, 1997.
110. Ryan, A.M. et al. Preclinical safety evaluation of rhuMAbVEGF, an antiangiogenic humanized mAb. *Toxicol Pathol*, 27, 78, 1999.
111. Kabbinavar, F. et al. Phase II, randomized trial comparing bevacizumab plus fluorouracil (FU)/leucovorin (LV) with FU/LV alone in patients with metastatic colorectal cancer. *J Clin Oncol*, 21, 60, 2003.
112. Hurwitz, H. et al. Bevacizumab plus irinotecan, fluorouracil, and leucovorin for metastatic colorectal cancer. *N Engl J Med*, 350, 2335, 2004.
113. See Genetench Inc. Web site at www.gene.com for press release on January 27, 2005: preliminary positive results from phase III trial of Avastin plus FOLFOX4 chemotherapy presented at ASCO GI meeting.
114. Willett, C.G. et al. Direct evidence that the VEGF-specific antibody bevacizumab has antivascular effects in human rectal cancer. *Nat Med*, 10, 145, 2004.
115. Kabbinavar, F.F. et al. Addition of bevacizumab to bolus fluorouracil and leucovorin in first-line metastatic colorectal cancer: results of a randomized Phase II trial. *J Clin Oncol*, 23, 3697, 2005.
116. Chen, H.X. et al. TRC Participating Investigators. Bevacizumab (BV) plus 5-FU/leucovorin (FU/LV) for advanced colorectal cancer (CRC) that progressed after standard chemotherapies: an NCI Treatment Referral Center trial (TRC-0301). *J Clin Oncol—2004 ASCO Annual Meeting Proceedings* (*Post-Meeting Edition*), 22 (Abstract 3515), 2004.
117. Johnson, D.H. et al. Randomized phase II trial comparing bevacizumab plus carboplatin and paclitaxel with carboplatin and paclitaxel alone in previously untreated locally advanced or metastatic non-small-cell lung cancer. *J Clin Oncol*, 22, 2184, 2004.
118. See Genetench Inc. Web site at www.gene.com for press release on May 13, 2005: interim analysis of phase III trial shows Avastin plus chemotherapy extends survival of patients with first-line non-squamous, non-small cell lung cancer.
119. Yang, J.C. et al. A randomized trial of bevacizumab, an anti-vascular endothelial growth factor antibody, for metastatic renal cancer. *N Engl J Med*, 349, 427, 2003.
120. Cobleigh, M.A. et al. A phase I/II dose-escalation trial of bevacizumab in previously treated metastatic breast cancer. *Semin Oncol*, 30, 117, 2003.

121. Miller, K.D. et al. Randomized phase III trial of capecitabine compared with bevacizumab plus capecitabine in patients with previously treated metastatic breast cancer. *J Clin Oncol*, 23, 792, 2005.
122. See Genetench Inc. Web site at www.gene.com for press release on May 13, 2005: avastin plus chemotherapy showed doubling of time without cancer progression in phase III trial of first-line metastatic breast cancer patients.
123. Rockwell, P. et al. In vitro neutralization of vascular endothelial growth factor activation of Flk-1 by a monoclonal antibody. *Mol Cell Differ*, 3, 91, 1995.
124. Prewett, M. et al. Antivascular endothelial growth factor receptor (fetal liver kinase 1) monoclonal antibody inhibits tumor angiogenesis and growth of several mouse and human tumors. *Cancer Res*, 59, 5209, 1999.
125. Kozin, S.V. et al. Vascular endothelial growth factor receptor-2-blocking antibody potentiates radiation-induced long-term control of human tumor xenografts. *Cancer Res*, 61, 39, 2001.
126. Bruns, C.J. et al. Effect of the vascular endothelial growth factor receptor-2 antibody DC101 plus gemcitabine on growth, metastasis and angiogenesis of human pancreatic cancer growing orthotopically in nude mice. *Int J Cancer*, 102, 101, 2002.
127. Sweeney, P. et al. Anti-vascular endothelial growth factor receptor 2 antibody reduces tumorigenicity and metastasis in orthotopic prostate cancer xenografts via induction of endothelial cell apoptosis and reduction of endothelial cell matrix metalloproteinase type 9 production. *Clin Cancer Res*, 8, 2714, 2002.
128. Hanahan, D., Bergers, G., and Bergsland, E. Less is more, regularly: metronomic dosing of cytotoxic drugs can target tumor angiogenesis in mice. *J Clin Invest*, 105, 1045, 2000.
129. Klement, G. et al. Continuous low-dose therapy with vinblastine and VEGF receptor-2 antibody induces sustained tumor regression without overt toxicity. *J Clin Invest*, 105, R15, 2000.
130. Man, S. et al. Antitumor effects in mice of low-dose (metronomic) cyclophosphamide administered continuously through the drinking water. *Cancer Res*, 62, 2731, 2002.
131. Zhang, L. et al. Combined anti-fetal liver kinase 1 monoclonal antibody and continuous low-dose doxorubicin inhibits angiogenesis and growth of human soft tissue sarcoma xenografts by induction of endothelial cell apoptosis. *Cancer Res*, 62, 2034, 2002.
132. Zimmermann, R.C. et al. Preovulatory treatment of mice with anti-VEGF receptor 2 antibody inhibits angiogenesis in corpora lutea. *Microvasc Res*, 62, 15, 2001.
133. Pauli, S.A. et al. The vascular endothelial growth factor (VEGF)/VEGF receptor 2 pathway is critical for blood vessel survival in corpora lutea of pregnancy in the rodent. *Endocrinology*, 146, 1301, 2004.
134. Hunt, S. Technology evaluation: IMC-1C11, ImClone Systems. *Curr Opin Mol Ther*, 3, 418, 2001.
135. Zhu, Z. et al. Inhibition of vascular endothelial growth factor-induced receptor activation with anti-kinase insert domain-containing receptor single-chain antibodies from a phage display library. *Cancer Res*, 58, 3209, 1998.
136. Zhu, Z. et al. Inhibition of vascular endothelial growth factor induced mitogenesis of human endothelial cells by a chimeric anti-kinase insert domain-containing receptor antibody. *Cancer Lett*, 136, 203, 1999.
137. Lu, D. et al. Identification of the residues in the extracellular region of KDR important for interaction with vascular endothelial growth factor and neutralizing anti-KDR antibodies. *J Biol Chem*, 275, 14321, 2000.
138. McLeod, D.S. et al. Localization of VEGF receptor-2 (KDR/Flk-1) and effects of blocking it in oxygen-induced retinopathy. *Invest Ophthalmol Vis Sci*, 43, 474, 2002.
139. Zimmermann, R.C. et al. Administration of antivascular endothelial growth factor receptor 2 antibody in the early follicular phase delays follicular selection and development in the rhesus monkey. *Endocrinology*, 143, 2496, 2002.
140. Fiedler, W. et al. Vascular endothelial growth factor, a possible paracrine growth factor in human acute myeloid leukemia. *Blood*, 89, 1870, 1997.
141. Zhu, Z. et al. Inhibition of human leukemia in an animal model with human antibodies directed against vascular endothelial growth factor receptor 2. Correlation between antibody affinity and biological activity. *Leukemia*, 17, 604, 2003.
142. Posey, J.A. et al. A phase I study of anti-kinase insert domain-containing receptor antibody, IMC-1C11, in patients with liver metastases from colorectal carcinoma. *Clin Cancer Res*, 9, 1323, 2003.

143. Lu, D. et al. Selection of high affinity human neutralizing antibodies to VEGFR2 from a large antibody phage display library for antiangiogenesis therapy. *Int J Cancer*, 97, 393, 2002.
144. Lu, D. et al. Tailoring in vitro selection for a picomolar affinity human antibody directed against vascular endothelial growth factor receptor 2 for enhanced neutralizing activity. *J Biol Chem*, 278, 43496, 2003.
145. Pavco, P.A. et al. Antitumor and antimetastatic activity of ribozymes targeting the messenger RNA of vascular endothelial growth factor receptors. *Clin Canc Res*, 5, 2094, 2000.
146. Kanno, S. et al. Roles of two VEGF receptors, Flt-1 and KDR, in the signal transduction of VEGF effects in human vascular endothelial cells. *Oncogene*, 19, 2138, 2000.
147. Wu, Y. et al. Inhibition of tumor growth and angiogenesis in a mouse model by a neutralizing anti-Flt-1 monoclonal antibody. *Proc Amer Assoc Cancer Res*, 42 (Abstract 4436), 2001.
148. Hattori, K. et al. Placental growth factor reconstitutes hematopoiesis by recruiting VEGFR1(+) stem cells from bone-marrow microenvironment. *Nat Med*, 8, 841, 2002.
149. Luttun, A. et al. Revascularization of ischemic tissues by PlGF treatment, and inhibition of tumor angiogenesis, arthritis and atherosclerosis by anti-Flt1. *Nat Med*, 8, 831, 2002.
150. Hiratsuka, S. et al. MMP9 induction by vascular endothelial growth factor receptor-1 is involved in lung-specific metastasis. *Cancer Cell*, 2, 289, 2002.
151. Lyden, D. et al. Impaired recruitment of bone-marrow-derived endothelial and hematopoietic precursor cells blocks tumor angiogenesis and growth. *Nat Med*, 7, 1194, 2001.
152. Wu, Y. et al. Monoclonal antibody against VEGFR-1 directly inhibits growth of human breast tumors. *Proc Amer Assoc Cancer Res*, 44 (Abstract 6340), 2003.
153. Wu, Y. et al. A fully human monoclonal antibody against vegfr-1 inhibits growth of human breast cancers. *Proc Amer Assoc Cancer Res*, 45 (Abstract), 2004.
154. Kaipainen, A. et al. Expression of the fms-like tyrosine kinase *FLT4* gene becomes restricted to endothelium of lymphatic vessels during development. *Proc Natl Acad Sci USA*, 92, 3566, 1995.
155. Partanen, T.A. et al. VEGF-C and VEGF-D expression in neuroendocrine cells and their receptor, VEGFR-3, in fenestrated blood vessels in human tissues. *FASEB J*, 14, 2087, 2000.
156. Jeltsch, M. et al. Hyperplasia of lymphatic vessels in VEGF-C transgenic mice. *Science*, 276, 1423, 1997.
157. Karkkainen, M.J. et al. Missense mutations interfere with VEGFR-3 signalling in primary lymphoedema. *Nat Genet*, 25, 153, 2000.
158. Karpanen, T. et al. Vascular endothelial growth factor C promotes tumor lymphangiogenesis and intralymphatic tumor growth. *Cancer Res*, 61, 1786, 2001.
159. Tsurusaki, T. et al. Vascular endothelial growth factor-C expression in human prostatic carcinoma and its relationship to lymph node metastasis. *Br J Cancer*, 80, 309, 1999.
160. Achen, M.G. et al. Localization of vascular endothelial growth factor-D in malignant melanoma suggests a role in tumour angiogenesis. *J Pathol*, 193, 147, 2001.
161. White, J.D. et al. Vascular endothelial growth factor-D expression is an independent prognostic marker for survival in colorectal carcinoma. *Cancer Res*, 62, 1669, 2002.
162. Andre, T. et al. VEGF, VEGF-B, VEGF-C and their receptors KDR, FLT-1 and FLT-4 during the neoplastic progression of human colonic mucosa. *Int J Cancer*, 86, 174, 2000.
163. Valtola, R. et al. VEGFR-3 and its ligand VEGF-C are associated with angiogenesis in breast cancer. *Am J Pathol*, 154, 1381, 1999.
164. Clarijs, R. et al. Induction of vascular endothelial growth factor receptor-3 expression on tumor microvasculature as a new progression marker in human cutaneous melanoma. *Cancer Res*, 62, 7059, 2002.
165. Kubo, H. et al. Involvement of vascular endothelial growth factor receptor-3 in maintenance of integrity of endothelial cell lining during tumor angiogenesis. *Blood*, 96, 546, 2000.
166. Kubo, H. et al. Blockade of vascular endothelial growth factor receptor-3 signaling inhibits fibroblast growth factor-2-induced lymphangiogenesis in mouse cornea. *Proc Natl Acad Sci USA*, 99, 8868, 2002.
167. Karpanen, T. and Alitalo, K. Lymphatic vessels as targets of tumor therapy? *J Exp Med*, 194, F37, 2001.
168. He, Y., Karpanen, T., and Alitalo, K. Role of lymphangiogenic factors in tumor metastasis. *Biochim Biophys Acta*, 1654, 3, 2004.

169. Detmar, M. and Hirakawa, S. The formation of lymphatic vessels and its importance in the setting of malignancy. *J Exp Med*, 196, 713, 2002.
170. Pytowski, B. et al. Complete and specific inhibition of adult lymphatic regeneration by a novel VEGFR-3 neutralizing antibody. *J Natl Cancer Inst*, 97, 14, 2005.
171. Robert, N. et al. Inhibition of VEGR-3 activation with the antagonistic antibody more potently suppresses lymph node and distant metastases than inactivation of VEGFR-2. *Cancer Res*, 66, 2650, 2006.
172. Persaud, K. et al. Involvement of the VEGF receptor 3 in tubular morphogenesis demonstrated with a human anti-human VEGFR-3 monoclonal antibody that antagonizes receptor activation by VEGF-C. *J Cell Sci*, 117, 2745, 2004.
173. Hynes, R.O. A reevaluation of integrins as regulators of angiogenesis. *Nat Med*, 8, 918, 2002.
174. Tucker, G.C. Alpha v integrin inhibitors and cancer therapy. *Curr Opin Investig Drugs*, 4, 722, 2003.
175. Brooks, P., Clark, R.A., and Cheresh, D.A. Requirement of vascular integrin alpha v beta 3 for angiogenesis. *Science*, 264, 569, 1994.
176. Stromblad, S. and Cheresh, D. Cell adhesion and angiogenesis. *Trends Cell Biol*, 6, 462, 1996.
177. Soldi, R. et al. Role of alpha v beta3 integrin in the activation of vascular endothelial growth factor receptor-2. *EMBO J*, 18, 882, 1999.
178. Bader, B. et al. Extensive vasculogenesis, angiogenesis, and organogenesis precede lethality in mice lacking all alpha v integrins. *Cell*, 95, 507, 1998.
179. Drake, C., Cheresh, D.A., and Little, C.D. An antagonist of integrin alpha v beta 3 prevents maturation of blood vessels during embryonic neovascularization. *J Cell Sci*, 108, 2655, 1995.
180. Brooks, P. et al. Anti-integrin alpha v beta 3 blocks human breast cancer growth and angiogenesis in human skin. *J Clin Invest*, 96, 1815, 1995.
181. Brooks, P. et al. Integrin alpha(v)beta(3) antagonists promote tumor regression by inducing apoptosis of angiogenic blood vessels. *Cell*, 79, 1157, 1994.
182. Posey, J.A. et al. A pilot trial of Vitaxin, a humanized anti-vitronectin receptor (anti alpha v beta 3) antibody in patients with metastatic cancer. *Cancer Biother Radiopharm*, 16, 125, 2001.
183. Patel, S.R. et al. Pilot study of Vitaxin—an angiogenesis inhibitor in patients with advanced leiomyosarcomas. *Cancer*, 92, 1347, 2001.
184. Faivre, S.J. et al. Safety profile and pharmacokinetic analysis of Medi-522, a novel humanized monoclonal antibody that targets $\alpha v\beta 3$ integrin receptor, in patients with refractory solid tumors. *Proc Amer Soc Clin Oncol*, 22 (Abstract 832), 2003.
185. Ho, S.K. et al. The effect of a chimeric anti-integrin $\alpha 5\beta 1$ antibody (M200) on the migration of HUVECs and human cancer cells. *Proc Amer Assoc Cancer Res*, 45 (Abstract 1452), 2004.
186. Ramakrishnan, V. et al. A function-blocking chimeric antibody, Eos200-4, against $\alpha 5\beta 1$ integrin inhibits angiogenesis in a monkey model. *Proc Amer Assoc of Cancer Res*, 44 (Abstract 3052), 2003.
187. Ricart, A. et al. A phase I dose-escalation study of anti-$\alpha 5\beta 1$ integrin monoclonal antibody (M200) in patients with refractory solid tumors. *Eur J Cancer Suppl*, 2, 8 (Abstract 166), 2004.
188. Nakada, M.T. et al. CNTO 95, a novel human anti-$\alpha v\beta 3/\alpha v\beta 5$ antibody demonstrates anti-inflammatory activity by inhibiting leukocyte transmigration in vivo. *Proc Amer Assoc Cancer Res*, 44 (Abstract 3036), 2003.
189. Trikha, M. et al. Inhibition of tumor growth and angiogenesis by CNTO 95, a fully-human monoclonal antibody to integrins $\alpha v\beta 3$ and avB5. *Proc Amer Assoc Cancer Res*, 44 (Abstract 754), 2003.
190. Nemeth, J.A. et al. Synergistic anti-tumor activity of paclitaxel with CNTO 95, a fully human monoclonal antibody that targets αv integrins. *Proc Amer Assoc Cancer Res*, 45 (Abstract 71), 2004.
191. Dejana, E. Endothelial cell–cell junctions: happy together. *Nat Rev Mol Cell Biol*, 5, 261, 2004.
192. Corada, M. et al. Monoclonal antibodies directed to different regions of vascular endothelial cadherin extracellular domain affect adhesion and clustering of the protein and modulate endothelial permeability. *Blood*, 97, 1679, 2001.
193. Liao, F. et al. Monoclonal antibody to vascular endothelial-cadherin is a potent inhibitor of angiogenesis, tumor growth, and metastasis. *Cancer Res*, 60, 6805, 2000.
194. Corada, M. et al. Vascular endothelial-cadherin is an important determinant of microvascular integrity in vivo. *Proc Natl Acad Sci USA*, 96, 9815, 1999.

195. Liao, F. et al. Selective targeting of angiogenic tumor vasculature by vascular endothelial-cadherin antibody inhibits tumor growth without affecting vascular permeability. *Cancer Res*, 62, 2567, 2002.
196. May, C. et al. Identification of a transiently exposed VE-cadherin epitope that allows for specific targeting of an antibody to the tumor neovasculature. *Blood*, 105, 4337, 2005.
197. Ran, S., Downes, A., and Thorpe, P.E. Increased exposure of anionic phospholipids on the surface of tumor blood vessels. *Cancer Res*, 62, 6132, 2002.
198. Ran, S. and Thorpe, P.E. Phosphatidylserine is a marker of tumor vasculature and a potential target for cancer imaging and therapy. *Int J Radiat Oncol Biol Phys*, 54, 1479, 2002.
199. Ran, S. et al. Antitumor effects of a monoclonal antibody that binds anionic phospholipids on the surface of tumor blood vessels in mice. *Clin Cancer Res*, 11, 1551, 2005.
200. Nicolo, G. et al. Expression of tenascin and of the ED-B containing oncofetal fibronectin isoform in human cancer. *Cell Differ Dev*, 32, 401, 1990.
201. Ebbinghaus, C. et al. Diagnostic and therapeutic applications of recombinant antibodies: targeting the extra-domain B of fibronectin, a marker of tumor angiogenesis. *Curr Pharm Des*, 10, 1537, 2004.
202. Pini, A. et al. Design and use of a phage display library. Human antibodies with subnanomolar affinity against a marker of angiogenesis eluted from a two-dimensional gel. *J Biol Chem*, 273, 21769, 1998.
203. Tarli, L. et al. A high-affinity human antibody that targets tumoral blood vessels. *Blood*, 94, 192, 1999.
204. Birchler, M. et al. Selective targeting and photocoagulation of ocular angiogenesis mediated by a phage-derived human antibody fragment. *Nat Biotechnol*, 17, 984, 1999.
205. Demartis, S. et al. Selective targeting of tumour neovasculature by a radiohalogenated human antibody fragment specific for the ED-B domain of fibronectin. *Eur J Nucl Med*, 28, 534, 2001.
206. Nilsson, F. et al. Targeted delivery of tissue factor to the ED-B domain of fibronectin, a marker of angiogenesis, mediates the infarction of solid tumors in mice. *Cancer Res*, 61, 711, 2001.
207. Borsi, L. et al. Selective targeted delivery of TNFalpha to tumor blood vessels. *Blood*, 102, 4384, 2003.
208. Carnemolla, B. et al. Enhancement of the antitumor properties of interleukin-2 by its targeted delivery to the tumor blood vessel extracellular matrix. *Blood*, 99, 1659, 2002.
209. Halin, C. et al. Enhancement of the antitumor activity of interleukin-12 by targeted delivery to neovasculature. *Nat Biotechnol*, 20, 264, 2002.
210. Ebbinghaus, C. et al. Engineered vascular-targeting antibody–interferon-gamma fusion protein for cancer therapy. *Int J Cancer* [Epub ahead of print], 2005.
211. Ellis, L.M. et al. Angiopoietins and their role in colon cancer angiogenesis. *Oncology* (*Huntingt*), 16 (4 Suppl 3), 31, 2002.
212. Karnani, P. and Kairemo, K. Targeting endothelial growth with monoclonal antibodies against Tie-1 kinase in mouse models. *Clin Cancer Res*, 9 (10 Pt 2), 3821S, 2003.
213. Matsuno, F. et al. Induction of lasting complete regression of preformed distinct solid tumors by targeting the tumor vasculature using two new anti-endoglin monoclonal antibodies. *Clin Cancer Res*, 5, 371, 1999.
214. Martinez, A., Zudaire, E., Portal-Nunez, S., Guedez, L., Libutti, S.K., Stetler-Stevenson, W.G., and Cuttitta, F. Proadrenomedullin NH2-terminal 20 peptide is a potent angiogenic factor, and its inhibition results in reduction of tumor growth. *Cancer Res*, 64, 6489, 2004.
215. Fukushi, J., Makagiansar, I.T., and Stallcup, W.B., NG2 proteoglycan promotes endothelial cell motility and angiogenesis via engagement of galectin-3 and alpha3beta1 integrin. *Mol Biol Cell*, 15, 3580, 2004.
216. Kennel, S.J., Lankford, T., Davern, S., Foote, L., Taniguchi, K., Ohizumi, I., Tsutsumi, Y., Nakagawa, S., Mayumi, T., and Mirzadeh, S. Therapy of rat tracheal carcinoma IC-12 in SCID mice: vascular targeting with [^{213}Bi]-MAb TES-23. *Eur J Cancer*, 38, 1278, 2002.
217. Carson-Walter, E.B., Watkins, D.N., Nanda, A., Vogelstein, B., Kinzler, K.W., and Croix, B. Cell surface tumor endothelial markers are conserved in mice and humans. *Cancer Res*, 61, 6649, 2001.

8 Targeting Fibroblast Growth Factor/Fibroblast Growth Factor Receptor System in Angiogenesis

Marco Rusnati and Marco Presta

CONTENTS

8.1 THE FIBROBLAST GROWTH FACTOR/FIBROBLAST GROWTH FACTOR RECEPTOR SYSTEM IN ANGIOGENESIS

Angiogenesis, the process of new blood vessel formation from preexisting ones, plays a key role in various physiological and pathological conditions, including embryonic development, wound repair, inflammation, and tumor growth.[1] The local, uncontrolled release of angiogenic growth factors contributes to neovascularization that takes place during angiogenesis-dependent diseases, including cancer.[2]

The 1980s saw for the first time the purification to homogeneity of proangiogenic proteins, the breakthrough coming as a result of the observation that endothelial cell (EC) growth factors showed a marked affinity for heparin.[3,4] This led to the identification, purification, and sequencing of the two prototypic heparin-binding angiogenic fibroblast growth factors (FGFs) 1 and 2. Since then, numerous inducers of angiogenesis have been identified, including the members of the vascular endothelial growth factor (VEGF) family, angiopoietins, transforming growth factor-α and -β (TGF-α and TGF-β), platelet-derived growth factor (PDGF), tumor necrosis factor-α (TNF-α), interleukins (ILs), and chemokines.[5]

So far, 23 structurally related members of the FGF family have been identified.[6] FGFs are pleiotropic factors acting on different cell types, including ECs, following interaction with heparan-sulfate proteoglycans (HSPGs) and tyrosine kinase (TK) FGF receptors (FGFRs) (see Section 8.1.2). Only a limited number among them have been investigated for their angiogenic potential, the bulk of experimental data referring to the prototypic FGF1 and FGF2.[7] Nevertheless, several experimental evidences point to a role for FGFs in tumor angiogenesis, inflammation, and angioproliferative diseases (discussed in Ref. 7). Thus, despite some limitations (see later and Table 8.1), the FGF/FGFR system may represent a target for antiangiogenic therapies.

8.1.1 THE "ANGIOGENIC PHENOTYPE"

FGFs induce a complex "proangiogenic phenotype" in cultured ECs (Figure 8.1) that recapitulates several aspects of the angiogenesis process in vivo, including the modulation of EC proliferation, migration, intercellular gap-junction communication, expression of proteases, integrins, and cadherin receptors (summarized in Ref. 8).

As stated earlier, FGFs exert their biological activities by binding to TK-FGFRs on the surface of ECs (see Section 8.1.2). Activation of TK-FGFRs by angiogenic FGFs (including FGF1, FGF2, and FGF4) leads to EC proliferation.[9] Recently, the FGF8b isoform has also been shown to stimulate EC proliferation in vitro.[10]

TABLE 8.1
FGF/FGFR Features That May Limit the Efficacy of Anti-FGF Strategies for the Development of Antiangiogenic Therapies

Multiple members of the FGF family: 23 members
Multiple members of the TK-FGFR family: 4 members
Multiple FGF isoforms: e.g., low and high molecular weight FGF2 and FGF5 isoforms
Multiple TK-FGFR splicing variants
Multiple mechanisms of action: extracellular, intracellular, autocrine, paracrine, intracrine
Multiple FGFRs: TK-FGFRs, HSPGs, integrins, gangliosides
Multiple signal transduction pathways: see Table 8.2, Table 8.3, and Figure 8.2
Complex FGF bioavailability: free, immobilized, bound to ECM or serum components
Cross-talk with other angiogenic growth factors: e.g., VEGF, HIV-1 Tat
EC heterogeneity: e.g., microvascular versus macrovascular endothelium

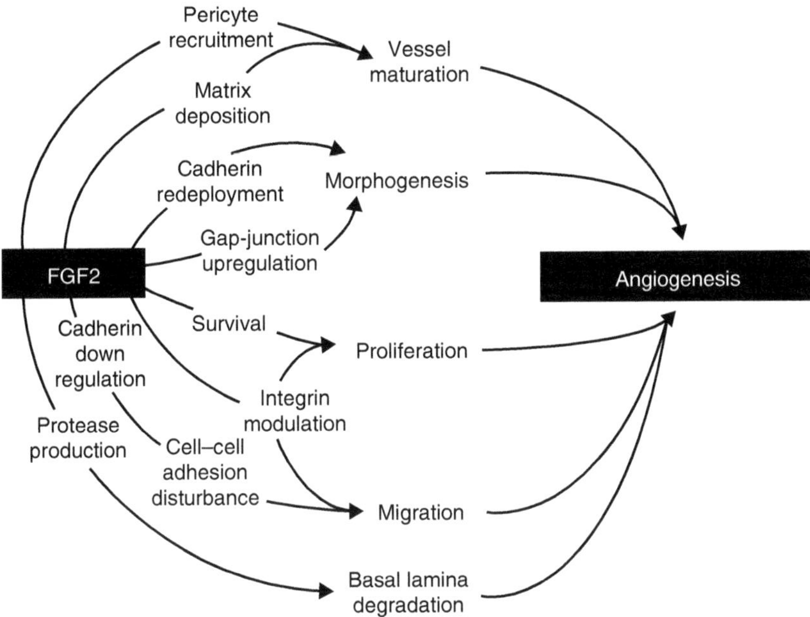

FIGURE 8.1 Events triggered by FGF/FGFR interaction in ECs, which contribute to the acquisition of the angiogenic phenotype in vitro and to neovascularization in vivo.

Extracellular matrix (ECM) degradation represents another important step during the first phases of the angiogenesis process. The plasmin–plasminogen activator (PA) system and matrix metalloproteinases (MMPs) cooperate in this degradation.[11] Urokinase-type PA (uPA) converts plasminogen into plasmin, a serine protease that degrades fibrin and other matrix proteins, and activates several MMPs, including stromelysin-1 (MMP-3), collagenase-1 (MMP-1), type IV collagenases (MMP-2 and MMP-9).[12] FGFs upregulate uPA and MMPs production in ECs.[13,14]

FGF2 stimulates chemotaxis and chemokinesis in ECs.[15,16] Interestingly, FGF2-mediated chemotaxis on fibronectin-coated substrata can be attenuated by $\alpha_v\beta_3$ integrin inhibitors,[17] underlying the tight cross-talk between FGFs and the integrin receptor system (see later). When cultured within or on the top of a permissive three-dimensional matrix, ECs invade the substratum and organize capillary-like structures with a hollow lumen.[18] FGF2 enhances this response in type I collagen[19] and fibrin gels[20] in a CD44[21]- and integrin[22]-dependent manner. In addition, FGF2 enhances EC reorganization on Matrigel.[23] The process requires $\alpha_6\beta_1$ integrin receptor engagement,[24] PECAM-1 activation,[25] and the activation of the proteolytic machinery consisting in the upregulation of MMP-2, MMP-9, and related inhibitors (TIMPs)[26] of uPA and its type-1 inhibitor (PAI-1).[27]

EC migration and proliferation are limited by lateral cell–cell adhesion and ECM interactions,[28] which, in turn, are mediated by cadherin and integrin receptors. As mentioned earlier, FGF2-mediated adhesion and migration of ECs onto type I collagen depends on integrin expression.[29] Interestingly, FGF2 regulates the expression of different integrins, including $\alpha_v\beta_3$[30–32] and cadherins,[28,33] and the production of various ECM components by ECs,[34] contributing to the maturation of the new blood vessels (Figure 8.1).

However, it should be pointed out that significant differences exist in the response to FGFs between large-vessel and microvascular endothelium, as well as between neovasculature and quiescent endothelium. In addition, a high degree of heterogeneity exists in the behavior of FGF-stimulated ECs isolated from different tissues or animal species.[35–37]

The angiogenic activity of different members of the FGF family has been demonstrated in vivo in different experimental models, including the chick embryo chorion-allantoid membrane (CAM),[38] the avascular rabbit[39] or mouse[40] cornea, and the subcutaneous Matrigel injection.[41] In these experimental models, a potent angiogenic response can be obtained by the delivery of FGFs as recombinant proteins, via retroviral, adenoviral, lentiviral, and adeno-associated viral vector transduction, or via implantation of FGF-overexpressing cell transfectants. The latter approach allows the continuous delivery of FGFs produced by a limited number of cells, thus mimicking more closely the in vivo situation.[42] Indeed, the release of 1.0 pg FGF2 per day from viable cells triggers an angiogenic response in the CAM assay quantitatively similar to that elicited by 1.0 μg of the recombinant molecule.[43] These considerations may affect the design of FGF-antagonist strategies.

8.1.2 FGF Receptors in ECs

FGFs establish a complex interaction with EC surface.[7] As stated earlier, FGFs interact with TK-FGFRs and HSPGs.[7] In addition, to trigger a full angiogenic response, FGFs may require the engagement of the integrin receptor $\alpha_v\beta_3$[44,45] and of cell surface associated gangliosides.[46]

The four members of the TK-FGFR family [TK-FGFR1 (*flg*), TK-FGFR2 (*bek*), TK-FGFR3, and TK-FGFR4] are encoded by distinct genes and their structural variability is remarkably increased by the existence of variants generated by alternative splicing of their RNA transcripts.[47] FGFR1 is expressed by ECs in vivo.[48–50] In vitro, ECs of different origin express TK-FGFR1[8,51] and, under some circumstances, TK-FGFR2,[52] whereas the expression of TK-FGFR3 or TK-FGFR4 has never been reported in endothelium. The interactions of the FGFs with TK-FGFRs occur with high affinity (K_d = 10–550 pM) and trigger the activation of complex signal transduction pathways (described later).

HSPGs are associated with the surface of ECs at densities ranging between 10^5 and 10^6 molecules per cells. They consist of a core protein and of glycosaminoglycan (GAG) chains represented by unbranched heparin-like anionic polysaccharides.[53] The interaction of HSPGs with FGFs (including FGF1, FGF2, FGF4, FGF7, and FGF8) occurs with low affinity (K_d = 2–200 nM) and is mediated by the negatively charged sulfated groups of the GAG chain,[54] which bind to basic amino acid motifs.[55] FGF/HSPG interaction modulates angiogenesis in vitro and in vivo[56] by direct activation of intracellular signaling,[57] by promoting FGF internalization,[58,59] and/or by presenting FGFs to TK-FGFRs in a proper conformation, thus facilitating the formation of productive HSPG/FGF/TK-FGFR ternary complexes.[56] In addition, HSPGs may act as a reservoir for extracellular FGFs, which are protected from degradation[60,61] and accumulate in the microenvironment to sustain a long-term stimulation of ECs.[62] Interestingly, FGF2 regulates the synthesis and release of protease/glycosidase, which digest HSPGs and induce the mobilization of free HSPG/HS chains.[63] In addition, ECM degradation can lead to mobilization of entrapped FGF2 with consequent activation of an angiogenic response.[64] The capacity of FGFs to complex HSPGs (as well as other ECM or serum components[7]) may modify their accessibility to neutralizing antibodies or antagonist compounds.

Integrins are transmembrane adhesion-receptor heterodimers composed of α and β subunits that mediate cell adhesion to a variety of adhesive proteins of the ECM.[65] Integrins also regulate the response of ECs to soluble growth factors, including FGF2.[66] In particular, $\alpha_v\beta_3$ integrin is expressed on ECs where it plays a central role in neovascularization. For this reason, $\alpha_v\beta_3$ has been considered a target for the development of antiangiogenic therapies.[67] Similar to classical adhesive proteins, FGF2 binds to $\alpha_v\beta_3$.[44] Consequently, immobilized FGF2 promotes EC adhesion and spreading, leading to uPA upregulation, cell migration, proliferation, and morphogenesis.[45] $\alpha_v\beta_3$/FGF2 interaction and EC adhesion to immobilized

FGF2 lead to the assembly of focal adhesion plaques containing $\alpha_v\beta_3$ and TK-FGFR1.[45] Consistently, a direct $\alpha_v\beta_3$/TK-FGFR1 interaction is required for a full response to FGF2.[68] EC adhesion and activation by immobilized FGF2 may have relevance in vivo. Indeed, as stated earlier, FGF2 accumulates as an immobilized protein in the ECM, mainly by binding to HSPGs. Relevant to this point, heparin-bound FGF2 retains its cell-adhesive capacity.[69] In addition, HSPGs bound to fibronectin can present FGF2 in a biologically active form.[70] Thus, HSPGs may facilitate the interaction of ECM components with FGF2, which, in turn, promote EC adhesion and activation.

Gangliosides are neuraminic acid (NeuAc)-containing glycosphingolipids. Under physiological conditions, gangliosides are mainly associated to the EC membrane, where they modulate cell growth, adhesion, and cell–cell interaction.[71] Gangliosides bind FGF1, FGF2, and FGF4 via negatively charged NeuAc residues.[46,72] Consistently, the ganglioside GM_1 expressed on the EC surface binds FGF2 acting as a functional FGF2 coreceptor.[46]

8.1.3 Signal Transduction Triggered by FGFRs

A schematic representation of the cross-talk among the various signal transduction pathways activated by the interaction of FGFs with their EC receptors is shown in Figure 8.2.

Briefly, signaling via TK-FGFRs is mediated by the direct binding of the growth factor to the extracellular domain of the receptor that causes receptor dimerization and autophosphorylation of specific tyrosine residues located in its intracytoplasmic tail. This, in turn, leads to the recruitment of intracellular signaling/adaptor proteins that bind to phosphorylated tyrosine residues on the activated receptor. Among the various signaling proteins

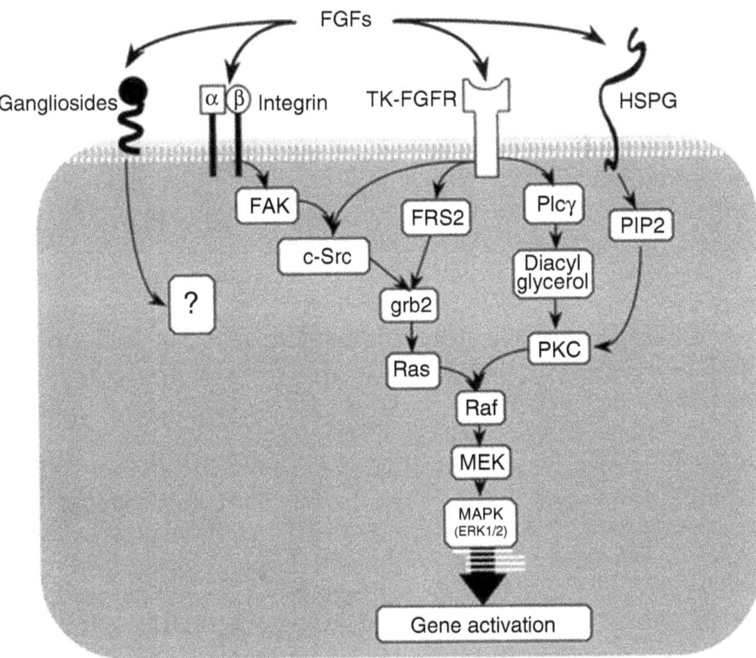

FIGURE 8.2 Schematic representation of the signal transduction pathways triggered by FGF2 in ECs. The interaction of FGF2 with EC integrins, TK-FGFRs, and HSPGs triggers complementary intracellular signals converging to the MAPK transduction pathway. No data are available about the possibility that FGF2/ganglioside interaction may directly activate intracellular second messengers.

activated by the FGF2/FGFR1 system are FRS2α, Shc, phospholipase-Cγ, STAT1, and Gab1, which eventually lead to the activation of the Ras/mitogen activated kinase (MAPK) and phosphatidylinositol-3 kinase (PI-3K)/Akt pathways (for further details see Ref. 73).

Unlike TK-FGFRs, integrins lack intrinsic TK activity. Indeed, an early event during integrin signaling is the tyrosine phosphorylation of the nonreceptor TK focal adhesion kinase (FAK),[74,75] which, in turn, leads to the activation of the GTPase RhoA and $pp60^{src}$.[76–78] In ECs, this signal transduction pathway can be activated on integrin engagement by adhesive proteins and leads to nuclear translocation of NF-κB[79] and to the activation of MAPKs.[45]

HSPGs, long viewed as silent coreceptors able to present FGFs to TK-FGFRs, have been found to trigger an intracellular response following FGF interaction (Figure 8.2). For instance, engagement of syndecan-4 by FGF2 leads to activation of phosphatidylinositol 4,5-bisphosphate and protein kinase C (PKC)-α,[57] which eventually leads to MAPK activation.[80]

Even though no data are available about the involvement of gangliosides in FGF signaling, ganglioside-rich lipid rafts have been implicated in the modulation of signal transduction and biological activity of different growth factors.[81] Indeed, specific ganglioside ligands (e.g., the GM_1-binding cholera toxin B subunit) may act as FGF2 antagonists in ECs.[46]

The complex signal transduction pathways activated by FGFR engagement are mirrored by the complexity of the angiogenic phenotype elicited by FGFs in ECs, raising the possibility that different intracellular signals may be responsible for the various steps of the angiogenesis process (Table 8.2 and Table 8.3). However, inhibition of a single response may be sufficient to hamper the whole angiogenic program.

8.1.4 Intracrine, Autocrine, and Paracrine Mechanisms of Action of FGFs

FGFs can exert their effects on ECs via a paracrine mode consequent to their release by tumor and stromal cells and by their mobilization from the ECM. On the other hand, FGFs may play an autocrine role in ECs (see Ref. 82 and references therein). Moreover, the single-copy human *fgf2* gene encodes multiple FGF2 isoforms with molecular weight ranging from 18 to 24 kDa, and experimental evidences point to different functions of FGF2 isoforms possibly related to differences in their subcellular localization and release.[83] Indeed, high molecular weight FGF2 isoforms contain a nuclear localization sequence and are mostly recovered in the nucleus whereas the 18 kDa FGF2 isoform is mostly cytosolic.[84] The constitutive overexpression of high molecular weight FGF2 isoforms leads to cell immortalization whereas 18 kDa FGF2 overexpression induces a transformed phenotype.[85] Taken together, these data suggest that endogenous FGFs produced by cells of the endothelial lineage may play important autocrine, intracrine, or paracrine roles in angiogenesis and in the pathogenesis of vascular lesions (Figure 8.3).

Finally, a further increase in the complexity of the action of FGFs on ECs is given by the intimate relationship that takes place during neovascularization among FGFs and other angiogenic growth factors, including VEGFs.[7] Actually, an intimate cross-talk exists among FGF2 and the different members of the VEGF family during angiogenesis, lymphangiogenesis, and vasculogenesis (discussed in Ref. 7). Several experimental evidences point to the possibility that FGF2 induces neovascularization indirectly by activation of the VEGF/VEGFR system and that inhibition of this system may affect FGF2-driven angiogenesis. Indeed (i) VEGFR2 antagonists inhibit both VEGF- and FGF2-induced angiogenesis in vitro and in vivo;[86] (ii) systemic administration of anti-VEGF neutralizing antibodies dramatically reduces FGF2-induced vascularization in mouse cornea;[40] (iii) VEGFR1-blocking antibodies or the

TABLE 8.2
Inhibition of FGF2-Mediated Intracellular Signaling by Chemical Inhibitors

Second Messenger	Inhibitor	FGF2 Activity Inhibited
TK-FGFR	SU5416	Survival,[232,233] angiogenesis,[234] motogenesis[a]
	SU6668	Proliferation[235,b]
	SU5402	Proliferation[232]
	Z24	Angiogenesis[234]
	PD173074	Morphogenesis, angiogenesis[236]
	CP-547,632	Proliferation, angiogenesis[237]
$ERK_{1/2}$	PD 098059	Proliferation,[45,238] survival,[239] uPA expression,[240] MMP3 expression,[241] migration,[242–244] CD13 expression,[245] morphogenesis,[245,246] angiogenesis,[245,244] survival, integrin activation,[246] Egr-1 expression,[247] KDR expression[248]
	U0126	Morphogenesis,[240] survival,[249] MMP3 expression,[241] motogenesis[a]
	Apigenin	Proliferation[250]
P38	SB203580	Morphogenesis[251]
PI-3K	LY294002	Survival,[249,252] CD13 expression, morphogenesis,[245] migration,[218,253] proliferation,[250,254] cytoskeleton organization,[255,256] motogenesis[a], FGF2 production[105]
	Antibodies	Proliferation[254]
	Apigenin	Proliferation[250]
PKC	Bis I	Survival[249]
	GO6983	Survival[249]
	GFX	KDR expression[248]
	Chelerythrine	Proliferation[257]
	H7	Proliferation,[257–260] survival[261]
	NSC 639366	Migration, uPA expression, angiogenesis[262]
	Calphostin C	Angiogenesis,[263] FGF2 production[106]
Ras	Manumycin A	CD13 expression,[245] morphogenesis,[245] proliferation[238]
	FTS	Proliferation[238]
	FPT inhibitor III	Proliferation[250]
Pan-TK	Tyrphostin[23]	Proliferation,[45] motogenesis[a]
	Genistein	Proliferation[264,265]
	Herbimycin A	Proliferation[264]
c-Src	PP1	Migration,[243] morphogenesis[238]
	PP2	Angiogenesis, morphogenesis, cytoskeleton organization[266,267]
PLC-γ	Antibodies	Proliferation[268]
PLC-α2	Aristolochic acid	Migration[269]
	ONO-RS-082	Migration[269]
$P70^{S6K}$	Rapamycin	Proliferation[250]
RhoA	C3	ICAM-1 expression[270]
Grb2	Grb2–Src homology 2 domain binding antagonist	Proliferation, migration, angiogenesis[271]
cAMP	Forskolin	Proliferation[272]
	8-bromo AMPc	Proliferation[272]
Akt	ML-9	Angiogenesis[273]
Ca^{2+} influx	CAI	Proliferation, adhesion, MMP-2 expression[274]
G-proteins	Pertussis toxin	Migration[269]

[a] Urbinati, C., personal communication, motogenesis refers to the capacity of an EC monolayer to repair a mechanical wound in response to FGF2.

[b] EC proliferation stimulated by FGF1.

TABLE 8.3
Inhibition of FGF2-Mediated Intracellular Signaling by Dominant Negative or Antisense Overexpression

Target	Inhibition Strategy	FGF2 Activity Inhibited
TK-FGFR	Dominant negative overexpression	Proliferation,[45] cytoskeleton organization,[275] migration, angiogenesis,[276] uPA expression[13]
Syndecan docking sites	Dominant negative overexpression	Proliferation, migration, morphogenesis[57]
FAK	Dominant negative overexpression	Angiogenesis[277]
c-Src	Dominant negative overexpression	Chemotaxis,[243] angiogenesis[266]
Rac	Dominant negative overexpression	Proliferation[278]
Ras	Dominant negative overexpression	CD13 expression,[245] angiogenesis[277]
Raf	Dominant negative overexpression	CD13 expression,[245] survival,[239,279] angiogenesis[277]
MEK	Dominant negative overexpression	CD13 expression,[245] proliferation, migration[280]
$ERK_{1/2}$	Dominant negative overexpression	CD13 expression,[245] FGF2 production[108]
SH2	Dominant negative overexpression	Cytoskeleton organization,[256] proliferation[281]
PKC	Dominant negative overexpression	Proliferation, morphogenesis[282]
c-FES	Dominant negative overexpression	Chemotaxis[283]
PI-3K	Dominant negative overexpression	Survival[284]
PAK	Dominant negative overexpression	Angiogenesis[277]
Akt	Dominant negative overexpression	Survival,[252] morphogenesis[87]
NF-κB	IkB-2A overexpression	Angiogenesis[285]
c-Fyn	Dominant negative overexpression	Morphogenesis[267]
c-jun	Antisense oligonucleotide treatment	FGF2 production[107]
JNK	Dominant negative overexpression	FGF2 production[108]
Ets-1	Dominant negative overexpression	Angiogenesis[286]
Egr-1	Neutralizing single-stranded DNA treatment	Proliferation[109]

expression of a dominant negative VEGFR1 cause a significant reduction of FGF2-induced cell extensions and capillary morphogenesis;[87] and (iv) a peptide encoded by exon 6 of VEGF inhibits FGF2-dependent proliferation and angiogenesis.[88] On the other hand, EC tube formation stimulated by VEGF in murine embryonic explants depends on endogenous

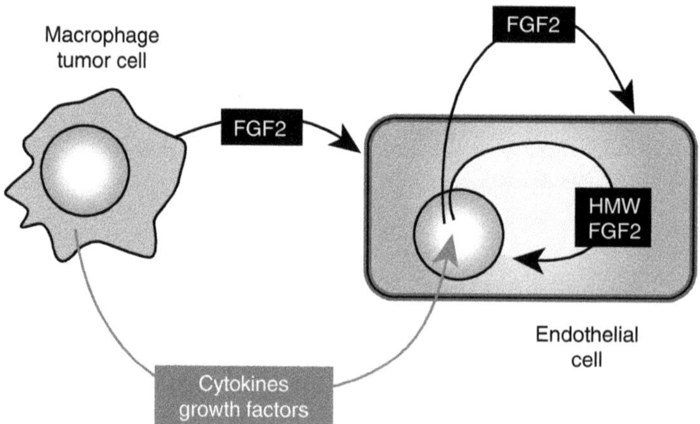

FIGURE 8.3 Different mechanisms of action of FGFs. FGFs released by producing cells (e.g., macrophages and tumor cells) activate ECs via a paracrine mode of action. Alternatively, cytokines/growth factors can stimulate ECs to produce FGFs, which, in turn, will act at the intracellular level (intracrine stimulation) or in an autocrine manner via an extracellular loop of stimulation.

FGF2.[89] In addition, FGF2 and VEGF may exert a synergistic effect in different angiogenesis models,[90–92] even though this may not be the case when the two factors are applied onto the CAM.[93]

8.2 INHIBITION OF THE FGF/FGFR SYSTEM

Theoretically, the angiogenic activity of FGFs can be neutralized at different levels (Figure 8.4): (i) by inhibiting FGF production and release; (ii) by inhibiting the expression of the different FGFRs in ECs; (iii) by sequestering FGFs in the extracellular environment or preventing their interaction with ECs; (iv) by interrupting the signal transduction pathways triggered by FGF/FGFRs interaction in ECs; and (v) by neutralizing FGF-induced effectors whose activity is essential in mediating the angiogenic potential of FGFs (like proteases). All these approaches have been challenged experimentally and are described later.

8.2.1 Inhibition of FGF Production

As mentioned earlier (see Section 8.1.4 and Figure 8.3), various cell types, including tumor cells and leukocytes, can produce FGFs (leading to paracrine EC activation) and

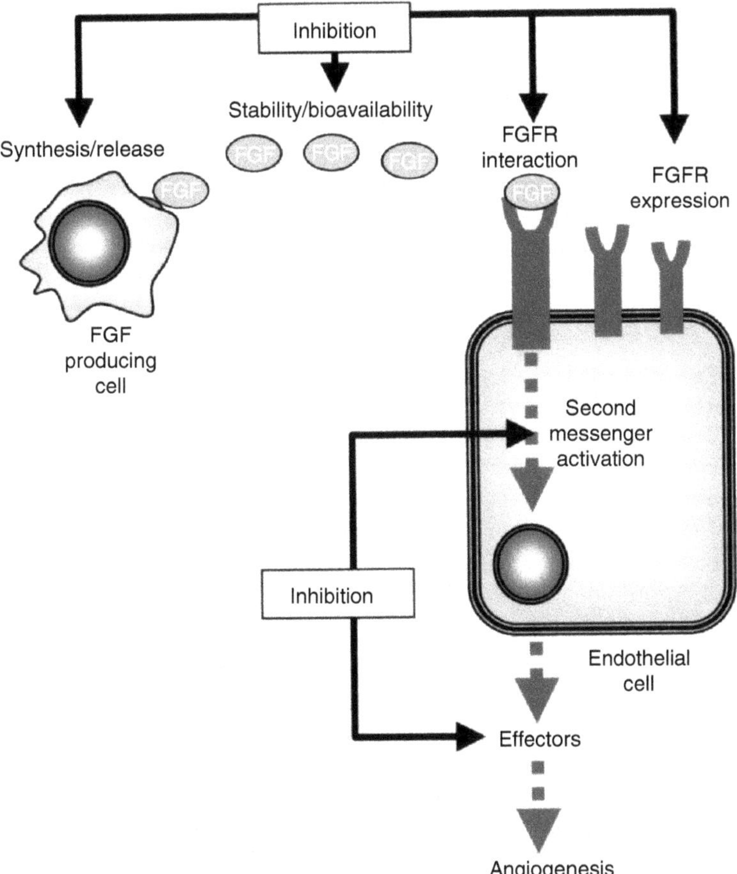

FIGURE 8.4 Anti-FGF strategies for the development of antiangiogenic therapies. Intracellular and extracellular FGF inhibitors can act on different targets. See Section 8.2 for further details.

different cytokines and growth factors that stimulate FGF synthesis in ECs (leading to autocrine/intracrine EC activation). In both cases, the inhibition of FGF production will lead to inhibition of neovascularization.

Liposome-mediated gene transfer of an antisense-oriented FGF2 cDNA into human melanomas leads to an arrest of the growth of subcutaneous tumors in nude mice as a result of blocked intratumoral angiogenesis.[94] In addition, inhibition of the production of angiogenic growth factors (including FGFs) is a common feature of cytotoxic chemotherapeutics. Inhibition of FGF2 production by tumors has been achieved with the novel taxane IDN 5109 (BAY59-8862),[95] with the microtubule-stabilizing drug docetaxel[96] and with the epidermal growth factor receptor TK inhibitor ZD1839 (Iressa).[97] Finally, doxycycline normalizes the urine levels of FGF2 in patients with pulmonary capillary hemangiomatosis.[98]

The production of FGFs in tumor cells can be inhibited also by specific inhibitors of second messengers, including JAK, PI-3K,[99] STAT1,[100] and STAT3[99] (Table 8.2 and Table 8.3), by endogenous cytokines like interferon (IFN)-β,[100] and by natural products (genistein,[101] fumagillin, and its analog TNP-470,[102] curcumin,[103] and the green tea component epigallocatechin-3-gallate[104]).

The latter compound is also able to inhibit the production of both FGF2 and FGF1 by ECs.[104] IL-1 produced by polymorphonuclear leukocytes induces FGF2 synthesis and consequent morphological changes in ECs[105] that are inhibited by the PI-3K inhibitor LY294002[105] (Table 8.2). Similarly, the PKC inhibitor calphostin C inhibits 17β-estradiol-dependent expression of FGF2 in ECs[106] (Table 8.2). In addition, a c-jun antisense cDNA inhibits TNF-α-dependent production of FGF2 and tube formation in ECs[107] (Table 8.3). Consistently, EC transfection with dominant negative mutants of extracellular regulated kinase (ERK) and JNK inhibits the production of FGF2 induced by IL-1, TNF-α, and oncostatin M[108] (Table 8.3). FGF2 expression and consequent neovascularization are inhibited also by DNA-cleaving deoxyribozymes (DNAzymes) targeting the multifunctional transcription factor early growth response-1.[109] Finally, antisense oligonucleotides that directly target FGF2 expression can block FGF2 production and cell proliferation in ECs stimulated by oncostatin M[110] or TNF-α.[111] Accordingly, anti-FGF2 antisense oligonucleotides inhibit angiogenesis in vivo,[94,112] including FGF2-dependent angiogenesis in Kaposi's sarcoma (KS) lesions.[113]

8.2.2 Inhibition of FGFR Expression

The blockage of the FGF/FGFR system can be achieved by hampering the expression of the various FGFRs on EC surface (see Section 8.1.2).

FGF2-dependent proliferation and migration of ECs are abolished by transfection of ECs with a TK-FGFR1 antisense cDNA.[80] Accordingly, liposome-mediated gene transfer of an episomal vector containing antisense-oriented TK-FGFR1 cDNA blocks intratumoral angiogenesis in human melanomas grown subcutaneously in nude mice.[94] In addition, the synthetic retinoid fenretinide (HPR) inhibits FGF2-induced angiogenesis in vivo and EC proliferation in vitro by reducing the expression of FGFR2 on the EC surface.[114] Finally, EC surface expression of TK-FGFR1 and TK-FGFR2 can be inhibited by antibodies directed against $\alpha_v\beta_3$ and $\alpha_5\beta_1$ integrins or by exposure to fibrin.[115]

Lead exposure causes HSPG downregulation leading to inhibition of EC responsiveness to FGF2.[116] In addition, antiangiogenic antithrombin inhibits EC proliferation by down-regulating the surface expression of perlecan HSPG.[117] Accordingly, the overexpression of perlecan antisense cDNA suppresses the autocrine and paracrine functions of FGF2 in fibroblasts,[118] suggesting that this approach can be applied also to ECs. Heparinase removes HSPGs from ECs, abolishing their capacity to migrate in response to FGF2,[80] and accordingly, the GAG 6–0-endosulfatase inhibits neovascularization induced in vivo by FGF2.[119]

EC morphogenesis on three-dimensional fibrin gel or Matrigel is suppressed by downregulation of $\alpha_v\beta_3$ expression obtained by specific DNAzymes,[120] raising the possibility that a similar inhibitory effect may be obtained also for FGF2-dependent activities.

Finally, specific inhibitors of the synthesis of complex gangliosides, including fumonisin B_1, D-threo-1-phenyl-2-decanoylamino-3-morpholino-1-propanol, and D-1-threo-1-phenyl-2-hexadecanoylamino-3-pyrrolidino-1-propanol, affect EC proliferation triggered by FGF2.[46]

8.2.3 Sequestration of FGFs in the Extracellular Environment and Inhibition of FGFR Interaction

Once released in the extracellular environment, FGFs interact with several partners that modulate their bioavailability, stability, local concentration, interaction with EC receptors, and intracellular fate.[7] The identification of these molecules and the biochemical characterization of their FGF-binding/antagonist capacity may allow the design of selective inhibitors. Since the bulk of data refer to FGF2, we focus on this member of the FGF family, even though many of the interactions that are described later may apply also to different FGFs.

8.2.3.1 Heparin-Like Polyanionic Compounds

FGFs bind to free heparin, a negatively charged GAG released in the blood stream during inflammation. At variance with HSPGs, which act as FGF coreceptors (see earlier), free heparin sequesters FGFs in the extracellular environment exerting an antagonist effect. However, due to its anticoagulant activity and its capacity to bind a wide array of growth factors, cytokines, enzymes and proteases, unmodified heparin cannot be used as an antiangiogenic drug.[121] This started a series of researches aimed at identifying heparin derivatives and heparin-like molecules endowed with a more specific FGF-antagonist activity and a more favorable therapeutic window (reviewed in Ref. 121). A list of polyanionic compounds able to bind FGFs and to modulate their binding and related biological effects in ECs is reported in Table 8.4.

It must be pointed out that polyanionic compounds can exert both inhibitory and costimulatory effects on FGF activity (Table 8.4). This apparently contradictory behavior may depend on: (i) the different FGF targeted or the biological assay used; (ii) the concentration of FGFs and medium composition;[122] (iii) the features of the EC line studied;[56,123,124] and (iv) the structural property of the polyanion. In this regard, heparin-derived oligosaccharides can inhibit or enhance FGF2 activity depending on the length of their saccharidic chains.[125] Taken together, these considerations call for extreme caution in the design of this class of compounds and in the evaluation of their biological activity.

Other considerations may limit the exploitability of polyanionic compounds as antiangiogenic FGF antagonists. FGFs can act at the intracellular level, therefore inaccessible to the majority of polyanionic compounds. Nevertheless, suramin can be internalized by ECs, suggesting that it may act also intracellularly.[126] In addition, phosphorothioate oligodeoxynucleotides inhibit EC stimulation by FGF2 possibly acting either with a sequence-specific antisense mechanism of action and as extracellular polyanionic FGF2-binders.[127]

Given the structural similarity among the various members of the FGF family and the heparin-binding capacity shared by a variety of angiogenic growth factors and cytokines, it may be difficult to envisage the design of selective polyanionic antagonists. Nevertheless, recent observations have shown the possibility to achieve a certain degree of specificity by selective sulfation of the *Escherichia coli* K5 polysaccharide.[128,129] It must be pointed out, however, that the "multitarget" activity of certain polyanionic compounds may increase their efficacy in vivo. Indeed, tumor angiogenesis and growth are often the result of the synergic

TABLE 8.4
Heparin-Like Polyanionic Compounds Affecting FGF Activity in ECs

Polyanionic Molecule	Targeted FGF	Affected Biological Activity
Sulfated malto-oligosaccharides	FGF2	Proliferation↓↑→[a], morphogenesis↓[123]
Sulfated beta-(1→4)-galacto-oligosaccharides	FGF2	Angiogenesis↓[287]
RG-13577 (nonsulfated aromatic compound)	FGF2	Proliferation↓, morphogenesis↓[288]
Heparin-derived oligosaccharides	FGF1	Proliferation↑↓,[122,b] angiogenesis↑[39]
	FGF2	Proliferation↓,[122] angiogenesis→[39]
Fucoidan	FGF2	Proliferation↓, migration↓,[289] morphogenesis↑, integrin expression↑[290]
	FGF1	Proliferation↑, motogenesis↑[289]
Suramin	FGF2	Motogenesis↓,[168] uPA expression→[13]
	FGF3	Angiogenesis↓,[112,291] proliferation↓[112]
Suramin derivatives	FGF2	Angiogenesis↓,[292–295] proliferation↓,[293,294,296] migration↓,[294] uPA expression↓,[294] FGF2-dependent tumorigenesis↓[292]
	FGF1	Proliferation↓[297]
PPS (pentosan polysulfate)	FGF1	Proliferation↑, migration↑[124]
	FGF2	Proliferation↑↓[a], migration↓,[124]
	FGF3	Angiogenesis↓[298]
	FGFs[c]	Proliferation↓[298]
TMPP (porphyrin analog)	FGF2	Morphogenesis↓[299]
K5 derivatives (chemically sulfated polysaccharides from *E. coli*)	FGF2	Proliferation↓, FGF2-dependent cell–cell interaction↓, morphogenesis↓, angiogenesis↓,[128,300] cell-adhesion↓[300]
Suleparoide (heparan-sulfate analog)	FGF2	Angiogenesis↓[301]
Undersulfated glycol-split heparins	FGF2	Proliferation↓, FGF2-dependent cell–cell interaction↓,[302,303] angiogenesis↓[303]
Synthetic sulfonic acid polymers	FGF2	FGF2-dependent cell–cell interaction↓,[304] proliferation↓,[304,305] angiogenesis, morphogenesis↓[305]
β-cyclodextrin polysulfate	FGF2	Angiogenesis↓[306]
ATA (aurintricarboxylic acid)	FGF2	Angiogenesis↓[307]
OS-EPS (oversulfated exopolysaccharide from Alteromonas infernos)	FGF2	Proliferation↑, morphogenesis↑[308]
PS-ODN (phosphorothioate oligodeoxynucleotides)	FGF2	Morphogenesis↓, angiogenesis↓[127]
Gangliosides	FGF2	Proliferation↓, uPA expression→,[72] angiogenesis↓↑[138]
Carrageenan	FGF2	Proliferation↓[203,309]

↑, ↓, and → indicate enhancement, inhibition, or lack of effect, respectively.

[a] Depending on the EC line used for the assay.

[b] Depending on FGF1 concentration.

[c] Conditioned media from different tumor cell lines whose mitogenic potential is neutralized after heparin-sepharose chromatography.

action of more than one angiogenic growth factor.[130,131] Relevant to this point, pentosan polysulfate (PPS) efficiently inhibits the biological activity of both HIV-1 Tat[132] and FGF2.[124] Accordingly, phase I and II clinical trials have shown that PPS may lead to stabilization of KS tumors,[133–135] a lesion in which HIV-1 Tat and FGF2 exert a synergistic effect.[130] Finally, it is also worth noting that polyanionic FGF-binders PNU 145156E[136] and heparin derivatives may exert beneficial effects preventing venous thromboembolism, a well-known complication in malignant disease.[137]

A peculiar class of polyanionic compounds is represented by gangliosides. As mentioned earlier (see Section 8.1.2), they act as functional FGF2 coreceptors when associated to the EC surface. Accordingly, the cholera toxin B subunit, which specifically binds to GM_1 ganglioside, inhibits FGF2-dependent proliferation of ECs.[46] However, gangliosides are shed in the microenvironment during tumor growth, and free gangliosides bind FGF1, FGF2, and FGF4 via negatively charged NeuAc residues.[46,72] In the extracellular environment, gangliosides sequester FGF2, inhibiting EC interaction and mitogenic activity.[72] However, as described earlier for heparin derivatives, also gangliosides can differently affect the biological activity of FGFs, depending on their structure, concentration, and combinations[138] (Table 8.4).

Finally, by the mechanism of action opposite to that shown for polyanionic compounds, protamine, an arginine-rich polypeptide, can disrupt FGF/FGFR interaction and inhibit FGF2-dependent proliferation of ECs[13] possibly by binding and masking TK-FGFRs and HSPGs.[139]

8.2.3.2 ECM Components

Several ECM components or their degradation products can affect FGF-driven angiogenesis (Table 8.5). Thrombospondin-1 (TSP-1) is a modular glycoprotein secreted by different cell types, including ECs. It is composed of multiple active domains that bind to soluble factors, cell receptors, and ECM components.[140] In particular, TSP-1 associates to HSPGs of the ECM and binds integrin receptors.[140] TSP-1 was the first endogenous inhibitor of angiogenesis to be identified and its effect is due, at least in part, to its capacity to bind FGF2.[141] The interaction is mediated by the COOH-terminal, antiangiogenic 140 kDa fragment of TSP-1. TSP-1 prevents the interaction of FGF2 with soluble heparin and with HSPGs and TK-FGFRs. Accordingly, TSP-1 inhibits the mitogenic and chemotactic activity of FGF2 in ECs. TSP-1 also prevents the accumulation of FGF2 in the ECM and favors the mobilization of matrix-bound FGF2, generating inactive TSP-1/FGF2 complexes.[142] These observations suggest that free TSP-1 can act as a scavenger for matrix-associated FGFs, affecting their location, bioavailability, and function, whereas ECM-associated TSP-1 may act as an "FGF decoy," sequestering the growth factors in an inactive form. Beside its well-characterized FGF-binding capacity, TSP may inhibit FGFs also by alternative mechanisms of action. For instance, TSP-1 may antagonize the binding of FGF2 to $\alpha_v\beta_3$ (Figure 8.5). In addition, through its binding to its CD36 receptor, TSP-1 may transduce antiangiogenic signals in ECs, as suggested by the observation that a modified peptide from TSP-1 inhibits EC responsiveness to FGF1 and FGF2 in a CD36-dependent manner.[143]

Fibstatin is a fibronectin fragment that binds FGF2 but not FGF1, FGF3, FGF6, or FGF12.[144] Fibstatin inhibits FGF2-dependent proliferation, migration, and tubulogenesis in ECs in vitro and angiogenesis and tumor growth in vivo.[144] Like other FGF-binding partners (see earlier), fibstatin is endowed with the capacity to bind heparin and integrin receptors, suggesting that multiple interactions are responsible for the antiangiogenic activity of this molecule.

8.2.3.3 Serum Proteins

A variety of serum components can affect FGF activity in ECs (Table 8.5). α_2-Macroglobulin (α_2M) is a 718 kDa homotetrameric protein present in human plasma where it acts as a broad-specific protease inhibitor. To exert its activity, α_2M undergoes major conformational changes that lead to the activated form α_2M*. Both α_2M and α_2M* bind a variety of cytokines and growth factors, including FGF1, FGF2, FGF4, and FGF6, but not FGF5, FGF7,

TABLE 8.5
Endogenous Inhibitors of FGFs

Localization	Molecule	Mechanism of Action
Intracellular	Homeobox gene GAX	Inhibition of NF-kB[310]
	Sprouty proteins	Inhibition of FGF signaling[311]
	Heat shock proteins (Hsp) 70 and 90	Downmodulation of signaling molecules including pAkt, c-Raf-1, and $ERK_{1/2}$[312]
ECM	Collagen I	Unknown[313]
	TSP-1	Sequestration of FGF2,[141] CD36 engagement,[143] integrin occupancy (?)
	Alphastatin (fibrinogen fragment)	Unknown[314]
	Endostatin (collagen fragment)	Cytoskeleton organization,[315] Shb activation[316]
	Fibstatin (fibronectin fragment)	Sequestration of FGF2[144]
Blood	CXCL13	Sequestration of FGF2[162]
	CXCL14	Unknown[163]
	PDGF	Sequestration of FGF2[153]
	α_2-macroglobulin	Sequestration of FGF2[145]
	PTX3	Sequestration of FGF2[150]
	Heparin	Sequestration of FGF2[56]
	Gangliosides	Sequestration of FGF2[72]
	PF4	Sequestration of FGF2,[156] HSPG occupancy,[158] unknown[157]
	Soluble form of the extracellular portion of FGFR1 (xcFGFR1)	Sequestration of FGF2,[166] formation of heterodimers with cellular TK-FGFR1[167]
	Histidine-rich glycoprotein	HSPGs occupancy,[158] tropomyosin engagement[317]
	Antithrombin	Downregulation of HSPGs in ECs[117]
	Thromboxane	Impairment of TK-FGFR1 internalization[318]
	Angiostatin (fragment of plasminogen)	Inhibition of ERK cascade[319]
	Prolactin (16 kDa fragment)	Unknown[320]
	Vitamin D3-binding protein	CD36 engagement[321]
	Ghrelin	Inhibition of TK/MAPK cascades[322]
	Lysophosphatidylcholine	Inhibition of ras/ERK cascades[280]
	Cleaved HMW kininogen	Tropomyosin engagement[323]
	IL-4	Alteration of cell cycle regulatory molecules[324]
	IP-10	Unknown[325]
	Pigment epithelium-derived factor	Fes-dependent inhibition of Fyn[326]
	Vasculostatin (fragment of brain angiogenesis inhibitor-1)	Unknown[327]
	Vasostatin	Unknown[328]
	Kininostatin (fragment of kininogen)	Inhibition of cyclin D1 expression[329]
	Kallistatin	Unknown[152]
	TGF-β1	Unknown[20]
	TIMP-2, 4	Inhibition of FGF-induced proteases[330]
	IFN-γ	Downregulation of FGFRs[331]
	IL-1	Downregulation of FGFRs[331]
	TNF-α, β	Unknown[332]
	Somatostatin	Unknown[333]
	Retinoids	Unknown[114]
Extracellular microenvironment	Heparan sulfate 6-0-endosulfatase	Desulfation of HSPGs in ECs[119]
	Heparinase	HSPG degradation[80]
	Semaphorin-3F	Inhibition of ERK cascade[334]

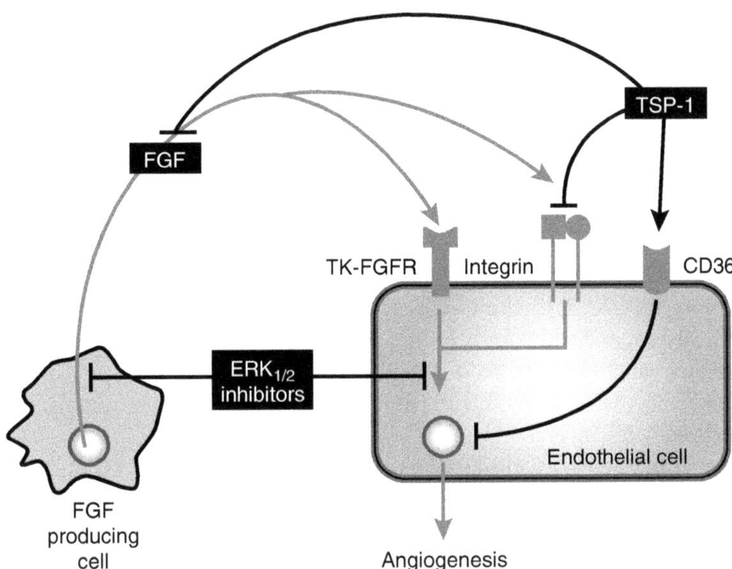

FIGURE 8.5 Multitarget activity of selected FGF inhibitors. Possible mechanism of action of antiangiogenic TSP-1 and $ERK_{1/2}$ inhibitors. See Section 8.3 for further details.

FGF9, and FGF10.[145] The binding of α_2M to FGF2 occurs with high affinity and is primarily hydrophobic in nature.[146] α_2M sequesters FGF2 in the extracellular environment and inhibits its cell interaction, protease-inducing activity,[147] and mitogenic capacity.[145] Interestingly, both TGF-β[146] and PDGF[147] compete with FGF2 for binding to α_2M. In addition, α_2M competes with ECM components for FGF2 interaction.[145]

Pentraxin 3 (PTX3) is a 45 kDa glycosylated protein predominantly assembled in 10–20 mer multimers.[148] Its COOH-terminal domain shares homology with the classic short-pentraxin C-reactive protein whereas its NH_2-terminal portion does not show significant homology with any other known protein.[149] PTX3 is synthesized and released by activated mononuclear phagocytes and ECs.[149] PTX3 acts as a soluble pattern recognition receptor with unique functions in various physio-pathological conditions. These functions rely, at least in part, on the capacity of PTX3 to bind different structures.[150] PTX3 binds FGF2, but not FGF1 and FGF4, with high affinity.[150] In ECs, PTX3 prevents the binding of FGF2 to cell surface TK-FGFRs and HSPGs, with a consequent inhibition of cell proliferation and migration, and inhibits FGF2-dependent neovascularization in the CAM assay. In addition, PTX3 overexpression in FGF2-transformed ECs inhibits FGF2-dependent proliferation and invasion in vitro and tumorigenesis in vivo.[150] PTX3 exists both as a free or ECM-immobilized molecule.[151] Relevant to this point, FGF2 and PTX3 retain their binding capacity independently of their free or immobilized status.[150] Thus, as described for TSP-1, free PTX3 may have access to ECM-bound FGF2 by acting as a scavenger for the stored growth factor, whereas ECM-associated PTX3 may act as an "FGF decoy," sequestering the growth factor in an inactive form.

Kallistatin, a serpin first identified as a specific inhibitor of tissue kallikrein, inhibits FGF2-induced proliferation, migration, and adhesion of cultured ECs. In addition, kallistatin attenuated FGF2-induced neovascularization in subcutaneously implanted Matrigel plugs in mice.[152]

PDGF-BB binds FGF2 with a 1:2 stoichiometry.[153] This interaction may contribute to the inhibitory effect exerted by PDGF-BB on FGF2-dependent neovascularization.[154] The heparin-binding C-X-C chemokine platelet factor 4 (PF4) is a well-known inhibitor of angiogenesis (see Ref. 155 and references therein). PF4 binds FGF1[156] and FGF2.[155] In ECs, PF4 inhibits FGF2 interaction with HSPGs and TK-FGFR1, FGF2 internalization, and mitogenic activity.[155] It is interesting to note that PF4 may inhibit FGF2 also by an intracellular mechanism of action.[157] Moreover, PF4 and the histidine-rich glycoprotein bind to and mask cell surface or ECM-associated HSPGs, hindering these receptors to FGF2 and FGF1 binding.[158] The observation that peptides from PF4 can be modified obtaining a significant increase in their FGF2-binding and antagonist activity suggests that peptides from FGF-binding proteins represent a potential class of antiangiogenic agents with defined modes of action.[159–161]

Similar to PF4, the CXCL13 chemokine (formerly known as B cell-attracting chemokine 1) binds to FGF2, displaces the growth factor from ECs, impairs the formation of functional FGF2 homodimers, and inhibits FGF2-dependent survival of ECs.[162] BRAK/CXCL14, a chemokine whose expression is absent in tumors, inhibits FGF2-dependent migration of ECs in vitro and angiogenesis in vivo.[163]

A soluble form of the extracellular portion of TK-FGFR1 (xcFGFR1) was identified in body fluids[164] and endothelial ECM,[165] which is able to bind FGF2, with an affinity that is lower than that of the intact receptor, but sufficient to prevent FGF2/TK-FGFR1 interaction.[166] Accordingly, xcFGFR1 inhibits signal transduction triggered by FGF1, FGF2, and FGF3 by forming heterodimers with cellular TK-FGFR1[167] and inhibits FGF2-dependent proliferation of ECs.[13] Similarly, neutralizing anti-TK-FGFR antibodies can effectively block FGF2-mediated angiogenesis in vivo.[168]

8.2.3.4 FGF-Based Peptides and Neutralizing Antibodies

A schematic representation of the FGF2-based peptides that have been so far discovered to exert FGF2-antagonist activity is reported in Figure 8.6. In detail, a structural analysis

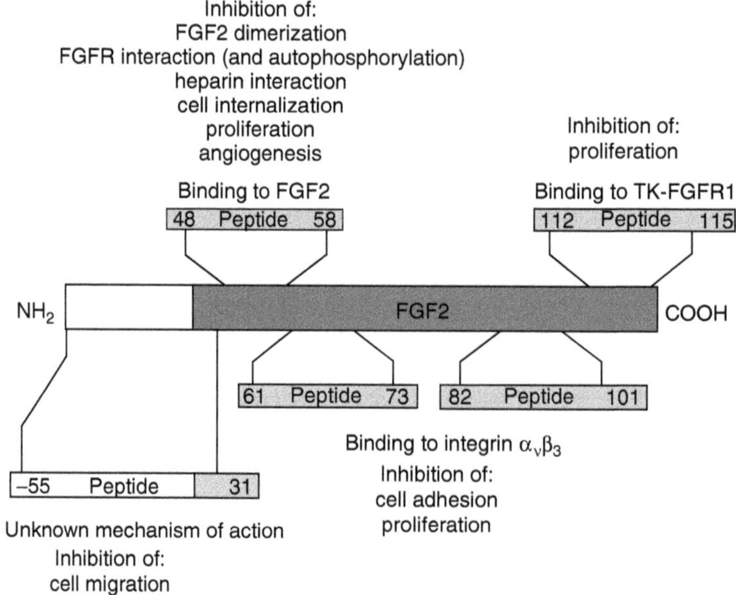

FIGURE 8.6 FGF2-related synthetic peptides as FGF2 antagonists. Grey box represents the 18 kDa FGF2 isoforms. White box represents the amino-terminal extension present in the high molecular weight FGF2 isoforms. The mechanism of action and the biological activities antagonized in ECs are shown for each peptide. See Section 8.2.3.4 for further details.

carried out on FGF2 identified a region encompassing residues 48–58 as involved in FGF2 dimerization. Accordingly, the derived peptide FREG-(48–58) prevents dimerization of the growth factor and its interaction with heparin and TK-FGFR1, thus inhibiting TK-FGFR phosphorylation, FGF2-dependent EC proliferation and migration in vitro, and angiogenesis in vivo.[169] Furthermore, a polyclonal antibody directed against FREG-(48–58) blocks FGF2 action in vitro.[169] On the other hand, the interaction of FGF2 with FGFR1 can be inhibited by peptides derived by the amino acid sequence FGF2 (112–155).[13]

Alternatively, antibodies directed against the amino acid sequences FGF2 (61–73) and FGF2 (82–101) and related peptides inhibit FGF2-dependent EC proliferation without affecting FGF2/TK-FGFR1 interaction.[170] Interestingly, both FGF2 regions contain a DGR sequence, which is the inverse of the integrin-recognition sequence RGD present in several cell-adhesive proteins. Consistently, the two DGR-containing FGF2 peptides inhibit $\alpha_v\beta_3$-mediated EC adhesion to FGF2.[44] In addition, RGD-containing tetra or eptapeptides affect FGF2-dependent EC adhesion[44] and proliferation.[170] Accordingly, RGD-peptidomimetics inhibit FGF2-dependent neovascularization and tumorigenesis.[171] Relevant to this point, monoclonal anti-$\alpha_v\beta_3$ antibodies inhibit FGF2/$\alpha_v\beta_3$ interaction and EC adhesion, proliferation, and uPA upregulation in vitro[44] and FGF2-mediated angiogenesis in vivo,[172] indicating that RGD- and DGR-containing peptides exert their FGF2-antagonist activity by preventing the interaction of the growth factor with $\alpha_v\beta_3$ integrin. A similar mechanism of action may be shared by disintegrins, a class of naturally occurring integrin antagonists that have been demonstrated to inhibit different aspects of FGF2 biology.[173]

A peptide derived from the amino-terminal extension of the high molecular weight 24 kDa FGF2 isoform and the first 31 amino acids from the canonic 18 kDa isoform inhibits FGF2-dependent migration of ECs without affecting FGF2/TK-FGFR1 interaction or $ERK_{1/2}$ activation.[174]

A different approach to block the activity of extracellular FGF2 has been developed by Plum and coworkers.[175] They demonstrated the capacity of a liposome-based peptide vaccine, targeting the heparin-binding domain of FGF2, to generate a specific anti-FGF2 antibody that was able to inhibit FGF2 binding to HSPGs and FGF2-dependent angiogenesis in vivo. Accordingly, anti-FGF immunization has been shown to inhibit tumor growth.[176–179]

8.2.4 Inhibition of Signal Transduction Triggered by FGFR Engagement

Theoretically, each of the intracellular signaling pathways activated by FGFs in ECs (see Section 8.1.3, Figure 8.2, and Ref. 73) might be considered a target for angiogenesis inhibitors. Accordingly, FGF activity can be efficiently inhibited both in vitro and in vivo by synthetic compounds (Table 8.2), selective dominant negative mutants, and antisense cDNAs (Table 8.3) targeting various signal transduction pathways triggered by FGFs. In addition, different endogenous inhibitors of angiogenesis have been shown to affect FGF signaling (Table 8.5). Among them, several cytokines modulate EC activation and neovascularization induced by FGF2. It is possible that these cytokines, by interacting with specific receptors on ECs, may interfere with the signal transduction pathways activated by the angiogenic growth factor and trigger a genetic program that affects EC responsiveness to angiogenic stimuli. However, the therapeutic exploitation of this approach is greatly limited by the fact that several of the second messengers activated by FGF2 during pathological neovascularization are implicated in various physiological processes. Their inhibition may thus induce undesired side effects.

8.2.5 Inhibition of Effectors/Biological Activities Induced by FGFs in ECs

As described earlier in Section 8.1.1 and Figure 8.1, FGFs induce in ECs a complex proangiogenic phenotype characterized by an increase in ECM degradation and in EC motility, proliferation, and morphogenesis. These processes are mediated by distinct effectors induced/activated by FGFs, and their blockage may result in the inhibition of FGF-dependent angiogenesis.

For instance, to degrade ECM, FGFs upregulate the production of several proteases (see Section 8.1.1). TIMPs inhibit FGF2 neovascularization[180,181] and an MMP-independent mechanism of inhibition of FGF-dependent angiogenesis has been proposed for TIMP-2.[180] In addition, the epidermal growth factor-like domain of murine uPA alone or fused to the Fc portion of human IgG acts as high-affinity urokinase receptor antagonist and inhibits FGF2-induced angiogenesis in vivo.[182] Accordingly, medroxyprogesterone acetate exerts an angiostatic effect by increasing the expression of PAI-1, thus counteracting the uPA-inducing activity of FGF2.[183]

The properties of neovasculature differ from those of quiescent endothelium. Vascular targeting agents exploit differences in cell proliferation, permeability, maturation, and reliance on tubulin cytoskeleton to induce selective blood vessel occlusion and destruction.[184] In particular, microtubule-destabilizing agents, including combretastatin-derived prodrugs and analogs, disrupt rapidly proliferating and immature tumor endothelium, leading to reduced tumor blood flow and hypoxia.[185–187] Interestingly, microtubule-destabilizing agents, for example, combretastatin A-4 and vinblastine, may also show a distinct antiangiogenic activity.[188,189] Accordingly, the *trans*-resveratrol derivative 3,5,4′-trimethoxystilbene acts as a microtubule-destabilizing agent endowed with both anti-FGF2 antiangiogenic activity and vascular targeting capacity.[190] Similarly, microtubule-stabilizing agents, including paclitaxel and taxane derivatives,[191,192] affect FGF2-triggered angiogenesis in vitro and in vivo. In addition, by preventing the formation of stress fibers, the antifungal polyether macrolide goniodomin A inhibits FGF2-induced migration and morphogenesis in ECs, leaving their proliferation unaffected.[193] These findings are of importance when considering that combining vascular targeting agents with angiogenesis inhibitors may result in additive or synergistic effects on tumor growth and vascularization.[194–196]

Finally, apoptosis-inducing agents can inhibit the action of FGF2, possibly counteracting its mitogenic activity. This is the case of betulinic acid, a proapoptotic mitochondria-damaging pentacyclic triterpenoid, which inhibits FGF2-induced EC invasion and tube formation.[197]

8.2.6 FGF Inhibitors with Undefined Mechanism of Action

8.2.6.1 Natural Products

Numerous bioactive plant compounds, often referred to as nutraceuticals, have been tested for their potential clinical application. Some of these compounds are currently under study for their anti-FGF, antiangiogenic potential, including curcumin from *Curcuma longa* and epigallocatechin-3-gallate from green tea. The *Gleditsia sinensis* fruit extract inhibits the angiogenic activity of FGF2 in vivo.[198] Citrus pectin inhibits the formation of the productive ternary complex heparin/FGF2/TK-FGFR1, probably by interacting directly with the growth factor and competing for heparin binding.[199] The antineoplastic compound aplidine, a new marine-derived depsipeptide, inhibits angiogenesis elicited by FGF2 in vivo and FGF2-dependent EC proliferation in vitro. Philinopside A, a novel sulfated saponin isolated from the sea cucumber *Pentacta quadrangulari*, and the Chinese folk medicine-derived phytochemical 11,11′-dideoxyverticillin from fungus *Shiraia bambusicola* are potent inhibitors of TK-FGFR1 activity.[200,201] Psammaplin A is a phenolic natural product isolated from a marine sponge that suppresses the invasion and tube formation of ECs stimulated by FGF2. Resveratrol, found in grapes and wine, inhibits FGF-driven angiogenesis in vitro and in vivo.[202] l-Carrageenan is a natural polysulfated carbohydrate that inhibits FGF2 mitogenic activity in ECs.[203] Finally, a naturally occurring agent isolated from cartilage, referred to as Neovastat (AE-941), inhibits FGF2-dependent angiogenesis in the CAM and Matrigel plug assays.[204]

8.2.6.2 Other Drugs

Interestingly, several drugs developed for the treatment of tumor-unrelated diseases have been shown to be endowed with FGF-antagonist activity. Spironolactone, a mineralcorticoid receptor antagonist mainly used in the treatment of heart failure, inhibits neovascularization triggered in vivo by FGF2.[205] Transilast, first developed as an antiallergic drug, inhibits FGF2-dependent EC proliferation.[206] Bisphosphonate drugs inhibit osteoclastic bone resorption and are widely used to treat skeletal complications. Zoledronic acid, a new generation bisphosphonate, inhibits FGF2-induced EC proliferation and in vivo neovascularization.[207] Cidofovir, approved for the treatment of cytomegalovirus retinitis in AIDS patients, inhibits FGF2-dependent tumorigenesis.[42] Indomethacin, a nonsteroidal anti-inflammatory drug, inhibits angiogenesis in vivo by affecting FGF2-induced EC proliferation.[208] Cerivastatin, an HMG-CoA reductase inhibitor used in the treatment of hypercholesterolemia-related diseases, inhibits EC locomotion in vitro and angiogenesis in vivo.[209] SR 25989, an esterified derivative of ticlopidine, inhibits FGF1-dependent healing of a mechanical wound in confluent endothelium.[210] Finally, the inhibitor of the secretory phospholipase-A2 (HyPE) prevents FGF2-dependent proliferation, migration, and morphogenesis in vitro.[211]

8.3 THE MULTITARGET OPTION

Different inhibitors of FGFs act with a multitarget mechanism of action (Figure 8.5 and Table 8.5). PF4 binds FGFs,[155,156] masks HSPGs,[158] or acts by an intracellular mechanism of action.[157] Similarly, TSP-1 sequesters FGF2 in an inactive form,[141,142] binds $\alpha_v\beta_3$[140] (possibly preventing FGF2 interaction), and inhibits FGF2 activity by a CD36-dependent mechanism of action.[143] RGD-containing peptides antagonize FGF2 mainly by competing for $\alpha_v\beta_3$ interaction.[44] However, their direct binding to integrin leads to a caspase-dependent apoptotic signal, which contributes to EC inhibition.[212] The blockage of $ERK_{1/2}$ activation leads to inhibition of FGF2 production and of FGF2-mediated biological activity in ECs (Table 8.2, Table 8.3, and Figure 8.5). The histidine-rich glycoprotein, already demonstrated to bind and mask HSPGs to FGF1 and FGF2, binds and transduces antiangiogenic signals through cell surface tropomyosin on ECs.[158] Curcuminoids can either inhibit FGF production by tumor cells[103] or prevent FGF2-dependent protease production in ECs.[213]

In tumors, FGF inhibitors with a multitarget mechanism of action as well as the combination of FGF antagonists and classic chemotherapeutic agents should prevent the development of drug resistance and decrease the dosage and related toxicity of each single drug, as shown for Neovastat used in combination with cisplatinum.[204]

Several antitumor agents are endowed with an intrinsic FGF-antagonist activity (Figure 8.7). For instance, the quinazoline-derived $\alpha 1$-adrenoreceptor antagonist doxazosin, used in the treatment of prostate cancer, inhibits FGF2-induced morphogenesis in ECs.[214] Thalidomide, used in the treatment of relapsing malignant gliomas, inhibits FGF2-induced EC proliferation.[215] The same effect is exerted by the antiestrogen tamoxifen used as adjuvant in the treatment of breast cancer[216] and by the functionally related medroxyprogesterone-acetate, which inhibits the release of uPA induced by FGF2 in ECs.[217] The topoisomerase I inhibitor topotecan possesses an indirect antitumor effect in vivo mediated by angiosuppression due, at least in part, to the inhibition of FGF2-induced EC migration.[218] Aplidine, which exerts a cytotoxic effect in tumor cells and is currently tested in early phase clinical trials,[219] possesses FGF2-antagonist activity (see Section 8.2.6.1). The same dual effect has been

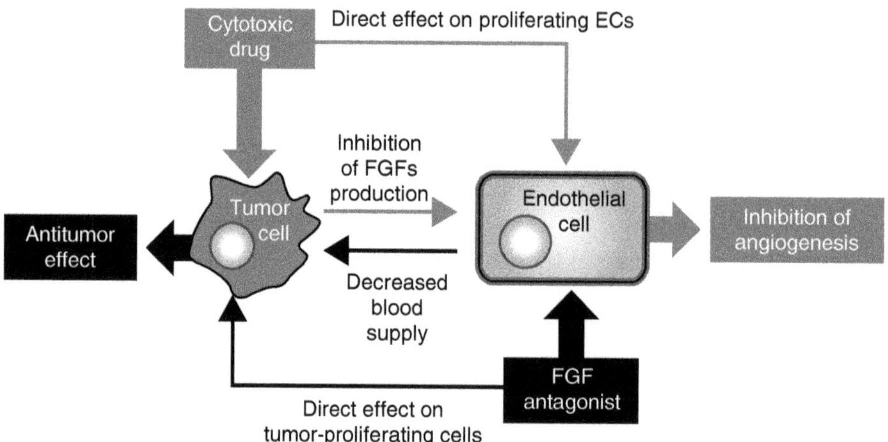

FIGURE 8.7 Multiple effects of FGF antagonists and antineoplastic drugs on tumor growth and neovascularization. FGF antagonists can affect tumor growth indirectly by decreasing blood supply and directly by blocking FGF-dependent tumor cell proliferation. On the other hand, cytotoxic drugs can inhibit EC proliferation and decrease the amount of FGF available to ECs by killing FGF-producing tumor cells.

demonstrated for Neovastat (AE-941).[204] The chemotherapic 6-methylmercaptopurine-riboside inhibits FGF2-dependent angiogenesis in vitro and in vivo.[220] A combination of tegafur and uracil (UFT), used in the treatment of a variety of malignant tumors, inhibits EC proliferation induced by FGF2.[221] Finally, the antimetabolite 6-thioguanine, used in the management of acute myelogenous leukemia, inhibits EC proliferation and angiogenesis triggered by FGF2.[220]

On the other hand, it must be pointed out that due to their pleiotropic nature, FGFs may contribute to cancer progression also by acting directly on tumor cells. Indeed, the coexpression of FGF7/KGF and its receptor TK-FGFR2 IIIb/KGFR correlates with high proliferative activity and poor prognosis in lung adenocarcinoma[222] and high levels of FGF8[223,224] or FGF17[225] are associated with less favorable prognosis in human prostate cancer. Thus, targeting the FGF/FGFR system in cancer may gain benefits not only in terms of inhibition of the neovascularization process but also by a direct inhibition of tumor cell proliferation (Figure 8.7). For instance, inhibition of the FGF/FGFR system in glioma cells by dominant negative TK-FGFR transfection[226] or in prostate cancer cells by *fgf2* gene knockout[227] results in inhibition of tumor growth by both angiogenesis-dependent and angiogenesis-independent mechanisms.

8.4 CONCLUSION

The bulk of experimental data summarized here clearly indicate that the FGF/FGFR system may represent a target for numerous antiangiogenic strategies in different pathological settings, including cancer. Accordingly, cancer clinical trials are in progress to assess the safety and efficacy of various compounds with a potential capacity to affect the FGF/FGFR system at different levels.[228,229] It must be pointed out, however, that the ability to interact with this system may not represent the main rationale that has lead to their testing. This is the case for various heparin derivatives that have been tested in cancer patients because of their antithrombotic effect rather than for their potential FGF-antagonist activity. Similarly, the humanized monoclonal anti-$\alpha_v\beta_3$ antibody vitaxin[230,231] has been investigated for its ability to affect the cell-adhesive function of this integrin receptor rather than its potential ability to act as a signaling FGFR. In addition, as stated earlier, numerous cytotoxic drugs can affect the FGF/FGFR system and angiogenesis. Novel strategies aimed at inhibiting multiple targets, including the FGF/FGFR system, may represent a more efficacious approach for the treatment of angiogenesis-dependent diseases, including cancer.

ACKNOWLEDGMENT

This work was supported by grants from MIUR (Centro di Eccellenza IDET, FIRB 2001, Cofin 2004), Fondazione Berlucchi, ISS (Progetto Oncotecnologico), and AIRC to M.P, and by grants from ISS (AIDS Project), AIRC, and ≪60%≫ to MR.

REFERENCES

1. Carmeliet, P. and Jain, R.K., Angiogenesis in cancer and other diseases, *Nature*, 407, 249, 2000.
2. Folkman, J., Angiogenesis in cancer, vascular, rheumatoid and other disease, *Nat Med*, 1, 27, 1995.
3. Shing, Y. et al., Heparin affinity: purification of a tumor-derived capillary endothelial cell growth factor, *Science*, 223, 1296, 1984.
4. Maciag, T. et al., Heparin binds endothelial cell growth factor, the principal endothelial cell mitogen in bovine brain, *Science*, 225, 932, 1984.
5. Carmeliet, P., Mechanisms of angiogenesis and arteriogenesis, *Nat Med*, 6, 389, 2000.
6. Itoh, N. and Ornitz, D.M., Evolution of the Fgf and Fgfr gene families, *Trends Genet*, 20, 563, 2004.
7. Presta, M. et al., Fibroblast growth factor/fibroblast growth factor receptor system in angiogenesis, *Cytokine Growth Factor Rev*, 16, 159, 2005.
8. Javerzat, S., Auguste, P., and Bikfalvi, A., The role of fibroblast growth factors in vascular development, *Trends Mol Med*, 8, 483, 2002.
9. Cross, M.J. and Claesson-Welsh, L., FGF and VEGF function in angiogenesis: signalling pathways, biological responses and therapeutic inhibition, *Trends Pharmacol Sci*, 22, 201, 2001.
10. Mattila, M.M. et al., FGF-8b increases angiogenic capacity and tumor growth of androgen-regulated S115 breast cancer cells, *Oncogene*, 20, 2791, 2001.
11. Liotta, L.A., Steeg, P.S., and Stetler-Stevenson, W.G., Cancer metastasis and angiogenesis: an imbalance of positive and negative regulation, *Cell*, 64, 327, 1991.
12. Hiraoka, N. et al., Matrix metalloproteinases regulate neovascularization by acting as pericellular fibrinolysins, *Cell*, 95, 365, 1998.
13. Rusnati, M. et al., A distinct basic fibroblast growth factor (FGF-2)/FGF receptor interaction distinguishes urokinase-type plasminogen activator induction from mitogenicity in endothelial cells, *Mol Biol Cell*, 7, 369, 1996.
14. Taraboletti, G. et al., Shedding of the matrix metalloproteinases MMP-2, MMP-9, and MT1-MMP as membrane vesicle-associated components by endothelial cells, *Am J Pathol*, 160, 673, 2002.
15. Terranova, V.P. et al., Human endothelial cells are chemotactic to endothelial cell growth factor and heparin, *J Cell Biol*, 101, 2330, 1985.
16. Stokes, C.L. et al., Chemotaxis of human microvessel endothelial cells in response to acidic fibroblast growth factor, *Lab Invest*, 63, 657, 1990.
17. Shono, T. et al., Inhibition of FGF-2-mediated chemotaxis of murine brain capillary endothelial cells by cyclic RGDfV peptide through blocking the redistribution of c-Src into focal adhesions, *Exp Cell Res*, 268, 169, 2001.
18. Montesano, R., Orci, L., and Vassalli, P., In vitro rapid organization of endothelial cells into capillary-like networks is promoted by collagen matrices, *J Cell Biol*, 97, 1648, 1983.
19. Montesano, R. et al., Basic fibroblast growth factor induces angiogenesis in vitro, *Proc Natl Acad Sci USA*, 83, 7297, 1986.
20. Pepper, M.S. et al., Transforming growth factor-beta 1 modulates basic fibroblast growth factor-induced proteolytic and angiogenic properties of endothelial cells in vitro, *J Cell Biol*, 111, 743, 1990.

21. Henke, C.A. et al., CD44-related chondroitin sulfate proteoglycan, a cell surface receptor implicated with tumor cell invasion, mediates endothelial cell migration on fibrinogen and invasion into a fibrin matrix, *J Clin Invest*, 97, 2541, 1996.
22. Takei, A. et al., Effects of fibrin on the angiogenesis in vitro of bovine endothelial cells in collagen gel, *In Vitro Cell Dev Biol Anim*, 31, 467, 1995.
23. Kumar, R. et al., Regulation of distinct steps of angiogenesis by different angiogenic molecules, *Int J Oncol*, 12, 749, 1998.
24. Davis, G.E. and Camarillo, C.W., Regulation of endothelial cell morphogenesis by integrins, mechanical forces, and matrix guidance pathways, *Exp Cell Res*, 216, 113, 1995.
25. Sheibani, N., Newman, P.J., and Frazier, W.A., Thrombospondin-1, a natural inhibitor of angiogenesis, regulates platelet-endothelial cell adhesion molecule-1 expression and endothelial cell morphogenesis, *Mol Biol Cell*, 8, 1329, 1997.
26. Schnaper, H.W. et al., Type IV collagenase(s) and TIMPs modulate endothelial cell morphogenesis in vitro, *J Cell Physiol*, 156, 235, 1993.
27. Schnaper, H.W. et al., Plasminogen activators augment endothelial cell organization in vitro by two distinct pathways, *J Cell Physiol*, 165, 107, 1995.
28. Underwood, P.A., Bean, P.A., and Gamble, J.R., Rate of endothelial expansion is controlled by cell:cell adhesion, *Int J Biochem Cell Biol*, 34, 55, 2002.
29. Hoying, J.B. and Williams, S.K., Effects of basic fibroblast growth factor on human microvessel endothelial cell migration on collagen I correlates inversely with adhesion and is cell density dependent, *J Cell Physiol*, 168, 294, 1996.
30. Klein, S. et al., Basic fibroblast growth factor modulates integrin expression in microvascular endothelial cells, *Mol Biol Cell*, 4, 973, 1993.
31. Collo, G. and Pepper, M.S., Endothelial cell integrin alpha 5 beta 1 expression is modulated by cytokines and during migration in vitro, *J Cell Sci*, 112 (Pt 4), 569, 1999.
32. Sepp, N.T. et al., Basic fibroblast growth factor increases expression of the alpha v beta 3 integrin complex on human microvascular endothelial cells, *J Invest Dermatol*, 103, 295, 1994.
33. Zhou, L. et al., Divergent effects of extracellular oxygen on the growth, morphology, and function of human skin microvascular endothelial cells, *J Cell Physiol*, 182, 134, 2000.
34. Gerritsen, M.E. et al., Branching out: a molecular fingerprint of endothelial differentiation into tube-like structures generated by Affymetrix oligonucleotide arrays, *Microcirculation*, 10, 63, 2003.
35. Garlanda, C. and Dejana, E., Heterogeneity of endothelial cells. Specific markers, *Arterioscler Thromb Vasc Biol*, 17, 1193, 1997.
36. McCarthy, S.A. et al., Heterogeneity of the endothelial cell and its role in organ preference of tumour metastasis, *Trends Pharmacol Sci*, 12, 462, 1991.
37. Chi, J.T. et al., Endothelial cell diversity revealed by global expression profiling, *Proc Natl Acad Sci USA*, 100, 10623, 2003.
38. Ribatti, D. et al., The chick embryo chorioallantoic membrane as a model for in vivo research on anti-angiogenesis, *Curr Pharm Biotechnol*, 1, 73, 2000.
39. Herbert, J.M., Laplace, M.C., and Maffrand, J.P., Effect of heparin on the angiogenic potency of basic and acidic fibroblast growth factors in the rabbit cornea assay, *Int J Tissue React*, 10, 133, 1988.
40. Seghezzi, G. et al., Fibroblast growth factor-2 (FGF-2) induces vascular endothelial growth factor (VEGF) expression in the endothelial cells of forming capillaries: an autocrine mechanism contributing to angiogenesis, *J Cell Biol*, 141, 1659, 1998.
41. Passaniti, A. et al., A simple, quantitative method for assessing angiogenesis and antiangiogenic agents using reconstituted basement membrane, heparin, and fibroblast growth factor, *Lab Invest*, 67, 519, 1992.
42. Liekens, S. et al., Inhibition of fibroblast growth factor-2-induced vascular tumor formation by the acyclic nucleoside phosphonate cidofovir, *Cancer Res*, 61, 5057, 2001.
43. Ribatti, D. et al., Cell-mediated delivery of fibroblast growth factor-2 and vascular endothelial growth factor onto the chick chorioallantoic membrane: endothelial fenestration and angiogenesis, *J Vasc Res*, 38, 389, 2001.
44. Rusnati, M. et al., Alpha v beta 3 integrin mediates the cell-adhesive capacity and biological activity of basic fibroblast growth factor (FGF-2) in cultured endothelial cells, *Mol Biol Cell*, 8, 2449, 1997.

45. Tanghetti, E. et al., Biological activity of substrate-bound basic fibroblast growth factor (FGF2): recruitment of FGF receptor-1 in endothelial cell adhesion contacts, *Oncogene*, 21, 3889, 2002.
46. Rusnati, M. et al., Cell membrane GM1 ganglioside is a functional coreceptor for fibroblast growth factor 2, *Proc Natl Acad Sci USA*, 99, 4367, 2002.
47. Johnson, D.E. and Williams, L.T., Structural and functional diversity in the FGF receptor multigene family, *Adv Cancer Res*, 60, 1, 1993.
48. Gonzalez, A.M. et al., Distribution of fibroblast growth factor (FGF)-2 and FGF receptor-1 messenger RNA expression and protein presence in the mid-trimester human fetus, *Pediatr Res*, 39, 375, 1996.
49. Arany, E. and Hill, D.J., Fibroblast growth factor-2 and fibroblast growth factor receptor-1 mRNA expression and peptide localization in placentae from normal and diabetic pregnancies, *Placenta*, 19, 133, 1998.
50. Yoon, S.Y., Tefferi, A., and Li, C.Y., Cellular distribution of platelet-derived growth factor, transforming growth factor-beta, basic fibroblast growth factor, and their receptors in normal bone marrow, *Acta Haematol*, 104, 151, 2000.
51. Ribatti, D. et al., New model for the study of angiogenesis and antiangiogenesis in the chick embryo chorioallantoic membrane: the gelatin sponge/chorioallantoic membrane assay, *J Vasc Res*, 34, 455, 1997.
52. Dell'Era, P. et al., Paracrine and autocrine effects of fibroblast growth factor-4 in endothelial cells, *Oncogene*, 20, 2655, 2001.
53. Lindahl, U. et al., More to "heparin" than anticoagulation, *Thromb Res*, 75, 1, 1994.
54. Pellegrini, L., Role of heparan sulfate in fibroblast growth factor signalling: a structural view, *Curr Opin Struct Biol*, 11, 629, 2001.
55. Eriksson, A.E. et al., Three-dimensional structure of human basic fibroblast growth factor, *Proc Natl Acad Sci USA*, 88, 3441, 1991.
56. Rusnati, M. and Presta, M., Interaction of angiogenic basic fibroblast growth factor with endothelial cell heparan sulfate proteoglycans. Biological implications in neovascularization, *Int J Clin Lab Res*, 26, 15, 1996.
57. Horowitz, A., Tkachenko, E., and Simons, M., Fibroblast growth factor-specific modulation of cellular response by syndecan-4, *J Cell Biol*, 157, 715, 2002.
58. Rusnati, M., Urbinati, C., and Presta, M., Internalization of basic fibroblast growth factor (bFGF) in cultured endothelial cells: role of the low affinity heparin-like bFGF receptors, *J Cell Physiol*, 154, 152, 1993.
59. Hsia, E., Richardson, T.P., and Nugent, M.A., Nuclear localization of basic fibroblast growth factor is mediated by heparan sulfate proteoglycans through protein kinase C signaling, *J Cell Biochem*, 88, 1214, 2003.
60. Gospodarowicz, D. and Cheng, J., Heparin protects basic and acidic FGF from inactivation, *J Cell Physiol*, 128, 475, 1986.
61. Coltrini, D. et al., Different effects of mucosal, bovine lung and chemically modified heparin on selected biological properties of basic fibroblast growth factor, *Biochem J*, 303 (Pt 2), 583, 1994.
62. Presta, M. et al., Basic fibroblast growth factor is released from endothelial extracellular matrix in a biologically active form, *J Cell Physiol*, 140, 68, 1989.
63. Vlodavsky, I. et al., Extracellular matrix-resident growth factors and enzymes: possible involvement in tumor metastasis and angiogenesis, *Cancer Metastasis Rev*, 9, 203, 1990.
64. Ribatti, D. et al., In vivo angiogenic activity of urokinase: role of endogenous fibroblast growth factor-2, *J Cell Sci*, 112 (Pt 23), 4213, 1999.
65. Ruegg, C. and Mariotti, A., Vascular integrins: pleiotropic adhesion and signaling molecules in vascular homeostasis and angiogenesis, *Cell Mol Life Sci*, 60, 1135, 2003.
66. Eliceiri, B.P., Integrin and growth factor receptor crosstalk, *Circ Res*, 89, 1104, 2001.
67. Kumar, C.C., Integrin alpha v beta 3 as a therapeutic target for blocking tumor-induced angiogenesis,*Curr Drug Targets*, 4, 123, 2003.
68. Sahni, A. and Francis, C.W., Stimulation of endothelial cell proliferation by FGF-2 in the presence of fibrinogen requires $\alpha_v\beta_3$, *Blood*, 104, 3635, 2004.

69. Presta, M. et al., Antiangiogenic activity of semisynthetic biotechnological heparins. Low-molecular-weight-sulfated *Escherichia coli* K5 polysaccharide derivatives as fibroblast growth factor antagonists, *Arterioscler Thromb Vasc Biol*, 25, 71, 2005.
70. Salmivirta, M., Heino, J., and Jalkanen, M., Basic fibroblast growth factor-syndecan complex at cell surface or immobilized to matrix promotes cell growth, *J Biol Chem*, 267, 17606, 1992.
71. Birkle, S. et al., Role of tumor-associated gangliosides in cancer progression, *Biochimie*, 85, 455, 2003.
72. Rusnati, M. et al., Interaction of fibroblast growth factor-2 (FGF-2) with free gangliosides: biochemical characterization and biological consequences in endothelial cell cultures, *Mol Biol Cell*, 10, 313, 1999.
73. Eswarakumar, V.P., Lax, I., and Schlessinger, J., Cellular signaling by fibroblast growth factor receptors, *Cytokine Growth Factor Rev*, 16, 139, 2005.
74. Schlaepfer, D.D., Hauck, C.R., and Sieg, D.J., Signaling through focal adhesion kinase, *Prog Biophys Mol Biol*, 71, 435, 1999.
75. Kumar, C.C., Signaling by integrin receptors, *Oncogene*, 17, 1365, 1998.
76. Palazzo, A.F. et al., Localized stabilization of microtubules by integrin- and FAK-facilitated Rho signaling, *Science*, 303, 836, 2004.
77. Zhai, J. et al., Direct interaction of focal adhesion kinase with p190RhoGEF, *J Biol Chem*, 278, 24865, 2003.
78. Sharma-Walia, N. et al., Kaposi's sarcoma-associated herpesvirus/human herpesvirus 8 envelope glycoprotein gB induces the integrin-dependent focal adhesion kinase-Src-phosphatidylinositol 3-kinase-rho GTPase signal pathways and cytoskeletal rearrangements, *J Virol*, 78, 4207, 2004.
79. Scatena, M. et al., NF-kappaB mediates alpha v beta 3 integrin-induced endothelial cell survival, *J Cell Biol*, 141, 1083, 1998.
80. Chua, C.C. et al., Heparan sulfate proteoglycans function as receptors for fibroblast growth factor-2 activation of extracellular signal-regulated kinases 1 and 2, *Circ Res*, 94, 316, 2004.
81. Miljan, E.A. and Bremer, E.G., Regulation of growth factor receptors by gangliosides, *Sci STKE*, 2002, RE15, 2002.
82. Gualandris, A. et al., Basic fibroblast growth factor overexpression in endothelial cells: anautocrine mechanism for angiogenesis and angioproliferative diseases, *Cell Growth Differ*, 7, 147, 1996.
83. Bikfalvi, A. et al., Differential modulation of cell phenotype by different molecular weight forms of basic fibroblast growth factor: possible intracellular signaling by the high molecular weight forms, *J Cell Biol*, 129, 233, 1995.
84. Florkiewicz, R.Z., Baird, A., and Gonzalez, A.M., Multiple forms of bFGF: differential nuclear and cell surface localization, *Growth Factors*, 4, 265, 1991.
85. Quarto, N. et al., Selective expression of high molecular weight basic fibroblast growth factor confers a unique phenotype to NIH 3T3 cells, *Cell Regul*, 2, 699, 1991.
86. Tille, J.C. et al., Vascular endothelial growth factor (VEGF) receptor-2 antagonists inhibit VEGF- and basic fibroblast growth factor-induced angiogenesis in vivo and in vitro, *J Pharmacol Exp Ther*, 299, 1073, 2001.
87. Kanda, S., Miyata, Y., and Kanetake, H., Fibroblast growth factor-2-mediated capillary morphogenesis of endothelial cells requires signals via Flt-1/vascular endothelial growth factor receptor-1: possible involvement of c-Akt, *J Biol Chem*, 279, 4007, 2004.
88. Jia, H. et al., Peptides encoded by exon 6 of VEGF inhibit endothelial cell biological responses and angiogenesis induced by VEGF, *Biochem Biophys Res Commun*, 283, 164, 2001.
89. Tomanek, R.J. et al., Vascular endothelial growth factor and basic fibroblast growth factor differentially modulate early postnatal coronary angiogenesis, *Circ Res*, 88, 1135, 2001.
90. Xue, L. and Greisler, H.P., Angiogenic effect of fibroblast growth factor-1 and vascular endothelial growth factor and their synergism in a novel in vitro quantitative fibrin-based 3-dimensional angiogenesis system, *Surgery*, 132, 259, 2002.
91. Castellon, R. et al., Effects of angiogenic growth factor combinations on retinal endothelial cells, *Exp Eye Res*, 74, 523, 2002.
92. Pepper, M.S. and Mandriota, S.J., Regulation of vascular endothelial growth factor receptor-2 (Flk-1) expression in vascular endothelial cells, *Exp Cell Res*, 241, 414, 1998.

93. Nico, B. et al., In vivo absence of synergism between fibroblast growth factor-2 and vascular endothelial growth factor, *J Hematother Stem Cell Res*, 10, 905, 2001.
94. Wang, Y. and Becker, D., Antisense targeting of basic fibroblast growth factor and fibroblast growth factor receptor-1 in human melanomas blocks intratumoral angiogenesis and tumor growth, *Nat Med*, 3, 887, 1997.
95. Cassinelli, G. et al., Cellular bases of the antitumor activity of the novel taxane IDN 5109 (BAY59–8862) on hormone-refractory prostate cancer, *Clin Cancer Res*, 8, 2647, 2002.
96. Hotchkiss, K.A. et al., Inhibition of endothelial cell function in vitro and angiogenesis in vivo by docetaxel (Taxotere): association with impaired repositioning of the microtubule organizing center, *Mol Cancer Ther*, 1, 1191, 2002.
97. Ciardiello, F. et al., Inhibition of growth factor production and angiogenesis in human cancer cells by ZD1839 (Iressa), a selective epidermal growth factor receptor tyrosine kinase inhibitor, *Clin Cancer Res*, 7, 1459, 2001.
98. Ginns, L.C. et al., Pulmonary capillary hemangiomatosis with atypical endotheliomatosis: successful antiangiogenic therapy with doxycycline, *Chest*, 124, 2017, 2003.
99. Jee, S.H. et al., Interleukin-6 induced basic fibroblast growth factor-dependent angiogenesis in basal cell carcinoma cell line via JAK/STAT3 and PI3-kinase/Akt pathways, *J Invest Dermatol*, 123, 1169, 2004.
100. Huang, S. et al., Stat1 negatively regulates angiogenesis, tumorigenicity and metastasis of tumor cells, *Oncogene*, 21, 2504, 2002.
101. Sasamura, H. et al., Inhibitory effect on expression of angiogenic factors by antiangiogenic agents in renal cell carcinoma, *Br J Cancer*, 86, 768, 2002.
102. Fujimoto, J. et al., Plausible novel therapeutic strategy of uterine endometrial cancer with reduction of basic fibroblast growth factor secretion by progestin and O-(chloroacetyl-carbamoyl) fumagillol (TNP-470; AGM-1470), *Cancer Lett*, 113, 187, 1997.
103. Shao, Z.M. et al., Curcumin exerts multiple suppressive effects on human breast carcinoma cells, *Int J Cancer*, 98, 234, 2002.
104. Sartippour, M.R. et al., Inhibition of fibroblast growth factors by green tea, *Int J Oncol*, 21, 487, 2002.
105. Lee, H.T. et al., FGF-2 induced by interleukin-1 beta through the action of phosphatidylinositol 3-kinase mediates endothelial mesenchymal transformation in corneal endothelial cells, *J Biol Chem*, 279, 32325, 2004.
106. Albuquerque, M.L., Akiyama, S.K., and Schnaper, H.W., Basic fibroblast growth factor release by human coronary artery endothelial cells is enhanced by matrix proteins, 17beta-estradiol, and a PKC signaling pathway, *Exp Cell Res*, 245, 163, 1998.
107. Yoshida, S. et al., Involvement of interleukin-8, vascular endothelial growth factor, and basic fibroblast growth factor in tumor necrosis factor alpha-dependent angiogenesis, *Mol Cell Biol*, 17, 4015, 1997.
108. Faris, M. et al., Inflammatory cytokines induce the expression of basic fibroblast growth factor (bFGF) isoforms required for the growth of Kaposi's sarcoma and endothelial cells through the activation of AP-1 response elements in the bFGF promoter, *Aids*, 12, 19, 1998.
109. Khachigian, L.M., Early growth response-1: blocking angiogenesis by shooting the messenger, *Cell Cycle*, 3, 10, 2004.
110. Wijelath, E.S. et al., Oncostatin M induces basic fibroblast growth factor expression in endothelial cells and promotes endothelial cell proliferation, migration and spindle morphology, *J Cell Sci*, 110 (Pt 7), 871, 1997.
111. Maier, J.A. et al., Tumor-necrosis-factor-induced fibroblast growth factor-1 acts as a survival factor in a transformed endothelial cell line, *Am J Pathol*, 149, 945, 1996.
112. Danesi, R. et al., Suramin inhibits bFGF-induced endothelial cell proliferation and angiogenesis in the chick chorioallantoic membrane, *Br J Cancer*, 68, 932, 1993.
113. Ensoli, B. et al., Block of AIDS-Kaposi's sarcoma (KS) cell growth, angiogenesis, and lesion formation in nude mice by antisense oligonucleotide targeting basic fibroblast growth factor. A novel strategy for the therapy of KS, *J Clin Invest*, 94, 1736, 1994.
114. Ribatti, D. et al., Inhibition of neuroblastoma-induced angiogenesis by fenretinide, *Int J Cancer*, 94, 314, 2001.

115. Tsou, R. and Isik, F.F., Integrin activation is required for VEGF and FGF receptor protein presence on human microvascular endothelial cells, *Mol Cell Biochem*, 224, 81, 2001.
116. Fujiwara, Y. and Kaji, T., Possible mechanism for lead inhibition of vascular endothelial cell proliferation: a lower response to basic fibroblast growth factor through inhibition of heparan sulfate synthesis, *Toxicology*, 133, 147, 1999.
117. Zhang, W. et al., Antiangiogenic antithrombin down-regulates the expression of the proangiogenic heparan sulfate proteoglycan, perlecan, in endothelial cells, *Blood*, 103, 1185, 2004.
118. Aviezer, D. et al., Suppression of autocrine and paracrine functions of basic fibroblast growth factor by stable expression of perlecan antisense cDNA, *Mol Cell Biol*, 17, 1938, 1997.
119. Wang, S. et al., QSulf1, a heparan sulfate 6-O-endosulfatase, inhibits fibroblast growth factor signaling in mesoderm induction and angiogenesis, *Proc Natl Acad Sci USA*, 101, 4833, 2004.
120. Cieslak, M. et al., DNAzymes to beta 1 and beta 3 mRNA down-regulate expression of the targeted integrins and inhibit endothelial cell capillary tube formation in fibrin and matrigel, *J Biol Chem*, 277, 6779, 2002.
121. Presta, M. et al., Heparin derivatives as angiogenesis inhibitors, *Curr Pharm Des*, 9, 553, 2003.
122. Barzu, T. et al., Heparin-derived oligosaccharides: affinity for acidic fibroblast growth factor and effect on its growth-promoting activity for human endothelial cells, *J Cell Physiol*, 140, 538, 1989.
123. Foxall, C. et al., Sulfated malto-oligosaccharides bind to basic FGF, inhibit endothelial cell proliferation, and disrupt endothelial cell tube formation, *J Cell Physiol*, 168, 657, 1996.
124. Herbert, J.M. et al., Activity of pentosan polysulphate and derived compounds on vascular endothelial cell proliferation and migration induced by acidic and basic FGF in vitro, *Biochem Pharmacol*, 37, 4281, 1988.
125. Ishihara, M. et al., Structural features in heparin which modulate specific biological activities mediated by basic fibroblast growth factor, *Glycobiology*, 4, 451, 1994.
126. Gagliardi, A.R., Taylor, M.F., and Collins, D.C., Uptake of suramin by human microvascular endothelial cells, *Cancer Lett*, 125, 97, 1998.
127. Kitajima, I., Unoki, K., and Maruyama, I., Phosphorothioate oligodeoxynucleotides inhibit basic fibroblast growth factor-induced angiogenesis in vitro and in vivo, *Antisense Nucleic Acid Drug Dev*, 9, 233, 1999.
128. Leali, D. et al., Fibroblast growth factor-2 antagonist activity and angiostatic capacity of sulfated *Escherichia coli* K5 polysaccharide derivatives, *J Biol Chem*, 276, 37900, 2001.
129. Urbinati, C. et al., Chemically sulfated *Escherichia coli* K5 polysaccharide derivatives as extracellular HIV-1 Tat protein antagonists, *FEBS Lett*, 568, 171, 2004.
130. Ensoli, B. et al., Synergy between basic fibroblast growth factor and HIV-1 Tat protein in induction of Kaposi's sarcoma, *Nature*, 371, 674, 1994.
131. Giavazzi, R. et al., Distinct role of fibroblast growth factor-2 and vascular endothelial growth factor on tumor growth and angiogenesis, *Am J Pathol*, 162, 1913, 2003.
132. Rusnati, M. et al., Pentosan polysulfate as an inhibitor of extracellular HIV-1 Tat, *J Biol Chem*, 276, 22420, 2001.
133. Marshall, J.L. et al., Phase I trial of orally administered pentosan polysulfate in patients with advanced cancer, *Clin Cancer Res*, 3, 2347, 1997.
134. Swain, S.M. et al., Phase I trial of pentosan polysulfate, *Invest New Drugs*, 13, 55, 1995.
135. Schwartsmann, G. et al., Phase II study of pentosan polysulfate (PPS) in patients with AIDS-related Kaposi's sarcoma, *Tumori*, 82, 360, 1996.
136. Sola, F. et al., The antitumor efficacy of cytotoxic drugs is potentiated by treatment with PNU 145156E, a growth-factor-complexing molecule, *Cancer Chemother Pharmacol*, 43, 241, 1999.
137. Zacharski, L.R. and Loynes, J.T., Low-molecular-weight heparin in oncology, *Anticancer Res*, 23, 2789, 2003.
138. Ziche, M. et al., Angiogenesis can be stimulated or repressed in vivo by a change in GM3:GD3 ganglioside ratio, *Lab Invest*, 67, 711, 1992.
139. Neufeld, G. and Gospodarowicz, D., Protamine sulfate inhibits mitogenic activities of the extracellular matrix and fibroblast growth factor, but potentiates that of epidermal growth factor, *J Cell Physiol*, 132, 287, 1987.
140. Bornstein, P. et al., Thrombospondin 2, a matricellular protein with diverse functions, *Matrix Biol*, 19, 557, 2000.

141. Taraboletti, G. et al., The 140-kilodalton antiangiogenic fragment of thrombospondin-1 binds to basic fibroblast growth factor, *Cell Growth Differ*, 8, 471, 1997.
142. Margosio, B. et al., Thrombospondin 1 as a scavenger for matrix-associated fibroblast growth factor 2, *Blood*, 102, 4399, 2003.
143. Dawson, D.W. et al., Three distinct D-amino acid substitutions confer potent antiangiogenic activity on an inactive peptide derived from a thrombospondin-1 type 1 repeat, *Mol Pharmacol*, 55, 332, 1999.
144. Bossard, C. et al., Antiangiogenic properties of fibstatin, an extracellular FGF-2-binding polypeptide, *Cancer Res*, 64, 7507, 2004.
145. Asplin, I.R. et al., Differential regulation of the fibroblast growth factor (FGF) family by alpha(2)-macroglobulin: evidence for selective modulation of FGF-2-induced angiogenesis, *Blood*, 97, 3450, 2001.
146. Mathew, S. et al., Characterization of the interaction between alpha2-macroglobulin and fibroblast growth factor-2: the role of hydrophobic interactions, *Biochem J*, 374, 123, 2003.
147. Dennis, P.A. et al., Alpha 2-macroglobulin is a binding protein for basic fibroblast growth factor, *J Biol Chem*, 264, 7210, 1989.
148. Bottazzi, B. et al., Multimer formation and ligand recognition by the long pentraxin PTX3. Similarities and differences with the short pentraxins C-reactive protein and serum amyloid P component, *J Biol Chem*, 272, 32817, 1997.
149. Breviario, F. et al., Interleukin-1-inducible genes in endothelial cells. Cloning of a new gene related to C-reactive protein and serum amyloid P component, *J Biol Chem*, 267, 22190, 1992.
150. Rusnati, M. et al., Selective recognition of fibroblast growth factor-2 by the long pentraxin PTX3 inhibits angiogenesis, *Blood*, 104, 92, 2004.
151. Salustri, A. et al., PTX3 plays a key role in the organization of the cumulus oophorus extracellular matrix and in in vivo fertilization, *Development*, 131, 1577, 2004.
152. Miao, R.Q. et al., Kallistatin is a new inhibitor of angiogenesis and tumor growth, *Blood*, 100, 3245, 2002.
153. Russo, K. et al., Platelet-derived growth factor-BB and basic fibroblast growth factor directly interact in vitro with high affinity, *J Biol Chem*, 277, 1284, 2002.
154. De Marchis, F. et al., Platelet-derived growth factor inhibits basic fibroblast growth factor angiogenic properties in vitro and in vivo through its alpha receptor, *Blood*, 99, 2045, 2002.
155. Perollet, C. et al., Platelet factor 4 modulates fibroblast growth factor 2 (FGF-2) activity and inhibits FGF-2 dimerization, *Blood*, 91, 3289, 1998.
156. Lozano, R.M. et al., Solution structure and interaction with basic and acidic fibroblast growth factor of a 3-kDa human platelet factor-4 fragment with antiangiogenic activity, *J Biol Chem*, 276, 35723, 2001.
157. Sulpice, E. et al., Platelet factor 4 inhibits FGF2-induced endothelial cell proliferation via the extracellular signal-regulated kinase pathway but not by the phosphatidylinositol 3-kinase pathway, *Blood*, 100, 3087, 2002.
158. Brown, K.J. and Parish, C.R., Histidine-rich glycoprotein and platelet factor 4 mask heparan sulfate proteoglycans recognized by acidic and basic fibroblast growth factor, *Biochemistry*, 33, 13918, 1994.
159. Hagedorn, M. et al., Domain swapping in a COOH-terminal fragment of platelet factor 4 generates potent angiogenesis inhibitors, *Cancer Res*, 62, 6884, 2002.
160. Hagedorn, M. et al., A short peptide domain of platelet factor 4 blocks angiogenic key events induced by FGF-2, *FASEB J*, 15, 550, 2001.
161. Jouan, V. et al., Inhibition of in vitro angiogenesis by platelet factor-4-derived peptides and mechanism of action, *Blood*, 94, 984, 1999.
162. Spinetti, G. et al., The chemokine CXCL13 (BCA-1) inhibits FGF-2 effects on endothelial cells, *Biochem Biophys Res Commun*, 289, 19, 2001.
163. Shellenberger, T.D. et al., BRAK/CXCL14 is a potent inhibitor of angiogenesis and a chemotactic factor for immature dendritic cells, *Cancer Res*, 64, 8262, 2004.
164. Hanneken, A. and Baird, A., Soluble forms of the high-affinity fibroblast growth factor receptor in human vitreous fluid, *Invest Ophthalmol Vis Sci*, 36, 1192, 1995.
165. Hanneken, A., Maher, P.A., and Baird, A., High affinity immunoreactive FGF receptors in the extracellular matrix of vascular endothelial cells—implications for the modulation of FGF-2, *J Cell Biol*, 128, 1221, 1995.

166. Bergonzoni, L. et al., Characterization of a biologically active extracellular domain of fibroblast growth factor receptor 1 expressed in *Escherichia coli*, *Eur J Biochem*, 210, 823, 1992.
167. Ueno, H. et al., A truncated form of fibroblast growth factor receptor 1 inhibits signal transduction by multiple types of fibroblast growth factor receptor, *J Biol Chem*, 267, 1470, 1992.
168. Schilling-Schon, A. et al., The role of endogenous growth factors to support corneal endothelial migration after wounding in vitro, *Exp Eye Res*, 71, 583, 2000.
169. Facchiano, A. et al., Identification of a novel domain of fibroblast growth factor 2 controlling its angiogenic properties, *J Biol Chem*, 278, 8751, 2003.
170. Presta, M. et al., Biologically active synthetic fragments of human basic fibroblast growth factor (bFGF): identification of two Asp-Gly-Arg-containing domains involved in the mitogenic activity of bFGF in endothelial cells, *J Cell Physiol*, 149, 512, 1991.
171. Kumar, C.C. et al., Inhibition of angiogenesis and tumor growth by SCH221153, a dual alpha(v)beta3 and alpha(v)beta5 integrin receptor antagonist, *Cancer Res*, 61, 2232, 2001.
172. Friedlander, M. et al., Definition of two angiogenic pathways by distinct alpha v integrins, *Science*, 270, 1500, 1995.
173. Yeh, C.H. et al., Rhodostomin, a snake venom disintegrin, inhibits angiogenesis elicited by basic fibroblast growth factor and suppresses tumor growth by a selective alpha(v) beta(3) blockade of endothelial cells, *Mol Pharmacol*, 59, 1333, 2001.
174. Ding, L. et al., Inhibition of cell migration and angiogenesis by the amino-terminal fragment of 24kD basic fibroblast growth factor, *J Biol Chem*, 277, 31056, 2002.
175. Plum, S.M. et al., Generation of a specific immunological response to FGF-2 does not affect wound healing or reproduction, *Immunopharmacol Immunotoxicol*, 26, 29, 2004.
176. Baird, A., Mormede, P., and Bohlen, P., Immunoreactive fibroblast growth factor (FGF) in a transplantable chondrosarcoma: inhibition of tumor growth by antibodies to FGF, *J Cell Biochem*, 30, 79, 1986.
177. Hori, A. et al., Suppression of solid tumor growth by immunoneutralizing monoclonal antibody against human basic fibroblast growth factor, *Cancer Research*, 51, 6180, 1991.
178. Talarico, D. et al., Protection of mice against tumor growth by immunization with an oncogene-encoded growth factor, *Proc Natl Acad Sci USA*, 87, 4222, 1990.
179. He, Q.M. et al., Inhibition of tumor growth with a vaccine based on xenogeneic homologous fibroblast growth factor receptor-1 in mice, *J Biol Chem*, 278, 21831, 2003.
180. Seo, D.W. et al., TIMP-2 mediated inhibition of angiogenesis: an MMP-independent mechanism, *Cell*, 114, 171, 2003.
181. Gatto, C. et al., BAY 12–9566, a novel inhibitor of matrix metalloproteinases with antiangiogenic activity, *Clin Cancer Res*, 5, 3603, 1999.
182. Min, H.Y. et al., Urokinase receptor antagonists inhibit angiogenesis and primary tumor growth in syngeneic mice, *Cancer Res*, 56, 2428, 1996.
183. Blei, F. et al., Mechanism of action of angiostatic steroids: suppression of plasminogen activator activity via stimulation of plasminogen activator inhibitor synthesis, *J Cell Physiol*, 155, 568, 1993.
184. Thorpe, P.E., Vascular targeting agents as cancer therapeutics, *Clin Cancer Res*, 10, 415, 2004.
185. Tozer, G.M. et al., Combretastatin A-4 phosphate as a tumor vascular-targeting agent: early effects in tumors and normal tissues, *Cancer Res*, 59, 1626, 1999.
186. Tozer, G.M. et al., Mechanisms associated with tumor vascular shut-down induced by combretastatin A-4 phosphate: intravital microscopy and measurement of vascular permeability, *Cancer Res*, 61, 6413, 2001.
187. Tozer, G.M. et al., The biology of the combretastatins as tumour vascular targeting agents, *Int J Exp Pathol*, 83, 21, 2002.
188. Vacca, A. et al., Antiangiogenesis is produced by nontoxic doses of vinblastine, *Blood*, 94, 4143, 1999.
189. Ahmed, B. et al., Vascular targeting effect of combretastatin A-4 phosphate dominates the inherent angiogenesis inhibitory activity, *Int J Cancer*, 105, 20, 2003.
190. Belleri, M. et al., Antiangiogenic and vascular-targeting activity of the microtubule-destabilizing trans-resveratrol derivative 3,5,4′-trimethoxystilbene, *Mol Pharmacol*, 67, 1451, 2005.
191. Belotti, D. et al., The microtubule-affecting drug paclitaxel has antiangiogenic activity, *Clin Cancer Res*, 2, 1843, 1996.

192. Taraboletti, G. et al., Antiangiogenic and antitumor activity of IDN 5390, a new taxane derivative, *Clin Cancer Res*, 8, 1182, 2002.
193. Abe, M. et al., Goniodomin A, an antifungal polyether macrolide, exhibits antiangiogenic activities via inhibition of actin reorganization in endothelial cells, *J Cell Physiol*, 190, 109, 2002.
194. Siemann, D.W. et al., Vascular targeting agents enhance chemotherapeutic agent activities in solid tumor therapy, *Int J Cancer*, 99, 1, 2002.
195. Siim, B.G. et al., Marked potentiation of the antitumour activity of chemotherapeutic drugs by the antivascular agent 5,6-dimethylxanthenone-4-acetic acid (DMXAA), *Cancer Chemother Pharmacol*, 51, 43, 2003.
196. Wildiers, H. et al., Combretastatin A-4 phosphate enhances CPT-11 activity independently of the administration sequence, *Eur J Cancer*, 40, 284, 2004.
197. Kwon, H.J. et al., Betulinic acid inhibits growth factor-induced in vitro angiogenesis via the modulation of mitochondrial function in endothelial cells, *Jpn J Cancer Res*, 93, 417, 2002.
198. Chow, L.M. et al., Anti-angiogenic potential of Gleditsia sinensis fruit extract, *Int J Mol Med*, 12, 269, 2003.
199. Liu, Y. et al., Citrus pectin: characterization and inhibitory effect on fibroblast growth factor-receptor interaction, *J Agric Food Chem*, 49, 3051, 2001.
200. Tong, Y. et al., Philinopside A, a novel marine-derived compound possessing dual anti-angiogenic and anti-tumor effects, *Int J Cancer*, 114, 843, 2005.
201. Tong, Y. et al., Anti-angiogenic effects of Shiraiachrome A, a compound isolated from a Chinese folk medicine used to treat rheumatoid arthritis, *Eur J Pharmacol*, 494, 101, 2004.
202. Brakenhielm, E., Cao, R., and Cao, Y., Suppression of angiogenesis, tumor growth, and wound healing by resveratrol, a natural compound in red wine and grapes, *FASEB J*, 15, 1798, 2001.
203. Hoffman, R., Burns, W.W., 3rd, and Paper, D.H., Selective inhibition of cell proliferation and DNA synthesis by the polysulphated carbohydrate l-carrageenan, *Cancer Chemother Pharmacol*, 36, 325, 1995.
204. Dupont, E. et al., Antiangiogenic and antimetastatic properties of Neovastat (AE-941), an orally active extract derived from cartilage tissue, *Clin Exp Metastasis*, 19, 145, 2002.
205. Klauber, N. et al., New activity of spironolactone. Inhibition of angiogenesis in vitro and in vivo, *Circulation*, 94, 2566, 1996.
206. Koyama, S. et al., Tranilast inhibits protein kinase C-dependent signalling pathway linked to angiogenic activities and gene expression of retinal microcapillary endothelial cells, *Br J Pharmacol*, 127, 537, 1999.
207. Wood, J. et al., Novel antiangiogenic effects of the bisphosphonate compound zoledronic acid, *J Pharmacol Exp Ther*, 302, 1055, 2002.
208. Pai, R. et al., Indomethacin inhibits endothelial cell proliferation by suppressing cell cycle proteins and PRB phosphorylation: a key to its antiangiogenic action? *Mol Cell Biol Res Commun*, 4, 111, 2000.
209. Vincent, L. et al., Cerivastatin, an inhibitor of 3-hydroxy-3-methylglutaryl coenzyme a reductase, inhibits endothelial cell proliferation induced by angiogenic factors in vitro and angiogenesis in in vivo models, *Arterioscler Thromb Vasc Biol*, 22, 623, 2002.
210. Klein-Soyer, C. et al., SR 25989 inhibits healing of a mechanical wound of confluent human saphenous vein endothelial cells which is modulated by standard heparin and growth factors, *J Cell Physiol*, 160, 316, 1994.
211. Chen, W. et al., Control of angiogenesis by inhibitor of phospholipase A2, *Chin Med Sci J*, 19, 6, 2004.
212. Aguzzi, M.S. et al., RGDS peptide induces caspase 8 and caspase 9 activation in human endothelial cells, *Blood*, 103, 4180, 2004.
213. Mohan, R. et al., Curcuminoids inhibit the angiogenic response stimulated by fibroblast growth factor-2, including expression of matrix metalloproteinase gelatinase B, *J Biol Chem*, 275, 10405, 2000.
214. Keledjian, K., Garrison, J.B., and Kyprianou, N., Doxazosin inhibits human vascular endothelial cell adhesion, migration, and invasion, *J Cell Biochem*, 94, 374, 2005.
215. Gelati, M. et al., Effects of thalidomide on parameters involved in angiogenesis: an in vitro study, *J Neurooncol*, 64, 193, 2003.

216. Gagliardi, A.R., Hennig, B., and Collins, D.C., Antiestrogens inhibit endothelial cell growth stimulated by angiogenic growth factors, *Anticancer Res*, 16, 1101, 1996.
217. Ashino-Fuse, H. et al., Medroxyprogesterone acetate, an anti-cancer and anti-angiogenic steroid, inhibits the plasminogen activator in bovine endothelial cells, *Int J Cancer*, 44, 859, 1989.
218. Nakashio, A., Fujita, N., and Tsuruo, T., Topotecan inhibits VEGF- and bFGF-induced vascular endothelial cell migration via downregulation of the PI3K-Akt signaling pathway, *Int J Cancer*, 98, 36, 2002.
219. Taraboletti, G. et al., Antiangiogenic activity of aplidine, a new agent of marine origin, *Br J Cancer*, 90, 2418, 2004.
220. Presta, M. et al., Anti-angiogenic activity of the purine analog 6-thioguanine, *Leukemia*, 16, 1490, 2002.
221. Basaki, Y. et al., UFT and its metabolites inhibit cancer-induced angiogenesis. Via a VEGF-related pathway, *Oncology* (*Huntingt*), 14, 68, 2000.
222. Yamayoshi, T. et al., Expression of keratinocyte growth factor/fibroblast growth factor-7 and its receptor in human lung cancer: correlation with tumour proliferative activity and patient prognosis, *J Pathol*, 204, 110, 2004.
223. Dorkin, T.J. et al., FGF8 over-expression in prostate cancer is associated with decreased patient survival and persists in androgen independent disease, *Oncogene*, 18, 2755, 1999.
224. Gnanapragasam, V.J. et al., FGF8 isoform b expression in human prostate cancer, *Br J Cancer*, 88, 1432, 2003.
225. Heer, R. et al., Fibroblast growth factor 17 is over-expressed in human prostate cancer, *J Pathol*, 204, 578, 2004.
226. Auguste, P. et al., Inhibition of fibroblast growth factor/fibroblast growth factor receptor activity in glioma cells impedes tumor growth by both angiogenesis-dependent and -independent mechanisms, *Cancer Res*, 61, 1717, 2001.
227. Polnaszek, N. et al., Fibroblast growth factor 2 promotes tumor progression in an autochthonous mouse model of prostate cancer, *Cancer Res*, 63, 5754, 2003.
228. Hagedorn, M. and Bikfalvi, A., Target molecules for anti-angiogenic therapy: from basic research to clinical trials, *Crit Rev Oncol Hematol*, 34, 89, 2000.
229. Ziche, M., Donnini, S., and Morbidelli, L., Development of new drugs in angiogenesis, *Curr Drug Targets*, 5, 485, 2004.
230. Patel, S.R. et al., Pilot study of vitaxin—an angiogenesis inhibitor-in patients with advanced leiomyosarcomas, *Cancer*, 92, 1347, 2001.
231. Posey, J.A. et al., A pilot trial of Vitaxin, a humanized anti-vitronectin receptor (anti alpha v beta 3) antibody in patients with metastatic cancer, *Cancer Biother Radiopharm*, 16, 125, 2001.
232. Heryanto, B., Lipson, K.E., and Rogers, P.A., Effect of angiogenesis inhibitors on oestrogen-mediated endometrial endothelial cell proliferation in the ovariectomized mouse, *Reproduction*, 125, 337, 2003.
233. Abdollahi, A. et al., SU5416 and SU6668 attenuate the angiogenic effects of radiation-induced tumor cell growth factor production and amplify the direct anti-endothelial action of radiation in vitro, *Cancer Res*, 63, 3755, 2003.
234. Wang, L.L. et al., Antitumor activities of a novel indolin-2-ketone compound, Z24: more potent inhibition on bFGF-induced angiogenesis and bcl-2 over-expressing cancer cells, *Eur J Pharmacol*, 502, 1, 2004.
235. Laird, A.D. et al., SU6668 is a potent antiangiogenic and antitumor agent that induces regression of established tumors, *Cancer Res*, 60, 4152, 2000.
236. Dimitroff, C.J. et al., Anti-angiogenic activity of selected receptor tyrosine kinase inhibitors, PD166285 and PD173074: implications for combination treatment with photodynamic therapy, *Invest New Drugs*, 17, 121, 1999.
237. Beebe, J.S. et al., Pharmacological characterization of CP-547,632, a novel vascular endothelial growth factor receptor-2 tyrosine kinase inhibitor for cancer therapy, *Cancer Res*, 63, 7301, 2003.
238. Klint, P. et al., Contribution of Src and Ras pathways in FGF-2 induced endothelial cell differentiation, *Oncogene*, 18, 3354, 1999.

239. Alavi, A. et al., Role of Raf in vascular protection from distinct apoptotic stimuli, *Science*, 301, 94, 2003.
240. Giuliani, R. et al., Role of endothelial cell extracellular signal-regulated kinase1/2 in urokinase-type plasminogen activator upregulation and in vitro angiogenesis by fibroblast growth factor-2, *J Cell Sci*, 112 (Pt 15), 2597, 1999.
241. Pintucci, G. et al., Induction of stromelysin-1 (MMP-3) by fibroblast growth factor-2 (FGF-2) in FGF-$2^{-/-}$ microvascular endothelial cells requires prolonged activation of extracellular signal-regulated kinases-1 and -2 (ERK-1/2), *J Cell Biochem*, 90, 1015, 2003.
242. Naik, M.U., Vuppalanchi, D., and Naik, U.P., Essential role of junctional adhesion molecule-1 in basic fibroblast growth factor-induced endothelial cell migration, *Arterioscler Thromb Vasc Biol*, 23, 2165, 2003.
243. Shono, T., Kanetake, H., and Kanda, S., The role of mitogen-activated protein kinase activation within focal adhesions in chemotaxis toward FGF-2 by murine brain capillary endothelial cells, *Exp Cell Res*, 264, 275, 2001.
244. Eliceiri, B.P. et al., Integrin alpha v beta 3 requirement for sustained mitogen-activated protein kinase activity during angiogenesis, *J Cell Biol*, 140, 1255, 1998.
245. Bhagwat, S.V. et al., The angiogenic regulator CD13/APN is a transcriptional target of Ras signaling pathways in endothelial morphogenesis, *Blood*, 101, 1818, 2003.
246. Kuzuya, M. et al., Induction of apoptotic cell death in vascular endothelial cells cultured in three-dimensional collagen lattice, *Exp Cell Res*, 248, 498, 1999.
247. Santiago, F.S. et al., Early growth response factor-1 induction by injury is triggered by release and paracrine activation by fibroblast growth factor-2, *Am J Pathol*, 154, 937, 1999.
248. Hata, Y., Rook, S.L., and Aiello, L.P., Basic fibroblast growth factor induces expression of VEGF receptor KDR through a protein kinase C and p44/p42 mitogen-activated protein kinase-dependent pathway, *Diabetes*, 48, 1145, 1999.
249. Langford, D. et al., Signalling crosstalk in FGF2-mediated protection of endothelial cells from HIV-gp120, *BMC Neurosci*, 6, 8, 2005.
250. Zubilewicz, A. et al., Two distinct signalling pathways are involved in FGF2-stimulated proliferation of choriocapillary endothelial cells: a comparative study with VEGF, *Oncogene*, 20, 1403, 2001.
251. Tanaka, K., Abe, M., and Sato, Y., Roles of extracellular signal-regulated kinase 1/2 and p38 mitogen-activated protein kinase in the signal transduction of basic fibroblast growth factor in endothelial cells during angiogenesis, *Jpn J Cancer Res*, 90, 647, 1999.
252. Gu, Q. et al., Basic fibroblast growth factor inhibits radiation-induced apoptosis of HUVECs. I. The PI3K/AKT pathway and induction of phosphorylation of BAD, *Radiat Res*, 161, 692, 2004.
253. Rieck, P.W., Cholidis, S., and Hartmann, C., Intracellular signaling pathway of FGF-2-modulated corneal endothelial cell migration during wound healing in vitro, *Exp Eye Res*, 73, 639, 2001.
254. Kay, E.P. et al., Fibroblast growth factor 2 uses PLC-gamma1 for cell proliferation and PI3-kinase for alteration of cell shape and cell proliferation in corneal endothelial cells, *Mol Vis*, 4, 22, 1998.
255. Lee, H.T. and Kay, E.P., FGF-2 induced reorganization and disruption of actin cytoskeleton through PI 3-kinase, Rho, and Cdc42 in corneal endothelial cells, *Mol Vis*, 9, 624, 2003.
256. Lu, L. et al., Role of the Src homology 2 domain-containing protein Shb in murine brain endothelial cell proliferation and differentiation, *Cell Growth Differ*, 13, 141, 2002.
257. Kent, K.C. et al., Requirement for protein kinase C activation in basic fibroblast growth factor-induced human endothelial cell proliferation, *Circ Res*, 77, 231, 1995.
258. Presta, M. et al., Basic fibroblast growth factor requires a long-lasting activation of protein kinase C to induce cell proliferation in transformed fetal bovine aortic endothelial cells, *Cell Regul*, 2, 719, 1991.
259. Presta, M. et al., Basic fibroblast growth factor: production, mitogenic response, and post-receptor signal transduction in cultured normal and transformed fetal bovine aortic endothelial cells, *J Cell Physiol*, 141, 517, 1989.
260. Presta, M., Maier, J.A., and Ragnotti, G., The mitogenic signaling pathway but not the plasminogen activator-inducing pathway of basic fibroblast growth factor is mediated through protein kinase C in fetal bovine aortic endothelial cells, *J Cell Biol*, 109, 1877, 1989.

261. Haimovitz-Friedman, A. et al., Protein kinase C mediates basic fibroblast growth factor protection of endothelial cells against radiation-induced apoptosis, *Cancer Res*, 54, 2591, 1994.
262. Takano, S. et al., A diaminoantraquinone inhibitor of angiogenesis, *J Pharmacol Exp Ther*, 271, 1027, 1994.
263. Hu, D.E. and Fan, T.P., Protein kinase C inhibitor calphostin C prevents cytokine-induced angiogenesis in the rat, *Inflammation*, 19, 39, 1995.
264. van Hinsbergh, V.W. et al., Genistein reduces tumor necrosis factor alpha-induced plasminogen activator inhibitor-1 transcription but not urokinase expression in human endothelial cells, *Blood*, 84, 2984, 1994.
265. Koroma, B.M. and de Juan, E., Jr., Phosphotyrosine inhibition and control of vascular endothelial cell proliferation by genistein, *Biochem Pharmacol*, 48, 809, 1994.
266. Kilarski, W.W., Jura, N., and Gerwins, P., Inactivation of Src family kinases inhibits angiogenesis in vivo: implications for a mechanism involving organization of the actin cytoskeleton, *Exp Cell Res*, 291, 70, 2003.
267. Tsuda, S. et al., Role of c-Fyn in FGF-2-mediated tube-like structure formation by murine brain capillary endothelial cells, *Biochem Biophys Res Commun*, 290, 1354, 2002.
268. Lee, H.T., Kim, T.Y., and Kay, E.P., Cdk4 and p27Kip1 play a role in PLC-gamma1-mediated mitogenic signaling pathway of 18 kDa FGF-2 in corneal endothelial cells, *Mol Vis*, 8, 17, 2002.
269. Sa, G. and Fox, P.L., Basic fibroblast growth factor-stimulated endothelial cell movement is mediated by a pertussis toxin-sensitive pathway regulating phospholipase A2 activity, *J Biol Chem*, 269, 3219, 1994.
270. Anwar, K.N. et al., RhoA/Rho-associated kinase pathway selectively regulates thrombin-induced intercellular adhesion molecule-1 expression in endothelial cells via activation of I kappa B kinase beta and phosphorylation of RelA/p65, *J Immunol*, 173, 6965, 2004.
271. Soriano, J.V. et al., Inhibition of angiogenesis by growth factor receptor bound protein 2-Src homology 2 domain bound antagonists, *Mol Cancer Ther*, 3, 1289, 2004.
272. D'Angelo, G., Lee, H., and Weiner, R.I., cAMP-dependent protein kinase inhibits the mitogenic action of vascular endothelial growth factor and fibroblast growth factor in capillary endothelial cells by blocking Raf activation, *J Cell Biochem*, 67, 353, 1997.
273. Forough, R. et al., Role of AKT/PKB signaling in fibroblast growth factor-1 (FGF-1)-induced angiogenesis in the chicken chorioallantoic membrane (CAM), *J Cell Biochem*, 94, 109, 2005.
274. Hoffmann, S. et al., Carboxyamido-triazole modulates retinal pigment epithelial and choroidal endothelial cell attachment, migration, proliferation, and MMP-2 secretion of choroidal endothelial cells, *Curr Eye Res*, 30, 103, 2005.
275. Cross, M.J. et al., Tyrosine 766 in the fibroblast growth factor receptor-1 is required for FGF-stimulation of phospholipase C, phospholipase D, phospholipase A(2), phosphoinositide 3-kinase and cytoskeletal reorganisation in porcine aortic endothelial cells, *J Cell Sci*, 113 (Pt 4), 643, 2000.
276. Lee, S.H., Schloss, D.J., and Swain, J.L., Maintenance of vascular integrity in the embryo requires signaling through the fibroblast growth factor receptor, *J Biol Chem*, 275, 33679, 2000.
277. Hood, J.D. et al., Differential alpha v integrin-mediated Ras-ERK signaling during two pathways of angiogenesis, *J Cell Biol*, 162, 933, 2003.
278. Mettouchi, A. et al., Integrin-specific activation of Rac controls progression through the G(1) phase of the cell cycle, *Mol Cell*, 8, 115, 2001.
279. Wary, K.K., Signaling through Raf-1 in the neovasculature and target validation by nanoparticles, *Mol Cancer*, 2, 27, 2003.
280. Rikitake, Y. et al., Lysophosphatidylcholine inhibits endothelial cell migration and proliferation via inhibition of the extracellular signal-regulated kinase pathway, *Arterioscler Thromb Vasc Biol*, 20, 1006, 2000.
281. Cross, M.J. et al., The Shb adaptor protein binds to tyrosine 766 in the FGFR-1 and regulates the Ras/MEK/MAPK pathway via FRS2 phosphorylation in endothelial cells, *Mol Biol Cell*, 13, 2881, 2002.

282. Murakami, M. et al., Protein kinase C (PKC) delta regulates PKC alpha activity in a Syndecan-4-dependent manner, *J Biol Chem*, 277, 20367, 2002.
283. Kanda, S. et al., The nonreceptor protein-tyrosine kinase c-Fes is involved in fibroblast growth factor-2-induced chemotaxis of murine brain capillary endothelial cells, *J Biol Chem*, 275, 10105, 2000.
284. Tan, J. and Hallahan, D.E., Growth factor-independent activation of protein kinase B contributes to the inherent resistance of vascular endothelium to radiation-induced apoptotic response, *Cancer Res*, 63, 7663, 2003.
285. Klein, S. et al., Alpha 5 beta 1 integrin activates an NF-kappa B-dependent program of gene expression important for angiogenesis and inflammation, *Mol Cell Biol*, 22, 5912, 2002.
286. Pourtier-Manzanedo, A. et al., Expression of an Ets-1 dominant-negative mutant perturbs normal and tumor angiogenesis in a mouse ear model, *Oncogene*, 22, 1795, 2003.
287. Kasbauer, C.W., Paper, D.H., and Franz, G., Sulfated beta-(1→4)-galacto-oligosaccharides and their effect on angiogenesis, *Carbohydr Res*, 330, 427, 2001.
288. Miao, H.Q. et al., Modulation of fibroblast growth factor-2 receptor binding, dimerization, signaling, and angiogenic activity by a synthetic heparin-mimicking polyanionic compound, *J Clin Invest*, 99, 1565, 1997.
289. Giraux, J.L. et al., Modulation of human endothelial cell proliferation and migration by fucoidan and heparin, *Eur J Cell Biol*, 77, 352, 1998.
290. Chabut, D. et al., Low molecular weight fucoidan and heparin enhance the basic fibroblast growth factor-induced tube formation of endothelial cells through heparan sulfate-dependent alpha 6 overexpression, *Mol Pharmacol*, 64, 696, 2003.
291. Bocci, G. et al., Inhibitory effect of suramin in rat models of angiogenesis in vitro and in vivo, *Cancer Chemother Pharmacol*, 43, 205, 1999.
292. Sola, F. et al., Endothelial cells overexpressing basic fibroblast growth factor (FGF-2) induce vascular tumors in immunodeficient mice, *Angiogenesis*, 1, 102, 1997.
293. Gagliardi, A.R. et al., Antiangiogenic and antiproliferative activity of suramin analogues, *Cancer Chemother Pharmacol*, 41, 117, 1998.
294. Takano, S. et al., Suramin, an anticancer and angiosuppressive agent, inhibits endothelial cell binding of basic fibroblast growth factor, migration, proliferation, and induction of urokinase-type plasminogen activator, *Cancer Res*, 54, 2654, 1994.
295. Ciomei, M. et al., New sulfonated distamycin A derivatives with bFGF complexing activity, *Biochem Pharmacol*, 47, 295, 1994.
296. Braddock, P.S. et al., A structure–activity analysis of antagonism of the growth factor and angiogenic activity of basic fibroblast growth factor by suramin and related polyanions, *Br J Cancer*, 69, 890, 1994.
297. Finch, P.W. et al., Inhibition of growth factor mitogenicity and growth of tumor cell xenografts by a sulfonated distamycin A derivative, *Pharmacology*, 55, 269, 1997.
298. Zugmaier, G., Lippman, M.E., and Wellstein, A., Inhibition by pentosan polysulfate (PPS) of heparin-binding growth factors released from tumor cells and blockage by PPS of tumor growth in animals, *J Natl Cancer Inst*, 84, 1716, 1992.
299. Aviezer, D. et al., Porphyrin analogues as novel antagonists of fibroblast growth factor and vascular endothelial growth factor receptor binding that inhibit endothelial cell proliferation, tumor progression, and metastasis, *Cancer Res*, 60, 2973, 2000.
300. Presta, M. et al., Antiangiogenic activity of semisynthetic biotechnological heparins: low-molecular-weight-sulfated *Escherichia coli* K5 polysaccharide derivatives as fibroblast growth factor antagonists, *Arterioscler Thromb Vasc Biol*, 25, 71, 2005.
301. Benelli, U. et al., The heparan sulfate suleparoide inhibits rat corneal angiogenesis and in vitro neovascularization, *Exp Eye Res*, 67, 133, 1998.
302. Casu, B. et al., Short heparin sequences spaced by glycol-split uronate residues are antagonists of fibroblast growth factor 2 and angiogenesis inhibitors, *Biochemistry*, 41, 10519, 2002.
303. Casu, B. et al., Undersulfated and glycol-split heparins endowed with antiangiogenic activity, *J Med Chem*, 47, 838, 2004.
304. Liekens, S. et al., Modulation of fibroblast growth factor-2 receptor binding, signaling, and mitogenic activity by heparin-mimicking polysulfonated compounds, *Mol Pharmacol*, 56, 204, 1999.

305. Liekens, S. et al., The sulfonic acid polymers PAMPS [poly(2-acrylamido-2-methyl-1-propanesulfonic acid)] and related analogues are highly potent inhibitors of angiogenesis, *Oncol Res*, 9, 173, 1997.
306. Sakairi, N. et al., Synthesis and biological evaluation of 2-amino-2-deoxy- and 6-amino-6-deoxy-cyclomaltoheptaose polysulfates as synergists for angiogenesis inhibition, *Bioorg Med Chem*, 4, 2187, 1996.
307. Gagliardi, A.R. and Collins, D.C., Inhibition of angiogenesis by aurintricarboxylic acid, *Anticancer Res*, 14, 475, 1994.
308. Matou, S. et al., Effect of an oversulfated exopolysaccharide on angiogenesis induced by fibroblast growth factor-2 or vascular endothelial growth factor in vitro, *Biochem Pharmacol*, 69, 751, 2005.
309. Hoffman, R. and Sykes, D., Inhibition of binding of basic fibroblast growth factor to low and high affinity receptors by carrageenans, *Biochem Pharmacol*, 45, 2348, 1993.
310. Patel, S., Leal, A.D., and Gorski, D.H., The homeobox gene Gax inhibits angiogenesis through inhibition of nuclear factor-kappaB-dependent endothelial cell gene expression, *Cancer Res*, 65, 1414, 2005.
311. Hanafusa, H. et al., Sprouty1 and Sprouty2 provide a control mechanism for the Ras/MAPK signalling pathway, *Nat Cell Biol*, 4, 850, 2002.
312. Kaur, G. et al., Antiangiogenic properties of 17-(dimethylaminoethylamino)-17-demethoxygeldanamycin: an orally bioavailable heat shock protein 90 modulator, *Clin Cancer Res*, 10, 4813, 2004.
313. Kroon, M.E. et al., Collagen type 1 retards tube formation by human microvascular endothelial cells in a fibrin matrix, *Angiogenesis*, 5, 257, 2002.
314. Staton, C.A. et al., Alphastatin, a 24-amino acid fragment of human fibrinogen, is a potent new inhibitor of activated endothelial cells in vitro and in vivo, *Blood*, 103, 601, 2004.
315. Dixelius, J. et al., Endostatin regulates endothelial cell adhesion and cytoskeletal organization, *Cancer Res*, 62, 1944, 2002.
316. Dixelius, J. et al., Endostatin-induced tyrosine kinase signaling through the Shb adaptor protein regulates endothelial cell apoptosis, *Blood*, 95, 3403, 2000.
317. Guan, X. et al., Histidine-proline rich glycoprotein (HPRG) binds and transduces anti-angiogenic signals through cell surface tropomyosin on endothelial cells, *Thromb Haemost*, 92, 403, 2004.
318. Ashton, A.W. et al., Thromboxane A2 receptor agonists antagonize the proangiogenic effects of fibroblast growth factor-2: role of receptor internalization, thrombospondin-1, and alpha (v) beta 3, *Circ Res*, 94, 735, 2004.
319. Redlitz, A., Daum, G., and Sage, E.H., Angiostatin diminishes activation of the mitogen-activated protein kinases ERK-1 and ERK-2 in human dermal microvascular endothelial cells, *J Vasc Res*, 36, 28, 1999.
320. Duenas, Z. et al., Inhibition of rat corneal angiogenesis by 16-kDa prolactin and by endogenous prolactin-like molecules, *Invest Ophthalmol Vis Sci*, 40, 2498, 1999.
321. Kanda, S. et al., Effects of vitamin D(3)-binding protein-derived macrophage activating factor (GcMAF) on angiogenesis, *J Natl Cancer Inst*, 94, 1311, 2002.
322. Baiguera, S. et al., Ghrelin inhibits in vitro angiogenic activity of rat brain microvascular endothelial cells, *Int J Mol Med*, 14, 849, 2004.
323. Zhang, J.C. et al., The antiangiogenic activity of cleaved high molecular weight kininogen is mediated through binding to endothelial cell tropomyosin, *Proc Natl Acad Sci USA*, 99, 12224, 2002.
324. Kim, J. et al., IL-4 inhibits cell cycle progression of human umbilical vein endothelial cells by affecting p53, p21(Waf1), cyclin D1, and cyclin E expression, *Mol Cells*, 16, 92, 2003.
325. Angiolillo, A.L. et al., Human interferon-inducible protein 10 is a potent inhibitor of angiogenesis in vivo, *J Exp Med*, 182, 155, 1995.
326. Kanda, S. et al., Pigment epithelium-derived factor inhibits fibroblast-growth-factor-2-induced capillary morphogenesis of endothelial cells through Fyn, *J Cell Sci*, 118, 961, 2005.
327. Kaur, B. et al., Vasculostatin, a proteolytic fragment of brain angiogenesis inhibitor 1, is an antiangiogenic and antitumorigenic factor, *Oncogene*, 24, 3632, 2005.
328. Pike, S.E. et al., Vasostatin, a calreticulin fragment, inhibits angiogenesis and suppresses tumor growth, *J Exp Med*, 188, 2349, 1998.
329. Guo, Y.L., Wang, S., and Colman, R.W., Kininostatin, an angiogenic inhibitor, inhibits proliferation and induces apoptosis of human endothelial cells, *Arterioscler Thromb Vasc Biol*, 21, 1427, 2001.

330. Lafleur, M.A. et al., Endothelial tubulogenesis within fibrin gels specifically requires the activity of membrane-type-matrix metalloproteinases (MT-MMPs), *J Cell Sci*, 115, 3427, 2002.
331. Norioka, K. et al., Interaction of interleukin-1 and interferon-gamma on fibroblast growth factor-induced angiogenesis, *Jpn J Cancer Res*, 85, 522, 1994.
332. Sato, N. et al., Actions of TNF and IFN-gamma on angiogenesis in vitro, *J Invest Dermatol*, 95, 85S, 1990.
333. Grant, M.B., Caballero, S., and Millard, W.J., Inhibition of IGF-I and b-FGF stimulated growth of human retinal endothelial cells by the somatostatin analogue, octreotide: a potential treatment for ocular neovascularization, *Regul Pept*, 48, 267, 1993.
334. Kessler, O. et al., Semaphorin-3F is an inhibitor of tumor angiogenesis, *Cancer Res*, 64, 1008, 2004.

9 Development of the VEGF Trap as a Novel Antiangiogenic Treatment Currently in Clinical Trials for Cancer and Eye Diseases, and Discovery of the Next Generation of Angiogenesis Targets

John S. Rudge, Gavin Thurston, S. Davis, Nicholas Papadopoulos, Nicholas W. Gale, Stanley J. Wiegand, and George D. Yancopoulos

CONTENTS

The hope that tumors can be controlled by directly targeting their vascular supply is finally becoming a reality. Blocking vascular endothelial growth factor (VEGF-A) is the best validated antiangiogenesis approach in cancer, with the most advanced clinical data having been generated with a humanized monoclonal antibody termed bevacizumab (Avastin®), which directly binds and blocks all isoforms of VEGF-A (1). Notwithstanding the promising data described to date, dose–response studies suggest that higher doses of bevacizumab may provide even greater benefit (2,3), implying that current bevacizumab regimens may not optimize VEGF inhibition and thus may not have yet demonstrated the maximum potential of VEGF blockade in cancer.

Not only do anti-VEGF approaches look promising in cancer, but blocking VEGF-A has also been shown to have marked efficacy in maintaining and improving vision in wet age-related macular degeneration (AMD), a disease marked by leaky and proliferating vessels that distort the retina, and these data suggest that VEGF blockade may provide benefit in other eye diseases involving vascular leak and proliferation (4). Efficacy in wet AMD has most notably been achieved using a modified fragment of the bevacizumab antibody, termed ranibizumab (Lucentis®), delivered via monthly intraocular injections (5). Although the most impressive clinical data in cancer and eye diseases have been achieved using the antibody-based approaches mentioned here, which have the advantages of providing very specific, high-affinity as well as extended blockade from a single dose, numerous other VEGF pathway blockers are being investigated in clinical trials. These include small molecule kinase inhibitors, which target VEGF receptors but tend to be less specific and require frequent administration, and aptamers, which bind directly to VEGF.

In this review, we focus on the development and status of a novel VEGF blocking agent, termed the VEGF Trap, that retains many of the advantages of a blocking antibody but may offer further potential (6). The VEGF Trap comprises portions of VEGF receptors that have been fused to the constant region of an antibody, resulting in a fully human biologic with exceedingly high affinity, which not only blocks all isoforms of VEGF-A, but also related VEGF family members such as placental growth factor (PlGF). The VEGF Trap also displays extended pharmacologic half-life, allowing long-term as well as very high-affinity blockade. The VEGF Trap has performed impressively in extensive animal studies of cancer and eye diseases, and initial clinical trials appear promising. The VEGF Trap may provide the opportunity to explore the potential of more complete VEGF blockade in cancer, as well as the opportunity for even longer-interval dosing regimens in eye diseases. At the end of this review we describe approaches in which we have used the VEGF Trap as a research tool in efforts to discover and validate the next generation of antiangiogenesis targets.

9.1 HISTORY OF VEGF, AND ELUCIDATION OF ITS REQUISITE ROLES DURING NORMAL DEVELOPMENT AND IN DISEASE SETTINGS

Initial studies by Dvorak and colleagues in 1986 (7,8) identified a protein in tumor ascites fluid that was capable of inducing vascular leak and permeability, which they termed vascular permeability factor (VPF). Independent efforts by Ferrara and colleagues to identify secreted factors that could promote tumor angiogenesis led to the discovery of a protein in bovine pituitary follicular cell conditioned medium with mitogenic properties for endothelial cells, which they termed VEGF (9,10). Upon sequencing and further studies, this VEGF protein was unexpectedly found to correspond to the VPF previously identified by the Dvorak lab. These findings set the stage for a concerted effort to define the role of VEGF/VPF (hereon VEGF) in cancer angiogenesis as well as other settings of vascular disease, which have led to the understanding that both of its initially realized actions—that is, promoting vascular permeability as well as vascular growth—appear critical to understanding its roles during normal biology as well as in disease.

About two decades of subsequent intensive investigation by numerous laboratories have revealed a great deal about VEGF and its actions. It is now clear that VEGF is perhaps the most critical vascular regulator during normal development as well as in many disease states, and that its dosage must be exquisitely regulated in a spatial and temporal manner to avoid vascular disaster. Disruption of both VEGF alleles in developing mice, which ablates all VEGF production, results in complete failure to develop even a primordial vasculature, demonstrating that VEGF is absolutely essential for the earliest stages of blood vessel development. Even more remarkably, disruption of even a single VEGF allele in developing mice, which decreases VEGF levels by half, also results in embryonic lethality due to severe vascular abnormalities (11,12), demonstrating the need for exquisite regulation of VEGF levels to form normal vessels. Reciprocally, modest increases in VEGF levels during development also lead to vascular disaster and lethality (13). VEGF continues to be critical during early postnatal growth and development, as evidenced by the lethality and major growth disturbances caused by conditional disruption of the VEGF gene or by administration of VEGF blockers (11,12,14–21). However, VEGF blockade in older animals is much less traumatic, profoundly affecting only those structures that continue to depend on ongoing vascular remodeling, such as that occurring in bone growth plates or during remodeling of the female reproductive organs (15,17,22). As discussed in greater detail later, vascular remodeling is absolutely required in a variety of pathologic settings such as during tumor growth and in other settings, providing major therapeutic opportunities for VEGF blockade in the adult setting such blockade can be tolerated.

9.2 VEGF ISOFORMS, FAMILY MEMBERS, AND RECEPTORS

Further study of the gene encoding human VEGF revealed eight exons separated by seven introns, which encode four isoforms of increasing size—$VEGF_{121}$, $VEGF_{165}$, $VEGF_{189}$, and $VEGF_{206}$ (subscripts refer to the number of amino acids comprising the isoform, with the VEGF isoforms varying in length at their carboxy-termini). The main purpose of these isoforms appears to relate to their bioavailability such that the 121 isoform is diffusible whereas the higher molecular weight isoforms remain bound to the extracellular matrix requiring cleavage to be released (23–25).

Due to the discovery of additional members of the VEGF family, VEGF is now often referred to as VEGF-A. Other members of the VEGF family were identified based on their homology to VEGF, as well as their ability to interact with a related set of cell surface receptors (26, see later). The first VEGF relative to be identified is known as PlGF, and until recently little was known about its normal function (27). Whereas mice lacking PlGF appear to undergo normal vascular development, recent findings indicate that adult mice lacking PlGF exhibit deficiencies in certain models of adult vascular remodeling, including in tumors and eye disease models, raising the interesting possibility that the activity of PlGF may be limited to these settings, and that blockade of PlGF may provide for enhanced efficacy when combined with VEGF blockade (28,29). Little is known about VEGF-B, and mice lacking VEGF-B are overtly healthy and fertile. VEGF-C and D seem to play more critical roles in the lymphatic vasculature than in the blood vasculature, showing specificity for a VEGF receptor (see later) expressed on this vasculature; administration of both these factors leads to lymphatic vessel hyperplasia (30–32).

Following rapidly on the heels of the discovery of VEGF came the identification of two closely related high-affinity receptors for VEGF—FLT1 (FMS-like tyrosine kinase), now termed VEGFR1 (33), and KDR or Flk1, now termed VEGFR2 (34–36). These high-affinity receptors share features of many other growth factor receptors, in that they contain an extracellular domain that binds and is dimerized by ligand and a cytoplasmic tyrosine kinase domain that can be regulated upon binding of ligand to the extracellular domain. VEGFR2 seems to be the receptor that mediates the major growth and permeability actions of VEGF, whereas VEGFR1 may have a negative role, either by acting as a decoy receptor or by suppressing signaling through VEGFR2. Thus, mice engineered to lack VEGFR2 fail to develop a vasculature and have very few endothelial cells (37), phenocopying mice lacking VEGF, whereas mice lacking VEGFR1 seem to have excess formation of endothelial cells, which abnormally coalesce into disorganized tubules (38). Mice engineered to express only a truncated form of VEGFR1, lacking its kinase domain, appear rather normal, consistent with the notion that the primary role of VEGFR1 may be that of a decoy receptor (39) and supporting only a minor role for the cytoplasmic kinase domain. The third member of this receptor family, initially called Flt-4 and now termed VEGFR3, does not bind to VEGF-A nor PlGF and instead binds to VEGF-C and VEGF-D and seems to mediate the actions of these latter two factors on the lymphatic vasculature (40).

In addition to these primary receptors, a number of potential accessory receptors for the VEGFs have been identified, though the requisite roles of these receptors in mediating VEGF responses have not been clearly elucidated. These potential accessory receptors include the neuropilins (41).

9.3 ROLE OF VEGF IN TUMOR ANGIOGENESIS SUPPORTS THE NOTION OF BLOCKING VEGF AS AN ANTITUMOR STRATEGY

One area of intense study after the discovery of VEGF was the analysis of VEGF expression levels in different tumor types using in situ hybridization. VEGF was found to be expressed in a number of different tumors such as renal, gastrointestinal, breast, ovarian, pancreatic, and lung but the variability in expression across the tumor and between different tumor types made the simple correlation between VEGF and severity of the cancer impossible (42–46). However, out of these studies came the interesting finding that one tumor type, renal cell carcinoma, had particularly high VEGF expression, which correlated with inactivation of the Von Hippel Lindau locus resulting in the loss of control of the tumor's oxygen sensor—hypoxia inducible factor (HIF) (47,48). The upregulation of VEGF in an attempt to reoxygenate the tumor through revascularization led to the belief that this tumor may either be highly sensitive to anti-VEGF therapy or highly refractory. Fortunately, the former seems to be the case (3).

Concomitant with the analysis of human tumors for VEGF expression came the development of animal models of cancer where the hypothesis that VEGF was required for tumor vasculature and thus tumor growth could be tested. In 1993, 4 years after their discovery of VEGF, Ferrara and colleagues demonstrated that a mouse monoclonal antibody to human VEGF (A.4.6.1) could inhibit the growth of several human tumor types in nude mice with inhibition ranging from 70% to >90% (49). Subsequent to this observation, a number of laboratories using different strategies to inhibit VEGF signaling have shown to a greater or lesser extent that inhibition of VEGF can have a major impact on tumor growth in mice. In addition to numerous studies using the VEGF blocking antibody, other strategies to block VEGF in tumor models included blocking antibodies targeting VEGFR2 (50), soluble VEGF receptors acting as circulating decoys to capture VEGF and preventing it from binding cell surface receptors (3,17,22,51); dominant-negative VEGF receptors expressed at high levels on tumor surfaces, small molecule inhibitors of VEGF receptor kinases, and other kinases (52); antisense oligonucleotides targeting VEGF and VEGF siRNA (53,54).

As the number of studies increased comparing the different modes of inhibiting VEGF, it became apparent that blocking tumor-derived VEGF without blocking stromal VEGF was not as efficacious, implicating stromal VEGF as a crucial player in tumor growth and angiogenesis. Thus, antibodies such as A.4.6.1, which only block human VEGF, did not fare as well in blocking human tumor growth in immunocompromised mice as reagents blocking both tumor and host stroma-derived VEGF (55,51).

9.4 EXPLOITING REGENERON'S TRAP TECHNOLOGY PLATFORM TO CREATE THE VEGF TRAP

The clinical promise of initial anti-VEGF approaches highlighted the need to optimize blockade of this pathway. Early studies indicated that one of the most effective ways to block the VEGF signaling pathway is to prevent VEGF from binding to its endogenous receptors by administering soluble decoy receptors (15,22). In particular, a soluble decoy receptor created by fusing the first three immunoglobulin-like (Ig) domains of VEGFR1 to the constant region (Fc portion) of human IgG1 resulted in a forced homodimer that acted as a very high-affinity blocking reagent with 5–20 pM binding affinity for VEGF. In tumor experiments this VEGFR1-Fc reagent was efficacious at approximately 500-fold lower concentration than a similar VEGFR2-Fc construct (56). Despite its high affinity, the VEGFR1-Fc was not a feasible clinical candidate because of its poor pharmacokinetic profile; in rodent studies, this protein had to be administered frequently and at very high (and impractical) doses to achieve efficacious levels. In addition, this agent seemed to have nonspecific toxicity effects that did not seem to be accounted for by its blocking of VEGF (56). We decided to exploit our Trap Technology Platform (57), which involves defining and fusing minimal binding units from different receptor components to generate chimeric fusion proteins that act as high-affinity soluble blockers, in an attempt to create a potent Trap for VEGF that might improve on the deficiencies of the VEGFR1-Fc. The result was a chimeric fusion protein containing a modified domain 2 of VEGFR1 and the third Ig domain of VEGFR2 fused to the Fc region of human IgG1, resulting in a fully human protein that we term VEGF Trap (Figure 9.1) (6).

The VEGF Trap has the advantage of being fully human and thus potentially nonimmunogenic, as well as being substantially smaller than previous fusion proteins and antibodies, raising the possibility that it might allow for improved tissue and tumor penetration. In addition, the VEGF Trap had greatly improved pharmacologic bioavailability as compared with the initial VEGFR1-Fc reagent, exhibiting about a 300-fold increase in the maximum concentration achieved in the circulation (i.e., C_{max}), as well as about a 1000-fold increase in total circulation exposure (i.e., AUC) (6) (Table 9.1). Importantly, the affinity of VEGF Trap binding to both mouse and human VEGF isoforms (subpicomolar binding affinity) was superior to that of the parental VEGFR1-Fc (~5–20 pM) (6). In addition, the VEGF Trap also bound PlGF with high affinity (Table 9.2).

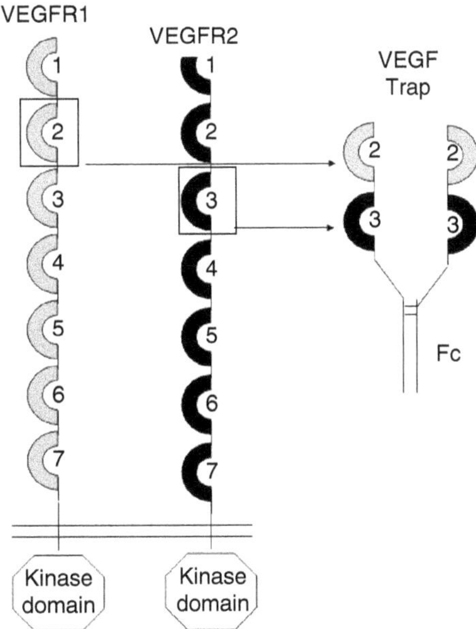

FIGURE 9.1 VEGF Trap is a fully human chimeric protein consisting of the second immunoglobulin domain of VEGF receptor 1 and the third immunoglobulin domain of VEGF receptor 2 fused to the Fc domain of human IgG1.

TABLE 9.1
Pharmacokinetic Properties of VEGF Trap Compared to VEGFR1(1–3 Ig)-Fc

Construct	p*I*	C_{max} (μg/mL)	AUC (μg × days/mL)
VEGFR1(1–3 Ig)-Fc	9.4	0.05	0.04
VEGF Trap	8.8	16	36.28

In order to determine if the improved pharmacologic bioavailability and high-affinity binding of VEGF Trap translated into superior performance in vivo, we first used a short-term and quantitative in vivo model of VEGF activity in which a single dose of VEGF induces a stereotypic reduction in blood pressure. In this acute assay model, we found that equivalent doses of VEGF Trap were indeed far superior to that of the parental VEGFR1-Fc (6).

9.5 VEGF TRAP WORKS IMPRESSIVELY IN PRECLINICAL TUMOR MODELS

Based on the above evidence suggesting that the VEGF Trap was a potent VEGF blocker that was efficacious in vivo, we moved to evaluate the VEGF Trap in tumor models. Initial studies confirmed the remarkable efficacy of the VEGF Trap. Not only did the VEGF Trap blunt tumor growth in the early models, but it could block tumor angiogenesis in many cases, resulting in completely avascular tumors (6). These initial studies inspired exploration of the VEGF Trap in multiple tumor models, in the laboratories of many different investigators, with impressive results in almost every case.

The following is a summary of the different preclinical tumor models in which VEGF Trap has been studied:

1. Multiple subcutaneous tumor models: Initial studies using VEGF Trap in murine tumor models showed that subcutaneously administered VEGF Trap (2.5 and 25 mg/kg) could significantly inhibit the growth of murine B16F10 melanoma ($p=0.01$), human A673 rhabdomyosarcoma ($p=0.06$), and rat C6 glioma ($p<0.0001$). At both doses of VEGF Trap, significant tumor necrosis was observed with an almost complete

TABLE 9.2
Binding Properties of VEGF Trap to Human and Mouse VEGF and PlGF Isoforms

	K_a (1/Ms)	K_d (1/s)[a]	*K*D (pmol/L)[b]
$hVEGF_{165}$	4.5×10^7	2.1×10^{-5}	0.46
$hVEGF_{121}$	1.69×10^7	8.8×10^{-6}	0.52
hPlGF2	1.0×10^7	2.57×10^{-4}	25.6
$mVEGF_{164}$	6.8×10^7	3.96×10^{-5}	0.58
mPlGF2	5.5×10^7	1.01×10^{-4}	1.8

K_a is equilibrium association constant.

[a] K_d is dissociation constant.

[b] *K*D is affinity constant.

blockade of tumor angiogenesis at the higher dose with the lower dose being slightly less effective (6).

2. Pancreatic tumor models: Four human pancreatic cell lines (BxPC3, T3M4, COLO-357, and PANC-1) were implanted into athymic nude mice and treated with VEGF Trap (25 mg/kg) twice weekly for 5–6 weeks. At the end of the treatment period, VEGF Trap-treated animals showed inhibition of tumor growth in all the models tested—COLO-357 and PANC-1 (97%), BxPC3 (92%), and T3M4 (89%) (58).
3. Wilms tumor model: Using an orthotopic human SK-NEP Wilms tumor xenograft model, Kandel and colleagues treated mice twice weekly with 500 μg of VEGF Trap and showed frank regression of these tumors resulting in an 81% reduction in tumor size. This correlated with histological analysis of the tumors where there were decreases in microvessel density (54%, $p = 0.037$), total vessel length (42%, $p = 0.01$), vessel ends (63%, $p < 0.004$), and branch points/nodes (80%, $p < 0.004$) with a 79.3% reduction in mean tumor weight ($p < 0.0002$). In addition, growth inhibition of lung micrometastases was also observed (59).
4. Ewing's sarcoma models: Analysis of VEGF Trap effects on six Ewing's sarcoma cell lines (EFST) revealed a significant correlation between microvessel density and the expression of VEGF but not other overexpressed growth factors. When VEGF Trap was administered twice weekly, it delayed subcutaneous tumor growth of EFST and RD-ES cells at high (25 mg/kg) and low (2.5 mg/kg) doses compared with controls ($p = 0.001$) (60).
5. Ovarian cancer and ovarian ascites model: VEGF Trap has also demonstrated activity in combination with the chemotherapeutic paclitaxel showing a 97.7% reduction ($p < 0.01$) in OVCAR-3 tumor volume and ascites. VEGF Trap alone resulted in a 55.7% ($p < 0.05$) reduction and paclitaxel alone resulted in a 54.8% ($p < 0.05$) reduction compared with controls ($p = 0.001$) (61).
6. Glioblastoma model: In a recent study using U87 glioblastoma, low-dose VEGF Trap plus fractionated radiotherapy was superior to either treatment modality alone in slowing tumor growth rate ($p < 0.04$). Tumor microvessel density in VEGF Trap- and radiation therapy-treated tumors was decreased to between 15% and 30% of control or radiation therapy-treated tumors ($p = 0.001$). FDG PET imaging revealed that the percentage of metabolically inactive tumor was higher in VEGF Trap-treated tumors than in untreated tumors (62).

In addition to the earlier published studies, recent unpublished temporal studies indicate that vascular regression can be seen in most tumors within hours of VEGF Trap treatment, resulting in marked and widespread hypoxia within the tumors. In addition, transcription profiling studies during these temporal studies have revealed a set of endothelial-specific genes that are rapidly and profoundly regulated in response to VEGF Trap treatment. Further studies on some of these genes have led to their identification as potential targets for new antiangiogenesis therapies (see later).

In summary, animal tumor studies have indicated that treatment with VEGF Trap effectively inhibited tumor growth of a wide variety of murine, rat, and human tumor cell lines implanted either subcutaneously or orthotopically in mice. VEGF Trap treatment inhibited the growth of tumors representing a variety of tumor types, including melanoma, glioma, rhabdomyosarcoma, ovarian, pancreatic, renal, and mammary tumor tissue, with a broad therapeutic index. Growth of small, established tumors was also inhibited. Histological analysis indicated that treatment with VEGF Trap resulted in the formation of largely avascular and necrotic tumors, demonstrating that tumor-induced angiogenesis was blocked. VEGF Trap was also active in blocking tumor growth in similar animal tumor models in combination with paclitaxel, docetaxel, or radiation, and was synergistic with 5-FU. VEGF

Trap as a single agent and in combination with paclitaxel also prevented the formation of ascites in mouse tumor models.

9.6 VEGF TRAP ENTERS CLINICAL TRIALS FOR CANCER

The results reported here in animal tumor models supported the exploration of the VEGF Trap in human studies. Initial clinical studies are promising (63–65) in that the VEGF Trap as a single agent has resulted in objective radiographic responses in several advanced cancer patients suffering from multiply treated chemotherapy-refractory disease, as well as long-term stable disease in patients. Similar data have been generated in patients treated with VEGF Trap in combination with various chemotherapeutic agents. The VEGF Trap is now entering an assortment of additional exploratory as well as potentially pivotal efficacy studies, both as a single agent and in combination with chemotherapy.

9.7 VEGF TRAP WORKS IMPRESSIVELY IN PRECLINICAL MODELS OF VASCULAR EYE DISEASES

In addition to the role of VEGF in tumor angiogenesis, a variety of studies have indicated that VEGF may play a key pathologic role in vascular eye diseases, in particular in diabetic edema and retinopathy settings, and in AMD, which are leading causes of vision loss and blindness. In these diseases, excess VEGF is thought to result in vascular leak, which contributes to abnormal swelling of the retina and resulting vision impairment, as well as to the abnormal growth of choroidal and retinal vessels that can destroy normal retinal architecture. Consistent with these possibilities, the VEGF Trap has demonstrated impressive efficacy in an assortment of animal models of these eye diseases.

Preclinical rodent studies have shown that VEGF Trap can inhibit choroidal (66) and corneal (67) neovascularization, as well as suppress vascular leak into the retina (68) and that the VEGF Trap can also promote the survival of corneal transplants by inhibiting associated neovascularization (69). The following summarizes some of the most important preclinical studies with the VEGF Trap in animal models of vascular eye diseases:

1. Choroidal neovascularization was induced in C57/BL6 mice and VEGF Trap (25 mg/kg) or human Fc was administered systemically 1 day before laser and on days 2, 5, 8, and 11 thereafter. The VEGF Trap-treated mice showed almost complete inhibition of pathological neovascularization compared with the controls. Intraocular administration of VEGF Trap also produced a statistically significant reduction in the mean area of neovascularization compared with controls ($p < 0.0001$) (66).
2. Corneal vascularization was induced in C57/BL6 mice by application of NaOH and mechanical debridement of the corneal epithelium. Treatment with VEGF Trap (12.5 mg/kg subcutaneously) inhibited corneal neovascularization even 4 weeks after treatment cessation. In addition, VEGF Trap also reduced corneal inflammation and edema (67).
3. In a mouse model of early diabetic vascular disease associated with abnormal leak, the VEGF Trap almost completely blocked the development of this abnormal leak, suggesting that it might be a useful therapeutic in settings of retinal edema associated with vision loss (68).
4. In a primate model of AMD in which choroidal neovascular lesions and vascular leak are induced by using a laser to create small lesions in the retinas of adult cynomolgus macaques, both systemically and intravitreally delivered VEGF Trap not only prevented development of vascular leak and neovascular membranes when administered

before laser lesion, but also induced regression when administered after lesions had developed (70).

5. In a mouse model of corneal transplantation, administration of the VEGF Trap improved long-term graft survival, apparently because of associated inhibition of transplant-associated neovascularization (69).

9.8 VEGF TRAP ENTERS CLINICAL TRIALS FOR VASCULAR EYE DISEASES

The results reported here in animal models supported the exploration of the VEGF Trap in human studies of vascular eye diseases. Initial clinical studies in human patients suffering from both AMD and diabetic edema and retinopathy appear quite promising, with evidence in early trials that the VEGF Trap can rapidly and impressively decrease retinal swelling and that these changes can be associated with improvement in visual acuity (71). The VEGF Trap is now entering more advanced clinical trials in vascular eye diseases.

9.9 NEXT GENERATION OF ANGIOGENESIS TARGETS: ANGIOPOIETINS AND Dll4

Despite the promise of anti-VEGF approaches in general and that of the VEGF Trap in particular, it is clear both from the animal studies and from the emerging human trials that tumors display varying degrees of responsivity to VEGF blockade. While some tumors might show marked regression or very long-term stabilization, other tumors can continue to grow even in the setting of anti-VEGF treatments. The realization that some tumors can be relatively resistant to anti-VEGF approaches raises the need for additional antiangiogenesis approaches that might be useful in such settings. Toward this end, as had already been noted earlier, we performed transcriptional profiling screens to identify endothelial-specific targets that are markedly regulated by either VEGF blockade or by excess VEGF activity, reasoning that such targets might prove interesting as new antiangiogenesis targets. Confirming the potential of such a screen, one target that was "rediscovered" was Angiopoietin-2. We had previously independently identified the Angiopoietins as key new angiogenic regulators that seemed to work in tandem with the VEGFs (72–77) and moreover obtained substantial data that Angiopoietin-2 was specifically induced in tumor vasculature and that it was important for tumor angiogenesis (78). A recent study employing Angiopoietin-2 blocking antibodies confirmed notable antitumor effects (79). Based on the confidence in these transcriptional profiling screens engendered by the reidentification of Angiopoietin-2, we explored additional potential targets identified by the screens. Among these targets we have reported the identification of delta-like ligand 4 (Dll4) (a ligand for the Notch family of receptors) as a gene that is markedly and specifically induced in tumor vasculature (80). Moreover, Dll4 is strikingly upregulated in VEGF overexpressing tumors and downregulated in tumors by VEGF blockade. Using VelociGene technology, which provides for a high-throughput approach to create mouse mutants for genes of interest (81), we found that mice lacking Dll4 exhibit profound vascular defects early in development (80). Remarkably, and as previously seen only for VEGF (see earlier), deletion of even just one of the two Dll4 alleles in developing embryos results in embryonic lethality due to vascular defects (80). All this evidence for a critical role for Dll4 in normal as well as tumor angiogenesis provided a rationale to develop blockers for Dll4. Recent testing in tumor models indicates that Dll4 may indeed prove to be an important new antiangiogenesis target, either alone or in combination with the VEGF Trap, or in settings of relative resistance to anti-VEGF therapies. In addition, the success with this screen in terms of yielding Dll4 as an exciting new antiangiogenesis targets has led us to rigorously pursue several additional similarly identified targets.

9.10 CONCLUSION

In summary, exploiting our Trap Technology Platform (57), we generated the VEGF Trap, a very potent VEGF blocking agent with excellent pharmacologic properties. This drug candidate has proven to be highly efficacious in a number of diverse preclinical tumor models. It dramatically inhibits the growth of a variety of types of tumors and can even cause frank tumor regression in some settings. In other preclinical cancer models, we have found that combination of VEGF Trap with a cytotoxic agent can result in potency far greater than that of either single agent. Furthermore, the VEGF Trap is also very effective in animal models of vascular eye diseases. The impressive efficacy in preclinical models of cancer and eye diseases provided rationale for advancement of the VEGF Trap into clinical trials, where it is producing promising initial results in both cancer and eye diseases.

In addition to its potential therapeutic value in cancer and vascular eye diseases, we have found that the VEGF Trap is also an invaluable research tool. Transcription profiling screens using the VEGF Trap as a tool have allowed for a number of strategies designed to identify new antiangiogenesis targets. Hopefully, these strategies are helping to identify the next generation of antiangiogenesis targets, which may work either alone or in combination with the VEGF Trap, or in settings of relative resistance to anti-VEGF therapies.

ACKNOWLEDGMENTS

The authors gratefully acknowledge the substantial contributions of our colleagues at Regeneron, in particular Jocelyn Holash, Susan D. Croll, Lillian Ho, Michelle Russell, Patricia Boland, Ray Leidich, Donna Hylton, Shelly Jiang, Sarah Nandor, Alexander Adler, Hua Jiang, Elena Burova, Irene Noguera, Ella Ioffe, Tammy Huang, Czeslaw Radziejewski, Calvin Lin, Jingtai Cao, Kevin Bailey, James Fandl, Tom Daly, Eric Furfine, Jesse Cedarbaum, Neil Stahl and our colleagues at Zanofi-aventis. In addition, we would like to acknowledge our collaborators on VEGF Trap studies, including Jessica Kandel, Darrell Yamashiro, Robert Jaffe, Donald McDonald, Murray Korc, Phyllis Wachsberger, Adam Dicker, Tony Adamis, C. Cursiefen, J.W. Streilein, and Peter Campochiaro. We regret if we have omitted key contributors.

REFERENCES

1. Ferrara, N., K.J. Hillan, et al. (2004). Discovery and development of bevacizumab, an anti-VEGF antibody for treating cancer. *Nat Rev Drug Discov* 3(5): 391–400.
2. Yang, J.C. (2004). Bevacizumab for patients with metastatic renal cancer: an update. *Clin Cancer Res* 10(18 Pt 2): 6367S–6370S.
3. Yang, J.C., L. Haworth, et al. (2003). A randomized trial of bevacizumab, an anti-vascular endothelial growth factor antibody, for metastatic renal cancer. *N Engl J Med* 349(5): 427–434.
4. Rothen, M., E. Jablon, et al. (2005). "Anti-macular degeneration agents." *Ophthalmol Clin North Am* 18(4): 561–567.
5. Heier, J.S., A.N. Antoszyk, et al. (2006). Ranibizumab for treatment of neovascular age-related macular degeneration a phase I/II multicenter, controlled, multidose study. *Ophthalmology* 113(4): 633–642.
6. Holash, J., S. Davis, et al. (2002). VEGF-Trap: a VEGF blocker with potent antitumor effects. *Proc Natl Acad Sci USA* 99(17): 11393–11398.
7. Dvorak, H.F., J.A. Nagy, et al. (1999). Vascular permeability factor/vascular endothelial growth factor and the significance of microvascular hyperpermeability in angiogenesis. *Curr Top Microbiol Immunol* 237: 97–132.
8. Senger, D.R., C.A. Perruzzi, et al. (1986). A highly conserved vascular permeability factor secreted by a variety of human and rodent tumor cell lines. *Cancer Res* 46(11): 5629–5632.

9. Ferrara, N. and W.J. Henzel (1989). Pituitary follicular cells secrete a novel heparin-binding growth factor specific for vascular endothelial cells. *Biochem Biophys Res Commun* 161(2): 851–858.
10. Leung, D.W., G. Cachianes, et al. (1989). Vascular endothelial growth factor is a secreted angiogenic mitogen. *Science* 246(4935): 1306–1309.
11. Carmeliet, P., V. Ferreira, et al. (1996). Abnormal blood vessel development and lethality in embryos lacking a single VEGF allele. *Nature* 380(6573): 435–439.
12. Ferrara, N., K. Carver-Moore, et al. (1996). Heterozygous embryonic lethality induced by targeted inactivation of the VEGF gene. *Nature* 380(6573): 439–442.
13. Miquerol, L., B.L. Langille, et al. (2000). Embryonic development is disrupted by modest increases in vascular endothelial growth factor gene expression. *Development* 127(18): 3941–3946.
14. Eremina, V., M. Sood, et al. (2003). Glomerular-specific alterations of VEGF-A expression lead to distinct congenital and acquired renal diseases. *J Clin Invest* 111(5): 707–716.
15. Ferrara, N., H. Chen, et al. (1998). Vascular endothelial growth factor is essential for corpus luteum angiogenesis. *Nat Med* 4(3): 336–340.
16. Fraser, H.M., S.E. Dickson, et al. (2000). Suppression of luteal angiogenesis in the primate after neutralization of vascular endothelial growth factor. *Endocrinology* 141(3): 995–1000.
17. Gerber, H.P., T.H. Vu, et al. (1999). VEGF couples hypertrophic cartilage remodeling, ossification and angiogenesis during endochondral bone formation. *Nat Med* 5(6): 623–628.
18. Hazzard, T.M., F. Xu, et al. (2002). Injection of soluble vascular endothelial growth factor receptor 1 into the preovulatory follicle disrupts ovulation and subsequent luteal function in Rrhesus monkeys. *Biol Reprod* 67(4): 1305–1312.
19. Ravindranath, N., L. Little-Ihrig, et al. (1992). Vascular endothelial growth factor messenger ribonucleic acid expression in the primate ovary. *Endocrinology* 131(1): 254–260.
20. Ryan, A.M., D.B. Eppler, et al. (1999). Preclinical safety evaluation of rhuMAbVEGF, an antiangiogenic humanized monoclonal antibody. *Toxicol Pathol* 27(1): 78–86.
21. Zimmermann, R.C., E. Xiao, et al. (2001). Short-term administration of antivascular endothelial growth factor antibody in the late follicular phase delays follicular development in the rhesus monkey. *J Clin Endocrinol Metab* 86(2): 768–772.
22. Gerber, H.P., K.J. Hillan, et al. (1999). VEGF is required for growth and survival in neonatal mice. *Development* 126(6): 1149–1159.
23. Houck, K.A., D.W. Leung, et al. (1992). Dual regulation of vascular endothelial growth factor bioavailability by genetic and proteolytic mechanisms. *J Biol Chem* 267(36): 26031–26037.
24. Keyt, B.A., L.T. Berleau, et al. (1996). The carboxyl-terminal domain (111–165) of vascular endothelial growth factor is critical for its mitogenic potency. *J Biol Chem* 271(13): 7788–7795.
25. Park, J.E., G.A. Keller, et al. (1993). The vascular endothelial growth factor (VEGF) isoforms: differential deposition into the subepithelial extracellular matrix and bioactivity of extracellular matrix-bound VEGF. *Mol Biol Cell* 4(12): 1317–1326.
26. Eriksson, U. and K. Alitalo (1999). Structure, expression and receptor-binding properties of novel vascular endothelial growth factors. *Curr Top Microbiol Immunol* 237: 41–57.
27. Maglione, D., V. Guerriero, et al. (1991). Isolation of a human placenta cDNA coding for a protein related to the vascular permeability factor. *Proc Natl Acad Sci USA* 88(20): 9267–9271.
28. Persico, M.G., V. Vincenti, et al. (1999). Structure, expression and receptor-binding properties of placenta growth factor (PlGF). *Curr Top Microbiol Immunol* 237: 31–40.
29. Carmeliet, P. (2000). Mechanisms of angiogenesis and arteriogenesis. *Nat Med* 6(4): 389–395.
30. Joukov, V., K. Pajusola, et al. (1996). A novel vascular endothelial growth factor, VEGF-C, is a ligand for the Flt4 (VEGFR-3) and KDR (VEGFR-2) receptor tyrosine kinases. *EMBO J* 15(7): 1751.
31. Orlandini, M., L. Marconcini, et al. (1996). Identification of a c-fos-induced gene that is related to the platelet-derived growth factor/vascular endothelial growth factor family. *Proc Natl Acad Sci USA* 93(21): 11675–11680.
32. Olofsson, B., M. Jeltsch, et al. (1999). Current biology of VEGF-B and VEGF-C. *Curr Opin Biotechnol* 10(6): 528–535.
33. de Vries, C., J.A. Escobedo, et al. (1992). The fms-like tyrosine kinase, a receptor for vascular endothelial growth factor. *Science* 255(5047): 989–991.

34. Shibuya, M., S. Yamaguchi, et al. (1990). Nucleotide sequence and expression of a novel human receptor-type tyrosine kinase gene (flt) closely related to the fms family. *Oncogene* 5(4): 519–524.
35. Terman, B.I., M. Dougher-Vermazen, et al. (1992). Identification of the KDR tyrosine kinase as a receptor for vascular endothelial cell growth factor. *Biochem Biophys Res Commun* 187(3): 1579–1586.
36. Millauer, B., S. Wizigmann-Voos, et al. (1993). High affinity VEGF binding and developmental expression suggest Flk-1 as a major regulator of vasculogenesis and angiogenesis. *Cell* 72(6): 835–846.
37. Shalaby, F., J. Rossant, et al. (1995). Failure of blood-island formation and vasculogenesis in Flk-1-deficient mice. *Nature* 376(6535): 62–66.
38. Fong, G.H., J. Rossant, et al. (1995). Role of the Flt-1 receptor tyrosine kinase in regulating the assembly of vascular endothelium. *Nature* 376(6535): 66–70.
39. Hiratsuka, S., O. Minowa, et al. (1998). Flt-1 lacking the tyrosine kinase domain is sufficient for normal development and angiogenesis in mice. *Proc Natl Acad Sci USA* 95(16): 9349–9354.
40. Taipale, J., T. Makinen, et al. (1999). Vascular endothelial growth factor receptor-3. *Curr Top Microbiol Immunol* 237: 85–96.
41. Soker, S., S. Takashima, et al. (1998). Neuropilin-1 is expressed by endothelial and tumor cells as an isoform-specific receptor for vascular endothelial growth factor. *Cell* 92(6): 735–745.
42. Ellis, L.M., Y. Takahashi, et al. (1998). Vessel counts and vascular endothelial growth factor expression in pancreatic adenocarcinoma. *Eur J Cancer* 34(3): 337–340.
43. Sowter, H.M., A.N. Corps, et al. (1997). Expression and localization of the vascular endothelial growth factor family in ovarian epithelial tumors. *Lab Invest* 77(6): 607–614.
44. Tomisawa, M., T. Tokunaga, et al. (1999). Expression pattern of vascular endothelial growth factor isoform is closely correlated with tumour stage and vascularisation in renal cell carcinoma. *Eur J Cancer* 35(1): 133–137.
45. Volm, M., R. Koomagi, et al. (1997). Prognostic value of vascular endothelial growth factor and its receptor Flt-1 in squamous cell lung cancer. *Int J Cancer* 74(1): 64–68.
46. Yoshiji, H., D.E. Gomez, et al. (1996). Expression of vascular endothelial growth factor, its receptor, and other angiogenic factors in human breast cancer. *Cancer Res* 56(9): 2013–2016.
47. Iliopoulos, O., A.P. Levy, et al. (1996). Negative regulation of hypoxia-inducible genes by the von Hippel-Lindau protein. *Proc Natl Acad Sci USA* 93(20): 10595–10599.
48. Lonser, R.R., G.M. Glenn, et al. (2003). von Hippel-Lindau disease. *Lancet* 361(9374): 2059–2067.
49. Kim, K.J., B. Li, et al. (1993). Inhibition of vascular endothelial growth factor-induced angiogenesis suppresses tumour growth in vivo. *Nature* 362(6423): 841–844.
50. Prewett, M., J. Huber, et al. (1999). Antivascular endothelial growth factor receptor (fetal liver kinase 1) monoclonal antibody inhibits tumor angiogenesis and growth of several mouse and human tumors. *Cancer Res* 59(20): 5209–5218.
51. Liang, W.C., X. Wu, et al. (2006). Cross-species vascular endothelial growth factor (VEGF)-blocking antibodies completely inhibit the growth of human tumor xenografts and measure the contribution of stromal VEGF. *J Biol Chem* 281(2): 951–961.
52. Smith, J.K., N.M. Mamoon, et al. (2004). Emerging roles of targeted small molecule protein-tyrosine kinase inhibitors in cancer therapy. *Oncol Res* 14(4 and 5): 175–225.
53. Grunweller, A. and R.K. Hartmann (2005). RNA interference as a gene-specific approach for molecular medicine. *Curr Med Chem* 12(26): 3143–3161.
54. Lu, P.Y., F.Y. Xie, et al. (2005). Modulation of angiogenesis with siRNA inhibitors for novel therapeutics. *Trends Mol Med* 11(3): 104–113.
55. Gerber, H.P., J. Kowalski, et al. (2000). Complete inhibition of rhabdomyosarcoma xenograft growth and neovascularization requires blockade of both tumor and host vascular endothelial growth factor. *Cancer Res* 60(22): 6253–6258.
56. Kuo, C.J., F. Farnebo, et al. (2001). Comparative evaluation of the antitumor activity of antiangiogenic proteins delivered by gene transfer. *Proc Natl Acad Sci USA* 98(8): 4605–4610.
57. Economides, A.N., L.R. Carpenter, et al. (2003). Cytokine traps: multi-component, high-affinity blockers of cytokine action. *Nat Med* 9(1): 47–52.
58. Fukasawa, M. and M. Korc (2004). Vascular endothelial growth factor-trap suppresses tumorigenicity of multiple pancreatic cancer cell lines. *Clin Cancer Res* 10(10): 3327–3332.

59. Huang, J., J.S. Frischer, et al. (2003). Regression of established tumors and metastases by potent vascular endothelial growth factor blockade. *Proc Natl Acad Sci USA* 100(13): 7785–7790.
60. Dalal, S., A.M. Berry, et al. (2005). Vascular endothelial growth factor: a therapeutic target for tumors of the Ewing's sarcoma family. *Clin Cancer Res* 11(6): 2364–2378.
61. Hu, L., J. Hofmann, et al. (2005). Vascular endothelial growth factor trap combined with paclitaxel strikingly inhibits tumor and ascites, prolonging survival in a human ovarian cancer model. *Clin Cancer Res* 11(19 Pt 1): 6966–6971.
62. Wachsberger, P.R., R. Burd, et al. (2007). VEGF Trap in combination with radiotherapy improves tumor control in U87 glioblastoma. *Int J Radiat Oncol Biol Phys*. In Press.
63. Dupont, J., M.L. Rothenberg, et al. (2005). Safety and pharmacokinetics of intravenous VEGF Trap in a phase I clinical trial of patients with advanced solid tumors. *Proc Am Soc Clin Oncol* 23: 3029.
64. Mulay, M., S.A. Limentani, et al. (2006). Safety and pharmacokinetics of intravenous VEGF Trap plus FOLFOX4 in a combination phase I clinical trial of patients with advanced solid tumors. *Proc Am Soc Clin Oncol* 24: Abstract.
65. Rixe, O., C. Verslype, et al. (2006). Safety and pharmacokinetics of intravenous VEGF Trap plus irinotecan, 5-fluorouracil, and leucovorin (I-LV5FU2) in a combination phase I clinical trial of patients with advanced solid tumors. *Proc Am Soc Clin Oncol* 24: Abstract.
66. Saishin, Y., Y. Saishin, et al. (2003). VEGF-TRAP (R1R2) suppresses choroidal neovascularization and VEGF-induced breakdown of the blood-retinal barrier. *J Cell Physiol* 195(2): 241–248.
67. Wiegand, S., J. Cao, et al. (2003). Long-lasting inhibition of corneal neovascularization following systemic administration of the VEGF Trap. *Invest Ophthalmol Vis Sci* 44: 829 (Abstract).
68. Qaum, T., Q. Xu, et al. (2001). VEGF-initiated blood-retinal barrier breakdown in early diabetes. *Invest Ophthalmol Vis Sci* 42(10): 2408–2413.
69. Cursiefen, C., J. Cao, et al. (2004). Inhibition of hemangiogenesis and lymphangiogenesis after normal-risk corneal transplantation by neutralizing VEGF promotes graft survival. *Invest Ophthalmol Vis Sci* 45(8): 2666–2673.
70. Wiegand, S.J., E. Zimmer, et al. (2005). VEGF trap both prevents experimental choroidal neovascularization and causes regression of established lesions in non-human primates. *Invest Ophthalmol Vis Sci* 46: E-Abstract 1411.
71. Shah, S.M., Q.D. Nguyen, et al. (2006). A double-masked, placebo-controlled, safety, and tolerability study of intravenous VEGF trap in patients with diabetic macular edema. *IOVS ARVO*: Abstract.
72. Davis, S., T.H. Aldrich, et al. (1996). Isolation of angiopoietin-1, a ligand for the TIE2 receptor, by secretion-trap expression cloning. *Cell* 87(7): 1161–1169.
73. Gale, N.W., G. Thurston, et al. (2002). Complementary and coordinated roles of the VEGFs and angiopoietins during normal and pathologic vascular formation. *Cold Spring Harb Symp Quant Biol* 67: 267–273.
74. Maisonpierre, P.C., C. Suri, et al. (1997). Angiopoietin-2, a natural antagonist for Tie2 that disrupts in vivo angiogenesis. *Science* 277(5322): 55–60.
75. Suri, C., P.F. Jones, et al. (1996). Requisite role of angiopoietin-1, a ligand for the TIE2 receptor, during embryonic angiogenesis. *Cell* 87(7): 1171–1180.
76. Valenzuela, D.M., J.A. Griffiths, et al. (1999). Angiopoietins 3 and 4: diverging gene counterparts in mice and humans. *Proc Natl Acad Sci USA* 96(5): 1904–1909.
77. Yancopoulos, G.D., S. Davis, et al. (2000). Vascular-specific growth factors and blood vessel formation. *Nature* 407(6801): 242–248.
78. Holash, J., P.C. Maisonpierre, et al. (1999). Vessel cooption, regression, and growth in tumors mediated by angiopoietins and VEGF. *Science* 284(5422): 1994–1998.
79. Oliner, J., H. Min, et al. (2004). Suppression of angiogenesis and tumor growth by selective inhibition of angiopoietin-2. *Cancer Cell* 6(5): 507–516.
80. Gale, N.W., M.G. Dominguez, et al. (2004). Haploinsufficiency of delta-like 4 ligand results in embryonic lethality due to major defects in arterial and vascular development. *Proc Natl Acad Sci USA* 101(45): 15949–15954.
81. Valenzuela, D.M., A.J. Murphy, et al. (2003). High-throughput engineering of the mouse genome coupled with high-resolution expression analysis. *Nat Biotechnol* 21(6): 652–659.

10 Proteinases and Their Inhibitors in Angiogenesis

Yves A. DeClerck, Khalid Bajou, and Walter E. Laug

CONTENTS

Extracellular proteases play an important role in many physiological and pathological conditions associated with tissue remodeling, including organogenesis, wound repair, osteoarthritis, tumor invasion, and angiogenesis. In angiogenesis, the predominant concept for many years has been that these proteases were critical to allow the migration of endothelial cells (EC) through the dense network of extracellular matrix (ECM) proteins and proteoglycans. However, more recently it has become clear that proteolytic processing of ECM proteins generates antiangiogenic molecules and that extracellular proteases affect the bioavailability of angiogenic factors. As a result, these proteases have been reported to have proangiogenic as well as antiangiogenic activities. Endogenous inhibitors tightly regulate the activity of extracellular proteases in the tumor environment and the balance between enzymes and inhibitors is responsible for tissue homeostasis. Some of these inhibitors have also been found to have paradoxically a proangiogenic activity. In this chapter, the role of extracellular proteases and their natural inhibitors in angiogenesis is reviewed, and the impact of this knowledge on cancer therapy is discussed.

Angiogenesis is typically described in two phases. The first phase consists of an activation process during which, after increased vascular permeability and the deposition of a fibrin matrix, EC dissociate, start to proliferate, migrate and invade, and ultimately form a new lumen. The second phase consists of a resolution process during which cells stop to proliferate and migrate, deposit a collagenous matrix (basement membrane), which replaces the transitional fibrin matrix, establish functional tight junctions, and recruit pericytes.[1] Extracellular proteolysis has been implicated in many of these processes, in particular degradation of the ECM and its remodeling. However, as discussed in this chapter, the role of extracellular proteolysis in angiogenesis goes far beyond this process. Evidence supporting a role for extracellular proteases in angiogenesis came as early as 1939 when Clark and Clark described the presence of extracellular proteolytic activity in EC during angiogenesis in transparent chambers placed in the rabbit ear.[2] Since then there have been numerous publications supporting a critical role for extracellular proteases in angiogenesis. This evidence comes in general from three lines of observation:

1. Expression of extracellular proteases is upregulated by angiogenic factors and by hypoxia.
2. Inhibition of proteolytic activity of EC is often associated with inhibition of angiogenesis.
3. In mice deficient in some proteases, there is a defect in angiogenesis.

However, the contribution of proteases to angiogenesis is complex and includes positive and negative regulatory elements. Several key concepts of the role of extracellular proteases in angiogenesis have emerged over the last decade: The first concept is that angiogenesis requires an appropriate proteolytic balance to occur. Total absence of extracellular proteolysis as well as excessive extracellular proteolysis will both lead to defective angiogenesis. EC need ECM proteins to invade and migrate.[3] The second concept is that what matters for angiogenesis is the proteolytic activity that occurs at the close interface between EC and the ECM. Many extracellular proteases are localized at the cell surface, either because of the presence of a transmembrane domain or a glycosyl phosphatidyl inositol (GPI) anchor or because they are actively recruited at the cell surface by binding to cell surface associated receptors. This mechanism provides a tight control of the extracellular proteolysis at the cell surface.[4] The third concept is that extracellular proteolysis does much more than degrade ECM proteins and has a profound effect on the tumor microenvironment. Proteolytic degradation of ECM proteins releases biologically active protein fragments that inhibit angiogenesis. It exposes a new cryptic domain (called matricryptic) to which EC adhere. It also releases and solubilizes numerous ECM-bound growth factors, enhancing their bioavailability. The fourth concept is that many extracellular proteases target proteins other than ECM proteins and act as sheddase solubilizing cell surface associated growth factors and growth factor receptors.[5] How these emerging concepts explain the complex and sometimes paradoxical effects of extracellular proteases and their inhibitors in angiogenesis is discussed here. Two classes of proteases, the metalloproteases and the serine proteases, have been primarily implicated in angiogenesis and these are discussed in further detail. The role of other extracellular proteases like cathepsins or heparinase in angiogenesis is less well understood and will not be addressed.

10.1 METALLOPROTEINASES IN ANGIOGENESIS

10.1.1 Metalloproteinases and Their Endogenous Inhibitors

Metalloproteinases consist of two major groups of metallo-dependent proteinases, the matrix metalloproteinases (MMP) and the *A* disintegrin and metalloproteinases (ADAM) (Figure 10.1). MMP are a family of 24 calcium-dependent neutral proteases, which share a conserved catalytic motif HEXXHXXGXXH containing three histidine residues (underlined) that bind a Zn^{2+} at the catalytic site and a glutamic acid residue (bold) that is essential for cleavage of a peptide bond.[6–8] These proteases have an N-terminal prodomain present in the

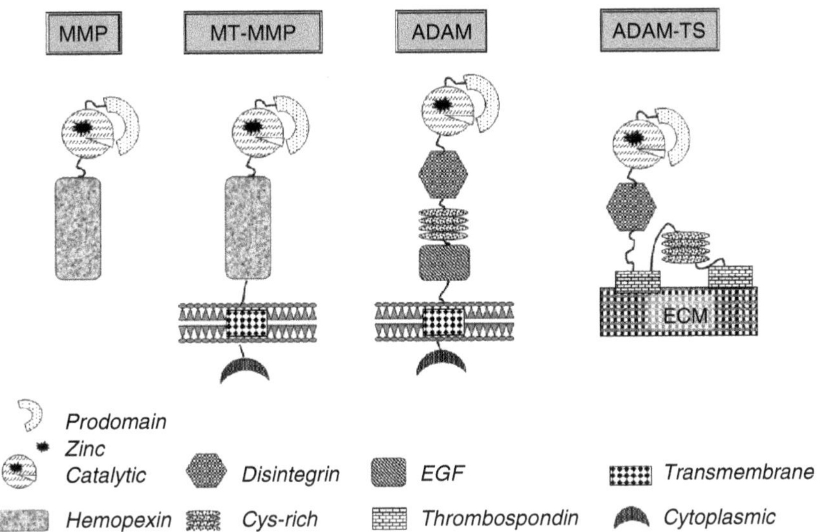

FIGURE 10.1 Comparison between the structures of MMP, MT-MMP, ADAM, and ADAM-TS. The figure illustrates the various domains of these metalloproteases responsible for their binding to the cell surface or to the ECM. (From DeClerck, Y.A., *Eur. J. Cancer*, 36, 1258, 2000. With permission.)

inactive form of the enzyme that is proteolytically removed by extracellular proteases or by intracellular furins upon activation. MMP are typically classified into four major groups according to their structure. The prototype is the matrilysins (MMP-7 and MMP-26), which consist of an N-terminal prodomain and a catalytic domain (Group 1). All other members of the family have a C-terminal hemopexin-like domain, which is involved in substrate recognition and several (MMP-1, MMP-3, MMP-8, MMP-10, MMP-11, MMP-12, MMP-13, MMP-19, MMP-20, MMP-21, MMP-27, and MMP-28) are soluble proteins (Group 2). Gelatinases (MMP-2 and MMP-9) form a separate and third group of soluble MMP characterized by several collagen-binding domains present between the hemopexin domain and the catalytic domain (Group 3). The fourth group consists of MMP designated membrane-type (MT) MMP and presents at the cell surface either through a transmembrane domain present in the C-terminal domain (MMP-14 or MT1-MMP, MMP-15, MMP-16, and MMP-24), the N-terminal domain (MMP-23A and MMP-23B), or a GPI anchor (MMP-17 and MMP-25) (Group 4). ADAM are proteins that contain a metalloproteinase-like domain but all do not have proteolytic activity.[9,10] The presence of a disintegrin-like domain binding to integrins gives them a unique property in mediating cell–cell and cell–ECM interactions. They are thus involved in many interactive processes like fertilization, cell fusion, and development. A unique property of some ADAM is their activity as sheddases of cell-surface receptors and adhesion molecules. ADAM-17, the TNF-α converting enzyme (TACE), processes the membrane-bound form of TNF-α[11] and Kuzbanian (ADAM-10) releases a soluble form of the Notch ligand, Delta, facilitating Notch signaling.[12] ADAM-TS [ADAM with thrombospondin (TS) motif] do not possess a transmembrane domain and are secreted proteins also known as aggrecanases, which bind to the ECM through a thrombospondin-like domain present in their C-terminal.[13,14]

In the extracellular milieu, the activity of metalloproteinases is controlled by several endogenous inhibitors including four tissue inhibitors of metalloproteinases (TIMPs), the nonspecific protease inhibitor α2-macroglobulin, and the membrane-anchored glycoprotein reversion-inducing cysteine-rich protein with Kazal motif (RECK). TIMP-1, 2, and 4 are soluble whereas TIMP-3 binds to the ECM.[15,16] TIMPs inhibit most MMP without selectivity except for TIMP-1, which is a poor inhibitor for most MT-MMP.[17,18] They also selectively inhibit some ADAM and ADAM-TS. For example, TIMP-3 selectively inhibits ADAM-17,

ADAM-TS 4 and 5,[19,20] and TIMP-1 and TIMP-3 but not TIMP-2 or TIMP-4 inhibit ADAM 10.[21] RECK is a recently reported cell-surface glycoprotein, which regulates the catalytic activities of MMP-2, MMP-9, and MT1-MMP.[22]

10.1.2 Evidence Supporting a Role for Metalloproteinases in Angiogenesis

Among the MMP specifically involved in angiogenesis are in particular the two gelatinases MMP-2 and MMP-9 and the membrane-associated MT1-MMP (MMP-14). Indirect evidence supporting their role comes first from the observation that the expression of these proteases in EC is stimulated by hypoxia or angiogenic factors like VEGF or bFGF and in conditions that promote EC tubulogenesis-like growth in three-dimensional matrices.[23–26] A second line of evidence comes from a series of observations in the presence of synthetic as well as natural MMP inhibitors. Several studies have shown that TIMPs and synthetic MMP inhibitors inhibit tubulogenesis and EC invasion in Matrigel and type I collagen matrices.[27–29] More direct evidence in support of their role has come from observations made in transgenic null mice. Mice deficient in MMP-2 have a normal phenotype and no vascular defect; however, tumors implanted in these mice show abnormal angiogenesis.[30] Consistently MMP-2-deficient EC have a growth defect in vivo when grown in the cornea. In the combined absence of MMP-9, MMP-2-deficient EC fail to outgrow fragments of aorta plated in collagen gels (aortic ring assay).[31–33] Evidence for a role for MMP-9 in angiogenesis is even more strongly obtained in MMP-9-deficient mice. MMP-9 null mice have an abnormal phenotype characterized by short femoral bones associated with a deficiency in angiogenesis in the growth plate and ossification of long bones.[34] When bred with transgenic mice carrying the SV40 T antigen transgene under the control of the rat insulin promoter (RIP-Tag2 mice), which develops pancreatic islet cell carcinomas or with mice carrying the human papilloma virus-16, expressed under the control of the keratin 14 promoter (K14-HPV-16 model) and developing spontaneous and invasive squamous cell carcinoma, these double transgenic mice show a delay in tumor formation associated with a lack of initiation of the angiogenic switch critical for tumor development.[35–37] When transplanted with human neuroblastoma tumors, immunodeficient MMP-9 null mice show a defect in tumor vascularization characterized by a lack of pericyte recruitment and an immature vasculature.[38] Interestingly, in these mice transplantation with bone marrow–derived cells from wild-type mice restores a normal angiogenic and tumorigenic phenotype indicating that bone marrow–derived inflammatory cells and not EC are the primary source of the enzyme.[36,39] Mice deficient in MT1-MMP (MMP-14) are normal at birth but die in a few weeks because of bone defects and suffer from deficient vascularization and reduced angiogenesis.[40] In the aortic ring assay, MMP-14-deficient EC fail to degrade type I collagen and to migrate when placed in a collagen matrix. Coculture of aortic rings from MMP-14-deficient mice with aortic rings from wild-type mice does not promote tubulogenesis in MMP-14$^{-/-}$ EC, indicating that in contrast to MMP-9 where the enzyme is not expressed by EC, expression of MMP-14 by EC is necessary for normal angiogenesis.[41] Evidence in support of a critical role for ADAM in tumor angiogenesis is so far limited. ADAM-15 deficient mice have a decreased angiogenic response to hypoxia in the retina but a normal angiogenic response to tumor-mediated angiogenesis.[42] In contrast, some ADAM-TS have been shown to have anti rather than proangiogenic activity. ADAM-TS 1 and 8 suppress bFGF-induced vascularization in the cornea and inhibit VEGF-induced angiogenesis in the chorioallantoid membrane (CAM) assay.[13]

10.1.3 Proangiogenic Activities of Metalloproteinases

Metalloproteinases promote angiogenesis by at least six specific mechanisms: (i) degradation of ECM proteins and promotion of EC migration (invasion), (ii) release and activation of matrix-bound growth factors, (iii) sheddase activity and solubilization of membrane-bound growth factors, (iv) exposure of matricryptic ligands, (v) modulation of vasculogenesis, and (vi) stimulation of vessel maturation.

Degradation of ECM proteins: Through their ability to degrade interstitial and basement membrane collagens and other ECM proteins, MMP promote the invasion of EC into a proteolytically modified ECM as demonstrated in a variety of in vitro and in vivo assays such as the formation of tubes in Matrigel, aortic ring assays in collagen gels, and CAM assays. The demonstration that these metalloproteases specifically act by their ability to degrade the ECM has however often been lacking. It has for example been recently shown that cells can migrate into the ECM using either a mesenchymal-type of movement that requires proteases or an amoeboid-type of movement that does not require ECM proteolysis.[43] Tumor cells use both migratory patterns to penetrate tissues and can switch from one to the other depending on their ability to degrade the ECM. Whether EC similarly can use both migratory mechanisms has not been presently demonstrated.[44,45]

Release and activation of growth factors: Many growth factors have heparin-binding domains, which are responsible for their sequestration by proteoglycans abundantly present in the ECM. On degradation of the ECM, these growth factors become soluble and diffuse through the extracellular space. As a result, their biological activity is increased. For example, in MMP-9-deficient mice there is a decrease in the levels of soluble VEGF and transplantation of these mice with MMP-9 expressing bone marrow cells is associated with an increase in VEGF release.[35]

Sheddase activity of metalloproteases: Several metalloproteases have specific proteolytic activity for membrane-associated growth factors, and as a consequence these growth factors are released in the extracellular milieu. This is well illustrated in the case of heparin-bound EGF (HB-EGF), the ligand for ErbB4. In tumor cells and uterine and mammary epithelial cells, MMP-7 is trapped at the cell surface by binding to CD44 (the receptor for hyaluronic acid) and thus becomes localized in close proximity of the membrane-associated precursor form of HB-EGF. This enables MMP-7 to cleave HB-EGF at a specific peptide bond located outside the transmembrane domain, releasing a soluble form of HB-EGF.[46] ADAM-17 (TACE), which processes the membrane-bound form of pro-TNFα, has a similar activity on HB-EGF. Some EC express MMP-7[47] and TACE,[48] but whether this mechanism plays a role in angiogenesis has not been demonstrated.

Exposure of matricryptic ligands: Collagens are cross-linked, polymeric proteins, which are denatured on the action of metalloproteinases. This activity unfolds the proteins and exposes cryptic domains that have a modulatory function on cell growth and survival. Several specific cryptic domains in the noncollagenous part of collagen IV (NC1 domain) have for example been identified and shown to have a modulatory activity on EC adhesion and migration and to protect EC from apoptosis.[49] These cryptic domains are exposed on ECM digestion by MMP-9.[50] Similarly, degradation of the laminin-5 γ2 chain by MMP-2 and MT1-MMP releases promigratory fragments and exposes epitopes that promote vasculogenic mimicry.[51]

Modulation of vasculogenesis: Vasculogenesis differs from angiogenesis because it consists of the formation of new blood vessels that arise not from existing vessels (as in angiogenesis) but from vascular precursor cells. These vascular precursor cells originate from hematopoietic precursor cells present in the bone marrow and circulate in the peripheral blood as endothelial progenitor cells (EPC). In tumors, EPC can associate with mature EC to form functional blood vessels.[52–54] EPC are actively recruited by the tumor and this process is controlled by MMP-9. The mobilization of EPC from the bone marrow into the blood circulation is stimulated by VEGF and the soluble form of Kit ligand.[55] MMP-9 cleaves the membrane form of Kit ligand, therefore releasing it in a soluble form. In mice deficient in MMP-9, VEGF-induced mobilization does not occur and there is a decreased level of soluble Kit ligand associated with a lack of mobilization of hematopoietic stem cells from the bone marrow into the circulation.[56] In our laboratory, we have recently demonstrated that in tumor-bearing MMP-9 null mice, there is a decrease in the presence of bone marrow–derived CD31 expressing cells in the tumor vasculature, suggesting therefore that MMP-9 plays a critical role in tumor vasculogenesis by controlling the release of EPC from the bone marrow and their recruitment in the tumor vasculature.[39]

Effect on vessel maturation: It is increasingly recognized that blood vessel maturation is a critical aspect of tumor angiogenesis. Often tumor-derived blood vessels have a disorganized architecture, are irregular in size and shape, and lack pericytes. As a result, the vascularization of tumors is deficient and includes areas of necrosis and hemorrhage typical of many solid tumors. Using a model of human neuroblastoma tumors orthotopically implanted in immunodeficient MMP-9-deficient mice, our laboratory has recently shown that in tumor-bearing MMP-9 null mice there is a lack of pericyte recruitment along EC, associated with a reduction in blood vessel size and shape.[38] This defect was corrected by transplanting mice with bone marrow cells from MMP-9 positive mice. We also observed that in more aggressive forms of neuroblastoma tumors, there is an increase in MMP-9 expressing inflammatory cells in conjunction with an increase in pericyte coverage of EC.[39] The data thus suggest that MMP-9 positively contributes to the establishment of a mature tumor vasculature.

10.1.4 Antiangiogenic Activity of Metalloproteinases

Whereas the bulk of in vitro experiments and preclinical studies in animals have supported a positive role for MMP and ADAM in angiogenesis, clinical trials with MMP inhibitors have generated disappointing and sometimes paradoxical results to the point that most of these trials have been terminated.[57] In one particular study with the compound BAY 12-9566,[58] acceleration rather than inhibition of tumor progression was in fact documented in patients with non-small cell lung carcinoma (NSCLC) treated with the MMP inhibitor, suggesting a paradoxical effect of MMP inhibition in tumor progression and angiogenesis. Three mechanisms by which metalloproteases may act as negative stimulators of angiogenesis have been so far identified.

Generation of soluble matricryptic ligands with antiangiogenic activity: The discovery of angiostatin as a critical negative regulator of angiogenesis and the observation that it is generated by proteolytic cleavage of plasminogen by MMP such as MMP-2, MMP-3, MMP-7, MMP-9, and MMP-12 provided the first suggestion that MMP could be a double-edged sword in angiogenesis.[59] Mice deficient in α1-integrin have increased levels of MMP-9 and MMP-7, which correlate with increased levels of angiostatin and decreased tumor vascularization. A paradoxical stimulation of angiogenesis associated with decreased levels of circulating angiostatin was then accomplished by treatment of these mice with the synthetic MMP inhibitor BB94.[60] Many other antiangiogenic proteolytic fragments derived from proteolytic cleavage of collagen XVIII (endostatin), α_1 type IV collagen (canstatin), α_2 type IV collagen (tumstatin), and α_3 type IV collagen (arresten) by MMP have since been reported and their generation explains the paradoxical antiangiogenic activity of MMP.[61,62] However, whether all MMP have equal activity in their ability to generate antiangiogenic matricryptic ligands is unclear. Interestingly MMP-9, which is among the most active metalloproteases in angiogenesis, is also the most active in generating antiangiogenic factors. MMP-9 null mice have lower levels of angiostatin and tumstatin and tumors in these mice exhibit increased rather than decreased vascularization.[63] These data illustrate the complex nature of MMP in tumor angiogenesis.

Interference with angiogenic factor: Both ADAM-TS 1 and ADAM-TS 8 suppress VEGF-induced angiogenesis in the corneal micropocket assay and the CAM assay by inhibiting VEGF-A 165 binding to its receptor VEGFR-2 and VEFGR-mediated signaling.[13] Whether this mechanism plays a role in tumor-mediated angiogenesis in vivo is unknown at this point.

Interference with the uPA/plasminogen system: MMP can also have an antiangiogenic activity by their ability to interfere with uPA-mediated plasminogen activation. For example, MMP-12 can release the receptor for uPA (uPAR). This sheddase activity results in a decrease in cell-associated PA activity in EC and an inhibition of their ability to form tubes in a fibrin matrix.[64]

10.1.5 TIMPs in Angiogenesis

As key regulators of MMP activity, TIMPs have been suspected earlier to have antiangiogenic activity. For example, the high content of cartilage extracts in TIMPs was considered to be responsible for the antiangiogenic activity of this biological material.[65] The activity of TIMPs in angiogenesis is however complex and includes stimulatory as well as inhibitory effects and the activity of some TIMPs does not require the inhibition of the catalytic action of MMP. For example, early on it was demonstrated that TIMP-2 inhibits the proliferation of bFGF-stimulated EC,[66] and recently it was shown that a similar effect can be achieved with an inactive mutant (Ala-TIMP-2). This antiproliferative activity of TIMP-2 involves binding to the $\alpha_3\beta_1$ integrin preventing integrin-mediated activation of receptor tyrosine kinases.[67] TIMP-3 has a known antiangiogenic activity.[27] Intratumoral delivery of TIMP-3 by adenovirus inhibits angiogenesis and tumors overexpressing TIMP-3 have a defect in pericyte recruitment.[68] Some of the antiangiogenic activity of TIMP-3 seems linked to its ability to bind to VEGF-R2, interfering thus with VEGF-mediated receptor activation.[69] The effect of TIMP-4 on tumor angiogenesis is unclear and depends on the method and timing of administration. When administered prior to tumor implantation, TIMP-4 has a strong antiangiogenic activity.[70] However, when administrated in the form of naked DNA in mice with established tumors, it promotes tumor formation and angiogenesis.[71] These data illustrate the complexity of the role of TIMP in angiogenesis.

10.1.6 Synthetic and Exogenous Metalloproteinase Inhibitors in Cancer Therapy

The scientific literature published over the last two decades supports the use of MMP inhibitors as anticancer agents. On the basis of these convincing data, in the late 1990s several MMP inhibitors developed by pharmaceutical companies were tested in phase I, II, and III clinical trials. These clinical trials have however been disappointing and most have been discontinued as they have shown no effect or in some cases, as discussed earlier, an acceleration of tumor progression.[57,72] The results of recent clinical trials with Marimastat, an orally available broad-spectrum inhibitor, in patients with prostate cancer and advanced NSCLC have been reported,[73,74] as well as the results of a phase I clinical trial of BMS-275291, a sheddase-sparing MMP inhibitor in advanced or metastatic cancers (Table 10.1).[75] Presently only one inhibitor, Neovastat (AE-941), a cartilage extract with both anti-MMP and anti-VEGF activities, is still tested in clinical trials.[76–78] XL-784, a small-molecular weight orally available inhibitor of ADAM-10 without anti-MMP1 activity, has been shown to be free of side effects when administered to healthy volunteers and is currently tested in clinical trials. An analysis of the failure of MMP inhibitors in the clinics, however, provides helpful lessons. Because of the abundance of convincing preliminary data in vitro and experiments in preclinical models, MMP inhibitors went into clinical trial very rapidly and before a more complete understanding of their effect in cancer progression and angiogenesis was obtained. As the complexity and paradoxical role of MMP in angiogenesis have been unraveled, it is now easier to understand why MMP inhibition may have failed in clinical trials such as in patients with NSCLC treated with BAY 12-9566. The absence of well-defined biological end points to measure the activity of these MMP inhibitors in treated patients has also made the determination of the maximum biologically effective dose not possible and phase I trials were

TABLE 10.1
Some Recent Clinical Trials with Extracellular Protease Inhibitors in Cancer

Compound	Company	Target	Clinical Trial	Results	Ref.
Marimastat	British Biotechnology	MMP (broad spectrum)	Phase I advanced NSCLC with carboplatin and paclitaxel	Well tolerated	74
Marimastat	British Biotechnology	MMP (broad spectrum)	Phase I/II biochemically relapsed prostate cancer	Toxicity at 20 (mg)	73
BMS-275291	Bristol-Myers Squibb	MMP (broad spectrum)	Phase I advanced and metastatic cancer	Disease stabilization in 12/44 patients	75
Neovastat	Aeterna	VEGF and MMP	Phase I/II NSCLC	Well tolerated; survival advantage at higher doses. No antitumor effect	77
Neovastat	Aeterna	VEGF and MMP	Phase II refractory renal cell carcinoma	Well tolerated; survival benefit	78
Neovastat	Aeterna	VEGF and MMP	Phase III NSCLC randomized with chemotherapy and radiation	Ongoing	—
XL-784	Exelixis	ADAM-10	Phase I healthy volunteer	No side effect	—
WX-UK1	Wilex	uPA	Phase III advanced malignancies in combination with capecitabine	Ongoing	—

thus primarily aimed at the determination of the maximal tolerable dose (MTD) as is the goal with cytotoxic anticancer agents. As a consequence the doses used in phase II studies were often toxic and associated with a reversible arthritis-like syndrome characterized by intractable joint pains and fibrotic nodules, which necessitated discontinuing the treatment in up to 30% of treated patients.[79] Thus, many patients who might have responded to a lower dose were not investigated. Most trials were also performed with broad-spectrum inhibitors, targeting therefore MMP with proangiogenic activities as well as MMP with antiangiogenic activities. Finally, most trials involved patients with advanced-stage disease where inhibition of angiogenesis was unlikely to be sufficient to have a significant effect on tumor progression. Animal experiments in the RIP-Tag2 model suggest for example that MMP inhibitors need to be employed before activation of the angiogenic switch has occurred.[80] Whether MMP inhibition could still be a valid target in the treatment of cancer remains an open question. The abundance of convincing preclinical data supporting a role for MMP in angiogenesis and the demonstrated effect of MMP inhibition in many preclinical models still suggest that if administered at the correct dose and at the correct time, MMP inhibitors may be a valuable component of the multidrug therapeutic arsenal used in cancer patients. For example, evidence that MMP-9 plays a stimulatory role in promoting the colonization of primary tumors by bone marrow–derived endothelial precursor cells and that there is an increase in circulating EPC postintensive chemotherapy[81] suggests that a short course of MMP inhibition may decrease the risk of EPC colonization of a primary tumor postintensive chemotherapy. Whether a selective versus a broad-spectrum MMP inhibition is a better approach in antiangiogenic therapies remains unclear. If MMP with antiangiogenic activity can be distinguished from MMP with proangiogenic activity, selective inhibition could be a promising approach. MMP-7 and MMP-12 in particular are the most active MMP in generating angiostatic molecules and therefore the development of MMP inhibitors with little or no activity against these MMP may be a promising approach.[82] However, in the case of MMP-9, antiangiogenic as well as proangiogenic activities have been reported, and selective inhibition of MMP-9 may have unpredictable results.

10.2 SERINE PROTEASES IN ANGIOGENESIS

10.2.1 Role of Plasminogen Activation in Angiogenesis

Control of plasminogen activation at the surface of EC: Among the serine proteases involved in angiogenesis is plasmin, a broad-spectrum protease with tryptic specificity that hydrolyzes many ECM proteins in addition to fibrin.[3] Plasmin also activates several proMMP and as such increases MMP-dependent proteolysis. Plasmin is generated from its precursor plasminogen by proteolytic cleavage by other serine proteases like urokinase-type (uPA) or tissue-type (tPA) plasminogen activators. uPA (or more precisely its inactive precursor pro-uPA) is secreted as a soluble protein and binds to a high-affinity GPI-anchored cell-surface receptor (uPAR) present on the surface of many cells, including EC.[83] Plasminogen can also bind to various cell-surface receptors such as annexin II,[84] α-enolase,[85,86] and amphoterin.[87,88] The presence of these receptors at the cell surface provides a mechanism for tightly controlling the amount of plasmin generated at the cell surface. tPA, which is present in a soluble form, can also be trapped at the cell surface through its ability to bind to annexin II and amphoterin.[88] The activation of plasminogen by PA is further controlled by specific PA inhibitors (PAI), in particular PAI-1 and PAI-2. Binding of PAI-1 to uPA–uPAR promotes the internalization of the complex in a caveolin-dependent endocytosis process in which uPA and PAI-1 are internally degraded by late endosomes and uPAR is recycled to the cell surface providing new binding sites for uPA.[89] This process is further controlled by the low-density lipoprotein receptor-related protein (LRP).[90] Quiescent

EC do not express uPA, uPAR, and PAI-1 but express these proteins during angiogenesis in vivo and on stimulation by angiogenic factors like VEGF and bFGF.[91,92] Hypoxia is also a major stimulus for uPAR and PAI-1.[93]

Angiogenic activity of plasmin: Evidence supporting a role for uPA in angiogenesis is multifaceted: (i) uPA and uPAR are expressed by EC and their expression is upregulated on stimulation by angiogenic factors, (ii) uPA and uPAR are abundantly expressed in tumors, (iii) downregulation of uPA and uPAR in EC inhibits tube formation and angiogenesis in various tubulogenesis assays, (iv) administration of endogenous or synthetic inhibitors of uPA activity and of uPA–uPAR interaction inhibits angiogenesis. Experiments in mice deficient in uPA, uPAR, or tPA, however, indicate that they are not required for vascular development as these mice have a normal phenotype. Mice deficient in either uPA or tPA have a normal embryonic and postnatal development and a normal angiogenic response in the cornea;[94] however, plasminogen-deficient mice exhibit a deficient response.[94,95] EC from the aortic rings of uPA-, tPA-, and uPAR-deficient mice form normal tubes. However, in mice deficient in uPA, there is an inhibition of angiogenesis postmyocardial infarction.[96] In contrast, in mice deficient in plasminogen, there is a deficit in angiogenesis as demonstrated by a delay in tube formation by EC cultured from aortic rings placed in a collagen gel.[97] Thus, altogether the data suggest that the activation of plasminogen is a necessary step for EC tubulogenesis. Whether it is necessary for tumor angiogenesis in vivo requires more complex experiments combining plasminogen-deficient mice and xenotransplanted or transgenic tumor models. In plasminogen-deficient mice, syngeneic transformed keratinocytes implanted on the skin fail to show a significant reduction in the angiogenic response,[98] and develop smaller and less-hemorrhagic Lewis lung carcinoma tumors.[99] The observation that plasmin can activate several proMMP also links the uPA-plasminogen system to the MMP. For example, although plasmin is a poor activator of MMP-9, it can activate proMMP-3 (stromelysin-1), which becomes a potent activator of MMP-9.[100]

Angiogenic activity of uPA–uPAR interaction: The critical role that uPA–uPAR interaction plays in angiogenesis has been well demonstrated. Angiogenesis in subcutaneous Matrigel plugs containing bFGF is inhibited by a fusion uPA protein in which the uPAR-binding site is replaced by the Fc portion of IgG.[101] Inhibition of uPA–uPAR interaction by a noncompetitive octamer peptide antagonist also inhibits neovascularization in a xenotransplanted prostate cancer model in mice.[102] We have recently shown that a 48 amino acid N-terminal fragment of uPA (uPA 1–48), which prevents uPA binding to its receptor uPAR, inhibits EC tubulogenesis in Matrigel and EC migration on vitronectin gradient. When administered to mice orthotopically implanted with human U87MG glioma cells, this peptide suppresses brain tumor growth through inhibition of angiogenesis.[103] The exact mechanism by which uPA and uPAR interactions promote angiogenesis is not entirely understood. uPA promotes plasmin-mediated proteolytic degradation of ECM glycoproteins, which are critical for cell migration and invasion. Although proteolytic degradation of these proteins is important, excessive degradation has a negative effect on EC attachment to ECM proteins, a critical step in migration and survival.[104] uPA binding to uPAR also has a cell adhesion function independent of its proteolytic activity. uPAR binds to the domain of vitronectin through its D3 domain and this binding is enhanced by uPA. Binding of uPA to uPAR thus promotes EC adhesion to vitronectin in an integrin-independent manner.[105] The consequence of uPA–uPAR binding to vitronectin on EC migration is however unclear. Some of the conflicting observations made on the role of uPA and uPAR on angiogenesis can be explained by its dose-dependent effect. At low concentrations, uPA acts as a promoter of angiogenesis by providing the sufficient amount of plasmin-mediated proteolysis to promote EC invasion and migration. Excessive uPA activity in contrast is likely to decrease the amount of ECM proteins necessary for migration and for protecting cells from undergoing apoptosis.

tPA in angiogenesis: The role of tPA in angiogenesis is less clear. In vitro, tPA has been implicated in EC tubulogenesis in Matrigel and type I collagen[106] but its role in EC migration has been controversial.[3] Angiostatin inhibits EC invasion by blocking tPA.[107] However, more definitive evidence supporting a role for tPA in angiogenesis is still lacking.

10.2.2 Membrane-Anchored Serine Proteases in Angiogenesis

Recently a new class of membrane-anchored serine proteases (SP) has been recognized.[108,109] These proteases are attached to the plasma membrane by either a C-terminal transmembrane domain (type I SP), a GPI link (GPI-anchored SP), or an N-terminal transmembrane domain (type II SP).[109,110] Fifteen SP have been described so far and several are expressed by EC, suggesting a possible role in angiogenesis.[111] Further studies, however, are required to determine their precise role in angiogenesis.

10.2.3 PAI-1 in Angiogenesis

The contribution of PAI-1 to tumor development and angiogenesis has long been controversial. Several clinical studies in human cancers indicated that high and not low levels of PAI were a strong negative prognosticator of cancer progression.[112] Higher levels of PAI-1 correlated with poorer prognosis in multiple cohorts of patients with breast and ovarian cancers[113,114] and were predictive of lack of response to agents like tamoxifen.[115] This paradoxical observation was recently explained by the discovery of a proangiogenic function of PAI-1. Mice deficient in PAI-1 and implanted with syngeneic malignantly transformed keratinocytes fail to develop a normal angiogenic response.[116] The mechanism by which PAI-1 promotes angiogenesis is complex and involves both its antiproteolytic activity and its modulatory activity on cell adhesion. It is also dose dependent as PAI-1 is proangiogenic at physiological concentrations and antiangiogenic at high, pharmacological concentrations.[97,117] PAI-1 can promote angiogenesis by two potential mechanisms. First, by protecting the ECM against plasmin-mediated proteolysis, PAI-1 maintains a scaffold required for EC migration. Second, PAI-1 is a modulator of EC adhesion to vitronectin. PAI-1 has in its N-terminal region, a domain that binds with high affinity to the somatomedin B-like domain of vitronectin. Such binding acts as a stabilizer of PAI-1. As a result the RGD-binding site of vitronectin, which is in close proximity to the somatomedin B-like domain and responsible for EC binding to vitronectin via integrin $\alpha v\beta 3$ and $\alpha v\beta 5$, becomes cryptic. By this mechanism PAI-1 inhibits integrin-mediated EC adhesion to vitronectin.[118–121] However, in contrast to other inhibitors of α_V-mediated integrin binding, which induce EC apoptosis, PAI-1 does not. Our laboratory has shown that the binding of PAI-1 to vitronectin in fact promotes EC migration from vitronectin toward fibronectin and EC adhesion to fibronectin via the integrin $\alpha_5\beta_1$.[122] We have also shown that in neuroblastoma tumors, vitronectin is more abundant in the perivascular space and fibronectin in the nonvascularized compartment of the tumor, suggesting that one of the proangiogenic effects of PAI-1 consists in promoting the migration of EC from vitronectin-rich perivascular areas of the tumor toward fibronectin-rich less-vascularized areas. Whether both antiproteolytic and antiadhesive functions of PAI-1 play an equal role in the proangiogenic activity of PAI-1 is controversial. Using adenoviral gene transfer of PAI-1 mutants in PAI-1-deficient mice, it was shown that a mutant with antiproteolytic activity but no inhibitory activity on vitronectin binding promoted angiogenesis, whereas a mutant without antiproteolytic activity but that inhibited vitronectin binding had no effect.[98] However, other studies have suggested that both functions play a role,[123] and in our laboratory we observed that the stimulatory activity of PAI-1 on EC migration from vitronectin toward fibronectin required the vitronectin binding and not the antiproteolytic function of PAI-1.[122]

10.2.4 Serine Protease Inhibitors in Antiangiogenic Therapy

Many synthetic inhibitors of serine proteases have been developed but their clinical use as systemic drugs has been limited because of their potent effect on hemostasis. One such inhibitor, WX-UK1, an oral low-molecular weight 3-aminophenylalanine serine protease inhibitor with activity against uPA and with in vitro and in vivo activities against carcinoma cells,[124,125] is currently in phase III trial in combination with capecitabine in patients with advanced malignancies (Table 10.1). Interfering with uPA–uPAR interaction has also been an attractive target for antiangiogenesis therapy because of its well-demonstrated role in angiogenesis. Compounds that interfere with the binding of uPA to uPAR include peptides, peptidomimetic, monoclonal antibodies, and recombinant uPA- or uPAR-derived proteins. All have been shown to have activity in vitro and in vivo,[103,126,127] but they have not entered clinical trials yet.

10.3 CONCLUSION AND PERSPECTIVES

The role of extracellular proteases in angiogenesis has been well supported by a large body of experimental data in vitro and in vivo, and by observations in human cancer. However, the role of these proteases is more complex than initially anticipated because of their activity on multiple substrates, their ability to unmask biologically active cryptic domains in the ECM, and to affect growth factor–ECM and cell–ECM interaction. With a more complete understanding of the mechanisms by which extracellular proteases and their inhibitors affect tumor angiogenesis, novel therapeutic opportunities will continue to be identified. Their validation in the clinic will require the design of better clinical trials where specific biological effects can be measured as end points and where the testing of inhibitors can be done in patients at earlier stages of cancer progression.

ACKNOWLEDGMENT

This work was supported by grants CA 084103-04 and CA 082989 from the National Institutes of Health and the T.J. Martell Foundation.

REFERENCES

1. Jain, R.K., Normalization of tumor vasculature: an emerging concept in antiangiogenic therapy, *Science*, 307, 58, 2005.
2. Clark, E.R. and Clark, E.L., Growth and behavior of epidermis as observed microscopically in observation chambers inserted in the ears of rabbits, *Am. J. Anat.*, 93, 171, 1953.
3. Pepper, M.S., Role of the matrix metalloproteinase and plasminogen activator–plasmin systems in angiogenesis, *Arterioscler. Thromb. Vasc. Biol.*, 21, 1104, 2001.
4. Vassalli, J.D. and Pepper, M.S., Tumour biology. Membrane proteases in focus, *Nature*, 370, 14, 1994.
5. McCawley, L.J. and Matrisian, L.M., Matrix metalloproteinases: they're not just for matrix anymore, *Curr. Opin. Cell Biol.*, 13, 534, 2001.
6. Egeblad, M. and Werb, Z., New functions for the matrix metalloproteinases in cancer progression, *Nat. Rev. Cancer*, 2, 161, 2002.
7. Matrisian, L.M., Matrix metalloproteinases, *Curr. Biol.*, 10, R692, 2000.
8. Nagase, H. and Woessner, J.F., Jr., Matrix metalloproteinases, *J. Biol. Chem.*, 274, 21491, 1999.
9. Wolfsberg, T.G. et al., ADAM, a novel family of membrane proteins containing a disintegrin and metalloprotease domain: multipotential functions in cell–cell and cell–matrix interactions, *J. Cell Biol.*, 131, 275, 1995.

10. Seals, D.F. and Courtneidge, S.A., The ADAMs family of metalloproteases: multidomain proteins with multiple functions, *Genes Dev.*, 17, 7, 2003.
11. Black, R.A. et al., A metalloproteinase disintegrin that releases tumour-necrosis factor-a from cells, *Nature*, 385, 729, 1997.
12. Primakoff, P. and Myles, D.G., The ADAM gene family: surface proteins with adhesion and protease activity, *Trends Genet.*, 16, 83, 2000.
13. Vazquez, F. et al., METH-1, a human ortholog of ADAMTS-1, and METH-2 are members of a new family of proteins with angio-inhibitory activity, *J. Biol. Chem.*, 274, 23349, 1999.
14. Nagase, H. and Kashiwagi, M., Aggrecanases and cartilage matrix degradation, *Arthritis Res. Ther.*, 5, 94, 2003.
15. Blavier, L. et al., Tissue inhibitors of matrix metalloproteinases in cancer, *Ann. NY Acad. Sci.*, 878, 108, 1999.
16. Brew, K., Dinakarpandian, D., and Nagase, H., Tissue inhibitors of metalloproteinases: evolution, structure and function, *Biochim. Biophys. Acta*, 1477, 267, 2000.
17. Will, H. et al., The soluble catalytic domain of membrane type 1 matrix metalloproteinase cleaves the propeptide of progelatinase A and initiates autoproteolytic activation—regulation by TIMP-2 and TIMP-3, *J. Biol. Chem.*, 271, 17119, 1996.
18. Zhao, H. et al., Differential inhibition of membrane type 3 (MT3)-matrix metalloproteinase (MMP) and MT1-MMP by tissue inhibitor of metalloproteinase (TIMP)-2 and TIMP-3 regulates pro-MMP-2 activation, *J. Biol. Chem.*, 279, 8592, 2004.
19. Kashiwagi, M. et al., TIMP-3 is a potent inhibitor of aggrecanase 1 (ADAM-TS4) and aggrecanase 2 (ADAM-TS5), *J. Biol. Chem.*, 276, 12501, 2001.
20. Amour, A. et al., TNF-a converting enzyme (TACE) is inhibited by TIMP-3, *FEBS Lett.*, 435, 39, 1998.
21. Amour, A. et al., The in vitro activity of ADAM-10 is inhibited by TIMP-1 and TIMP-3, *FEBS Lett.*, 473, 275, 2000.
22. Takahashi, C. et al., Regulation of matrix metalloproteinase-9 and inhibition of tumor invasion by the membrane-anchored glycoprotein RECK, *Proc. Natl. Acad. Sci. USA*, 95, 13221, 1998.
23. Chan, V.T. et al., Membrane-type matrix metalloproteinases in human dermal microvascular endothelial cells: expression and morphogenetic correlation, *J. Invest. Dermatol.*, 111, 1153, 1998.
24. Ben Yosef, Y. et al., Regulation of endothelial matrix metalloproteinase-2 by hypoxia/reoxygena-oxygenation, *Circ. Res.*, 90, 784, 2002.
25. Haas, T.L., Davis, S.J., and Madri, J.A., Three-dimensional type I collagen lattices induce coordinate expression of matrix metalloproteinases MT1-MMP and MMP-2 in microvascular endothelial cells, *J. Biol. Chem.*, 273, 3604, 1998.
26. Kondo, S. et al., Connective tissue growth factor increased by hypoxia may initiate angiogenesis in collaboration with matrix metalloproteinases, *Carcinogenesis*, 23, 769, 2002.
27. Anand-Apte, B. et al., Inhibition of angiogenesis by tissue inhibitor of metalloproteinase-3, *Invest. Ophthalmol. Vis. Sci.*, 38, 817, 1997.
28. Koike, T. et al., Inhibited angiogenesis in aging: a role for TIMP-2, *J. Gerontol. A Biol. Sci. Med. Sci.*, 58, B798–B805, 2003.
29. Robinet, A. et al., Elastin-derived peptides enhance angiogenesis by promoting endothelial cell migration and tubulogenesis through upregulation of MT1-MMP, *J. Cell Sci.*, 118, 343, 2005.
30. Itoh, T. et al., Reduced angiogenesis and tumor progression in gelatinase A-deficient mice, *Cancer Res.*, 58, 1048, 1998.
31. Kato, T. et al., Diminished corneal angiogenesis in gelatinase A-deficient mice, *FEBS Lett.*, 508, 187, 2001.
32. Masson, V. et al., Contribution of host MMP-2 and MMP-9 to promote tumor vascularization and invasion of malignant keratinocytes, *FASEB J.*, 19, 234, 2005.
33. Masson, V. et al., Mouse aortic ring assay: a new approach of the molecular genetics of angiogenesis, *Biol. Proceed. Online*, 4, 24, 2002.
34. Vu, T.H. et al., MMP-9/gelatinase B is a key regulator of growth plate angiogenesis and apoptosis of hypertrophic chondrocytes, *Cell*, 93, 411, 1998.
35. Bergers, G. et al., Matrix metalloproteinase-9 triggers the angiogenic switch during carcinogenesis, *Nat. Cell Biol.*, 2, 737, 2000.

36. Coussens, L.M. et al., MMP-9 supplied by bone marrow–derived cells contributes to skin carcinogenesis, *Cell*, 103, 481, 2000.
37. Coussens, L.M. et al., Inflammatory mast cells up-regulate angiogenesis during squamous epithelial carcinogenesis, *Genes Dev.*, 13, 1382, 1999.
38. Chantrain, C.F. et al., Stromal matrix metalloproteinase-9 regulates the vascular architecture in neuroblastoma by promoting pericyte recruitment, *Cancer Res.*, 64, 1675, 2004.
39. Jodele, S. et al., The contribution of bone marrow–derived cells to the tumor vasculature in neuroblastoma is matrix metalloproteinase-9 dependent, *Cancer Res.*, 65, 3200, 2005.
40. Holmbeck, K. et al., MT1-MMP: a tethered collagenase, *J. Cell Physiol.*, 200, 11, 2004.
41. Chun, T.H. et al., MT1-MMP-dependent neovessel formation within the confines of the three-dimensional extracellular matrix, *J. Cell Biol.*, 167, 757, 2004.
42. Horiuchi, K. et al., Potential role for ADAM15 in pathological neovascularization in mice, *Mol. Cell Biol.*, 23, 5614, 2003.
43. Wolf, K. et al., Compensation mechanism in tumor cell migration: mesenchymal-amoeboid transition after blocking of pericellular proteolysis, *J. Cell Biol.*, 160, 267, 2003.
44. Friedl, P. and Brocker, E.B., The biology of cell locomotion within three-dimensional extracellular matrix, *Cell Mol. Life Sci.*, 57, 41, 2000.
45. Friedl, P., Prespecification and plasticity: shifting mechanisms of cell migration, *Curr. Opin. Cell Biol.*, 16, 14, 2004.
46. Yu, W.H. et al., CD44 anchors the assembly of matrilysin/MMP-7 with heparin-binding epidermal growth factor precursor and ErbB4 and regulates female reproductive organ remodeling, *Genes Dev.*, 16, 307, 2002.
47. Nagashima, Y. et al., Expression of matrilysin in vascular endothelial cells adjacent to matrilysin-producing tumors, *Int. J. Cancer*, 72, 441, 1997.
48. Goddard, D.R., Bunning, R.A., and Woodroofe, M.N., Astrocyte and endothelial cell expression of ADAM 17 (TACE) in adult human CNS, *Glia*, 34, 267, 2001.
49. Petitclerc, E. et al., New functions for non-collagenous domains of human collagen type IV. Novel integrin ligands inhibiting angiogenesis and tumor growth in vivo, *J. Biol. Chem.*, 275, 8051, 2000.
50. Hangai, M. et al., Matrix metalloproteinase-9-dependent exposure of a cryptic migratory control site in collagen is required before retinal angiogenesis, *Am. J. Pathol.*, 161, 1429, 2002.
51. Seftor, R.E. et al., Cooperative interactions of laminin 5 gamma2 chain, matrix metalloproteinase-2, and membrane type-1-matrix/metalloproteinase are required for mimicry of embryonic vasculogenesis by aggressive melanoma, *Cancer Res.*, 61, 6322, 2001.
52. Reyes, M. et al., Origin of endothelial progenitors in human postnatal bone marrow, *J. Clin. Invest.*, 109, 337, 2002.
53. Jain, R.K., Molecular regulation of vessel maturation, *Nat. Med.*, 9, 685, 2003.
54. Lyden, D. et al., Impaired recruitment of bone-marrow-derived endothelial and hematopoietic precursor cells blocks tumor angiogenesis and growth, *Nat. Med.*, 7, 1194, 2001.
55. Heissig, B. et al., Role of c-kit/Kit ligand signaling in regulating vasculogenesis, *Thromb. Haemost.*, 90, 570, 2003.
56. Heissig, B. et al., Recruitment of stem and progenitor cells from the bone marrow niche requires MMP-9 mediated release of kit-ligand, *Cell*, 109, 625, 2002.
57. Coussens, L.M., Fingleton, B., and Matrisian, L.M., Matrix metalloproteinase inhibitors and cancer: trials and tribulations, *Science*, 295, 2387, 2002.
58. Erlichman, C. et al., Phase I study of the matrix metalloproteinase inhibitor, BAY 12–9566, *Ann. Oncol.*, 12, 389, 2001.
59. O'Reilly, M.S. et al., Regulation of angiostatin production by matrix metalloproteinase-2 in a model of concomitant resistance, *J. Biol. Chem.*, 274, 29568, 1999.
60. Pozzi, A. et al., Elevated matrix metalloprotease and angiostatin levels in integrin alpha 1 knockout mice cause reduced tumor vascularization, *Proc. Natl. Acad. Sci. USA*, 97, 2202, 2000.
61. Maeshima, Y. et al., Tumstatin, an endothelial cell-specific inhibitor of protein synthesis, *Science*, 295, 140, 2002.
62. Colorado, P.C. et al., Anti-angiogenic cues from vascular basement membrane collagen, *Cancer Res.*, 60, 2520, 2000.

63. Pozzi, A., LeVine, W.F., and Gardner, H.A., Low plasma levels of matrix metalloproteinase 9 permit increased tumor angiogenesis, *Oncogene*, 21, 272, 2002.
64. Koolwijk, P. et al., Proteolysis of the urokinase-type plasminogen activator receptor by metalloproteinase-12: implication for angiogenesis in fibrin matrices, *Blood*, 97, 3123, 2001.
65. Moses, M.A., Sudhalter, J., and Langer, R., Identification of an inhibitor of neovascularization from cartilage, *Science*, 248, 1408, 1990.
66. Murphy, A.N., Unsworth, E.J., and Stetler Stevenson, W.G., Tissue inhibitor of metalloproteinases-2 inhibits bFGF-induced human microvascular endothelial cell proliferation, *J. Cell Physiol.*, 157, 351, 1993.
67. Seo, D.W. et al., TIMP-2 mediated inhibition of angiogenesis: an MMP-independent mechanism, *Cell*, 114, 171, 2003.
68. Spurbeck, W.W. et al., Enforced expression of tissue inhibitor of matrix metalloproteinase-3 affects functional capillary morphogenesis and inhibits tumor growth in a murine tumor model, *Blood*, 100, 3361, 2002.
69. Qi, J.H. et al., A novel function for tissue inhibitor of metalloproteinases-3 (TIMP3): inhibition of angiogenesis by blockage of VEGF binding to VEGF receptor-2, *Nat. Med.*, 9, 407, 2003.
70. Wang, M.S. et al., Inhibition of tumor growth and metastasis of human breast cancer cells transfected with tissue inhibitor of metalloproteinase 4, *Oncogene*, 14, 2767, 1997.
71. Jiang, Y. et al., Stimulation of mammary tumorigenesis by systemic tissue inhibitor of matrix metalloproteinase 4 gene delivery, *Cancer Res.*, 61, 2365, 2001.
72. Nelson, A.R. et al., Matrix metalloproteinases: biologic activity and clinical implications, *J. Clin. Oncol.*, 18, 1135, 2000.
73. Rosenbaum, E. et al., Marimastat in the treatment of patients with biochemically relapsed prostate cancer: a prospective randomized, double-blind, phase I/II trial, *Clin. Cancer Res.*, 11, 4437, 2005.
74. Goffin, J.R. et al., Phase I trial of the matrix metalloproteinase inhibitor marimastat combined with carboplatin and paclitaxel in patients with advanced non-small cell lung cancer, *Clin. Cancer Res.*, 11, 3417, 2005.
75. Rizvi, N.A. et al., A phase I study of oral BMS-275291, a novel nonhydroxamate sheddase-sparing matrix metalloproteinase inhibitor, in patients with advanced or metastatic cancer, *Clin. Cancer Res.*, 10, 1963, 2004.
76. Dredge, K., AE-941 *Curr. Opin. Investig. Drugs*, 5, 668, 2004.
77. Latreille, J. et al., Phase I/II trial of the safety and efficacy of AE-941 (Neovastat) in the treatment of non-small-cell lung cancer, *Clin. Lung Cancer*, 4, 231, 2003.
78. Batist, G. et al., Neovastat (AE-941) in refractory renal cell carcinoma patients: report of a phase II trial with two dose levels, *Ann. Oncol.*, 13, 1259, 2002.
79. Wojtowicz-Praga, S. et al., Phase I trial of marimastat, a novel matrix metalloproteinase inhibitor, administered orally to patients with advanced lung cancer, *J. Clin. Oncol.*, 16, 2150, 1998.
80. Bergers, G. et al., Effects of angiogenesis inhibitors on multistage carcinogenesis in mice, *Science*, 284, 808, 1999.
81. Bertolini, F. et al., Maximum tolerable dose and low-dose metronomic chemotherapy have opposite effects on the mobilization and viability of circulating endothelial progenitor cells, *Cancer Res.*, 63, 4342, 2003.
82. Zucker, S., Cao, J., and Chen, W.T., Critical appraisal of the use of matrix metalloproteinase inhibitors in cancer treatment, *Oncogene*, 19, 6642, 2000.
83. Ploug, M. et al., Cellular receptor for urokinase plasminogen activator. Carboxyl-terminal processing and membrane anchoring by glycosyl-phosphatidylinositol, *J. Biol. Chem.*, 266, 1926, 1991.
84. Hajjar, K.A., Cellular receptors in the regulation of plasmin generation, *Thromb. Haemost.*, 74, 294, 1995.
85. Miles, L.A. et al., Plasminogen receptors: the sine qua non of cell surface plasminogen activation, *Front Biosci.*, 10, 1754, 2005.
86. Lopez-Alemany, R. et al., Inhibition of cell surface mediated plasminogen activation by a monoclonal antibody against alpha-enolase, *Am. J. Hematol.*, 72, 234, 2003.
87. Parkkinen, J. and Rauvala, H., Interactions of plasminogen and tissue plasminogen activator (t-PA) with amphoterin. Enhancement of t-PA-catalyzed plasminogen activation by amphoterin, *J. Biol. Chem.*, 266, 16730, 1991.

88. Huttunen, H.J. and Rauvala, H., Amphoterin as an extracellular regulator of cell motility: from discovery to disease, *J. Intern. Med.*, 255, 351, 2004.
89. Conese, M. and Blasi, F., Urokinase/urokinase receptor system: internalization/degradation of urokinase–serpin complexes: mechanism and regulation, *Biol. Chem. Hoppe Seyler*, 376, 143, 1995.
90. Czekay, R.P. et al., Direct binding of occupied urokinase receptor (uPAR) to LDL receptor-related protein is required for endocytosis of uPAR and regulation of cell surface urokinase activity, *Mol. Biol. Cell*, 12, 1467, 2001.
91. Pepper, M.S. et al., Vascular endothelial growth factor (VEGF) induces plasminogen activators and plasminogen activator inhibitor-1 in microvascular endothelial cells, *Biochem. Biophys. Res. Commun.*, 181, 902, 1991.
92. Pepper, M.S. et al., Vascular endothelial growth factor (VEGF)-C synergizes with basic fibroblast growth factor and VEGF in the induction of angiogenesis in vitro and alters endothelial cell extracellular proteolytic activity, *J. Cell. Physiol.*, 177, 439, 1998.
93. Kroon, M.E. et al., Urokinase receptor expression on human microvascular endothelial cells is increased by hypoxia: implications for capillary-like tube formation in a fibrin matrix, *Blood*, 96, 2775, 2000.
94. Vogten, J.M. et al., The role of the fibrinolytic system in corneal angiogenesis, *Angiogenesis*, 6, 311, 2003.
95. Oh, C.W., Hoover-Plow, J., and Plow, E.F., The role of plasminogen in angiogenesis in vivo, *J. Thromb. Haemost.*, 1, 1683, 2003.
96. Heymans, S. et al., Inhibition of plasminogen activators or matrix metalloproteinases prevents cardiac rupture but impairs therapeutic angiogenesis and causes cardiac failure, *Nat. Med.*, 5, 1135, 1999.
97. Devy, L. et al., The pro- or antiangiogenic effect of plasminogen activator inhibitor 1 is dose dependent, *FASEB J.*, 16, 147, 2002.
98. Bajou, K. et al., The plasminogen activator inhibitor PAI-1 controls in vivo tumor vascularization by interaction with proteases, not vitronectin. Implications for antiangiogenic strategies, *J. Cell Biol.*, 152, 777, 2001.
99. Bugge, T.H. et al., Growth and dissemination of Lewis lung carcinoma in plasminogen-deficient mice, *Blood*, 90, 4522, 1997.
100. Ramos-DeSimone, N. et al., Activation of matrix metalloproteinase-9 (MMP-9) via a converging plasmin/stromelysin-1 cascade enhances tumor cell invasion, *J. Biol. Chem.*, 274, 13066, 1999.
101. Min, H.Y. et al., Urokinase receptor antagonists inhibit angiogenesis and primary tumor growth in syngeneic mice, *Cancer Res.*, 56, 2428, 1996.
102. Guo, Y. et al., A peptide derived from the nonreceptor binding region of urokinase plasminogen activator (uPA) inhibits tumor progression and angiogenesis and induces tumor cell death in vivo, *FASEB J.*, 14, 1400, 2000.
103. Bu, X. et al., Species-specific urokinase receptor ligands reduce glioma growth and increase survival primarily by an antiangiogenesis mechanism, *Lab. Invest.*, 84, 667, 2004.
104. Plow, E.F. et al., Plasminogen and cell migration in vivo, *Fibrinolysis Proteol.*, 13, 49, 1999.
105. Wei, Y. et al., Identification of the urokinase receptor as an adhesion receptor for vitronectin, *J. Biol. Chem.*, 269, 32380, 1994.
106. Sato, Y. et al., Indispensable role of tissue-type plasminogen activator in growth factor-dependent tube formation of human microvascular endothelial cells in vitro, *Exp. Cell Res.*, 204, 223, 1993.
107. Stack, M.S. et al., Angiostatin inhibits endothelial and melanoma cellular invasion by blocking matrix-enhanced plasminogen activation, *Biochem. J.*, 340, 77, 1999.
108. Aimes, R.T. et al., Endothelial cell serine proteases expressed during vascular morphogenesis and angiogenesis, *Thromb. Haemost.*, 89, 561, 2003.
109. Hooper, J.D. et al., Type II transmembrane serine proteases. Insights into an emerging class of cell surface proteolytic enzymes, *J. Biol. Chem.*, 276, 857, 2001.
110. Szabo, R. et al., Type II transmembrane serine proteases, *Thromb. Haemost.*, 90, 185, 2003.
111. Netzel-Arnett, S. et al., Membrane anchored serine proteases: a rapidly expanding group of cell surface proteolytic enzymes with potential roles in cancer, *Cancer Metastasis Rev.*, 22, 237, 2003.
112. Foekens, J.A. et al., Plasminogen activator inhibitor-1 and prognosis in primary breast cancer, *J. Clin. Oncol.*, 12, 1648, 1994.

113. Chambers, S.K., Ivins, C.M., and Carcangiu, M.L., Plasminogen activator inhibitor-1 is an independent poor prognostic factor for survival in advanced stage epithelial ovarian cancer patients, *Int. J. Cancer*, 79, 449, 1998.
114. Foekens, J.A. et al., The urokinase system of plasminogen activation and prognosis in 2780 breast cancer patients, *Cancer Res.*, 60, 636, 2000.
115. Foekens, J.A. et al., Urokinase-type plasminogen activator and its inhibitor PAI-1: predictors of poor response to tamoxifen therapy in recurrent breast cancer, *J. Natl. Cancer Inst.*, 87, 751, 1995.
116. Bajou, K. et al., Absence of host plasminogen activator inhibitor 1 prevents cancer invasion and vascularization, *Nat. Med.*, 4, 923, 1998.
117. Bajou, K. et al., Host-derived plasminogen activator inhibitor-1 (PAI-1) concentration is critical for in vivo tumoral angiogenesis and growth, *Oncogene*, 23, 6986, 2004.
118. Chapman, H.A., Plasminogen activators, integrins, and the coordinated regulation of cell adhesion and migration, *Curr. Opin. Cell Biol.*, 9, 714, 1997.
119. Deng, G. et al., Is plasminogen activator inhibitor-1 the molecular switch that governs urokinase receptor-mediated cell adhesion and release? *J. Cell Biol.*, 134, 1563, 1996.
120. Loskutoff, D.J. et al., Regulation of cell adhesion by PAI-1, *APMIS*, 107, 54, 1999.
121. Kjoller, L. et al., Plasminogen activator inhibitor-1 represses integrin- and vitronectin-mediated cell migration independently of its function as an inhibitor of plasminogen activation, *Exp. Cell Res.*, 232, 420, 1997.
122. Isogai, C. et al., Plasminogen activator inhibitor-1 promotes angiogenesis by stimulating endothelial cell migration toward fibronectin, *Cancer Res.*, 61, 5587, 2001.
123. Stefansson, S. et al., Inhibition of angiogenesis in vivo by plasminogen activator inhibitor-1, *J. Biol. Chem.*, 276, 8135, 2001.
124. Ertongur, S. et al., Inhibition of the invasion capacity of carcinoma cells by WX-UK1, a novel synthetic inhibitor of the urokinase-type plasminogen activator system, *Int. J. Cancer*, 110, 815, 2004.
125. Setyono-Han, B. et al., Suppression of rat breast cancer metastasis and reduction of primary tumour growth by the small synthetic urokinase inhibitor WX-UK1, *Thromb. Haemost.*, 93, 779, 2005.
126. Reuning, U. et al., Urokinase-type plasminogen activator (uPA) and its receptor (uPAR): development of antagonists of uPA/uPAR interaction and their effects in vitro and in vivo, *Curr. Pharm. Des.*, 9, 1529, 2003.
127. Mazar, A.P., The urokinase plasminogen activator receptor (uPAR) as a target for the diagnosis and therapy of cancer, *Anticancer Drugs*, 12, 387, 2001.
128. DeClerck, Y.A., Interactions between tumour cells and stromal cells and proteolytic modification of the extracellular matrix by metalloproteinases in cancer, *Eur. J. Cancer*, 36, 1258, 2000.

11 Prostaglandins and COX-2: Role in Antiangiogenic Therapy

Kristin Hennenfent, Daniel Morgensztern, and Ramaswamy Govindan

CONTENTS

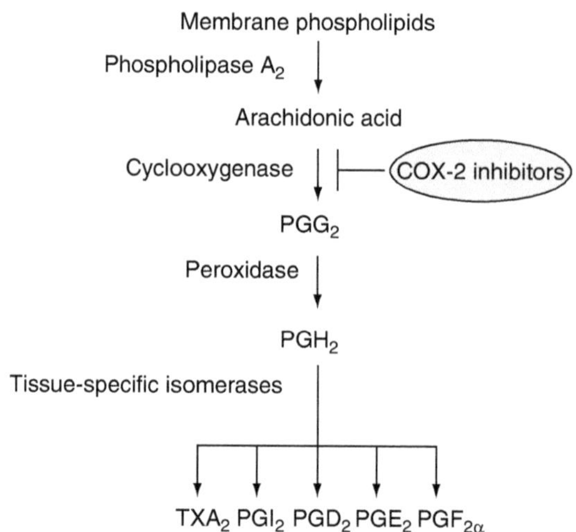

FIGURE 11.1 The prostaglandin pathway. Once the arachidonic acid is liberated from the membrane phospholipids by the phospholipase A_2, it is converted to prostaglandin G_2 (PGG_2) by cyclooxygenase-1 and Cyclooxygenase-2. PGG_2 is subsequently converted to prostaglandin H_2 (PGH_2) by peroxidase and then undergoes the action of tissue-specific isomerases to originate multiple prostanoids including prostaglandin I_2 (PGI_2), prostaglandin D_2 (PGD_2), prostaglandin E_2 (PGE_2), prostaglandin $F_{2\alpha}$ ($PGF_{2\alpha}$), and thromboxane A_2 (TXA_2).

Several lines of evidence suggest that cyclooxygenase-2 (COX-2) is a rational target for anticancer therapy. Epidemiologic data have shown that chronic use of nonsteroidal anti-inflammatory agents (NSAIDs) reduce the risk of a number of solid malignancies, including colorectal and esophageal carcinoma.[1] NSAIDs inhibit cyclooxygenase, a family of enzymes critical for arachidonic acid metabolism (Figure 11.1).[2] The cyclooxygenase enzymes convert arachidonic acid to prostaglandin H_2 (PGH_2) and subsequently thromboxane (TXA_2) and prostaglandins PGE_2, PGF_2, PGD_2, and PGI_2.[3] The constitutive cyclooxygenase-1 (COX-1) isozyme is present in most normal tissues to control normal physiologic functions, including maintenance of the gastrointestinal mucosa, regulation of renal blood flow, and platelet aggregation. In contrast, the COX-2 enzyme isozyme inducible by cytokines and growth factors is often present in inflammatory conditions and cancer.[4]

Conventional NSAIDs inhibit both COX-1 and COX-2. Long-term use for chronic inflammatory conditions such as osteoarthritis has been associated with gastrointestinal and renal toxicity due to chronic inhibition of the COX-1-derived prostaglandins. This finding led to the emergence of selective COX-2 inhibitors over the past decade for the treatment of these conditions (Figure 11.2). At therapeutic doses in humans, these pharmacologic agents do not lower protective gastrointestinal prostaglandins and are less likely to be associated with the severe toxicities associated with chronic use of traditional NSAIDs.[5]

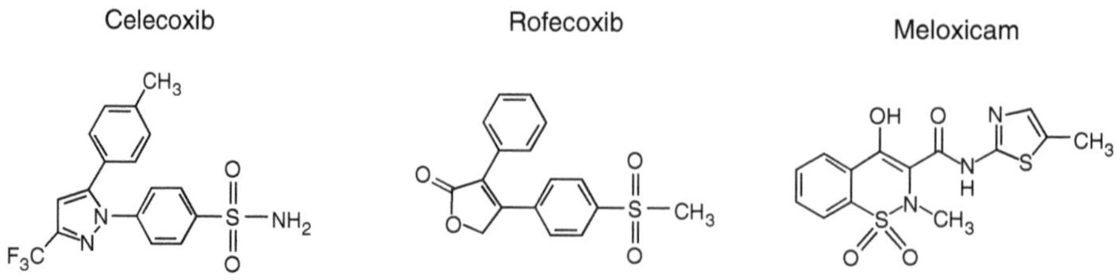

FIGURE 11.2 Selected cyclooxygenase-2 inhibitors.

Human solid tumors, including colorectal, breast, prostate, and lung carcinomas, express COX-2, whereas normal quiescent vascular endothelial cells solely express COX-1.[2,6] COX-2 overexpression has been linked to increased tumor growth and angiogenesis, thereby suggesting the potential role of COX-2 inhibitors for both chemoprevention and therapeutic treatment.[2] Celecoxib and rofecoxib have demonstrated potent antiangiogenic effects in animal models both in vitro and in vivo.[2,7,8] In addition, celecoxib effectively decreases the number and size of adenomatous polyps, known angiogenesis-dependent lesions, in patients with familial adenomatous polyposis.[9] The role of COX-2 in tumor angiogenesis and the rationale for targeting COX-2 to treat human malignancies are discussed in this chapter.

11.1 ROLE OF PROSTAGLANDINS AND COX-2 IN TUMOR ANGIOGENESIS

Within the tumor microenvironment, the maintenance of a vascular supply is essential for tumor growth and metastasis.[10] COX-2 triggers prostaglandin synthesis and results in new capillary (further explanation follows later). In experimental corneal angiogenesis, Leahy and colleagues demonstrated that COX-2 expression was present in branching capillaries, but absent in pre-existing limbal vessels surrounding the cornea.[2] Moreover, in chronic arthritis, COX-2 is localized to the expanding endothelium, macrophages, and synovial fibroblasts of the inflamed joint in vivo.[11] In experimental models using mice null for COX-2, tumor growth and vascular density were dramatically reduced in COX-$2^{-/-}$, but not COX-$1^{-/-}$ mice.[12] COX-2 is expressed in the human tumor vasculature.[6] These observations suggest that COX-2 plays an important role in tumor angiogenesis.

COX-2 is a key regulator of tumor angiogenesis through multiple mechanisms including inducing eicosanoid prostaglandin products (e.g., TXA_2, PGE_2, PGI_2), increasing production of vascular endothelial growth factor (VEGF), upregulating the antiapoptotic protein Bcl-2 (mechanism described later), upregulation of matrix metalloproteinases (MMP), and activating the epidermal growth factor receptor (EGFR) for downstream angiogenic effects. However, the major role of COX-2 in tumor angiogenesis is thought to stem from stimulation of eicosanoid prostaglandin products, specifically TXA_2, PGE_2, and PGI_2, which subsequently stimulate the expression of proangiogenic factors.[3,13] Defining the signaling mechanisms by which prostanoids stimulate tumorigenesis is an active area of investigation.

11.1.1 Influence of Eicosanoid Prostaglandin Products in Tumor Angiogenesis

Eicosanoid products of arachidonic acid metabolism are associated with new blood vessel formation at various points in the complex pathway. In vitro, TXA_2 stimulates endothelial cell motility and capillary tube-like structures, while basic fibroblast growth factor (bFGF) increases TXB_2 production, the active metabolite of TXA_2.[2,14,15] Inhibition of its synthesis suppresses endothelial cell migration stimulated by proangiogenic factors, such as bFGF or VEGF.[16] Likewise, local delivery of a TXA_2-antagonist inhibited blood vessel growth in a mouse model.[15] Notably, COX-2 inhibition of endothelial cell migration can be reversed with TXA_2-mimetic agent.[15]

Prostaglandin E_2 (PGE_2), the most abundant prostanoid detected in most epithelial malignancies, induces angiogenesis by enhancing VEGF production through interaction with its receptor (EP). In fact, in $EP_3^{-/-}$ receptor xenograft models, there was a substantial reduction in tumor angiogenesis.[17] In $APC^{\Delta 716}$ mice (a xenograft model of familial adenomatous polyposis (FAP)) homozygous deletion of the EP_2 receptor caused a significant reduction in the number and size of intestinal polyps through antiangiogenic activities.[18] On activation of the EP_2 receptor, PGE_2 stimulates VEGF production via protein kinase A and ERK2/JNK1 signaling pathways, an effect that is suppressed by an EP_2 selective receptor antagonist.[19,20] A significant correlation between COX-2 expression and tumor vascularization ($p = 0.007$) as well as between COX-2 and microvessel density ($p = 0.007$) was demonstrated in patients with head and neck cancers.[21] In addition to TXA_2 and PGE_2, prostacyclin (PGI_2) is released when endothelial cells are exposed to bFGF or VEGF and then regulates endothelial sprouting and VEGF-induced vascular permeability.[22–24]

11.1.2 Prosurvival Factors and Tumor Angiogenesis

Beyond the impact of increased VEGF production on tumor angiogenesis, COX-2-mediated resistance to apoptosis is thought to contribute to the increased tumorigenic potential of COX-2 overexpressing cells.[14] Increased production of Bcl-2, a family of anti-apoptotic proteins, is correlated with COX-2 overexpression.[25] Bcl-2 stimulates angiogenic activity and its expression is upregulated by both COX-2 and PGE_2.[26] Notably, Bcl-2-expressing human vascular endothelial cells spawn a persistent angiogenic state and are refractory to the apoptotic effects of thrombospondin-1, an endogenous angiogenesis inhibitor.[26,27] Selective COX-2 inhibitors induce apoptosis in colorectal carcinoma cell lines.[28,29] Downregulation of Bcl-2 by COX-2 inhibitors may account for these findings.[30]

11.1.3 Induction of Matrix Metalloproteinases and Tumor Angiogenesis

Matrix metalloproteinases, specifically MMP-2 and MMP-9, are zinc-dependent endopeptidases that may play a role in tumor and vascular invasion.[31] Recent data from in vitro mouse models null for gelatinase A (MMP-2) and gelatinase B (MMP-9) suggest that these proteins may be associated with tumor neovascularization.[32] These effects, however, have not been tested in vivo. PGE_2 interacts with the EP4 receptor to promote MMP-2 expression, an effect that can be negated by PGE_2 antagonism.[33] Even though the currently available MMP inhibitors did not show any benefit in clinical trials, in preclinical models the MMP inhibitor (AG3340) demonstrated reduction in the number of blood vessels in human non-small cell lung cancer (NSCLC) by up to 77% in a dose-dependent manner as evidenced by CD31 staining.[34]

11.1.4 EGFR-Mediated Tumor Angiogenesis

The EGFR is activated and potentiated by PGE_2.[35,36] EGFR activation subsequently upregulates proangiogenic factors, including VEGF and interleukin-8 (IL-8).[37–39] Selective inhibition of either COX-2 or EGFR may potentially mitigate these effects.

11.2 TARGETING ANGIOGENESIS WITH SELECTIVE COX-2 INHIBITORS

Available preclinical evidence suggests that COX-2 inhibitors will suppress carcinogenesis. COX-2 inhibitors commonly used for pain management, including celecoxib (Celebrex, Pfizer, Inc., New York, NY) and rofecoxib (Vioxx, Merck & Co., Whitehouse Station, NJ), are among the pharmacologic agents evaluated as antitumor agents with antiangiogenic properties. In numerous pharmacological studies involving animal models, treatment with selective COX-2 inhibitors reduced the growth and metastases of a variety of experimental tumors.[40] However, tumor regression has rarely been demonstrated, thereby suggesting that the usefulness of selective COX-2 inhibitors as single agents will likely be limited. The utility of these agents, in all likelihood, will remain in combination with standard antineoplastic modalities. In fact, selective COX-2 inhibitors are evaluated with concurrent chemotherapy and radiotherapy in patients with cancers of the lung, colon, breast, esophagus, pancreas, liver, cervix, and brain.[41]

11.2.1 Exploring Preclinical Data

11.2.1.1 COX-2 Inhibitors with Radiation Therapy

In response to cellular injury by ionizing radiation, normal cells upregulate certain proteins, including COX-2 and PGE_2, in an effort to survive fatal damage.[42] COX-2 and PGE_2 are produced in response to phospholipase A_2 activation by radiation-induced injury in a dose-dependent manner. Furuta and Barkley reported that indomethacin improved the antitumor effects of radiation on murine fibrosarcoma while sparing normal tissues.[43] Subsequently, greater synergistic effects were noted when a selective COX-2 inhibitor, SC-236, was administered for the same malignancy.[42,44] These findings have been replicated in human glioma xenografts.[45] In a preclinical study NS-398, a specific COX-2 inhibitor functioned as a radiation sensitizer preferentially in colon cancer xenografts that overexpressed COX-2.[46] These findings suggest a potential usefulness of COX-2 inhibitors to augment radiation therapy in patients with epithelial malignancies.

11.2.1.2 COX-2 Inhibitors with Chemotherapy

The incremental benefits of COX-2 inhibitors with conventional chemotherapeutic agents on tumor regression have been demonstrated in a number of preclinical and clinical models. Witters and colleagues reported that combination therapy with docetaxel and the COX-2 inhibitor SC-236 more effectively inhibited the growth of MCF-7 human breast cancer cells than either agent when used alone.[47] Hida and Muramatsu reported the additive effects of the specific COX-2 inhibitor nimesulide when used in combination with docetaxel, irinotecan, cisplatin, and etoposide on cultured NSCLC tumor cell and normal lung epithelial cell lines.[48] In this model, the addition of nimesulide at clinically achievable concentrations reduced the IC_{50} of the chemotherapeutic agents up to 77%. The greatest benefit was achieved in those cell lines with the highest expression of COX-2. Interestingly, the sensitivity of normal epithelial cells to the chemotherapeutic agents did not change in the presence of nimesulide. Thus, concurrent use of a selective COX-2 inhibitor may improve chemosensitivity to conventional chemotherapeutic agents.

The taxanes and vinca alkaloids, microtubule-targeting chemotherapeutic agents, are believed to stimulate COX-2 transcription and thereby decrease their intrinsic efficacy.[49] This finding provides additional rationale for combination therapy with selective COX-2 inhibitors and conventional chemotherapy. In addition to the previously discussed findings by Witters and colleagues in $HER2/neu^{+}$ breast cancer cell lines treated with docetaxel, other reports of enhanced in vitro chemosensitivity by lung and colon cancer cell lines to doxorubicin, vincristine, etoposide, and paclitaxel when given in combination with either a selective or nonselective inhibitor of COX-2 support these findings.[50–52]

In addition to the synergistic effects observed in vitro, there is ample evidence to suggest these effects occur in animal models of human malignancies. In the experimental Lewis model of metastasizing murine lung cancer and HT-29 colon carcinoma, celecoxib inhibited tumor growth and the number and size of metastatic lung lesions in a dose-dependent fashion.[6] Combination therapy with celecoxib and cyclophosphamide retarded tumor growth in the Lewis model by 97%. In contrast, monotherapy with celecoxib and cyclophosphamide produced an 84% and 35% tumor growth inhibition, respectively.[53] Similar synergistic effects were demonstrated in the colon cancer xenograft models; concurrent administration of celecoxib and fluorouracil inhibited tumor growth by 80%, compared with 35% for fluorouracil and 69% for celecoxib as single agents.[53]

11.2.2 Impact of COX-2 Inhibitors in Cancer Therapy: What Have We Learned from Clinical Trials?

11.2.2.1 COX-2 Inhibitors with Radiation Therapy

Based on preclinical information, clinical studies have begun to investigate the role of COX-2 inhibitors as radiosensitizers in the treatment of human malignancies, including NSCLC and rectal cancer. Carbone and colleagues initiated a phase II trial of celecoxib 400 mg twice daily in combination with concurrent weekly chemoradiation (paclitaxel 50 mg/m^2, carboplatin AUC 2, and 63 Gy chest radiation) for stage III NSCLC.[54] The authors reported that serum/plasma levels of VEGF declined at 2, 5, and 7 months following treatment in 8 of the 9 patients enrolled at the time of data presentation. These findings suggest that celecoxib may cause a time-dependent reduction in circulating angiogenic markers. Prostaglandin levels are measured to more fully understand the role of COX-2 inhibition in this patient population. In a retrospective analysis, Adler and colleagues explored the radiosensitizing activity of neoadjuvant celecoxib (200 mg twice daily), oxaliplatin, and capecitabine in combination with radiation therapy (5040 Gy over 5.5 weeks) for the treatment of advanced, operable rectal cancer.[55] At the time of preliminary reporting, 10 patients had been enrolled and treated; 9 patients underwent surgery; and 5 (50%) had downstaging of their disease. This regimen was well tolerated; no dose reductions or treatment delays resulted from drug toxicities. A prospective phase II evaluation of this therapeutic regimen is underway.

11.2.2.2 COX-2 Inhibitors with Chemotherapy

Several recent clinical studies have investigated the antitumor activity of the selective COX-2 inhibitors with standard chemotherapeutic agents or targeted therapy in the treatment of a number of solid tumors, including carcinoma of the breast, lung, and gastrointestinal tract (Table 11.1).

11.2.3 Breast Cancer

11.2.3.1 Neoadjuvant Therapy

Chow and Toi evaluated the addition of celecoxib to the combination consisting of 5-florouracil, epirubicin, and cyclophosphamide (FEC) in the neoadjuvant treatment of locally advanced breast cancer. A total of 31 patients were treated with three cycles of chemotherapy with (FEC-C) or without celecoxib 400 mg po bid. Both regimens were well tolerated. The response rate for the 16 patients treated with the FEC-C regimen was 81% compared with 62.5% in the 15 patients who were treated with FEC alone. Both the clinical complete response (cCR) and pathological complete response (pCR) were higher in patients treated with the addition of celecoxib (18.8% and 12.5% compared with 6.3% and 6.3%). In patients treated with FEC-C, response rates were associated with a significantly higher level of COX-2 gene expression compared with nonresponders.[56]

11.2.3.2 Adjuvant Therapy

The National Cancer Institute of the Canada Clinical Trials Group (NCIC CTG) has recently opened accrual to a study comparing the efficacy of anastrazole and exemestane in the adjuvant setting. After completing 5 years of therapy, patients will be randomized to celecoxib 400 mg po bid or placebo for an additional 3 years.

TABLE 11.1
Clinical Studies with the Combination of Celecoxib and Systemic Therapy

Author	Tumor	Patients	Additional Therapy	Response *n* (%)
Chow[56]	Breast (neoadjuvant)	16	5-FU/Epi/Cy (FEC)	2 pCR (12.5), 3 cCR (19), 10 cPR (62.5)
Dang[57]	Breast (>second line)	11	Trastuzumab	No clinical responses
Altorki[84]	Lung (neoadjuvant)	28	Carb/Pac	5 CR (18), 14 PR (50)
Milella[59]	Lung (neoadjuvant)	19	Gem/Cis	3 CR (16%), 7 PR (37%), 2 pCR (10%)
Chaplen[60]	Lung (first line)	19	Doc	1 CR (5), 3 PR (14)
Burton[61]	Lung (first line)	44	Carb/Gem	9 RR (20%)
Nishiyama[62]	Lung (first line)	27	Carb/Pac[a]	1 CR (4%), 11 PR (41%)
Csiki[57]	Lung (second line)	13	Doc	2 PR (15%)
Gadgeel[64]	Lung (second line)	20	Doc	2 PR (10)
Nugent[65]	Lung (second line)	22	Doc	1 PR (4.5)
Stani[85]	Lung (second line)	43	Pac	1 CR (2), 10 PR (23)
Keresztes[67]	Lung (second line)	51	Irin/Doc or Irin/Gem	2 PR (4)
O'Byrne[68]	Lung (second line)	31	Gefitinib[b]	1 CR (3%), 2 PR (6%)
Gadgeel[69]	Lung (second line)	10	Gefitinib	2 PR (20%)
Vervenne[72]	Esophageal (first line)	14	Gefitinib	No clinical responses
Govindan[70]	Esophageal (first line)	18	Irin/cisplatin/RT	7 pCR
Blanke[73]	Colorectal (first line)	18	IFL	5 PR (28)
El-Rayes[74]	Colorectal (first line)	38	Cap/Irin	16 PR (41)
Maiello[75]	Colorectal (first line)	27	FOLFIRI	1 CR (4%), 9 PR (33%)
Kobrossy[76]	Pancreas (first line)	22	Gem/Cis	2 PR (9)
Kerr[77]	Pancreas (first line)	17	Gem/Irin	3 PR (18%)
Millela[78]	Pancreas (second line)	17	5-FU	2 PR (12)
Kasimis[79]	Prostate (first line)	27	Doc	PSA 11/27 (41), soft tissue 6/18 (33)
Rini[80]	Renal (first line)	20	Interferon	2 PR (10)
Pannullo[81]	Brain (refractory or relapsed)	29	Temozolamide	5 PR (17)

Cap represents capecitabine, Carb represents carboplatin, Cy represents cyclophosphamide, Doc represents docetaxel, Epi represents epirubicin, Irin represents irinotecan, LV represents leucovorin, Oxa represents oxaliplatin, Pac represents paclitaxel, Cis represents cisplatin, RT represents radiation therapy.

[a] Meloxican.

[b] Rofecoxib.

11.2.3.3 Metastatic Disease

Based on preclinical studies indicating a link between the overexpression of Her-2/neu and the activity of COX-2, Dang and colleagues evaluated the combination of celecoxib and trastuzumab in patients with metastatic breast cancer overexpressing Her-2/neu, who had progressive disease on trastuzumab. Despite good tolerability, there were no responses among the 11 evaluable patients.[57]

11.2.4 Lung Cancer

11.2.4.1 Neoadjuvant Therapy

Altorki and colleagues treated 29 patients with stages I to IIIA NSCLC with neoadjuvant therapy consisting of carboplatin, paclitaxel, and celecoxib. Grade 3 or 4 neutropenia was observed in 18 patients (62%). Of the 28 patients evaluated for response, 5 patients achieved CR (17%), 14 patients had partial response PR (48%), and 8 patients had stable disease SD (28%). Furthermore, treatment with celecoxib 400 mg twice daily was sufficient to normalize the increase in PGE_2 levels found in NSCLC patients after treatment with paclitaxel and carboplatin.[58]

Milella and colleagues treated 19 stage III NSCLC patients with a combination of fixed-dose gemcitabine, cisplatin, and celecoxib as induction therapy. Three patients achieved clinical CR (16%), 7 patients achieved PR (37%), and 7 patients had SD (37%). Of the 9 patients who underwent subsequent surgery, 2 patients had pathologic CR. Grade 3 neutropenia occurred in 6 patients.[59]

11.2.4.2 First-Line Therapy for Metastatic Disease

Chaplen and colleagues evaluated the combination of celecoxib and weekly docetaxel in advanced NSCLC patients older than 70 years or with poor performance status. Four of the 19 evaluable patients achieved tumor response including one who had a complete response. The median time to progression and survival time for the elderly patients were 3.9 months and 8.6 months, respectively. This therapy was well tolerated with no grade 3 or 4 toxicities.[60]

The combination of celecoxib with carboplatin and gemcitabine was evaluated as first-line therapy in 49 patients with NSCLC. Response rates were seen in 9 out of the 44 evaluable patients (20%) whereas 19 patients had disease stabilization for more than 4 months. Overall survival was 31.2% at one year and 12.5% at two years.[61]

Nishiyama and colleagues evaluated the addition of meloxicam to the combination of paclitaxel and carboplatin as the first-line therapy in 27 patients with NSCLC. The overall response rate was 44% with 1 CR (4%) and 11 PR (41%). The most common significant toxicity was grade 3 or 4 neutropenia, which occurred in 6 patients.[62]

11.2.4.3 Second-Line Therapy in Metastatic Disease

Csiki and colleagues reported the preliminary results of a phase II trial of combination therapy with celecoxib (400 mg twice daily) and taxotere 75 mg/m^2 intravenously thrice weekly in 15 patients with advanced NSCLC after cisplatin failure. The majority of the enrolled patients had previously failed a prior taxane-containing regimen. Of the 13 evaluable patients, 2 patients (15.4%) achieved a partial response and 3 patients (23%) had disease stabilization. Pre- and postcelecoxib intratumoral PGE_2 levels demonstrated a marked decline after therapy (100.7 to 18.1 ng/g, respectively). These data provide clinical evidence that celecoxib abrogates intratumoral COX-2 resulting in intratumoral PGE_2 changes.[63] Two similar studies were subsequently performed. Gadgeel and colleagues used a combination of docetaxel and celecoxib to treat 20 patients with NSCLC progressing after a cisplatin-based therapy. Two patients had a confirmed PR (10%) and the median time to progression was 10 weeks. Median overall survival had not been reached but at the time of presentation it was greater than 6 months. Although the nonhematologic toxicities were mild, 14 patients developed grade 3 or 4 neutropenia and 5 patients (25%) had neutropenic fever.[64] Nugent and colleagues treated 27 patients with the same combination. Of the 22 evaluable patients, 1 had PR (4.5%) and 18 had SD (82%). The Kaplan–Meier estimates of median TTP and median OS were 19.6 weeks and 39.3 weeks, respectively. Three patients had neutropenic fever, with one death resulting from sepsis.[65] Stani and colleagues reported a 25% response rate with the combination of paclitaxel and celecoxib as second-line therapy.[66]

Keresztes and colleagues evaluated the addition of celecoxib to a combination of irinotecan with docetaxel or gemcitabine as second-line therapy in NSCLC. Celecoxib was associated with increased hematologic toxicity without any increase in clinical efficacy. Of the 51 patients treated with either regimen in combination with celecoxib, only 2 (4%) responded to therapy. In contrast, responses were seen in 4 out of 46 patients (9%) treated without the addition of celecoxib.[67]

The addition of rofecoxib to gefitinib as second-line therapy in 31 patients with NSCLC produced responses in 3 patients, including one complete response and two partial responses. Furthermore, 9 patients achieved stable disease for a combined disease control rate of 39%.[68] Two patients (20%) achieved PR and 3 patients (30%) out of 10 patients achieved SD in a small prospective study using celecoxib and gefitinib in the second- or third-line setting.[69]

11.2.5 Esophageal Cancer

The Hoosier Oncology Group evaluated the addition of celecoxib to the combination of 5-fluoracil, cisplatin, and radiation therapy in 31 patients with resectable esophageal cancer. Of the 22 patients who underwent surgery, 5 patients achieved pathologic CR.[70] The combination of celecoxib, irinotecan, cisplatin, and radiation was evaluated in patients with locally advanced esophageal cancer. Of the 18 enrolled patients, 14 completed the therapy and underwent complete resection. Pathologic CR was achieved in 7 patients.[71] Lastly, the combination of celecoxib and gefitinib as first-line therapy for advanced esophageal cancer revealed good tolerability, but no radiologic responses in 14 evaluable patients.[72]

11.2.6 Colorectal Cancer

The addition of celecoxib to combination of IFL (irinotecan, 5-fluoracil, and leucovorin) in 23 patients with previously untreated metastatic colorectal cancer produced partial responses in 5 of the 18 patients evaluated for response (28%). Significant neutropenia was seen in 9 patients, with 2 of those developing neutropenic fever. The most common grade 3 or 4 nonhematologic toxicities were diarrhea and nausea, each of them occurring in 5 patients.[73]

El-Rayes evaluated the combination of celecoxib, irinotecan, and capecitabine in 38 patients with metastatic colorectal cancer. In an intention to treat analysis, 41% of the patients achieved objective response with a median duration of 6.5 months. The therapy was well tolerated with no grade 4 toxicities.[74]

Maiello and colleagues evaluated the addition of celecoxib to the folinic acid, 5-fluorouracil, and irinotecan (FOLFIRI) regimen in a phase II study. Whereas 11 out of the 29 (38%) evaluable patients treated with FOLFIRI achieved PR (38%), 9 out of 27 patients treated with the addition of celecoxib achieved PR (33%) and 1 patient had CR (4%).[75]

11.2.7 Pancreatic Cancer

Addition of celecoxib to the combination of cisplatin and fixed-dose ratio gemcitabine failed to produce a significant impact on the survival of previously untreated patients with metastatic pancreatic cancer.[76] In a recent phase II trial, the addition of celecoxib to the combination of gemcitabine and irinotecan was evaluated in 20 patients with metastatic pancreatic cancer. Of the 17 patients evaluable for response, 3 patients achieved PR (18%) and 12 patients had SD (70%). Serological response with decrease in CA 19-9 and CEA was observed in 12 out of 14 and 8 out of 10 patients, respectively. The median and one-year OS were 13 months and 64%, respectively. Thirteen patients reported an improvement in the quality of life.[77] In patients who had progressed on gemcitabine, the combination of infusional 5-fluorouracil and celecoxib produced two partial responses in 17 patients. Therapy was well tolerated and responses lasted 68 weeks and 23 weeks, indicating that this combination is feasible and further studies may be indicated.[78]

11.2.8 Prostate Cancer

The combination of docetaxel and celecoxib cancer has been recently evaluated as first-line therapy in 28 patients with hormone-resistant prostate cancer. Of the 27 evaluable patients,

11 (41%) had PSA response with normalization of the levels in 3 patients. Furthermore, radiologic responses were seen in 6 out of 18 patients with soft tissue mass, including 3 patients with complete response.[79]

11.2.9 Renal Cell Carcinoma

Rini and colleagues evaluated the combination of interferon-alpha and celecoxib in patients with metastatic renal cell carcinoma. Although there was no increase in the toxicity compared with the expected side effects from interferon alone, only 2 out of the 20 treated patients achieved PR.[80]

11.2.10 Brain

Pannullo and colleagues used the combination of temozolamide and celecoxib in 36 patients with refractory or relapsed glioblastoma multiforme. A total of 36 patients were treated with temozolomide at a loading dose of 200 mg/m^2 followed by 9 doses of 90 mg/m^2 bid for 5 days every 28 days. Celecoxib was given to a maximum dose of 480 mg/m^2 for 10 days. This combination was well tolerated and the hematologic side effects did not recur following dose reduction. Of the 29 patients evaluated for response after two cycles, 5 patients achieved PR (17%) and 21 (72.5%) achieved SD. One patient achieved CR after seven cycles. The 6-month PFS and OS were 35% and 83%, respectively.[81]

11.2.11 Melanoma

Wilson and Allan studied the efficacy of celecoxib in 24 patients with surgically incurable and recurrent metastatic melanoma. Tumor regression was observed in 5 patients, including 2 CR and 3 mixed PR/SD. Both patients achieving CR had small volume lung metastases and 1 patient had no previous systemic therapy.[82]

The Hellenic Cooperative Oncologic Group evaluated the combination of celecoxib and temozolomide in 52 patients with advanced melanoma. Among the 50 evaluable patients, there were 11 responders (22%), including 5 CR (10%) and 6 PR (12%). An additional 20 patients had stable disease. Therapy was well tolerated with only one patient discontinued from the study due to severe toxicity. The most common side effects included nausea and vomiting (75%), fatigue (46.5%), thrombocytopenia (33%), anemia (27.5%), and neutropenia (17.5%).[83]

In summary, several small phase trials evaluating the combination of celecoxib and chemotherapy or targeted therapy have been recently reported. To date, although this form of therapy is usually well tolerated, there has been no study demonstrating that the addition of celecoxib provides a clear advantage in terms of response rate or survival.

11.3 FUTURE DIRECTIONS

Significant advances have been made in understanding COX-2 biology as well as its role in carcinogenesis over the past decade, but the preliminary results from several small phase II studies have been very unimpressive. However, these studies are flawed for many reasons: small sample size, lack of control arm, comparison only with historic controls, to name a few. The optimal dose of COX-2 inhibitors for cancer therapy is unknown. No dose escalating studies using celecoxib or rofecoxib to define the optimal tolerable dose for cancer therapy have been performed to the best of our knowledge. It is unfortunate that the clinical research in this area was hampered by simultaneous activation of several small studies instead of few large phase III studies that would have identified definitely the role of COX-2 inhibitors in cancer therapy. The recent recognition that COX-2 inhibitors are associated with increased risk of cardiovascular events certainly dampened further investigations using this class of agents. The role of COX-2 inhibitors in cancer therapy remains very uncertain at the moment.

REFERENCES

1. Thun MJ HS, Patrono C. Nonsteroidal antiinflammatory drugs as anticancer agents: mechanistic, pharmacologic, and clinical issues. *J Natl Cancer Inst* 2002; 94:252–266.
2. Leahy KM OR, Yu W, Zweifel BS, Koki AT, Masferrer JL. Cyclooxygenase-2 inhibition by celecoxib reduces proliferation and induces apoptosis in angiogenic endothelial cells *in vivo*. *Cancer Res* 2002; 62:625–631.
3. Gately S LW. Multiple roles of COX-2 in tumor angiogenesis: a target for antiangiogenic therapy. *Semin Oncol* 2004; 31:2–11.
4. Williams CS MM, Dubois RN. The role of cyclooxygenase in inflammation, cancer, and development. *Oncogene* 1999; 18.
5. Silverstein FE FG, Goldstein JL, Simon LS, Pincus T, Whelton A, Makuch R, Eisen G, Agrawal NM, Stenson WF, Burr AM, Zhao WW, Kent JD, Lefkowith JB, Verburg KM, Geis GS. Gastrointestinal toxicity with celecoxib vs nonsteroidal anti-inflammatory drugs for osteoarthritis and rheumatoid arthritis: the CLASS study: a randomized controlled trial. *JAMA* 2000; 284:1247–1255.
6. Masferrer JL LK, Koki AT, Zweifel BS, Settle SL, Woerner BM, Edwards DA, Flickinger AG, Moore RJ, Seibert K. Antiangiogenic and antitumor activities of cyclooxygnase-2 inhibitors. *Cancer Res* 2000; 60:1306–1311.
7. Dicker AP WT, Grant DS. Targeting angiogenic processes by combination rofecoxib and ionizing radiation. *Am J Clin Oncol* 2001; 24:438–442.
8. Harris RE AG, Abou-Issa H. Chemoprevention of breast cancer in rats by celecoxib, a cyclooxygenase 2 inhibitor. *Cancer Res* 2000; 60:2101–2103.
9. Steinbach G LP, Phillips RK. The effect of celecoxib, a cyclooxygenase-2 inhibitor in familial adenomatous polyposis. *N Engl J Med* 2000; 342:1946–1952.
10. Folkman J. Tumor angiogenesis: therapeutic implications. *N Engl J Med* 1971; 285:1182–1186.
11. Sano H EA. *In vivo* cyclooxygenase expression in synovial tissues of patients with rheumatoid arthritis and osteoarthritis and rats with adjuvant and streptococcal cell wall arthritis. *J Clin Invest* 1992; 89:97–108.
12. Williams CS TM, Reese J. Host cyclooxygenase-2 modulates carcinoma growth. *J Clin Invest* 2000; 105:1589–1594.
13. Tsujii M KS, DuBois RN. Cyclooxygenase regulates angiogenesis induced by colon cancer cells. *Cell* 1998; 93:705–716.
14. Tsujii M DR. Alterations in cellular adhesion and apoptosis in endothelial cells overexpressing prostaglandin endoperoxide synthase 2. *Cell* 1995; 83:493–501.
15. Daniel DA LH, Morrow JD, Crews BC, Marnett LJ. Thromboxane A2 is a mediator of cyclooxygenase-2 dependent endothelial migration and angiogenesis. *Cancer Res* 1999; 59:4574–4577.
16. Nie D LM, Zacharek A. Thromboxane A2 regulation of endothelial cell migration, angiogenesis, and tumor metastasis. *Biochem Biophys Res Commum* 2000; 267:245–251.
17. Amano H HI, Endo H, Kitasato H, Yamashina S, Maruyama T, Kobayashi M, Satoh K, Narita M, Sugimoto Y. Host prostablandin E2-EP3 signaling regulates tumor-associated angiogenesis and tumor growth. *J Exp Med* 2003; 197:221–232.
18. Sonoshita M TK, Sasaki N, Sugimoto Y, Ushikubi F, Narumiya S, Oshima M, Taketo MM. Acceleration of intestinal polyposis through prostaglandin receptor EP2 in APC716 knockout mice. *Nat Med* 2001; 7:1048–1051.
19. Inoue H TM, Shimoyama Y. Regulation by PGE(2) of the production of interleukin-6, macrophage colony stimulating factor, and vascular endothelial growth factor in human synovial fibroblasts. *Br J Pharmacol* 2002; 136:287–295.

20. Cheng T CW, Wen R. Prostaglandin E2 induces vascular endothelial growth factor and basic fibroblast growth factor mRNA expression in cultured rat Muller cells. *Invest Opthalmol Vis Sci* 1998; 39:581–591.
21. Gallo O FA, Magnelli L, Sardi I, Vannacci A, Boddi V, Chiarugi V, Masini E. Cyclooxygenase-2 pathway correlates with VEGF expression in head and neck cancer: implications for tumor angiogenesis and metastasis. *Neoplasia* 2001; 3:53–61.
22. Zachary I. Signaling mechanisms mediating vascular protective actions of vascular endothelial growth factor. *Am J Physiol Cell Physiol* 2001; 280:C1375–C1386.
23. Kuwashima L GJ, Glaser BM. Stimulation of endothelial cell prostacyclin release by retina-derived factors. *Invest Opthalmol Vis Sci* 1988; 29:1213–1220.
24. Murohara T HJ, Silver M. Vascular endothelial growth factor/vascular permeability factor enhances vascular permeability via nitric oxide and prostacyclin. *Circulation* 1998; 97:99–107.
25. Liu CH CS, Narko K. Overexpression of cyclooxygenase-2 is sufficient to induce tumorigenesis in transgenic mice. *J Biol Chem* 2001; 276:18563–18569.
26. Nor JE CJ, Mooney DJ. Vascular endothelial growth factor (VEGF)-mediated angiogenesis is associated with enhanced endothelial cell survival and induction of Bcl-2 expression. *Am J Pathol* 1999; 154:375–384.
27. Nor JE CJ, Liu J. Up-regulation of Bcl-2 in microvascular endothelial cells enhances intratumoral angiogenesis and accelerates tumor growth. *Cancer Res* 2001; 61:2183–2188.
28. Arico S PS, Bauvy C. Celecoxib induces apoptosis by inhibiting 3-phosphoinositide-dependent protein kinase-1 activity in the human colon cancer TH-29 cell line. *J Biol Chem* 2002; 277.
29. Elder DJ HD, Playle LC. The MEK/ERK pathway mediates COX-2-selective NSAID-induced apoptosis and induced COX-2 protein expression in colorectal carcinoma cells. *Int J Cancer* 2002; 99:323–327.
30. Sheng H SJ, Morrow JD. Modulation of apoptosis and Bcl-2 expression by prostaglandin E2 in human colon cancer cells. *Cancer Res* 1998; 58:362–366.
31. Nelson AR RB, Rothenberg ML. Matrix metalloproteinases: biologic activity and clinical implications. *J Clin Oncol* 2000; 18:1135–1149.
32. Itoh T TM, Yoshida H. Reduced angiogenesis and tumor progression in gelatinase A-deficient mice. *Cancer Res* 1998; 58:1048–1051.
33. Dohadwala M BR, Luo J. Autocrine/paracrine prostaglandin E2 production by non-small cell lung cancer cells regulates matrix metalloproteinase-2 and CD44 in cyclooxygenase-2-dependent invasion. *J Biol Chem* 2002; 277:50828–50833.
34. Shalinsky DR ZH, McDermott CD. AG3340, a selective MMP inhibitor, has broad antiangiogenic activity across oncology and opthalmology models in vivo. *Proc Am Assoc Cancer Res* 1999 (abstr); 40:66.
35. Pai R SB, Szabo IL. Prostaglandin E2 transactivates EGF receptor: a novel mechanism for promoting colon cancer growth and gastrointestinal hypertrophy. *Nat Med* 2002; 8:289–293.
36. Buchanan FG WD, Bargiacchi F, DuBois RN. Prostaglandin E2 regulates cell migration via the intracellular activation of the epidermal growth factor receptor. *J Biol Chem* 2003; 278:35451–35457.
37. Russell KS SD, Polverini PJ. Neuregulin activation of ErbB receptors in vascular endothelium leads to angiogenesis. *Am J Physiol* 1999; 277:H2205–H2211.
38. Kedar D BC, Killion JJ. Blockade of the epidermal growth factor receptor signaling inhibits angiogenesis leading to regression of human renal cell carcinoma growth orthotopically in nude mice. *Clin Cancer Res* 2002; 8:3592–3600.
39. Pore N LS, Haas-Kogan DA. PTEN mutation and epidermal growth factor receptor activation regulate vascular endothelial growth factor (VEGF) mRNA expression in human glioblastoma cells by transactivating the proximal VEGF promoter. *Cancer Res* 2003; 63:236–241.
40. Subbaramaiah K DA. Cyclooxygenase 2: a molecular target for cancer prevention and treatment. *Trends Pharmacol Sci* 2003; 24:96–102.
41. Gasparini G LR, Sarmiento R, Morabito A. Inhibitors of cyclo-oxygenase 2: a new class of anticancer agents? *Lancet Oncol* 2003; 4:605–615.
42. Kishi K PS, Petersen C. Preferential enhancement of tumor radioresponse by a cyclooxygenase-2 inhibitor. *Cancer Res* 2000; 60:1326–1331.

43. Furuta Y HN, Barkley T. Increase in radioresponse of murine tumors by treatment with indomethacin. *Cancer Res* 1988; 48:3008–3013.
44. Milas L KK, Hunter N. Enhancement of tumor response to gamma-radiation by an inhibitor of cyclooxygenase-2 enzyme. *J Natl Cancer Inst* 1999; 91:1501–1504.
45. Petersen C PS, Milas L. Enhancement of intrinsic tumor cell radiosensitivity induced by a selective cyclooxygenase-2 inhibitor. *Clin Cancer Res* 2000; 6:2513–2520.
46. Pyo H CH, Amorino GP. A selective cyclooxygenase-2 inhibitor, NS-398, enhances the effect of radiation in vitro and in vivo preferentially on the cells that express cyclooxygenase-2. *Clin Cancer Res* 2001; 7:2998–3005.
47. Witters L LK, Sachdeva K, Lipton A, Bremer M. Inhibition of a HER-2/neu transfected human breast cancer cell line with a COX-2 inhibitor. *Proc Am Soc Clin Oncol* 2001 (abstr); 20.
48. Hida T KK, Muramatsu H. Cyclooxygenase-2 inhibitor induces apoptosis and enhances cytotoxicity of various anticancer agents in non-small cell lung cancer cell lines. *Clin Cancer Res* 2000; 6:2006–2011.
49. Subbaramaiah K HJ, Norton L. Microtubule-interfering agents stimulated the transcription of cyclooxygenase-2. Evidence for involvement of ERK1/2 and p38 mitogen-activated protein kinase pathways. *J Biol Chem* 2000; 275:14838–14845.
50. Nokihara H NY, Shinohara T. Critical role of cyclooxygenase-2 in sensitivity of human lung cancer cells to anticancer agents. *Proc Am Assoc Cancer Res* 1999 (abstr); 40:26.
51. Chodkiewicz C GJ, Ling YH. Cyclooxygenase-2 (COX-2) expression and in vitro and in vivo sensitivity to paclitaxel in non small cell lung cancer (NSCLC). *Proc Am Assoc Cancer Res* 2000 (abstr); 41:867.
52. Singh MS SM. Potentiation of selective COX-2 inhibitor nimesulide against colon and lung cancer cell lines. *Proc Am Assoc Cancer Res* 2001 (abstr); 49:926.
53. Moore RJ ZB, Heuvelman DM. Enhanced antitumor activity by co-adminstration of celecoxib and the chemotherapeutic agents cyclophosphamide and 5-FU. *Proc Am Assoc Cancer Res* 2002 (abstr):409.
54. Carbone D CH, Csiki I, Dang T, Campbell N, Garcia B, Morrow J, Saha D, Johnson DH, Sandler A. Serum/plasma VEGF level changes with cyclooxygease-2 (COX-2) inhibition in combined modality therapy in stage III non-small cell lung cancer (NSCLC): preliminary results of a phase II trial (THO-0059). *Proc Am Soc Clin Oncol* 2002 (abstr):1270.
55. Adler WM LB, Heywood G, Wong G, Lee F. CEX (celecoxib, oxaliplatin, capecitabine) and XRT: a novel neoadjuvant approach for locally advanced rectal cancer. *Proc Am Soc Clin Oncol* 2005 (abstr):3744.
56. Chow L WC, Toi M. Combination therapy of cytotoxic agents and COX-2 inhibitor in primary systemic therapy of breast cancer patients: a phase II study on efficacy and cardiac toxicity. *J Clin Oncol (Meeting Abstracts)* 2005; 23:3146.
57. Dang CT, Dannenberg AJ, Subbaramaiah K, et al. Phase II study of celecoxib and trastuzumab in metastatic breast cancer patients who have progressed after prior trastuzumab-based treatments. *Clin Cancer Res* 2004; 10:4062–4067.
58. Altorki NK, Keresztes RS, Port JL, et al. Celecoxib, a selective cyclo-oxygenase-2 inhibitor, enhances the response to preoperative paclitaxel and carboplatin in early-stage non-small-cell lung cancer. *J Clin Oncol* 2003; 21:2645–2650.
59. Milella M, Ceribelli A, Gelibter A, et al. Celecoxib combined with fixed dose-rate gemcitabine (FDR-Gem)/CDDP as induction chemotherapy for stage III non-small cell lung cancer (NSCLC). *J Clin Oncol (Meeting Abstracts)* 2005; 23:7324.
60. Chaplen RA, Kalemkerian GP, Wozniak A, et al. Celecoxib (CEL) and weekly docetaxel (DOC) in elderly or PS2 patients (pts) with advanced non-small cell lung cancer (NSCLC). *J Clin Oncol (Meeting Abstracts)* 2004; 22:7102.
61. Burton JD, El-Sayah D, Cherry M, et al. Results of a phase I/II trial of carboplatin/gemcitabine plus celecoxib for first-line treatment of stage IIIB/IV non-small cell lung cancer (NSCLC). *J Clin Oncol (Meeting Abstracts)* 2005; 23:7250.
62. Nishiyama O, Taniguchi H, Kondoh Y, et al. Meloxicam, a selective cyclooxygenase-2 (COX-2) inhibitor, enhances the response to carboplatin and weekly paclitaxel in advanced non-small cell lung cancer (NSCLC). *J Clin Oncol (Meeting Abstracts)* 2005; 23:7312.

63. Csiki I DT, Gonzalez A, Sandler A, Carbone D, Choy H, Campbell N, Garcia B, Morrow J, Johnson DH. Cyclooxygenase-2 (COX-2) inhibition + docetaxel in recurrent non-small cell lung cancer (NSCLC): preliminary results of a phase II trial (THO-0054). *J Clin Oncol* 2002 (abstr):1187.
64. Gadgeel S, Thatai L, Kraut M, et al. Phase II study of celecoxib and docetaxel in non-small cell lung cancer (NSCLC) patients with progression after platinum-based therapy. *Proc Am Soc Clin Oncol* 2003; 22:684 (abstr 2749).
65. Nugent F, Graziano N, Levitan R, et al. Docetaxel and COX-2 inhibition with celecoxib in relapsed/refractory non-small cell lung cancer (NSCLC): promising progression-free survival in a phase II study. *Proc Am Soc Clin Oncol* 2003; 22:671 (abstr 2697).
66. Stani SC, Carillio G, Meo S, et al. Phase II study of celecoxib and weekly paclitaxel in the treatment of pretreated advanced non-small cell lung cancer (NSCLC). *J Clin Oncol (Meeting Abstracts)* 2004; 22:7337.
67. Keresztes RS, Socinski M, Bonomi P, et al. Phase II randomized trial of irinotecan/docetaxel (ID) or irinotecan/gemcitabine (IG) with or without celecoxib (CBX) in 2nd-line treatment of non-small-cell lung cancer (NSCLC). *J Clin Oncol (Meeting Abstracts)* 2004; 22:7137.
68. O'Byrne K CL, Dunlop D, Ranson M, Danson S, Botwood N, Carbone D. Combination therapy with gefitinib (ZD1839) and rofecoxib in platinum-pretreated relapsed non-small cell lung cancer (NSCLC). *Proc Am Soc Clin Oncol* 2004 (abstr):7101.
69. Gadgeel SM, Shehadeh NJ, Ruckdeschel JC, et al. Gefitinib and celecoxib in patients with platinum refractory non-small cell lung cancer (NSCLC). *J Clin Oncol (Meeting Abstracts)* 2004; 22:7094.
70. Govindan R, McLeod H, Mantravadi P, et al. Cisplatin, fluorouracil, celecoxib, and RT in resectable esophageal cancer: preliminary results. *Oncology (Williston Park)* 2004; 18:18–21.
71. Enzinger P, Mamon H, Bueno R, et al. Phase II cisplatin, irinotecan, celecoxib and concurrent radiation therapy followed by surgery for locally advanced esophageal cancer. *Proc Am Soc Clin Oncol* 2003; 22:361 (abstr 1451).
72. Vervenne WL, Bollen JM, Bergman JJGHM, et al. Evaluation of the antitumor activity of gefitinib (ZD1839) in combination with celecoxib in patients with advanced esophageal cancer. *J Clin Oncol (Meeting Abstracts)* 2004; 22:4054.
73. Blanke C, Benson III A, Dragovich T, et al. A phase II trial of celecoxib (CX), irinotecan (I), 5-fluoracil (5FU), and leucovorin (LCV) in patients (pts) with unresectable or metastatic colorectal cancer (CRC). *Proc Am Soc Clin Oncol* 2002; 21:127a (abstr 505).
74. El-Rayes BF, Zalupski MM, Shields AF, et al. A phase II trial of celecoxib, irinotecan, and capecitabine in metastatic colorectal cancer. *J Clin Oncol (Meeting Abstracts)* 2005; 23:3677.
75. Maiello E, Giuliani F, Romito S, et al. Phase II randomized study of FOLFIRI vs FOLFIRI + celecoxib in first line treatment of advanced colorectal cancer (ACRC). *J Clin Oncol (Meeting Abstracts)* 2005; 23:3656.
76. Kobrossy B, El-Rayes BF, Shields AF, et al. A phase II study of gemcitabine by fixed-dose rate infusion, cisplatin, and celecoxib in metastatic pancreatic cancer. *J Clin Oncol (Meeting Abstracts)* 2004; 22:4120.
77. Kerr S, Campbell C, Legore K, et al. Phase II trial of gemcitabine and irinotecan plus celecoxib in advanced adenocarcinoma of the pancreas. *J Clin Oncol (Meeting Abstracts)* 2005; 23:4155.
78. Milella M, Gelibter A, Di Cosimo S, et al. Pilot study of celecoxib and infusional 5-fluorouracil as second-line treatment for advanced pancreatic carcinoma. *Cancer* 2004; 101:133–138.
79. Kasimis B, Cogswell J, Hwang S, et al. High dose celecoxib (C) and docetaxel (D) in patients (pts) with hormone resistant prostate cancer (HRPC). Results of an ongoing phase II trial. *J Clin Oncol (Meeting Abstracts)* 2005; 23:4704.
80. Rini B, Weinberg V, Cadman B, et al. A phase II trial of interferon-alpha and celecoxib in metastatic renal carcinoma: clinical and anti-angiogenic effects. *J Clin Oncol (Meeting Abstracts)* 2004; 22:4604.
81. Pannullo S, Hariharam S, Serventi J, et al. Phase I/II trial of a twice-daily regimen of temozolomide and celecoxib for treatment of relapsed/refractory glioblastoma multiforme and anaplastic astrocytoma. *J Clin Oncol (Meeting Abstracts)* 2004; 22:1549.
82. Wilson KS, Allan SJ. Clinical activity of celecoxib (CXB) in metastatic malignant melanoma (MMM). *J Clin Oncol (Meeting Abstracts)* 2005; 23:7555.
83. Gogas H, Polyzos A, Tsoutsos D, et al. Temozolomide in combination with celecoxib in patients with advanced melanoma. A phase II study of the Hellenic Cooperative Oncology Group. *J Clin Oncol (Meeting Abstracts)* 2005; 23:7554.

84. Altorki NK KR, Port JL. Celecoxib, a selective cyclooxygenase-2 inhibitor, enhances the response to preoperative paclitaxel and carboplatin in early-stage non-small cell lung cancer. *J Clin Oncol* 2003; 21:2645–2650.
85. Stani S, Carillio G, Meo S, et al. Phase II study of celecoxib and weekly paclitaxel in the treatment of pretreated advanced non-small cell lung cancer (NSCLC). *Proc Am Soc Clin Oncol* 2004; 23:697 (abstr 7337).

12 Integrins, Adhesion, and Coadhesion Inhibitors in Angiogenesis

Abebe Akalu and Peter C. Brooks

CONTENTS

The critical importance of communication between cells and the noncellular microenvironment in regulating development and homeostasis has been appreciated for decades. Both cell–cell and cell–matrix interactions are known to regulate tissue morphogenesis, differentiation, and organ architecture. New molecular insight into bidirectional communication between components of the extracellular matrix (ECM) and cells has led to a more complete understanding of the ability of cells to sense their local surroundings, transfer this molecular information from outside the cell to the cells interior, interpret this information, and ultimately initiate appropriate cellular responses. As the flow of molecular information from outside to inside is clearly of great importance, the transfer of biochemical information from cells to the ECM and thus, the ability of cells to alter their local environment, is also of great significance. A key molecular component that functions to integrate and connect this bidirectional communication system is a conserved family of cell adhesion receptors termed integrins (1–5). Integrins represent a large family of heterodimeric integral membrane proteins composed of noncovalently associated α and β chains. Not surprisingly, members of this family have been identified in nearly all cells from such divergent species as sponges to man (1–5).

Given the importance of cell–ECM interactions in controlling either directly or indirectly nearly all cellular behavior, it is not surprising that cell–ECM interactions play critical roles in regulating a number of pathological processes including aberrant angiogenesis, tumor growth, and metastasis. This chapter focuses primarily on cell adhesive interactions in controlling angiogenesis or the development of new blood vessels from preexisting vessels. In particular, we discuss the general biochemical and molecular mechanisms by which cells initiate contact with the ECM and cooperatively interact with other regulatory factors to control neovascularization. In addition, we examine how selective targeting of these adhesive events may lead to the development of novel inhibitors for controlling malignant tumor progression. New basic, translational, and clinical research on angiogenesis has lead to the recent development and approval of some of the first new antiangiogenic compounds such as Avastin (bevacizumab) and Macugen (pegaptanib sodium), drugs that target vascular endothelial growth factor (VEGF) (6,7). With continued work and a more detailed molecular understanding of the roles that cellular interactions with the ECM play in angiogenesis, additional inhibitors that selectively target adhesive processes are likely to make their way into the clinical testing. In fact, several drugs targeting integrin receptors are currently evaluated in various stages of clinical trials including, Vitaxin and Ciligatide, inhibitors of integrin $\alpha_v\beta_3$ and M200, an inhibitor of $\alpha_5\beta_1$ (8–10).

12.1 GENERAL COMPOSITION AND STRUCTURAL ORGANIZATION OF THE ECM

To begin to understand the significance of adhesive cellular events in regulating pathological processes such as tumor angiogenesis, molecular insight into the composition and structural organization of the ECM is essential. Evidence from both ECM and integrin mutant or null mice suggests a crucial role for bidirectional communication between cells and ECM (5,11). Alterations in the expression and functions of ECM molecules such as collagen, fibronectin, and laminin have been shown to impact angiogenesis, since mice deficient in or harboring specific mutations in these proteins exhibit an array of defects including vascular abnormalities (5,11). While an exhaustive discussion of the components, functions, and detailed structural organization of the ECM would be well beyond the scope of this chapter, we briefly summarize some of the major structural components and features of the ECM with regard to specific ligands for integrin receptors. Moreover, we discuss these ECM molecules in terms of their importance in regulating bidirectional signals required for new blood vessel formation.

In simplistic terms, the ECM can be thought of as being organized into two major interconnected compartments, which include the thin sheetlike network of the basement membrane (BM) and the loosely arranged interstitial matrix (Figure 12.1). In general, BMs are composed of two major sheetlike structures of multidomain glycoprotein composed predominately of collagen type-IV and laminin, which are interconnected with a variety of proteoglycans such as nidogen and perlecan (12–14). Distinct types of collagen including type-VIII, XV, and XVIII have also been suggested to be present as well as several matricellular proteins. Though the specific composition and the structural integrity of the BM can vary

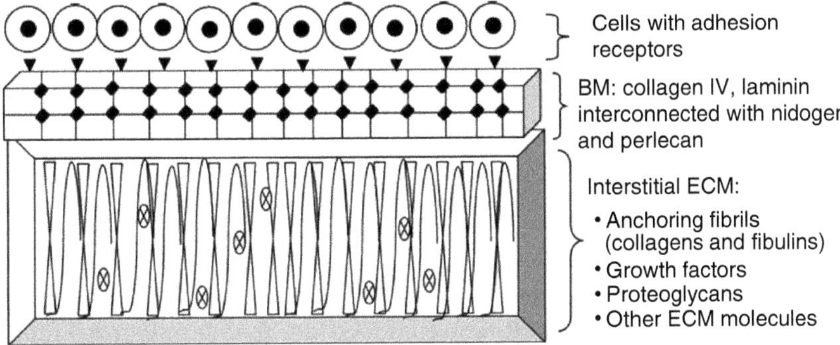

FIGURE 12.1 General structure and function of the extracellular matrix. A schematic diagram of the general structural organization of the ECM. The major components of the basement membrane (BM) include collagen type-IV, laminin, nidogen, and perlecan. The major components of the interstitial matrix include an array of fiber-forming collagens, fibronectin, vitronectin, and numerous proteoglycans.

between different tissue compartments and during distinct developmental and pathological processes, in general they are found underlying all epithelial sheets and blood vessels. Moreover, BMs not only provide mechanical support for cells and tissues but also provide critical binding sites for other ECM proteins, integrin receptors, growth factors, and cell surface proteoglycans. Integrin-mediated cellular interactions with these BM components are known to regulate an array of signaling pathways including PI3 kinase/Akt, MAPK/ERK cascades, and the JNK pathways, which play crucial roles in angiogenesis including modulation of cell adhesion, migration, invasion, proliferation, survival, and differentiation (15–18).

In addition to the BM, the surrounding interstitial matrix is also of great importance in regulating angiogenesis. The interstitial matrix can be thought of as a loosely organized gel-like composition of collagenous and noncollagenous glycoprotein, proteoglycans, and matricellular proteins. Some of the major ubiquitous components of the interstitial matrix include distinct forms of collagen such as type-I, II, and III as well as fibronectin, fibrinogen, thrombospondin, vitronectin, and elastin (12–14). Multiple protein-binding sites within these ECM molecules provide an elaborate interconnecting mesh-type structure, which allows the creation of unique microenvironments that control different cellular processes. For example, studies have shown that the interstitial matrix can function as a reservoir of regulatory molecules including angiogenic growth factors, cytokines, motility factors, proteolytic enzymes, and protease inhibitors (12–14). In this regard, studies have led to the identification of two VEGF-binding domains within fibronectin, which function to enhance the biological activity of VEGF (19). New studies have indicated that specific matrix metalloproteinases (MMPs) may cleave matrix-immobilized VEGF, thereby releasing soluble fragments that play distinct roles in vascular patterning (20). For example, cleaved VEGF was shown to stimulate capillary dilation of existing blood vessels whereas the MMP-resistant VEGF promoted outgrowth and branching of blood vessels (20). Thus, the coordinated storage and temporal release of these regulatory factors can significantly influence distinct aspects of angiogenesis and tumor progression. In addition to proteolytic release of growth factors, a number of antiangiogenic fragments of ECM proteins may be released such as tumstatin and endostatin as well as soluble proangiogenic fragments from laminin (14,21). As one might expect, this complex gel-like network of proteins is known to regulate a wide diversity of signaling pathways as well as providing important binding sites for a multitude of integrins. Thus, the cooperative interactions of proteases, ECM components, integrins, and growth factors function together to regulate the balance of angiogenesis inducers and inhibitors, which ultimately control new blood vessel development.

12.2 ANGIOGENESIS AND CELL ADHESION RECEPTORS

The field of vascular biology has grown extensively over the past decade, due in part to the realization of the importance that neovascularization has in both normal as well as pathological processes. Elegant studies in the field of vascular biology have revealed at least three general mechanisms by which new blood vessels form. The first, vasculogenesis, is the process by which functional blood vessels develop from precursor cells called angioblasts, which are derived from the primitive mesoderm (22,23). The second mechanism is called arteriogenesis and can be defined as the process of activation, dilation, and remodeling of preexisting small nonfunctional vessels into functional ones (24). Interestingly, the recent findings concerning the roles of MMP-cleaved VEGF in modulating vessel dilatation may shed new light on the molecular mechanisms regulating arteriogenesis (20,24).

Angiogenesis (Figure 12.2) is the process by which new blood vessels sprout from preexisting vessels (25,26). Angiogenesis is regulated by controlling the balance and functional bioavailability of inhibitors and inducers by a series of interconnected molecular events. Shifts in these mechanisms in favor of inducers (angiogenic switch) ultimately result in the formation of new blood vessels. Interestingly, recent studies have suggested that endothelial progenitor cells may also contribute to angiogenesis by incorporating into the growing vessel (27,28).

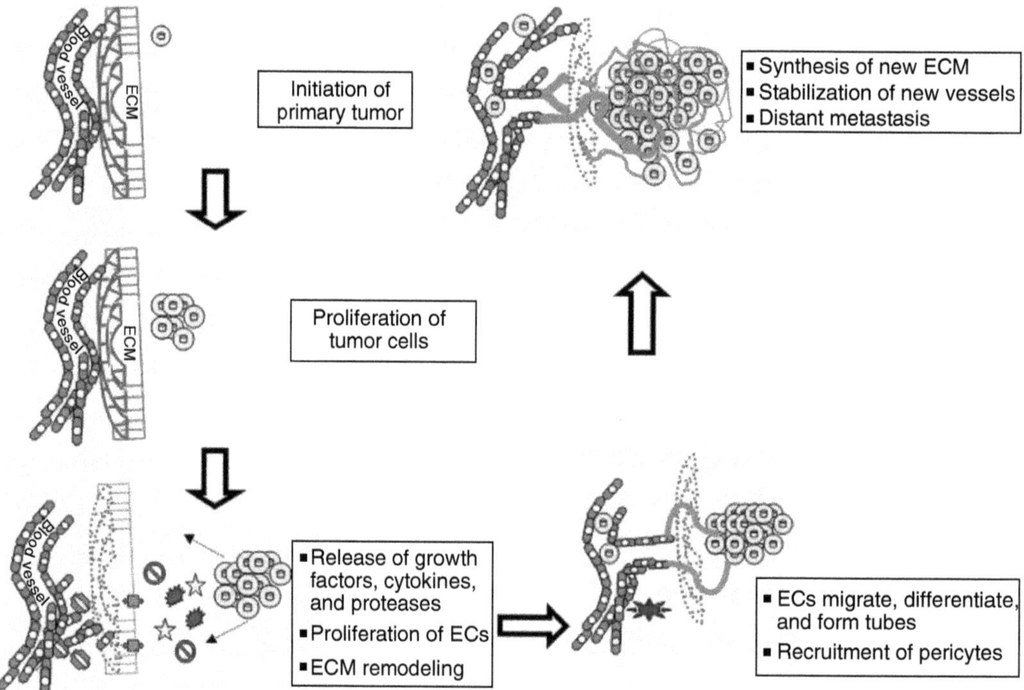

FIGURE 12.2 The angiogenic cascade. Schematic overview of major events involved in angiogenesis. Angiogenesis involves a complex series of interconnected steps, which can be organized into three stages including an initiation phase, an invasive and proliferative phase, and a maturation phase. The initiation phase is characterized by release of a variety of proangiogenic factors such as growth factor and cytokines and the downregulation of endogenous inhibitors of angiogenesis. The invasive and proliferative phase involves proteolytic remodeling of ECM components and local invasion, migration, and proliferation of activated vascular cells. Finally, the maturation phase involves the secretion of new ECM components, recruitment of stabilizing accessory cells, and new cell–cell and cell–ECM interactions. ECs: endothelial cells.

However, to what extent this process contributes to pathological angiogenesis still remains to be determined. Due in large part to space limitations and the focus of this chapter, we do not discuss vasculogenesis and arteriogenesis, but instead, focus on exciting new advances in the molecular understanding of angiogenesis. In particular, we discuss the expanding body of work that demonstrates unique molecular cooperativity between cell adhesion molecules and other regulatory factors that control neovascularization, with a specific focus on integrins and the importance of these molecules as potential therapeutic targets.

The critical importance of blood vessel formation for normal physiological processes such as trophoblast implantation, embryonic development, and wound healing has been appreciated for some time (29–33). However, it was not until the early 1970s that pioneering work by Dr. Judah Folkman suggested that angiogenesis may also play an important role in the regulation of tumor growth (32,33). These early studies stimulated interest within the scientific community that aberrant neovascularization may contribute to a number of pathological processes such as diabetic retinopathy, rheumatoid arthritis, psoriasis, tumor growth, and metastasis (29–33). Recent efforts have now been focused on the identification of molecules that regulate angiogenesis. Some of the more widely studied of these angiogenesis regulatory factors can be grouped into at least five categories, including growth factors and their receptors, cell adhesion molecules, proteolytic enzymes, ECM components, and transcription factors (34,35). Though our chapter is primarily focused on adhesion receptors and ECM molecules, our discussion is not intended to suggest that these are the only factors involved in angiogenesis, nor that they are necessarily the most important, but rather we use these molecules as examples to illustrate the important concept of molecular cooperativity and cellular adhesion in the regulation of neovascularization.

Angiogenesis is a complex series of interconnected molecular, cellular, and biochemical events that ultimately result in the formation of new blood vessels. In general, angiogenesis can be organized into three steps or stages (Figure 12.2), including an initiation phase where angiogenic cytokines and growth factors are released from a variety of sources (29,31,33,36). Growth factors are thought to initiate signal transduction pathways that ultimately activate endothelial cells, stromal cells, and pericytes (37,38). Vascular cell activation results in the acquisition of an invasive phenotype, which promotes cell–cell disassociation, production of matrix-remodeling proteases, and alterations in the expression of cell surface adhesion receptors (29–38). These activated vascular cells remodel BM components, interact with altered matrix molecules, and invade the local interstitium. Once in the interstitial compartment, the vascular cells again remodel their microenvironment and interact with these modified interstitial ECM components. Vascular cell interactions with these remodeled interstitial components lead to morphogenesis and cellular reorganization into tubelike structures. Importantly, new studies have emerged concerning the roles of integrin signaling and ECM components in the regulation of canalization and endothelial tube formation, thus expanding the mechanisms by which integrins regulate angiogenesis (39). In the final maturation phase of angiogenesis, vascular cells express new matrix components, undergo cell–cell interactions with accessory cells such as pericytes, and differentiate into functional blood vessels. These newly formed vessels can then provide nutrients, exchange gases, and remove metabolic wastes for proper maintenance of tissues and organs. Importantly, angiogenesis is not a process in which individual molecules function in isolation, but rather a series of events that functions cooperatively to form new vessels. In this regard, we focus our discussion of angiogenesis in terms of molecular cooperativity and cell adhesion and examine important new insights into the regulation of angiogenesis by altering adhesive functions.

Cell adhesion molecules can be grouped into at least four major families including cadherins, selectins, immunoglobulin (Ig) supergene family members, and integrins (40–43). Interestingly, recent evidence has suggested that certain cell surface receptor tyrosine kinases and various proteoglycans may also contribute to the regulation of cell adhesion, however much less is known concerning the functional impact of these molecules on angiogenesis. Cadherins, selectins, and Ig family members predominately mediate cell–cell interactions, whereas members of the integrin family primarily facilitate cell–matrix interactions (40–43). Integrin receptors, however, can also promote cell–cell interactions (40–43). Exciting new studies have provided evidence that the view of cell adhesion molecules as simply molecular glue may be only a part of their diverse functions during the angiogenic cascade. For example, integrins not only facilitate physical associations between other cells and the ECM, but transmit biochemical information from outside the cell to the interior and from the inside of the cell to the outside (44). Thus, integrins allow cells to sense their immediate extracellular surroundings and in turn, respond to changes in their dynamic microenvironment. These integrin-mediated signaling pathways have been shown to involve changes in intracellular pH, fluxes in calcium concentrations, protein phosphorylation, and regulation of gene expression (44,45). These biochemical changes activate signaling molecules including kinases, phosphatases, and adapter proteins, which in turn regulate processes such as cell adhesion, migration, invasion, gene expression, and differentiation (43–45). Moreover, integrins can bind to a variety of other cell surface and secreted molecules including certain growth factors and their receptors, proteolytic enzymes, and TM4 proteins. In fact, $\alpha_v\beta_3$ was shown to bind to partially activated MMP-2, thereby localizing proteolytic activity to close ECM contact sites (46,47).

Cadherins, selectins, and members of the Ig superfamily are also thought to contribute to signal transduction pathways (48–50). For example, cadherins interact with intracellular accessory proteins called catenins, which help potentiate signaling (48–50). Interestingly, studies show that cell adhesion molecules such as integrins and cadherins may function cooperatively to regulate complex cellular behavior such as migration (48–50). The development of new blood vessels is known to require adhesive processes. In fact, function-blocking antibodies directed to specific adhesion molecules block endothelial tube formation in vitro and angiogenesis in vivo (39–50). While the well-established functions of cell adhesion molecules in promoting adhesive processes play active roles in neovascularization, cooperative interactions with other regulatory molecules such as growth factors and proteolytic enzymes also contribute to the angiogenic cascade.

12.3 STRUCTURAL MODIFICATION OF THE ECM IN CONTROLLING INVASIVE CELLULAR BEHAVIOR

While cellular interactions with native or intact ECM molecules play important regulatory roles in angiogenesis, invasive cellular behavior can also be profoundly influenced by structural and biochemical modification of the ECM (51). In this regard, it is becoming increasingly clear that during many pathological processes such as tumor growth, invasion, metastasis, and angiogenesis, the composition, structural integrity, and biochemical characteristics of the ECM can be significantly modified. Inflammatory cells, stromal fibroblasts, tumor cells, and endothelial cells can all secrete new ECM proteins in variable quantities, thereby significantly altering the ratios of ECM components. These changes in the ratio and composition of ECM molecules can in turn alter the potential pool of integrin ligands within the local microenvironment. Biochemical modifications of existing ECM molecules also occur. For example, ECM proteins have been suggested to be susceptible to biochemical alterations such as glycation (52–54). In fact, studies have shown that glycation of ECM proteins such as collagen and fibronectin can alter the ability of these proteins to polymerize and in addition, can modify specific integrin-binding sites. Studies have demonstrated that glycation of the RGD integrin recognition motif can significantly alter integrin binding resulting in altered cell adhesion (52–54). Given the importance of integrins in controlling signaling pathways and gene expression, it is clear that alterations of either the composition or biochemical characteristics of ECM proteins would result in altered integrin-mediated communication between cells and the ECM.

Finally, one of the most well-studied processes that impacts cellular behavior includes proteolytic remodeling of the ECM. Matrix-degrading enzymes play important roles in regulating invasive cellular processes such as angiogenesis, tumor invasion, and metastasis. In this regard, mice deficient in MMPs such as MMP-2 and MMP-9 have exhibited defects in tumor growth and angiogenesis (55–57). Interestingly, MMP-9 but not MMP-2-deficient mice exhibited reduced exposure of the HUIV26 cryptic collagen epitope known to bind to integrin $\alpha_v\beta_3$ and regulate angiogenesis (58). Moreover, other studies have indicated that naturally occurring tissue inhibitor of metalloproteinases (TIMPs), synthetic inhibitors of MMPs, as well as synthetic inhibitors of serine proteases such as urokinase-type plasminogen activator (uPA) can also inhibit angiogenesis and tumor growth (59,60). However, much work still remains concerning the mechanisms by which these proteolytic enzymes contribute to invasive cell behavior, since early clinical studies of many of these protease inhibitors exhibited little therapeutic efficacy in clinical trials. The limited efficacy of these antiproteolytic strategies is not likely due to the lack of functional significance of ECM degradation, but rather may be associated with the timing of the treatment, dosing, and specificity of the particular drugs. In this regard, studies have suggested that proteases as well as growth factors may play distinct roles at different stages of tumor development, some functioning primarily during early stages and others at later stages. Thus, it is critical to have an in-depth understanding of the temporal expression of a particular therapeutic target during tumor progression to optimize clinical benefit.

Given the importance of cellular communication with the ECM in controlling invasive cell behavior, it is not surprising that structural and biochemical remodeling of the ECM can significantly impact angiogenesis and tumor invasion. Recent work has provided convincing evidence that proteolytic enzymes may regulate invasive cellular behavior by altering ECM barriers, thereby creating a less restrictive microenvironment through which tumor and endothelial cells can migrate (59,60). Other studies have suggested that proteolytic enzymes may not simply destroy ECM molecules, but in addition, selectively cleave or alter their three-dimensional structure thereby exposing cryptic epitopes, which may serve as integrin ligands that regulate motility, proliferation, and gene expression (61,62). Studies have suggested that MMP-mediated exposure of cryptic epitopes within collagen type-IV (HUIV26 and HUI77) and laminin-5 can facilitate adhesion and migration in vitro and angiogenesis and tumor growth in vivo (63–65). While proteolytic enzymes may regulate invasive cellular behavior by exposing active matrix-immobilized cryptic ECM epitopes (MICEE), a variety of additional mechanisms also exist. For example, studies have suggested that proteolytic enzymes such as MMP-9 may facilitate the release of matrix-sequestered angiogenic factors such as VEGF, thereby regulating the angiogenic switch (66). Moreover, the ability of certain MMPs to cleave or not to cleave matrix-bound VEGF may modulate its functional activity from facilitating outgrowth and branching to dilation of existing vessels (20). In addition, activated MMP-2 has been shown to bind integrins such as $\alpha_v\beta_3$ helping to localize proteolytic activity to close cell–ECM sites, thereby potentiating ECM remodeling. It is also possible that this $\alpha_v\beta_3$-mediated localization of MMP-2 could promote release of matrix-sequestered growth factors. Localization of proteolytic enzymes to the cell surface is not only restricted to MMP-2, but other studies have suggested that additional secreted MMPs including MMP-1 and MMP-9 may be localized to the cell surface through integrins as well as other proteins such as CD44 (67,68). Moreover, serine proteases such as uPA are known to bind to the cell through specific cell surface receptors. To this end, the uPA receptor (uPAR) has been shown to associate with a number of integrins including $\alpha_3\beta_1$, $\alpha_5\beta_1$, $\alpha_v\beta_3$, and $\alpha_{II}\beta\beta_3$ (69–71). It is clear from these limited examples that proteolytic enzymes can function cooperatively with integrin receptors to modulate ECM remodeling and thereby regulate angiogenesis. Thus, cooperative interactions between integrins, ECM components, proteases, and growth factors represent potential areas for therapeutic intervention for controlling angiogenesis and invasive cell behavior.

12.4 ANGIOGENESIS INHIBITORS FROM THE EXTRACELLULAR MATRIX

Recently, many laboratories have been focused on the generation and release of endogenous angiogenesis inhibitors. A great number of these naturally occurring inhibitors appear to be derived from fragments of ECM proteins (72,73). For example, our laboratory as well as others has shown that specific fragments of collagen type-IV can inhibit angiogenesis and tumor growth in a number of models (73–76). Studies have demonstrated that α_1(IV)NC1 (arrestin), α_2(IV)NC1 (canstatin), α_3(IV)NC1 (tumstatin), and α_6(IV)NC1 all inhibit angiogenesis (73–76). Importantly, these angiogenesis inhibitors have been shown to bind to integrin receptors including $\alpha_v\beta_3$ and a number of β_1-containing integrins (73–76). Many other inhibitors of angiogenesis are also thought to be derived from ECM proteins including endostatin the NC domains from collagen type-XVIII, restin the NC domain from collagen type-XV, and endorepellin a fragment from the ECM protein perlecan. Importantly, generation of these antiangiogenic ECM fragments may be facilitated by proteolytic enzymes such as MMPs. Thus, the possibility exists that dramatically reducing MMPs with broad spectrum MMP inhibitors such as those that have recently been examined in clinical trials may alter the endogenous levels of naturally occurring angiogenesis regulator, thereby shifting the balance of inhibitors and inducers in vivo.

In addition to the variety of angiogenesis inhibitors described earlier, many other examples of either small peptides or domains from ECM proteins including collagen type-I, fibronectin, fibrinogen, laminin, and thrombospondin have been reported (73–76). While an in-depth description and analysis of the potential mechanisms of action of these inhibitors would be beyond the focus of our discussion, the general descriptions of these ECM-derived inhibitors are useful in emphasizing the critical importance of adhesive interactions in controlling angiogenesis. It also clear from the brief descriptions that integrins play a central role in the function of these inhibitors given the facts that the majority of these angiogenesis inhibitors specifically interact with one or more integrins.

12.5 INTEGRIN STRUCTURE AND FUNCTIONS AS KEY INTEGRATORS IN THE BIDIRECTIONAL COMMUNICATION BETWEEN CELLS AND THE ECM

As can be appreciated from the discussions earlier, the bidirectional communication between cells and their local microenvironment plays crucial roles in regulating angiogenesis. The integrin family of cell adhesion receptors represents one of the most well-studied families of molecules, which mediate cell–ECM interactions. In the 1980s, pioneering work carried out by a number of investigators, notably Drs. Hynes, Ruoslahti, and Springer, began to examine the functional importance of cell–ECM interactions in regulating complex cellular processes (77–79). These early studies paved the way to the isolation and eventually cloning of some of the first integrins described such as the fibronectin receptor $\alpha_5\beta_1$ and the platelet fibrinogen receptor $\alpha_{II}\beta\beta_3$ (79,80). Sequence analysis in conjunction with functional studies rapidly leads to the realization that these interesting molecules represented members of a conserved group of integral membrane proteins. Early studies on these receptors lead to the discovery that these heterodimers connect ECM molecules such as fibronectin to the cytoskeletal (80). These observations provided important evidence for the concept that cells communicate

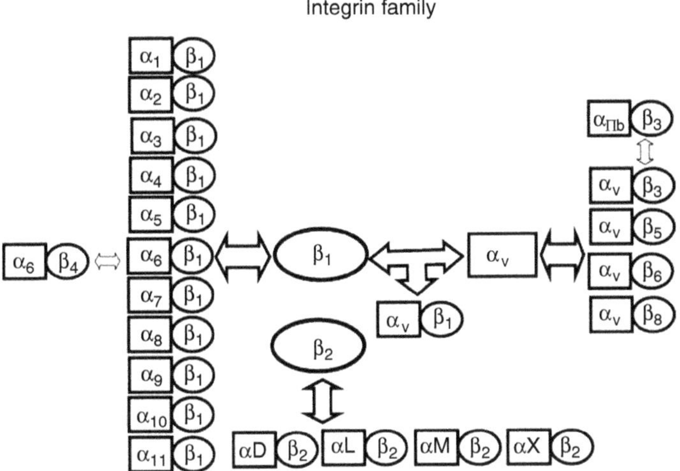

FIGURE 12.3 Schematic diagram of major integrin heterodimers. The integrin family of cell surface adhesion receptors comprises at least 18α and 8β subunits, which assemble to form at least 24 different heterodimers with distinct and overlapping functions and binding specificities.

bidirectionally with the local microenvironment. Given the unique features of these cell surface receptors, the term "integrin" was coined to help describe the ability of these proteins to facilitate connections between cells and the ECM.

The explosion of adhesion-related studies in the late 1980s and early 1990s resulted in the establishment of the fascinating field of integrin biology. To date at least 18α and 8β subunits have been identified resulting in ~24 integrin heterodimers, each with their own unique tissue distribution and functional specificities (Figure 12.3). In general, integrins are organized into three domains including an extracellular cation-dependent ligand-binding domain, a transmembrane region, and an intracellular tail known to interact with cytoskeletal components (Figure 12.4). The α and β subunits are derived from a single gene product and some of the α subunits are proteolytically cleaved during maturation resulting in two chains (heavy and light) held together by disulfide bonds. The N-terminal region of the α subunit is thought to be organized in a seven-bladed β-propeller structure, which contains residues crucial for ligand binding. In approximately half of the known integrins, a 200 amino acid I(inset)-domain is present, which has homology with von Willebrand factor A (81–85). This I-domain contains metal ion–dependent adhesion sites (MIDAS) that are thought to coordinate divalent cation binding, which is needed for efficient binding of negatively charged residues within integrin ligands (77,81–85). Importantly, some of the many function-blocking integrin antagonists are likely directed to this critical MIDAS region. Thus, an in-depth molecular understanding of the structure and function of the MIDAS likely contributes to the development and design of more effective integrin inhibitors.

Recent x-ray studies have revealed a more detailed molecular understanding of these receptors and led to the first crystal structure of the extracellular domain of the integrin $\alpha_v\beta_3$ (86). This recently solved structure has revealed an extracellular head domain organized into a seven-bladed β-propeller-like structure. Cation-dependent ligand binding to this globular head domain is thought to be dependent in part on molecular contacts between the extracellular regions of both α and β chains. In particular, studies have suggested that acidic residues such as aspartic acid (D) and glutamic acid (E) within certain ECM ligands coordinate with MIDAS along with oxygen atoms to form a coordination sphere (77,81–86). These cation-dependent binding events then initiate distinct conformational changes, which

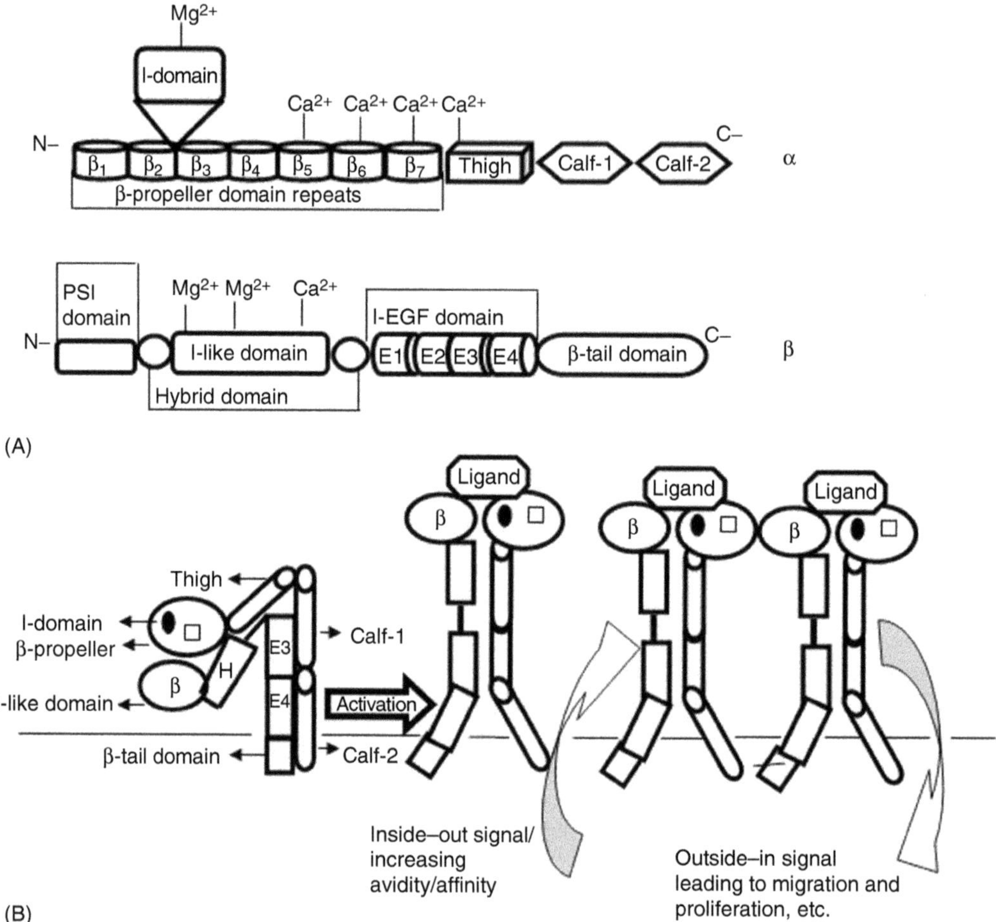

FIGURE 12.4 Schematic overview of integrin architecture and activation states. (A) Organization of extracellular domains of α and β integrin subunits. The α subunit contains an I-domain, which is a major ligand-binding region, a seven-bladed β-propeller domain, a thigh domain, and two calf domains. The β subunit contains an N-terminal PSI domain, an I-like domain inserted in the hybrid domain, a region containing a series of tandem EGF repeats and a C-terminal β-tail domain. (B) Schematic of the proposed arrangement of domains within integrins and their potential activation states. Inactive unoccupied integrins are thought to be bent at the "knee" region, with the head domain in close proximity to the plasma membrane. Inside–out signaling can trigger conformational changes leading to receptor activation, separation of the leg regions, extension of the head domain, and enhanced affinity and ligand binding. β1–7, β-propeller domain; E1–4, epidermal growth factors (EGF); PSI, plexin/semaphorin/integrin; H, hybrid domain; N–, N-terminal; C–, C-terminal.

are mechanically transmitted throughout the extracellular domain, the transmembrane domain and eventually to the cytoplasmic tails, which in turn are thought to physically separate to allow assembly of regulatory proteins such as talin, focal adhesion kinase (Fak), and integrin-linked kinase (ILK) (87–89).

The integrin globular head domain is next followed by leglike structures for both α and β chains which are further organized into subdomains. The leg region of the α chain contains a region termed the thigh domain and two Ig-like β-sandwich structures, called the Calf-1 and Calf-2 domains. The β chain is somewhat different and contains a 54 amino acid plexin, semaphorin, integrin (PSI) domain and in some cases an I-like domain, which may contain metal-binding sequences similar to the I-domains in the α subunit (81,85,86). This region is followed by four tandem EGF-like repeats ending with a structure that has been suggested to resemble cystain-C and termed β-T domain (βTD). Interestingly, both the α and β leg regions contain kneelike joints called α and β genu, which allow considerable flexibility to the leg or stalklike domains of the integrin.

The unique molecular structure and flexibility of the integrin molecules likely contribute to their ability to transmit as well as translate biomechanical binding into distinct signaling events.

Thus, a detailed understanding of the molecular events regulating cellular interactions with the ECM will no doubt provide important new insight into the development of highly selective approaches for disrupting cell–ECM interactions.

12.6 CELL ADHESION AND INTEGRIN CONFORMATION: PHYSICAL CONNECTIONS WITH TALIN

Exciting new studies have begun to shed light on some of the earliest molecular events regulating the ability of cells to interact with the ECM. In fact, new work has suggested that some of the earliest events may not involve integrins, but rather involve rapid (seconds or less) weak interactions of cells to the local environment through hyaluronan (90). These transient interactions may serve to tether the cells to the ECM to provide time to initiate more stable integrin-mediated interactions (90). Studies in the fields of integrin biology and cell adhesion have led to the identification of at least three types of stable adhesive structures including focal adhesions, fibrillar adhesions, and focal complex adhesions (91,92). These structures are formed during close physical contact of cells with the ECM and result in the organization of focal plaques containing a variety of distinct proteins including integrins and a number of cytosolic proteins (91,92). These fascinating adhesive structures have been predominately studied in vitro and thus their functional significance in the three-dimensional in vivo environment remains to be clarified. Recent evidence has suggested that the formation and composition of these adhesive structures can vary extensively between two-dimensional and three-dimensional matrices (91,92). Thus, more detailed studies as to the functional relevance of these adhesive structures in the in vivo environment are needed.

Though early integrin studies focused on their ability to connect cells with the ECM, it quickly became apparent that integrins are not simply molecular glue, but despite the lack of intrinsic enzymatic activity, they are multifunctional molecules capable of signal transduction (Figure 12.5). The cytoplasmic tails of both the α and β chains have been shown to interact with a variety of proteins including talin, α-actinin, Fak, Src, and ILK. Recent studies have confirmed the importance of talin–integrin interactions in focal adhesion assembly and subsequent cellular adhesion to the ECM (93,94). Microinjection of antibodies directed to talin or downregulated expression of talin disrupted integrin-dependent focal adhesion assembly and cell–ECM interactions (93,94). In this regard, new studies have provided key evidence that the N-terminal headlike region of talin has a four-point-one ezrin, radixin, moesin (FERM) domain that is thought to contain binding sites for β-integrins, Fak, and actin, thus providing a unique scaffold for protein–protein interactions and assembly of a signaling complex (93,94). Integrin-binding sites have also been identified in the C-terminal region of talin. Integrin-binding sites within talin may be conformation dependent, since structural changes including calpain cleavage of talin have been suggested to regulate integrin–talin interactions (93,94). Moreover, it has also been suggested that a physical scissorlike separation of the α and β integrin tails occurs following integrin binding, which may result in enhanced talin interactions. These structural alterations in both integrins

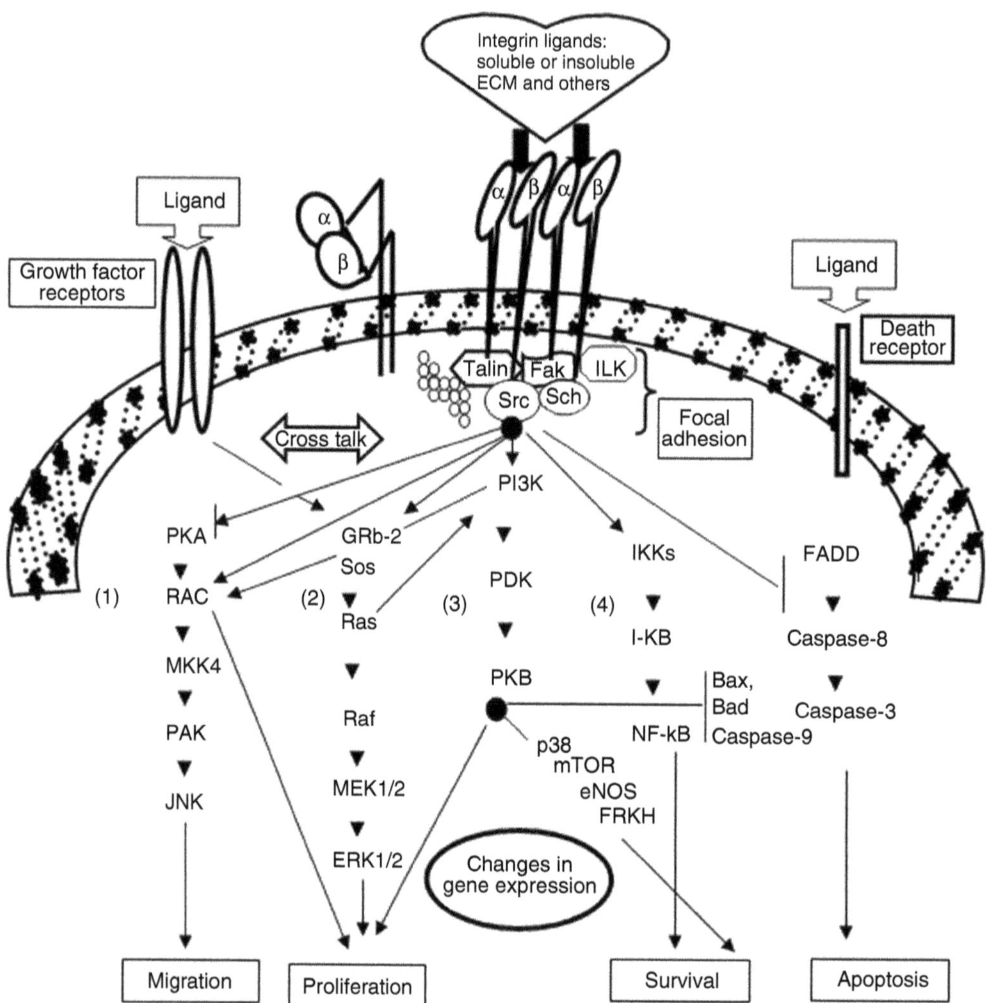

FIGURE 12.5 Schematic diagram depicting integrin-mediated signaling. Integrin binding can regulate a number of signaling pathways that may impact angiogenesis, including Rac-GTPase, MAP kinase, PI3 kinase, JNK, and NF-κB cascades. Due to the complexity of the cascades, not all components from each pathway are listed. These signaling pathways have been suggested to regulate crucial processes such as cellular proliferation, migration, apoptosis, survival, and gene expression. PI3K, phosphatidylinositol 3-kinase; MEKK-4, MAPK/ERK kinase kinase-4; JNK, the Jun-NH_2-terminal kinase; MAPK, mitogen-activated protein kinase; ERK1/2, extracellular regulated kinase 1/2; ILK, integrin-linked kinase; PKB, protein kinase B; PKA, protein kinase A; NF-κB, nuclear factor-κB; FADD, Fas-associated protein with death domain; BAX, Bcl-2-associated X protein; mTOR, mammalian target of rapamycin; PDK1, phosphoinositide-dependent protein kinase 1; eNOS, endothelial nitric oxide synthase; I-κB, inhibitor of κB; FRKH, fork head.

and talin are crucial for establishing a molecular link between the ECM and the intracellular compartment. Given these facts, it is not surprising that disruption of these connections could significantly impact angiogenesis, tumor growth, and metastasis. Thus, the possibility exists that small cell permeable compounds might be developed that could target interactions between talin and the integrin tails. However, these approaches might suffer from the lack of specificity given the apparent commonality of the talin interactions with most integrin heterodimers.

12.7 INTEGRIN–LIGAND INTERACTIONS AND SIGNALING

As mentioned earlier, integrins do not possess intrinsic enzymatic activity, thus their ability to facilitate bidirectional signal transduction depends on the recruitment of accessory molecules into signaling complexes (Figure 12.5). The ability of integrins to facilitate signal transduction is not the major focus of our discussion and therefore, readers interested in a more comprehensive review of integrin signaling are referred to a number of excellent reviews on the subject (95–98). Here, we briefly examine only some of the more general integrin-signaling concepts critical to understanding the roles of integrins in angiogenesis and mechanisms by which disruption of integrin-mediated adhesive events may impact neovascularization.

As discussed earlier, integrins are multifunctional receptors with distinct and overlapping binding specificities. Early studies first identified the RGD tripeptide sequence within fibronectin as an important integrin recognition motif for the fibronectin receptor $\alpha_5\beta_1$ (99,100). Additional studies have provided evidence that a second synergy site within fibronectin contributes to the ability of $\alpha_5\beta_1$ to bind to fibronectin. Further work has confirmed the RGD tripeptide as a critical ligand for the interactions of a number of α_v integrins such as $\alpha_v\beta_1$, $\alpha_v\beta_3$, and $\alpha_v\beta_5$. The capacity of integrins to recognize and bind to their respective ligands can depend on their activation state (81,90,101). For example, intracellular signaling events can lead to conformational changes within the cytoplasmic tails that cause integrin activation and enhanced ligand binding through a process termed inside–out signaling (81,90,101). In contrast, signaling events and ligand binding initiated outside the cell can transmit molecular information to the cell's interior in a process called outside–in signaling (81,90,101). To this end, new data concerning the three-dimensional structure of integrins have contributed to a better understanding of these events and the concept that integrins may exist in at least two and possibly three activity states (81,90). For example, data from the recent crystal structure and electron rotary shadowing studies suggest that integrin $\alpha_v\beta_3$ can exist in a bent, inactive, or low-affinity state characterized by the headlike domain in a conformation close to the plasma membrane, thus limiting access to ECM ligands. These findings were in good agreement with the conformation of the ectodomain observed in the presence of inhibitory cations such as Ca^+, which is known to reduce ligand binding. In the presence of soluble ligand or activating cations such as Mg^+ and Mn^+, the integrin appeared to adopt an extended conformation in which the head domain is physically extended away from the plasma membrane (81,90,101). Given the facts that integrins exhibit a variety of binding affinities to distinct ECM components, it is likely that a dynamic equilibrium exists between fully activated, intermediate, and inactive states.

Genetic evidence from several integrin studies has confirmed their importance in angiogenesis and blood vessel formation. In this regard, recent evidence has suggested that the conditional knockout of Fak, a key component of integrin-signaling complexes, resulted in numerous vascular defects in mice (102,103). Moreover, the ability of integrins to differentially regulate endothelial cell responses to angiogenic growth factors such as bFGF and VEGF also exemplifies the importance of integrin signaling in controlling neovascularization. To this end, previous studies have suggested that bFGF-induced angiogenesis can be inhibited by antagonists of $\alpha_v\beta_3$, but not by antagonists of $\alpha_v\beta_5$ (104). Studies have also indicated that bFGF signaling in association with $\alpha_v\beta_3$ integrin engagement may lead to enhanced resistance of endothelial cell death by a mechanism involving P21-activated protein kinase-1 (PAK1)–mediated phosphorylation of Raf-1 on serine 338 and 339 (105,106). In contrast, $\alpha_v\beta_5$ binding in association with stimulation of endothelial cells with VEGF leads to Src-mediated phosphorylation of tyrosine 340 and 341, which was associated with enhanced resistance to endothelial cell apoptosis mediated by Fas (105,106). These interesting findings suggest that integrin-mediated signaling can differentially modulate the response of cells to specific apoptotic pathways. Finally, whereas numerous studies have indicated that antagonists of $\alpha_v\beta_3$ can inhibit angiogenesis in a wide array of animal models, intriguing new studies have indicated that mice deficient in β_3 integrins are associated with elevated levels of VEGF receptor-2 (VEGFR-2) and enhanced tumor-associated angiogenesis (107,108). Thus, it is clear that integrin-mediated signals required for angiogenesis are complex and more in-depth studies are needed to clarify the roles of individual integrins during new blood vessel formation.

As can be appreciated from the examples described earlier, integrins impact a variety of signaling cascades. Some of the most well-studied processes known to be influenced by integrin signaling include regulation of cytoskeletal organization, cell adhesion, migration, proliferation, apoptosis, and gene expression. In particular, studies have shown that integrin engagement can regulate the activity of Rho GTPases such as Rac, Rho, and Cdc42, which are known to impact events required for angiogenesis including extension of lamipodia, filipoda, and cell spreading and motility (95–98). In addition, integrin-binding events regulate signaling involved in controlling cell survival and apoptosis such as the PI3 kinase/Akt pathway (95–98). Recent evidence also suggests that regulation of the endogenous angiogenesis inhibitor thrombospondin-1 (TSP-1) may be controlled by signaling events that involve PI3 kinase activation, which ultimately leads to regulation of c-Myc, a transcription factor known to control expression of TSP-1 (109). Interestingly, studies have indicated that integrin binding may impact p53 activation as well as $p21^{CIP1}$ levels thereby regulating apoptosis and cell cycle control (110,111). Moreover, integrin binding influences the Ras/MAP kinase/ERK pathways, ultimately regulating proliferation and gene expression. In this regard, the response of vascular cells to many angiogenic growth factors depends on distinct integrin-mediated interactions with the ECM. Thus, it can be appreciated from this rather limited discussion of integrin-signaling events that key components of these cascades may provide important new therapeutic targets for controlling angiogenesis.

12.8 INTEGRINS AS KEY THERAPEUTIC TARGETS FOR CONTROLLING ANGIOGENESIS

As discussed earlier, a plethora of molecular, biochemical, and cellular evidence clearly supports that notion that integrin receptors represent an important set of clinically relevant therapeutic targets for the control of angiogenesis. A wide array of integrin receptors have been suggested to play key roles in angiogenesis including $\alpha_1\beta_1$, $\alpha_2\beta_1$, $\alpha_3\beta_1$, $\alpha_4\beta_1$, $\alpha_5\beta_1$, $\alpha_6\beta_4$, $\alpha_v\beta_3$, and $\alpha_v\beta_5$ (112–114). The possibility exists that altering the expression or function of these adhesion receptors could significantly impact angiogenesis. Given what we currently understand concerning the multitude of functions that integrins control, a number of strategies could be devised to alter expression, ligand binding, and activation state of these integrins including siRNA oligos, small molecule inhibitors, peptide memetics, and antibodies. Many of these approaches are currently developed and tested in both animal models and human clinical trials for the control of angiogenesis and tumor progression. In fact, encouraging clinical data have been described concerning the evaluation of Vitaxin, a humanized Mab directed to $\alpha_v\beta_3$ integrin (8,115–117). Moreover, Mab CNTO 95 a new Mab directed to α_v will also be evaluated in clinical trials for treatment of human tumors (118). In addition, studies are currently underway to assess the effects of Mab M200 directed to the fibronectin receptor $\alpha_5\beta_1$ in human clinical trials. Interestingly, a number of other integrin antagonists directed to such integrins as $\alpha_{II}\beta\beta_3$, $\alpha_4\beta_7$, and $\alpha_4\beta_1$ have also been studied in clinical trials for a variety of other noncancer indications (119,120). Thus, it is possible that some of these previously evaluated antagonists may be useful in the treatment of angiogenic diseases such as those targeting $\alpha_4\beta_1$ integrin.

Clinical success in utilizing integrin antagonists lays in optimizing the functional design and dosing of these inhibitors. In this regard, key features to consider when optimizing the effectiveness of an integrin antagonist include a molecular understanding of the particular functions that the integrin inhibitors are disrupting. It is also important to identify the particular cell type in which the integrin targets are expressed, given that binding specificity as well as functions can vary for a particular integrin, depending on the specific cell type in which it is expressed. This concept is of particular importance given the emerging appreciation of the variety of cell types that contribute to the regulation of angiogenesis. For example, recent studies have indicated that cells such as endothelial progenitors, inflammatory cells, stromal fibroblasts, pericytes, and tumor cells all contribute either directly or indirectly to the formation of new blood vessels (121–123). In this regard, integrin antagonists may specifically block binding to a particular ECM component, whereas others may inhibit binding of integrins to proteolytic enzymes and still others may disrupt integrin associations with growth factors or growth factor receptors (124–126). Importantly, depending on the concentration and affinities of the particular integrin-binding molecule, the reagent could change the integrin conformation and either activate or inhibit integrin function. Thus, great care should be used when considering the appropriate doses of an integrin inhibitor to be examined clinically. Moreover, blocking or activating one integrin may alter the function of another integrin which may lead to complicating side effects, thus an understanding of potential integrin cross talk should also be considered. Importantly, angiogenesis inhibitors such as endrepellin, tumstatin, canstatin, arrestin, and endostatin may bind directly to a number of different integrins expressed in different cellular compartments, thereby initiating unique signaling cascades, which may alter angiogenesis in unexpected ways. Thus, an investigation of the redundancy of integrin binding by the antagonists should also be carried out.

12.9 TARGETING COLLAGEN-BINDING INTEGRINS ($\alpha_1\beta_1$ AND $\alpha_2\beta_1$) FOR THE TREATMENT OF ANGIOGENESIS

As mentioned, several β_1 integrins have been suggested to play roles in regulating angiogenesis. Studies with α_1 integrin null mice suggest that these mice are associated with reduced tumor growth and angiogenesis (127). Moreover, studies of mice-harboring mutation in several collagen molecules, which represent ligands for certain β_1 integrins, have also exhibited defects in vascular development (128,129). To this end, recent studies have indicated that the collagen-binding integrins $\alpha_1\beta_1$ and $\alpha_2\beta_1$ contribute to the regulation of VEGF-induced angiogenesis. Stimulation with VEGF was shown to upregulate expression of $\alpha_1\beta_1$ and $\alpha_2\beta_1$ in endothelial cells (130). Treatment of mice with function-blocking antibodies directed to either α_1 or α_2 integrin partially inhibited angiogenesis in vivo (130). These function-blocking antibodies were shown to disrupt endothelial cell adhesion to collagen, suggesting the possibility that distinct β_1 integrin-signaling cascades mediated by either $\alpha_1\beta_1$ or $\alpha_2\beta_1$ play roles in VEGF-dependent angiogenesis. It is interesting to note that studies have shown collagen-binding integrins can also regulate cell cycle transition and proliferation by modulating the expression of cyclin-dependent kinase inhibitors including $p21^{CIP1}$ and $p27^{KIP1}$ (131). Interestingly, endorepellin, an angiogenesis inhibitor derived from the matrix protein, perlecan can bind to integrin $\alpha_2\beta_1$ (132). This unique integrin-binding event has shown to initiate a signaling cascade involving cyclic AMP and activation of protein kinase A (PKA) which resulted in disruption of Fak–cytoskeletal interactions, ultimately inhibiting endothelial cell migration (132). These interesting experimental findings suggest that either blocking $\alpha_2\beta_1$ interactions with collagen or initiating unique signaling cascades via endorepellin binding to $\alpha_2\beta_1$ may inhibit angiogenesis. A similar paradigm has been shown with integrin $\alpha_v\beta_3$. For example, studies have shown that blocking $\alpha_v\beta_3$ interactions with ECM proteins can induce endothelial cell apoptosis and inhibit angiogenesis by mechanisms involving altered regulation of p53 DNA-binding activity. In addition, rather than inhibiting physical association of $\alpha_v\beta_3$ with the ECM, functional binding of $\alpha_v\beta_3$ to the endogenous angiogenesis inhibitor tumstatin initiates a signaling cascade, which acts to induce endothelial cell–specific apoptosis and inhibits angiogenesis (133,134). These findings support the concept that distinct signaling events from either unligated or ligand-bound integrins may impact neovascularization. Taken together, these studies suggest that integrin antagonists directed to the collagen-binding integrins $\alpha_1\beta_1$ and $\alpha_2\beta_1$ may inhibit angiogenesis by disrupting cellular adhesion, motility, and perhaps proliferation.

12.10 TARGETING COLLAGEN/LAMININ-BINDING INTEGRIN ($\alpha_3\beta_1$) FOR THE TREATMENT OF ANGIOGENESIS

In addition to the major collagen-binding integrins, recent studies have implicated integrin $\alpha_3\beta_1$ as a potential regulator of angiogenesis (135–137). Integrin $\alpha_3\beta_1$ has been suggested to be expressed in endothelial cells and angiogenic blood vessels and bound to laminin as well as collagen (135). Moreover, a peptide derived from the N-terminal domain of the endogenous angiogenesis inhibitor TSP-1 binds to $\alpha_3\beta_1$ and modulates endothelial cell proliferation and angiogenesis in vivo (136,137). Additionally, the antagonist of $\alpha_3\beta_1$ was shown to reverse the endothelial cell antimigratory activity of an inhibitor derived from TSP-1. These findings provide further evidence that $\alpha_3\beta_1$ may play a role in the response of endothelial cells to angiogenesis inhibitors. Interestingly, $\alpha_3\beta_1$ has also been shown to associate with uPAR, which is thought to facilitate signaling cascades involved in proliferation and motility (69). Importantly, the uPA–uPAR system has been shown to play a role in angiogenesis (138). Interestingly, specific NC1 domains from collagen type-IV including the angiogenesis inhibitor canstatin were also shown to interact with $\alpha_3\beta_1$ (74). As has been observed with other integrins, $\alpha_3\beta_1$ can also bind to a number of non-ECM ligands. In this regard, intriguing new studies suggest that TIMP-2 can bind to $\alpha_3\beta_1$ (139). This unique integrin interaction was shown to inhibit angiogenesis by a mechanism independent from the ability of TIMP-2 to inhibit protease activity (139). Given the importance of endothelial cell interactions with both laminin and collagen in conjunction with the ability of $\alpha_3\beta_1$ to bind to a number of novel antiangiogenic molecules, the possibility exists that integrin $\alpha_3\beta_1$ may represent an important angiogenesis regulator and potential therapeutic target for controlling neovascularization.

12.11 TARGETING FIBRONECTIN-BINDING INTEGRINS ($\alpha_4\beta_1$ AND $\alpha_5\beta_1$) FOR THE TREATMENT OF ANGIOGENESIS

Receptors capable of binding fibronectin may regulate new blood vessel formation including both $\alpha_4\beta_1$ and $\alpha_5\beta_1$ (140,141). In fact, studies with α_4 and α_5 null mice suggested a role for these integrin in vascular development and blood vessel formation (142,143). Integrin $\alpha_4\beta_1$ has been suggested to bind to fibronectin as well as serve as a counterreceptor for the cell adhesion molecule VCAM-1. As VCAM-1 is thought to play a role in angiogenesis, recent studies suggest that specific antagonists of $\alpha_4\beta_1$ can inhibit angiogenesis in the chick CAM model as well as a model of muscle vascularization (141,142). As was observed with integrin $\alpha_3\beta_1$, $\alpha_4\beta_1$ may be capable of binding TSP-1 and TSP-2 (141). In this regard, $\alpha_4\beta_1$ has been suggested to facilitate large vessel endothelial cell adhesion and proliferation on TSP-1 (141). These data imply that $\alpha_4\beta_1$ may play a role in angiogenesis that is independent from the known role of VCAM in neovascularization.

Studies have documented the critical role of fibronectin in controlling a number of events important to the development of new blood vessels including endothelial cell adhesion, migration, invasion, and proliferation (144,145). Moreover, studies have suggested that integrin-mediated interactions with distinct domains within fibronectin differentially modulate expression of proteolytic enzymes such as MMPs (146). Studies have also indicated that the fibronectin receptor $\alpha_5\beta_1$ is highly expressed on angiogenic blood vessels and that $\alpha_5\beta_1$-mediated interactions with fibronectin can initiate signaling pathways thought to regulate cell survival and apoptosis (147,148). In particular, $\alpha_5\beta_1$-mediated signaling has been suggested to regulate the relative levels of Bcl-2 and Bax (147,148). Given our previous discussion concerning the mechanisms by which integrins might regulate endothelial cell behavior, it is not surprising that $\alpha_5\beta_1$ appears to play an important role in regulating angiogenesis. Antagonists of $\alpha_5\beta_1$ have been shown to inhibit endothelial cell function in vitro and angiogenesis in vivo (149,150). Given the strong supporting evidence for a role of $\alpha_5\beta_1$ in controlling neovascularization, a Mab termed M200 directed to the $\alpha_5\beta_1$ will be evaluated in clinical trials for the treatment of malignant tumors.

Interestingly, the endogenous angiogenesis inhibitor endostatin, the NC domain of collagen type XVIII, was also shown to interact with $\alpha_5\beta_1$ (72,75). Endostatin has been suggested to disrupt Wnt signaling (151,152). However, whether modulation of Wnt signaling and binding of $\alpha_5\beta_1$ play a role in the antiangiogenic activity of endostatin in vivo remains to be fully clarified. Finally, as was observed with $\alpha_3\beta_1$ and $\alpha_4\beta_1$, $\alpha_5\beta_1$ also associates with the uPAR. Thus, the ability of $\alpha_5\beta_1$ to regulate signaling mechanism dependent on uPAR may also impact the angiogenesis. Given the interconnected signaling pathways controlled by integrins, it is not surprising that binding and signaling through one integrin might impact either positively or negatively the functional activity of a second integrin. In fact, studies have shown that this type of activity can occur and has been termed in one case, transdominant integrin inhibition. In this regard, recent studies have suggested the ability of $\alpha_v\beta_3$ to facilitate migration and angiogenesis may be regulated in part, by ligation of the fibronectin receptor $\alpha_5\beta_1$ (150,153). Antagonists of $\alpha_5\beta_1$ suppressed $\alpha_v\beta_3$-dependent focal contact formation and migration and this activity appeared to depend on PKA (150,153). Taken together, it is clear that the potential mechanisms by which $\alpha_5\beta_1$ may impact angiogenesis are many, and additional studies are needed to clarify the numerous roles of $\alpha_5\beta_1$ in angiogenesis.

12.12 TARGETING LAMININ-BINDING INTEGRIN ($\alpha_6\beta_4$) FOR THE TREATMENT OF ANGIOGENESIS

Whereas cellular interactions with laminin are thought to play a role in angiogenesis, the roles of specific laminin-binding integrins in new blood vessel growth are not completely understood. Interestingly, several integrin-binding peptides derived from laminin were shown to have both proangiogenic and antiangiogenic activity (154–156). Moreover, the expression of the laminin-binding integrin $\alpha_6\beta_4$, in contrast to $\alpha_3\beta_1$, may be restricted to more mature blood vessels or transiently expressed during angiogenesis since little $\alpha_6\beta_4$ was detected on proliferating blood vessels. Integrin $\alpha_6\beta_4$ may also regulate VEGF expression and signaling (157). Moreover, in studies of α_6 or β_4 null mice, investigators observed little disruption of normal vascular development. Based largely on these studies, it was assumed that $\alpha_6\beta_4$ might not play a significant role in angiogenesis. However, recent evidence now suggests that $\alpha_6\beta_4$ may indeed play a role in angiogenesis. In this regard, mice were generated that contained a mutation within the signaling portion of the β_4 subunit such that $\alpha_6\beta_4$ was expressed yet did not signal properly (158). Interestingly, these β_4 mutant mice exhibited a significantly reduced angiogenic response to bFGF in Matrigel implants as well as reduced ischemia-associated retinal neovascularization (158). In addition, xenograft tumors grown in these mice were associated with significantly reduced blood vessel counts. Interestingly, $\alpha_6\beta_4$-mediated regulation of endothelial cell migration and invasion was shown to be controlled in part by mechanisms involving translocation of phosphorylated Erk and NF-κB to the nucleus (158). Thus, though the exact role of $\alpha_6\beta_4$ in angiogenesis is not completely understood, these studies shed new light on a possible role for $\alpha_6\beta_4$ in distinct temporal phases of new blood vessel formation.

12.13 TARGETING α_v INTEGRINS ($\alpha_v\beta_3$ AND $\alpha_v\beta_5$) FOR THE TREATMENT OF ANGIOGENESIS

Perhaps the most widely studied integrins thought to play roles in angiogenesis belong to the α_v integrin subfamily. Studies have suggested that α_v integrins such as $\alpha_v\beta_3$ and $\alpha_v\beta_5$ are upregulated in activated endothelial cells and on angiogenic blood vessels in vivo (159–161). Several angiogenic growth factors such as bFGF and VEGF have been suggested to stimulate expression of $\alpha_v\beta_3$ and $\alpha_v\beta_5$. In addition to the traditional roles of these integrins in facilitating endothelial cell adhesion and migration, a number of other non-ECM proteins may also bind $\alpha_v\beta_3$ and regulate angiogenesis. For example, $\alpha_v\beta_3$ can bind to activated MMP-2, thereby localizing proteolytic activity to the cell–ECM interface thereby facilitating ECM remodeling (46,47). The serine protease thrombin may also interact with $\alpha_v\beta_3$ through an RGD-dependent mechanism (162). In this regard, studies have suggested that thrombin can promote angiogenesis in vivo (163). Moreover, other studies have indicated that $\alpha_v\beta_3$ can associate with the uPAR, which is thought to modulate endothelial cell migration and angiogenesis (164–166). Interestingly, studies have suggested that $\alpha_v\beta_3$-mediated interactions with the cell adhesion molecule L1 may stimulate angiogenesis through a mechanism involving cooperative signaling with VEGFR-2 (167). Recent work has also suggested that members of the CCN protein family such as CY61 may also bind to $\alpha_v\beta_3$ and play a role in the ability of CY61 to regulate angiogenesis (168). Finally, evidence was provided that $\alpha_v\beta_3$ may contain a binding site for thyroid hormone, which may facilitate angiogenesis through MAP kinase signaling pathway (169). Thus, it is clear that α_v integrins may regulate angiogenesis by multiple mechanisms.

Antagonists of $\alpha_v\beta_5$ have been shown to inhibit cellular adhesion and migration on specific ECM components in vitro including vitronectin (170–172). Interestingly, $\alpha_v\beta_5$-dependent tumor cell migration and metastasis required stimulation of the insulin-like growth factor signaling pathway and recruitment of α-actinin to cytoplasmic tails of β_5 integrins (173). Moreover, while antagonists of $\alpha_v\beta_5$ could inhibit angiogenesis induced by VEGF, it had little effect on angiogenesis induced by bFGF in the chick CAM and corneal micropocket assays (104). Taken together, these studies suggest that angiogenesis can be controlled in part, by at least two distinct pathways dependent on two different α_v integrins (104–106). As has been observed with integrin $\alpha_v\beta_5$, antagonists of $\alpha_v\beta_3$ have been shown to inhibit endothelial cell adhesion and migration on specific ECM components. Studies have also suggested that antagonists of $\alpha_v\beta_3$ including Mabs, peptides, and nonpeptide memetics exhibit antiangiogenic and antitumor activity in several animal models (90,97,104–106). Whereas antagonists of $\alpha_v\beta_3$ have clearly been shown to inhibit angiogenesis, tumor growth, and metastasis, mice lacking α_v or β_3 exhibited minimal embryonic vascular defects (107,108,174). However, some defects in the vasculature were observed in the brain and intestine. Moreover, additional studies suggested that mice defective in β_3 integrin actually exhibited enhanced vascularization of xenograft tumors indicating that the role of $\alpha_v\beta_3$ integrin in angiogenesis and vascular development is not completely understood and additional mechanistic studies are clearly needed.

Given the recent wealth of evidence that $\alpha_v\beta_3$ plays a significant role in angiogenesis, much attention has been focused on defining potential mechanisms by which $\alpha_v\beta_3$ contributes to new blood vessel growth. Studies have demonstrated that in addition to facilitating endothelial cell adhesion and migration, $\alpha_v\beta_3$ may regulate endothelial cell survival and apoptosis, since Mabs and peptide antagonists of $\alpha_v\beta_3$ induced apoptosis in tumor-associated blood vessels in vivo (106,175). Consistent with these findings, studies in melanoma demonstrated that $\alpha_v\beta_3$-mediated binding to denatured collagen resulted in enhanced survival and reduced apoptosis in vitro (176). Inhibiting $\alpha_v\beta_3$ could also induce apoptosis of human melanoma tumors growing in full thickness human skin (177). Taken together, these studies indicate that integrin $\alpha_v\beta_3$ may regulate apoptosis in both tumor cells as well as endothelial cells. The regulation of apoptosis by $\alpha_v\beta_3$ may depend on a mechanism involving suppressing p53 activity as well as altering levels of the cyclin-dependent kinase inhibitor $p21^{CIP1}$. Consistent with this hypothesis, antagonist of $\alpha_v\beta_3$ failed to inhibit retinal neovascularization in p53 null mice suggesting a key role for p53 in the function of $\alpha_v\beta_3$ during angiogenesis (110). In addition to modulating the activity of p53 and $p21^{CIP1}$, $\alpha_v\beta_3$-mediated cellular interactions with a cryptic epitope in collagen significantly increased the relative Bcl-2:Bax ratio in melanoma cells (177). To this end, a high Bcl-2 to Bax ratio is known to be associated with suppression of apoptosis and increased survival. Finally, intriguing new studies suggest that unligated $\alpha_v\beta_3$ may function by actively initiating death signals associated with recruitment and activation of caspase 8 (178).

As mentioned previously, integrins are well known to modulate the activity of growth factor signaling pathways including VEGF. In β_3 null mice the relative levels of the VEGF receptor-2/Flk1 were enhanced leading to increased VEGF receptor signaling (108). In other studies, VEGF receptors were shown to colocalize with $\alpha_v\beta_3$ in both tumor and endothelial cells (179). Recent studies have indicated that endothelial cells may be able to adhere to and spread on $VEGF_{189}$ and $VEGF_{165}$ but not $VEGF_{121}$ via interactions with integrins $\alpha_3\beta_1$ and $\alpha_v\beta_3$ (180). Taken together, these findings imply that integrins such as $\alpha_v\beta_3$ may be able to associate with VEGF and/or VEGF receptors. Finally, recent work has suggested that expression and signaling mediated by $\alpha_v\beta_3$ in tumor cells may directly regulate the expression of VEGF by a mechanism involving p66 Shc (181). Thus, as can be appreciated from these limited examples of functional cooperation between $\alpha_v\beta_3$ and VEGF, it is possible that disruption of $\alpha_v\beta_3$ function in either endothelial or tumor cells can significantly alter angiogenesis both in vitro and in vivo. Given the complexity of the interconnected signaling events, in conjunction with integrin cross talk, more detailed molecular studies are needed to clarify the role of $\alpha_v\beta_3$ in angiogenesis in relationship growth factor signaling.

As has been discussed earlier, soluble angiogenesis inhibitors from the ECM can also bind to $\alpha_v\beta_3$ and result in distinct signaling pathways that regulate angiogenesis. For example, tumstatin as well as canstatin both have been shown to bind to $\alpha_v\beta_3$ and inhibit angiogenesis. Endostatin may also bind to $\alpha_v\beta_3$ as well as $\alpha_5\beta_1$. Moreover, PEX, a naturally occurring angiogenesis inhibitor derived from the C-terminal hemopexin-like domain of MMP-2 also binds to $\alpha_v\beta_3$ and inhibits angiogenesis by disrupting localization of active MMP-2 at the cell–ECM interface. Interestingly, a novel compound has recently been developed that blocked the ability of $\alpha_v\beta_3$ to bind to MMP-2 yet did not impact the ability to bind to its natural ECM ligand vitronectin (124,126). Importantly, this unique $\alpha_v\beta_3$-binding antagonists inhibited angiogenesis in vivo (124,126). These studies suggest the capacity of $\alpha_v\beta_3$ to regulate angiogenesis by modulating proteolytic activity independently from ECM binding. Interestingly, fastatin, the fourth FAS1 domain of the ECM protein Big-h3, which inhibits angiogenesis and induces apoptosis, also binds to $\alpha_v\beta_3$ in an RGD-dependent manner (182). Finally, recent studies have utilized nanoparticles targeting $\alpha_v\beta_3$ to selectively deliver a mutant form of Raf, which inhibits signaling to angiogenic blood vessels, thereby inhibiting angiogenesis (183). Moreover, the chemotherapeutic agent, Pacitaxel conjugated with an integrin-targeting ligand was also shown to inhibit angiogenesis and tumor growth in vivo (184). Taken together, these studies illustrate still another potential use of selective integrin targeting for the control of abnormal blood vessel formation. Given the large number of angiogenesis inhibitors that appear to bind to $\alpha_v\beta_3$, it will be interesting to determine whether these binding events initiate unique antiproliferative or apoptotic signals or whether the antiangiogenic activity of these molecules results from blocking natural ECM ligand bindings and thereby mimicking the unligated state of the integrin. Importantly, unligated integrin $\alpha_v\beta_3$ has been suggested to induce apoptosis by mechanisms involving caspase-8 recruitment to the integrin and subsequent activation (178).

12.14 TARGETING MATRIX-IMMOBILIZED CRYPTIC ECM EPITOPES FOR THE TREATMENT OF ANGIOGENESIS

As can be appreciated from our discussions earlier, direct targeting of integrins represents an important therapeutic target for the treatment of many neovascular diseases. However, given the importance of integrin signaling in controlling angiogenesis, indirect targeting of integrin function may also represent a clinically useful strategy for controlling aberrant neovascularization. Recent studies have suggested that cryptic epitopes within ECM proteins including laminin, fibronectin, and collagen can play crucial roles in regulating invasive cellular behavior (185). In this regard, studies have indicated that a cryptic epitope exposed by MMP-mediated remodeling of laminin-5 can promote tumor cell migration and invasion (65). Moreover, expression of specific chains of laminin-5 has been suggested to correlate with more aggressive malignant epithelial tumors (186). Given the importance of integrins and in particular $\alpha_3\beta_1$ in binding laminin, it is possible that integrin-mediated cellular interactions with the cryptic epitope of laminin-5 may contribute to the regulation of angiogenesis. Thus, it is possible that targeting cryptic epitopes in the ECM, instead of the cell surface integrins, may effectively inhibit adhesive cellular processes without directly targeting a specific subset of cells. Moreover, these cryptic ECM epitopes likely represent more stable targets as compared with cell surface epitopes, in which the expression may change due to mutations. Interestingly, an alternatively spliced form of fibronectin, containing a region called the EDB domain, was shown to be specifically associated with angiogenic- and tumor-associated blood vessels, while little if any was detected surrounding normal vessels (187,188). Whereas the functional role of this region of fibronectin is not completely understood, this domain may allow targeting of angiogenic blood vessels with cytotoxic agents that may inhibit angiogenesis and tumor growth.

Finally, studies have identified MICEE within collagen (HUIV26 and HU177) that play critical roles in angiogenesis and tumor growth (58,63,64). Exposure of the HUIV26 cryptic epitope within collagen type-IV was restricted to the ECM surrounding angiogenic blood vessels and invasive fronts of malignant tumors, but not detected in normal nonproliferating tissues (58,63,64,189). Studies have indicated that these cryptic collagen epitopes can be specifically exposed following MMP-mediated remodeling of collagen in vitro and in vivo. Interestingly, MMP-9 but not MMP-2 null mice exhibited significantly reduced levels of the HUIV26 cryptic epitope surrounding angiogenic retinal vessels as compared with wild-type controls (58). These findings suggest a role for MMP-9 in exposure of the HUIV26 epitope during retinal angiogenesis. The HUIV26 cryptic collagen epitope is recognized by integrin $\alpha_v\beta_3$ (63). Function-blocking antibodies directed to the HUIV26 cryptic collagen epitope potently inhibited angiogenesis and tumor growth in a number of animal models (58,63,64). These data suggest that either targeting integrin receptors or specific integrin ECM ligands may represent clinically useful approaches for the treatment of neovascular diseases.

12.15 CONCLUSION

In recent years, the focused attention on the molecular mechanisms regulating angiogenesis has led to unique insight into the functional roles of regulatory molecules, signaling pathways, and various cell types that contribute to new blood vessel development. With this new appreciation for the complexity and molecular cooperativity between angiogenesis regulatory factors, as well as the contributions that cells other than endothelial cells added to the angiogenic cascade, many new targets and therapeutic strategies have been proposed for clinical intervention in neovascular disorders. In this regard, over the last decade a number of important new endogenous angiogenesis inhibitors have been identified. The vast majority of these molecules have proven to be fragments of the ECM proteins, including endostatin, tumstatin, canstatin, arrestin, restin, thrombospondin, and endorepellin to name just a few. A striking common feature of all these angiogenesis regulators is the fact that they all bind to various members of the integrin family of cell adhesion receptors and their antiangiogenic activity may depend, in part, on their ability to bind integrins.

Interestingly, in addition to the numerous soluble integrin-binding angiogenesis inhibitors identified recently, a growing body of evidence is emerging that selectively exposed MICEE, including HUIV26 and HU177, represent acellular solid-state regulatory elements which play important roles in facilitating angiogenesis, tumor growth, and metastasis. As was observed with the soluble angiogenesis inhibitors, these MICEE also bind directly to integrin receptors. These findings in conjunction with the known importance of the bidirectional communication between cells and the ECM provide convincing evidence that integrins, ECM, and adhesive processes represent a rapidly expanding set of new antiangiogenic targets. The intriguing concept that the acellular ECM and their integrin receptors do not simply represent an insoluble scaffold or molecular glue, but rather a largely untapped source of potential for the design of clinically useful compounds for the treatment of human diseases, which has been driven in large part by the greater understanding of the molecular mechanisms by which the ECM and integrins control normal and pathological events.

ACKNOWLEDGMENT

We would like to thank Jennifer Roth for her help in preparing this manuscript. This work was supported in part by a grant from the NIH ROICA9164.

REFERENCES

1. Stupack, D.G. and Cheresh, D.A., Integrins and angiogenesis, *Curr. Opin. Devel. Biol.*, 64, 207, 2004.
2. Guo, W. and Giancotti, G., Integrin signaling during tumor progression, *Nat. Rev. Mol. Cell Biol.*, 5, 816, 2004.
3. French-Constabt, C. and Colognato, H., Integrins: versatile integrators of extracellular signals, *Trends Cell Biol.*, 14, 678, 2004.
4. DeMali, K.A., Wennerberg, K., and Burridge, K., Integrin signaling to the actin cytoskeleton, *Curr. Opin. Cell Biol.*, 15, 572, 2003.
5. Hynes, R.O., Integrins: bidirectional, allosteric signaling machines, *Cell*, 110, 673, 2002.
6. Moshfeghi, A.A. and Puliafito, C.A., Pegaptanib sodium for the treatment of neovascular age-related macular degeneration, *Expert Opin. Investig. Drugs*, 14, 671, 2005.
7. Yang, J.C. et al., A randomized trial of bevacizumab, an anti-vascular endothelial growth factor antibody, for metastatic renal cancer, *N. Engl. J. Med.*, 349, 427, 2003.
8. Mikecz, K., Vitaxin applied molecular evolution, *Curr. Opin. Investig. Drugs*, 2, 199, 2000.
9. Smith, J.W., Cilengitide Merck, *Curr. Opin. Investig. Drugs*, 6, 741, 2003.
10. Bhaskar, V. et al., *M200, an integrin* $\alpha_5\beta_1$ *antibody, promotes cell death in proliferating endothelial cells.* Protein Design Labs, Inc. Poster Presentation, AACR, 2003.
11. Hirsch, E., Brancaccio, M., and Altruda, F., Tissue-specific KO of ECM proteins, *Methods Mol. Biol.*, 139, 147, 2000.
12. Akalu, A. and Brooks, P.C., Matrix, extracellular and interstitial, *Encycl. Mol. Cell Biol. Mol. Med.*, 8. Second Edition, Edited by Robert A. Myers. Wiley-VCH Verlag GMbH & Co., KGaA, 45, 2005.

13. Mott, J.D. and Werb, Z., Regulation of matrix biology by matrix metalloproteinases, *Curr. Opin. Cell Biol.*, 16, 558, 2004.
14. Kalluri, R., Basement membranes: structure, assembly and role in tumor angiogenesis, *Nat. Rev. Cancer*, 3, 422, 2003.
15. Chen, C.S., Tan, J., and Tien, J., Mechanotransduction at cell–matrix and cell–cell contacts, *Annu. Rev. Biomed. Eng.*, 6, 275, 2004.
16. Giancotti, F.G. and Tarone, G., Positional control of cell fate through joint integrin/receptor protein kinase signaling, *Annu. Rev. Cell Dev. Biol.*, 19, 173, 2003.
17. Stupack, D.G., Integrins as a distinct subtype of dependence receptors, *Cell Death Diff.*, 8, 1021, 2005.
18. Brader, S. and Eccles, S.A., Phosphoinositide 3-kinase signaling pathways in tumor progression, invasion and angiogenesis, *Tumori*, 90, 2, 2004.
19. Wijelath, E.S. et al., Novel vascular endothelial growth factor binding domains of fibronectin enhance vascular endothelial growth factor biological activity, *Cir. Res.*, 91, 25, 2002.
20. Lee, S. et al., Processing of VEGF-A by matrix metalloproteinases regulate bioavailability and vascular patterning in tumors, *J. Cell Biol.*, 169, 681, 2005.
21. Pounce, M.L., Nomizu, M., and Kleinman, H.K., An angiogenic laminin site and its antagonists bind through the $\alpha_v\beta_3$ and $\alpha_5\beta_1$ integrins, *FASEB J.*, 15, 1389, 2001.
22. Flamme, I., Frolich, T., and Riasu, W., Molecular mechanisms of vasculogenesis and embryonic angiogenesis, *J. Cell Physiol.*, 173, 206, 1997.
23. Tyagi, S.C., Vasculogenesis and angiogenesis: extracellular matrix remodeling in coronary collateral arteries and the ischemic heart, *J. Cell Biochem.*, 65, 388, 1997.
24. Scholz, D., Cai, W.J., and Schaper, W., Arteriogenesis, a new concept of vascular adaptation in occlusive disease, *Angiogenesis*, 4, 247, 2001.
25. Noden, D.M., Embryonic origins and assembly of blood vessels, *Ann. Rev. Respir. Dis.*, 140, 1097, 1989.
26. Risau, W., Mechanisms of angiogenesis, *Nature*, 368, 1997.
27. Rafii, S., Lyden, D., and Benezra, R., Vascular and haematopoietic stem cells: novel targets for anti-angiogenesis therapy, *Nat. Rev. Cancer*, 2, 826, 2002.
28. Dignat-George, J. et al., Circulating endothelial cells: realities and promises in vascular disorders, *Pathophysiol. Haemost. Thromb.*, 5, 495, 2003.
29. Blood, C.H. and Zetter, B.R., Tumor interactions with the vasculature: angiogenesis and tumor metastasis, *Biochem. Acta*, 1032, 89, 1990.
30. Cockerill, G.W., Gamble, J.R., and Vadas, M.A., Angiogenesis: models and modulators, *Int. Rev. Cyt.*, 159, 113, 1995.
31. D'Amore, P.A. and Thompson, R.W., Mechanisms of angiogenesis, *Ann. Rev. Physiol.*, 49, 453, 1987.
32. Folkman, J., The role of angiogenesis in tumor growth, *Cancer Biol. Semin.*, 3, 65, 1992.
33. Paku, S. and Paweletz, N., First steps of tumor-related angiogenesis, *Lab. Invest.*, 65, 334, 1991.
34. Fox, S.B., Gatter, K.C., and Harris, A.L., Tumor angiogenesis, *J. Pathol.*, 179, 232, 1996.
35. Auerbach, W. and Auerbach, R., Angiogenesis inhibition: a review, *Pharmac. Ther.*, 63, 265, 1994.
36. Sholley, M.M. et al., Mechanisms of neovascularization: vascular sprouting can occur without proliferation of endothelial cells, *Lab. Invest.*, 51, 624, 1984.
37. Leek, D., Harris, A.L., and Lewis, C.E., Cytokine networks in solid human tumors: regulation of angiogenesis, *Leuk. Biol.*, 55, 423, 1994.
38. Sunderkotter, C. et al., Macrophages and angiogenesis, *J. Leuk. Biol.*, 55, 410, 1994.
39. Davis, G.E., Bayless, K.J., and Mavila, A., Molecular basis of endothelial cell morphogenesis in three-dimensional extracellular matrices, *Anat. Rec.*, 268, 252, 2002.
40. Brooks, P.C., Cell adhesion molecules in angiogenesis, *Cancer Met. Rev.*, 15, 187, 1996.
41. Brooks, P.C., Role of integrins in angiogenesis, *Eur. J. Cancer*, 32A, 2423, 1996.
42. Stad, R.K. and Buurman, W.A., Current views on structure and function of endothelial adhesion molecules, *Cell Adhes. Commun.*, 2, 261, 1994.
43. Hynes, R.O., Integrins: versatility, modulation and signaling in cell adhesion, *Cell*, 69, 11, 1992.
44. Yamada, K.M. and Miyamoto, S., Integrin transmembrane signaling and cytoskeletal control, *Curr. Opin. Cell Biol.*, 7, 681, 1995.

45. Sastry, S.K. and Horwitz, A.F., Integrin cytoplasmic domains: mediators of cytoskeletal linkages and extra- and intracellular initiated transmembrane signaling, *Curr. Opin. Cell Biol.*, 5, 819, 1993.
46. Brooks, P.C. et al., Localization of matrix metalloproteinase MMP-2 to the surface of invasive cells by interaction with integrin $\alpha_v\beta_3$, *Cell*, 85, 683, 1996.
47. Brooks, P.C. et al., Disruption of angiogenesis by PEX, a noncatalytic metalloproteinase fragment with integrin binding activity, *Cell*, 92, 391, 1998.
48. Yamada, K.M. and Geiger, B., Molecular interactions in cell adhesion complexes, *Curr. Opin. Cell Biol.*, 9, 76, 1997.
49. Lampugnani, M.G. and Gejana, E., Interendothelial junctions: structure, signaling and functional roles, *Curr. Opin. Cell Biol.*, 9, 674, 1997.
50. Bischoff, J., Approaches to studying cell adhesion in angiogenesis, *Trends Cell Biol.*, 6, 69, 1995.
51. Seiki, M., Koshikawa, N., and Yana, I., Role of pericellular proteolysis by membrane-type 1 matrix metalloproteinase in cancer invasion and angiogenesis, *Cancer Met. Rev.*, 22, 129, 2003.
52. Paul, R.G. and Bailey, A.J., The effect of advanced glycation end-product formation upon cell–matrix interactions, *Int. J. Biochem. Cell Biol.*, 31, 653, 1999.
53. McCarthy, A.D. et al., Advanced glycation end products interfere with integrin-mediated osteoblastic attachment to a type-I collagen matrix, *Int. J. Biochem. Cell. Biol.*, 36, 840, 2004.
54. Sakata, N. et al., Possible involvement of altered RGD sequence in reduced adhesive and spreading activities of advanced glycation end product-modified fibronectin to vascular smooth muscle cells, *Connect. Tissue Res.*, 41, 213, 2000.
55. Itoh, T. et al., Reduced angiogenesis and tumor progression in gelatinase A-deficient mice, *Cancer Res.*, 58, 1048, 1998.
56. Lambert, V. et al., MMP-2 and MMP-9 synergize in promoting choroidal neovascularization, *FASEB J.*, 17, 2290, 2003.
57. Masson, V. et al., Contribution of host MMP-2 and MMP-9 to promote tumor vascularization and invasion of malignant keratinocytes, *FASEB J.*, 19, 234, 2005.
58. Hangai, M. et al., Matrix metalloproteinase-9-dependent exposure of a cryptic migratory control site in collagen is required before retinal angiogenesis, *Am. J. Pathol.*, 161, 1429, 2002.
59. Rundhaug, J.E., Matrix metalloproteinases and angiogenesis, *J. Cell Mol. Med.*, 9, 267, 2005.
60. Noel, A. et al., Membrane associated proteases and their inhibitors in tumor angiogenesis, *J. Clin. Pathol.*, 57, 577, 2004.
61. Davis, G.E. et al., Regulation of tissue injury responses by the exposure of matricryptic sites within extracellular matrix molecules, *Am. J. Pathol.*, 156, 1489, 2000.
62. Schenk, S. and Quaranta, V., Tales from the cryptic sites of the extracellular matrix, *Trends Cell Biol.*, 13, 366, 2003.
63. Xu, J. et al., Generation of monoclonal antibodies to cryptic collagen sites by using subtractive immunization, *Hybridoma*, 19, 375, 2000.
64. Xu, J. et al., Proteolytic exposure of a cryptic site within collagen type IV is required for angiogenesis and tumor growth in vivo, *J. Cell Biol.*, 154, 1069, 2001.
65. Giannelli, G. et al., Induction of cell migration by matrix metalloprotease-2 cleavage of laminin-5, *Science*, 277, 225, 1997.
66. Bergers, G. et al., Matrix metalloproteinase-9 triggers the angiogenic switch during carcinogenesis, *Nat. Cell Biol.*, 2, 737, 2000.
67. Dumin, J.A. et al., Pro-collagenase-1 (matrix metalloproteinase-1) binds the $\alpha_2\beta_1$ integrin upon release from keratinocytes migrating on type-I collagen, *J. Biol. Chem.*, 31, 29368, 2001.
68. Yu, Q. and Stamenkovic, I., Localization of metalloproteinase 9 to the surface provides a mechanism for CD44-mediated tumor invasion, *Genes Dev.*, 1, 35, 1999.
69. Wei, Y. et al., Urokinase receptors promote β_1 integrin function through interactions with integrin $\alpha_3\beta_1$, *Mol. Biol. Cell*, 12, 2975, 2001.
70. Gellert, G.C., Goldfarb, R.H., and Kitson, R.P., Physical association of uPAR with the av integrin on the surface of human NK cells, *Biochem. Biophys. Res. Commun.*, 315, 1025, 2004.
71. Wei, Y. et al., Regulation of $\alpha_5\beta_1$ integrin conformation and function by urokinase receptor binding, *J. Cell Biol.*, 168, 501, 2005.
72. Ortega, N. and Werb, Z., New functional roles for non-collagenous domains of basement membrane collagens, *J. Cell Sci.*, 115, 4201, 2002.

73. Bix, G. and Lozzo, R.V., Matrix revolutions: "tails" of basement-membrane components with angiostatic functions, *Trends Cell Biol.*, 15, 52, 2005.
74. Petitclerc, E. et al., New functions for non-collagenous domains of human collagen type IV. Novel integrin ligands inhibiting angiogenesis and tumor growth in vivo, *J. Biol. Chem.*, 275, 8051, 2000.
75. Nyberg, P., Xie, L., and Kalluri, R., Endogenous inhibitors of angiogenesis, *Cancer Res.*, 65, 3967, 2005.
76. Roth, J.M. et al., Recombinant α_2(IV)NC1 domain inhibits tumor cell–extracellular matrix interactions, induces cellular senescence, and inhibits tumor growth in vivo, *Am. J. Pathol.*, 166, 901, 2005.
77. Xiong, J.-P. et al., Integrins, cations and ligands: making the connection, *J. Thromb. Haemost.*, 1, 1642, 2003.
78. Hynes, R.O., The emergence of integrins: a personal and historical perspective, *Matrix Biol.*, 23, 333, 2004.
79. Pytela, R., Piershbacher, M.D., and Ruosahti, E., Identification and isolation of a 140 kd cell surface glycoprotein with properties expected of a fibronectin receptor, *Cell*, 40, 191, 1985.
80. Bennett, J.S., Vilaire, G., and Cines, D.B., Identification of the fibrinogen receptor on human platelets by photoaffinity labeling, *J. Biol. Chem.*, 257, 8049, 1982.
81. Arnaout, M.A., Integrin structure: new twists and turns in dynamic cell adhesion, *Immunol. Rev.*, 186, 125, 2002.
82. Mould, A.P. and Humphries, M.J., Regulation of integrin function through conformational complexity: not simply a knee-jerk reaction, *Curr. Opin. Cell Biol.*, 16, 544, 2004.
83. Adair, B.D. et al., Three-dimensional EM structure of the ectodomain of integrin $\alpha_v\beta_3$ in a complex with fibronectin, *J. Cell Biol.*, 168, 1109, 2005.
84. De Melker, A.A. and Sonnenberg, A., Integrins: alternative splicing as a mechanism to regulate signaling events, *Bio. Essays*, 21, 499, 1999.
85. Shimaoka, M. and Springer, T.A., Therapeutic antagonists and conformational regulation of integrin function, *Nat. Rev. Drug Dis.*, 2, 703, 2003.
86. Xiong, J.P. et al., Crystal structure of the extracellular segment of integrin $\alpha_v\beta_3$, *Science*, 294, 339, 2001.
87. Carragher, N.O. and Frame, M.C., Focal adhesion and actin dynamics: a place where kinases and proteases meet to promote invasion, *Trends Cell Biol.*, 14, 241, 2004.
88. Grashoff, C. et al., Integrin-linked kinase: integrin's mysterious partner, *Curr. Opin. Cell Biol.*, 16, 565, 2004.
89. Critchley, D.R., Cytoskeletal proteins talin and vinculin in integrin-mediated adhesion, *Biochem. Soc. Trans.*, 32, 831, 2004.
90. Zaidel-Bar, R. et al., Hierarchical assembly of cell–matrix adhesion complexes, *Biochem. Soc. Trans.*, 32, 416, 2004.
91. Humphries, M.J. et al., Mechanisms of integration of cells and extracellular matrices by integrins, *Biochem. Soc. Trans.*, 32, 822, 2004.
92. Yamada, K.M., Pankov, R., and Cukierman, E., Dimensions and dynamics in integrin function, *Braz. J. Med. Biol. Res.*, 36, 959, 2003.
93. Nayal, A., Webb, D.J., and Horwitz, A.F., Talin: an emerging focal point of adhesion dynamics, *Curr. Opin. Cell Biol.*, 16, 94, 2004.
94. Calderwood, D.A. and Ginsberg, M.H., Talin forges the links between integrins and actin, *Nat. Cell Biol.*, 5, 694, 2003.
95. Kuphal, S., Bauer, R., and Bosserhoff, A.-K., Integrin signaling in malignant melanoma, *Cancer Met. Rev.*, 24, 195, 2005.
96. Parise, L.V., Lee, J.W., and Juliano, R.L., New aspects of integrin signaling in cancer, *Semin. Cancer Biol.*, 10, 407, 2000.
97. Miranti, C.K. and Brugge, J.S., Sensing the environment: a historical perspective on integrin signal transduction, *Nat. Cell Biol.*, 4, E83, 2002.
98. Katsumi, A. et al., Integrins in mechanotransduction, *J. Biol. Chem.*, 279, 12001, 2004.
99. Ruoslahti, E., The RGD story: a personal account, *Matrix Biol.*, 22, 459, 2003.
100. Takagi, J., Structural basis for ligand recognition by RGD (Arg–Gly–Asp)-dependent integrins, *Biochem. Soc. Trans.*, 32, 403, 2004.

101. Kim, M., Carmen, C.V., and Springer, T.A., Bidirectional transmembrane signaling by cytoplasmic domain separation in integrins, *Science*, 301, 1720, 2003.
102. Shen, T. et al., Conditional knockout of focal adhesion kinase in endothelial cells reveals its role in angiogenesis and vascular development in late embryogenesis, *J. Cell Biol.*, 169, 941, 2005.
103. McLean, G.W. et al., The role of focal-adhesion kinase in cancer—a new therapeutic opportunity, *Nat. Rev. Cancer*, 5, 505, 2005.
104. Friedlander, M. et al., Definition of two angiogenic pathways by distinct alpha v integrins, *Science*, 270, 1500, 1995.
105. Hood, J.D. et al., Differential alpha v integrin-mediated Ras-ERK signaling during two pathways of angiogenesis, *J. Cell Biol.*, 162, 933, 2003.
106. Alavi, A. et al., Role of Raf in vascular protection from distinct apoptotic stimuli, *Science*, 301, 94, 2003.
107. Taverna, D. et al., Increased primary tumor growth in mice null for beta3- or beta3/beta5-integrins or selectins, *Proc. Natl. Acad. Sci. USA*, 101, 763, 2004.
108. Reynolds, A.R. et al., Elevated Flk1 (vascular endothelial growth factor receptor 2) signaling mediates enhanced angiogenesis in beta3-integrin-deficient mice, *Cancer Res.*, 64, 8643, 2004.
109. Watnick, R.S. et al., Ras modulates Myc activity to repress thrombospondin-1 expression and increase tumor angiogenesis, *Cancer Cell*, 3, 219, 2003.
110. Stromblad, S. et al., Suppression of p53 activity and p21WAF1/CIP1 expression by vascular cell integrin $\alpha_v\beta_3$ during angiogenesis, *J. Clin. Invest.*, 98, 426, 1996.
111. Bao, W. et al., Cell attachment to the extracellular matrix induces proteasomal degradation of p21(CIP1) via Cdc42/Rac1 signaling, *Mol. Cell Biol.*, 22, 4587, 2002.
112. Iivanainen, E. et al., Endothelial cell–matrix interactions, *Microsc. Res. Tech.*, 60, 13, 2003.
113. Hwang, R. and Varner, J., The role of integrins in tumor angiogenesis, *Hematol. Oncol. Clin. North Am.*, 18, 991, 2004.
114. Ruegg, C. and Mariotti, A., Vascular integrins: pleiotropic adhesion and signaling molecules in vascular homeostasis and angiogenesis, *Cell Mol. Life Sci.*, 60, 1135, 2003.
115. Gutheil, J.C. et al., Targeted antiangiogenic therapy for cancer using Vitaxin: a humanized monoclonal antibody to the integrin $\alpha_v\beta_3$, *Clin. Cancer Res.*, 6, 3056, 2000.
116. Patel, S.R. et al., Pilot study of Vitaxin—an angiogenesis inhibitor—in patients with advanced leiomyosarcomas, *Cancer*, 92, 1347, 2001.
117. Posey, J.A. et al., A pilot trial of Vitaxin, a humanized anti-vitronectin receptor (anti-alpha v beta 3) antibody in patients with metastatic cancer, *Cancer Biother. Radiopharm.*, 2, 125, 2001.
118. Trikha, M. et al., CNTO 95, a fully human monoclonal antibody that inhibits av integrins, has antitumor and antiangiogenic activity in vivo, *Int. J. Cancer*, 110, 326, 2004.
119. Boersma, E. and Westerhout, C.M., Intravenous glycoprotein IIb/IIIa inhibitors in acute coronary syndromes: lessons from recently conducted randomized clinical trials, *Curr. Opin. Investig. Drugs*, 5, 313, 2004.
120. Hijazi, Y. et al., Pharmacokinetics, safety, and tolerability of R411, a dual alpha4beta1–alpha4beta7 integrin antagonist after oral administration at single and multiple once-daily ascending doses in healthy volunteers, *J. Clin. Pharmacol.*, 44, 1368, 2004.
121. Jung, Y.D. et al., The role of the microenvironment and intercellular cross-talk in tumor angiogenesis, *Semin. Cancer Biol.*, 12, 105, 2002.
122. Betsholtz, C., Lindblom, P., and Gerhardt, H., Role of pericytes in vascular morphogenesis, *EXS*, 94, 115, 2005.
123. Moldovan, L. and Moldovan, N.I., Role of monocytes and macrophages in angiogenesis, *EXS*, 94, 127, 2005.
124. Boger, D.L. et al., Identification of a novel class of small-molecule antiangiogenic agents through the screening of combinatorial libraries which function by inhibiting the binding and localization of proteinase MMP2 to integrin $\alpha_v\beta_3$, *J. Am. Chem. Soc.*, 123, 1280, 2001.
125. Furlan, F. et al., The soluble D2D3 (88–274) fragment of the urokinase receptor inhibits monocyte chemotaxis and integrin-dependent cell adhesion, *J. Cell Sci.*, 117, 2909, 2004.
126. Silletti, S. et al., Disruption of matrix metalloproteinase 2 binding to integrin $\alpha_v\beta_3$ by an organic molecule inhibits angiogenesis and tumor growth in vivo, *Proc. Natl. Acad. Sci. USA*, 98, 119, 2001.

127. Pozzi, A. et al., Elevated matrix metalloprotease and angiostatin levels in integrin α_1 knockout mice cause reduced tumor vascularization, *Proc. Natl. Acad. Sci. USA*, 97, 2202, 2000.
128. Merneros, A.G. and Olsen, B.R., Physiological role of collagen XVIII and endostatin, *FASEB J.*, 19, 716, 2005.
129. Gould, D.B. et al., Mutations in Col4a1 cause perinatal cerebral hemorrhage and porencephaly, *Science*, 308, 1167, 2005.
130. Senger, D.R. et al., Angiogenesis promoted by vascular endothelial growth factor: regulation through $\alpha_1\beta_1$ and $\alpha_2\beta_1$ integrins, *Proc. Natl. Acad. Sci. USA*, 94, 13612, 1997.
131. Henriet, P. et al., Contact with fibrillar collagen inhibits melanoma cell proliferation by up-regulating p27KIP1, *Proc. Natl. Acad. Sci. USA*, 97, 10026, 2000.
132. Bix, G. et al., Endorepellin causes endothelial cell disassembly of actin cytoskeleton and focal adhesions through $\alpha_2\beta_1$ integrin, *J. Cell Biol.*, 166, 97, 2004.
133. Maeshima, Y. et al., Tumstatin, an endothelial cell-specific inhibitor of protein synthesis, *Science*, 295, 140, 2002.
134. Sudhakar, A. et al., Human tumstatin and human endostatin exhibit distinct antiangiogenic activities mediated by $\alpha_v\beta_3$ and $\alpha_5\beta_1$ integrins, *Proc. Natl. Acad. Sci. USA*, 100, 4766, 2003.
135. Kreidberg, J.A., Functions of $\alpha_3\beta_1$ integrin, *Curr. Opin. Cell Biol.*, 12, 548, 2000.
136. Chandrasekaran, L. et al., Cell contact-dependent activation of $\alpha_3\beta_1$ integrin modulates endothelial cell responses to thrombospondin-1, *Mol. Biol. Cell*, 11, 2885, 2000.
137. Short, S.M. et al., Inhibition of endothelial cell migration by thrombospondin-1 type-1 repeats is mediated by β_1 integrins, *J. Cell Biol.*, 168, 643, 2005.
138. Weidle, U.H. and Konig, B., Urokinase receptor antagonists: novel agents for the treatment of cancer, *Expert Opin. Investig. Drugs*, 7, 391, 1998.
139. Seo, D.W. et al., TIMP-2 mediated inhibition of angiogenesis: an MMP-independent mechanism, *Cell*, 114, 171, 2003.
140. Calzada, M.J. et al., Alpha4beta1 integrin mediates selective endothelial cell responses to thrombospondins 1 and 2 in vitro and modulates angiogenesis in vivo, *Circ. Res.*, 94, 462, 2004.
141. Taverna, D. and Hynes, R.O., Reduced blood vessel formation and tumor growth in alpha 5-integrin-negative teratocarcinomas and embryoid bodies, *Cancer Res.*, 61, 5255, 2001.
142. Yang, J.T., Rayburn, H., and Hynes, R.O., Cell adhesion events mediated by alpha 4 integrins are essential in placental and cardiac development, *Development*, 121, 549, 1995.
143. Yang, J.T., Rayburn, H., and Hynes, R.O., Embryonic mesodermal defects in alpha 5 integrin-deficient mice, *Development*, 119, 1093, 1993.
144. George, E.L. et al., Defects in mesoderm, neural tube and vascular development in mouse embryos lacking fibronectin, *Development*, 119, 1079, 1993.
145. Georges-Labouesse, E.N. et al., Mesodermal development in mouse embryos mutant for fibronectin, *Dev. Dyn.*, 207, 145, 1996.
146. Tremble, P., Chiquet-Ehrismann, R., and Werb, Z., The extracellular matrix ligands fibronectin and tenascin collaborate in regulating collagenase gene expression in fibroblasts, *Mol. Biol. Cell*, 5, 439, 1994.
147. Lee, B.H. and Ruoslahti, E., $\alpha_5\beta_1$ integrin stimulates Bcl-2 expression and cell survival through Akt, focal adhesion kinase, and Ca(2+)/calmodulin-dependent protein kinase IV, *J. Cell Biochem.*, 95, 1214, 2005.
148. Zhang, Z. et al., The $\alpha_5\beta_1$ integrin supports survival of cells on fibronectin and up-regulates Bcl-2 expression, *Proc. Natl. Acad. Sci. USA*, 92, 6161, 1995.
149. Kim, S. et al., Regulation of angiogenesis in vivo by ligation of integrin 5β1 with the central cell-binding domain of fibronectin, *Am. J. Pathol.*, 156, 1345, 2000.
150. Kim, S., Harris, M., and Varner, J.A., Regulation of integrin $\alpha_v\beta_3$-mediated endothelial cell migration and angiogenesis by integrin $\alpha_5\beta_1$ and protein kinase A, *J. Biol. Chem.*, 275, 33920, 2000.
151. Hanai, J. et al., Endostatin is a potential inhibitor of Wnt signaling, *J. Cell Biol.*, 158, 529, 2002.
152. Abdollahi, A. et al., Endostatin's antiangiogenic signaling network, *Mol. Cell*, 13, 649, 2004.
153. Boudreau, N.J. and Varner, J.A., The homeobox transcription factor Hox D3 promotes integrin $\alpha_5\beta_1$ expression and function during angiogenesis, *J. Biol. Chem.*, 279, 4862, 2004.
154. Grant, D.S. and Kleinman, H.K., Regulation of capillary formation by laminin and other components of the extracellular matrix, *EXS*, 79, 317, 1997.

155. Grant, D.S., The role of basement membrane in angiogenesis and tumor growth, *Pathol. Res. Pract.*, 190, 854, 1994.
156. Tejindervir, S.H. et al., Endothelial expression of the $\alpha_6\beta_4$ integrin is negatively regulated during angiogenesis, *J. Cell Sci.*, 116, 3771, 2003.
157. Mercurio, A.M. et al., Autocrine signaling in carcinoma: VEGF and the α6β4 integrin, *Semin. Cancer Biol.*, 14, 115, 2004.
158. Nikolopoulos, S.N. et al., Integrin β_4 signaling promotes tumor angiogenesis, *Cancer Cell*, 6, 471, 2004.
159. Brooks, P.C., Clark, R.A., and Cheresh, D.A., Requirement of vascular integrin $\alpha_v\beta_3$ for angiogenesis, *Science*, 264, 569, 1994.
160. Brooks, P.C., Cell adhesion molecules in angiogenesis, *Cancer Met. Rev.*, 15, 187, 1996.
161. Gasparini, G. et al., Vascular integrin $\alpha_v\beta_3$: a new prognostic indicator in breast cancer, *Clin. Cancer Res.*, 4, 2625, 1998.
162. Bar-Shavit, R. et al., The involvement of thrombin RGD in metastasis: characterization of a cryptic adhesive site, *Isr. J. Med. Sci.*, 31, 86, 1995.
163. Caunt, M. et al., Thrombin induces neoangiogenesis in the chick chorioallantoic membrane, *J. Thromb. Haemost.*, 1, 2097, 2003.
164. Wei, Y. et al., Regulation of integrin function by the urokinase receptor, *Science*, 273, 1551, 1996.
165. Xue, W. et al., Urokinase-type plasminogen activator receptors associate with β_1 and β_3 integrins of fibrosarcoma cells: dependence on extracellular matrix components, *Cancer Res.*, 57, 1682, 1997.
166. Ossowski, L. and Aguirre-Ghiso, J.A., Urokinase receptor and integrin partnership: coordination of signaling for cell adhesion, migration and growth, *Curr. Opin. Cell Biol.*, 12, 613, 2000.
167. Hall, H. and Hubbell, J.A., Matrix-bound sixth Ig-like domain of cell adhesion molecule L1 acts as an angiogenic factor by ligating $\alpha_v\beta_3$-integrin and activating VEGF-R2, *Microvasc. Res.*, 68, 169, 2004.
168. Brigstock, D.R., Regulation of angiogenesis and endothelial cell function by connective tissue growth factor (CTGF) and cysteine-rich 61 (CYR61), *Angiogenesis*, 5, 153, 2002.
169. Bergh, J.J. et al., Integrin $\alpha_v\beta_3$ contains a cell surface receptor site for thyroid hormone that is linked to activation of mitogen-activated protein kinase and induction of angiogenesis, *Endocrinology*, 146, 2864, 2005.
170. Kumar, C.C. et al., Inhibition of angiogenesis and tumor growth by SCH221153, a dual $\alpha_v\beta_3$ and $\alpha_v\beta_5$ integrin receptor antagonist, *Cancer Res.*, 61, 2232, 2001.
171. Nisato, R.E. et al., $\alpha_v\beta_3$ and $\alpha_v\beta_5$ integrin antagonists inhibit angiogenesis in vitro, *Angiogenesis*, 6, 105, 2003.
172. Zhou, Q. et al., Contortrostatin, a homodimeric disintegrin, binds to integrin $\alpha_v\beta_5$, *Biochem. Biophys. Res. Commun.*, 267, 350, 2000.
173. Brooks, P.C. et al., Insulin-like growth factor receptor cooperates with integrin $\alpha_v\beta_5$ to promote tumor cell dissemination in vivo, *J. Clin. Invest.*, 99, 1390, 1997.
174. McCarty, J.H. et al., Defective associations between blood vessels and brain parenchyma lead to cerebral hemorrhage in mice lacking av integrins, *Mol. Cell Biol.*, 22, 7667, 2002.
175. Brooks, P.C. et al., Integrin $\alpha_v\beta_3$ antagonists promote tumor regression by inducing apoptosis of angiogenic blood vessels, *Cell*, 79, 1157, 1994.
176. Montgomery, A.M., Reisfeld, R.A., and Cheresh, D.A., Integrin $\alpha_v\beta_3$ rescues melanoma cells from apoptosis in three-dimensional dermal collagen, *Proc. Natl. Acad. Sci. USA*, 91, 8856, 1994.
177. Petitclere, E. et al., Integrin $\alpha_v\beta_3$ promotes M21 melanoma growth in human skin by regulating tumor cell survival, *Cancer Res.*, 59, 2724, 1999.
178. Stupack, D.G. et al., Apoptosis of adherent cells by recruitment of caspase-8 to unligated integrins, *J. Cell Biol.*, 155, 459, 2001.
179. Rawling, N.G. et al., Localization of integrin $\alpha_v\beta_3$ and vascular endothelial growth factor receptor-2 (KDR/Flk-1) in cutaneous and oral melanomas of dog, *Histol. Histopathol.*, 3, 819, 2003.
180. Hutchings, H., Ortega, N., and Plouet, J., Extracellular matrix-bound vascular endothelial growth factor promotes endothelial cell adhesion, migration, and survival through integrin ligation, *FASEB J.*, 17, 1520, 2003.

181. De, S. et al., VEGF-integrin interplay controls tumor growth and vascularization, *Proc. Natl. Acad. Sci. USA*, 102, 7589, 2005.
182. Nam, J.O. et al., Regulation of tumor angiogenesis by fastatin, the fourth FAS1 domain of big-β_3, via $\alpha_v\beta_3$ integrin, *Cancer Res.*, 65, 4153, 2005.
183. Hood, J.D. et al., Tumor regression by targeted gene delivery to the neovasculature, *Science*, 296, 2404, 2002.
184. Chen, X. et al., Synthesis and biological evaluation of dimeric RGD peptide-paclitaxel conjugated as a model for integrin-targeted drug delivery, *J. Med. Chem.*, 48, 1098, 2005.
185. Bellon, G., Martiny, L., and Robinet, A., Matrix metalloproteinase and matrikines in angiogenesis, *Crit. Rev. Oncol. Hematol.*, 49, 203, 2004.
186. Kiyoshima, K. et al., Overexpression of laminin-5 gamma-2 chain and its prognostic significance in urothelial carcinoma of urinary bladder: association with expression of cyclooxygenase 2, epidermal growth factor, and human epidermal growth factor 2, *Hum. Pathol.*, 36, 522, 2005.
187. Ebbinghaus, C. et al., Diagnostic and therapeutic applications of recombinant antibodies: targeting the extra-domain B of fibronectin, a marker of tumor angiogenesis, *Curr. Pharm. Des.*, 10, 1537, 2004.
188. Menrad, A. and Menssen, H.D., ED-B fibronectin as a target for antibody-based cancer treatments, *Expert. Opin. Ther. Targets*, 9, 491, 2005.
189. Lobov, I.B., Brooks, P.C., and Lang, R.A., Angiopoietin-2 displays VEGF-dependent modulation of capillary structure and endothelial cell survival in vivo, *Proc. Natl. Acad. Sci. USA*, 99, 11205, 2002.

13 Conventional Therapeutics with Antiangiogenic Activity

Christian Hafner, Thomas Vogt, and Albrecht Reichle

CONTENTS

In search of new strategies for the treatment of advanced and metastasized cancer, stroma-targeted therapies have attracted the attention of scientists and clinicians in the last years. The microenvironment is crucial for tumor survival and growth, and the tumor cells can influence stroma functions and therefore favor invasive and metastatic behavior. Consequently, the tumor–stroma interaction has evolved into a promising new field for molecular-targeted approaches in oncology. Stroma-targeted therapy strategies represent a new dogma in oncology. Established cytotoxic chemotherapy schedules had been developed to kill as many tumor cells as possible by "maximum tolerated doses" of chemotherapeutic drugs. As a consequence, other tissues with a high proliferation rate were also affected through these therapies, resulting in severe and dose-limiting side effects like bone marrow insufficiency, gastrointestinal toxicity, and hair loss. Furthermore, the duration of response usually is limited, since the tumors develop drug resistance after repeated cycles of therapy and the patients experience a relapse. Therefore, the overall benefit regarding survival and quality of life remains only modest in many cancers.[1]

In contrast, stroma-targeted approaches seem to have some major advantages. Acquired drug resistance might be delayed or even circumvented over long periods of treatment,[2] since the stroma is genetically more stable than the tumor.[3] Due to lower doses of the applied drugs, the observed side effects of stroma-targeted therapies are usually milder than in conventional chemotherapies, and the quality of life for the patients can therefore be improved. In addition, the concept emerges that the tumor-associated "activated" stroma is different from normal stroma, opening the gate for specific intervention with a sufficient therapeutic index. The aim of stroma-targeted therapies is often to control and stabilize disease and to prolong progression-free survival rates rather than to achieve high response rates, which is also very different to many established high-dose chemotherapy schedules.

Within the stroma, the endothelial cell is a critical component. Tumors cannot grow beyond a critical size (about 100 to 200 μm) or metastasize to other organs without adequate nutritional support through blood vessels. Increased angiogenesis has therefore been identified as a hallmark of cancer.[4] Metabolic stress (hypoglycemia, low pH, low pO_2), concomitant inflammatory cells, genetic mutations (activation of oncogenes or deletion of tumor-suppressor genes controlling the transcription of angiogenic factors), and mechanical factors (e.g., pressure caused by the mass of cancer cells) can shift the highly sensitive balance between pro- and antiangiogenic factors toward angiogenesis.[5]

Some conventional therapeutics, which were initially developed for other indications like metabolic and rheumatic disorders, have revealed also remarkable efficiency against cancer cells (Table 13.1). Beside direct antitumor effects, the inhibition of angiogenesis and further effects on the tumor–stroma interaction are important mechanisms underlying these observations. In this chapter, we focus on several conventional drugs with antiangiogenic activity, including PPARγ agonists, COX-2 inhibitors, thalidomide, and mammalian target of rapamycin (mTOR) antagonists. Furthermore, it has turned out that low-dose continuous application of conventional chemotherapeutic drugs, referred to as metronomic chemotherapy, targets especially the endothelial cells in tumors. Metronomic scheduling of conventional

TABLE 13.1
Selection of Clinical Trials Using Conventional Drugs with Antiangiogenic Activity for Cancer Therapy

Cancer	Drug	Reference
Angiosarcoma	Pioglitazone, rofecoxib, metronomic trofosfamide	98
Breast cancer	Troglitazone	239
Breast cancer	Metronomic methotrexate, metronomic cyclophosphamide	195
Cervical cancer	Celecoxib	240
Colorectal cancer	Rofecoxib	16
Duodenal polyps	Celecoxib	64
Glioblastoma multiforme	Rofecoxib, metronomic temozolamide	221
Melanoma	Pioglitazone, rofecoxib, metronomic trofosfamide	206
Pancreatic cancer	Celecoxib, metronomic 5-fluorouracil	241
Pancreatic cancer	Celecoxib, thalidomide, metronomic irinotecan (*case report*)	222
Prostate cancer	Celecoxib	54
Renal cell carcinoma	CCI-779	160
Soft-tissue sarcoma	Pioglitazone, rofecoxib, metronomic trofosfamide	206

chemotherapeutics is therefore discussed as an effective component of antiangiogenic therapies. The combination of different conventional drugs with antiangiogenic activity or the additional use of "biologicals" like vascular endothelial growth factor (VEGF) antibodies or tyrosine kinase inhibitors may be even more efficient.

13.1 COX-2 INHIBITORS

13.1.1 Expression and Function of COX-2

Two cyclooxygenase (COX) enzyme isoforms can be found, named COX-1 and COX-2. COX-2 represents the inducible isoform of the COX enzyme.[6] COX-2 modulates the conversion of arachidonic acid to prostaglandin H_2, which is further converted to prostaglandin D_2, prostaglandin E_2 (PGE_2), prostaglandin F_2, prostaglandin I_2 (prostacyclin), and thromboxane A_2 (TXA_2). PGE_2 is a major downstream player of COX-2.[7,8] COX-1 is constitutively expressed in a variety of tissues. In contrast, COX-2 is lowly or not expressed in most tissues,[9] but can be induced in inflammation and tumors.[10] COX-2 expression and enzymatic activity have been reported to be upregulated by ultraviolet B light, hypoxia, TNF-α, bacterial lipopolysaccharide (LPS), IL-2, and IL-1β.[11–13] Upregulation of COX-2 is also known from many cancers.[9]

Selective COX-2 inhibitors were initially developed for pain therapy, for example, in rheumatoid arthritis. However, a multitude of in vitro and in vivo studies have revealed that these drugs bear the potential for anticancer efficacy as well. Celecoxib, etoricoxib, rofecoxib, lumiracoxib, valdecoxib, and parecoxib are members of this drug family. However, valdecoxib and rofecoxib have already been withdrawn from the market. Rofecoxib, which is together with celecoxib the most frequently investigated COX-2 inhibitor for cancer therapy, revealed an increased rate of cardiovascular side effects in patients with long-term application (>18 months) of the drug. Valdecoxib followed rofecoxib because of an unfavorable toxicity profile including also cardiovascular events and severe skin reactions. Other COX-2 inhibitors are currently under observation. In our opinion, COX-2 inhibitors nevertheless represent an attractive drug class for the palliative treatment of cancer irrespective of an increased rate of cardiovascular side effects, since the ethical assessment of side effects in these patients is substantially different compared with noncancer patients. A separate approval of COX-2 inhibitors for the use in palliative cancer therapy is therefore imaginable.

13.1.2 Antiangiogenic Mechanisms of COX-2 Inhibitors

COX-2 inhibitors have been successfully used in animal models and preliminary clinical trials of lung cancer,[14] breast cancer,[15] colorectal cancer,[16] gastric cancer,[17] liver cancer,[18] prostate cancer,[19,20] and further cancer entities. The fact that the inhibition of tumor growth by COX-2 inhibitors was more efficient in vivo than in vitro suggested also indirect antineoplastic effects of COX-2 inhibitors mediated by stromal cells. Among these effects on the tumor stroma, angiogenesis has turned out to be a major target of COX-2 inhibitors (Figure 13.1). Tumor-related endothelial cells express COX-2, whereas normal endothelium expresses COX-1.[21] This observation opens the gate for a certain tumor selectivity.[22] COX-2 and its prostaglandin derivatives are fundamentally involved in tumor angiogenesis.[23, 24]

COX-2 inhibitors are able to shift the balance between the proangiogenic VEGF and the antiangiogenic endostatin serum levels in favor of endostatin.[25] Selective COX-2 inhibitors reduced the diameter of tumor vessels in a rat model of pulmonary metastases from colorectal cancer.[26] When COX-2 is overexpressed, the endothelial cell migration and the formation of tubes are significantly enhanced.[27] Likewise, pharmacologic blocking of COX-2 can inhibit neovascularization.[21,28] The proangiogenic effects of COX-2 are mediated by the eicosanoid

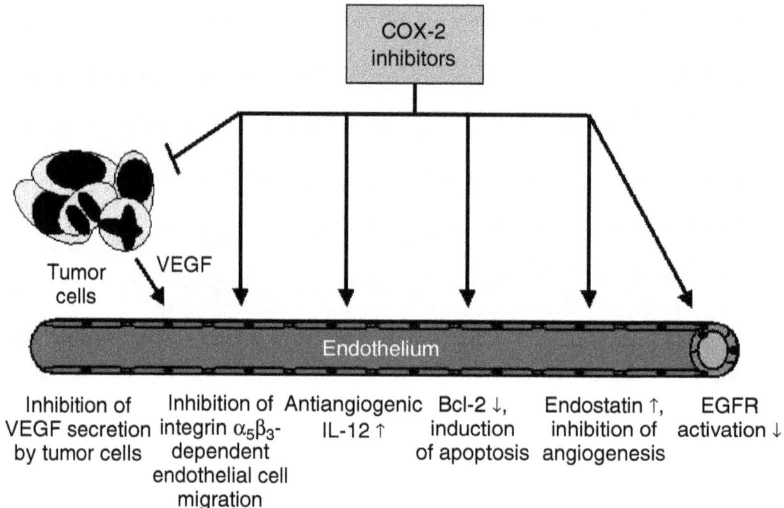

FIGURE 13.1 Antiangiogenic mechanisms of COX-2 inhibitors.

products of this enzyme. PGE_2 enhances the production and release of VEGF.[29] Fibroblasts are also an important source for VEGF release induced by COX-2.[30] The selective COX-2 inhibitor celecoxib can decrease the expression of VEGF in pancreatic cancer cell lines. This leads to a reduced angiogenesis and metastasis in a mouse model, and was mediated by the suppression of Sp1 transcription factor activity, suggesting further COX-2-independent anti-angiogenic mechanisms of COX-2 inhibitors.[31] Celecoxib rather than rofecoxib demonstrated marked antiproliferative effects in a model of prostate carcinoma cells which were correlated with a decreased microvessel density, although these cells did not express COX-2. Therefore, COX-2 inhibitors seem to possess some COX-2-independent anticancer efficacy which may vary between the different drugs.

COX-2 has been found to counteract the apoptosis of endothelial cells by the upregulation of antiapoptotic molecules like Bcl-2 and the activation of Akt.[32–35] Activation of the PGE_2 receptor EP_2 also increases integrin $\alpha_5\beta_3$-dependent endothelial cell migration, adhesion, and spreading.[35–37] Accordingly, COX-2 inhibitors can decrease endothelial migration. Further antiangiogenic effects of COX-2 inhibitors might be mediated by the modulation of matrix metalloproteinases. PGE_2 controls the expression of MMP-2 and the release of MMP-9.[38–40] Both matrix metalloproteinases are key regulators in angiogenesis and show a fatal cooperation during tumor angiogenesis.[41] The downregulation of the antiangiogenic IL-12[42,43] and the transactivation of the epidermal growth factor receptor (EGFR)[44] have also been reported to contribute to the antiangiogenic effects of COX-2 inhibitors. Activation of EGFR in endothelial cells enhances the production of proangiogenic molecules like VEGF, IL-8, and trefoil peptide pS2.[45–47]

Since further eicosanoid derivatives of COX-2 are involved into angiogenesis, the blocking of these derivatives by COX-2 inhibitors has also been supposed to contribute to antiangiogenesis. TXA_2, for example, modulates endothelial cell migration and therefore angiogenesis,[48] and prostaglandin I_2 is a well-known regulator of VEGF-dependent vascular permeability.[49]

Although antiangiogenic effects of COX-2 inhibitors seem to be very important for the antitumor efficacy of this drug class, further effects on the tumor stroma are involved into this action like modulation of immune functions[43,50,51] and extracellular matrix homeostasis.[52]

13.1.3 COX-2 Inhibitors in Clinical Trials

COX-2 inhibitors have already been tested as single agents or, more often, in combination with other antiangiogenic or cytostatic drugs in clinical trials. The first results indicate that these drugs might be effective both in cancer preventive strategies and in the treatment of metastasized and advanced tumors. Celecoxib and also diclofenac, a nonspecific COX inhibitor, could significantly inhibit the growth of neuroblastomas in rats.[53] Rofecoxib reduced the microvessel density in patients with liver metastases of colorectal cancer, although apoptosis and proliferation of the cancer cells were not affected compared with the control group.[16] Celecoxib showed induction of apoptosis and antiangiogenic effects in animal models of breast cancer.[15] In patients with recurrent prostate cancer, celecoxib was effective to delay or prevent disease progression, which was monitored by the measurement of prostate specific antigen (PSA) levels.[54] However, the proportion of antiangiogenic mechanism contributing to the antitumor effects, compared with direct effects on tumor cells and other stroma-dependent effects like modulation of the immune system, is unknown. Especially in patients with advanced disease, the use of COX-2 inhibitors in combination with other drugs seems to be more promising in terms of tumor mass reduction and avoidance of drug resistance (which is discussed later in this chapter).

Many studies have reported successful application of COX-2 inhibitors for the prevention of tumor development. Celecoxib seems to have possible chemopreventive effects for oral carcinomas in mice.[55] Oral carcinomas often arise multifocally according to the "field cancerization" theory. Pharmacological prevention and early treatment is therefore of special interest in tumors with field cancerization. Interestingly, celecoxib could prevent the development of esophageal inflammation–metaplasia–adenocarcinoma sequence in rats.[56] Likewise, COX-2 inhibitors were effective in the treatment of prostatic intraepithelial neoplasia.[20] In colorectal adenoma and cancer, the long-term use of nonsteroidal anti-inflammatory drugs (NSAIDs) decreased the risk for the development of these lesions.[57] In addition, the long-term use of NSAIDs was correlated with a decreased risk for the development of stomach, breast, lung, esophagus, bladder, prostate, and ovary cancer,[58–60] whereas in pancreatic cancer NSAIDs seem to act not preventively[61] or might even increase the risk for these tumors in females.[62] Patients with familial adenomatous polyposis (FAP), who received 400 mg of celecoxib twice daily for 6 months, showed a significant reduction in the number of colorectal polyps.[63] Similar results were obtained for duodenal polyps.[64] Celecoxib prevented tumor-induced wasting in mouse models of colon carcinoma and head and neck cancer, suggesting advantageous effects of this drug also in palliative therapy situations when no response or prolongation of progression-free survival can be achieved.[65]

13.2 THALIDOMIDE

13.2.1 History

Thalidomide [α-(N-phtalimido)glutarimide] is a synthetic derivative of glutamic acid. The drug is a racemic mixture of S(−) and R(+) enantiomers. It was initially developed as a sedative and antiemetic drug. In the mid-1950s, thalidomide was very popular because of its obvious lack of toxicity and therefore frequently used by pregnant women. In 1961, a relationship between the use of thalidomide in pregnancy and the occurrence of severe limb defects (dysmelia) in newborns was reported.[66] As a consequence, thalidomide was withdrawn from the market. A few years later, the beneficial use of thalidomide for the treatment of erythema nodosum leprosum was reported in dermatology, suggesting some anti-inflammatory activity of this drug.[67] This observation was the starting point for further investigations of thalidomide application in various diseases including cancer. It has now turned out that thalidomide has remarkable antineoplastic activity. Currently, several clinical trials have already been finished or are on the way, investigating thalidomide for different cancer entities like multiple myeloma, malignant melanoma, prostate cancer, glioma, or renal cell cancer.[68–72] Synthetic analogs of thalidomide, also referred to as immunomodulatory drugs (IMiDs), have been developed to reduce the side effects of thalidomide and to enhance the anti-inflammatory and anticancer effects. Indeed, IMiDs revealed significantly more anti-angiogenic potential than thalidomide in in vitro assays.[73] Among those IMiDs, CC-4047 and CC-5013 have already entered clinical trials. Other analogs like CPS11 and CPS49 represent promising candidates which have been tested successfully in preclinical models.[74]

13.2.2 Antiangiogenic Mechanisms of Thalidomide

The anticancer effects partly rely on the antiangiogenic potential of thalidomide (Figure 13.2). It has not been clarified yet whether the teratogenic effects of thalidomide are linked to the antiangiogenic potential. Folkman and colleagues demonstrated in a rabbit cornea micro-pocket assay that thalidomide inhibits angiogenesis induced by basic fibroblast growth factor (bFGF).[75] Interestingly, high plasma levels of bFGF predicted responsiveness in patients with multiple myeloma, although no correlation with a prolonged progression-free survival was found.[76] Furthermore, the production of proangiogenic factors like VEGF, TNF-α, IL-6 and IL-8 is suppressed by thalidomide.[77–79] Thalidomide and the immunomodulatory analog CC-4047 decreased the upregulation of IL-6 and VEGF secretion in cultures of bone marrow stromal cells, interfering the interaction between multiple myeloma cells and the stroma.[80] Treatment of nude mice bearing hepatocellular carcinoma with thalidomide resulted in significantly decreased levels of VEGF mRNA and microvessel density in the tumors and circulating TNF-α.[81] Thalidomide decreased also the stability of COX-2 mRNA in murine monocytes, resulting in a reduced expression of COX-2 protein and consequently in decreased synthesis of prostaglandins.[82] Since prostaglandins (e.g., PGE_2) have proangiogenic properties (see earlier), this interference of thalidomide with the COX-2 pathway might contribute to the antiangiogenic effects as well. Inhibition of NF-κB activation through the suppression of IκB kinase activity has been identified as a further potential mechanism of thalidomide.[83] NF-κB is a well-known key regulator of genes involved in inflammation and angiogenesis. Recently it has been shown that thalidomide exerts antiangiogenic action by the activation of neutral sphingomyelinase and, subsequently, an increase of ceramide levels. The elevated

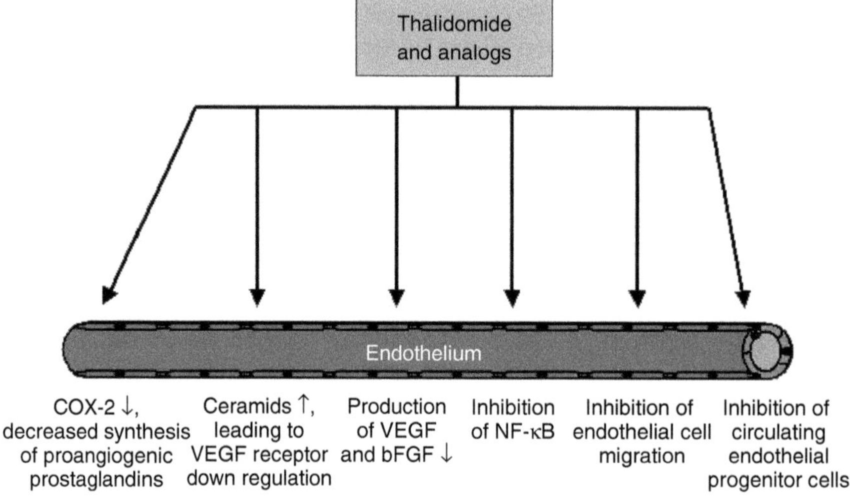

FIGURE 13.2 Antiangiogenic mechanisms of thalidomide and analogs.

ceramide levels in turn lead to downregulation of VEGF receptors.[84] Another study implicates that thalidomide might inhibit circulating endothelial progenitor cells (EPCs) in patients with multiple myeloma.[85] These bone marrow derived cells are thought to contribute to angiogenesis in tumors.[86] The thalidomide analog lenalidomide was found to inhibit endothelial cell migration in vitro by the inhibition of growth factor induced Akt phosphorylation.[87] Therefore, a variety of mechanisms seems to contribute to the antiangiogenic action of thalidomide and analogs. Further mechanisms might be detected in the future.

However, it has to be pointed out that the antineoplastic effects of thalidomide are not solely based on its antiangiogenic activity. Further direct antitumor effects like induction of apoptosis and growth arrest of tumor cells have been reported,[88] as well as further indirect antitumor effects like stimulation of the immune system[89,90] or inhibition of metastatic tumor spread through downregulation of adhesion molecules on endothelial cells.[91]

13.2.3 Thalidomide and Analogs in Clinical Trials

Thalidomide is an effective drug for the treatment of patients with refractory multiple myeloma.[92] Response rates of 25%–33% can be achieved in these patients. The anticancer activity of thalidomide in multiple myeloma was significantly linked to a decrease of bone marrow microvessel density in responders.[93] The combination of thalidomide with dexamethasone and cyclophosphamide seems to be even more effective in refractory multiple myeloma.[94,95] Interestingly, thalidomide was also successfully used for the treatment of metastasized hepatic epithelioid hemangioendothelioma, a vascular tumor with unpredictable malignant potential.[96] Likewise, thalidomide produced encouraging results in the treatment of HIV-infected patients with Kaposi's sarcoma.[97] Tumors deriving from a vascular origin might be particularly sensitive to antiangiogenic therapies.[98]

Although thalidomide is well established for the treatment of multiple myeloma, this drug is now increasingly studied in a variety of solid tumors.[79] Thalidomide was effective in the treatment of glioblastoma multiforme, especially in the combination with additionally administered temozolomide[99] and carmustine.[100] Promising results were also reported from malignant melanoma. In a phase II study, 38 patients with advanced metastasized melanoma (but no brain metastases) received temozolomide and thalidomide. An objective tumor response was observed in 32% of the patients with a median survival of 9.5 months.[101] This combination might be also effective in melanoma patients with brain metastases[102] and has been reported to be favorable in terms of toxicity.[103] Thalidomide was also studied in metastatic renal cell carcinoma as a single agent or in combination with IL-2 or interferon α-2b.[78,104] The response to thalidomide was associated with a marked reduction of C-reactive protein (CRP) and IL-6 serum levels, underlining the anti-inflammatory potential of this drug. In androgen-independent metastatic prostate cancer, the combination of thalidomide plus docetaxel was superior to docetaxel alone in terms of PSA decline, median progression-free survival, and overall survival.[105] The available data on thalidomide indicate that this drug is a promising candidate for the treatment of various hematological and solid advanced cancers, especially when used in a combination with other antineoplastic drugs. The development of thalidomide derivatives (IMiDs) with improved antiangiogenic and anticancer activity, but a more favorable toxicity profile, might further increase the attractiveness of this drug class for oncologists in the future.

13.3 PPARγ AGONISTS

13.3.1 The PPAR Family

Peroxisome proliferator-activated receptors (PPARs) are members of the large superfamily of nuclear hormone receptors, comprising about 75 proteins in the mammalian proteome.[106]

Three different isoforms exist, PPARα, PPARβ/δ, and PPARγ1/2. PPARγ is mainly expressed in adipocytes and cells of the immune system.[107] Interestingly, all isoforms can heterodimerize with the 9-*cis*-retinoic acid receptor (RXR). Endogenous ligands of PPARs include eicosanoid derivatives and long-chain polyunsaturated fatty acids like gamma-linolenic acid and arachidonic acid. Modified oxidized lipids can activate PPARγ as well. The most potent endogenous ligand for PPARγ is probably 15-deoxy-Δ 12,14-prostaglandin J2.[108] Activation of PPARγ leads to the transcriptional regulation of gene expression. These genes are involved in the regulation of lipid homoeostasis and glucose metabolism.

Synthetic PPARγ agonists have been developed in addition to the natural ligands. The thiazolidinediones (also referred to as glitazones) including troglitazone, pioglitazone, ciglitazone, and rosiglitazone are the best known members. Thiazolidinediones have initially been developed for the treatment of type 2 diabetes.[109] However, troglitazone has been withdrawn from the market because of severe liver toxicity. In addition to the glitazones, other drugs like indometacin, ibuprofen, flufenamic acid, and fenoprofen represent also PPARγ ligands.[110] Similar to the COX-2 inhibitors, PPARγ agonists exert both receptor dependent and independent antineoplastic effects. Moreover, some authors have suggested that thiazolidinedione-mediated anticancer effects might not be mediated mainly by PPARγ.[111–113] And also similar to COX-2 inhibitors, PPARγ agonists target both the tumor cell[114] and the tumor stroma. PPARγ agonists clearly show antineoplastic effects beyond their original "metabolic" indication.[110]

13.3.2 Antiangiogenic Effects of PPARγ Agonists

PPARγ agonists can target different cell types in the peritumoral stroma. For example, they can activate iNKT cells and enhance the maturation of dendritic cells, inhibit the transcription of proinflammatory cytokines, and regulate NK cells.[115–117] Furthermore, PPARγ agonists modulate the expression of several matrix metalloproteinases and influence the adhesion of tumor cells to the extracellular matrix.[118,119] However, we will focus on the antiangiogenic effects of this drug class (Figure 13.3). These antiangiogenic effects enhance the direct antitumor effects of PPARγ agonists in tumor therapy. PPARγ plays a crucial role for physiological angiogenesis. PPARγ knock-out mice embryos die on day 10, as a result of disrupted terminal differentiation of the trophoblast and placental vascularization.[120] PPARγ

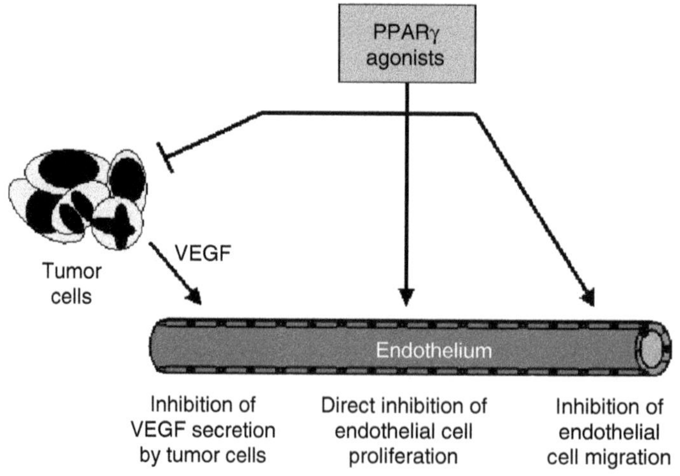

FIGURE 13.3 Antiangiogenic mechanisms of PPARγ agonists.

expression in proliferating tumor endothelia seems to be increased,[121] which might favor specific effects of these drugs in tumor therapy.

Direct and indirect antiangiogenic effects of PPARγ agonists on endothelial cells have been identified. Thiazolidinediones inhibit endothelial cell proliferation in a dose-dependent manner. Furthermore, rosiglitazone decreased the VEGF secretion of several tumor cell lines, which results also in inhibition of the tumor endothelium.[121] In contrast, in smooth muscle cells PPARγ activation leads to increased VEGF levels.[122,123]

PPARγ agonists could efficiently block neovascularization in an in vivo chick embryo model.[121] Since leptin-induced migration of endothelial cells is also affected by PPARγ agonists, this mechanism might contribute to the antiangiogenic effects.[124] The size of tumors in mice was significantly decreased with rosiglitazone treatment, although these tumors (prostate cancer cell line LNCaP) expressed PPARγ only at low levels and are in vitro relatively resistant to glitazones.[121] This suggests that the antiangiogenic effects contribute considerably to the therapeutic effects at least in this tumor entity. The antiangiogenic efficacy was underlined by the observation of significantly decreased microvessel density and endothelial cell proliferation in the treated mice. Apart from antiangiogenic effects, PPARγ agonists might also prevent the development of metastases, since the invasion of tumor cells through the endothelium is disturbed by these drugs.[121]

13.4 mTOR INHIBITORS

13.4.1 Inhibition of the mTOR Pathway

Rapamycin (sirolimus) is a lipophilic macrocyclic lactone and represents a natural fungicide isolated from the soil bacteria *Streptomyces hygrosopicus*. It is used as an immunosuppressive drug in organ transplant patients. The antitumor activity of this drug was recognized early in preclinical studies.[125] CCI-779 and RAD001 (everolimus) are analogs of rapamycin and currently tested in phase I/II trials. Rapamycin forms a complex with FK binding protein (FKBP12). This complex interacts with the mTOR, a serine/threonine kinase, and pivotal regulator of cell growth and proliferation. The function of mTOR is inhibited by rapamycin,[126] resulting in a cell cycle arrest in many different cell types. Inhibitors of mTOR prevent progression of dividing cells from the G_1 to S phase. The major effects of mTOR inhibitors are therefore decreased cell proliferation and induction of apoptosis.[127,128]

The activity of mTOR is influenced by growth factors (e.g., insulin-like growth factor) and their receptors (e.g., insulin-like growth factor receptor [IGFR], ErbB family receptors, estrogen receptor).[129,130] Furthermore, the nutrient concentration (glucose and amino acids)[131,132] is a regulator of mTOR activity. A high concentration of nutrients causes upregulation of mTOR signaling, while starvation of the cells dramatically decreases mTOR activity.[133,134] Protein kinase B (Akt/PKB) and phosphatidylinositol 3′-kinase (PI3K) are involved into the signaling cascade upstream to mTOR.[135] Due to this involvement, the mTOR signaling pathway is also called the IGFR–PI3K–Akt–mTOR pathway. Interestingly, cells with PTEN mutations show an enhanced sensitivity to mTOR inhibitors.[136] Under serum-free conditions, p53 deficient tumor cells respond to rapamycin with apoptosis, whereas p53 wild-type cells show cytostasis but not apoptosis.[137] The intracellular concentration of ATP has been found to influence mTOR activity. Thus, mTOR can act as an ATP sensor.[138]

The inhibition of mTOR results in a 5%–20% reduction in total protein synthesis of a cell.[135,139] This reduction is mediated by a block of translational pathways and results in dephosphorylation of the eukaryotic translation initiation factor, 4E binding protein-1 (4E-BP1),[140] and inhibition of the 40S ribosomal protein p70 S6 kinase (S6K1), which affects the ribosomal biogenesis.[135,141] Since p70 S6 kinase promotes antiapoptotic signals, mTOR antagonists increase cell death.[142] A multitude of other mechanisms have been reported to be affected by the inhibition of mTOR: cytoskeletal actin organization, membrane traffic, protein degradation, protein phosphatases, cyclins, cyclin-dependent kinases, cyclin-dependent kinase inhibitors, transcription of RNA polymerases I/II/III, translation of rRNA and tRNA, hypoxia-inducible factor-1α (HIF-1α), phosphorylation of retinoblastoma protein (pRb), CLIP-170, and protein kinase C signaling.[135,141,143–145] Altogether these interactions result in remarkable inhibition of tumor cell growth.[146] Antineoplastic effects of this drug are known for various cell lines and animal models including breast cancer, pancreatic cancer, prostate cancer, melanoma, renal cell cancer, glioblastoma, multiple myeloma, and leukemia.[126,147–151]

13.4.2 Antiangiogenic Effects of mTOR Inhibitors

Some antitumor effects of mTOR antagonists are obviously based on antiangiogenic effects (Figure 13.4). For example, loss of TSC1/2 results in enhanced production of VEGF through the mTOR signaling pathway.[152] This suggests otherwise that inhibition of mTOR can block angiogenesis. Rapamycin indeed shows remarkable antiangiogenic activity, which is especially associated with a decrease of VEGF and a decreased response of endothelial cells to stimulation by VEGF. These effects resulted in inhibition of angiogenesis and growth of metastases in in vivo mouse models.[153] Hypoxia-mediated proliferation is increased in endothelial cells which overexpress mTOR. Thus, inhibition of mTOR by rapamycin abrogated hypoxia-mediated proliferation of vascular cells and angiogenesis.[154] In a model of human renal cell cancer pulmonary metastasis, rapamycin significantly reduced VEGF-A expression.[147] Another study demonstrated that rapamycin can cause thrombosis of tumor vessels and apoptosis of tumor endothelial cells in a pancreatic cancer mouse model.[155] The latter effect could be confirmed in human umbilical vein endothelial cells in vitro.[155,156] Interestingly, CCI-779 also inhibited angiogenesis in a multiple myeloma xenograft model.[151] In mesenchymal cells, rapamycin has been shown to suppress the expression of VEGF by silencing the PDGFRα-p70S6K pathway.[157] Rapamycin might also have an effect on the quality of the blood circulation within a tumor.[158]

In the meantime, several clinical trials confirmed that rapamycin can be an efficient drug for (antiangiogenic) tumor therapy. A multitude of phase-I/II trials are on the way or already

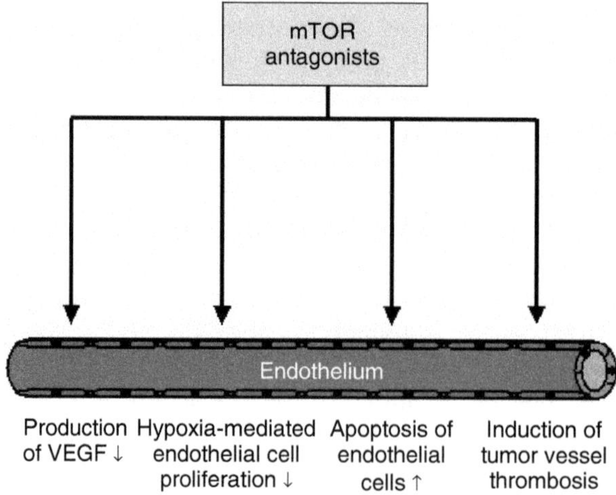

FIGURE 13.4 Antiangiogenic mechanisms of mTOR antagonists.

finished, including nonsmall-cell lung cancer, metastatic breast cancer, metastatic renal cell cancer, anaplastic astrocytoma, uterine cancer, cervical cancer, soft-tissue sarcoma, and mesothelioma.[159–161]

13.5 METRONOMIC CHEMOTHERAPY

13.5.1 The Rationale of Metronomic Dosing Schedules

More than 50 years after introduction of cytostatic chemotherapy, these drugs are a major component in cancer therapies. Conventional chemotherapeutic drugs are applied in repeated cycles at the maximum tolerated dose. Killing as many malignant cells as possible by maximum tolerated doses of chemotherapy has evolved into a kind of dogma in oncology. Breaks between the cycles are necessary for the patients to recover from the often severe side effects, including myelosuppression, gastrointestinal toxicity, mucositis, and hair loss. Although the initially observed responses to therapy can be remarkable, their duration is very limited in most cases. After repeated cycles of therapy, the development of drug resistant clones often is observed and the patients experience a relapse. A multitude of strategies have been established in the last decades to avoid or overcome this drug resistance: maximum doses were steadily increased, supportive treatment was optimized to impair side effects, and the drugs were used in combinations. Nevertheless, the overall benefit in terms of survival and quality of life remains modest in most cancers.

In contrast to conventional chemotherapy, the so-called "metronomic" chemotherapy is administered frequently (e.g., daily), without prolonged drug-free breaks, at doses significantly below the maximum tolerated dose[162] (Figure 13.5). Therefore, metronomically scheduled chemotherapy decreases toxicity and improves the quality of live of patients. Despite the fact that the cumulative dose of the drug is much lower than in conventional schedules, metronomic regimens are obviously effective in cancer therapy.[163] Recent studies have shown that these effects might be preferentially based on antiangiogenic effects.[164] Since tumor endothelial cells show a dividing proportion at any given time[165] and dividing cells are the target of cytotoxic drugs, every conventional chemotherapy probably has some

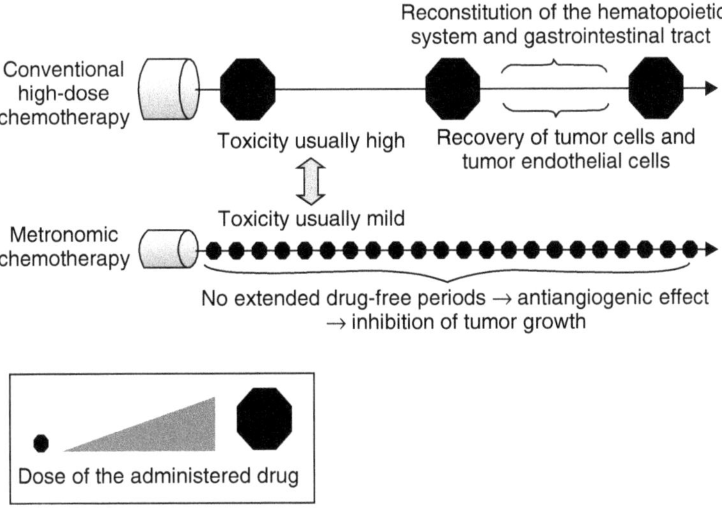

FIGURE 13.5 Comparison of conventional high-dose chemotherapy and low-dose metronomic chemotherapy.

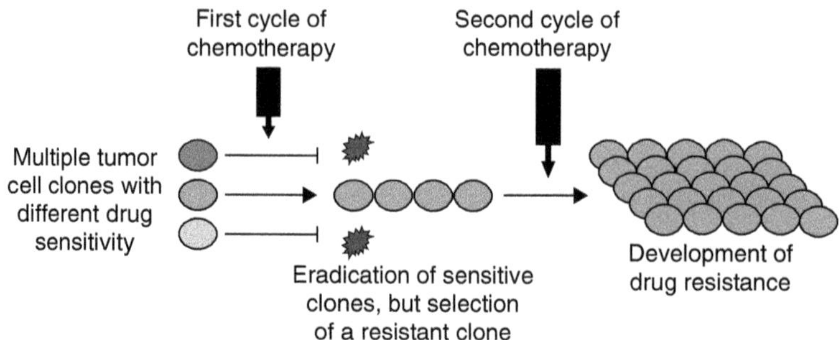

FIGURE 13.6 Tumors contain multiple different subclones due to their genetic instability. Application of repeated cycles of chemotherapy leads to the selection of drug resistant clones, which in turn can cause a relapse or progression of disease.

antiangiogenic effects.[166] In conventional chemotherapy regimen the damage of the tumor vessels can be repaired during drug-free breaks, analogous to the recovery of the bone marrow. A dosing schedule was therefore established to enhance the antiangiogenic efficacy by targeting especially the proliferating endothelial cells in tumors.[167] The peritumoral or intratumoral endothelial cells are the major target of metronomic chemotherapy.[168] In contrast to tumor cells which acquire rapidly resistance to cytotoxic drugs due to their genetic instability and selection of resistant clones (Figure 13.6), endothelial cells lack this genetic instability and might therefore be less susceptible for the development of drug resistance. The fact that in patients receiving conventional chemotherapy the myelosuppression of the bone marrow is always observed even when the tumors become resistant to the same chemotherapeutic drugs might be explained by the different genetic stability of these compartments.[169] The observation that drug resistant tumor cells respond to this drug when the drug is administered in metronomic dosing schedules, in contrast to conventional schedules, confirms this antiangiogenic concept.[167] The rationale of metronomic chemotherapy is also based on theoretical models, demonstrating that the minimization of total tumor burden, rather than complete eradication, is probably the more practical target. Metronomic dosing is thought to be the best way to achieve this target.[170] In addition, mathematical considerations predict a targeting bias toward the endothelial compartment of a tumor when chemotherapy is administered in a metronomic regimen.

13.5.2 Antiangiogenic Mechanisms of Metronomic Chemotherapy

The molecular mechanisms of metronomic chemotherapy are not fully understood in detail yet. In the last years, some studies have revealed that chemotherapeutic drugs selectively inhibit the proliferation of human endothelial cells at ultralow (picomolar) concentrations, while they inhibit nonendothelial cells at 10^4- to 10^5-fold higher concentrations.[171] It has turned out that proliferation, motility, and tube formation of endothelial cells are affected at picomolar levels of metronomic administered chemotherapeutic drugs.[172–175] In vitro studies suggest that a broad spectrum of chemotherapeutic drugs are feasible for metronomic chemotherapy.[176] If metronomic chemotherapy acts specifically against endothelial cells, severe side effects due to "collateral" damage to all endothelial cells of the organism might be feared. However, previous preclinical and clinical studies have not observed the occurrence of those side effects. A possible selectivity of metronomic chemotherapy on the tumor endothelium might be based on the observation that the tumor endothelium is activated compared with the normal endothelium.[164,177] This can be explained by a modified peritumoral cytokine milieu. The increased proliferation index of the tumor endothelium at the site of newly forming blood vessels might open the gate for a sufficient therapeutic index of metronomic chemotherapy.[165]

The antiangiogenic effect of metronomic chemotherapy on endothelial cells might be mediated by thrombospondin 1 (TSP-1), because the expression of TSP-1 in endothelial cells is increased during metronomic chemotherapy.[177,178] TSP-1 suppresses angiogenesis by the inhibition of endothelial cell proliferation, increased apoptosis, and displacement of VEGF from endothelial cells.[179–181] Antiangiogenic mechanisms of metronomic chemotherapy comprise also the decreased expression of HIF-1α protein, which is an important positive regulatory checkpoint for many proangiogenic genes.[182]

A significant proportion of endothelial cells in newly forming blood vessels is probably derived from the bone marrow. This proportion contributes to postnatal physiological and pathological neovascularization, including cancer.[183–185] Circulating endothelial precursor cells, derived from the bone marrow, are indeed involved in tumor angiogenesis.[86] EPCs are mobilized from the bone marrow by erythropoietin, cytokines, growth factors such as VEGF, and under ischemic conditions.[186–188] After entering the peripheral blood circulation, they can migrate to sites of ongoing angiogenesis like tumors.[164] Animals treated with a cycle of cyclophosphamide at maximum tolerable dose show a decrease of circulating EPCs in the blood. However, in drug-free periods between the cycles, the number of circulating EPCs rapidly increases again. The administration of metronomic scheduled cyclophosphamide decreased the number and viability of circulating EPCs permanently with consecutive inhibition of the tumors.[189] Modulation of the number of circulating EPCs in the blood might therefore also contribute to the antiangiogenic effects of metronomic chemotherapy.

In addition to the inhibition of tumor angiogenesis, some other mechanisms have been suggested to be responsible for the antitumor efficacy of metronomic chemotherapy, including effects on the immune system[190,191] and direct effects on tumor cells.[164,192]

13.5.3 Metronomic Chemotherapy in Clinical Trials

Metronomic temozolomide in combination with radiotherapy was successfully used for the treatment of children with high-risk brain tumors including medulloblastomas.[193] Temozolomide inhibits angiogenesis in vitro at low doses corresponding to the plasma concentrations achieved by metronomic scheduling. This indicates that the antitumor activity of metronomic temozolomide is partly based on its antiangiogenic activity.[194]

Patients with metastatic breast cancer received a combination of metronomic cyclophosphamide and methotrexate.[195] An overall clinical benefit of 31% was observed in these heavily pretreated patients. This was accompanied by a significant decrease of the median serum VEGF levels for at least 2 months, although both responders and nonresponders showed similar reductions in serum VEGF.[195] Another study identified a subset of patients with more aggressive tumors and less response to metronomic treatment with cyclophosphamide and methotrexate. These patients were characterized by increased pretreatment serum levels of the proto-oncogene HER2/neu extracellular domain.[196] Metronomic cyclophosphamide combined with low-dose dexamethasone demonstrated efficacy and tolerability in the treatment of patients with hormone-refractory prostate carcinoma, indicated by the reduction of PSA.[197] A clinical benefit was also observed in patients with hormone-refractory metastatic prostate cancer treated with continuous oral low-dose cyclophosphamide.[198] Low-dose oral etoposide, given continuously over prolonged periods to elderly patients with refractory non-Hodgkin's lymphoma, was linked to a low incidence of serious adverse effects, an excellent subjective tolerance, and a response rate of 65%.[199] Metronomic trofosfamide might be a therapeutic option for the palliative treatment of heavily pretreated patients with advanced ovarian carcinoma[200] and metastatic soft-tissue sarcoma.[201]

Metronomic therapy was found to be at least equal or even superior in terms of tumor growth inhibition when directly compared with conventional dosing schedules of the same chemotherapeutic drug.[163,167,189] However, patients with large tumor masses can profit by a modified schedule including initial high-dose chemotherapy which induces a strong mass reduction of the tumors, and a subsequent switch to long-term metronomic chemotherapy for stabilization of this response.[202] Orally available drugs should be preferred for metronomic chemotherapy, because the therapy should be practicable for an outpatient setting. The administration of metronomic chemotherapy in animal models or clinical trials has usually been reported to be significantly less toxic than conventional high-dose chemotherapy.[98,197,198,200,201,203–206] The low toxicity compared with conventional high-dose chemotherapy improves the acceptance of the therapy by the patients and the quality of the patients' life.

13.6 COMBINATION OF ANTIANGIOGENIC DRUGS

13.6.1 Mechanisms of Resistance to Antiangiogenic Drugs

The above mentioned drugs have demonstrated antitumor efficiency in vitro and in vivo in various cancer entities. Some drugs have already entered clinical trials for patients with advanced disease, and the first reports from these studies are at least partly encouraging. However, when these drugs are used as single agents for therapy, the tumors seem to develop rapidly escape mechanisms, despite of theoretical considerations regarding the genetic stability of the stroma. Several mechanisms might be responsible for this escape: induction of apoptosis of tumor endothelial cells is an important mechanism of stroma-targeted therapies.[155,167] Chronic antiangiogenic treatment can lead to the selection of tumor clones which have the ability to survive better under hypoxic and nutrient-starved conditions. Epigenetic changes contribute to the development of resistance in endothelial cells. Epigenetic mechanisms may comprise antiapoptotic survival gene expression or induction of genes which modify the metabolism of administered drugs. Since the tumor cell population is very heterogeneous, the single clones may show variable capabilities to survive in relatively hypoxic conditions[207] and consequently vary in their dependence on blood vessel supply.[192,208] Hypoxia has already been shown to mediate the selection of cells with reduced apoptotic potential in solid tumors.[209] High levels of the antiapoptotic VEGF and other endothelial survival factors in the vicinity of tumor vessels which might be produced by the tumors themselves probably can antagonize these antiapoptotic effects.[208] Some antiangiogenic effects of stroma-targeted conventional drugs are based on the inhibition of proangiogenic growth factors such as VEGF. Proangiogenic growth factors are redundantly expressed by both tumor and stromal cells.[5] If one proangiogenic component is blocked, this can lead to a compensatory overexpression of alternative proangiogenic factors.[210] The tumor cell itself is an important source for the production of growth factors. Considering the genetic instability of the tumor cells, tumor clones might be selected during antiangiogenic therapies which produce compensatory vascular growth factors.[211] These factors would not be affected by the current therapy and allow tumor cells to overcome growth inhibition. Some studies seem to confirm this considerations, since higher expression levels of VEGF, Ang-2, transforming growth factor α, PDGF-α, and bFGF are found in patients with advanced disease compared with initial stages.[212] A further possible escape mechanism in stroma-targeted therapies has been demonstrated recently.[213] Antiangiogenic therapy with metronomic topotecan and anti-VEGF antibodies induced the development of strikingly remodeled vessels. These modified blood vessels showed significant increases in diameter and proliferation of mural cells. In addition, the vessels revealed remarkable expression of PDGF-β, which enhances vascular integrity via stromal cell recruitment,[214] and ephrinB2, which is required for the proper assembly of stromal cells into blood vessels.[215] This study indicates that improved vascular stability might be a characteristic escape mechanism for tumor vessels under chronic anti-angiogenic treatment.

13.6.2 Advantages of Combination Therapies

The combination of conventional therapeutics with antiangiogenic activity is thought to be superior for the circumvention of escape mechanisms, since different pathways are targeted simultaneously. First results from clinical studies seem to confirm this hypothesis. In this context, the combination of PPARγ agonists, COX-2 inhibitors, and metronomically scheduled trofosfamide has shown promising results. Patients with advanced angiosarcomas received this triple therapy.[98] These patients had been pretreated and did not respond to various first line therapies. However, the application of pioglitazone, rofecoxib, and metronomic trofosfamide led to remarkable responses and progression-free survival times (Figure 13.7). Moreover, the toxicity of the triple therapy was low and the treatment could be performed in an outpatient setting. Similar encouraging results of this triple combination were observed for endemic Kaposi sarcoma,[216] chemorefractory malignant melanoma,

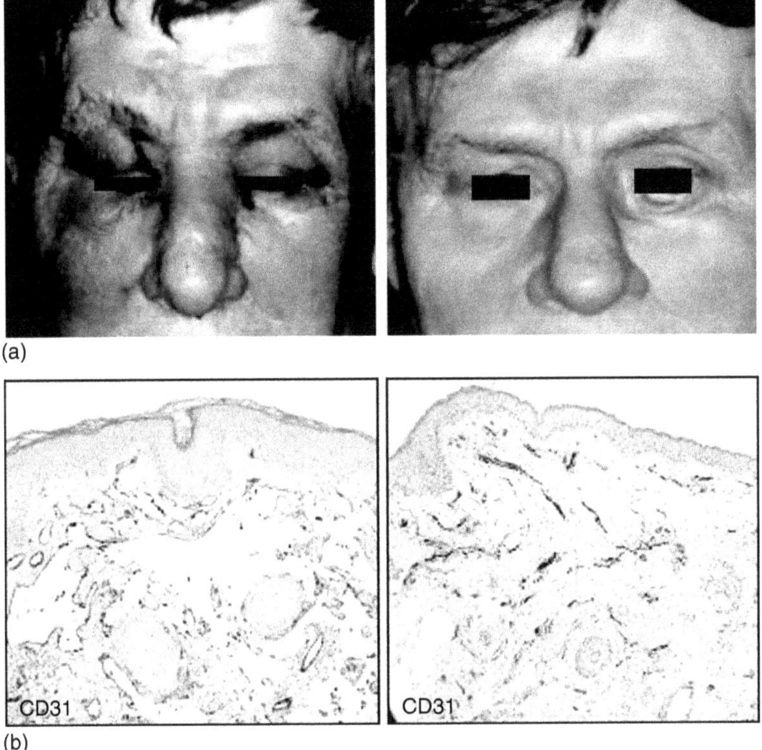

FIGURE 13.7 (See color insert following page 558.) (a) Patient before (*left*) and after 2 months (*right*) of treatment with an antiangiogenic therapy (pioglitazone, rofecoxib, and metronomic trofosfamide). A complete clinical response was observed with a duration of 6 months. (b) Biopsies of the facial skin before (*left*) and 14 days after begining of the antiangiogenic treatment with pioglitazone and rofecoxib alone (*right*) show signs of regression with loss of the luminal architecture after administration of the biomodulating drugs. However, CD31 immunohistochemistry still shows a positive staining for residual endothelial cells. (From Vogt, T. et al., *Cancer*, 98, 2251–2256, 2003. With permission.)

soft-tissue sarcoma,[206] and chemorefractory multisystem Langerhans' cell histiocytosis.[217] In a new multicenter phase-II study, the combination of metronomic trofosfamide and pioglitazone/rofecoxib was superior to the monotherapy with metronomic trofosfamide.[218] To our knowledge, this combination therapy including biomodulators (pioglitazone and rofecoxib) and metronomic chemotherapy (trofosfamide) is the first to demonstrate that a novel, completely orally administered combined biomodulator/metronomic chemotherapy regimen may be active and well tolerated in patients with chemorefractory advanced malignancies (melanomas, sarcomas, especially vascular tumors). The novel regimen showed disease control in about one-third of all patients with advanced chemorefractory disease.[98,206] There is growing evidence that anti-inflammatory therapy with PPARγ agonists and COX-2 inhibitors posses activity against early neoplastic progression.[219] Data on the ability of anti-inflammatory agents (e.g., rofecoxib and pioglitazone) to control the progression of advanced-stage disease are scarcer.[206,218,220] Stroma-targeted therapies can also attenuate tumor-associated CRP levels.[218] CRP levels may be useful to identify subgroups of patients with significantly different progression-free survival rates, regardless of the tumor type. The recently published trials on combined therapy with biomodulators (COX-2 inhibitors and PPARγ agonists) and metronomic chemotherapy did not show whether antiangiogenic effects, immunomodulatory or direct antiproliferative effects on tumor cells are most responsible for the observed responses.

In breast cancer, a combination of metronomic cyclophosphamide and metronomic methotrexate proved efficiency.[195] In a pilot study, patients with glioblastoma multiforme responded well to a combination of metronomic temozolomide and rofecoxib, especially when the tumors exhibited a high degree of vascularization.[221] Celecoxib, thalidomide, and metronomic irinotecan have been successfully applied in pancreatic carcinoma.[222]

In vitro studies identified that the combination of different conventional therapeutics can exhibit synergistic and superadditive antitumor effects. Rapamycin combined with CC-5013, the immunomodulatory analog of thalidomide, revealed strong synergistic effects in multiple myeloma. The combination of these two drugs could even overcome drug resistance in multiple myeloma cell lines resistant to conventional chemotherapeutic drugs.[223] Thalidomide also increased the expression of PPARγ in nonsmall cell lung cancer cells, suggesting possible synergistic effects of these two drug classes.[224] Several studies report that the combination of COX-2 inhibitors and PPARγ agonists is especially advantageous, as demonstrated in human and rat breast cancer cells,[225,226] where these drugs showed even superadditive effects. An explanation for this observation is probably the linkage of both pathways.[227–231] Whereas PPARγ agonists can decrease the expression of COX-2, some COX-2 inhibitors like sulindac can activate PPARγ.

13.6.3 Future Aspects

Stroma-targeted therapies with conventional drugs are promising for the palliative treatment of patients with advanced tumors in terms of disease stabilization, prolongation of progression-free survival, and tolerability of therapy related side effects. However, available results from previous studies indicate that a complete eradication of cancer will not be possible with these therapeutics. The combination with antiangiogenic biologicals like antibodies against proangiogenic factors, tyrosine kinase inhibitors, or endogenous angiogenesis inhibitors might be more effective and even eradicate metastasized tumors without an increase of toxicity.[232] The FDA has already approved bevacizumab (Avastin®), a recombinant humanized antibody against VEGF for the treatment of metastatic colorectal cancer (in combination with 5-fluorouracil) in 2004. This represents the first approval of a "biological" with exclusively antiangiogenic activity. A large set of new angiogenesis inhibitors are currently under way or have already entered clinical tests (e.g., TNP-470, endostatin, angiostatin). Since upregulation of receptor tyrosine kinase pathways like PDGF/PDGFR and EphB/ephrin-B2 has been shown to be fundamentally involved in escape mechanisms[213] during antiangiogenic therapy, specific tyrosine kinase inhibitors might perfectly supplement stroma-targeted strategies. Some tyrosine kinase inhibitors like imatinib (Glivec®) are already approved and in clinical use. Each cancer entity and each antiangiogenic combination therapy will probably induce a specific escape mechanism. Array analysis is a promising tool to detect the involved pathways in each case for the development of specific counteractive strategies.

Moreover, additional classes of conventional drugs might bear an antiangiogenic potential. For example, captopril shows inhibition of angiogenesis, resulting in inhibition of tumor growth in vivo.[233,234] Heparin can also inhibit angiogenesis, cause regression of tumors, and prevent metastases.[235,236] Other results indicate that the antiangiogenic activity is restricted to the low-molecular-weight-fraction (LMW) of heparin.[237] LMW heparin derivatives antagonize the fibroblast growth factor pathway and provide the basis for the development of novel therapeutics with a high antiangiogenic potential.[238]

Conventional therapeutics with antiangiogenic potential are 1000-fold approved for their safety and tolerability, and exhibit a manageable toxicity profile. According to their remarkable antitumor efficacy, these drugs represent promising components of antiangiogenic cancer therapies in the future, rather in combination therapies than as single agents.

REFERENCES

1. Leaf, C., Why we're losing the war on cancer (and how to win it). *Fortune*, 149, 76–82, 84–86, 88, 2004.
2. Boehm, T. et al., Antiangiogenic therapy of experimental cancer does not induce acquired drug resistance. *Nature*, 390, 404–407, 1997.
3. Graham, C.H. et al., Rapid acquisition of multicellular drug resistance after a single exposure of mammary tumor cells to antitumor alkylating agents. *J Natl Cancer Inst*, 86, 975–982, 1994.
4. Hanahan, D. and Weinberg, R.A., The hallmarks of cancer. *Cell*, 100, 57–70, 2000.
5. Carmeliet, P. and Jain, R.K., Angiogenesis in cancer and other diseases. *Nature*, 407, 249–257, 2000.
6. Hla, T. and Neilson, K., Human cyclooxygenase-2 cDNA. *Proc Natl Acad Sci USA*, 89, 7384–7388, 1992.
7. Yang, V.W. et al., Size-dependent increase in prostanoid levels in adenomas of patients with familial adenomatous polyposis. *Cancer Res*, 58, 1750–1753, 1998.
8. Fukuda, R., Kelly, B., and Semenza, G.L., Vascular endothelial growth factor gene expression in colon cancer cells exposed to prostaglandin E2 is mediated by hypoxia-inducible factor 1. *Cancer Res*, 63, 2330–2334, 2003.
9. Zha, S. et al., Cyclooxygenases in cancer: progress and perspective. *Cancer Lett*, 215, 1–20, 2004.
10. Dubois, R.N. et al., Cyclooxygenase in biology and disease. *FASEB J*, 12, 1063–1073, 1998.
11. Fosslien, E., Molecular pathology of cyclooxygenase-2 in neoplasia. *Ann Clin Lab Sci*, 30, 3–21, 2000.
12. Chen, C.C. et al., Tumor necrosis factor-alpha-induced cyclooxygenase-2 expression via sequential activation of ceramide-dependent mitogen-activated protein kinases, and IkappaB kinase 1/2 in human alveolar epithelial cells. *Mol Pharmacol*, 59, 493–500, 2001.
13. Liu, X.H. et al., Cycloxygenase-2 suppresses hypoxia-induced apoptosis via a combination of direct and indirect inhibition of p53 activity in a human prostate cancer cell line. *J Biol Chem*, 280, 3817–3823, 2004.
14. Diperna, C.A. et al., Cyclooxygenase-2 inhibition decreases primary and metastatic tumor burden in a murine model of orthotopic lung adenocarcinoma. *J Thorac Cardiovasc Surg*, 126, 1129–1133, 2003.
15. Basu, G.D. et al., Cyclooxygenase-2 inhibitor induces apoptosis in breast cancer cells in an in vivo model of spontaneous metastatic breast cancer. *Mol Cancer Res*, 2, 632–642, 2004.
16. Fenwick, S.W. et al., The effect of the selective cyclooxygenase-2 inhibitor rofecoxib on human colorectal cancer liver metastases. *Gastroenterology*, 125, 716–729, 2003.
17. Hu, P.J. et al., Chemoprevention of gastric cancer by celecoxib in rats. *Gut*, 53, 195–200, 2004.
18. Kern, M.A. et al., Cyclooxygenase-2 inhibitors suppress the growth of human hepatocellular carcinoma implants in nude mice. *Carcinogenesis*, 25, 1193–1199, 2004.

19. Narayanan, B.A. et al., Suppression of *N*-methyl-*N*-nitrosourea/testosterone-induced rat prostate cancer growth by celecoxib: effects on cyclooxygenase-2, cell cycle regulation, and apoptosis mechanism(s). *Clin Cancer Res*, 9, 3503–3513, 2003.
20. Narayanan, B.A. et al., Regression of mouse prostatic intraepithelial neoplasia by nonsteroidal anti-inflammatory drugs in the transgenic adenocarcinoma mouse prostate model. *Clin Cancer Res*, 10, 7727–7737, 2004.
21. Masferrer, J.L. et al., Antiangiogenic and antitumor activities of cyclooxygenase-2 inhibitors. *Cancer Res*, 60, 1306–1311, 2000.
22. Raut, C.P. et al., Celecoxib inhibits angiogenesis by inducing endothelial cell apoptosis in human pancreatic tumor xenografts. *Cancer Biol Ther*, 3, 2004.
23. Form, D.M. and Auerbach, R., PGE2 and angiogenesis. *Proc Soc Exp Biol Med*, 172, 214–218, 1983.
24. Tatsuguchi, A. et al., Cyclooxygenase-2 expression correlates with angiogenesis and apoptosis in gastric cancer tissue. *Hum Pathol*, 35, 488–495, 2004.
25. Ma, L., del Soldato, P., and Wallace, J.L., Divergent effects of new cyclooxygenase inhibitors on gastric ulcer healing: shifting the angiogenic balance. *Proc Natl Acad Sci USA*, 99, 13243–13247, 2002.
26. Kobayashi, H. et al., JTE-522, a selective COX-2 inhibitor, interferes with the growth of lung metastases from colorectal cancer in rats due to inhibition of neovascularization: a vascular cast model study. *Int J Cancer*, 112, 920–926, 2004.
27. Tsujii, M. et al., Cyclooxygenase regulates angiogenesis induced by colon cancer cells. *Cell*, 93, 705–716, 1998.
28. Majima, M. et al., Cyclo-oxygenase-2 enhances basic fibroblast growth factor-induced angiogenesis through induction of vascular endothelial growth factor in rat sponge implants. *Br J Pharmacol*, 130, 641–649, 2000.
29. Cheng, T. et al., Prostaglandin E2 induces vascular endothelial growth factor and basic fibroblast growth factor mRNA expression in cultured rat Muller cells. *Invest Ophthalmol Vis Sci*, 39, 581–591, 1998.
30. Williams, C.S. et al., Host cyclooxygenase-2 modulates carcinoma growth. *J Clin Invest*, 105, 1589–1594, 2000.
31. Wei, D. et al., Celecoxib inhibits vascular endothelial growth factor expression in and reduces angiogenesis and metastasis of human pancreatic cancer via suppression of Sp1 transcription factor activity. *Cancer Res*, 64, 2030–2038, 2004.
32. Nor, J.E. et al., Up-regulation of Bcl-2 in microvascular endothelial cells enhances intratumoral angiogenesis and accelerates tumor growth. *Cancer Res*, 61, 2183–2188, 2001.
33. Nor, J.E. et al., Vascular endothelial growth factor (VEGF)-mediated angiogenesis is associated with enhanced endothelial cell survival and induction of Bcl-2 expression. *Am J Pathol*, 154, 375–384, 1999.
34. Dimmeler, S. and Zeiher, A.M., Akt takes center stage in angiogenesis signaling. *Circ Res*, 86, 4–5, 2000.
35. Gately, S. and Li, W.W., Multiple roles of COX-2 in tumor angiogenesis: a target for antiangiogenic therapy. *Semin Oncol*, 31, 2–11, 2004.
36. Dormond, O. et al., Prostaglandin E2 promotes integrin alpha V beta 3-dependent endothelial cell adhesion, rac-activation, and spreading through cAMP/PKA-dependent signaling. *J Biol Chem*, 277, 45838–45846, 2002.
37. Ruegg, C., Dormond, O., and Mariotti, A., Endothelial cell integrins and COX-2: mediators and therapeutic targets of tumor angiogenesis. *Biochim Biophys Acta*, 1654, 51–67, 2004.
38. Dohadwala, M. et al., Autocrine/paracrine prostaglandin E2 production by non-small cell lung cancer cells regulates matrix metalloproteinase-2 and CD44 in cyclooxygenase-2-dependent invasion. *J Biol Chem*, 277, 50828–50833, 2002.
39. Callejas, N.A. et al., Expression of cyclooxygenase-2 promotes the release of matrix metalloproteinase-2 and -9 in fetal rat hepatocytes. *Hepatology*, 33, 860–867, 2001.
40. Vaday, G.G. et al., Transforming growth factor-beta suppresses tumor necrosis factor alpha-induced matrix metalloproteinase-9 expression in monocytes. *J Leukoc Biol*, 69, 613–621, 2001.
41. Masson, V. et al., Contribution of host MMP-2 and MMP-9 to promote tumor vascularization and invasion of malignant keratinocytes. *FASEB J*, 19, 234–236, 2005.

42. Voest, E.E. et al., Inhibition of angiogenesis in vivo by interleukin 12. *J Natl Cancer Inst*, 87, 581–586, 1995.
43. Stolina, M. et al., Specific inhibition of cyclooxygenase 2 restores antitumor reactivity by altering the balance of IL-10 and IL-12 synthesis. *J Immunol*, 164, 361–370, 2000.
44. Pai, R. et al., Prostaglandin E2 transactivates EGF receptor: a novel mechanism for promoting colon cancer growth and gastrointestinal hypertrophy. *Nat Med*, 8, 289–293, 2002.
45. Kedar, D. et al., Blockade of the epidermal growth factor receptor signaling inhibits angiogenesis leading to regression of human renal cell carcinoma growing orthotopically in nude mice. *Clin Cancer Res*, 8, 3592–3600, 2002.
46. Rodrigues, S. et al., Trefoil peptides as proangiogenic factors in vivo and in vitro: implication of cyclooxygenase-2 and EGF receptor signaling. *FASEB J*, 17, 7–16, 2003.
47. Pore, N. et al., PTEN mutation and epidermal growth factor receptor activation regulate vascular endothelial growth factor (VEGF) mRNA expression in human glioblastoma cells by transactivating the proximal VEGF promoter. *Cancer Res*, 63, 236–241, 2003.
48. Daniel, T.O. et al., Thromboxane A2 is a mediator of cyclooxygenase-2-dependent endothelial migration and angiogenesis. *Cancer Res*, 59, 4574–4577, 1999.
49. Murohara, T. et al., Vascular endothelial growth factor/vascular permeability factor enhances vascular permeability via nitric oxide and prostacyclin. *Circulation*, 97, 99–107, 1998.
50. Nataraj, C. et al., Receptors for prostaglandin E(2) that regulate cellular immune responses in the mouse. *J Clin Invest*, 108, 1229–1235, 2001.
51. Lang, S. et al., Impaired monocyte function in cancer patients: restoration with a cyclooxygenase-2 inhibitor. *FASEB J*, 17, 286–288, 2003.
52. Baratelli, F.E. et al., Prostaglandin E2-dependent enhancement of tissue inhibitors of metalloproteinases-1 production limits dendritic cell migration through extracellular matrix. *J Immunol*, 173, 5458–5466, 2004.
53. Johnsen, J.I. et al., Cyclooxygenase-2 is expressed in neuroblastoma, and nonsteroidal anti-inflammatory drugs induce apoptosis and inhibit tumor growth in vivo. *Cancer Res*, 64, 7210–7215, 2004.
54. Pruthi, R.S., Derksen, J.E., and Moore, D., A pilot study of use of the cyclooxygenase-2 inhibitor celecoxib in recurrent prostate cancer after definitive radiation therapy or radical prostatectomy. *BJU Int*, 93, 275–278, 2004.
55. Wang, Z., Fuentes, C.F., and Shapshay, S.M., Antiangiogenic and chemopreventive activities of celecoxib in oral carcinoma cell. *Laryngoscope*, 112, 839–843, 2002.
56. Oyama, K. et al., A COX-2 inhibitor prevents esophageal inflammation–metaplasia–adenocarcinoma sequence in rats. *Carcinogenesis*, 26, 565–570, 2005.
57. Garcia Rodriguez, L.A. and Huerta-Alvarez, C., Reduced incidence of colorectal adenoma among long-term users of nonsteroidal antiinflammatory drugs: a pooled analysis of published studies and a new population-based study. *Epidemiology*, 11, 376–381, 2000.
58. Thun, M.J., Henley, S.J., and Patrono, C., Nonsteroidal anti-inflammatory drugs as anticancer agents: mechanistic, pharmacologic, and clinical issues. *J Natl Cancer Inst*, 94, 252–266, 2002.
59. Moran, E.M., Epidemiological and clinical aspects of nonsteroidal anti-inflammatory drugs and cancer risks. *J Environ Pathol Toxicol Oncol*, 21, 193–201, 2002.
60. Harris, R.E., Beebe-Donk, J., and Schuller, H.M., Chemoprevention of lung cancer by non-steroidal anti-inflammatory drugs among cigarette smokers. *Oncol Rep*, 9, 693–695, 2002.
61. Jacobs, E.J. et al., Aspirin use and pancreatic cancer mortality in a large United States cohort. *J Natl Cancer Inst*, 96, 524–528, 2004.
62. Schernhammer, E.S. et al., A prospective study of aspirin use and the risk of pancreatic cancer in women. *J Natl Cancer Inst*, 96, 22–28, 2004.
63. Steinbach, G. et al., The effect of celecoxib, a cyclooxygenase-2 inhibitor, in familial adenomatous polyposis. *N Engl J Med*, 342, 1946–1952, 2000.
64. Phillips, R.K. et al., A randomised, double blind, placebo controlled study of celecoxib, a selective cyclooxygenase 2 inhibitor, on duodenal polyposis in familial adenomatous polyposis. *Gut*, 50, 857–860, 2002.
65. Davis, T.W. et al., Inhibition of cyclooxygenase-2 by celecoxib reverses tumor-induced wasting. *J Pharmacol Exp Ther*, 308, 929–934, 2004.

66. Lenz, W. and Knapp, K., [Thalidomide embryopathy.] *Dtsch Med Wochenschr*, 87, 1232–1242, 1962.
67. Sampaio, E.P. et al., The influence of thalidomide on the clinical and immunologic manifestation of erythema nodosum leprosum. *J Infect Dis*, 168, 408–414, 1993.
68. Singhal, S. et al., Antitumor activity of thalidomide in refractory multiple myeloma. *N Engl J Med*, 341, 1565–1571, 1999.
69. Eisen, T. et al., Continuous low dose thalidomide: a phase II study in advanced melanoma, renal cell, ovarian and breast cancer. *Br J Cancer*, 82, 812–817, 2000.
70. Macpherson, G.R. et al., Current status of thalidomide and its role in the treatment of metastatic prostate cancer. *Crit Rev Oncol Hematol*, 46 Suppl, S49–57, 2003.
71. Fine, H.A. et al., Phase II trial of the antiangiogenic agent thalidomide in patients with recurrent high-grade gliomas. *J Clin Oncol*, 18, 708–715, 2000.
72. Stebbing, J. et al., The treatment of advanced renal cell cancer with high-dose oral thalidomide. *Br J Cancer*, 85, 953–958, 2001.
73. Dredge, K. et al., Novel thalidomide analogues display anti-angiogenic activity independently of immunomodulatory effects. *Br J Cancer*, 87, 1166–1172, 2002.
74. Kumar, S. et al., Antimyeloma activity of two novel N-substituted and tetraflourinated thalidomide analogs. *Leukemia*, 19, 1253–1261, 2005.
75. D'Amato, R.J. et al., Thalidomide is an inhibitor of angiogenesis. *Proc Natl Acad Sci USA*, 91, 4082–4085, 1994.
76. Neben, K. et al., High plasma basic fibroblast growth factor concentration is associated with response to thalidomide in progressive multiple myeloma. *Clin Cancer Res*, 7, 2675–2681, 2001.
77. Sampaio, E.P. et al., Thalidomide selectively inhibits tumor necrosis factor alpha production by stimulated human monocytes. *J Exp Med*, 173, 699–703, 1991.
78. Kedar, I., Mermershtain, W., and Ivgi, H., Thalidomide reduces serum C-reactive protein and interleukin-6 and induces response to IL-2 in a fraction of metastatic renal cell cancer patients who failed IL-2-based therapy. *Int J Cancer*, 110, 260–265, 2004.
79. Sleijfer, S., Kruit, W.H., and Stoter, G., Thalidomide in solid tumours: the resurrection of an old drug. *Eur J Cancer*, 40, 2377–2382, 2004.
80. Gupta, D. et al., Adherence of multiple myeloma cells to bone marrow stromal cells upregulates vascular endothelial growth factor secretion: therapeutic applications. *Leukemia*, 15, 1950–1961, 2001.
81. Zhang, Z.L., Liu, Z.S., and Sun, Q., Effects of thalidomide on angiogenesis and tumor growth and metastasis of human hepatocellular carcinoma in nude mice. *World J Gastroenterol*, 11, 216–220, 2005.
82. Fujita, J. et al., Thalidomide and its analogues inhibit lipopolysaccharide-mediated induction of cyclooxygenase-2. *Clin Cancer Res*, 7, 3349–3355, 2001.
83. Keifer, J.A. et al., Inhibition of NF-kappa B activity by thalidomide through suppression of IkappaB kinase activity. *J Biol Chem*, 276, 22382–22387, 2001.
84. Yabu, T. et al., Thalidomide-induced anti-angiogenic action is mediated by ceramide through depletion of VEGF receptors, and antagonized by sphingosine-1-phosphate. *Blood*, 106, 125–134, 2005.
85. Zhang, H. et al., Circulating endothelial progenitor cells in multiple myeloma: implications and significance. *Blood*, 105, 3286–3294, 2005.
86. Lyden, D. et al., Impaired recruitment of bone-marrow-derived endothelial and hematopoietic precursor cells blocks tumor angiogenesis and growth. *Nat Med*, 7, 1194–1201, 2001.
87. Dredge, K. et al., Orally administered lenalidomide (CC-5013) is anti-angiogenic in vivo and inhibits endothelial cell migration and Akt phosphorylation in vitro. *Microvasc Res*, 69, 56–63, 2005.
88. Hideshima, T. et al., Thalidomide and its analogs overcome drug resistance of human multiple myeloma cells to conventional therapy. *Blood*, 96, 2943–2950, 2000.
89. Haslett, P.A. et al., Thalidomide costimulates primary human T lymphocytes, preferentially inducing proliferation, cytokine production, and cytotoxic responses in the $CD8^+$ subset. *J Exp Med*, 187, 1885–1892, 1998.
90. Davies, F.E. et al., Thalidomide and immunomodulatory derivatives augment natural killer cell cytotoxicity in multiple myeloma. *Blood*, 98, 210–216, 2001.
91. Geitz, H., Handt, S., and Zwingenberger, K., Thalidomide selectively modulates the density of cell surface molecules involved in the adhesion cascade. *Immunopharmacology*, 31, 213–221, 1996.

92. Juliusson, G. et al., Frequent good partial remissions from thalidomide including best response ever in patients with advanced refractory and relapsed myeloma. *Br J Haematol*, 109, 89–96, 2000.
93. Kumar, S. et al., Effect of thalidomide therapy on bone marrow angiogenesis in multiple myeloma. *Leukemia*, 18, 624–627, 2004.
94. Garcia-Sanz, R. et al., The oral combination of thalidomide, cyclophosphamide and dexamethasone (ThaCyDex) is effective in relapsed/refractory multiple myeloma. *Leukemia*, 18, 856–863, 2004.
95. Weber, D. et al., Thalidomide alone or with dexamethasone for previously untreated multiple myeloma. *J Clin Oncol*, 21, 16–19, 2003.
96. Mascarenhas, R.C. et al., Thalidomide inhibits the growth and progression of hepatic epithelioid hemangioendothelioma. *Oncology*, 67, 471–475, 2004.
97. Little, R.F. et al., Activity of thalidomide in AIDS-related Kaposi's sarcoma. *J Clin Oncol*, 18, 2593–2602, 2000.
98. Vogt, T. et al., Antiangiogenetic therapy with pioglitazone, rofecoxib, and metronomic trofosfamide in patients with advanced malignant vascular tumors. *Cancer*, 98, 2251–2256, 2003.
99. Baumann, F. et al., Combined thalidomide and temozolomide treatment in patients with glioblastoma multiforme. *J Neuro Oncol*, 67, 191–200, 2004.
100. Fine, H.A. et al., Phase II trial of thalidomide and carmustine for patients with recurrent high-grade gliomas. *J Clin Oncol*, 21, 2299–2304, 2003.
101. Hwu, W.J. et al., Phase II study of temozolomide plus thalidomide for the treatment of metastatic melanoma. *J Clin Oncol*, 21, 3351–3356, 2003.
102. Hwu, W.J. et al., Treatment of metastatic melanoma in the brain with temozolomide and thalidomide. *Lancet Oncol*, 2, 634–635, 2001.
103. Danson, S. et al., Randomized phase II study of temozolomide given every 8 h or daily with either interferon alfa-2b or thalidomide in metastatic malignant melanoma. *J Clin Oncol*, 21, 2551–2557, 2003.
104. Hernberg, M. et al., Interferon alfa-2b three times daily and thalidomide in the treatment of metastatic renal cell carcinoma. *J Clin Oncol*, 21, 3770–3776, 2003.
105. Dahut, W.L. et al., Randomized phase II trial of docetaxel plus thalidomide in androgen-independent prostate cancer. *J Clin Oncol*, 22, 2532–2539, 2004.
106. Robinson-Rechavi, M. et al., How many nuclear hormone receptors are there in the human genome? *Trends Genet*, 17, 554–556, 2001.
107. Fajas, L. et al., The organization, promoter analysis, and expression of the human PPAR gamma gene. *J Biol Chem*, 272, 18779–18789, 1997.
108. Forman, B.M. et al., 15-Deoxy-delta 12,14-prostaglandin J2 is a ligand for the adipocyte determination factor PPAR gamma. *Cell*, 83, 803–812, 1995.
109. Yki-Jarvinen, H., Thiazolidinediones. *N Engl J Med*, 351, 1106–1118, 2004.
110. Grommes, C., Landreth, G.E., and Heneka, M.T., Antineoplastic effects of peroxisome proliferator-activated receptor gamma agonists. *Lancet Oncol*, 5, 419–429, 2004.
111. Palakurthi, S.S. et al., Anticancer effects of thiazolidinediones are independent of peroxisome proliferator-activated receptor gamma and mediated by inhibition of translation initiation. *Cancer Res*, 61, 6213–6218, 2001.
112. Mueller, E. et al., Effects of ligand activation of peroxisome proliferator-activated receptor gamma in human prostate cancer. *Proc Natl Acad Sci USA*, 97, 10990–10995, 2000.
113. Mueller, E. et al., Terminal differentiation of human breast cancer through PPAR gamma. *Mol Cell*, 1, 465–470, 1998.
114. Elstner, E. et al., Ligands for peroxisome proliferator-activated receptor gamma and retinoic acid receptor inhibit growth and induce apoptosis of human breast cancer cells in vitro and in BNX mice. *Proc Natl Acad Sci USA*, 95, 8806–8811, 1998.
115. Ricote, M. et al., The peroxisome proliferator-activated receptor (PPAR gamma) as a regulator of monocyte/macrophage function. *J Leukoc Biol*, 66, 733–739, 1999.
116. Zhang, X. et al., Peroxisome proliferator-activated receptor-gamma and its ligands attenuate biologic functions of human natural killer cells. *Blood*, 104, 3276–3284, 2004.
117. Szatmari, I. et al., Activation of PPAR gamma specifies a dendritic cell subtype capable of enhanced induction of iNKT cell expansion. *Immunity*, 21, 95–106, 2004.

118. Sunami, E. et al., Decreased synthesis of matrix metalloproteinase-7 and adhesion to the extracellular matrix proteins of human colon cancer cells treated with troglitazone. *Surg Today*, 32, 343–350, 2002.
119. Galli, A. et al., Antidiabetic thiazolidinediones inhibit invasiveness of pancreatic cancer cells via PPAR gamma independent mechanisms. *Gut*, 53, 1688–1697, 2004.
120. Barak, Y. et al., PPAR gamma is required for placental, cardiac, and adipose tissue development. *Mol Cell*, 4, 585–595, 1999.
121. Panigrahy, D. et al., PPAR gamma ligands inhibit primary tumor growth and metastasis by inhibiting angiogenesis. *J Clin Invest*, 110, 923–932, 2002.
122. Yamakawa, K. et al., Peroxisome proliferator-activated receptor-gamma agonists increase vascular endothelial growth factor expression in human vascular smooth muscle cells. *Biochem Biophys Res Commun*, 271, 571–574, 2000.
123. Jozkowicz, A. et al., Ligands of peroxisome proliferator-activated receptor-gamma increase the generation of vascular endothelial growth factor in vascular smooth muscle cells and in macrophages. *Acta Biochim Pol*, 47, 1147–1157, 2000.
124. Goetze, S. et al., Leptin induces endothelial cell migration through Akt, which is inhibited by PPAR gamma-ligands. *Hypertension*, 40, 748–754, 2002.
125. Douros, J. and Suffness, M., New antitumor substances of natural origin. *Cancer Treat Rev*, 8, 63–87, 1981.
126. Huang, S. and Houghton, P.J., Inhibitors of mammalian target of rapamycin as novel antitumor agents: from bench to clinic. *Curr Opin Investig Drugs*, 3, 295–304, 2002.
127. Wiederrecht, G.J. et al., Mechanism of action of rapamycin: new insights into the regulation of G1-phase progression in eukaryotic cells. *Prog Cell Cycle Res*, 1, 53–71, 1995.
128. Hosoi, H. et al., Rapamycin causes poorly reversible inhibition of mTOR and induces p53-independent apoptosis in human rhabdomyosarcoma cells. *Cancer Res*, 59, 886–894, 1999.
129. Oldham, S. and Hafen, E., Insulin/IGF and target of rapamycin signaling: a TOR de force in growth control. *Trends Cell Biol*, 13, 79–85, 2003.
130. Carraway, H. and Hidalgo, M., New targets for therapy in breast cancer: mammalian target of rapamycin (mTOR) antagonists. *Breast Cancer Res*, 6, 219–224, 2004.
131. Jacinto, E. and Hall, M.N., Tor signalling in bugs, brain and brawn. *Nat Rev Mol Cell Biol*, 4, 117–126, 2003.
132. Beugnet, A. et al., Regulation of targets of mTOR (mammalian target of rapamycin) signalling by intracellular amino acid availability. *Biochem J*, 372, 555–566, 2003.
133. Peng, T., Golub, T.R., and Sabatini, D.M., The immunosuppressant rapamycin mimics a starvation-like signal distinct from amino acid and glucose deprivation. *Mol Cell Biol*, 22, 5575–5584, 2002.
134. Proud, C.G., Regulation of mammalian translation factors by nutrients. *Eur J Biochem*, 269, 5338–5349, 2002.
135. Huang, S. and Houghton, P.J., Targeting mTOR signaling for cancer therapy. *Curr Opin Pharmacol*, 3, 371–377, 2003.
136. Shi, Y. et al., Enhanced sensitivity of multiple myeloma cells containing PTEN mutations to CCI-779. *Cancer Res*, 62, 5027–5034, 2002.
137. Huang, S. et al., p53/p21 (CIP1) cooperate in enforcing rapamycin-induced G(1) arrest and determine the cellular response to rapamycin. *Cancer Res*, 61, 3373–3381, 2001.
138. Dennis, P.B. et al., Mammalian TOR: a homeostatic ATP sensor. *Science*, 294, 1102–1105, 2001.
139. Dutcher, J.P., Mammalian target of rapamycin inhibition. *Clin Cancer Res*, 10, 6382S–6387S, 2004.
140. Dilling, M.B. et al., 4E-binding proteins, the suppressors of eukaryotic initiation factor 4E, are down-regulated in cells with acquired or intrinsic resistance to rapamycin. *J Biol Chem*, 277, 13907–13917, 2002.
141. Hidalgo, M. and Rowinsky, E.K., The rapamycin-sensitive signal transduction pathway as a target for cancer therapy. *Oncogene*, 19, 6680–6686, 2000.
142. Harada, H. et al., p70S6 kinase signals cell survival as well as growth, inactivating the pro-apoptotic molecule BAD. *Proc Natl Acad Sci USA*, 98, 9666–9670, 2001.
143. Schmelzle, T. and Hall, M.N., TOR, a central controller of cell growth. *Cell*, 103, 253–262, 2000.

144. Rosenwald, I.B. et al., Eukaryotic translation initiation factor 4E regulates expression of cyclin D1 at transcriptional and post-transcriptional levels. *J Biol Chem*, 270, 21176–21180, 1995.
145. Hudson, C.C. et al., Regulation of hypoxia-inducible factor 1alpha expression and function by the mammalian target of rapamycin. *Mol Cell Biol*, 22, 7004–7014, 2002.
146. Rao, R.D., Buckner, J.C., and Sarkaria, J.N., Mammalian target of rapamycin (mTOR) inhibitors as anti-cancer agents. *Curr Cancer Drug Targets*, 4, 621–636, 2004.
147. Luan, F.L. et al., Rapamycin is an effective inhibitor of human renal cancer metastasis. *Kidney Int*, 63, 917–926, 2003.
148. Gibbons, J.J. et al., The effect of CCI-779, a novel macrolide antitumor agent, on the growth of human tumor cells in vitro and in nude mouse xenograft in vivo. *Proc Am Assoc Cancer Res*, 40, 301, 2000.
149. Koehl, G.E. et al., Rapamycin protects allografts from rejection while simultaneously attacking tumors in immunosuppressed mice. *Transplantation*, 77, 1319–1326, 2004.
150. Yu, K. et al., mTOR, a novel target in breast cancer: the effect of CCI-779, an mTOR inhibitor, in preclinical models of breast cancer. *Endocr Relat Cancer*, 8, 249–258, 2001.
151. Frost, P. et al., In vivo antitumor effects of the mTOR inhibitor CCI-779 against human multiple myeloma cells in a xenograft model. *Blood*, 104, 4181–4187, 2004.
152. El-Hashemite, N. et al., Loss of Tsc1 or Tsc2 induces vascular endothelial growth factor production through mammalian target of rapamycin. *Cancer Res*, 63, 5173–5177, 2003.
153. Guba, M. et al., Rapamycin inhibits primary and metastatic tumor growth by antiangiogenesis: involvement of vascular endothelial growth factor. *Nat Med*, 8, 128–135, 2002.
154. Humar, R. et al., Hypoxia enhances vascular cell proliferation and angiogenesis in vitro via rapamycin (mTOR)-dependent signaling. *FASEB J*, 16, 771–780, 2002.
155. Bruns, C.J. et al., Rapamycin-induced endothelial cell death and tumor vessel thrombosis potentiate cytotoxic therapy against pancreatic cancer. *Clin Cancer Res*, 10, 2109–2119, 2004.
156. Bagli, E. et al., Luteolin inhibits vascular endothelial growth factor-induced angiogenesis; inhibition of endothelial cell survival and proliferation by targeting phosphatidylinositol 3′-kinase activity. *Cancer Res*, 64, 7936–7946, 2004.
157. Tsutsumi, N. et al., Essential role of PDGFRalpha-p70S6K signaling in mesenchymal cells during therapeutic and tumor angiogenesis in vivo: role of PDGFRalpha during angiogenesis. *Circ Res*, 94, 1186–1194, 2004.
158. Koehl, G. et al., Rapamycin treatment at immunosuppressive doses affects tumor blood vessel circulation. *Transplant Proc*, 35, 2135–2136, 2003.
159. Dutcher, J.P., Mammalian target of rapamycin (mTOR) inhibitors. *Curr Oncol Rep*, 6, 111–115, 2004.
160. Atkins, M.B. et al., Randomized phase II study of multiple dose levels of CCI-779, a novel mammalian target of rapamycin kinase inhibitor, in patients with advanced refractory renal cell carcinoma. *J Clin Oncol*, 22, 909–918, 2004.
161. Chan, S., Targeting the mammalian target of rapamycin (mTOR): a new approach to treating cancer. *Br J Cancer*, 91, 1420–1424, 2004.
162. Hanahan, D., Bergers, G., and Bergsland, E., Less is more, regularly: metronomic dosing of cytotoxic drugs can target tumor angiogenesis in mice. *J Clin Invest*, 105, 1045–1047, 2000.
163. Man, S. et al., Antitumor effects in mice of low-dose (metronomic) cyclophosphamide administered continuously through the drinking water. *Cancer Res*, 62, 2731–2735, 2002.
164. Kerbel, R.S. and Kamen, B.A., The anti-angiogenic basis of metronomic chemotherapy. *Nat Rev Cancer*, 4, 423–436, 2004.
165. Eberhard, A. et al., Heterogeneity of angiogenesis and blood vessel maturation in human tumors: implications for antiangiogenic tumor therapies. *Cancer Res*, 60, 1388–1393, 2000.
166. Miller, K.D., Sweeney, C.J., and Sledge, G.W., Jr., Redefining the target: chemotherapeutics as antiangiogenics. *J Clin Oncol*, 19, 1195–1206, 2001.
167. Browder, T. et al., Antiangiogenic scheduling of chemotherapy improves efficacy against experimental drug-resistant cancer. *Cancer Res*, 60, 1878–1886, 2000.
168. Lennernas, B. et al., Antiangiogenic effect of metronomic paclitaxel treatment in prostate cancer and non-tumor tissue in the same animals: a quantitative study. *APMIS*, 112, 201–209, 2004.
169. Kerbel, R.S. et al., Continuous low-dose anti-angiogenic/metronomic chemotherapy: from the research laboratory into the oncology clinic. *Ann Oncol*, 13, 12–15, 2002.

170. Hahnfeldt, P., Folkman, J., and Hlatky, L., Minimizing long-term tumor burden: the logic for metronomic chemotherapeutic dosing and its antiangiogenic basis. *J Theor Biol*, 220, 545–554, 2003.
171. Wang, J. et al., Paclitaxel at ultra low concentrations inhibits angiogenesis without affecting cellular microtubule assembly. *Anticancer Drugs*, 14, 13–19, 2003.
172. Hirata, S. et al., Inhibition of in vitro vascular endothelial cell proliferation and in vivo neovascularization by low-dose methotrexate. *Arthritis Rheum*, 32, 1065–1073, 1989.
173. Bocci, G., Nicolaou, K.C., and Kerbel, R.S., Protracted low-dose effects on human endothelial cell proliferation and survival in vitro reveal a selective antiangiogenic window for various chemotherapeutic drugs. *Cancer Res*, 62, 6938–6943, 2002.
174. Grant, D.S. et al., Comparison of antiangiogenic activities using paclitaxel (taxol) and docetaxel (taxotere). *Int J Cancer*, 104, 121–129, 2003.
175. Klement, G. et al., Differences in therapeutic indexes of combination metronomic chemotherapy and an anti-VEGFR-2 antibody in multidrug-resistant human breast cancer xenografts. *Clin Cancer Res*, 8, 221–232, 2002.
176. Drevs, J. et al., Antiangiogenic potency of various chemotherapeutic drugs for metronomic chemotherapy. *Anticancer Res*, 24, 1759–1763, 2004.
177. Bocci, G. et al., Thrombospondin 1, a mediator of the antiangiogenic effects of low-dose metronomic chemotherapy. *Proc Natl Acad Sci USA*, 100, 12917–12922, 2003.
178. Hamano, Y. et al., Thrombospondin-1 associated with tumor microenvironment contributes to low-dose cyclophosphamide-mediated endothelial cell apoptosis and tumor growth suppression. *Cancer Res*, 64, 1570–1574, 2004.
179. de Fraipont, F. et al., Thrombospondins and tumor angiogenesis. *Trends Mol Med*, 7, 401–407, 2001.
180. Lawler, J., Thrombospondin-1 as an endogenous inhibitor of angiogenesis and tumor growth. *J Cell Mol Med*, 6, 1–12, 2002.
181. Gupta, K. et al., Binding and displacement of vascular endothelial growth factor (VEGF) by thrombospondin: effect on human microvascular endothelial cell proliferation and angiogenesis. *Angiogenesis*, 3, 147–158, 1999.
182. Rapisarda, A. et al., Schedule-dependent inhibition of hypoxia-inducible factor-1 alpha protein accumulation, angiogenesis, and tumor growth by topotecan in U251-HRE glioblastoma xenografts. *Cancer Res*, 64, 6845–6848, 2004.
183. Asahara, T. et al., Isolation of putative progenitor endothelial cells for angiogenesis. *Science*, 275, 964–967, 1997.
184. Asahara, T. et al., Bone marrow origin of endothelial progenitor cells responsible for postnatal vasculogenesis in physiological and pathological neovascularization. *Circ Res*, 85, 221–228, 1999.
185. Murayama, T. et al., Determination of bone marrow–derived endothelial progenitor cell significance in angiogenic growth factor-induced neovascularization in vivo. *Exp Hematol*, 30, 967–972, 2002.
186. Takahashi, T. et al., Ischemia- and cytokine-induced mobilization of bone marrow–derived endothelial progenitor cells for neovascularization. *Nat Med*, 5, 434–438, 1999.
187. Hattori, K. et al., Vascular endothelial growth factor and angiopoietin-1 stimulate postnatal hematopoiesis by recruitment of vasculogenic and hematopoietic stem cells. *J Exp Med*, 193, 1005–1014, 2001.
188. Heeschen, C. et al., Erythropoietin is a potent physiologic stimulus for endothelial progenitor cell mobilization. *Blood*, 102, 1340–1346, 2003.
189. Bertolini, F. et al., Maximum tolerable dose and low-dose metronomic chemotherapy have opposite effects on the mobilization and viability of circulating endothelial progenitor cells. *Cancer Res*, 63, 4342–4346, 2003.
190. Ben-Efraim, S., Immunomodulating anticancer alkylating drugs: targets and mechanisms of activity. *Curr Drug Targets*, 2, 197–212, 2001.
191. Matar, P. et al., Th2/Th1 switch induced by a single low dose of cyclophosphamide in a rat metastatic lymphoma model. *Cancer Immunol Immunother*, 50, 588–596, 2002.
192. Garber, K., Could less be more? Low-dose chemotherapy goes on trial. *J Natl Cancer Inst*, 94, 82–84, 2002.
193. Sterba, J., Pavelka, Z., and Slampa, P., Concomitant radiotherapy and metronomic temozolomide in pediatric high-risk brain tumors. *Neoplasma*, 49, 117–120, 2002.

194. Kurzen, H. et al., Inhibition of angiogenesis by non-toxic doses of temozolomide. *Anticancer Drugs*, 14, 515–522, 2003.
195. Colleoni, M. et al., Low-dose oral methotrexate and cyclophosphamide in metastatic breast cancer: antitumor activity and correlation with vascular endothelial growth factor levels. *Ann Oncol*, 13, 73–80, 2002.
196. Sandri, M.T. et al., Serum levels of HER2 ECD can determine the response rate to low dose oral cyclophosphamide and methotrexate in patients with advanced stage breast carcinoma. *Anticancer Res*, 24, 1261–1266, 2004.
197. Glode, L.M. et al., Metronomic therapy with cyclophosphamide and dexamethasone for prostate carcinoma. *Cancer*, 98, 1643–1648, 2003.
198. Nicolini, A. et al., Oral low-dose cyclophosphamide in metastatic hormone refractory prostate cancer (MHRPC). *Biomed Pharmacother*, 58, 447–450, 2004.
199. Niitsu, N. and Umeda, M., Evaluation of long-term daily administration of oral low-dose etoposide in elderly patients with relapsing or refractory non-Hodgkin's lymphoma. *Am J Clin Oncol*, 20, 311–314, 1997.
200. Gunsilius, E. et al., Palliative chemotherapy in pretreated patients with advanced cancer: oral trofosfamide is effective in ovarian carcinoma. *Cancer Invest*, 19, 808–811, 2001.
201. Kollmannsberger, C. et al., Phase II study of oral trofosfamide as palliative therapy in pretreated patients with metastatic soft-tissue sarcoma. *Anticancer Drugs*, 10, 453–456, 1999.
202. Pietras, K. and Hanahan, D., A multitargeted, metronomic, and maximum-tolerated dose "Chemo-Switch" regimen is antiangiogenic, producing objective responses and survival benefit in a mouse model of cancer. *J Clin Oncol*, 23, 939–952, 2005.
203. Rozados, V.R. et al., Metronomic therapy with cyclophosphamide induces rat lymphoma and sarcoma regression, and is devoid of toxicity. *Ann Oncol*, 15, 1543–1550, 2004.
204. Emmenegger, U. et al., A comparative analysis of low-dose metronomic cyclophosphamide reveals absent or low-grade toxicity on tissues highly sensitive to the toxic effects of maximum tolerated dose regimens. *Cancer Res*, 64, 3994–4000, 2004.
205. Strumberg, D. et al., Phase II trial of continuous oral trofosfamide in patients with advanced colorectal cancer refractory to 5-fluorouracil. *Anticancer Drugs*, 8, 293–295, 1997.
206. Reichle, A. et al., Pioglitazone and rofecoxib combined with angiostatically scheduled trofosfamide in the treatment of far-advanced melanoma and soft tissue sarcoma. *Cancer*, 101, 2247–2256, 2004.
207. Yu, J.L. et al., Heterogeneous vascular dependence of tumor cell populations. *Am J Pathol*, 158, 1325–1334, 2001.
208. Kerbel, R.S. et al., Possible mechanisms of acquired resistance to anti-angiogenic drugs: implications for the use of combination therapy approaches. *Cancer Metastasis Rev*, 20, 79–86, 2001.
209. Graeber, T.G. et al., Hypoxia-mediated selection of cells with diminished apoptotic potential in solid tumours. *Nature*, 379, 88–91, 1996.
210. Relf, M. et al., Expression of the angiogenic factors vascular endothelial cell growth factor, acidic and basic fibroblast growth factor, tumor growth factor beta-1, platelet-derived endothelial cell growth factor, placenta growth factor, and pleiotrophin in human primary breast cancer and its relation to angiogenesis. *Cancer Res*, 57, 963–969, 1997.
211. Rak, J. et al., What do oncogenic mutations have to do with angiogenesis/vascular dependence of tumors? *Cancer Res*, 62, 1931–1934, 2002.
212. Eggert, A. et al., High-level expression of angiogenic factors is associated with advanced tumor stage in human neuroblastomas. *Clin Cancer Res*, 6, 1900–1908, 2000.
213. Huang, J. et al., Vascular remodeling marks tumors that recur during chronic suppression of angiogenesis. *Mol Cancer Res*, 2, 36–42, 2004.
214. Korff, T. et al., Blood vessel maturation in a 3-dimensional spheroidal coculture model: direct contact with smooth muscle cells regulates endothelial cell quiescence and abrogates VEGF responsiveness. *FASEB J*, 15, 447–457, 2001.
215. Oike, Y. et al., Regulation of vasculogenesis and angiogenesis by EphB/ephrin-B2 signaling between endothelial cells and surrounding mesenchymal cells. *Blood*, 100, 1326–1333, 2002.
216. Coras, B. et al., Antiangiogenic therapy with pioglitazone, rofecoxib, and trofosfamide in a patient with endemic kaposi sarcoma. *Arch Dermatol*, 140, 1504–1507, 2004.

217. Reichle, A. et al., Anti-inflammatory and angiostatic therapy in chemorefractory multisystem Langerhans' cell histiocytosis of adults. *Br J Haematol*, 128, 730–732, 2005.
218. Reichle, A. et al., Metronomic chemotherapy plus/minus antiinflammatory treatment in far-advanced melanoma: a randomized multi-institutional phase II trial. *Proc Am Soc Clin Oncol*, A3139, 2005.
219. Giovannucci, E., The prevention of colorectal cancer by aspirin use. *Biomed Pharmacother*, 53, 303–308, 1999.
220. Lundholm, K. et al., Evidence that long-term COX-treatment improves energy homeostasis and body composition in cancer patients with progressive cachexia. *Int J Oncol*, 24, 505–512, 2004.
221. Tuettenberg, J. et al., Continuous low-dose chemotherapy plus inhibition of cyclooxygenase-2 as an antiangiogenic therapy of glioblastoma multiforme. *J Cancer Res Clin Oncol*, 131, 31–40, 2005.
222. Hada, M. and Mizutari, K., [A case report of metastatic pancreatic cancer that responded remarkably to the combination of thalidomide, celecoxib and irinotecan]. *Gan To Kagaku Ryoho*, 31, 1407–1410, 2004.
223. Raje, N. et al., Combination of the mTOR inhibitor rapamycin and CC-5013 has synergistic activity in multiple myeloma. *Blood*, 104, 4188–4193, 2004.
224. DeCicco, K.L. et al., The effect of thalidomide on non-small cell lung cancer (NSCLC) cell lines: possible involvement in the PPAR gamma pathway. *Carcinogenesis*, 25, 1805–1812, 2004.
225. Badawi, A.F. et al., Inhibition of rat mammary gland carcinogenesis by simultaneous targeting of cyclooxygenase-2 and peroxisome proliferator-activated receptor gamma. *Cancer Res*, 64, 1181–1189, 2004.
226. Michael, M.S., Badr, M.Z., and Badawi, A.F., Inhibition of cyclooxygenase-2 and activation of peroxisome proliferator-activated receptor-gamma synergistically induces apoptosis and inhibits growth of human breast cancer cells. *Int J Mol Med*, 11, 733–736, 2003.
227. Li, M.Y. et al., PPAR gamma pathway activation results in apoptosis and COX-2 inhibition in HepG2 cells. *World J Gastroenterol*, 9, 1220–1226, 2003.
228. Sabichi, A.L. et al., Peroxisome proliferator-activated receptor-gamma suppresses cyclooxygenase-2 expression in human prostate cells. *Cancer Epidemiol Biomarkers Prev*, 13, 1704–1709, 2004.
229. Han, S. et al., Control of COX-2 gene expression through peroxisome proliferator-activated receptor gamma in human cervical cancer cells. *Clin Cancer Res*, 9, 4627–4635, 2003.
230. Grau, R., Iniguez, M.A., and Fresno, M., Inhibition of activator protein 1 activation, vascular endothelial growth factor, and cyclooxygenase-2 expression by 15-deoxy-delta12,14-prostaglandin J2 in colon carcinoma cells: evidence for a redox-sensitive peroxisome proliferator-activated receptor-gamma-independent mechanism. *Cancer Res*, 64, 5162–5171, 2004.
231. Wick, M. et al., Peroxisome proliferator-activated receptor-gamma is a target of nonsteroidal anti-inflammatory drugs mediating cyclooxygenase-independent inhibition of lung cancer cell growth. *Mol Pharmacol*, 62, 1207–1214, 2002.
232. Folkman, J., Endogenous angiogenesis inhibitors. *APMIS*, 112, 496–507, 2004.
233. Crawford, S.E. et al., Captopril suppresses post-transplantation angiogenic activity in rat allograft coronary vessels. *J Heart Lung Transplant*, 23, 666–673, 2004.
234. Volpert, O.V. et al., Captopril inhibits angiogenesis and slows the growth of experimental tumors in rats. *J Clin Invest*, 98, 671–679, 1996.
235. Folkman, J. et al., Angiogenesis inhibition and tumor regression caused by heparin or a heparin fragment in the presence of cortisone. *Science*, 221, 719–725, 1983.
236. Folkman, J. and Shing, Y., Control of angiogenesis by heparin and other sulfated polysaccharides. *Adv Exp Med Biol*, 313, 355–364, 1992.
237. Norrby, K., Heparin and angiogenesis: a low-molecular-weight fraction inhibits and a high-molecular-weight fraction stimulates angiogenesis systemically. *Haemostasis*, 23 Suppl 1, 141–149, 1993.
238. Presta, M. et al., Antiangiogenic activity of semisynthetic biotechnological heparins: low-molecular-weight-sulfated *Escherichia coli* K5 polysaccharide derivatives as fibroblast growth factor antagonists. *Arterioscler Thromb Vasc Biol*, 25, 71–76, 2005.
239. Burstein, H.J. et al., Use of the peroxisome proliferator-activated receptor (PPAR) gamma ligand troglitazone as treatment for refractory breast cancer: a phase II study. *Breast Cancer Res Treat*, 79, 391–397, 2003.

240. Ferrandina, G. et al., Celecoxib modulates the expression of cyclooxygenase-2, ki67, apoptosis-related marker, and microvessel density in human cervical cancer: a pilot study. *Clin Cancer Res*, 9, 4324–4331, 2003.
241. Milella, M. et al., Pilot study of celecoxib and infusional 5-fluorouracil as second-line treatment for advanced pancreatic carcinoma. *Cancer*, 101, 133–138, 2004.

14 Vascular Disrupting Agents

David J. Chaplin, Graeme J. Dougherty, and Dietmar W. Siemann

CONTENTS

Over the last 30 years, there has been considerable interest in developing therapeutics that prevent angiogenesis, the process of blood vessel formation. This drive to discover so-called antiangiogenic agents was stimulated by the work of Folkman and colleagues in the early 1970s [1–4]. Based on this work it is now widely accepted that for any tumor to exceed a size of 1–2 mm^3 requires a change to a proangiogenic phenotype, leading to the formation of new blood vessels. Several proangiogenic growth factors and their relevant receptors have subsequently been identified, including vascular endothelial growth factor (VEGF), and have led to the development of novel antiangiogenic therapeutics. Recently, proof of principle for antiangiogenics which target the VEGF pathway has been established in the clinic with the humanized anti-VEGF antibody bevacizumab (Avastin) [5]. Several small molecule therapeutics that target the VEGF receptor tyrosine kinase are also in late stage clinical development.

An alternative and potentially complementary approach to targeting the neovasculature associated with tumors is not simply to interfere with its development but to disrupt its function when it has already been formed. Vascular disrupting agents (VDAs) differ from antiangiogenic agents in that, rather than preventing new blood vessel formation, they are designed to cause a rapid and selective vascular shutdown in tumors, which occurs over a period of minutes to hours. The resulting ischemia produces rapid and extensive tumor cell kill (Figure 14.1). VDAs are expected to be used in intermittent doses, to synergize with conventional treatments, rather than chronically over months or years, as is required for the antiangiogenic therapies. VDAs divide into two main classes, the biologics which use antibodies and peptides to deliver toxins, procoagulant and proapoptotic effectors to tumor endothelium and the small molecules, which do not specifically localize to tumor endothelium but exploit the known differences between tumor and normal endothelial cells to induce selective vascular dysfunction.

The VDA approach has several advantages:

- Each vessel provides the nutrition for, and facilitates removal of waste products of metabolism from, many thousands of tumor cells.

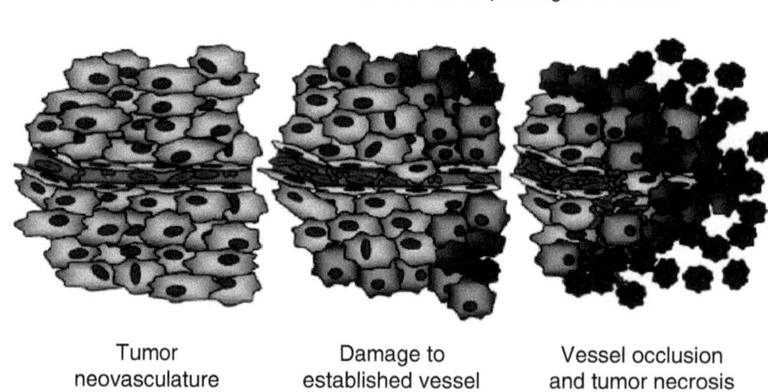

FIGURE 14.1 (See color insert following page 558.) Proposed principle of action for VDAs. Treatment with such agents induces endothelial cell dysfunction, which leads to partial occlusion of the vessel. The reduction in blood flow leads to tumor cells farthest from the vessel becoming increasingly hypoxic. As damage to the vessel progresses, coagulation events are initiated and blockage of the vessel ultimately occurs. The loss of vessel patency results in widespread tumor cell necrosis. (From Siemann, D., Chaplin, D., and Horsman, M. *Cancer*, 100, 2491, 2004.)

- Since a vessel is much like any channel involved in transportation, it only has to be damaged at one point to block both upstream and downstream function.
- It is not necessary to kill endothelial cells, a change of shape or function could also be effective.
- Endothelial cell is adjacent to the blood stream; therefore, delivery problems of any therapeutic are minimized.
- A surrogate marker of biological activity, i.e., blood flow, is readily measurable in the clinic.

As mentioned previously, in contrast to the focus on antiangiogenic approaches to therapy, there has until recently been relatively little effort afforded to the development of VDAs. In this chapter, we detail the background behind the development of these agents, their therapeutic potential when used in combination with conventional and antiangiogenic agents, and the current clinical status of the VDAs in development.

Formulation of the VDA Concept

In 1982, Juliana Denekamp first proposed that the higher proliferation rate of endothelial cells in tumors than normal tissues might enable them to be selectively targeted [6]. Until this time, most investigators considered the relatively poorly formed and inadequate vascular networks in tumors to be a problem which limited the effectiveness of conventional therapies. Many tumor cells would be hypoxic and thus radiation resistant and many cells would be located distant from vessels and lack an optimal supply of nutrients to facilitate proliferation, thus limiting the delivery and effectiveness of cell cycle specific chemotherapy. In a series of papers, Denekamp advocated that tumor vasculature, rather than being a problem, represented an opportunity for therapeutic approaches and should be considered the "Achilles' heel" of solid tumors [6–8].

Effects of Vascular Disruption on Tumor Cell Survival

One of the questions raised about disrupting vasculature within tumors is how long blood flow needs to be shut down to induce significant tumor cell kill. The initial study investigating the effects of induced ischemia within tumors was performed by Denekamp and colleagues in 1983 [9]. The ischemia was achieved by applying D-shaped metal clamps across the base of subcutaneously implanted tumors. Marked tumor regression, delayed growth, and long-term tumor control were seen, with the magnitude of the response being proportional to the duration of clamping. Vessel occlusion for at least 15 h was necessary to achieve local cure of the tumor. One potential issue with this study that was not addressed was the actual kinetics of tumor cell death and the potential effects of cooling likely to have occurred in the tissue during the clamping period. A subsequent study by Chaplin and Horsman investigated these issues in different tumor models [10]. In this study, it was shown that a 2 h period of ischemia reduced survival by over 99% when the tumor temperature was maintained at 37°C; similar results were obtained for other tumor types. The rate of cell death was much reduced if the tumors were allowed to cool during the ischemic insult. It was also shown in this study that a 6 h period of ischemia induced in C3H mammary tumors maintained at 37°C could result in the local control of three of the seven tumors treated. Similar kinetics of tumor cell death was also reported for the CaNT tumor following ischemic insult under conditions where the tumor temperature was maintained [11]. These authors attributed the decreased rate of cell death in tumors allowed to cool to a reduced metabolic process and showed a slower acidification of extracellular pH in these tumors. The cell killing effect of a 2 h period of ischemia on three different tumor lines is shown in Figure 14.2.

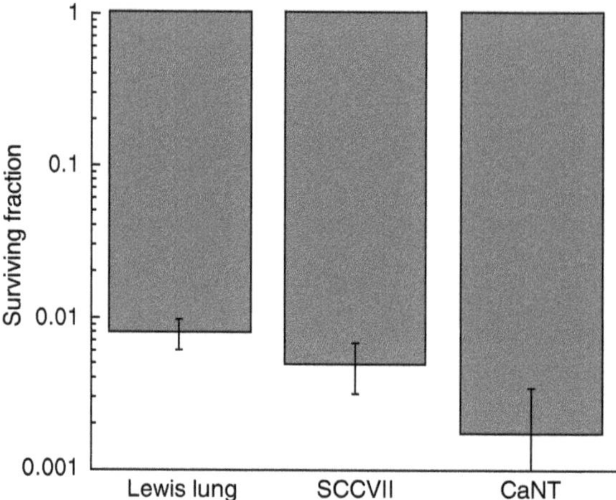

FIGURE 14.2 Clonogenic cell survival in three different tumor types: Lewis lung carcinoma, SCCVII, and carcinoma NT following a 2 h period of ischemia. Ischemia was artificially induced using a metal clamp and tumor temperature maintained at 37°C. Full details have been described previously. (From Chaplin, D.J. and Horsman, M.R. *Radiother Oncol*, 30, 59, 1994.)

14.1 SMALL MOLECULE VDAs

Interestingly, many early cancer therapies induced histological changes in animal tumors consistent with vascular disruption. These hallmarks of VDA therapy are necrosis, hemorrhage, and thrombosis. Such therapies included the tubulin depolymerizing agents colchicine and podophyllotoxin [12–16], bacterial products (so-called Coleys toxins) [12,17], lead colloids [18], and trivalent arsenicals [19].

At present there are three distinct classes of small molecule VDAs undergoing clinical development:

- Tubulin depolymerizing agents
- Synthetic flavonoids
- N-cadherin inhibitors

The chemical structures of some of these compounds are shown in Figure 14.3.

14.1.1 Tubulin Depolymerizing Agents—Development

As mentioned previously, early studies in the 1930s had indicated that colchicine could induce rapid hemorrhagic necrosis in experimental tumors. The drawback was that these effects were only observed at, or close to, the maximum tolerated dose (MTD). Despite these early hints of vascular disrupting activity, efforts to discover other tubulin depolymerizing agents focused only on their ability to inhibit cell proliferation in vitro. These studies led to the discovery and development of many compounds including the vinca alkaloids vincristine and vinblastine. Although these compounds exert their antitumor action, at doses used clinically, through their antiproliferative action, VDA-like activity is observed at doses close to the MTD in animals [20,21]. Based on the knowledge that tubulin depolymerizing agents, as a class, could elicit vascular disruption selectively in tumors at high doses, a drug discovery effort was initiated in the mid 1990s to identify tubulin interacting agents which could elicit such effects at doses much below the MTD and therefore elicit vascular effects at potential clinically achievable doses. One of the first groups of agents to be identified was the combretastatins,

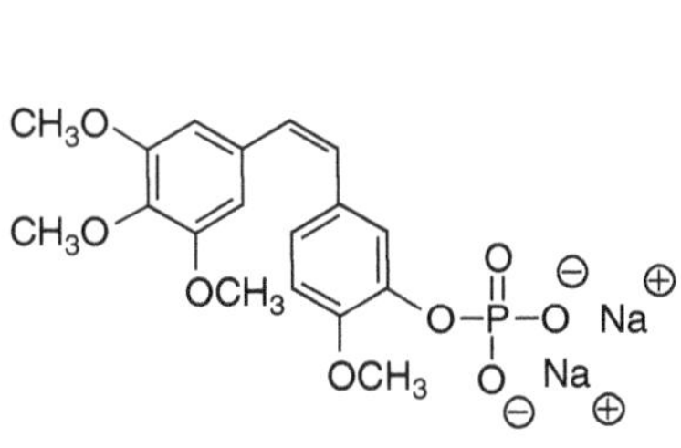

CA4P

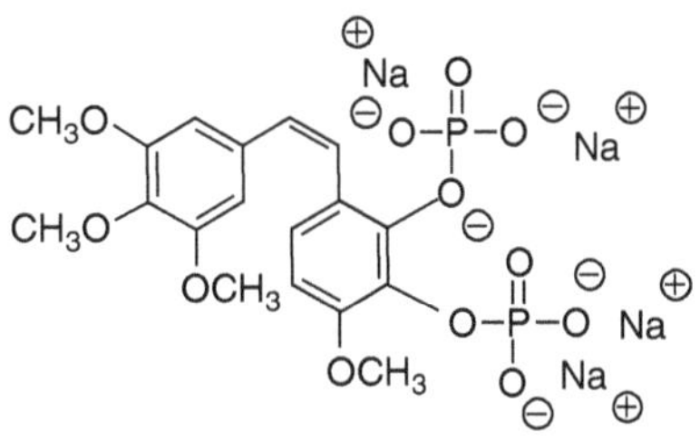

OXi4503 (CA1P)

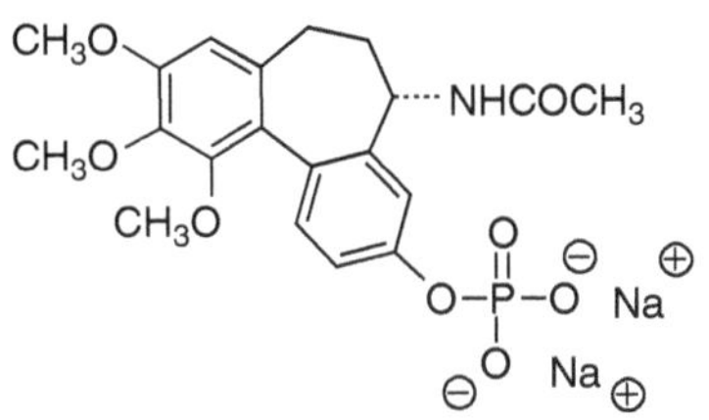

ZD6126

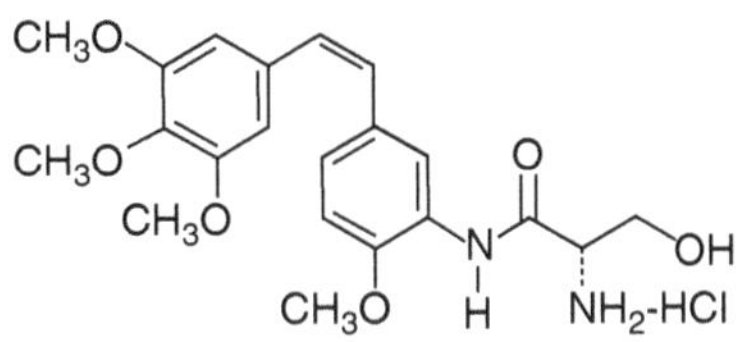

AVE8062A

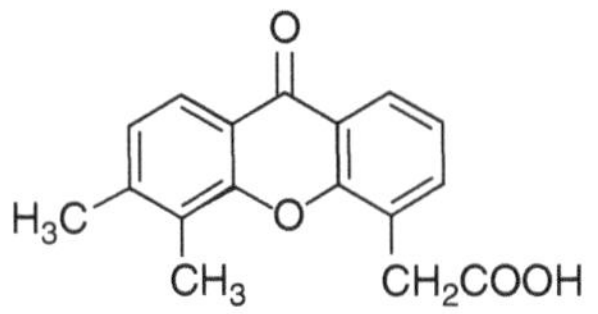

DMXAA

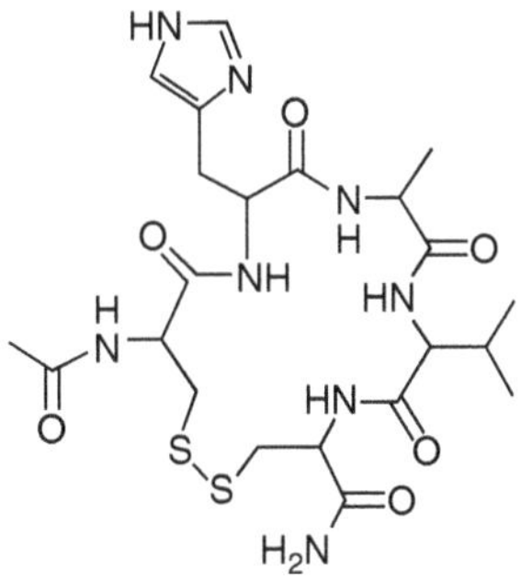

Exherin

FIGURE 14.3 Structures of small molecule VDAs in clinical development.

namely CA4 and CA1 [22]. Currently three combretastatin derivatives are in clinical development, the phosphate prodrug of CA4 (OXi2021), OXi4503, and AVE8062. Two other noncombretastatin-based tubulin depolymerizing agents developed as VDAs have entered into clinical trial, ZD6126 and MN-029.

The change from solely antiproliferative agents to VDAs for this subgroup of tubulin destabilizing agents is believed to center on the duration of their interaction with the target protein and their pharmacokinetic behavior. More reversible binding and rapid plasma clearance will result in a reduction in the severity of the long-term effects on tubulin assembly, such as inhibition of cell division processes, but still retain the acute effects, e.g., cell morphological changes. As will be discussed later, rapid changes in the shape of immature endothelial cells are implicated in their VDA action. The reason such effects are not achieved with low doses of the vinca alkaloids or colchicines is because they are strongly antiproliferative in vivo by the time any VDA action is seen; dose limiting antiproliferative action limits further dose escalations.

14.1.1.1 CA4P

The combretastatins were originally derived from the bark of the African willow tree *Combretum caffrum* by Pettit and colleagues [23]. Numerous synthetic analogs of combretastatins have now been made [24–33]. The compound which has been the subject of most study in this class is combrestatin A4 phosphate (CA4P). This compound was shown to shut down blood flow in tumors at doses below 1/10th the MTD in mice [34]. Moreover, this blood flow shutdown was shown to be selective for the tumor tissue [35,36]. CA4P represented the first small molecular weight agent that could induce selective vascular disruption in tumors over a wide dose range. At the time that its vascular activity was discovered little was known about the molecule which could clearly explain why it induced vascular shutdown.

Studies have shown that CA4P can induce selective cytotoxic/antiproliferative effects against dividing, compared with quiescent, endothelial cells in culture [34]. However, the rapidity of vascular shutdown (~10 min) observed in an isolated rat tumor model suggests that more immediate changes are responsible for the drug effects seen [34,35]. Work has shown that CA4P and related compounds can have dramatic effects on the three-dimensional shape of newly formed endothelial cells, but less effect on quiescent endothelial cells [37,38]. It is believed that the reason newly formed endothelial cells are more sensitive than more mature cells is that the latter have a more highly developed actin cytoskeleton which maintains the cell shape despite depolymerization of the tubulin cytoskeleton [39]. Recent studies have indicated that morphological changes in immature endothelial cells in vitro also involve signaling through the small GTPase Rho and Rho kinase which drives stress fiber formation, focal adhesion assembly, and redistribution of junctional proteins [40]. Clearly, rapid changes in endothelial cell shape in vivo would dramatically alter capillary blood flow, expose basement membrane, and as a result, induce hemorrhage and coagulation. The sensitivity of the immature tumor vasculature to CA4P probably relates to not only differences between newly formed and mature endothelial cells, but also to specific characteristics of the tumor microcirculation such as high vascular permeability, high interstitial fluid pressure, procoagulant status, vessel tortuosity, and heterogeneous blood flow distribution [41].

CA4P has been shown to reduce perfusion in isolated tumor models where clotting factors and circulating blood cells are absent, further supporting the idea that the first key step in vascular disruption is a direct effect on endothelial cells. However, in tumors in vivo, in the presence of clotting factors and blood cells, one would expect the vascular disrupting activity to be amplified. In normal tissue inflammation neutrophil adhesion to damaged endothelium will produce additional damaging oxidizing elements through the action of neutrophilic myeloperoxidase. Following CA4P treatment it has been shown that neutrophils are recruited to the endothelium [42] and into the tumor [43]. This neutrophil recruitment probably contributes to the vascular effects of CA4P and other VDAs.

14.1.1.2 AVE8062 (AC7700)

AVE8062, previously termed AC7700, is a CA4P derivative originally developed by the Ajinomoto Company in Japan. The compound is a tubulin depolymerizing agent which, like CA4P, has been shown to elicit blood flow changes at doses well below the MTD [44,45]. These blood flow changes, as with CA4P, occur in a matter of minutes and lead to extensive necrosis. Although little data have been published to date, as regards mechanism, it is to be assumed that, as with CA4P, endothelial cell shape changes play a key role in its action. In addition, it has recently been shown that AVE8062 can induce constriction of the host arterioles which feed the tumor [46]. This finding led to the suggestion that, at least with this compound, vasoconstriction could play a major role in its mechanism of action. It is of interest that tubulin depolymerizing agents as a class can, in addition to their direct effects on endothelial cell morphology, induce arteriolar vasoconstriction, even in endothelial cell denuded vessels [47]. In addition, it has been established that concomitant use of vasodilators with CA4P in rats can eliminate vasoconstriction without altering the blood flow shutdown effects in the tumor, further indicating the independence of the two effects [48]. These data indicate that, at least for CA4P, vasoconstriction is not involved in its action as a VDA.

14.1.1.3 OXi4503

OXi4503 is the diphosphate prodrug of combretastatin A1 (CA1) originally synthesized by Pettit and Lippert [49]. Preclinical work in a variety of experimental tumor systems has shown rapid and extensive blood flow shutdown leading to necrosis. OXi4503 is more effective at producing tumor growth delay than CA4P [50]. These increased antitumor effects include a more efficient killing of the viable rim that remains after solid tumor treatment with CA4P. OXi4503 is also seen to be more efficacious against metastatic liver deposits of HT-29 and DLD-1 colon tumors [51]. In a recent study by Hua and colleagues, it was shown that in a preclinical xenograft model of breast cancer OXi4503 induced tumor regressions (including complete regression) [52]. In vitro metabolic comparisons of CA1 (OXi4500) and CA4 strongly suggest that once dephosphorylated, OXi4503 is metabolized to a more reactive species than CA4P, presumably an orthoquinone [53]. The ability of OXi4503 to induce direct cytotoxicity via an orthoquinone intermediate, while retaining the VDA activity mediated through the interaction with tubulin, could account for the tumor responses seen with this agent when used alone. This work identified OXi4503 as a preclinical developmental candidate with more potent antivascular and antitumor effects than CA4P when used as a single agent. OXi4503 entered Phase I clinical trials in 2005.

14.1.1.4 ZD6126

ZD6126 is a colchinol-based compound that is a close structural analog of colchicine. Similar to the other agents discussed earlier, ZD6126 was shown to induce extensive necrosis in preclinical tumor models, including human tumor xenografts. ZD6126 possesses a wide therapeutic index for antitumor effects [54]. Clear evidence of rapid and selective blood flow shutdown within tumors has been observed in both rat and mouse models [55,56].

14.1.1.5 MN-029

MN-029 is currently in Phase I clinical trial. This compound is a benzimidazole carbamate-based tubulin depolymerizing agent that binds to a site on tubulin distinct from colchicines and the vincas. Early related members of this chemical class (e.g., Mebendazole) are used as anthelminthics.

14.1.2 Tubulin Depolymerizing Agents—Clinical Experience

Three Phase I clinical trials with CA4P have been completed. These trials explored the safety and pharmacokinetics (PK) of three different IV administration schedules of CA4P as a single agent (i.e., single dose schedule every 21 days, weekly for three consecutive weeks every 28 days, and daily for five consecutive days every 21 days) [57–61]. The results from all the Phase I trials have recently been reviewed [62]. Doses as high as 114 mg/m^2 (weekly schedule) have been studied. Dose limiting toxicities (DLTs) have been identified in all three trials, resulting in an MTD in the range of 60–68 mg/m^2. DLTs have included dyspnea, myocardial ischemia, reversible neurological events, and tumor pain. In general, across all three studies, most frequently reported adverse events after a single intravenous injection of CA4P were mild (Grade 1 or 2) nausea, vomiting, headache, fatigue, and tumor pain. CA4P did not produce significant clinical chemistry or hematologic toxicity associated with other commonly used tubulin-binding agents such as the vinca alkaloids and taxanes, as well as colchicine. The other common toxicities associated with traditional cytotoxic drugs, such as stomatitis and alopecia, were not reported in these clinical trials.

In order to establish whether CA4P was reducing tumor perfusion, all three studies evaluated blood flow, using either dynamic contrast—MRI (DCE-MRI) or positron emission tomography (PET). The former involved the iv injection of the contrast agent gadolinium diethylenetriaminepentaacetate (Gd-DTPA). The later used ^{15}O-labelled water as the tracer ($H_2{}^{15}O$). Significant blood flow reductions were observed in the majority of patients receiving doses at or above 52 mg/m^2 [62]. Interestingly, very few patients were evaluated for blood flow changes below this dose. Although patients receiving 10 mg/m^2 or less had no discernable blood flow changes detected using PET, no data clearly established a dose response between 10 and 52 mg/m^2 [63]. Recent PET studies have indicated significant tumor blood flow reductions at doses down to 30 mg/m^2 [62]. The blood flow changes seen in the tumor tissue appear to be selective, with no change in normal muscle perfusion seen [60].

In terms of tumor response in the Phase I studies, a complete response was seen in an anaplastic thyroid cancer and a partial response in a sarcoma. In addition, several minor responses in kidney and pancreas were observed as well as prolonged (>12 months) disease stabilization in three patients, one with colon cancer and two with medullary thyroid cancer [62].

Based on the Phase I data, several Phase I/II clinical trials, in which CA4P is combined with chemotherapy, radiotherapy, or antibody-based therapeutics, have been initiated. These studies are currently using CA4P at doses of 30–60 mg/m^2, which are known to elicit VDA-like activity.

Three Phase I clinical trials with ZD6126, using similar dosing regimes to those used with CA4P, were initiated and recently completed. Although the full results have not yet been published, some data have been reported [64,65]. In the once every 21 days dosing regime a maximum dose of 112 mg/m^2 was reached. The most common adverse events were pain, constipation, fatigue, and dyspnea. Dose escalation above 112 mg/m^2 was not pursued because of a variety of adverse events at this dose, including abdominal pain and hypertension. No tumor responses were reported. However, blood flow data, using DCE-MRI, established that at doses of 80 mg/m^2 and higher, ZD6126 treatment caused a 36%–72% decrease in tumor blood flow in all patients studied, whereas no significant changes in muscle or spleen flow were observed [66].

A summary of the ongoing clinical trials with AVE8062 was presented at the recent AACR meeting [67]. In the dx5 and weekly studies, the occurrence of four potentially drug related vascular events (i.e., myocardial ischemia (MI), transient asymptomatic hypotension, transient cerebral ischemia, asymptomatic ventricular tachycardia) without residual clinical deficits, led to a voluntary interruption of all trials. No vascular event was observed in the Q21d schedule up to a 22 mg/m^2 dose, thus this trial was resumed after restricting eligibility criteria and increasing cardiovascular monitoring. No cardiac effects or tumor blood flow changes have been observed to date using this protocol.

There has been considerable focus on potential cardiovascular side effects of the tubulin-binding VDAs based around their known vascular mechanism of action and the events seen in Phase I, particularly with the recent data on AVE8062. Most work to understand any cardiovascular issues has been done around CA4P. In the reported Phase I studies of CA4P, three cases of reversible myocardial effects were reported and these contributed to the establishment of the current MTD. In these patients, hypertension was evident before the myocardial effects occurring. It is of interest that tubulin depolymerizing agents as a class can, in addition to their direct effects on endothelial cell morphology, induce arteriolar vasoconstriction even in endothelial cell denuded vessels [47]. In addition, it has been established that concomitant use of vasodilators with CA4P in rats can eliminate vasoconstriction without altering the blood flow shutdown effects in the tumor, further indicating the independence of the two effects [48]. If, as is now believed, the vasoconstrictive actions of CA4P contribute to the DLT, including cardiovascular ones, there is a possibility that the prophylactic use of vasodilators may not only alter its toxicity profile but also enable higher dose levels to be attained. This approach requires evaluation in a clinical setting. However, using the doses of CA4P chosen for Phase I/II, cardiovascular effects, in terms of changes in blood pressure and heart rate, have been mild and reversible [68].

14.1.3 Flavonoids: DMXAA—Development

The interest in this nontubulin-interacting class of VDAs began in the 1980s with the finding that flavone acetic acid (FAA), a drug originally synthesized as a nonsteroidal anti-inflammatory agent, had significant antitumor effects in a wide variety of experimental tumor models in mice [69]. Evidence that this drug could elicit rapid hemorrhagic necrosis in experimental solid tumors was obtained by several investigators [70–72]. Other studies confirmed that the drug elicited a rapid reduction in tumor blood flow [71,73,74]. The mechanism of action of the drug was attributed, in part, to its ability to induce the production of the cytokine tumor necrosis factor alpha (TNFα) in tumor cells and the tumor-associated host cells [75]. Interestingly, the identification of the vascular action of FAA came after the compound had entered clinical trials. The lack of response seen in the trials when administered as a single agent [76–78], together with the fact that no tumor necrosis was seen in biopsies following treatment [79], led to the discontinuation of the clinical program. The lack of effect of FAA was attributed at the time to its inability to induce significant TNF production in cells from human, rather than rodent, origin. This led to a search for other structurally related agents which could overcome this issue. The current lead compound 5,6-dimethylxanthenone 4-acetic acid (DMXAA, AS1404) was the result of research carried out by Denny and Baguley at the University of Auckland in New Zealand [80]. This compound has been shown to induce TNFα mRNA in both murine and human cells [81]. The mechanism of action of DMXAA is not thought just to involve TNF production, since it has a rapid and direct proapototic effect on endothelial cells in vitro [82]. Subsequent activation of platelets causes the release of serotonin which itself is known to reduce tumor blood flow [83]. The current understanding of the mechanism of action of DMXAA has been discussed in recent reviews [84,85].

DMXAA has shown single agent antitumor activity, including long-term regressions, in some tumor models [86]. In contrast to the tubulin-binding VDAs discussed earlier, in mice the therapeutic index for DMXAA is rather narrow [87]. The fact that in the clinic blood flow reductions have been observed at doses below the MTD may however indicate that VDA activity is clinically achievable with this agent.

14.1.4 Flavonoids: DMXAA—Clinical Experience

As alluded to previously, the clinical experience with FAA was disappointing in terms of response however, the initial Phase I evaluations were planned before the knowledge of its vascular effects, so no measures of blood flow were incorporated in the design.

DMXAA entered clinical trials in 1996 and was administered as a 20 min intravenous infusion [88,89]. In these studies, DMXAA was used in doses ranging from 6 to 4900 mg/m^2. Toxicities observed included tremor, confusion, slurred speech, anxiety, and urinary incontinence. In addition, transient prolongation of cardiac QTc interval was seen in some of the patients administered doses at or above 2000 mg/m^2.

In terms of direct response, only two unconfirmed PRs were observed, at doses of 1100 and 1300 mg/m^2. However, in terms of biological activity, over the dose range of 500–4900 mg/m^2 9/16 patients had significant reductions in tumor blood flow 24 h after the first dose of DMXAA [90]. In addition, dose dependent increases in the serotonin metabolite 5-hydroxyindoleacetic acid, a potential marker of vascular disruption with DMXAA in the mouse studies, were observed at doses exceeding 650 mg/m^2. These later two findings indicate that DMXAA can act as a VDA in man at clinically achievable doses. Based on these findings, together with recent data showing no significant QTc changes at doses up to 1200 mg/m^2, three Phase II studies have been initiated in prostate, ovarian, and NSCLC, where DMXAA is incorporated in taxane-based chemotherapy regimes.

14.1.5 N-Cadherin Antagonists

The endothelial cell adhesion molecules, neural- and vascular endothelial-cadherin, play essential roles in the formation of stable and fully functional blood vessels [91]. In particular, N-cadherin has been shown to be directly involved in regulating adhesion between the endothelial cells and pericytes of blood vessels. Exherin (ADH-1) is a novel cyclic pentapeptide which is an N-cadherin antagonist containing the cadherin cell adhesion recognition sequence His-Ala-Val. Injection of Exherin into tumor-bearing mice has been shown to induce vascular damage and extensive tumor necrosis [92]. Data from the first Phase I study with Exherin have recently been reported [93]. Thirty-three patients were treated, Exherin was administered as a bolus iv injection, doses from 50 to 840 mg/m^2 were evaluated, and tumors characterized for their expression of N-cadherin. No responses or changes in blood flow were reported in patients with N-cadherin negative tumors. Three patients with N-cadherin positive tumors showed some evidence of antitumor activity, including a PR in an esophageal cancer. Although the MTD had not been reached, the most common adverse events at the doses used were nausea, fatigue, and hot flushes. Some cardiovascular effects were seen but were thought to be related, at least in part, to prior cardiac history and injection rate. Based on the Phase I data, a Phase II trial has been initiated, focused on patients with N-cadherin positive tumors.

14.2 BIOLOGY-BASED APPROACHES TO SELECTIVE VASCULAR DISRUPTION

The promising results that have been obtained using small molecule VDAs have encouraged the search for additional approaches by which selective vascular damage can be induced. Among the strategies that appear to possess the greatest clinical potential are a number that employ biological agents such as monoclonal antibodies (mAbs), cytokines or other soluble ligands, aptamers, and viral or nonviral vectors to preferentially target therapeutic agents or processes to a tumor site, effectively minimizing the damage inflicted elsewhere.

Experimental validation, demonstrating the potential of biologic vascular disrupting strategies, was provided in a seminal study by Burrows and Thorpe [94]. Their approach targeted the MHC class II antigen, which was specifically induced on tumor endothelium using a neuroblastoma cell line genetically engineered to express interferon-γ. High affinity mAbs directed against MHC class II determinants conjugated to the A chain of the toxin ricin were shown to preferentially localize to tumor following intravenous administration, triggering endothelial cell death on endocytosis, producing vascular occlusion and ultimately massive tumor necrosis. The effect was dramatic, leading to the eradication of even large tumors in this model system.

14.2.1 Identification of Differentially Expressed Targetable Structures on Tumor Endothelium

Although the work of Burrows and Thorpe [94,95] is certainly encouraging, the further development and future clinical application of such biologic vascular disrupting therapies are, of course, predicated on the presence, on tumor-associated endothelial cells, of differentially expressed surface structures that possess both the distribution and characteristics necessary to function in a targeting capacity. Molecules orientated toward the luminal surface of blood vessels are particularly attractive in this regard as they are obviously in direct contact with the circulation and therefore more readily accessible to intravenously administered agents. Although it is appreciated that angiogenesis is a normal physiologic process, which in adults occurs during wound healing and in the course of the menstrual cycle, the poorly regulated nature of the angiogenic process within tumors, as well as the exposure of tumor-associated endothelial cells to a variety of stimuli that are unique to the tumor microenvironment, such as oxidative stress, hypoxia, low glucose levels, and low pH (as a result of the production of lactate during anaerobic glycolysis), produces a vascular network that is clearly abnormal, from both a structural and functional perspective. Such studies lend credence to the belief that differentially expressed surface structures suitable for targeting purposes will indeed exist on endothelial cells and may even be abundant [96].

Initial efforts to identify such determinants focused largely on cultured endothelial cells. The basic approach employed by most workers involved simply exposing endothelial cell monolayers to a variety of stimuli that they might reasonably be expected to encounter within the tumor microenvironment and then, at various time points thereafter, comparing the stimulated and unstimulated populations for gene expression using various biochemical and molecular techniques. Among these, the generation of panels of mAbs, serial analysis of gene expression (SAGE), and microarray analysis have been particularly informative. Irrespective of which screening technique was used, in general, the process was carried out in two stages. In the first, markers exhibiting apparent endothelial specificity are identified, with promising candidates moving forward to a second round in which a broad range of normal and malignant tissues are tested to define those that are differentially expressed on tumor endothelium.

The process of angiogenesis is initiated and controlled by a network of cytokines that interact with a number of constitutively expressed or inducible cell surface receptors that in turn trigger endothelial cell proliferation, migration, differentiation, and spatial reorganization. These downstream events too are dependent on the appropriate expression and functional activity of a broad range of structurally distinct surface molecules that mediate the highly coordinated and temporally regulated series of cell–cell or cell–matrix interactions that control these complex processes.

While the endothelial cells present in normal vessels are essentially dormant, dividing on average only once every 2 or 3 years during adult life, a large proportion of the endothelial cells associated with tumor vessels appear to be in cycle, turning over as often as every 1 to 2 days [97]. Thus, from a tumor-targeting perspective, cell surface molecules that are specifically upregulated on rapidly proliferating endothelial cells would seem worthy of further consideration. cDNA array analysis comparing resting endothelial cells with those stimulated with various proangiogenic cytokines commonly produced within the tumor microenvironment, such as VEGF or fibroblast growth factor (FGF), has identified a number of such molecules, including the receptor tyrosine kinases Tie-1 and Tie-2 (angiopoietin-1 receptors), and various adhesion proteins, including the integrin subunits αv, $\alpha 5$, $\beta 1$, and $\beta 5$, neural (N)-cadherin and vascular endothelial (VE)-cadherin [98–100]. Importantly, in most cases, the differential expression of these molecules within tumors and other sites of active angiogenesis has been confirmed using immunohistochemistry, in situ hybridization or other approaches

[98,101,102]. Moreover, functional studies have demonstrated a direct role for certain of these molecules in the angiogenic process. Thus, mAbs and other inhibitors directed against $\alpha v\beta 1$, $\alpha 1\beta 1$, and $\alpha 2\beta 2$ have been shown to block the formation of new vessels in response to VEGF in a number of model systems [103–105]. Although encouraging, it is important to keep in mind that differential expression on actively proliferating endothelial cells is often relative rather than absolute and some potential targeting proteins, including the integrins in particular, are fairly widely distributed and can be readily detected on many different cell types, although perhaps at a somewhat lower level.

Publicly accessible SAGE databases are a valuable source of genetic information and it should come as no surprise that they have been successfully mined for endothelial-specific genes using *in silico* approaches [106]. In one recent study, Ho et al. [107] used virtual subtraction to identify a subset of genes that exhibited evidence of differential expression on endothelial cells. These served as a starting point for further high throughput microarray analysis ultimately leading to the identification of 64 "pan-endothelial" markers including several that were shown by in situ hybridization to be expressed at sites of active angiogenesis in embryonic tissues.

While ease of identification and study has focused initial attention on proteins differentially expressed on proliferating versus resting endothelial cells, a number of molecules that are induced at later stages in the angiogenic process have also been identified and there seems no reason why these too could not be exploited in a tumor-targeting context. Thus, using a differential display strategy Prols et al. [100] identified two genes, osteopontin and PC4, that are upregulated when rat endothelial cells organize into vascular tubes in a 3D in vitro culture system. In a similar but more recent study, Gerritsen et al. [108] used Affymatrix oligonucleotide array analysis to identify genes upregulated when endothelial cells were induced to form tube-like structures in three different well-defined in vitro culture systems. This powerful approach yielded over a 1000 genes that showed at least a twofold increase in expression. Since molecules present on the cell surface are likely to prove more useful in targeting and imaging studies, established bioinformatic algorithms were applied to the data set to identify proteins containing putative transmembrane domains. In order to favor identification of markers differentially expressed on tumor endothelium rather than endothelial cells in general, the resultant subset was compared to a list of genes upregulated in colon tumor tissue and genes coexpressed in any of the six established colon tumor cell lines excluded from the final list. This systematic approach yielded 24 candidate genes with apparent tumor vascular specificity, including 3, gp34, stannicalcin-1 (STC-1), and GA733-1 that were expressed at a 10-fold or greater level in colon tumor tissue versus normal colon. Subsequent in situ hybridization confirmed that STC-1 was indeed differentially expressed on tumor endothelium [108].

Genes induced by stimuli unique to the tumor microenvironment are potentially the most useful in the development of tumor-targeted vascular disrupting therapies. Efforts to identify such determinants have involved analysis of differential gene expression in endothelial cells exposed in vitro to conditions characteristic of the tumor microenvironment, including tumor-conditioned media, cocktails of proangiogenic cytokines, hypoxia, oxidative stress, glucose deprivation, and acidosis. DELTA4, a previously characterized member of the Notch family of cell surface signaling molecules, was shown, for example, to be induced on endothelial cells in response to hypoxia [109] and VEGF [110]. In vivo studies confirmed that while DELTA4 is expressed on the dorsal aorta, umbilical artery and heart during embryonic development, it is almost undetectable in adult tissues except at sites of angiogenesis [109]. Reflecting their role in the maintenance of adequate tissue oxygenation, endothelial cells are particularly responsive to hypoxia. In addition to genes previously known to be regulated by oxygen tension, screening studies have identified several novel transcripts that show elevated expression under hypoxic conditions, including CXCR4, CD24, Dell, PLOD1 and PLOD2, claudin-3, and tetranectin. Among the other endothelial cell-associated cell surface proteins upregulated by hypoxia are the sortilin-related receptor-1 protein (SORL1) and the leucine-rich repeat-containing protein 19 (LRRC19).

Although convenient from a practical perspective, cultured endothelial cells may not constitute the ideal starting material with which to identify targeting moieties induced by the tumor microenvironment. When maintained in vitro, endothelial cells are no longer exposed to the complex, heterogeneous, and interactive spectrum of signals found in the in vivo situation. Tumor-conditioned media are a poor substitute and in the absence of such microenvironmental cues, even recently derived cells may no longer respond in an appropriate fashion to additional insults such as hypoxia. Moreover, endothelial cells are clearly heterogeneous in nature and HUVEC, which are often used in studies of this type, may not behave in the same way as the cells that line particular tumor vessels.

In response to such concerns, various techniques have been developed that allow the direct identification of antigens expressed on tumor-associated endothelial cells in vivo. One simple approach that has been much explored is to physically isolate such cells from disaggregated tumor tissue and use the material obtained to generate panels of mAbs that can be screened for reactivity against various normal and malignant tissues. While such efforts have resulted in identification of numerous pan-endothelial antibodies, reagents that define tumor-specific markers have been more difficult to obtain. In part, this finding may reflect the relative paucity of such determinants in comparison to the large number of competing highly immunogenic molecules present on the surface of endothelial cells.

In an effort to overcome this problem, Schnitzer and colleagues have developed a novel subtractive proteomic mapping approach that permits even rare differentially expressed proteins present on the endothelial cell surface to be identified [111,112]. The basic strategy involves in vivo labeling of the luminal surface of endothelial cells with silica particles, facilitating the subsequent isolation of a luminal plasma membrane fraction, which is then analyzed by 2D gel electrophoresis or multidimensional mass spectroscopy to produce a high-resolution proteomic map. Proteins unique to endothelial cells in a particular tissue can be readily identified, isolated, and sequenced using established techniques and subjected to bioinformatic analysis. Using this very powerful approach, Schnitzer's group defined a number of proteins that appear to be specifically upregulated on endothelial cells within the lungs of rats bearing metastatic breast adenocarcinomas [111]. Although not present on normal lung, several of these proteins are, however, found on endothelial cells in other tissues, precluding their use to tumor targeting (e.g., endoglin, which is constitutively expressed on certain brain endothelial cells) [113]. Others were already known from previous studies to exhibit elevated expression on tumor endothelium. However, two proteins, aminopeptidase-P and annexin A1, do appear to be differentially expressed on lung tumor-associated endothelium. Emphasizing the importance of such in vivo studies, a comparison of the luminal plasma membrane proteome of endothelial cells in normal rat lung with that generated from cultured rat lung microvascular endothelial cells revealed that over 40% of proteins expressed in vivo cannot be detected on cells maintained in vitro [112].

Another technique that has been widely used to identify determinants differentially expressed on the luminal surface of endothelial cells within tumors in vivo is phage display. In this approach, large and complex libraries of phage bearing short random peptide sequences are injected intravenously and tumor tissue removed after sufficient time has elapsed for the particles to distribute widely throughout the body. Phage with affinity for proteins and other molecules expressed on tumor endothelium will be retained within tumor tissue and can be recovered following surgical resection, expanded and subjected to further rounds of enrichment. Ultimately, the peptide sequence displayed by phage exhibiting high affinity for tumor endothelium can be determined and used to characterize the corresponding ligand or incorporated into various other targeting entities. This simple but powerful approach has proven useful in both human [114] and animal [115,116] studies. Interestingly, among the large number of putative tumor endothelial targets that have been identified in this way are some that were quite unexpected including the ubiquitous RNA binding protein nucleolin [117] which was also identified in the proteomic screening studies of Oh et al. [111]. Subsequent studies have revealed that although nucleolin is restricted solely to the nucleus in nonproliferating cells, it can be readily detected on the surface of actively growing cells [117]. The tumor homing peptide F3 identified in phage display studies bound to surface expressed nucleolin and was subsequently internalized and transported to the nucleus. Although not yet explored, this process clearly presents unique therapeutic opportunities.

Even in adults, tumors are not the only site of active angiogenesis. The process occurs during the course of the menstrual cycle and in wound healing and is characteristic of a number of pathologic conditions including diabetic retinopathy, coronary vascular disease, primary pulmonary hypertension, and Kawasaki disease. Thus, targeting determinants associated with angiogenesis per se, including those induced on proliferating endothelial cells or involved in later stages of the new vessel formation such as endothelial cell migration or tube formation, whereas certainly a good starting point, may not achieve the specificity for tumors that safety considerations demand. On the positive side, since the process occurs in nearly all solid tumors, "pan-angiogenesis" markers are likely to be shared by many malignancies irrespective of histology or location, greatly facilitating their exploitation in a therapeutic setting. In contrast, target structures induced in response to the unique spectrum of signals generated within a particular tumor microenvironment, whereas potentially more specific for malignant tissue, are likely to vary from tumor to tumor, depending for example on the degree of hypoxia or the particular cytokine milieu produced within the mass. If so, therapy may need to be individually tailored, based, for example, on the results of phenotypic analysis of biopsied tissue.

14.2.2 Targeting Therapy to Differentially Expressed Endothelial Cell Determinants

A broad range of ligands have been explored in the context of tumor-targeted vascular disrupting therapy. Of these, mAbs remain perhaps the most attractive from a practical perspective. They can be rapidly generated against almost any antigen using a broad range of well-established technologies. Dissociation constants in the picomolar range are achievable and various engineered derivatives including Fab fragments, single chain antibodies (scFv), mini-antibodies, and so on, can be selected to achieve desired pharmacokinetic properties.

Physiologic ligands such as VEGF that are recognized and bound by differentially expressed receptors have also been successfully conjugated to various effectors and used to target tumor vasculature. Peptides too have proven effective in some circumstances although in vivo stability issues and the difficulty implicit in generating high-affinity reagents against certain defined target structures remains a challenge.

Aptamers have great but largely unexplored potential. They can, however, be produced in a rapid and cost effective manner using automated procedures and selected to possess high specificity and subnanomolar affinities, characteristics that might facilitate the future development of a clinically useful reagent.

14.2.3 Effector Molecules of Use in Tumor-Targeted Vascular Disrupting Therapy

Perhaps the most straightforward way to induce vascular occlusion within tumors is to simply kill a proportion of the endothelial cells that line tumor-associated blood vessels exposing the basement membrane and inducing intravascular coagulation. There is no shortage of effectors that can be conjugated to targeting ligands and used to achieve this goal including toxins, cytotoxic agents, cytokines, radioisotopes, and so on.

For example, deglycosylated ricin-A immunoconjugates directed against human endoglin (CD105), that exhibited weak cross reactivity with determinants present on murine endothelial cells, induced long-lasting and complete regression of MCF-7 xenografts in the majority of immunocompromised hosts following intravenous administration [118]. In another similar study, Tsunoda et al. [119] demonstrated that an immunoconjugate consisting of the cytotoxic agent neocarzinostatin linked to TES-23 (a mAb that recognizes an epitope present on an isoform of the adhesion protein CD44 that is differentially expressed on tumor-associated endothelium) induced marked regression of established KMT-17 fibrosarcomas. This was an event that was accompanied by substantial hemorrhagic necrosis characteristic of an antivascular effect.

In promising recent studies, Thorpe and colleagues have demonstrated that anionic phospholipids, principally phosphatidylserine, which in viable cells are normally restricted to the inner surface of the plasma membrane, are differentially expressed on the outer surface on tumor-associated vascular endothelial cells [120]. An unconjugated IgG3 mAb that binds anionic phospholipids in a beta2-glycoprotein I-dependent manner, designated 3G4, was shown to localize specifically to tumors in vivo and reduce the growth of both syngeneic murine tumors and orthotopic human tumor xenografts following systemic administration [120]. This activity was associated with the induction of substantial vascular damage and a reduction in both vascular density and tumor plasma volume. Interestingly from a mechanistic point of view, evidence was obtained suggesting that treatment with mAb 3G4 promoted the binding of monocytes to tumor endothelium and enhanced infiltration of macrophages into the tumor site [120]. Treatment of mice with subtoxic concentrations of docetaxel enhanced the percentage of tumor vessels that express anionic phospholipids on their outer surface from 35% to 60% and in the case of orthotopic MDA-MB-435 human breast tumor xenografts dramatically enhanced the degree of local and metastatic tumor control that can be achieved with mAb 3G4 [121].

Immunoliposomes have also been used to induce tumor-targeted vascular disruption and offer some advantages over immunoconjugates, most notably with respect to the large drug payloads they can carry. In one study, unilamellar liposomes derivatized with scFv directed against the ED-B domain of the extracellular matrix protein fibronectin, which is expressed at sites of active angiogenesis in tumors and elsewhere, were shown to accumulate within tumor tissue and substantially inhibit the growth of subcutaneous F9 teratocarcinomas when loaded with the cytotoxic agent arabinofuranosylcytosine [122].

There have also been some encouraging results using cytokines or other soluble mediators to target toxins and other effector molecules to their corresponding cognate receptors differentially expressed on tumor-associated vascular endothelium. For example, Olson et al. [123] have shown that a conjugate in which a truncated form of diphtheria toxin was chemically linked to the proangiogenic cytokine VEGF-A greatly inhibited the growth of subcutaneous tumors following intraperitoneal administration. This response was associated with substantial hemorrhagic necrosis consistent a vascular-mediated effect. There was no evidence of hemorrhage outside the tumor tissue and the therapy was apparently well tolerated.

In a more recent study, a similar VEGF-diphtheria toxin conjugate was shown to delay the appearance and dramatically inhibit the growth of spontaneous breast tumors in the well characterized C3(1)/SV40 TAg transgenic mouse model [124]. Survival was also increased and other than some temporary weight loss, no long-lasting toxicity was evident [124].

Rather than killing endothelial cells, tumor vessels can be selectively occluded by targeting the procoagulant protein tissue factor (TF) to determinants differentially expressed on tumor-associated endothelial cells. Although TF lacks coagulant activity when free in the circulation, it becomes a powerful inducer of the clotting cascade when presented bound to the surface of cells. Various antibodies have been used to target TF to the vasculature of tumors including those directed against the cell-adhesion molecule VCAM-1 [125], the oncofetal ED-B domain of fibronectin [126], and prostate-specific membrane antigen [127]. All of these reagents have shown potent antitumor activity characterized by the induction of widespread vessel thrombosis and massive central necrosis in one or more model systems.

Fusion proteins that target TF to necrotic areas of tumors, fibronectin determinants present in the basement membrane of tumor vessels, or integrins expressed on the luminal surface of tumor vessels, have also been generated and tested for their antitumor activity [128]. Although all three fusion proteins induced microregional thrombosis and caused massive tumor necrosis, substantial inhibition of tumor growth was only achieved if medium to large vessels were occluded. TF targeted to integrins on the luminal surface of tumor endothelial cells was apparently less effective in this regard [128].

14.2.4 Gene Therapy Approaches

Despite the challenging developmental and regulator obstacles that will need to be overcome to produce a clinically viable treatment, there is increasing interest in the possibility of using gene therapy-based approaches to trigger tumor vascular occlusion. Such enthusiasm is driven in large part by the specificity for tumor sites that could potentially be achieved through the use of molecular approaches. Certainly, the range of genes that might prove useful in the context of vascular-targeted therapy is large. It includes toxin and other genes that simply kill transduced target cells, as well as genes that activate prodrugs or sensitize cells to the cytotoxic effects of chemotherapeutic agents or ionizing radiation. Based on the results of antibody targeting studies, genes that regulate complement activation or trigger the coagulation cascade also have potential as do genes that control endothelial cell shape and adhesion to the basement membrane. As with other approaches, the challenge, of course, is to develop practical means of targeting such genes specifically to the tumor vasculature to minimize as much as possible the damage inflicted on various normal tissues. In general terms, efforts to achieve this goal have focused on four main approaches: *transductional targeting*, *transcriptional targeting*, *posttranscriptional targeting*, *and functional targeting*. The potential approaches to gene therapy-based targeting are illustrated in Figure 14.4.

14.2.5 Transductional Targeting

The term "transductional targeting" refers to the use of vectors and techniques designed to achieve selective delivery of a DNA construct to a desired target cell type and not to other cells that the vector might come into contact with.

Early efforts to target genes to vascular endothelial cells in vivo employed physical means such as vessel in-dwell procedures, catheterization, or direct injection to restrict vectors to a particular region or tissue. Although reasonable transduction efficiencies were obtained in some studies [129] such approaches do not readily lend themselves to the treatment of most malignant diseases.

Although several pathogenic viruses infect endothelial cells at certain stages in the disease process, any tropism for the vasculature is at best relative, precluding their use in a therapeutic setting. The viral vectors commonly used in cancer and other gene therapy applications including retroviruses, adenoviruses, and adeno-associated viruses recognize widely distributed counter-receptors and, as such, infect many different cell types and tissues following systemic administration. While adenoviral and retroviral vectors are capable of infecting endothelial cells, serotype-2-based adeno-associated viruses (AAV-2) are not very efficient in this regard. Thus high levels of gene expression are observed in smooth muscle cells but not endothelial cells when vessels are perfused with an AAV-2 vector [130]. Although AAV-2

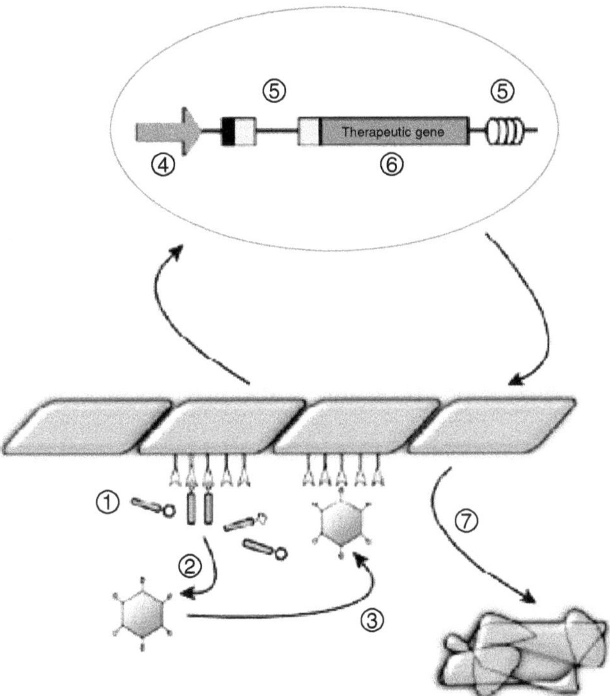

FIGURE 14.4 Targeting gene therapy to tumor vasculature. In vitro or in vivo phage display and other techniques can be used to identify ligands that recognize counter-receptors differentially expressed on tumor-associated vascular endothelial cells (1). Promising ligands may be incorporated into appropriate vectors (2) to promote selective transduction of the desired cell population (3). Further specificity can be achieved by the inclusion within the construct delivered by such vectors of additional regulatory elements that control expression of a therapeutic gene. Examples include transcriptional (4) and post-transcriptional (5) control elements that restrict gene expression to tumor endothelial cells. Additional benefit may be gained by employing therapeutic genes that are differentially active within the tumor microenvironment (6). Although no one technique may possess sufficient specificity, the inclusion of multiple independent control elements within a particular vector may help ensure selective destruction of tumor-associated endothelial cells (7).

particles are to some extent bound and sequestered by extracellular matrix components associated with the surface of endothelial cells, the observed poor transduction efficiencies have been mainly attributed to the rapid proteasome-mediated degradation of viral particles by endothelial cells [131,132].

Given the broad distribution of the corresponding viral counter-receptors, it is evident that the targeting of therapeutic genes to tumor vasculature will require both abrogation of binding to the natural receptors while simultaneously redirecting infection to alternative surface structures differentially expressed on the desired endothelial cell population.

In the case of adenoviruses, infection is mediated by an initial interaction between the coxsackie and adenovirus receptor protein (CAR) on target cells and determinants present on the knob portion of the viral fiber coat protein. Subsequent viral uptake requires the involvement of coexpressed cell surface integrins (principally αvβ3 and αvβ5) that recognize an Arg-Gly-Asp/RGD motif in the viral penton base protein [133–136]. Pseudotyping approaches that result in the replacement of the Ad5 fiber protein, which recognizes and binds to CAR, with the fiber protein from the subgroup B virus Ad16, which uses CD46 as a receptor, produces viral particles that exhibit a greatly enhanced ability to transduce CD46-positive vascular cells both in vitro and in venous explants [137]. More importantly from a targeting perspective, however, studies using appropriately engineered adenoviral vectors and target cells deficient in both CAR and the required integrins have demonstrated that infection can, in at least some circumstances, be achieved through binding to a single alternative cell surface protein that does not normally function as a adenoviral receptor [138]. Subsequently, numerous groups have shown that adenoviral tropism can be redirected to particular cell types, including endothelial cells, by introducing peptide sequences with affinity for differentially expressed target structures directly into the Ad5 fiber protein of viral particles in which both binding to CAR and interaction of the penton base protein with heparin are compromised by mutation. Appropriate peptide sequences can be readily identified by phage display and the cloning steps required to generate the chimeric fiber protein are relatively straightforward. Emphasizing the potential power of this approach in the context of vascular targeting, Nicklin et al. [139] have generated a number of recombinant adenoviruses in which 12-mer peptides with affinity for endothelial cells derived by phage display were incorporated into the H1-loop of the Ad5 fiber protein. These engineered vectors efficiently transduced endothelial cells but remained poor at infecting a broad range of other cell types [139].

The same basic approach has been employed with equal success in the targeting of AAV-2 [140–142]. In one example, selective transduction of endothelial cells both in vitro and in vivo was achieved by incorporating 7-mer or 12-mer peptide sequences with affinity for endothelial cells, once again selected by phage display, directly into the viral capsid protein [143]. A possible advantage of AAV-2 over other vector systems in the development of vascular-targeted therapies relates to the ability to circumvent the requirement for phage display studies by directly generating AAV-2 libraries in which random peptide sequences are incorporated into the capsid protein [144].

14.2.6 Transcriptional Targeting

The term "transcriptional targeting" refers to the use of approaches in which expression of a therapeutic gene in a particular cell type is transcriptionally regulated, most often by placing it downstream of a defined promoter or enhancer element. As indicated earlier, numerous genes have been identified that exhibit either specificity for endothelial cells in general or unique expression in endothelial cells within tumors or other sites of active angiogenesis. The promoters that drive such genes are obvious candidates for use in the transcriptional targeting of endothelial cells. Indeed, a number have already been tested in cultured human endothelial cells or in various animal models and some promising results obtained. Among the elements that appear to exhibit the highest and most consistent degree of endothelial cell specificity are those associated with the genes encoding endoglin [145], Flt-1 [146,147], Tie 2 [148,149], ICAM-2 [150–152], vascular endothelial-cadherin, [153] and prepro-endothelin-1 [154–158]. Interestingly, detailed analysis of these and other putative promoter elements located 5′ of genes differentially expressed in endothelial cells did not reveal any common structural features or evidence of a characteristic arrangement of particular transcription factor binding sites.

Although a combination of deletion analysis and site-directed mutagenesis can potentially be used to improve both the specificity and activity of promising promoter/enhancer sequences, the complex nature of most naturally occurring regulatory elements can make studies of this kind a long and frustrating experience. Such considerations have encouraged the development of alternative approaches that involve the generation of entirely synthetic promoters. In one recent study, Dai et al. [159] described a very promising approach based on earlier work on skeletal muscle carried out by Li et al. [160] and Edelman et al. [161], in which short oligonucleotides corresponding to a broad range of established transcription factor binding sites are ligated together in a random fashion and then cloned adjacent to a minimal

ICAM-2 promoter upstream of a green fluorescent protein (GFP) indicator gene in a self-inactivating human immunodeficiency virus type 1 (HIV-1)-based vector. In initial experiments, the resultant library of synthetic promoter elements was screened on a Rhesus monkey choroidal endothelial cell line. Cells expressing the highest levels of GFP were sorted on the FACS, expanded, and PCR ultimately performed to recover the corresponding synthetic promoter element. Although at an early stage in development, rescreening on a panel of human cell lines has identified several elements that exhibit both strong activity and promising specificity for endothelial cells [159].

14.2.7 Functional Targeting

"Functional targeting" is the term used to describe a novel approach in which the differential presence of certain initiating signals (e.g., cytokines and other soluble mediators, extracellular matrix proteins, etc.) within the tumor microenvironment can be exploited to induce a biological effect (i.e., apoptosis) in a desired target cell population. Since a response is only triggered at sites where the required initiating signal is found, expression of the effector molecule in other tissues should not induce a harmful response. This strategy has been explored in the context of vascular-targeting and it has been shown that the differential production of VEGF within the tumor microenvironment could potentially be exploited to induce selective destruction of the tumor vasculature [162]. The basic approach employs a chimeric effector protein in which the extracellular ligand-binding domain of the VEGF receptor VEGFR-2/Flk-1 [163] is fused in frame to the cytoplasmic "death domain" of the proapoptotic protein Fas [162]. Preliminary studies have confirmed that rather than triggering cellular proliferation, migration, and the other events normally induced in endothelial cells on binding of VEGF, the interaction of VEGF with the chimeric receptor instead induced cell death. Importantly, with regard to potential specificity for tumor vasculature, even cells expressing high levels of the chimeric receptor remain viable and proliferate normally if VEGF is present below a certain critical level [162]. Although initial efforts have focused on VEGF, the same general strategy could presumably be applied other differentially expressed multivalent molecules produced within tumors.

14.2.8 Posttranscriptional Targeting

Although posttranscriptional approaches have not yet been exploited in the context of vascular targeting this is likely to be an area that will prove fruitful in the future. The expression and functional activity of many endothelial genes are regulated by posttranscriptional processes such as mRNA degradation and alternative pre-mRNA splicing. Theoretically, the regulatory motifs normally located at the 3′ end of mRNA transcripts that control turnover rate could be incorporated into vectors to control the time that a therapeutic gene persists within a transduced cell. Similarly, one could conceivably exploit alternative splicing in a vascular-targeting context by building mini-gene constructs in which expression of a therapeutic gene depends on an alternative splicing event that occurs preferentially within a particular endothelial cell population. The potential of this latter approach has already been validated in early proof-of-principle studies [164,165].

14.3 POTENTIAL THERAPEUTIC APPLICATION OF VDAs IN CANCER TREATMENT

14.3.1 VDAs—Single Agent Antitumor Efficacies

As discussed earlier, both small molecule and biological VDAs now have been shown to exploit differences between tumor and normal tissue endothelia to induce selective occlusion

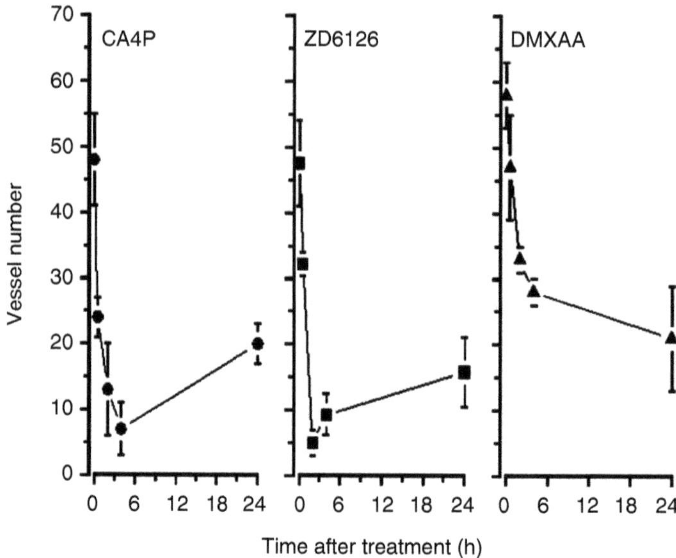

FIGURE 14.5 Impact of VDA treatment on patent number of tumor blood vessels. Response of KHT sarcomas to single dose treatments with CA4P, ZD6126, and DMXAA is illustrated. (Redrawn from Siemann, D., Chaplin, D., and Horsman, M. *Cancer*, 100, 2491, 2004.)

of tumor vessels [87,166]. Both types of agents have demonstrated potent antivascular and antitumor efficacy in transplanted and spontaneous rodent tumors, orthotopically transplanted tumors, and human tumor xenografts [34,36,54,80,167].

In vivo treatments with VDAs result in the rapid onset of physiological events leading to dose-dependent vascular shutdown in tumors that occur in minutes to hours (Figure 14.5). The suppression of functional vessels typically lasts for 24 h and gives rise to histological findings of wide-spread tumor necrosis (Figure 14.6) and loss of viable tumor cells (Figure 14.7); consequences of VDA treatments that have been widely recognized [168]. The potential for vascular damaging therapies may be most significant in large tumor masses; that is, those tumors that are typically proven to be particularly difficult to control with conventional anticancer therapies. Preclinical studies with VDAs CA4P and ZD6126 suggested that larger tumors might be particularly responsive to VDAs [169–171]. When the impact of tumor size on ZD6126 treatment efficacy was examined in detail in six rodent tumor and human tumor xenograft models, the results showed a strong tumor size-dependent treatment response in all models, with the greatest efficacy of ZD6126 noted in larger tumors [171]. This was also reflected in the impact of VDA treatment on tumor vascularity and necrosis. When a dose of ZD6126 was given to small tumors, only minimal loss of patent blood vessels and a limited increase in tumor necrosis were observed. In contrast, large tumors treated with the same dose of the VDA showed >90% reduction in patent blood vessels and 80%–90% necrosis.

The extensive necrosis and viable rim of neoplastic cells at the periphery (Figure 14.6) are characteristic features of tumors treated with VDAs [87,172,173]. These residual areas of tumor tissue, believed to survive VDA treatment because their nutritional support is derived from vasculature in the adjacent normal tissue (Figure 14.8), can act as a source of tumor regrowth. One consequence of the viable rim is that single dose treatments with VDAs typically do not lead to significant growth delays in preclinical tumor models; a finding widely reported in numerous tumor growth delay studies conducted with a variety of VDAs [174]. However, repeated, multiple dose treatments with such agents can result in measurable growth delays [54,173,175,176]. Thus, in general, it is highly unlikely that VDA treatment

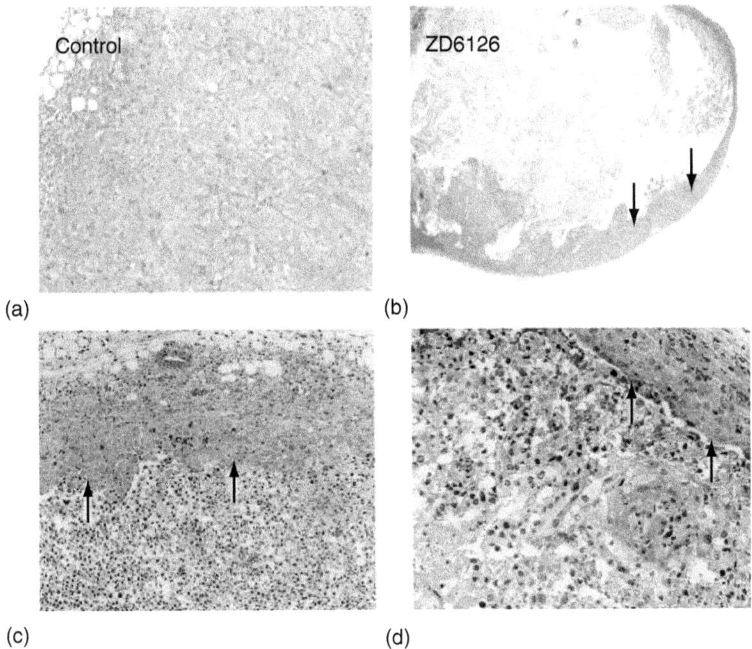

FIGURE 14.6 (See color insert following page 558.) Tumor histology following VDA exposure. Treatment with VDAs leads to extensive tumor necrosis leaving viable tumor cells only in areas adjacent to normal tissue. Twenty-four hours after treatment with ZD6126 extensive necrosis is seen at the tumor core with viable cells only at the tumor periphery (*arrows*). (From Siemann, D.W. *Horizons in Cancer Therapeutics: From Bench to Bedside*, 2002.)

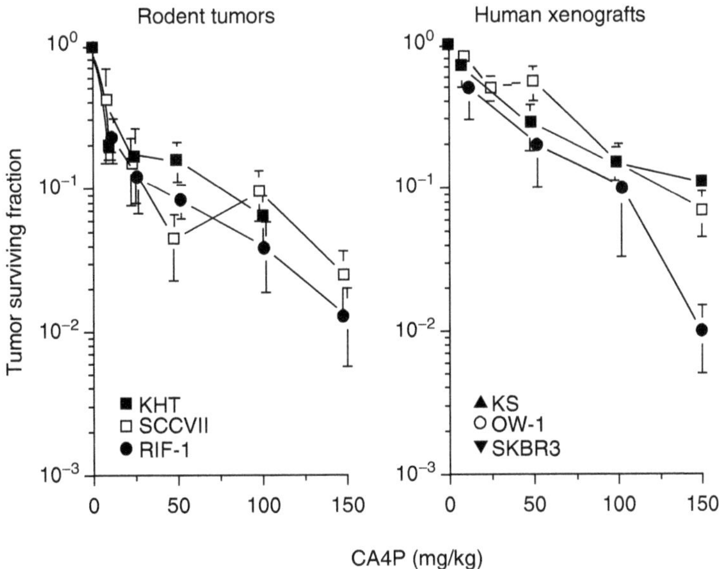

FIGURE 14.7 Clonogenic tumor cell survival determined 24 h after treatment with a range of doses of CA4P. Results for several rodent (*left panel*) and human (*right panel*) tumor models are shown. (Redrawn and modified from Siemann, D.W. et al., *Int J Cancer*, 99, 1, 2002.)

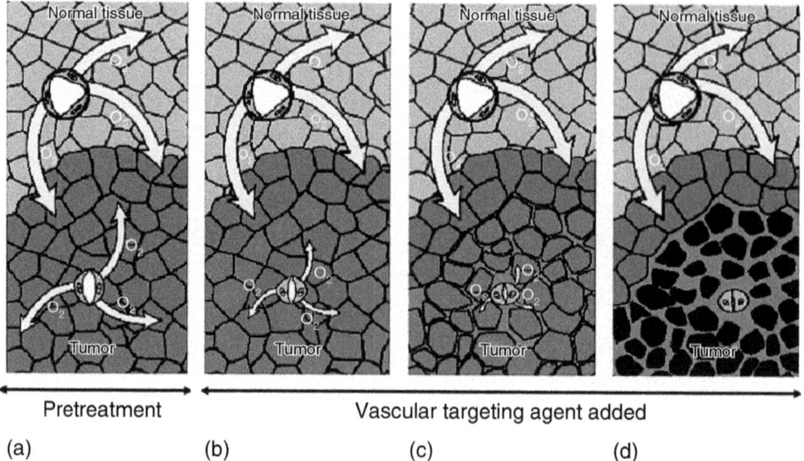

FIGURE 14.8 (See color insert following page 558.) Postulated mechanism of VDA treatment leading to a viable rim of tumor cells. (a) The tumor receives nutritive support from host and tumor vasculature. (b) VDA treatment compromises the patency of tumor neovessels but not that of the mature host vasculature. (c) Tumor cells undergo hypoxia and starvation stress. (d) With the loss of the tumor vessel the supported tumor cells undergo necrosis. However, tumor cells supplied by the host vasculature are relatively unaffected by the treatment. (From Siemann, D.W. *Horizons in Cancer Therapeutics: From Bench to Bedside*, 2002).

alone will eradicate the entire tumor mass. Still, the destruction of large tumor areas, particularly in the central and typically most radiation and chemotherapy treatment-resistant regions of the tumors, is clearly highly desirable.

14.3.2 Combining VDAs with Conventional Anticancer Therapies

In general, greater antitumor effects may be achieved when agents that have different mechanisms of action, different cellular targets, and nonoverlapping toxicities are combined. In addition to these fundamental differences, the concept of combining vascular targeting therapies with conventional anticancer therapies is based on two principles. First, even with aggressive chemotherapy and radiotherapy, a significant number of cancer patients treated with curative intent fail. Second, VDA treatment, though capable of inducing extensive tumor necrosis, is not able to eliminate the entire tumor mass. Yet by applying these two treatment strategies in concert it may be possible to overcome some of the shortcomings of both.

Cells comprising the viable rim of tumor tissue that survives VDA treatment are likely to be in a state of high proliferation and excellent nutrition. These factors, coupled with their ready accessibility to systemically administered agents, make the surviving tumor cells susceptible to killing by radiation and anticancer drugs. Conversely, by causing the destruction of large areas of the interior of tumors, VDA treatments reduce or eliminate tumor cells residing in environments that typically lead to resistance to conventional anticancer therapies. The strategy of combining VDAs with conventional cytotoxic modalities therefore represents a treatment strategy that holds significant therapeutic potential.

14.3.3 VDAs and Anticancer Drugs

In general, marked enhancements in antitumor activities have been observed when VDAs were combined with standard chemotherapy [54,174,177–183] (Figure 14.9). Typically these enhanced responses were obtained by administering the VDA postchemotherapy.

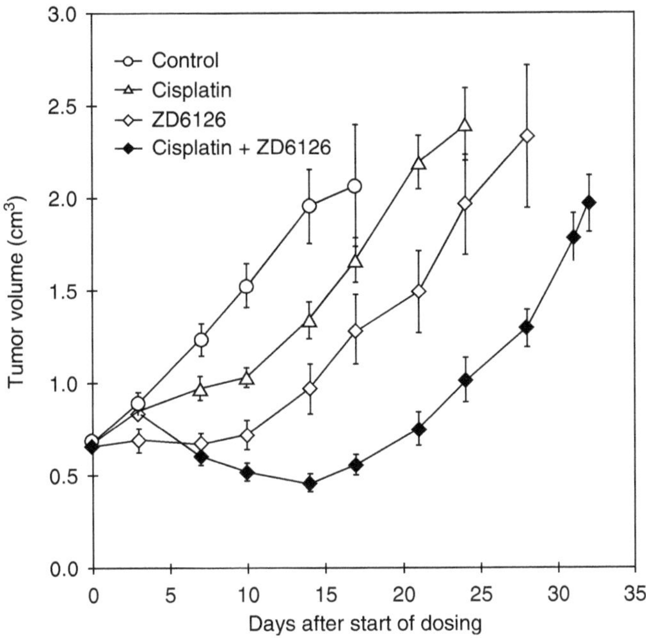

FIGURE 14.9 Response of Calu-6-bearing nude mice to treatment with cisplatin and or ZD6126. (From Blakey, D.C. et al., *Clin Cancer Res*, 8, 1974, 2002.)

The rationale for such a sequence was usually based on a desire to avoid any possible pharmacologic interference by the VDA on the action of the anticancer drug. Indeed, those preclinical investigations that have reported on the issue of timing and sequencing of agents in particular have concluded that VDA-chemotherapy combinations may be less effective when the VDA is administered just before the anticancer drug [174,183]. A possible explanation for this observation is that under these conditions some parts of the tumors may experience transient reductions in blood flow sufficient to impair chemotherapeutic agent delivery. Longer separations may once again be effective if the treatment interval is sufficiently long to allow expression of ischemic tumor cell death in those areas where vessels were damaged sufficiently to lead to necrosis, whereas areas experiencing only transient vascular changes would have returned to normal leaving the tumor cells susceptible to the action of the chemotherapeutic agent. Still, administering the VDA postchemotherapy may provide the most appropriate scheme for combining VDAs and anticancer drugs. Indeed a wide variety of tumor models now have been evaluated and most have shown clear evidence that the VDA-chemotherapy combination can result in significantly improved tumor responses under such treatment conditions [54,174,177–179,181–183]. The observed enhanced treatment responses, while most likely due to the two therapies acting against different tumor cell populations, could, at least in part, be the consequence of trapping the chemotherapeutic agent within the tumor. However, the relative contribution of these factors to the net effect is unclear and indeed may vary for different VDAs.

Demonstrating improved tumor responses through the combination of VDAs and chemotherapy will only be of benefit if such a combined modality treatment does not enhance the response of critical normal tissues. Results from preclinical investigations addressing this question indicate that the combination of VDAs and anticancer drugs can improve the antitumor efficacy without a concomitant increase in general host toxicity [54,177–183].

Data on chemotherapeutic agent specific side effects are more limited but the absence of enhanced bone marrow toxicity is encouraging [183].

It is clear that the combination of VDAs with clinically active anticancer drugs holds great promise. The possibility of using two agents against different cellular targets, while expressing different mechanisms of action and toxicity profiles, ascribes to the very principle of combined modality therapy. Preclinical investigations have demonstrated that combining VDAs with conventional chemotherapy is not only sensible, but provides an approach that offers complimentary antitumor effects. Given their mode of action, these agents may prove to be excellent companions for conventional anticancer drugs. Further evaluations of the therapeutic potential of VDAs in an adjuvant setting clearly are warranted.

14.3.4 VDAs and Radiotherapy

Combining radiation treatments with strategies aimed at compromising the tumor vasculature may improve radiotherapy outcomes by capitalizing on principles of enhanced antitumor efficacy, nonoverlapping toxicities, or spatial cooperation.

The tumor vasculature plays a key role in the response of tumors to radiation. Because tumor vasculature is morphologically and functionally abnormal [184], the associated blood flow is often irregular, sluggish, and intermittent [185,186]. As a consequence oxygen-deficient regions are a common occurrence in solid tumors [185,186]. The presence of such hypoxic cells in tumors can have a detrimental effect on the ability to control human malignancies treated with curative intent [187–190]. VDA treatments impact the central core of tumors leaving a viable rim of cells near the periphery. This viable rim of tumor cells presumably survives because it derives its nutritional support from nearby normal tissue blood vessels that are typically nonresponsive to VDAs (Figure 14.8). The surviving cells are most likely to be well oxygenated and thus radiation sensitive. Thus, a logical rationale for combining a VDA with radiation is that the two treatments interact in a complimentary fashion at the tumor microregional level, i.e., the former reducing or eliminating the poorly oxygenated and hence radioresistant tumor cell subpopulations, the latter destroying cells not affected by the VDA.

A number of studies have investigated the combination of VDAs and radiotherapy [170,191–193]. Most involved the use of single doses of radiation and VDA treatments, although a few have examined the efficacy of VDAs in fractionated radiation dose regimes [174,194–197]. Evaluations of the critical issues of timing and sequencing indicated that administering the VDA just before irradiation provided little or no antitumor benefit [192,195]. A possible explanation for this finding is that the vascular shutdown induced by the VDAs may have rendered some tumor cells hypoxic at the time of irradiation and that these cells later reoxygenated and survived. That such a scenario can occur is demonstrated by the finding that the hypoxia-selective bioreductive drug tirapazamine can further improve the efficacy of radiation and DMXAA [195] under such conditions. As was the case for chemotherapy, tumor responses were significantly enhanced when the VDAs were administered within a few hours after the radiotherapy (Figure 14.10) [195,196,198], but the mechanisms underlying these positive interactions have not been clearly delineated. A likely explanation is that the two therapies are acting independently against different tumor cell subpopulations [168] though one investigation did indicate that VDAs may increase the extent of radiation-induced apoptosis [199].

Since clinically most radiotherapy is delivered using daily fractionated dose treatments, an important issue to be addressed is how to incorporate VDA exposures into such a setting. Given the vascular and physiologic effects of VDAs most investigators who have evaluated the efficacy of VDAs in fractionated radiotherapy protocols have chosen to deliver the VDA after the last radiation fraction at the end of each week of treatment. Generally, this approach was taken in the belief that such an approach would minimize any possible negative consequences

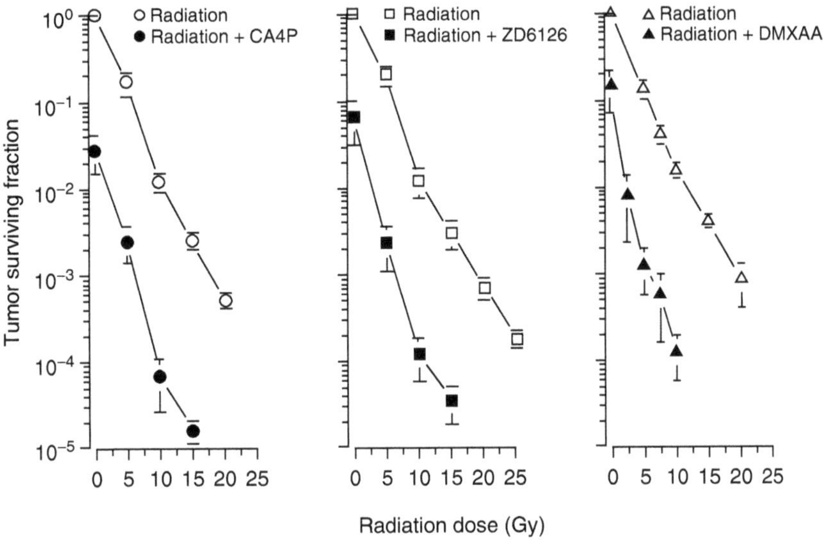

FIGURE 14.10 Clonogenic cell survival in KHT sarcomas treated with CA4P, ZD6126, or DMXAA 1 h after a range of single doses of radiotherapy. (From Siemann, D., Chaplin, D., and Horsman, M., *Cancer*, 100, 2491, 2004.)

VDA treatment might have on radiation efficacy. When this was done, both CA4P and ZD6126 significantly enhanced the tumor response to fractionated radiotherapy [174,200]. However, when it comes to combining VDAs with fractionated radiotherapy, the issue of timing and sequencing has not been fully resolved. Indeed at least one preclinical study has shown that a VDA may be administered successfully more often during the course of fractionated radiation [195]. Clearly additional evaluation of this topic is required.

The observed tumor size-dependent efficacy of VDAs may have particular significance for combined modality treatment strategies as it is well established that advanced neoplastic disease is typically difficult to control. Studies which examined the importance of tumor size in combined radiotherapy–VDA treatments demonstrated that while small tumors were most radiation sensitive, larger tumors were consistently most responsive to VDA treatment. When VDA and radiation treatments were combined, the antitumor efficacy was enhanced at all tumor sizes, but most significantly when larger tumors were treated [171]. These findings suggest that the VDA–radiotherapy combination may be particularly effective when applied to advanced tumors.

A number of studies have now investigated the combination of VDAs and radiotherapy and the consensus reached is that significant enhancement of the radiation response can be achieved when VDAs are included in the treatment regimen. Importantly, VDAs have shown little effect on the radiation response of relevant normal tissues. VDAs have shown no influence on the radiation response of early responding normal tissue such as skin [195,198,201], nor on late responding normal tissues such as bladder and lung [200]. Taken together, these findings support the notion that combining VDAs with radiotherapy may yield a therapeutic benefit.

14.3.5 Combining VDAs with Antiangiogenic Treatment Strategies

Two primary approaches to targeting the tumor blood vessel network have been pursued [202]. The approach that is the focus of this chapter (vascular disruption) emphasizes agents that disrupt established tumor blood vessels leading to rapid vascular collapse, vessel congestion, and extensive tumor necrosis. An alternative approach, and indeed the approach that has received by far the greater scientific attention, is directed at interfering with a tumor's ability to form a blood vessel network (inhibition of angiogenesis). Since both the initiation of new vessel formation and the integrity of the existing blood vessel network are critical to a tumor's growth and survival, a two-pronged attack comprised of a combination of antiangiogenic and vascular disrupting strategies would seem like a logical approach.

There are additional reasons for considering combining these two classes of agents. Evidence suggests that antiangiogenic therapies may be especially well suited for treating micrometastatic disease or early-stage cancer [203–205]. The use of such agents in cancer patients likely will mean long-term therapy because on treatment cessation, surviving dormant tumor cells may reinitiate proangiogenic activities and progress. On the other hand, the effects of VDA exposure manifest themselves within hours after treatment [87]. These agents are applied intermittently and are very efficacious against large bulky tumors [169–171]. Although both antiangiogenic and vascular disrupting approaches have shown significant antitumor effects in preclinical studies, achieving tumor cures with either, when used as monotherapy, is likely to be extremely difficult. In antiangiogenic approaches, the complexity of angiogenesis, with its multiple redundant or alternative pathways, strongly implies that disabling a single target will ultimately be insufficient to fully impair this process [206,207]. When vascular disrupting strategies are applied, a thin viable rim of tumor cells inevitably survives at the periphery near the normal tissue as discussed in detail previously.

Preclinical evaluations of the combination of VDAs and antiangiogenic agents now are beginning to emerge. This strategy was first examined in a setting which combined ZD6474, a selective inhibitor of VEGFR2 associated tyrosine kinase, with the microtubulin disrupting agent ZD6126 in several human tumor xenograft models [176,208]. In these studies, the antiangiogenic agent was administered on a daily basis, whereas the VDA was given intermittently during the course of treatment. The results showed that although each agent alone was capable of inducing a tumor response, when ZD6474 and ZD6126 were combined the resultant tumor growth delay was markedly increased (Figure 14.11). In Kaposi's sarcoma model this combination also significantly increased host survival and resulted in

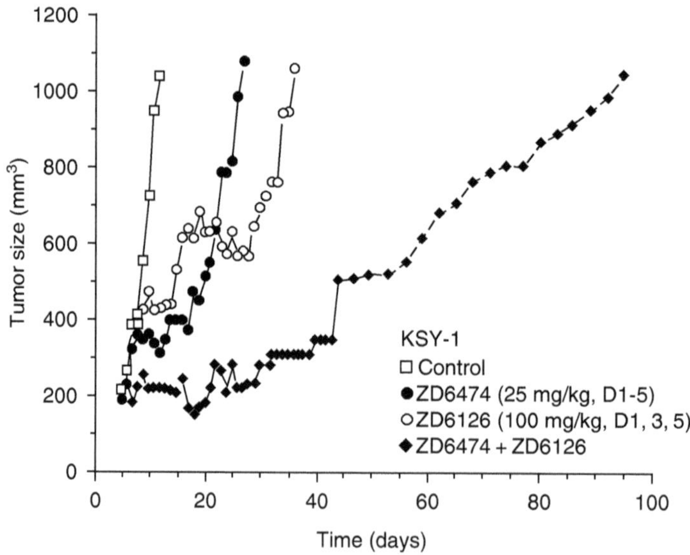

FIGURE 14.11 Response of Kaposi's sarcoma xenografts to therapy combining the antiangiogenic agent ZD6474 and the VDA ZD6126. The median tumor responses of the various treatment groups are shown. (From Siemann, D.W. and Shi, W., *Int J Radiat Oncol Biol Phys*, 60, 1233, 2004.)

some long-term tumor cures [176]. A likely explanation for the effectiveness of this combination is that while the VDA (ZD6126) significantly "debulks" the tumor mass, the angiogenesis inhibitor (ZD6474) impairs the rapid tumor cell regrowth from the surviving viable rim by interfering with the reestablishment of the tumor vasculature. While this particular example used an inhibitor of VEGFR2 associated kinase in combination with a tubulin depolymerizing agent, combinations of other antiangiogenic and vascular disrupting therapies are also likely to be very efficacious. Indeed, recent studies in xenografts have demonstrated that avastin applied in conjunction with CA4P or OXi4503 results in enhanced tumor responses [209].

Antiangiogenic and vascular disrupting therapies target different components of the expanding tumor blood vessel network. Given that different targets and mechanisms of action are involved, the application of an antiangiogenic agent in conjunction with one that disrupts established tumor blood vessels is likely to lead to complimentary antitumor effects. Such a combination therapy has the potential to have catastrophic consequences for a tumor's blood vessel support network and hence its growth and survival. The combined application of antiangiogenic and vascular disruption therapies may also capitalize on principles of nonoverlapping toxicities and spatial cooperation.

14.4 POTENTIAL THERAPEUTIC APPLICATIONS OF VDAs IN OTHER DISEASES

It is now well established that abnormal development of new blood vessels contributes to the pathology of several other nonmalignant diseases. In particular, the decreasing visual acuity associated with diseases of the eye, such as wet age-related macular degeneration (AMD), myopic macular degeneration (MMD), and diabetic retinopathy, is associated with abnormal neovascularization of the choroidal and retinal compartments of the eye. As a result, there is considerable interest in the use of antiangiogenic agents, and potentially VDAs, in these conditions. This interest has led to the clinical evaluation of several approaches, particularly antiangiogenic strategies, which target the VEGF pathway. The success of these approaches in slowing down the process of vision loss in patients with AMD has recently led to the approval of Macugen for treatment of this disease. Several other agents are in late stage clinical trial and encouraging results have been reported.

Initial work with VDAs has focused to date mainly around CA4P. Studies by Griggs and colleagues looked at the effect of CA4P in 12 day old mouse pups where abnormal retinal neovascularization had been induced following exposure to a high-oxygen environment [210]. After 5 days of CA4P treatment they evaluated the effect on both the abnormal neovascularization associated with the retinopathy and the continued neovascularization associated with the normal retinal development in these young mice. They concluded that "CA4P permitted the development of normal retinal vasculature while inhibiting aberrant neovascularization." This study was followed by one from Nambu and colleagues using two models of ocular neovascularization one inducing retinal neovascularization the other choroidal neovascularization [211]. Systemically administered CA4P was active in both models and was shown not only to block development, but also to induce regression of the neovasculature. The authors conclude, "Therefore CA4P shows potential for both prevention and treatment of ocular neovascularization." Based on these data, CA4P has moved into two clinical trials in the field of ophthalmology, one Phase I/II study in AMD and another, larger, multinational Phase II trial in MMD.

14.5 CONCLUSION

Although the importance of tumor vasculature as a potential therapeutic target for drug development was first alluded to over 80 years ago, it is only over the last 10 to 15 years that a concerted effort has been made to develop such therapeutics. With several compounds now in clinical trial and many more due to enter the clinical evaluation phase, it is of interest that the most widely studied group of compounds are the tubulin depolymerizing drugs and that their development stems from early evidence that colchicines induced hemorrhagic necrosis in tumors.

Several ligand-based VDA approaches have demonstrated impressive effects in animal models and several are expected to enter clinical trials shortly. Conceptually, the ability to target unique antigenic determinants expressed on tumor endothelium is very appealing. It has become clear that even the ideal VDA would not be curative because a small but significant number of tumor cells reside adjacent to normal tissue vessels. However, the combination of VDAs with many conventional and emerging therapeutic strategies, including antiangiogenics, has produced impressive results in preclinical models.

Although originally developed for use in oncology it has become evident that VDAs may have application in other diseases, which are characterized by abnormal neovascularization in particular certain ocular disorders such as wet AMD and MMD. Positive results in preclinical models have led to the initiation of two clinical trials evaluating the use of CA4P in both AMD and MMD.

REFERENCES

1. Folkman, J. Tumor angiogenesis: therapeutic implications. *N Engl J Med*, 285, 1182, 1971.
2. Folkman, J. Anti-angiogenesis: new concept for therapy of solid tumors. *Ann Surg*, 175, 409, 1972.
3. Folkman, J. Tumor angiogenesis: role in regulation of tumor growth. *Symp Soc Dev Biol*, 30, 43, 1974.
4. Folkman, J. Tumor angiogenesis: a possible control point in tumor growth. *Ann Intern Med*, 82, 96, 1975.
5. Hurwitz, H. et al. Bevacizumab plus irinotecan, fluorouracil, and leucovorin for metastatic colorectal cancer. *N Engl J Med*, 350, 2335, 2004.
6. Denekamp, J. Endothelial cell proliferation as a novel approach to targeting tumour therapy. *Br J Cancer*, 45, 136, 1982.
7. Denekamp, J. Vascular endothelium as the vulnerable element in tumours. *Acta Radiologica Oncol*, 23, 217, 1984.
8. Denekamp, J. Endothelial cell attack as a novel approach to cancer therapy. *Cancer Topics*, 6, 6, 1986.
9. Denekamp, J., Hill, S.A., and Hobson, B. Vascular occlusion and tumour cell death. *Eur J Cancer Clin Oncol*, 19, 271, 1983.
10. Chaplin, D.J. and Horsman, M.R. The influence of tumour temperature on ischemia-induced cell death: potential implications for the evaluation of vascular mediated therapies. *Radiother Oncol*, 30, 59, 1994.
11. Parkins, C.S. et al. Ischaemia induced cell death in tumors: importance of temperature and pH. *Int J Radiat Oncol Biol Phys*, 29, 499, 1994.
12. Boyland, E. and Boyland, M.E. Studies in tissue metabolism. IX. The action of colchicine and B. typhosus extract. *Biochem J*, 31, 454, 1937.
13. Ludford, R.J. Colchicine in the experimental chemotherapy of cancer. *J Natl Cancer Inst*, 6, 89, 1945.
14. Ludford, R.J. Factors determining the action of colchicine on tumour growth. *Br J Cancer*, 2, 75, 1948.
15. Seed, L., Slaughter, D.P., and Limarzi, L.R. Effect of colchicine on human carcinoma. *Surgery*, 7, 696, 1940.
16. Leiter, J. et al. Damage induced in sarcoma 37 with podophyllin, podophyllotoxin alpha-peltatin, beta-peltatin, and quercetin. *J Natl Cancer Inst*, 10, 1273, 1950.
17. Andervont, H.B. Effect of colchicine and bacterial products on transplantable and spontaneous tumors in mice. *J Natl Cancer Inst*, 1, 361, 1940.

18. Mottram, J.C. Observations on the combined action of colloidal lead and radiation on tumours. *Brit Med J*, 1, 928, 1928.
19. Leiter, J. et al. Damage induced in sarcoma 37 with chemical agents. II. Trivalent and pentavalent arsenicals. *J Natl Cancer Inst*, 13, 365, 1952.
20. Hill, S.A. et al. Vinca alkaloids: anti-vascular effects in a murine tumour. *Eur J Cancer*, 9, 1320, 1993.
21. Hill, S.A. et al. The effect of vinca alkaloids on tumour blood flow. *Adv Exp Med Biol*, 345, 417, 1994.
22. Chaplin, D.J. et al. Antivascular approaches to solid tumour therapy: evaluation of tubulin binding agents. *Br J Cancer*, 27 (Suppl), S86, 1996.
23. Pettit, G.R., Cragg, G.M., and Singh, S.B. Antineoplastic agents, 122. Constituents of Combretum caffrum. *J Nat Prod*, 50, 386, 1987.
24. Aleksandrzak, K., McGown, A.T., and Hadfield, J.A. Antimitotic activity of diaryl compounds with structural features resembling combretastatin A-4. *Anticancer Drugs*, 9, 545, 1998.
25. Hatanaka, T. et al. Novel B-ring modified combretastatin analogues: syntheses and antineoplastic activity. *Bioorg Med Chem Lett*, 8, 3371, 1998.
26. Ohsumi, K. et al. Novel combretastatin analogues effective against murine solid tumors: design and structure–activity relationships. *J Med Chem*, 41, 3022, 1998.
27. Pettit, G.R. et al. Antineoplastic agents. 379. Synthesis of phenstatin phosphate. *J Med Chem*, 41, 1688, 1998.
28. Lawrence, N.J. et al. Antimitotic and cell growth inhibitory properties of combretastatin A-4-like ethers. *Bioorg Med Chem Lett*, 11, 51, 2001.
29. Hadimani, M.B. et al. Synthesis, in vitro, and in vivo evaluation of phosphate ester derivatives of combretastatin A-4. *Bioorg Med Chem Lett*, 13, 1505, 2003.
30. Liou, J.P. et al. Concise synthesis and structure–activity relationships of combretastatin A-4 analogues, 1-aroylindoles and 3-aroylindoles, as novel classes of potent antitubulin agents. *J Med Chem*, 47, 4247, 2004.
31. Perez-Melero, C. et al. A new family of quinoline and quinoxaline analogues of combretastatins. *Bioorg Med Chem Lett*, 14, 3771, 2004.
32. Sun, L. et al. Examination of the 1,4-disubstituted azetidinone ring system as a template for combretastatin A-4 conformationally restricted analogue design. *Bioorg Med Chem Lett*, 14, 2041, 2004.
33. Prinz, H. Recent advances in the field of tubulin polymerization inhibitors. *Expert Rev Anticancer Ther*, 2, 695, 2002.
34. Dark, G.G. et al. Combretastatin A-4, an agent that displays potent and selective toxicity toward tumor vasculature. *Cancer Res*, 57, 1829, 1997.
35. Tozer, G.M. et al. Combretastatin A-4 phosphate as a tumor vascular-targeting agent: early effects in tumors and normal tissues. *Cancer Res*, 59, 1626, 1999.
36. Murata, R., Overgaard, J., and Horsman, M.R. Comparative effects of combretastatin A-4 disodium phosphate and 5,6-dimethylxanthenone-4-acetic acid on blood perfusion in a murine tumour and normal tissues. *Int J Radiat Biol*, 77, 195, 2001.
37. Watts, M.E. et al. Effects of novel and conventional anti-cancer agents on human endothelial permeability: influence of tumour secreted factors. *Anticancer Res*, 17, 71, 1997.
38. Galbraith, S.M. et al. Effects of combretastatin A4 phosphate on endothelial cell morphology in vitro and relationship to tumour vascular targeting activity in vivo. *Anticancer Res*, 21, 93, 2001.
39. Chaplin, D.J. and Dougherty, G.J. Tumour vasculature as a target for cancer therapy. *Br J Cancer*, 80 (Suppl 1), 57, 1999.
40. Kanthou, C. and Tozer, G.M. The tumor vascular targeting agent combretastatin A-4-phosphate induces reorganization of the actin cytoskeleton and early membrane blebbing in human endothelial cells. *Blood*, 99, 2060, 2002.
41. Tozer, G.M. et al. The biology of the combretastatins as vascular targeting agents. *Int J Exp Pathol*, 83, 21, 2002.
42. Brooks, A.C. et al. The vascular targeting agent combretastatin A-4-phosphate induces neutrophil recruitment to endothelial cells in vitro. *Anticancer Res*, 23, 3199, 2003.
43. Parkins, C.S. et al. Determinants of anti-vascular action by combretastatin A-4 phosphate: role of nitric oxide. *Br J Cancer*, 83, 811, 2000.

44. Hori, K. et al. Antitumor effects due to irreversible stoppage of tumor tissue blood flow: evaluation of a novel combretastatin A-4 derivative, AC7700. *Jap J Cancer Res*, 90, 1026, 1999.
45. Hori, K., Saito, S., and Kubota, K. A novel combretastatin A-4 derivative, AC7700, strongly stanches tumour blood flow and inhibits growth of tumours developing in various tissues and organs. *Br J Cancer*, 86, 1604, 2002.
46. Hori, K. and Saito, S. Induction of tumour blood flow stasis and necrosis: a new function for epinephrine similar to that of combretastatin A-4 derivative AVE8062 (AC7700). *Br J Cancer*, 90, 549, 2004.
47. Platts, S.H. et al. Alteration of microtubule polymerization modulates arteriolar vasomotor tone. *Am J Physiol*, 277, H100, 1999.
48. Honess, D.J., Hylands, F., Chaplin, D.J., and Tozer, G.M. Comparison of strategies to overcome the hypertensive effect of combretastatin-A-4-phosphate in a rat model. *Br J Cancer*, 86, S118, 2002.
49. Pettit, G.R. and Lippert, J.W. Antineoplastic agents 429. Syntheses of the combretastatin A-1 and combretastatin B-1 prodrugs. *Anti-Cancer Drug Des*, 15, 203, 2000.
50. Hill, S.A. et al. Schedule dependence of combretastatin A4 phosphate in transplanted and spontaneous tumour models. *Int J Cancer*, 102, 70, 2002.
51. Holwell, S.E. et al. Anti-tumor and anti-vascular effects of the novel tubulin-binding agent combretastatin A-1 phosphate. *Anticancer Res*, 22, 3933, 2002.
52. Hua, J. et al. Oxi4503, a novel vascular targeting agent: effects on blood flow and antitumor activity in comparison to combretastatin A-4 phosphate. *Anticancer Res*, 23, 1433, 2003.
53. Kirwan, I.G. et al. Comparative preclinical pharmacokinetic and metabolic studies of the combretastatin prodrugs combretastatin A4 phosphate and A1 phosphate. *Clin Cancer Res*, 10, 1446, 2004.
54. Blakey, D.C. et al. Antitumor activity of the novel vascular targeting agent ZD6126 in a panel of tumor models. *Clin Cancer Res*, 8, 1974, 2002.
55. Robinson, S.P. et al. Tumour dose response to the antivascular agent ZD6126 assessed by magnetic resonance imaging. *Br J Cancer*, 88, 1592, 2003.
56. McIntyre, D.J. et al. Single dose of the antivascular agent, ZD6126 (N-acetylcolchinol-O-phosphate), reduces perfusion for at least 96 h in the GH3 prolactinoma rat tumor model. *Neoplasia*, 6, 150, 2004.
57. Dowlati, A. et al. A phase I pharmacokinetic and translational study of the novel vascular targeting agent combretastatin a-4 phosphate on a single-dose intravenous schedule in patients with advanced cancer. *Cancer Res*, 62, 3408, 2002.
58. Stevenson, J.P. et al. Phase I trial of the antivascular agent combretastatin A4 phosphate on a 5-day schedule to patients with cancer: magnetic resonance imaging evidence for altered tumor blood flow. *J Clin Oncology*, 21, 4428, 2003.
59. Rustin, G.J. et al. Phase I clinical trial of weekly combretastatin A4 phosphate: clinical and pharmacokinetic results. *J Clin Oncol*, 21, 2815, 2003.
60. Galbraith, S.M. et al. Combretastatin A4 phosphate has tumor antivascular activity in rat and man as demonstrated by dynamic magnetic resonance imaging. *J Clin Oncol*, 21, 2831, 2003.
61. Anderson, H.L. et al. Assessment of pharmacodynamic vascular response in a phase I trial of combretastatin A4 phosphate. *J Clin Oncol*, 21, 2823, 2003.
62. Young, S.L. and Chaplin, D.J. Combretastatin A4 phosphate: background and current clinical status. *Expert Opin Investig Drugs*, 13, 1171, 2004.
63. West, C.M. and Price, P. Combretastatin A4 phosphate. *Anticancer Drugs*, 15, 179, 2004.
64. Gadgeel, S.M. et al. A dose-escalation study of the novel vascular-targeting agent ZD6126, in patients with solid tumors. Abstract 438. *Am Soc Clin Oncol Annu Meet*, 2002.
65. Tolcher, A.W. et al. Phase I, pharmacokinetic, and DCE-MRI correlative study of AVE8062A, an antivascular combretastatin analogue, administered weekly for 3 weeks every 28-days. Abstract 834. *Am Soc Clin Oncol Annu Meet*, 2003.
66. Evelhoch, J.L. et al. Magnetic resonance imaging measurements of the response of murine and human tumors to the vascular-targeting agent ZD6126. *Clin Cancer Res*, 10, 3650, 2004.
67. Sessa, C. et al. A pharmacokinetic and DCE-MRI-dynamic phase I study of the antivascular combretastatin analogue AVE8062A administered every 3 weeks. Abstract 5827. *96th Annual Meeting of the American Association for Cancer Research*. Anaheim, CA, 2005.

68. Rustin, G.J. et al. A Phase Ib trial of combretastatin A-4 phosphate (CA4P) in combination with carboplatin or paclitaxel chemotherapy in patients with advanced cancer. Abstract 3013. *Am Soc Clin Oncol Annu Meet*, 2005.
69. Corbett, T.H. et al. Activity of flavone acetic acid (NSC-347512) against solid tumors of mice. *Invest New Drugs*, 4, 207, 1986.
70. Smith, G.P. et al. Flavone acetic acid (NSC 347512) induces haemorrhagic necrosis of mouse colon 26 and 38 tumours. *Eur J Cancer Clin Oncol*, 23, 1209, 1987.
71. Bibby, M.C. et al. Reduction of tumor blood flow by flavone acetic acid: a possible component of therapy. *J Natl Cancer Inst*, 81, 216, 1989.
72. Hill, S.A., Williams, K.B., and Denekamp, J. A comparison of vascular-mediated tumor cell death by the necrotizing agents GR63178 and flavone acetic acid. *Int J Radiat Oncol Biol Phys*, 22, 437, 1992.
73. Evelhoch, J.L. et al. Flavone acetic acid (NSC 347512)-induced modulation of murine tumor physiology monitored by in vivo nuclear magnetic resonance spectroscopy. *Cancer Res*, 48, 4749, 1988.
74. Hill, S.A., Williams, K.B., and Denekamp, J. Vascular collapse after flavone acetic acid: a possible mechanism of it's anti-tumour action. *Eur J Cancer Clin Oncol*, 25, 1419, 1989.
75. Mahadevan, V. et al. Role of tumor necrosis factor in flavone acetic acid-induced tumor vasculature shutdown. *Cancer Res*, 50, 5537, 1990.
76. Kerr, D.J. et al. Phase I and pharmacokinetic study of flavone acetic acid. *Cancer Res*, 47, 6776, 1987.
77. Weiss, R.B. et al. Phase I and clinical pharmacology study of intravenous flavone acetic acid (NSC 347512). *Cancer Res*, 48, 5878, 1988.
78. Kerr, D.J. et al. Phase II trials of flavone acetic acid in advanced malignant melanoma and colorectal carcinoma. *Br J Cancer*, 60, 104, 1989.
79. O'Reilly, S.M. et al. Flavone acetic acid (FAA) with recombinant interleukin-2 (rIL-2) in advanced malignant melanoma: I. Clinical and vascular studies. *Br J Cancer*, 67, 1342, 1993.
80. Rewcastle, G.W. et al. Potential antitumor agents. 61. Structure–activity relationships for in vivo colon 38 activity among disubstituted 9-oxo-9H-xanthene-4-acetic acids. *J Med Chem*, 34, 217, 1991.
81. Ching, L.M. et al. Induction of tumor necrosis factor-alpha messenger RNA in human and murine cells by the flavone acetic acid analogue 5,6-dimethylxanthenone-4-acetic acid (NSC 640488). *Cancer Res*, 54, 870, 1994.
82. Ching, L.M. et al. Induction of endothelial cell apoptosis by the antivascular agent 5,6-dimethylxanthenone-4-acetic acid. *Br J Cancer*, 86, 1937, 2002.
83. Baguley, B.C., Zhuang, L., and Kestell, P. Increased plasma serotonin following treatment with flavone-8-acetic acid, 5,6-dimethylxanthenone-4-acetic acid, vinblastine and colchicine: relation to vascular effects. *Oncol Res*, 9, 55, 1997.
84. Baguley, B.C. Antivascular therapy of cancer: DMXAA. *Lancet Oncol*, 4, 141, 2003.
85. Tozer, G.M., Kanthou, C., and Baguley, B.C. Disrupting tumour blood vessels. *Nat Rev Cancer*, 5, 423, 2005.
86. Zhao, L. et al. Improvement of the antitumor activity of intraperitoneally and orally administered 5,6-dimethylxanthenone-4-acetic acid by optimal scheduling. *Clin Cancer Res*, 9, 6545, 2003.
87. Siemann, D., Chaplin, D., and Horsman, M. Vascular-targeting therapies for treatment of malignant disease. *Cancer*, 100, 2491, 2004.
88. Rustin, G.J. et al. 5,6-dimethylxanthenone-4-acetic acid (DMXAA), a novel antivascular agent: phase I clinical and pharmacokinetic study. *Br J Cancer*, 88, 1160, 2003.
89. Jameson, M.B. et al. Clinical aspects of a phase I trial of 5,6-dimethylxanthenone-4-acetic acid (DMXAA), a novel antivascular agent. *Br J Cancer*, 88, 1844, 2003.
90. Galbraith, S.M. et al. Effects of 5,6-dimethylxanthenone-4-acetic acid on human tumor microcirculation assessed by dynamic contrast-enhanced magnetic resonance imaging. *J Clin Oncol*, 20, 3826, 2002.
91. Blaschuk, O.W. and Rowlands, T.M. Cadherins as modulators of angiogenesis and the structural integrity of blood vessels. *Cancer Metast Rev*, 19, 1, 2000.
92. Lepekhin, E. et al. Early and long term effects of exherin on tumor vasculature. Abstract 883. *Am Soc Clin Oncol Annu Meet*, 2003.

93. Jonker, D.J. et al. A phase I study of the novel molecularly targeted vascular targeting agent exherinTM (ADH-1), shows activity in some patients with refractory solid tumors stratified according to N-cadherin expression. Abstract 3038. *Am Soc Clin Oncol Annu Meet*, 2005.
94. Burrows, F.J. and Thorpe, P.E. Eradication of large solid tumors in mice with an immunotoxin directed against tumor vasculature. *Proc Natl Acad Sci USA*, 90, 8996, 1993.
95. Burrows, F. and Thorpe, P. Vascular targeting—a new approach to the therapy of solid tumors. *Pharmacol Ther*, 64, 155, 1994.
96. Denekamp, J. Vascular attack as a therapeutic strategy for cancer. *Cancer Metast Rev*, 9, 267, 1990.
97. Denekamp, J. Review article: angiogenesis, neovascular proliferation and vascular pathophysiology as targets for cancer therapy. *Br J Radiol*, 66, 181, 1993.
98. Zhang, H. et al. Transcriptional profiling of human microvascular endothelial cells in the proliferative and quiescent state using cDNA arrays. *Angiogenesis*, 3, 211, 1999.
99. Kozian, D. and Augustin, H. Rapid identification of differentially expressed endothelial cell genes by RNA display. *Biochem Biophys Res Commun*, 209, 1068, 1995.
100. Prols, F., Loser, B., and Marx, M. Differential expression of osteopontin, PC4, and CEC5, a novel mRNA species, during in vitro angiogenesis. *Exp Cell Res*, 239, 1, 1998.
101. Fabbri, M. et al. A functional monoclonal antibody recognizing the human alpha 1-integrin I-domain. *Tissue Antigens*, 48, 47, 1996.
102. Enenstein, J. and Kramer, R. Confocal microscopic analysis of integrin expression on the microvasculature and its sprouts in the neonatal foreskin. *J Invest Dermatol*, 103, 381, 1994.
103. Brooks, P. et al. Integrin alpha v beta 3 antagonists promote tumor regression by inducing apoptosis of angiogenic blood vessels. *Cell*, 79, 1157, 1994.
104. Brooks, P. et al. Antiintegrin alpha v beta 3 blocks human breast cancer growth and angiogenesis in human skin. *J Clin Invest*, 96, 1815, 1995.
105. Senger, D. et al. Angiogenesis promoted by vascular endothelial growth factor: regulation through alpha1beta1 and alpha2beta1 integrins. *Proc Natl Acad Sci USA*, 94, 13612, 1997.
106. Huminiecki, L. and Bicknell, R. In silico cloning of novel endothelial-specific genes. *Genome Res*, 10, 1796, 2000.
107. Ho, M. et al. Identification of endothelial cell genes by combined database mining and microarray analysis. *Physiol Genomics*, 13, 249, 2003.
108. Gerritsen, M. et al. In silico data filtering to identify new angiogenesis targets from a large in vitro gene profiling data set. *Physiol Genomics*, 10, 13, 2002.
109. Mailhos, C. et al. Delta4, an endothelial specific notch ligand expressed at sites of physiological and tumor angiogenesis. *Differentiation*, 69, 135, 2001.
110. Liu, Z. et al. Regulation of Notch1 and Dll4 by vascular endothelial growth factor in arterial endothelial cells: implications for modulating arteriogenesis and angiogenesis. *Mol Cell Biol*, 23, 14, 2003.
111. Oh, P. et al. Subtractive proteomic mapping of the endothelial surface in lung and solid tumours for tissue-specific therapy. *Nature*, 429, 629, 2004.
112. Durr, E. et al. Direct proteomic mapping of the lung microvascular endothelial cell surface in vivo and in cell culture. *Nat Biotechnol*, 22, 985, 2004.
113. Matsubara, S. et al. Analysis of endoglin expression in normal brain tissue and in cerebral arteriovenous malformations. *Stroke*, 31, 2653, 2000.
114. Arap, W. et al. Steps toward mapping the human vasculature by phage display. *Nat Med*, 8, 121, 2002.
115. Kolonin, M., Pasqualini, R., and Arap, W. Molecular addresses in blood vessels as targets for therapy. *Curr Opin Chem Biol*, 5, 308, 2001.
116. Trepel, M., Arap, W., and Pasqualini, R. In vivo phage display and vascular heterogeneity: implications for targeted medicine. *Curr Opin Chem Biol*, 6, 399, 2002.
117. Christian, S. et al. Nucleolin expressed at the cell surface is a marker of endothelial cells in angiogenic blood vessels. *J Cell Biol*, 163, 871, 2003.
118. Matsuno, F. et al. Induction of lasting complete regression of preformed distinct solid tumors by targeting the tumor vasculature using two new anti-endoglin monoclonal antibodies. *Clin Cancer Res*, 5, 371, 1999.

119. Tsunoda, S. et al. Specific binding of TES-23 antibody to tumour vascular endothelium in mice, rats and human cancer tissue: a novel drug carrier for cancer targeting therapy. *Br J Cancer*, 81, 1155, 1999.
120. Ran, S. et al. Antitumor effects of a monoclonal antibody that binds anionic phospholipids on the surface of tumor blood vessels in mice. *Clin Cancer Res*, 11, 1551, 2005.
121. Huang, X., Bennett, M., and Thorpe, P. A monoclonal antibody that binds anionic phospholipids on tumor blood vessels enhances the antitumor effect of docetaxel on human breast tumors in mice. *Cancer Res*, 65, 4408, 2005.
122. Marty, C. et al. Cytotoxic targeting of F9 teratocarcinoma tumours with anti-ED-B fibronectin scFv antibody modified liposomes. *Br J Cancer*, 87, 106, 2002.
123. Olson, T. et al. Targeting the tumor vasculature: inhibition of tumor growth by a vascular endothelial growth factor-toxin conjugate. *Int J Cancer*, 73, 865, 1997.
124. Wild, R. et al. VEGF-DT385 toxin conjugate inhibits mammary adenocarcinoma development in a transgenic mouse model of spontaneous tumorigenesis. *Breast Cancer Res Treat*, 85, 161, 2004.
125. Ran, S. et al. Infarction of solid Hodgkin's tumors in mice by antibody-directed targeting of tissue factor to tumor vasculature. *Cancer Res*, 58, 4646, 1998.
126. Nilsson, F. et al. Targeted delivery of tissue factor to the ED-B domain of fibronectin, a marker of angiogenesis, mediates the infarction of solid tumors in mice. *Cancer Res*, 61, 711, 2001.
127. Liu, C. et al. Prostate-specific membrane antigen directed selective thrombotic infarction of tumors. *Cancer Res*, 62, 5470, 2002.
128. Hu, P. et al. Comparison of three different targeted tissue factor fusion proteins for inducing tumor vessel thrombosis. *Cancer Res*, 63, 5046, 2003.
129. Nabel, E., Plautz, G., and Nabel, G. Gene transfer into vascular cells. *J Am College Cardiol*, 17, 189B, 1991.
130. Richter, M. et al. Adeno-associated virus vector transduction of vascular smooth muscle cells in vivo. *Physiol Genomics*, 2, 117, 2000.
131. Nicklin, S. et al. Efficient and selective AAV2-mediated gene transfer directed to human vascular endothelial cells. *Mol Ther*, 4, 174, 2001.
132. Pajusola, K. et al. Cell-type-specific characteristics modulate the transduction efficiency of adeno-associated virus type 2 and restrain infection of endothelial cells. *J Virol*, 76, 11530, 2002.
133. Wickham, T. et al. Integrins alpha v beta 3 and alpha v beta 5 promote adenovirus internalization but not virus attachment. *Cell*, 73, 309, 1993.
134. Greber, U. et al. Stepwise dismantling of adenovirus 2 during entry into cells. *Cell*, 75, 477, 1993.
135. Wang, K. et al. Adenovirus internalization and infection require dynamin. *J Virol*, 72, 3455, 1998.
136. Bergelson, J. Receptors mediating adenovirus attachment and internalization. *Biochem Pharmacol*, 57, 975, 1999.
137. Gaggar, A., Shayakhmetov, D., and Lieber, A. CD46 is a cellular receptor for group B adenoviruses. *Nat Med*, 9, 1408, 2003.
138. Wickham, T. et al. Targeted adenovirus-mediated gene delivery to T cells via CD3. *J Virol*, 71, 7663, 1997.
139. Nicklin, S. et al. In vitro and in vivo characterisation of endothelial cell selective adenoviral vectors. *J Gene Med*, 6, 300, 2004.
140. Liu, L. et al. Incorporation of tumor vasculature targeting motifs into moloney murine leukemia virus env escort proteins enhances retrovirus binding and transduction of human endothelial cells. *J Virol*, 74, 5320, 2000.
141. Masood, R. et al. Retroviral vectors bearing IgG-binding motifs for antibody-mediated targeting of vascular endothelial growth factor receptors. *Int J Mol Med*, 8, 335, 2001.
142. Nicklin, S. and Baker, A. Tropism-modified adenoviral and adeno-associated viral vectors for gene therapy. *Curr Gene Ther*, 2, 273, 2002.
143. White, S. et al. Targeted gene delivery to vascular tissue in vivo by tropism-modified adeno-associated virus vectors. *Circulation*, 109, 513, 2004.
144. Muller, O. et al. Random peptide libraries displayed on adeno-associated virus to select for targeted gene therapy vectors. *Nat Biotechnol*, 21, 1040, 2003.

145. Graulich, W. et al. Cell type specificity of the human endoglin promoter. *Gene*, 227, 55, 1999.
146. Nicklin, S. et al. Analysis of cell-specific promoters for viral gene therapy targeted at the vascular endothelium. *Hypertension*, 38, 65, 2001.
147. Reynolds, P. et al. Combined transductional and transcriptional targeting improves the specificity of transgene expression in vivo. *Nat Biotechnol*, 19, 838, 2001.
148. Korhonen, J. et al. Endothelial-specific gene expression directed by the tie gene promoter in vivo. *Blood*, 86, 1828, 1995.
149. De Palma, M., Venneri, M., and Naldini, L. In vivo targeting of tumor endothelial cells by systemic delivery of lentiviral vectors. *Hum Gene Ther*, 14, 1193, 2003.
150. Cowan, P. et al. Targeting gene expression to endothelial cells in transgenic mice using the human intercellular adhesion molecule 2 promoter. *Transplantation*, 62, 155, 1996.
151. Velasco, B. et al. Vascular gene transfer driven by endoglin and ICAM-2 endothelial-specific promoters. *Gene Ther*, 8, 897, 2001.
152. Richardson, T., Kaspers, J., and Porter, C. Retroviral hybrid LTR vector strategy: functional analysis of LTR elements and generation of endothelial cell specificity. *Gene Ther*, 11, 775, 2004.
153. Dancer, A. et al. Expression of thymidine kinase driven by an endothelial-specific promoter inhibits tumor growth of Lewis lung carcinoma cells in transgenic mice. *Gene Ther*, 10, 1170, 2003.
154. Jager, U., Zhao, Y., and Porter, C. Endothelial cell-specific transcriptional targeting from a hybrid long terminal repeat retrovirus vector containing human prepro-endothelin-1 promoter sequences. *J Virol*, 73, 9702, 1999.
155. Mavria, G., Jager, U., and Porter, C. Generation of a high titre retroviral vector for endothelial cell-specific gene expression in vivo. *Gene Ther*, 7, 368, 2000.
156. Varda-Bloom, N. et al. Tissue-specific gene therapy directed to tumor angiogenesis. *Gene Ther*, 8, 819, 2001.
157. Cho, J. et al. Development of an efficient endothelial cell specific vector using promoter and 5′ untranslated sequences from the human preproendothelin-1 gene. *Exp Mol Med*, 35, 269, 2003.
158. Greenberger, S. et al. Transcription-controlled gene therapy against tumor angiogenesis. *J Clin Invest*, 113, 1017, 2004.
159. Dai, C., McAninch, R., and Sutton, R. Identification of synthetic endothelial cell-specific promoters by use of a high-throughput screen. *J Virol*, 78, 6209, 2004.
160. Li, X. et al. Synthetic muscle promoters: activities exceeding naturally occurring regulatory sequences. *Nat Biotechnol*, 17, 241, 1999.
161. Edelman, G. et al. Synthetic promoter elements obtained by nucleotide sequence variation and selection for activity. *Proc Natl Acad Sci USA*, 97, 3038, 2000.
162. Carpenito, C. et al. Exploiting the differential production of angiogenic factors within the tumor microenvironment in the design of a novel vascular-targeted gene therapy-based approach to the treatment of cancer. *Int J Radiat Oncol Biol Phys*, 54, 1473, 2002.
163. Cross, M. et al. VEGF-receptor signal transduction. *Trends Biochem Sci*, 28, 488, 2003.
164. Hayes, G. et al. Alternative splicing as a novel of means of regulating the expression of therapeutic genes. *Cancer Gene Ther*, 9, 133, 2002.
165. Hayes, G. et al. Molecular mechanisms regulating the tumor-targeting potential of splice-activated gene expression. *Cancer Gene Ther*, 11, 797, 2004.
166. Thorpe, P.E. Vascular targeting agents as cancer therapeutics. *Clin Cancer Res*, 10, 415, 2004.
167. Huang, X. et al. Tumor infarction in mice by antibody-directed targeting of tissue factor to tumor vasculature. *Science*, 275, 547, 1997.
168. Siemann, D.W. Therapeutic strategies that selectively target and disrupt established tumor vasculature. *Hematol Oncol Clin North Am*, 18, 1023, 2004.
169. Landuyt, W. et al. In vivo antitumor effect of vascular targeting combined with either ionizing radiation or anti-angiogenesis treatment. *Int J Radiat Oncol Biol Phys*, 49, 443, 2001.
170. Siemann, D.W. and Shi, W. Targeting the tumor blood vessel network to enhance the efficacy of radiation therapy. *Semin Radiat Oncol*, 13, 53, 2003.
171. Siemann, D.W. and Rojiani, A.M. The vascular disrupting agent ZD6126 shows increased antitumor efficacy and enhanced radiation response in large advanced tumors. *Int J Radiat Oncol Biol Phys*, in press, 62, 846, 2005.

172. Grosios, K. et al. In vivo and in vitro evaluation of combretastatin A-4 and its sodium phosphate prodrug. *Br J Cancer*, 81, 1318, 1999.
173. Davis, P.D. et al. ZD6126: a novel vascular-targeting agent that causes selective destruction of tumor vasculature. *Cancer Res*, 62, 7247, 2002.
174. Chaplin, D.J., Pettit, G.R., and Hill, S.A. Anti-vascular approaches to solid tumour therapy: evaluation of combretastatin A4 phosphate. *Anticancer Res*, 19, 189, 1999.
175. Li, L., Rojiani, A.M., and Siemann, D.W. Preclinical evaluations of therapies combining the vascular targeting agent combretastatin A-4 disodium phosphate and conventional anticancer therapies in the treatment of Kaposi's sarcoma. *Acta Oncol*, 41, 91, 2002.
176. Siemann, D.W. and Shi, W. Efficacy of combined antiangiogenic and vascular disrupting agents in treatment of solid tumors. *Int J Radiat Oncol Biol Phys*, 60, 1233, 2004.
177. Grosios, K. et al. Combination chemotherapy with combretastatin A-4 phosphate and 5-fluorouracil in an experimental murine colon adenocarcinoma. *Anticancer Res*, 20, 229, 2000.
178. Cliffe, S. et al. Combining bioreductive drugs (SR 4233 or SN 23862) with the vasoactive agents flavone acetic acid or 5,6-dimethylxanthenone acetic acid. *Int J Radiat Oncol Biol Phys*, 29, 373, 1994.
179. Lash, C.J. et al. Enhancement of the anti-tumour effects of the antivascular agent 5,6-dimethylxanthenone-4-acetic acid (DMXAA) by combination with 5-hydroxytryptamine and bioreductive drugs. *Br J Cancer*, 78, 439, 1998.
180. Pruijn, F.B. et al. Mechanisms of enhancement of the antitumour activity of melphalan by the tumour-blood-flow inhibitor 5,6-dimethylxanthenone-4-acetic acid. *Cancer Chemother Pharmacol*, 39, 541, 1997.
181. Siim, B.G. et al. Marked potentiation of the antitumour activity of chemotherapeutic drugs by the antivascular agent 5,6-dimethylxanthenone-4-acetic acid (DMXAA). *Cancer Chemother Pharmacol*, 51, 43, 2003.
182. Nelkin, B.D. and Ball, D.W. Combretastatin A-4 and doxorubicin combination treatment is effective in a preclinical model of human medullary thyroid carcinoma. *Oncol Rep*, 8, 157, 2001.
183. Siemann, D.W. et al. Vascular targeting agents enhance chemotherapeutic agent activities in solid tumor therapy. *Int J Cancer*, 99, 1, 2002.
184. Konerding, M.A., Miodonski, A.J., and Lametschwandtner, A. Microvascular corrosion casting in the study of tumor vascularity: a review. *Scanning Microsc*, 9, 1233, 1995.
185. Vaupel, P. et al. Blood flow, oxygen consumption, and tissue oxygenation of human breast cancer xenografts in nude rats. *Cancer Res*, 47, 3496, 1987.
186. Vaupel, P., Kallinowski, F., and Okunieff, P. Blood flow, oxygen consumption and tissue oxygenation of human tumors. *Adv Exp Med Biol*, 277, 895, 1990.
187. Hockel, M. et al. Intratumoral pO2 predicts survival in advanced cancer of the uterine cervix. *Radiother Oncol*, 26, 45, 1993.
188. Brizel, D.M. et al. Pretreatment oxygenation profiles of human soft tissue sarcomas. *Int J Radiat Oncol Biol Phys*, 30, 635, 1994.
189. Nordsmark, M., Overgaard, M., and Overgaard, J. Pretreatment oxygenation predicts radiation response in advanced squamous cell carcinoma of the head and neck. *Radiother Oncol*, 41, 31, 1996.
190. Knocke, T.H. et al. Intratumoral pO2-measurements as predictive assay in the treatment of carcinoma of the uterine cervix. *Radiother Oncol*, 53, 99, 1999.
191. Horsman, M.R. and Murata, R. Vascular targeting effects of ZD6126 in a C3H mouse mammary carcinoma and the enhancement of radiation response. *Int J Radiat Oncol Biol Phys*, 57, 1047, 2003.
192. Siemann, D.W. and Horsman, M.R. Enhancement of radiation therapy by vascular targeting agents. *Curr Opin Investig Drugs*, 3, 1660, 2002.
193. Siemann, D.W. and Horsman, M.R. Targeting the tumor vasculature: a strategy to improve radiation therapy. *Expert Rev Anticancer Ther*, 4, 321, 2004.
194. Nishiguchi, I., Willingham, V., and Milas, L. Tumor necrosis factor as an adjunct to fractionated radiotherapy in the treatment of murine tumors. *Int J Radiat Oncol Biol Phys*, 18, 555, 1990.
195. Wilson, W.R. et al. Enhancement of tumor radiation response by the antivascular agent 5,6-dimethylxanthenone-4-acetic acid. *Int J Radiat Oncol Biol Phys*, 42, 905, 1998.
196. Siemann, D.W. and Rojiani, A.M. Enhancement of radiation therapy by the novel vascular targeting agent ZD6126. *Int J Radiat Oncol Biol Phys*, 53, 164, 2002.

197. Lew, Y.S. et al. Synergistic interaction with arsenic trioxide and fractionated radiation in locally advanced murine tumor. *Cancer Res*, 62, 4202, 2002.
198. Murata, R. et al. Improved tumor response by combining radiation and the vascular-damaging drug 5,6-dimethylxanthenone-4-acetic acid. *Radiat Res*, 156, 503, 2001.
199. Kimura, H. et al. Fluctuations in red cell flux in tumor microvessels can lead to transient hypoxia and reoxygenation in tumor parenchyma. *Cancer Res*, 56, 5522, 1996.
200. Horsman, M.R., Murata, R., and Overgaard, J. Combination studies with combretastatin and radiation: effects in early and late responding tissues. *Radiother Oncol*, 63 (Suppl 1), S50, 2002.
201. Murata, R. et al. Interaction between combretastatin A-4 disodium phosphate and radiation in murine tumors. *Radiother Oncol*, 60, 155, 2001.
202. Siemann, D.W. et al. Differentiation and definition of vascular-targeted therapies. *Clin Cancer Res*, 11, 416, 2005.
203. O'Reilly, M.S. et al. Angiostatin: a novel angiogenesis inhibitor that mediates the suppression of metastases by a Lewis lung carcinoma. *Cell*, 79, 315, 1994.
204. Yoon, S.S. et al. Mouse endostatin inhibits the formation of lung and liver metastases. *Cancer Res*, 59, 6251, 1999.
205. Liao, F. et al. Monoclonal antibody to vascular endothelial-cadherin is a potent inhibitor of angiogenesis, tumor growth, and metastasis. *Cancer Res*, 60, 6805, 2000.
206. Ellis, L.M. et al. Vascular endothelial growth factor in human colon cancer: biology and therapeutic implications. *Oncologist*, 5 (Suppl 1), 11, 2000.
207. Fidler, I.J. and Ellis, L.M. The implications of angiogenesis for the biology and therapy of cancer metastasis. *Cell*, 79, 185, 1994.
208. Wedge, S.R., Kendrew, J., Ogilvie, D.J., Hennequin, L.F., Brave, A.J., Ryan, A.J. et al. Combination of the VEGF receptor tyrosine kinase inhibitor ZD6474 and vascular-targeting agent ZD6126 produces an enhanced anti-tumor response. *Proc Am Assoc Cancer Res*, 43, 1081, 2002.
209. Siemann, D.W. and Shi, W. Dual Targeting of Tumor Vasculature: Combining Avastin with CA4P or OXi4503. *Clin Cancer Res*, 11, 8965, 2005.
210. Griggs, J. et al. Inhibition of proliferative retinopathy by the anti-vascular agent combretastatin-A4. *Am J Pathol*, 160, 1097, 2002.
211. Nambu, H. et al. Combretastatin A-4 phosphate suppresses development and induces regression of choroidal neovascularization. *Invest Ophthalmol Vis Sci*, 44, 3650, 2003.
212. Siemann, D.W. Vascular Targeting Agents, in *Horizons in Cancer Therapeutics: From Bench to Bedside*, 3, 4, 2002.

15 Vascular and Hematopoietic Stem Cells as Targets for Antiangiogenic Therapy

Carlos Almeida Ramos, Hans-Georg Kopp, and Shahin Rafii

CONTENTS

The processes underlying formation of new blood vessels have traditionally been divided into two categories: vasculogenesis and angiogenesis [1]. In vasculogenesis, new vessels form by the differentiation of mesenchymal cells (angioblasts) into endothelial cells, a process that occurs in parallel with the formation of blood islands, which are co-derived from hematopoietic progenitors. Very likely, an earlier precursor (hemangioblast) gives rise to both populations of cells. In angiogenesis, endothelial cells lining previously formed vessels proliferate and form sprouts, which give rise to new vascular structures. Whereas there is good evidence that angiogenesis occurs both during development and after birth, until a few years ago it was widely accepted that vasculogenesis, in the broader sense of an undifferentiated cell type giving rise to endothelial cells, took place only in the embryo. Recent findings have challenged this dogma [2], however. There is now evidence that supports the idea that circulating vascular progenitors are involved in blood vessel formation after birth, a process that has been referred to as neoangiogenesis [3]. Moreover, it has also been newly recognized that hematopoietic cells contribute to the maintenance and initiation of these processes, and that co-mobilization of both vascular and hematopoietic elements may be an essential part of neoangiogenesis [4].

Several studies have indicated that circulating angiogenic cells (endothelial cell precursors, or EPCs) and their progeny, most likely derived from the bone marrow (BM), functionally contribute to revascularization of ischemic tissues and vascularization of tumors. Processes in which this mechanism has been shown to play an important role include wound healing [2,5–9], myocardial ischemia [10–15], cerebral ischemia [16], limb ischemia [17–21], endothelialization of vascular grafts [22–25], atherosclerosis [26], retinal neovascularization [27,28], and tumor growth [3,6,29–34]. Moreover, while vascular endothelial growth factor receptor 2 (VEGFR2)-positive myelomonocytic cells have been previously implicated in neoangiogenesis, there is recent evidence that other hematopoietic lineages can also contribute to these processes [35–40].

In light of these findings, similar to what happens in the embryo, postnatal hematopoiesis and neoangiogenesis seem to be intimately interconnected. Understanding the mechanisms by which EPCs and hematopoietic cells (stem, progenitors, and differentiated lineages) cooperate provides the basis for new targeted therapies and monitoring strategies for malignancies, cardiovascular disease, and aberrant inflammatory processes. In this chapter, we review briefly the contribution of endothelial progenitors and pro-angiogenic hematopoietic elements to the vascularization of tumors and ischemic tissues. We address initially the identification of EPCs and the characteristics of pro-angiogenic HPCs. We then discuss some of the pathways involved in the recruitment of pro-angiogenic elements from the bone marrow. Finally, we deal briefly with the mechanisms responsible for homing of these cells to their ultimate destinations.

15.1 IDENTITY OF CIRCULATING ENDOTHELIAL PROGENITOR CELLS

Although there are substantial data supporting the idea that bone marrow can remotely give rise to vascular progenitors that are able to incorporate in sites of new vessel formation [2,29], there is still significant controversy regarding the exact identity of EPCs. These have been loosely defined as cells that are mobilized from the bone marrow in response to various physiological stimuli including chemotherapy, cytokines, ischemia, or the angiogenic switch that accompanies tumor growth. Phenotypically, EPCs have been identified as circulating cells expressing various combinations of antigens traditionally associated with hematopoietic stem or progenitor cells, such as c-Kit, Sca-1, and CD133; as well as endothelial cells, like VEGFR2 and Vascular Endothelial (VE)-cadherin; or both, as with CD34. For instance, a $CD34^+$ $CD133^+$ $VEGFR2^+$ cell found at a frequency of about 1:10,000 nucleated cells in the peripheral blood of G-CSF treated individuals and in cord blood has been proposed as a bona fide circulating EPC [22,41].

Alternatively, functional characteristics that have been used to define EPCs include the ability to generate cells that display in vitro aspects of mature endothelial cells (such as low-density lipoprotein uptake and endothelial nitric oxide expression), especially when coupled with high proliferative and clonogenic ability and serial replating potential [42] (the so called high proliferative potential-endothelial cell, HPP-EC, described both in cord and adult blood). Furthermore, the ability to integrate in blood vessel endothelium, as assessed by several tracking techniques (such as FISH and GFP or beta-galactosidase positivity), is also often used to demonstrate the presence of endothelial precursors in a specific population of cells.

At this point, it is unclear whether EPCs are derived from undifferentiated bone marrow cells (a putative adult hemangioblast) or from differentiated cells (such as monocytes) that acquire an endothelial cell phenotype [43]. Additionally unresolved is whether bone marrow-derived EPCs are hematopoietic cells or part of another resident subset. A population of multipotent adult progenitor cells (MAPCs), capable of differentiating into endothelial cells [30,44], has been obtained from bone marrow after removal of $CD45^+$ and Glycophorin A^+ cells, a procedure that should eliminate all hematopoietic elements. A similarly unsettled matter is whether these circulating EPCs are indeed derived solely from bone marrow precursors or whether they represent sloughed off endothelial cells. Recent studies have shown that the vascular wall also contains HPP-ECs [45,46]. Finally, it remains to be clearly established whether the different cell populations that have been described represent the same cell or correspond to precursors for distinct subsets of vascular wall constituents [35].

15.2 ENDOTHELIAL PROGENITORS CONTRIBUTE TO TUMOR NEOANGIOGENESIS IN HUMANS

The analysis of the origin of endothelial cells in solid tumors and lymphomas that developed after transplantation has shown that marrow-derived endothelial precursors can incorporate into tumor neovessels, with a rate varying between 0.5% and 12%, the highest being in lymphomas [47]. Another study, also done after bone marrow transplantation, demonstrated contribution of marrow-derived cells to vessels in regenerating organs, but also at low levels [48].

An important factor that could explain the low levels of incorporation seen after bone marrow transplantation is the likely low number of transplanted true repopulating endothelial progenitors. In studies where an appreciably higher number of tagged bone marrow cells were transplanted, there was significantly more incorporation into tumor vessels, presumably because there was a higher degree of engraftment of endothelial progenitors into the bone marrow [29,49]. Indeed, most of the bone marrow vasculature after irradiation and bone marrow transplantation is host-derived both in humans [50] and in mice [51]. If this vasculature is the origin of circulating endothelial precursors, the low degree of endothelial chimerism could be the determining factor for the low percentage of donor-derived cells in endothelial beds. It is thus likely that the true contribution of bone marrow–derived endothelial precursors is underestimated by these reports.

Other reasons why these studies report low levels of contribution from bone marrow could be variations in the extent of tumor ischemia and vascular injury, as well as differences in tumor growth rate. Emerging data show that the incorporation of circulating endothelial cells is highest during early phases of the angiogenic switch, when tumors transform from dormant to rapidly growing states [29]. Finally, one cannot ignore the competition of local angiogenic processes, which not only may include the recruitment of pre-existent endothelial cells from nearby vessels, but possibly also vascular mimicry by the tumor cells [52].

In summary, under permissive conditions, circulating EPCs, likely bone marrow–derived, have the capacity to contribute to the wall of neovessels of specific tumors that undergo rapid growth. However, the co-recruitment of specific subsets of hematopoietic cells may be essential for the proper incorporation of circulating and locally derived endothelial cells [35,53].

15.3 CO-MOBILIZED HEMATOPOIETIC CELLS REGULATE TUMOR ANGIOGENESIS

Our group and others have found evidence of hematopoietic cells regulating tumor angiogenesis as well as contributing to vascular regeneration in hind limb ischemia in mice. In order to avoid confusion and to emphasize the fact that these are truly hematopoietic cells contributing indirectly to neoangiogenesis, the term "hemangiocyte" has been coined to name this population of CXCR4-positive bone marrow–derived cells that can be found perivascularly in neoangiogenic areas [54].

Pro-angiogenic hematopoietic cells are thought to support angiogenesis both during embryonic development and postnatally by delivering bioavailable angiogenic factors, including VEGFs, matrix metalloproteinases (MMPs), and angiopoietins to the forming vessels [55–61]. Studies performed with MMP-9 deficient mice, which have impaired tumor growth, have established the importance of recruitment of myeloid cells for neoangiogenesis [56].

Transplantation of wild-type MMP-9$^{+/+}$ bone marrow resulted in enhanced tumor growth in MMP-9$^{-/-}$ mice by delivering MMP-9-producing hematopoietic cells to the tumor vasculature. Our group has also shown that VEGFR1^{+} hematopoietic cells contribute to tumor angiogenesis and metastasis [29]. In addition, Gr1^{+}CD11b^{+} cells have been shown to contribute to tumor neoangiogenesis by releasing MMP-9 [62]. Activated VEGFR1^{+} myeloid cells can indeed release angiogenic factors such as VEGF, platelet-derived growth factor (PDGF), and brain-derived neurotrophic factor (BDNF) enhancing vessel formation and stability [56,63,64]. We and others have demonstrated that inhibition of either VEGFR1 or VEGFR2 signaling is insufficient to induce tumor regression and necrosis, and that inhibition of both pathways is needed for this to occur [29,55].

These data suggest that co-recruitment of VEGFR1^{+} hematopoietic cells conveys signals that support incorporation and differentiation of VEGFR2^{+} endothelial cells into functional neovessels. Therefore, co-mobilization of pro-angiogenic hematopoietic cells and EPCs from the bone marrow may be the key event regulating neoangiogenesis. This suggests that traditional angiogenic factors not only support the mobilization of endothelial progenitors, but also induce mobilization of hematopoietic cells to the tumor vasculature. Co-recruitment of various lineages including myelomonocytic cells, progenitors, and possibly hematopoietic stem cells (HSCs) likely provides necessary signals for sprouting and stabilization of endothelial cells.

15.4 ANGIOGENIC FACTORS PROMOTE MOBILIZATION OF HEMATOPOIETIC CELLS AND ENDOTHELIAL PRECURSOR CELLS

In the mid-1980s, it was reported that several tumors produced a protein that rapidly increased microvascular permeability [65]. Soon after, it was realized that this protein secreted by tumor cells was an endothelial cell mitogen and it was named VEGF [66,67]. This characteristic made it a good candidate for a factor stimulating angiogenesis in tumors, an essential process for their growth [68]. It was later further recognized that several conditions associated with vascular trauma and ischemia, including cardiac ischemia [69], resulted in VEGF elevation within hours of the insult. Interestingly, in parallel with the rise in VEGF, rapid mobilization of endothelial and hematopoietic progenitors has been observed as a sequel of vascular trauma [70]. Similarly, tumor implantation has been shown to result in rapid mobilization of the same cells to the peripheral circulation.

The microenvironment where stem and progenitor cells dwell, the so called “niche,” is thought to be critical for regulating self-renewal and cell fate decisions. Yet, the molecular mechanisms governing recruitment to exit these niches are not well known. Under steady-state conditions, most stem cells are quiescent in their niche [71]. Whether a stem cell remains in its niche or is recruited into other areas of the bone marrow or the circulation is likely dictated by the bioavailability of stem cell-active cytokines, which are bound to the extracellular matrix or tethered to the membrane of stromal cells. Stress, such as bone marrow ablation by cytotoxic agents, switches on sequences of events whereby HSCs are recruited from their niches to reconstitute hematopoiesis.

Our group has shown that release of angiogenic factors by a tumor mass results in upregulation of MMP-9 within the bone marrow. Matrix metalloproteinases promote the release of extracellular matrix-bound or cell-surface bound cytokines [72], including, in the case of MMP-9, the stem cell-active cytokine soluble Kit-ligand (sKitL). This, in turn, may direct stem and progenitor cell recruitment [73]. Increased bioactive sKitL was shown to promote HSC cell cycling and enhance their motility, the latter being needed for the rapid recruitment of circulating endothelial and hematopoietic progenitors. G-CSF, another well-known progenitor mobilizing agent, induces expression of other proteolytic enzymes, such as elastase, cathepsin G, and MMP-2 (apart from MMP-9), which are likely important for recruitment [74]. Conflicting data from other groups about the role of proteases in stem-cell trafficking point to the possibility of redundant protease activity as well as to variability among different mouse strains in mobilization capacity [75,76].

It is intriguing to think that the conversion of KitL from a membrane-bound adhesion/survival-promoting molecule to a soluble survival/motogenic factor by MMP-9 is a critical event in regulating EPC mobilization. Because modulating the bioavailability of local cytokines may change the EPC fate, regulators of enzyme activity leading to proteolytic cleavage may be critical elements in this process. It remains to be determined whether alterations in the cytokine repertoire influence the differentiation status of EPCs before being launched to the peripheral circulation, and whether differentiation of EPCs takes place during their sojourn in the peripheral circulation or after they have arrived at the tumor vasculature.

Although MMP-9 activation is required, the actual molecular transducer that promotes mobilization of endothelial and hematopoietic progenitors is not entirely known. Our group and others have shown that VEGFR2 activation leads to MMP-9 activation and mobilization of VEGFR2$^+$ EPCs [73]. Within the VEGF family of cytokines, VEGF-A is the most potent mediator of the angiogenic switch [77,78]. VEGF-A exerts its effect through interaction with two tyrosine kinase receptors, VEGF receptor 1 (VEGFR1, Flt-1) and VEGF receptor 2 (VEGFR2, Flk-1, KDR). Placental growth factor (PlGF), a member of the VEGF family, signals mainly through VEGFR1 [79]. Although VEGFR1 and VEGFR2 convey signals that regulate angiogenesis, their role in the regulation of hematopoiesis is less clear.

Mice deficient in VEGFR2 display profound defects in vasculogenesis and hematopoiesis [80–82]. Because of early lethality of embryos, it is difficult to assess whether VEGFR2 expression is absolutely essential for HSC proliferation or migration during adulthood. Similarly, as VEGFR1$^{-/-}$ mice die from vascular disorganization in early embryogenesis, the role of VEGFR1 in the regulation of hematopoiesis during late embryonic development has been difficult to evaluate [83,84].

Several lines of evidence suggest that VEGFR1 may play an essential role in regulating specific aspects of adult hematopoiesis. Functional VEGFR1 has been shown to be expressed on mature myelomonocytic cells, conveying signals that support their migration in vitro [85–87]. In *Drosophila*, VEGF supports the motility of hemocytes [88], which are ancestral homologues of murine and human hematopoietic cells. Based on these data, it is likely that expression of VEGF receptors, in particular VEGFR1, apart from its role in classical angiogenic processes, also regulates hematopoiesis by increasing motility and recruitment of HSCs. Therefore, suppression of hematopoiesis through VEGFR1 inhibition may have two ramifications. On the one hand, blocking VEGFR1 may inhibit tumor angiogenesis and growth. On the other hand, it may introduce unwanted untoward marrow toxicity.

15.5 STROMAL CELL-DERIVED FACTOR 1 (SDF-1, CXCL-12) AND CXCR4 PLAY AN IMPORTANT ROLE IN THE HOMING OF MOBILIZED CELLS TO SITES OF NEOANGIOGENESIS

SDF-1/CXCL-12 [89] is a chemokine that is thought to exclusively signal through its cognate receptor CXCR4. Targeted deletion of either SDF-1 or CXCR4 results in an identical phenotype, which is characterized by embryonic lethality due to disturbed hematopoiesis as well as abnormal neuronal and cardiovascular development [90–93]. The SDF-1/CXCR4 signaling axis is highly conserved among species, and HSCs seem to be exclusively responsive to SDF-1 among all chemokines [94]. Indeed, SDF-1/CXCR4 interaction is thought to be responsible for the retention of HSCs in their marrow niches, a notion that is corroborated by the finding that the CXCR4 antagonist AMD3100 can be used to mobilize hematopoietic stem and progenitor cells to the peripheral circulation both in mice and humans [95].

The fact that the bone marrow represents a hypoxic microenvironment [96] may contribute to SDF-1 expression, which has shown to be regulated by hypoxia through upregulated Hypoxia Induced Factor (HIF-1), not only within the bone marrow, but also in peripheral ischemic tissue [97]. With endothelial cells expressing high amounts of SDF-1 on their surface, a haptotactic gradient is created where by CXCR4-positive cells, which include hematopoietic progenitors, adhere, roll, and eventually transmigrate to home to the periendothelial niche, where SDF-1 is highly expressed by perivascular myofibroblasts, thereby creating a retention signal for pro-angiogenic hematopoietic cells [98–100]. Bone marrow endothelial and stromal cells have even been shown to be able to take up SDF-1 on the luminal surface to secrete it into the bone marrow hematopoietic compartment [101].

How do these findings relate to malignant or ischemic disease? In the case of malignant tumors, SDF-1 expression in organs like bone and lungs mediates, for example, CXCR4-expressing breast or lung tumor cell chemotaxis, thereby allowing the formation of metastases to these organs [102,103]. Accordingly, inhibition of CXCR4 using monoclonal antibodies or small molecule inhibitors has been shown to diminish metastatic spread [102]. However, when the dynamic growth and angiogenesis in xenotransplanted Colon38 and Panco2 cells were examined closely, the response to CXCR4 inhibiting agents was shown to be mediated by impaired angiogenesis and to be independent of CXCR4-expression by the tumor itself [104]. Instead, the authors found tumor angiogenesis to be severely impaired—to an extent that was unexpected given the comparably small amount of bone marrow-derived endothelial cells that could be detected in the tumor vasculature. As previously mentioned, our group has recently shown that CXCR4$^+$ hematopoietic cells comprise a population of pro-angiogenic cells, which have been termed "hemangiocytes" to emphasize the fact that they are functionally pro-angiogenic, but truly hematopoietic instead of endothelial in origin [54]. Moreover, we demonstrated that SDF-1 can be released by platelets in response to hematopoietic cytokines and thereby create chemotactic gradients for hemangiocytes homing to ischemic tissues. Other groups have published similar results in models of arterial injury, where platelet-derived SDF-1 was necessary to enable hematopoietic progenitor cells to home and adhere to the injured vessel wall [105]. Taken together, these results not only confirm the importance of SDF-1/CXCR4 signaling in ischemic and malignant disease, but also point out the fact that targeting this signaling pathway is of potential worth in non-CXCR4 expressing tumors.

15.6 CONCLUSION

Circulating endothelial progenitor cells are recruited to sites of neoangiogenesis during tumor growth or ischemia, where they differentiate into mature endothelial cells. The co-mobilization of hematopoietic cells occurs in parallel with this process and provides cues for the proper assembly of new vessels. Classic angiogenic factors (such as VEGFs), by signaling through specific receptors, are involved in recruitment of EPCs and pro-angiogenic hematopoietic cells, likely by activation of bone marrow stromal proteases, such as MMP-9. These modulate the bioavailability of stem cell-active cytokines and potentially sever interactions between progenitors and their microenvironment allowing their release from the bone marrow. Homing interactions (such as the SDF-1/CXCR4 axis) are important not only for keeping these cells in their bone marrow niches, but also to guide them to sites of neoangiogenesis. Each of these mechanisms offers potential therapeutic targets, some already in development, as described in the previous chapters.

Cellular therapy with populations containing stem or progenitor cells is being evaluated in several trials of myocardial ischemia [106]. VEGF inhibition with monoclonal antibodies has shown promising results in metastatic tumors [107,108]. A soluble VEGFR is currently in clinical trials [109]. Small molecule inhibitors of intracellular kinases involved in angiogenic axes are currently being developed, with exciting results in some type of tumors [110]. MMP inhibitors are evaluated as potential targets [111]. CXCR4 inhibitors have been shown to have antitumor activity in animal models [112–114] and they are being evaluated in metastatic osteosarcoma [115]. The CXCR4 antagonists AMD3100 and ALX40-4C are currently used for stem cell mobilization [95,116]. Further characterization of the molecular pathways involved in recruitment of vessel precursors will provide us with new targets for antiangiogenesis strategies for inhibition of tumor growth, as well as for stimulating angiogenesis in order to accelerate the recovery of ischemic tissues.

REFERENCES

1. Risau, W., Mechanisms of angiogenesis, *Nature*, 386, 671, 1997.
2. Asahara, T. et al., Isolation of putative progenitor endothelial cells for angiogenesis, *Science*, 275, 964, 1997.
3. Moore, M.A., Putting the neo into neoangiogenesis, *J Clin Invest*, 109, 313, 2002.
4. Rafii, S. et al., Vascular and haematopoietic stem cells: novel targets for anti-angiogenesis therapy? *Nat Rev Cancer*, 2, 826, 2002.
5. Rafii, S. and Lyden, D., Therapeutic stem and progenitor cell transplantation for organ vascularization and regeneration, *Nat Med*, 9, 702, 2003.
6. Asahara, T. et al., Bone marrow origin of endothelial progenitor cells responsible for postnatal vasculogenesis in physiological and pathological neovascularization, *Circ Res*, 85, 221, 1999.
7. Asahara, T. et al., VEGF contributes to postnatal neovascularization by mobilizing bone marrow-derived endothelial progenitor cells, *EMBO J*, 18, 3964, 1999.
8. Crisa, L. et al., Human cord blood progenitors sustain thymic T-cell development and a novel form of angiogenesis, *Blood*, 94, 3928, 1999.
9. Carmeliet, P. and Luttun, A., The emerging role of the bone marrow–derived stem cells in (therapeutic) angiogenesis, *Thromb Haemost*, 86, 289, 2001.
10. Orlic, D. et al., Bone marrow cells regenerate infarcted myocardium, *Nature*, 410, 701, 2001.
11. Orlic, D. et al., Mobilized bone marrow cells repair the infarcted heart, improving function and survival, *Proc Natl Acad Sci USA*, 98, 10344, 2001.
12. Kocher, A.A. et al., Neovascularization of ischemic myocardium by human bone-marrow-derived angioblasts prevents cardiomyocyte apoptosis, reduces remodeling and improves cardiac function, *Nat Med*, 7, 430, 2001.
13. Shintani, S. et al., Mobilization of endothelial progenitor cells in patients with acute myocardial infarction, *Circulation*, 103, 2776, 2001.
14. Jackson, K.A. et al., Regeneration of ischemic cardiac muscle and vascular endothelium by adult stem cells, *J Clin Invest*, 107, 1395, 2001.
15. Edelberg, J.M. et al., Young adult bone marrow–derived endothelial precursor cells restore aging-impaired cardiac angiogenic function, *Circ Res*, 90, E89, 2002.
16. Zhang, Z.G. et al., Bone marrow–derived endothelial progenitor cells participate in cerebral neovascularization after focal cerebral ischemia in the adult mouse, *Circ Res*, 90, 284, 2002.
17. Iwaguro, H. et al., Endothelial progenitor cell vascular endothelial growth factor gene transfer for vascular regeneration, *Circulation*, 105, 732, 2002.
18. Kalka, C. et al., Transplantation of ex vivo expanded endothelial progenitor cells for therapeutic neovascularization, *Proc Natl Acad Sci USA*, 97, 3422, 2000.
19. Schatteman, G.C. et al., Blood-derived angioblasts accelerate blood-flow restoration in diabetic mice, *J Clin Invest*, 106, 571, 2000.
20. Murohara, T., Therapeutic vasculogenesis using human cord blood-derived endothelial progenitors, *Trends Cardiovasc Med*, 11, 303, 2001.
21. Murohara, T. et al., Transplanted cord blood-derived endothelial precursor cells augment postnatal neovascularization, *J Clin Invest*, 105, 1527, 2000.
22. Shi, Q. et al., Evidence for circulating bone marrow–derived endothelial cells, *Blood*, 92, 362, 1998.

23. Bhattacharya, V. et al., Enhanced endothelialization and microvessel formation in polyester grafts seeded with CD34(+) bone marrow cells, *Blood*, 95, 581, 2000.
24. Kaushal, S. et al., Functional small-diameter neovessels created using endothelial progenitor cells expanded ex vivo, *Nat Med*, 7, 1035, 2001.
25. Maeda, M. et al., Progenitor endothelial cells on vascular grafts: an ultrastructural study, *J Biomed Mater Res*, 51, 55, 2000.
26. Sata, M. et al., Hematopoietic stem cells differentiate into vascular cells that participate in the pathogenesis of atherosclerosis, *Nat Med*, 8, 403, 2002.
27. Otani, A. et al., Bone marrow derived stem cells target retinal astrocytes and can promote or inhibit retinal angiogenesis, *Nat Med*, 29, 29, 2002.
28. Grant, M.B. et al., Adult hematopoietic stem cells provide functional hemangioblast activity during retinal neovascularization, *Nat Med*, 8, 607, 2002.
29. Lyden, D. et al., Impaired recruitment of bone-marrow-derived endothelial and hematopoietic precursor cells blocks tumor angiogenesis and growth, *Nat Med*, 7, 1194, 2001.
30. Reyes, M. et al., Origin of endothelial progenitors in human postnatal bone marrow, *J Clin Invest*, 109, 337, 2002.
31. Gehling, U.M. et al., In vitro differentiation of endothelial cells from AC133-positive progenitor cells, *Blood*, 95, 3106, 2000.
32. Marchetti, S. et al., Endothelial cells genetically selected from differentiating mouse embryonic stem cells incorporate at sites of neovascularization in vivo, *J Cell Sci*, 115, 2075, 2002.
33. Davidoff, A.M. et al., Bone marrow–derived cells contribute to tumor neovasculature and, when modified to express an angiogenesis inhibitor, can restrict tumor growth in mice, *Clin Cancer Res*, 7, 2870, 2001.
34. Ito, H. et al., Endothelial progenitor cells as putative targets for angiostatin, *Cancer Res*, 59, 5875, 1999.
35. De Palma, M. et al., Tie2 identifies a hematopoietic lineage of proangiogenic monocytes required for tumor vessel formation and a mesenchymal population of pericyte progenitors, *Cancer Cell*, 8, 211, 2005.
36. Conejo-Garcia, J.R. et al., Tumor-infiltrating dendritic cell precursors recruited by a beta-defensin contribute to vasculogenesis under the influence of Vegf-A, *Nat Med*, 10, 950, 2004.
37. Riboldi, E. et al., Cutting edge: proangiogenic properties of alternatively activated dendritic cells, *J Immunol*, 175, 2788, 2005.
38. Schruefer, R. et al., Human neutrophils promote angiogenesis by a paracrine feedforward mechanism involving endothelial interleukin-8, *Am J Physiol Heart Circ Physiol*, 288, H1186, 2005.
39. Puxeddu, I. et al., Human peripheral blood eosinophils induce angiogenesis, *Int J Biochem Cell Biol*, 37, 628, 2005.
40. Heissig, B. et al., Low-dose irradiation promotes tissue revascularization through VEGF release from mast cells and MMP-9-mediated progenitor cell mobilization, *J Exp Med*, 202, 739, 2005.
41. Peichev, M. et al., Expression of VEGFR-2 and AC133 by circulating human CD34(+) cells identifies a population of functional endothelial precursors, *Blood*, 95, 952, 2000.
42. Ingram, D.A. et al., Identification of a novel hierarchy of endothelial progenitor cells using human peripheral and umbilical cord blood, *Blood*, 104, 2752, 2004.
43. Rohde, E. et al., Blood monocytes mimic endothelial progenitor cells, *Stem Cells*, 24, 357, 2006.
44. Jiang, Y. et al., Pluripotency of mesenchymal stem cells derived from adult marrow, *Nature*, 418, 41, 2002.
45. Ingram, D.A., Caplice, N.M., and Yoder, M.C., Unresolved questions, changing definitions, and novel paradigms for defining endothelial progenitor cells, *Blood*, 106, 1525, 2005.
46. Ingram, D.A. et al., Vessel wall-derived endothelial cells rapidly proliferate because they contain a complete hierarchy of endothelial progenitor cells, *Blood*, 105, 2783, 2005.
47. Peters, B.A. et al., Contribution of bone marrow–derived endothelial cells to human tumor vasculature, *Nat Med*, 11, 261, 2005.
48. Jiang, S. et al., Transplanted human bone marrow contributes to vascular endothelium, *Proc Natl Acad Sci USA*, 101, 16891, 2004.
49. Ruzinova, M.B. et al., Effect of angiogenesis inhibition by Id loss and the contribution of bone-marrow-derived endothelial cells in spontaneous murine tumors, *Cancer Cell*, 4, 277, 2003.

50. Kvasnicka, H.M. et al., Mixed chimerism of bone marrow vessels (endothelial cells, myofibroblasts) following allogeneic transplantation for chronic myelogenous leukemia, *Leuk Lymphoma*, 44, 321, 2003.
51. Kopp, H.G., Ramos, C.A., and Rafii, S., Contribution of endothelial progenitors and pro-angiogenic hematopoietic cells to vascularization of tumor and ischemic tissue, *Curr Opin Hematol*, 13, 175, 2006.
52. Folberg, R., Hendrix, M.J., and Maniotis, A.J., Vasculogenic mimicry and tumor angiogenesis, *Am J Pathol*, 156, 361, 2000.
53. Kaplan, R.N. et al., VEGFR1-positive haematopoietic bone marrow progenitors initiate the pre-metastatic niche, *Nature*, 438, 820, 2005.
54. Jin, D.K. et al., Cytokine-mediated deployment of SDF-1 induces revascularization through recruitment of CXCR4(+) hemangiocytes, *Nat Med*, 12, 557, 2006.
55. Luttun, A., Carmeliet, G., and Carmeliet, P., Vascular progenitors: from biology to treatment, *Trends Cardiovasc Med*, 12, 88, 2002.
56. Coussens, L.M. et al., MMP-9 supplied by bone marrow–derived cells contributes to skin carcinogenesis, *Cell*, 103, 481, 2000.
57. Coussens, L.M. et al., Inflammatory mast cells up-regulate angiogenesis during squamous epithelial carcinogenesis, *Genes Dev*, 13, 1382, 1999.
58. Hiratsuka, S. et al., MMP-9 induction by vascular endothelial growth factor receptor-1 is involved in lung-specific metastasis, *Cancer Cell*, 2, 289, 2002.
59. Pipp, F. et al., VEGFR-1-selective VEGF homologue PlGF is arteriogenic: evidence for a monocyte-mediated mechanism, *Circ Res*, 92, 378, 2003.
60. Cursiefen, C. et al., VEGF-A stimulates lymphangiogenesis and hemangiogenesis in inflammatory neovascularization via macrophage recruitment, *J Clin Invest*, 113, 1040, 2004.
61. Rafii, S. et al., Contribution of marrow-derived progenitors to vascular and cardiac regeneration, *Semin Cell Dev Biol*, 13, 61, 2002.
62. Yang, L. et al., Expansion of myeloid immune suppressor Gr+CD11b+ cells in tumor-bearing host directly promotes tumor angiogenesis, *Cancer Cell*, 6, 409, 2004.
63. Takakura, N. et al., A role for hematopoietic stem cells in promoting angiogenesis, *Cell*, 102, 199, 2000.
64. Donovan, M.J. et al., Brain derived neurotrophic factor is an endothelial cell survival factor required for intramyocardial vessel stabilization, *Development*, 127, 4531, 2000.
65. Senger, D.R. et al., Tumor cells secrete a vascular permeability factor that promotes accumulation of ascites fluid, *Science*, 219, 983, 1983.
66. Leung, D.W. et al., Vascular endothelial growth factor is a secreted angiogenic mitogen, *Science*, 246, 1306, 1989.
67. Keck, P.J. et al., Vascular permeability factor, an endothelial cell mitogen related to PDGF, *Science*, 246, 1309, 1989.
68. Folkman, J. et al., Induction of angiogenesis during the transition from hyperplasia to neoplasia, *Nature*, 339, 58, 1989.
69. Hashimoto, E. et al., Rapid induction of vascular endothelial growth factor expression by transient ischemia in rat heart, *Am J Physiol*, 267, H1948, 1994.
70. Gill, M. et al., Vascular trauma induces rapid but transient mobilization of VEGFR2(+)AC133(+) endothelial precursor cells, *Circ Res*, 88, 167, 2001.
71. Cheng, T. et al., Hematopoietic stem cell quiescence maintained by p21cip1/waf1, *Science*, 287, 1804, 2000.
72. Vu, T.H. and Werb, Z., Matrix metalloproteinases: effectors of development and normal physiology, *Genes Dev*, 14, 2123, 2000.
73. Heissig, B. et al., Recruitment of stem and progenitor cells from the bone marrow niche requires MMP-9 mediated release of kit-ligand, *Cell*, 109, 625, 2002.
74. McQuibban, G.A. et al., Matrix metalloproteinase activity inactivates the CXC chemokine stromal cell-derived factor-1, *J Biol Chem*, 276, 43503, 2001.
75. Levesque, J.P. et al., Characterization of hematopoietic progenitor mobilization in protease-deficient mice, *Blood*, 104, 65, 2004.
76. Roberts, A.W. et al., Genetic influences determining progenitor cell mobilization and leukocytosis induced by granulocyte colony-stimulating factor, *Blood*, 89, 2736, 1997.

77. Carmeliet, P. and Jain, R.K., Angiogenesis in cancer and other diseases, *Nature*, 407, 249, 2000.
78. Hanahan, D. and Folkman, J., Patterns and emerging mechanisms of the angiogenic switch during tumorigenesis, *Cell*, 86, 353, 1996.
79. Carmeliet, P. et al., Synergism between vascular endothelial growth factor and placental growth factor contributes to angiogenesis and plasma extravasation in pathological conditions, *Nat Med*, 7, 575, 2001.
80. Kabrun, N. et al., Flk-1 expression defines a population of early embryonic hematopoietic precursors, *Development*, 124, 2039, 1997.
81. Shalaby, F. et al., Failure of blood-island formation and vasculogenesis in Flk-1-deficient mice, *Nature*, 376, 62, 1995.
82. Shalaby, F. et al., A requirement for Flk1 in primitive and definitive hematopoiesis and vasculogenesis, *Cell*, 89, 981, 1997.
83. Fong, G.H. et al., Role of the Flt-1 receptor tyrosine kinase in regulating the assembly of vascular endothelium, *Nature*, 376, 66, 1995.
84. Fong, G.H. et al., Increased hemangioblast commitment, not vascular disorganization, is the primary defect in flt-1 knock-out mice, *Development*, 126, 3015, 1999.
85. Sawano, A. et al., Flt-1, vascular endothelial growth factor receptor 1, is a novel cell surface marker for the lineage of monocyte-macrophages in humans, *Blood*, 97, 785, 2001.
86. Clauss, M. et al., The vascular endothelial growth factor receptor Flt-1 mediates biological activities. Implications for a functional role of placenta growth factor in monocyte activation and chemotaxis, *J Biol Chem*, 271, 17629, 1996.
87. Barleon, B. et al., Migration of human monocytes in response to vascular endothelial growth factor (VEGF) is mediated via the VEGF receptor flt-1, *Blood*, 87, 3336, 1996.
88. Cho, N.K. et al., Developmental control of blood cell migration by the Drosophila VEGF pathway, *Cell*, 108, 865, 2002.
89. Zlotnik, A. and Yoshie, O., Chemokines: a new classification system and their role in immunity, *Immunity*, 12, 121, 2000.
90. Nagasawa, T. et al., Defects of B-cell lymphopoiesis and bone-marrow myelopoiesis in mice lacking the CXC chemokine PBSF/SDF-1, *Nature*, 382, 635, 1996.
91. Tachibana, K. et al., The chemokine receptor CXCR4 is essential for vascularization of the gastrointestinal tract, *Nature*, 393, 591, 1998.
92. Zou, Y.R. et al., Function of the chemokine receptor CXCR4 in haematopoiesis and in cerebellar development, *Nature*, 393, 595, 1998.
93. Ma, Q. et al., Impaired B-lymphopoiesis, myelopoiesis, and derailed cerebellar neuron migration in CXCR4- and SDF-1-deficient mice, *Proc Natl Acad Sci USA*, 95, 9448, 1998.
94. Wright, D.E. et al., Hematopoietic stem cells are uniquely selective in their migratory response to chemokines, *J Exp Med*, 195, 1145, 2002.
95. Broxmeyer, H.E. et al., Rapid mobilization of murine and human hematopoietic stem and progenitor cells with AMD3100, a CXCR4 antagonist, *J Exp Med*, 201, 1307, 2005.
96. Asosingh, K. et al., Role of the hypoxic bone marrow microenvironment in 5T2MM murine myeloma tumor progression, *Haematologica*, 90, 810, 2005.
97. Ceradini, D.J. et al., Progenitor cell trafficking is regulated by hypoxic gradients through HIF-1 induction of SDF-1, *Nat Med*, 10, 858, 2004.
98. Mohle, R. et al., The chemokine receptor CXCR-4 is expressed on $CD34^+$ hematopoietic progenitors and leukemic cells and mediates transendothelial migration induced by stromal cell-derived factor-1, *Blood*, 91, 4523, 1998.
99. Kopp, H.G. et al., The bone marrow vascular niche: home of HSC differentiation and mobilization, *Physiology (Bethesda)*, 20, 349, 2005.
100. Grunewald, M. et al., VEGF-induced adult neovascularization: recruitment, retention, and role of accessory cells, *Cell*, 124, 175, 2006.
101. Dar, A. et al., Chemokine receptor CXCR4-dependent internalization and resecretion of functional chemokine SDF-1 by bone marrow endothelial and stromal cells, *Nat Immunol*, 6, 1038, 2005.
102. Muller, A. et al., Involvement of chemokine receptors in breast cancer metastasis, *Nature*, 410, 50, 2001.

103. Burger, J.A. and Kipps, T.J., CXCR4: A key receptor in the cross talk between tumor cells and their microenvironment, *Blood*, 107, 1761, 2006.
104. Guleng, B. et al., Blockade of the stromal cell-derived factor-1/CXCR4 axis attenuates in vivo tumor growth by inhibiting angiogenesis in a vascular endothelial growth factor-independent manner, *Cancer Res*, 65, 5864, 2005.
105. Massberg, S. et al., Platelets secrete stromal cell-derived factor 1α and recruit bone marrow-derived progenitor cells to arterial thrombi in vivo, *J Exp Med*, 203, 1221, 2006.
106. Povsic, T.J. and Peterson, E.D., Stem cells for the ischaemic heart, *Expert Opin Biol Ther*, 6, 427, 2006.
107. Goldberg, R.M., Advances in the Treatment of Metastatic Colorectal Cancer 10.1634/theoncologist.10-90003-40, *Oncologist*, 10, 40, 2005.
108. de Gramont, A. and Van Cutsem, E., Investigating the potential of bevacizumab in other indications: metastatic renal cell, non-small cell lung, pancreatic and breast cancer, *Oncology*, 69 Suppl 3, 46, 2005.
109. Hu, L. et al., Vascular endothelial growth factor trap combined with paclitaxel strikingly inhibits tumor and ascites, prolonging survival in a human ovarian cancer model, *Clin Cancer Res*, 11, 6966, 2005.
110. Schoffski, P. et al., Emerging role of tyrosine kinase inhibitors in the treatment of advanced renal cell cancer: a review, *Ann Oncol*, 17, 1185, 2006.
111. Sang, Q.X. et al., Matrix metalloproteinase inhibitors as prospective agents for the prevention and treatment of cardiovascular and neoplastic diseases, *Curr Top Med Chem*, 6, 289, 2006.
112. Tamamura, H. et al., T140 analogs as CXCR4 antagonists identified as anti-metastatic agents in the treatment of breast cancer, *FEBS Lett*, 550, 79, 2003.
113. Smith, M.C. et al., CXCR4 regulates growth of both primary and metastatic breast cancer, *Cancer Res*, 64, 8604, 2004.
114. Takenaga, M. et al., A single treatment with microcapsules containing a CXCR4 antagonist suppresses pulmonary metastasis of murine melanoma, *Biochem Biophys Res Commun*, 320, 226, 2004.
115. Kim, S.Y. et al., Inhibition of murine osteosarcoma lung metastases using the CXCR4 antagonist, CTCE-9908., *AACR Meeting Abstracts*, 2005, 60, 2005.
116. Hendrix, C.W. et al., Safety, pharmacokinetics, and antiviral activity of AMD3100, a selective CXCR4 receptor inhibitor, in HIV-1 infection, *J Acquir Immune Defic Syndr*, 37, 1253, 2004.

16 Genetic Strategies for Targeting Angiogenesis

Anita Tandle, Mijung Kwon, and Steven K. Libutti

CONTENTS

Angiogenesis is the process of the formation of new blood vessels from preexisting capillaries and is a tightly regulated biological process. Angiogenesis is normally observed only transiently under physiological conditions such as embryogenesis, wound healing, and reproductive functions in adults [1,2]. The process of angiogenesis is composed of two main phases, an activation phase and a resolution phase [3]. The phase of activation encompasses increased vascular permeability, basement membrane degradation, cell migration, extracellular matrix (ECM) invasion, and EC proliferation. The phase of resolution leads to the establishment of blood flow in the newly formed vessels. The process is governed by a tight balance between pro- and antiangiogenic growth factors. This balance, however, is lost in tumors. Unregulated angiogenesis may lead to several angiogenic diseases including cancer and is thought to be indispensable for cancer growth and metastasis [4]. As a result, tumor angiogenesis appears distinctly different from physiological angiogenesis with its aberrant vascular structure, altered endothelial cell–pericyte interactions, abnormal blood flow, increased permeability, and delayed maturation [5].

Tumor growth has been hypothesized to have two phases, which are separated by the angiogenic switch. The first is defined as an avascular phase, which corresponds to small occult lesions that are no more than 1–2 mm in diameter. These tumors stay dormant by reaching a steady state between proliferation and apoptosis [6,7]. It is believed that only a small subset of dormant tumors enters the second phase, a vascular phase, in which exponential tumor growth occurs. Within a given microenvironment, the angiogenic response is determined by a net balance between pro- and antiangiogenic regulators released from activated ECs, monocytes, smooth muscle cells, and platelets [8]. The principal growth factors driving angiogenesis are VEGF, bFGF, Ang-1, PDGF, and integrins [1]. The angiogenic inhibitors include naturally occurring inhibitors endostatin, thrombospondin, angiostatin, fumagillin, inhibitors of matrix metalloproteinases (MMPs), and TNF-α. In certain tumors, the angiogenic switch also involves downregulation of endogenous antiangiogenic factors such as thrombospondin-1 [9]. Considering the importance of angiogenesis for the growth and progression of dormant lesions, therapeutic approaches targeting the angiogenic switch may provide long-term, effective control of cancers and their metastases.

The antiangiogenic inhibitors have been divided into two broad categories: direct and indirect angiogenesis inhibitors [10,11]. Direct angiogenesis inhibitors (e.g., angiostatin, endostatin, vitaxin) prevent ECs from responding to various proangiogenic factors, whereas indirect angiogenesis inhibitors (e.g., anti-VEGF monoclonal antibody) generally inhibit the expression of proangiogenic factors from tumor cells, or the expression of proangiogenic receptors on ECs.

Given the notion that angiogenesis is regulated by a complex balance between pro- and antiangiogenic factors, identifying the molecular pathways involving these factors and using the knowledge to develop genetic strategies for targeting tumor angiogenesis will assist in the development of better therapeutic modalities. The recent development of microarray and gene profiling techniques has allowed us to understand the mechanism of action of these factors to a great extent [12,13]. In the current chapter, we have attempted to provide an overview of the recently discovered molecular pathways for pro- and antiangiogenic factors. The chapter also focuses on preclinical tumor-targeted therapy to inhibit angiogenesis. Finally, it addresses the need for in-depth research required to effectively translate antiangiogenic therapy to clinic.

STRATEGIES FOR TARGETING ANGIOGENESIS

Most cancer deaths result from metastases [14]. Most conventional treatment modalities can do very little to improve survival once the tumor has spread beyond the primary stage. The rationale for antiangiogenic tumor therapy has been supported by a growing body of literature in the field. The fact that tumors require new vasculature to grow and that inhibition of the growth of tumor vasculature or antiangiogenesis could be an effective anticancer strategy has been the subject of intense research in last decade. Antiangiogenic agents can reduce the vascular density and ultimately the tumor burden [15]. Hence, systemic as well as tumor-targeted antiangiogenic therapy has a great potential in treating advanced human cancers. Antiangiogenic therapy has potential advantages over traditional modes of cancer treatment: (a) it is less toxic than conventional chemotherapy; (b) there is less development of traditional resistance; and (c) it controls tumor growth independent of the growth fraction of the tumor cell type [16]. Antiangiogenic agents work through different mechanisms, such as inhibition of EC proliferation, migration, and EC apoptosis. When an angiogenesis inhibitor induces EC apoptosis in a microvessel, tumor cells supported by that vessel subsequently undergo apoptosis. It has been established that EC apoptosis precedes tumor cell apoptosis [17].

However, to see the full benefits of antiangiogenic therapy, the identification of molecular targets and strategies to inhibit these targets should be carefully elucidated. The following section describes the development of newer molecular techniques to identify the genetic targets and strategies to inhibit them for effective antiangiogenic therapy.

16.1 UTILIZATION OF MICROARRAY TECHNIQUES FOR IDENTIFICATION OF MOLECULAR TARGETS

16.1.1 Identification of Downstream Mediators of Antiangiogenic Factors

Endostatin, the 20 kDa C-terminal fragment of collagen XVIII, is a member of a group of endogenous antiangiogenic proteins that are generated by proteolytic processing [18,19]. Endostatin inhibits EC proliferation, migration, invasion, and tube formation, and induces apoptosis [20]. It was the first endogenous angiogenesis inhibitor to enter clinical trials, and the preliminary results were promising [21]. Although several binding partners to endostatin have been identified, the detailed mechanism for the antiangiogenic activity of endostatin has been elusive. Using cDNA microarray along with RT–PCR and phosphorylation analysis, Abdollahi et al. have recently shown that a 4 h endostatin treatment on human microvascular ECs results in the induction of a large cluster of genes known to affect EC growth and development [22]. Interestingly, known proangiogenic factors (such as Ids, HIF-1α, NF-κB, ETS1, STATs, ephrins, AP-1, and thrombin receptors) are all downregulated, and antiangiogenic factors (such as thrombospondin-1, kininogen, AT-III, chromogranin A, precursor of vasostatin, and mapsin) are all upregulated. In addition, upstream and downstream of these key regulatory elements, endostatin downregulates a cascade of interdependent genes such as the VEGF family, Bcl-2, LDH-A, MMPs, TNF-α, COX-2, and $\alpha_v\beta_3$. Furthermore, endostatin exposure results in the dephosphorylation of many proteins involved in angiogenic cell signaling such as Id1, JNK, NF-κB, and Bcl-2, and the phosphorylation of proteins including cyclin D. Perhaps the most important lesson from these studies would be the realization that angiogenic and antiangiogenic proteins are shown to be regulated in a way that promotes the antiangiogenic action of endostatin, enabling the prediction of the antiangiogenic behaviors of new agents that are yet to be studied. Another important message is that certain pathways thought to be distinct, in fact, work together to trigger an orchestrated antiangiogenic response. For instance, while thrombospondin-1 itself is a potent antiangiogenic factor [23], endostatin is shown to upregulate thrombospondin-1 expression. Additionally, Id1, a downregulator of thrombospondin [24], is suppressed by endostatin.

Our laboratory has pursued a comparative approach to identify common pathways involved in the response of ECs to angiogenesis inhibitors. Using cDNA microarray and real-time RT–PCR techniques (TaqMan assay), we have examined the effects of two different antiangiogenic reagents (endostatin as an example of an endogenous protein and fumagillin as an example of an exogenous compound) on the gene expression profiles of human umbilical vein endothelial cells (HUVECs) in order to elucidate commonly affected pathways [25]. Fumagillin (a natural metabolite from *Aspergillus fumigatus*) and its synthetic analog TNP-470 have been shown to block angiogenesis both in vitro and in vivo by directly inhibiting ECs [26]. Each reagent was incubated with HUVECs for up to 8 h, and interestingly, a majority of gene expression changes were observed as early as 1 and 2 h following treatment [25]. The genes that showed these early expression changes are involved in cell proliferation, gene transcription, and matrix formation, and many of them have unknown functions. Four genes (*DOC1*, *KLF4*, *TC-1*, and *Id1*) that showed a similar expression profile for both reagents were selected, and the specificity of their effects was validated by TaqMan assay. KLF4 and Id1 are known to be a transcription factor and a regulator of transcription, respectively [27,28], and the function of DOC1 and TC-1 is currently unknown [29,30].

DOC1, KLF4, and TC-1 revealed the same gene expression profile as seen in the microarray data, whereas Id1 did not show any significant changes by TaqMan assay [25]. Since the TaqMan assay is more sensitive and specific than the microarray technique, it is valuable to verify the microarray data with TaqMan assay results. Importantly, no significant changes in expression for these three genes occurred over the same time course of treatment in fibroblasts, indicating that these changes are unique responses of ECs to endostatin and fumagillin. Furthermore, we have shown by small interfering RNA (siRNA) technique that KLF4 and TC-1 fail to demonstrate upregulation in expression in response to endostatin treatment when DOC1 was silenced, suggesting that DOC1 may be upstream of KLF4 and TC-1 in mediating the antiangiogenic response to endostatin. In contrast, the abrogation of the KLF4 and TC-1 response by DOC1 silencing was not observed following fumagillin treatment. Collectively, these data suggest that the DOC1, KLF4, and TC-1 genes might play an important role in the early responsive pathways for endostatin and fumagillin, where the interactions among these genes can be different depending on the antiangiogenic reagent, which may lead to affects on different downstream pathways.

We have also investigated the effects of the tumor-derived cytokine, endothelial monocyte-activating polypeptide-II (EMAP-II), on the gene expression profile of HUVECs [31]. EMAP-II was first detected in the supernatants of murine tumor cells by virtue of its ability to stimulate endothelial-dependent coagulation in vitro [32]. The biologically active, mature 23 kDa form of EMAP-II was shown to be produced from inactive pro-EMAP-II [33]. Purified EMAP-II has pleiotropic effects on ECs, monocytes, and neutrophils and has been shown to inhibit EC proliferation and induces apoptosis [34,35]. Gene expression changes from HUVECs treated with EMAP-II for 0.5, 1, 2, 4, and 8 h were analyzed by cDNA microarray [31]. We observed changes in 69 genes based on a twofold cutoff threshold (log ratio $\geq$2 or $\leq$0.5). Then, we selected 10 of the 69 genes, based on their function and expression pattern, to validate the gene expression changes with TaqMan assay. Among these, the gene expression changes in ADM, DOC1, ICAM-1, Id1, KIT, SOCS3, and TNFAIP3 were confirmed by TaqMan analysis. Importantly, gene expression profiles for these 10 genes from fibroblasts treated with EMAP-II were considerably different from those of HUVECs, suggesting that changes in mRNA expression for these genes may be endothelial cell-specific. In addition, silencing of *DOC1* gene expression by siRNA to DOC1 abrogated the modulatory effect of EMAP-II on ADM, KLF4, SOCS3, and TNFAIP3, suggesting that DOC1 might play a role in mediating some of the effects of EMAP-II on ECs. This observation was intriguing because we previously showed that DOC1 silencing also abrogated changes in KLF4 and TC-1 expression in response to endostatin treatment, which suggest that DOC1 might be a common mediator of antiangiogenic reagents such as endostatin and EMAP-II.

An antiangiogenic antithrombin fragment has been shown to downregulate the expression of perlecan, the proangiogenic heparan sulfate proteoglycan, in ECs [36]. Perlecan was previously shown to be an essential coreceptor for bFGF-induced angiogenesis [37]. Significantly, the inhibition of bFGF-induced EC proliferation by the antithrombin fragment can be overcome through the upregulation of perlecan expression by TGF-β1 [36]. Most recently, it has been demonstrated that blocking the heparin-binding site of cleaved or latent antithrombin by complexing with a high-affinity heparin pentasaccharide abolishes the serpin's ability to inhibit proliferation, migration, capillary-like tube formation, bFGF signaling, and perlecan gene expression in bFGF-stimulated HUVECs, indicating that the heparin-binding site of antithrombin is of crucial importance for mediating the serpin's antiangiogenic activity [38].

Another laboratory has studied thrombin-induced changes in gene expression by microarray analysis in human microvascular endothelial cells (HMEC-1) and HUVECs [39]. Thrombin induces the expression of a set of at least 65 genes including thrombospondin-1. Increase in thrombospondin-1 expression has been linked to disease status such as tumor progression, atherosclerosis, and wound healing. Changes in thrombospondin-1 mRNA expression correlated with an increase in the extracellular thrombospondin-1 protein concentration [39]. Interestingly, analysis of the promoter region of thrombospondin-1 and other genes of similar expression profile identified from the microarray predicted an EBOX/EGRF transcription model. It was subsequently shown that the expression of each family member, MYC and EGR1, respectively, correlated with thrombospondin-1 expression, demonstrating that a transcriptional regulatory pathway for thrombospondin-1 induction in response to thrombin in ECs could be delineated by microarray analysis.

Microarray analysis for ECs treated with the inflammatory cytokines TNF-α and IFN-γ has revealed the upregulation of many antiangiogenic molecules, such as IFN-inducible protein-10, tryptophanyl-tRNA synthetase, and TIMP-1, and the downregulation of many proangiogenic molecules including VEGFR-2, collagen type IV, chemokine SDF-1 (stromal cell-derived factor), and chemokine receptor CXCR4 [40]. SDF-1 and CXCR4 were previously shown to mediate EC tube formation [41]. The reduced expression of both proteins by TNF-α and IFN-γ results in the inhibition of EC tube formation, suggesting that the downregulation of proangiogenic SDF-1 and CXCR4 could play an important role in mediating antiangiogenic responses to these inflammatory cytokines. Using oligonucleotide microarray analysis, bone morphogenetic protein-2 and macrophage inhibitory cytokine-1, two multifunctional cytokines of the TGF-β family, are shown to be upregulated in ECs following *N*-(4-hydroxyphenyl)retinamide (fenretinide; 4HPR) treatment [42]. The treatment with 4HPR is shown to inhibit and prevent tumor growth under diverse conditions and is known to be antiangiogenic. Both bone morphogenetic protein-2 and macrophage inhibitory cytokine-1 inhibit EC growth, migration, and invasion in vitro and suppress angiogenesis in the matrigel plug assay in vivo. In addition, blocking antibodies to bone morphogenetic protein-2 reverses the suppressive effects of 4HPR both in vitro and in vivo. These results suggest that the TGF-β family proteins play an important role in mediating antiangiogenic activities of 4HPR.

Hyperthermia combined with radiotherapy or chemotherapy has become a promising method for cancer treatment [43]. Although the biological effects of elevated temperatures have been extensively studied, the molecular mechanisms underlying the beneficial outcome in patients are not well understood. Roca et al. examined the effect of hyperthermia on in vitro and in vivo angiogenesis [44]. Heat treatment of HUVECs affects their differentiation into capillary-like structures in two models of in vitro angiogenesis and inhibits the formation of new blood vessels in the CAM assay. To elucidate the molecular mechanisms of this process, they performed cDNA microarray analysis for HUVECs subjected to heat shock. Plasminogen activator inhibitor-1 (PAI-1), a protein involved in the control of extracellular matrix degradation, was shown to be specifically upregulated on heat treatment. They subsequently demonstrated that anti-PAI-1 neutralizing antibodies reversed the heat shock effect on in vitro capillary formation and in vivo vessel formation. They also showed that heat treatment of murine mammary adenocarcinomas resulted in inhibition of tumor growth, associated with a reduction in microvessel counts and an increase in PAI-1 expression. The data therefore suggest that upregulation of endogenous PAI-1 in ECs is sufficient to inhibit angiogenesis both in vitro and in vivo and that heat-mediated PAI-1 induction is an important mechanism by which hyperthermia exerts its antitumor activity.

Cline et al. performed microarray analysis on ECs treated with six angiogenesis inhibitors including endostatin, TNP-470, thrombospondin-mimetic peptide, PEDF, retinoic acid, and IFN-α to predict possible synergistic antiangiogenic activity among these reagents [45]. Interestingly, the expression profiles of thrombospondin-mimetic peptide [46,47] and TNP-470, for example, were very similar, whereas endostatin had a dramatically different profile. They subsequently demonstrated in an in vivo tumor model that both tumor growth and angiogenesis were markedly inhibited when endostatin and either thrombospondin-mimetic peptide or TNP-470, at doses that were ineffective when used alone, were combined. In contrast, animals treated with both thrombospondin-mimetic peptide and TNP-470 showed a modest effect on both tumor growth and angiogenesis. Therefore, these results suggest that microarray technology could be useful in predicting the possible synergy among antiangiogenic reagents, allowing investigators to maximize their antitumor efficacy even in the absence of a complete understanding of their mechanism of action.

16.1.2 Identification of Negative Regulators of Angiogenesis

Microarray techniques have also been used to identify negative feedback regulators of angiogenesis. Although negative feedback regulation is one of the most crucial physiological mechanisms with which organisms are endowed, no such regulators have been identified for the regulation of angiogenesis so far. Since many proangiogenic factors have multiple effects on EC behavior, it is reasonable to suspect that these factors might induce negative regulators of angiogenesis. For example, it was previously shown that the dose–response curve for the effect of VEGF on the migration of ECs exhibited a bell-shaped pattern, which may suggest that the loss of responsiveness to higher concentration of VEGF is mediated by VEGF-induced expression of endogenous inhibitor(s) [48]. Surveying VEGF-inducible genes in human ECs by microarray analysis, a gene of unknown function, *KIAA1036*, was identified [49]. Surprisingly, the recombinant protein of KIAA1036 (subsequently named vasohibin) inhibited angiogenesis in vivo (shown by CAM and mouse corneal micropocket assay) as well as angiogenesis in vitro (shown by proliferation, migration, and tube formation). This inhibitory effect was selective for EC types. Importantly, antisense inhibition of KIAA1036 expression in ECs restored their responsiveness to a higher concentration of VEGF. As VEGF induced KIAA1036 mRNA in a dose-dependent manner, these data therefore suggest that the induced vasohibin inhibited the effect of VEGF at higher concentrations, behaving as a negative feedback regulator.

Using cDNA microarray analysis, Down syndrome critical region (DSCR)-1 gene has also been identified as a responsive gene to VEGF and thrombin treatment in ECs [50]. DSCR-1 was known to be a negative feedback regulator of calcium–calcineurin–NF-AT signaling [51]. Constitutive expression of DSCR-1 in HUVECs markedly impaired NF-AT nuclear localization, proliferation, and tube formation. In addition, overexpression of DSCR-1 attenuated neovascularization in matrigel plug assay and reduced B16 melanoma tumor growth in mice. Therefore, these findings suggest that DSCR-1 is a novel angiogenesis inhibitor that is induced in ECs by proangiogenic factors such as VEGF and thrombin, and inhibits angiogenesis in an autocrine manner.

16.1.3 Role of Transcription Factors in Angiogenesis

Angiogenesis results from a series of complex events such that gene expression during the process has to be tightly regulated by transcription factors. Overexpression of transcription factors in ECs followed by microarray analysis has been performed to identify the downstream targets and subsequent phenotype. Homeobox (Hox) genes are one of the major regulatory genes that play a role in organogenesis and maintaining differentiated tissue function [52]. Hox genes are also shown to be expressed in adult cells including vascular endothelium and regulate expression of genes involved in cell–cell, cell–extracellular matrix interaction [53,54]. It has been shown that HoxD3 induces an angiogenic phenotype and promotes EC migration and invasion via upregulation of integrin $\alpha_v\beta_3$ and urokinase-type plasminogen activator [55]. In addition, HoxB3 plays a role in angiogenesis by increasing expression of ephrinA1 that facilitates capillary morphogenesis [56]. In contrast, Myers et al. have shown that HoxD10 expression is higher in quiescent endothelium as compared with tumor-associated angiogenic endothelium [57]. Microarray analysis of HoxD10-overexpressing human endothelial cells (HMEC-1) revealed a pattern of gene expression consistent with a nonangiogenic phenotype. The expression of RhoC, β_4 integrin, α_3 integrin, and MMP-14 was decreased, whereas that of PAI-1 was increased in HoxD10-overexpressing HMEC-1 cells. Sustained expression of HoxD10 blocked EC migration and inhibited neovascularization in the CAM assay. Furthermore, HoxD10-transfected HMEC-1 directly implanted into nude mice failed to form new vessels. These data therefore suggest that HoxD10 can inhibit angiogenesis by directly modulating gene expression in the endothelium, and that HoxD10 plays a role in maintaining a nonangiogenic state in the endothelium.

Hematopoietically expressed homeobox (HEX) gene is a member of the nonclustered homeobox genes and is shown to play an essential role in the endoderm for normal development of the forebrain, liver, and thyroid gland [58]. HEX is also demonstrated to play some roles in embryonic vascular development [59]. Overexpression of HEX in HUVECs completely abrogated the response of HUVECs to VEGF with regard to proliferation, migration, and invasion and inhibited network formation of HUVECs on matrigel assay [60]. Subsequent cDNA microarray analysis revealed that the expression of VEGFR-1, VEGFR-2, neuropilin-1, TIE-1, TIE-2, and α_v integrin was decreased, whereas that of endoglin was increased in HEX-overexpressing HUVECs. These changes in gene expression were verified by TaqMan and western blot analysis. Thus, these results suggest that HEX may act as a negative regulator of angiogenesis by modulating the expression of angiogenesis-related genes.

Hypoxia-inducible factor 1 (HIF-1) plays an essential role in angiogenesis by activating transcription of genes encoding angiogenic growth factors such as VEGF under hypoxic conditions [61]. To determine whether HIF-1 also mediates cell-autonomous responses to hypoxia, Manalo et al. compared gene expression profiles in arterial ECs cultured under nonhypoxic versus hypoxic conditions [62]. Microarray analysis was also performed with arterial ECs infected with adenovirus encoding a constitutively active form of HIF-1α (deletion in amino acids 392–520 and substitutions in Pro567Thr and Pro658Gln) [63], in nonhypoxic conditions. More than half of genes upregulated by both hypoxia and a constitutively active HIF-1α can be placed into one of six categories: cytokines/growth factors, receptors, other signal transduction proteins, oxidoreductases, collagens/modifying enzymes, and transcription factors. VEGF-C was upregulated by HIF-1 in hypoxic ECs. In addition, one of the largest functional categories of genes regulated by HIF-1 encodes transcription factors, indicating that HIF-1 may be at the top of a hierarchy of hypoxia-regulated gene expression in endothelial cells. As suggested by gene expression profiles, ECs exposed to hypoxia or expressing a constitutively active HIF-1α exhibited major functional changes in these cells. Hypoxic ECs demonstrated an increased matrigel invasion and accelerated tube formation compared with those treated with nonhypoxic conditions. Infection of ECs with a constitutively active HIF-1α under nonhypoxic conditions was sufficient to show the similar responses induced by hypoxia. Therefore, these findings indicate that HIF-1 mediates autonomous EC activation.

Several members of the ETS family of transcription factors have been shown to play a role in angiogenesis [64,65]. Dominant-negative ETS1 inhibits motility of cancer cells and in vivo angiogenesis [66,67]. Although the mechanism by which interferon's (IFNs) function on ECs is elusive, they are known to be antiangiogenic. SP100, a constituent member of nuclear body (also called PML body), is one of the proteins whose expression is potently induced by IFNs [68]. It has also been shown that SP100 interacts with ETS1, reduces ETS1-DNA binding, decreases the transcriptional activity of ETS1, and inhibits MDA-MB-231 cancer cell invasion [68]. Thus, to test the hypothesis that ETS1 activity is negatively modulated by SP100 in ECs, Yordy et al. examined phenotype changes following the infection of HUVECs with adenovirus-expressing ETS1 and/or SP100 [69]. SP100 modulates ETS1-dependent morphological changes in HUVEC network formation and inhibits ETS1-induced migration and invasion. Subsequent cDNA microarray analysis from these infected cells revealed that, although SP100 modulated only a subset of genes responsive to ETS1 expression, the majority of differentially regulated genes showed reciprocal expression in response to ETS1 and SP100. Specifically, genes that are downregulated by ETS1 and upregulated by SP100 have antimigratory or antiangiogenic properties. Collectively, these data suggest that SP100 negatively modulates ETS1-dependent angiogenic processes in ECs.

16.1.4 Tumor–Host Interactions

Although in vitro studies to identify the molecular pathways mediating pro- and antiangiogenic factors have given us great insights in terms of the development of better anticancer therapy, it has also been shown that the translation from in vitro results to in vivo studies is not always successful. Perhaps one of the main reasons for this is because tumor-associated ECs are different from those of normal quiescent vessels. For example, compared with normal microvascular ECs, endothelial cells purified from renal cell carcinoma showed an increased survival and enhanced proadhesive and angiogenic properties [70]. Subsequent microarray analysis revealed that compared with normal ECs, ECs derived from tumor demonstrated an increased expression of genes involved in survival such as IAP and Bcl-2 families; in cell adhesion such as integrins, ICAM-1, CD44, and NCAM; and in angiogenesis such as VEGFR-2, angiopoietin-1, angiogenin, and VEGF-D. Consequently, in contrast to normal ECs, tumor-derived ECs were resistant to apoptosis, proadhesive for renal carcinoma cells, and able to grow and organize capillary-like structures in the absence of serum. Investigating the differences in gene expression between normal and tumor-associated ECs, therefore, could be relevant for the development of anticancer therapy.

It is now well accepted that tumor growth and invasion result from an interaction between the tumor and host microenvironment in which the host actively participates in the induction, selection, and expansion of the tumor cells [71]. Tumor and host stroma exchange enzymes and cytokines that modify the surrounding extracellular matrix and stimulate the migration, invasion, and proliferation of each other. The stromal compartment consists of many cell types including immune cells, inflammatory cells, fibroblasts, and vascular cells including ECs, pericytes, and smooth muscle cells. Since in many cases solid tumors are dependent on angiogenesis for their continued growth and invasion, tumor angiogenesis is a typical example of such tumor–host interactions [2]. Elucidation of molecular pathways involved in tumor–host interactions may, therefore, lead us to develop better anticancer therapies.

To this end, our laboratory has recently proposed a promising strategy to identify these pathways [72]. Specifically, MC38 murine colon adenocarcinoma cells were transduced with either an empty retrovirus or a retrovirus encoding human TIMP-2, and stable clones were subsequently developed. Although TIMP-2 was originally shown to inhibit MMP activity [73], it was recently reported that TIMP-2 has also a potent antiangiogenic activity independent of MMP-inhibition function [74]. TIMP-2 overexpression resulted in substantial inhibition of tumor growth in syngenic mouse xenografts, but not in in vitro cell culture, suggesting that tumor–host interactions are responsible for the TIMP-2-induced inhibition of tumor growth [72]. In addition, MC38/TIMP-2 tumors demonstrated a significantly reduced microvascular density compared with MC38/null tumor, which indicates that TIMP-2 overexpression results in the inhibition of tumor angiogenesis. To identify the genes involved in tumor–host interactions, cDNA microarray analysis was performed for both in vitro culture and in vivo tumor, and the results were compared. Thirteen candidate genes, which showed persistent expression differences in vivo that were not present in vitro, were selected. Among these, Ptpn16, the murine analog for human MKP-1, was chosen for further study based on the known relationship of this gene with TIMP-2-related processes such as angiogenesis. Specifically, MKP-1 is a dual specificity phosphatase, which has been implicated in the regulation of the MAPK pathway [75,76], and MAPK is known to play an important role in angiogenesis [77,78]. p38 MAPK, for instance, is shown to mediate the angiogenic response to VEGF and bFGF [79,80]. Further, p38 MAPK is known to be dephosphorylated and thus inactivated by MKP-1 [81]. To examine the possible relationship among these molecules, protein expression patterns were analyzed by a layered protein scanning technique, a multireplica protein-blotting for tissue array [82]. MC38/TIMP-2 tumor displayed a nearly threefold increase in MKP-1 expression, 34% decrease in p38 phosphorylation, and no changes in total p38 expression compared with MC38/null tumor [72]. However, the level of MKP-1 expression from null- and TIMP-2-transduced tumor cells grown in vitro was similar. These data, therefore, suggest that there is a link between MKP-1 upregulation and TIMP-2-induced inhibition of angiogenesis in vivo, although the cell of origin of increased MKP-1 expression in vivo cannot be definitively demonstrated from these data. When orthovanadate, a phosphatase inhibitor [83,84], was used to treat animals bearing MC38/TIMP-2 tumor, the growth of MC38/TIMP-2 tumor was accelerated, and TIMP-2-induced dephosphorylation of p38 was completely reversed [72]. Furthermore, the addition of TIMP-2 to the endothelial culture medium resulted in a twofold increase in the amount of MKP-1 bound to MAPK, supporting the hypothesis that MKP-1-dependent MAPK inactivation mediates the antiangiogenic effects of TIMP-2 in vivo. Collectively, these findings suggest that MKP-1 may be a new molecular target for anticancer therapy, and that these experimental approaches might be useful as a general strategy to elucidate molecular pathways associated with the host response to tumor.

Although tumor xenograft models have been useful to give us insights to develop anticancer therapy, they certainly have their limitations [8]. Since xenografts are often grown subcutaneously, they do not develop in common sites for human tumors, thus may not be physiologically relevant models to human tumors. In addition, since xenografts grow very fast, it may exaggerate antiangiogenic and antitumor responses, as compared with human epithelial malignancies, which typically develop over a period of 5–15 years. These notions have recently been illustrated by the study of Ruzinova et al. Id proteins are transcription factors [85], and Id1 and Id3 are shown to be required for neurogenesis, angiogenesis, and vascularization of tumor xenografts [28]. To compare the consequences of Id loss in spontaneous tumors with those of xenografts, Ruzinova et al. have crossed $Id1^{-/-}$ $Id3^{+/-}$ mice with $Pten^{+/-}$ mice [24]. Homozygous mutation of the Pten tumor suppressor gene has been implicated in a high percentage of human tumors, and $Pten^{+/-}$ mice exhibit a severe lymphocytic hyperplasia and high tumor incidence including uterine carcinomas, prostate intraepithelial neoplasias, and pheochromocytomas [86,87]. Significantly, in $Pten^{+/-}$ spontaneous tumors, the consequence of angiogenesis inhibition by Id loss turns out to vary in severity depending on tumor type, with some tumors growing to a large size despite hypoxic stress, suggesting that the response of spontaneously arising tumors to angiogenic stress is somewhat different from that observed in tumor xenografts [24]. To examine differences in gene expression between wild type and mutant Id1 tissues in these tumors, cDNA microarray analysis was performed by comparing gene expression in hyperplastic lymphoid tissue from $Pten^{+/-}$ $Id1^{+/+}$ and $Pten^{+/-}$ $Id1^{-/-}$ mice. Several proangiogenic genes, including α6 and β4 integrins, FGFR-1, and MMP-2 as well as members of ephrin and IGF2 families, were downregulated in $Id1^{-/-}$ ECs. In addition, treatment of blocking antibody to FGFR-1 or α6 integrin and peptide inhibitor of MMP-2 resulted in the reduced EC invasion and tube formation in matrigel plug assay, producing a phenotype very similar to that of $Id1^{-/-}$ $Id3^{+/-}$ mice. These results therefore suggest that these molecules are coordinately regulated by Id1, and that Id1 regulates several angiogenic pathways in vivo. Furthermore, these experimental strategies may be very useful for the development of new antiangiogenic therapies.

In the vasculature, ECs, pericytes, and vascular smooth muscle cells are closely associated with each other. Pericytes and vascular smooth muscle cells support and maintain the mature vasculature, but are also crucial for normal blood vessel development. To examine how communication between pericytes and ECs influences vascular development [89], used an in vitro model of TGF-β-stimulated differentiation of 10T1/2 cells into pericytes [88]. Microarray analysis was performed to identify genes that are differentially expressed when 10T1/2 cells differentiate to pericytes in vitro [89]. Microarray analysis revealed that an entire angiogenic program of genes, including VEGF, IL-6, VEGF-C, HB-EGF, CTGF, tenascin C, integrin α_5, and Eph receptor A2, was expressed during the differentiation process. In the presence of neutralizing antibodies to VEGF and IL-6, the differentiation of 10T1/2 cells was considerably reduced, indicating that the expression of these genes plays an important role in this differentiation process. When HUVECs and 10T1/2 cells were cocultured in a two-dimensional culture system, HUVECs formed cord-like structures and 10T1/2 cells that were in contact with HUVECs differentiated into pericytes. Importantly, the formation of cord-like structures was disrupted when neutralizing antibodies to VEGF and IL-6 were added to the coculture system, suggesting that factors produced by pericytes may be responsible for recruiting ECs and promoting angiogenesis. Therefore, further understanding of the genes that are expressed during the differentiation of mesenchymal precursors to pericytes may provide a better strategy for the development of anticancer therapy.

16.2 STRATEGIES TO INHIBIT ANGIOGENESIS

Recent progress in molecular genetic and computational technology has enabled us to analyze differential gene expression profiles of thousands of genes collectively in response to a single stimulus. As discussed in the earlier section, DNA microarray technology has identified a number of molecular targets important in angiogenesis. The next question is how to use this vast information to build on methods to target angiogenesis. The following section discusses different strategies to inhibit angiogenesis using the information gathered on molecular targets. Some of the strategies include use of neutralizing antibodies to angiogenic growth factors, soluble receptors, antisense oligonucleotide, or small interfering RNA to inhibit gene transcription and gene therapy.

16.2.1 Use of Neutralizing Antibodies and Soluble Receptors

A number of studies have demonstrated that neutralizing antibodies to angiogenic growth factors and their receptors can impair tumor-associated neovascularization [90]. The neutralizing antibody binds to its target and prevents binding to the cellular target, which may be a cell surface receptor, and blocks the receptor from activating downstream signaling pathways. The best example of a neutralizing antibody to one of the most important angiogenic growth factors, vascular endothelial growth factor (VEGF), is Bevacizumab (BV). The antibody is in Phase I/II/III clinical trials [91,92]. A multicenter randomized Phase II trial in advanced colorectal cancer showed statistically significant improvement in overall response in groups treated with low-dose BV combined with 5-fluorouracil (5-FU) and leucovorin (LV) [93]. This led Hurwitz et al. to conduct the trial of BV in metastatic colorectal cancer, the first antiangiogenic agent in Phase III clinical trial [92]. A statistically significant improvement in overall survival, progression-free survival, and response rate was seen leading to the FDA approving this agent for the treatment of Stage IV colon cancer in combination with chemotherapy.

A similar strategy has been used to block VEGF receptors. DC101, a monoclonal antibody to VEGFR-2, can inhibit binding of VEGF to its receptor and prevent signaling in vitro as well as strongly inhibit angiogenesis and subsequent subcutaneous tumor growth in variety of tumor models [94,95].

The blocking of the receptor for a newly identified ephrin target can inhibit tumor angiogenesis in animal models [22,96]. The ephrin ligands and Eph receptors constitute a large family of signaling molecules that are widely expressed in many tissues. The EphrinA1 ligand and its EphA2 receptor are also present on vascular endothelium [97]. The EphrinA1 plays a role in the inflammatory angiogenesis induced by tumor necrosis factor-α (TNF-α) [98]. A dominant-negative EphA2 receptor blocked the formation of capillary endothelial tubes in vitro. Further studies have shown that soluble EphA2-Fc and EphA3-Fc receptor constructs inhibit tumor angiogenesis and growth in vivo [96]. Recent studies have shown that blockade of the EphA/EphB4 receptor specifically inhibits VEGF-induced angiogenesis [99,100].

The development of antibodies against these growth factors would generate a new tool against angiogenesis. Some of the points to be considered in treatment scheduling are the right dose, timing, and sequence of different combinations of treatment.

16.2.2 Gene Therapy-Mediated Delivery of Angiogenesis Inhibitors

Several lessons learned from early clinical trials in antiangiogenic therapy would seem to support a role for antiangiogenic gene transfer strategies in the future. These lessons include:

(1) Genetic stability—ECs are much more genetically stable than tumor cells and are therefore less likely to accumulate mutations that confer early drug resistance [101]. Gene therapy strategies that result in constitutive expression of an antiangiogenic protein would be expected to be more effective in this setting than a gene therapy approach targeting a tumor cell that might quickly develop an escape mechanism.

(2) Low, continuous dosing—Constitutive expression of an antiangiogenic protein even at lower concentrations than bolus doses may be more effective than the intermittent peaks associated with repeated delivery of a recombinant protein. Some evidence for this thinking has been demonstrated in a mouse model [102].

(3) Angiogenic switch—Understanding the genetic and epigenetic events that transform a normal cell into a cancer cell has been one of the major advances in the field of tumor biology. However, in addition to these changes that occur during transformation, the induction of tumor vasculature, called the angiogenic switch, has increasingly become recognized as a critical step in tumor propagation and progression [103]. From this perspective, the body may harbor many in situ tumors yet the tumors do not progress to lethal tumors unless there is an imbalance between a tumor's proangiogenic output and the body's total angiogenic defense [104]. Gene therapy offers a strategy whereby

an individual could boost their endogenous angiogenic defenses and tip the balance favorably.

(4) Cost of production—The production of functional proteins can be expensive and the availability of some of the recombinant proteins may become scarce. For example, production of two of the most well known and widely studied angiogenesis inhibitors—angiostatin and endostatin—was recently halted [105]. Gene therapy offers the opportunity for patients to become their own source of production, an endogenous "factory" for antiangiogenic protein production.

(5) Multiple pathways—The ability to inhibit multiple angiogenic pathways could be made easier utilizing gene transfer strategies, which could achieve prolonged, sustained levels of multiple therapeutic agents rather than repeated systemic boluses of numerous antiangiogenic agents [106].

16.2.3 Methods for Gene Transfer

16.2.3.1 Antisense RNA and siRNA

Antisense oligodeoxynucleotides (ODNs) are synthetic molecules that block mRNA translation, thereby inhibiting the action of the gene. The phosphorothioate ODNs are the most extensively studied and have become the "first generation" of antisense molecules [107]. Some of the ODNs are in Phase I/II clinical trials for cancer therapy. However, the use of antisense oligos against angiogenesis is still at preclinical level. Use of ODNs against VEGF, its receptors and Ang-1, suppresses tumor growth and angiogenesis and increases apoptosis [108,109]. Angiogenesis and tumorigenicity (as measured by microvessel density and tumor volume, respectively) of human esophageal squamous cell carcinoma can be effectively inhibited by VEGF-165 antisense RNA [110]. VEGF-mediated neovascularization can also be inhibited by a combination of antisense oligonucleotides to VEGFR-1 and VEGFR-2 [108]. Simultaneously targeting VEGF production with antisense oligonucleotide and VEGF receptor signaling with receptor tyrosine kinase inhibitors enhances the anticancer efficacy of either therapy alone [111]. Expression of antisense RNA to integrin subunit $\beta 3$ inhibits microvascular EC capillary tube formation, with the extent of downregulation correlating with the extent of tube formation inhibition [112]. The main hurdle to take antisense oligos to the clinic is the delivery to target cells and their uptake. The use of tumor-associated receptors for the uptake of ODNs has shown considerable promise for the development of tumor-specific antisense therapy.

The ability of small double-stranded RNA to suppress the expression of a gene corresponding to its own sequence is called RNA interference (RNAi). RNAi is a conserved surveillance system that responds to double-stranded RNA by silencing mRNAs with homology to the double-stranded RNA trigger. The application of siRNA to silence gene expression has profound implications for the treatment of human diseases including cancer and has largely displaced efforts with antisense and ribozymes. The ability of siRNA to mediate gene-specific posttranscriptional silencing in mammalian cells will undoubtedly revolutionize functional genomics, as well as drug target identification and validation. siRNA oligonucleotides targeted to either subunit of the $\alpha 6\beta 4$ (a laminin adhesion receptor) integrin reduced cell surface expression of this integrin and resulted in decreased invasion of breast carcinoma cells [113]. Knock-down of Her-2/neu expression by siRNA is associated with increased expression of the antiangiogenic factor TSP-1 and decreased expression of VEGF in human breast or ovarian cancer. This strategy may be a useful therapeutic strategy for Her-2/neu-overexpressing breast or ovarian cancer [114].

The treatment of microvascular ECs with endostatin has identified ADM, NF-κB, HIF-1α, ephrins as new antiangiogenic targets [22]. Adrenomedullin (ADM) is a biologically active peptide released from the vascular wall, which increases blood flow through its vasorelaxant effects and prevents platelet activation by stimulation of nitric oxide synthesis [115]. ADM has been reported to act as a growth factor in tumor cells, to be involved in survival from apoptosis, and in angiogenesis. Additionally, an elevation of ADM levels is observed in hypoxic conditions [115]. Cytokine-induced ADM expression can be abolished by siRNA to HIF-1α [116]. Microarray work from our laboratory has identified KLF4 as a new target [31]. Platelet-derived growth factor BB (PDGF-BB) has been shown to be an extremely potent negative regulator of smooth muscle cell (SMC) differentiation [117]. A recent study suggests that KLF4 may be a key effector of PDGF-BB phenotypic switching of SMC [118]. A small interfering RNA to KLF4 partially blocked PDGF-BB-induced SMC gene repression.

Potential drawbacks for the use of siRNA clinically include such limitations as, chemically synthesized siRNA is expensive, requires high transfection efficiency, and the gene silencing effects are transient in nature. To overcome these limitations, viral systems can be used to deliver siRNA via short hairpin constructs (shRNA), thus providing more efficient and stable gene silencing. These vectors employ the U6, H1, or tRNA polymerase III promoters to direct transcription of small RNA hairpins, which are then processed by cellular enzymes into functional siRNA [119]. These can be used to generate stable knock-down cell lines. Many researchers have developed vectors like adenovirus, adeno-associated virus (AAV), retrovirus, and lentivirus expressing siRNA. Retroviruses, the transgene-delivery vector of choice for many experimental gene therapy studies, have been engineered to deliver and stably express therapeutic siRNA within cells, both in vitro and in vivo.

Brummelkamp et al. developed a vector system, named pSUPER, which directs the synthesis of siRNA in mammalian cells [120]. The authors have shown that siRNA expression mediated by this vector results in persistent and specific downregulation of gene expression, leading to functional inactivation of the targeted gene over longer periods of time.

VEGF carries out multifaceted functions in tumor development. DNA-vector-based RNAi, in which RNAi sequences targeting VEGF isoforms, has potential applications in isoform-specific knock-down of VEGF [121]. Rubinson et al. and Hemann et al. have demonstrated that RNAi delivered by retroviral and lentiviral vectors can silence genes in mice [122,123]. With these reports and creation of transgenic mice expressing siRNA, it is now possible to "knock-down" diseased genes in vivo and test these concepts [124]. However, some of the concerns in applying siRNA as a therapeutic modality are the targeting should be specific, it should not target any other gene, and it should not result in the initiation of nonspecific interferon responses and gene silencing in normal tissues [125]. Gene silencing in normal tissues can be avoided by targeting viral vectors expressing siRNA specifically to tumor cells. The strategies to target viral vectors to tumor tissues are discussed in the "targeted gene therapy" section of this chapter.

16.2.3.2 DNA Vaccines

DNA vaccine can be a simple plasmid DNA injected either intramuscularly or intradermally, and engineered to express an antigen, which serves as a target for the immune system. Direct injection of free DNA into certain tissues, particularly muscle, has been shown to produce surprisingly high levels of gene expression, and the simplicity of this approach has led to its adoption in a number of clinical protocols. Intratumoral administration of naked plasmid DNA encoding mouse endostatin inhibits renal carcinoma growth [126]. Intramuscular administration of the endostatin gene could significantly retard the growth of metastatic brain tumors [127]. A single intramuscular administration of the endostatin gene resulted in secretion of endostatin for up to 2 weeks and the inhibition of systemic angiogenesis [128]. Injection of the endostatin gene also inhibited both the growth of primary tumors and the development of metastatic lesions.

DNA vaccines can result in the generation of an adaptive immune response comprising antigen-specific CD4$^+$ cells and antibodies as well as CD8$^+$ cells. In recent years, a number of tumor vaccination strategies have been developed [129]. Most of these rely on the identification of tumor antigens that can be recognized by the immune system. DNA vaccination represents one such approach for the induction of both humoral and cellular immune responses against tumor antigens. Studies in animal models have demonstrated the feasibility of using DNA vaccination to elicit protective antitumor immune responses. However, most tumor antigens expressed by cancer cells in humans are weakly immunogenic, and therefore require the development of strategies to potentiate DNA vaccine efficacy in the clinical setting. Different strategies to increase immunogenecity of DNA vaccines include codon optimization of encoding DNA, linkage of a tumor antigen to either a microbial or viral epitope, and use of immunostimulatory molecules including cytokines [129]. Multiple clinical trials using tumor-associated antigen peptides have reported tumor regression, however, complete response has rarely been reported [130]. One of the possible reasons of modest clinical efficacy could be loss or downregulation of HLA class I molecules on the tumor cells [131]. This problem could be overcome with the development of effective vaccine against tumor angiogenesis, because ECs are genetically stable, do not show downregulation of HLA class I molecules, and are critically involved in the progression of a variety of tumors [132].

Recent reports have shown that vaccination using cDNA or recombinant protein of mouse VEGFR-2 is associated with significant antitumor effects in mouse tumor model of melanoma, colon, and non-small-cell lung carcinoma [133,134]. Niethammer et al. have shown an effective antitumor immune response was induced against subcutaneous tumors by an orally administered DNA vaccine encoding murine VEGFR-2 carried by attenuated *Salmonella typhimurium* [133]. Prolonged antitumor effects were demonstrated since C57BL/6J mice challenged subcutaneously with MC38 colon carcinoma cells 10 months after their last vaccination revealed a marked decrease in tumor growth in all experimental animals compared with controls. Vaccination could protect against spontaneous pulmonary as well as established metastases and prolonged the life span of treated animals [135]. There was a marked increase in T-cell activation markers in splenocytes from successfully vaccinated C57BL/6J mice, indicating involvement of cytotoxic T-cells (CTLs). Microscopic evaluation revealed an association of CD8$^+$ T-cells with vessel structures throughout the tumor tissue or matrigel sections of animals immunized with pcDNA3.1-VEGFR-2. ECs are genetically stable and do not downregulate MHC-class I and II antigens—an event that frequently occurs in solid tumors and severely impairs T-cell-mediated antitumor responses. The downside of the study is there was a statistically significant delay in wound healing.

Wada et al. checked epitope peptides derived from human VEGFR-2 [132]. The authors identified the epitope peptides of human VEGFR-2 showing that CTLs induced with these peptides have potent and specific cytotoxicity against not only peptide-pulsed target cells but also ECs endogenously expressing VEGFR-2 in the HLA class I–restricted fashion. The efficacy of the vaccine was checked in a unique mouse model system, where CTL recognizes tumor ECs expressing VEGFR-2 in the HLA-restricted fashion. These results strongly suggest that VEGFR-2 could be a promising target of immunologic therapy using cellular immunity and support the definitive rationale of the clinical development of this strategy against a broad range of cancers. Vaccination using these epitope peptides derived from VEGFR-2 is now in the process of Phase I clinical study [132].

Several characteristics of DNA vaccines make them an attractive strategy for therapy against cancer. Any tumor EC-specific antigen can be used, it is relatively cheap to make large amounts of vaccine, no immune responses toward antigen are seen, and they are safer compared to viral vactors [129]. Microarray techniques have generated a variety of new targets, which could be tried in the form of DNA vaccines in preclinical models.

However, in comparison to naked DNA, complexes of plasmid DNA with liposomes are relatively more stable with higher potency for transfection [136].

16.2.3.3 Cationic Liposomes

Liposomes are microscopic spherical vesicles of phospholipids and cholesterol. Recently, liposomes have been evaluated as delivery systems for drugs and have been loaded with a great variety of agents such as small drug molecules, proteins, nucleotides, and even plasmids. The advantages of using liposomes as drug/gene carriers are that they can be injected intravenously and when they are modified with lipids, which render their surface more hydrophilic, their circulation time in the bloodstream can be increased significantly. In addition, they lack immunogenicity after in vivo administration, repeated administrations can be done, they have a low cost of synthesis, and they can be made in safer formulations [137]. DNA can be engineered to provide specific or long-term gene expression, replication, or integration. The inverted terminal repeats from adenovirus or AAV have been used to prolong gene expression [137]. Plasmid DNA can be efficiently encapsulated between two bilamellar vesicles created using lipids, DOTAP and cholesterol to give robust gene expression in variety of tissues [138]. Liposomes can be targeted to the tumor cells by conjugating them to specific molecules like antibodies, proteins, and small peptides [139]. Tissue specific promoters (TSPs) have been used for the production of gene expression exclusively in the target cells.

Systemic, liposome-mediated administration of angiostatin suppressed the growth of melanoma tumors in mice [140]. Similar findings were observed by Chen et al. with angiostatin and endostatin [141]. Angiogenic ECs averaged 15–33-fold more uptake of cationic liposomes than corresponding normal ECs [142]. Thus, the encapsulation of antineoplastic drugs into cationic liposomes is a promising tool to improve selective drug delivery by targeting tumor vasculature. Strieth et al. used cationic liposomes encapsulating paclitaxel to treat melanomas in the dorsal skinfold chamber model. Tumor growth was significantly retarded after treatment with liposomes containing paclitaxel, compared with the treatment with paclitaxel. Analysis of intratumoral microcirculation revealed a reduced functional vessel density in tumors after application of liposomal paclitaxel. At the end of the observation time, vessel diameters were significantly smaller in animals treated with paclitaxel encapsulated in cationic liposomes. This resulted in a significantly reduced blood flow in vessel segments and increased EC apoptosis. Remarkable microcirculatory changes indicate that encapsulation of paclitaxel in cationic liposomes resulted in a mechanistic switch from tumor cell toxicity to an antivascular therapy. Cationic liposomes have been used successfully to target either a chemotherapeutic drug or an angiogenesis inhibitor to tumor vasculature to achieve vascular-targeted treatment [143–145]. Kondo et al. used liposomes modified with peptide, identified on phage-displayed library for membrane-type 1 matrix metalloproteinase (MT1-MMP) [146]. MT1-MMP is a key enzyme for the activation of MMP-2, which is closely related to tumor angiogenesis as well as to invasion and metastasis of cancer cells and is located on tumor cells and angiogenic ECs [147]. Liposomes modified with MT1-MMP homing peptide for EC can strongly suppress tumor growth compared with unmodified liposomes [146]. Janssen et al. showed that the coupling of cyclic RGD-peptides to the surface of PEG-liposomes can target to tumor endothelium [148]. Tumor vessel-targeted liposomes could be used to efficiently deliver therapeutic doses of chemotherapy [149]. Thus, use of the DNA: liposome complexes for delivery of specific DNA might avoid problems associated with degradation and removal of naked DNA. However, the following points should be considered while designing DNA: liposome complexes for clinical use, plasmid design, plasmid DNA preparation, delivery vehicle formulation, route of administration, detection of gene expression, and the dosing and administration schedule [137].

The disadvantages of nonviral delivery systems such as liposome systems have been the low levels of delivery and gene expression produced. Some of these problems can be overcome using viral vectors.

16.2.3.4 Nanoparticles

Nanoparticles (NPs) are submicron-sized polymeric colloidal particles with a therapeutic agent of interest encapsulated within their polymeric matrix or adsorbed or conjugated onto the surface [150]. The NPs are mostly composed of lipids that cross-link to form liquid crystal structures that self-assemble through polymerization. They have received considerable attention due to their smaller size, which offers a greater cellular uptake over microparticles. NPs can be used to provide targeted (cellular/tissue) delivery of drugs, to improve oral bioavailability, to sustain drug/gene effect in target tissue, to solubilize drugs for intravascular delivery, and to improve the stability of therapeutic agents against enzymatic degradation (nucleases and proteases), especially of protein, peptide, and nucleic acid drugs [151]. They are less immunogenic than viral vectors and hence it may be feasible to deliver therapeutic genes repeatedly to angiogenic blood vessels for long-term treatment of diseases. They can be conjugated to a biospecific ligand, which could direct them to the target tissue or organ [152]. The synthetic polyester-based NPs, PLGA, escape rapidly from the degradative endo-lysosomal compartment to the cytoplasmic compartment thus protecting the therapeutic agent from degradation due to lysosomal enzymes [150]. A cationic NP coupled to an integrin $\alpha_v\beta_3$-targeting ligand can deliver genes selectively to angiogenic blood vessels in tumor-bearing mice [153]. Systemic injection of the NP attached to mutant Raf gene (ATP^{μ}-*Raf*), which blocks endothelial signaling and angiogenesis in response to multiple growth factors, into mice resulted in apoptosis of the tumor-associated endothelium, ultimately leading to tumor cell apoptosis and sustained regression of established primary and metastatic tumors. A combination of radiotherapy and antiangiogenic agent can be delivered successfully to murine tumors in an animal model [154].

Recent advances in nanomaterials have produced a new class of fluorescent labels by conjugating semiconductor quantum dots (Qdots) with biorecognition molecules [155]. Qdots are small (<10 nm) inorganic nanocrystals that possess unique luminescent properties; their fluorescence emission is stable and tuned by varying the particle size or composition. Researchers have identified peptides homing specifically to tumor blood vessels and tumor lymphatic vessels [156,157]. In a recent report, studies have shown that these peptides specifically direct Qdots to blood vessels or lymphatic vessels in tumors [158]. The other tested organs such as brain, heart, kidney, or skin did not contain detectable Qdots, however, accumulation was seen in both the liver and spleen, in addition to the targeted tissues. Adding polyethylene glycol to the Qdot coating reduced the accumulation in the liver and spleen by about 95%, without noticeably altering Qdot accumulation in tumor tissue.

16.2.3.5 Nanocell

A nanocell is a composite vehicle particle of 80–120 nm, consisting of a solid biodegradable polymer core of nanoparticle surrounded by a lipid membrane. A chemotherapeutic agent is bound chemically to the inner core of the particle and an antiangiogenic agent is trapped within the lipid envelope. Disruption of this envelope inside a tumor would result in a rapid deployment of the agent, leading to vascular collapse and the intratumoral trapping of the chemotherapeutic agent [159,160]. The subsequent slow release of the cytotoxic agent from the nanoparticle should then kill the tumor cells. Using doxorubicin and combretastatin-A4 the authors have demonstrated that the staged release of the two drugs using the new delivery vehicle improved the survival time further than when the drugs were delivered simultaneously [159]. The delivery vehicles tended to accumulate in the tumors, rather than in other body tissues, killing both endothelial and cancer cells. The effect of the sequential delivery of these two drugs on tumor growth was dramatic.

16.2.3.6 Viral Vectors

The basic concept of viral vectors is to harness the innate ability of viruses to deliver genetic material into the infected cell. Delivery of genes expressing inhibitors of angiogenesis using viral vectors represents a more effective strategy, as it can produce stable and higher quantities of gene product compared with the systemic infusion of antiangiogenic DNA/protein. A number of viral vectors have been studied to check for the efficient delivery of a therapeutic molecule/gene without generating systemic toxicity. There are five main classes of clinically applicable viral vectors: adenoviruses, AAVs, retroviruses, lentiviruses, and herpes simplex-1 viruses (HSV-1s) [161,162]. The main difference between these vectors is that retroviruses and lentiviruses can integrate into the genome, whereas the other three classes predominantly persist as extrachromosomal episomes. Although the advantage of chromosomal integration is long-term transgene expression, these vectors can infect dividing cells only. The nonintegrated viral vectors can infect nondividing cells, but are not a favorable choice to bring about stable genetic change.

Adenovirus is a double-stranded DNA virus. For the majority of adenoviruses binding occurs via a cell surface receptor called Coxsackie-adenovirus receptor (CAR). In addition to binding to CAR, efficient virus internalization requires an interaction between the viral penton base and the cellular integrin αv receptor [163]. Adenoviruses can be produced in high titers and can efficiently deliver a therapeutic gene.

We have demonstrated inhibition of tumor growth by adenovirus expressing endostatin [164]. We have also examined the adenoviral delivery of endostatin in a C3(1)/Tag transgenic mouse model of mammary cancer [165,166]. Lesion progression in this model follows a very predictable time course with low-grade MIN lesions progressing to high-grade MIN leading to invasive carcinomas. Mice treated with Ad-mEndo (an adenovirus expressing secretable murine endostatin) for 12 weeks generated high levels of circulating endostatin compared with controls. By 20–21 weeks of age, the mice treated with Ad-mEndo had an approximately 50% reduction in the average tumor burden, compared with the control animals. A statistically significant positive correlation ($P = 0.04$) was found between levels of endostatin and survival in the group of mice treated with Ad-mEndo. Elevation of systemic levels of endostatin was associated with a significant decrease in levels of VEGF in the preinvasive mammary lesions.

Replication-defective adenoviral vectors are indeed particularly well suited for cancer gene therapy as they lead to a transient, but robust, expression of the transgene, and efficient in vivo gene transfer has been reported especially in the liver after systemic injection. However, they do not integrate into the host genome and the gene expression is transient. Moreover, adenovirus elicits an immune response resulting in an elimination of vector expressing cells. Despite these apparent drawbacks, the adenovirus remains a popular vector for gene therapy due to its high gene transfer efficiency and high level of expression in a wide variety of cell types. This has made the adenovirus an attractive vector for clinical trials.

One such effort has involved a replication-deficient adenoviral vector called TNFerade. This vector expresses human TNF-α and contains a radiation-inducible Egr-1 promoter. TNFerade was recently evaluated in a Phase I, dose escalation trial in patients with treatment refractory solid tumors [167]. TNFerade was injected intratumorally weekly once for 6 weeks with concomitant radiation. TNFerade-related toxicities included fever, injection site pain, and chills but dose-limiting toxicities were not seen. Overall, 21 of 30 patients demonstrated an objective response (5 CRs, 9 PRs) and, interestingly, in patients with synchronous lesions, a more favorable tumor response was seen in lesions treated with TNFerade + radiation as compared with lesions treated with radiation alone (4 of 5 patients). In a Phase I trial of

TNFerade in patients with soft tissue sarcoma in the extremities, of the 13 evaluable patients, 11 patients (85%) showed objective or pathological tumor responses (2 complete and 9 partial), and 1 had stable disease [168]. Partial responses were achieved though some of these tumors were very large (up to 675 cm^2). Of the 11 patients who underwent surgery, 10 (91%) showed a pathological complete response/partial response. Currently, Phase II randomized trials with TNFerade are open for patients with rectal cancer and unresectable pancreatic cancer. A single arm Phase II trial is open for patients with locally advanced esophageal cancer.

Adenovirus can be targeted to pulmonary endothelium by complexing it with a bispecific antibody to viral particle and angiotensin-converting enzyme [169]. Bilamellar cationic liposomes can also be used to encapsulate adenovirus. The encapsulated adenovirus can transduce CAR negative cells and is resistant to the neutralizing anti-adenoviral antibodies, allowing the readministration of the adenovirus [170].

Retroviruses are a class of enveloped viruses containing a single-stranded RNA molecule as the genome. Retrovirus vectors have been used in the majority of human gene transfers. They are able to efficiently integrate permanently into the human genome where they provide the basis for permanent expression of up to 8–9 kb of foreign DNA. Simple retroviruses, such as murine leukemia virus (MLV), and the vectors derived from them, require cell division for infection and thus possess a degree of inherent specificity for the rapidly dividing cells of neoplastic tissue [171]. Though transgene expression is usually adequate in vitro, prolonged expression is difficult to attain in vivo. In addition, retroviruses are inactivated by complement proteins and inflammatory IFN, specifically IFN-α and IFN-γ [172,173].

We studied a retroviral construct expressing mouse endostatin under the control of EF1α promoter in both subcutaneous and intraperitoneal cancer models [174]. The parental cell line (NMuLi—an epithelioid nonparenchymal cell line) or NEF-null cells developed rapidly growing tumors and mice were sacrificed 63 days after injection when the tumor volumes were 2400 and 2700 mm^3, respectively. The mean tumor volumes were less than 30 mm^3 in all four groups given an injection of NEF-Endo clones. After 122 days of follow-up, mice given an injection of different NEF-Endo clones had tumor volumes of 70 to 419 mm^3. Mice given an intraperitoneal injection of parental NMuLi or NEF-null cells had median survival times of 58 and 56 days, respectively, and all mice were dead by day 123. At this same time point, only 3 (9%) of the 32 mice receiving intraperitoneal NEF-Endo clones had died. Autopsies of mice that died following the injecion of wild type or null tumor clones revealed massive peritoneal tumor deposits and ascites. Surviving mice injected with endostatin expressing clones revealed only occasional, small peritoneal deposits at necropsy.

We have also evaluated retroviral delivery of interferon-gamma-inducible protein-10 (IP-10) to human melanoma xenografts in mice [175]. IP-10 is a chemokine that, in addition to being a chemoattractant for stimulated T lymphocytes, has been demonstrated to have antiangiogenic properties in vivo [176]. Intratumor injection of IP-10 reduced tumor growth via an angiostatic mechanism [177]. IP-10 combined with gemcitabine has shown significantly synergistic antitumor effect in hepatocellular carcinoma [178]. We observed, by 28 days after tumor implantation, animals bearing subcutaneous parental or null-transduced A375 melanoma had tumor volumes of 1100 ± 185 and 848 ± 227 mm^3, compared with tumor volumes in animals injected with IP-10 clone 2, were 47 ± 31 and clone 11, 124 ± 37 mm^3, respectively [175]. The size of subcutaneous tumors resulting from the injection of a mixed population of cells containing 50% each of null- and IP-10-transduced cells was also significantly smaller than tumors consisting of null-transduced cells at day 40 indicating gene delivery to every tumor cell is not required for efficacy. The histopathological analysis showed IP-10-transduced tumors with central areas of necrosis surrounded by a thin rim of viable tumor cells and a fibrotic capsule with only rare, isolated microvessels. However, more studies are needed to understand the antiangiogenic properties and clinical utility of IP-10.

A major shortcoming of retrovirus-derived vectors is their tendency to revert to replication-competent retrovirus (RCR), which could lead to fatal neoplasms. With the use of the latest packaging cell lines and vectors, the risk of RCR-generation has been drastically reduced. Currently, the greatest safety concern of using retroviral vectors is related to the risk of malignant transformation following oncogene activation due to random retroviral genomic integration and is discussed in more detail later.

The AAV is a nonpathogenic human parvovirus. Productive AAV infection requires helper functions that can be supplied by coinfection with helper viruses, such as adenovirus and herpesvirus. AAV can also replicate in cells that have been put under stress, such as irradiation or treatment with genotoxic agents. In the absence of a permissive environment that supports AAV replication, the viral DNA becomes integrated into the host chromosomal genome to establish a latent infection [179]. AAV combines some of the advantages of both the adenoviral and retroviral vectors. AAV vectors based on the serotype 2 capsid have been the most commonly used for gene therapy studies and have demonstrated transduction in a large number of cell types and experimental model systems [180]. The broad host range, low level of immune response, and longevity of gene expression observed with these vectors have enabled the initiation of a number of clinical trials using this gene delivery system [181]. One of the major limitations for the use of AAV as a gene delivery vehicle is the relatively small packaging capacity and low transduction efficiencies of recombinant AAV vectors. The AAV vectors do not contain any viral coding regions, and therefore, there is no toxicity associated with gene expression. However, a single injection of AAV vector elicits a strong humoral immune response against the viral capsid, which interferes with readministration of the vector [182]. AAV tropism can be genetically engineered by use of phage display-derived peptides to generate vectors that are selective for the vasculature [183]. The journal *Nature* recently reported concerns that there is a possibility that recombinant AAV vectors may cause or contribute to cancer in gene therapy subjects [184]. However, because of infrequent integration efficiency of AAV, the risk of cancer in current AAV trials is negligible [185].

Lentiviral vectors represent a new vector system that can achieve permanent integration of the gene into nondividing cells. Gene transfer can be achieved in very quiescent cells, nondividing or terminally differentiated cells such as neurons. Lentiviral vectors are especially useful in transducing cells, which lack receptors for adenoviruses. A broad tissue tropism for lentivirus can be achieved using variety of viral envelopes [186]. So far, lentiviral vectors expressing matrix metalloproteinase-2 (MMP-2), angiostatin, and endostatin have been developed [186,187]. However, lentivirus has low transduction efficiency for ECs and may result in significant vector-associated cytotoxicity [186].

Human herpes simplex viruses are a class of large DNA viruses with double-stranded genomes capable of accommodating a large amount of foreign DNA. The herpes simplex virus (HSV) thymidine kinase gene (tk) therapy with gancyclovir forms the basis of a widely used strategy for suicide gene therapy [188]. These vectors have been applied to multiple gene therapy approaches, including neurological diseases, spinal nerve injury, glioblastoma, and even pain therapy [189]. The advantages in using HSV include its wide host range, its ability to accommodate large genes, and its ability to establish long-lived asymptomatic infections in neuronal cells. The large packaging capacity of HSV-1 amplicons (up to a theoretical 152 kb) may be very useful for delivering complex genes and regulatory sequences or multiple copies of the transgene [190]. However, the virus's ability to replicate lytically in the brain, under some circumstances causing encephalitis, has led to fears about its potential safety for ultimate use in humans.

Conditionally replicating vaccinia viruses have been shown to target tumors in vivo after systemic delivery resulting in antitumor response [191]. Vaccinia virus has a large genome that is able to be engineered for the insertion and simultaneous expression of multiple genes.

It reliably infects a large variety of tumors and expresses high levels of the gene(s) of interest. Although naturally cytopathic, it can be engineered to be noncytopathic but still infectious [191]. Our earlier work has shown that by using recombinant vaccinia virus, tumor vasculature resisitant to TNF-α therapy can be made sensitized to the therapy by delivering EMAP-II [192]. EMAP-II is a tumor-derived cytokine with potent effects on ECs, including induction of tissue factor, upregulation of TNF receptor 1, and upregulation of E-selectin and P-selectin [193]. EMAP-II inhibits EC proliferation but has little effect on tumor cell or fibroblast proliferation [193]. Inhibition appears to be mediated through pro-apoptotic pathways. Release of EMAP-II in melanoma cells appears to render the tumor-associated vasculature sensitive to TNF [194]. We used these findings to develop a gene transfer strategy using EMAP-II to sensitize tumors to TNF-α [192]. We constructed a recombinant vaccinia virus encoding the human EMAP-II gene and transfected a human melanoma cell line previously insensitive to TNF-α treatment. Vaccinia virus expressing a reporter gene showed approximately 100 times higher gene expression in tumor tissues than those of other tissues, providing evidence of tumor-selective delivery of the foreign gene. EMAP-II expressing human melanoma cells implanted in nude mice demonstrated significant tumor regression after treatment with systemic TNF-α. Control groups remained insensitive to TNF-α therapy [192]. A recent report showed that a regional hyperthermia can improve vaccinia virus targeting to tumors, thereby enhancing the antitumor response [195].

16.2.3.7 Phage Vectors

Recently, bacteriophage vectors have been developed, which combine desirable properties of both viral and nonviral systems minimizing the typical drawbacks of both the systems [196]. They can be an attractive alternative to existing animal viral vectors because they lack intrinsic tropism for mammalian cells and can be produced in bacteria in large titers. Phage particles are extremely stable under a variety of harsh conditions of pH, DNase, and proteolytic enzymes. The most significant advantage is the genetic flexibility of filamentous phage, which allows a wide variety of proteins, antibodies, and peptides to be displayed on the phage coat, thus allowing phage to be targeted genetically to cell surface receptors [196]. Phage vectors engineered to express specific ligands can be targeted to tumor cells. Bacteriophage containing fibroblast growth factor (FGF) as a ligand can be used to express a reporter gene in FGF receptor positive cells but not to the receptor negative cells [197]. Using this approach, long-term transgene expression was established, indicating that with the appropriate targeted tropism, phage vectors can be targeted to mammalian cells. However, multivalent phage vectors are much more efficient than the monovalent phages [198]. These observations have been confirmed by another group using phage displaying single-chain antibody against the HER2 receptor [199]. The main concern about phage vectors is the low transduction efficiency. Currently, phage vector transduction efficiency (1%–4%) is considerably lower than most viral vectors; however, it can be improved by certain genotoxic treatments [200]. Thus, using specific targets on the activated tumor endothelium, it is possible to target gene therapy specifically to the tumor ECs. Much work is needed to show the efficacy of this novel vector system using different preclinical models.

16.2.3.8 Safety of Viral Vectors

A recent review of over 600 clinical trials using gene therapy with the enrollment of thousands of patients worldwide has been completed [190]. The successes and setbacks were reviewed [190]. The use of viral vectors, although very impressive, presents some challenges. The most important point to be considered in using viral vectors is to have them replication-deficient. The more genes that are removed from the virus, the more replication defective the vector will be, and there is less chance of recombination to generate the infectious parental virus.

Next the immune response of the host against adenoviral proteins is the major hurdle to the efficient and safe use of adenoviral vectors. Replication-defective Ad vectors are designed by replacing crucial adenoviral coding regions. In the first generation of Ad vectors, the E1 gene was replaced with the transgene. Although E1-deleted viruses are defective for replication, in some cell types they can produce virus proteins that serve as foreign antigens to induce a cellular immune response. More specifically, a recent study using E1/E4-deleted adenovirus for the correction of ornithine transcarbamylase deficiency regrettably caused the death of a nonterminal participant, due to a cytokine cascade, disseminated intravascular coagulation, acute respiratory distress, and multiorgan failure [201]. Many attempts have been made to reduce immunogenicity by engineering the second generation of Ad vectors that are additionally deleted in other viral transcription units, such as E2 and E4 [190]. Adenoviruses require helper viruses for propagation, generating a problem in the purification of a helper-free virus. An important step toward a third generation of Ad vectors is the development of high-capacity, helper-dependent vectors based on the Cre/*loxP*-system of site-specific DNA excision [202].

Major concerns in the use of retroviral vectors are the possibility of vector mobilization and recombination with defective (endogenous) retroviruses in the target cell. This prompted the development of self-inactivating vectors [203]. In these vectors LTR-driven transcription is prevented in transduced cells by deletion in the U3, and transgene expression is driven instead by an internal promoter, allowing the use of regulated and tissue specific promoters. For retrovirus infections, integration into the host genomes is a necessary step in the virus life cycle, which can sometimes cause life-threatening conditions. The first true successes of gene therapy in clinical trials were reported in 2000 and 2002, in which physicians successfully treated children suffering from SCIDs by retrovirally transducing hematopoietic stem cells with the gamma-c gene [104]. Unfortunately, 2 of the 10 children subsequently developed leukemia-like conditions, eventually attributed to retroviral vector integration in proximity to the LMO2 proto-oncogene promoter, leading to aberrant transcription and expression of LMO2 [204]. These cases garnered enormous attention from scientists, regulators, and the general public [205]. Reaction varied but some countries even imposed a general moratorium on trials involving retroviral gene transfer. Though in most countries trials have now been allowed to resume, the reaction has thrown the field of gene therapy into recession and has discouraged many scientists from starting new clinical trials [205].

Another problem using viral vectors to deliver a therapeutic gene is that introducing a large amount of virus into general circulation can cause widespread inflammation and organ damage or failure [201]. Thus, extensive preclinical evaluations are required to evaluate various safety-related issues. Tumor-targeted gene expression using viral vectors could help to decrease some of the viral-related toxicity.

16.2.4 Strategies to Achieve Tumor Targeting

The systemic delivery of a gene product can lead to nonspecific uptake by various different tissues and hence result in systemic toxicity because of transgene expression. Local delivery can circumvent this problem; however, many times the tumor is not accessible for local injection. Other considerations in using targeted gene delivery are, although VEGF is an important player in tumor angiogenesis, it also helps to maintain normal bone marrow microvascular endothelial function [206]. Therefore, care should be taken not to disrupt the physiological role played by this important angiogenic growth factor. Additionally, systemic administration of a synthetic analog of fumagillin can inhibit endometrial maturation and complete failure of embryonic growth in pregnant mice [207]. Thus antiangiogenic effects on normal functions can be avoided using targeted gene therapy. The pharmacological aspects of targeting cancer gene therapy to ECs have been reviewed previously [208].

Targeting tumor vasculature by gene therapy represents an ideal strategy, as tumor blood vessels are easily accessible to systemically applied vectors. The development of methods to achieve effective targeted gene expression and to enable appropriate regulation of gene expression may help to maximize therapeutic efficacy and minimize systemic toxicity for physiological angiogenesis. Targeted viral vectors can be constructed in several ways. The first approach is pseudotyping, in which one species of virus is made to incorporate the envelope protein of another virus. AAV genome with its inverted terminal repeats can be packaged in the capsids of different serotypes, which enables transduction with broad specificity [209]. The second approach is to genetically modify the viral capsid protein to incorporate a small peptide coding for a specific receptor and hence allow targeting by ligand–receptor internalization. Tissue specific targeting can also be achieved by conjugating specific antibodies to receptors onto the viral capsids [210].

The added advantages of systemic targeted gene therapy are: (a) it can prevent recurrence of distant metastases after surgery or radiotherapy; and (b) it can be used in combination with conventional chemotherapy, vaccine therapy, immunotherapy, or in combination with other types of gene therapy, for example, delivery of tumor suppressor genes [16]. However, efforts to influence antiangiogenic therapy by gene delivery have been hampered by a lack of targeting vectors specific for ECs in diseased tissues. The next section reviews the different modes of tissue targeting, vector availability, and specific targets on tumor endothelium.

16.2.4.1 Transcriptional Targeting

Transcriptional targeting can be achieved by the use of an expression cassette, which is switched on using TSPs [211]. The choice is either to use the gene promoter with a high expression in tumor cells/tumor ECs or to achieve such a level of expression by inducible promoters. Some of the promoters with high expression in tumors with minimal expression in the normal tissues are CXCR4, Cyclooxygenase-2 (COX-2), survivin (a novel member of the inhibitor of apoptosis protein family), and pre-proendothelin-1 (PPE-1), a precursor protein for endothelin-1. The human CXCR4 gene is expressed at high levels in many types of cancers, but is repressed in the liver. Thus, the CXCR4 promoter is shown to have a "tumor-on" and "liver-off" status in vitro and in vivo, and CXCR4 may prove to be a good candidate TSP for cancer gene therapy approaches for melanoma and breast cancers [212]. COX-2, a key enzyme in the synthesis of prostaglandins and thromboxanes, is highly upregulated in tumor cells, stromal cells, and angiogenic ECs during tumor progression [213]. Endothelial cell COX-2 promotes integrin $\alpha_v\beta_3$-mediated EC adhesion, spreading, migration, and angiogenesis [214]. COX-2 promoter can direct expression of caspases in COX-2-overexpressing cancer cells and induces apoptosis with no expression in normal cells [215]. Similarly, an adenovirus containing the survivin promoter showed high levels of reporter expression in tumor cells in contrast to normal cells [216].

PPE-1 is the precursor protein for endothelin-1, a potent vasoconstrictor and smooth muscle cell mitogen, and is synthesized by ECs. The murine PPE-1 promoter contains a hypoxia-responsive element that increases its expression under hypoxic conditions, such as those seen in tumors [217,218]. Greenberger et al. used a chimeric death receptor, composed of Fas and TNF receptor 1 under the control of PPE-1 promoter, which resulted in specific apoptosis of ECs in vitro and sensitization of cells to the proapoptotic effect of TNF-α [219]. Recently, a conditionally replicating adenovirus capable of targeting dividing ECs has been reported [220]. The adenovirus uses two promoters with the regulatory elements of VEGFR-2 and endoglin genes, which have been shown to be highly overexpressed in angiogenic ECs. However, the clinical utility of this virus needs to be tested.

Hallahan et al. used an adenovirus expressing TNF-α under the control of radiation-inducible Egr-1 promoter (Ad5.Egr-TNF) [221]. Combined treatment with Ad5.Egr-TNF and 5000 cGy (rad) resulted in increased intratumoral TNF-α production and increased tumor control, without an increase in normal tissue damage.

16.2.4.2 Transductional Targeting or Receptor Targeting

Transductional targeting involves the chemical or genetic modification of a vector, redirecting its tropism to a new target expressed preferentially on the target cell. This can be achieved either by direct targeting or indirect targeting. In direct targeting, the cell-specific targeting of the vector is mediated by a ligand that is directly inserted into the viral capsid [222]. In ligand-mediated targeting, some points to consider are there should be a good internalization site, the ligand should be structure independent, not too large to avoid the destabilization of the entire capsid, the ligand should be cell-type specific, and the ligand–receptor complex should be internalized in a way that allows an efficient transport of the virus and the release of the viral DNA in the cell nucleus [223]. The arginine–glycine–aspartic acid (RGD) containing peptide ligands targeted to integrin receptors expressed on activated ECs have been studied recently [224].

In indirect targeting the interaction between the viral vector and the target cell is mediated by an associated molecule (e.g., a glycoside molecule or a bispecific antibody), which is bound to the viral surface and interacts with a specific cell surface molecule [223,225]. Tumor cell targeting using single-chain or bispecific antibodies has been done. A bispecific antibody containing Fab arms of αIIbβ3 integrin and AAV capsid antibodies could target AAV to cells, which are not normally permissive for AAV infection [226]. Everts et al. demonstrated a drug targeting strategy for the selective delivery of dexamethasone to activated ECs, using an E-selectin-directed drug-Ab conjugate [227]. The dexamethasone-Ab conjugate did not bind to resting ECs.

Reynolds et al. used an adenoviral system combining transductional and transcriptional targeting for achieving cell-specific transgene expression in pulmonary endothelium [228]. The combination of transductional targeting to a pulmonary endothelial marker (angiotensin-converting enzyme, ACE) and an endothelial-specific promoter (VEGFR-1) resulted in a synergistic, 300,000-fold improvement in the selectivity of transgene expression for lung versus the usual site of vector sequestration, the liver. Thus the combined approach is much more significant than either approach alone.

16.2.4.3 Combination Therapy

Studies have shown advantages to combining antiangiogenic agents with each other, as well as with chemotherapy and radiation [229]. Combining agents that target a number of different and synergistic pathways may yield improved antitumor effects and long-term suppression of metastatic progression [230]. The antiangiogenic therapy can normalize tumor vasculature, improving drug delivery by pruning the immature and inefficient vessels [231]. For example, the combination of a VEGFR-2 tyrosine kinase inhibitor and low-dose endostatin reduced tumor growth more efficiently than monotherapy with either agent [232]. Whereas VEGFR-2 inhibitor specifically inhibits VEGF signaling, low-dose endostatin is able to inhibit a broader spectrum of diverse angiogenic pathways directly in the endothelium.

Recently, Kerbel and Kamen introduced a concept of metronomic chemotherapy. Metronomic chemotherapy represents a strategy wherein conventional cytotoxic agents are used at noncytotoxic doses on a frequent or continuous schedule, with no extended interruptions [233]. Three main purposes are served in this model. When delivered via the metronomic method, conventional agents can be: (1) antiangiogenic; (2) pro-apoptotic; and (3) immunostimulatory. Three conventional agents have recently been shown to have angiogenic inhibitory qualities, namely docetaxel and paclitaxel (*taxanes*) and vinflunine, a vinca alkaloid [234,235]. In preclinical models, metronomic chemotherapy can be effective in treating tumors in which cancer cells have developed resistance to the same chemotherapeutic drug. The efficacy can be further increased by combination with angiogenesis inhibitors, which can increase the levels of the endogenous inhibitors [233].

A new concept, treating cancer as a chronic manageable disease with angiogenesis inhibitors, is now emerging. Since tumor progression depends crucially on the balance between the in situ tumor's total angiogenic output and an individual's total angiogenic defense, a beneficial long-term balance may be achieved [104]. In order to have long-term tumor-free survival by using antiangiogenic therapy, the factors controlling tumor neovasculature need to be systemically maintained at stable therapeutic levels.

16.3 FUTURE CONSIDERATIONS AND RECOMMENDATIONS

Inhibition of tumor growth by depriving the tumor of its blood supply seems to be a very practical approach to cancer therapy. However, there are some pitfalls to this approach:

1. Long-term effects of angiogenesis inhibition on normal processes are still largely obscure.
2. It is not easy to extrapolate the knowledge generated in the laboratory to the clinical situation, where we face lot of heterogeneity. Tumors expressing the same proangiogenic growth factor like VEGF may respond differently to anti-VEGF therapy [91]. Past experiences with chemotherapy teach us that combination therapy works better than monotherapy. With few exceptions, the scheduling and timing of combination therapy has not been worked out for angiogenesis inhibitors.
3. Although we know that small tumors might give us more favorable outcome, most of the clinical trials are being done on previously untreated and bulky tumors.
4. One of the current challenges in the clinical application of therapy monitoring is to determine the time frame during therapy, which is most appropriate for response classification. Often a tumor shows initial response to most therapies, even if they are subsequently unsuccessful.
5. Future clinical trials are likely to build on past experience with stricter entry criteria, supportive care guidelines, and the use of surrogate markers. The format of a typical cancer study has developed over many years to evaluate cytotoxic chemotherapy regimens, where tumor regression is the expected outcome for an active agent. In contrast, antiangiogenic agents, like many other types of targeted drugs, are more likely to elicit a cytostatic effect, rather than tumor shrinkage. In addition, while cytotoxic agents are typically used at doses at or near the maximum tolerated dose (MTD), the optimal dose for targeted agents may be at doses far below the MTD. Evaluation of these agents will therefore require refinements in trial design to robustly detect "soft" end points such as stable disease, coupled with the development and validation of improved markers for assessing biological activity.
6. Genetic alterations, which decrease the vascular dependence of tumor cells, can influence the therapeutic response of tumors to therapy [91]. Tumor-targeted therapy with a single agent is more likely to develop resistance. The key strategy should involve either a combination of potent agents or a single agent, which targets more than one pathway, for example, a tyrosine kinase inhibitor, which can inhibit VEGFR-dependent tumor angiogenesis and EGFR-dependent tumor cell proliferation and survival [236].
7. Future research should identify new, more specific tumor endothelial markers and in parallel, efforts are required to identify novel effectors [237]. Angiogenesis-related promoters also need additional investigation because they raise the possibility of restricting gene expression more completely to sites of tumor angiogenesis.

8. Another consideration can be combining antiangiogenesis with vascular targeting to destroy new vessels with established tumor vasculature [238]. The two approaches may be useful in combination, for example, destroy the tumor vasculature using a vascular targeting agent to elicit tumor collapse and then inhibit regrowth by long-term administration of an antiangiogenic agent [14].
9. Repopulation of surviving tumor cells during cancer chemo- and radiation therapy is an important cause of treatment failure [239]. There is evidence that the repopulation of tumor cells limits the effectiveness of radiation therapy and that tumor-cell repopulation might accelerate during a course of radiotherapy [239]. Repopulation probably depends on the activation of signaling pathways that stimulate the proliferation of tumor cells, and many molecular-targeted agents have been developed that inhibit these pathways.
10. Mining of databases linking sequence and functional information plays an increasingly significant role in optimization of mammalian expression cassette design [240].

16.4 CONCLUSION

Studies of the processes involved in tumor angiogenesis have led to the identification of promising new targets and therapies for the treatment of human cancers. This is reflected in increasing number of antiangiogenic inhibitors entering the clinic. Tumor-targeted gene therapy could be used across many tumor types with tolerable systemic toxicity. As better vectors are developed, combination strategies continue to evolve, and an increased understanding of the complex role that endogenous angiogenesis inhibitors play in tumor growth and progression takes place; antiangiogenic gene therapy will certainly be evaluated in future clinical trials.

A concept that cancer might be treated as a chronic manageable disease in the future has recently been proposed [104]. This notion originally comes from the fact that many of us carry in situ tumors, which do not develop into invasive cancer. It has been proposed that it is an angiogenic switch, which converts in situ tumors into lethal cancer, and that this angiogenic switch depends on the balance between the in situ tumor's total angiogenic output and the body's total angiogenic defense generated by endogenous angiogenesis inhibitors. This notion was further emphasized based on the observation that Down syndrome patients, who have an elevated level of circulating endostatin due to an extra copy of the gene, rarely develop solid tumors and that a polymorphism in endostatin predisposes for the development of prostate cancer [241,242]. These observations suggest that modulation of the angiogenic defense could affect cancer progression in humans. Understanding the molecular pathways responsible for pro- and antiangiogenic responses will, therefore, not only benefit the field in the development of new anticancer therapies but also, possibly, in the development of cancer prevention strategies.

REFERENCES

1. Folkman, J. and Klagsbrun, M. Angiogenic factors. *Science, 235:* 442–447, 1987.
2. Folkman, J. What is the evidence that tumors are angiogenesis dependent? *J Natl Cancer Inst, 82:* 4–6, 1990.
3. Pepper, M.S. Role of the matrix metalloproteinase and plasminogen activator-plasmin systems in angiogenesis. *Arterioscler Thromb Vasc Biol, 21:* 1104–1117, 2001.
4. Hanahan, D.A. Flanking attack on cancer. *Nat Med, 4:* 13–14, 1998.
5. Benjamin, L.E. and Keshet, E. Conditional switching of vascular endothelial growth factor (VEGF) expression in tumors: induction of endothelial cell shedding and regression of hemangioblastoma-like vessels by VEGF withdrawal. *Proc Natl Acad Sci USA, 94:* 8761–8766, 1997.

6. Holmgren, L., O'Reilly, M.S., and Folkman, J. Dormancy of micrometastases: balanced proliferation and apoptosis in the presence of angiogenesis suppression. *Nat Med, 1:* 149–153, 1995.
7. Gimbrone, M.A., Jr., Leapman, S.B., Cotran, R.S., and Folkman, J. Tumor dormancy in vivo by prevention of neovascularization. *J Exp Med, 136:* 261–276, 1972.
8. Carmeliet, P. and Jain, R.K. Angiogenesis in cancer and other diseases. *Nature, 407:* 249–257, 2000.
9. Watnick, R.S., Cheng, Y.N., Rangarajan, A., Ince, T.A., and Weinberg, R.A. Ras modulates Myc activity to repress thrombospondin-1 expression and increase tumor angiogenesis. *Cancer Cell, 3:* 219–231, 2003.
10. Folkman, J. Angiogenesis inhibitors: a new class of drugs. *Cancer Biol Ther, 2:* S127–133, 2003.
11. Wary, K.K. Molecular targets for anti-angiogenic therapy. *Curr Opin Mol Ther, 6:* 54–70, 2004.
12. Costouros, N.G., Lorang, D., Zhang, Y., Miller, M.S., Diehn, F.E., Hewitt, S.M., Knopp, M.V., Li, K.C., Choyke, P.L., Alexander, H.R., and Libutti, S.K. Microarray gene expression analysis of murine tumor heterogeneity defined by dynamic contrast-enhanced MRI. *Mol Imaging, 1:* 301–308, 2002.
13. Zogakis, T.G., Costouros, N.G., Kruger, E.A., Forbes, S., He, M., Qian, M., Feldman, A.L., Figg, W.D., Alexander, H.R., Liu, E.T., Kohn, E.C., and Libutti, S.K. Microarray gene expression profiling of angiogenesis inhibitors using the rat aortic ring assay. *Biotechniques, 33:* 664–666, 668, 670, 2002.
14. Bicknell, R. The realisation of targeted antitumour therapy. *Br J Cancer, 92 Suppl 1:* S2–S5, 2005.
15. Kerbel, R. and Folkman, J. Clinical translation of angiogenesis inhibitors. *Nat Rev Cancer, 2:* 727–739, 2002.
16. Folkman, J. Antiangiogenic gene therapy. *Proc Natl Acad Sci USA, 95:* 9064–9066, 1998.
17. Folkman, J. Angiogenesis and apoptosis. *Semin Cancer Biol, 13:* 159–167, 2003.
18. O'Reilly, M.S., Boehm, T., Shing, Y., Fukai, N., Vasios, G., Lane, W.S., Flynn, E., Birkhead, J.R., Olsen, B.R., and Folkman, J. Endostatin: an endogenous inhibitor of angiogenesis and tumor growth. *Cell, 88:* 277–285, 1997.
19. Ferreras, M., Felbor, U., Lenhard, T., Olsen, B.R., and Delaisse, J. Generation and degradation of human endostatin proteins by various proteinases. *FEBS Lett, 486:* 247–251, 2000.
20. Dixelius, J., Cross, M.J., Matsumoto, T., and Claesson-Welsh, L. Endostatin action and intracellular signaling: beta-catenin as a potential target? *Cancer Lett, 196:* 1–12, 2003.
21. Thomas, J.P., Arzoomanian, R.Z., Alberti, D., Marnocha, R., Lee, F., Friedl, A., Tutsch, K., Dresen, A., Geiger, P., Pluda, J., Fogler, W., Schiller, J.H., and Wilding, G. Phase I pharmacokinetic and pharmacodynamic study of recombinant human endostatin in patients with advanced solid tumors. *J Clin Oncol, 21:* 223–231, 2003.
22. Abdollahi, A., Hahnfeldt, P., Maercker, C., Grone, H.J., Debus, J., Ansorge, W., Folkman, J., Hlatky, L., and Huber, P.E. Endostatin's antiangiogenic signaling network. *Mol Cell, 13:* 649–663, 2004.
23. Jimenez, B., Volpert, O.V., Crawford, S.E., Febbraio, M., Silverstein, R.L., and Bouck, N. Signals leading to apoptosis-dependent inhibition of neovascularization by thrombospondin-1. *Nat Med, 6:* 41–48, 2000.
24. Ruzinova, M.B., Schoer, R.A., Gerald, W., Egan, J.E., Pandolfi, P.P., Rafii, S., Manova, K., Mittal, V., and Benezra, R. Effect of angiogenesis inhibition by Id loss and the contribution of bone-marrow-derived endothelial cells in spontaneous murine tumors. *Cancer Cell, 4:* 277–289, 2003.
25. Mazzanti, C.M., Tandle, A., Lorang, D., Costouros, N., Roberts, D., Bevilacqua, G., and Libutti, S.K. Early genetic mechanisms underlying the inhibitory effects of endostatin and fumagillin on human endothelial cells. *Genome Res, 14:* 1585–1593, 2004.
26. Ingber, D., Fujita, T., Kishimoto, S., Sudo, K., Kanamaru, T., Brem, H., and Folkman, J. Synthetic analogues of fumagillin that inhibit angiogenesis and suppress tumour growth. *Nature, 348:* 555–557, 1990.
27. Yet, S.F., McA'Nulty, M.M., Folta, S.C., Yen, H.W., Yoshizumi, M., Hsieh, C.M., Layne, M.D., Chin, M.T., Wang, H., Perrella, M.A., Jain, M.K., and Lee, M.E. Human EZF, a Kruppel-like zinc finger protein, is expressed in vascular endothelial cells and contains transcriptional activation and repression domains. *J Biol Chem, 273:* 1026–1031, 1998.
28. Lyden, D., Young, A.Z., Zagzag, D., Yan, W., Gerald, W., O'Reilly, R., Bader, B.L., Hynes, R.O., Zhuang, Y., Manova, K., and Benezra, R. Id1 and Id3 are required for neurogenesis, angiogenesis and vascularization of tumour xenografts. *Nature, 401:* 670–677, 1999.

29. Mok, S.C., Wong, K.K., Chan, R.K., Lau, C.C., Tsao, S.W., Knapp, R.C., and Berkowitz, R.S. Molecular cloning of differentially expressed genes in human epithelial ovarian cancer. *Gynecol Oncol, 52:* 247–252, 1994.
30. Chua, E.L., Young, L., Wu, W.M., Turtle, J.R., and Dong, Q. Cloning of TC-1 (C8orf4), a novel gene found to be overexpressed in thyroid cancer. *Genomics, 69:* 342–347, 2000.
31. Tandle, A.T., Mazzanti, C., Alexander, H.R., Roberts, D.D., and Libutti, S.K. Endothelial monocyte activating polypeptide-II induced gene expression changes in endothelial cells. *Cytokine, 30:* 347–358, 2005.
32. Kao, J., Ryan, J., Brett, G., Chen, J., Shen, H., Fan, Y.G., Godman, G., Familletti, P.C., Wang, F., Pan, Y.C., et al. Endothelial monocyte-activating polypeptide II. A novel tumor-derived polypeptide that activates host-response mechanisms. *J Biol Chem, 267:* 20239–20247, 1992.
33. Kao, J., Houck, K., Fan, Y., Haehnel, I., Libutti, S.K., Kayton, M.L., Grikscheit, T., Chabot, J., Nowygrod, R., Greenberg, S., et al. Characterization of a novel tumor-derived cytokine. Endothelial-monocyte activating polypeptide II. *J Biol Chem, 269:* 25106–25119, 1994.
34. Berger, A.C., Alexander, H.R., Wu, P.C., Tang, G., Gnant, M.F., Mixon, A., Turner, E.S., and Libutti, S.K. Tumour necrosis factor receptor I (p55) is upregulated on endothelial cells by exposure to the tumour-derived cytokine endothelial monocyte-activating polypeptide II (EMAP-II). *Cytokine, 12:* 992–1000, 2000.
35. Schwarz, M.A., Kandel, J., Brett, J., Li, J., Hayward, J., Schwarz, R.E., Chappey, O., Wautier, J.L., Chabot, J., Lo Gerfo, P., and Stern, D. Endothelial-monocyte activating polypeptide II, a novel antitumor cytokine that suppresses primary and metastatic tumor growth and induces apoptosis in growing endothelial cells. *J Exp Med, 190:* 341–354, 1999.
36. Zhang, W., Chuang, Y.J., Swanson, R., Li, J., Seo, K., Leung, L., Lau, L.F., and Olson, S.T. Antiangiogenic antithrombin down-regulates the expression of the proangiogenic heparan sulfate proteoglycan, perlecan, in endothelial cells. *Blood, 103:* 1185–1191, 2004.
37. Iozzo, R.V. and San Antonio, J.D. Heparan sulfate proteoglycans: heavy hitters in the angiogenesis arena. *J Clin Invest, 108:* 349–355, 2001.
38. Zhang, W., Swanson, R., Izaguirre, G., Xiong, Y., Lau, L.F., and Olson, S.T. The heparin-binding site of antithrombin is crucial for antiangiogenic activity. *Blood, 106:* 1621–1628, 2005.
39. McLaughlin, J.N., Mazzoni, M.R., Cleator, J.H., Earls, L., Perdigoto, A.L., Brooks, J.D., Muldowney, J.A., 3rd, Vaughan, D.E., and Hamm, H.E. Thrombin modulates the expression of a set of genes including thrombospondin-1 in human microvascular endothelial cells. *J Biol Chem, 280:* 22172–22180, 2005.
40. Salvucci, O., Basik, M., Yao, L., Bianchi, R., and Tosato, G. Evidence for the involvement of SDF-1 and CXCR4 in the disruption of endothelial cell-branching morphogenesis and angiogenesis by TNF-alpha and IFN-gamma. *J Leukoc Biol, 76:* 217–226, 2004.
41. Salvucci, O., Yao, L., Villalba, S., Sajewicz, A., Pittaluga, S., and Tosato, G. Regulation of endothelial cell branching morphogenesis by endogenous chemokine stromal-derived factor-1. *Blood, 99:* 2703–2711, 2002.
42. Ferrari, N., Pfeffer, U., Dell'Eva, R., Ambrosini, C., Noonan, D.M., and Albini, A. The transforming growth factor-beta family members bone morphogenetic protein-2 and macrophage inhibitory cytokine-1 as mediators of the antiangiogenic activity of *N*-(4-hydroxyphenyl) retinamide. *Clin Cancer Res, 11:* 4610–4619, 2005.
43. Overgaard, J., Gonzalez Gonzalez, D., Hulshof, M.C., Arcangeli, G., Dahl, O., Mella, O., and Bentzen, S.M. Randomised trial of hyperthermia as adjuvant to radiotherapy for recurrent or metastatic malignant melanoma. European Society for Hyperthermic Oncology. *Lancet, 345:* 540–543, 1995.
44. Roca, C., Primo, L., Valdembri, D., Cividalli, A., Declerck, P., Carmeliet, P., Gabriele, P., and Bussolino, F. Hyperthermia inhibits angiogenesis by a plasminogen activator inhibitor 1-dependent mechanism. *Cancer Res, 63:* 1500–1507, 2003.
45. Cline, E.I., Bicciato, S., DiBello, C., and Lingen, M.W. Prediction of in vivo synergistic activity of antiangiogenic compounds by gene expression profiling. *Cancer Res, 62:* 7143–7148, 2002.
46. Westphal, J.R. Technology evaluation: ABT-510, Abbott. *Curr Opin Mol Ther, 6:* 451–457, 2004.
47. Reiher, F.K., Volpert, O.V., Jimenez, B., Crawford, S.E., Dinney, C.P., Henkin, J., Haviv, F., Bouck, N.P., and Campbell, S.C. Inhibition of tumor growth by systemic treatment with thrombospondin-1 peptide mimetics. *Int J Cancer, 98:* 682–689, 2002.

48. Kanno, S., Oda, N., Abe, M., Terai, Y., Ito, M., Shitara, K., Tabayashi, K., Shibuya, M., and Sato, Y. Roles of two VEGF receptors, Flt-1 and KDR, in the signal transduction of VEGF effects in human vascular endothelial cells. *Oncogene, 19:* 2138–2146, 2000.
49. Watanabe, K., Hasegawa, Y., Yamashita, H., Shimizu, K., Ding, Y., Abe, M., Ohta, H., Imagawa, K., Hojo, K., Maki, H., Sonoda, H., and Sato, Y. Vasohibin as an endothelium-derived negative feedback regulator of angiogenesis. *J Clin Invest, 114:* 898–907, 2004.
50. Minami, T., Horiuchi, K., Miura, M., Abid, M.R., Takabe, W., Noguchi, N., Kohro, T., Ge, X., Aburatani, H., Hamakubo, T., Kodama, T., and Aird, W.C. Vascular endothelial growth factor- and thrombin-induced termination factor, Down syndrome critical region-1, attenuates endothelial cell proliferation and angiogenesis. *J Biol Chem, 279:* 50537–50554, 2004.
51. Klee, C.B., Ren, H., and Wang, X. Regulation of the calmodulin-stimulated protein phosphatase, calcineurin. *J Biol Chem, 273:* 13367–13370, 1998.
52. Botas, J. Control of morphogenesis and differentiation by HOM/Hox genes. *Curr Opin Cell Biol, 5:* 1015–1022, 1993.
53. Boudreau, N. and Bissell, M.J. Extracellular matrix signaling: integration of form and function in normal and malignant cells. *Curr Opin Cell Biol, 10:* 640–646, 1998.
54. Patel, C.V., Sharangpani, R., Bandyopadhyay, S., and DiCorleto, P.E. Endothelial cells express a novel, tumor necrosis factor-alpha-regulated variant of HOXA9. *J Biol Chem, 274:* 1415–1422, 1999.
55. Boudreau, N., Andrews, C., Srebrow, A., Ravanpay, A., and Cheresh, D.A. Induction of the angiogenic phenotype by Hox D3. *J Cell Biol, 139:* 257–264, 1997.
56. Myers, C., Charboneau, A., and Boudreau, N. Homeobox B3 promotes capillary morphogenesis and angiogenesis. *J Cell Biol, 148:* 343–351, 2000.
57. Myers, C., Charboneau, A., Cheung, I., Hanks, D., and Boudreau, N. Sustained expression of homeobox D10 inhibits angiogenesis. *Am J Pathol, 161:* 2099–2109, 2002.
58. Martinez Barbera, J.P., Clements, M., Thomas, P., Rodriguez, T., Meloy, D., Kioussis, D., and Beddington, R.S. The homeobox gene Hex is required in definitive endodermal tissues for normal forebrain, liver and thyroid formation. *Development, 127:* 2433–2445, 2000.
59. Liao, W., Ho, C.Y., Yan, Y.L., Postlethwait, J., and Stainier, D.Y. Hhex and scl function in parallel to regulate early endothelial and blood differentiation in zebrafish. *Development, 127:* 4303–4313, 2000.
60. Nakagawa, T., Abe, M., Yamazaki, T., Miyashita, H., Niwa, H., Kokubun, S., and Sato, Y. HEX acts as a negative regulator of angiogenesis by modulating the expression of angiogenesis-related gene in endothelial cells in vitro. *Arterioscler Thromb Vasc Biol, 23:* 231–237, 2003.
61. Pugh, C.W. and Ratcliffe, P.J. The von Hippel-Lindau tumor suppressor, hypoxia-inducible factor-1 (HIF-1) degradation, and cancer pathogenesis. *Semin Cancer Biol, 13:* 83–89, 2003.
62. Manalo, D.J., Rowan, A., Lavoie, T., Natarajan, L., Kelly, B.D., Ye, S.Q., Garcia, J.G., and Semenza, G.L. Transcriptional regulation of vascular endothelial cell responses to hypoxia by HIF-1. *Blood, 105:* 659–669, 2005.
63. Kelly, B.D., Hackett, S.F., Hirota, K., Oshima, Y., Cai, Z., Berg-Dixon, S., Rowan, A., Yan, Z., Campochiaro, P.A., and Semenza, G.L. Cell type-specific regulation of angiogenic growth factor gene expression and induction of angiogenesis in nonischemic tissue by a constitutively active form of hypoxia-inducible factor 1. *Circ Res, 93:* 1074–1081, 2003.
64. Lelievre, E., Lionneton, F., Soncin, F., and Vandenbunder, B. The ETS family contains transcriptional activators and repressors involved in angiogenesis. *Int J Biochem Cell Biol, 33:* 391–407, 2001.
65. Sato, Y., Teruyama, K., Nakano, T., Oda, N., Abe, M., Tanaka, K., and Iwasaka-Yagi, C. Role of transcription factors in angiogenesis: Ets-1 promotes angiogenesis as well as endothelial apoptosis. *Ann NY Acad Sci, 947:* 117–123, 2001.
66. Delannoy-Courdent, A., Mattot, V., Fafeur, V., Fauquette, W., Pollet, I., Calmels, T., Vercamer, C., Boilly, B., Vandenbunder, B., and Desbiens, X. The expression of an Ets1 transcription factor lacking its activation domain decreases uPA proteolytic activity and cell motility, and impairs normal tubulogenesis and cancerous scattering in mammary epithelial cells. *J Cell Sci, 111 (Pt 11)*: 1521–1534, 1998.

67. Nakano, T., Abe, M., Tanaka, K., Shineha, R., Satomi, S., and Sato, Y. Angiogenesis inhibition by transdominant mutant Ets-1. *J Cell Physiol, 184:* 255–262, 2000.
68. Yordy, J.S., Li, R., Sementchenko, V.I., Pei, H., Muise-Helmericks, R.C., and Watson, D.K. SP100 expression modulates ETS1 transcriptional activity and inhibits cell invasion. *Oncogene, 23:* 6654–6665, 2004.
69. Yordy, J.S., Moussa, O., Pei, H., Chaussabel, D., Li, R., and Watson, D.K. SP100 inhibits ETS1 activity in primary endothelial cells. *Oncogene, 24:* 916–931, 2005.
70. Bussolati, B., Deambrosis, I., Russo, S., Deregibus, M.C., and Camussi, G. Altered angiogenesis and survival in human tumor-derived endothelial cells. *Faseb J, 17:* 1159–1161, 2003.
71. Liotta, L.A. and Kohn, E.C. The microenvironment of the tumour–host interface. *Nature, 411:* 375–379, 2001.
72. Feldman, A.L., Stetler-Stevenson, W.G., Costouros, N.G., Knezevic, V., Baibakov, G., Alexander, H.R., Jr., Lorang, D., Hewitt, S.M., Seo, D.W., Miller, M.S., O'Connor, S., and Libutti, S.K. Modulation of tumor–host interactions, angiogenesis, and tumor growth by tissue inhibitor of metalloproteinase 2 via a novel mechanism. *Cancer Res, 64:* 4481–4486, 2004.
73. Stetler-Stevenson, W.G., Krutzsch, H.C., and Liotta, L.A. TIMP-2: identification and characterization of a new member of the metalloproteinase inhibitor family. *Matrix Suppl, 1:* 299–306, 1992.
74. Seo, D.W., Li, H., Guedez, L., Wingfield, P.T., Diaz, T., Salloum, R., Wei, B.Y., and Stetler-Stevenson, W.G. TIMP-2 mediated inhibition of angiogenesis: an MMP-independent mechanism. *Cell, 114:* 171–180, 2003.
75. Keyse, S.M. Protein phosphatases and the regulation of MAP kinase activity. *Semin Cell Dev Biol, 9:* 143–152, 1998.
76. Keyse, S.M. Protein phosphatases and the regulation of mitogen-activated protein kinase signalling. *Curr Opin Cell Biol, 12:* 186–192, 2000.
77. Berra, E., Pages, G., and Pouyssegur, J. MAP kinases and hypoxia in the control of VEGF expression. *Cancer Metastasis Rev, 19:* 139–145, 2000.
78. Pages, G., Milanini, J., Richard, D.E., Berra, E., Gothie, E., Vinals, F., and Pouyssegur, J. Signaling angiogenesis via p42/p44 MAP kinase cascade. *Ann NY Acad Sci, 902:* 187–200, 2000.
79. Rousseau, S., Houle, F., Landry, J., and Huot, J. p38 MAP kinase activation by vascular endothelial growth factor mediates actin reorganization and cell migration in human endothelial cells. *Oncogene, 15:* 2169–2177, 1997.
80. Tanaka, K., Abe, M., and Sato, Y. Roles of extracellular signal-regulated kinase 1/2 and p38 mitogen-activated protein kinase in the signal transduction of basic fibroblast growth factor in endothelial cells during angiogenesis. *Jpn J Cancer Res, 90:* 647–654, 1999.
81. Kiemer, A.K., Weber, N.C., Furst, R., Bildner, N., Kulhanek-Heinze, S., and Vollmar, A.M. Inhibition of p38 MAPK activation via induction of MKP-1: atrial natriuretic peptide reduces TNF-alpha-induced actin polymerization and endothelial permeability. *Circ Res, 90:* 874–881, 2002.
82. Englert, C.R., Baibakov, G.V., and Emmert-Buck, M.R. Layered expression scanning: rapid molecular profiling of tumor samples. *Cancer Res, 60:* 1526–1530, 2000.
83. Charles, C.H., Sun, H., Lau, L.F., and Tonks, N.K. The growth factor-inducible immediate-early gene 3CH134 encodes a protein-tyrosine-phosphatase. *Proc Natl Acad Sci USA, 90:* 5292–5296, 1993.
84. Lin, W.W. and Hsu, Y.W. Cycloheximide-induced cPLA(2) activation is via the MKP-1 down-regulation and ERK activation. *Cell Signal, 12:* 457–461, 2000.
85. Ruzinova, M.B. and Benezra, R. Id proteins in development, cell cycle and cancer. *Trends Cell Biol, 13:* 410–418, 2003.
86. Di Cristofano, A., Kotsi, P., Peng, Y.F., Cordon-Cardo, C., Elkon, K.B., and Pandolfi, P.P. Impaired Fas response and autoimmunity in Pten +/− mice. *Science, 285:* 2122–2125, 1999.
87. Di Cristofano, A., De Acetis, M., Koff, A., Cordon-Cardo, C., and Pandolfi, P.P. Pten and p27KIP1 cooperate in prostate cancer tumor suppression in the mouse. *Nat Genet, 27:* 222–224, 2001.
88. Gazit, D., Ebner, R., Kahn, A.J., and Derynck, R. Modulation of expression and cell surface binding of members of the transforming growth factor-beta superfamily during retinoic acid-induced osteoblastic differentiation of multipotential mesenchymal cells. *Mol Endocrinol, 7:* 189–198, 1993.

89. Kale, S., Hanai, J., Chan, B., Karihaloo, A., Grotendorst, G., Cantley, L., and Sukhatme, V.P. Microarray analysis of in vitro pericyte differentiation reveals an angiogenic program of gene expression. *Faseb J, 19:* 270–271, 2005.
90. Davidoff, A.M. and Nathwani, A.C. Antiangiogenic gene therapy for cancer treatment. *Curr Hematol Rep, 3:* 267–273, 2004.
91. Glade-Bender, J., Kandel, J.J., and Yamashiro, D.J. VEGF blocking therapy in the treatment of cancer. *Expert Opin Biol Ther, 3:* 263–276, 2003.
92. Hurwitz, H., Fehrenbacher, L., Novotny, W., Cartwright, T., Hainsworth, J., Heim, W., Berlin, J., Baron, A., Griffing, S., Holmgren, E., Ferrara, N., Fyfe, G., Rogers, B., Ross, R., and Kabbinavar, F. Bevacizumab plus irinotecan, fluorouracil, and leucovorin for metastatic colorectal cancer. *N Engl J Med, 350:* 2335–2342, 2004.
93. Kabbinavar, F., Hurwitz, H.I., Fehrenbacher, L., Meropol, N.J., Novotny, W.F., Lieberman, G., Griffing, S., and Bergsland, E. Phase II, randomized trial comparing bevacizumab plus fluorouracil (FU)/leucovorin (LV) with FU/LV alone in patients with metastatic colorectal cancer. *J Clin Oncol, 21:* 60–65, 2003.
94. Witte, L., Hicklin, D.J., Zhu, Z., Pytowski, B., Kotanides, H., Rockwell, P., and Bohlen, P. Monoclonal antibodies targeting the VEGF receptor-2 (Flk1/KDR) as an anti-angiogenic therapeutic strategy. *Cancer Metastasis Rev, 17:* 155–161, 1998.
95. Prewett, M., Huber, J., Li, Y., Santiago, A., O'Connor, W., King, K., Overholser, J., Hooper, A., Pytowski, B., Witte, L., Bohlen, P., and Hicklin, D.J. Antivascular endothelial growth factor receptor (fetal liver kinase 1) monoclonal antibody inhibits tumor angiogenesis and growth of several mouse and human tumors. *Cancer Res, 59:* 5209–5218, 1999.
96. Brantley, D.M., Cheng, N., Thompson, E.J., Lin, Q., Brekken, R.A., Thorpe, P.E., Muraoka, R.S., Cerretti, D.P., Pozzi, A., Jackson, D., Lin, C., and Chen, J. Soluble Eph A receptors inhibit tumor angiogenesis and progression in vivo. *Oncogene, 21:* 7011–7026, 2002.
97. Bicknell, R. and Harris, A.L. Novel angiogenic signaling pathways and vascular targets. *Annu Rev Pharmacol Toxicol, 44:* 219–238, 2004.
98. Pandey, A., Shao, H., Marks, R.M., Polverini, P.J., and Dixit, V.M. Role of B61, the ligand for the Eck receptor tyrosine kinase, in TNF-alpha-induced angiogenesis. *Science, 268:* 567–569, 1995.
99. Martiny-Baron, G., Korff, T., Schaffner, F., Esser, N., Eggstein, S., Marme, D., and Augustin, H.G. Inhibition of tumor growth and angiogenesis by soluble EphB4. *Neoplasia, 6:* 248–257, 2004.
100. Cheng, N., Brantley, D.M., Liu, H., Lin, Q., Enriquez, M., Gale, N., Yancopoulos, G., Cerretti, D.P., Daniel, T.O., and Chen, J. Blockade of EphA receptor tyrosine kinase activation inhibits vascular endothelial cell growth factor-induced angiogenesis. *Mol Cancer Res, 1:* 2–11, 2002.
101. Kerbel, R.S., Yu, J., Tran, J., Man, S., Viloria-Petit, A., Klement, G., Coomber, B.L., and Rak, J. Possible mechanisms of acquired resistance to anti-angiogenic drugs: implications for the use of combination therapy approaches. *Cancer Metastasis Rev, 20:* 79–86, 2001.
102. Kisker, O., Becker, C.M., Prox, D., Fannon, M., D'Amato, R., Flynn, E., Fogler, W.E., Sim, B.K., Allred, E.N., Pirie-Shepherd, S.R., and Folkman, J. Continuous administration of endostatin by intraperitoneally implanted osmotic pump improves the efficacy and potency of therapy in a mouse xenograft tumor model. *Cancer Res, 61:* 7669–7674, 2001.
103. Bergers, G. and Benjamin, L.E. Tumorigenesis and the angiogenic switch. *Nat Rev Cancer, 3:* 401–410, 2003.
104. Folkman, J. and Kalluri, R. Cancer without disease. *Nature, 427:* 787, 2004.
105. Marx, J. Angiogenesis. A boost for tumor starvation. *Science, 301:* 452–454, 2003.
106. Liau, G., Su, E.J., and Dixon, K.D. Clinical efforts to modulate angiogenesis in the adult: gene therapy versus conventional approaches. *Drug Discov Today, 6:* 689–697, 2001.
107. Stephens, A.C. and Rivers, R.P. Antisense oligonucleotide therapy in cancer. *Curr Opin Mol Ther, 5:* 118–122, 2003.
108. Marchand, G.S., Noiseux, N., Tanguay, J.F., and Sirois, M.G. Blockade of in vivo VEGF-mediated angiogenesis by antisense gene therapy: role of Flk-1 and Flt-1 receptors. *Am J Physiol Heart Circ Physiol, 282:* H194–204, 2002.
109. Shim, W.S., Teh, M., Mack, P.O., and Ge, R. Inhibition of angiopoietin-1 expression in tumor cells by an antisense RNA approach inhibited xenograft tumor growth in immunodeficient mice. *Int J Cancer, 94:* 6–15, 2001.

110. Gu, Z.P., Wang, Y.J., Li, J.G., and Zhou, Y.A. VEGF165 antisense RNA suppresses oncogenic properties of human esophageal squamous cell carcinoma. *World J Gastroenterol, 8:* 44–48, 2002.
111. Shi, W. and Siemann, D.W. Simultaneous targeting of VEGF message and VEGF receptor signaling as a therapeutic anticancer approach. *Anticancer Res, 24:* 213–218, 2004.
112. Dallabrida, S.M., De Sousa, M.A., and Farrell, D.H. Expression of antisense to integrin subunit beta 3 inhibits microvascular endothelial cell capillary tube formation in fibrin. *J Biol Chem, 275:* 32281–32288, 2000.
113. Lipscomb, E.A., Dugan, A.S., Rabinovitz, I., and Mercurio, A.M. Use of RNA interference to inhibit integrin (alpha6beta4)-mediated invasion and migration of breast carcinoma cells. *Clin Exp Metastasis, 20:* 569–576, 2003.
114. Yang, G., Cai, K.Q., Thompson-Lanza, J.A., Bast, R.C., Jr., and Liu, J. Inhibition of breast and ovarian tumor growth through multiple signaling pathways by using retrovirus-mediated small interfering RNA against Her-2/neu gene expression. *J Biol Chem, 279:* 4339–4345, 2004.
115. Julian, M., Cacho, M., Garcia, M.A., Martin-Santamaria, S., de Pascual-Teresa, B., Ramos, A., Martinez, A., and Cuttitta, F. Adrenomedullin: a new target for the design of small molecule modulators with promising pharmacological activities. *Eur J Med Chem*, 2005.
116. Frede, S., Freitag, P., Otto, T., Heilmaier, C., and Fandrey, J. The proinflammatory cytokine interleukin 1beta and hypoxia cooperatively induce the expression of adrenomedullin in ovarian carcinoma cells through hypoxia inducible factor 1 activation. *Cancer Res, 65:* 4690–4697, 2005.
117. Owens, G.K., Kumar, M.S., and Wamhoff, B.R. Molecular regulation of vascular smooth muscle cell differentiation in development and disease. *Physiol Rev, 84:* 767–801, 2004.
118. Liu, Y., Sinha, S., McDonald, O.G., Shang, Y., Hoofnagle, M.H., and Owens, G.K. Kruppel-like factor 4 abrogates myocardin-induced activation of smooth muscle gene expression. *J Biol Chem, 280:* 9719–9727, 2005.
119. Devroe, E. and Silver, P.A. Therapeutic potential of retroviral RNAi vectors. *Expert Opin Biol Ther, 4:* 319–327, 2004.
120. Brummelkamp, T.R., Bernards, R., and Agami, R. A system for stable expression of short interfering RNAs in mammalian cells. *Science, 296:* 550–553, 2002.
121. Zhang, L., Yang, N., Mohamed-Hadley, A., Rubin, S.C., and Coukos, G. Vector-based RNAi, a novel tool for isoform-specific knock-down of VEGF and anti-angiogenesis gene therapy of cancer. *Biochem Biophys Res Commun, 303:* 1169–1178, 2003.
122. Rubinson, D.A., Dillon, C.P., Kwiatkowski, A.V., Sievers, C., Yang, L., Kopinja, J., Rooney, D.L., Ihrig, M.M., McManus, M.T., Gertler, F.B., Scott, M.L., and Van Parijs, L. A lentivirus-based system to functionally silence genes in primary mammalian cells, stem cells and transgenic mice by RNA interference. *Nat Genet, 33:* 401–406, 2003.
123. Hemann, M.T., Fridman, J.S., Zilfou, J.T., Hernando, E., Paddison, P.J., Cordon-Cardo, C., Hannon, G.J., and Lowe, S.W. An epi-allelic series of p53 hypomorphs created by stable RNAi produces distinct tumor phenotypes in vivo. *Nat Genet, 33:* 396–400, 2003.
124. Hasuwa, H., Kaseda, K., Einarsdottir, T., and Okabe, M. Small interfering RNA and gene silencing in transgenic mice and rats. *FEBS Lett, 532:* 227–230, 2002.
125. Caplen, N.J., Parrish, S., Imani, F., Fire, A., and Morgan, R.A. Specific inhibition of gene expression by small double-stranded RNAs in invertebrate and vertebrate systems. *Proc Natl Acad Sci USA, 98:* 9742–9747, 2001.
126. Szary, J. and Szala, S. Intra-tumoral administration of naked plasmid DNA encoding mouse endostatin inhibits renal carcinoma growth. *Int J Cancer, 91:* 835–839, 2001.
127. Oga, M., Takenaga, K., Sato, Y., Nakajima, H., Koshikawa, N., Osato, K., and Sakiyama, S. Inhibition of metastatic brain tumor growth by intramuscular administration of the endostatin gene. *Int J Oncol, 23:* 73–79, 2003.
128. Blezinger, P., Wang, J., Gondo, M., Quezada, A., Mehrens, D., French, M., Singhal, A., Sullivan, S., Rolland, A., Ralston, R., and Min, W. Systemic inhibition of tumor growth and tumor metastases by intramuscular administration of the endostatin gene. *Nat Biotechnol, 17:* 343–348, 1999.
129. Pavlenko, M., Leder, C., and Pisa, P. Plasmid DNA vaccines against cancer: cytotoxic T-lymphocyte induction against tumor antigens. *Expert Rev Vaccines, 4:* 315–327, 2005.
130. Restifo, N.P. and Rosenberg, S.A. Use of standard criteria for assessment of cancer vaccines. *Lancet Oncol, 6:* 3–4, 2005.

131. Fonteneau, J.F., Le Drean, E., Le Guiner, S., Gervois, N., Diez, E., and Jotereau, F. Heterogeneity of biologic responses of melanoma-specific CTL. *J Immunol, 159:* 2831–2839, 1997.
132. Wada, S., Tsunoda, T., Baba, T., Primus, F.J., Kuwano, H., Shibuya, M., and Tahara, H. Rationale for antiangiogenic cancer therapy with vaccination using epitope peptides derived from human vascular endothelial growth factor receptor 2. *Cancer Res, 65:* 4939–4946, 2005.
133. Niethammer, A.G., Xiang, R., Becker, J.C., Wodrich, H., Pertl, U., Karsten, G., Eliceiri, B.P., and Reisfeld, R.A. A DNA vaccine against VEGF receptor 2 prevents effective angiogenesis and inhibits tumor growth. *Nat Med, 8:* 1369–1375, 2002.
134. Li, Y., Wang, M.N., Li, H., King, K.D., Bassi, R., Sun, H., Santiago, A., Hooper, A.T., Bohlen, P., and Hicklin, D.J. Active immunization against the vascular endothelial growth factor receptor flk1 inhibits tumor angiogenesis and metastasis. *J Exp Med, 195:* 1575–1584, 2002.
135. Reisfeld, R.A., Niethammer, A.G., Luo, Y., and Xiang, R. DNA vaccines suppress tumor growth and metastases by the induction of anti-angiogenesis. *Immunol Rev, 199:* 181–190, 2004.
136. Barron, L.G., Uyechi, L.S., and Szoka, F.C., Jr. Cationic lipids are essential for gene delivery mediated by intravenous administration of lipoplexes. *Gene Ther, 6:* 1179–1183, 1999.
137. Templeton, N.S. Cationic liposome-mediated gene delivery in vivo. Biosci Rep, *22:* 283–295, 2002.
138. Templeton, N.S., Lasic, D.D., Frederik, P.M., Strey, H.H., Roberts, D.D., and Pavlakis, G.N. Improved DNA: liposome complexes for increased systemic delivery and gene expression. *Nat Biotechnol, 15:* 647–652, 1997.
139. Marty, C., Odermatt, B., Schott, H., Neri, D., Ballmer-Hofer, K., Klemenz, R., and Schwendener, R.A. Cytotoxic targeting of F9 teratocarcinoma tumours with anti-ED-B fibronectin scFv antibody modified liposomes. *Br J Cancer, 87:* 106–112, 2002.
140. Rodolfo, M., Cato, E.M., Soldati, S., Ceruti, R., Asioli, M., Scanziani, E., Vezzoni, P., Parmiani, G., and Sacco, M.G. Growth of human melanoma xenografts is suppressed by systemic angiostatin gene therapy. *Cancer Gene Ther, 8:* 491–496, 2001.
141. Chen, Q.R., Kumar, D., Stass, S.A., and Mixson, A.J. Liposomes complexed to plasmids encoding angiostatin and endostatin inhibit breast cancer in nude mice. *Cancer Res, 59:* 3308–3312, 1999.
142. Thurston, G., McLean, J.W., Rizen, M., Baluk, P., Haskell, A., Murphy, T.J., Hanahan, D., and McDonald, D.M. Cationic liposomes target angiogenic endothelial cells in tumors and chronic inflammation in mice. *J Clin Invest, 101:* 1401–1413, 1998.
143. Strieth, S., Eichhorn, M.E., Sauer, B., Schulze, B., Teifel, M., Michaelis, U., and Dellian, M. Neovascular targeting chemotherapy: encapsulation of paclitaxel in cationic liposomes impairs functional tumor microvasculature. *Int J Cancer, 110:* 117–124, 2004.
144. Schmitt-Sody, M., Strieth, S., Krasnici, S., Sauer, B., Schulze, B., Teifel, M., Michaelis, U., Naujoks, K., and Dellian, M. Neovascular targeting therapy: paclitaxel encapsulated in cationic liposomes improves antitumoral efficacy. *Clin Cancer Res, 9:* 2335–2341, 2003.
145. Blezinger, P., Yin, G., Xie, L., Wang, J., Matar, M., Bishop, J.S., and Min, W. Intravenous delivery of an endostatin gene complexed in cationic lipid inhibits systemic angiogenesis and tumor growth in murine models. *Angiogenesis, 3:* 205–210, 1999.
146. Kondo, M., Asai, T., Katanasaka, Y., Sadzuka, Y., Tsukada, H., Ogino, K., Taki, T., Baba, K., and Oku, N. Anti-neovascular therapy by liposomal drug targeted to membrane type-1 matrix metalloproteinase. *Int J Cancer, 108:* 301–306, 2004.
147. Butler, G.S., Butler, M.J., Atkinson, S.J., Will, H., Tamura, T., van Westrum, S.S., Crabbe, T., Clements, J., d'Ortho, M.P., and Murphy, G. The TIMP2 membrane type 1 metalloproteinase "receptor" regulates the concentration and efficient activation of progelatinase A. A kinetic study. *J Biol Chem, 273:* 871–880, 1998.
148. Janssen, A.P., Schiffelers, R.M., ten Hagen, T.L., Koning, G.A., Schraa, A.J., Kok, R.J., Storm, G., and Molema, G. Peptide-targeted PEG-liposomes in anti-angiogenic therapy. *Int J Pharm, 254:* 55–58, 2003.
149. Pastorino, F., Brignole, C., Marimpietri, D., Cilli, M., Gambini, C., Ribatti, D., Longhi, R., Allen, T.M., Corti, A., and Ponzoni, M. Vascular damage and anti-angiogenic effects of tumor vessel-targeted liposomal chemotherapy. *Cancer Res, 63:* 7400–7409, 2003.
150. Panyam, J. and Labhasetwar, V. Biodegradable nanoparticles for drug and gene delivery to cells and tissue. *Adv Drug Deliv Rev, 55:* 329–347, 2003.
151. Moghimi, S.M., Hunter, A.C., and Murray, J.C. Long-circulating and target-specific nanoparticles: theory to practice. *Pharmacol Rev, 53:* 283–318, 2001.

152. Luo, D., Han, E., Belcheva, N., and Saltzman, W.M. A self-assembled, modular DNA delivery system mediated by silica nanoparticles. *J Control Release*, *95:* 333–341, 2004.
153. Hood, J.D., Bednarski, M., Frausto, R., Guccione, S., Reisfeld, R.A., Xiang, R., and Cheresh, D.A. Tumor regression by targeted gene delivery to the neovasculature. *Science*, *296:* 2404–2407, 2002.
154. Li, L., Wartchow, C.A., Danthi, S.N., Shen, Z., Dechene, N., Pease, J., Choi, H.S., Doede, T., Chu, P., Ning, S., Lee, D.Y., Bednarski, M.D., and Knox, S.J. A novel antiangiogenesis therapy using an integrin antagonist or anti-Flk-1 antibody coated 90Y-labeled nanoparticles. *Int J Radiat Oncol Biol Phys*, *58:* 1215–1227, 2004.
155. Chan, W.C., Maxwell, D.J., Gao, X., Bailey, R.E., Han, M., and Nie, S. Luminescent quantum-dots for multiplexed biological detection and imaging. *Curr Opin Biotechnol*, *13:* 40–46, 2002.
156. Porkka, K., Laakkonen, P., Hoffman, J.A., Bernasconi, M., and Ruoslahti, E. A fragment of the HMGN2 protein homes to the nuclei of tumor cells and tumor endothelial cells in vivo. *Proc Natl Acad Sci USA*, *99:* 7444–7449, 2002.
157. Laakkonen, P., Porkka, K., Hoffman, J.A., and Ruoslahti, E. A tumor-homing peptide with a targeting specificity related to lymphatic vessels. *Nat Med*, *8:* 751–755, 2002.
158. Akerman, M.E., Chan, W.C., Laakkonen, P., Bhatia, S.N., and Ruoslahti, E. Nanocrystal targeting in vivo. *Proc Natl Acad Sci USA*, *99:* 12617–12621, 2002.
159. Sengupta, S., Eavarone, D., Capila, I., Zhao, G., Watson, N., Kiziltepe, T., and Sasisekharan, R. Temporal targeting of tumour cells and neovasculature with a nanoscale delivery system. *Nature*, *436:* 568–572, 2005.
160. Mooney, D. Cancer: one step at a time. *Nature*, *436:* 468–469, 2005.
161. Thomas, C.E., Ehrhardt, A., and Kay, M.A. Progress and problems with the use of viral vectors for gene therapy. *Nat Rev Genet*, *4:* 346–358, 2003.
162. Tandle, A., Blazer, D.G., 3rd, and Libutti, S.K. Antiangiogenic gene therapy of cancer: recent developments. *J Transl Med*, *2:* 22, 2004.
163. Wickham, T.J., Mathias, P., Cheresh, D.A., and Nemerow, G.R. Integrins alpha v beta 3 and alpha v beta 5 promote adenovirus internalization but not virus attachment. *Cell*, *73:* 309–319, 1993.
164. Feldman, A.L., Restifo, N.P., Alexander, H.R., Bartlett, D.L., Hwu, P., Seth, P., and Libutti, S.K. Antiangiogenic gene therapy of cancer utilizing a recombinant adenovirus to elevate systemic endostatin levels in mice. *Cancer Res*, *60:* 1503–1506, 2000.
165. Green, J.E., Shibata, M.A., Yoshidome, K., Liu, M.L., Jorcyk, C., Anver, M.R., Wigginton, J., Wiltrout, R., Shibata, E., Kaczmarczyk, S., Wang, W., Liu, Z.Y., Calvo, A., and Couldrey, C. The C3(1)/SV40 T-antigen transgenic mouse model of mammary cancer: ductal epithelial cell targeting with multistage progression to carcinoma. *Oncogene*, *19:* 1020–1027, 2000.
166. Calvo, A., Feldman, A.L., Libutti, S.K., and Green, J.E. Adenovirus-mediated endostatin delivery results in inhibition of mammary gland tumor growth in C3(1)/SV40 T-antigen transgenic mice. *Cancer Res*, *62:* 3934–3938, 2002.
167. Senzer, N., Mani, S., Rosemurgy, A., Nemunaitis, J., Cunningham, C., Guha, C., Bayol, N., Gillen, M., Chu, K., Rasmussen, C., Rasmussen, H., Kufe, D., Weichselbaum, R., and Hanna, N. TNFerade biologic, an adenovector with a radiation-inducible promoter, carrying the human tumor necrosis factor alpha gene: a phase I study in patients with solid tumors. *J Clin Oncol*, *22:* 592–601, 2004.
168. Mundt, A.J., Vijayakumar, S., Nemunaitis, J., Sandler, A., Schwartz, H., Hanna, N., Peabody, T., Senzer, N., Chu, K., Rasmussen, C.S., Kessler, P.D., Rasmussen, H.S., Warso, M., Kufe, D.W., Gupta, T.D., and Weichselbaum, R.R. A Phase I trial of TNFerade biologic in patients with soft tissue sarcoma in the extremities. *Clin Cancer Res*, *10:* 5747–5753, 2004.
169. Reynolds, P.N., Zinn, K.R., Gavrilyuk, V.D., Balyasnikova, I.V., Rogers, B.E., Buchsbaum, D.J., Wang, M.H., Miletich, D.J., Grizzle, W.E., Douglas, J.T., Danilov, S.M., and Curiel, D.T. A targetable, injectable adenoviral vector for selective gene delivery to pulmonary endothelium in vivo. *Mol Ther*, *2:* 562–578, 2000.
170. Yotnda, P., Chen, D.H., Chiu, W., Piedra, P.A., Davis, A., Templeton, N.S., and Brenner, M.K. Bilamellar cationic liposomes protect adenovectors from preexisting humoral immune responses. *Mol Ther*, *5:* 233–241, 2002.
171. Logg, C.R. and Kasahara, N. Retrovirus-mediated gene transfer to tumors: utilizing the replicative power of viruses to achieve highly efficient tumor transduction in vivo. *Methods Mol Biol*, *246:* 499–525, 2004.

172. Rollins, S.A., Birks, C.W., Setter, E., Squinto, S.P., and Rother, R.P. Retroviral vector producer cell killing in human serum is mediated by natural antibody and complement: strategies for evading the humoral immune response. *Hum Gene Ther*, *7:* 619–626, 1996.
173. Ghazizadeh, S., Carroll, J.M., and Taichman, L.B. Repression of retrovirus-mediated transgene expression by interferons: implications for gene therapy. *J Virol*, *71:* 9163–9169, 1997.
174. Feldman, A.L., Alexander, H.R., Hewitt, S.M., Lorang, D., Thiruvathukal, C.E., Turner, E.M., and Libutti, S.K. Effect of retroviral endostatin gene transfer on subcutaneous and intraperitoneal growth of murine tumors. *J Natl Cancer Inst*, *93:* 1014–1020, 2001.
175. Feldman, A.L., Friedl, J., Lans, T.E., Libutti, S.K., Lorang, D., Miller, M.S., Turner, E.M., Hewitt, S.M., and Alexander, H.R. Retroviral gene transfer of interferon-inducible protein 10 inhibits growth of human melanoma xenografts. *Int J Cancer*, *99:* 149–153, 2002.
176. Angiolillo, A.L., Sgadari, C., Taub, D.D., Liao, F., Farber, J.M., Maheshwari, S., Kleinman, H.K., Reaman, G.H., and Tosato, G. Human interferon-inducible protein 10 is a potent inhibitor of angiogenesis in vivo. *J Exp Med*, *182:* 155–162, 1995.
177. Arenberg, D.A., Kunkel, S.L., Polverini, P.J., Morris, S.B., Burdick, M.D., Glass, M.C., Taub, D.T., Iannettoni, M.D., Whyte, R.I., and Strieter, R.M. Interferon-gamma-inducible protein 10 (IP-10) is an angiostatic factor that inhibits human non-small cell lung cancer (NSCLC) tumorigenesis and spontaneous metastases. *J Exp Med*, *184:* 981–992, 1996.
178. Mei, K., Tian, L., Wei, Y.Q., Li, J., Wen, Y.J., Kan, B., and Deng, H.X. [Antitumor effects of interferon-gamma-inducible protein 10 combined with gemcitabine]. *Ai Zheng*, *24:* 397–402, 2005.
179. McCarty, D.M., Young, S.M., Jr., and Samulski, R.J. Integration of adeno-associated virus (AAV) and recombinant AAV vectors. *Annu Rev Genet*, *38:* 819–845, 2004.
180. Buning, H., Nicklin, S.A., Perabo, L., Hallek, M., and Baker, A.H. AAV-based gene transfer. *Curr Opin Mol Ther*, *5:* 367–375, 2003.
181. Smith-Arica, J.R. and Bartlett, J.S. Gene therapy: recombinant adeno-associated virus vectors. *Curr Cardiol Rep*, *3:* 43–49, 2001.
182. Peden, C.S., Burger, C., Muzyczka, N., and Mandel, R.J. Circulating anti-wild-type adeno-associated virus type 2 (AAV2) antibodies inhibit recombinant AAV2 (rAAV2)-mediated, but not rAAV5-mediated, gene transfer in the brain. *J Virol*, *78:* 6344–6359, 2004.
183. White, S.J., Nicklin, S.A., Buning, H., Brosnan, M.J., Leike, K., Papadakis, E.D., Hallek, M., and Baker, A.H. Targeted gene delivery to vascular tissue in vivo by tropism-modified adeno-associated virus vectors. *Circulation*, *109:* 513–519, 2004.
184. Giles, J. Protests win reprieve for renowned medical lab. *Nature*, *423:* 573, 2003.
185. Kay, M.A. and Nakai, H. Looking into the safety of AAV vectors. *Nature*, *424:* 251, 2003.
186. Shichinohe, T., Bochner, B.H., Mizutani, K., Nishida, M., Hegerich-Gilliam, S., Naldini, L., and Kasahara, N. *Development* of lentiviral vectors for antiangiogenic gene delivery. *Cancer Gene Ther*, *8:* 879–889, 2001.
187. Pfeifer, A., Kessler, T., Silletti, S., Cheresh, D.A., and Verma, I.M. Suppression of angiogenesis by lentiviral delivery of PEX, a noncatalytic fragment of matrix metalloproteinase 2. *Proc Natl Acad Sci USA*, *97:* 12227–12232, 2000.
188. Latchman, D.S. Herpes simplex virus vectors for gene therapy. *Mol Biotechnol*, *2:* 179–195, 1994.
189. Burton, E.A., Fink, D.J., and Glorioso, J.C. Gene delivery using herpes simplex virus vectors. *DNA Cell Biol*, *21:* 915–936, 2002.
190. Verma, I.M. and Weitzman, M.D. Gene therapy: twenty-first century medicine. *Annu Rev Biochem*, *74:* 711–738, 2005.
191. Peplinski, G.R., Tsung, K., and Norton, J.A. Vaccinia virus for human gene therapy. *Surg Oncol Clin N Am*, *7:* 575–588, 1998.
192. Gnant, M.F., Berger, A.C., Huang, J., Puhlmann, M., Wu, P.C., Merino, M.J., Bartlett, D.L., Alexander, H.R., Jr., and Libutti, S.K. Sensitization of tumor necrosis factor alpha-resistant human melanoma by tumor-specific in vivo transfer of the gene encoding endothelial monocyte-activating polypeptide II using recombinant vaccinia virus. *Cancer Res*, *59:* 4668–4674, 1999.
193. Berger, A.C., Alexander, H.R., Tang, G., Wu, P.S., Hewitt, S.M., Turner, E., Kruger, E., Figg, W.D., Grove, A., Kohn, E., Stern, D., and Libutti, S.K. Endothelial monocyte activating polypeptide II induces endothelial cell apoptosis and may inhibit tumor angiogenesis. *Microvasc Res*, *60:* 70–80, 2000.

194. Wu, P.C., Alexander, H.R., Huang, J., Hwu, P., Gnant, M., Berger, A.C., Turner, E., Wilson, O., and Libutti, S.K. In vivo sensitivity of human melanoma to tumor necrosis factor (TNF)-alpha is determined by tumor production of the novel cytokine endothelial-monocyte activating polypeptide II (EMAPII). *Cancer Res*, *59:* 205–212, 1999.
195. Chang, E., Chalikonda, S., Friedl, J., Xu, H., Phan, G.Q., Marincola, F.M., Alexander, H.R., and Bartlett, D.L. Targeting vaccinia to solid tumors with local hyperthermia. *Hum Gene Ther*, *16:* 435–444, 2005.
196. Larocca, D., Burg, M.A., Jensen-Pergakes, K., Ravey, E.P., Gonzalez, A.M., and Baird, A. Evolving phage vectors for cell targeted gene delivery. *Curr Pharm Biotechnol*, *3:* 45–57, 2002.
197. Larocca, D., Witte, A., Johnson, W., Pierce, G.F., and Baird, A. Targeting bacteriophage to mammalian cell surface receptors for gene delivery. *Hum Gene Ther*, *9:* 2393–2399, 1998.
198. Larocca, D., Jensen-Pergakes, K., Burg, M.A., and Baird, A. Receptor-targeted gene delivery using multivalent phagemid particles. *Mol Ther*, *3:* 476–484, 2001.
199. Poul, M.A. and Marks, J.D. Targeted gene delivery to mammalian cells by filamentous bacteriophage. *J Mol Biol*, *288:* 203–211, 1999.
200. Burg, M.A., Jensen-Pergakes, K., Gonzalez, A.M., Ravey, P., Baird, A., and Larocca, D. Enhanced phagemid particle gene transfer in camptothecin-treated carcinoma cells. *Cancer Res*, *62:* 977–981, 2002.
201. Davis, J.J. and Fang, B. Oncolytic virotherapy for cancer treatment: challenges and solutions. *J Gene Med*, *7:* 1380–1389, 2005.
202. Hardy, S., Kitamura, M., Harris-Stansil, T., Dai, Y., and Phipps, M.L. Construction of adenovirus vectors through Cre-lox recombination. *J Virol*, *71:* 1842–1849, 1997.
203. Yu, S.F., von Ruden, T., Kantoff, P.W., Garber, C., Seiberg, M., Ruther, U., Anderson, W.F., Wagner, E.F., and Gilboa, E. Self-inactivating retroviral vectors designed for transfer of whole genes into mammalian cells. *Proc Natl Acad Sci USA*, *83:* 3194–3198, 1986.
204. Hacein-Bey-Abina, S., Von Kalle, C., Schmidt, M., McCormack, M.P., Wulffraat, N., Leboulch, P., Lim, A., Osborne, C.S., Pawliuk, R., Morillon, E., Sorensen, R., Forster, A., Fraser, P., Cohen, J.I., de Saint Basile, G., Alexander, I., Wintergerst, U., Frebourg, T., Aurias, A., Stoppa-Lyonnet, D., Romana, S., Radford-Weiss, I., Gross, F., Valensi, F., Delabesse, E., Macintyre, E., Sigaux, F., Soulier, J., Leiva, L.E., Wissler, M., Prinz, C., Rabbitts, T.H., Le Deist, F., Fischer, A., and Cavazzana-Calvo, M. LMO2-associated clonal T cell proliferation in two patients after gene therapy for SCID-X1. *Science*, *302:* 415–419, 2003.
205. Cavazzana-Calvo, M., Thrasher, A., and Mavilio, F. The future of gene therapy. *Nature*, *427:* 779–781, 2004.
206. Kong, H.L. and Crystal, R.G. Gene therapy strategies for tumor antiangiogenesis. *J Natl Cancer Inst*, *90:* 273–286, 1998.
207. Klauber, N., Rohan, R.M., Flynn, E., and D'Amato, R.J. Critical components of the female reproductive pathway are suppressed by the angiogenesis inhibitor AGM-1470. *Nat Med*, *3:* 443–446, 1997.
208. Sedlacek, H.H. Pharmacological aspects of targeting cancer gene therapy to endothelial cells. *Crit Rev Oncol Hematol*, *37:* 169–215, 2001.
209. Rabinowitz, J.E., Rolling, F., Li, C., Conrath, H., Xiao, W., Xiao, X., and Samulski, R.J. Cross-packaging of a single adeno-associated virus (AAV) type 2 vector genome into multiple AAV serotypes enables transduction with broad specificity. *J Virol*, *76:* 791–801, 2002.
210. Khare, P.D., Liao, S., Hirose, Y., Kuroki, M., Fujimura, S., Yamauchi, Y., and Miyajima-Uchida, H. Tumor growth suppression by a retroviral vector displaying scFv antibody to CEA and carrying the iNOS gene. *Anticancer Res*, *22:* 2443–2446, 2002.
211. Wagner, E., Kircheis, R., and Walker, G.F. Targeted nucleic acid delivery into tumors: new avenues for cancer therapy. *Biomed Pharmacother*, *58:* 152–161, 2004.
212. Zhu, Z.B., Makhija, S.K., Lu, B., Wang, M., Kaliberova, L., Liu, B., Rivera, A.A., Nettelbeck, D.M., Mahasreshti, P.J., Leath, C.A., 3rd, Yamaoto, M., Alvarez, R.D., and Curiel, D.T. Transcriptional targeting of adenoviral vector through the CXCR4 tumor-specific promoter. *Gene Ther*, *11:* 645–648, 2004.
213. Ruegg, C., Dormond, O., and Mariotti, A. Endothelial cell integrins and COX-2: mediators and therapeutic targets of tumor angiogenesis. *Biochim Biophys Acta*, *1654:* 51–67, 2004.

214. Dormond, O., Bezzi, M., Mariotti, A., and Ruegg, C. Prostaglandin E2 promotes integrin alpha Vbeta 3-dependent endothelial cell adhesion, rac-activation, and spreading through cAMP/PKA-dependent signaling. *J Biol Chem, 277:* 45838–45846, 2002.
215. Godbey, W.T. and Atala, A. Directed apoptosis in Cox-2-overexpressing cancer cells through expression-targeted gene delivery. *Gene Ther, 10:* 1519–1527, 2003.
216. Zhu, Z.B., Makhija, S.K., Lu, B., Wang, M., Kaliberova, L., Liu, B., Rivera, A.A., Nettelbeck, D.M., Mahasreshti, P.J., Leath, C.A., Barker, S., Yamaoto, M., Li, F., Alvarez, R.D., and Curiel, D.T. Transcriptional targeting of tumors with a novel tumor-specific survivin promoter. *Cancer Gene Ther, 11:* 256–262, 2004.
217. Yanagisawa, M., Kurihara, H., Kimura, S., Goto, K., and Masaki, T. A novel peptide vasoconstrictor, endothelin, is produced by vascular endothelium and modulates smooth muscle Ca2+ channels. *J Hypertens Suppl, 6:* S188–191, 1988.
218. Hu, J., Discher, D.J., Bishopric, N.H., and Webster, K.A. Hypoxia regulates expression of the endothelin-1 gene through a proximal hypoxia-inducible factor-1 binding site on the antisense strand. *Biochem Biophys Res Commun, 245:* 894–899, 1998.
219. Greenberger, S., Shaish, A., Varda-Bloom, N., Levanon, K., Breitbart, E., Goldberg, I., Barshack, I., Hodish, I., Yaacov, N., Bangio, L., Goncharov, T., Wallach, D., and Harats, D. Transcription-controlled gene therapy against tumor angiogenesis. *J Clin Invest, 113:* 1017–1024, 2004.
220. Savontaus, M.J., Sauter, B.V., Huang, T.G., and Woo, S.L. Transcriptional targeting of conditionally replicating adenovirus to dividing endothelial cells. *Gene Ther, 9:* 972–979, 2002.
221. Hallahan, D.E., Mauceri, H.J., Seung, L.P., Dunphy, E.J., Wayne, J.D., Hanna, N.N., Toledano, A., Hellman, S., Kufe, D.W., and Weichselbaum, R.R. Spatial and temporal control of gene therapy using ionizing radiation. *Nat Med, 1:* 786–791, 1995.
222. Walther, W. and Stein, U. Cell type specific and inducible promoters for vectors in gene therapy as an approach for cell targeting. *J Mol Med, 74:* 379–392, 1996.
223. Buning, H., Ried, M.U., Perabo, L., Gerner, F.M., Huttner, N.A., Enssle, J., and Hallek, M. Receptor targeting of adeno-associated virus vectors. *Gene Ther, 10:* 1142–1151, 2003.
224. Nakamura, T., Sato, K., and Hamada, H. Effective gene transfer to human melanomas via integrin-targeted adenoviral vectors. *Hum Gene Ther, 13:* 613–626, 2002.
225. Miller, A.D. Cell-surface receptors for retroviruses and implications for gene transfer. *Proc Natl Acad Sci USA, 93:* 11407–11413, 1996.
226. Bartlett, J.S., Kleinschmidt, J., Boucher, R.C., and Samulski, R.J. Targeted adeno-associated virus vector transduction of nonpermissive cells mediated by a bispecific F(ab′gamma) 2 antibody. *Nat Biotechnol, 17:* 181–186, 1999.
227. Everts, M., Kok, R.J., Asgeirsdottir, S.A., Melgert, B.N., Moolenaar, T.J., Koning, G.A., van Luyn, M.J., Meijer, D.K., and Molema, G. Selective intracellular delivery of dexamethasone into activated endothelial cells using an E-selectin-directed immunoconjugate. *J Immunol, 168:* 883–889, 2002.
228. Reynolds, P.N., Nicklin, S.A., Kaliberova, L., Boatman, B.G., Grizzle, W.E., Balyasnikova, I.V., Baker, A.H., Danilov, S.M., and Curiel, D.T. Combined transductional and transcriptional targeting improves the specificity of transgene expression in vivo. *Nat Biotechnol, 19:* 838–842, 2001.
229. Scappaticci, F.A. Mechanisms and future directions for angiogenesis-based cancer therapies. *J Clin Oncol, 20:* 3906–3927, 2002.
230. Emmenegger, U., Man, S., Shaked, Y., Francia, G., Wong, J.W., Hicklin, D.J., and Kerbel, R.S. A comparative analysis of low-dose metronomic cyclophosphamide reveals absent or low-grade toxicity on tissues highly sensitive to the toxic effects of maximum tolerated dose regimens. *Cancer Res, 64:* 3994–4000, 2004.
231. Jain, R.K. Normalization of tumor vasculature: an emerging concept in antiangiogenic therapy. *Science, 307:* 58–62, 2005.
232. Abdollahi, A., Lipson, K.E., Sckell, A., Zieher, H., Klenke, F., Poerschke, D., Roth, A., Han, X., Krix, M., Bischof, M., Hahnfeldt, P., Grone, H.J., Debus, J., Hlatky, L., and Huber, P.E. Combined therapy with direct and indirect angiogenesis inhibition results in enhanced antiangiogenic and antitumor effects. *Cancer Res, 63:* 8890–8898, 2003.
233. Kerbel, R.S. and Kamen, B.A. The anti-angiogenic basis of metronomic chemotherapy. *Nat Rev Cancer, 4:* 423–436, 2004.

234. Vacca, A., Ribatti, D., Iurlaro, M., Merchionne, F., Nico, B., Ria, R., and Dammacco, F. Docetaxel versus paclitaxel for antiangiogenesis. *J Hematother Stem Cell Res*, *11:* 103–118, 2002.
235. Kruczynski, A. and Hill, B.T. Vinflunine, the latest Vinca alkaloid in clinical development. A review of its preclinical anticancer properties. *Crit Rev Oncol Hematol*, *40:* 159–173, 2001.
236. Wedge, S.R., Ogilvie, D.J., Dukes, M., Kendrew, J., Chester, R., Jackson, J.A., Boffey, S.J., Valentine, P.J., Curwen, J.O., Musgrove, H.L., Graham, G.A., Hughes, G.D., Thomas, A.P., Stokes, E.S., Curry, B., Richmond, G.H., Wadsworth, P.F., Bigley, A.L., and Hennequin, L.F. ZD6474 inhibits vascular endothelial growth factor signaling, angiogenesis, and tumor growth following oral administration. *Cancer Res*, *62:* 4645–4655, 2002.
237. Thorpe, P.E., Chaplin, D.J., and Blakey, D.C. The first international conference on vascular targeting: meeting overview. *Cancer Res*, *63:* 1144–1147, 2003.
238. Neri, D. and Bicknell, R. Tumour vascular targeting. *Nat Rev Cancer*, *5:* 436–446, 2005.
239. Kim, J.J. and Tannock, I.F. Repopulation of cancer cells during therapy: an important cause of treatment failure. *Nat Rev Cancer*, *5:* 516–525, 2005.
240. Ill, C.R. and Chiou, H.C. Gene therapy progress and prospects: recent progress in transgene and RNAi expression cassettes. *Gene Ther*, *12:* 795–802, 2005.
241. Zorick, T.S., Mustacchi, Z., Bando, S.Y., Zatz, M., Moreira-Filho, C.A., Olsen, B., and Passos-Bueno, M.R. High serum endostatin levels in Down syndrome: implications for improved treatment and prevention of solid tumours. *Eur J Hum Genet*, *9:* 811–814, 2001.
242. Iughetti, P., Suzuki, O., Godoi, P.H., Alves, V.A., Sertie, A.L., Zorick, T., Soares, F., Camargo, A., Moreira, E.S., di Loreto, C., Moreira-Filho, C.A., Simpson, A., Oliva, G., and Passos-Bueno, M.R. A polymorphism in endostatin, an angiogenesis inhibitor, predisposes for the development of prostatic adenocarcinoma. *Cancer Res*, *61:* 7375–7378, 2001.

17 Identification of New Targets Using Expression Profiles

Thomas R. Burkard, Zlatko Trajanoski, Maria Novatchkova, Hubert Hackl, and Frank Eisenhaber

CONTENTS

When large-scale expression profiling methods were first published about a decade ago, their introduction into the lab praxis was accompanied with high expectations of getting system-wide understanding of gene interactions. Although this hope has not yet materialized, expression profiling tools have become an indispensable part of the methodical arsenal in biological laboratories. In this review, we describe the experimental variants of gene expression profiling. The computational methods for postexperimental expression profile analysis, both for expression value treatment and gene/protein function prediction from sequence, are considered in detail since they do not receive the necessary attention in many practical applications. Finally, biologically significant results from expression profiling studies with special emphasis on angiogenic systems including the identification of new targets are assessed.

The end of the twentieth century has seen life sciences entering a qualitatively new era of their development. The major change is connected with progress in research technologies, the introduction of high-throughput methods for the generation of large amounts of uniform, quantitative data describing the status and the development of biological systems. This trend is associated with a new, essential, and creative role of computational biology for handling this data and for interpreting it in terms of biological functions. DNA sequencing was in the first wave and the complete sequencing of genomes of all major model organisms and human (beginning with that of *Haemophilus influencae* in 1995) was the major highlight of this development.

In the late 1990s, expression profiling was seen as a continuation of this process with the promise of simultaneous, large-scale, and high-throughput characterization of the transcriptional status of all genes, especially in contrast, at that time, to the usual single-gene northern blot technology. The euphoria that systematic understanding of networks and pathway interactions is just in reach has vanished due to several reasons. First, high-throughput expression profiling methods did not reach the necessary level of sensitivity to monitor simultaneously frequently occurring and low-abundant mRNAs with sufficient accuracy. The hope of mathematical network modeling has often moved away instead of a semi-quantitative evaluation of expression data (up to a distinction of only "high" and "low" expression) and the determination of sets of genes that are markedly transcriptionally regulated. That is, expression profiling is used just as one type of screen for target identification. Second, the transcriptional status is only a facet of a complex biological reality that has many additional regulation mechanisms at the translational, proteome, physiological, or population level. After the euphoria has settled, expression profiling tools have become an important part of the laboratory repertoire that complements other techniques and approaches.

Application of expression profiling in the angiogenic context is an especially demanding task since angiogenesis is a process that involves many cell types, tissues, and both local and organism-level regulation mechanisms. Most often, the focus is on the transcriptional status of endothelial cells (ECs), a material that is not easy to collect from sample organisms and where isolated in vitro culturing substantially influences intracellular processes [1,2].

In the first part of this chapter, we present an overview of experimental methods applicable for monitoring gene expression profiles as well as a review of computational approaches used for the analysis and the biological interpretation of the resulting complex data. The second part of this treatise summarizes achievements of these techniques with an emphasis on angiogenesis research but also a few other cases that appear especially methodically significant.

17.1 EXPRESSION PROFILING METHODS: DATA CREATION

A variety of expression profiling methods are available, which take a snapshot of the currently expressed genes at a given time, state, environment, genetic background, and treatment in one or numerous cells. The older and simpler methods, differential display and subtractive hybridization, are methods that lack the sensitivity to detect relative differences in the transcript abundance between two samples. Compared with other technologies, these two methods belong to the group of low-throughput processes together with differential hybridization. On the other hand, they have the advantage not to require specialized equipment and, therefore, can be applied in a standard biological laboratory. The state of the art for large-scale gene expression profiling is represented by serial analysis of gene expression (SAGE) [3], massive parallel signature sequencing (MPSS) [4], oligo- and cDNA-microarrays [5,6], and real-time polymerase chain reaction (RT-PCR). With these approaches, it is possible to determine the absolute or relative abundance of RNA in a medium- or high-throughput manner. Any of these methods has their own advantages and limitations. SAGE and MPSS profit from the possibility to quantify transcript numbers in absolute terms without prior knowledge of the transcriptome. Oligo- and cDNA-microarray are the most commonly used technology. Arrays are able to measure the expression level from several hundreds or thousands of entities on one single glass slide. Since the scientific community in this field is large, a variety of analytical tools are available. Finally, RT-PCR has an outstandingly low detection limit of only 1 transcript in 1000 cells and is often applied for validation of array measurements.

If sufficiently complete, the transcript abundance data generated by gene expression profiling methods described above are very useful for the determination of transcriptional networks. Nevertheless, it should be noted that the transcriptional status is only one aspect of the cellular physiology. Although an essentially linear relationship between the amount of transcripts and protein abundance is assumed in most studies for the extraction of biological conclusions, this preposition is not true in many cases, for example, in instances of posttranscriptional regulation or translational control.

In addition to the problem of the protein/transcript ratio, the diversity of proteins originating from a single transcript must be taken into account. Posttranslational modifications (intein exclusion, proteolytic processing, phosphorylation, acetylation, myristylation, glycosylaton, etc.) have a great influence on the functional properties of proteins but their occurrence cannot be measured with gene expression profiling methods. Therefore, transcription profiling results would need to be complemented with proteome analysis methods such as LC/MS/MS and protein arrays.

Several efforts using genomic and proteomic methods have tried to shed light in the assumption of protein/transcript correlation. Gygi et al. [7] showed that for a set of 106 *Saccharomyces cerevisiae* genes the Pearson product moment correlation coefficient between the SAGE transcript expression and protein expression was 0.935. This high level was biased by few highly abundant species. For transcripts represented by less than 10 copies per cell (69% of the 106 gene set), the coefficient dropped to 0.356. This effect might be a general observation or a problem of higher inaccuracies in measuring low abundances. Another study by Ideker et al. [8] compared the protein/transcript ratios of *S. cerevisiae* $wt + gal$ and $wt - gal$. A moderate correlation ($r = 0.61$) was observed for 289 expression pairs. Interestingly, the change of ribosomal-proteins on the transcript level was not passed on to the protein level. On the other hand, Kern et al. [9] observed that there is a significant correlation between protein and gene expression in acute myeloid leukemia for several markers.

For a final conclusion, the accuracy of the quantitative measuring methods of transcripts and especially of proteins has to improve considerably. The limits of all involved methods have to be considered exhaustively for an appropriate comparison to avoid influences of all kind of biases on the comparison. Most likely, a general correlation is not possible and the proteins have to be divided into two groups of highly transcript-sensitive and -insensitive ones. The insensitive proteins might be further separated into subgroups depending on different regulatory mechanisms (e.g., codon bias) probably associated with cellular localization or functional groups/processes (metabolism, transcription factor, transport, etc.). The expression level might be an additional but important parameter.

17.1.1 Differential Display Is a Simple but Limited Method

Differential display is one of the simplest expression profiling methods. It determines the abundance of an RNA species through comparison of two probes. Extracted total RNAs of the sample and the control are separately reverse transcribed to cDNA and, thereafter, are amplified by PCR using arbitrary primers. Control and sample RNAs are separated on a denaturing polyacryl gel by electrophoresis and, after visualization, the discrete bands are compared side by side. Quantitative differences can be estimated by the intensities of the bands in the gel. cDNA that is abundant in only one part is extracted, further amplified with PCR and sequenced [10,11]. The classical differential display uses radioactive nucleotides to detect mRNAs but this disadvantage can be overcome by fluorescence techniques [12].

A more advanced variation allows a separation of the RNA fraction into smaller ones. Therefore, the reverse transcription uses an oligo-dT primer with two additional nucleotides. Due to one more hydrogen bond, primers with G and C are more efficient than those containing A and T [13,14]. Depending on the nucleotide variations, many different RNA subjects can be separated.

The advantage of differential display lies in the simplicity by using only standard procedures. It allows simultaneous determination of up- and downregulated genes in two or more samples of about 5–10 ng of total RNA. Its main purpose is to identify uncharacterized and not yet sequenced mRNA species easily. On the other side, the classical differential display often generates false-negative results during PCR amplification and cloning [15]. The workload can grow considerably with the number of different nucleotide variations in the primers. Finally, differential display is more or less only a qualitative method.

17.1.2 cDNA Libraries and Differential Hybridization

cDNA libraries are constructed from reverse transcribed mRNA of a tissue of interest typically with oligo-dT primers. The resulting cDNAs are cloned into bacteria and, therefore, each bacterial colony represents one gene of a tissue [16]. Nonnormalized libraries preserve the relative abundance of the mRNA levels [16]. An expression profile can be determined by randomly picking out clones and sequencing an expressed sequence tag (EST) of ~300 bp. This method is biased toward strongly abundant expressed transcripts; it is labor-intensive and expensive. To identify rare mRNA species, methods to construct normalized libraries exist, which contains each mRNA species at a similar level [17].

Differential hybridization relies on the hybridization of cDNA of a tissue of interest to a cDNA library with fixed colony positions. The cDNA library is transferred to filter membranes and is treated with alkali to release the cDNA and to denature it. mRNAs of different tissues of interest are subjected to reverse transcription with radioactive nucleotides and hybridized to fixed cDNAs on the filters. Comparison of the x-ray films reveals the relative abundance of a gene in the tissues [18].

17.1.3 Subtractive Hybridization

The principle of this method is based on the fact that single-stranded cDNA (ss-cDNA) can be separated from double-stranded cDNA (ds-cDNA) with a hydroxylapatite column or the avidin-biotin method [19]. After reverse transcription, cDNA obtained from the sample mRNA is hybridized to the mRNA of the control fraction. A part of the sample cDNA finds no complementary RNA in the control and remains as ss-cDNA, which can be separated by chromatography with hydroxylapatite columns. To increase the purity, further rounds of hybridization with control mRNA and chromatography remove the slower hybridizing sequences. A sample-overrepresented cDNA library is constructed from the ss-cDNA fraction and specific clones are sequenced. In typical cases of successful application of this technique, less than 5% of the initial cDNA sequences are isolated as single-stranded [20].

Similarly to differential display, this technique is mainly used to identify the abundance of an mRNA species in one tissue compared with another. It is an inexpensive method, which can be carried out in a basic molecular biology laboratory. Compared with differential display, it has a lower rate of false-negatives [15] but determines only the abundance of an mRNA species in one sample. RNAs with only slight differences in expression are hard to identify (for reliability, a 5- to 10-fold difference in abundance is required) [20]. The fact that high abundant transcripts are easier to hybridize than low-abundant ones represents another problem. Due to the subtractive nature of the procedure, relative expression values cannot be measured.

17.1.4 Serial Analysis of Gene Expression Is a Good Choice for Acquiring Transcript Counts

Serial analysis of gene expression (SAGE) [3] is a method based on sequencing concatemer clones of 9–10 bp sequence tags instead of whole EST libraries. The tags have a unique position within the mRNA species and, typically, contain enough information to describe the whole mRNA [1]. Many variations of SAGE are now available that can work with a reduced amount of necessary raw material [21] or that generate sequence tags with increased length [22].

Shortly, RNA of purified cells is captured by hybridization of the poly-A tail to oligo-T magnetic beads. ds-cDNA is constructed and cleaved with the endonuclease *Nla*III. The bead-bound part is processed further [21]. In another approach, cDNA with biotinylated primers is synthesized. The cDNAs are cleaved and attached to streptavidin beads [3]. The tags are divided into two parts and are ligated each to two types of adaptors. The adapters contain sites for the restriction enzyme *Bsm*F1, which produces an overhang of 14-base pairs (bp) composed of 10 tag-specific and 4 nonspecific bp. The blunt-ended fragments are pooled and ligated to form di-tags. PCR is used to amplify the resulting di-tags with primers for the adapters. Finally, the di-tags are released by the restriction enzyme *Nla*III, concatenated, cloned, and sequenced. To create a typical library of 100,000 tags, preparation, screening, and sequencing of ~3000 concatemer clones are needed [21]. The frequency distribution is calculated by dividing the count of a tag by the total of sequenced tags. In the typical case, each 9–10 bp long tag entity has enough information to describe sufficiently a unique transcript in the analyzed cells.

The transcript numbers generated by SAGE are measured effectively in absolute terms and, thus, are easy to compare with different datasets [23]. It should be noted that sensitivity of transcript detection and accuracy of transcript occurrence measurements depend greatly on mRNA abundance in the sample and the number of sequenced tags. For example, there is a 37% chance to miss a transcript that occurs in the sample with the fraction of $p = 1{:}100{,}00$ in the mRNA pool if 100,000 tags are sequenced [1]. If the transcript is 10 times more abundant ($p = 1{:}100{,}00$), there is still a 22% chance that the tag is found not more than 7 times or a 14% chance that the tag occurs at least 15 times. Thus, the systematic measurement error ranges at least between the 0.7- and 1.5-fold of the final number for rare transcripts. Statistical estimates show that SAGE yields good quantitative measures of transcript occurrences for $p = 1{:}100{,}00$, qualitative estimates (high/low expression) for transcript occurrences with $p = 1{:}100{,}00$ and no reliable results for more rare transcripts [1].

SAGE needs no prior knowledge about the expressed transcripts. This is especially interesting for organisms without a sequenced genome. On the contrary, SAGE is an expensive and labor-intensive method (a large number of clones must be purified and sequenced). It is biased as a result of possible sampling and sequencing errors, nonuniqueness and nonrandomness of tag sequences [1,24]. Further, a link to a gene or genomic entry is needed for each tag in a sequence database. SAGE can result in a different distribution of expression values than microarray experiments. Therefore, the clustering methods based on Pearson correlation and Euclidean distance, used in microarray analysis, are not appropriate for SAGE profiles. Cai et al. considered that Poisson-based distances appear more suitable, which have been incorporated into a clustering algorithm for SAGE analysis [25].

17.1.5 Massive Parallel Signature Sequencing Arrays Transcript Tags on Microbeads

Massive parallel signature sequencing (MPSS) [4] is a combination of arraying sequences fixed on a bead and sequencing 16–20 bp long signatures. In the first step, transcripts are cloned on microbeads. Therefore, a complex conjugate mixture of transcripts and 32-oligomere microbead-specific tags are generated. The number of different tags has to be at least 100-fold higher than that of the transcript entities. A representative part is subjected to PCR and hybridized to tag-specific microbeads. Beads with sequences attached are sorted out with a fluorescence-activated cell sorter (FACS). These microbeads are arrayed in a fixed position within a flow cell. After cutting the sequence, it is ligated with an initiating adapter, which contains a type II restriction site for *Bbv*1. As in SAGE experiments, *Bbv*1 cuts at a specific distance away from its recognition site with a 4 bp overhang. This overhang is sequenced by ligating it to an adapter, which is specific for one nucleotide, at a certain position. Each microbead is attached to four adapters, which are specific for the 4 bp sequence. In 16 cycles, phycoerythrin-labeled decoder probes are hybridized to the adapters, scanned with CCD detectors, and washed off. The fluorescent images of the arrayed microbeads allow reconstructing the sequence. The cycle of *Bbv*1 cutting, adaptor ligation and decoder hybridization and imaging is repeated until enough sequence information is achieved. Analyzing a statistical significant amount of microbeads enables the measurement of the expression profile of the transcripts.

Similar to SAGE, MPSS requires no prior knowledge of the transcripts. The measured absolute transcript counts allow direct comparison between experiments. Due to cloning on the microbeads and parallel sequencing, the workload is lower than in the case of SAGE. The disadvantage of the technology is the requirement of specialized equipment only available through Lynx Genetics [23].

17.1.6 Oligo- and cDNA-Microarrays Are the Most Widely Used Gene Expression Profiling Methods in Genomic Scale

Microarray technology [6,5] is the most widely used method to measure gene expression profiles in a highly parallel manner. All different microarray platforms have in common that a nucleotide sequence is arrayed on a solid carrier, which is hybridized with a labeled nucleotide mixture of interest (with a radioactive marker or a fluorescent dye). The expression of each gene is measured by scanning each spot on the chip. Due to the fact that the gene position on the carrier is exactly known, the expression can be instantly assigned to a specific gene.

The fabrication of chips can be generally divided into spotted and in situ synthesized microarrays. Up to 780.000 oligonucleotides can be in situ synthesized on a single silicon wafer with photolithographic methods. The production of these high density chips remains in commercial hands like Affymetrix [5] and NimbleGen [26,27]. Further, an ink-jet system is developed by Rosetta Inpharmatics, which is licensed by Agilent [28]. These systems avoid handling of large cDNA and oligo-libraries as in the case of spotting methods. On the contrary, the relative short length of in situ synthesized oligonucleotides (20–25 mers) can result in a reduced selectivity and sensitivity [29]. Advances in this technology by NimbleGen use a maskless array synteziser (MAS) method, which reduces costs and allow oligonucleotide lengths up to 70 mers combined with higher flexibility in customized chip design [26,27].

Spotted chips [6] are based on arrayed cDNA clones of an EST library or on presynthesized oligonucleotides. Many resources of cDNA libraries, PCR primers, and oligonucleotide (50–70 mer) sets of various organisms are reviewed by Lyons [30]. A density of ~30,000 genes per chip can be reached. Contrary to in situ synthesis, it is possible to use not yet sequenced clones, which enables the identification of completely uncharacterized mRNA species [29]. Due to the higher flexibility of spotted elements and the lower cost than in situ techniques, it is widely used in the academic research field. ESTs and oligos are printed with high-throughput robots on poly-lysin, amino silanes, or amino-reactive silanes coated glass slide, which have a low inherent fluorescence. The surface coating enhances hydrophobicity and the adherence of the spotted probes [31]. The arrayed DNA is typically fixed by ultraviolet irradiation to the chip surface. A variety of currently available slides and printing systems are summarized by Affara [32]. To minimize the spot size and to improve the reproducibility the diameter of the pin tip, the hydrophobicity of the carrier surface, the humidity, the buffer types, and the temperature must be optimized [32].

Total RNA is extracted from the samples of interest. The RNA must be purified from proteins, lipids, and carbohydrate, which influence the hybridization negatively. If insufficient amounts of RNA are available, the signal can be amplified to continue the experiment with standard labeling and hybridization techniques. Different amplification methods are reviewed by Livesey [33]. One example of signal amplification consists in marking the reverse transcribed cDNA with specific tags. These are recognized by antibodies or proteins that are linked to an enzyme, which breaks down a tyramide-fluorophore dye that is added to the system. This technique has a 10- to 100-fold lower detection level with representative and reproducible results [34]. Sample amplification is divided into linear (T7) [35] and exponential (PCR) methods. Standardized linear T7 protocols allow amplification of nanogram quantities but are suffering from the transcript shortening resulting in a 3′ bias [35,36].

Due to the high reproducibility of in situ syntheses of oligonucleotide chips, it is possible to compare accurately between chips hybridized only with one sample [29]. In contrast, spotted microarray techniques are relative methods and need hybridization of Cy3/Cy5 labeled sample and reference. During direct Cy3/Cy5 labeling, the two fluorescents are incorporated with a different rate. To resolve this issue, amino-allyl modified nucleotides are incorporated in the first reverse transcription step followed by a coupling of the dyes to the reactive amino groups of the cDNA [37]. Nevertheless, some dye-bias is persistent but it can be addressed by the dye-swapping technique, which is explained below. After labeling, the cDNA is denatured and hybridized to the chip in specific chambers. Unmatched cDNA is removed with stringent washing solutions. The last step in the experimental procedure consists of scanning of the slides in a chip reader. A variety of available scanners are summarized by Affara [32].

17.1.7 Real-Time Reverse Transcription Polymerase Chain Reaction Is the Preferred Method to Detect Expression of Low-Abundant Genes

Real-time reverse transcription polymerase chain reaction (RT-PCR) is a powerful method to identify the expression profile of low-abundant transcripts (100-fold more sensitive than microarrays). Real-time RT-PCR has a detection range of 0.001–100 transcripts per cell [38,39]. Detailed information about the sequences of interest is needed. For each transcript, two unique primers of 20–24 nucleotides in length with a melting temperature of 60°C ± 2°C, a guanine-cysteine content of 45%–55%, and a PCR amplicon length of 60–150 bp have to be designed. Splice variants can be assessed if the amplicon reaches over an exon–exon boundary. The PCR (with its cycling temperature profile for primer annealing, DNA synthesizing, and denaturing) is performed in an optical 384-well plate using a fluorescence dye to monitor the dsDNA synthesis. A large RT-PCR profiling study measured the expression level of more than 1400 low-abundant transcription factors (known and putative) [38].

17.1.8 Proteomic Methods Allow Insights into Posttranscriptional and Posttranslational Expression Profiles

Transcripts are the blue prints for their corresponding proteins, which are the main players of cellular processes beside functional RNA. It is widely assumed that gene expression correlates with the translated products, but to ultimately derive the protein amounts, turn-over, and posttranslational modification, quantitative proteomic methods are needed. Due to the diverse physico-biological properties of proteins and the high dynamic range from one to millions of copies, it is nearly impossible to determine reliable and accurate amounts of proteins among others with the state-of-the-art methodology [40,41]. Nevertheless, ambitious technologies aim at solving these problems. Isotope-coded-affinity-tag (ICAT) uses a combination of liquid chromatography and mass spectrometry (LC/MS) to measure protein levels quantitatively in a complex mixture [42]. Therefore, light and heavy ICATs are used as an internal standard. Relative expression values are obtained by binding the different ICATs to two different cell states. LC separates the proteins of the combined samples but the isotopes remain close together. Fragments of the co-eluted isotopes are separated by MS with an 8 Da mass difference, which can be used to obtain the relative abundance.

As the different "omic" technologies and the bioinformatic tools for data integration evolve further, the most powerful approaches will be an appropriate combination of genomic, transcriptomic, proteomic, and metabolomic research. This will lead to an extensive understanding of the globular mechanism of cells and give complex insights into development and diseases.

17.1.9 To Obtain Meaningful Results, a Thoughtful Experimental Design Is Mandatory for Large-Scale Expression Profiling

Since large-scale expression studies are labor-intensive and costly, a significant time should be spent for planning the experiments to maximize the informative results and minimize result bias. Randomization, replications, and local control should be considered [43]. If differences in personnel, equipment, reagents, and times, which can leave a signature on the gene expression profile, cannot be controlled by a good design, randomization, and experiment repetitions are the best way to deal with them. Measurements of only one single sample do not allow the inference to the general behavior of the population of the respective biological systems. Therefore, biological (at least two samples for one set of measurements) and technical (one sample divided among at least two sets of measurement) replicates are needed [44]. Since biological variations are generally higher in any expression study than the technical noise, independent biological replicates are strongly preferred to generalize the results. Local control is used for arranging experimental material in relation to extraneous sources of variability. For example, if two different sample probes are to be compared (mice, tissues, etc.), the groups should be analyzed equally distributed over time to minimize a possible day-to-day difference (e.g., environmental influences, operator condition), which could otherwise bias the expression profile, if one group is investigated solely on 1 day and the other on a different day.

Pooling biological samples might provide a form of "biological averaging" by achieving more accurate results with fewer measurements. A danger lies in a possible unrecognized camouflage of an outlier within the pool that cannot be identified but could influence the expression profile significantly in a wrong way. Therefore, pooling should be avoided, unless cost issues or material amounts do not allow it [43].

Ruijter et al. [45] investigated pair-wise comparisons of SAGE libraries and concluded that the most efficient way to set up a SAGE study without knowledge of transcript abundance is to compile two SAGE libraries of equal size. A major drawback is that SAGE libraries represent only one experimental measurement lacking information of biological variation and experimental precision and, therefore, depend on simulations and assumptions of these parameters.

Two-color microarray studies rely on appropriate labeling and hybridization strategies. Several reviews address these concerns [43,46,47]. Issues regarding the aim of the experiment, availability of RNA and chips as well as the experimental processes have to be considered. Two principal comparison routines are available: direct and indirect. If two samples are hybridized to the same chip, these can be compared directly. In contrast, indirect comparison between two slides needs a common reference (universal or biological relevant). Often a combination is the most practical way to go. Single-factor, loop, or multifactorial design has to be chosen to address the biological question correctly [48,49]. For example, if treated or diseased tissues should be compared with healthy samples, the clear design choice is to compare all atypical once to a common reference (the healthy). To confront the systematic color bias of the two fluorescent dyes used in cDNA microarrays, dye-swapping should be used wherever possible. Therefore, two chips have to be hybridized with vice versa labeled samples.

17.2 COMPUTATIONAL ANALYSIS OF EXPRESSION PROFILING DATA: BIOLOGICAL SIGNIFICANCE OF THE EXPRESSED GENES

Designing the experiment and obtaining the expression values are the first part of an expression profiling study, resulting only in a plain list of differentially expressed tags/transcripts/proteins under specific conditions. Computational methods can help to correct the raw expression data from systematic and random bias, to determine the true expression status of genes, to derive new insights and hypotheses with regard to function of genes and biomolecular networks, processes, and mechanisms. Due to the large scale of SAGE, MPSS, and microarray experiments, these important analyses typically take a significant if not the largest time of the whole study. Since expression profiling studies are typically initiated in laboratories without tradition of computational biology research, human resources (with respect to number and qualification of researchers), time ranges, and capital investments in an appropriate computer and biological software environment dedicated to this effort are typically grossly underestimated. In addition to allocating appropriate resources, it might be useful to consider co-operation with experienced specialists.

Normalization of data minimizes biases of the expression values. There are general statistical considerations specialized for the different profiling applications [48]. Several tools integrate different normalization algorithms. ArrayNorm [50] is an academic, freely available, stand-alone software package; the Bioconductor project [51] is another alternative. Some normalization functions are also available as part of the Genesis suite [52]. Several public and commercial expression profile databases are available for sharing the measured profiles within the community. These function as a central repository to retrieve data for reanalyzes and comparisons. Clustering algorithms group genes or samples with similar expression profiles together. Co-expressed genes could share possibly a common regulatory mechanism. Groups of samples may identify specific cancer types. De novo prediction of functional domains can identify novel molecular and cellular functions of proteins. Especially for pharmacological treatment, transmembrane proteins, receptors, and enzymes with small binding pockets are of enhanced interest. Finally, pathway mapping and network reconstruction together with extensive literature enquiry might give an insight into the global interactions and the molecular atlas of the cell.

17.2.1 Repositories of Expression Data: Differences to Other Tissues and Health States Can Be Uncovered through Comparison

Expression databases (Table 17.1) provide access to raw and sometimes also processed versions of a multitude of measured gene expression profiles. Such deposits allow to reanalyze and validate experiments as well as to compare data with other related topics. Data sharing in publicly available databases is, therefore, an important issue that is, especially in microarray studies, mandatory for submission in some journals. Since experimental design, treatment, tissue type, normalization, and many other factors influence the expression profiles, these parameters have to be stored together with the measured and normalized data. Minimal information about a microarray experiment (MIAME) is a standard proposed for describing chip datasets by the Microarray Gene Expression Data Society to ensure that data can be easily interpreted and results can be independently verified [53].

An analysis of the expression datasets in the database GEO reveals interesting trends. As shown in Figure 17.1, there are two techniques that are most popular in expression profiling – in situ hybridization microarrays (including Affymetrix chips) and cDNA spotted

TABLE 17.1
Gene Expression Databases Accessible through the Internet

Database	Database Type	Internet Location
GEO [137,138]	Repository for expression data	http://www.ncbi.nlm.nih.gov/projects/geo/
ArrayExpress [139,140]	Repository for microarrays	http://www.ebi.ac.uk/arrayexpress/
SAGE map [141]	Repository for SAGE	http://www.ncbi.nlm.nih.gov/projects/SAGE/
GNF SymAtlas [142]	Normal tissues (Affymetrix)	http://symatlas.gnf.org/SymAtlas/
Riken Expression Array Database [143]	Normal tissues (cDNA microarray)	http://read.gsc.riken.go.jp/fantom2/
SAGE Genie [144]	Normal tissues and cancers (SAGE)	http://cgap.nci.nih.gov/SAGE

GEO is a public repository of high-throughput molecular abundance data. GEO is dedicated mainly to gene expression data but can even store proteomic abundances. Approximately 1300 platform entries and over 40,000 public samples have been stored till June 2005: ~38,500 cDNA-, ~1100 genomic-, and ~600 SAGE-samples [137,138]. ArrayExpress is a public microarray repository of well annotated raw and normalized data, which is stored in the minimum information about a microarray experiment (MIAME) standard [139,140]. SAGEmap is a repository for SAGE expression data, which allows the mapping of the tags to the corresponding transcripts [141]. Tissue-specific transcript expression of a normal physiological status in mouse and human is extensively profiled with Affymetrix- and cDNA-microarrays and stored in databases. The expression profiles of normal tissues provide the base-line for identification of transcript candidates of diseased and cancerous states through comparison [145,142,143]. The SAGE Genie of the Cancer Genome Anatomy Project is helpful in comparing SAGE results with numerous nonpathological tissue and cancers [144].

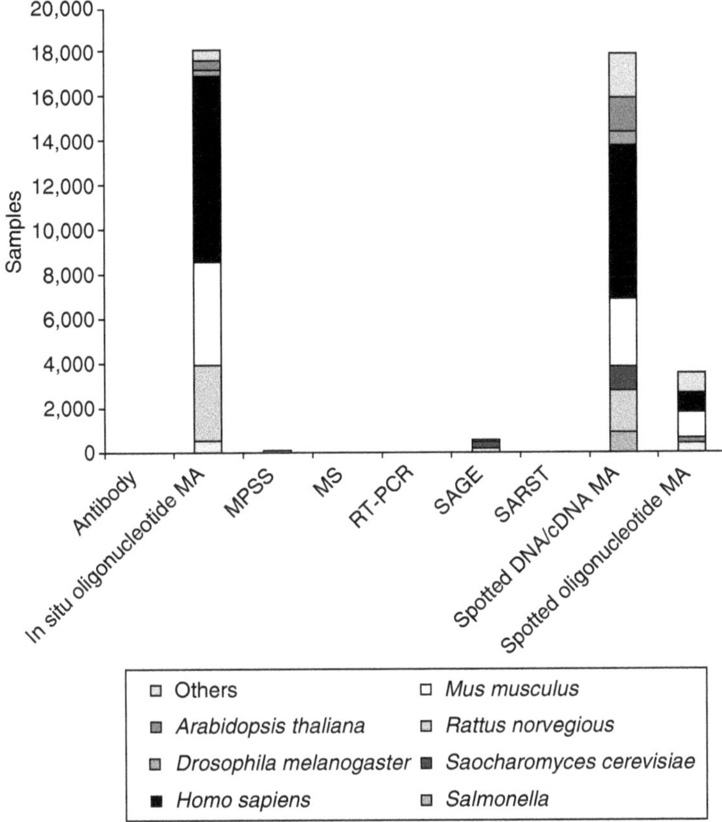

FIGURE 17.1 Gene Expression Omnibus (GEO) sample distribution. The prevalent method for expression profiling stored in the largest expression database GEO (http://www.ncbi.nlm.nih.gov/projects/geo/) is the microarray (MA) technology. According to the sample number, in situ hybridized and cDNA spotted chips are approximately equally popular. The dedication of these two technologies differs in the sample origin. In situ hybridized chips are mainly used for the species *Homo sapiens, Mus musculus, and Rattus norvegicus*. In contrast, the flexibility of cDNA microarrays allows a much broader range of ~70 different organism origins.

microarrays. Whereas the first type of arrays can be applied only in manufacturer-supplied customized forms, the second is technologically simpler and cheaper to produce and, therefore, the method of choice for researchers working on more esoteric model organisms then human, rat, or mouse. Figure 17.2 illustrates that expression profiling is still far from reaching a saturation phase, the number of new datasets coming into public databases each quarter is still growing.

Comparison of absolute as well as relative expression values can be carried out between datasets of various sources. Often, it is an extremely difficult task due to differences in methods, experimental designs, and systematic biases. Relative measurements already implicate comparisons intrinsically. For a "counting" method like SAGE and for in situ oligo-microarrays, comparison with public data is easy. Such studies can reveal specific marker genes, differences in mechanisms between healthy and diseased states, and show variations among sample groups (e.g., cancer subtypes). If similar sources are compared, hypotheses for upcoming new experiments can be derived and, hopefully, an expansion of the present mechanistic knowledge is achieved.

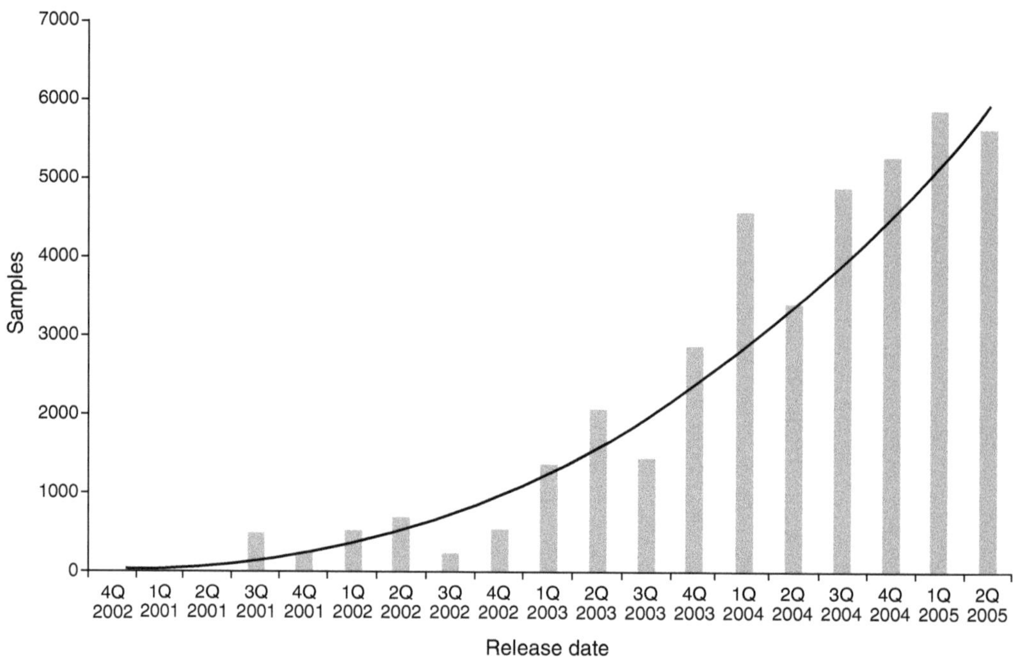

FIGURE 17.2 Incremental change of dataset numbers in GEO. This diagram illustrates the number of incoming datasets for GEO per quarter. Apparently, the stream of new expression profiles grows in accordance to a power law (with exponent of ~2.5).

17.2.2 Clustering Reveals Distinct Patterns in Expression Profiles

To find similarities and differences between biologically comparable datasets, algorithms clustering genes with similar expression status are used. Generally, it is possible to group genes by their co-expressed genes, samples by similar expression profiles, or to combine both. Clusters of co-expressed genes can share similar cellular functions and are likely to be transcriptionally co-regulated. For example, groups of samples can identify cancer subclasses by their expression profiles [54,55].

Several public and commercial tools are available on the market that have implemented a large variety of clustering algorithms and distance measurements (e.g., Genesis [52], TM4 [56]). It should be noted that there are more than 30 papers that offer different mathematical forms of clustering expression profiles but it is not really clear whether mathematical sophistication is always good for the application. None of the clustering algorithms can claim for itself that its special analytical form does reflect aspects of the biological system better than the others and, therefore, they should be similarly good for the purpose. Even more, cluster assignments of genes and experiments that change dramatically on application of specific clustering algorithms are rather not informative from the biological point of view. It should also not be forgotten that some datasets contain considerable noise (error margins of ~100% are not rare) and clustering of such data can result in artifacts.

Typically, the clustered expression profiles are visualized by coloring them in accordance with their log2(ratio). This analytical form of the ratio between a sample and a reference state is used because of its simplicity for interpretation (a twofold change increases or decreases the log2(ratio) by one unit). Mostly, repression of the genes in comparison with the reference is visualized by green (negative log2(ratio)) and overexpressed genes are colored red (positive log2(ratio)). The higher the expression or repression the brighter the colors are. Black represents equal amounts.

For measuring the similarity of expression profiles, each gene is characterized by an n-dimensional vector defining n as the number of samples/experiments. Various distance metrics (Euclidean distances, correlation distances, semi-metric distances, etc.) are proposed to identify a closely related group [57]. The correct choice of the appropriate distance metric between the vectors is a difficult question and might depend on the data preprocessing (e.g., normalization). For SAGE experiments, it is shown that Poisson-based distances fit well [25]. Generally, the observed biological effects should be stable with regard to minor aspects of the distance metric.

Provided with a distance measurement, supervised and unsupervised clustering methods can group the genes by their similarities. Unsupervised approaches are used to find co-regulated genes or related samples. For supervised methods (e.g., supported vector machine), which can create classifiers to predict some behavior, additional knowledge for some profiles must be provided like functional classes or healthy and diseased states [58]. Three different unsupervised methods are described shortly to give an insight into the mechanism behind the most important algorithms.

Hierarchical clustering [59], one of the most common unsupervised methods, joins consecutively the nearest clusters/genes together until all are combined in one large group. As a result, the clustered genes are placed into a dendrogram that visualizes the closeness of the transcriptional status among the genes. Single-, complete-, and average-linkage clustering corresponds to joining of growing clusters regarding the minimum, maximum, and average distances between them [57]. Comparing the minimum distances results in loose clusters, whereas compact clusters are achieved with complete-linkage clustering. It is a drawback of the simple hierarchical clustering that an incorrect intermediate assignment cannot be reversed and, thus, the expression vector defining a cluster might no longer represent the genes in it.

K-means clustering [60] is an iterative method that requires the initial knowledge of the cluster number K. The genes are randomly assigned to the K clusters. Through iterative gene shuffling between the clusters, the within-cluster dispersion is minimized and the inter-cluster distances are maximized. It stops if no improvement can be achieved anymore or if a specified number of iterations is exceeded. The method is computing-intensive but some early incorrect assignment can be modified at later iterations. A hierarchical ordering of clusters is not achieved and, therefore, no dendrogram is generated.

Principal component analysis (PCA) [61] reduces the dimensionality of the expression data through eigenvector calculations and data projection into the space spanned by the most important eigenvectors. Therefore, redundant properties of the expression profile are filtered out and major distinct patterning that describes the data separation of genes best might become clear. PCA results are visualized in the 3D-space with small clouds of data points that correspond to clustered genes or experiments.

17.2.3 Molecular Role of a Gene Product Can Be Identified through Sequence Analysis

Essentially, the basic element of expression data has a composite structure: On the one hand, it comprises a vector of n real numbers describing the expression status for n conditions/experiments/time points. Quantitative and qualitative methods for analyzing the real value part of the data have been considered in the previous chapters. On the other hand, a sequence tag is associated with the vector of expression values. Without the knowledge of the identity of the gene represented by the EST and the biological function of the respective gene, interpretation of the expression data in terms of biological mechanisms and processes is impossible. It should be noted that protein function requires a hierarchical description with molecular function, cellular function, and phenotypic function [62].

The first step is to allocate the corresponding gene and protein if available. With the knowledge of complete genomes, sequence comparisons can identify genes from sufficiently long sequence tags. For instance rounds of Megablast searches [63] against various nucleotide databases of decreasing trust-levels (in the order of RefSeq [64,65], FANTOM [66], UniGene [67], nr GenBank, and TIGR Mouse Gene Index [68] or other organism-specific databases) can be initially applied. If not all nucleic sequences can be assigned, the routine should be repeated with blastn [69], and finally against the genome of the whole organism. Long stretches (greater than 100) of nonspecific nucleotides have to be excluded. For the obtained genes, the respective gene product can be retrieved: This is either a protein sequence if the expressed RNA is translated to a protein or a functional RNA species (microRNA, etc.).

The molecular and cellular role of novel protein targets derived from an expression study is of special interest. A first insight of the processes involved can be obtained from gene ontology (GO) terms [70]. With the GenPept/RefSeq accession numbers, GO numbers for molecular function, biological process, and cellular component can be derived from the gene ontology database (Gene Ontology Consortium). If considered in context with the expression cluster identity, it is possible to observe groups of co-regulated proteins, which are part of a unique process or share a molecular function. Such groups of genes can be visualized with Genesis [52].

Detailed de novo function prediction for the target proteins can reveal new aspects of their molecular and cellular function. This analysis step is especially important for sequences that entered the databases after large-scale sequencing efforts and that have remained experimentally uncharacterized. A large variety of different sequence analysis tools are available. Sometimes, a homology search with the blast tool [69] (with low e-value cut-off less than e-10) finds an almost full-length homologue with a described function that can be transferred to the protein of interest.

In-depth sequence-analytic studies involve a three-step procedure. This approach is based on the assumption that proteins consist of linear sequence modules that have their own structural and functional characteristics. The function of the whole protein is a superposition of the segments' functions. The sequence modules can represent globular domains or nonglobular segments such as fibrillar segments with secondary structure, transmembrane helical segments, or polar, flexible regions without inherent structural preferences.

The first step involves the detection of the nonglobular part of the protein sequence, which houses many membrane-embedded segments, localization, and posttranslational modification sequence signals. Typically, nonglobular segments are characterized by some type of amino acid type compositional bias. Therefore, the principle procedure is to begin with analyses of compositional biases. They are found as low complexity regions with SEG [71] and compositional biased stretches with SAPS [72], Xnu, Cast [73], and GlobPlot 1.2 [74]. N-terminal localization signals can be studied with SignalP [75] and Sigcleave [76], the C-terminal PTS1 signal for peroxisomal targeting via the PEX5 mechanism is checked for with the predictor of Neuberger et al. [77–79]. A number of quite accurate predictors test the capacity of the query protein for lipid posttranslational modifications such as GPI lipid anchors [80–82], myristoyl [83], or prenyl (farnesyl or geranylgeranyl) [84] anchors. Membrane-embedded regions are recognized via the occurrence of strongly hydrophobic stretches. Standard predictors such as HMMTOP [85], TOPPRED, DAS-TMfilter [86], and SAPS [72] find transmembrane helical regions. Generally, several transmembrane prediction methods should map to the same loci to obtain reliable results. Secondary structure prediction methods (COILS [87], SSCP [88,89], Predator [90], NPS@ [91]) can generate information about secondary structural elements in nonglobular parts of proteins.

The second step involves comparisons of the query sequence with libraries of known domains. Since domain definitions of even the same domain type do slightly differ among authors, it is necessary to check all available for the given query sequence. Known sequence domains of the repositories PFAM and SMART [92] are characterized by hidden Markov models (HMM), which can be searched against a sequence with HMMER both in a global and local search mode. Further domain prediction method and databases are RPS-BLAST (CDD) [93], IMPALA [94], and PROSITE [95].

The third and last step involves analysis of query protein segments (with at least 50 amino acid residues long except for domains with many cysteines) that are not covered by hits produced by any of the prediction tools for nonglobular regions and known domains. Most likely, these sequence segments represent not yet characterized globular domains. It is necessary to collect protein sequence segments that are significantly similar to the respective part of the query. This can be done with tools from the BLAST/PSI-BLAST suite [96–98], HMMsearch, or SAM [99,100]. There are several strategies to collect as complete as possible sequence families; for example, each identified homologous protein can be resubmitted to an additional PSI-BLAST in an attempt to find more homologues. Through multiple alignments of the conserved residue stretches, sequence analysis of the found proteins, and extensive literature search, a new functional domain can be characterized. This sequence of steps has, for example, led to the characterization of the brix [101], GACKIX [102], and ProflAP [103] domains. Further, prediction of tertiary structure can sometimes help to answer additional questions (e.g., small pockets for therapeutics). PDBblast can identify the structure through homology to proteins with known tertiary structure. Several de novo structure prediction methods are available on the net. An excellent meta-server combining various structural prediction method can be found at bioinfo.pl [104].

Finally, the different sequence-analytic findings for the various query segments need to be synthesized for a description of the function of the whole protein. Overlaps of predicted features have to be resolved by assessing significances of hits, the amino acid compositional status of the respective segment, and expected false-positive prediction rates of algorithms. With the entirety of the sequence-analytic methods, some functional conclusion at least in very general terms is possible for most sequences. All these academic prediction tools for de novo sequence prediction are integrated in the user-friendly ANNOTATOR/NAVIGATOR environment, a novel protein sequence analysis system, which is in development at the IMP Bioinformatics group. A single experienced researcher can reasonably classify 50–100 sequence targets per day in this environment.

17.2.4 Network Reconstruction Gives Insight into the Transcriptional Mechanisms of Small Genomes

Expression profiling is, due to the wealth of data, a good source for reconstruction of gene networks. Generally, these methods determine genes, which affect the activity of other ones. Genes are represented in the networks as a node and the impact on another gene is shown in a connective arc. The arcs can receive a label, like up- or downregulation [105,106]. Networks can describe as diverse conditions as the influence of one gene expression on another expression [106] or the influence of a deletion mutant on other gene expressions [105], and even the mentioning of two genes in the same paper as in literature-based networks [107,108]. The construction of these interactions relies among others on Boolean networks, Bayesian networks, or biological-meaningful decision trees. Further, it is possible to integrate additional knowledge like transcription factor binding of chromatin-immunoprecipitation (ChIP) or computational prediction of transcription factor binding [109].

Unfortunately, most of these studies are limited to small genomes and dataset. Especially, the well-characterized yeast *S. cerevisiae* is a preferred model system for network reconstruction. This is no surprise since the accuracy of the genome sequence, the established methodologies of measuring precise and reproducible expression profiles and the wealth of other biological data available invite for such analyses. It will take a while until methods generate similar biologically reasonable insight for more complex organisms.

Several procedures allow prediction of regulatory elements of genes important for the regulatory network, which is implicated in the expression profile. One way is to cluster the genes by their expression profile and search for transcription factor binding sites of public databases (e.g., TRANSFAC) that are specific for a cluster [110]. A different approach is a de novo prediction of regulatory oligonucleotides based on the genome-wide expression profiles and the promoter sequences. The web-based tool, regulatory element detection using correlation with expression (REDUCE), tries to explain the log-ratio of a gene expression by the number of occurrences of the motif in its promoter. The regression coefficient associated with the motif occurrence can be directly interpreted as the change in concentration of the active transcription factor in the nucleus [111].

17.3 VALIDATION OF EXPRESSION PROFILES: TESTING THE HYPOTHESIS

In large-scale expression profiling, there are many sources of errors beginning from wrong sequencing, tag assignment, arraying, labeling biases, difficulties in managing large clone sets and datasets, over reagent, color, equipment, personal and day-by-day biases, and sample specific errors. Therefore, an identified novel target gene of a large-scale expression study must always be validated with a second different method to exclude a wrong measurement of a specific method and to be sure to address the right target. Especially northern-blotting and RT-PCR are the preferred methods to verify DNA microarray expression data [37]. Expression profile analysis-based hypotheses have to be investigated in further follow-up studies using appropriate techniques such as RNAi, overexpression, or knockout experiments.

17.4 SUCCESS STORIES WITH EXPRESSION PROFILING IN ANGIOGENESIS AND OTHER BIOLOGICAL PROCESSES

The original idea of using large-scale gene expression profiling for the elucidation of system biological properties (e.g., gene networks) of cells has found its realization only in a few studies and these focused mainly on simple model organisms. At the beginning, the problems with accuracy and reproducibility of large-scale expression measurements favored two other types of applications, namely (i) the usage of microarrays for the identification of a handful or even a single target that are most upregulated in certain physiological conditions and (ii) the numerical overall comparison of large sets of expression values between different physiological states for diagnostic purposes (e.g., for the differentiation between cancer development stages). In both cases, even gross errors of expression values for a limited number of genes do not affect the outcome dramatically. With the technology improving, the original idea of understanding gene networks has revived and it has become possible to delineate groups of genes in expression profiles that are together responsible for certain cellular processes and explain the molecular mechanism behind physiological phenomena.

The following sections highlight some selected recent expression profiling studies, which have led to the identification (i) of only a few but critical targets, (ii) of group of transcripts that are involved in common pathways and processes, and (iii) of genes that could be placed into a network construct. We put special emphasis on attempts to study expression profiles of cells/tissues with a role in angiogenic processes but, if appropriate, we also present other examples.

17.4.1 Identification of Targets with Prominent Expression Regulation

Roca et al. [112] investigated the mechanisms underlying the inhibition of in vitro and in vivo angiogenesis during hyperthermia. This treatment often complements radiotherapy or chemotherapy of cancer [113,114]. First, an inhibition of angiogenesis was observed in in vitro ECs on a three-dimensional preparation, which were exposed to temperatures 41°C–45°C. Culture medium treatment with the angiogenic factor VEGF-A_{165} could not neutralize this inhibition. The same observation was found in the in vivo chick embryo chorioallantoic membrane assay. To understand the molecular mechanism, a microarray experiment was performed. The results showed a marked upregulation of the plasminogen activator inhibitor 1 (PAI-1) [112]. PAI-1 is responsible for the extracellular matrix homeostasis during angiogenesis [115]. Finally, the hypothesis of the PAI-1 involvement was further manifested by the indication that anti-PAI-1 antibodies antagonize the hypothermia effects on angiogenesis of in vitro and in vivo systems [112].

The conceptual role of expression profiling in this work is characteristic for many studies: This work started with a phenotypic observation in in vitro and in vivo systems. The investigation has relevance for real therapy. Further, this led to the characterization of the effects at the transcriptional level with microarrays. The hypothesis of PAI-1 involvement in the process of angiogenic inhibition during hypothermia evolved from this experiment and it was proven with additional experiments.

Watanabe et al. [116] searched with cDNA microarrays for vascular endothelial growth factor (VEGF)-inducible genes in human umbilical vein endothelial cells (HUVECs). Ninety-seven genes were upregulated more than twofold by VEGF. This group contained 11 uncharacterized products. In their follow-up study, they concentrated on KIAA1036, which was named vasohibin. The recombinant protein-inhibited VEGF- and FGF2-induced angiogenesis, which indicated that vasohibin secreted into the extracellular space acts on ECs. Further, this endothelial cell-specific transcript inhibited tumor angiogenesis in vivo in transfected Lewis lung carcinoma cells. Therefore, the first evidence of a feedback inhibitory factor involved in angiogenesis was found by large-scale microarray techniques combined with appropriate experiments to prove the hypothesis.

Tubulogenesis by epithelial cells regulates kidney, lung, and mammary development, whereas that by ECs regulates vascular development. Albig and Schiemann [117] studied gene expression differences of endothelial tubulogenesis and of epithelial tubulogenesis with the microarray technology. The regulator of G protein signaling 4 (RGS4), which was not previously associated with tubulogenesis, was found to be upregulated greater than eightfold in tubulating cells compared to the control samples, thereby implicating RGS4 as a potential regulator of tubulogenesis.

Zhang et al. [118] wanted to shed light into the molecular mechanism of the phenotypic observation that antithrombin is transformed to a potent angiogenic inhibitor by limited proteolysis or mild heating [119]. Treatment of bFGF-induced and control HUVECs with latent antiangiogenic antithrombin repressed the key proangiogenic heparan sulfate proteoglycan, perlecan, significantly compared with HUVECs treated with native antithrombin. This observation made with the cDNA microarray technology was confirmed on the mRNA and protein level with semiquantitative RT-PCR, northern blotting, and immunoblotting analyses. Finally, it was shown that the exposure to TGF-beta1, another stimulator of perlecan expression, overcomes the inhibitory effect of antiangiogenic antithrombin [118].

Yuan et al. [120] investigated the role of ephrinB2 and its receptor ephB4, which have to be localized in the membranes of adjacent cells for interaction, in inflammatory angiogenesis. cDNA membrane microarray experiments identified a group of 13 upregulated genes after ephrinB2 stimulation of HUVEC. Syntenin, which binds syndecan, is a member of this gene group. Subsequently, RT-PCR and northern blotting confirmed the increased expression of syndecan-1. Further experiments showed that ephrin stimulated upregulation of syndecan-1, which exhibited in vitro antiangiogenic and in vivo angiogenic effects. Due to the fact that the enzyme heparanase, which converts soluble syndecan-1 from an inhibitor into a potent activator of angiogenesis, is expressed in inflammatory environments, controversial effects were observed [120]. This study emphasizes that the observations of simplified in vitro studies cannot always be generalized for in vivo systems even if the same expression is measured. Due to additional, probably unmeasured signals and proteins, further rational inspiration and literature research can help to identify missing links of the original expression profiles. This is especially important if only a limited number of genes were profiled. Therefore, it is critical to keep in mind that a gene expression profile mirrors only a snapshot of the transcripts at a specific time in a specific environment, which might not be able to explain effects caused at higher levels of regulations, for example, posttranslation modifications. Further, genes that are not arrayed slip through the measurement in the case of DNA chips.

The Nanog story nicely illustrates the difficulties in proper interpretation of expression data. Microarray-based studies of expression profiles focused on embryonic stem cells have generated a long list of putative important genes involved in the core stem cell properties [121]. Only a year later, Nanog was identified as a key gene with important roles in pluripotency and self-renewal of embryonic stem cells [122]. The vast amount of uncharacterized genes of the original experiment left this protein unrecognized until it was reidentified with cDNA library screening methods followed by the molecular characterization of the homeodomain containing transcriptional regulator, which proved to be a key player of self-renewal [122]. Yet another year later, a microarray study [123] revealed with hierarchical clustering that Nanog is also highly expressed in carcinoma in situ (CIS), the common precursor of testicular germ cell tumors, and might play an important role in malignant behavior. Comparison of the CIS expression profile showed common patterns with embryonic stem cells supporting the hypothesis of early fetal origin of CIS cells.

Expression analysis studies that focus only on the handful of topregulated genes have received considerable support from large-scale expression profiling and lead to many research results with biological and medical significance. At the same time, it might be considered a waste of effort if the rest of the expression profile was not studied with appropriate scrutiny. Interestingly, most single-factor identification experiments in angiogenesis research did not use basic profile analyses methods such as normalization and clustering. Instead, they concentrated on a single differentially regulated factor and performed further experiments to verify the hypotheses. A higher success rate in identification of novel targets and mechanisms might be achieved if the profiles of large-scale expression studies are analyzed rigorously with the goal of deriving hypothesis from these results. Naturally, this effort is combined with huge work of computational analysis (both with respect to expression value studies and gene function prediction with sequence-analytic methods together with literature research). We think that such a hypothesis-driven strategy might prevent some trial-and-error approaches and, therefore, redirects the sequel experiments into the right direction reducing time- and resource-consumption of follow-up studies.

17.4.2 Identification of Groups of Target Genes Responsible for Cellular Mechanisms

In the consideration of expression studies, quality aspects of sample preparation are often overlooked. This issue is especially important for angiogenesis-related investigations since ECs form only single layers on the base membrane in vessels that intimately penetrate other tissues. The pure extraction of endothelial cell from human tissue samples in a form useful for expression profiling from in vivo material was first achieved by St. Croix et al. [2]. As an application, they examined differential gene expression in normal and tumor ECs.

Purified human ECs were subjected to SAGE profiling (with sequencing of approximately 100,000 tags). On the one hand, 93 pan-endothelial markers were extracted with similarly high expression in normal and tumor ECs. On the other hand, 33 tags (named normal endothelial markers; NEM) and 46 tags were identified to be specifically elevated in normal endothelium and in tumor endothelium, respectively. Despite the methodical progress, the work of St. Croix et al. left several issues open.

1. Surprisingly, the second group of 46 tags was called tumor endothelial markers (TEMs), although the authors could not bring up any biological argument justifying specificity of the ECs extracted from tumor tissue for the tumor state. Not surprisingly, an in-depth sequence-analytic study of the TEMs [1] showed that they are generally highly expressed during wound healing, corpus luteum formation, bone restructuring, and so on. Thus, the TEMs are in fact markers of restructuring tissues and the angiogenesis therein and, thus, describe common features of the respective cellular processes.
2. Many tags were left without link to genes and many genes missed a functional characterization. Novatchkova and Eisenhaber [1] carried out a case study of the identification of molecular mechanism from gene expression profiles. A careful strategy of tag-to-gene mapping was applied, which led to only nine tags without a link to a gene. With a variety of different sequence analysis algorithms, putative nonglobular domains and globular domains were assigned to the protein sequence. This information of the protein architecture together with homology searches and target-specific scientific literature links led to molecular and cellular function and mechanistic insights of the involved PEMs and TEMs.

Novatchkova and Eisenhaber found that most TEM targets identified are relatively downstream of the regulatory cascades leading to angiogenesis. They are involved in more general features such as extracellular matrix remodeling, migration, adhesion, and cell–cell communication typical for migrating cells in restructuring tissues rather than specific for angiogenesis initiation and control. Therefore, these genes should be termed more correctly as angiogenic endothelial markers. Apparently, also the abundance of transcripts is important. The identification of genes that are mainly endpoints of regulation cascades mirrors the preferred registration of highly abundant transcripts in the original SAGE study. Further, it was shown that in vitro endothelial expression studies are very different from in vivo profiles and, therefore, hardly representative for the in vivo situation.

SP100 interacts with the proangiogenic transcription factor ETS1 and represses, thereby, the DNA-binding and transcriptional activity [124,125]. Since SP100 is also expressed in ECs, Yordy et al. [126] tested the hypothesis that SP100 alters the ETS1 response in ECs. It was shown that the antiangiogenic interferons IFN-α, IFN-γ, and TNF-γ induce SP100 in ECs. To characterize the negative phenomenological modulation of SP100 on ETS1 on the molecular level, gene expression profiles were generated with microarrays. With hierarchical clustering, 1000 ESTs were grouped and the complexity was further reduced with a self-organizing tree algorithm. This resulted in 129 ESTs, which are reciprocally regulated between the SP100 and ETS1 experiments. Gene ontology terms were assigned. This strategy led to the conclusion that only a part of ETS1 regulated genes can be modulated by SP100 but genes that show reciprocal expression in response to ETS1 and SP100 have antimigratory or antiangiogenic properties.

Nagawaka et al. [127] investigated the influence of hematopoietically expressed homeobox (HEX), which is transiently upregulated in ECs during vascular formation in embryos [128]. Therefore, HUVECs were transfected with the transcription factor HEX and the effects on the transcriptome were compared relative to not transfected cells with cDNA microarray. The repression of several important factors including VEGFR-1, VEGFR-1, TIE-1, TIE-2, and neuropilin-1 was observed. In contrast, mRNA levels of endoglin, ephrin B2, urokinase-type plasminogen activator, tissue-type plasminogen activator, plasminogen activator inhibitor-1, TIMP-1, TIMP-2, and TIMP-3 were upregulated greater than twofold [127]. This expression profile mirrors a repression of angiogenesis induced by HEX in an in vitro study and assigns, thereby, an antiangiogenic role to HEX in this environment.

Angiogenesis is an important process for tumor growth [129]. Nevertheless, it was reported that tumors can grow without new vessel formation by taking advantage of preexisting ones [130]. Hu et al. [131] investigated the transcriptional differences between angiogenic and nonangiogenic non-small-cell lung cancer with cDNA microarrays. These two phenomenological characteristics could be distinguished by a group of 62 genes. It was found that nonangiogenic tumors have more upregulated genes involved in mitochondrial metabolism. Thus, intracellular respiration appears more effective in nonangiogenic tumors.

Thijssen et al. [132] followed an interesting approach to profile the important communication and interaction between tumor and ECs. Therefore, a xenograft tumor was constructed by inoculation of mice with human tumor cells. Quantitative RT-PCR (qRT-PCR) was used to measure the expression level of 16 angiogenesis-related genes. To distinguish between the mRNA origins, primers specific for mouse and human were constructed in low homology regions. bFGF expression was for instance 300-fold higher in nontumor cells. The real power of this method was shown by monitoring the changes in the two compartments in response to different antitumor treatments.

The combination of VEGF and hepatocyte growth factor (HGF) induces a more robust angiogenesis than by each factor alone [133]. Gerritsen et al. [134] investigated the underlying mechanisms of this additive affect by measuring the response of VEGF and HGF as well as the combination of both with Affymetrix microarrays. Relative log ratios were obtained by comparing the induced profile with the basal one. Only a small fraction of the enhanced genes were upregulated in both datasets after VEGF or HGF stimulation. The synergistic interaction differentially regulated twice as many genes as for each growth factor alone, including many totally different genes. This study emphasizes the importance to investigate not only the response of a single-factor but also the synergistic combination of two or more to identify the important targets for an efficient antiangiogenic cancer therapy.

17.4.3 Identification of Gene Networks

The reconstruction of gene networks from expression profiles is the ultimate goal; although, it remains a matter of the future with respect to angiogenesis. In this section, we describe a few examples that highlight the principal possibilities. It should be noted that most network reconstruction studies were carried out on yeast and other model organisms with small gene numbers. Even under these conditions, the level of system understanding and the biological significance of results remain limited.

Ideker et al. [8] investigated the galactose utilization in yeast by an integrated approach of genomic and proteomic data. Therefore, the genome sequence and the identity genes, proteins, and other players in pathways were known at the beginning and an initial qualitative (topology) model of the galactose related part of the network could be constructed a priori. The authors introduced several perturbations in yeast cultures and recorded the genomic and proteomic response. Analysis of the expression profiles allowed improving the model of the network. This process of perturbation and integration of its resulting influences into the model was iteratively repeated. This approach led to a number of refinements of the galactose-utilization model with regard to network interactions and new regulators.

Basso et al. [135] claim to have reconstructed a genetic network from large-scale expression profiles in the mammalian cells. For a substantial range of absolute expression, a large set of 336 human B-cell phenotypes were profiled with microarrays and an algorithm for the reconstruction of accurate cellular networks (ARACN) was applied. The network indicates a topology of few highly connected hubs and a high number of less connected nodes with a hierarchical control mechanism, which is already known for phylogenetically earlier organisms [135,136]. Further the gene named BYSL was biochemically identified to be part of the proto-oncogene MYC network, which might have important cellular roles. The interpretation approach might be especially interesting for investigation of tumor subtype differentiation [135].

Gao et al. [109] analyzed transcription networks by combining genome-wide expression data of ~750 expression pattern and the genome-wide promoter occupancies for 113 transcription factors. Surprisingly, roughly half of the transcription factor targets, which were predicted by the ChIP analysis, were not regulated by them. Therefore, the regulatory coupling strength was calculated between all transcription factors and genes, which might enhance the specificity of target predictions. The importance of coupling was further manifested by enrichment of specific gene ontology categories and significant change in mRNA expression of transcription factor deletion mutants in the coupled test group compared with the bound transcription factor but uncoupled group. This study highlights the fact that binding of a transcription factor is not sufficient to predict the regulation of the gene.

17.5 CONCLUSION

This review has considered both experimental and computational biology methodologies used for gene expression profiling as well as applications of these techniques to angiogenesis. Although the expectations from expression profiling techniques within the scientific community were initially high (especially with respect to the system-wide view and analysis), the additional insight produced by the respective research efforts is limited and our understanding of molecular processes remains fragmentary. Nobody has ever calculated the costs that were funneled into expression profiling in general and for angiogenesis and tumor analysis specifically and analyzed possible outcomes of alternative investment scenarios. Nevertheless, it is widely accepted that expression profiling methods have become an indispensable part of the arsenal of modern research methods. And indeed, in carefully selected cases, these techniques will generate the decisive data.

We think that, for some applications, currently available gene expression profiling methods might be simply insufficiently mature. We expect their considerable technical sophistication in coming years with emerging solutions for improving accuracy, reproducibility, sensitivity, and reduced requirements in sample size. Single-cell expression profiling will certainly become a standard. In addition to technical improvements for expression measuring, the consequent application of available computational biology approaches (for the postexperimental treatment of data) can extract more value from available expression profiles already now. Normalization and clustering of expression vectors should become a standard as well as sequence-based function prediction for insufficiently characterized genes and their protein products. Most studies have not invested enough for these purposes and the respective principal investigators should be encouraged to look after appropriate collaborations.

Finally, expression profiling should be viewed in context of cellular hierarchies and the different methods that generate information about them. For proper biological interpretation, it will be important to integrate genome sequence data, gene and protein functional annotation, gene expression profiles, and protein occurrence data from proteomics experiments.

ACKNOWLEDGMENTS

This work was supported by Boehringer Ingelheim and by the Gen-AU BIN Program.

REFERENCES

1. Novatchkova, M. and Eisenhaber, F., Can molecular mechanisms of biological processes be extracted from expression profiles? Case study: endothelial contribution to tumor-induced angiogenesis, *Bioessays*, 23, 1159, 2001.
2. St. Croix, B. et al., Genes expressed in human tumor endothelium, *Science*, 289, 1197, 2000.
3. Velculescu, V.E. et al., Serial analysis of gene expression, *Science*, 270, 484, 1995.
4. Brenner, S. et al., Gene expression analysis by massively parallel signature sequencing (MPSS) on microbead arrays, *Nat Biotechnol*, 18, 630, 2000.
5. Lipshutz, R.J. et al., Using oligonucleotide probe arrays to access genetic diversity, *Biotechniques*, 19, 442, 1995.
6. Schena, M. et al., Quantitative monitoring of gene expression patterns with a complementary DNA microarray, *Science*, 270, 467, 1995.
7. Gygi, S.P. et al., Correlation between protein and mRNA abundance in yeast, *Mol Cell Biol*, 19, 1720, 1999.
8. Ideker, T. et al., Integrated genomic and proteomic analyses of a systematically perturbed metabolic network, *Science*, 292, 929, 2001.
9. Kern, W. et al., Correlation of protein expression and gene expression in acute leukemia, *Cytometry B Clin Cytom*, 55, 29, 2003.
10. Liang, P. and Pardee, A.B., Differential display of eukaryotic messenger RNA by means of the polymerase chain reaction, *Science*, 257, 967, 1992.
11. Welsh, J. et al., Arbitrarily primed PCR fingerprinting of RNA, *Nucleic Acids Res*, 20, 4965, 1992.
12. Ito, T. et al., Fluorescent differential display: arbitrarily primed RT-PCR fingerprinting on an automated DNA sequencer, *FEBS Lett*, 351, 231, 1994.
13. Liang, P. et al., Differential display using one-base anchored oligo-dT primers, *Nucleic Acids Res*, 22, 5763, 1994.
14. Mou, L. et al., Improvements to the differential display method for gene analysis, *Biochem Biophys Res Commun*, 199, 564, 1994.
15. Carulli, J.P. et al., High throughput analysis of differential gene expression, *J Cell Biochem Suppl*, 30–31, 286, 1998.
16. Sim, G.K. et al., Use of a cDNA library for studies on evolution and developmental expression of the chorion multigene families, *Cell*, 18, 1303, 1979.
17. Patanjali, S.R., Parimoo, S., and Weissman, S.M., Construction of a uniform-abundance (normalized) cDNA library, *Proc Natl Acad Sci USA*, 88, 1943, 1991.
18. Gergen, J.P., Stern, R.H., and Wensink, P.C., Filter replicas and permanent collections of recombinant DNA plasmids, *Nucleic Acids Res*, 7, 2115, 1979.
19. Sive, H.L. and St John, T., A simple subtractive hybridization technique employing photoactivatable biotin and phenol extraction, *Nucleic Acids Res*, 16, 10937, 1988.
20. Byers, R.J. et al., Subtractive hybridization–genetic takeaways and the search for meaning, *Int J Exp Pathol*, 81, 391, 2000.
21. Velculescu, V.E., Vogelstein, B., and Kinzler, K.W., Analysing uncharted transcriptomes with SAGE, *Trends Genet*, 16, 423, 2000.
22. Saha, S. et al., Using the transcriptome to annotate the genome, *Nat Biotechnol*, 20, 508, 2002.
23. Pollock, J.D., Gene expression profiling: methodological challenges, results, and prospects for addiction research, *Chem Phys Lipids*, 121, 241, 2002.
24. Tuteja, R. and Tuteja, N., Serial analysis of gene expression (SAGE): unraveling the bioinformatics tools, *Bioessays*, 26, 916, 2004.
25. Cai, L. et al., Clustering analysis of SAGE data using a Poisson approach, *Genome Biol*, 5, R51, 2004.

26. Singh-Gasson, S. et al., Maskless fabrication of light-directed oligonucleotide microarrays using a digital micromirror array, *Nat Biotechnol*, 17, 974, 1999.
27. Nuwaysir, E.F. et al., Gene expression analysis using oligonucleotide arrays produced by maskless photolithography, *Genome Res*, 12, 1749, 2002.
28. Shoemaker, D.D. et al., Experimental annotation of the human genome using microarray technology, *Nature*, 409, 922, 2001.
29. Schulze, A. and Downward, J., Navigating gene expression using microarrays—a technology review, *Nat Cell Biol*, 3, E190, 2001.
30. Lyons, P., Advances in spotted microarray resources for expression profiling, *Brief Funct Genomic Proteomic*, 2, 21, 2003.
31. Duggan, D.J. et al., Expression profiling using cDNA microarrays, *Nat Genet*, 21, 10, 1999.
32. Affara, N.A., Resource and hardware options for microarray-based experimentation, *Brief Funct Genomic Proteomic*, 2, 7, 2003.
33. Livesey, F.J., Strategies for microarray analysis of limiting amounts of RNA, *Brief Funct Genomic Proteomic*, 2, 31, 2003.
34. Karsten, S.L. et al., An evaluation of tyramide signal amplification and archived fixed and frozen tissue in microarray gene expression analysis, *Nucleic Acids Res*, 30, E4, 2002.
35. Eberwine, J. et al., Analysis of gene expression in single live neurons, *Proc Natl Acad Sci USA*, 89, 3010, 1992.
36. Baugh, L.R. et al., Quantitative analysis of mRNA amplification by in vitro transcription, *Nucleic Acids Res*, 29, E29, 2001.
37. Richter, A. et al., Comparison of fluorescent tag DNA labeling methods used for expression analysis by DNA microarrays, *Biotechniques*, 33, 620, 2002.
38. Czechowski, T. et al., Real-time RT-PCR profiling of over 1400 Arabidopsis transcription factors: unprecedented sensitivity reveals novel root- and shoot-specific genes, *Plant J*, 38, 366, 2004.
39. Horak, C.E. and Snyder, M., Global analysis of gene expression in yeast, *Funct Integr Genomics*, 2, 171, 2002.
40. Zhang, H., Yan, W., and Aebersold, R., Chemical probes and tandem mass spectrometry: a strategy for the quantitative analysis of proteomes and subproteomes, *Curr Opin Chem Biol*, 8, 66, 2004.
41. Linscheid, M.W., Quantitative proteomics, *Anal Bioanal Chem*, 381, 64, 2005.
42. Gygi, S.P. et al., Quantitative analysis of complex protein mixtures using isotope-coded affinity tags, *Nat Biotechnol*, 17, 994, 1999.
43. Bolstad, B.M. et al., Experimental design and low-level analysis of microarray data, *Int Rev Neurobiol*, 60, 25, 2004.
44. Churchill, G.A., Fundamentals of experimental design for cDNA microarrays, *Nat Genet*, 32 Suppl, 490, 2002.
45. Ruijter, J.M., Van Kampen, A.H., and Baas, F., Statistical evaluation of SAGE libraries: consequences for experimental design, *Physiol Genomics*, 11, 37, 2002.
46. Armstrong, N.J. and van de Wiel, M.A., Microarray data analysis: from hypotheses to conclusions using gene expression data, *Cell Oncol*, 26, 279, 2004.
47. Yang, Y.H. and Speed, T., Design issues for cDNA microarray experiments, *Nat Rev Genet*, 3, 579, 2002.
48. Quackenbush, J., Microarray data normalization and transformation, *Nat Genet*, 32 Suppl, 496, 2002.
49. Yang, I.V. et al., Within the fold: assessing differential expression measures and reproducibility in microarray assays, *Genome Biol*, 3, research0062, 2002.
50. Pieler, R. et al., ArrayNorm: comprehensive normalization and analysis of microarray data, *Bioinformatics*, 20, 1971, 2004.
51. Gentleman, R.C. et al., Bioconductor: open software development for computational biology and bioinformatics, *Genome Biol*, 5, R80, 2004.
52. Sturn, A., Quackenbush, J., and Trajanoski, Z., Genesis: cluster analysis of microarray data, *Bioinformatics*, 18, 207, 2002.
53. Brazma, A. et al., Minimum information about a microarray experiment (MIAME)-toward standards for microarray data, *Nat Genet*, 29, 365, 2001.

54. Wessels, L.F. et al., A protocol for building and evaluating predictors of disease state based on microarray data, *Bioinformatics*, 21, 3755, 2005.
55. 't Veer, L.J. et al., Expression profiling predicts outcome in breast cancer, *Breast Cancer Res*, 5, 57, 2003.
56. Saeed, A.I. et al., TM4: a free, open-source system for microarray data management and analysis, *Biotechniques*, 34, 374, 2003.
57. Quackenbush, J., Computational analysis of microarray data, *Nat Rev Genet*, 2, 418, 2001.
58. Brazma, A. and Vilo, J., Gene expression data analysis, *FEBS Lett*, 480, 17, 2000.
59. Eisen, M.B. et al., Cluster analysis and display of genome-wide expression patterns, *Proc Natl Acad Sci USA*, 95, 14863, 1998.
60. Tavazoie, S. et al., Systematic determination of genetic network architecture, *Nat Genet*, 22, 281, 1999.
61. Raychaudhuri, S., Stuart, J.M., and Altman, R.B., Principal components analysis to summarize microarray experiments: application to sporulation time series, *Pac Symp Biocomput*, 455, 2000.
62. Bork, P. et al., Predicting function: from genes to genomes and back, *J Mol Biol*, 283, 707, 1998.
63. Zhang, Z. et al., A greedy algorithm for aligning DNA sequences, *J Comput Biol*, 7, 203, 2000.
64. Pruitt, K.D. et al., Introducing RefSeq and LocusLink: curated human genome resources at the NCBI, *Trends Genet*, 16, 44, 2000.
65. Pruitt, K.D., Tatusova, T., and Maglott, D.R., NCBI Reference Sequence (RefSeq): a curated non-redundant sequence database of genomes, transcripts and proteins, *Nucleic Acids Res*, 33 Database Issue, D501, 2005.
66. Okazaki, Y. et al., Analysis of the mouse transcriptome based on functional annotation of 60,770 full-length cDNAs, *Nature*, 420, 563, 2002.
67. Schuler, G.D., Pieces of the puzzle: expressed sequence tags and the catalog of human genes, *J Mol Med*, 75, 694, 1997.
68. Quackenbush, J. et al., The TIGR gene indices: reconstruction and representation of expressed gene sequences, *Nucleic Acids Res*, 28, 141, 2000.
69. Altschul, S.F. et al., Basic local alignment search tool, *J Mol Biol*, 215, 403, 1990.
70. Ashburner, M. et al., Gene ontology: tool for the unification of biology. The Gene Ontology Consortium, *Nat Genet*, 25, 25, 2000.
71. Wootton, J.C. and Federhen, S., Analysis of compositionally biased regions in sequence databases, *Methods Enzymol*, 266, 554, 1996.
72. Brendel, V. et al., Methods and algorithms for statistical analysis of protein sequences, *Proc Natl Acad Sci USA*, 89, 2002, 1992.
73. Promponas, V.J. et al., CAST: an iterative algorithm for the complexity analysis of sequence tracts. Complexity analysis of sequence tracts, *Bioinformatics*, 16, 915, 2000.
74. Linding, R. et al., GlobPlot: Exploring protein sequences for globularity and disorder, *Nucleic Acids Res*, 31, 3701, 2003.
75. Bendtsen, J.D. et al., Improved prediction of signal peptides: SignalP 3.0, *J Mol Biol*, 340, 783, 2004.
76. von Heijne, G., A new method for predicting signal sequence cleavage sites, *Nucleic Acids Res*, 14, 4683, 1986.
77. Neuberger, G. et al., Hidden localization motifs: naturally occurring peroxisomal targeting signals in non-peroxisomal proteins, *Genome Biol*, 5, R97, 2004.
78. Neuberger, G. et al., Prediction of peroxisomal targeting signal 1 containing proteins from amino acid sequence, *J Mol Biol*, 328, 581, 2003.
79. Neuberger, G. et al., Motif refinement of the peroxisomal targeting signal 1 and evaluation of taxon-specific differences, *J Mol Biol*, 328, 567, 2003.
80. Eisenhaber, B., Bork, P., and Eisenhaber, F., Prediction of potential GPI-modification sites in proprotein sequences, *J Mol Biol*, 292, 741, 1999.
81. Eisenhaber, B. et al., Glycosylphosphatidylinositol lipid anchoring of plant proteins. Sensitive prediction from sequence- and genome-wide studies for Arabidopsis and rice, *Plant Physiol*, 133, 1691, 2003.
82. Eisenhaber, B. et al., A sensitive predictor for potential GPI lipid modification sites in fungal protein sequences and its application to genome-wide studies for *Aspergillus nidulans*, *Candida*

albicans, *Neurospora crassa*, *Saccharomyces cerevisiae* and *Schizosaccharomyces pombe*, *J Mol Biol*, 337, 243, 2004.

83. Maurer-Stroh, S., Eisenhaber, B., and Eisenhaber, F., N-terminal N-myristoylation of proteins: prediction of substrate proteins from amino acid sequence, *J Mol Biol*, 317, 541, 2002.
84. Maurer-Stroh, S. and Eisenhaber, F., Refinement and prediction of protein prenylation motifs, *Genome Biology*, 6, R55, 2005.
85. Tusnady, G.E. and Simon, I., Principles governing amino acid composition of integral membrane proteins: application to topology prediction, *J Mol Biol*, 283, 489, 1998.
86. Cserzo, M. et al., On filtering false positive transmembrane protein predictions, *Protein Eng*, 15, 745, 2002.
87. Lupas, A., Van Dyke, M., and Stock, J., Predicting coiled coils from protein sequences, *Science*, 252, 1162, 1991.
88. Eisenhaber, F. et al., Prediction of secondary structural content of proteins from their amino acid composition alone. I. New analytic vector decomposition methods, *Proteins*, 25, 157, 1996.
89. Eisenhaber, F., Frommel, C., and Argos, P., Prediction of secondary structural content of proteins from their amino acid composition alone. II. The paradox with secondary structural class, *Proteins*, 25, 169, 1996.
90. Frishman, D. and Argos, P., Incorporation of non-local interactions in protein secondary structure prediction from the amino acid sequence, *Protein Eng*, 9, 133, 1996.
91. Combet, C. et al., NPS@: network protein sequence analysis, *Trends Biochem Sci*, 25, 147, 2000.
92. Letunic, I. et al., SMART 4.0: towards genomic data integration, *Nucleic Acids Res*, 32 Database issue, 142, 2004.
93. Marchler-Bauer, A. et al., CDD: a database of conserved domain alignments with links to domain three-dimensional structure, *Nucleic Acids Res*, 30, 281, 2002.
94. Schaffer, A.A. et al., IMPALA: matching a protein sequence against a collection of PSI-BLAST-constructed position-specific score matrices, *Bioinformatics*, 15, 1000, 1999.
95. Sigrist, C.J. et al., PROSITE: a documented database using patterns and profiles as motif descriptors, *Brief Bioinform*, 3, 265, 2002.
96. Schaffer, A.A. et al., Improving the accuracy of PSI-BLAST protein database searches with composition-based statistics and other refinements, *Nucleic Acids Res*, 29, 2994, 2001.
97. Altschul, S.F. and Koonin, E.V., Iterated profile searches with PSI-BLAST—a tool for discovery in protein databases, *Trends Biochem Sci*, 23, 444, 1998.
98. Altschul, S.F. et al., Gapped BLAST and PSI-BLAST: a new generation of protein database search programs, *Nucleic Acids Res.*, 25, 3389, 1997.
99. Wistrand, M. and Sonnhammer, E.L., Improved profile HMM performance by assessment of critical algorithmic features in SAM and HMMER, *BMC Bioinformatics*, 6, 99, 2005.
100. Wistrand, M. and Sonnhammer, E.L., Improving profile HMM discrimination by adapting transition probabilities, *J Mol Biol*, 338, 847, 2004.
101. Eisenhaber, F., Wechselberger, C., and Kreil, G., The Brix domain protein family—a key to the ribosomal biogenesis pathway? *Trends Biochem Sci*, 26, 345, 2001.
102. Novatchkova, M. and Eisenhaber, F., Linking transcriptional mediators via the GACKIX domain super family, *Curr Biol*, 14, R54, 2004.
103. Kurzbauer, R. et al., Crystal structure of the p14/MP1 scaffolding complex: how a twin couple attaches mitogen-activated protein kinase signaling to late endosomes, *Proc Natl Acad Sci USA*, 101, 10984, 2004.
104. Ginalski, K. et al., 3D-Jury: a simple approach to improve protein structure predictions, *Bioinformatics*, 19, 1015, 2003.
105. Rung, J. et al., Building and analysing genome-wide gene disruption networks, *Bioinformatics*, 18 Suppl 2, 202, 2002.
106. Soinov, L.A., Krestyaninova, M.A., and Brazma, A., Towards reconstruction of gene networks from expression data by supervised learning, *Genome Biol*, 4, R6, 2003.
107. Jenssen, T.K. et al., A literature network of human genes for high-throughput analysis of gene expression, *Nat Genet*, 28, 21, 2001.
108. Stapley, B.J. and Benoit, G., Biobibliometrics: information retrieval and visualization from co-occurrences of gene names in Medline abstracts, *Pac Symp Biocomput*, 529, 2000.

109. Gao, F., Foat, B.C., and Bussemaker, H.J., Defining transcriptional networks through integrative modeling of mRNA expression and transcription factor binding data, *BMC Bioinformatics*, 5, 31, 2004.
110. Matys, V. et al., TRANSFAC: transcriptional regulation, from patterns to profiles, *Nucleic Acids Res*, 31, 374, 2003.
111. Roven, C. and Bussemaker, H.J., REDUCE: an online tool for inferring cis-regulatory elements and transcriptional module activities from microarray data, *Nucleic Acids Res*, 31, 3487, 2003.
112. Roca, C. et al., Hyperthermia inhibits angiogenesis by a plasminogen activator inhibitor 1-dependent mechanism, *Cancer Res*, 63, 1500, 2003.
113. Overgaard, J. et al., Randomised trial of hyperthermia as adjuvant to radiotherapy for recurrent or metastatic malignant melanoma. European Society for Hyperthermic Oncology, *Lancet*, 345, 540, 1995.
114. Falk, M.H. and Issels, R.D., Hyperthermia in oncology, *Int J Hyperthermia*, 17, 1, 2001.
115. Johnsen, M. et al., Cancer invasion and tissue remodeling: common themes in proteolytic matrix degradation, *Curr Opin Cell Biol*, 10, 667, 1998.
116. Watanabe, K. et al., Vasohibin as an endothelium-derived negative feedback regulator of angiogenesis, *J Clin Invest*, 114, 898, 2004.
117. Albig, A.R. and Schiemann, W.P., Identification and characterization of regulator of G protein signaling 4 (RGS4) as a novel inhibitor of tubulogenesis: RGS4 inhibits mitogen-activated protein kinases and vascular endothelial growth factor signaling, *Mol Biol Cell*, 16, 609, 2005.
118. Zhang, W. et al., Antiangiogenic antithrombin down-regulates the expression of the proangiogenic heparan sulfate proteoglycan, perlecan, in endothelial cells, *Blood*, 103, 1185, 2004.
119. O'Reilly, M.S. et al., Antiangiogenic activity of the cleaved conformation of the serpin antithrombin, *Science*, 285, 1926, 1999.
120. Yuan, K. et al., Syndecan-1 up-regulated by ephrinB2/EphB4 plays dual roles in inflammatory angiogenesis, *Blood*, 104, 1025, 2004.
121. Ramalho-Santos, M. et al., "Stemness": transcriptional profiling of embryonic and adult stem cells, *Science*, 298, 597, 2002.
122. Chambers, I. et al., Functional expression cloning of Nanog, a pluripotency sustaining factor in embryonic stem cells, *Cell*, 113, 643, 2003.
123. Almstrup, K. et al., Embryonic stem cell-like features of testicular carcinoma in situ revealed by genome-wide gene expression profiling, *Cancer Res*, 64, 4736, 2004.
124. Vandenbunder, B. et al., Complementary patterns of expression of c-ets 1, c-myb and c-myc in the blood-forming system of the chick embryo, *Development*, 107, 265, 1989.
125. Yordy, J.S. et al., SP100 expression modulates ETS1 transcriptional activity and inhibits cell invasion, *Oncogene*, 23, 6654, 2004.
126. Yordy, J.S. et al., SP100 inhibits ETS1 activity in primary endothelial cells, *Oncogene*, 24, 916, 2005.
127. Nakagawa, T. et al., HEX acts as a negative regulator of angiogenesis by modulating the expression of angiogenesis-related gene in endothelial cells in vitro, *Arterioscler Thromb Vasc Biol*, 23, 231, 2003.
128. Thomas, P.Q., Brown, A., and Beddington, R.S., Hex: a homeobox gene revealing peri-implantation asymmetry in the mouse embryo and an early transient marker of endothelial cell precursors, *Development*, 125, 85, 1998.
129. Folkman, J., Angiogenesis in cancer, vascular, rheumatoid and other disease, *Nat Med*, 1, 27, 1995.
130. Pezzella, F. et al., Non-small-cell lung carcinoma tumor growth without morphological evidence of neo-angiogenesis, *Am J Pathol*, 151, 1417, 1997.
131. Hu, J. et al., Gene expression signature for angiogenic and nonangiogenic non-small-cell lung cancer, *Oncogene*, 24, 1212, 2005.
132. Thijssen, V.L. et al., Angiogenesis gene expression profiling in xenograft models to study cellular interactions, *Exp Cell Res*, 299, 286, 2004.
133. Xin, X. et al., Hepatocyte growth factor enhances vascular endothelial growth factor-induced angiogenesis in vitro and in vivo, *Am J Pathol*, 158, 1111, 2001.
134. Gerritsen, M.E. et al., Using gene expression profiling to identify the molecular basis of the synergistic actions of hepatocyte growth factor and vascular endothelial growth factor in human endothelial cells, *Br J Pharmacol*, 140, 595, 2003.

135. Basso, K. et al., Reverse engineering of regulatory networks in human B cells, *Nat Genet*, 37, 382, 2005.
136. Jeong, H. et al., The large-scale organization of metabolic networks, *Nature*, 407, 651, 2000.
137. Barrett, T. et al., NCBI GEO: mining millions of expression profiles—database and tools, *Nucleic Acids Res*, 33, D562, 2005.
138. Edgar, R., Domrachev, M., and Lash, A.E., Gene Expression Omnibus: NCBI gene expression and hybridization array data repository, *Nucleic Acids Res*, 30, 207, 2002.
139. Parkinson, H. et al., ArrayExpress—a public repository for microarray gene expression data at the EBI, *Nucleic Acids Res*, 33, D553, 2005.
140. Brazma, A. et al., ArrayExpress—a public repository for microarray gene expression data at the EBI, *Nucleic Acids Res*, 31, 68, 2003.
141. Lash, A.E. et al., SAGEmap: a public gene expression resource, *Genome Res*, 10, 1051, 2000.
142. Su, A.I. et al., Large-scale analysis of the human and mouse transcriptomes, *Proc Natl Acad Sci USA*, 99, 4465, 2002.
143. Bono, H. et al., Systematic expression profiling of the mouse transcriptome using RIKEN cDNA microarrays, *Genome Res*, 13, 1318, 2003.
144. Liang, P., SAGE Genie: a suite with panoramic view of gene expression, *Proc Natl Acad Sci USA*, 99, 11547, 2002.
145. Shyamsundar, R. et al., A DNA microarray survey of gene expression in normal human tissues, *Genome Biol*, 6, R22, 2005.

Part III

Translating Angiogenesis Inhibitors to the Clinic

18 Clinical Trial Design and Regulatory Issues

Ramzi N. Dagher and Richard Pazdur

CONTENTS

Historically, drugs developed for cancer treatment have largely been cytotoxic agents with significant toxicities related to effects on rapidly dividing cells. These drugs have mechanisms of action that involve major disruptions in DNA and RNA synthesis or function (1,2). Early stages of development have focused on identification of maximally tolerated doses and signals of activity in refractory populations. Later stages of development have usually focused on evaluation of safety and efficacy in patient populations mainly defined by clinical and histologic criteria.

The explosion in knowledge regarding disease pathogenesis and the role that intercellular signaling and intracellular processes play in malignant transformation provide the potential for development of drugs that target specific disease pathways or components of a pathway. Processes such as angiogenesis, apoptosis, and the biological basis for metastatic behavior are all examples (3,4). An understanding of these processes provides an opportunity to explore new therapeutic classes and offers challenges in defining patient populations using molecular diagnostic techniques that extend beyond our current clinical and histologic armamentarium (5). The more favorable toxicity profiles of these novel agents and their availability in oral formulations enable oncologists to implement clinical trial designs not used with intravenously administered cytotoxic drugs.

This chapter provides a summary of regulatory considerations and their impact on oncology drug development. An overview of FDA's role in the drug development process, requirements for demonstration of safety and effectiveness, and discussion of end points used in the evaluation of oncology drugs are discussed. Clinical trial design considerations relevant to targeted therapies, novel approaches for defining patient populations, and the role of diagnostic tests are also discussed. Finally, FDA initiatives to promote evaluation of drug therapies in pediatric populations are briefly summarized.

18.1 FDA OVERSIGHT OF CLINICAL TRIALS

Investigational drugs must be administered under an investigational new drug (IND) application submitted to the FDA. The regulations describe two parties involved in IND submission, the *sponsor*, who is responsible for reporting to the FDA, and the *investigator*, who performs the trial. The sponsor may be a pharmaceutical company, an academic institution, or an individual (e.g., the sponsor/investigator). Sponsors are to select only investigators "qualified by training and experience as appropriate experts to investigate the drug."

18.1.1 Initial IND Submission

When an IND is submitted to the FDA, a team of scientific reviewers evaluate the safety data from animals or other sources, evaluate the proposed phase 1 study, and determine whether patients would be exposed to an unreasonable and significant risk. These issues are discussed individually in the later sections.

18.1.2 Need to Submit an IND Application

All studies of nonapproved drugs must be done under an IND. For approved drugs, however, some studies require an IND and others are exempt from the IND requirement. To determine that an IND is not needed, the investigator and sponsor must find that the study meets all of the five exemption requirements: the study (i) is not intended to support approval of a new indication or a significant change in the product labeling, (ii) is not intended to support a significant change in advertising, (iii) does not involve a route of administration or dosage level or use in a patient population or other factor *that significantly increases the risks* associated with the use of the drug product, (iv) is conducted in compliance with institutional review board (IRB) and informed consent regulations, and (v) will not be used to promote unapproved indications.

A recent cancer drug guidance clarifies the FDA's interpretation of the IND exemption regulations with regard to the criteria that constitute a significant increase in risk from use of approved cancer drugs in clinical studies. Oncologists frequently use cancer drugs in doses and in combinations not yet described in the label. Such "off-label" use, when safety has been demonstrated by published data or past clinical experience, is not considered an increased risk, and would not require an IND for study. The cancer IND exemption guidance provides examples to clarify the FDA interpretation (6).

18.1.3 IND Application Process

The IND process spans the entire period of drug investigation. It includes the initial IND application and later IND amendments to provide safety reports or additional protocols. The initial IND application usually consists of a phase 1 clinical protocol and data to support the safety of the proposal. The latter would include in vitro animal or human evidence describing drug toxicity and allowing prediction of a safe starting dose. Manufacturing data describing the composition, manufacture, and control of the drug substance and drug product are also included. After FDA receives the initial IND, sponsors are required to wait for 30 days before initiating the proposed study unless they request and receive a waiver of the 30-day review period from FDA. A multidisciplinary team of FDA reviewers, including oncologists, animal toxicologists, chemists, and clinical pharmacologists, determine whether the study is safe to proceed. The FDA may put an IND "on hold" if the agency believes subjects would be exposed to unreasonable risk of injury or if there is insufficient information to assess the risks. The most common reason for a hold is insufficient information to support the safety of the proposed dose or regimen.

The FDA frequently meets with sponsors and investigators in pre-IND meetings to review proposed IND plans and to clarify IND requirements. The FDA has provided guidance on the design of preclinical studies required to support the proposed phase 1 study. For oncology drugs, at least two studies are usually needed, one in a rodent and one in a nonrodent species. Animal studies should use the same schedule of administration proposed for the phase 1 clinical study. The starting dose for investigational drugs used in human studies is usually one-tenth of the mouse STD_{10} (dose where 10% of animals have severe toxicity) calculated on a milligrams per meter squared basis provided this dose does not cause irreversible toxicity in nonrodents. If this dose causes irreversible toxicity, one-sixth of the highest dose that does not produce irreversible toxicity is selected for the starting dose (7).

18.1.3.1 Phase 1 Trial Design

The FDA has accepted a variety of phase 1 trial designs for cancer drugs. In the 1980s and early 1990s, the modified Fibonacci scheme was commonly used. Pharmacologically guided dosing was evaluated in the early 1990s with limited success, but was difficult logistically. Beginning in the early and mid 1990s, the FDA allowed investigators to use a variety of new methods for accelerating dose escalation (8).

18.1.4 FDA Involvement in Clinical Trial Design

The FDA meets frequently with commercial IND sponsors throughout drug development. A critical FDA role in drug development is to meet with sponsors to provide advice on the design of phase 3 (and sometimes phase 2) clinical trials that will support marketing applications. The multidisciplinary FDA team attending these meetings includes oncologists, statisticians, clinical pharmacologists, and often external expert consultants. FDA chemists also meet to discuss manufacturing and quality control issues. Recent legislation allows sponsors to submit protocols subsequent to these meetings and request a "Special Protocol Assessment" that provides for a binding agreement on protocol design (9). After the clinical trials have been conducted and trial results are available, sponsors again meet with FDA in "pre-NDA" meetings to discuss whether a new drug application (NDA) may be warranted and, if so, to discuss details of an NDA submission.

18.1.5 FDA and the Drug Approval Process

After clinical trials have been completed and an NDA has been submitted, the FDA verifies data quality and judges whether trial results demonstrate that the drug is safe and effective for the proposed use. After approval, the FDA continues to evaluate drug safety and regulate drug marketing.

The package insert describes clinical trial results from data that have been reviewed and validated by FDA review teams. Regulations require that NDAs contain all relevant information about manufacturing, preclinical pharmacology and toxicology, human pharmacokinetics and bioavailability, clinical data, and statistical analyses. NDA applicants must submit financial disclosure information about investigators. The FDA review of the NDA involves a multidisciplinary team of chemists, toxicologists, clinical pharmacologists, oncologists, statisticians, microbiologists, site inspectors, and a project manager. The FDA reviewers evaluate the primary data, available in the form of case report forms or electronic data, verify analyses, and where appropriate perform additional analyses. The FDA field inspectors verify that information on case report forms is supported by source data such as hospital charts. This review process leads to a high level of confidence in the information that supports NDA approval and that is described in the package insert.

Applications are prioritized for review according to their importance. Based on the Prescription Drug User Fee Act (PDUFA), the FDA performs NDA review with either a 6 month or a 10 month goal. Applications representing a significant improvement compared to marketed products are assigned *priority* status and a 6 month review goal; whereas, *standard* applications have a 10 month review goal.

FDA routinely seeks external advice on the design, analysis, and interpretation of clinical trials. Consultants are screened to exclude conflict of interest. Individual consultants advise FDA during the design of clinical trials and early stages of NDA review. After initial NDA review, the FDA presents selected NDAs to the Oncologic Drugs Advisory Committee (ODAC). This group is composed of oncologists, statisticians, patient advocates, consumer representatives, and a nonvoting industry representative. At the public meetings of ODAC, the NDA applicant summarizes the results in an initial presentation, FDA presents review findings, ODAC discusses the issues, and ODAC votes on questions submitted by FDA. The FDA is not obligated to adhere to the advice provided.

18.2 REQUIREMENTS FOR DEMONSTRATION OF SAFETY AND EFFICACY

The requirements for demonstration of safety and efficacy before marketing approval stem from two congressional actions. The Food, Drug, and Cosmetic Act of 1938 first outlined requirements for submission of marketing applications with safety information. In 1962, the Kefauver-Harris amendments stipulated demonstration of effectiveness and an assessment of the risk–benefit relationship. For approval in a specific indication, sufficient information must be provided to define an appropriate population for treatment with the drug. The product label, which represents a licensing agreement between the sponsor and the federal government, must also provide adequate information to enable safe and effective use.

In oncology, clinical benefit supporting an efficacy claim has been generally predicated on demonstration of prolongation of life, a better life, or an established surrogate for at least one of these (10,11). An example of an established surrogate may be blood pressure or blood cholesterol. In the oncology setting, durable complete response in some hematologic malignancies has been considered an established surrogate since it is associated with reduction in cancer-related mortality and morbidities such as infections and bleeding.

In 1992, subpart H was added to the NDA regulations, allowing for accelerated approval (AA) of drugs for serious or life-threatening diseases where a drug demonstrates an advantage over available therapy or a benefit in a setting where no available therapy exists (12). Approval is based on a surrogate end point "reasonably likely to predict clinical benefit." The sponsor must study the drug further to demonstrate clinical benefit in subsequent or ongoing clinical trials. In oncology, one strategy used in the development of some anticancer agents (especially for solid tumors) has been to demonstrate an objective response of meaningful magnitude and duration for accelerated approval in a refractory population. To demonstrate clinical benefit, sponsors may evaluate the drug in randomized trials in less refractory patients examining clinical benefit end points such as survival (13). An alternative strategy has been to evaluate response and time-to-progression (TTP) on an interim basis in large randomized trials.

The same trial then provides confirmation of clinical benefit in the same population based on a final analysis of survival. Table 18.1 provides a comparison of regular approval and accelerated approval.

TABLE 18.1
Regular Approval and Accelerated Approval

Regular Approval	Accelerated Approval
Since 1962 amendments to Food Drug and Cosmetic Act	Since 1992 subpart H
Must demonstrate clinical benefit	Surrogate reasonably likely to predict benefit
All disease settings	Serious or life-threatening diseases only
Comparison to available therapy not required	Must demonstrate an advantage over available therapy or effect in a population with no available therapy
Voluntary postmarketing commitments	Confirmation of clinical benefit in ongoing or subsequent trials required

18.3 ONCOLOGY END POINTS

In oncology clinical trials, selection of appropriate end points is a complex decision influenced by the natural history of the disease, stage of disease, the therapies evaluated, and goals of the clinical trial. The following is a discussion of efficacy end points commonly examined in oncology clinical trials.

Overall survival (OR), defined as time from randomization to time of death from any cause, is considered an optimal efficacy end point in randomized clinical cancer trials.

Survival is assessed daily and is easily documented through direct patient contact (office visit, hospitalization) or verbal contact by phone. Date of death is usually confirmed with little difficulty and is independent of causality. For patients who are lost to follow-up before documentation of death, censoring is usually undertaken at the time of last documented contact.

Disease-free survival (DFS), defined as time from randomization to disease recurrence or death owing to disease progression, is also frequently examined in cancer clinical trials. In some cases, DFS is more difficult to document as an end point compared with OS since it requires careful follow-up to detect disease recurrence, and the cause of death can be difficult to ascertain in cancer patients. Cancer patients often have comorbid conditions that may confound the interpretation of DFS. Furthermore, cancer patients often die outside of a hospital setting, and autopsies are not routinely performed.

TTP is defined as the time from randomization to time of progressive disease or death. Potential advantages associated with the use of TTP include a smaller sample size and shorter follow-up than is necessary when a survival end point is used. Furthermore, evaluation of TTP findings is not obscured by secondary therapies if a crossover effect exists. This end point may be enhanced by correlating radiographic changes with delay in onset of symptoms or delay of symptom worsening. On the other hand, use of TTP is associated with a number of limitations. First, many oncology drug trials are not blinded, potentially introducing bias into the designation of progression events. Second, patients must be evaluated on a regular basis with uniform assessment techniques and evaluation schedules for TTP findings to be interpretable. Third, determination of a magnitude of difference considered clinically meaningful can be problematic since most measurements are performed every 2–3 months and differences in TTP may be of similar magnitude. Finally, handling of missing data and censoring decisions can be difficult.

Time-to-treatment failure (TTF) is often defined as a composite end point including time from randomization to documentation of death, progression, withdrawal due to toxicity, patient refusal to continue treatment, or introduction of new therapy. The composite nature of TTF allows for a potential tradeoff of efficacy for toxicity reduction, and hence TTF is not often used as a primary end point of randomized clinical cancer trials.

Objective tumor response (OR), defined in the case of solid tumors as a reduction in tumor size over baseline, provides initial evidence of a treatment's biologic activity. In contrast to end points such as survival or TTP which are influenced both by the treatment effect and the natural history of the disease, tumor size reduction can usually be attributed entirely to treatment. Limitations associated with interpretation of response data include variation in the criteria used for determination of response and individual reader subjectivity. For evaluation of solid tumors, the Response Evaluation Criteria in Solid Tumors (RECIST) guidelines developed by a collaboration of the European Organization for Research and Treatment of Cancer, the National Cancer Institute of the United States, and the National Cancer Institute of Canada are used commonly in clinical trials (14).

Anticancer effects may also result in improvements in cancer-related symptoms, or in the case of therapies evaluated in the adjuvant setting, a delay in onset of symptoms.

These findings are most convincing when correlated with objective evidence of an antitumor effect such as radiographic response or prolonged survival. FDA has included information on symptom improvement in the product label in some cases, with examples such as gemcitabine for the treatment of pancreatic cancer (with an associated survival benefit) and photofrin for obstructive esophageal or nonsmall cell lung cancer (15).

18.4 TARGETED THERAPIES VERSUS CONVENTIONAL CYTOTOXINS

Although several cytotoxic drugs have identifiable mechanisms of action and some drugs classified as "targeted" probably have effects that are mediated through multiple processes, clinical trials examining targeted agents may allow for novel trial designs. Cancer clinical trials have not historically used blinding with a placebo due, in part, to the difficulty in maintaining the blind given the toxicity and schedule differences between the drugs or drug combinations that are compared. With orally administered agents associated with limited toxicities, randomized blinded placebo-controlled trials have been used. These trials compare an investigational drug to placebo alone or may use an "add on" design. In the latter case, the investigational drug is added to an available therapy and compared with the same therapy plus placebo.

Oral availability and limited toxicity may also allow for chronic administration schedules. Preclinical toxicity studies should evaluate schedules of administration that extend beyond the typical 28-day toxicity study if chronic administration schedules are planned. Oral agents with limited toxicity profiles and negative preclinical screening for genotoxicity may allow the possibility of healthy subjects to evaluate some components of clinical development, such as oral bioavailability and dose finding studies. For this to be possible, several safety concerns must be addressed. In addition to lack of genotoxicity, one must be assured of limited exposure and strict criteria for determination of dose-limiting toxicity.

Novel mechanisms of action may result in benefits not usually evaluated using standard end points. Targeted therapies may reduce the rate of tumor growth rather than causing tumor shrinkage. These benefits may not be evaluated adequately using response rates. Disease stabilization is optimally evaluated in a randomized trial examining time-to-event end points such as TTP or DFS. These end points assume greater clinical relevance if they are associated with an improvement in tumor-related symptoms or delay in onset of symptoms. Evaluation of end points that examine symptoms is predicated on the ability to minimize bias using blinding and randomization, to use validated measurement tools, and to prospectively address plans for multiplicity of end points and the potential for missing data.

18.5 TARGETED SUBGROUPS

Targeted subgroups may be useful in selecting patients who may be more likely to benefit from a specific therapy or who may be at greater risk to develop toxicities. The development

TABLE 18.2
To Target or Not to Target Prospectively

Advantages	Disadvantages
Greater treatment effect	May exclude patients who would benefit due to an unrecognized mechanism of action
Potentially smaller sample size	May limit potential market
Target population could "redefine disease" from histopathologic to molecular criteria	

of assays to define subgroups should be established early in drug development. Ideally, the same assay would be used in early drug development, later development, and in the postmarketing phase. Several trial design models may be applied to address these issues as outlined later. In addition, Table 18.2 outlines potential advantages and disadvantages of prospectively evaluating a targeted subgroup in efficacy studies.

A retrospective design is seldom useful for regulatory decision-making without prospective verification. In this model (Figure 18.1) all patients are treated with drug and all patients are evaluated for efficacy. Some patients respond and others do not. Retrospective comparison of pharmacogenomic markers suggests that responders have a higher frequency of a certain marker. From a regulatory perspective, this would be considered a retrospective subgroup analysis. Such findings would be more credible if the subgroup had been prospectively identified and the analysis adjusted for multiplicity. Generally, the retrospective subgroup analysis would be considered exploratory.

A prospective design (Figure 18.2) assumes an absence of benefit in the marker-negative population. In this setting, because the safety and efficacy would only be evaluated in the marker-positive subset, the target assay must be available at the time of drug approval. This design

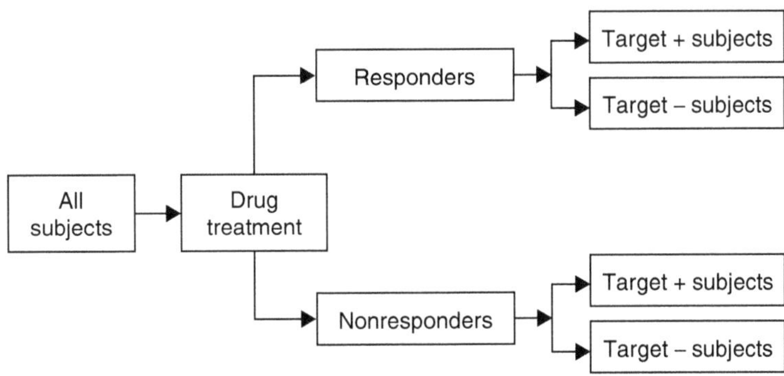

FIGURE 18.1 Retrospective. This design outlines the retrospective discovery process. This design is seldom useful for regulatory decisions without prospective verification. All patients are treated with drug and evaluated for efficacy. Some patients respond and others do not. Retrospective comparison of pharmacogenomic markers suggests that responders have a higher frequency of a certain marker. From a regulatory standpoint, this would be considered a retrospective subgroup analysis. Such a finding would be more credible if the subgroup had been identified prospectively. From a statistical standpoint, the analysis would have to be adjusted for multiplicity. Mistakes have been made over the years when investigators have relied on posthoc subgroup analyses. One example was the evaluation of aspirin for preventing stroke. A subgroup analysis strongly suggested that this effect did not apply to women. Years later, this conclusion was shown to be erroneous. Again, these sorts of data would generally be exploratory, which we might expect to be included in a safe harbor policy.

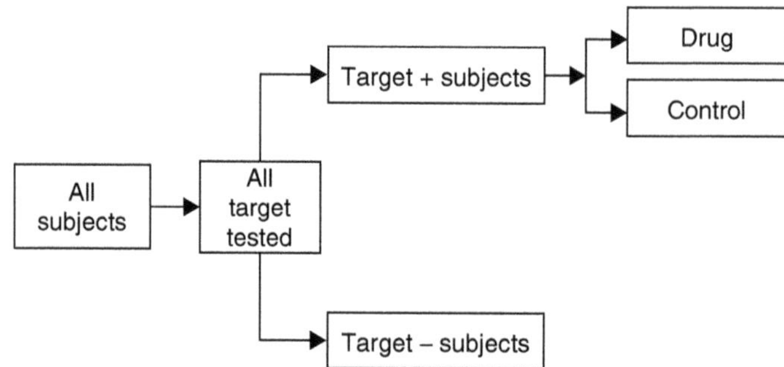

FIGURE 18.2 Prospective, screened: no expected effect in target-negative patients. This design describes the situation where we assume there will be no benefit in the marker-negative population. In this setting, because safety and efficacy would only be evaluated in the marker-positive subset, the target assay must be available at the time the drug is approved. This design might be used to test the proof-of-concept in phase 2 studies to screen for activity in the population most likely to respond. It might be used in the first stage of the two-stage Simon phase 2 design. If no activity is found even in the marker-positive enriched population, then the trial would not proceed. The development of Herceptin is an example where this approach was used. Only patients with tumors that were at least 2-plus in the Her2 expression assay were studied. In retrospect, however, we do not definitely know if patients with lesser expression of Her2 might not also have derived benefit from Herceptin. Clinical testing of all groups in early studies might have provided a more secure basis for determining who would benefit.

may be used to test a proof-of-concept to screen for activity in subpopulations most likely to respond. It might be useful in the first stage of a two-stage Simon design. If no activity is observed even in the marker-positive enriched population, then the trial would not proceed further. This approach was used in the development of herceptin. Only patients with tumors that were at least 2-plus in the Her2 expression assay were studied.

Figure 18.3 provides an alternative prospective design that introduces a stratification element. Unless the predictive value of the target test is nearly 100%, or unless lack of activity in the marker-negative group has been demonstrated in phase 2 trials, this clinical trial design

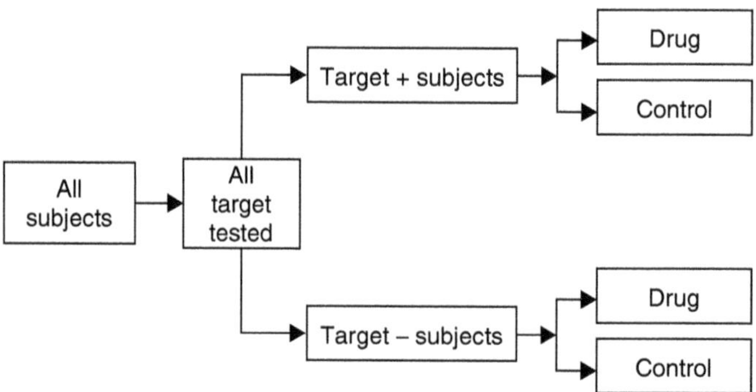

FIGURE 18.3 Prospective, stratified: (1) possible effect in target-patients or (2) pretreatment selection not possible. Unless the predictive value of the target test is nearly 100%, or unless lack of activity in the marker-negative group has been demonstrated in phase 2 trials, FDA will probably recommend this clinical trial design for phase 3 clinical trials. In this design, all patients have target testing at baseline and all patients are treated according to a stratified randomization. This design allows an estimate of efficacy and safety in the marker-positive and marker-negative population. This design also allows determination of the utility of the target assay. A useful adaptation of this design would be to include interim analyses that examine efficacy in the marker-negative subgroup and allow this arm to be dropped if it shows inadequate response.

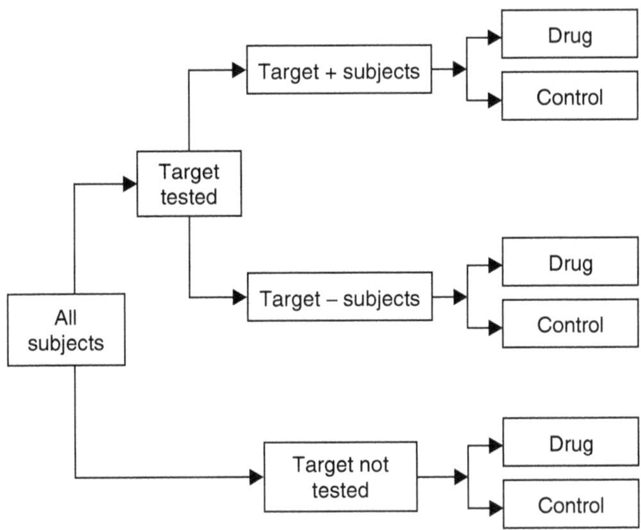

FIGURE 18.4 Prospective, stratified, explicit study of unselected population. This is another study design, probably more useful in areas outside oncology. In this model, one group of patients undergoes pharmacogenetic testing with stratified randomization and another group is not tested. This design provides a comparison of efficacy and safety in a controlled environment where the pharmacogenomic marker can be determined and an overall evaluation of safety and efficacy in a more general population. Perhaps this design could be useful when the assay would not be widely available, especially for a relatively nontoxic therapy and in settings where there are no other treatments available. This example has been used in international breast cancer trials of hormonal therapies. Because the measurement of hormone receptor status may not be possible or unknown in certain geographic areas there are basically two subgroups—those tested for HR and another not tested. Those tested for HR status are either positive or negative and the treatment effect can be examined in either population.

would be most appropriate for phase 3 clinical trials. In this design, all patients have target testing at baseline and all patients are treated according to a stratified randomization. This design allows an estimate of efficacy and safety in the marker-positive and marker-negative population. It also allows determination of the utility of the target assay. An adaptation of this design would include an interim analysis that examines efficacy in the marker-negative subgroup and would allow this arm to be discontinued if there is an inadequate response.

Finally, Figure 18.4 provides a prospective, stratified evaluation of tested subjects, but another group of untested patients is also evaluated. This design provides a comparison of efficacy and safety in a controlled environment where the marker can be evaluated, and an overall evaluation of safety and efficacy in a more general population can be performed. This design may be useful when an assay would not be widely available. International trials of hormonal therapies for breast cancer are examples of this design. Because the measurement of hormone receptor status may not be widely available in certain geographic regions, patients are either receptor status positive, receptor status negative, or receptor status untested/unknown. The treatment effect can be examined in the overall population and in these subgroups.

18.6 DIAGNOSTIC TESTS AND VGDS

Regulatory review of assays is performed at FDA by the Center for Devices and Radiologic Health (CDRH). During drug development, CDRH is frequently consulted about assay development. We encourage investigators and developers to do the same, either during meetings with the Office of Oncology Drug Products or in separate discussions with CDRH. The FDA pharmacogenomics guidance that has been developed includes input from CDRH on assay development (16).

This guidance outlines the concept of voluntary genomic data submissions (VGDS) and provides the FDA and drug sponsors with a framework for submission and review of genomic data without regulatory penalty. This process will benefit both the FDA and sponsors as we prepare for future submissions that include genomic data. Sponsors will gain experience in genomic data submission. The FDA will develop the capability to evaluate data across submissions with the potential for discovering general principles that will benefit the public health.

For information that the FDA and the sponsor consider to be exploratory given the current state of scientific knowledge, the FDA will agree to use the information only for nonregulatory purposes. Information from individual applications will be held in the same confidence that FDA provides for proprietary information for the drug applications. Furthermore, FDA will work with sponsors to find a format for data submission that will not be burdensome.

18.7 FDA PEDIATRIC INITIATIVES

There has been general recognition of the lack of adequate labeling of drug products for children. The small pediatric oncology market compared with the adult oncology market makes pediatric drug development less financially attractive and likely contributes to the lack of pediatric labeling information (17). In order to encourage evaluation of new therapies in pediatric populations and to optimize submission of pediatric data to support labeling of commercially available drugs, the FDA has undertaken two initiatives that can be summarized as follows.

A voluntary program, commonly referred to as "pediatric exclusivity," was outlined in the Food and Drug Modernization Act (FDAMA) of 1997 (18). Under this program, a commercial sponsor may receive a 6 month extension of existing exclusivity by submitting data from pediatric clinical trials in support of an NDA or a labeling supplement. The general design of studies to be conducted and data to be submitted are outlined in a pediatric-written request issued by FDA. This program applies to drug products including those with orphan designation. However, biological therapies such as monoclonal antibodies and vaccines are excluded. The Best Pharmaceuticals for Children Act (BPCA) subsequently provided some refinements to the program, including a framework for making summaries of medical and pharmacology reviews available to the public. It also established an Office of Pediatric Therapeutics and defined the membership of a Pediatric Subcommittee of the Oncology Drugs Advisory Committee (19).

A mandatory program applying to indications where the disease in adults is similar to the adult disease was previously known as the "Pediatric Rule" and was subsequently formally legislated in the Pediatric Research Equity Act (PREA) (20). This mandatory requirement for sponsors to submit data from pediatric studies when an adult indication is granted applies to a specific drug and indication under review. In order to avoid the potential delay in drug development and in providing access to therapies for life-threatening disease that can arise when the adult data support approval and pediatric studies are still ongoing, a deferral for submission of the pediatric data may be granted. Both drug and biologic products are subject to the PREA. However, products with orphan designation are excluded from this requirement.

18.8 CONCLUSION

Expanding knowledge regarding cancer pathogenesis provides for development of drugs that target specific pathways. Potentially limited toxicity profiles and availability of oral formulations allow for clinical trial designs previously difficult to implement with conventional intravenous cytotoxic drugs. End points used in the evaluation of anticancer drugs include survival, time-to-tumor progression, objective response rate, and symptom benefit or delay in symptom deterioration. Biomarkers may help "target" or define subgroups of patients who may more likely benefit or more likely experience an adverse event. Trial designs aimed at defining efficacy or safety in a subpopulation should prospectively examine these subgroups. Concurrent development and validation of target assays are essential.

ACKNOWLEDGMENT

The views expressed are the result of independent work and do not necessarily represent the views or findings of the U.S. Food and Drug Administration or the United States.

REFERENCES

1. Balis FM, Holcenberg JS, and Blaney SM. Chapter 9. General Principles of Chemotherapy, pages 237–308, in *Principles and Practice of Pediatric Oncology*, Pizzo and Poplack editors, 4th edition, Lippincott Williams & Wilkins, Philadelphia, PA, 2002.
2. Teicher BA, O'Dywer PJ, Johnson SW, Hamilton TC, Allegra CJ, and Grem JL. Chapters 4–6. *Cancer Principles and Practice of Oncology*, DeVita, Hellman and Rosenberg editors, 6th edition, Lippincott Williams & Wilkins, Philadelphia, PA, 2002.
3. Arbuck SG, Dancey J, Pluda JM et al. New targets for cancer chemotherapy. *Cancer Chemother Biol Response Modif* 19:237–288, 2001.
4. Murgo AJ, Dancey J, Eckhardt SG et al. New targets for cancer chemotherapy. *Cancer Chemother Biol Response Modif* 20:239–272, 2002.
5. Fox E, Curt GA, and Balis FM. Clinical trial design for target-based therapy. *Oncologist* 7:401–409, 2002.
6. Guidance for Industry: IND Exemptions for Studies of Lawfully Marketed Drug or Biological Products for the Treatment of Cancer, September 2003. Available at: http://www.fda.gov/cder/guidance/549.htm
7. DeGeorge J, Ahn C, Andrews P et al. Regulatory considerations for preclinical development of anticancer drugs. *Cancer Chemother Pharmacol* 41:173–185, 1998.
8. Simon R, Freidlin B, Rubinstein L et al. Accelerated titration designs for phase I clinical trials in oncology. *J Natl Cancer Inst* 89:1138–1147, 1997.
9. Guidance for Industry: Special Protocol Assessment, May 2002. Available at http://www.fda.gov/cder/guidance/3764.htm
10. O'Shaughnessy JA, Wittes RE, Burke G et al. Commentary concerning demonstration of safety and efficacy of investigational anticancer agents in clinical trials. *J Clin Oncol* 9:2225–2232, 1991.
11. Johnson JR, Williams G, and Pazdur R. Endpoints and United States Food and Drug Administration approval of oncology drugs. *J Clin Oncol* 21:1404–1411, 2003.
12. 21 Code of Federal Regulations, Parts 314.500 to 314.530.
13. Dagher R, Johnson J, Williams G, Keegan P, and Pazdur R. Accelerated approval of oncology products: a decade of experience. *J NCI* 96:1500–1509, 2004.
14. Therasse P, Arbuck SG, Eisenhauer EA et al. New guidelines to evaluate the response to treatment in solid tumors: European Organization for Research and Treatment of Cancer, National Cancer Institute of the United States, National Cancer Institute of Canada. *J Natl Inst* 92:205–216, 2000.
15. Williams G, Pazdur R, and Temple R. Assessing tumor-related signs and symptoms to support cancer drug approval. *J Biopharm Stat* 14:5–21, 2004.
16. Guidance for Industry: Pharmacogenomic Data Submissions. http://www.fda.gov/cder/guidance/index/htm
17. Hirschfeld S, Ho PTC, Smith M et al. Regulatory approvals of pediatric oncology drugs: previous experience and new initiatives. *J Clin Oncol* 21:1066–1073, 2003.

18. The Food and Drug Modernization Act, 21 USC 355a. Public Law 105–115, 1997.
19. Best Pharmaceuticals for Children Act, amended to the Federal Food, Drug and Cosmetic Act. Public Law 107–109, 1789, 2002.
20. Pediatric Research Equity Act. Public Law 108–155, 2003.

19 Surrogate Markers for Antiangiogenic Cancer Therapy

Darren W. Davis

CONTENTS

19.1 RATIONALE FOR SURROGATE MARKERS

In recent years, there has been significant progress in the development of angiogenesis inhibitors. Antiangiogenic agents that are currently being developed may be more effective against solid tumors and less toxic than cytotoxic chemotherapy. As a result of the early clinical trials of angiogenesis inhibitors, investigators are beginning to appreciate the complexity of targeting angiogenesis and the realization that developing optimal therapeutic benefit will be more challenging than originally thought. New methods and surrogate markers are crucial for the successful clinical development of these new drugs. This chapter introduces methods used to measure the pharmacodynamic effects of angiogenesis inhibitors and summarizes the clinical data on the development of surrogate markers in serum, plasma, and urine (Figure 19.1). The next three chapters provide details on the development of methods and markers for imaging, tissue analysis, and blood-based markers including circulating endothelial cells.

Preclinical studies suggest that angiogenesis inhibitors are cytostatic (growth-delaying) rather than cytotoxic. Traditionally, the maximum tolerated dose of a cytotoxic agent (determined by its dose-limiting toxicity) provided an estimation of the active dose range for subsequent clinical studies (1). However, for antiangiogenic agents it is uncertain whether doses associated with clinical toxicity correlate with an antiangiogenic effect. Because there is an uncertain relationship between toxicity and response, determining the optimal biological dose (OBD) of angiogenesis inhibitors depends on a different logic than conventional cytotoxic chemotherapy. Several clinical studies have shown that angiogenesis inhibitors are optimal at doses well below the maximum dose studied (2,3). Thus, clinical investigation of angiogenesis inhibitors has accentuated the need to develop new methods and biological markers to monitor antiangiogenic activity. Advantages and disadvantages of these methods are summarized in Table 19.1.

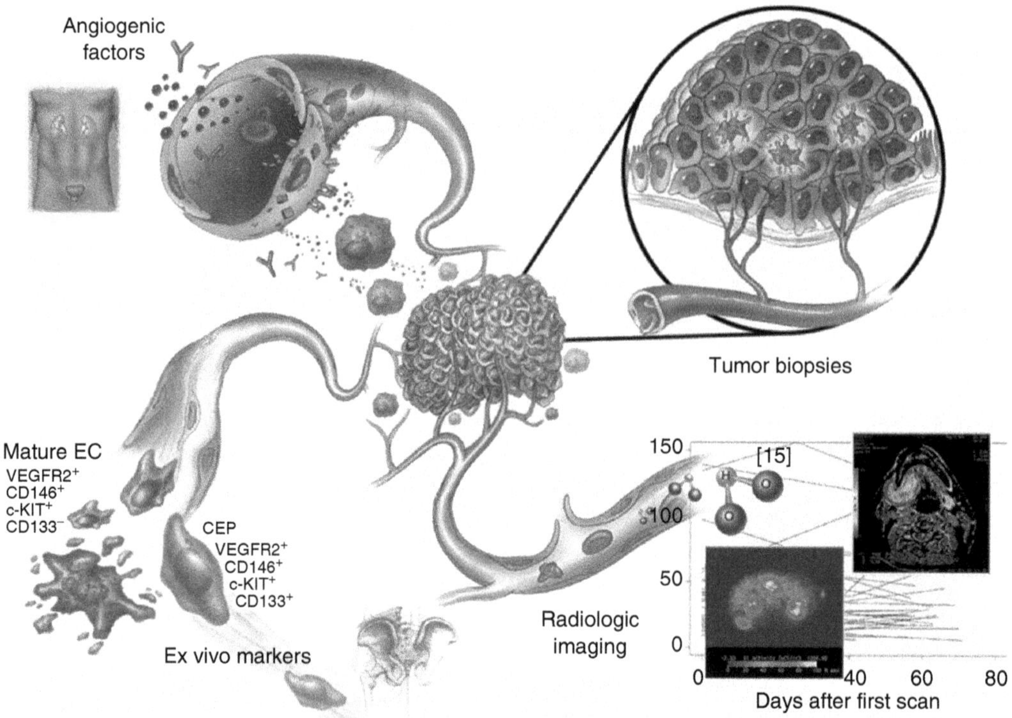

FIGURE 19.1 (See color insert following page 558.) Novel assays to measure surrogate markers for antiangiogenic cancer therapy. The complex biology of angiogenesis inhibitors has accentuated the need for developing technologies that can be used to assess the effects of biological markers. A compilation of data from multiple assays including measuring angiogenic factors in serum, plasma, and urine; tumor biopsy analysis; radiologic imaging; and, recently, ex vivo analyses of isolated peripheral blood cells (circulating endothelial cells) may facilitate defining the OBD for subsequent clinical studies of angiogenesis inhibitors. (From Davis et al., *British J. Cancer*, 89, 14, 2003. With permission.)

19.2 SERUM, PLASMA, AND URINE FACTORS

The most commonly used surrogate assays measure specific serum, plasma, and urine levels of angiogenic factors that are hypothesized to be direct or indirect targets of the antiangiogenic agent. Proteins thought to be important mediators of angiogenesis can be measured in serum, plasma, and urine by using enzyme-linked immunosorbent assays. A significant change in the

TABLE 19.1
Comparison of Methods Used to Monitor Surrogate Markers of Antiangiogenic Activity

Method	Advantages	Disadvantages
Angiogenic factors	Noninvasive (urine/serum/plasma), quantitative, multiple time points	Effects following inhibition of receptor unknown, interpatient variability, tumor specific
PET, MRI, CT, US	Assessment of tumor size, blood flow and volume, determine disease specific response	Expensive, not routinely available, effects on tumor blood flow not validated
Biopsy analysis	Provides proof of concept, specific cell types, quantitative	Invasive, effects of tumor heterogeneity unknown
Ex vivo markers	Noninvasive, feasible, multiple time points, quantitative	Difficult to detect rare cell populations, requires immediate analysis

level of angiogenic proteins after initiation of treatment might provide an early indication of antiangiogenic activity before clinically demonstrable reductions in tumor size. Elevated levels of angiogenic growth factors, proteases, and endothelial adhesion molecules have been detected in sera of patients with malignant disease (4). These important promoters of tumor angiogenesis include VEGF (5), bFGF (6), urokinase-type plasminogen activator and its soluble receptor (7), E-selectin and vascular cell adhesion molecule-1 (VCAM-1) (8), and von Willebrand's factor (vWF) (9). Intriguingly, in a study of 65 patients treated with radiation therapy, urine VEGF and MMP were shown to be strong predictors of patient 1 year progression free survival (10). Of the many mediators of angiogenesis, VEGF and bFGF are thought to play the most important roles and thus have been frequently measured as potential surrogate markers of antiangiogenic activity in Phase I and II clinical trials.

In a Phase I dose-escalation study of SU5416, a small-molecule inhibitor of the VEGF receptor-2 tyrosine kinase, urinary VEGF, and bFGF levels was found to vary one- to two-fold during the course of therapy but did not correlate with response or treatment dose (11). Interestingly, the baseline urine VEGF levels were significantly higher in the 4 patients that clinically responded than in the 18 nonresponders. However, bFGF levels were not significantly affected. The effects of angiogenesis inhibitors on the level of angiogenic factors may be drug or tumor-type specific and may be dependent on the level of proteins detectable in different bodily fluids. For example, urine, serum, and plasma levels of bFGF and VEGF were measured in patients with renal cell carcinoma enrolled in a Phase II study of razoxane (12), an antiangiogenic topoisomerase II inhibitor derived from the chelating agent EDTA. The levels of urinary VEGF significantly increased in patients who developed progressive disease but not in those with stable disease. However, no significant increases were observed in serum VEGF after one cycle of therapy, although baseline levels of serum VEGF were significantly higher in patients who subsequently developed progressive disease than in those with stable disease. Other mediators of angiogenesis including serum VCAM-1, vWF, and urokinase plasminogen activator-soluble receptor hypothesized to be indirect targets of razoxane were significantly higher in progressive disease patients than in stable disease patients before and after one cycle of treatment (12). In a Phase II study of the antiangiogenic agent thalidomide, 16 of 36 patients with recurrent, high-grade gliomas that had radiographic and clinically stable disease had either stable or decreased serum bFGF levels (13). Interestingly, serum bFGF levels significantly correlated to survival; patients whose serum bFGF significantly decreased (compared with baseline) survived approximately twice as long (43 weeks) than in those patients whose levels increased beyond the third week of therapy. In a Phase I study of endostatin, 2 of 25 patients experienced minor anticancer effects, but their serum VEGF, bFGF, VCAM-1, and E-selectin levels did not significantly change after treatment (14).

An important family of proteinases essential for the degradation of the extracellular matrix (ECM) and the basement membrane during angiogenesis include the family of matrix metalloproteinases (MMPs) (15). Proteolytic activation of the MMPs is regulated by the tissue inhibitor metalloproteinase (TIMP) family of proteins (16). MMPs also increase the bioavailability of factors such as bFGF that are essential for the ECM (17). In patients, high levels of MMP expression have been shown to correlate directly with metastasis (15,18) and poor prognosis (19,20). Therefore, inhibition of MMPs has become an important target for antiangiogenesis therapy, and the proteins themselves have become surrogate markers of antiangiogenic activity.

Patients enrolled in a Phase I study of MM1270, a nonspecific inhibitor of MMPs, had a significant increase in plasma levels of MMP-2, MMP-9, TIMP-1, TIMP-2, and bFGF after one cycle of treatment (21). Stable disease was observed in 19 of 92 patients. Other indirect surrogate markers that were measured but failed to demonstrate any significant change after treatment included VEGF, VCAM-1, soluble urokinase plasminogen activator receptor, and cathepsin B and H (proteases involved in the degradation of the ECM). In another study, BAY12-9566, an MMP inhibitor that targets MMP-2, MMP-3, and MMP-9, changes in the plasma levels of TIMP-2 were significantly related to dose (3). Ironically, despite achieving biologically relevant plasma concentrations of BAY12-9566, changes in the levels of MMP-2 and MMP-9 were not significantly affected by the dose. It is possible that negative feedback loops become activated, such that increasing inhibition of the enzyme results in further production. Interestingly, in a Phase I study with COL-3, changes in plasma MMP-2 were significantly related to dose when compared with patients with progressive disease versus those with stable disease (22). The direct effect on MMP-2 is not surprising because COL-3, which was derived from tetracycline, is a competitive inhibitor of MMP-2 (23). Serum levels of VEGF and bFGF were also measured but failed to demonstrate any significant correlations.

TABLE 19.2
Summary of Changes in Angiogenic Factors and Correlation with Outcome

Agent[Ref]	Patients	Outcome	PD Marker	Source	Time Point, Correlation
SU5416[11]	22	3 SD, 1 MR	VEGF	U	Pre > in responders[a]
			bFGF	U	None
Razoxane[12]	35	11 SD	VEGF	S, U	S—pre PD > SD[a]; U—post PD > SD[a]
			bFGF	S, U	S—post PD > SD[a]; U—pre PD > SD[a]
			VCAM-1	S, U	S—post PD > SD[a]; U—post PD > SD[a]
			vWF	S	Pre and post PD > SD[a]
			uPAsr	P	Pre and post PD > SD[a]
Thalidomide[13]	16	2 PR, 2 MR, 12 SD	VEGF	U	No change
			bFGF	S, U	S—decrease with survival[a]; U—no change
Endostatin[14]	25	1 SD, 1 MR	VEGF	S	No change
			bFGF	S	No change
			VCAM-1	S	No change
			E-selectin	S	No change
MM1270[21]	75	19 SD	VEGF	P	No change
			bFGF	P	Increase with AUC[a]
			VCAM-1	P	No change
			MMP-2	P	Increase with AUC[a]
			MMP8	P	No change
			MMP9	P	Increase with C^{a}_{max}
			TIMP1	P	Increase with AUC[a]
			TIMP2	P	Increase with AUC[a]
BAY 12-9566[3]	21	No response	MMP-2	P	No change
			MMP9	P	No change
			TIMP2	P	Increase with dose[a]
COL-3[22]	35	8 SD	VEGF	S	No change
			bFGF	S	No change
			MMP-2	P	Post PD > SD[a]
			MMP9	P	No change
TNP-470[24]	12	No response	bFGF	S	No change
			Thm	S	No change
			E-selectin	U	No change
Bevacizumab[25]	52	11 PR, 25 SD	VEGF	P	Pre, None

[a] Denotes significant change; Pre—before treatment; Post—after treatment; P—plasma; S—serum; U—urine; SD—stable disease; MR—minor response; PR—partial response; AUC—area under concentration curve; Thm—thrombomodulin.

Other surrogate markers include specific target-related proteins associated with the mechanism(s) of action of the angiogenesis inhibitor or the disease under study. For example, in a Phase I study of tetrathiomolybdate (2), an anticopper agent developed for Wilson's disease, serum ceruloplasmin, was used as a surrogate marker to monitor total body copper. The goal of the study was to reduce ceruloplasmin to 20% of baseline value. Five of six patients with stable disease reached the target range. In another study, TNP-470, an analogue of the antibiotic fumagillin, prostate specific antigen (PSA), was measured to monitor antitumor activity in patients with progressive androgen-independent prostate cancer (24). Surprisingly, a reproducible transient increase in PSA concentration was observed in some patients treated and rechallenged with TNP-470. However, the biological effect on PSA did not appear to be influenced by dose. Furthermore, urine bFGF levels revealed no significant relationship to dose or baseline serum bFGF concentrations.

To date, only a few angiogenic factors evaluated in clinical studies have proven useful for prognosis rather than monitoring response (Table 19.2). Even limited studies with bevacizumab have failed to demonstrate significant changes in plasma VEGF levels (25). Validation of angiogenic factors from homogeneous types of cancer and with agents known to have biological activity may facilitate the identification of surrogate markers. Further studies are needed to refine methods and develop specific markers that may be more sensitive to the effects of angiogenesis inhibitors.

19.3 RADIOLOGIC IMAGING

Routine radiographic imaging techniques (magnetic resonance imaging, dynamic computed tomography, and three-dimensional ultrasound) are useful for conventional measurement of disease and evaluating the effect of a biologic agent on tumor size (11,12,22). However, changes in tumor size may occur long after initiation of therapy or in some cases not at all. In addition, the effects of antiangiogenesis inhibitors on disease stabilization may be attributable to continuous, low doses of the drug. Therefore, developing surrogate markers by employing noninvasive imaging would facilitate defining the OBD, especially if the antiangiogenic effects were optimal at intermediate (low) doses or in specific cancers. Emerging radiologic techniques include positron emission tomography (PET), magnetic resonance imaging, dynamic computed tomography, and three-dimensional ultrasound. Each of these techniques is used to assess changes in tumor blood flow, vascular permeability, and in some cases metabolism (measured by PET). For example, three-dimensional ultrasound depicting vascularity revealed a 4.4-fold decrease in tumor blood flow in a rib metastasis from renal cell carcinoma after 8 weeks of tetrathiomolybdate therapy, and dynamic computed tomography scan imaging confirmed that the lesions with decreased blood flow were stable (2). In a Phase I study of endostatin, PET imaging was used to monitor changes in tumor blood flow [^{15}O] (H_2O) and tumor metabolism [^{18}F] (FDG-fluorodeoxyglucose) 28 and 56 days after endostatin therapy (14). It was hypothesized that the antiangiogenic effect of endostatin would decrease tumor blood flow and metabolism. Interestingly, the study did reveal that endostatin decreased tumor blood flow with maximal effects between 180 and 300 mg/m^2/day. However, tumor metabolism appeared to have a complex relationship with tumor blood flow, increasing at a dose of approximately 180 mg/m^2/day before decreasing at doses >300 mg/m^2/day. Analysis of the tumor blood flow data using a quadratic polynomial model showed that endostatin-induced changes in tumor blood flow were significantly related to dose at 56 days. Importantly, the 95% confidence interval identified for blood flow overlapped the OBD (250 mg/m^2/day) determined from the tumor biopsy studies. Thus, integration of surrogate marker data along with imaging studies was crucial for assessing the OBD of recombinant human endostatin. The use of imaging to monitor antiangiogenic activity is discussed in more detail in Chapter 20.

19.4 TUMOR BIOPSIES

Perhaps the most direct approach for determining the biological activity of antiangiogenic agents involves the analysis of tumor biopsies before and at specified time points after the initiation of treatment. The ability to measure molecular changes such as phosphorylation of receptor tyrosine kinases associated with drug–target interactions before and after therapy can provide early proof of whether the biologic agent has successfully reached its intended target. In addition, analysis of tumor biopsies permits the direct measurement of apoptosis in endothelial cells, an important end point of antiangiogenic therapy, and surrounding tumor cells. For example, analysis of the effects of SU5416 and SU6668 in preclinical studies suggested that these small molecules had great potential, as they induced significant levels of apoptosis in endothelial cells, consequently inhibiting tumor growth (26,27). However, analysis of tumor biopsy specimens obtained from Phase I/II studies demonstrated that these small molecules did not significantly inhibit phosphorylation of their primary target, VEGF receptor-2, or significantly increase apoptosis in tumor-associated endothelial cells (27,28). These data may explain the lack of clinical activity observed in the clinical studies but also demonstrate that the VEGF receptor remains an adequate therapeutic target. Quantitative analysis of biomarkers in tissues has become an important correlative end point in clinical studies. Chapter 21 discusses the latest developments of methods and biomarkers for analyzing skin and tumor biopsies to monitor the effects of angiogenesis inhibitors.

19.5 EX VIVO MARKERS

The heterogeneity of tumor biology and drug delivery to solid tumors may lead to variability in the results, making the interpretation of tumor biopsy and imaging data extremely challenging. Consequently, other creative strategies are developed to monitor surrogate markers of antiangiogenic activity in isolated peripheral blood cells (29,30). For example, a cytokine release assay was used to measure the effect of MM1270 on release of tumor necrosis factor-α from ex vivo-stimulated peripheral blood cells (21). MMP inhibitors have been shown to regulate TNF-α activity (31), and hence, it was hypothesized that MM1270 treatment would inhibit TNF-α release from stimulated whole blood cultures. Although there was slight inhibition of TNF-α release during MM1270 treatment, the results were not statistically significant, and there was no relationship to dose. Moreover, attempts to demonstrate apoptosis in endothelial cells isolated from patients after exposure to angiogenesis inhibitors have not been successful (14). Unfortunately, ex vivo analyses remain complex and their utility may depend on several factors including the inhibitor's relative specificity for tumor-associated endothelial cells, the sensitivity of the assay to detect effects on rare cell populations, and limited exposure time of the drug to have an effect on peripheral blood cells before they are isolated.

The use of flow cytometry to quantify activated circulating endothelial cells from the peripheral blood of cancer patients has provided another approach to assessing the effects of antiangiogenic activity (32). Recently, work has shown that VEGF-dependent mobilization of bone marrow derived CEPs can contribute to tumor neovascularization (29). Clinical studies have shown that angiogenesis inhibitors decrease the number of CEPs and increase the number of mature circulating endothelial cells (CECs) by causing shedding or damage to the endothelium (33,34). In fact, increases in mature CECs have been observed as early as 6 h in patients treated with ZD6126 (35). The feasibility of using circulating endothelial cells to measure biomarkers of early antiangiogenic activity is an encouraging approach. Additional methods and development of blood-based markers for antiangiogenic activity are discussed in Chapter 22.

19.6 CONCLUSION AND PERSPECTIVES

Angiogenesis inhibitors have great potential as cancer therapeutics (bevacizumab). However, the conventional end points of toxicity and response are inadequate for assessing these agents in clinical studies. The specificity and cytostatic nature of angiogenesis inhibitors requires the development of new methods and surrogate markers to identify doses and schedules with

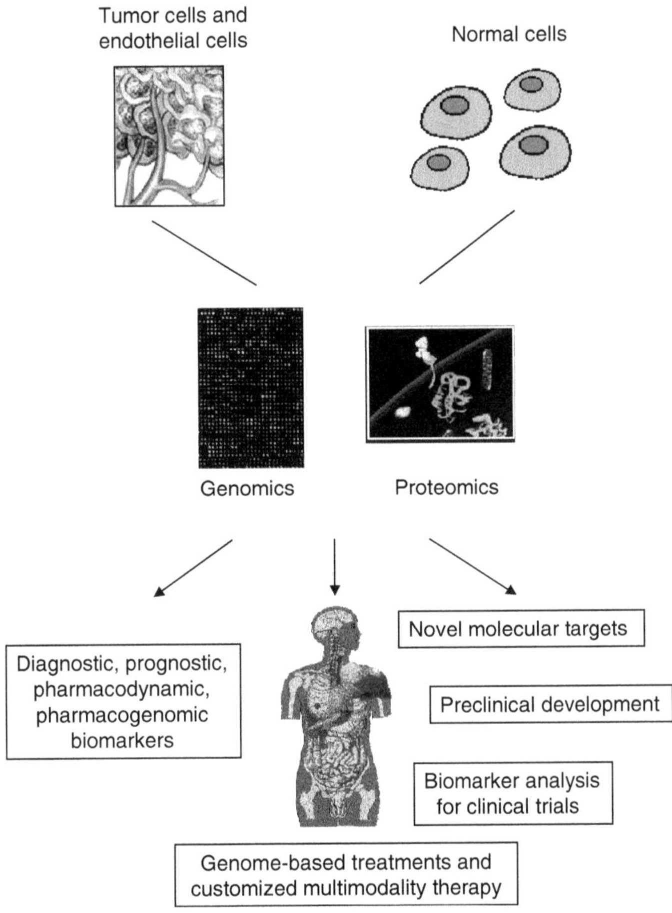

FIGURE 19.2 (See color insert following page 558.) Genomics and proteomics in the development and optimization of markers for antiangiogenic cancer therapy. Microarray expression profiling of endothelial cells from normal and tumor tissue can be used to develop novel molecular targets that effect angiogenesis and metastasis. Genome and proteome analyses should complement each other to better understand the crosstalk between tumor cells and tumor-associated endothelial cells and the response of the tumor microenvironment to antiangiogenic agents. In addition, these technologies are crucial for the discovery of molecular biomarkers that will become increasingly important for genome-based cancer treatments, especially to determine the optimal biological dose of angiogenesis inhibitors in early clinical studies. Further, combining selective genome-based therapeutics with genome derived biomarkers has the potential to optimize multimodality (angiogenesis inhibitors + cytotoxic agents) therapy and customize cancer treatment in a patient-specific fashion. (From Davis et al., *Biotechniques*, 34, 1048, May 2003. With permission.)

optimal antiangiogenic activity. In addition, a better understanding of the molecular mechanisms of angiogenesis inhibitors and the consequential effects on surrounding tumor and stromal cells is necessary for successful translation into the clinic. Recent genomic studies have shown that the tumor endothelial cells have a unique gene expression pattern compared to normal endothelial phenotypes (36,37). Thus, genomic and proteomic-based technologies will be needed to develop better molecular targets, diagnostic and prognostic markers that can be used to assess the optimal activity of angiogenesis inhibitors (Figure 19.2).

Although exploratory studies of angiogenic factors in plasma, serum, and urine as surrogate markers have been somewhat disappointing, some tumor types may produce higher levels and more quantifiable factors than other tumors. Angiogenic factors should continue to be studied, especially in more homogeneous patient populations. Further, it is anticipated that the combination of antiangiogenic therapy with traditional cytotoxic agents will offer optimal therapeutic benefit in the management of metastatic disease. Paradoxically, the addition of cytotoxic therapies will introduce toxicities that clinicians hoped to avoid by using angiogenesis inhibitors. Validation of those surrogates suggested to be prognostic indicators of clinical response, for example, urine VEGF, becomes critically important for monitoring combination therapy using cytotoxic and antiangiogenic drugs.

Quantitative analysis of drug–target interactions in tumor biopsies is essential for early proof-of-concept, especially during Phase I/II studies. The effects of an angiogenesis inhibitor on signal transduction are likely evident within hours of exposure to inhibitors of tyrosine kinase receptors (27). Confirming target inhibition and measuring apoptotic endothelial cells in tumor biopsy specimens obtained early after initiation of therapy may provide an assessment of optimal antiangiogenic activity (38). Indeed, automated quantitative analysis of apoptosis in tumor cells 48 h after initiation of treatment has proven useful for predicting clinical response to cytotoxic therapies in breast cancer (39). Although many of the assays discussed in this chapter are still in the stage of development and the surrogate markers must be scrupulously validated, a combination of these types of assays may provide the best strategy to determine the OBD of antiangiogenic agents.

REFERENCES

1. Simon, R., Freidlin, B., Rubinstein, L., Arbuck, S.G., Collins, J., and Christian, M.C. Accelerated titration designs for phase I clinical trials in oncology. *J Natl Cancer Inst*, 89: 1138–1147, 1997.
2. Brewer, G.J., Dick, R.D., Grover, D.K., LeClaire, V., Tseng, M., Wicha, M., Pienta, K., Redman, B.G., Jahan, T., Sondak, V.K., Strawderman, M., LeCarpentier, G., and Merajver, S.D. Treatment of metastatic cancer with tetrathiomolybdate, an anticopper, antiangiogenic agent: Phase I study. *Clin Cancer Res*, 6: 1–10, 2000.
3. Rowinsky, E.K., Humphrey, R., Hammond, L.A., Aylesworth, C., Smetzer, L., Hidalgo, M., Morrow, M., Smith, L., Garner, A., Sorensen, J.M., Von Hoff, D.D., and Eckhardt, S.G. Phase I and pharmacologic study of the specific matrix metalloproteinase inhibitor BAY 12-9566 on a protracted oral daily dosing schedule in patients with solid malignancies. *J Clin Oncol*, 18: 178–186, 2000.
4. Dirix, L.Y., Vermeulen, P.B., Pawinski, A., Prove, A., Benoy, I., De Pooter, C., Martin, M., and Van Oosterom, A.T. Elevated levels of the angiogenic cytokines basic fibroblast growth factor and vascular endothelial growth factor in sera of cancer patients. *Br J Cancer*, 76: 238–243, 1997.
5. Takahashi, A., Sasaki, H., Kim, S.J., Tobisu, K., Kakizoe, T., Tsukamoto, T., Kumamoto, Y., Sugimura, T., and Terada, M. Markedly increased amounts of messenger RNAs for vascular endothelial growth factor and placenta growth factor in renal cell carcinoma associated with angiogenesis. *Cancer Res*, 54: 4233–4237, 1994.
6. Fujimoto, K., Ichimori, Y., Yamaguchi, H., Arai, K., Futami, T., Ozono, S., Hirao, Y., Kakizoe, T., Terada, M., and Okajima, E. Basic fibroblast growth factor as a candidate tumor marker for renal cell carcinoma. *Jpn J Cancer Res*, 86: 182–186, 1995.

7. Xu, Y., Hagege, J., Doublet, J.D., Callard, P., Sraer, J.D., Ronne, E., and Rondeau, E. Endothelial and macrophage upregulation of urokinase receptor expression in human renal cell carcinoma. *Hum Pathol*, 28: 206–213, 1997.
8. Banks, R.E., Gearing, A.J., Hemingway, I.K., Norfolk, D.R., Perren, T.J., and Selby, P.J. Circulating intercellular adhesion molecule-1 (ICAM-1), E-selectin and vascular cell adhesion molecule-1 (VCAM-1) in human malignancies. *Br J Cancer*, 68: 122–124, 1993.
9. Gadducci, A., Baicchi, U., Marrai, R., Del Bravo, B., Fosella, P.V., and Facchini, V. Pretreatment plasma levels of fibrinopeptide-A (FPA), D-dimer (DD), and von Willebrand factor (vWF) in patients with ovarian carcinoma. *Gynecol Oncol*, 53: 352–356, 1994.
10. Chan, L.W., Moses, M.A., Goley, E., Sproull, M., Muanza, T., Coleman, C.N., Figg, W.D., Albert, P.S., Menard, C., and Camphausen, K. Urinary VEGF and MMP levels as predictive markers of 1-year progression-free survival in cancer patients treated with radiation therapy: a longitudinal study of protein kinetics throughout tumor progression and therapy. *J Clin Oncol*, 22: 499–506, 2004.
11. Stopeck, A., Sheldon, M., Vahedian, M., Cropp, G., Gosalia, R., and Hannah, A. Results of a Phase I dose-escalating study of the antiangiogenic agent, SU5416, in patients with advanced malignancies. *Clin Cancer Res*, 8: 2798–2805, 2002.
12. Braybrooke, J.P., O'Byrne, K.J., Propper, D.J., Blann, A., Saunders, M., Dobbs, N., Han, C., Woodhull, J., Mitchell, K., Crew, J., Smith, K., Stephens, R., Ganesan, T.S., Talbot, D.C., and Harris, A.L. A phase II study of razoxane, an antiangiogenic topoisomerase II inhibitor, in renal cell cancer with assessment of potential surrogate markers of angiogenesis. *Clin Cancer Res*, 6: 4697–4704, 2000.
13. Fine, H.A., Figg, W.D., Jaeckle, K., Wen, P.Y., Kyritsis, A.P., Loeffler, J.S., Levin, V.A., Black, P.M., Kaplan, R., Pluda, J.M., and Yung, W.K. Phase II trial of the antiangiogenic agent thalidomide in patients with recurrent high-grade gliomas. *J Clin Oncol*, 18: 708–715, 2000.
14. Herbst, R.S., Mullani, N.A., Davis, D.W., Hess, K.R., McConkey, D.J., Charnsangavej, C., O'Reilly, M.S., Kim, H.W., Baker, C., Roach, J., Ellis, L.M., Rashid, A., Pluda, J., Bucana, C., Madden, T.L., Tran, H.T., and Abbruzzese, J.L. Development of biologic markers of response and assessment of antiangiogenic activity in a clinical trial of human recombinant endostatin. *J Clin Oncol*, 20: 3804–3814, 2002.
15. Liotta, L.A. and Stetler-Stevenson, W.G. Metalloproteinases and cancer invasion. *Semin Cancer Biol*, 1: 99–106, 1990.
16. Gomez, D.E., Alonso, D.F., Yoshiji, H., and Thorgeirsson, U.P. Tissue inhibitors of metalloproteinases: structure, regulation and biological functions. *Eur J Cell Biol*, 74: 111–122, 1997.
17. Haro, H., Crawford, H.C., Fingleton, B., Shinomiya, K., Spengler, D.M., and Matrisian, L.M. Matrix metalloproteinase-7-dependent release of tumor necrosis factor-alpha in a model of herniated disc resorption. *J Clin Invest*, 105: 143–150, 2000.
18. Brown, P.D., Bloxidge, R.E., Stuart, N.S., Gatter, K.C., and Carmichael, J. Association between expression of activated 72-kilodalton gelatinase and tumor spread in non-small-cell lung carcinoma. *J Natl Cancer Inst*, 85: 574–578, 1993.
19. Murray, G.I., Duncan, M.E., O'Neil, P., Melvin, W.T., and Fothergill, J.E. Matrix metalloproteinase-1 is associated with poor prognosis in colorectal cancer. *Nat Med*, 2: 461–462, 1996.
20. Vihinen, P. and Kahari, V.M. Matrix metalloproteinases in cancer: prognostic markers and therapeutic targets. *Int J Cancer*, 99: 157–166, 2002.
21. Levitt, N.C., Eskens, F.A., O'Byrne, K.J., Propper, D.J., Denis, L.J., Owen, S.J., Choi, L., Foekens, J.A., Wilner, S., Wood, J.M., Nakajima, M., Talbot, D.C., Steward, W.P., Harris, A.L., and Verweij, J. Phase I and pharmacological study of the oral matrix metalloproteinase inhibitor, MMI270 (CGS27023A), in patients with advanced solid cancer. *Clin Cancer Res*, 7: 1912–1922, 2001.
22. Rudek, M.A., Figg, W.D., Dyer, V., Dahut, W., Turner, M.L., Steinberg, S.M., Liewehr, D.J., Kohler, D.R., Pluda, J.M., and Reed, E. Phase I clinical trial of oral COL-3, a matrix metalloproteinase inhibitor, in patients with refractory metastatic cancer. *J Clin Oncol*, 19: 584–592, 2001.
23. Seftor, R.E., Seftor, E.A., De Larco, J.E., Kleiner, D.E., Leferson, J., Stetler-Stevenson, W.G., McNamara, T.F., Golub, L.M., and Hendrix, M.J. Chemically modified tetracyclines inhibit human melanoma cell invasion and metastasis. *Clin Exp Metastasis*, 16: 217–225, 1998.

24. Logothetis, C.J., Wu, K.K., Finn, L.D., Daliani, D., Figg, W., Ghaddar, H., and Gutterman, J.U. Phase I trial of the angiogenesis inhibitor TNP-470 for progressive androgen-independent prostate cancer. *Clin Cancer Res*, 7: 1198–1203, 2001.
25. Kindler, H.L., Friberg, G., Singh, D.A., Locker, G., Nattam, S., Kozloff, M., Taber, D.A., Karrison, T., Dachman, A., Stadler, W.M., and Vokes, E.E. Phase II trial of bevacizumab plus gemcitabine in patients with advanced pancreatic cancer. *J Clin Oncol*, 23: 8033–8040, 2005.
26. Shaheen, R.M., Davis, D.W., Liu, W., Zebrowski, B.K., Wilson, M.R., Bucana, C.D., McConkey, D.J., McMahon, G., and Ellis, L.M. Antiangiogenic therapy targeting the tyrosine kinase receptor for vascular endothelial growth factor receptor inhibits the growth of colon cancer liver metastasis and induces tumor and endothelial cell apoptosis. *Cancer Res*, 59: 5412–5416, 1999.
27. Davis, D.W., Takamori, R., Raut, C.P., Xiong, H.Q., Herbst, R.S., Stadler, W.M., Heymach, J.V., Demetri, G.D., Rashid, A., Shen, Y., Wen, S., Abbruzzese, J.L., and McConkey, D.J. Pharmacodynamic analysis of target inhibition and endothelial cell death in tumors treated with the vascular endothelial growth factor receptor antagonists SU5416 or SU6668. *Clin Cancer Res*, 11: 678–689, 2005.
28. Heymach, J.V., Desai, J., Manola, J., Davis, D.W., McConkey, D.J., Harmon, D., Ryan, D.P., Goss, G., Quigley, T., Van den Abbeele, A.D., Silverman, S.G., Connors, S., Folkman, J., Fletcher, C.D., and Demetri, G.D. Phase II study of the antiangiogenic agent SU5416 in patients with advanced soft tissue sarcomas. *Clin Cancer Res*, 10: 5732–5740, 2004.
29. Lyden, D., Hattori, K., Dias, S., Costa, C., Blaikie, P., Butros, L., Chadburn, A., Heissig, B., Marks, W., Witte, L., Wu, Y., Hicklin, D., Zhu, Z., Hackett, N.R., Crystal, R.G., Moore, M.A., Hajjar, K.A., Manova, K., Benezra, R., and Rafii, S. Impaired recruitment of bone-marrow-derived endothelial and hematopoietic precursor cells blocks tumor angiogenesis and growth. *Nat Med*, 7: 1194–1201, 2001.
30. Reyes, M., Dudek, A., Jahagirdar, B., Koodie, L., Marker, P.H., and Verfaillie, C.M. Origin of endothelial progenitors in human postnatal bone marrow. *J Clin Invest*, 109: 337–346, 2002.
31. Gearing, A.J., Beckett, P., Christodoulou, M., Churchill, M., Clements, J., Davidson, A.H., Drummond, A.H., Galloway, W.A., Gilbert, R., Gordon, J.L., et al. Processing of tumour necrosis factor-alpha precursor by metalloproteinases. *Nature*, 370: 555–557, 1994.
32. Mancuso, P., Burlini, A., Pruneri, G., Goldhirsch, A., Martinelli, G., and Bertolini, F. Resting and activated endothelial cells are increased in the peripheral blood of cancer patients. *Blood*, 97: 3658–3661, 2001.
33. Beaudry, P., Force, J., Naumov, G.N., Wang, A., Baker, C.H., Ryan, A., Soker, S., Johnson, B.E., Folkman, J., and Heymach, J.V. Differential effects of vascular endothelial growth factor receptor-2 inhibitor ZD6474 on circulating endothelial progenitors and mature circulating endothelial cells: implications for use as a surrogate marker of antiangiogenic activity. *Clin Cancer Res*, 11: 3514–3522, 2005.
34. Bertolini, F., Mingrone, W., Alietti, A., Ferrucci, P.F., Cocorocchio, E., Peccatori, F., Cinieri, S., Mancuso, P., Corsini, C., Burlini, A., Zucca, E., and Martinelli, G. Thalidomide in multiple myeloma, myelodysplastic syndromes and histiocytosis. Analysis of clinical results and of surrogate angiogenesis markers. *Ann Oncol*, 12: 987–990, 2001.
35. Beerepoot, L.V., Radema, S.A., Witteveen, E.O., Thomas, T., Wheeler, C., Kempin, S., and Voest, E.E. Phase I clinical evaluation of weekly administration of the novel vascular-targeting agent, ZD6126, in patients with solid tumors. *J Clin Oncol*, 24: 1491–1498, 2006.
36. St Croix, B., Rago, C., Velculescu, V., Traverso, G., Romans, K.E., Montgomery, E., Lal, A., Riggins, G.J., Lengauer, C., Vogelstein, B., and Kinzler, K.W. Genes expressed in human tumor endothelium. *Science*, 289: 1197–1202, 2000.
37. Kobayashi, T., Kim, S., Zhang, W., Yamaguchi, M., Ueno, S., Morikawa, J., and Shiku, H. [Gene expression profiling of de novo CD5-positive diffuse large B-cell lymphoma]. *Rinsho Ketsueki*, 44: 144–148, 2003.
38. Davis, D.W., Shen, Y., Mullani, N.A., Wen, S., Herbst, R.S., O'Reilly, M., Abbruzzese, J.L., and McConkey, D.J. Quantitative analysis of biomarkers defines an optimal biological dose for recombinant human endostatin in primary human tumors. *Clin Cancer Res*, 10: 33–42, 2004.
39. Davis, D.W., Buchholz, T.A., Hess, K.R., Sahin, A.A., Valero, V., and McConkey, D.J. Automated quantification of apoptosis after neoadjuvant chemotherapy for breast cancer: early assessment predicts clinical response. *Clin Cancer Res*, 9: 955–960, 2003.

20 Noninvasive Surrogates

Bruno Morgan and Mark A. Horsfield

CONTENTS

Imaging departments today bear little relation to those of 25 years ago, with the old equipment replaced by advanced scanning equipment, which use computed x-ray tomography, ultrasound, magnetic resonance, and positron-emitting agents. Despite these advances in medical imaging technology, the organization and function of most clinical cancer imaging departments remain largely unchanged. The emphasis is still on a surgical approach to disease, concentrating on the localization, size, shape, and appearance of tumor lesions. The pressure for change has been limited by the lack of innovative pharmacological approaches to cancer therapy, with surgical resection still giving the best chance of cure in most solid tumors.[1] As increasing knowledge of molecular medicine has been applied to the study of cancer, treatment approaches and the "imaging" questions that these approaches bring have changed dramatically. There is an increasing need to study biological processes in vivo at both the preclinical and clinical stages of pharmacological development and application. This is particularly true for treatments based on tumor angiogenesis.

Currently, there are no clinical paradigms that require information on angiogenesis status to decide diagnosis and management. However, this book demonstrates that angiogenesis is important for the prognosis and treatment of cancer and this is reflected in the fact that several antiangiogenesis strategies are currently in clinical development. It is likely that, in future, antiangiogenesis treatment protocols will be specific to the histology and stage of the disease.[2] It has always been recognized that angiogenesis inhibitors may cause tumor stasis, preventing further growth, rather than reducing tumor size.[3] This is important since, with standard chemotherapy, it has been demonstrated that it is probably the duration of the stable disease phase, and not tumor shrinkage, that has the major effect in prolonging survival of patients.[4]

Tumor-size measurements taken from CT scans have shown some correlation with clinical outcome in chemotherapy[5] and are useful for monitoring treatment where rapid size changes are expected. However, the possible absence of tumor shrinkage, and the fact that cross-sectional imaging of tumors requires a 30% reduction in size to reliably gauge the response to treatment,[6,7] makes evaluation of the potential efficacy of angiogenesis inhibitors in phase I/II trials problematic. The traditional end point of oncology trials therefore needs to change, away from the short-term goal of improvement in tumor size and patient's well-being, to goals that may be related to a lack of progression of disease and a failure to metastasize. New trial designs are therefore needed.[8]

This has led to the search for new surrogate end points as markers of efficacy. Angiogenesis is a local tissue phenomenon, so attempts to measure surrogate markers in the blood, while promising, may be nonspecific.[9] Biopsies with histopathological staining are invasive, may not be representative of the whole tumor, and only give information at a single time point.[10] Several imaging modalities potentially offer a noninvasive in vivo test, with the ability both to measure tumors spatially and to map their changes over time. Functional and molecular information is increasingly emphasized, and the in vivo characterization and measurement of biologic processes at the cellular and molecular level are often described as "molecular imaging."[11] Measuring vascular or other physiological changes can be described as "functional" imaging. Such molecular or functional imaging could very quickly show whether the drug is working at a mechanistic/molecular level. Imaging is already changing the study of drug development in the preclinical phase, but is just beginning to have an impact in the clinic. Clinical studies are particularly important, as animal studies do not directly translate to humans, and most preclinical studies are carried out in non-wild-type tumors with different locations, maturity, and stages of development.

Traditionally, the purpose of a phase I trial is to establish pharmacokinetics and potential toxicity in humans, and imaging is simply used as a measure of tumor size to determine response. The extra emphasis on a molecular/mechanistic approach to drug development has increased the importance of imaging in early clinical trials, with the usage of up to four different modalities.[12] The questions in early trials were not just whether the agent works in the broadest sense, but also does it have an effect through the intended mechanism (e.g., changes in vascularity) in vivo; does it affect the intended receptor in vivo; and is it getting to the right place in sufficient quantities and times in the studied dose regime[13]? This is important for early clinical studies in advanced cancer, since it may be difficult to detect clinical efficacy by traditional methods such as survival.[14]

In reality, radiologists are no strangers to imaging biologic processes. Imaging of the uptake of isotope-labeled substances specific to thyroid (iodine) and bone (^{99m} technetium diphosphonates) has been available for more than 30 years. Furthermore, when performing CT or MRI scans with injected contrast media, radiologists have been using different contrast-enhancement patterns in the liver to differentiate benign lesions from malignant ones for several years. However, further development of functional or molecular imaging is more problematic. Imaging studies investigating the molecular basis of cancer have been used extensively in the preclinical setting,[15] but translation of this work from bench to bedside is more difficult. First, radiology departments are often focused on high turnover, clinical service work with little budget or poor organizational systems for research. Radiologists are only broadly aware of the revolution that is taking place and may have insufficient knowledge and training.[16] Furthermore, using complex imaging in clinical trials to study biological processes requires more time from the patient, testing their commitment.[17] One main opportunity arises, however, because imaging has become an integral part of a patient's cancer journey with increasing sophistication of standard imaging tests. This means that the technology for many advanced functional imaging tests is often available in imaging departments.

This chapter looks at the role of clinical imaging in the development of antiangiogenesis treatments. Instead of giving an exhaustive list of techniques and approaches for imaging, we concentrate on the questions that are asked of imaging by oncologists and pharmaceutical companies, the tools and approaches available now, and the potential for the future.

20.1 GENERAL IMAGING ISSUES

The "switch" to angiogenesis involves many processes and factors that may be potential treatment targets. These range from genetic alterations to the consequences of angiogenic factors, such as vascular endothelial growth factor (VEGF), causing changes in vascular permeability, proliferation, and maturity.[18–20] Vascular changes may result in increased circulation or, more commonly, the presence of inadequate vascular networks and lymphatics leading to hypoxia and high interstitial pressure.[21] Hypoxia can be chronic "diffusion-limited" tumor hypoxia or "perfusion-limited" hypoxia due to a dynamic process in which the vessels periodically open and close.[22] This hypoxia has been recognized for many years as hypodense (necrotic) centers on contrast-enhanced CT scans (Figure 20.1). In neck lymph nodes, these areas have been correlated to hypoxia using oxygen-sensitive electrodes and have been shown to be a marker of poor prognosis.[23]

There are therefore several aspects of the angiogenic process that are amenable to imaging. First are the direct processes that may be manifested by overexpression of cell surface markers, falling into the realms of molecular imaging. These processes may also lead to changes in circulating angiogenic factors.[9] Less directly, there will be changes in the vasculature itself, including the vascular permeability to macromolecules (and therefore contrast agents), the perfusion of the tissue, and the maturity of the vessels in the tumor, which may change in response to pharmacologic manipulation. Ultimately, failure of tissue perfusion will lead to cell death by necrosis and apoptosis, which suggests further potential

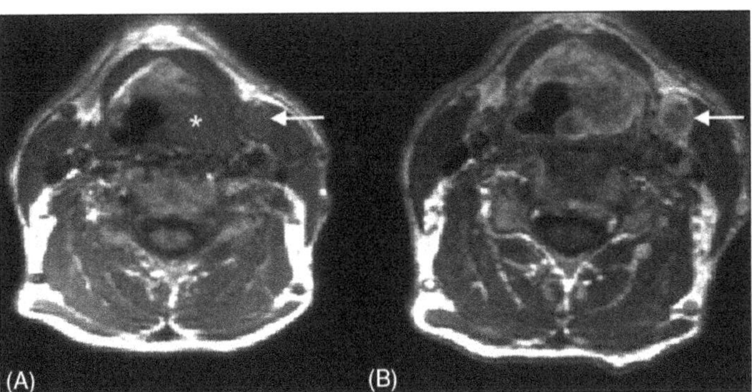

FIGURE 20.1 Axial CT images before (A) and after (B) the administration of a contrast agent, which shows a supraglottic laryngeal carcinoma (*). The administration of contrast shows the rim-enhancing effect of an adjacent lymph node (arrow) around a hypodense "necrotic" center, known to be associated with tumor hypoxia.

imaging targets including hypoxia and apoptosis markers as well as tumor volume. Antiangiogenic treatment may therefore have a range of effects on the vascular characteristics of a tumor but, ultimately, all should be amenable to imaging of some kind.

There are three major mechanisms by which pharmacological targeting of tumor blood vessels could be achieved:

1. True angiogenesis inhibition
2. Vascular targeting
3. Nonselective antiangiogenic effects such as those proposed for some chemotherapeutic agents at low dose[24]

Hence, any imaging test that has been shown to correlate with a successful clinical outcome may be used in clinical studies of antiangiogenesis agents, and, further, any successful cancer treatment may affect vasculature and therefore cause alteration in imaging-derived vascular parameters.

Angiogenesis inhibitors may not result in substantial reductions of tumor volume. Other problems with imaging the response to antiangiogenic treatments are that they may inhibit only a single positive factor, and individual tumors can express several angiogenic factors, which may lead to partial or complete resistance of the tumor vessels to therapy.[25] This diversity encourages the use of combinations of these agents.[18] Furthermore, advanced cancer may not be the ultimate indication, but is likely to be the target in early treatment trials. The efficacy of treatment could vary between patients and different tumor types, and the heterogeneity of delivery of drugs to solid tumors may lead to further variability in response.[13] Without knowledge of what to target, phase I trials might therefore miss potential efficacy by including too diverse a group of patients and tumors.[26] For these reasons, the aims of a single-agent phase I clinical trial are to demonstrate a potentially useful biological effect and to direct future choice of combination therapies, method of delivery, and target tumor types.

Phase I trial design is made more difficult by the lack of toxicity of these drugs, such that toxicity-based selection of dose for further development may not be optimal. Although conventional imaging techniques are still needed, since tumor-size monitoring will remain an important response variable,[19] methods to demonstrate biologic activity before reaching maximum-tolerated dose, or even to show an optimal dose well below the maximum-tolerated dose, would greatly enhance the utility of such studies.

There is currently no proven method of imaging the angiogenic process. The reasons for this are easy to understand: angiogenesis is a complex process involving many steps, defying a simple single method approach. Furthermore, if the individual molecular processes are to be studied, the method has to be sensitive to microscopic changes or nanomolar concentrations of naturally occurring substances or deliverable imaging contrast agents.

Therefore, for imaging to be successful, it needs to be established as a "surrogate" or "biomarker" for the activity of the drug. A biological marker (biomarker) is defined as an objective measurement indicating a pharmacological response to a therapeutic intervention. A surrogate end point is a biomarker that is intended to substitute for a clinical end point, a characteristic or variable that reflects the patient's well-being.[27] It should be noted that in a dose-escalating trial, even if a potential biomarker shows a correlation between dose and efficacy, this may be purely a side effect or even toxicity of the drug and does not imply cause and effect.

Although many imaging tests are in the early stages of validation as biomarkers, there are, as yet, no studies validating "functional imaging" as surrogate end points of clinical efficacy for antiangiogenic treatments.

An ideal biomarker indicates the presence of a target disease in an accurate and reproducible manner and is closely linked to success or failure of the therapeutic effect of

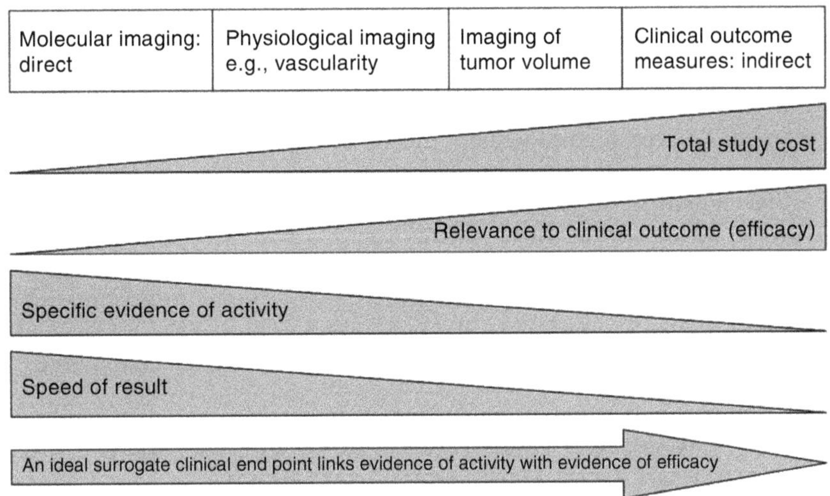

FIGURE 20.2 The progression of direct to indirect tests to monitor drug therapy.

the product that is evaluated. This provides a quicker trial result, since "true" end points include parameters such as 5 year survival and helps to avoid confounding factors, incidental to the treated disease, developing. Another advantage is that fewer patients are required: each patient can act as their own control.[28] Figure 20.2 charts the progression of tests demonstrating specific activity of a drug through to true end points demonstrating the broader clinical efficacy of treatment. An ideal test shows both mechanistic activity and clinical efficacy.

There is therefore a balance between the importance of a specific biomarker and true clinical end points. On the one hand, it is useful to know whether treatments are promising in early therapeutic trials, even if advanced tumors are not the ultimate target, since this avoids rejecting a potentially useful therapy.[29] On the other hand, in combination treatment, antiangiogenic treatment may be expected to reduce vasculature, which may compromise the effectiveness of radiotherapy or chemotherapy. In this case, biomarkers showing specific drug activity are less helpful, and true indirect clinical end points such as tumor-size response or delayed time to progression are required.[30,31]

20.2 TYPES OF IMAGING

Imaging can be performed by a variety of modalities, including x-ray computed tomography (CT), magnetic resonance imaging (MRI), radioisotope imaging (single-photon emission computerized tomography [SPECT] and positron emission tomography [PET]), ultrasound, and optical imaging. One fundamental aspect of all imaging modalities is their resolution, which impacts on the ability to separate tissues that are different either by structure or function in a spatial and temporal manner. Resolution is related to the ratio of signal to noise or the ratio of the information returned from the tissue to the random variation in that information due to measurement imperfections. All in vivo imaging techniques have their own strengths and weaknesses because of the different types of information returned and therefore have varying limits of spatial and temporal resolution. In some cases, the imaging modality alone provides information that is relevant for studying angiogenesis, for example, Doppler ultrasound, which provides blood flow measures, or MRI using diffusion or spectroscopy techniques. However, with other techniques a contrast agent (or probe, possibly radioactively labeled) is required. In these cases, the attributes of the agent largely dictate what information can be gained. The imaging test determines at what concentration, speed, and spatial resolution the agent can be studied.

When a contrast agent is used, it may provide direct information related to a specific aspect of angiogenesis or indirect (downstream) information related to a consequence of successful therapy. Tests performed without the use of contrast media or probes are generally indirect.

The multiple potential targets and treatment strategies suggest that the most useful approach may be an indirect measure of angiogenesis. The most common current indirect indicators of angiogenesis are changes in metabolism and vascularity. These represent the expected downstream consequences of depriving the tumor of blood supply and are particularly useful if there is uncertainty about the exact nature of the mechanisms of action of the drug and the need to test drugs with different mechanisms in combination.

20.3 INDIRECT MEASURES: VASCULARITY AND BLOOD FLOW AND INTRODUCTION TO IMAGING METHODS

The vascularity of a tumor can be measured in terms of the blood volume (the volume of the intravascular space compared with the volume of the tumor) and perfusion (rate of blood flow into the tumor). A further aspect is the permeability of the vasculature, which is related to the ease with which substances can pass from the intravascular to the extravascular extracellular (interstitial) space. The permeability depends on the molecular weight (size) of the substance, and angiogenesis promoters such as VEGF have been shown to increase this permeability to macromolecules.[20] There is debate about the exact mechanism of changes in macromolecular permeability that may impact on interpretation of imaging results. One possibility is that macromolecules extravasate predominantly by an opening of the junctions between adjacent endothelial cells.[32] However, Dvorak and colleagues argue that, although it is likely that very small hydrophilic molecules up to 3 nm in diameter pass through intact interendothelial cell junctions, in response to VEGF-A, macromolecules up to 50 to 70 nm in diameter cross the endothelium predominantly by means of a transendothelial cell pathway that involves vesiculovacuolar organelles.[20,33,34] There are therefore potentially two distinct mechanisms for the leakage of macromolecules up to 3 nm in diameter.

20.3.1 Magnetic Resonance Imaging with Contrast Agents

Studies of the vasculature with MRI normally include rapid injection of a contrast agent, often referred to as dynamic contrast-enhanced MRI (DCE-MRI). Consequent changes in the image brightness are then used to detect and characterize lesions. DCE-MRI is already finding routine clinical application in MR mammography.

The MR image is created from the nuclei of hydrogen atoms (protons) that are mainly in water. Applying both a large static magnetic field and a series of radio-frequency pulses causes "excitation" of the protons and generates the signal. The image (a spatial location map of the proton signal) is created by applying magnetic field gradients along different directions. Although the signal intensity is largely dependent on the water concentration (or proton density), the image can be made sensitive to two different ways the signal changes or "relaxes": the time constants that govern these two relaxation processes are called the T_1 and T_2 relaxation times. When the image is sensitized to one of the relaxation processes, this is called either T_1-weighted or T_2-weighted imaging.

Contrast media are available that can cause both T_1 and T_2 to change, with a consequent change in signal intensity. The presence of contrast agent is indicated by either signal hyperintensity (with T_1-weighted imaging) or hypointensity (with T_2-weighted imaging). In normal clinical use in the brain, just a single image is acquired some time (typically 5 min) after contrast injection to show the distribution of the agent and to confirm opening of the blood–brain barrier. Outside the brain, however, where even in healthy tissue the contrast agent leaks from the vasculature, measuring the time course of the signal change in the tissue can be much more revealing.

Using contrast media of varying molecular weights and magnetic properties, MRI can be used to measure blood volume, perfusion, and blood vessel permeability.[35] Large molecular weight contrast agents will stay within the intravascular space, and by collecting MR images continuously as the contrast is injected, both blood volume and perfusion can be estimated. Very small molecules, such as water, will leak rapidly into the interstitial space, again providing a guide to perfusion in the dynamic phase and the size of the interstitial space. Small to intermediate molecular weight agents, however, are neither freely diffusible nor do they remain purely in the blood pool, and the degree of signal change will be related to both flow and permeability parameters to varying degrees. Temporal analysis of the enhancement pattern for intermediate size agents can help elucidate the separate components of blood volume, perfusion, and permeability, but for low molecular weight compounds such as the standard gadolinium (Gd) chelates, available for use in humans, the enhancement pattern seen often results from an inseparable combination of flow, blood volume, and permeability.[36]

Although nonspecific, all these factors are related to angiogenesis[19] and microvascular density, malignancy, and prognosis have all been correlated with enhancement parameters.[37–40] Correlations are not reliable, however, probably due to the variable effects of malignancy on vascularity and vascular permeability.[41] In tumors, where permeability is often very high, contrast enhancement mainly depends on perfusion regardless of the contrast agent used.[42] However, for a particular tumor, as the molecular weight of the contrast agent increases, kinetic parameters derived from DCE-MRI change in magnitude and spatial heterogeneity, suggesting that the utility of the measurement depends on optimizing the size of the agent.[43] Several studies have shown that successful therapies may result in changes in parameters derived from DCE-MRI data in animals and humans, which may prove a more accurate and earlier indication of response than standard clinical and imaging parameters.[44,45]

The signal intensity from T_2-weighted MRI depends on inherent tissue properties and requires MRI sequences that are insensitive to local magnetic field inhomogeneity (spin-echo sequences). Other types of scan (T_2^*-weighted imaging) are sensitized to any local magnetic field inhomogeneity and show a reduction in signal intensity in regions of poor field uniformity. Standard Gd chelates in high concentration cause shortening of the T_2^* relaxation time. Such concentrations are found in the vascular tree after bolus injection, causing a decrease in observed signal intensity. This T_2^* effect reduces dramatically as leakage into the extravascular space occurs. Therefore, Gd chelates are sometimes considered as extravascular agents with T_1-weighted imaging and as intravascular agents for T_2^* sequences.

Gd chelates are used routinely in clinical practice with T_2^*-weighted imaging for cerebral perfusion studies, where the blood–brain barrier prevents leakage into the extravascular space. In tumors, breakdown of the blood–brain barrier, which is essential for standard contrast enhancement, makes the T_2^* effect less consistent and more difficult to interpret quantitatively. Despite this, promising results have been obtained in brain tumors with hybrid scanning that combines T_2^* (perfusion-sensitive) and T_1 (leakage-sensitive) acquisitions.[46]

Other contrast agents are now becoming available for use in humans, and these are based around iron oxide particles in a dextran coating, giving a strong T_2^* change even at low concentrations. These agents, called superparamagnetic iron oxide particles (SPIOs) or ultrasmall superparamagnetic iron oxide particles (USPIOs), may have potential as blood pool markers when used in conjunction with a dynamic MRI scan.[47,48] They are already proving useful in the clinic as specific lymph-node markers. In a trial of 80 patients with presurgical, clinical stage T1, T2, or T3 prostate cancer, high-resolution MRI with highly lymphotropic superparamagnetic nanoparticles allowed the detection of small and otherwise undetectable lymph-node metastases.[49]

These developments, and the fact that contrast-enhanced MRI is often performed routinely in cancer patients, have led to increased interest in their use to study the effects of treatment. This has met with varying degrees of success, with some clinical studies showing that DCE-MRI with standard Gd chelates (using a variety of imaging and analysis methods) can successfully be used to assess different types of therapy.

20.3.2 Quantification of DCE-MRI

As stated earlier, in cases where the permeability of the vessel walls is relatively low, the rate of leakage into the interstitial space is determined by this permeability. However, when permeability is high, delivery of contrast agent by the vasculature can be the rate-limiting step, and then the rate of enhancement is more indicative of tissue perfusion. However, in the general case of standard Gd chelates, enhancement profiles result from an intractable combination of perfusion and permeability.[36] As well as perfusion and permeability, the enhancement profile depends on the volume of tissue to which the contrast agent has access, the extravascular extracellular space: the higher its volume fraction, the slower the contrast agent equilibrates between the blood and the tissue.

Figure 20.3 shows a schematic enhancement curve, indicating the key features. First, it should be noted that this is a plot of MR signal intensity vs. time. In MRI, the signal intensity is not a physical parameter, but depends on the type of imaging sequence used and is on an arbitrary scale. Furthermore, if two such curves were collected from the same patient without any real change in the tumor characteristics, the intensity range seen could be markedly different even when using the same scanner and pulse sequence. We must therefore find ways to evaluate these enhancement curves without reference to the signal intensities directly. To make the assessment independent of the arbitrary intensity scale, this must be "normalized" in some way. Two simple methods of normalization are subtracting or dividing by the

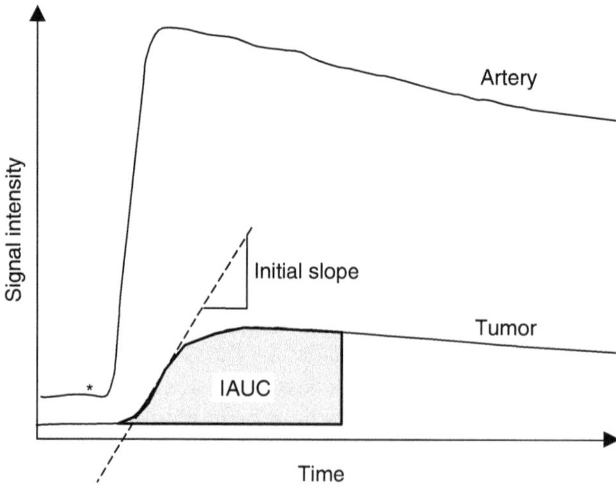

FIGURE 20.3 Schematic showing the signal-intensity time course in a tumor and in an artery that feeds the tumor. Asterisk denotes timing of contrast injection. The change in signal intensity in the artery is typically much greater than that in the tumor. The shaded area represents the initial area under the enhancement curve (IAUC), evaluated out to a particular time point (typically 60 or 180 s) after contrast arrival in the tumor, after subtracting the precontrast signal intensity.

baseline (precontrast) signal. The disadvantage of the first method is that it does not take into account the "scale" of the signal-intensity changes, whereas for the second method, baseline signal intensity may change in follow-up studies due to physiological reasons, thereby changing the apparent enhancement. For example in a trial of cancer therapy, the treatment may increase edema in a tumor. This will increase the T_1 parameter of a tissue thereby lowering the precontrast signal intensity. If signal intensities of the enhancement curve are divided by this lower value, there will be an apparent increase in enhancement. Other methods of "normalization" could be against an intensity standard included in the image field of view (such as a water-containing vial), or against the intensities seen in the artery that feeds the tissue. The latter approach has the advantage of also countering any variation in the dose and timing of the contrast agent injection (the arterial input function [AIF]). Another major hurdle to overcome when attempting quantitative assessment is that the changes in signal intensity we observe in T_1-weighted MR images are not proportional to the concentration of contrast agent.[50] A typical response curve is shown in Figure 20.4, and it should be noted that the exact form of the response depends very much on the exact pulse sequence used, and can vary from one MRI scanner to the next even with the same nominal pulse sequence implementation. True quantification requires the signal intensities to be converted to R_1 values ($R_1 = 1/T_1$) since, in the case of standard Gd chelates the change in R_1 is proportional to the contrast agent concentration. The calculation of R_1 from signal intensities is possible, although it is challenging to devise methods that can quantify concentration of the Gd-based contrast agent over the wide range seen in slowly enhancing tumors and in the feeding arteries.

Simple methods of analyzing the data include measuring peak enhancement or the peak slope of the enhancement curve. As the concentration in the feeding artery is constantly changing, the peak enhancement depends on different physiological parameters at different times. The peak enhancement after a minute depends mainly on the perfusion and permeability of the vasculature, whereas after 10 min the peak enhancement may depend more on the extravascular extracellular space. Peak slope can also be a problem as, in rapidly enhancing tumors, it may be difficult to define, and, in heterogeneous tumors, there be more than one component to the initial slope. Perhaps the simplest semiquantitative approach to assessment of contrast dynamics is to evaluate the area under the enhancement curve out to a certain fixed time after contrast injection. This measurement has no direct physiologic meaning but is a robust measurement that depends on the vascularity of the tumor in its broadest sense. This is often termed the initial area under the enhancement curve (IAUC). Unless the signal intensities have been converted to R_1 values as earlier, it is usual to divide the IAUC by the area under the AIF curve out to the same time point. An ideal analysis method must be able to take the AIF into account although in follow-up studies, where relative changes from pretreatment values are important, there may be less need to measure the AIF.

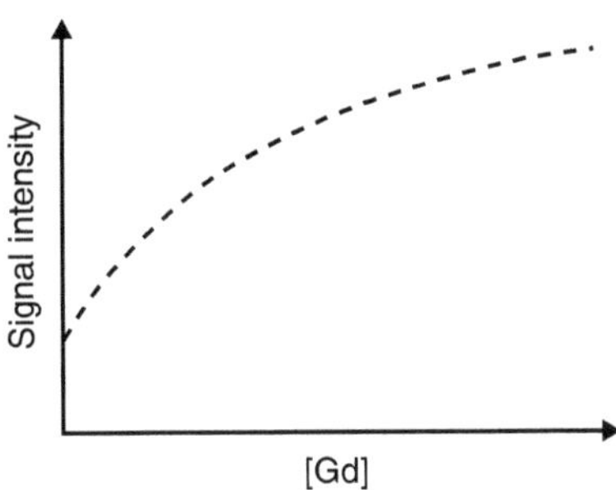

FIGURE 20.4 Nonlinear relationship between signal intensity and contrast agent concentration [Gd] for a typical T_1-weighted MRI pulse sequence. The signal intensity saturates at high concentrations, making quantification of [Gd] difficult in the artery that feeds the tumor.

A more sophisticated approach attempts to quantify the perfusion/permeability in terms of a nonspecific leakage rate constant (K^{trans}) and also the extravascular extracellular space volume fraction (v_e). Currently, there is no complete consensus of the best method, although many studies use similar methods to estimate the transfer constant for Gd chelate as it passes into the interstitial space including K^{trans}, K_i, K_{ep}, and K_{21}. The imaging community is working to develop uniformity of imaging and analysis protocols,[36] but this will remain difficult until successful treatments are available to assess the utility of different imaging tests in different tumors and organ systems.

The process of calculating K^{trans} and v_e from a tissue enhancement curve and an AIF curve is one of deconvolution, a mathematical process that enables any variability in the AIF to be accounted for in a rigorous way, with the result that K^{trans} and v_e are parameters that are interpretable in a physically meaningful way, independent of the technique used.

Under certain idealized circumstances of a very short, tight bolus injection of contrast, the value of K^{trans} is numerically the same as the ratio of the initial upslope of the tissue enhancement curve to the instantaneous contrast agent concentration in the plasma. However, as the signal intensities change rapidly in the artery, certain tumors can also enhance rapidly; acquiring images fast enough to measure these slopes and peaks accurately is difficult.

The choice of pulse sequence used in DCE-MRI is usually a compromise between spatial resolution, time resolution, and spatial coverage. Achieving higher spatial resolution requires a longer scan time for any individual image and for DCE-MRI, where multiple images are sequentially acquired, the time resolution is compromised. Good time resolution is needed to fully capture the dynamic of contrast uptake, but also with increased scan times there is the risk of patient movement during the acquisition which can result in artifacts and unusable images. The choice of pulse sequence thus depends on the expected rate of uptake, which depends on the type of tumor and also the tumor locations, since anatomical areas that are easily immobilized (such the head, limbs, or breast) are more suitable for high-resolution imaging.

In this chapter, all studies quoted have different approaches, with different spatial and temporal resolutions and different sensitivities to tumor heterogeneity. Figure 20.5 shows

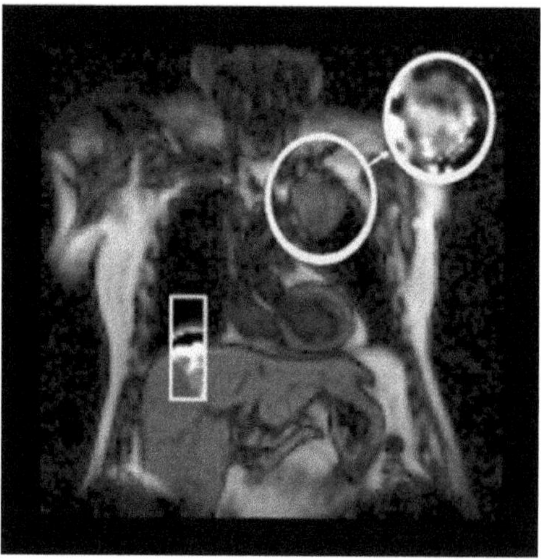

FIGURE 20.5 A coronal MR image showing a tumor in the apex of the left lung. A pixel map of enhancement parameter K^{trans} for this region is shown in the displaced circle with higher intensity indicating higher vascularity/permeability. The square region shows a pixel map of enhancement (K^{trans}) in an area of breathing-related movement with accompanying artifacts at the junction between liver and lung.

how a map of enhancement can be made by calculating K^{trans} for each pixel for an immobile area in the lung apex, and also how problems occur in a moving area such as the lung–liver interface. For such mobile areas, averaging techniques may be more appropriate. In addition, for follow-up studies, tumor growth or shrinkage makes comparison of these "parameter maps" difficult. There is no doubt, however, that such parameter maps increase the available information, and the heterogeneity and variability of microcirculation within the tumor rather than the average value over the tumor may be the main factor influencing therapeutic outcome in "staging" scans.[51] The vascularity, vascular permeability, and interstitial pressures may be very different in areas such as the tumor rim and tumor core. This means that the mechanism of contrast enhancement (and, in particular, the rate-limiting step of enhancement) will be different: for example, contrast enhancement in the tumor core may depend mainly on interstitial pressure due to poor venous and lymphatic drainage whereas in the hypervascular rim enhancement may be related to flow. Different treatments may therefore have different effects in different parts of the tumor.[43]

To make measurements quantitative, it is necessary to assess the tracer concentration in both the tissue of interest, and in the artery that feeds the tissue. There are several methods of estimating this AIF for PET and MRI, none of which are completely satisfactory. The lack of agreement about data acquisition and analysis methods makes comparison of results between groups difficult.

Different methods lend themselves to different agents with different mechanisms of action, and too much standardization may stifle development. An example of this is shown in Figure 20.6, where a steep enhancement curve is not fitted by the standard two-compartment model used in the estimation of K^{trans}, although curve b, more typical of a liver metastasis, does. The steep part of the enhancement curve is likely to represent the contribution of the tracer in the blood pool to enhancement, not taken into account in the standard analysis of K^{trans}, making the analysis sensitive to changes in the blood volume. It is possible to make more complex models to account for this,[52] but higher quality data are required. Furthermore, in all models, assumptions are still made about the distribution of the contrast agent or isotope, which may not prove true in tumors.

20.3.3 Clinical Trials Using DCE-MRI

In phase I studies, conducted in Leicester and Freiburg, we have shown that DCE-MRI can provide useful information in the clinical study of an angiogenesis inhibitor.[53,54] The agent (PTK787/ZK222584) is a small molecule inhibitor of VEGF tyrosine kinases. In a subcohort of 25 patients with liver metastases from colorectal carcinoma, we showed significant reductions in enhancement rates as early as day 2 after start of treatment, which may persist for months. These changes correlated significantly with increasing dose and plasma levels, as dose was escalated. Furthermore, the degree of enhancement reduction correlated well with changes in tumor size and clinical response. In these types of study, enhancement can be measured using a variety of methods. In our study, we measured the bidirectional transfer constant K_i[55] (which is proportional to the other common parameter K^{trans}[36]), which reflects tumor vascular permeability and flow. Similar changes have been seen in glioblastoma multiforme with the same agent.[56] These studies have helped in the selection of the dose and tumor types to be used in phase II and III studies. A similar trial using an anti-VEGF antibody HuMV833 shows a similar magnitude of changes in contrast-enhancement parameters.[57]

Although there is a clear relationship between the measured enhancement parameters and treatment with the drug, the exact mechanism is not clear. The mechanism therefore has to be inferred, based on our knowledge of preclinical data and knowledge of the various factors that can affect the imaging test.[58] In preclinical studies with other anti-VEGF agents, changes

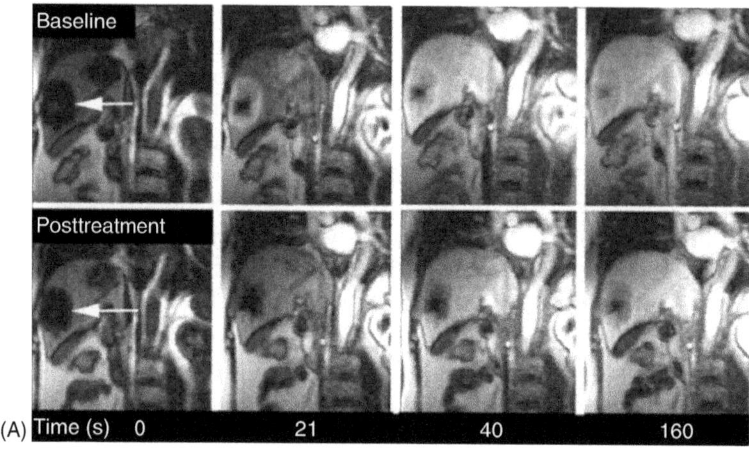

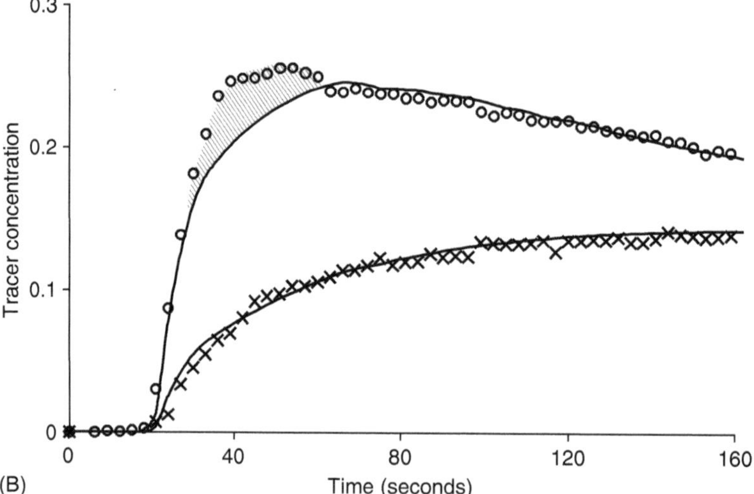

FIGURE 20.6 (A) MR Images showing enhancement of a liver metastasis (arrow) over time after contrast media injection; before (baseline) and after treatment with an angiogenesis inhibitor. (B) Graphs of the contrast-enhancement curve over time ○ = baseline, × = posttreatment. The solid lines represent the best curve-fit obtained by a two-compartment model (K^{trans}). The shaded area shows data that are not fitted well by the model.

in endothelial cell survival and vessel density may not occur for some days after initiation of therapy.[59] However, some studies show more rapid reductions in the density of immature vessels.[60] K^{trans} reflects not only tumor vascularity and blood flow, but also vascular permeability. The effects observed may therefore be due to acute changes in permeability, vascularity, or even other factors such as the action of VEGF on nitric oxide production.[61] It seems likely, however, that specific targeting of immature vessels contributes to rapid enhancement changes and the normalization of the vasculature as proposed by Jain.[62]

Similar results have been obtained with the vascular targeting agent, combretastatin A4 phosphate (CA4P).[63,64] DCE-MRI studies were performed to examine changes in parameters related to blood flow and vascular permeability (K^{trans}) and the initial area under the contrast medium concentration–time curve (IAUC) during a 24 h period after treatment with CA4P. Eighteen patients in a phase I trial received escalating doses, and significant reductions in tumor K^{trans} after treatment were seen.

A similar study of 16 patients treated with 5,6-dimethylxanthenone-4-acetic acid (DMXAA), an agent that causes vascular shutdown in preclinical models, in a dose-escalating trial[65] showed that 9 of 16 patients had significant reductions in IAUC 24 h after treatment.

In the last two studies, although there was some evidence of a dose-related response, the correlations were not clear-cut, presumably due to variation in response in each dose cohort. The PTK787/ZK222584 trial only studied the dose–response for a similar tumor type (colorectal liver metastases) and although the dose–response exists for all tumor types, it is not as strong.[66] Furthermore, establishing a strong dose correlation requires starting with a dose that has little efficacy.

Effects on blood flow have also been observed with classical chemotherapy agents. DCE-MRI measurements using K^{trans} have been made in 16 patients receiving preoperative chemoradiotherapy.[67] K^{trans} and VEGF levels correlated before treatment. Eight responsive tumors had higher pretreatment K^{trans} values than nonresponsive tumors, and showed a marked reduction in K^{trans} at the end of treatment suggesting the possibility of antiangiogenic action. In a separate, similar trial from the same center, taxane-based chemotherapy showed no effect.[68] However, in separate trials of preoperative chemotherapy, reductions in enhancement parameters have been seen in a variety of tumors with docetaxel, with some relation to efficacy.[69] These trials concentrated on high temporal resolution scans with analysis of the "first-pass" gradient of the enhancement curve rather than K^{trans} or IAUC. To know if chemotherapy regimes affect vascularity as seen on DCE-MRI is important, not only to judge DCE-MRI results from chemotherapy in combination with antiangiogenesis agents,[70] but also to assess possible antiangiogenic effects of low-dose metronomic chemotherapy.

Initial DCE-MRI results may also predict response to treatment, generally with increased enhancement parameters predicting a good response.[51,67] This may be related to many factors, including tumor oxygenation, access of chemotherapy, and potential correlation with angiogenesis.

Cerebral contrast-enhanced T_2^*-weighted imaging has also proved useful in the brain in 24 patients undergoing treatment with carboplatin and thalidomide for malignant gliomas. Cerebral blood volume (CBV) maps created for the tumors before and after treatment showed marked reduction in patients treated with thalidomide and carboplatin in comparison with carboplatin alone. These changes correlated with efficacy after 1 year.[70]

20.3.4 MRI Measures of Tumor Blood Flow without Contrast Media

For some time, MRI angiography methods have been available that can measure flow velocity in large blood vessels, but these are of no use in the tumor microvasculature. A newer technique, MRI arterial spin labeling (ASL) is a perfusion imaging technique that involves exciting protons (spin tagging) in a well-defined vessel that feeds the organ, and then recording the consequent signal change in that organ/tumor. This effectively measures tissue perfusion using arterial water as the probe. ASL has shown correlation with contrast-enhanced methods of cerebral blood flow in brain tumors.[71] Although this technique has the advantage of using no exogenous contrast agent, allowing multiple studies to be performed sequentially, the signal to noise is considerably poorer than contrast-enhanced methods. The available signal to noise has been improved with the introduction of clinical MRI scanners operating at a higher magnetic field of 3 T (compared with 1.5 T in more common usage).[72] In investigational studies, the ability to do multiple sequential measurements in the same day may allow assessment of the perfusion changes over time, particularly in combination with pharmacokinetic measurements.

A less direct method of assessing vasculature uses intrinsic blood oxygenation level dependent (BOLD) contrast MRI. The BOLD technique uses a heavily T_2^*-weighted image that can depict changes in blood oxygenation. As an isolated study it is difficult to quantify, but changes in oxygenation can be detected by comparison with baseline. Signal intensities can be affected by vasodilatation by pharmacologic agents or even mental activity (functional brain imaging).[73] These techniques may be useful for mapping vascular maturity, since immature vessels do not have smooth muscle activity and do not respond to vasodilators. Bold contrast MRI has been used for mapping vascular maturation using the response of mature vessels to hypercapnia (inhalation of air vs. air 5% and CO_2) and the response of all vessels to hyperoxia (air and 5% CO_2 vs. oxygen and 5% CO_2 (carbogen)).[74–76] This may help predict response to antivascular therapy.[77] These techniques can be used in the clinic, although breathing carbogen can be unpleasant.[75]

20.3.5 Computed Tomography

CT has the advantage of wide availability and, like T_1-weighted MRI, shows enhancement of the image with increasing concentration of contrast agent, although the volumes of contrast agent used are much higher with CT. Iodinated contrast agents in CT show similar distribution to MRI media, that is, they are extravascular extracellular agents with no specific uptake mechanism. However, the mechanism of contrast in CT is different: the image brightness depends on the degree of attenuation of the x-ray beam, and is related to the density of the tissue or contrast agent. With iodinated compounds, the degree of enhancement (i.e., increase in tissue attenuation) is proportional to the concentration of iodine, making quantification straightforward, particularly when measuring arterial contrast enhancement during bolus injections.[78,79] As for DCE-MRI, "functional" contrast-enhanced CT techniques have shown increases in tissue perfusion that may reflect malignancy and stage.[80] Furthermore, in a trial of 35 patients contrast-enhanced CT parameters were shown to correlate with microvascular density and VEGF expression in lung adenocarcinoma.[81] CT is a faster and easier procedure to perform than MRI, with fewer potential artifacts and a higher spatial resolution (typically around 0.5 mm). Clinical MRI has a resolution in the order of 1–2 mm, although preclinical MRI can achieve resolutions of a few tens of microns. This makes CT a more robust technique that potentially allows automated analysis. Indeed, in one example, a retrospective trial was possible in 130 patients with primary lung carcinoma showing correlations with VEGF expression and microvascular density based on relatively straightforward acquisition parameters.[82]

CT has the weakness of potentially poor anatomical coverage, which is solved to some extent by multislice spiral technology.[83] Generally, CT contrast media are safe, but have a worse side effect profile than standard MRI contrast agents. In particular, the "hot flush" that many patients experience can make reliable multiple breath-holding protocols for dynamic-enhancement studies problematic.

The chief weakness of CT compared with MRI, and MRI compared with nuclear medicine, is the low sensitivity for detecting current clinical contrast agents or labeled probe. Whereas CT contrast agents are often used in the millimolar concentration range, and MRI agents range from millimolar to micromolar, nuclear medicine agents are down to true "tracer" picomolar concentrations. One must therefore be careful not to overwhelm the system under investigation, and the development of new, more targeted CT agents is difficult due to potential toxicity.

Theoretically, since CT and MRI standard contrast agents have similar pharmacokinetics, any DCE-MRI findings should be translatable to CT. This was demonstrated in a study of the VEGF-specific antibody bevacizumab in human rectal cancer, where a rapid antivascular effect was shown by dynamic contrast-enhanced CT.[84]

Despite the ease of using CT in clinical applications, there is the concern about the associated radiation exposure. The increased risks of radiation from a CT scan may seem trivial for many cancer patients; however, there are now strict regulations in Europe concerning techniques used for research that involve radiation, but are not of direct benefit to the patient.[85] Currently, tumor perfusion studies are not of proven benefit to the patient, so these regulations make MRI easier to organize in clinical trials, unless CT studies can be linked with standard clinical CT protocols.

20.3.6 Radionuclide Imaging

SPECT and PET utilize compounds labeled with radioisotopes as molecular probes. Both techniques have considerably poorer spatial resolution in comparison to CT and MRI, but much better sensitivity to low concentrations of tracer. While SPECT uses gamma ray emitters, PET uses positron emitters, with the annihilation of the emitted positron and an electron producing two photons (gamma rays of 512 keV energy) that travel in almost exactly opposite directions. Detection of both these photons allows the location of the original positron emission to be determined to within a few millimeters. PET has advantages over SPECT: it has better spatial resolution, greater sensitivity to radiopharmaceuticals, easier quantification of tissue radiopharmaceutical concentration, and hence biologically important radiopharmaceuticals are easier to manufacture.[86] Since most SPECT applications (such as labeled annexin to image apoptosis) can be adapted to PET imaging techniques, this chapter concentrates on PET applications.

A wide variety of simple positron-emitting atoms can be created, such as isotopes of oxygen, nitrogen, carbon, fluorine (chemically similar to hydrogen), and numerous others, without changing their chemical and biological properties. Since very low concentrations of probe are required, pharmacological effect is not usually a concern. Unfortunately, PET is not widely available since, in addition to the scanner, on-site (or nearby) radiochemistry facilities and a cyclotron are needed to generate the short-lived isotopes, making it an expensive technique. Quantification of the tracer concentration is difficult, due to variable attenuation of photons from the deep structures, which can make follow-up studies difficult.[87]

Despite the expense, PET imaging in oncology is becoming standard in some areas. Its main applications use the probe 18-fluorodeoxyglucose (^{18}FDG) as an indirect marker of metabolically active cancer cells. This is a glucose analog, which is transported into cells and undergoes hexokinase-mediated phosphorylation. The end product, FDG-6-PO_4, is not a significant substrate for subsequent reactions and is retained in the cell in proportion to rate of glycolysis. Increased metabolism is a biomarker for the presence of a tumor, since many tumors have high levels of glucose utilization via glycolysis rather than oxidative metabolism (Warburg effect).[88] The relative specificity for FDG uptake by tumors has lead to PET becoming a standard tool in the staging of lung cancer, particularly in combination with CT scanning which improves spatial localization. Figure 20.7 shows an example of fused CT/PET imaging in a case of non-small cell lung cancer, with no evidence of distant spread.

^{18}FDG-PET is often used during treatment of tumors such as lymphoma, and early work has shown changes related to prognosis after one cycle of chemotherapy.[89] Dramatic responses have been seen in patients with advanced gastrointestinal stromal tumors within days of the first dose of the signal transduction inhibitor, imatinib (Gleevec).[90] In a study of 57 patients with non-small cell lung cancer, reduction of metabolic activity after one cycle of chemotherapy as shown by ^{18}FDG-PET was closely correlated with final outcome of therapy.[91]

Using radiotracers, such as $H_2{}^{15}O$, ^{11}CO or $C^{15}O$, and dynamic phase (monitoring the concentration over time) ^{18}FDG-PET, blood flow and blood volume estimations can be made. Water provides perfusion information and CO, which binds to hemoglobin, giving blood volume information. To calculate flow, tissue and arterial tracer concentration measurements

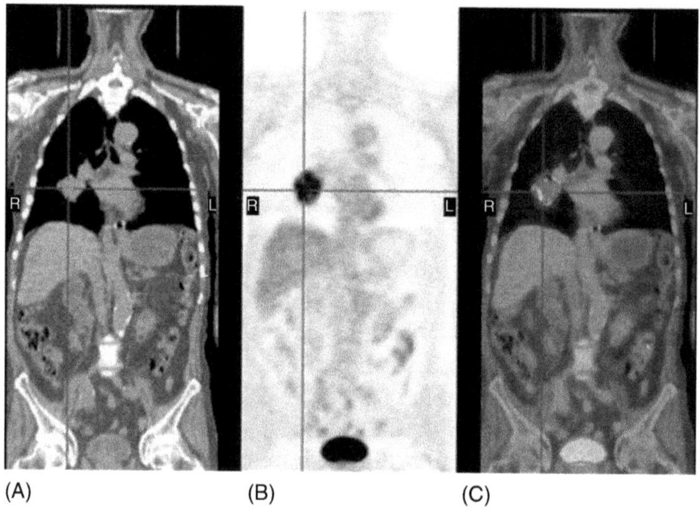

FIGURE 20.7 (See color insert following page 558.) A fused CT/PET study showing: (A) a coronal CT image showing a lung mass, (B) a coronal PET image showing increased ^{18}FDG uptake in lung, and (C) a fused CT/PET image showing that the increased ^{18}FDG uptake is directly related to the lung mass.

need to be made. Methodology (and problems encountered) is similar to that described for quantifying DCE-MRI enhancement and is largely based on continuing developments to the Fick and Kety principles in 1870 and 1951.[92] Measurement of arterial tracer concentration is difficult for both PET and DCE-MRI. Unlike DCE-MRI, however, where arterial tracer concentrations can be estimated based on injection rate, injected volume, and patient body weight, radionuclide studies require that the tracer activity is measured during the scan, since it changes over time due to rapid radioactive decay. Measurement of arterial tracer activity can be determined directly by arterial sampling, or by measuring the signal from a region of interest over the left ventricle or larger arteries, such as the aorta.[93]

PET perfusion studies are complex and may take as long as 3 h. The half-life of "non"-^{18}FDG tracers is short: for ^{15}O it is 123 s, requiring on-site cyclotron facility. Like CT and MRI, the image is made of voxels—cuboid volumes, each of which has signal intensity—and smaller voxels imply better spatial resolution. Partial volume effects occur when voxels are too large to capture the details of signal intensities that change frequently in small-imaged volumes. These partial volume effects may be significant if the tumor size is of in the order of (or less than) the resolution of the scanner (~2 cm). Partial volume effects are compounded by a phenomenon called "spill over" or "spill in" of signal counts from surrounding structures with high blood flow, such as the heart and aorta, or within areas of relatively high flow, such as liver.[94]

As for MRI, PET has been used to measure the effects of CA4P on tumor and normal tissue perfusion and blood volume in humans. Significant dose-dependent reductions were seen in tumor perfusion and tumor blood volume within 30 min after dosing, although by 24 h there was evidence of tumor vascular recovery.[95] Interestingly, the twofold decrease seen in humans was not nearly as dramatic as the eightfold reduction in tumor perfusion seen in rats at 1 h, or the 100-fold decrease at 6 h. This again emphasizes the need to confirm preclinical findings in the clinic.

Herbst et al.[96] imaged primary and metastatic lesions serially using $H_2{}^{15}O$-PET and ^{18}FDG-PET to assess changes in tumor blood flow and metabolism during treatment with human recombinant endostatin. They showed measurable effects on tumor blood flow and metabolism even in the absence of demonstrable anticancer effects. The data suggest that there is a complex, possibly nonlinear, relationship between tumor blood flow, tumor metabolism, and endostatin dose. In a study of 35 patients with locally advanced breast cancer, ^{18}FDG and ^{15}O-water PET imaging before and after 2 months of chemotherapy were used to assess metabolism and perfusion. Although both resistant and responsive tumors had an average decline in metabolic rate over the course of chemotherapy, resistant tumors had an average increase in blood flow. Patients whose tumors failed to show a decline in blood flow after 2 months of therapy had poorer disease-free and overall survival.[97] These studies clearly show that measured perfusion is not necessarily coupled with metabolism or response.

20.3.7 Ultrasound

Ultrasound imaging is inexpensive, quick to perform, and a mainstay in obstetrics and the diagnosis of disease. Ultrasound uses pulses of high-frequency sound waves (usually between 3 and 20 MHz), which are transmitted into the body and reflected by the different structures. These echoes are detected by a piezoelectric crystal, which can turn the reflected sound waves into an electrical voltage. The resolution of traditional ultrasound depends on the frequency used, with higher frequency giving better resolution, but poorer depth penetration, a problem for high-resolution clinical imaging. High-frequency ultrasound may be useful in accessible human tumors such as ocular melanoma and skin tumors. Imaging deep structures is also compromised by poor accessibility to certain anatomical regions (e.g., those that are behind bone) and operator dependence. Blood flow can be measured by using the Doppler shift in the echo frequencies caused by movement of the blood. Using pulsed Doppler, this information can be displayed as a waveform of vascular flow velocity at a certain position (Figure 20.8), whereas color Doppler gives an image of mean blood flow velocities. Power Doppler, on the other hand, shows a map of blood flow amplitude, which is useful for assessing flow in small vessels. Ultrasound therefore has the potential to provide effective, low cost, sequential monitoring of vascular changes associated with malignant tumors and their response to treatment.

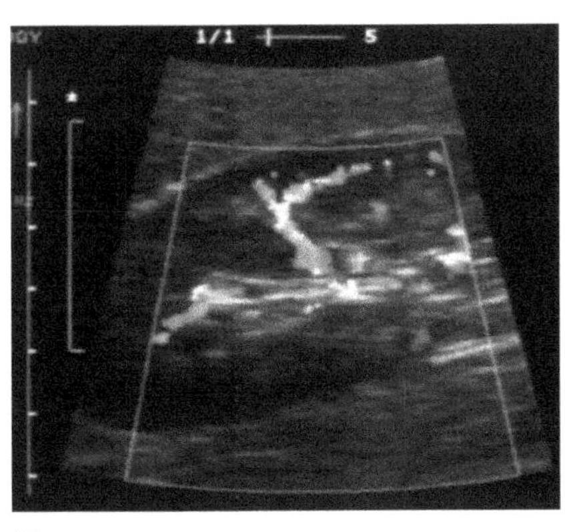

(A)

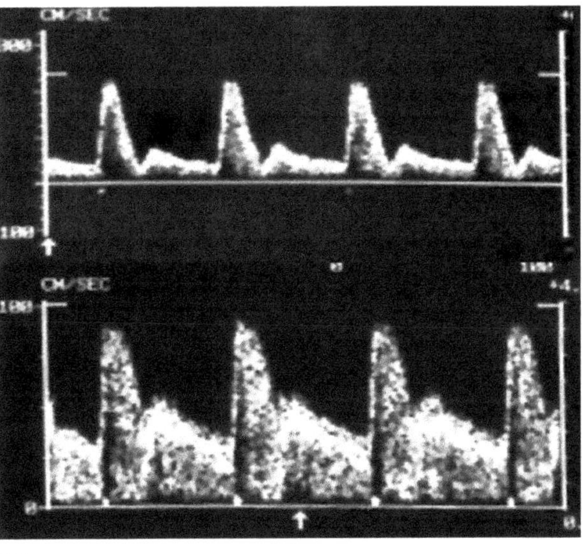

(B)

FIGURE 20.8 (A) Color Doppler image of the renal vessels. **(See color insert following page 558.)** Red represents flow towards, and blue away from the ultrasound probe. Color intensity is related to flow velocity. (B) Pulsed Doppler trace showing flow velocity patterns with systolic (peak) and diastolic flow. The upper trace shows a high resistance pattern with relatively low diastolic flow compared to the lower trace. Changes in vascular resistance may be of use in monitoring treatment.

Noninvasive monitoring of antiangiogenic therapy has been performed by serial power Doppler and color Doppler ultrasound imaging of preclinical tumors, showing reduction in vascularity with treatment by antivascular and anti-VEGF therapies.[98] Color flow Doppler has also been used to characterize superficial solid tumors in patients. In a study of 67 patients with melanomas before surgical excision, high-frequency sonography and color Doppler sonography parameters correlated with tumor aggressiveness.[99] In a further study, tumor vascularity index was evaluated with power Doppler US in 44 patients with advanced hepatocellular carcinoma treated with 200–300 mg/day thalidomide. The pretreatment vascularity index was significantly higher in responders than in nonresponders.[100]

The development of ultrasound contrast agents relies on one of the main disadvantages of ultrasound: the fact that ultrasound waves do not travel well through air. Any air/soft tissue interface causes strong echogenicity, and deeper structures cannot be seen. Ultrasound therefore cannot be used to "see" though lung, and gas-filled bowel loops, can prevent successful abdominal imaging. Ultrasound contrast media use "microbubbles" of air surrounded by a polymer shell, which are intensely echogenic. This improves the image of any vascular structure and enhances Doppler studies allowing smaller vessel sizes down to 40 μm to be discriminated. Increasing the energy of the ultrasound pulse, or selecting a particular "harmonic" frequency, can also destroy these microbubbles. This allows imaging the reappearance of the microbubbles, reflecting flow into the imaged area, and quantification of perfusion.[101]

Correlations between ultrasound-derived enhancement parameters and microvascular density have been demonstrated in animals.[101] In human breast tumors, enhancement parameters have been shown to be different in carcinomas and benign lesions after intravenous injection of microbubbles.[102] Contrast-enhanced power Doppler ultrasonography (PDUS) has also been used in determining the angiogenic status of 21 patients with renal cell carcinoma. The color pixel ratios of selected images were calculated as the ratio of the number of pixels showing power Doppler signals to the total number of pixels within the lesion. A significant correlation was found between color pixel ratio and microvascular density.[103]

Thirty-five consecutive patients with pathologically confirmed, nonresectable pancreatic carcinoma were examined with contrast-enhanced US before systemic chemotherapy. The median time to progression and median survival was longer in patients who had avascular tumors compared with patients who had vascular tumors.[104] In 15 patients, follow-up examinations after stereotactic, single-dose radiotherapy were performed using contrast-enhanced ultrasound showing a significant reduction of the arterial vascularization in treated tumors ($p < 0.05$).[105]

20.4 DISCUSSION OF BLOOD FLOW IMAGING

The results described earlier have caused great excitement in the field of drug development, because they offer the hope not only of establishing a "proof of concept" of drug activity with relatively few patients, but also of aiding dose selection for phase II trials without relying on dose-limiting toxicity. There is a problem, however, in that although a positive result is reassuring, many promising agents have not revealed positive results using PET and MRI.[106] In addition, initial encouraging findings using these tests are no guarantee of later success. Positive results in combination therapy in the presence of toxicity may give encouragement for other regimes to be explored. In evaluating all biological agents, it must be recognized that they affect not only their primary target but also the activity of other kinases, some known and some possibly unknown. The exact mechanism of enhancement reduction is also unclear and may be different for different agents and at different times. A positive result from an indirect test therefore may not relate to the expected activity of the agent. Further there is a danger that efficacious treatments could be dismissed because DCE-MRI with standard contrast media is insensitive to their mode of action, or their onset of action is too slow. The development of new contrast media and isotope probes will considerably aid understanding of the mechanisms of enhancement reductions. Furthermore, the fact that endostatin has produced measurable effects on tumor blood flow using PET but not MRI[12] in the absence of tumor regression, and that, as previously stated, measures of metabolism do not always couple with measures of perfusion, provides evidence that different tumor imaging methods may be required as end points in different situations.

There is also no consensus about how MRI or PET scans should be performed and how the data should be analyzed. Although there is a wealth of experience in animal models, these sometimes do not help in the planning of human trials, since different tumor types are often studied, with imaging protocols that are not feasible for clinical trials due to potentially toxic, unlicensed agents or clinically impractical imaging protocols. Translational imaging studies comparing effects in animals and humans using similar regimes do, however, provide information helpful in planning and interpreting clinical studies.[63,107] Reproducibility studies are required not only to judge the numbers required to obtain significant results in trials but also, possibly more importantly, to judge the significance of changes in the individual patient. Reproducibility varies depending on methods employed, but often shows coefficient of variation of ~14%–20%. This implies that such studies should be sensitive to treatment changes of ~15% if cohort studies of 10 patients are used. The intrapatient repeatability, which is an indicator of the significance on an individual patient's response, is generally higher, of the order of 30%–40%.[108–111] Technique refinement should improve these values in future.

Due to the dynamic nature of MRI contrast enhancement and PET studies, care must be taken not to interpret "reductions in enhancement" as an indication that the drug is delivered less effectively to the tissues. In both PET and MRI studies, a reduction in enhancement may simply relate to a delay in achieving maximum tissue concentrations of the tracer, due to reducing the perfusion of a tumor or vascular permeability. The potential maximum concentration of the tracer in the extravascular space may never be achieved due to limited clinical imaging times, ranging from 5 to 20 min, and the fact that the tracer concentration declines, either by renal excretion or the short half-life of isotopes, during this time. This is important since, during treatment, "steady-state" plasma levels of a pharmaceutical compound should be achieved, and delays in achieving peak tissue concentration, even of several hours, should not be significant. In addition, it has been suggested that blocking VEGF signaling "normalizes" the tumor vasculature by selective destruction of immature blood vessels. A further treatment effect includes lowering the interstitial fluid pressure creating a hydrostatic pressure gradient across the vascular wall. This induced pressure gradient may actually lead to better delivery of molecules into tumors. Thus, anti-VEGF therapy may paradoxically improve the access of therapeutic agents to cancer cells.[112]

20.5 INDIRECT TESTS NOT MEASURING BLOOD FLOW

20.5.1 MR Spectroscopy

By altering the way in which the signal from hydrogen is measured, the slightly different resonance frequency of some common metabolites allows their concentration to be estimated (^{1}H-MR spectroscopy). With more specialist MRI equipment, metabolites containing phosphorus, such as adenosine triphosphate, can also be measured (^{31}P-MR spectroscopy).

Measuring the levels of different molecules in vivo has considerable appeal, although progress has been slow due to the poor sensitivity of the technique and therefore the limited range of molecules that can be studied. Considerable improvement in clinical results has been possible with use of increasing magnetic field strength in commercially available MRI platforms. Elevated choline levels are detectable by ^{1}H-MRS in cancer, and correlate with

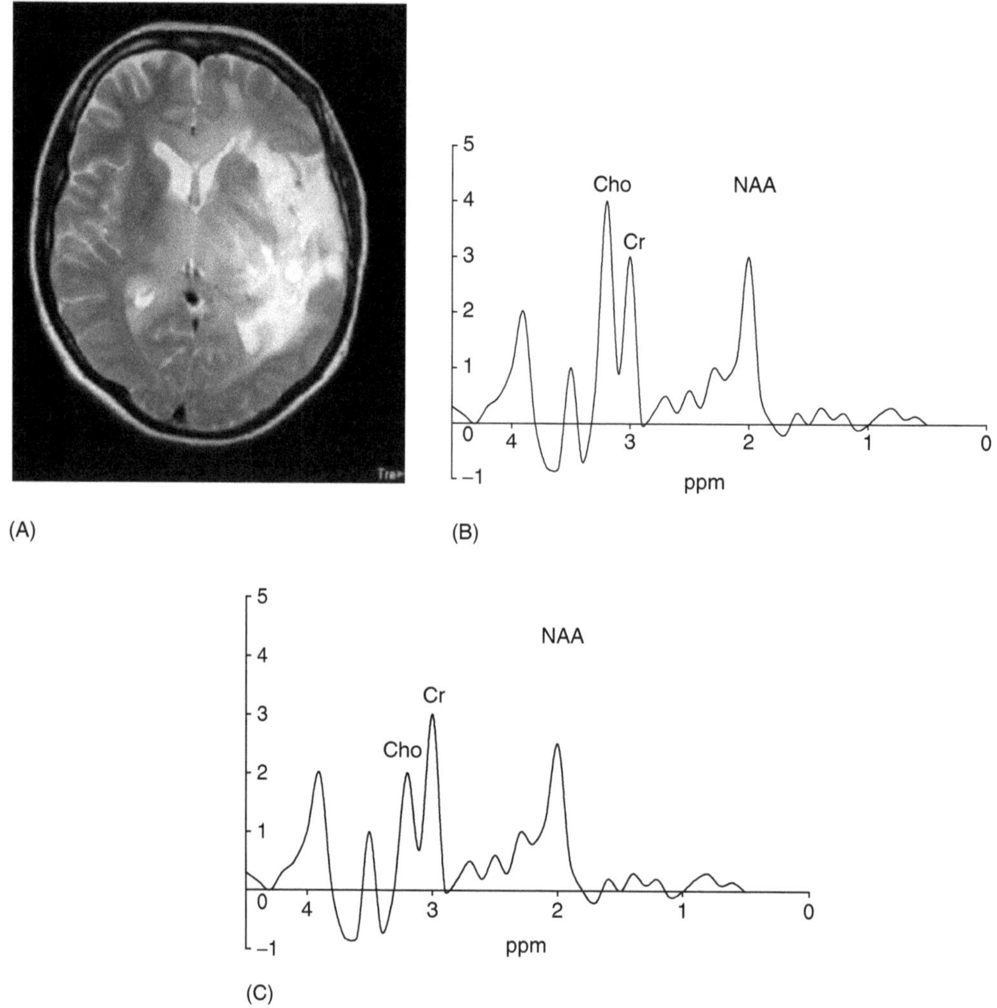

FIGURE 20.9 (A) A T_2-weighted MR image of a brain lesion after radiotherapy for a brain tumor. Black box represents a region of interest over a brain lesion for study by ^{1}H-MRI spectroscopy. (B) The expected spectrum for a malignancy with high choline (Cho) level in comparison with creatine (Cr). *N*-acetyl aspartate (NAA), an axonal marker, is low in concentration. (C) The spectrum for radiation necrosis returned in this case. Both the choline and *N*-acetyl aspartate are low compared with creatine.

malignancy and cell proliferation in brain tumors.[113] This can help diagnose malignancy, and has particular clinical value in distinguishing radiation necrosis from recurrent tumor in the brain (Figure 20.9). Although not specifically related to angiogenesis, any technique that shows a measurable difference between benign and malignant tissue could be adapted as a potential test for response to treatment. Garwood and colleagues have shown that ^{1}H-MRS can demonstrate changes as early as 1 day in neoadjuvant breast cancer therapy, which are correlated with response after 6 weeks.[114,115]

20.5.2 Diffusion-Weighted MRI

By using the magnetic field gradients, the MRI signal can be made sensitive to water motion at the microscopic level. Images that are sensitized in this way are used extensively in the clinical investigation of stroke, as they are extremely sensitive to the changes in water mobility that occur in acute ischemia because of cytotoxic edema due to membrane pump failure. There is also increasing interest in this technique as a method of monitoring apoptosis, which is a demonstrated result of antiangiogenesis treatment.[116]

Whereas water mobility decreases in acute ischemia, it increases in the case of extracellular edema. This is consistent with the observation that treatment of tumors can cause an initial decrease in measured water diffusion with a subsequent increase.[117] These techniques have been shown to be of value in a combination trial demonstrating the value of adding Taxol to radiation therapy.[118]

20.5.3 Radionuclide Imaging

Although ^{18}FDG-PET scans are those most commonly used in clinical oncology, there is increasing use of other agents with indirect mechanisms of measurement but acting as more specific indicators. Agents are available that are sensitive to programed cell death (apoptosis), due to affinity to phosphatidyl serine which is externalized on the cell wall early in the apoptosis pathway (^{99m}Tc-labeled annexin)[119] and proliferation by [^{18}F]-fluorothymidine (^{18}FLT-PET).[120] These agents can also show some perfusion information in the dynamic phase. Detection of programed cell death (apoptosis) by imaging is potentially interesting for assessing malignant and benign disorders, since apoptosis mediates tumor cell and angiogenic vascular endothelial cell regression.

Hypoxia in tumor tissue is also an important prognostic indicator of response to either chemotherapy or radiation therapy. Therefore, detection of hypoxia in advance of such interventions is of importance in optimizing the use and outcome of different therapeutic modalities. Furthermore, many antiangiogenic therapies alter oxygen levels in tumors. Misonidazole molecules bind in inverse proportion to oxygen levels and [^{18}F]-fluoromisonidazole (^{18}FMISO) or more recently ^{60}copper diacetyl-bis(*N*-methyl-thiosemicarbazone) (^{60}Cu-ATSM) can be used to study hypoxia and changes in oxygen status.[121] ^{18}FMISO has been used to quantify hypoxia in the rat glioma by PET, and may provide functional information about the results of antiangiogenic therapy.[122] In 14 patients with biopsy-proved cervical cancer, ^{60}Cu-ATSM-PET, before initiation of radiotherapy and chemotherapy, showed that the frequency of locoregional nodal metastasis was greater in hypoxic tumors. Tumor ^{18}FDG uptake did not correlate with ^{60}Cu-ATSM-PET uptake, showing measurement of hypoxia is independent of metabolism as measured by ^{18}FDG-PET.[123] Similar correlations have been found in lung cancer.[124]

^{18}FLT acts as a marker for proliferation and has the potential to be used as a specific agent for assessing disease activity in various stages of different malignancies. As cytotoxic chemotherapeutic agents affect cell division earlier and more prominently than glucose metabolism, ^{18}FLT-PET may prove to be superior to ^{18}FDG-PET for assessing response to treatment.[125]

20.6 SPECIFIC (DIRECT) IMAGING IN DEVELOPMENT

As well as observing downstream effects of successful treatment, whether specific to angiogenesis (blood flow) or simply related to successful treatment at a cellular level, there is interest in imaging specifically to document the effect of treatments on their intended site of action. This section relies mainly on preclinical in vivo data to speculate about what may be achieved in human trials in the future. Imaging will almost certainly rely on contrast media or other probes which, to be successful in humans, will need to be imaged at very low concentrations. Because of this, PET imaging is at the forefront. Due to the ability to label molecules with isotopes such as oxygen and carbon for PET imaging, it is possible to label just about any specific marker. PET studies also have the potential advantage that the treatment agent can be directly labeled. This allows direct imaging of drug delivery by "microdosing," and chemotherapeutic agents, such as ^{18}F-fluorouracil, have been synthesized to assess their pharmacokinetics and metabolism. The concentration of ^{18}F-fluorouracil in metastatic colorectal cancer has been correlated with patient survival.[126] Labeled VEGF and other mediators of angiogenesis can also be used to predict response to anti-VEGF treatment.[127] For studying angiogenesis, there has been some work in labeling integrins, specific to endothelial markers in angiogenesis, and endothelial growth factor receptors including a Her2/neu agent.[128,129] Direct labeling of the actual therapeutic agent can provide crucial information necessary for trial design and optimal dosing.[130] In a study of 20 patients with progressive solid tumors treated with various doses of the anti-VEGF antibody HuMV833, the agent was labeled with ^{124}iodine. PET showed antibody distribution and clearance were markedly heterogeneous between and within patients and between and within individual tumors.[57] This suggests future trial designs for this type of agent that use defined tumor types and potentially intrapatient dose escalation.

Molecular imaging by MRI has been thoroughly reviewed elsewhere.[11] The main problem is developing a contrast agent, which can be "seen" by MRI at nanomolar concentrations and that can be linked to specific probes. What works in animals may not be helpful in humans due to long development times and potential toxicity. Preclinical imaging with MRI scanners with much smaller access bores allows much higher magnet strengths to be achieved (typically six times that of a standard clinical scanner) giving greater sensitivity to low concentrations, or better spatial resolution in the range of 10–100 μm rather than millimeters.

Nanoparticles composed of a perfluorocarbon emulsion coated with a layer of lipid have been developed.[131] Linked to the lipid layer of each nanoparticle are up to 90,000 molecules of Gd-DTPA, enough to enable detection at low concentrations. Into the lipid outer layer, hundreds of homing molecules can be added, such as antibodies, peptides, or peptidomimetics. By targeting a protein alpha v beta 3-integrin, it is possible to detect the immature blood vessels that characterize angiogenesis in vivo in preclinical models.[132]

An exciting property of MRI contrast media is that they are not imaged directly but by their effect on surrounding water. This means they have the potential to be activated by chemical reactions in the body, an effect has been used in imaging gene expression in vivo in preclinical models. Where a gene transfer is attempted by a vector, a technique that may be used to modify angiogenesis in the future, transduction efficiency of the vector can be tested by the inclusion of a marker enzyme with the vector. The marker enzyme's effect could be to activate the MRI contrast agent. Such systems and further different approaches have been designed in preclinical models.[133]

The high sensitivity of ultrasound to microbubble contrast means that high-frequency ultrasound systems can be designed to be sensitive to a single microbubble. As well as microbubbles of air, perfluorocarbon nanospheres, similar to that used in MRI, have been developed. Vectorization of these contrast agents, in particular with a specific alpha v beta 3-integrin monoclonal antibody, directed at endothelium in tumor vessels, has already been accomplished in preclinical models.[134,135] Since it is possible to focus ultrasound energy to destroy these spheres, targeted drug delivery under ultrasound guidance may also be possible.[136]

Optical imaging is based on the use of molecules that may affect or emit radiation in the visible or near-visible spectrum in a variety of ways including scattering, absorption, and fluorescence. These "chromophores" or "fluorophores" may be intrinsic to the tissue, or may be administered.[137] Optical imaging is currently limited to research, but with endoscopic imaging technology, fluorescent and bioluminescent probes could be seen in clinically relevant sites in humans. The inability of light to pass from deep tissues is the biggest obstacle, although the use of near-infrared (NIR) light emitters and recent advances in laser technology and photon detection has improved this condition.[138] The main intrinsic mechanisms of NIR light attenuation in tissue are scattering due to variations of the cellular organelles, and absorption mainly due to oxy- and deoxyhemoglobin and some lipids and water. The combination of multiple NIR light measurements through tissue at several projections allows tomographic techniques to be used. The interpretation of these types of image may be helped by registration with CT or MRI images.

Administered contrast agents may be specific to anatomical, physiological, biochemical, or molecular function. Optical tomography, using intrinsic hemoglobin concentration,[139] and separately with extrinsic indocyanine green, a light absorber,[140] has been successfully used to detect breast lesions in a clinical setting.

Although these techniques do not immediately lend themselves to human studies, the approach may be useful. Whether they are translatable to humans remains to be seen, and depends on the toxicity of the agents and the ability to achieve satisfactory imaging resolution and signal to noise.

20.7 CONCLUSION AND PERSPECTIVES

No review can expect to be complete in a continually evolving field. As a review of clinically available imaging techniques, this chapter only hints at what may be developed in the future. What is clear is that imaging tests are available that can give useful information to aid development of antiangiogenesis strategies in humans. The good news is that when changes are seen in the clinic, they are almost always rapid and there are few cases where imaging "too early" has failed to see a response. Many studies either show or suggest a relationship between dose and response or efficacy, although human trials are always confounded by heterogeneity in patients and tumor types. It is clear that positive results may not always show correlation with each other or with clinical outcome, and there should not be overreliance on the accuracy of any one technique. Although perfusion and glucose metabolism are sometimes "coupled" in untreated tumors, studies in which both parameters have been measured before and after treatment show that perfusion and glucose metabolism may not change in parallel in response to therapy.[97,141]

However, when imaging is "successful," there is a danger of putting too much weight on cases where imaging is positive, ignoring tumor types where there is no imaging response. This is particularly true in comparisons of MRI enhancement effects in angiogenesis inhibition. It is reasonable to assume that treatment will have a bigger (or more rapid) effect on metastatic lesions, with high proportions of immature strongly angiogenic blood vessels, than a primary tumor. This, however, may take attention away from a more subtle but clinically significant response in the primary tumor with its larger proportion of mature vessels and better perfusion. Variations in the effect of angiogenesis treatments may also be due to differing levels of natural antiangiogenic agents. In one case, removal of a primary colorectal tumor resulted in an increase in metabolic activity in its liver metastasis with a concomitant drop in levels of angiostatin and endostatin in urine and plasma, respectively.[142] These circulating inhibitors of angiogenesis have been shown to affect the growth of distant micrometastatic disease in patients with cancer,[143] and the level of these factors may well affect the degree of response to be expected from biomarker studies. Variations in effect could also be caused by several other confounding factors.[144] The old adage therefore applies: treat the patient, not the images.

Many advances in the past have been made because of the observation that something works, without the need to discover the mechanism. Now, drugs are designed to have an effect on specific mechanisms. Unfortunately, understanding of these mechanisms is incomplete, and designing drugs to work perfectly in the test tube is no guarantee of ultimate success.

Furthermore, lack of understanding of the specific mechanism is no guarantee that it will not work for other reasons.[145] Though it will always be important to progress understanding of mechanisms of action for both imaging and treatment, a more pragmatic approach is needed in the interim. Whether an imaging test is valuable depends on whether it can be established as a surrogate end point or biomarker for the desired effect, and therefore answer key questions for drug development rather than simply providing interesting data.[146] These questions include

1. Did imaging help to assess whether the mechanistic goals were achieved?
2. Did imaging assist dose selection for phase II?
3. Did imaging provide assistance for schedule selection for phase II?
4. Can imaging select subpopulations enriched for response?

To achieve further progress, a huge multidisciplinary effort is required.[147] Radiologists clearly have to learn about molecular biology, but also clinical oncologists, molecular biologists, scientists, and particularly the pharmaceutical companies need to understand imaging. A multidisciplinary approach is essential to achieve validation and standardization of imaging methodology and to draw up guidelines to ensure consistent and standardized reporting on findings.[148–150] Comparison studies to determine which imaging methods work best (alone or in combination) should be instituted. The pharmaceutical companies could play a key role in developing advanced contrast agents whose main clinical role may be in the assessment of novel anticancer agents. Pharmaceutical companies must also take a translational "bench-to-bedside" approach to imaging: preclinical development of angiogenesis inhibitors should include developing imaging approaches suitable for use in subsequent clinical trials.

With further cooperation and progress, these imaging techniques, as surrogate end points for efficacy of biological agents, may become as commonplace as CT scans for drug development, and may even become standard imaging tests for all oncology patients.

ACKNOWLEDGMENTS

Parts of this chapter and Figure 20.5, Figure 20.6, Figure 20.8, and Figure 20.9 have been reprinted from *Hematology/Oncology Clinics*, Volume 18 (5), Morgan B, Horsfield MA, and Steward WP. The role of imaging in the clinical development of antiangiogenic agents, discussed in *Angiogenesis and Anti-Angiogenesis Therapy*, pages 1183–1206, 2004, has been published with permission from Elsevier.

REFERENCES

1. Bailar, J.C. and Gornik, H.L., Cancer undefeated, *New England Journal of Medicine*, 336, 1569, 1997.
2. Bergers, G., et al., Effects of angiogenesis inhibitors on multistage carcinogenesis in mice, *Science*, 284, 808, 1999.
3. O'Reilly, M.S., et al., Angiostatin induces and sustains dormancy of human primary tumors in mice, *Nature Medicine*, 2, 689, 1996.
4. Takahashi, Y., et al., Prolonged stable disease effects survival in patients with solid gastric tumor: Analysis of phase II studies of doxifluridine, *International Journal of Oncology*, 17, 285, 2000.
5. Buyse, M., et al., Relation between tumour response to first-line chemotherapy and survival in advanced colorectal cancer: A meta-analysis, *Lancet*, 356, 373, 2000.
6. James, K., et al., Measuring response in solid tumors: Unidimensional versus bidimensional measurement, *Journal of the National Cancer Institute*, 91, 523, 1999.

7. Therasse, P., et al., New guidelines to evaluate the response to treatment in solid tumors, *Journal of the National Cancer Institute*, 92, 205, 2000.
8. Korn, E.L., et al., Clinical trial designs for cytostatic agents: Are new approaches needed? *Journal of Clinical Oncology*, 19, 265, 2001.
9. Drevs, J., et al., Soluble markers for the assessment of biological activity with PTK787/ZK 222584 (PTK/ZK), a vascular endothelial growth factor receptor (VEGFR) tyrosine kinase inhibitor in patients with advanced colorectal cancer from two phase I trials, *Annals of Oncology*, 16, 558, 2005.
10. Kerbel, R.S., Tumor angiogenesis: Past, present and the near future, *Carcinogenesis*, 21, 505, 2000.
11. Weissleder, R. and Mahmood, U., Molecular imaging, *Radiology*, 219, 316, 2001.
12. Thomas, J.P., et al., Phase I pharmacokinetic and pharmacodynamic study of recombinant human endostatin in patients with advanced solid tumors, *Journal of Clinical Oncology*, 21, 223, 2003.
13. Jain, R.K., Barriers to drug-delivery in solid tumors, *Scientific American*, 271, 58, 1994.
14. Augustin, H.G., Translating angiogenesis research into the clinic: The challenges ahead, *British Journal of Radiology*, 76, S3, 2003.
15. Evelhoch, J.L., et al., Applications of magnetic resonance in model systems: Cancer therapeutics, *Neoplasia*, 2, 152, 2000.
16. Knopp, M.V., et al., Dynamic contrast-enhanced magnetic resonance imaging in oncology, *Topics in Magnetic Resonance Imaging*, 12, 301, 2001.
17. Hillman, B.J. and Neiman, H.L., Translating molecular imaging research into radiologic practice: Summary of the proceedings of the American College of Radiology Colloquium, April 22–24, 2001, *Radiology*, 222, 19, 2002.
18. Kerbel, R.S., Clinical trials of antiangiogenic drugs: Opportunities, problems, and assessment of initial results, *Journal of Clinical Oncology*, 19, 45S, 2001.
19. Carmeliet, P. and Jain, R.K., Angiogenesis in cancer and other diseases, *Nature*, 407, 249, 2000.
20. Dvorak, H.F., Vascular permeability factor/vascular endothelial growth factor: A critical cytokine in tumor angiogenesis and a potential target for diagnosis and therapy, *Journal of Clinical Oncology*, 20, 4368, 2002.
21. Raghunand, N., Gatenby, R.A., and Gillies, R.J., Microenvironmental and cellular consequences of altered blood flow in tumours, *British Journal of Radiology*, 76, S11, 2003.
22. Hermans, R., Estimation of tumour oxygenation levels with dynamic contrast-enhanced magnetic resonance imaging, *Radiotherapy and Oncology*, 57, 1, 2000.
23. Lartigau, E., et al., Oxygenation of head and neck tumors, *Cancer*, 71, 2319, 1993.
24. Miller, K.D., Sweeney, C.J., and Sledge, G.W., Redefining the target: Chemotherapeutics as antiangiogenics, *Journal of Clinical Oncology*, 19, 1195, 2001.
25. Kerbel, R.S., A cancer therapy resistant to resistance, *Nature*, 390, 335, 1997.
26. Betensky, R.A., Louis, D.N., and Cairncross, J.G., Influence of unrecognized molecular heterogeneity on randomized clinical trials, *Journal of Clinical Oncology*, 20, 2495, 2002.
27. Atkinson, A.J., et al., Biomarkers and surrogate endpoints: Preferred definitions and conceptual framework, *Clinical Pharmacology and Therapeutics*, 69, 89, 2001.
28. Smith, J.J., Sorensen, A.G., and Thrall, J.H., Biomarkers in imaging: Realizing radiology's future, *Radiology*, 227, 633, 2003.
29. Castro, M., The simpleton's error in drug development, *Journal of Clinical Oncology*, 20, 4606, 2002.
30. Hurwitz, H.I., et al., Bevacizumab in combination with fluorouracil and leucovorin: An active regimen for first-line metastatic colorectal cancer, *Journal of Clinical Oncology*, 23, 3502, 2005.
31. Siemann, D.W., Therapeutic strategies that selectively target and disrupt established tumor vasculature, *Hematology–Oncology Clinics of North America*, 18, 1023, 2004.
32. McDonald, D.M., Thurston, G., and Baluk, P., Endothelial gaps as sites for plasma leakage in inflammation, *Microcirculation*, 6, 7, 1999.
33. Dvorak, A.M., et al., The vesiculo-vacuolar organelle (VVO): A distinct endothelial cell structure that provides a transcellular pathway for macromolecular extravasation, *Journal of Leukocyte Biology*, 59, 100, 1996.
34. Dvorak, A.M. and Feng, D., The vesiculo-vacuolar organelle (VVO). A new endothelial cell permeability organelle, *Journal of Histochemistry and Cytochemistry*, 49, 419, 2001.

35. Su, M.Y., et al., Tumor characterization with dynamic contrast-enhanced MRI using MR contrast agents of various molecular weights, *Magnetic Resonance in Medicine*, 39, 259, 1998.
36. Tofts, P.S., et al., Estimating kinetic parameters from dynamic contrast-enhanced T-1-weighted MRI of a diffusable tracer: Standardized quantities and symbols, *Journal of Magnetic Resonance Imaging*, 10, 223, 1999.
37. Buadu, L.D., et al., Breast lesions: Correlation of contrast medium enhancement patterns on MR images with histopathologic findings and tumor angiogenesis, *Radiology*, 200, 639, 1996.
38. Hawighorst, H., et al., Angiogenic activity of cervical carcinoma: Assessment by functional magnetic resonance imaging-based parameters and a histomorphological approach in correlation with disease outcome, *Clinical Cancer Research*, 4, 2305, 1998.
39. Mayr, N.A., et al., Prediction of tumor control in patients with cervical cancer: Analysis of combined volume and dynamic enhancement pattern by MR imaging, *American Journal of Roentgenology*, 170, 177, 1998.
40. Stomper, P.C., et al., Angiogenesis and dynamic MR imaging gadolinium enhancement of malignant and benign breast lesions, *Breast Cancer Research and Treatment*, 45, 39, 1997.
41. Hulka, C.A., et al., Dynamic echo-planar imaging of the breast: Experience in diagnosing breast carcinoma and correlation with tumor angiogenesis, *Radiology*, 205, 837, 1997.
42. de Lussanet, Q.G., et al., Gadopentetate dimeglumine versus ultrasmall super-paramagnetic iron oxide for dynamic contrast-enhanced MR imaging of tumor angiogenesis in human colon carcinoma in mice, *Radiology*, 229, 429, 2003.
43. de Lussanet, Q.G., et al., Dynamic contrast-enhanced MR imaging kinetic parameters and molecular weight of dendritic contrast agents in tumor angiogenesis in mice, *Radiology*, 235, 65, 2005.
44. Reddick, W.E., Taylor, J.S., and Fletcher, B.D., Dynamic MR imaging (DEMRI) of microcirculation in bone sarcoma, *Journal of Magnetic Resonance Imaging*, 10, 277, 1999.
45. Pham, C.D., et al., Magnetic resonance imaging detects suppression of tumor vascular permeability after administration of antibody to vascular endothelial growth factor, *Cancer Investigation*, 16, 225, 1998.
46. Barbier, E.L., Lamalle, L., and Decorps, M., Methodology of brain perfusion imaging, *Journal of Magnetic Resonance Imaging*, 13, 496, 2001.
47. Enochs, W.S., et al., Improved delineation of human brain tumors on MR images using a long-circulating, superparamagnetic iron oxide agent, *Journal of Magnetic Resonance Imaging*, 9, 228, 1999.
48. Rydland, J., et al., New intravascular contrast agent applied to dynamic contrast enhanced MR imaging of human breast cancer, *Acta Radiologica*, 44, 275, 2003.
49. Harisinghani, M.G., et al., Noninvasive detection of clinically occult lymph-node metastases in prostate cancer, *New England Journal of Medicine*, 348, 2491, 2003.
50. Evelhoch, J.L., Key factors in the acquisition of contrast kinetic data for oncology, *Journal of Magnetic Resonance Imaging*, 10, 254, 1999.
51. DeVries, A.F., et al., Tumor microcirculation evaluated by dynamic magnetic resonance imaging predicts therapy outcome for primary rectal carcinoma, *Cancer Research*, 61, 2513, 2001.
52. Buckley, D.L., Uncertainty in the analysis of tracer kinetics using dynamic contrast-enhanced T-1-weighted MRI, *Magnetic Resonance in Medicine*, 47, 601, 2002.
53. Morgan, B., et al., Dynamic contrast-enhanced magnetic resonance imaging as a biomarker for the pharmacological response of PTK787/ZK 222584, an inhibitor of the vascular endothelial growth factor receptor tyrosine kinases, in patients with advanced colorectal cancer and liver metastases: Results from two phase I studies, *Journal of Clinical Oncology*, 21, 3955, 2003.
54. Thomas, A.L., et al., Phase I study of the safety, tolerability, pharmacokinetics, and pharmacodynamics of PTK787/ZK 222584 administered twice daily in patients with advanced cancer, *Journal of Clinical Oncology*, 23, 4162, 2005.
55. Larsson, H.B.W., et al., Myocardial perfusion modeling using MRI, *Magnetic Resonance in Medicine*, 35, 716, 1996.
56. Yung, W.K.A., Friedman, H., Conrad, C., Reardon, D., Provenzale, J., and Jackson, E. A phase I trial of single-agent PTK 787/ZK 222584 (PTK/ZK), an oral VEGFR tyrosine kinase inhibitor, in patients with recurrent glioblastoma multiforme, *Proceedings of the American Society of Clinical Oncology*, 22, 395(abstract), 2003.

57. Jayson, G.C., et al., Molecular imaging and biological evaluation of HuMV833 anti-VEGF antibody: Implications for trial design of antiangiogenic antibodies, *Journal of the National Cancer Institute*, 94, 1484, 2002.
58. Ellis, L.M., Antiangiogenic therapy: More promise and, yet again, more questions, *Journal of Clinical Oncology*, 21, 3897, 2003.
59. Bruns, C.J., et al., Vascular endothelial growth factor is an in vivo survival factor for tumor endothelium in a murine model of colorectal carcinoma liver metastases, *Cancer*, 89, 488, 2000.
60. Benjamin, L.E., et al., Selective ablation of immature blood vessels in established human tumors follows vascular endothelial growth factor withdrawal, *Journal of Clinical Investigation*, 103, 159, 1999.
61. He, H., et al., Vascular endothelial growth factor signals endothelial cell production of nitric oxide and prostacyclin through Flk-1/KDR activation of c-Src, *Journal of Biological Chemistry*, 274, 25130, 1999.
62. Jain, R.K., Normalization of tumor vasculature: An emerging concept in antiangiogenic therapy, *Science*, 307, 58, 2005.
63. Galbraith, S.M., et al., Combretastatin A4 phosphate has tumor antivascular activity in rat and man as demonstrated by dynamic magnetic resonance imaging, *Journal of Clinical Oncology*, 21, 2831, 2003.
64. Stevenson, J.P., et al., Phase I trial of the antivascular agent combretastatin A4 phosphate on a 5-day schedule to patients with cancer: Magnetic resonance imaging evidence for altered tumor blood flow, *Journal of Clinical Oncology*, 21, 4428, 2003.
65. Galbraith, S.M., et al., Effects of 5,6-dimethylxanthenone-4-acetic acid on human tumor microcirculation assessed by dynamic contrast-enhanced magnetic resonance imaging, *Journal of Clinical Oncology*, 20, 3826, 2002.
66. Morgan, B., et al., Dynamic contrast enhanced magnetic resonance imaging as a surrogate marker of efficacy in a phase I trial of a VEGF receptor tyrosine kinase inhibitor, *Proceedings of the American Association of Cancer Research*, 42, 587(abstract), 2001.
67. George, M.L., et al., Non-invasive methods of assessing angiogenesis and their value in predicting response to treatment in colorectal cancer, *British Journal of Surgery*, 88, 1628, 2001.
68. Lankester, K.J., et al., Conventional cytotoxic chemotherapy agents do not have acute antivascular effects, as measured by dynamic contrast enhanced MRI (DCE-MRI), *Proceedings of the American Society of Clinical Oncology*, 22, 588(abstract), 2003.
69. Wolfe, W., et al., Response to anticancer treatment with docetaxel administered every 3 weeks and weekly is associated with functional assessment of changes in tumoral blood flow/perfusion, *Proceedings of AACR*, 44, 5343(abstract), 2003.
70. Cha, S., et al., Dynamic contrast-enhanced T2*-weighted MR imaging of recurrent malignant gliomas treated with thalidomide and carboplatin, *American Journal of Neuroradiology*, 21, 881, 2000.
71. Warmuth, C., Gunther, M., and Zimmer, C., Quantification of blood flow in brain tumors: Comparison of arterial spin labeling and dynamic susceptibility-weighted contrast-enhanced MR imaging, *Radiology*, 228, 523, 2003.
72. Wang, B., Gao, Z.Q., and Yan, X., Correlative study of angiogenesis and dynamic contrast-enhanced magnetic resonance imaging features of hepatocellular carcinoma, *Acta Radiologica*, 46, 353, 2005.
73. Keogan, M.T. and Edelman, R.R., Technologic advances in abdominal MR imaging, *Radiology*, 220, 310, 2001.
74. Eberhard, A., et al., Heterogeneity of angiogenesis and blood vessel maturation in human tumors: Implications for antiangiogenic tumor therapies, *Cancer Research*, 60, 1388, 2000.
75. Taylor, N.J., et al., BOLD MRI of human tumor oxygenation during carbogen breathing, *Journal of Magnetic Resonance Imaging*, 14, 156, 2001.
76. Neeman, M., et al., In vivo BOLD contrast MRI mapping of subcutaneous vascular function and maturation: Validation by intravital microscopy, *Magnetic Resonance in Medicine*, 45, 887, 2001.
77. McDonald, D.M. and Choyke, P.L., Imaging of angiogenesis: From microscope to clinic, *Nature Medicine*, 9, 713, 2003.
78. Brix, G., et al., Regional blood flow capillary permeability, and compartmental volumes: Measurement with dynamic CT—Initial experience, *Radiology*, 210, 269, 1999.
79. Miles, K.A., Perfusion CT for the assessment of tumour vascularity: Which protocol? *British Journal of Radiology*, 76, S36, 2003.

80. Miles, K.A., Tumour angiogenesis and its relation to contrast enhancement on computed tomography: A review, *European Journal of Radiology*, 30, 198, 1999.
81. Tateishi, U., et al., Tumor angiogenesis and dynamic CT in lung adenocarcinoma: Radiologic–pathologic correlation, *Journal of Computer-Assisted Tomography*, 25, 23, 2001.
82. Tateishi, U., et al., Contrast-enhanced dynamic computed tomography for the evaluation of tumor angiogenesis in patients with lung carcinoma, *Cancer*, 95, 835, 2002.
83. Dawson, P., Dynamic contrast-enhanced functional imaging with multi-slice CT, *Academic Radiology*, 9, S368, 2002.
84. Willett, C.G., et al., Direct evidence that the VEGF-specific antibody bevacizumab has antivascular effects in human rectal cancer, *Nature Medicine*, 10, 649, 2004.
85. Anonymous, The ionising radiation (medical exposure) regulations 2000, in *Health and Safety (Statutory Instruments)*, The Stationery Office, Ed., 2000.
86. Alavi, A., Kung, J.W., and Zhuang, H.M., Implications of PET based molecular imaging on the current and future practice of medicine, *Seminars in Nuclear Medicine*, 34, 56, 2004.
87. Pomper, M.G., Functional and metabolic imaging, in *Cancer, Principles and Practice of Oncology*, DeVita, V.T., Hellman, S., and Rosenberg, S.A., Eds., Lippincott Williams & Wilkins, Philadelphia, 2001.
88. Gillies, R.J., et al., MRI of the tumor microenvironment, *Journal of Magnetic Resonance Imaging*, 16, 430, 2002.
89. Kostakoglu, L., et al., PET predicts prognosis after 1 cycle of chemotherapy in aggressive lymphoma and Hodgkin's disease, *Journal of Nuclear Medicine*, 43, 1018, 2002.
90. Joensuu, H., et al., Effect of the tyrosine kinase inhibitor STI571 in a patient with a metastatic gastrointestinal stromal tumor, *New England Journal of Medicine*, 344, 1052, 2001.
91. Weber, W.A., et al., Positron emission tomography in non-small-cell lung cancer: Prediction of response to chemotherapy by quantitative assessment of glucose use, *Journal of Clinical Oncology*, 21, 2651, 2003.
92. Anderson, H. and Price, P., Clinical measurement of blood flow in tumours using positron emission tomography: A review, *Nuclear Medicine Communications*, 23, 131, 2002.
93. Bacharach, S.L., Libutti, S.K., and Carrasquillo, J.A., Measuring tumor blood flow with (H_2O)-O-15: Practical considerations, *Nuclear Medicine and Biology*, 27, 671, 2000.
94. Laking, G.R. and Price, P.M., Positron emission tomographic imaging of angiogenesis and vascular function, *British Journal of Radiology*, 76, S50, 2003.
95. Anderson, H.L., et al., Assessment of pharmacodynamic vascular response in a phase I trial of combretastatin A4 phosphate, *Journal of Clinical Oncology*, 21, 2823, 2003.
96. Herbst, R.S., et al., Development of biologic markers of response and assessment of antiangiogenic activity in a clinical trial of human recombinant endostatin, *Journal of Clinical Oncology*, 20, 3804, 2002.
97. Mankoff, D.A., et al., Changes in blood flow and metabolism in locally advanced breast cancer treated with neoadjuvant chemotherapy, *Journal of Nuclear Medicine*, 44, 1806, 2003.
98. Drevs, J., et al., Effects of PTK787/ZK 222584, a specific inhibitor of vascular endothelial growth factor receptor tyrosine kinases, on primary tumor, metastasis, vessel density, and flood flow in a murine renal cell carcinoma model, *Cancer Research*, 60, 4819, 2000.
99. Lassau, N., et al., Prognostic value of angiogenesis evaluated with high-frequency and color Doppler sonography for preoperative assessment of melanomas, *American Journal of Roentgenology*, 178, 1547, 2002.
100. Hsu, C., et al., Effect of thalidomide in hepatocellular carcinoma: Assessment with power Doppler US and analysis of circulating angiogenic factors, *Radiology*, 235, 509, 2005.
101. Lassau, N., et al., Evaluation of contrast-enhanced color Doppler ultrasound for the quantification of angiogenesis in vivo, *Investigative Radiology*, 36, 50, 2001.
102. Huber, S., et al., Effects of a microbubble contrast agent on breast tumors: Computer-assisted quantitative assessment with color Doppler US—Early experience, *Radiology*, 208, 485, 1998.
103. Kabakci, N., et al., Echo contrast-enhanced power Doppler ultrasonography for assessment of angiogenesis in renal cell carcinoma, *Journal of Ultrasound in Medicine*, 24, 747, 2005.
104. Masaki, T., et al., Noninvasive assessment of tumor vascularity by contrast-enhanced ultrasonography and the prognosis of patients with nonresectable pancreatic carcinoma, *Cancer*, 103, 1026, 2005.

105. Krix, M., et al., Monitoring of liver metastases after stereotactic radiotherapy using low-Ml contrast-enhanced ultrasound—Initial results, *European Radiology*, 15, 677, 2005.
106. Scappaticci, F.A., Mechanisms and future directions for angiogenesis-based cancer therapies, *Journal of Clinical Oncology*, 20, 3906, 2002.
107. Lee, L., et al., Biomarkers for the assessment of pharmacologic activity for a vascular endothelial growth factor (VEGF) receptor inhibitor, PTK787/ZK 222584 (PTK/ZK): Comparison of preclinical data with results in phase I studies, *Clinical Cancer Research*, 9, 6203S, 2003.
108. Galbraith, S.M., et al., Reproducibility of dynamic contrast-enhanced MRI in human muscle and tumours: Comparison of quantitative and semi-quantitative analysis, *NMR in Biomedicine*, 15, 132, 2002.
109. Jackson, A., et al., Reproducibility of quantitative dynamic contrast-enhanced MRI in newly presenting glioma, *British Journal of Radiology*, 76, 153, 2003.
110. Evelhoch, J.L., et al., Magnetic resonance imaging measurements of the response of murine and human tumors to the vascular-targeting agent ZD6126, *Clinical Cancer Research*, 10, 3650, 2004.
111. Morgan, B., Higginson, A., and Horsfield, M., Reproducibility of a single slice, single region method for monitoring the effect of treatment of metastases using dynamic contrast enhanced MRI, *Proceedings of International Society for Magnetic Resonance in Medicine*, 13, 2099(abstract), 2005.
112. Tong, R.T., et al., Vascular normalization by vascular endothelial growth factor receptor 2 blockade induces a pressure gradient across the vasculature and improves drug penetration in tumors, *Cancer Research*, 64, 3731, 2004.
113. Aboagye, E.O. and Bhujwalla, Z.M., Malignant transformation alters membrane choline phospholipid metabolism of human mammary epithelial cells, *Cancer Research*, 59, 80, 1999.
114. Bolan, P.J., et al., In vivo quantification of choline compounds in the breast with H-1 MR spectroscopy, *Magnetic Resonance in Medicine*, 50, 1134, 2003.
115. Meisamy, S., et al., Neoadjuvant chemotherapy of locally advanced breast cancer: Predicting response with in vivo H-1 MR spectroscopy—A pilot study, *Radiology*, 233, 424, 2004.
116. Ross, B.D., et al., Evaluation of cancer therapy using diffusion magnetic resonance imaging, *Molecular Cancer Therapeutics*, 2, 581, 2003.
117. Chenevert, T.L., et al., Diffusion magnetic resonance imaging: An early surrogate marker of therapeutic efficacy in brain tumors, *Journal of the National Cancer Institute*, 92, 2029, 2000.
118. Chinnaiyan, A.M., et al., Combined effect of tumor necrosis factor-related apoptosis-inducing ligand and ionizing radiation in breast cancer therapy, *Proceedings of the National Academy of Sciences of the United States of America*, 97, 1754, 2000.
119. Blankenberg, F.G. and Strauss, H.W., Nuclear medicine applications in molecular imaging, *Journal of Magnetic Resonance Imaging*, 16, 352, 2002.
120. Shields, A.F., et al., Carbon-11-thymidine and FDG to measure therapy response, *Journal of Nuclear Medicine*, 39, 1757, 1998.
121. Rajendran, J.G. and Krohn, K.A., Imaging hypoxia and angiogenesis in tumors, *Radiologic Clinics of North America*, 43, 169, 2005.
122. Rasey, J.S., et al., Determining hypoxic fraction in a rat glioma by uptake of radiolabeled fluoromisonidazole, *Radiation Research*, 153, 84, 2000.
123. Dehdashti, F., et al., Assessing tumor hypoxia in cervical cancer by positron emission tomography with ^{60}Cu-ATSM: Relationship to therapeutic response—A preliminary report, *International Journal of Radiation Oncology, Biology, Physics*, 55, 1233, 2003.
124. Dehdashti, F., et al., In vivo assessment of tumor hypoxia in lung cancer with Cu-60-ATSM, *European Journal of Nuclear Medicine and Molecular Imaging*, 30, 844, 2003.
125. Shields, A.F., et al., Imaging proliferation in vivo with [F-18] FLT and positron emission tomography, *Nature Medicine*, 4, 1334, 1998.
126. Moehler, M., et al., F-18-labeled fluorouracil positron emission tomography and the prognoses of colorectal carcinoma patients with metastases to the liver treated with 5-fluorouracil, *Cancer*, 83, 245, 1998.
127. Collingridge, D.R., et al., The development of [I-124] iodinated-VG76e: A novel tracer for imaging vascular endothelial growth factor in vivo using positron emission tomography, *Cancer Research*, 62, 5912, 2002.

128. Goldenberg, D.M. and Nabi, H.A., Breast cancer imaging with radiolabeled antibodies, *Seminars in Nuclear Medicine*, 29, 41, 1999.
129. Haubner, R., et al., Noninvasive imaging of alpha(v)beta(3) integrin expression using F-18-labeled RGD-containing glycopeptide and positron emission tomography, *Cancer Research*, 61, 1781, 2001.
130. Propper, D.J., et al., Use of positron emission tomography in pharmacokinetic studies to investigate therapeutic advantage in a phase I study of 120-hour intravenous infusion XR5000, *Journal of Clinical Oncology*, 21, 203, 2003.
131. Lanza, G.M., et al., Molecular imaging and targeted drug delivery with a novel, ligand-directed paramagnetic nanoparticle technology, *Academic Radiology*, 9, S330, 2002.
132. Sipkins, D.A., et al., Detection of tumor angiogenesis in vivo by alpha(v)beta(3)-targeted magnetic resonance imaging, *Nature Medicine*, 4, 623, 1998.
133. Bremer, C. and Weissleder, R., Molecular imaging—In vivo imaging of gene expression: MR and optical technologies, *Academic Radiology*, 8, 15, 2001.
134. Ellegala, D.B., et al., Imaging tumor angiogenesis with contrast ultrasound and microbubbles targeted to alpha(v)beta(3), *Circulation*, 108, 336, 2003.
135. Leong-Poi, H., et al., Noninvasive assessment of angiogenesis by ultrasound and microbubbles targeted to alpha(v)-integrins, *Circulation*, 107, 455, 2003.
136. Unger, E.C., et al., Acoustically active lipospheres containing paclitaxel—A new therapeutic ultrasound contrast agent, *Investigative Radiology*, 33, 886, 1998.
137. Weissleder, R. and Ntziachristos, V., Shedding light onto live molecular targets, *Nature Medicine*, 9, 123, 2003.
138. Mahmood, U., et al., Near-infrared optical imaging of protease activity for tumor detection, *Radiology*, 213, 866, 1999.
139. Pogue, B.W., et al., Quantitative hemoglobin tomography with diffuse near-infrared spectroscopy: Pilot results in the breast, *Radiology*, 218, 261, 2001.
140. Ntziachristos, V., et al., Concurrent MRI and diffuse optical tomography of breast after indocyanine green enhancement, *Proceedings of the National Academy of Sciences of the United States of America*, 97, 2767, 2000.
141. Tateishi, U., et al., Lung tumors evaluated with FDG-PET and dynamic CT: The relationship between vascular density and glucose metabolism, *Journal of Computer Assisted Tomography*, 26, 185, 2002.
142. Peeters, C.F.J.M., et al., Decrease in circulating anti-angiogenic factors (angiostatin and endostatin) after surgical removal of primary colorectal carcinoma coincides with increased metabolic activity of liver metastases, *Surgery*, 137, 246, 2005.
143. Feldman, A.L., et al., A prospective analysis of plasma endostatin levels in colorectal cancer patients with liver metastases, *Annals of Surgical Oncology*, 8, 741, 2001.
144. Allen, J. and Bergsland, K., Angiogenesis in colorectal cancer: Therapeutic implications and future directions, *Hematology–Oncology Clinics of North America*, 18, 1087, 2004.
145. Rajan, T.V., The myth of mechanism, *Scientist*, 15, 6, 2001.
146. Collins, J.M., Functional imaging in phase I studies: Decorations or decision making? *Journal of Clinical Oncology*, 21, 2807, 2003.
147. Leach, M.O., et al., The assessment of antiangiogenic and antivascular therapies in early-stage clinical trials using magnetic resonance imaging: Issues and recommendations, *British Journal of Cancer*, 92, 1599, 2005.
148. Tatum, J.L. and Hoffman, J.M., Angiogenesis Imaging Methodology, AIM for Clinical Trials. The Role of Imaging in Clinical Trials of Anti-Angiogenesis Therapy in Oncology Program, http://imaging.cancer.gov/reportsandpublications/AngiogenesisImagingMethodsforClinicalTrials, 2005.
149. Anonymous, DCE-MRI Group Consensus Recommendations, NCI CIP MR Workshop on Translational Research in Cancer–Tumor Response, http://imaging.cancer.gov/reportsandpublications/ReportsandPresentations/MagneticResonance, 2004.
150. Kothari, M., et al., Imaging in antiangiogenesis trial: A clinical trials radiology perspective, *British Journal of Radiology*, 76, S92, 2003.

21 Pharmacodynamic Markers in Tissues

Darren W. Davis

CONTENTS

In recent years, there has been significant progress in the development of angiogenesis inhibitors, particularly those that target the vascular endothelial growth factor receptor-2 (VEGFR-2) (1,2). The rapid emergence of hundreds of molecular agents against numerous angiogenesis targets offers greater anticancer efficacy with fewer side effects (3). Despite these recent advances, assessing the effects of angiogenesis inhibitors individually or in combination, or combined with conventional therapies, has created significant challenges for basic scientists and clinical investigators to effectively integrate antiangiogenic agents into clinical practice (4). Thus, better strategies are needed to understand the pharmacodynamic effects of agents that target angiogenesis by different mechanisms (5).

Targeting tumor vascular endothelial cells to inhibit angiogenesis offers several therapeutic advantages compared with conventional cytotoxic therapies (6–8). Essentially, angiogenesis inhibitors can be classified as direct, indirect, and multitargeted (Figure 21.1). Direct angiogenesis inhibitors target endothelial cells involved in tumor vessel formation to prevent their recruitment, proliferation, or migration in response to pro-angiogenic proteins (9,10). In contrast, indirect angiogenesis inhibitors block the production or prevent binding of angiogenic factors secreted by stromal or tumor cells, and inhibit other angiogenic signaling pathways (11). A third type, multitargeted inhibitors, combines the effects of both

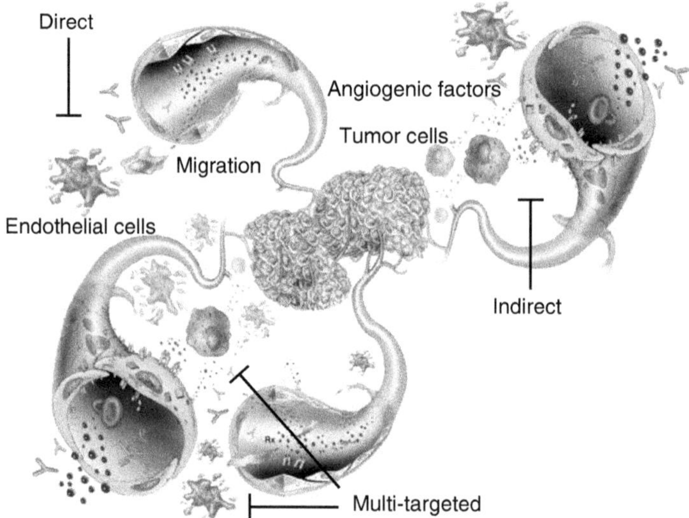

FIGURE 21.1 (See color insert following page 558.) Three major types of angiogenesis inhibitors in clinical development. Antiangiogenic agents can be classified as (1) direct inhibitors that target endothelial cells to prevent them from proliferating or migrating, for example, endostatin, (2) indirect inhibitors that block the production and prevent binding of endothelial growth factors or downstream angiogenic signaling pathways, for example, bevacizumab, gefitinib, and (3) multitargeted inhibitors that combine the effects of both direct and indirect inhibitors to block signal transduction pathways, in tumor-associated endothelial cells and tumor cells or other angiogenesis signaling pathways, for example, cyclo-oxygenase 2 inhibitors, sunitinib malate. Multitargeted agents may offer the best therapeutic approach for inhibiting angiogenesis in advanced disease.

direct and indirect mechanisms to block multiple signal transduction pathways on both tumor-associated endothelial and tumor cells (12,13). The mechanisms of resistance to angiogenesis inhibitors remain unknown; however, it is likely that during tumor progression, multiple endothelial growth factors are overexpressed thereby circumventing the effects of these agents (14). Currently, the optimal therapeutic strategy for using angiogenesis inhibitors remains unknown.

Clinical studies have shown that certain surrogate biomarkers (e.g., microvessel density) of angiogenesis have prognostic value in human tumors (15). However, to date, no surrogate markers that predict or correlate with clinical outcome have been identified or validated. The most direct approach for determining the biological activity of antiangiogenic agents is to implement correlative tissue-based analyses in clinical studies (16). Although tumor biopsies are limited by cancer type and accessibility, data obtained from correlative studies may help address important questions about the use of antiangiogenic agents in clinical practice. (1) Did the agent hit its intended target? (2) Was the target inhibition transient or stable to induce apoptosis in tumor-associated endothelial cells? (3) What biomarkers indicate which patients are most likely or least likely to respond? (4) What are the molecular mechanisms of action and resistance of single agents and their combinations? (5) What is the optimal biological dose or schedule? This chapter addresses the questions mentioned previously, discusses the development of reliable assays for quantifying pharmacodynamic markers in tissues, the effects of different angiogenesis inhibitors on various markers and their correlation with clinical outcome, and issues that pose challenges for incorporating tumor tissue analysis into clinical trials.

21.1 QUANTITATIVE ANALYSIS OF PHARMACODYNAMIC MARKERS IN TISSUES

Investigators typically rely on immunohistochemistry assays to measure the pharmacodynamic effects of targeted therapies in tissues. The majority of these studies use chromogenic or immunoperoxidase staining, which are semiquantitative in nature and have other limitations (17). In contrast, immunofluorescence detection methods can provide simultaneous labeling of multiple proteins in one sample (18). It was not until the development of angiogenesis inhibitors that the need for a technique to detect apoptosis in endothelial cells became apparent. Multi-immunofluorescent labeling is particularly useful for the detection and analysis of protein expression patterns or apoptosis in specific cell types, for example, tumor-associated endothelial cells, which require three fluorochromes to visualize the total cell nuclei, endothelial cells, and Terminal deoxynucleotidyl Transferase-dUTP Nick End Labeling (TUNEL)-positive cells (Figure 21.2). This dual-labeling technique was considered to be a significant milestone and became widely used to measure the efficacy of antiangiogenic

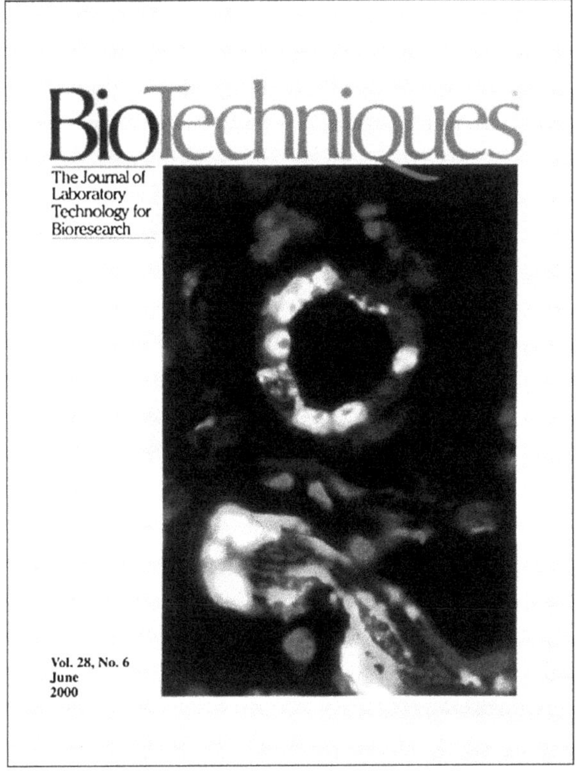

FIGURE 21.2 (See color insert following page 558.) Immunofluorescent double staining of apoptotic endothelial cells following DC101 and paclitaxel therapy in an orthotopic human bladder cancer model. Tumor endothelial cells were stained with anti-CD31 (red) and apoptotic cells were detected by TUNEL-FITC (green). This highly sensitive method facilitates identification and quantification of apoptotic endothelial cells in situ. Co-localization of red and green turns yellow and indicates apoptotic endothelial cells. [Immunofluorescent image appears with permission of Eaton Publishing, Westborough, MA; Cover, *BioTechniques* Vol. 28, No. 6 (June 2000).]

agents in both preclinical and clinical studies (5,19–24). Although immunofluorescent techniques can facilitate visualization of specific cell types by eye as a result of co-localization of different fluorochromes, manual quantification is limited to enumerating "positive" and "negative" cells in random microscopic fields using a categorical score and may not be able to detect subtle, but significant changes (25). Recent research efforts have focused on the development of immunofluorescent-based automated assays to quantify protein expression patterns and apoptosis in tumor-associated endothelial cells in situ (26,27).

Various platform technologies have been developed to facilitate automated, quantitative in situ measurements (28). Most of these systems are designed for standard immunohistochemistry assays using chromogenic substrates. Measuring the pharmacodynamic effects of antiangiogenic therapies requires the ability to distinguish between tumor-associated endothelial cells, tumor cells, and other cell types. One platform technology capable of quantifying multiple fluorochromes in fixed tissue specimens is the laser scanning cytometer (LSC). The LSC platform is an automated analysis system described as a cross between a flow and a static image cytometer. Lasers are used to simultaneously excite different fluorochromes in cellular specimens that emit discrete wavelengths detected by a set of photomultiplier tubes. Together these features permit the ability to generate high-content stoichiometric data on heterogeneous populations of large numbers of cells. Thus, the LSC is used much like a flow cytometer to obtain multicolor immunofluorescence intensity information on fixed specimens using a continuous scale (18,29).

Several clinical studies have incorporated LSC-mediated analysis to determine drug–target interactions, effects on downstream signaling pathways, and levels of apoptosis in xenografts, human skin, and tumor tissues (29–31). Because the LSC is a platform technology, many different applications can be developed to exploit its inherent capabilities. Recent research efforts have been focused on developing specific tissue-based applications using LSC technology in an attempt to standardize the methodology for consistent data generation so that the data can be compared between different tissue specimens and antiangiogenic therapies. Because LSC-mediated data acquisition is automated, the process requires a systematic interactive approach to maintain high quality control standards and ensure consistent data generation (Figure 21.3). Pharmacodynamic data generated using this type of process to analyze biomarkers in entire tissue cross-sections have consistently provided data that correlate with clinical outcome (29,32) (Proc Am Soc Clin Oncol. 23: 2005, Abstr 3006). The next few sections summarize the most recent pharmacodynamic data obtained from tissues using standard immunohistochemistry and LSC-mediated methods. A summary of antiangiogenic clinical studies that have incorporated analysis of tumor and skin tissues, their effects on various pathways, and correlation with clinical outcome are shown in Table 21.1 and Table 21.2, respectively.

21.2 PHARMACODYNAMIC ANALYSIS OF DIRECT ANGIOGENESIS INHIBITORS

Aberrant expression of cell-surface RTKs, for example, VEGFR-2, plays a pivotal role in angiogenesis and the progression of cancer. Drugs that target RTKs are designed to block the intrinsic enzymatic activity that catalyzes the transfer of the gamma-phosphate of ATP to tyrosine residues in protein substrates (33). Inhibiting phosphorylation of these tyrosine residues prevents downstream signaling events, which affect cellular function (e.g., proliferation, differentiation, migration, or apoptosis (34)). Thus, the ability to measure phosphorylation status and signal transduction pathways has become an important pharmacodynamic endpoint in clinical studies (Table 21.1).

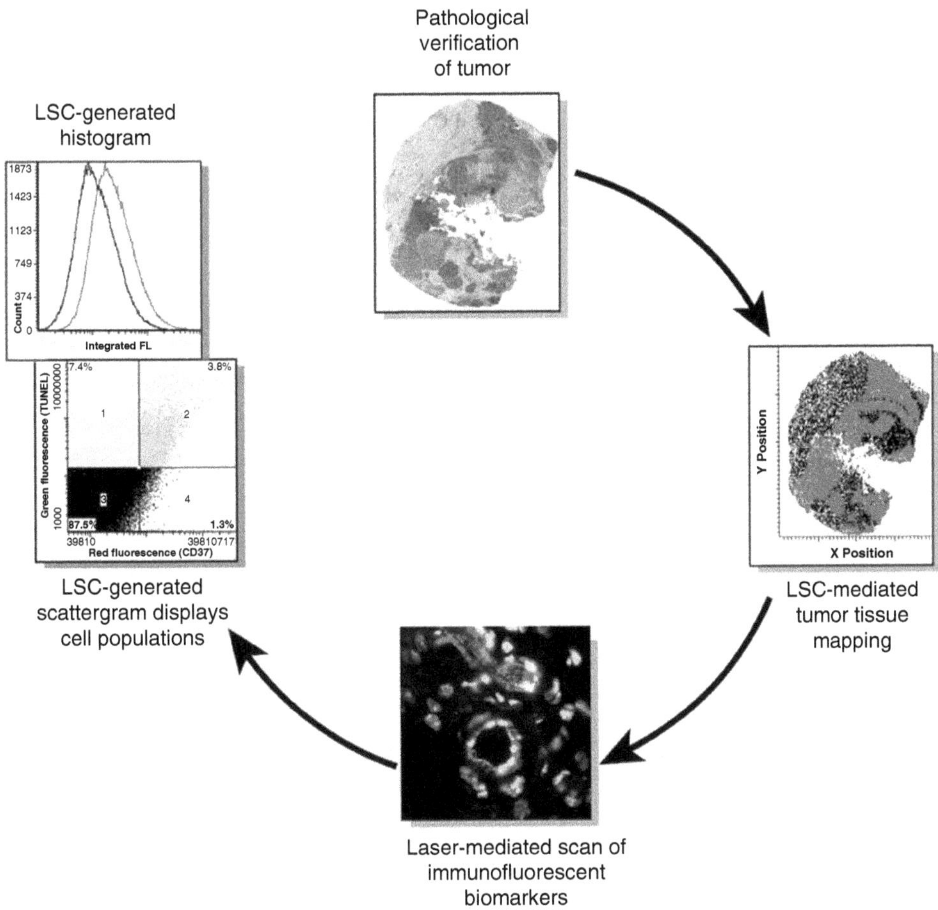

FIGURE 21.3 (See color insert following page 558.) Quantitative analysis of pharmacodynamic effects in tissues using LSC technology. Pathological verification of biopsy samples is essential for mapping tumor regions and excluding normal and necrotic regions from the analysis. Lasers detect individual cells within the mapped region of interest based on immunofluorescence staining. LSC-generated scattergrams display the percentage of cell populations based on user-defined gating using controls, for example, apoptotic endothelial cells. Alternatively, protein expression levels, for example, phosphorylated VEGF receptor-2, measured by mean fluorescent intensity may be determined as shown in the histogram. [Immunofluorescent image appears with permission of Eaton Publishing, Westborough, MA; Cover, *BioTechniques* Vol. 28, No. 6 (June 2000).]

21.2.1 SU5416

One of the first small molecule inhibitors of the VEGFR-2 to enter clinical studies was SU5416 (35,36). The compound is a competitive inhibitor of the ATP-binding site in the VEGFR-2 tyrosine kinase domain ($Ki = 0.16\ \mu mol/L$). It displays 20-fold selectivity for VEGFR-2 compared with the structurally related platelet-derived growth factor receptor (PDGFR), and it displays almost no activity against the receptors for epidermal growth factor or basic fibroblast growth factor (37). Because the antitumor activity of SU5416 did not meet clinical expectations, however, new clinical strategies were implemented to evaluate the potential biological activity in the tumor. Dowlati and coworkers conducted a novel Phase I study of SU5416, referred to as "dose de-escalation design" that monitored

TABLE 21.1
Analysis of Pharmacodynamic Markers in Tumor Tissues

Agent[Reference]	Target	PD Markers	Change on Therapy	Correlation	Response
Direct angiogenesis inhibitors					
SU5416[Dowlati] Phase I, advanced solid tumors	VEGFR-2 TKI	MVD	5/19 Increase	None	4 SD
SU5416[Davis, Haymach] Phase II, advanced solid tumors	VEGFR-2 TKI	pVEGFR-2	No change	With lack of response	None
		EC + TUNEL	No change	With lack of response	
		TC + TUNEL	No change	With lack of response	
Endostatin[Mundhenke] Phase I, advanced solid tumors	Endogenous Angiogenesis inhibitor	MVD	No change	None	SD
		CD31 + Ki67	No change	None	
		vWF + TUNEL	Rare event	None	
Endostatin[Herbst, Davis] Phase I, advanced solid tumors	Endogenous Angiogenesis inhibitor	MVD	Decrease*	Optimal dose and response	1PR, 1SD
		EC + TUNEL	Increase*	Optimal dose and response	
		BCL-2	No change	None	
		HIF-1	No change	None	
		TC + TUNEL	No change	None	
Indirect angiogenesis inhibitors					
ZD1839[Daneshmand] (Iressa, Gefitinib) Phase I, metastatic colorectal cancer	EGFR TKI	EGFR	7/10 No change, 3 decrease	None	None
		pEGFR	1/16 Decrease	None	
		pAKT	2/10 Decrease	None	
		pERK	5/9 Decrease in tumor fibroblasts	None	
		p27	2/9 Increase in nucleus	None	
		Beta-Catenin	2/9 No change	None	
		TUNEL	6/10 Increase, 2 decrease	None	
		Ki67	8/10 Decrease*	With tumor burden	
ZD1839[Baselga] (Iressa, Gefitinib) Phase II, metastatic breast cancer	EGFR TKI	EGFR	No change	None	12 SD
		pEGFR	Decrease*	None	
		pMAPK	Decrease*	None	
		Ki67	No change	None	
		pAKT	No change	None	
		p27	No change	None	
		TGF-a	No change	None	
		TUNEL	Increase*	None	
ZD1839[Rothenberg] (Iressa, Gefitinib) Phase II, recurrent colorectal adenocarcinoma	EGFR TKI	EGFR	Increase*	None	1 PR, 20 SD
		pEGFR	Increase* (3/19 decrease)	None	

TABLE 21.1 (continued)
Analysis of Pharmacodynamic Markers in Tumor Tissues

Agent[Reference]	Target	PD Markers	Change on Therapy	Correlation	Response
		pAKT	Increase*	None	
		pMAPK	No change	None	
		Ki67	No change	None	
ZD1839[Cohen] (Iressa, Gefitinib) Phase II, metastatic squamous cell carcinoma	EGFR TKI	EGFR	No change	None	1 CR, 4 PR, 20 SD
		ERK	Increase*	None	
		pERK	No change	None	
ZD1839[Gasparini] (Iressa, Gefitinib)	EGFR TKI	ERBB2	No change	None	2 PR, 7 SD
		EGFR	No change	None	
GW72016[Bacus] (Lapatinib)	ErbB1 and ErbB2 TKI	pErbB1	Decrease	response	4 PR, 11 SD
		ErbB1	No change	None	
		pErbB2	Decrease	Pre higher in responders	
		ErbB2	No change	Pre higher in responders	
		pAKT	Decrease	None	
		pErk1/2	Decrease	None	
		IGF-1R	Not measured	Pre higher in responders	
		p70 S6 Kinase	Not measured	Pre higher in responders	
		Cyclin D	Decrease	None	
		TGF-a	Decrease	Pre higher in responders	
		TUNEL	Increase	Pre higher in responders	
Multitargeted inhibitors					
SU6668[Davis] advanced solid tumors	VEGFR-2, PDGFR TKI	pVEGFR-2	1/8 Decrease	With lack of response	None
		EC and TC pPDGFR	1/6 TC Decrease	With lack of response	
		EC TUNEL	No change	With lack of response	
		TC TUNEL	No change	With lack of response	

* Significant.

pharmacodynamic endpoints at the maximum tolerated dose (MTD) (38). If a biological effect was shown, dose de-escalation to a predefined dose level would occur to determine if the lower dose induced the same pharmacodynamic effect as the higher dose. From the 19 patients enrolled in the study, only 5 had pre- and posttreatment biopsies available. All 5 patients displayed an increase in microvessel density, the only pharmacodynamic end point measured in the tumor tissue. These data were consistent with an overall increase in blood flow assessed by dynamic contrast-enhanced magnetic resonance imaging and the overall lack of clinical activity in this study (38). In fact, in a separate Phase II study,

TABLE 21.2
Analysis of Pharmacodynamic Markers in Skin Tissues

Agent[Reference]	Target	PD Markers	Change on Therapy	Correlation	Response
Direct angiogenesis inhibitors					
Endostatin[Mundhenke]	Endogenous	MVD	No change	None	SD
Phase I, advanced	angiogenesis	CD31 + Ki67	Slight decrease	None	
solid tumors	inhibitor	vWF + TUNEL	Rare event	None	
Indirect angiogenesis inhibitors					
ZD1839[Baselga, Albanell]	EGFR TKI	EGFR	No change	None	19 SD
(Iressa, Gefitinib)		pEGFR	Decrease*	None	
Phase I, advanced		pMAPK	Decrease*	None	
solid tumors		Ki67	Decrease*	None	
		p27	Increase*	None	
		TGF-a	No change	None	
ZD1839[Baselga]	EGFR TKI	EGFR	No change	None	12 SD
(Iressa, Gefitinib)		pEGFR	Decrease*	None	
Phase II, advanced		pMAPK	Decrease*	None	
solid tumors		pSTAT3	Increase*	None	
		p27	Increase*	None	
		Ki67	Decrease*	None	
EMD 72000[Vanhoefer]	EGFR mAB	EGFR	No change	None	5 PR, 6 SD, 1 MR
advanced solid		p-EGFR	Decrease*	None	
tumors		pMAPK	Decrease*	None	
		pSTAT3	Increase*	None	
		TGF-alpha	No change	None	
		p27	Increase*	None	
		Ki67	Decrease*	None	

it was confirmed that SU5416 failed to significantly inhibit VEGFR-2 phosphorylation using LSC-mediated quantitative analysis (31). Importantly, levels of tumor-associated endothelial cell and tumor-cell apoptosis did not significantly increase following treatment with SU5416 (29). Together, these data suggest that SU5416-mediated receptor inhibition was transient in nature and insufficient to trigger significant levels of endothelial cell death. Further pharmacodynamic studies with clinically active compounds are needed to determine whether the dose de-escalation strategy can benefit the clinical development of antiangiogenic therapies.

21.2.2 Endostatin

Endostatin is a COOH-terminal cleavage fragment of collagen XVIII that can function as an endogenous inhibitor of tumor angiogenesis (7,39). However, endostatin's receptor has not been isolated, and its mechanism of action remains obscure. Two separate Phase I studies were conducted that incorporated tissue-based analyses to assess the pharmacodynamic effects of endostatin. In one study, the goal was to determine the effects of endostatin on blood vessels in tumors and wound sites (40). Blood vessel formation in skin tissue was evaluated by creating a wound site on the arm with a punch biopsy device followed by a second overlapping biopsy after a 7-day interval. This sequential biopsy procedure was performed before and after 3 weeks initiation of endostatin treatment. Mundhenke and coworkers used dual-labeling immunohistochemical techniques to detect endothelial cell proliferation, mature blood vessels, and endothelial-cell apoptosis. No significant differences were detected between pre- and posttreatment samples of tumors or skin wounds for any of the markers. Interestingly, apoptotic endothelial and tumor cells were identified but the frequency was rare. Low levels of apoptosis were observed in a Phase I dose-finding study (15–600 mg/m^2) of endostatin and, therefore, it was apparent that more sensitive quantitative methods were required (26,41). Using LSC-mediated quantitative analysis, it was demonstrated that the average level of apoptosis significantly increased from 0.2% to 1.1% after endostatin therapy (30). Although the levels of apoptosis were low, a fivefold increase is comparable to tumor cell lines exposed to cytotoxic agents in vitro. In addition, significant decreases in microvessel density were observed following endostatin treatment. Effects on the vasculature were quantified using an innovative technique to enumerate microvessels in an entire tissue cross-section (27). Other parameters quantified by LSC that did not show significant changes with dose were BCL-2 and hypoxia-inducible factor-1α. Intriguingly, the changes in endothelial-cell apoptosis and microvessel density were consistent with significant decreases in blood flow measured by positron emission tomography. Together, these data strongly suggested that endostatin had optimal biological activity at doses of 250 mg/m^2 in this cohort of patients (30).

21.3 PHARMACODYNAMIC ANALYSIS OF INDIRECT ANGIOGENESIS INHIBITORS

21.3.1 ZD1839 (Iressa, Gefitinib)

ZD1839 was the first in a new class of small, molecular-targeted therapies against EGFR to gain market approval (based on two Phase II studies) for nonsmall cell lung cancer (42,43). Although these pivotal Phase II trials did not incorporate correlative tissue studies, several clinical studies suggest that ZD1839 exerts biological activity in tumor tissues, albeit not through direct inhibition of EGFR phosphorylation. In a Phase I, metastatic colorectal cancer trial, EGFR was detected by immunohistochemistry in all 16 pretreatment tumor biopsies (44). Total EGFR levels remained unchanged in 7 of 10 patients after treatment with ZD1839, whereas the other 3 demonstrated a decrease. Interestingly, 2 of the patients whose EGFR levels decreased displayed a large increase in apoptosis. Phosphorylation of EGFR was also measured; however, it was reported that detection was reproducible in only 1 patient and that this patient displayed complete inhibition after treatment with ZD1839. Other downstream markers of the EGFR signaling pathway were measured including phosphorylation of AKT and extracellular receptor kinase (ERK), $p27^{kip1}$ and Beta-Catenin expression. Phosphorylated-AKT and ERK (in tumor cells) were decreased in 2 and 1 patient, respectively. However, analysis of cellular sub-types revealed that phosphorylated-ERK was lower in tumor stromal fibroblasts in 5 of 9 patients after ZD1839 therapy. In addition, analysis of the cyclin-dependent protein kinase $p27^{Kip1}$ was not detected in 7 of 9 patients, in part because EGFR activation promotes degradation of $p27^{Kip1}$ (45). Interestingly, levels of $p27^{Kip1}$ increased in 2 patients whose tumors also displayed an increase in apoptosis. ZD1839 treatment did not affect Beta-Catenin expression in 3 of the paired samples that were evaluated. The proliferation index, measured by Ki67, was the only marker in this study that significantly correlated with change in tumor burden. Intriguingly, the 2 patients that had the largest decrease in Ki67 also showed the largest increase in apoptosis. These data suggested that ZD1839 affected downstream markers (e.g., Ki67), but failed to demonstrate the ability of the compound to directly block activation of EGFR. In fact, Daneshmand and coworkers concluded that it is likely other patients had phosphorylated EGFR and AKT but at levels that were below the detection limit of standard immunohistochemistry methods.

Serial skin biopsies have been analyzed as potential surrogate tissues for monitoring the biologic effects of antiangiogenic agents (Table 21.2). A Phase I study of ZD1839 in advanced solid malignancies incorporated skin, but not tumor, biopsies to determine effects on EGFR signaling (46,47). Surprisingly, levels of phosphorylated-EGFR expression were completely inhibited. No changes in total EGFR expression were observed after treatment. Other downstream markers in the EGFR network were affected by ZD1839 including phosphorylated-Ras-mitogen-activated protein kinase (MAPK) and STAT3, Ki67, $p27^{kip1}$, and apoptosis. Although significant changes were observed in almost all of the markers when comparing pre- and posttreatment skin biopsies, none of the changes correlated with dose or clinical response. However, because the inhibitory effects in the skin were significant at doses well below the one producing unacceptable toxicity, the investigators concluded that pharmacodynamic assessments may be useful for selecting optimal biological doses rather than an MTD for efficacy and safety in clinical trials. Other studies by the same group of investigators have incorporated both skin and tumor tissues into the same trial. Sequential (pre- and posttreatment) skin and tumor biopsies from 27 and 16 patients, respectively, in a Phase II study of metastatic breast cancer demonstrated complete inhibition of EGFR phosphorylation in both normal and malignant tissues following therapy with ZD1839 (48). However, the downstream consequences of receptor inhibition were distinct in skin and tumor samples. Phosphorylation of MAPK was inhibited in both skin and tumor, whereas ZD1839 treatment increased p27 and decreased Ki67 in skin, whereas no change was observed in tumors. In addition, ZD1839 did not inhibit phosphorylation of AKT in tumors. Total EGFR and TGF-α expression were not affected following ZD1839 treatment. Taken together, the comparison of the data in the skin and the tumor revealed that the effects of ZD1839 were identical at the level of receptor phosphorylation but not in the downstream markers p27 and Ki67. Thus, Baselga and coworkers concluded that the lack of significant clinical activity of ZD1839 was due to the lack of EGFR dependence in the tested population. Furthermore, because inhibition of MAPK phosphorylation also occurred in EGFR negative tumor biopsies, the investigators concluded that this could be an indication that immunohistochemistry fails to detect low EGFR levels in some tumors, or that ZD1839 may be inhibiting other members of the ErbB receptor family. Because there were no complete or partial responses in this cohort of patients, it is possible that ZD1839 transiently blocked activation of EGFR, failing to inhibit proliferation, and induce apoptosis in tumor cells. Indeed, recent studies have shown that sustained RTK target inhibition is critical for induction of apoptosis and subsequent clinical response (29).

In contrast, other clinical studies have shown that ZD1839 failed to inhibit its primary target and downstream effects on signaling pathways. A randomized Phase II, recurrent colorectal adenocarcinoma study demonstrated that total or activated EGFR, activated Akt, activated MAPkinase, or Ki67 in tumors did not decrease following 1 week of ZD1839 treatment (49). Surprisingly, significant increases in total EGFR, phosphorylated-EGFR, and phosphorylated-AKT were observed following treatment. In fact, only 3 patients displayed a decrease in phosphorylated EGFR. Interestingly, the investigators noted a trend toward decreased posttreatment levels of activated Akt and Ki67 in patients with a PFS higher than the median (1.9 months); however, these levels did not reach significance. In another Phase II study of metastatic squamous cell carcinoma of the head and neck, no statistically significant difference in EGFR expression following treatment with ZD1839 (50) was observed. Interestingly, total ERK expression was higher in posttreatment samples; however, phosphorylated-ERK did not increase. No correlation between biomarkers and clinical response was observed in this study. In contrast to the other studies, development of skin toxicity was a statistically significant predictor of response and improved outcome. Although this study did not evaluate pharmacodynamic markers in the skin, it is likely that molecular changes would be expected to precede the morphological and histological changes in the skin. Furthermore, in a Phase I study of metastatic breast cancer with two partial responses and seven stable diseases, pre- and posttreatment analyses of ERBB2 and EGFR levels were not statistically significant between the subgroups of patients responding to ZD1839 treatment (51). Perhaps the ability to detect significant changes in phosphorylated-EGFR is due to the sensitivity of standard immunohistochemical methods or that ZD1839 inhibits other members of the ErbB receptor family.

21.3.2 EMD72000

EMD72000 is a monoclonal antibody directed at the extracellular domain of the EGFR. A Phase I study of EMD7200 in advanced solid malignancies incorporated serial skin biopsies to analyze the effects on the EGFR signaling pathway (52). No changes in total EGFR expression were observed. However, phosphorylation of EGFR and MAPK decreased and phosphorylation of STAT3 and $p27^{Kip1}$ increased as expected as a result of EGFR inhibition in the skin. Rates of proliferation decreased and apoptosis were not measured. Thus, similar to the findings of ZD1839, significant changes were observed in almost all of the markers when comparing pre- and posttreatment skin biopsies; however, none of the changes correlated with dose or clinical response.

Both studies of ZD1839 and EMD7200 in skin tissues suggest that the pharmacodynamic changes confirmed inhibition in EGFR signaling at doses well below the MTD. However, because these particular studies did not evaluate tumor biopsies in parallel, it is impossible to know whether the changes observed in skin are representative of the biological effects in the tumor. For example, complete inhibition in EGFR phosphorylation in the skin was observed in patients who did not respond to therapy. Moreover, these data did not correlate with adverse skin reactions even though EGFR promotes keratinocyte survival (53). In contrast, skin toxicity was a statistically significant predictor of response and improved outcome in other ZD1839 studies (50). In fact, other studies have shown that skin may be useful for monitoring the potential toxicity of molecular-targeted therapies (40). Given that there were no major clinical responses observed after ZD1839 or EMD7200 treatment, it is unlikely that complete inhibition in EGFR signaling occurred in the tumor or that tumor-cell survival in these studies is independent of EGFR. Ironically, clinical development of ZD1839 was hindered following disappointing results in December 2004 from a pivotal Phase III survival trial that compared ZD1839 with placebo in patients with advanced nonsmall cell lung cancer who had failed one or more lines of chemotherapy (42,43). Clearly, more sensitive methods are needed to identify relevant biomarkers that indicate which patients are most likely or least likely to respond.

To address these complications, our research efforts have focused on developing pharmacodynamic assays using LSC technology to determine the correlation between skin and tumor, and clinical outcome. Unpublished results from a large cohort of patients treated with a small molecule RTK inhibitor that targets both EGFR and VEGFR-2 indicate that the skin can be used to confirm drug–target inhibition. However, the pharmacodynamic changes in skin are not always consistent with the changes observed in tumor suggesting that heterogeneity, the tissue microenvironment, and compensatory survival mechanisms may become activated by negative feedback loops. Importantly, the changes observed in tumor are consistent with the lack of clinical activity associated with the investigational agent. Further studies are warranted to determine the potential value of using skin as a surrogate marker for monitoring the pharmacodynamic effects of antiangiogenic therapies.

21.3.3 GW572016 (Lapatinib)

Lapatinib is an orally active small molecule that reversibly inhibits the EGFR tyrosine kinases ErbB1 and ErbB2. Because ErbB2-containing heterodimers exert potent mitogenic signals, simultaneously interrupting both ErbB1 and ErbB2 signaling is an appealing therapeutic approach. This dual mechanism of inhibition potently blocks activation of Erk1/2 and Akt in ErbB1- and ErbB2-expressing tumor cell lines and xenografts (54,55). It is thought that Lapatinib elicits antiangiogenic or cytotoxic antitumor effects depending on the cell type (54,56). A clinical pilot study was conducted to assess the biological effects of Lapatinib on various tumor growth and survival pathways in patients with advanced ErbB1- and ErbB2-overexpressing solid malignancies (57). Sequential tumor biopsies from 33 patients were evaluated by standard immunohistochemistry. Tumors from all 4 patients with partial responses displayed elevated baseline levels of phosphorylated-ErbB2. These same patients displayed an inhibition of cyclin D and the EGFR ligand TGF-α following treatment with Lapatinib. However, phosphorylated levels of both ErbB1 and ErbB2 varied after initiation of treatment in the responding patients. Interestingly, no pattern of a biologic effect on ErbB-MAPK-Erk1/2 or ErbB-PI3K-AKT signaling pathways was observed in nonresponding patients.

21.4 PHARMACODYNAMIC ANALYSIS OF MULTITARGETED ANGIOGENESIS INHIBITORS

21.4.1 SU6668

SU6668 is a second-generation synthetic derivative of SU5416 (58) that displays somewhat less activity against VEGFR-2 (Ki = 2.1 μmol/L) but also inhibits the receptors for basic fibroblast growth factor (Ki = 1.2 μmol/L) and PDGF (Ki = 8 nmol/L) (59). Proof of concept studies in xenografts demonstrated that SU6668 significantly reduced phosphorylation of VEGFR-2 and PDGFR-β in tumors by 50% and 92%, respectively, and induced apoptosis in tumor-associated endothelial cells and tumor cells at the highest dose (29). Intriguingly, levels of phosphorylation of both receptors rebounded greater than baselines at 24 h. In a Phase I trial of SU6668 (60), the ability of this small molecule to inhibit its putative targets was measured by synchronized immunofluorescence and LSC-mediated quantitative analysis of phosphorylated VEGFR-2 (29). Levels of phosphorylated VEGFR-2 and PDGFR-β also decreased significantly (50%) 6 h after therapy in 1 of 6 primary human tumors treated with SU6668, but these effects were not associated with increased endothelial or tumor-cell apoptosis. Quantitative analysis of specific cell types revealed strong inhibition in phosphorylation of PDGFR-β in tumor cells compared with endothelial cells. Together, these data suggested that SU5416 and SU6668 displayed biological activity in xenografts; however, neither drug produced marked biological activity in primary human tumors. The lack of apoptosis in these tumors confirmed that any effect on RTK inhibition was transient in nature (29).

21.4.2 SU11248 (Sunitinib Malate, Sutent)

SU11248 is an oral multitargeted receptor tyrosine kinase inhibitor that has shown antiangiogenic and antitumor activities in several in vitro and in vivo tumor models (61–63). These effects were associated with the inhibition of receptor tyrosine kinase signaling by KIT, PDGFRs, all three isoforms of the vascular endothelial growth factor receptors (VEGFR-1, VEGFR-2, VEGFR-3), Fms-like tyrosine kinase-3 receptor (FLT3), and the receptor encoded by the RET proto-oncogene (62,64,65). The effects of SU11248 on angiogenesis inhibition were investigated in a Phase I/II trial with paired tumor biopsies obtained at baseline and 11 days after initiation of treatment (66). Synchronized immunofluorescence was used to measure activated levels of PDGFR-β and VEGFR-2. Interestingly, levels of phosphorylated PDGFR-β in tumor cells and tumor-associated endothelial cells decreased 18% and 42%, respectively, in patients who responded to SU11248 therapy (67). In contrast, phosphorylation of PDGFR-β in tumor cells and tumor-associated endothelial cells increased by approximately 10% and 23%, respectively, in patients who progressed on therapy. Importantly, tumors from patients with clinical benefit displayed an overall ten- and sixfold ($P < 0.05$) significant increase from baseline in endothelial and tumor-cell apoptosis, respectively, whereas tumors from patients with progressive disease exhibited little or no change in endothelial or tumor-cell apoptosis (66,67).

21.5 CHALLENGES AND PERSPECTIVES

There are many challenges to successfully incorporating tissue analysis in the design of a clinical study. Acquiring the tissue alone requires the commitment of the sponsor, scientists, oncologists, interventional radiologists, committees, and patients. Standardization of tumor sampling and tissue procurement is critical to ensure that quality tumor tissue is being evaluated. A lack of quantitative standardization among different assays may lead to unintentional interpretation and variability between laboratories. Other technical issues that may impact interpretation of pharmacodynamic data are intra- and intertumor heterogeneity, tissue microenvironment (e.g., skin versus tumor), compensatory mechanisms, and timing of biopsies after initiation of therapy and after the last dose. Although tremendous effort is needed to obtain tissues and standardize assays for biomarker analysis, information obtained from pharmacodynamic studies in tissues should facilitate development and optimal therapeutic use of angiogenesis inhibitors.

Pharmacodynamic analysis of tumor tissues can provide direct proof of whether an antiangiogenic agent affected its intended target and downstream consequences on signal transduction and tumor-associated endothelial-cell apoptosis. It is worth emphasizing that few tissue-based studies have attempted to link target or pathway inhibition with endothelial-cell or tumor-cell apoptosis. Some agents may elicit transient target inhibition and fail to induce apoptosis (29). Therefore, measuring pharmacodynamic end points that include target or pathway inhibition linked to cellular fate, for example, apoptosis, may provide better evidence of the biological effects of the drug in the tumor and correlation with clinical outcome (Figure 21.4). Several studies have demonstrated that skin may serve as a surrogate tissue to confirm drug–target inhibition, signal transduction, and kinetics in clinical studies. However, analysis of biomarkers in tumor tissues may better represent the biological effects of a targeted therapy as tumor cells often respond differently compared with normal cells. More quantitative studies are needed to identify reliable biomarkers and their correlation between the effects in skin, tumor, and clinical outcome. Furthermore, pharmacodynamic studies in tumor tissue may also identify the genomic and proteomic profile of the population with the greatest chance to benefit from treatment. For example, the therapeutic activity of Herceptin would likely have been missed if patients had not been preselected based on their HER2 status. Another promising surrogate source that could potentially be used to assess the effects of targeted agents is the circulating tumor cell or endothelial cell. These cells may better represent the tumor microenvironment and are now being routinely studied for monitoring the effects of angiogenesis inhibitors (68,69). Our ongoing research efforts are aimed at developing sensitive assays to analyze the pharmacodynamic effects of "rare" circulating cells, for example, of epithelial origin.

Clearly, there is a need for better strategies to assess the effects of antiangiogenic agents early in clinical development. For example, in a Phase I trial of bevacizumab, no objective responses were observed out of 25 patients (70). It was not until a series of randomized Phase II and III trials over a period of more than 5 years that the clinical activity of bevacizumab was established. However, it is generally not practical to perform large randomized trials for drugs without evidence of biological activity early in their development, and therefore, many

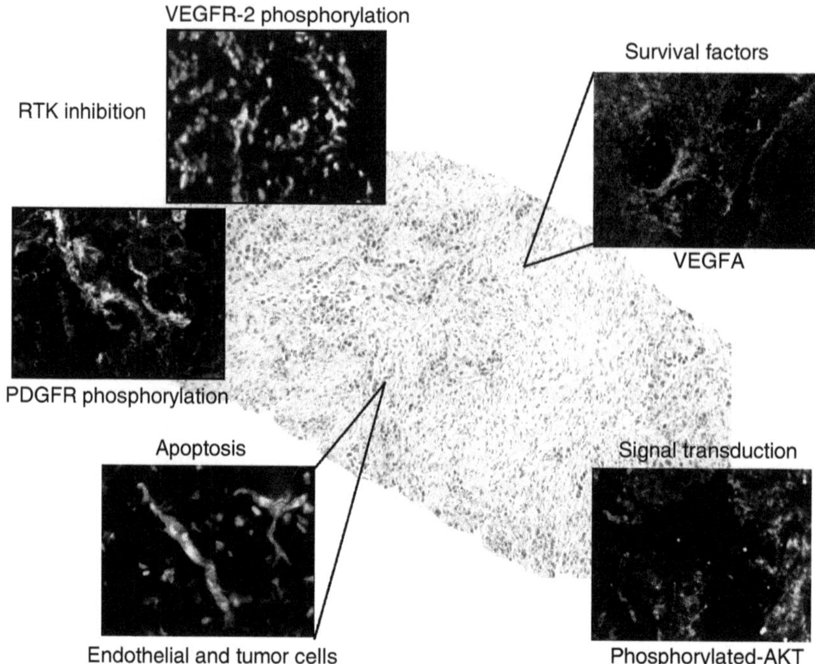

FIGURE 21.4 (See color insert following page 558.) Pharmacodynamic analysis of antiangiogenic therapies in tumor tissues. Correlative tissue studies may help determine the pharmacodynamic effects of angiogenesis inhibitors on receptor tyrosine kinase phosphorylation, growth factors, signal transduction, and apoptosis early in clinical studies. Immunofluorescence detection permits the analysis of biomarkers in specific cell types, for example, phosphorylation of PDGFR-β in endothelial cells. Measuring end points that include target or pathway inhibition linked to apoptosis may provide better evidence of the biological effects of the drug in the tumor and correlation with clinical outcome. (Red = endothelium, green = protein expression or TUNEL, yellow = Co-localization of endothelium and protein or TUNEL.)

promising drugs with therapeutic activity may never be developed. Given the large number of antiangiogenic therapies entering clinical testing, it is crucial that early studies incorporate correlative end points to determine optimal dosing and scheduling for subsequent Phase II and III trials. Ultimately, clinical development of antiangiogenic therapies would benefit if a biomarker was identified early and actually known to correlate with clinical outcome in specific types of cancer.

REFERENCES

1. Kabbinavar, F., Hurwitz, H.I., Fehrenbacher, L., Meropol, N.J., Novotny, W.F., Lieberman, G., Griffing, S., and Bergsland, E. Phase II, randomized trial comparing bevacizumab plus fluorouracil (FU)/leucovorin (LV) with FU/LV alone in patients with metastatic colorectal cancer. *J Clin Oncol*, 21: 60–65, 2003.
2. Demetri, G.D., van Oosterom, A.T., Garrett, C.R., Blackstein, M.E., Shah, M.H., Verweij, J., McArthur, G., Judson, I.R., Heinrich, M.C., Morgan, J.A., Desai, J., Fletcher, C.D., George, S., Bello, C.L., Huang, X., Baum, C.M., and Casali, P.G. Efficacy and safety of sunitinib in patients with advanced gastrointestinal stromal tumour after failure of imatinib: a randomised controlled trial. *Lancet*, 368: 1329–1338, 2006.
3. Ferrara, N. and Kerbel, R.S. Angiogenesis as a therapeutic target. *Nature*, 438: 967–974, 2005.
4. Dancey, J.E. and Chen, H.X. Strategies for optimizing combinations of molecularly targeted anticancer agents. *Nat Rev Drug Discov*, 5: 649–659, 2006.

5. Davis, D.W., McConkey, D.J., Abbruzzese, J.L., and Herbst, R.S. Surrogate markers in antiangiogenesis clinical trials. *Br J Cancer*, 89: 8–14, 2003.
6. Yancopoulos, G.D., Davis, S., Gale, N.W., Rudge, J.S., Wiegand, S.J., and Holash, J. Vascular-specific growth factors and blood vessel formation. *Nature*, 407: 242–248, 2000.
7. Boehm, T., Folkman, J., Browder, T., and O'Reilly, M.S. Antiangiogenic therapy of experimental cancer does not induce acquired drug resistance. *Nature*, 390: 404–407, 1997.
8. Kerbel, R.S. A cancer therapy resistant to resistance. *Nature*, 390: 335–336, 1997.
9. Bold, G., Altmann, K.H., Frei, J., Lang, M., Manley, P.W., Traxler, P., Wietfeld, B., Bruggen, J., Buchdunger, E., Cozens, R., Ferrari, S., Furet, P., Hofmann, F., Martiny-Baron, G., Mestan, J., Rosel, J., Sills, M., Stover, D., Acemoglu, F., Boss, E., Emmenegger, R., Lasser, L., Masso, E., Roth, R., Schlachter, C., and Vetterli, W. New anilinophthalazines as potent and orally well absorbed inhibitors of the VEGF receptor tyrosine kinases useful as antagonists of tumor-driven angiogenesis. *J Med Chem*, 43: 3200, 2000.
10. Keck, P.J., Hauser, S.D., Krivi, G., Sanzo, K., Warren, T., Feder, J., and Connolly, D.T. Vascular permeability factor, an endothelial cell mitogen related to PDGF. *Science*, 246: 1309–1312, 1989.
11. Ciardiello, F. and Tortora, G. A novel approach in the treatment of cancer: targeting the epidermal growth factor receptor. *Clin Cancer Res*, 7: 2958–2970, 2001.
12. Kerbel, R.S., Yu, J., Tran, J., Man, S., Viloria-Petit, A., Klement, G., Coomber, B.L., and Rak, J. Possible mechanisms of acquired resistance to anti-angiogenic drugs: implications for the use of combination therapy approaches. *Cancer Metastasis Rev*, 20: 79–86, 2001.
13. Kerbel, R. and Folkman, J. Clinical translation of angiogenesis inhibitors. *Nat Rev Cancer*, 2: 727–739, 2002.
14. Folkman, J. Looking for a good endothelial address. *Cancer Cell*, 1: 113–115, 2002.
15. Longo, R., Sarmiento, R., Fanelli, M., Capaccetti, B., Gattuso, D., and Gasparini, G. Anti-angiogenic therapy: rationale, challenges and clinical studies. *Angiogenesis*, 5: 237–256, 2002.
16. Dowlati, A., Haaga, J., Remick, S.C., Spiro, T.P., Gerson, S.L., Liu, L., Berger, S.J., Berger, N.A., and Willson, J.K. Sequential tumor biopsies in early phase clinical trials of anticancer agents for pharmacodynamic evaluation. *Clin Cancer Res*, 7: 2971–2976, 2001.
17. Rimm, D.L. What brown cannot do for you. *Nat Biotechnol*, 24: 914–916, 2006.
18. Davis, D.W., McConkey, D.J., Zhang, W., and Herbst, R.S. Antiangiogenic tumor therapy. *Biotechniques*, 34: 1048–1050, 1052, 1054 passim, 2003.
19. Bruns, C.J., Liu, W., Davis, D.W., Shaheen, R.M., McConkey, D.J., Wilson, M.R., Bucana, C.D., Hicklin, D.J., and Ellis, L.M. Vascular endothelial growth factor is an in vivo survival factor for tumor endothelium in a murine model of colorectal carcinoma liver metastases. *Cancer*, 89: 488–499, 2000.
20. Bruns, C.J., Shinohara, H., Harbison, M.T., Davis, D.W., Nelkin, G., Killion, J.J., McConkey, D.J., Dong, Z., and Fidler, I.J. Therapy of human pancreatic carcinoma implants by irinotecan and the oral immunomodulator JBT 3002 is associated with enhanced expression of inducible nitric oxide synthase in tumor-infiltrating macrophages. *Cancer Res*, 60: 2–7, 2000.
21. Inoue, K., Slaton, J.W., Davis, D.W., Hicklin, D.J., McConkey, D.J., Karashima, T., Radinsky, R., and Dinney, C.P. Treatment of human metastatic transitional cell carcinoma of the bladder in a murine model with the anti-vascular endothelial growth factor receptor monoclonal antibody DC101 and paclitaxel. *Clin Cancer Res*, 6: 2635–2643, 2000.
22. Raut, C.P., Takamori, R.K., Davis, D.W., Sweeney-Gotsch, B., O'Reilly, M.S., and McConkey, D.J. Direct effects of recombinant human endostatin on tumor cell IL-8 production are associated with increased endothelial cell apoptosis in an orthotopic model of human pancreatic cancer. *Cancer Biol Ther*, 3: 679–687, 2004.
23. Shaheen, R.M., Davis, D.W., Liu, W., Zebrowski, B.K., Wilson, M.R., Bucana, C.D., McConkey, D.J., McMahon, G., and Ellis, L.M. Antiangiogenic therapy targeting the tyrosine kinase receptor for vascular endothelial growth factor receptor inhibits the growth of colon cancer liver metastasis and induces tumor and endothelial cell apoptosis. *Cancer Res*, 59: 5412–5416, 1999.
24. Shaheen, R.M., Tseng, W.W., Davis, D.W., Liu, W., Reinmuth, N., Vellagas, R., Wieczorek, A.A., Ogura, Y., McConkey, D.J., Drazan, K.E., Bucana, C.D., McMahon, G., and Ellis, L.M. Tyrosine kinase inhibition of multiple angiogenic growth factor receptors improves survival in mice bearing

colon cancer liver metastases by inhibition of endothelial cell survival mechanisms. *Cancer Res*, 61: 1464–1468, 2001.

25. Berger, A.J., Davis, D.W., Tellez, C., Prieto, V.G., Gershenwald, J.E., Johnson, M.M., Rimm, D.L., and Bar-Eli, M. Automated quantitative analysis of activator protein-2alpha subcellular expression in melanoma tissue microarrays correlates with survival prediction. *Cancer Res*, 65: 11185–11192, 2005.
26. Herbst, R.S., Mullani, N.A., Davis, D.W., Hess, K.R., McConkey, D.J., Charnsangavej, C., O'Reilly, M.S., Kim, H.W., Baker, C., Roach, J., Ellis, L.M., Rashid, A., Pluda, J., Bucana, C., Madden, T.L., Tran, H.T., and Abbruzzese, J.L. Development of biologic markers of response and assessment of antiangiogenic activity in a clinical trial of human recombinant endostatin. *J Clin Oncol*, 20: 3804–3814, 2002.
27. Davis, D.W., Inoue, K., Dinney, C.P., Hicklin, D.J., Abbruzzese, J.L., and McConkey, D.J. Regional effects of an antivascular endothelial growth factor receptor monoclonal antibody on receptor phosphorylation and apoptosis in human 253J B-V bladder cancer xenografts. *Cancer Res*, 64: 4601–4610, 2004.
28. Cregger, M., Berger, A.J., and Rimm, D.L. Immunohistochemistry and quantitative analysis of protein expression. *Arch Pathol Lab Med*, 130: 1026–1030, 2006.
29. Davis, D.W., Takamori, R., Raut, C.P., Xiong, H.Q., Herbst, R.S., Stadler, W.M., Heymach, J.V., Demetri, G.D., Rashid, A., Shen, Y., Wen, S., Abbruzzese, J.L., and McConkey, D.J. Pharmacodynamic analysis of target inhibition and endothelial cell death in tumors treated with the vascular endothelial growth factor receptor antagonists SU5416 or SU6668. *Clin Cancer Res*, 11: 678–689, 2005.
30. Davis, D.W., Shen, Y., Mullani, N.A., Wen, S., Herbst, R.S., O'Reilly, M., Abbruzzese, J.L., and McConkey, D.J. Quantitative analysis of biomarkers defines an optimal biological dose for recombinant human endostatin in primary human tumors. *Clin Cancer Res*, 10: 33–42, 2004.
31. Heymach, J.V., Desai, J., Manola, J., Davis, D.W., McConkey, D.J., Harmon, D., Ryan, D.P., Goss, G., Quigley, T., Van den Abbeele, A.D., Silverman, S.G., Connors, S., Folkman, J., Fletcher, C.D., and Demetri, G.D. Phase II study of the antiangiogenic agent SU5416 in patients with advanced soft tissue sarcomas. *Clin Cancer Res*, 10: 5732–5740, 2004.
32. Davis, D.W., Buchholz, T.A., Hess, K.R., Sahin, A.A., Valero, V., and McConkey, D.J. Automated quantification of apoptosis after neoadjuvant chemotherapy for breast cancer: early assessment predicts clinical response. *Clin Cancer Res*, 9: 955–960, 2003.
33. Hubbard, S.R. Structural analysis of receptor tyrosine kinases. *Prog Biophys Mol Biol*, 71: 343–358, 1999.
34. Bernatchez, P.N., Soker, S., and Sirois, M.G. Vascular endothelial growth factor effect on endothelial cell proliferation, migration, and platelet-activating factor synthesis is Flk-1-dependent. *J Biol Chem*, 274: 31047–31054, 1999.
35. Glade-Bender, J., Kandel, J.J., and Yamashiro, D.J. VEGF blocking therapy in the treatment of cancer. *Expert Opin Biol Ther*, 3: 263–276, 2003.
36. Stadler, W. and Wilding, G. Angiogenesis inhibitors in genitourinary cancers. *Crit Rev Oncol Hematol*, 46 Suppl: S41–S47, 2003.
37. Fong, T.A., Shawver, L.K., Sun, L., Tang, C., App, H., Powell, T.J., Kim, Y.H., Schreck, R., Wang, X., Risau, W., Ullrich, A., Hirth, K.P., and McMahon, G. SU5416 is a potent and selective inhibitor of the vascular endothelial growth factor receptor (Flk-1/KDR) that inhibits tyrosine kinase catalysis, tumor vascularization, and growth of multiple tumor types. *Cancer Res*, 59: 99–106, 1999.
38. Dowlati, A., Robertson, K., Radivoyevitch, T., Waas, J., Ziats, N.P., Hartman, P., Abdul-Karim, F.W., Wasman, J.K., Jesberger, J., Lewin, J., McCrae, K., Ivy, P., and Remick, S.C. Novel Phase I dose de-escalation design trial to determine the biological modulatory dose of the antiangiogenic agent SU5416. *Clin Cancer Res*, 11: 7938–7944, 2005.
39. O'Reilly, M.S., Boehm, T., Shing, Y., Fukai, N., Vasios, G., Lane, W.S., Flynn, E., Birkhead, J.R., Olsen, B.R., and Folkman, J. Endostatin: an endogenous inhibitor of angiogenesis and tumor growth. *Cell*, 88: 277–285, 1997.
40. Mundhenke, C., Thomas, J.P., Wilding, G., Lee, F.T., Kelzc, F., Chappell, R., Neider, R., Sebree, L.A., and Friedl, A. Tissue examination to monitor antiangiogenic therapy: a phase I clinical trial with endostatin. *Clin Cancer Res*, 7: 3366–3374, 2001.

41. Herbst, R.S., Hess, K.R., Tran, H.T., Tseng, J.E., Mullani, N.A., Charnsangavej, C., Madden, T., Davis, D.W., McConkey, D.J., O'Reilly, M.S., Ellis, L.M., Pluda, J., Hong, W.K., and Abbruzzese, J.L. Phase I study of recombinant human endostatin in patients with advanced solid tumors. *J Clin Oncol*, 20: 3792–3803, 2002.
42. Fukuoka, M., Yano, S., Giaccone, G., Tamura, T., Nakagawa, K., Douillard, J.Y., Nishiwaki, Y., Vansteenkiste, J., Kudoh, S., Rischin, D., Eek, R., Horai, T., Noda, K., Takata, I., Smit, E., Averbuch, S., Macleod, A., Feyereislova, A., Dong, R.P., and Baselga, J. Multi-institutional randomized phase II trial of gefitinib for previously treated patients with advanced non-small-cell lung cancer (The IDEAL 1 Trial) [corrected]. *J Clin Oncol*, 21: 2237–2246, 2003.
43. Kris, M.G., Natale, R.B., Herbst, R.S., Lynch, T.J., Jr. Prager, D., Belani, C.P., Schiller, J.H., Kelly, K., Spiridonidis, H., Sandler, A., Albain, K.S., Cella, D., Wolf, M.K., Averbuch, S.D., Ochs, J.J., and Kay, A.C. Efficacy of gefitinib, an inhibitor of the epidermal growth factor receptor tyrosine kinase, in symptomatic patients with non-small cell lung cancer: a randomized trial. *JAMA*, 290: 2149–2158, 2003.
44. Daneshmand, M., Parolin, D.A., Hirte, H.W., Major, P., Goss, G., Stewart, D., Batist, G., Miller, W.H., Jr., Matthews, S., Seymour, L., and Lorimer, I.A. A pharmacodynamic study of the epidermal growth factor receptor tyrosine kinase inhibitor ZD1839 in metastatic colorectal cancer patients. *Clin Cancer Res*, 9: 2457–2464, 2003.
45. Busse, D., Doughty, R.S., Ramsey, T.T., Russell, W.E., Price, J.O., Flanagan, W.M., Shawver, L.K., and Arteaga, C.L. Reversible G(1) arrest induced by inhibition of the epidermal growth factor receptor tyrosine kinase requires up-regulation of p27(KIP1) independent of MAPK activity. *J Biol Chem*, 275: 6987–6995, 2000.
46. Albanell, J., Rojo, F., Averbuch, S., Feyereislova, A., Mascaro, J.M., Herbst, R., LoRusso, P., Rischin, D., Sauleda, S., Gee, J., Nicholson, R.I., and Baselga, J. Pharmacodynamic studies of the epidermal growth factor receptor inhibitor ZD1839 in skin from cancer patients: histopathologic and molecular consequences of receptor inhibition 10.1200/JCO.20.1.110. *J Clin Oncol*, 20: 110–124, 2002.
47. Baselga, J., Rischin, D., Ranson, M., Calvert, H., Raymond, E., Kieback, D.G., Kaye, S.B., Gianni, L., Harris, A., Bjork, T., Averbuch, S.D., Feyereislova, A., Swaisland, H., Rojo, F., and Albanell, J. Phase I safety, pharmacokinetic, and pharmacodynamic trial of ZD1839, a selective oral epidermal growth factor receptor tyrosine kinase inhibitor, in patients with five selected solid tumor types 10.1200/JCO.2002.03.100. *J Clin Oncol*, 20: 4292–4302, 2002.
48. Baselga, J., Albanell, J., Ruiz, A., Lluch, A., Gascon, P., Guillem, V., Gonzalez, S., Sauleda, S., Marimon, I., Tabernero, J.M., Koehler, M.T., and Rojo, F. Phase II and tumor pharmacodynamic study of gefitinib in patients with advanced breast cancer 10.1200/JCO.2005.08.326. *J Clin Oncol*, 23: 5323–5333, 2005.
49. Rothenberg, M.L., LaFleur, B., Levy, D.E., Washington, M.K., Morgan-Meadows, S.L., Ramanathan, R.K., Berlin, J.D., Benson, A.B., 3rd, and Coffey, R.J. Randomized phase II trial of the clinical and biological effects of two dose levels of gefitinib in patients with recurrent colorectal adenocarcinoma. *J Clin Oncol*, 23: 9265–9274, 2005.
50. Cohen, E.E., Rosen, F., Stadler, W.M., Recant, W., Stenson, K., Huo, D., and Vokes, E.E. Phase II trial of ZD1839 in recurrent or metastatic squamous cell carcinoma of the head and neck. *J Clin Oncol*, 21: 1980–1987, 2003.
51. Gasparini, G., Sarmiento, R., Amici, S., Longo, R., Gattuso, D., Zancan, M., and Gion, M. Gefitinib (ZD1839) combined with weekly epirubicin in patients with metastatic breast cancer: a phase I study with biological correlate. *Ann Oncol*, 16: 1867–1873, 2005.
52. Vanhoefer, U., Tewes, M., Rojo, F., Dirsch, O., Schleucher, N., Rosen, O., Tillner, J., Kovar, A., Braun, A.H., Trarbach, T., Seeber, S., Harstrick, A., and Baselga, J. Phase I study of the humanized antiepidermal growth factor receptor monoclonal antibody EMD72000 in patients with advanced solid tumors that express the epidermal growth factor receptor 10.1200/JCO.2004.05.114. *J Clin Oncol*, 22: 175–184, 2004.
53. Jost, M., Huggett, T.M., Kari, C., Boise, L.H., and Rodeck, U. Epidermal growth factor receptor-dependent control of keratinocyte survival and Bcl-xL expression through a MEK-dependent pathway. *J Biol Chem*, 276: 6320–6326, 2001.
54. Xia, W., Mullin, R.J., Keith, B.R., Liu, L.H., Ma, H., Rusnak, D.W., Owens, G., Alligood, K.J., and Spector, N.L. Anti-tumor activity of GW572016: a dual tyrosine kinase inhibitor blocks

EGF activation of EGFR/erbB2 and downstream Erk1/2 and AKT pathways. *Oncogene*, 21: 6255–6263, 2002.

55. Cockerill, S., Stubberfield, C., Stables, J., Carter, M., Guntrip, S., Smith, K., McKeown, S., Shaw, R., Topley, P., Thomsen, L., Affleck, K., Jowett, A., Hayes, D., Willson, M., Woollard, P., and Spalding, D. Indazolylamino quinazolines and pyridopyrimidines as inhibitors of the EGFr and C-erbB-2. *Bioorg Med Chem Lett*, 11: 1401–1405, 2001.
56. Rusnak, D.W., Lackey, K., Affleck, K., Wood, E.R., Alligood, K.J., Rhodes, N., Keith, B.R., Murray, D.M., Knight, W.B., Mullin, R.J., and Gilmer, T.M. The effects of the novel, reversible epidermal growth factor receptor/ErbB-2 tyrosine kinase inhibitor, GW2016, on the growth of human normal and tumor-derived cell lines in vitro and in vivo. *Mol Cancer Ther*, 1: 85–94, 2001.
57. Spector, N.L., Xia, W., Burris, H., 3rd, Hurwitz, H., Dees, E.C., Dowlati, A., O'Neil, B., Overmoyer, B., Marcom, P.K., Blackwell, K.L., Smith, D.A., Koch, K.M., Stead, A., Mangum, S., Ellis, M.J., Liu, L., Man, A.K., Bremer, T.M., Harris, J., and Bacus, S. Study of the biologic effects of lapatinib, a reversible inhibitor of ErbB1 and ErbB2 tyrosine kinases, on tumor growth and survival pathways in patients with advanced malignancies. *J Clin Oncol*, 23: 2502–2512, 2005.
58. Laird, A.D., Vajkoczy, P., Shawver, L.K., Thurnher, A., Liang, C., Mohammadi, M., Schlessinger, J., Ullrich, A., Hubbard, S.R., Blake, R.A., Fong, T.A., Strawn, L.M., Sun, L., Tang, C., Hawtin, R., Tang, F., Shenoy, N., Hirth, K.P., McMahon, G., and Cherrington, J.M. SU6668 is a potent antiangiogenic and antitumor agent that induces regression of established tumors. *Cancer Res*, 60: 4152–4160, 2000.
59. Sun, L., Tran, N., Tang, F., App, H., Hirth, P., McMahon, G., and Tang, C. Synthesis and biological evaluations of 3-substituted indolin-2-ones: a novel class of tyrosine kinase inhibitors that exhibit selectivity toward particular receptor tyrosine kinases. *J Med Chem*, 41: 2588–2603, 1998.
60. Xiong, H.Q., Herbst, R., Faria, S.C., Scholz, C., Davis, D., Jackson, E.F., Madden, T., McConkey, D., Hicks, M., Hess, K., Charnsangavej, C.A., and Abbruzzese, J.L. A phase I surrogate endpoint study of SU6668 in patients with solid tumors. *Invest New Drugs*, 22: 459–466, 2004.
61. Osusky, K.L., Hallahan, D.E., Fu, A., Ye, F., Shyr, Y., and Geng, L. The receptor tyrosine kinase inhibitor SU11248 impedes endothelial cell migration, tubule formation, and blood vessel formation in vivo, but has little effect on existing tumor vessels. *Angiogenesis*, 7: 225–233, 2004.
62. Mendel, D.B., Laird, A.D., Xin, X., Louie, S.G., Christensen, J.G., Li, G., Schreck, R.E., Abrams, T.J., Ngai, T.J., Lee, L.B., Murray, L.J., Carver, J., Chan, E., Moss, K.G., Haznedar, J.O., Sukbuntherng, J., Blake, R.A., Sun, L., Tang, C., Miller, T., Shirazian, S., McMahon, G., and Cherrington, J.M. In vivo antitumor activity of SU11248, a novel tyrosine kinase inhibitor targeting vascular endothelial growth factor and platelet-derived growth factor receptors: determination of a pharmacokinetic/pharmacodynamic relationship. *Clin Cancer Res*, 9: 327–337, 2003.
63. Schueneman, A.J., Himmelfarb, E., Geng, L., Tan, J., Donnelly, E., Mendel, D., McMahon, G., and Hallahan, D.E. SU11248 maintenance therapy prevents tumor regrowth after fractionated irradiation of murine tumor models. *Cancer Res*, 63: 4009–4016, 2003.
64. Abrams, T.J., Lee, L.B., Murray, L.J., Pryer, N.K., and Cherrington, J.M. SU11248 inhibits KIT and platelet-derived growth factor receptor beta in preclinical models of human small cell lung cancer. *Mol Cancer Ther*, 2: 471–478, 2003.
65. O'Farrell, A.M., Abrams, T.J., Yuen, H.A., Ngai, T.J., Louie, S.G., Yee, K.W., Wong, L.M., Hong, W., Lee, L.B., Town, A., Smolich, B.D., Manning, W.C., Murray, L.J., Heinrich, M.C., and Cherrington, J.M. SU11248 is a novel FLT3 tyrosine kinase inhibitor with potent activity in vitro and in vivo. *Blood*, 101: 3597–3605, 2003.
66. Davis, D.W., McConkey, D.J., Heymach, J.V., Desai, J., George, S., Jackson, J., Bello, C.L., Baum, C.M., Shalinsky, D.R., and Demetri, G.D. Pharmacodynamic analysis of target receptor tyrosine kinase activity and apoptosis in GIST tumors responding to therapy with SU11248. *Proc Am Soc Clin Oncol*, 23: 3006, 2005.
67. Davis, D.W., Heymach, J.V., Desai, J., George, S., Deprimo, S.E., Bello, C.L., Baum, C.M., and Demetri, G.D. Correlation of receptor tyrosine kinase (RTK) activity and apoptosis with response to sunitinib treatment in patients with gastrointestinal stromal tumor (GIST). EORTC-NCI-AACR Abstract 57, 2006.

68. Beaudry, P., Force, J., Naumov, G.N., Wang, A., Baker, C.H., Ryan, A., Soker, S., Johnson, B.E., Folkman, J., and Heymach, J.V. Differential effects of vascular endothelial growth factor receptor-2 inhibitor ZD6474 on circulating endothelial progenitors and mature circulating endothelial cells: implications for use as a surrogate marker of antiangiogenic activity. *Clin Cancer Res*, 11: 3514–3522, 2005.
69. Bertolini, F., Mingrone, W., Alietti, A., Ferrucci, P.F., Cocorocchio, E., Peccatori, F., Cinieri, S., Mancuso, P., Corsini, C., Burlini, A., Zucca, E., and Martinelli, G. Thalidomide in multiple myeloma, myelodysplastic syndromes and histiocytosis. Analysis of clinical results and of surrogate angiogenesis markers. *Ann Oncol*, 12: 987–990, 2001.
70. Gordon, M. et al. Phase I trial of recombinant humanized monoclonal anti-vascular endothelial growth factor (anti-VEGF MAB) in patients (PTS) with metastatic cancer (Meeting abstract). *Proc Am Soc Clin Oncol*, 17, 1998.

22 Blood-Based Biomarkers for VEGF Inhibitors

Amado J. Zurita, Hua-Kang Wu, and John V. Heymach

CONTENTS

In recent years, there has been significant progress in the clinical development of antiangiogenic therapy, particularly inhibitors of the VEGF pathway. Despite this progress, however, the biological activity of these agents remains difficult to assess because they do not typically lead to objective responses as judged by tumor shrinkage when used as monotherapy. Furthermore, we do not yet have validated methods for identifying which patients are most likely to respond to treatment, selecting the optimal dose, or determining whether the intended molecular target has been effectively inhibited. Ideally such methods should be noninvasive and practical for routine clinical care. In this chapter we review blood-based biomarkers currently under investigation for VEGF pathway inhibitors. These include circulating proangiogenic factors and receptors (i.e., soluble VEGFR-1 and VEGFR-2); markers of hypoxia and endothelial damage; and cellular populations in the peripheral blood such as circulating endothelial cells (CECs). Several of these markers are promising but further preclinical and clinical studies will be needed to determine their potential utility in the clinical setting.

22.1 NEED FOR BIOMARKERS FOR VEGF ACTIVITY

Over the past decade, antiangiogenic therapy has moved from theory to clinical practice. Bevacizumab, a monoclonal antibody directed against VEGF, has been demonstrated to provide clinical benefit when combined with chemotherapy for colorectal, lung, and breast cancer.[1–3] Furthermore, multitargeted inhibitors blocking the VEGFR pathway such as sunitinib and sorafenib have demonstrated significant clinical activity in chemotherapy-refractory tumors such as renal cell carcinoma.[4] Despite these advances, the biologic activity of these and newer agents remains difficult to assess. This is because these agents generally appear to be primarily cytostatic, and when they are used as monotherapy, only a minority of patients have an objective complete or partial response as judged by tumor shrinkage. As an illustration of this point, in a phase I trial of bevacizumab, no objective responses were observed out of 25 patients.[5] The clinical activity of bevacizumab was subsequently established in a series of randomized phase II and III trials over a period of more than 5 years. Given the large number of targeted agents now entering clinical testing as well as the chronic shortage of patients enrolled into clinical trials and the escalating costs of drug development, it is generally not feasible to perform large randomized trials for drugs without evidence of biological activity early in their development. Therefore, there remains a real risk that drugs that could ultimately benefit patients may not be developed, and even that many patients might be treated with less effective drug doses or schedules in phase II or III trials because of a lack of correlates of activity in earlier clinical testing. Surrogate biomarkers are therefore clearly needed to advance the clinical development of VEGF inhibitors. These markers may serve some, or all, of the following uses:

1. Assessment of whether an agent has the expected biological activity
2. Optimization of dosing
3. Identification of patients most likely, or least likely, to benefit from a given treatment
4. Monitoring of response to treatment and evaluation of potential mechanisms of resistance

The therapeutic efficacy of conventional chemotherapy is limited in part by the emergence of drug resistance by rapidly mutating tumor cells. Because antiangiogenic agents are directed against tumor endothelium, which is presumed to be diploid and genetically stable, it was initially thought that tumors may not develop resistance to antiangiogenic therapy[6,7] as they do to cytotoxic chemotherapy. More recent studies have, however, suggested at least two possible mechanisms by which tumors may initially have or acquire resistance, or decreased sensitivity, to VEGF pathway inhibitors.

22.1.1 Incomplete Target Inhibition

Drugs may not be present at sufficiently high concentrations, or for a sufficiently long time, to cause adequate inhibition of VEGF receptor signaling. As an example, in clinical trials of SU5416 and SU6668, two of the earliest VEGFR tyrosine kinase inhibitors (TKIs) in clinical development, we previously found that VEGFR and other key targets were not appreciably inhibited in posttreatment tumor biopsies.[8,9] This provided a potential explanation for the lack of clinical activity observed in these trials. For other TKIs such as gefitinib and imatinib, incomplete target inhibition has been shown to result from genetic changes, for example secondary mutations in EGFR or BCR-ABL, or epigenetic changes reducing the intracellular concentrations of the inhibitor.[10–12]

22.1.2 Bypass of the VEGF Pathway through Expression of Additional Angiogenic Factors

Genetic mutations or the activation of pathways such as hypoxia-inducible factor-1 (HIF-1) may lead to the expression of additional angiogenic factors (or the loss of angiogenic inhibitors) by malignant cells in the tumor.[13,14] In turn, these changes may promote the proliferation and survival of tumor endothelial cells even in the presence of VEGF blockade. It has been observed, for example, that advanced stage breast cancers express a greater number of proangiogenic factors than early stage cancers.[15] It is also known that factors such as basic fibroblast growth factor (bFGF) can promote endothelial cell survival despite a VEGF blockade.[16]

Understanding mechanisms of resistance to VEGF inhibitors therefore requires biomarkers capable of quantitatively assessing target inhibition as well as qualitatively assessing the profile of expressed pro- or antiangiogenic factors contributing to bypassing the VEGF pathway.

22.2 SOLUBLE MARKERS IN SERUM, PLASMA, AND URINE

22.2.1 Circulating Angiogenic Growth Factors, Inhibitors, and Related Vascular Molecules

Tumor angiogenesis is regulated by a balance between pro- and antiangiogenic factors and their influence on the tumor endothelium. In addition to the number and identity of the factors influencing such angiogenic balance, their relative concentrations determine pro- or antiangiogenic states. A number of different molecules implicated in angiogenesis can be detected in circulation and other biologic fluids and serve as biomarkers for monitoring anti-VEGF therapies.

VEGF-A (also called VEGF), one of the most potent proangiogenic growth factors, is expressed by nearly all cancer cell types and certain tumor stromal cells.[17–19] VEGF stimulates endothelial cell differentiation, proliferation, migration, survival, and expression of adhesion molecules, and mediates vascular permeability, being sometimes associated to the development malignant effusions.[17] VEGF can also mobilize bone marrow–derived endothelial progenitor cells.[20]

VEGF expression can be triggered during the early stages of neoplastic transformation by environmental stimuli (e.g., hypoxia or low pH) or by genetic mutations/activated oncogenes (e.g., K-ras, p53, VHL, HER2/ErbB2, WNT, EGFR, or bcr-abl), and often persists during tumor progression.[21–23] In addition, isoforms of VEGF are bound to the extracellular matrix and can be released by proteolytic enzymes produced by cancer or stromal cells.[17] Conventional anticancer therapy itself (e.g., radiation therapy) may increase VEGF production or accumulation.[24] Overexpression of VEGF occurs in most human tumors, and has been associated with tumor stage, grade, progression, and poor prognosis in several systems, including carcinomas of the colon-rectum, stomach, pancreas, breast, prostate, lung, and malignant melanoma.[17,25,26]

Plasma and serum levels of VEGF have been actively investigated as biomarkers of activity of VEGF inhibitors and clinical benefit. In preclinical models, rapid increases of plasma VEGF were detected both in nontumor bearing and tumor-bearing mice after single injections of anti-VEGFR-2 monoclonal antibodies.[27] These increases were induced in a dose-dependent manner, with maximum values being reached when doses previously determined to be optimal for therapy were used. In contrast, small-molecule antagonists of VEGFR-2 such as vatalanib (PTK787/ZK 222584) and SU5416, and a monoclonal antibody targeting VEGFR-1 (MF-1) did not cause any significant variation in plasma concentrations of VEGF, suggesting that therapy-induced increases in plasma VEGF might be restricted to antibody-mediated therapies that block or displace the VEGF ligand from the external domain of VEGFR-2.[27]

Several clinical trials of anti-VEGF agents have tested the value and significance of circulating levels of VEGF. Some of these data are preliminary and require further analysis. Bevacizumab, a humanized VEGF-A-specific antibody, has been shown to increase both serum and plasma total VEGF concentrations at different time points (30 days after therapy discontinuation,[28] on weeks 5 and 13,[29] and after only 12 days of therapy).[30] Interestingly, free serum VEGF concentrations decreased to undetectable levels even with low doses of bevacizumab in one of the studies.[28] In contrast to bevacizumab, VEGF concentrations are not homogeneously modulated in one direction by the variety of small-molecule receptor TKIs targeting the VEGF receptors under development. The TKI with the most data available is sunitinib malate (SU11248), which has shown to consistently induce on-therapy increases in VEGF plasma levels that are rapidly reversible to baseline when the therapy is stopped (in renal cell carcinoma, imatinib-resistant gastrointestinal stromal tumor [GIST], and several phase I clinical trials in patients with different types of solid tumors).[4,31–35] These studies suggest that plasma VEGF is a useful pharmacodynamic marker of exposure to sunitinib in diverse tumor types. Similarly, semaxanib (SU5416) also induced elevations in both urinary and plasma VEGF,[8,36] but produced the opposite effect when given in combination with low doses of IFN-α.[37] Unfortunately, data for other TKIs are yet very limited. In patients with advanced colorectal cancer, vatalanib caused a rise in plasma VEGF in the first cycle of therapy and a subsequent decrease.[38] An increment $\geq$150% from baseline VEGF correlated with clinical benefit defined as nonprogressive disease.[38] Another TKI in early clinical testing, AZD2171, triggered acute increases in VEGF that conversely were not dose-dependent.[39] In contrast, AG-013736, which has a similar spectrum of TK inhibition to sunitinib, induced no variation in VEGF levels in a phase I trial in patients with different solid tumor types.[40]

The mechanisms responsible for these fluctuations in serum or plasma VEGF are not known, but may be dependent on drug class. Interestingly, in preclinical models, a rise in VEGF was observed after treatment of either tumor-bearing or normal mice with a VEGFR-2 blocking monoclonal antibody, but not with small-molecule VEGFR-2 inhibitors,[27] implying that VEGF blockade at different levels elicits varied responses. Possible mechanisms for the changes observed in this study include induction of hypoxia (either in the tumor or normal tissues), increased release of VEGF from existing stores (i.e., platelets), displacement of receptor-bound VEGF, or alterations in VEGF clearance. Bevacizumab-induced elevations in VEGF levels are thought to be the result of an increase in VEGF synthesis/distribution and a decrease in VEGF clearance caused by formation of complexes between VEGF and the mAb. These studies highlight the complexities in VEGF regulation and suggest that plasma VEGF may serve as a pharmacodynamic marker of antiangiogenic therapy, but also that the observed effects are likely to depend on the specific drug and its mechanism of activity.

22.2.2 Soluble VEGFR-1 and VEGFR-2 (sVEGFR-1 and sVEGFR-2)

VEGF-A binds to two receptor tyrosine kinases, VEGFR-1 (Flt-1) and VEGFR-2 (KDR, flk-1). Of the two, it is generally agreed that VEGFR-2 is the major mediator of the mitogenic, angiogenic, and permeability-enhancing effects of VEGF-A.[41,42] The role of VEGFR-1 is more complex. Under some circumstances, VEGFR-1 can function as a "decoy" receptor that sequesters VEGF and prevents its interaction with VEGFR-2.[43]

However, VEGFR-1 also has significant roles in hematopoiesis and in the recruitment of "proangiogenic" monocytes and other bone marrow–derived cells.[44,45] In addition, VEGFR-1 is involved in the induction of matrix metalloproteinases (MMPs) and in the paracrine release of growth factors from endothelial cells.[46,47] Thus the VEGFR-1-selective ligands VEGF-B and placental-like growth factor (PlGF) may have a role in these processes. VEGFR-1 is also expressed by a number of tumor cells and may mediate a growth-promoting effect via an endothelial cell-independent pathway.[48]

Naturally occurring, soluble forms of VEGFR-1 and VEGFR-2 (sVEGFRs) have been described.[49,50] sVEGFR-1 can act as a natural endogenous inhibitor or modulator of VEGF and PlGF function, and has been studied as a surrogate marker for disease progression and inhibition of angiogenesis in different tumor types.[51–53] sVEGFR-1 also plays an important role in the pathogenesis of preeclampsia, with increased levels predicting the subsequent development of the disease.[54,55]

Even though its functional significance is still unclear, sVEGFR-2 has been found elevated in cancer patients and even to correlate with disease progression.[56–58] Preliminary experiences in assessing changes in serum or plasma levels as a surrogate for exposure to anti-VEGF agents suggest a clear potential value for sVEGFR-2. Most of the available results come from clinical trials using sunitinib. sVEGFR-2 and VEGF appear to change in a reciprocal manner during treatment; sunitinib induced consistent drops in plasma sVEGFR-2, accompanied by a rise in VEGF, with both proteins returning toward baseline levels when the drug was discontinued.[4,35] Similar observations were obtained for AZD2171.[39]

22.2.3 Other Angiogenic Regulatory Molecules

Circulating serum and plasma levels of FGF-2 (bFGF), PlGF, hepatocyte growth factor (HGF), thrombospondin-1 (TSP-1), soluble Tie-2, and interleukin-8 (IL-8) have been explored for their potential as surrogates of angiogenesis and to monitor response to VEGF-targeted therapies. Elevations in plasma levels of PlGF have been reported after therapy with bevacizumab and sunitinib.[4,30] Interestingly, PlGF is thought to play a role in pathological angiogenesis, without affecting quiescent vessels.[59]

22.2.4 Markers of Endothelial Cell Damage and Hemostasis

VEGF inhibition increases markedly the thrombo-embolic risk in cancer patients.[60] This phenomenon may be related to a reduced release of fibrinolytic components and/or an increased release of procoagulants,[61–63] a reduced synthesis and release of NO,[63] which inhibits platelet function, or endothelial cell dysfunction.[64] Markers of endothelial cell function/damage that can be measured in the laboratory are the von Willebrand factor (vWF), soluble thrombomodulin, soluble tissue factor (sTF), and soluble E-selectin. An increase in plasma levels of vWF, E-selectin, and sTF was reported during a phase I study investigating the feasibility of the combination of cisplatin-gemcitabine with the VEGFR-TKI SU5416. In this study, a particularly high number of thromboembolic events were observed.[65]

22.2.5 Markers of Hypoxia

Hypoxia promotes the production of proangiogenic factors, many of which are regulated by the HIF-1α pathway.[13] HIF-1α has been implicated in resistance to several treatment modalities, including chemotherapy and radiotherapy.[66] VEGF inhibitors may induce tumor hypoxia[67] and consequently the role of the HIF-1α pathway in the response to angiogenesis inhibitors is an area of active investigation. Hypoxia-regulated proteins like osteopontin[68–70] and galectin-1[71] hold promise for monitoring the effects of VEGF-targeted therapies.[70,71]

In summary, certain circulating angiogenic growth factors and vascular molecules appear to have a role as pharmacodynamic biomarkers of exposure to and effect of anti-VEGF therapies. More studies, however, are needed to expand and validate these initial findings and methodologies.

22.3 CIRCULATING ENDOTHELIAL CELLS AND OTHER BLOOD-BASED CELLULAR MARKERS IN PERIPHERAL BLOOD

22.3.1 Rationale

It has been recognized for more than three decades that cells bearing characteristics of endothelial cells can be detected in peripheral blood.[72–75] These cells, termed circulating endothelial cells (CECs), were initially thought to reflect vascular damage. Consistent with this hypothesis, the majority of CECs appear to arise from blood vessel walls and levels of CECs are increased after vascular damage caused by myocardial infection, sepsis, infection, sickle cell anemia, and other stimuli.[76–79]

In recent years, however, it has been demonstrated that a subset of CECs derived from bone marrow, circulating endothelial precursors or progenitors (CEPs), can differentiate into mature endothelial cells and contribute to neovascularization in murine models and in humans, although their contributions to tumor vasculature appear to be highly variable.[20,80–83] In humans, mature CECs and CEPs are distinguished from other peripheral blood mononuclear cells (PBMC) based on surface markers such as VEGF receptor-2 (KDR) and CD133.[84]

CECs (including both mature CECs and CEPs) have been investigated as potential biomarkers for antiangiogenic therapy in both preclinical models and in clinical studies of patients treated with antiangiogenic therapy.[35,85–90] The rationale for investigating these cells as a potential biomarker is severalfold. First, CECs are known to be increased in cancer patients, and are known to be associated with disease progression.[91,92] In addition, these cells are known to be mobilized in response to VEGF in both murine models[20,93] and in humans,[94–96] and to express molecular targets for many of the angiogenesis inhibitors in development including VEGFR-1 and VEGFR-2.[84,97] Finally, CECs may serve as a marker of damage to tumor vasculature.

Distinguishing between CEPs and mature CECs is therefore important if they are to be used as a marker because one would expect VEGFR inhibitors to decrease the number of CEPs by inhibiting their mobilization, but increase the number of sloughed vessel wall derived CECs (Figure 22.1). We and other investigators have observed increases in mature CECs in patients treated with antiangiogenic or vascular targeting agents.[98,99] This is in agreement with results in murine cancer models.[85] These studies support the hypothesis that measuring changes in CEC number and apoptosis may be informative.

22.3.2 Methods for CEC Detection

A variety of different methods have been used to detect CECs. Most recent studies have employed either immunobead capture, using antibodies directed against an endothelial antigen such as CD146,[79] or multicolor flow cytometry using a panel of antibodies to endothelial, hematopoietic, and progenitor cells (reviewed in Ref. 100). Flow cytometry-based methods have significant advantages in that they permit multiparametric analyses and high-speed of measurement. There are, however, significant methodologic issues related to differences in the antibodies and markers employed, cell preparation, viable cell staining, and gating strategies that make it difficult to compare results between different investigators and studies. Furthermore, no single antigen identified thus far is entirely specific to endothelium. As an example, one widely used antigen for detecting endothelial cells, CD146, is also

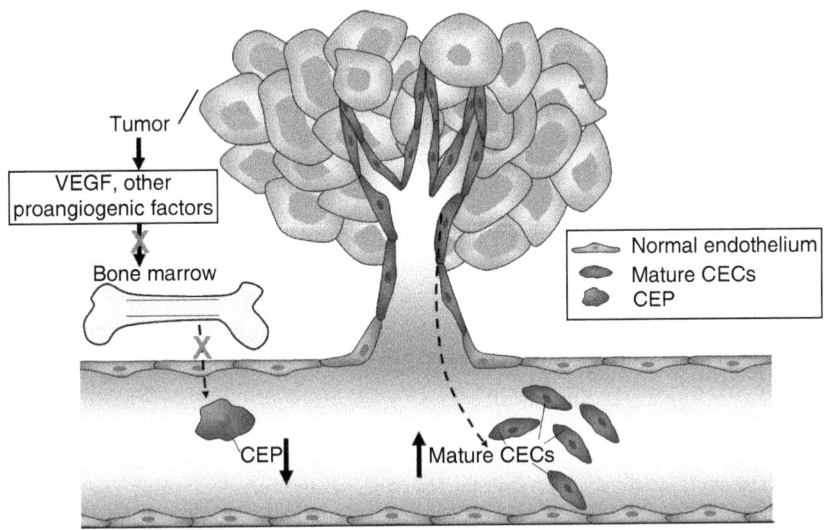

FIGURE 22.1 (See color insert following page 558.) Potential changes in circulating endothelial progenitors (CEPs) and mature CECs in response to treatment with a VEGF pathway inhibitor or other antiangiogenic agents. Angiogenesis inhibitors may decrease the mobilization of bone marrow–derived CEPs induced by VEGF and other proangiogenic cytokines, but at least transiently increase the shedding of mature endothelial cells by preferentially targeting fragile tumor endothelium (two-compartment model).[87]

detectable on activated lymphocytes and other cellular populations.[101] VEGFR-2 (KDR) and CD34 are expressed on both CEPs and hematopoietic stem cells (HSCs).[102] A variety of definitions of these different populations have been proposed based on different markers.[84,90,100,102,103] Most commonly, CD133 has been used as a marker of CEPs, with CD31, CD146, VEGFR-2, and VE-Cadherin as common endothelial markers, and CD45 and other leukocyte markers used to identify hematopoietic cells. Furthermore, in some studies there is a distinction made between viable and apoptotic CECs.[85,88,90] Clearly, standardization of these different methodologies would be an important step forward for the field so that results from different investigators, and from different studies, could be compared.

Published results of CECs and CEPs counts vary substantially in the literature, which undoubtedly is due in part to differences in the detection methods employed.[100] Studies done using immunobeads typically yield mean values of 10 cells/mL, whereas flow cytometry usually reports much greater values, often up to 1000-fold higher (Table 22.1).[100,104,105] Notably, cancer patients have consistently been reported to have increased levels of CECs.

TABLE 22.1
Detection of CECs and CEPs in Cancer Patients

				CEC/mL (Mean)		CEP/mL (Mean)	
Study	**Method**	**Cancer Type**	**Patients**	**Controls**	**Patients**	**Controls**	**Patients**
Mancuso[92]	FC	Breast/lymphoma	76	7,900	39,000	<500	<500
Beerepoot[91]	IB	Multiple tumor types	95	121 ± 16	399 ± 36	NA	NA
Zhang[114]	FC	Multiple myeloma	31	<1,000	>7,500	NA	NA
Wierzbowska[115]	FC	Acute myeloid leukemia	46	2,400	29,400	100	700
Cortelezzi[116]	FC	Myelodysplastic syndrome	50	12,300	0	250	0

CEC, circulating endothelial cell; CEP, circulating endothelial progenitor; FC, flow cytometry; IB, immunobead.

22.3.3 Preclinical Studies of CECs/CEPs

Tumor vasculature is typically disordered and abnormal, with evidence of co-existing vascular endothelial dysfunction.[106] Recent evidence has clearly indicated that tumor vasculature may be formed not only through angiogenic sprouting from local endothelium, but also through postnatal vasculogenesis via CEPs.[80,81] This involves mobilization from bone marrow, migration, and homing to sites of new vessel growth. These processes appear to be stimulated by high local levels of VEGF produced by tumors, as well as SDF-1 and other cytokines, several of which are hypoxia-regulated.[20,96,107] Lyden et al.[82] demonstrated that transplantation and engraftment of β-galactosidase-positive wild-type bone marrow or VEGF mobilized cells into lethally irradiated Id-mutant mice is sufficient to reconstitute tumor angiogenesis. In contrast to wild-type mice, Id-mutants fail to support the growth of tumors because of impaired angiogenesis. Tumor analysis demonstrated uptake of bone marrow–derived VEGFR-2-positive EPCs into vessels, which were surrounded by VEGFR-1-positive myeloid cells. Defective angiogenesis in Id-mutant mice is associated with impaired VEGF-induced mobilization and proliferation of the bone marrow precursor cells. Inhibition of both VEGFR-1 and VEGFR-2 signaling was needed to block tumor angiogenesis and induce necrosis. The mechanism for the underlying angiogenesis defect was believed to be due, in part, to impaired recruitment of EPCs to the tumor. Both B6RV2 lymphoma and Lewis lung carcinoma cells, which fail to grow tumors in Id-deficient mice, were able to form fully vascularized tumors in mice that had received wild-type bone marrow transplants.

More recently, Shaked and colleagues reported that the levels of CECs and CEPs vary greatly among animals with different genetic constitution, but correlate well with the degree of tumor angiogenesis or the response to antiangiogenic therapy. Moreover, they showed that antiangiogenic therapy dosing could be studied by monitoring the suppression of CECs and CEPs.[108] Recent work from our laboratory has demonstrated that VEGF inhibitors can have differential effects on mature CECs and on CEPs in murine lung cancer models, causing a decrease in CEPs and a concomitant rise in the mature CECs that is associated with a decrease in tumor angiogenesis.[87] This rise is not observed after treatment of nontumor bearing mice, suggesting that the rise is due to a shedding of tumor endothelium. Treatment of tumor-bearing mice with the vascular targeting agent ZD6126 also increased the number of mature CECs. This led to the proposed "two-compartment model" depicted in Figure 22.1. Preliminary results from several clinical studies of antiangiogenic or vascular targeting agents have been consistent with this model.[35,89,90] This illustrates that assessing both mature CECs and CEPs may provide insight into the mechanism of action of antiangiogenic agents. Moreover, a rise in mature CECs may serve as an early surrogate biomarker of the biologic activity of VEGF pathway inhibitors.

22.3.4 Clinical Results: Changes in CECs in Response to Treatment

CECs have been evaluated after treatment with angiogenesis inhibitors in several clinical studies. It is difficult to draw definitive conclusions from these studies thus far because they have generally been of modest size, and employed different detection methods and phenotypic markers. Nevertheless, they do suggest that CECs may be a useful marker that merits further investigation in larger trials. Several of these studies are discussed below.

Mancuso and colleagues investigated the correlation between CEC kinetics and clinical outcome in patients with advanced breast cancer receiving metronomic chemotherapy, using multicolor flow cytometry.[90] They observed that patients who experienced clinical benefit, defined as clinical response or stable disease, had an increased fraction of apoptotic CECs compared with those with progressive disease. Univariate and multivariate analyses indicated that elevated levels of CEC values after treatment were associated with a longer progression-free survival ($P = 0.001$) and an improved overall survival ($P = 0.005$). They suggested that the tumor vasculature is the most likely source of apoptotic CECs, and that the kinetics of CEC change over time may be critical to assess.

CECs were also recently investigated in 18 patients enrolled in a phase I study of the vascular targeting agent ZD6126[89] using immunobead isolation with antibodies directed against CD146. At week 1, a rise in CD146 + PBMCs was observed, increasing from 372 ± 109 CEC/mL to 524 ± 136 CEC/mL by 2–8 h after infusion.

In a study of six rectal cancer patients treated with the VEGF monoclonal antibody bevacizumab, a decrease in the number of progenitor (CD133+) cells was observed after 3 days of treatment.[109] There was also a decrease in the number of viable CECs, which the authors defined as CD45-CD31^{bright} cells. In a subsequent study, the same authors found that the kinetics of distinct CEC populations changed differently over time after bevacizumab treatment.[110]

We have also investigated changes in CECs in 16 patients with imatinib-refractory gastrointestinal stromal tumors (GIST) treated with the multitargeted tyrosine kinase inhibitor sunitinib.[35] CECs were assessed by multicolor flow cytometry. At baseline, patients with GIST had higher levels of CECs than normal controls. Patients who had clinical benefit, defined as a partial tumor response or stable disease, had a significant rise in mature CECs during the first 2 weeks of treatment, while patients with progressive disease had no significant change. Interestingly, levels of mature CECs returned to near baseline levels after 1–2 months of treatment in several patients who were followed longitudinally.

All these studies illustrate that CECs change in a dynamic manner after treatment with an angiogenesis inhibitor, and that the kinetics of change, as well as the specific population of cells that are changing, are critical parameters to assess. The development and standardization of robust, reproducible, and multiparametric analytical methods is clearly an important unmet need in the field.

22.4 OTHER CELLULAR POPULATIONS IN PERIPHERAL BLOOD AS POTENTIAL BIOMARKERS

Even though it was initially assumed that the contribution of bone marrow–derived cells to angiogenesis was through CEPs, it has recently been appreciated that circulating populations of cells with clear hematopoietic origin may contribute to the process as well. Common to them is their commitment to the myeloid lineage in a more or less advanced stage of differentiation.

Most available data are related to myelomonocytic cells that home to areas of vascular injury and contribute to angiogenesis by secreting angiogenic growth factors and, in some cases, by providing structural support to the adjacent endothelial cells by establishing themselves as mural cells or pericyte precursors. As an example, in preclinical models, a population of bone marrow–derived progenitor perivascular cells (PPCs) expressing PDGFR-β have been identified.[111] These cells have the ability to differentiate into pericytes and regulate vessel stability and vascular survival in tumors. A population of hemangiocytes expressing VEGFR-1 and CXCR-4 have also been found to be critical for cytokine-mediated revascularization.[112] Much more scant is the evidence implicating cells of the myeloid lineage in directly contributing to the vasculature, since most reports on this topic were hampered by technical limitations precluding the precise definition of the location of these cells in the vessel wall. However, a recent study shows that myeloid progenitor cells may directly integrate into blood vessels as endothelial cells.[113]

These results suggest that distinct populations of hematopoietic cells may functionally contribute to angiogenesis. It is not known, however, whether these populations may serve as a target or as a biomarker after treatment with antiangiogenic agents, or both. As a step toward identifying new potential cellular biomarkers for the VEGF pathway, we screened peripheral blood for cellular populations the VEGF binding and the expression of VEGF receptors. We found that the majority of VEGF binding cells in peripheral blood were VEGFR-1+ cells coexpressing the monocyte marker CD14 (CECs also expressed VEGFR-1 and VEGFR-2, but were present at lower levels).[35] Next, we monitored changes in CEC and monocyte numbers during treatment with sunitinib, which inhibits VEGFR-1, -2, and -3 and similar TK receptors, in patients with metastatic GIST and correlated these findings with clinical benefit. Mean monocyte numbers decreased by 54% from baseline to day 14 of treatment ($P < 0.001$; Wilcoxon rank-sum test), with a greater decrease occurring in the progressive disease group compared to the clinical benefit group (58% versus 48%; $P = 0.03$; Wilcoxon rank-sum test). Monocyte levels rebounded toward baseline during a rest period, and then repeated the pattern when sunitinib was restarted. This suggests that monocytes may be a pharmacodynamic marker for sunitinib activity and clinical benefit in this population. We are currently assessing specific monocyte populations to determine whether distinct cellular types in peripheral blood that may serve as more specific biomarkers of activity can be identified.

22.5 FUTURE DIRECTIONS

Due in part to recent randomized phase III trials demonstrating that angiogenesis inhibitors can benefit patients with different types of solid tumors,[1,2] there has been an explosion in the number and variety of these agents in development or clinical testing. These advances, while welcomed, present a new set of challenges and obstacles for the field. First, there is a major unmet need for validated, standardized, noninvasive markers for evaluating the activity of these agents, which are typically cytostatic when given as monotherapy, and for assessing the degree of target inhibition caused by treatment. This is needed both for selecting drugs to advance in clinical development, and for choosing the optimal dose for a given patient. Second, it is necessary to identify the mechanisms by which tumors become resistant to these agents, and to develop markers to monitor this resistance. It is clear from the initial clinical data that benefits from these agents are typically modest for most patients, and tumor progression eventually occurs. It is likely, therefore, that in the future more potent inhibitors will be developed and combinations of different angiogenesis inhibitors will need to be employed. Biomarkers can play a critical role in the rational design of such combination regimens. Third, it is critical to develop predictive markers to guide the selection of the most appropriate agents, or combination of agents, before initiating therapy. For all these reasons, biomarkers are likely to play an increasingly important role in the clinical development of VEGF inhibitors and other antiangiogenic agents. A growing body of research suggests that both soluble markers (e.g., sVEGFR-2) in plasma or serum and cellular markers (CECs, CEPs) have the potential to address some if not all of these issues and merit further investigation in larger clinical studies.

REFERENCES

1. Hurwitz H, et al., Bevacizumab plus irinotecan, fluorouracil, and leucovorin for metastatic colorectal cancer, *N Engl J Med*, 350, 2335–42, 2004.
2. Sandler AB, et al., Randomized phase II/III Trial of paclitaxel (P) plus carboplatin (C) with or without bevacizumab (NSC # 704865) in patients with advanced non-squamous non-small cell lung cancer (NSCLC): An Eastern Cooperative Oncology Group (ECOG) Trial-E4599, *J Clin Oncol (2005 ASCO Annual Meeting Proceedings)*, 23, abstr LBA4, 2005.

3. Miller KD, et al., E2100: A randomized phase III trial of paclitaxel versus paclitaxel plus bevacizumab as first-line therapy for locally recurrent or metastatic breast cancer, *Proc Am Soc Clin Oncol*, 23, 2005.
4. Motzer RJ, et al., Activity of SU11248, a multitargeted inhibitor of vascular endothelial growth factor receptor and platelet-derived growth factor receptor, in patients with metastatic renal cell carcinoma, *J Clin Oncol*, 24, 16–24, 2006.
5. Gordon M, et al., Phase I trial of recombinant humanized monoclonal anti-vascular endothelial growth factor (anti-VEGF MAB) in patients (PTS) with metastatic cancer (meeting abstract), *Proc Am Soc Clin Oncol*, 17, 1998.
6. Boehm T, et al., Antiangiogenic therapy of experimental cancer does not induce acquired drug resistance, *Nature*, 390, 404–7, 1997.
7. Kerbel RS, et al., Possible mechanisms of acquired resistance to anti-angiogenic drugs: implications for the use of combination therapy approaches, *Cancer Metastasis Rev*, 20, 79–86, 2001.
8. Heymach JV, et al., Phase II study of the antiangiogenic agent SU5416 in patients with advanced soft tissue sarcomas, *Clin Cancer Res*, 10, 5732–40, 2004.
9. Davis DW, et al., Pharmacodynamic analysis of target inhibition and endothelial cell death in tumors treated with the vascular endothelial growth factor receptor antagonists SU5416 or SU6668, *Clin Cancer Res*, 11, 678–89, 2005.
10. Gorre ME and Sawyers CL, Molecular mechanisms of resistance to STI571 in chronic myeloid leukemia, *Curr Opin Hematol*, 9, 303–7, 2002.
11. Gorre ME, et al., Clinical resistance to STI-571 cancer therapy caused by BCR-ABL gene mutation or amplification, *Science*, 293, 876–80, 2001.
12. Kobayashi S, et al., EGFR mutation and resistance of non-small-cell lung cancer to gefitinib, *N Engl J Med*, 352, 786–92, 2005.
13. Semenza GL, Targeting HIF-1 for cancer therapy, *Nat Rev Cancer*, 3, 721–32, 2003.
14. Yamakawa M, et al., Hypoxia-inducible factor-1 mediates activation of cultured vascular endothelial cells by inducing multiple angiogenic factors, *Circ Res*, 93, 664–73, 2003.
15. Relf M, et al., Expression of the angiogenic factors vascular endothelial cell growth factor, acidic and basic fibroblast growth factor, tumor growth factor beta-1, platelet-derived endothelial cell growth factor, placenta growth factor, and pleiotrophin in human primary breast cancer and its relation to angiogenesis, *Cancer Res*, 57, 963–9, 1997.
16. Sweeney CJ, et al., The antiangiogenic property of docetaxel is synergistic with a recombinant humanized monoclonal antibody against vascular endothelial growth factor or 2-methoxyestradiol but antagonized by endothelial growth factors, *Cancer Res*, 61, 3369–72, 2001.
17. Dvorak HF, Vascular permeability factor/vascular endothelial growth factor: a critical cytokine in tumor angiogenesis and a potential target for diagnosis and therapy, *J Clin Oncol*, 20, 4368–80, 2002.
18. Ferrara N, VEGF as a therapeutic target in cancer, *Oncology*, 69 Suppl 3, 11–6, 2005.
19. Fukumura D, et al., Tumor induction of VEGF promoter activity in stromal cells, *Cell*, 94, 715–25, 1998.
20. Asahara T, et al., VEGF contributes to postnatal neovascularization by mobilizing bone marrow-derived endothelial progenitor cells, *EMBO J*, 18, 3964–72, 1999.
21. Kerbel RS, et al., Establishing a link between oncogenes and tumor angiogenesis, *Mol Med*, 4, 286–95, 1998.
22. Ferrara N, et al., Discovery and development of bevacizumab, an anti-VEGF antibody for treating cancer, *Nat Rev Drug Discov*, 3, 391–400, 2004.
23. Carmeliet P and Jain RK, Angiogenesis in cancer and other diseases, *Nature*, 407, 249–57, 2000.
24. Wachsberger P, Burd R, and Dicker AP, Tumor response to ionizing radiation combined with antiangiogenesis or vascular targeting agents: exploring mechanisms of interaction, *Clin Cancer Res*, 9, 1957–71, 2003.
25. Poon RT, Fan ST, and Wong J, Clinical implications of circulating angiogenic factors in cancer patients, *J Clin Oncol*, 19, 1207–25, 2001.
26. Kuroi K and Toi M, Circulating angiogenesis regulators in cancer patients, *Int J Biol Markers*, 16, 5–26, 2001.

27. Bocci G, et al., Increased plasma vascular endothelial growth factor (VEGF) as a surrogate marker for optimal therapeutic dosing of VEGF receptor-2 monoclonal antibodies, *Cancer Res*, 64, 6616–25, 2004.
28. Gordon MS, et al., Phase I safety and pharmacokinetic study of recombinant human anti-vascular endothelial growth factor in patients with advanced cancer, *J Clin Oncol*, 19, 843–50, 2001.
29. Yang JC, et al., A randomized trial of bevacizumab, an anti-vascular endothelial growth factor antibody, for metastatic renal cancer, *N Engl J Med*, 349, 427–34, 2003.
30. Willett CG, et al., Surrogate markers for antiangiogenic therapy and dose-limiting toxicities for bevacizumab with radiation and chemotherapy: continued experience of a phase I trial in rectal cancer patients, *J Clin Oncol*, 23, 8136–9, 2005.
31. Faivre S, et al., Safety, pharmacokinetic, and antitumor activity of SU11248, a novel oral multi-target tyrosine kinase inhibitor, in patients with cancer, *J Clin Oncol*, 24, 25–35, 2006.
32. Manning WC, et al., Pharmacokinetic and pharmacodynamic evaluation of SU11248 in a phase I clinical trial of patients (pts) with imatinib-resistant gastrointestinal stromal tumor (GIST), in *Proc Am Soc Clin Oncol*, 22, 2003 (abstract 768).
33. O'Farrell A-M, et al., Analysis of biomarkers of SU11248 action in an exploratory study in patients with advanced malignancies, in *Proc Am Soc Clin Oncol*, 22, 2003 (abstract 939).
34. Rosen L, et al., Phase I trial of SU011248, a novel tyrosine kinase inhibitor in advanced solid tumors, in *Proc Am Soc Clin Oncol,* 22, 2003 (abstract 765).
35. Norden-Zfoni A, et al., Blood-based biomarkers of SU11248 activity and clinical outcome in patients with metastatic imatinib-resistant gastrointestinal stromal tumor (GIST), *Clin Cancer Res,* in press, 2007.
36. Dowlati A, et al., Novel Phase I dose de-escalation design trial to determine the biological modulatory dose of the antiangiogenic agent SU5416, *Clin Cancer Res*, 11, 7938–44, 2005.
37. Lara PN, Jr., et al., SU5416 plus interferon alpha in advanced renal cell carcinoma: a phase II California Cancer Consortium Study with biological and imaging correlates of angiogenesis inhibition, *Clin Cancer Res*, 9, 4772–81, 2003.
38. Drevs J, et al., Soluble markers for the assessment of biological activity with PTK787/ZK 222584 (PTK/ZK), a vascular endothelial growth factor receptor (VEGFR) tyrosine kinase inhibitor in patients with advanced colorectal cancer from two phase I trials, *Ann Oncol*, 16, 558–65, 2005.
39. Drevs J, et al., Phase I clinical evaluation of AZD2171, a highly potent VEGF receptor tyrosine kinase inhibitor, in patients with advanced tumors, *J Clin Oncol, 2005 ASCO Annual Meeting Proceedings. Part I of II (June 1 Supplement)*, 23, 3002, 2005.
40. Rugo HS, et al., Phase I trial of the oral antiangiogenesis agent AG-013736 in patients with advanced solid tumors: pharmacokinetic and clinical results, *J Clin Oncol*, 23, 5474–83, 2005.
41. Waltenberger J, et al., Different signal transduction properties of KDR and Flt1, two receptors for vascular endothelial growth factor, *J Biol Chem*, 269, 26988–95, 1994.
42. Seetharam L, et al., A unique signal transduction from FLT tyrosine kinase, a receptor for vascular endothelial growth factor VEGF, *Oncogene*, 10, 135–47, 1995.
43. Park JE, et al., Placenta growth factor. Potentiation of vascular endothelial growth factor bioactivity, in vitro and in vivo, and high affinity binding to Flt-1 but not to Flk-1/KDR, *J Biol Chem*, 269, 25646–54, 1994.
44. Gerber HP and Ferrara N, The role of VEGF in normal and neoplastic hematopoiesis, *J Mol Med*, 81, 20–31, 2003.
45. Hattori K, et al., Placental growth factor reconstitutes hematopoiesis by recruiting VEGFR1(+) stem cells from bone-marrow microenvironment, *Nat Med*, 8, 841–9, 2002.
46. Hiratsuka S, et al., MMP9 induction by vascular endothelial growth factor receptor-1 is involved in lung-specific metastasis, *Cancer Cell*, 2, 289–300, 2002.
47. Luttun A, et al., Revascularization of ischemic tissues by PlGF treatment, and inhibition of tumor angiogenesis, arthritis and atherosclerosis by anti-Flt1, *Nat Med*, 8, 831–40, 2002.
48. Hicklin DJ and Ellis LM, Role of the vascular endothelial growth factor pathway in tumor growth and angiogenesis, *J Clin Oncol*, 23, 1011–27, 2005.
49. Kendall RL and Thomas KA, Inhibition of vascular endothelial cell growth factor activity by an endogenously encoded soluble receptor, *Proc Natl Acad Sci USA*, 90, 10705–9, 1993.

50. Ebos JM, et al., A naturally occurring soluble form of vascular endothelial growth factor receptor 2 detected in mouse and human plasma, *Mol Cancer Res*, 2, 315–26, 2004.
51. Lamszus K, et al., Levels of soluble vascular endothelial growth factor (VEGF) receptor 1 in astrocytic tumors and its relation to malignancy, vascularity, and VEGF-A, *Clin Cancer Res*, 9, 1399–405, 2003.
52. Toi M, et al., Significance of vascular endothelial growth factor (VEGF)/soluble VEGF receptor-1 relationship in breast cancer, *Int J Cancer*, 98, 14–8, 2002.
53. Harris AL, et al., Soluble Tie2 and Flt1 extracellular domains in serum of patients with renal cancer and response to antiangiogenic therapy, *Clin Cancer Res*, 7, 1992–7, 2001.
54. Maynard SE, et al., Excess placental soluble fms-like tyrosine kinase 1 (sFlt1) may contribute to endothelial dysfunction, hypertension, and proteinuria in preeclampsia, *J Clin Invest*, 111, 649–58, 2003.
55. Levine RJ, et al., Circulating angiogenic factors and the risk of preeclampsia, *N Engl J Med*, 350, 672–83, 2004.
56. Gora-Tybor J, Blonski JZ, and Robak T, Circulating vascular endothelial growth factor (VEGF) and its soluble receptors in patients with chronic lymphocytic leukemia, *Eur Cytokine Netw*, 16, 41–6, 2005.
57. Wierzbowska A, et al., Circulating VEGF and its soluble receptors sVEGFR-1 and sVEGFR-2 in patients with acute leukemia, *Eur Cytokine Netw*, 14, 149–53, 2003.
58. Verstovsek S, et al., Clinical relevance of VEGF receptors 1 and 2 in patients with chronic myelogenous leukemia, *Leuk Res*, 27, 661–9, 2003.
59. Luttun A, et al., Genetic dissection of tumor angiogenesis: are PlGF and VEGFR-1 novel anti-cancer targets? *Biochim Biophys Acta*, 1654, 79–94, 2004.
60. Ferrara N and Kerbel RS, Angiogenesis as a therapeutic target, *Nature*, 438, 967–74, 2005.
61. Pepper MS, et al., Synergistic induction of t-PA by vascular endothelial growth factor and basic fibroblast growth factor and localization of t-PA to Weibel-Palade bodies in bovine microvascular endothelial cells, *Thromb Haemost*, 86, 702–9, 2001.
62. Ma L, et al., In vitro procoagulant activity induced in endothelial cells by chemotherapy and antiangiogenic drug combinations: modulation by lower-dose chemotherapy, *Cancer Res*, 65, 5365–73, 2005.
63. Yang R, et al., Effects of vascular endothelial growth factor on hemodynamics and cardiac performance, *J Cardiovasc Pharmacol*, 27, 838–44, 1996.
64. Baffert F, et al., Cellular changes in normal blood capillaries undergoing regression after inhibition of VEGF signaling, *Am J Physiol Heart Circ Physiol*, 290, H547–59, 2006.
65. Kuenen BC, et al., Potential role of platelets in endothelial damage observed during treatment with cisplatin, gemcitabine, and the angiogenesis inhibitor SU5416, *J Clin Oncol*, 21, 2192–8, 2003.
66. Koukourakis MI, et al., Endogenous markers of two separate hypoxia response pathways (hypoxia inducible factor 2 alpha and carbonic anhydrase 9) are associated with radiotherapy failure in head and neck cancer patients recruited in the CHART randomized trial, *J Clin Oncol*, 24, 727–35, 2006.
67. Franco M, et al., Targeted anti-vascular endothelial growth factor receptor-2 therapy leads to short-term and long-term impairment of vascular function and increase in tumor hypoxia, *Cancer Res*, 66, 3639–48, 2006.
68. Le QT, et al., An evaluation of tumor oxygenation and gene expression in patients with early stage non-small cell lung cancers, *Clin Cancer Res*, 12, 1507–14, 2006.
69. Overgaard J, et al., Plasma osteopontin, hypoxia, and response to the hypoxia sensitiser nimorazole in radiotherapy of head and neck cancer: results from the DAHANCA 5 randomised double-blind placebo-controlled trial, *Lancet Oncol*, 6, 757–64, 2005.
70. Le QT, et al., Identification of osteopontin as a prognostic plasma marker for head and neck squamous cell carcinomas, *Clin Cancer Res*, 9, 59–67, 2003.
71. Le QT, et al., Galectin-1: a link between tumor hypoxia and tumor immune privilege, *J Clin Oncol*, 23, 8932–41, 2005.
72. Bouvier C, et al., Circulating endothelium as an indication of vascular injury, *Throm Diath Haemorrh Suppl*, 40, 163–8, 1907.

73. Weber G, et al., Circulating endothelial-like cells in arterial peripheral blood of hypercholesterolemic rabbits, *Artery*, 5, 29–36, 1979.
74. Sbarbati R, et al., Immunologic detection of endothelial cells in human whole blood, *Blood*, 77, 764–9, 1991.
75. George F, et al., Cytofluorometric detection of human endothelial cells in whole blood using S-Endo 1 monoclonal antibody, *J Immunol Methods*, 139, 65–75, 1991.
76. Solovey A, et al., Circulating activated endothelial cells in sickle cell anemia, *N Engl J Med*, 337, 1584–90, 1997.
77. Lin Y, et al., Origins of circulating endothelial cells and endothelial outgrowth from blood, *J Clin Invest*, 105, 71–7, 2000.
78. George F, et al., Demonstration of Rickettsia conorii-induced endothelial injury in vivo by measuring circulating endothelial cells, thrombomodulin, and von Willebrand factor in patients with Mediterranean spotted fever, *Blood*, 82, 2109–16, 1993.
79. George F, et al., Rapid isolation of human endothelial cells from whole blood using S-Endo1 monoclonal antibody coupled to immuno-magnetic beads: demonstration of endothelial injury after angioplasty, *Thromb Haemost*, 67, 147–53, 1992.
80. Asahara T, et al., Isolation of putative progenitor endothelial cells for angiogenesis, *Science*, 275, 964–7, 1997.
81. Asahara T, et al, Bone marrow origin of endothelial progenitor cells responsible for postnatal vasculogenesis in physiological and pathological neovascularization, *Circ Res*, 85, 221–8, 1999.
82. Lyden D, et al., Impaired recruitment of bone-marrow-derived endothelial and hematopoietic precursor cells blocks tumor angiogenesis and growth, *Nat Med*, 7, 1194–201, 2001.
83. Peters BA, et al., Contribution of bone marrow–derived endothelial cells to human tumor vasculature, *Nat Med*, 11, 261–2, 2005.
84. Peichev M, et al., Expression of VEGFR-2 and AC133 by circulating human CD34(+) cells identifies a population of functional endothelial precursors, *Blood*, 95, 952–8, 2000.
85. Monestiroli S, et al., Kinetics and viability of circulating endothelial cells as surrogate angiogenesis marker in an animal model of human lymphoma, *Cancer Res*, 61, 4341–4, 2001.
86. Shaked Y, et al., Optimal biologic dose of metronomic chemotherapy regimens is associated with maximum antiangiogenic activity, *Blood*, 106, 3058–61, 2005.
87. Beaudry P, et al., Differential effects of vascular endothelial growth factor receptor-2 inhibitor ZD6474 on circulating endothelial progenitors and mature circulating endothelial cells: implications for use as a surrogate marker of antiangiogenic activity, *Clin Cancer Res*, 11, 3514–22, 2005.
88. Schuch G, et al., Endostatin inhibits the vascular endothelial growth factor-induced mobilization of endothelial progenitor cells, *Cancer Res*, 63, 8345–50, 2003.
89. Beerepoot LV, et al., Phase I clinical evaluation of weekly administration of the novel vascular-targeting agent, ZD6126, in patients with solid tumors, *J Clin Oncol*, 24, 1491–8, 2006.
90. Mancuso P, et al., Circulating endothelial-cell kinetics and viability predict survival in breast cancer patients receiving metronomic chemotherapy, *Blood*, 108, 452–9, 2006.
91. Beerepoot LV, et al., Increased levels of viable circulating endothelial cells are an indicator of progressive disease in cancer patients, *Ann Oncol*, 15, 139–45, 2004.
92. Mancuso P, et al., Resting and activated endothelial cells are increased in the peripheral blood of cancer patients, *Blood*, 97, 3658–61, 2001.
93. Takahashi T, et al., Ischemia- and cytokine-induced mobilization of bone marrow–derived endothelial progenitor cells for neovascularization, *Nat Med*, 5, 434–8, 1999.
94. Kalka C, et al., Vascular endothelial growth factor(165) gene transfer augments circulating endothelial progenitor cells in human subjects, *Circ Res*, 86, 1198–202, 2000.
95. Kalka C, et al., VEGF gene transfer mobilizes endothelial progenitor cells in patients with inoperable coronary disease, *Ann Thorac Surg*, 70, 829–34, 2000.
96. Hattori K, et al., Vascular endothelial growth factor and angiopoietin-1 stimulate postnatal hematopoiesis by recruitment of vasculogenic and hematopoietic stem cells, *J Exp Med*, 193, 1005–14, 2001.
97. Reyes M, et al., Origin of endothelial progenitors in human postnatal bone marrow, *J Clin Invest*, 109, 337–46, 2002.

98. Radema SA, et al., Clinical evaluation of the novel vascular-targeting agent, ZD6126: assessment of toxicity and surrogate markers of vascular damage, *Proc Am Soc Clin Oncol*, 21, 2002 (abstract 439).
99. Heymach J, et al., Circulating endothelial cells as a surrogate marker of antiangiogenic activity in patients treated with endostatin, *Proc Am Soc Clin Oncol*, 22, 979, 2003.
100. Khan SS, Solomon MA, and McCoy JP Jr., Detection of circulating endothelial cells and endothelial progenitor cells by flow cytometry, *Cytometry Part B: Clinical Cytometry*, 64B, 1–8, 2005.
101. Duda DG, et al., Differential CD146 expression on circulating versus tissue endothelial cells in rectal cancer patients: implications for circulating endothelial and progenitor cells as biomarkers for antiangiogenic therapy, *J Clin Oncol*, 24, 1449–53, 2006.
102. Ziegler BL, et al., KDR receptor: a key marker defining hematopoietic stem cells, *Science*, 285, 1553–8, 1999.
103. Burger PE, et al., Fibroblast growth factor receptor-1 is expressed by endothelial progenitor cells, *Blood*, 100, 3527–35, 2002.
104. Goon PK, Boos CJ, and Lip GY, Circulating endothelial cells: markers of vascular dysfunction, *Clin Lab*, 51, 531–8, 2005.
105. Shaffer RG, et al., Flow cytometric measurement of circulating endothelial cells: the effect of age and peripheral arterial disease on baseline levels of mature and progenitor populations, *Cytometry B Clin Cytom*, 70, 56–62, 2006.
106. Jain RK, Normalizing tumor vasculature with anti-angiogenic therapy: a new paradigm for combination therapy, *Nat Med*, 7, 987–9, 2001.
107. Ceradini DJ, et al., Progenitor cell trafficking is regulated by hypoxic gradients through HIF-1 induction of SDF-1, *Nat Med*, 10, 858–64, 2004.
108. Shaked Y, et al., Genetic heterogeneity of the vasculogenic phenotype parallels angiogenesis; implications for cellular surrogate marker analysis of antiangiogenesis, *Cancer Cell*, 7, 101–11, 2005.
109. Willett CG, et al., Direct evidence that the VEGF-specific antibody bevacizumab has antivascular effects in human rectal cancer, *Nat Med*, 10, 145–7, 2004.
110. Duda DG, et al., Differential circulation kinetics during antiangiogenic therapy of four distinct blood cell populations expressing endothelial markers., *J Clin Oncol, 2006 ASCO Annual Meeting Proceedings Part I*, 24, 3038, 2006.
111. Song S, et al., PDGFRbeta + perivascular progenitor cells in tumours regulate pericyte differentiation and vascular survival, *Nat Cell Biol*, 7, 870–9, 2005.
112. Jin DK, et al., Cytokine-mediated deployment of SDF-1 induces revascularization through recruitment of CXCR4 + hemangiocytes, *Nat Med*, 12, 557–67, 2006.
113. Bailey AS, et al., Myeloid lineage progenitors give rise to vascular endothelium, *Proc Natl Acad Sci USA*, 103, 13156–61, 2006.
114. Zhang H, et al., Circulating endothelial progenitor cells in multiple myeloma: implications and significance, *Blood*, 105, 3286–94, 2005.
115. Wierzbowska A, et al., Circulating endothelial cells in patients with acute myeloid leukemia, *Eur J Haematol*, 75, 492–7, 2005.
116. Cortelezzi A, et al., Endothelial precursors and mature endothelial cells are increased in the peripheral blood of myelodysplastic syndromes, *Leuk Lymphoma*, 46, 1345–51, 2005.

Part IV

Treatment of Specific Cancers with Angiogenesis Inhibitors

23 Antiangiogenic Therapy for Colorectal Cancer

Paulo M. Hoff and Everardo D. Saad

CONTENTS

Adenocarcinoma of the colon and rectum is a growing health problem around the world and accounts for most cases of colorectal cancer. Approximately two-thirds of such tumors arise in the colon and one-third in the rectum. According to recent estimates, nearly one million people are diagnosed with colorectal cancer each year, and almost half a million die from this disease worldwide.[1] Colorectal cancer is associated with the adoption of a so-called Westernized lifestyle, and the disease is more common in developed areas of the world. Fortunately, the relatively long course of colorectal tumorigenesis allows for early detection and effective interventions that aim at reducing morbidity and mortality.[2] Unfortunately, only a small proportion of eligible individuals currently undergo screening for colorectal cancer.[3] In the United States, where the lifetime risk of colorectal cancer is currently 1 in 17,[4] it was estimated that approximately 145,000 cases would be diagnosed in 2005.[5] Although only one in five patients with colorectal cancer has metastatic disease at diagnosis,[5] systemic recurrence will develop in a significant proportion of patients who have colorectal cancer that invades through the mucosa or the regional lymph nodes (i.e., tumor-node-metastasis stages II and III) at diagnosis. For patients with such recurrences, systemic therapy offers the potential to prolong survival and improve the quality of life.[6,7]

During the past decade, systemic therapy for colorectal cancer has evolved significantly. Two novel chemotherapeutic agents, irinotecan and oxaliplatin, have been shown to improve on the previous standard of fluorouracil (5-FU) and leucovorin.[8–11] More recently, antiangiogenic agents and other molecularly targeted therapies have been investigated in clinical trials among patients with colorectal cancer. In this chapter, we provide a brief overview of the state of the art regarding systemic therapy for advanced colorectal cancer, discuss the role of angiogenesis in colorectal cancer biology, present recent findings from studies of antiangiogenic agents, and discuss future perspectives relating to antiangiogenic treatment for colorectal cancer.

23.1 CURRENT CHEMOTHERAPY FOR ADVANCED COLORECTAL CANCER

In patients with advanced colorectal cancer, systemic chemotherapy prolongs survival.[6,7] Data on quality of life are of a lower level of evidence, but it is generally believed that early treatment is associated with a prolongation of the symptom-free period.[12] For nearly four decades, the thymidylate synthase inhibitor 5-FU remained the only chemotherapeutic agent with clinically significant, albeit modest, activity for the treatment of patients with advanced colorectal cancer. A variety of regimens containing 5-FU have been developed through the years. In these regimens, 5-FU may be delivered alone or with leucovorin, a biochemical modulator that enhances the activity of 5-FU.[13,14] It has also been shown that the activity of 5-FU is enhanced when it is administered through continuous infusion.[15] Because of practical issues relating to the use of portable infusion pumps and central venous catheters, bolus schedules have become most popular in the United States. Two of the most commonly used regimens of bolus schedules are known as (1) the Roswell Park regimen, in which 5-FU and leucovorin are delivered weekly for 6 consecutive weeks in 8-week cycles,[16] and (2) the Mayo Clinic regimen, in which 5-FU is given daily for 5 consecutive days in 28-day cycles.[17] In France and some other countries, the regimen developed by de Gramont et al.[18] was shown to increase response rates and the time to tumor progression, with less toxicity, when compared with the Mayo Clinic regimen.[18] According to this regimen, leucovorin is followed by a regimen of continuous infusion of 5-FU for 2 consecutive days every 2 weeks.

One of the novel chemotherapeutic agents, irinotecan, is an inhibitor of topoisomerase I, a nuclear enzyme that is important for maintaining the three-dimensional structure of DNA during its replication and RNA synthesis; irinotecan leads to DNA strand breaks and cell death by apoptosis.[19] In combination with 5-FU and leucovorin, irinotecan was shown to prolong the progression-free and overall survival of patients with metastatic colorectal cancer.[8,9] The regimen with infusional 5-FU,[8] known as FOLFIRI, seems to be less toxic than the one with a 5-FU bolus.[9] In addition, irinotecan prolongs overall survival when used in the second-line treatment of patients whose disease progressed after treatment with 5-FU.[20]

Another novel chemotherapeutic agent, oxaliplatin, is a platinum analog that works primarily by cross-linking DNA, leading to apoptosis, and is the only platinum compound with clinical activity in colorectal cancer.[21] Oxaliplatin is thought to partially reverse colorectal tumor resistance to 5-FU, because patients in whom previous treatment with 5-FU has failed are more likely to respond to oxaliplatin and 5-FU than to oxaliplatin alone.[22] Several combinations of 5-FU, leucovorin, and oxaliplatin have been developed and used in the treatment of patients with advanced colorectal cancer. In one of these regimens, developed by de Gramont's group, oxaliplatin was combined with the bimonthly 5-FU/leucovorin regimen, and this treatment came to be known as FOLFOX4.[10] In a recent phase III trial conducted in the United States, FOLFOX4 as the first-line treatment was shown to prolong the response and the time to tumor progression in patients with metastatic colorectal cancer.[11] Second-line treatment with FOLFOX4 also prolonged the time to tumor progression in patients previously treated with 5-FU and irinotecan in comparison with 5-FU/leucovorin or oxaliplatin.[23,24]

Another chemotherapeutic agent that deserves mention is capecitabine, an oral prodrug of 5-FU. Capecitabine is converted to 5-FU after three enzymatic reactions that take place in the liver and tumor tissue.[25] There is experimental evidence that the final reaction, which gives rise to 5-FU, occurs preferentially in colorectal tumor, when compared with adjacent normal tissue.[26] The toxicity profile of capecitabine is considered more favorable than that of intravenous 5-FU.[27] Capecitabine also produces more objective responses, with a time to tumor progression and overall survival similar to those seen in patients who receive the Mayo Clinic regimen.[28]

Bevacizumab, a monoclonal antibody against vascular endothelial growth factor (VEGF), also known as vascular permeability factor, is a new treatment for advanced colorectal cancer that is discussed in more detail below.[29] Before the advent of bevacizumab, the standard treatment for patients with advanced colorectal cancer who had an adequate performance status for chemotherapy consisted of 5-FU combined with either oxaliplatin or irinotecan. On tumor progression, 5-FU/leucovorin/oxaliplatin, 5-FU/leucovorin/irinotecan, or single-agent irinotecan were the most frequent options for second-line chemotherapy. From the standpoint of efficacy, the order in which oxaliplatin and irinotecan are used in the first- and second-line treatment does not seem to matter,[30] and a recent study suggested that survival is prolonged when both agents and 5-FU are used at some point during treatment.[31] Capecitabine may also be used as a substitute for 5-FU and leucovorin, despite the current lack of definitive results from large comparative trials of its combination with oxaliplatin or irinotecan. Recently, cetuximab, a chimeric monoclonal antibody against the epidermal growth factor receptor (EGFR), was shown to possess activity as a single agent in patients with refractory colorectal cancer.[32] In combination with irinotecan, cetuximab was shown to reverse clinical resistance to irinotecan and also to prolong the time to tumor progression,[33] thus representing a second- or third-line option in some cases. In selected patients, chemotherapy may be integrated into a multidisciplinary approach that includes the surgical resection of liver metastases.[34]

23.2 ANGIOGENESIS IN COLORECTAL CANCER

Several lines of evidence point to the key role played by angiogenesis in colorectal tumorigenesis and progression. In particular, microvessel density is gradually increased during the progression from normal mucosa to adenoma and adenocarcinoma.[35] Indeed, in a multivariate analysis, microvessel density was shown to be an adverse prognostic factor and to correlate with hematogenous metastasis in patients with colorectal cancer.[36–38]

A variety of mediators of angiogenesis, including proangiogenic molecules and their receptors, have been found in colorectal tumors. Foremost among these is VEGF, which is expressed in human colon cancer cells and in nearby endothelial cells; the latter also express VEGF receptors (VEGFRs).[39] The expression of VEGF and VEGFR-2 in colon cancer correlates with neovascularization and proliferation and is higher in metastatic than in nonmetastatic tumors.[40] VEGF, VEGFR-1, and VEGFR-2 are also expressed in colorectal cancer liver metastases, in which they are considered key regulators of growth.[41] In a mouse model of colon cancer liver metastasis, the anti-VEGF monoclonal antibody A.4.6.1 inhibited tumor growth in a dose-dependent fashion, which points to the potential therapeutic value of the strategy.[41]

Many other direct or indirect proangiogenic molecules have also been found in human colorectal cancer. These molecules include fibroblast growth factor,[42] platelet-derived growth factor[43] and platelet-derived growth factor receptors,[44] platelet-derived endothelial cell growth factor,[45,46] angiopoietin-2,[47] interleukins 8[48] and 15,[49] and cyclooxygenases,[50] among others. Additionally, noteworthy is the putative role played by EGFR in angiogenesis, as suggested by the finding that treatment with EGFR inhibitors leads to the reduced expression of VEGF and microvessel counts in colorectal tumor models.[51,52] The roles of these and other molecules, as well as the interplay of their activity, are currently actively investigated, given the fact that angiogenesis is an important target for antineoplastic drug development.

23.3 THERAPEUTIC STRATEGIES TARGETING VEGF

Theoretically, therapeutic agents may target many of the molecules and processes that constitute the VEGF signaling pathway. Hence, dominant-negative VEGFR mutants, antibodies against VEGF and VEGFRs, soluble VEGFRs, VEGF toxin conjugates, small-molecule tyrosine kinase inhibitors, VEGFR-1 ribozymes, and antisense oligonucleotides have all been tested in preclinical models against a variety of tumor types.[29,53] Of these strategies, monoclonal antibodies against VEGF and tyrosine kinase inhibitors targeting VEGFRs are the ones in more advanced phases of clinical development.[54]

23.3.1 Bevacizumab

Ferrara et al.[29] recently reviewed the preclinical and clinical development of bevacizumab, the humanized version of the murine antibody A.4.6.1, which their[29] group at Genentech first described.[55,56] Bevacizumab was generated by site-directed mutagenesis of a normal human immunoglobulin G framework, on which the complementarity-determining regions and several framework residues were changed from human to murine. Similar to A.4.6.1, bevacizumab binds with high affinity to all known VEGF-A isoforms and is active in a variety of preclinical models.[29] The activity of bevacizumab in colorectal cancer has been documented in a phase I clinical trial in patients with primary rectal cancer.[57] In this trial, six patients received an infusion of bevacizumab (5 mg/kg), which was followed by flexible sigmoidoscopy 12 days later. Tumor regression was observed in only one patient, but functional computed tomography indicated significant reductions in tumor perfusion. In addition, microvessel density and interstitial fluid pressure were decreased, demonstrating that bevacizumab has direct antivascular effects in human tumors and lending support to the hypothesis that antiangiogenic therapy may work by normalizing the structurally and functionally abnormal tumor vasculature.[58] Such normalization may in turn facilitate the activity of other antineoplastic agents, including chemotherapy.

In phase I trials conducted in patients with solid tumors, bevacizumab was administered alone or in combination with chemotherapy.[59,60] When used as a single agent in doses of up to 10 mg/kg on days 0, 28, 35, and 42, bevacizumab was well tolerated but produced no objective responses.[59] In this trial, there were only two episodes of grade 3 or 4 hemorrhage among 25 patients, but the relationship between this adverse event and bevacizumab initially remained unclear. In a second phase I trial, bevacizumab in combination with four different chemotherapy regimens was administered at a weekly dose of 3 mg/kg for 8 consecutive weeks.[60] No unexpected toxicity was recorded, and adverse events were attributed to the chemotherapy components of the regimens.

To explore the efficacy of bevacizumab in colorectal cancer, two randomized phase II trials of 5-FU/leucovorin with or without bevacizumab were initiated in previously untreated patients with metastatic colorectal cancer.[61,62] In the first trial, 104 patients were randomized to receive the Roswell Park regimen or the same chemotherapy combined with bevacizumab, at a dose of either 5 or 10 mg/kg administered every 2 weeks.[61] The median time to progression, the primary end point in the study, was 5.2 months in the control arm, 9.0 months in the 5 mg/kg bevacizumab arm ($P = 0.005$, compared with the control), and 7.2 months in the 10 mg/kg bevacizumab arm ($P = 0.217$, compared with the control). The response rates were 17%, 40% ($P = 0.029$), and 24% ($P = 0.434$), respectively. The adverse events attributed to bevacizumab were thrombosis, epistaxis, and proteinuria. In the second trial, 209 patients aged $\geq$65 years, with an Eastern Cooperative Oncology Group (ECOG) performance status of 1 or 2, serum albumin level $\geq$3.5 g/dL, or prior abdominal/pelvic radiotherapy, were randomized to receive the Roswell Park regimen with a placebo or bevacizumab, at a dose of 5 mg/kg every 2 weeks.[62] The median survival duration, the primary end point in this trial,

TABLE 23.1
Efficacy Results from Randomized Trials of Bevacizumab in Advanced Colorectal Cancer

First Author	Line of Treatment	Therapeutic Regimens	*N*	Response Rate (%)	Progression-Free Survival (Months)	Overall Survival (Months)
Kabbinavar[61]	First	5-FU/leucovorin	36	17	5.2	13.8
		5-FU/leucovorin/bevacizumab 5 mg/kg every 2 weeks	35	40*	9.0*	21.5
		5-FU/leucovorin/bevacizumab 10 mg/kg every 2 weeks	33	24	7.2	16.1
Kabbinavar[62]	First	5-FU/leucovorin/placebo	105	15.2	5.5	12.9
		5-FU/leucovorin/bevacizumab 5 mg/kg every 2 weeks	104	26	9.2*	16.6
Hurwitz[63,64]	First	Irinotecan/5-FU/leucovorin/placebo	411	34.8	6.2	15.6
		Irinotecan/5-FU/leucovorin/ bevacizumab 5 mg/kg every 2 weeks	402	44.8*	10.6*	20.3*
		5-FU/leucovorin/bevacizumab 5 mg/kg every 2 weeks	110	40	8.8	18.3
Giantonio[67]	Second	5-FU/leucovorin/oxaliplatin	289	9.2	5.5	10.7
		5-FU/leucovorin/oxaliplatin/ bevacizumab 10 mg/kg every 2 weeks	290	21.8	7.4*	12.5*
		Bevacizumab 10 mg/kg 2 weeks	243	3	3.5	10.2

5-FU, fluorouracil.

* Statistically significant difference from control arm.

was 16.6 months for the group treated with bevacizumab and 12.9 months for the group given a placebo ($P = 0.16$). Median progression-free survival duration was significantly prolonged, from 5.5 months with chemotherapy plus placebo to 9.2 months with chemotherapy plus bevacizumab ($P = 0.0002$). Grade 3 hypertension was more common in the bevacizumab arm (16% versus 3%) but was well controlled with oral medication. The results of these and other randomized trials of bevacizumab in colorectal cancer are shown in Table 23.1.

These phase II trials were followed by a phase III trial comparing the irinotecan/ 5-FU/leucovorin (IFL) regimen combined with bevacizumab or a placebo in the first-line treatment of advanced colorectal cancer.[63] The study initially enrolled 923 patients, who were randomized to one of three arms: IFL plus placebo, IFL plus bevacizumab (5 mg/kg every 2 weeks), and 5-FU/leucovorin plus bevacizumab. An interim analysis confirmed the safety of IFL plus bevacizumab, leading to discontinuation of the 5-FU/leucovorin/bevacizumab arm, the results of which have been subsequently reported.[64] Of the original 923 patients who underwent randomization, 402 received IFL plus bevacizumab and 411 received IFL plus placebo. As shown in Figure 23.1, median overall survival duration, the primary end point in the study, was 20.3 months in the IFL-plus-bevacizumab arm compared with 15.6 months in the IFL-plus-placebo arm ($P < 0.001$). The hazard ratio for death was 0.66, translating into a 34% reduction in the relative risk of mortality. Progression-free survival duration was also prolonged, from 6.2 months in patients treated without bevacizumab to 10.6 months in patients who received the antibody (hazard ratio, 0.54; $P < 0.001$), as depicted in Figure 23.2. Interestingly, response rates were significantly increased by the addition of bevacizumab, from 34.8% to 44.8% ($P = 0.004$), despite the fact that patients derived clinical benefit from bevacizumab regardless of whether an objective response was achieved.[65]

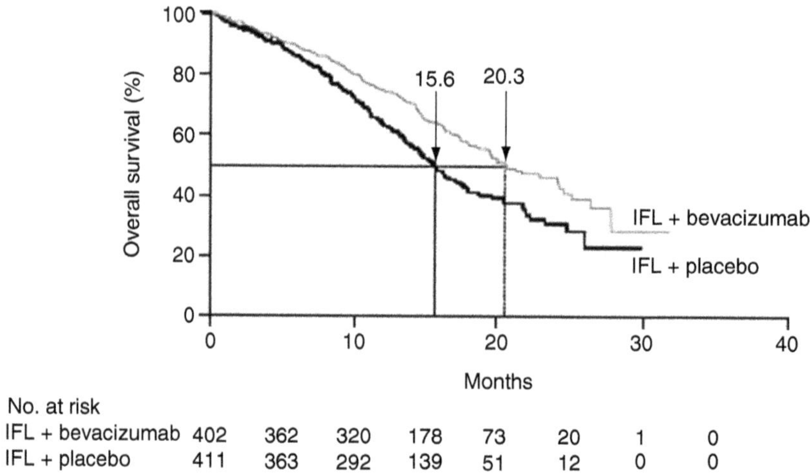

FIGURE 23.1 Kaplan–Meier estimates of overall survival. (Reproduced with permission from Hurwitz H, et al.

Adverse events associated with the use of bevacizumab in the phase III trial included a significantly increased risk of grade 3 hypertension and a small increase in the risk of grade 4 leukopenia and diarrhea; there were no significant differences in adverse events leading to hospitalization, study discontinuation, or 60-day mortality rate. Of note, gastrointestinal perforation occurred in six patients (1.5%) treated with IFL plus bevacizumab. The rates of proteinuria and bleeding were similar in both groups. Venous or arterial thrombotic events occurred in 19.4% of patients treated with chemotherapy plus bevacizumab and in 16.2% of patients given chemotherapy alone ($P = 0.26$). This landmark study was the first to show a survival advantage associated with treatment with an antiangiogenic agent, and its results led to the approval of bevacizumab by the Food and Drug Administration (FDA) in February 2004.

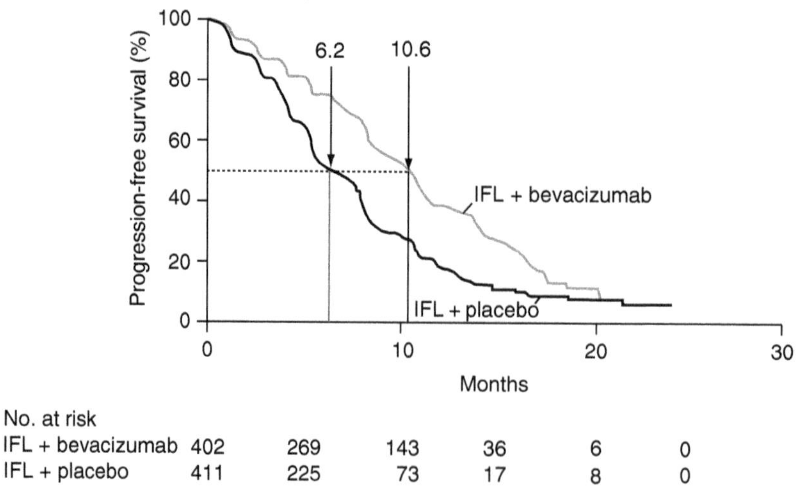

FIGURE 23.2 Kaplan–Meier estimates of progression-free survival. (Reproduced with permission from Hurwitz H, et al.

Bevacizumab is approved for use in combination with 5-FU-based chemotherapy for the first-line treatment of patients with metastatic colorectal cancer.

A recent report of the patients randomized before the interim analysis of this phase III trial showed that the addition of bevacizumab to the Roswell Park regimen produced superior results, in comparison with those seen among patients treated with IFL plus placebo.[64] In addition, in a combined analysis of these patients and those enrolled in the previous phase II trials discussed above,[61,62,66] data from 241 patients in a control group receiving either 5-FU/leucovorin or IFL and from 249 patients receiving 5-FU/leucovorin plus bevacizumab showed that the median survival duration was 17.9 months in the latter group compared with 14.6 months in the former.[66] This survival advantage corresponds to a hazard ratio for death of 0.74 ($P = 0.008$), translating into a 26% reduction in mortality for patients treated with bevacizumab as the first-line treatment. Further, the hazard ratio for disease progression was 0.63 ($P < 0.001$) and the response rate was increased from 24.5% to 34.1% ($P = 0.019$) with the addition of bevacizumab.

Bevacizumab has also been combined with the FOLFOX4 regimen, and results of second-line treatment were recently announced by the ECOG.[67] The ECOG study 3200 was a phase III trial in which patients who had been previously treated with 5-FU and irinotecan, either alone or in combination, were randomized to receive one of three treatment regimens: FOLFOX4, FOLFOX4 plus bevacizumab, or single-agent bevacizumab, which was administered at the dose of 10 mg/kg every 2 weeks. The trial enrolled 829 patients, and the bevacizumab arm was closed after the accrual of 243 patients. The main results of this study are shown in Table 23.1. Median overall survival duration, the primary end point in the study, was 10.7 months in the FOLFOX4 arm compared with 12.5 months in the FOLFOX4-plus-bevacizumab arm ($P = 0.0024$). The hazard ratio for death was 0.76, translating into a 24% reduction in the risk of mortality. Progression-free survival duration was also prolonged, from 5.5 months in patients treated with FOLFOX4 to 7.4 months in patients who also received bevacizumab (hazard ratio, 0.64; $P = 0.0003$). As seen in other studies, response rates were also increased by approximately 10%, in this case from 9.2% to 21.8%. Patients randomized to receive bevacizumab had a median overall survival duration of 10.2 months, a median progression-free survival duration of 3.5 months, and a response rate of 3%. The toxicity profile of bevacizumab confirmed that the addition of this monoclonal antibody increases the risk of hypertension and bleeding. Bowel perforation occurred in approximately 1% of patients given bevacizumab. Interestingly, the incidence of grade 3/4 sensory neuropathy, presumably caused by oxaliplatin, was increased from 9% in the FOLFOX4 arm to 16% with the addition of bevacizumab.

Other studies of bevacizumab for the treatment of advanced colorectal cancer are currently ongoing. In a randomized phase II trial, bevacizumab was given in combination with cetuximab, with or without irinotecan, in an attempt to improve on the results suggesting that irinotecan reverses clinical resistance to cetuximab.[33] The preliminary results thus far suggest that the combination of cetuximab and bevacizumab is safe and that the addition of irinotecan improves the response rate and the time to tumor progression, in comparison with treatment with the monoclonal antibodies alone.[68] In another phase II trial, bevacizumab has been combined with capecitabine and oxaliplatin.[69] The preliminary efficacy results are encouraging, but toxicity issues required dose reduction for capecitabine, and the final results are awaited.

Bevacizumab is also undergoing clinical trials in the adjuvant setting. Given the superiority of oxaliplatin-based regimens in the adjuvant treatment of patients with stage II or III colon cancer,[70,71] ongoing studies are investigating whether the addition of bevacizumab will improve on the results obtained with chemotherapy alone. It is also anticipated that the treatment of localized rectal cancer will benefit from the addition of bevacizumab, and clinical trials are currently ongoing.

23.4 OTHER ANTIANGIOGENIC AGENTS

23.4.1 Vatalanib

Vatalanib, also known as PTK787/ZK222584, is an orally administered small molecule that potently inhibits the tyrosine kinase activity of VEGFRs.[54,72] Vatalanib also inhibits other class III kinases, such as the platelet-derived growth factor receptor beta, and Kit, but at higher concentrations than it takes to inhibit the tyrosine kinase activity of VEGFRs. Vatalanib inhibits the VEGF-induced autophosphorylation of all three VEGFRs and consequent endothelial cell proliferation, migration, and survival. In preclinical models, the oral administration of vatalanib inhibited the growth of a variety of human tumor xenografts, including Ls174T and HT-29 colon carcinoma cell lines.[72] In humans, dynamic contrast-enhanced magnetic resonance imaging can be used to view changes in the contrast-enhanced parameters of tumor lesions and thereby assess the pharmacodynamic effects of vatalanib. Results from two phase I studies of 26 patients with liver metastases from colorectal cancer who were treated with vatalanib doses ranging from 50 to 2000 mg once daily suggested that dynamic contrast-enhanced magnetic resonance imaging may be a useful biomarker for evaluating response and defining doses for further clinical development.[73] Phase I trials have also shown that adverse effects of vatalanib are manageable and include lightheadedness, ataxia, nausea, vomiting, and hypertension. Disease stabilization for at least 6 months has been seen in heavily pretreated patients receiving vatalanib at higher doses.[74]

Phase I/II studies of vatalanib in patients with advanced colorectal cancer have shown that the drug may be safely combined with the FOLFOX4[75] and FOLFIRI[76] regimens. The main adverse event attributed to vatalanib has been central nervous system toxicity, which was observed with daily doses of 1500 and 2000 mg; no dose-limiting toxicity was seen with the dose of 1250 mg daily.[75] Two large multinational phase III trials of vatalanib in combination with FOLFOX4 have been launched. These studies, named Colorectal Oral Novel Therapy for the Inhibition of Angiogenesis and Retarding of Metastases (CONFIRM) trials, are comparing the FOLFOX4 regimen with vatalanib or placebo in the first-line (CONFIRM-1) or second-line (CONFIRM-2) treatment of patients with advanced colorectal cancer.

Preliminary results from CONFIRM-1 have recently been reported.[77] A total of 1168 patients with advanced colorectal cancer who had received no prior chemotherapy for metastatic disease were randomized to treatment with FOLFOX4 plus placebo or the same chemotherapy plus vatalanib, administered at a dose of 1250 mg once daily in a continuous fashion. Results of the analysis of overall survival, the primary end point in this trial, are expected in 2006. The analysis of progression-free survival by the study investigators showed a 17% reduction in the risk of progression (hazard ratio, 0.83; $P = 0.026$). However, a central review of the data led to a slightly higher hazard ratio (0.88), which lost statistical significance ($P = 0.118$). Response rates did not differ between either arm of the study, and the only toxicity attributed to vatalanib was hypertension. Given these results, the current role of vatalanib in the first-line treatment of patients with advanced colorectal cancer is unclear. CONFIRM-2, which is currently ongoing, is evaluating the potential survival benefit that results from the addition of vatalanib to the FOLFOX4 regimen in more than 800 patients whose disease progressed after first-line therapy consisting of 5-FU and irinotecan. The final results of this study are expected in 2006.

23.4.2 Newer Agents

The success of bevacizumab served as a proof of principle that therapies directed against VEGF and its receptor may work against colorectal cancer, generating great excitement toward the development of newer antiangiogenic agents. Besides bevacizumab and vatalanib, several other promising agents have entered in clinical trials. Most are tyrosine kinase inhibitors such as SU11248, ZD6474, ZD2171, AMG 706, and several others. However, other ways of blocking angiogenesis such as antibodies directed against the VEGF receptors as well as a VEGF-trap have been proposed and are already in clinical trials.

Besides therapies directed against VEGF and VEGF receptors, several other angiogenic factors have been the target of novel agents. Interestingly, even therapies not commonly thought to be antiangiogenic may actually exert some of its effects by blocking the release of proangiogenic factors. For example, the chimeric monoclonal antibody cetuximab, which is directed against the epidermal growth factor receptor, has been found to have an effect in the production of VEGF and IL-8, known angiogenic factors. Certainly, many more antiangiogenic agents will find their way into clinical testing against colorectal cancer.

23.5 CONCLUSION

Over the past decade, colorectal cancer has evolved from a relatively chemoresistant disease, for which only one modestly effective chemotherapeutic agent (5-FU) was available, to one that is amenable to several sequential, more effective systemic therapies. Furthermore, colorectal cancer has served as a platform for the development of novel, molecularly targeted agents. Among the targeted agents that have been pursued thus far, the greatest success has been seen for the antiangiogenic antibody bevacizumab. The clinical trials of bevacizumab have provided further proof of the hypothesis that antiangiogenic therapy should be a "form of cancer therapy worthy of serious exploration."[78] This hypothesis, put forward by Dr. Folkman three decades ago, continues to serve as a framework for the development of novel agents and combinations for the treatment of patients with colorectal cancer and other malignancies. Antiangiogenic cancer therapy is still in its early phases, and much work remains to be done to refine its efficacy, minimize its toxicity, and understand the long-term consequences of its use. Theoretically, the blocking of angiogenesis can also be of benefit in earlier phases in the natural history of solid tumors. However, given the efficacy of bevacizumab and the promising activity of other agents in the advanced setting, it is likely that antiangiogenic agents will be a component of various standard regimens for the systemic therapy of advanced neoplasms in the next several years.

REFERENCES

1. Stewart BJ, Kleihues, P. World cancer report. Lyon: IARC Press, 2003.
2. Fearon ER, Vogelstein B. A genetic model for colorectal tumorigenesis. *Cell* 1990; 61:759–67.
3. Yeazel MW, Church TR, Jones RM, et al. Colorectal cancer screening adherence in a general population. *Cancer Epidemiol Biomarkers Prev* 2004; 13:654–7.
4. Jemal A, Tiwari RC, Murray T, et al. Cancer statistics, 2004. *CA Cancer J Clin* 2004; 54:8–29.
5. Jemal A, Murray T, Ward E, et al. Cancer statistics, 2005. *CA Cancer J Clin* 2005; 55:10–30.
6. Simmonds PC. Palliative chemotherapy for advanced colorectal cancer: systematic review and meta-analysis. Colorectal Cancer Collaborative Group. *BMJ* 2000; 321:531–5.
7. Jonker DJ, Maroun JA, Kocha W. Survival benefit of chemotherapy in metastatic colorectal cancer: a meta-analysis of randomized controlled trials. *Br J Cancer* 2000; 82:1789–94.
8. Douillard JY, Cunningham D, Roth AD, et al. Irinotecan combined with fluorouracil compared with fluorouracil alone as first-line treatment for metastatic colorectal cancer: a multicentre randomised trial. *Lancet* 2000; 355:1041–7.
9. Saltz LB, Cox JV, Blanke C, et al. Irinotecan plus fluorouracil and leucovorin for metastatic colorectal cancer. Irinotecan Study Group. *N Engl J Med* 2000; 343:905–14.
10. de Gramont A, Figer A, Seymour M, et al. Leucovorin and fluorouracil with or without oxaliplatin as first-line treatment in advanced colorectal cancer. *J Clin Oncol* 2000; 18:2938–47.

11. Goldberg RM, Sargent DJ, Morton RF, et al. A randomized controlled trial of fluorouracil plus leucovorin, irinotecan, and oxaliplatin combinations in patients with previously untreated metastatic colorectal cancer. *J Clin Oncol* 2004; 22:23–30.
12. Expectancy or primary chemotherapy in patients with advanced asymptomatic colorectal cancer: a randomized trial. Nordic Gastrointestinal Tumor Adjuvant Therapy Group. *J Clin Oncol* 1992; 10:904–11.
13. Modulation of fluorouracil by leucovorin in patients with advanced colorectal cancer: evidence in terms of response rate. Advanced Colorectal Cancer Meta-Analysis Project. *J Clin Oncol* 1992; 10:896–903.
14. Thirion P, Michiels S, Pignon JP, et al. Modulation of fluorouracil by leucovorin in patients with advanced colorectal cancer: an updated meta-analysis. *J Clin Oncol* 2004; 22:3766–75.
15. Efficacy of intravenous continuous infusion of fluorouracil compared with bolus administration in advanced colorectal cancer. Meta-Analysis Group in Cancer. *J Clin Oncol* 1998; 16:301–8.
16. Petrelli N, Herrera L, Rustum Y, et al. A prospective randomized trial of 5-fluorouracil versus 5-fluorouracil and high-dose leucovorin versus 5-fluorouracil and methotrexate in previously untreated patients with advanced colorectal carcinoma. *J Clin Oncol* 1987; 5:1559–65.
17. Poon MA, O'Connell MJ, Moertel CG, et al. Biochemical modulation of fluorouracil: evidence of significant improvement of survival and quality of life in patients with advanced colorectal carcinoma. *J Clin Oncol* 1989; 7:1407–18.
18. de Gramont A, Bosset JF, Milan C, et al. Randomized trial comparing monthly low-dose leucovorin and fluorouracil bolus with bimonthly high-dose leucovorin and fluorouracil bolus plus continuous infusion for advanced colorectal cancer: a French intergroup study. *J Clin Oncol* 1997; 15:808–15.
19. Sinha BK. Topoisomerase inhibitors. A review of their therapeutic potential in cancer. *Drugs* 1995; 49:11–9.
20. Cunningham D, Pyrhonen S, James RD, et al. Randomised trial of irinotecan plus supportive care versus supportive care alone after fluorouracil failure for patients with metastatic colorectal cancer. *Lancet* 1998; 352:1413–8.
21. Culy CR, Clemett D, Wiseman LR. Oxaliplatin. A review of its pharmacological properties and clinical efficacy in metastatic colorectal cancer and its potential in other malignancies. *Drugs* 2000; 60:895–924.
22. deBraud F, Munzone E, Nole F, et al. Synergistic activity of oxaliplatin and 5-fluorouracil in patients with metastatic colorectal cancer with progressive disease while on or after 5-fluorouracil. *Am J Clin Oncol* 1998; 21:279–83.
23. Rothenberg ML, Oza AM, Bigelow RH, et al. Superiority of oxaliplatin and fluorouracil-leucovorin compared with either therapy alone in patients with progressive colorectal cancer after irinotecan and fluorouracil–leucovorin: interim results of a phase III trial. *J Clin Oncol* 2003; 21:2059–69.
24. Kemeny N, Garay CA, Gurtler J, et al. Randomized multicenter phase II trial of bolus plus infusional fluorouracil/leucovorin compared with fluorouracil/leucovorin plus oxaliplatin as third-line treatment of patients with advanced colorectal cancer. *J Clin Oncol* 2004; 22:4753–61.
25. Dooley M, Goa KL. Capecitabine. *Drugs* 1999; 58:69–76; discussion 77–8.
26. Schuller J, Cassidy J, Dumont E, et al. Preferential activation of capecitabine in tumor following oral administration to colorectal cancer patients. *Cancer Chemother Pharmacol* 2000; 45:291–7.
27. Hoff PM, Ansari R, Batist G, et al. Comparison of oral capecitabine versus intravenous fluorouracil plus leucovorin as first-line treatment in 605 patients with metastatic colorectal cancer: results of a randomized phase III study. *J Clin Oncol* 2001; 19:2282–92.
28. Van Cutsem E, Hoff PM, Harper P, et al. Oral capecitabine vs intravenous 5-fluorouracil and leucovorin: integrated efficacy data and novel analyses from two large, randomised, phase III trials. *Br J Cancer* 2004; 90:1190–7.
29. Ferrara N, Hillan KJ, Gerber HP, Novotny W. Discovery and development of bevacizumab, an anti-VEGF antibody for treating cancer. *Nat Rev Drug Discov* 2004; 3:391–400.
30. Tournigand C, Andre T, Achille E, et al. FOLFIRI followed by FOLFOX6 or the reverse sequence in advanced colorectal cancer: a randomized GERCOR study. *J Clin Oncol* 2004; 22:229–37.

31. Grothey A, Sargent D, Goldberg RM, Schmoll HJ. Survival of patients with advanced colorectal cancer improves with the availability of fluorouracil–leucovorin, irinotecan, and oxaliplatin in the course of treatment. *J Clin Oncol* 2004; 22:1209–14.
32. Saltz LB, Meropol NJ, Loehrer PJ, Sr., Needle MN, Kopit J, Mayer RJ. Phase II trial of cetuximab in patients with refractory colorectal cancer that expresses the epidermal growth factor receptor. *J Clin Oncol* 2004; 22:1201–8.
33. Cunningham D, Humblet Y, Siena S, et al. Cetuximab monotherapy and cetuximab plus irinotecan in irinotecan-refractory metastatic colorectal cancer. *N Engl J Med* 2004; 351:337–45.
34. Leonard GD, Brenner B, Kemeny NE. Neoadjuvant chemotherapy before liver resection for patients with unresectable liver metastases from colorectal carcinoma. *J Clin Oncol* 2005; 23:2038–48.
35. Bossi P, Viale G, Lee AK, Alfano R, Coggi G, Bosari S. Angiogenesis in colorectal tumors: microvessel quantitation in adenomas and carcinomas with clinicopathological correlations. *Cancer Res* 1995; 55:5049–53.
36. Tanigawa N, Amaya H, Matsumura M, et al. Tumor angiogenesis and mode of metastasis in patients with colorectal cancer. *Cancer Res* 1997; 57:1043–6.
37. Takahashi Y, Tucker SL, Kitadai Y, et al. Vessel counts and expression of vascular endothelial growth factor as prognostic factors in node-negative colon cancer. *Arch Surg* 1997; 132:541–6.
38. Vermeulen PB, Van den Eynden GG, Huget P, et al. Prospective study of intratumoral microvessel density, p53 expression and survival in colorectal cancer. *Br J Cancer* 1999; 79:316–22.
39. Brown LF, Berse B, Jackman RW, et al. Expression of vascular permeability factor (vascular endothelial growth factor) and its receptors in adenocarcinomas of the gastrointestinal tract. *Cancer Res* 1993; 53:4727–35.
40. Takahashi Y, Kitadai Y, Bucana CD, Cleary KR, Ellis LM. Expression of vascular endothelial growth factor and its receptor, KDR, correlates with vascularity, metastasis, and proliferation of human colon cancer. *Cancer Res* 1995; 55:3964–8.
41. Warren RS, Yuan H, Matli MR, Gillett NA, Ferrara N. Regulation by vascular endothelial growth factor of human colon cancer tumorigenesis in a mouse model of experimental liver metastasis. *J Clin Invest* 1995; 95:1789–97.
42. New BA, Yeoman LC. Identification of basic fibroblast growth factor sensitivity and receptor and ligand expression in human colon tumor cell lines. *J Cell Physiol* 1992; 150:320–6.
43. Anzano MA, Rieman D, Prichett W, Bowen-Pope DF, Greig R. Growth factor production by human colon carcinoma cell lines. *Cancer Res* 1989; 49:2898–904.
44. Craven RJ, Xu LH, Weiner TM, et al. Receptor tyrosine kinases expressed in metastatic colon cancer. *Int J Cancer* 1995; 60:791–7.
45. Takebayashi Y, Akiyama S, Akiba S, et al. Clinicopathologic and prognostic significance of an angiogenic factor, thymidine phosphorylase, in human colorectal carcinoma. *J Natl Cancer Inst* 1996; 88:1110–7.
46. Takahashi Y, Bucana CD, Liu W, et al. Platelet-derived endothelial cell growth factor in human colon cancer angiogenesis: role of infiltrating cells. *J Natl Cancer Inst* 1996; 88:1146–51.
47. Ahmad SA, Liu W, Jung YD, et al. Differential expression of angiopoietin-1 and angiopoietin-2 in colon carcinoma. A possible mechanism for the initiation of angiogenesis. *Cancer* 2001; 92:1138–43.
48. Brew R, Southern SA, Flanagan BF, McDicken IW, Christmas SE. Detection of interleukin-8 mRNA and protein in human colorectal carcinoma cells. *Eur J Cancer* 1996; 32A:2142–7.
49. Kuniyasu H, Ohmori H, Sasaki T, et al. Production of interleukin 15 by human colon cancer cells is associated with induction of mucosal hyperplasia, angiogenesis, and metastasis. *Clin Cancer Res* 2003; 9:4802–10.
50. Tsujii M, Kawano S, Tsuji S, Sawaoka H, Hori M, DuBois RN. Cyclooxygenase regulates angiogenesis induced by colon cancer cells. *Cell* 1998; 93:705–16.
51. Ciardiello F, Bianco R, Damiano V, et al. Antiangiogenic and antitumor activity of anti-epidermal growth factor receptor C225 monoclonal antibody in combination with vascular endothelial growth factor antisense oligonucleotide in human GEO colon cancer cells. *Clin Cancer Res* 2000; 6:3739–47.
52. Ciardiello F, Caputo R, Bianco R, et al. Inhibition of growth factor production and angiogenesis in human cancer cells by ZD1839 (Iressa), a selective epidermal growth factor receptor tyrosine kinase inhibitor. *Clin Cancer Res* 2001; 7:1459–65.

53. Hicklin DJ, Ellis LM. Role of the vascular endothelial growth factor pathway in tumor growth and angiogenesis. *J Clin Oncol* 2005; 23:1011–27.
54. Hoff PM. Future directions in the use of antiangiogenic agents in patients with colorectal cancer. *Semin Oncol* 2004; 31:17–21.
55. Kim KJ, Li B, Winer J, et al. Inhibition of vascular endothelial growth factor-induced angiogenesis suppresses tumour growth in vivo. *Nature* 1993; 362:841–4.
56. Presta LG, Chen H, O'Connor SJ, et al. Humanization of an anti-vascular endothelial growth factor monoclonal antibody for the therapy of solid tumors and other disorders. *Cancer Res* 1997; 57:4593–9.
57. Willett CG, Boucher Y, di Tomaso E, et al. Direct evidence that the VEGF-specific antibody bevacizumab has antivascular effects in human rectal cancer. *Nat Med* 2004; 10:145–7.
58. Jain RK. Normalizing tumor vasculature with anti-angiogenic therapy: a new paradigm for combination therapy. *Nat Med* 2001; 7:987–9.
59. Gordon MS, Margolin K, Talpaz M, et al. Phase I safety and pharmacokinetic study of recombinant human anti-vascular endothelial growth factor in patients with advanced cancer. *J Clin Oncol* 2001; 19:843–50.
60. Margolin K, Gordon MS, Holmgren E, et al. Phase Ib trial of intravenous recombinant humanized monoclonal antibody to vascular endothelial growth factor in combination with chemotherapy in patients with advanced cancer: pharmacologic and long-term safety data. *J Clin Oncol* 2001; 19:851–6.
61. Kabbinavar F, Hurwitz HI, Fehrenbacher L, et al. Phase II, randomized trial comparing bevacizumab plus fluorouracil (FU)/leucovorin (LV) with FU/LV alone in patients with metastatic colorectal cancer. *J Clin Oncol* 2003; 21:60–5.
62. Kabbinavar FF, Schulz J, McCleod M, et al. Addition of bevacizumab to bolus fluorouracil and leucovorin in first-line metastatic colorectal cancer: results of a randomized phase II trial. *J Clin Oncol* 2005; 23:3697–705.
63. Hurwitz H, Fehrenbacher L, Novotny W, et al. Bevacizumab plus irinotecan, fluorouracil, and leucovorin for metastatic colorectal cancer. *N Engl J Med* 2004; 350:2335–42.
64. Hurwitz HI, Fehrenbacher L, Hainsworth JD, et al. Bevacizumab in combination with fluorouracil and leucovorin: an active regimen for first-line metastatic colorectal cancer. *J Clin Oncol* 2005; 23:3502–8.
65. Mass RD, Sarkar S, Holden SN, Hurwitz H. Clinical benefit from bevacizumab (BV) in responding (R) and non-responding (NR) patients (pts) with metastatic colorectal cancer (mCRC). *J Clin Oncol* 2005; 23:249s (Supplement, abstract 3514).
66. Kabbinavar FF, Hambleton J, Mass RD, Hurwitz HI, Bergsland E, Sarkar S. Combined analysis of efficacy: the addition of bevacizumab to fluorouracil/leucovorin improves survival for patients with metastatic colorectal cancer. *J Clin Oncol* 2005; 23:3706–12.
67. Giantonio BJ, Catalano, PJ, Meropol, NJ, et al. High-dose bevacizumab improves survival when combined with FOLFOX4 in previously treated advanced colorectal cancer: results from the Eastern Cooperative Oncology Group (ECOG) study 3200. *J Clin Oncol* 2005; 23:1s (Supplement, abstract 2).
68. Saltz LB, Lenz, H-J, Hochster, H, et al. Randomized phase II trial of cetuximab/bevacizumab/irinotecan (CBI) versus cetuximab/bevacizumab (CB) in irinotecan-refractory colorectal cancer. *J Clin Oncol* 2005; 23:248s (Supplement, abstract 3508).
69. Fernando N, Yu D, Morse M, et al. A phase II study of oxaliplatin, capecitabine and bevacizumab in the treatment of metastatic colorectal cancer. *J Clin Oncol* 2005; 23:260s (Supplement, abstract 3556).
70. Andre T, Boni C, Mounedji-Boudiaf L, et al. Oxaliplatin, fluorouracil, and leucovorin as adjuvant treatment for colon cancer. *N Engl J Med* 2004; 350:2343–51.
71. Wolmark N, Wieand S, Kuebler JP, et al. A phase III trial comparing FULV to FULV + oxaliplatin in stage II or III carcinoma of the colon: results of NSABP Protocol C-07. *J Clin Oncol* 2005; 23:246s (Supplement, late breaking abstract 3500).
72. Wood JM, Bold G, Buchdunger E, et al. PTK787/ZK 222584, a novel and potent inhibitor of vascular endothelial growth factor receptor tyrosine kinases, impairs vascular endothelial growth factor-induced responses and tumor growth after oral administration. *Cancer Res* 2000; 60:2178–89.

73. Morgan B, Thomas AL, Drevs J, et al. Dynamic contrast-enhanced magnetic resonance imaging as a biomarker for the pharmacological response of PTK787/ZK 222584, an inhibitor of the vascular endothelial growth factor receptor tyrosine kinases, in patients with advanced colorectal cancer and liver metastases: results from two phase I studies. *J Clin Oncol* 2003; 21:3955–64.
74. Thomas AL, Morgan B, Drevs J, et al. Vascular endothelial growth factor receptor tyrosine kinase inhibitors: PTK787/ZK 222584. *Semin Oncol* 2003; 30:32–8.
75. Steward WP, Thomas A, Morgan B, et al. Expanded phase I/II study of PTK787/ZK 222584 (PTK/ZK), a novel, oral angiogenesis inhibitor, in combination with FOLFOX-4 as first-line treatment for patients with metastatic colorectal cancer. *J Clin Oncol* 2004; 22:259 (Supplement, abstract 3556).
76. Trarbach T, Schleucher N, Tewes M, et al. Phase I/II study of PTK787/ZK 222584 (PTK/ZK), a novel, oral angiogenesis inhibitor in combination with FOLFIRI as first-line treatment for patients with metastatic colorectal cancer (CRC). *J Clin Oncol* 2005; 23:272s (Supplement, abstract 3605).
77. Hecht JR, Trarbach T, Jaeger E, et al. A randomized, double-blind, placebo-controlled, phase III study in patients (Pts) with metastatic adenocarcinoma of the colon or rectum receiving first-line chemotherapy with oxaliplatin/5-fluorouracil/leucovorin and PTK787/ZK 222584 or placebo (CONFIRM-1). *J Clin Oncol* 2005; 23:2s (Supplement, late breaking abstract 3).
78. Folkman J. Tumor angiogenesis: therapeutic implications. *N Engl J Med* 1971; 285:1182–6.

24 Combined Modality Therapy of Rectal Cancer

Christopher G. Willett and Dan G. Duda

CONTENTS

In 2005, 40,340 new cases of rectal cancer were diagnosed in the United States [1]. The mainstay of therapy of this malignancy was surgery [2]. For patients with early stage tumors (lesions confined to the rectal wall without lymph node metastases), 5 year survival is excellent with recent series reporting 80% or greater cure rates [3]. In contrast, local and systemic failures pose significant challenges to patients undergoing potentially curative resection for more advanced staged tumors [4]. Treatment strategies of preoperative or postoperative chemotherapy and radiation therapy have been employed to prevent local and systemic failure and improve survival for these patients. Over the past 15 years, randomized trials have demonstrated statistically significant improvements in local control, freedom from distant metastases, and survival with radiation therapy and concurrent and maintenance 5-fluorouracil (5-FU)-based chemotherapy [5–12]. In a large Intergroup trial, the 7 year disease-free survival of 1695 patients with stage II and III rectal cancer undergoing resection and postoperative radiation therapy with concurrent and maintenance 5-FU-based chemotherapy was only 50% [11]. Despite the best contemporary adjuvant therapy, local recurrence and systemic failure remain important challenges, particularly in the treatment of patients with more advanced tumors. Innovative therapies should be pursued to improve on these outcomes.

24.1 5-FLUOROURACIL AND EXTERNAL BEAM RADIATION THERAPY

For patients with resected rectal cancers that have extended through the bowel wall or involve regional lymph nodes, external beam radiation therapy (EBRT) with concurrent and maintenance 5-FU prolongs disease-free survival and overall survival when administered as an adjuvant therapy [5–12]. In addition, the administration of 5-FU intravenous continuous infusion (225 mg/m^2/24 h) during the entire course of radiation therapy as opposed to bolus 5-FU at the beginning and end of radiation prolongs disease-free survival and overall survival in patients with Stage II or III rectal cancer [8]. Based on these and other studies, postoperative chemotherapy and radiation therapy is recommended for patients with resected but advanced stage rectal cancer.

In the past 10 years, there has been increasing interest in the use of preoperative radiation therapy and 5-FU-based chemotherapy for patients with clinical stage T3 and T4 or node-positive tumors [10,12]. Potential advantages to this approach (versus postoperative radiation therapy and chemotherapy) include higher rates of local control, improved tolerance, and enhanced sphincter preservation for patients with distal lesions. Surgery is typically performed 4–8 weeks following completion of radiation therapy and chemotherapy. Investigators from Germany recently reported the results of a randomized prospective phase III trial evaluating the efficacy of neoadjuvant radiation therapy and chemotherapy versus standard postoperative radiation therapy and chemotherapy [12]. In this study, 421 patients were randomly assigned to receive preoperative radiation therapy and chemotherapy and 402 patients to receive postoperative radiation therapy and chemotherapy. The overall 5 year survival rates were 76% and 74%, respectively. The 5 year cumulative incidence of local relapse was 6% for patients assigned to preoperative chemoradiotherapy and 13% in the postoperative-treatment group. Grade 3 or 4 acute toxic effects occurred in 27% of the patients in the preoperative-treatment group, as compared with 40% of the patients in the postoperative-treatment group; the corresponding rates of long-term toxic effects were 14% and 24%, respectively. The investigators conclude that preoperative chemoradiotherapy, as compared with postoperative chemoradiotherapy, improved local control and was associated with reduced toxicity but did not improve overall survival. Current and future clinical trials in rectal cancer are now employing preoperative strategies.

24.2 VASCULAR ENDOTHELIAL GROWTH FACTOR

Increased levels of vascular endothelial growth factor (VEGF) expression have been found in the tumors and sera of patients with localized as well as metastatic colon and rectal cancer [13–15]. High VEGF expression has been associated with disease progression and inferior survival. In a study of 79 consecutive patients with resected and adjuvantly treated (chemoradiation) node-positive rectal cancer, patients with tumors exhibiting VEGF overexpression were at statistically higher risk for the development of local recurrence and metastases [16]. Thus, inhibition of VEGF is a logical target in the treatment of patients with rectal and colon cancer.

A randomized phase II study (AVF0780g) first studied the efficacy and safety of anti-VEGF antibody, bevacizumab, combined with 5-FU and leucovorin in patients with metastatic colon and rectal cancer [17]. Patients were randomized to bevacizumab (5 or 10 mg/kg) and 5-FU and leucovorin or 5-FU and leucovorin only. Patients receiving bevacizumab and 5-FU and leucovorin had statistically improved response rates and time to disease progression compared with patients receiving 5-FU and leucovorin only. No additional toxicity was observed with the addition of bevacizumab to 5-FU and leucovorin chemotherapy.

Based on these data, a phase III trial was conducted, which randomized 813 patients with previously untreated metastatic colorectal cancer to receive irinotecan, 5-fluorouracil, and leucoverin (IFL) (402 patients) or to receive IFL plus placebo (411 patients) [18]. The median duration of survival was 20.3 months in the group given IFL plus bevacizumab as compared with 15.6 months in the group given IFL plus placebo ($p < 0.001$). The median duration of progression-free survival was 10.6 months in the group given IFL plus bevacizumab as compared with 6.2 months in the group given IFL plus placebo, $p < 0.0001$; the corresponding rates of response were 44.8% and 34.8% ($p = 0.004$). The median duration of response was 10.4 months in the group given IFL plus bevacizumab, as compared with 7.1 months in the group given IFL plus placebo. Grade 3 hypertension was more common during treatment with IFL plus bevacizumab than with IFL plus placebo (11% versus 2.3%) but was easily managed. The investigators concluded that the addition of bevacizumab to fluorouracil-based combination chemotherapy resulted in statistically significant and clinically meaningful improvement in survival among patients with metastatic colorectal cancer.

In another trial, previously treated patients (5-FU/irinotecan) with advanced metastatic colon and rectal cancer who received bevacizumab in combination with the proven second line therapy (an oxaliplatin regimen-FOLFOX4) had a survival benefit compared with patients who received FOLFOX4 alone [19]. Addition of bevacizumab increased the response rates (21.8% versus 9.2%, $p = 0.0001$), prolonged progression-free survival (7.2 months versus 4.8 months, $p < 0.0001$), and overall survival (12.9 months versus 10.8 months, $p < 0.001$). The common side effects for this regimen were occurrence of neuropathy, hypertension, bleeding, and bowel perforation. Whether combining bevacizumab with FOLFOX4 or other chemotherapy regimens are the best options for first line therapy for colorectal cancer is under investigation. Finally, a Treatment Referral Center (TRC) was established by NCI for patients with advanced metastatic colon and rectal cancer in the third line setting (who had disease progression after irinotecan- and oxaliplatin-based chemotherapy), where no standard treatment options were available. Bevacizumab was added to a 5-FU/LV regimen in these patients. While response data are not mature, bevacizumab does not seem to confer a survival advantage in this third line setting [20].

While VEGF inhibition has been demonstrated to be beneficial for patients with metastatic colon and rectal cancer and enhanced their survival by several months, the magnitude of the effect of anti-VEGF therapy in patients with localized and nonmetastatic disease is not known. This issue is under investigation in several phase I–II clinical trials that combine bevacizumab with cytotoxic regimens for localized rectal cancer. Recent experimental studies in human tumor xenografts as models of primary tumors have demonstrated that VEGF blockade serves as a potent and nontoxic enhancer of radiation therapy and reduces tumor interstitial pressure—a known barrier to drug delivery to tumors, and in some cases, reduces tumor hypoxia—a known barrier to radiation therapy [21–24]. The vascular normalization paradigm as proposed by Jain and colleagues in 2001 may explain these observations. In this paradigm, antiangiogenic agents transiently normalize the abnormal tumor vessels (formed as a result of excessive local production of angiogenic factors) [25,26]. With structural and functional remodeling of the tumor blood vessels, concentration of oxygen and penetration of macromolecules are improved and tumor's response to chemotherapy and radiation therapy is enhanced (See Chapter 3).

To test the hypothesis that inhibition of VEGF is safe and results in clinical benefit and enhancement of radiation therapy response, we initiated a neoadjuvant phase I/II trial with bevacizumab in combination with 5-FU and radiation therapy in patients with T3 or T4 rectal cancer [27,28]. Bevacizumab was delivered as a 90 min infusion on day 1 of each cycle (Figure 24.1). The dose was escalated in successive cohorts of six patients, beginning at 5 mg/kg followed by 10 mg/kg. Infusional 5-FU was administered over 24 h each day at a fixed dose of 225 mg/m^2 throughout each treatment week of cycles 2–4. External beam irradiation was administered during cycles 2–4 for a total dose of 50.4 Gy in 28 fractions over 5.5 weeks. The primary clinical objective of this study was to determine the maximum tolerated dose (MTD) of bevacizumab when delivered concurrently with 5-FU and

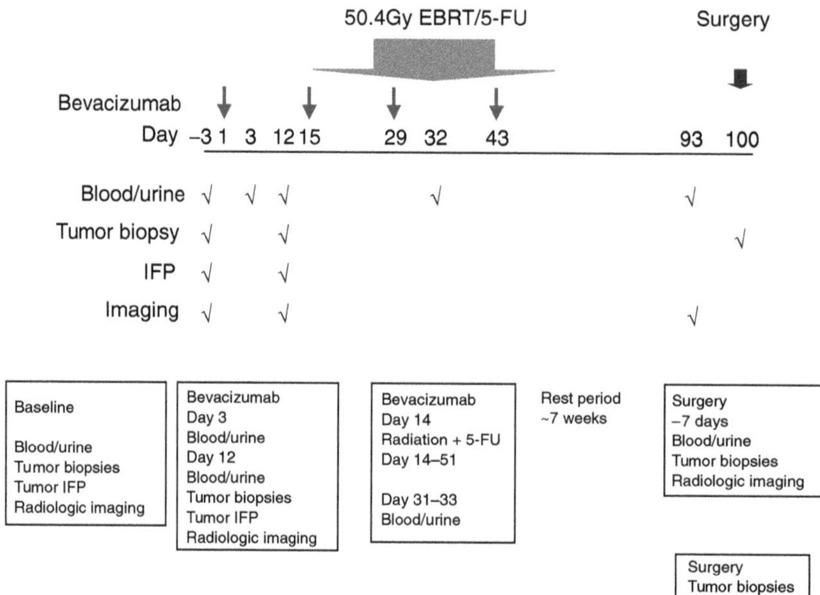

FIGURE 24.1 Schematic representation of the treatment schedule.

EBRT in patients with T3 or T4 rectal cancer before surgery. In parallel, a major goal of this study was to clarify through correlative studies, the mechanisms by which bevacizumab inhibits angiogenesis and improves the outcome of other therapeutic modalities in the treatment of this malignancy.

We have completed the phase I portion of the study [27,28]. The first six patients treated with the combination of bevacizumab at the 5 mg/kg dose level with chemotherapy and radiation therapy tolerated this treatment without difficulty. All six patients underwent surgery without complication. In contrast, two of five patients in the second cohort who were given "high-dose" bevacizumab (10 mg/kg) with chemotherapy and radiation therapy experienced Grade 3–4 dose-limiting toxicity (DLT) of diarrhea and colitis during the combined treatment. Following recovery from toxicity, these patients were able to resume and complete radiation therapy and 5-FU treatment. Because of these DLTs, only five patients were enrolled at the 10 mg/kg dose. All the patients underwent surgery. Of note, one patient on high-dose bevacizumab experienced a pulmonary embolus day 1 postoperatively and recovered completely with anticoagulation. Another patient in this cohort developed ileostomy obstruction with stent-related ileal perforation 10 days following resection requiring laparotomy and ileostomy revision.

At surgery, patients on the 5 mg/kg bevacizumab showed minimal residual disease, consistent with Manard Grade 3–4 regression. Of interest, pathologic evaluation of the surgical specimens for staging following completion of all therapy in the patients receiving 10 mg/kg bevacizumab showed two complete pathological responses when compared with incomplete pathological response in the 5 mg/kg bevacizumab group.

The study design of this trial permitted a unique opportunity to evaluate the effect of bevacizumab alone (cycle 1) on rectal cancer before its concurrent administration (cycles 2–4) with radiation therapy and chemotherapy. Correlative studies were undertaken to clarify the mechanism of action of bevacizumab on rectal cancer. Twelve days following the first bevacizumab infusion, patients underwent repeat flexible sigmoidoscopy with tumor biopsy,

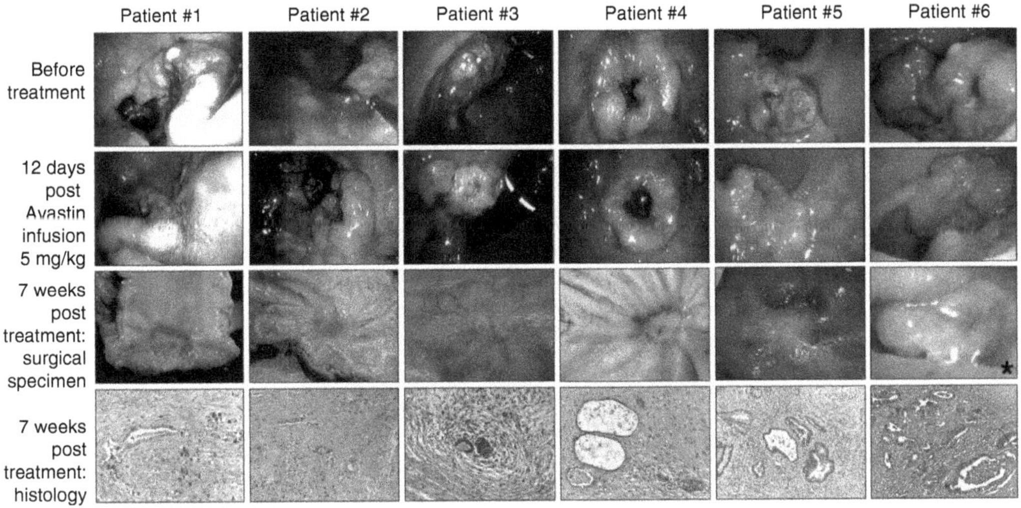

FIGURE 24.2 (See color insert following page 558.) Sigmoidoscopy surgical and histologic evaluation of rectal cancer patients after receiving bevacizumab (day 12) and then at surgery after bevacizumab with chemoradiation. (From Willet, C.G., Boucher, Y., di Tomaso, E., et al. *Nat Med.*, 10, 145–147, 2004. With permission.)

tumor interstitial pressure measurement, perfusion CT scan to measure blood flow, PET FDG scan, and analysis of blood and urine for a number of angiogenesis markers. At day 12, 1 of 11 patients exhibited a partial clinical response (Data from patients 1–6 is shown in Figure 24.2). At this early time point, a number of antivascular effects induced by bevacizumab were observed. Tumor interstitial pressure measurements were statistically lower following bevacizumab administration (Figure 24.3). These data are consistent with preclinical data and support the normalization hypothesis. Tumor vascular density measurements by immunohistochemistry and blood flow parameters by perfusion CT also dropped following bevacizumab infusion (Figure 24.4 through Figure 24.8). In contrast, FDG-uptake in the tumors measured on PET scans remained constant (Figures 24.6 and 24.7). Despite these decreases in tumor vascular density and blood flow, tumor metabolism as assessed by FDG activity was unchanged, thus supporting the normalization hypothesis. Further support of the normalization hypothesis has been the observation of increased alpha-smooth muscle actin (α-SMA)-positive pericyte coverage in tumor vessels following bevacizumab administration (Figure 24.8, right panel).

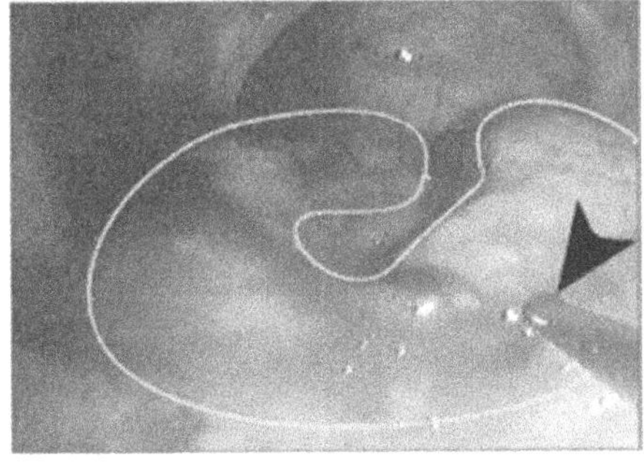

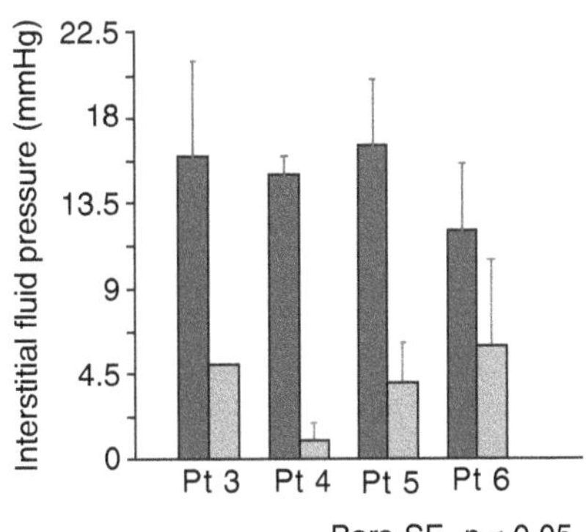

FIGURE 24.3 (See color insert following page 558.) Interstitial fluid pressure in rectal cancer after 1 BV infusion. (From Willet, C.G., Boucher, Y., di Tomaso, E., et al. *Nat Med.*, 10, 145–147, 2004. With permission.)

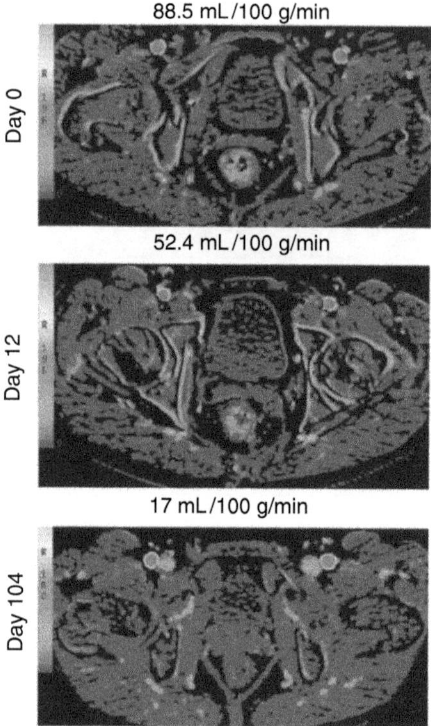

FIGURE 24.4 (See color insert following page 558.) Representative CT scans of rectal cancer after 1 BV infusion and presurgery. (From Willet, C.G., Boucher, Y., di Tomaso, E., et al. *Nat Med.*, 10, 145–147, 2004. With permission.)

24.3 CONCLUSION

A dose escalation phase I trial has demonstrated that the combination of bevacizumab at the 5 mg/kg dose with chemotherapy and radiation therapy was well tolerated by patients with rectal cancers. In addition, results from an array of correlative studies undertaken in the first 11 patients have shown the bevacizumab has antivascular effects and support

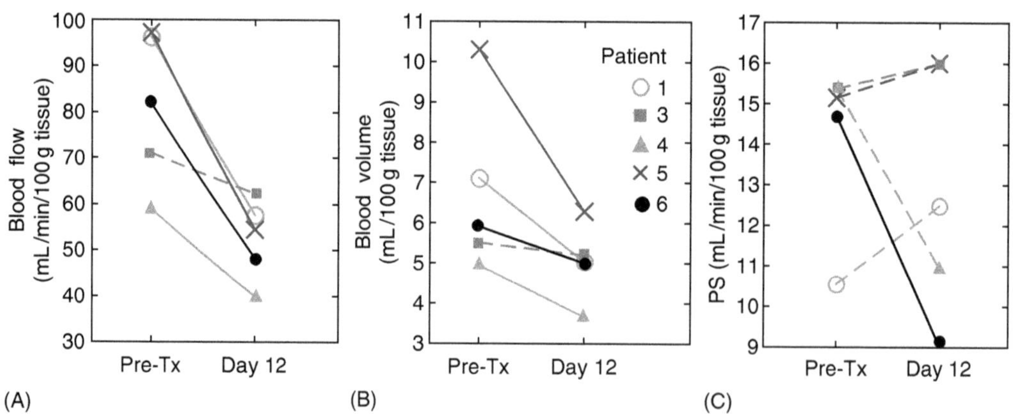

FIGURE 24.5 Functional vascular parameters calculated on CT scans. (From Willet, C.G., Boucher, Y., di Tomaso, E., et al. *Nat Med.*, 10, 145–147, 2004. With permission.)

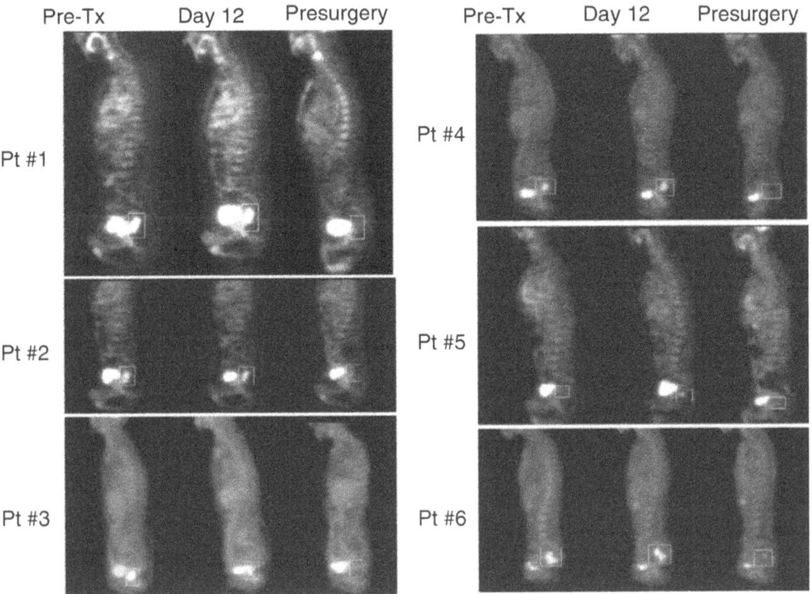

FIGURE 24.6 PET scans of FDG uptake in rectal cancer after 1 BV (5 mg/kg) infusion and at the end of the combined treatment. Tumor is outlined in box, posterior to bladder. (From Willet, C.G., Boucher, Y., di Tomaso, E., et al. *Nat Med.*, 10, 145–147, 2004. With permission.)

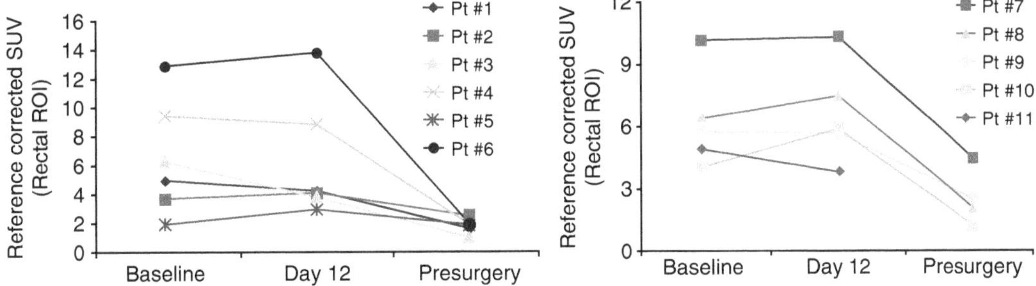

FIGURE 24.7 Tumor FDG uptake measured by PET in rectal cancer patients after 1 BV infusion and presurgery. (From Willet, C.G., Boucher, Y., di Tomaso, E., et al. *Nat Med.*, 10, 145–147, 2004. With permission.)

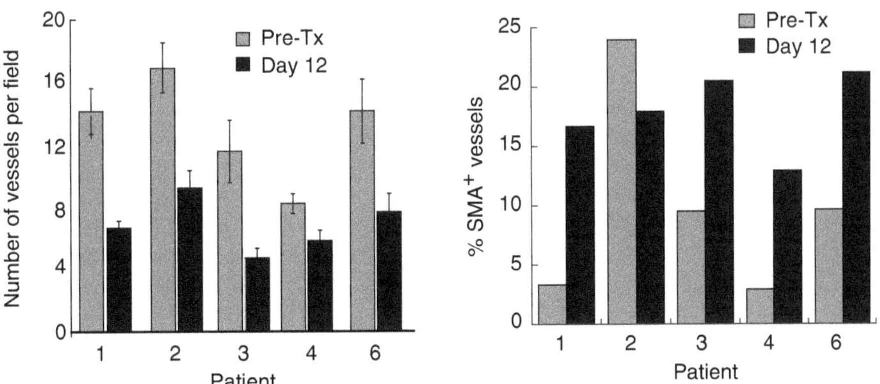

FIGURE 24.8 Tumor vascular parameters after 1 BV infusion quantified using immuohistochemistry in biopsy tissue sections. Left panel, microvascular density; right panel, pericyte coverage. (From Willet, C.G., Boucher, Y., di Tomaso, E., et al. *Nat Med.*, 10, 145–147, 2004. With permission.)

the normalization hypothesis. Continued accrual of patients in the phase II portion of the study will further elucidate the mechanisms of action and efficacy of bevacizumab in rectal cancer.

ACKNOWLEDGMENTS

This research was supported by funds from the National Cancer Institute R21 CA099237 to Christopher G. Willett, MD and P01 CA 80124 to Rakesh K. Jain, PhD, and the AACR to Dan G. Duda, DMD, PhD. We thank all Steele Laboratory collaborators—Drs. Jain, Boucher, di Tomaso, Kozin, Munn and Tong; we also thank our clinical collaborators, especially Drs. Blaszkowsky, Chung, Clark, Fishman, Lauwers, Ryan and Sahani.

REFERENCES

1. http://www.nci.gov/cancertopics/types/colon-and-rectal.
2. Kapiteijn, E., Marijnen, C.A.M., Nagtegaal, I.D., et al. Preoperative radiotherapy combined with total mesorectal excision for resectable rectal cancer. *New Engl J Med* 345:638–646 (2001).
3. Willett, C.G., Lewandrowski, K., Donnelly, S., et al. Are there patients with stage I rectal carcinoma at risk for failure after abdominoperineal resection? *Cancer* 69(7):1651–1655 (1992).
4. Rich, T., Gunderson, L.L., Lew, R., et al. Patterns of recurrence of rectal cancer after potentially curative surgery. *Cancer* 34:1317–1329 (1974).
5. Gastrointestinal Tumor Study Group. Prolongation of the disease-free interval in surgically treated rectal carcinoma. *New Engl J Med* 312(23):1465–1472 (1985).
6. Fisher, B., Wolmark, N., Rockette, H., et al. Postoperative adjuvant chemotherapy or radiation therapy for rectal cancer: Results from NSABP protocol R-01. *J Nat Cancer Inst* 80(1):21–29 (1988).
7. Gastrointestinal Tumor Study Group. Radiation therapy and fluorouracil with or without semustine for the treatment of patients with surgical adjuvant adenocarcinoma of the rectum. *J Clin Oncol* 10(4):549–558 (1992).
8. O'Connell, M.J., Martenson, J.A., Wieand, H.S., et al. Improving adjuvant therapy for rectal cancer by combining protracted-infusion fluorouracil with radiation therapy after curative surgery. *New Engl J Med* 331(8):502–507 (1994).
9. Hyams, D.M., Mamounas, E.P., Petrelli, N., et al. A clinical trial to evaluate the worth of preoperative multimodality therapy in patients with operable carcinoma of the rectum: a progress report of National Surgical Breast and Bowel Project Protocol R-03. *Dis Colon Rectum* 40(2):131–139 (1997).
10. Wolmark, N., Wieand, H.S., Hyams, D.M., et al. Randomized trial of postoperative adjuvant chemotherapy with or without radiotherapy for carcinoma of the rectum: National surgical adjuvant breast and bowel project protocol R-02. *J Natl Cancer Inst* 92(5):388–396 (2000).
11. Tepper, J.E., O'Connell, M.J., Niedzwiecki, D., et al. Final report of INT 0114—adjuvant therapy in rectal cancer. *ASCO Proceedings* 20:489 (2001).
12. Sauer, R., Becker, H., Hohenberger, W., et al. Preoperative versus postoperative chemoradiotherapy for rectal cancer. *New Engl J Med* 351:1731–1740 (2004).
13. Chin, K.F., Greenman, J., Gardiner, E., Kumar, H., Topping, K., and Monson, J. Preoperative serum vascular endothelial growth factor can select patients for adjuvant treatment after curative resection in colorectal cancer. *Br J Cancer* 83(11):1425–1431 (2000).
14. Hyodo, I., Doi, T., Endo, H., Hosokawa, Y., Nishikawa, Y., Tanimizu, M., Jinno, K., and Kotani, Y. Clinical significance of plasma vascular endothelial growth factor in gastrointestinal cancer. *Eur J Cancer* 34(13):2041–2045 (1998).
15. Nanashima, A., Ito, M., Sekine, I., Naito, S., Yamaguchi, H., Nakagoe, T., and Ayabe, H. Significance of angiogenic factors in liver metastatic tumors originating from colorectal cancers. *Dig Dis Sci* 43(12):2634–2640 (1998).

16. Cascinus, S., Graziano, F., Catalano, V., et al. Vascular Endothelial Growth Factor (VEGF), p53, and BAX Expression in node positive rectal cancer. *ASCO Proc* 20:595 (2001).
17. Kabbinavar, F., Hurwitz, H., Fehrenbacher, L., et al. Phase II trial comparing rhuMAB VERGF (recombinant humanized monocolonal antibody to vascular endothelial cell growth factor) plus 5-fluoroouracil/leucovorin (FU/LV) to FU/LV) alone in patients with metastatic colorectal cancer. *J Clin Oncol* 21:60–65 (2003).
18. Hurwitz, H., Fehrenbacher, L., Novotny, W., et al. Bevacizumab plus irinotecan, fluorouracil, and leucovorin for metastatic colorectal cancer. *New Engl J Med* 350:2335–2342 (2004).
19. Giatonio, B.J., Catalano, P., Maropol, N., et al. High dose Bevacizumab improves survival when combined with FOLFOX4 in previously treated advanced colorectal cancer: Results from the Eastern Cooperative Group (ECOG) study 3200. Proc ASCO: #2 (2005).
20. http://www.cancer.gov/newscenter/pressreleases/bevacizumab.
21. Lee, C.G., Heijn, M., di Tomaso, E., et al. Anti-vascular endothelial growth factor treatment augments tumor radiation response under normoxic or hypoxic conditions. *Cancer Res* 60:5565–5570 (2000).
22. Kozin, S.V., Boucher, Y., Hicklin, D.J., Bohlen, P., Jain, R.K., and Suit, H.D. Vascular endothelial growth factor receptor-2-blocking antibody potentiates radiation-induced long-term control of human tumor xenograts. *Cancer Res* 61:39–44 (2001).
23. Winkler, F., Kozin, S., Tong, R., Chae, S., Booth, M., Garkavstev, I., Xu, L., Hicklin, D.J., Fukumura, D., di Tomaso, E., Munn, L.L., and Jain, R.K., Kinetics of vascular normalization by VEGFR2 blockade governs brain tumor response to radiation: Role of oxygenation, Angiopoietin-1, and matrix metallproteinases, *Cancer Cell* 6:553–562 (2004).
24. Tong, R.T., Boucher, Y., Kozin, S.V., Winkler, F., Hicklin, D.J., and Jain, R.K., Vascular normalization by VEGFR2 blockade induces a pressure gradient across the vasculature and improves drug penetration in tumors, *Cancer Res* 64:3731–3736 (2004).
25. Jain, R.K. Normalizing tumor vasculature with anti-angiogenic therapy: A new paradigm for combination therapy. *Nat Med* 7:987–989 (2001).
26. Jain, R.K. Normalization of tumor vasculature: An emerging concept in anti-angiogenic therapy. *Science* 307:58–62 (2005).
27. Willett, C.G., Boucher, Y., di Tomaso, E., et al. Direct evidence that the VEGF-specific antibody Bevacizumab has antivascular effects in human rectal cancer. *Nat Med* 10:145–147 (2004).
28. Willett, C.G., Boucher, Y., Duda, D.G., et al. Surrogate markers for antiangiogenic therapy and dose limiting toxicities for Bevacizumab with radio-chemotherapy: Continued experience of a Phase I trial in rectal cancer patients. *J Clin Oncol* 23:8136–8139 (2005).

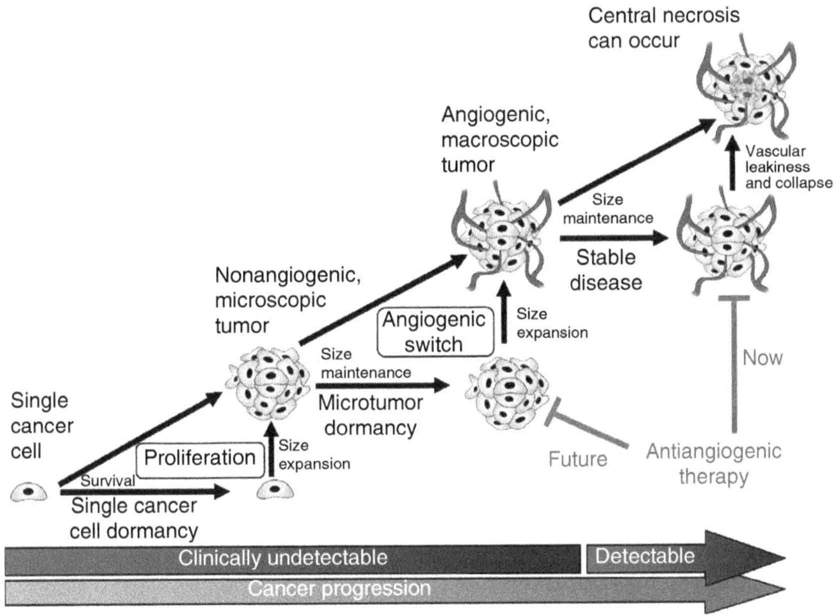

FIGURE 1.1

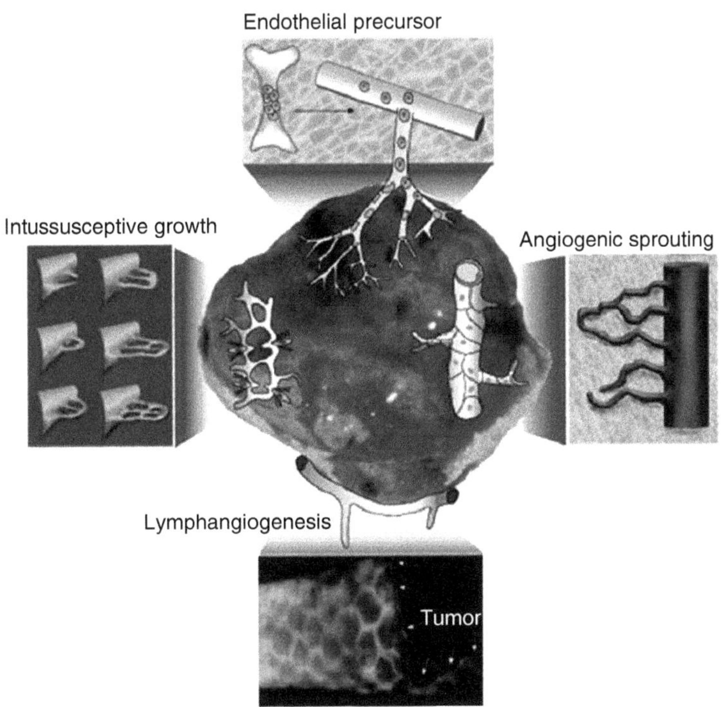

FIGURE 3.2

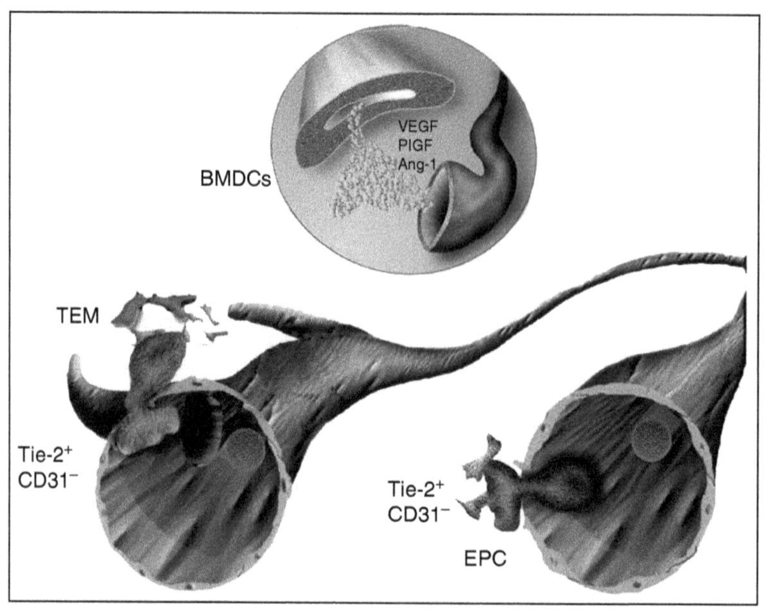

FIGURE 3.3

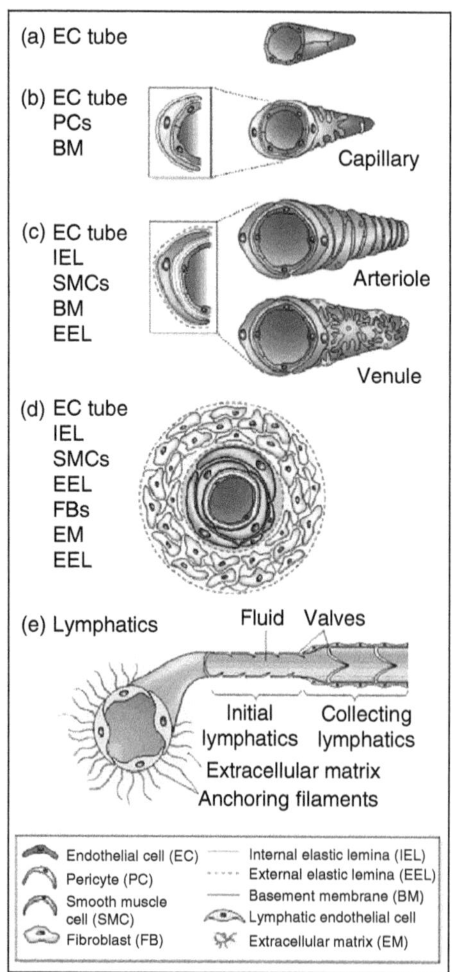

FIGURE 3.4

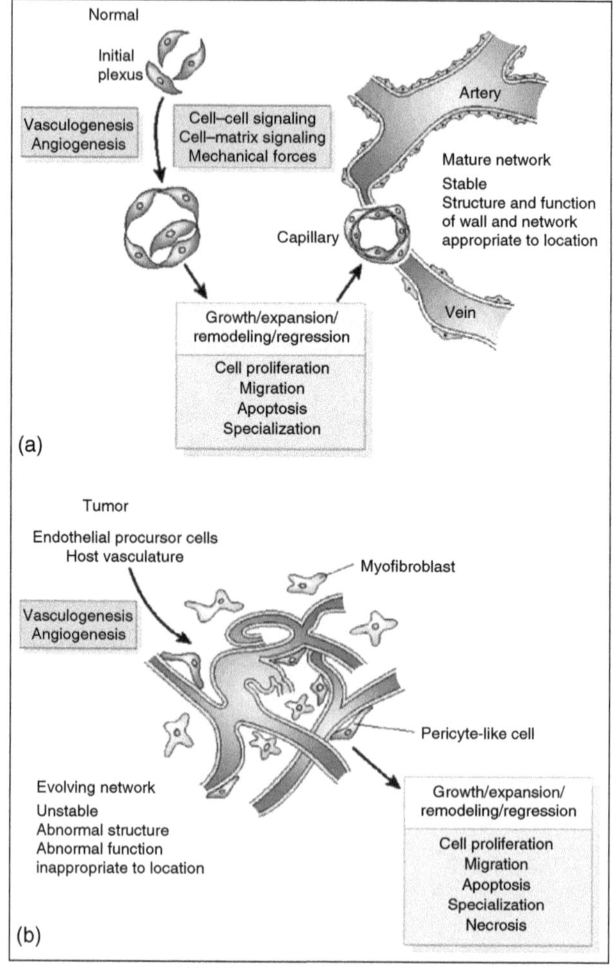

FIGURE 3.5

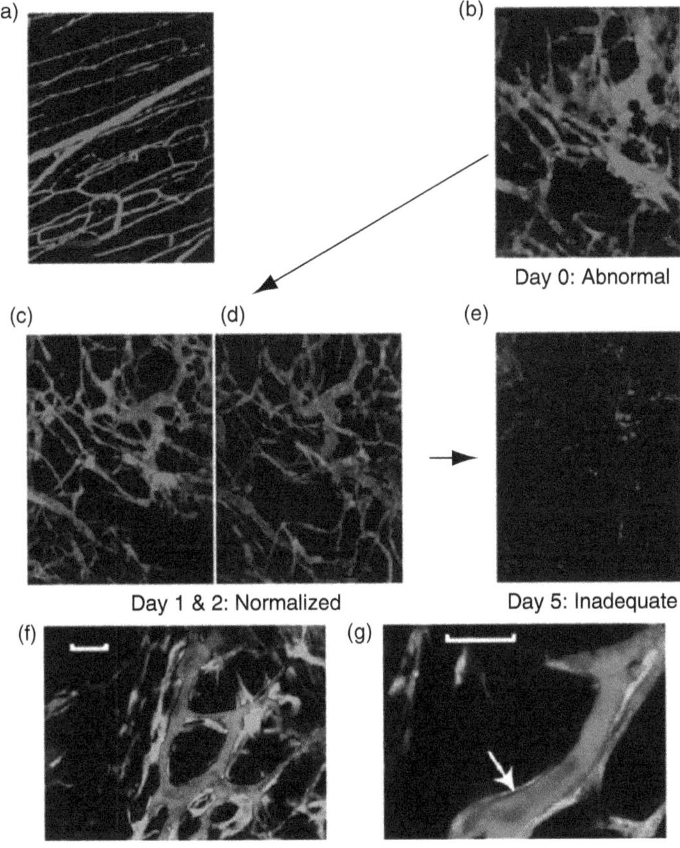

FIGURE 3.6

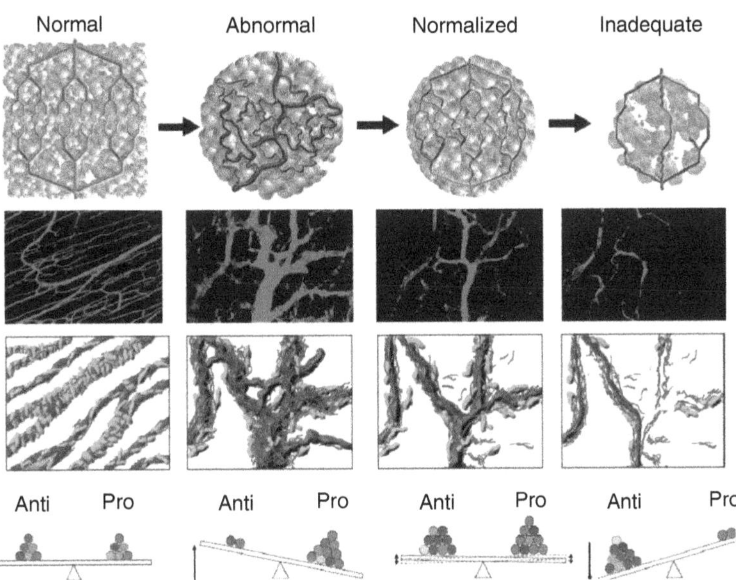

FIGURE 3.9

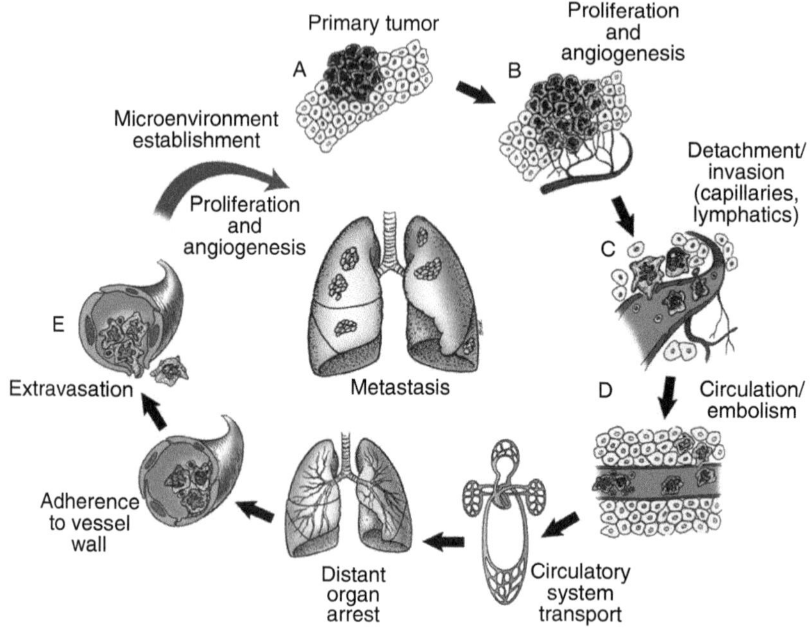

FIGURE 5.1

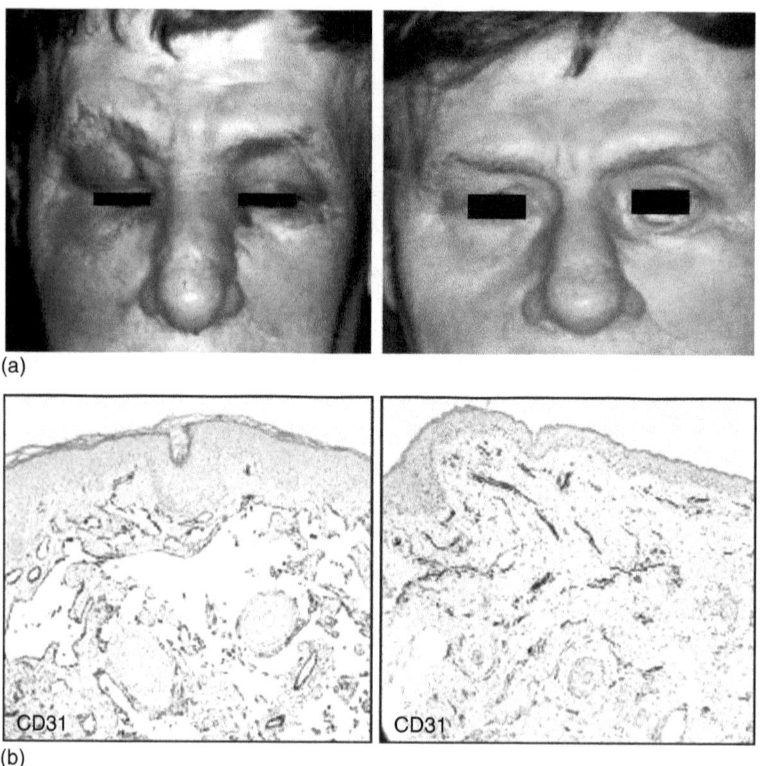

FIGURE 13.7

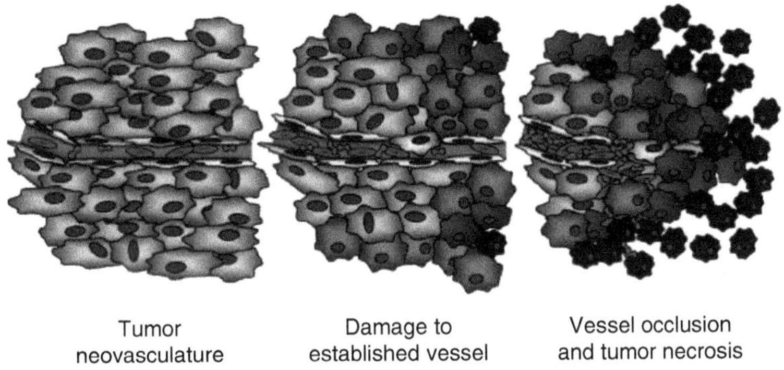

FIGURE 14.1

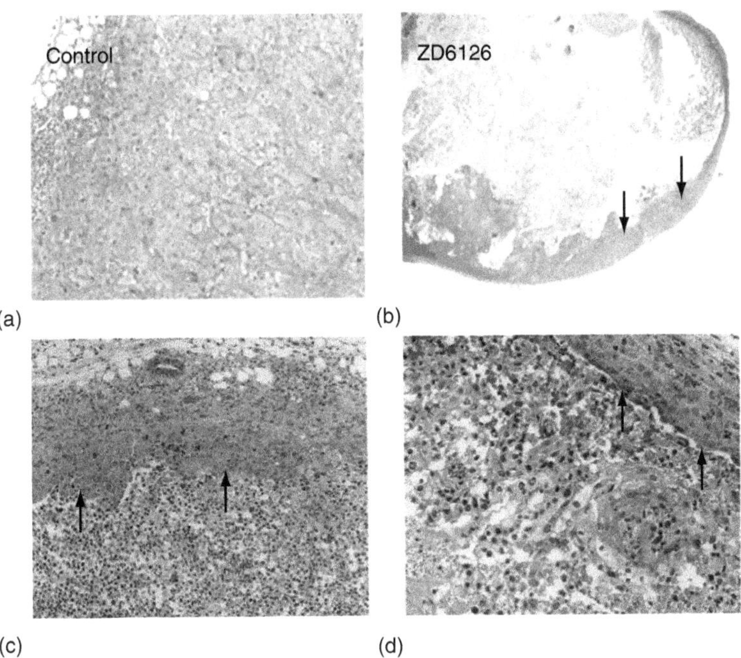

FIGURE 14.6

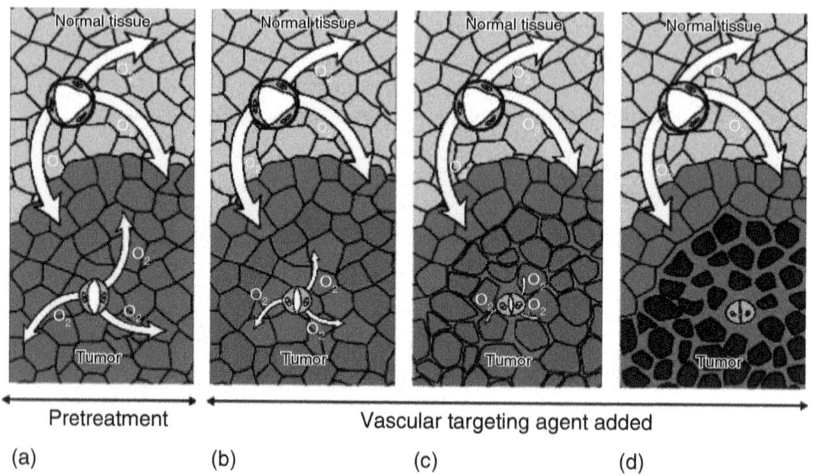

FIGURE 14.8

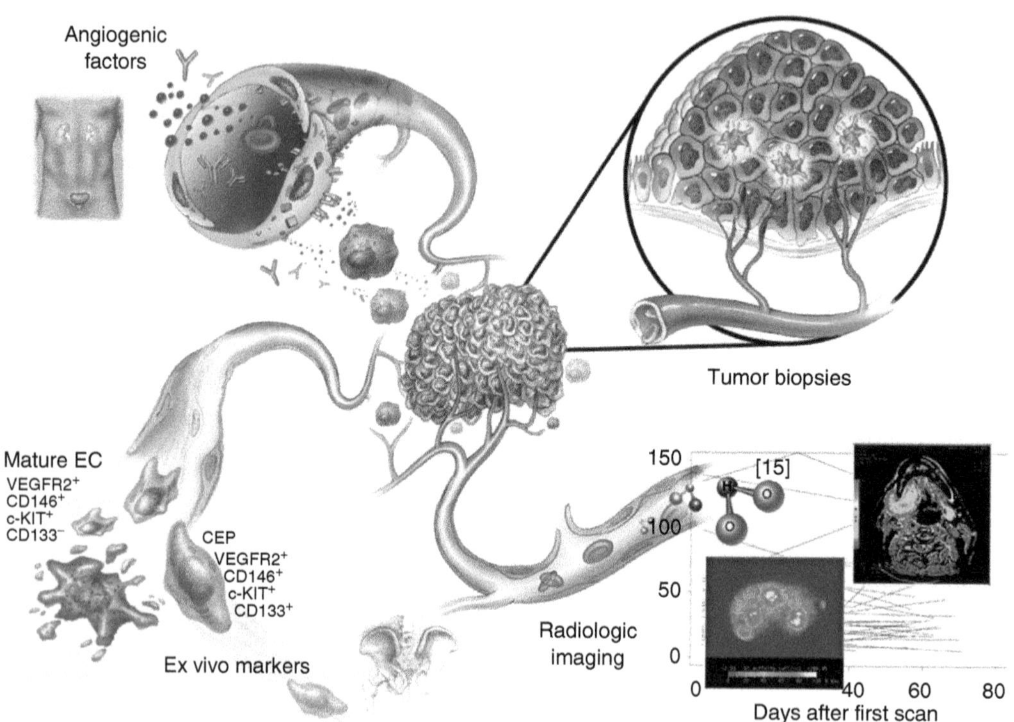

FIGURE 19.1

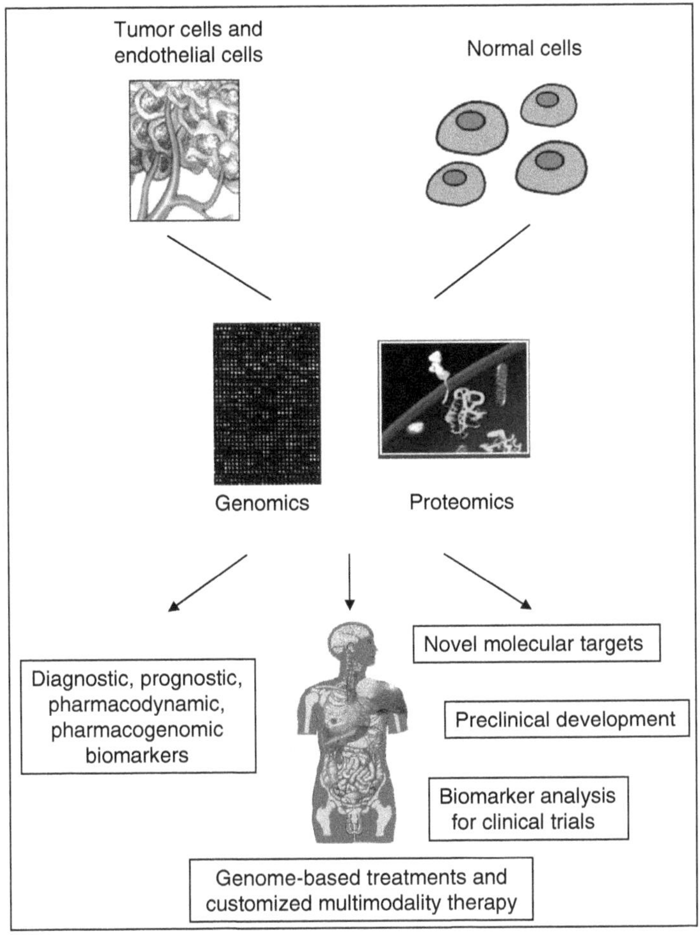

FIGURE 19.2

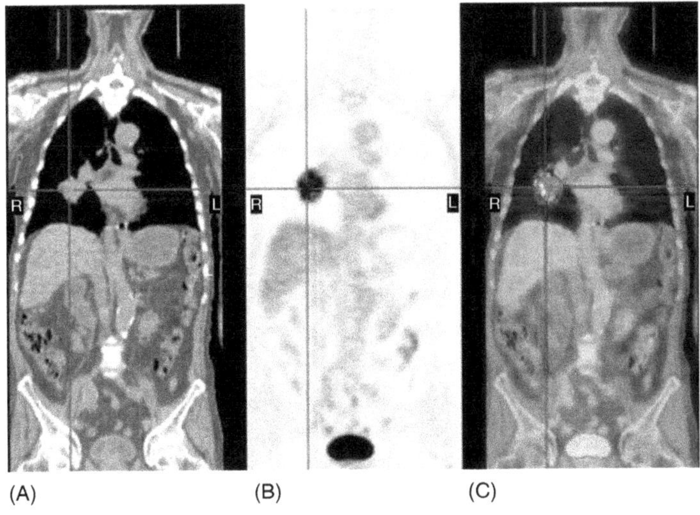

FIGURE 20.7

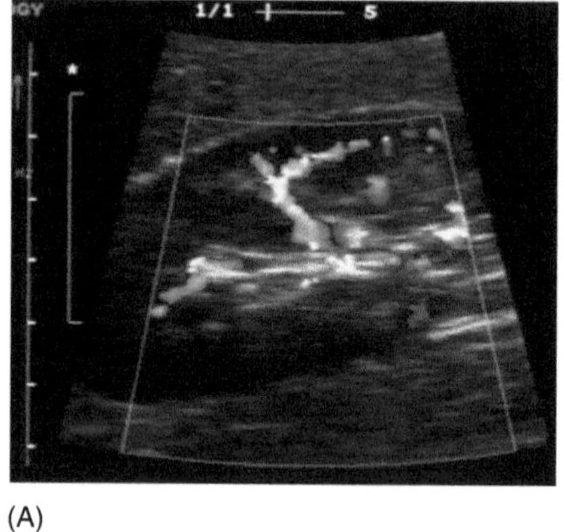

(A)

FIGURE 20.8A

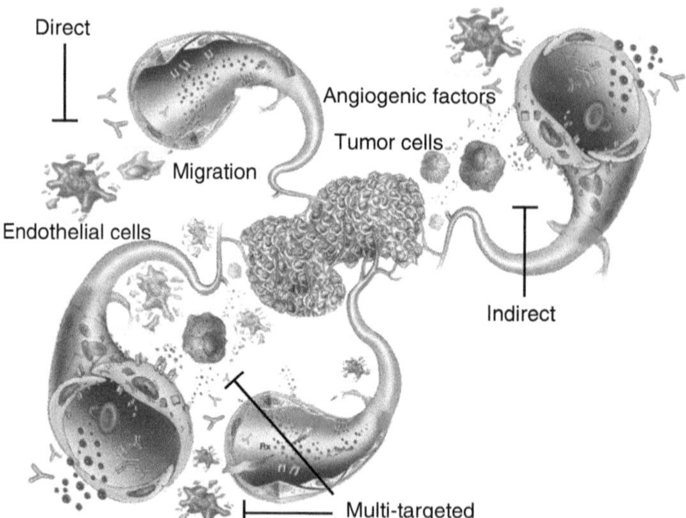

FIGURE 21.1

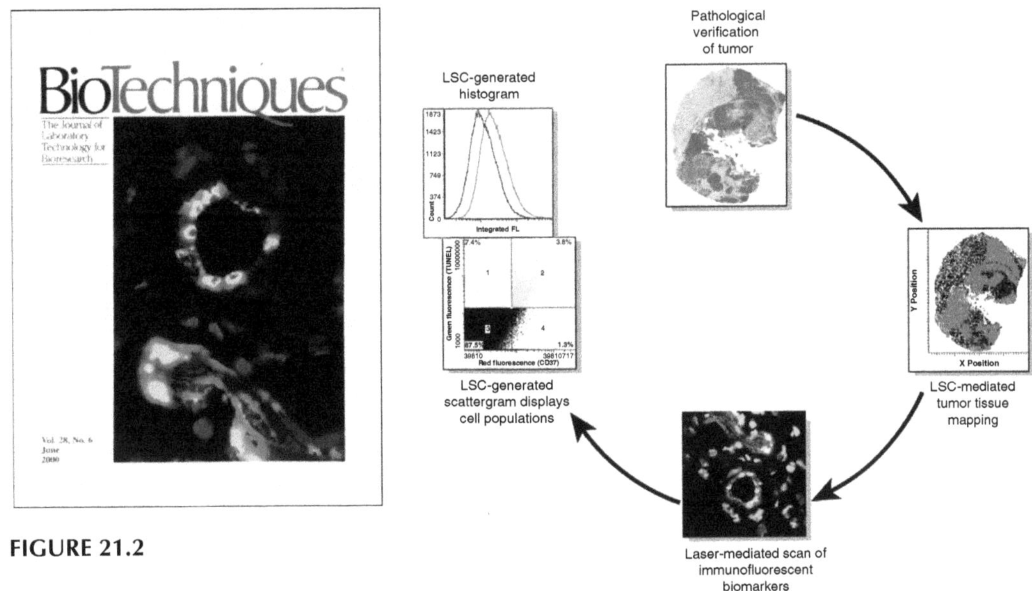

FIGURE 21.2

FIGURE 21.3

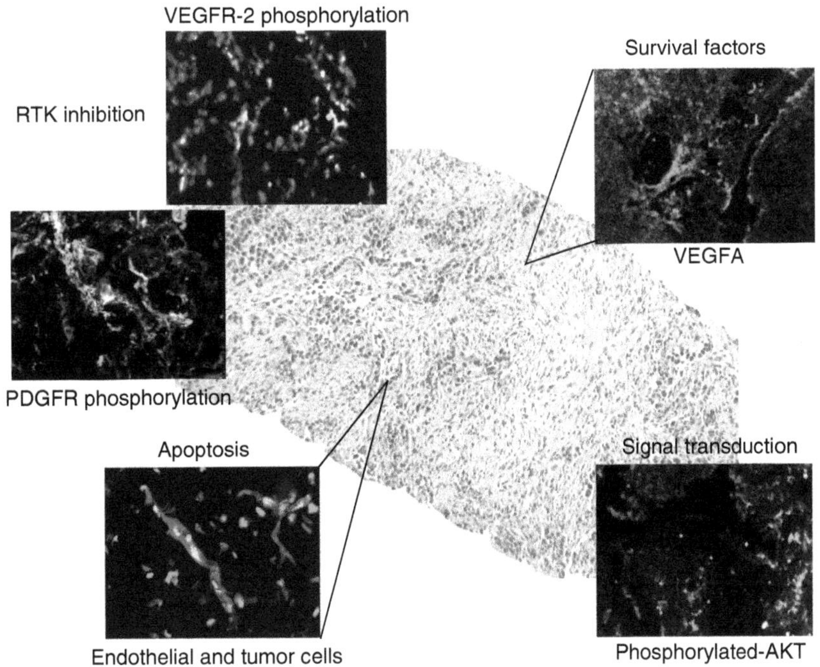

FIGURE 21.4

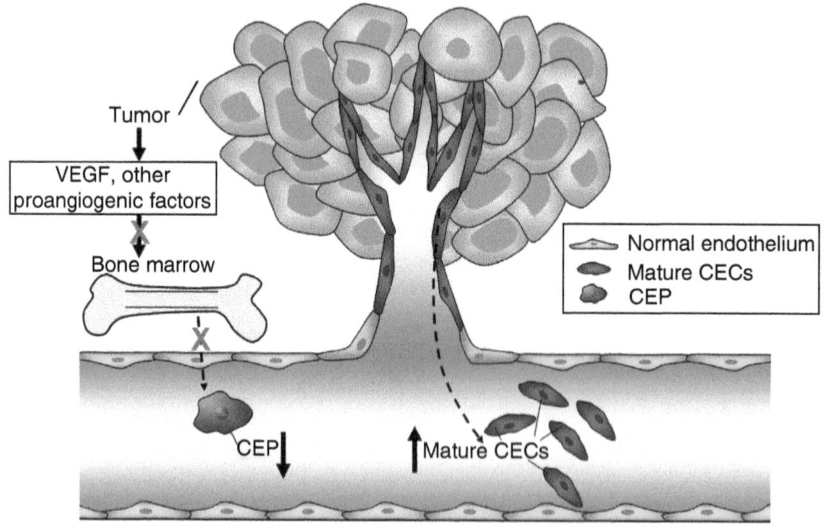

FIGURE 22.1

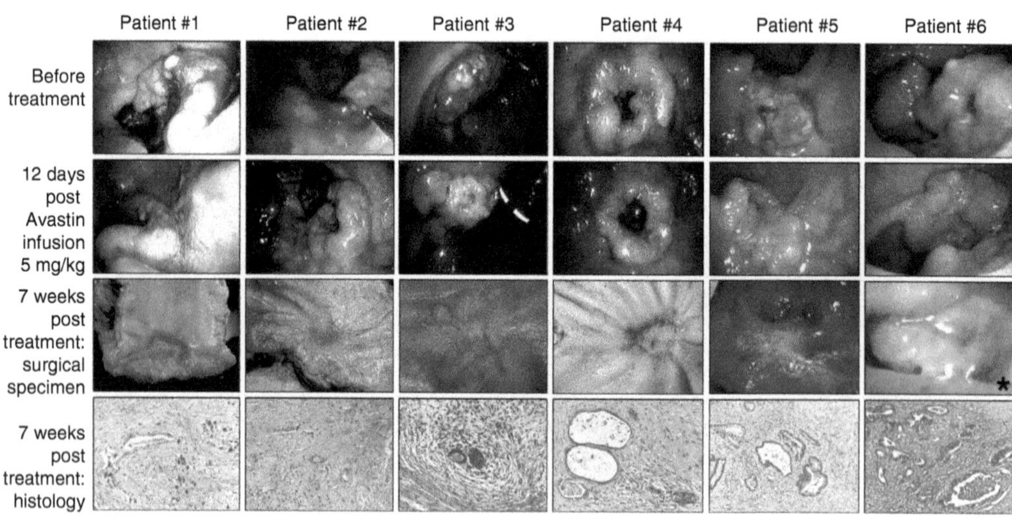

FIGURE 24.2

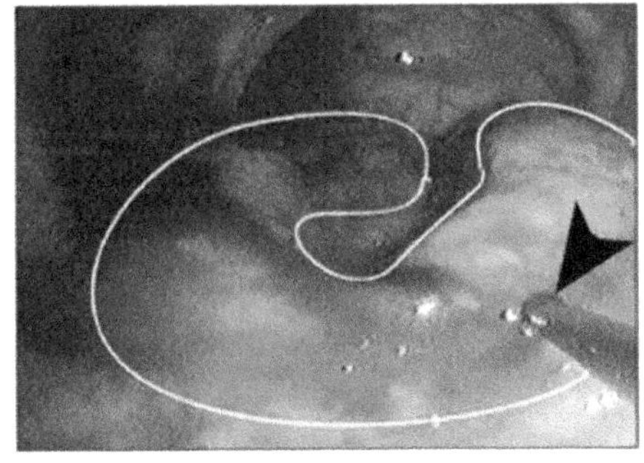

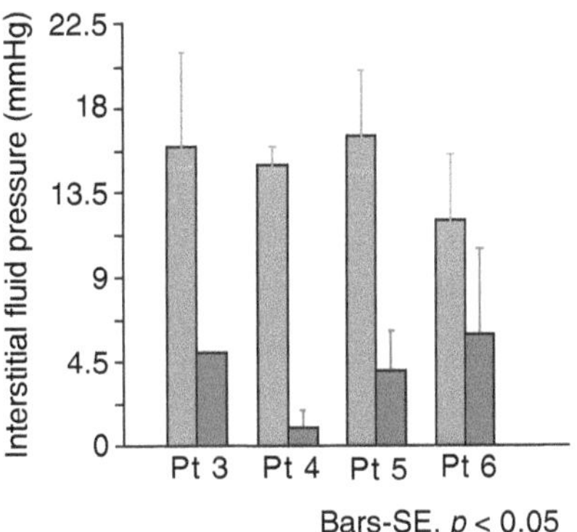

FIGURE 24.3

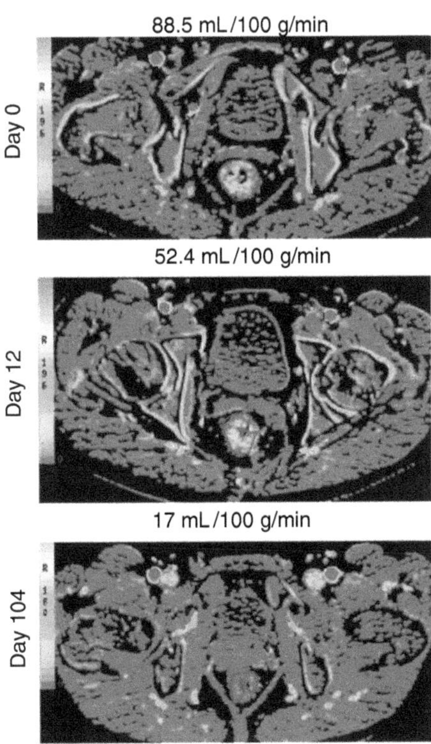

FIGURE 24.4

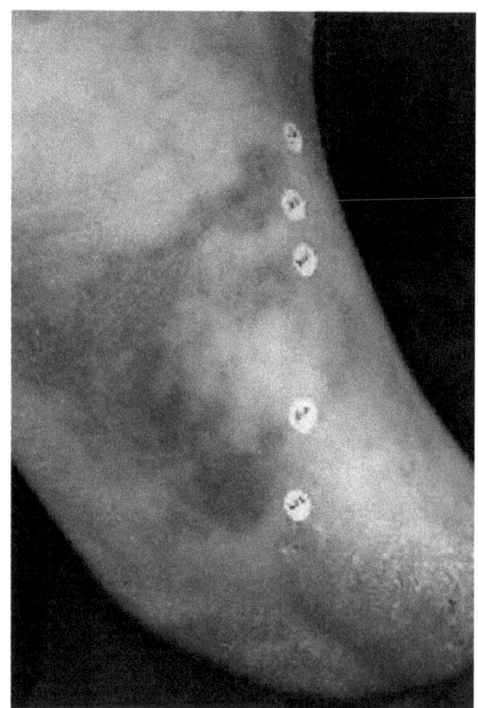

FIGURE 30.1

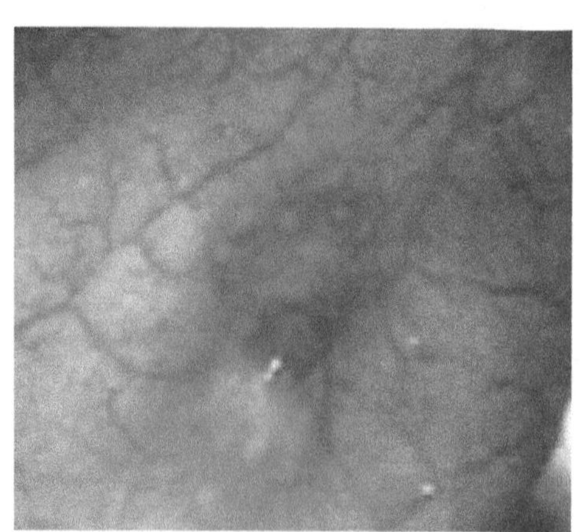

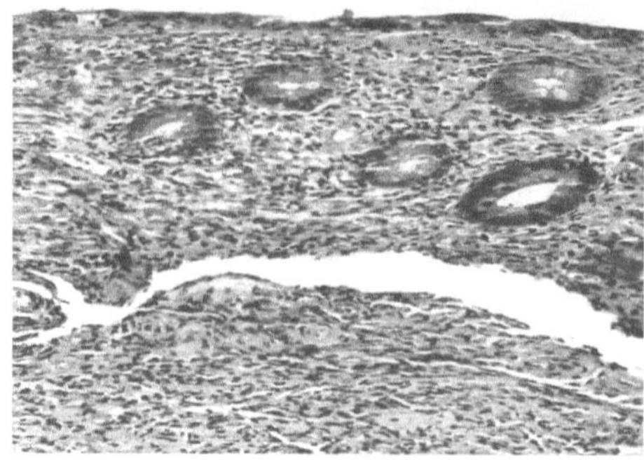

FIGURE 30.2

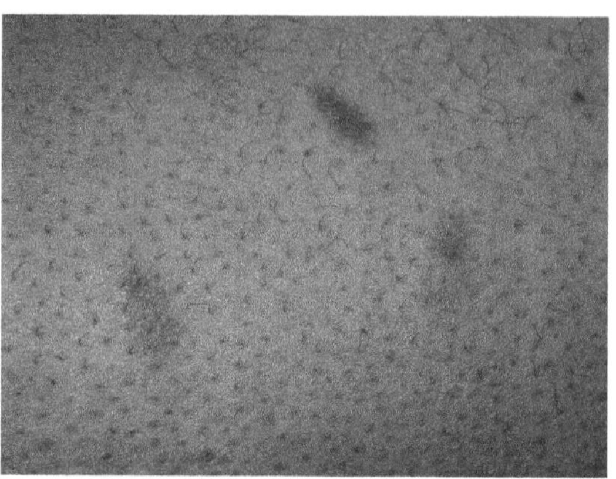

FIGURE 30.3

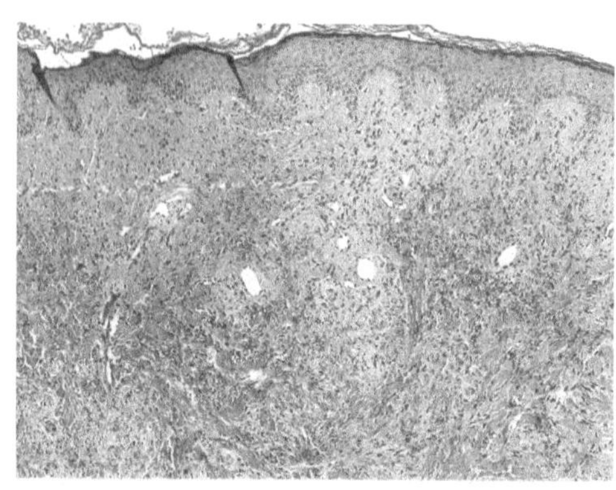

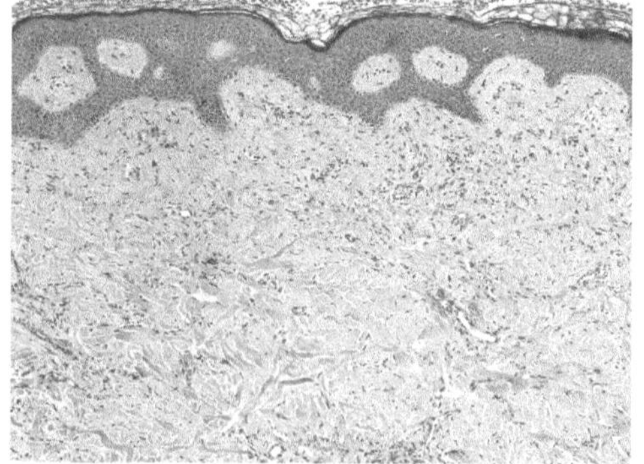

FIGURE 30.4

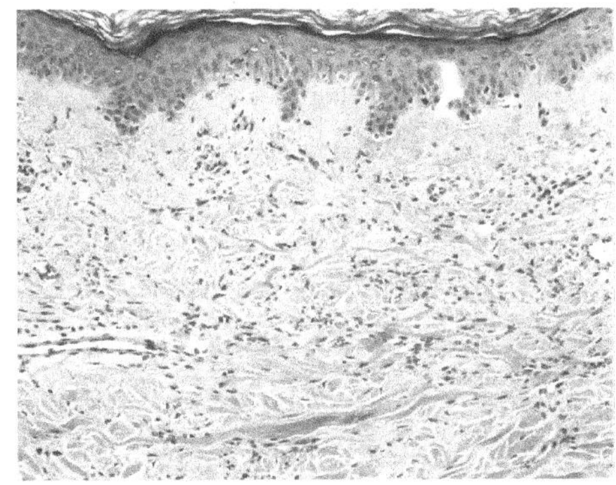

FIGURE 30.5

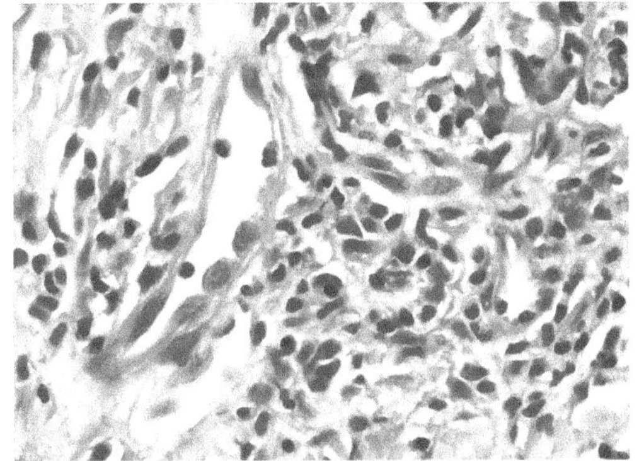

FIGURE 30.6

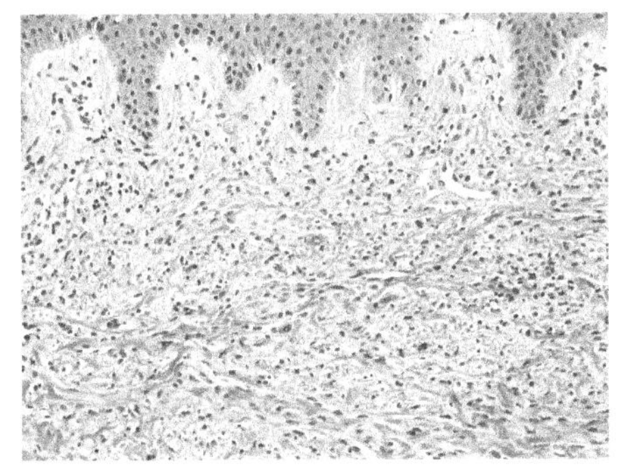

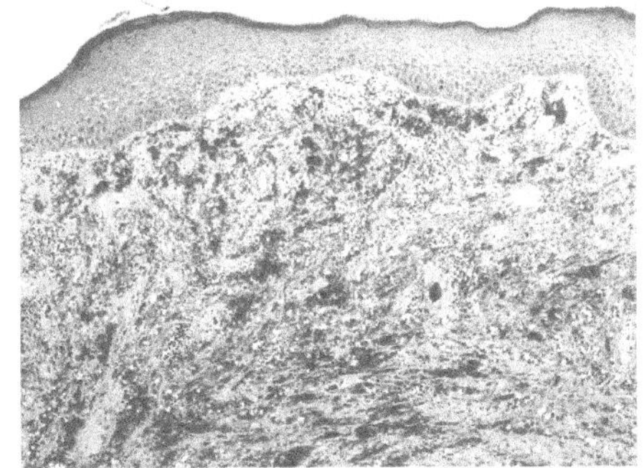

FIGURE 30.7

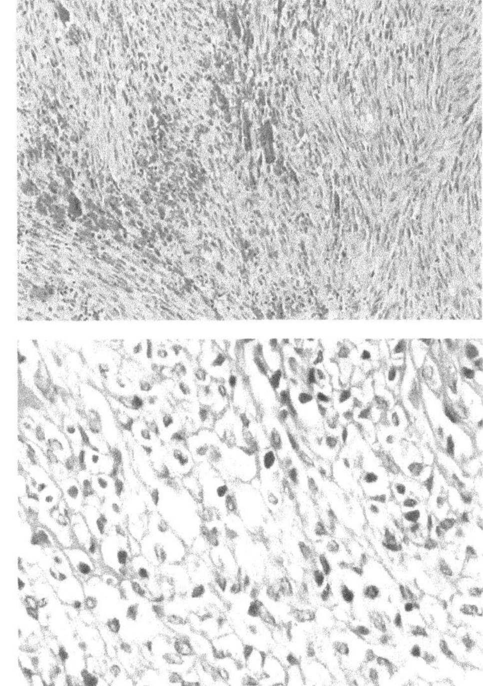

FIGURE 30.8

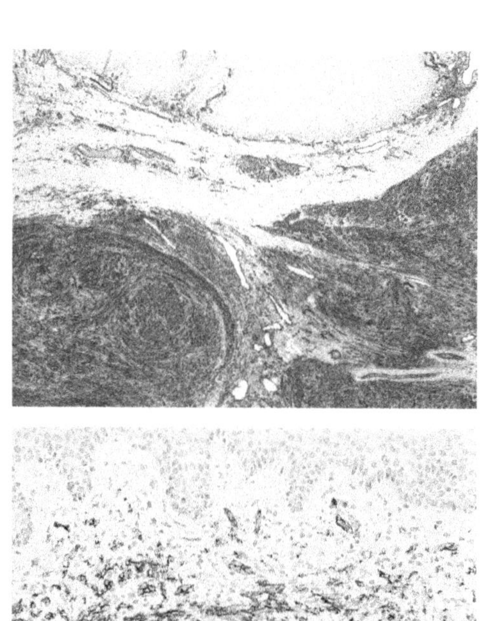

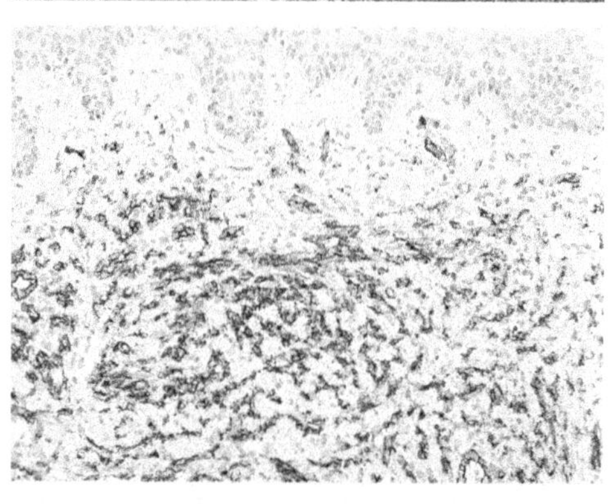

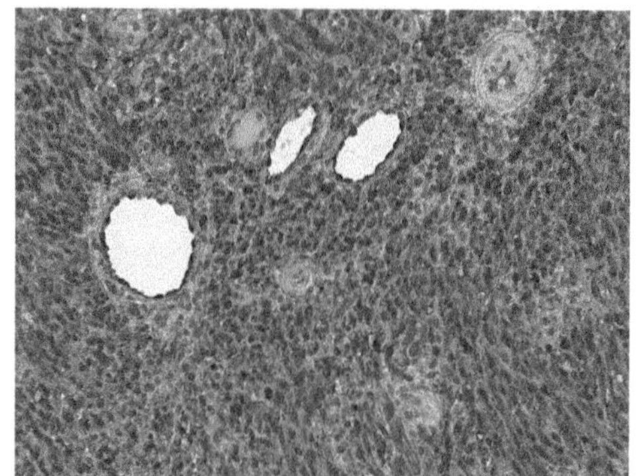

FIGURE 30.9

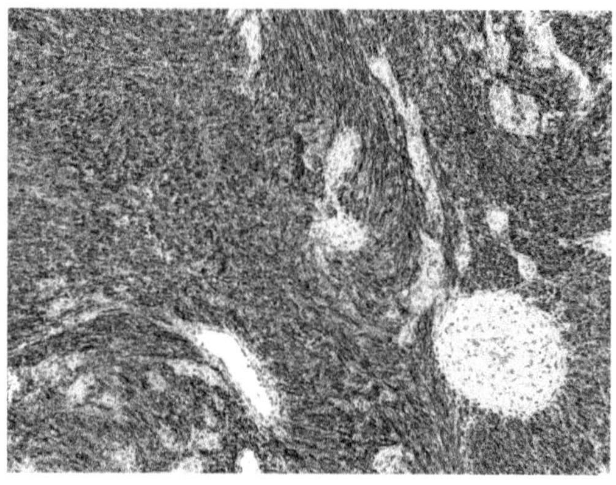

FIGURE 30.10

FIGURE 30.11

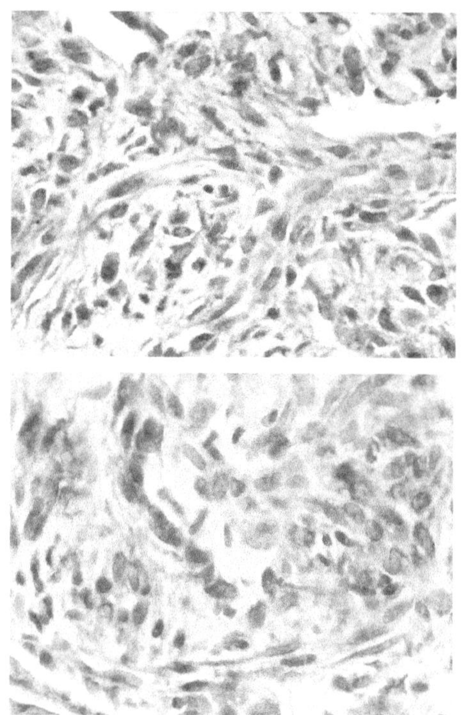

FIGURE 30.12

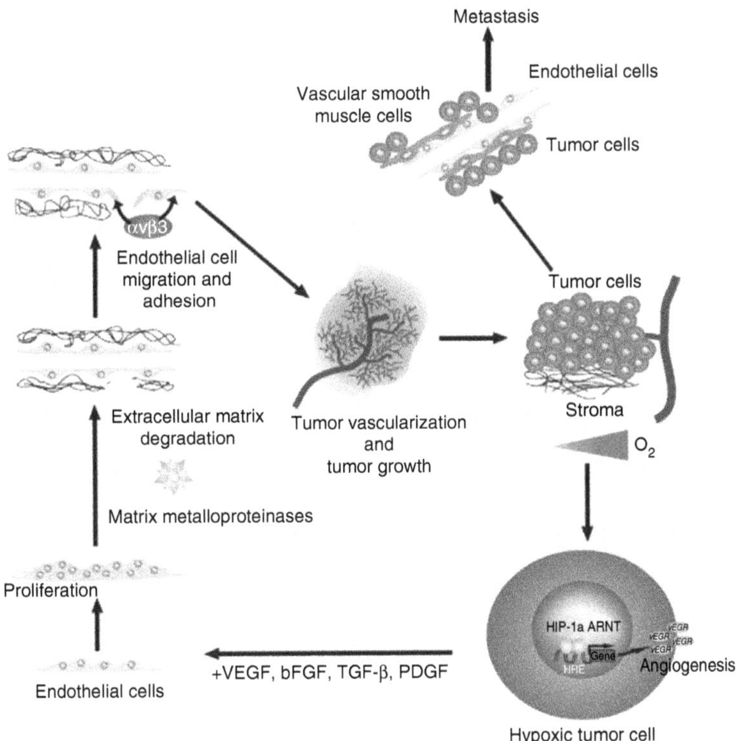

FIGURE 31.1

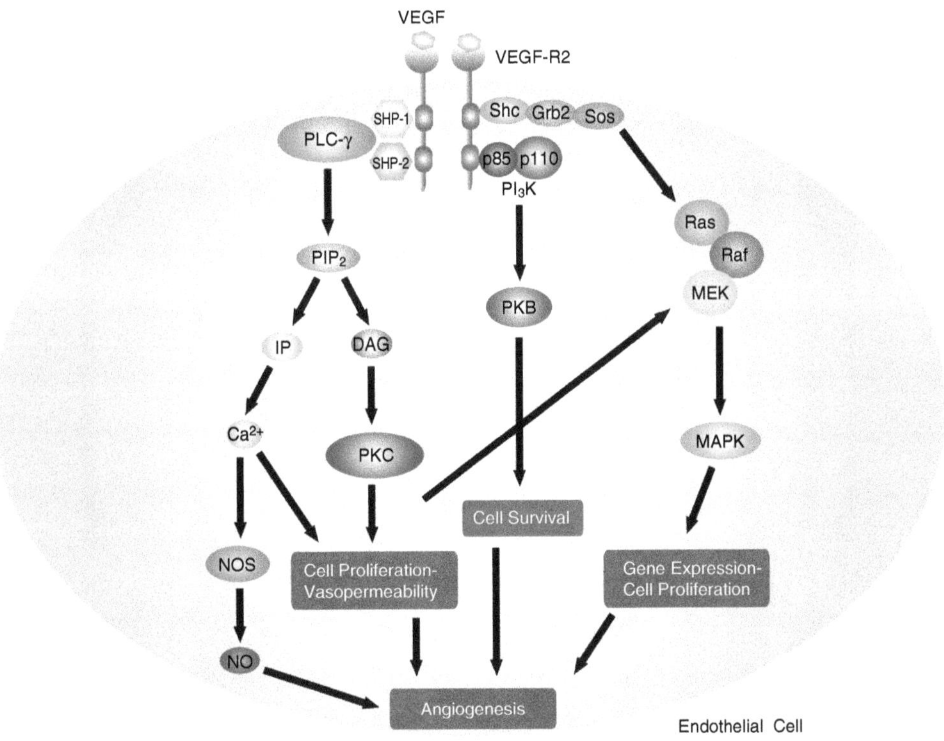

FIGURE 31.2

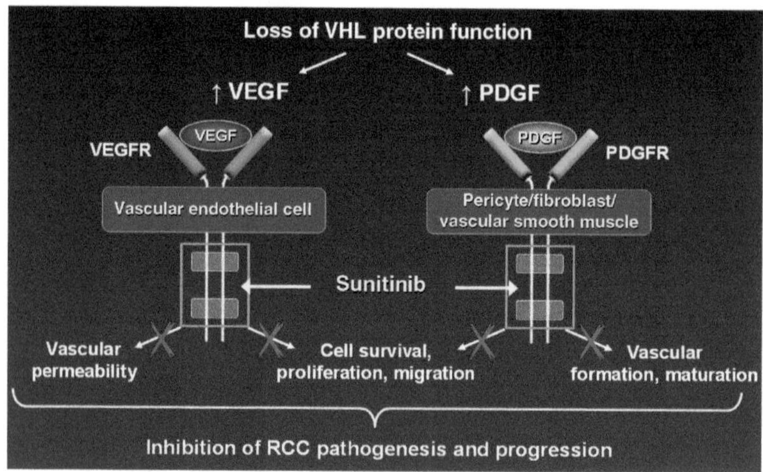

FIGURE 32.2

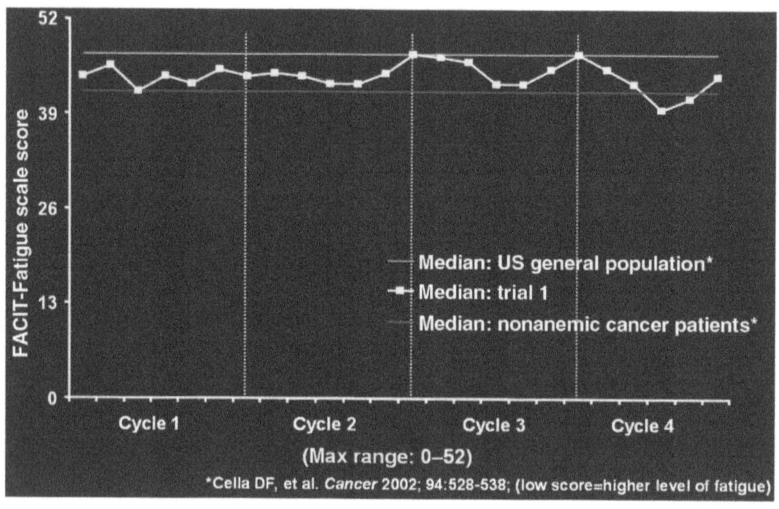

FIGURE 32.5

25 Antiangiogenic Therapy for Breast Cancer

Pablo M. Bedano, Brian P. Schneider, Kathy D. Miller, and George W. Sledge, Jr.

CONTENTS

Angiogenesis plays a central role in both local tumor growth as well as distant metastasis in breast cancer (1). The inhibition of angiogenesis is an attractive therapeutic target, with the potential for widespread activity and low toxicity, as well as the potential for synergy with classic cytotoxic therapy and radiotherapy. In this chapter, we explore the role of angiogenesis in breast cancer and recent attempts to quantify angiogenesis and tumor vasculature. We also discuss clinical trial results, potential mechanisms of resistance, and the special challenges posed by the development of antiangiogenic agents.

25.1 PRECLINICAL EVIDENCE

There is extensive preclinical data to support the role of angiogenesis as a key event in the transformation of mammary hyperplasia into malignancy (2,3). Transfection of tumor cells with angiogenic stimulatory peptides has been shown to increase tumor growth, invasiveness, and metastasis, whereas the transfection of tumor cells with inhibitors of angiogenesis decreases growth and metastasis (4). The expression of the matrix metalloproteinase (MMP) increases with the progression from benign to invasive breast cancer, and is associated with increasing histological tumor grade. Microscopic metastases are growth restricted and remain dormant until they undergo an angiogenic switch, presumably as a result of further mutation. This angiogenic switch often results in increased expression of MMPs (4). Hypoxia is a key signal for the induction of angiogenesis, associated with the expression of hypoxia-inducible factors (HIF-1 and HIF-2). The expression of HIF-1α has also been shown to increase during the progression from normal breast tissue to invasive ductal carcinoma. HIF-1α expression is higher in poorly differentiated lesions than in well-differentiated lesions, and it is associated with increased expression of the estrogen receptor and vascular endothelial growth factor (VEGF) (5). Similarly, the expression of carbonic anhydrase IX, an HIF-1α dependent enzyme, is associated with worse relapse-free survival (RFS) and overall survival (OS) in patients with invasive breast cancer (6,7).

25.2 CLINICAL EVIDENCE

Clinicopathologic correlation has also confirmed the central role of angiogenesis in breast cancer progression. Fibrocystic lesions with the highest vascular density are associated with a greater risk of breast cancer (8). Two distinct vascular patterns have been described in association with ductal carcinoma in situ (DCIS): a diffuse increase in stromal vascularity between ductal lesions and a dense rim of microvessels adjacent to the basement membrane of individual ducts (9,10). Microvessel density (MVD), a marker of angiogenesis, has been shown to be highest in aggressive DCIS (10) and is associated with increased VEGF expression (11).

Weidner et al. (12) assessed tumor vasculature in primary breast tumors, specifically staining endothelial cells with antibodies directed against factor VIII-associated antigen, scanning at low power to identify the areas of greatest MVD (or "hot spot"), and then meticulously counting microvessels in a single field at 200-fold magnification. In a preliminary study of 49 unselected patients, mean MVD was 101 in patients who developed metastasis as compared with 45 in those who remained disease free ($p = 0.003$ in univariate analysis). The prognostic significance of tumor MVD was then confirmed in a blinded study of consecutive patients by the same investigators (13). Tumor MVD was the only significant predictor of OS and RFS among the node-negative subset in this follow-up study.

Investigation in this area has flourished with other groups modifying the technique to use different endothelial antibodies (14,15) and counting strategies (16). These studies have generally, though not uniformly, validated the poor prognosis and early relapse associated with increasing MVD (Table 25.1). Differences in sample size, technique, methods, and interobserver variability likely account for the discrepancies (15,17–20). A recent meta-analysis of all these studies showed that MVD was predictive of poor survival in invasive breast cancer, especially in the node-negative subset, although the statistical link was weak, with a global risk ratio (RR) of 1.54 for OS and RFS (21).

Multiple angiogenic factors are commonly expressed by invasive human breast cancers; at least six different proangiogenic factors have been identified, with the 121-amino acid isoform of VEGF predominating (22). Carriers of the C936T allele in the VEGF gene were more frequent among controls (29.4%) than among a cohort of breast cancer patients (17.6%), implying a protective effect for carriers of the variant polymorphism (23). A recent report also demonstrated that this same polymorphism is associated with an improved metastasis-free survival time in patients that have a low-grade breast cancer (24). VEGF expression has also correlates with risk and outcomes in breast cancer. Several studies have found an inverse correlation between VEGF expression and OS in both node-positive and node-negative breast cancer (25,26). Increased VEGF expression has been associated with impaired response to tamoxifen or chemotherapy in patients with advanced breast cancer (27). Recently, VEGF

TABLE 25.1
Negative Prognostic Value of Tumor Microvessel Density

					High versus Low MVD (*p* Value)	
Investigator	**Pts.**	**Nodal Status**	**Antibody**	**Counting Method**	**RFS**	**OS**
Weidner 1991 (12)	49	N−/N+	FVIII	"hot spot"	0.003	NR
Weidner 1992 (13)	165	N−/N+	FVIII	"hot spot"	0.001	0.001
Hall 1992 (187)	87	N−/N+	FVIII	"hot spot"	NS	NS
Bosari 1992 (188)	88	N−	FVIII	"hot spot"	0.01	NR
	32	N+			NS	NR
Van Hoef 1993 (189)	93	N−	FVIII	"hot spot"	NS	NS
Obermair 1994 (190)			FVIII	"hot spot"		
Toi 1995	328	N−/N+	FVIII	"hot spot"	0.00001	NR
Ogawa 1995 (192)	155	N−/N+	FVIII	"hot spot"	0.025	0.01
Axelsson 1995 (193)	220	N−/N+	FVIII	"hot spot"	NS	NS
Obermair 1995 (190)	230	N−	FVIII	"hot spot"	<0.05	NR
Inada 1995 (191)	110	N−	FVIII	"hot spot"	<0.05	<0.05
Costello 1995 (194)			FVIII	"hot spot"		
Morphopoulas 1996 (195)	160	N−/N+	FVIII	"hot spot"	NS	NS
Karaiossifidi 1996 (196)	52	N−	FVIII	"hot spot"	<0.05	NR
Ozer 1997 (197)	35	N−	FVIII	"hot spot"	0.034	NR
Tynninen 1999 (198)	84	N−/N+	FVIII	"hot spot"	NS	NR
Medri 2000 (199)	378	N−	FVIII	"hot spot"	NS	NS
Vincent-Salomon 2001 (200)	685	N−/N+	FVIII	"hot spot"	NS	NS
Fox 1994 (201)	109	N−	CD31	Chalkley	0.01	0.028
Gasparini 1994 (202)	254	N−	CD31	"hot spot"	0.0004	0.047
Bevilacqua 1995 (203)	211	N−	CD31	"hot spot"	0.001	0.044
Fox 1995 (204)	211	N−/N+	CD31	Chalkley	NR	<0.05
Gasparini 1998 (205)	531	N−/N+	CD31	"hot spot"	<0.05	<0.05
Heimann 1996 (206)	167	N−	CD34	"hot spot"	0.018	NR
Kumar 1997 (207)	106	N−/N+	CD105	"hot spot"	0.0362	0.0029
			CD34		NS	NS
Hansen 2000 (208)	836	N−/N+	CD34	Chalkley	<0.05	NR
Guidi AJ 2000 (209)	110	N+	FVIII	"hot spot"	0.006	0.004
Gunel 2002 (210)	42	N−/N+	FVIII	"hot spot"	NS	NS
Fridman 2000 (211)	216	N−/N+	FVIII	"hot spot"	NS	NS
Tas 2000 (212)	120	N−/N+	CD34	"hot spot"	NR	0.01
Kato 2001 (213)	377	N−/N+	FVIII	"hot spot"	0.002	NS
Guidi AJ 2002 (214)	577	N+	FVIII	"hot spot"	NS	NS

Chalkley, chalkley counting as a measure of angiogenesis; FVIII, factor VIII related antigen; N+, lymph-node positive; N−, lymph-node negative; NR, not reported; NS, not significant; OS, overall survival; Pts, patients; RFS, relapse-free survival.

expression has been successfully quantified via immunohistochemistry in breast cancer tumor specimens (28). The expression and intensity of expression were found to correlate with significantly inferior outcomes to treatment of breast cancer.

25.3 SURROGATE END POINTS OF ANGIOGENESIS

As described earlier, there is a wealth of data supporting a central role for angiogenesis in breast cancer growth and metastasis. At present, however, the development of angiogenesis as a therapeutic target is limited by a lack of biologic predictors of therapeutic benefit.

Such biologic markers (e.g., estrogen receptor for hormone sensitivity and HER-2 for trastuzumab therapy) have played an important role in the development of therapeutics targeting specific subpopulations of breast cancer. Antiangiogenic therapy would clearly benefit from the development of surrogate markers that could guide therapy without repeated tissue samples. Though as yet no clear standard has emerged, the search for reliable surrogates of antiangiogenic activity has focused on two main areas: soluble factors and imaging of tumor vasculature.

VEGF has been measured in sera and it is typically detected at higher levels in patients with breast cancer than in healthy volunteers (29). Higher serum concentrations have also been found in those patients with stage III as compared with stage I or II breast cancer (30,31). Serum concentrations of VEGF may primarily reflect platelet count rather than tumor burden, limiting the interpretation of these results (32–34). VEGF can also be measured in urine with urinary levels unaffected by platelet counts (35). Though previous attempts to correlate single pretreatment plasma VEGF levels with response to therapy (36) and the ability of serial measurements to predict response (37) in patients with metastatic breast cancer have been disappointing, neither of these preliminary studies directly assessed angiogenesis.

Vascular cell adhesion molecule-1 (VCAM-1) levels have been increased in patients with breast cancer as compared to patients with benign breast disease (38). In one study, serum VCAM-1 levels were more tightly correlated with MVD than serum VEGF levels. Initial levels in metastatic disease were not found to be predictive of response, but they quickly rose in patients whose disease progressed but decreased or remained stable in responding patients (37).

Other surrogate markers that have been studied include the gelatinases (MMP-2 and MMP-9), whose overall expression is associated with grade and stage of breast cancer (39), and the fibroblast growth factor (FGF), which is commonly produced by breast cancers and can be measured in tumor cytosol (40). Similarly to VEGF, FGF levels are increased in patients with breast cancer (29,41) and may predict survival (42,43).

Imaging techniques have been used to measure tumor vasculature with varying success. Assessment of tumor blood flow by color Doppler ultrasonography (CDUS) has had inconsistent results, and is hampered by the technical limitations of traditional clinical ultrasound (2–10 MHz) (44). Low-velocity flow within the central region of a tumor and increased peripheral perfusion is common. Traditional ultrasound cannot readily distinguish these regional variations because the low-velocity flow approximates the tissue motion velocity. Modifications to clinical ultrasound, including laser Doppler flow measurements (45), use of encapsulated micro bubbles as contrast agents (46), and high-frequency Doppler (20–100 MHz) (44) to improve quantitative measurement of the tumor have not been proven in the clinic.

Positron emission tomography (PET) uses small doses of radiopharmaceuticals to depict and quantify biochemistry rather than measure static anatomic sites. The most commonly used agent in clinical PET imaging is 2-(^{18}F) fluoro-2-deoxyglucose (FDG). As a glucose analogue, FDG measures the uptake and phosphorylation of glucose as an indicator of metabolic activity. Sequential PET imaging using FDG predicted response to primary chemotherapy in some small pilot trials (47–49); however, FDG uptake has not correlated with VEGF expression in most studies (50,51). Radiotracers other than FDG for imaging breast cancer include ^{99m}technetium sestambibi, ^{201}thallium (TL), and ^{15}O-labeled water. ^{201}TL and ^{99mt}technetium sestambibi are taken up so avidly by most tissues that levels predominantly reflect relative blood flow. While ^{99m}Tc sestambibi and ^{201}TL have gained widespread acceptance for assessment of myocardial perfusion, increased uptake in tumors has also been demonstrated (52,53). Yoon et al. (54) evaluated ^{99m}Tc sestambibi uptake and washout in 31 patients with untreated primary breast cancer. Both early and late tumor-to-normal breast ratios correlated well with tumor MVD. ^{15}O-labeled water has a short half-life (2 min) mandating cyclotron production near the PET facility. However ^{15}O-labeled water can quantify perfusion (millimeters/gram/minute) in all body tissues, including breast cancer. A small pilot study using ^{15}O-labeled water found increased blood flow associated with the tumor compared with surrounding normal breast tissue but did not attempt any correlation with histology or MVD (55). Absolute tumor-associated blood volume can be calculated using ^{11}C carbon monoxide, either as and inhaled gas or mixed with whole blood ex vivo (56). ^{15}O-labeled oxygen has been used to measure changes in tumor oxygen metabolism during therapy (57). Several novel PET imaging strategies are currently in development to quantify tumor hypoxia, proliferation index, apoptosis, receptor/ligand interactions, and angiogenesis-related gene expression (58).

Magnetic resonance imaging (MRI) has been used to evaluate tumor-associated vasculature with either intrinsic or contrast-enhanced methods. Intrinsic methods use pulse sequences that detect the motion of water in the vascular bed (59,60) or the distortion of the magnetic field by deoxyhemoglobin (61,62). Intrinsic contrast methods rely solely on intravascular red blood cells and are not affected by changes in vascular permeability but have lower contrast-to-noise ratio limiting resolution. Dynamic contrast-enhanced MR using small molecular contrast media that equilibrate quickly between blood and the extravascular space is dependent on the tumor MVD, vascular permeability (K_{21}), and the transfer constant (K^{trans} = permeability-surface area product per unit volume of tissue). Such quantitative MRI parameters have been correlated with tumor MVD and VEGF expression in several studies (Table 25.2), although there is a high degree of inconsistency between MRI enhancements and vascular density (63). To date there is no validation of the prognostic value of MRI enhancement parameters in predicting RFS or OS (64).

TABLE 25.2
Contrast-Enhanced MRI Parameters Compared with Angiogenesis in Breast Cancer

Investigator	No. Tumors (B:M)	MR Parameter	VEGF	MVD	Comparison with MVD (Unless Otherwise Noted) r	p
Buada (215)	20:51	Steepest slope	—	CD34	0.83	0.001
Stomper (216)	23:25	Max. amplitude	—	FVIII	NR	0.02
Buckley (217)	0:40	Enhancement at 1 min	—	FVIII	0.47	0.002
Hulka (218)	36:21	Extraction flow product	—	FVIII	0.36	<0.01
Ikeda (219)	29:69	K^{trans}	—	NR	0.89	<0.01
Knopp (220)	20:7	K_{21}	IHC	CD31	K_{21}/VEGF, 0.52 MVD/VEGF(+) MVD/VEGF(−), 0.71	0.05 NS 0.07
Matsubayashi (221)	4:31	Rim enhancement	IHC	ND	VEGF, NR	0.008

B, benign; IHC, immunohistochemistry; K^{trans}, permeability-surface area product per unit volume of tissue; K_{21}, vascular permeability; M, malignant; MVD, microvessel density; ND, not done; NR, not reported; NS, not significant; p, level of statistical significance; r, correlation coefficient; VEGF, vascular endothelial growth factor.

TABLE 25.3
Association with MVD[a]

	Diagnosis	Surgery
Serum factors		
VCAM-1	0.0990	0.3462
bFGF	0.1384	0.0567
MMP-2	0.5714	0.0472
MMP-9	0.4317	0.8018
CDUS parameters		
Peak systolic velocity	0.2275	0.9072
End diastolic velocity	0.2377	0.9873
RI	0.9565	0.0698
TAM velocity	0.3009	0.8644
Mean blood flow	0.3591	0.4198
Maximum systolic diameter	0.0362	0.1959
Maximum diastolic diameter	0.1680	0.5814
PET parameters		
[F-18]FDG ratio	0.0662	0.4082
[F-18]FDG UI	0.0831	0.3305
[F-15]H_2O ratio	0.6495	0.6797
[F-15]H_2O UI	0.7841	0.7801
[C-11]CO ratio	0.5523	0.3186
[C-11]UI	0.9330	0.3107

[a] αβλε 0 Univariate *p* values for association with average MVD at time of diagnosis (initial biopsy) and definitive surgery.

We recently published the results of a trial in which we attempted to correlate a number of surrogate markers of angiogenesis with MVD and response to primary chemotherapy (65). Seventy patients with newly diagnosed stage II or III breast cancer were treated with sequential doxorubicin 75 mg/m^2 q2 weeks × 3 and docetaxel 40 mg/m^2 weekly × 6; treatment order was randomly assigned. Potential noninvasive markers of angiogenesis were obtained pretreatment, at crossover and completion of chemotherapy, and were correlated with MVD and pathologic response. Serum angiogenic factors collected included VCAM-1, bFGF, MMP-2, and MMP-9, and imaging with color Doppler ultrasound (CDUS) as well as PET with FDG, ^{15}O-labeled water, and (C-11) carbon monoxide. None of the measured serum proteins or imaging parameters consistently correlated with MVD (Table 25.3). Patient's demographics, tumor characteristics, and MVD at diagnosis did not predict pathologic response; only lower serum bFGF concentration at diagnosis was associated with pathologic complete response (pCR) (Table 25.4), which conflicts with the results of Linderholm et al. (66) in which patients with higher tumor cytosolic bFGF levels enjoyed a lower recurrence rate and improved OS. Given the small number of patients with pCR (12.8%) and lack of significant associations in univariate analysis, multivariate analysis was not conducted.

25.4 TARGETING ANGIOGENESIS

The challenges of developing antiangiogenic agents and measuring their efficacy are not unique to the treatment of breast cancer. Familiarity with the distinct biologic nature of these agents is essential for the interpretation of clinical trials. Unlike cytotoxic agents that are measured by

TABLE 25.4
Association with Complete Pathologic Response[a]

	Diagnosis	Change during Treatment
Age	0.790	NA
Tumor characteristics		
Initial tumor size	0.479	NA
ER	0.9368	NA
HER-2	0.0744	NA
MVD at diagnosis	0.2676	NA
Serum factors		
VCAM-1	0.0848	0.2028
bFGF	0.0227	0.2003
MMP-2	0.5481	0.3841
MMP-9	0.5062	0.2544
CDUS parameters		
Peak systolic velocity	0.6188	0.9229
End diastolic velocity	0.5941	0.9679
RI	0.3517	0.5300
TAM velocity	0.9877	0.7063
Mean blood flow	0.6692	0.6215
Maximum systolic diameter	0.5104	0.7894
Maximum diastolic diameter	0.5941	0.7456
PET parameters		
[F-18]FDG ratio	0.3364	0.7396
[F-18]FDG UI	0.6613	0.8676
[O-15]H_2O ratio	0.9824	0.6580
[O-15]H_2O UI	0.8533	0.9935
[C-11]CO ratio	0.6508	0.4024
[C-11]CO UI	0.3821	0.4642

[a] Univariate p values for association with complete pathologic response.

their ability to kill tumors directly, trials designed to evaluate these agents may require a paradigm shift. Because of the differences in mechanism of action, these agents must be delivered in a way that obtains optimal biological effect, rather than seeking the maximally tolerated dose. It may also be important to deliver chronic rather than intermittent therapy and strive to achieve induction of tumor dormancy rather than tumor cell kill as a therapeutic goal. Thus, measuring response rates in a traditional manner may not be appropriate and surrogate end points may be more predictive of therapeutic efficacy. Soluble measures of angiogenesis and imaging technology, as previously discussed, may provide better parameters to assess response, although no validated surrogate markers are as of yet available.

25.4.1 Antiangiogenic Activity of Existing Agents

The use of antiangiogenic agents may not be a novel concept in the treatment of breast cancer (67). Tamoxifen, initially thought to be merely a competitive inhibitor of estradiol, may also have estrogen-independent mechanisms of action (68). Tamoxifen inhibits VEGF- and FGF-stimulated angiogenesis in the chick chorioallantoic membrane model. This effect was not reversed by excess estradiol, suggesting that the antiangiogenic mechanism is not dependent on estradiol concentration or estrogen receptor content (69,70). Treatment with

TABLE 25.5
Criteria for Antiangiogenic Activity

- Differential cytotoxicity
- Alters endothelial cell function
- Critical mechanistic effects
- Inhibition of angiogenesis in vivo

tamoxifen resulted in a more than 50% decrease in the endothelial density of viable tumor and an increase in the extent of necrosis in MCF-7 tumors growing in nude mice (71). The inhibition of angiogenesis was detected before measurable effects on tumor volume (72). In a study using differential display technology to assess gene expression in tumor and normal breast tissue from two patients, brief treatment with tamoxifen resulted in down regulation of CD36, a glycoprotein receptor for matrix proteins thrombospondin-1 and collagen types I and IV (73).

Several chemotherapeutic agents used routinely in breast cancer treatment have known antiangiogenic activity (74). Unless the effect is purely related to endothelial cell death, a defined mechanistic effect specific to the process of angiogenesis should be measurable before a chemotherapeutic is accepted as antiangiogenic (Table 25.5). Maximal antiangiogenic activity typically requires prolonged exposure to low drug concentrations, exactly counter to maximum tolerated doses administered when optimal tumor cell kill is the goal (75). Many reports have confirmed the importance of dose and schedule in preclinical models. The combination of low, frequent dose chemotherapy plus an agent that specifically targets the endothelial cell compartment (TNP-470 and anti VEGF-2) controlled tumor growth much more effectively than the cytotoxic agent alone (76–78).

Thus far, few clinical trials have tested antiangiogenic dosing schedules of chemotherapy, so-called metronomic therapy (79). Metronomic dosing implements a frequent dosing schema at doses much lower than the maximum tolerated dosage. Preclinical data suggest that the mechanism responsible for the antiangiogenic effect is the induction of increased plasma levels of thrombospondin-1 (a potent and endothelial specific inhibitor of angiogenesis) (80). A phase II study of low-dose methotrexate (2.5 mg twice daily for 2 days each week) and cyclophosphamide (50 mg daily) in patients with previously treated metastatic breast cancer found an overall response rate of 19% (an additional 13% of patients had stable disease for 6 months or more). Serum VEGF levels decreased in all patients remaining on therapy for at least 2 months, but did not correlate with response (81). These studies suggest that activated endothelial cells may be more sensitive, or even selectively sensitive to protracted low-dose chemotherapy compared with other types of normal cells. This creates a potential therapeutic window. The Dana-Farber/Harvard Cancer Center (Boston, MA) is currently leading a phase II randomized trial of metronomic low-dose cyclophosphamide and methotrexate with or without bevacizumab (Avastin; Genentech, South San Francisco, CA) in women with metastatic breast cancer. Burnstein et al. presented the results of a randomized phase II study of metronomic chemotherapy with and without bevacizumab in patients with metastatic breast cancer (Burstein HJ, Spigel D, Kindsvogel K, et al. Metronomic chemotherapy with and without bevacizumab for advanced breast cancer: a randomized phase II study. Proceedings of the 28th Annual San Antonio Breast Cancer Symposium, San Antonio, 2005 [Abstr. 4]). Fifty-five women were enrolled, accrual to the metronomic therapy alone arm was discontinued early in the study since clinical activity was felt to be minimal. The clinical response rate (partial response and stable disease) in the 34 women who received the three agents was 70% vs. 48% in the 21 women who received metronomic therapy alone. The time to progression (TTP) also favored the triple therapy arm, with a median TTP of 5.5 months vs. 2 months in the metronomic therapy alone. Toxicity associated with the triple therapy arm consisted in a 21% incidence of grade 2/3 hypertension.

25.5 ANGIOGENESIS INHIBITORS IN BREAST CANCER

Our burgeoning understanding of angiogenesis has fostered the development of agents targeting specific steps in the angiogenic cascade, many of which have been tested in clinical trials. A detailed list of agents in clinical development can be obtained from the National Cancer Institute website (http://cancernet.nci.nih.gov). While the number of phase I and II trials has grown rapidly, few have been reported in the peer-reviewed literature; the first phase III clinical trial using a pure antiangiogenic agent in metastatic breast cancer has only been recently reported. Rather than an exhaustive review of all the agents currently in clinical testing, we have conceptually grouped agents into several categories: growth factor/receptor antagonists, which thwart signaling of proangiogenic growth factors; protease inhibitors, which either inhibit or otherwise interfere with the action of proteases critical for invasion; endothelial toxins, which specifically target endothelial antigens and natural inhibitors, which stimulate or mimic substances known to naturally inhibit angiogenesis. The clinical experience with representative categories is reviewed, with emphasis on Bevacizumab, which is the most matured of the agents targeting angiogenesis.

25.5.1 Proangiogenic Growth Factor/Receptor Antagonists

VEGF is the best studied angiogenic factor, with proven significance in breast cancer. There are many possible ways to target VEGF, and many inhibitors have entered the clinic, though not always, for patients with breast cancer. Several potential points of attack on VEGF are under examination: ligand sequestration, attacks on the external membrane receptor, inhibition of the internal (tyrosine kinase) portion of the receptor, inhibition of the VEGF receptor message, inhibition of downstream intermediate, and direct inhibition of upstream regulators of VEGF.

Bevacizumab is a humanized monoclonal antibody directed against the VEGF-A ligand, recognizing all isoforms. It has been approved by the U.S. Food and Drug Administration for the first-line treatment of metastatic colorectal cancer when given in combination with fluorouracil and irinotecan, based on an OS benefit over standard chemotherapy alone (82). Bevacizumab has also demonstrated promising activity in renal cell carcinoma (83,84).

In breast cancer, phase II studies in women who had previously progressed on at least an anthracycline- or taxane-based regimen revealed clinical activity (85). Therapy was generally well tolerated; no significant bleeding episodes were noted, although it should be mentioned that patients were screened for intracranial metastases, and those with brain metastases were excluded. This was based on phase I experience in which hemorrhage from unrecognized intracranial metastases occurred in a patient with hepatocellular carcinoma (86). Phase II trials have also combined bevacizumab with a variety of other agents, including Vinorelbine (87) and docetaxel (88) in the refractory metastatic setting. A phase III trial testing bevacizumab has also recently been reported. This study randomly assigned 462 patients with anthracycline- and taxane-refractory, metastatic breast cancer to receive capecitabine with or without bevacizumab (89). Hypertension requiring treatment (17.9% versus 0.5%), proteinuria (22.3% versus 7.4%), and thromboembolic events (7.4% versus 5.6%) were more frequent in patients receiving bevacizumab. The primary end point of this study was the time to progression (TTP) and there was no evidence of improvement in the bevacizumab arm. The combination, however, significantly increased the response rate (9.1% versus 19.8%; $p = 0.001$). Although attempts to correlate VEGF RNA overexpression (by in situ hybridization) and response rate in this study were unsuccessful, the sample size was too small for a definitive conclusion (90).

Bevacizumab has also been tested in the neoadjuvant setting. In one study, women were treated with docetaxel with and without bevacizumab. Eligible patients had locally unresectable breast cancer with or without distant metastasis. Patients whose disease responded underwent definitive surgery followed by four cycles of doxorubicine/cyclophosphamide and tamoxifen (if hormone receptor-positive). There were 5 complete clinical responses and 24 partial responses, and the therapy was generally well tolerated (91). The National Cancer Institute also recently reported the preliminary results of a neoadjuvant pilot study evaluating bevacizumab (with several correlative end points) in women with inflammatory breast cancer. Patients received bevacizumab alone for the first cycle followed by six cycles of bevacizumab with doxorubicin and docetaxel. After completion of chemotherapy 8 of 13 patients experienced a confirmed partial response. There was also evidence of a decrease in vascular permeability on dynamic contrast-enhanced MRI after the first cycle of bevacizumab monotherapy (92).

There is preclinical and clinical rationale to support the combination of bevacizumab with trastuzumab. HER-2 appears to play a role in the regulation of VEGF (93). An invitro study demonstrated increased HIF-1α and VEGF mRNA expression in HER-2 overexpressing cell lines (94). In another preclinical study, experiments have shown reduction in xenograft volume using a combination of trastuzumab and bevacizumab compared with single-agent control (95). In a cohort of 611 patients with primary breast cancer and a median follow-up of greater than 50 months, there was a significant positive association between HER-2 and VEGF expression (96). A recently reported phase I trial has evaluated the tolerability of the combination of bevacizumab with trastuzumab. In this trial, it was determined that the co-administration of these two humanized monoclonal antibodies did not alter the pharmacokinetics of either agent. Clinical responses were observed in five of nine patients, including one patient with prior disease progression on chemotherapy plus trastuzumab (97).

There is also rationale to support simultaneous blockade of VEGF and epidermal growth factor receptor (EGFR) pathways. The EGFR also appears to regulate VEGF (98,99) and several studies have demonstrated that blockade of the EGFR resulted in an antiangiogenic effect (100,101). Furthermore, data have suggested that an increased production of VEGF represents one mechanism by which tumor cells escape anti-EGFR monoclonal antibody therapy (102). One study tested the strategy of combining bevacizumab and erlotinib (EGFR tyrosine kinase inhibitor) in metastatic breast cancer. This combination has demonstrated activity and circulating endothelial cells and tumor cells may correlate with response (103).

A large, international phase III trial led by the Eastern Cooperative Oncology Group (E2100) compared paclitaxel with or without bevacizumab in chemotherapy-naïve patients with metastatic breast cancer. Women in this study received paclitaxel at 90 mg/m^2 weekly for 3 or 4 weeks with or without bevacizumab at 10 mg/kg every 2 weeks. This trial completed accrual in May of 2004, enrolling over 700 women. In April of 2005 an interim analysis demonstrated a significant improvement in progression-free survival (PFS) (10.97 versus 6.11 months, $p = 0.001$) and a doubling in the response rate (28.2 versus 14.2, $p = 0.0001$), with preliminary improvement in OS. Toxicity did not exceed predefined protocol limits for severe bleeding, hypertension, or thromboembolic events.

The importance of E2100 cannot be overestimated. This proof-of-concept trial is the first phase III trial to show benefit from the use of an antiangiogenic agent in breast cancer. The improvement in PFS seen with the addition of VEGF-targeting therapy to a standard metastatic chemotherapy surpasses that seen with any chemotherapeutic agent in the last generation of clinical trials, and suggests that a landmark has been reached in the treatment of breast cancer. The Eastern Cooperative Oncology Group has recently initiated a pilot adjuvant trial (E2104) evaluating bevacizumab in combination with dose-dense doxorubicin and cyclophosphamide followed by paclitaxel in women with node-positive breast cancer. It is anticipated that this trial will be followed by a large proof-of-concept adjuvant trial, where it is hoped that the PFS benefits seen in E2100 will translate to an increased cure rate in the setting of microscopic metastatic breast cancer.

Monoclonal antibodies have also been developed against the VEGF receptor type II (VEGF-R2) (104,105). Another ligand sequestration approach has been the use of soluble recombinant Flt-1 receptor. This approach has only been used in preclinical studies to date (106).

Many receptor tyrosine kinase inhibitors have been developed, targeting the internal membrane tyrosine kinase portion of VEGF-R1 and VEGF-R2. SU011248 is an inhibitor of receptor tyrosine kinases for VEGF-R1 and VEFG-R2, platelet-derived growth factor receptor (PDGFR), c-kit, and Flt-3. SU011248 has demonstrated preclinical activity in breast cancer models (107,108). Furthermore, physiologic imaging in a breast cancer model during treatment with SU011248 revealed that [^{11}C] carbon monoxide and [^{18}F] fluoromethane imaging may be a useful biologic surrogate for this agent (109). Preliminary data of a phase II study of SU011248 (Sutent) in women with anthracycline and taxane-resistant breast cancer has revealed a partial response in 4/23 women (Miller KD, Burstein HJ, Elias AD, et al. Phase II study of SU11248, a multitargeted receptor tyrosine kinase inhibitor (TKI), in patients (pts) with previously treated metastatic breast cancer (MBC). Proc Am Soc Clin Oncol 23: 16, 2005 [Abstr. 563]). The most common toxicities observed in 22 of the patients included diarrhea (32%), nausea (27%), fatigue (23%), hypertension (14%), headache (9%) and rash (5%). Laboratory data available for 37 patients showed grade 3 events of neutropenia (6 pts), thrombocytopenia (2 pts), and AST increase (1 pt), with one grade 4 event of ALT increase, but no treatment-related severe adverse events were reported.

PTK787 is a pan-VEGF, PDGFR, c-kit, and c-Fms receptor tyrosine kinase inhibitor. It inhibited the growth of a broad panel of carcinomas in rodent models, with histological examination revealing inhibition of microvessel formation (110,111). Patients with a variety of advanced cancers have received this agent and it has generally been well tolerated. The Hoosier Oncology Group has recently activated a phase I/II study of PTK787 in combination with trastuzumab in patients with newly diagnosed HER-2 overexpressing, locally recurrent, or metastatic breast cancer. Two randomized, double blind, phase III trials in patients with metastatic colorectal cancer are also ongoing.

ZD6474 is an inhibitor of VEGFR-2 and the EGFR tyrosine kinase. In a cohort of 7,12-dimethylbenz[a]anthracene-treated rats, there was inhibition of the formation of atypical ductal hyperplasia and carcinoma in situ by more than 95% and no invasive disease when they received ZD6474 (112). A phase II trial of this agent in women with previously treated metastatic breast cancer was recently reported. It was generally well tolerated, with 26% of patients experiencing a rash (but none worst than grade 2). There were no objective responses, and one patient had stable disease (113). This was, however, a heavily pretreated population with a median of four prior chemotherapeutic agents.

Inhibition of the VEGF receptor message has been attempted both with ribozyme (catalytic RNA) (114,115) and with antisense VEGF-108 (116). Angiozyme is a synthetic ribosome that cleaves the messenger RNA for the VEGF flt-1 receptor. Preclinical studies confirmed inhibition of both primary tumor growth and metastasis (117,118). In a trial on healthy volunteers, toxicity was minimal, with headaches and somnolence as the main complaints (115). A phase II trial in breast cancer has been reported with this agent and regrettably showed no evidence of clinical activity (119).

25.5.2 Protease Inhibitors

The inhibitors of the matrix metalloproteins (MMPs) currently under study can be grouped into three categories: (a) collagen peptidomimetics and nonpeptidomimetics, (b) tetracycline derivatives, and (c) bisphosphonates. Marimastat is a peptidomimetic that chelates the zinc atom of the active site of the MMPs; it has been the most studied MMP inhibitor. It inhibits a broad spectrum of the MMPs and has activity in multiple human xenograft models (120). Phase I trials have identified musculoskeletal syndromes including arthralgia/arthritis, tendonitis, and bursitis as the dose limiting toxicities with doses ranging from 5 to 10 mg twice daily in patients with advanced cancer (121). Although it is unlikely that MMP inhibitors induce a substantial clinical response in patients with well established bulky tumors, chronic therapy with an MMP inhibitor might in theory delay or eliminate tumor response after surgical excision or systemic chemotherapy. Two completed trials have evaluated marimastat in breast cancer. A pilot feasibility study of this drug evaluated patients with high-risk, node-negative or node-positive breast cancer. Marimastat was given either as a single agent following completion of adjuvant chemotherapy or concurrent with tamoxifen. Arthralgia and arthritis were the most common reported toxicities. Six patients (19%) who received the 5 mg (twice daily) and 11 patients (35%) who received the 10 mg (twice daily) dose discontinued marimastat therapy due to toxicity. In this study, plasma levels measured were rarely within the range for biologic activity (40–200 ng/mL). These findings were discouraging as the toxicity prohibited the maintenance of plasma levels within the test range (122). E2196 was a phase III trial of 190 patients with metastatic breast cancer who had responding or stable disease after six to eight cycles of first-line chemotherapy for metastatic disease. Patients were randomly assigned to receive marimastat or a placebo after the completion of chemotherapy. There was no significant difference in median PFS (4.7 versus 3.1 months; $p = 0.16$) or OS (26.6 versus 24.7 months; $p = 0.86$). Musculoskeletal toxicity was again highly prevalent. Surprisingly, higher trough plasma marimastat levels, at month 1 or 3, were associated with a greater risk for progression and death (123). A number of other trials have indicated no benefit for both peptidomimetic and nonpeptidomimetic MMP inhibitors when used either in combination with chemotherapy or sequentially after first-line chemotherapy in multiple other cancers (124–127).

Other MMP inhibitors have also been examined in breast cancer. BMS-275291 is an orally available peptidomometic MMP inhibitor that contains a mercapto-acyl zinc-binding group. It has antiangiogenic activity in vitro; it inhibited growth of B16F10 murine melanoma (128) and reduced tumor size and metastasis in the rat HOSP-1 mammary carcinoma (129). BMS-275291 inhibits a broad range of MMPs without inhibiting the sheddases, related metalloproteinases, which regulate proinflammatory cytokine and cytokine receptor-shedding from cell surfaces, and preserves the homeostasis of tumor necrosis factor-α (TNF-α), thought to play an important role in the musculoskeletal toxicity associated with marimastat. Combinations of this agent with both taxanes and platinating agents have been studied in the clinic setting (130). We conducted a pilot study to evaluate the role of BMS-275291 in patients with operable breast cancer who had completed adjuvant chemotherapy, to determine the safety of this agent and the feasibility of incorporating as part of adjuvant breast cancer therapy. Patients were randomized to receive BMS-275291 (1,200 mg/day; $n = 38$) or a placebo ($n = 19$) for up to 1 year (131). Main adverse events consisted of arthralgias and skin rash, which led to discontinuation of therapy in 21% of patients receiving BMS-275291 compared with 11% of patients receiving placebo. This rate of arthralgia and myalgia was sufficient to require early termination of the trial.

The available clinical data with MMP inhibitors do not support a role for their use in patients with breast cancer (or with any other malignancies). Whether the failure of these agents is a function of inadequate drug levels, unexpected toxicity, or a more basic failure of the underlying hypothesis is uncertain.

25.5.3 Endothelial Toxins

TNP-470 (AGM-1470), a synthetic derivation of fungillin, was shown to inhibit endothelial cells with low toxicity in animal models (132,133) and inhibited tumor growth and metastasis in preclinical models (134,135), so it was one of the first antiangiogenic agents to enter into clinical studies (136). Phase I studies of TNP-470 found dose-reversible neurological toxicity; only one objective response was reported, although several patients had stable disease (137,138). The half-life of TNP-470 is short, with practically no drug detectable an hour after treatment, suggesting that alternate dosing schedules may be needed for maximum activity (139,140). Antibodies against $\alpha_v\beta_3$ integrin inhibit angiogenesis and tumor growth both in vivo and in vitro (141). Vitaxin, a humanized monoclonal antibody against $\alpha_v\beta_3$ integrin, was well tolerated and showed some activity in a phase I trial, and phase II trials are ongoing (142,143). The RDG (Arg-Gly-Asp) epitope is critical for the function of many of the β_1 integrins, and it is the same epitope that recognizes $\alpha_v\beta_3$ integrin in its extracellular matrix ligands. This has led to the development of RDG-containing peptides that can selectively inhibit the vitronectin receptors. EMD-121974 (Cilengitide) is a cyclic Arg-Gly-Asp peptide with antiangiogenic activity that has been shown to synergize with radioimmunotherapy in breast cancer xenografts (144). Antibodies directed against endoglin have reduced tumor growth in mice xenografts, and antiendoglin antibodies complexed to deglycosylated ricin-A chain produce complete remission of preformed tumors in immunocompromised mice (145,146).

25.5.4 Natural Inhibitors of Angiogenesis

The administration of natural inhibitors of angiogenesis as therapeutics is also an area of intense research. A phase I trial of recombinant human angiostatin in patients with refractory solid tumors found no dose limiting toxicities or changes in coagulation factors. No objective responses were reported, although some patients had measurable decreases in urine bFGF and VEGF levels (147). Treatment with angiostatin has been found to increase sensitivity to radiation (148), and prolonged treatment with angiostatin after radiation therapy was effective in preventing growth of metastasis in the Lewis lung carcinoma animal model (149). Phase I studies of endostatin have also been reported. Patients with refractory solid tumors received daily bolus infusion ranging from 15 to 300 mg/m^2 with no apparent toxicity. Pharmacokinetics was linear, and treatment had no effect on wound healing (150,151). Correlative studies found dose-dependent decreases in tumor-associated blood flow with oxygen-15 PET imaging, increased tumor apoptosis, and decreased peripheral blood endothelial colony-forming precursors. A nonestrogenic metabolite of estradiol with antitumor and antiendothelial cell activity is 2-methoxyestradiol (2-ME$_2$) (152,153). The first phase I trial of 2-ME evaluated 31 patients with previously treated metastatic breast cancer. Seventeen patients had stable disease after the first treatment period and the therapy was well tolerated (154). A phase I study of 2-ME plus docetaxel in patients with metastatic breast cancer has also been performed. The overall response rate was 20% (including one complete response) and an additional 40% of patients had stable disease. Concurrent therapy with docetaxel and 2-ME was well tolerated and did not alter the pharmacokinetics of docetaxel or 2-ME (155). In these studies, however, 2-ME serum levels were well below those required for activity based on preclinical models. A new formulation with improved bioavailability has now entered clinical trials.

25.6 RESISTANCE TO INHIBITORS OF ANGIOGENESIS

Because normal endothelial cells are genetically stable, antiangiogenic therapy was initially theorized to be a treatment "resistant to resistance" (156). Initial xenograft studies supported these theoretical predictions: widespread activity, limited toxicity, and no resistance (157).

TABLE 25.6
Mechanisms of Resistance to Angiogenesis Inhibition

- Endothelial-cell heterogeneity
- Angiogenic-factor heterogeneity
- Tumor-cell heterogeneity
- Impact of the tumor microenvironment
- Compensatory responses to treatment
- Angiogenesis-independent tumor growth
- Pharmacokinetic resistance

For a time it was argued that disease control, if not outright cure, was close at hand. Regrettably, a growing body of laboratory and clinical data suggests that the problem of resistance continues. In support of this contention, we have proposed several theoretical and (often) substantiated mechanisms of acquired and de novo resistance to antiangiogenic therapies (Table 25.6).

One initial reason for belief in the "resistance to resistance" theory was that the endothelium was genetically stable in contradiction to the surrounding tumor, which was known to have frequent mutations (158). Mounting evidence now supports the concept that there is significant heterogeneity to the endothelium of vessels involved in angiogenesis, as well as sporadic mutations that they share with the parent tumor. Developing endothelium is dynamic and capable of differential gene expression based on the physiologic requirements and microenvironment of the associated tissue. A recent study found that 15%–85% of the endothelial cells of the vasculature in non-Hodgkin's lymphoma tumors carried chromosomal abnormalities that were also harbored in the lymphoma cells (159).

Beyond tumor-related mutational events (admittedly rare), there is growing evidence for the role of DNA polymorphisms for both pro- and antiangiogenic factors. Interindividual variability in drug response may be related to variations in host genes important in drug metabolism or transport. In addition to conferring a better or worse response to therapy, these polymorphisms may also play an important role in the development of cancers and development of early metastasis (160). This is plausible in a number of ways, such as less effective DNA repair mechanisms, more efficient ability to create tumor blood supply, and a quicker or slower ability to metabolize anticancer drugs, among others. A variety of polymorphisms in genes known to be important in the angiogenesis pathway have been shown to correlate with the likelihood of developing breast cancer (23) as well as serving as important prognostic markers (24,161). Collectively, these data suggest that considerable angiogenic heterogeneity is hardwired into individual patients, and that polymorphic genes may play an important functional role in angiogenesis and breast cancer.

Invasive cancers commonly express multiple angiogenic factors, and from a clinical standpoint, this heterogeneity occurs at an early point in time. Multiple different proangiogenic factors are found in primary breast tumors; genetic instability may result in modulation of both the amount and type of proangiogenic factors expressed in an individual tumor (162). Tumor heterogeneity may imply more than just heterogeneity of proangiogenic factors. It has been shown that disruption of p53 in tumor cells reduces sensitivity to antiangiogenic metronomic therapy (163).

Preclinical tumor models have demonstrated that orthotropic tumors have higher VEGF expression as well as more robust growth when compared with subcutaneous tumors (164). Culture systems have also demonstrated that the medium conditioned by malignant cells provides a more stable environment for tumor and vessel growth than those conditioned by nonmalignant cells (165). As many pro- and antiangiogenic factors are contained in or released from the extracellular matrix, differential sensitivity based on site of disease may be anticipated. For example, treatment with the metalloproteinase inhibitor batimastat had different effects on tumor progression and growth depending on the site of tumor implantation (166). The tumor microenvironment may also affect the delivery of the drug (167).

It seems reasonable to expect VEGF production to increase in response to treatment with the pure antiangiogenics as well. Indeed, VEGF levels increase after therapy with doxorubicin and a VEGFR tyrosine kinase inhibitor (168). Hypoxia may be chronic due to consumption/diffusion limitations or periodic resulting from transient reductions in tumor blood flow (so-called cyclic hypoxia) (169). It is reasonable to expect that these compensatory responses will be invoked by human cancers undergoing antiangiogenic attack.

Vessel cooption, growth by intussusceptions, vascular mimicry, and vasculogenesis may decrease a tumor's dependence on classical angiogenesis. Multiple studies from tumor models have demonstrated the ability of tumors to rely on these alternative methods to obtain the necessary blood supply when classical angiogenesis is not permitted (170–179). Recent data also suggest that inflammatory breast cancer relies almost entirely on vasculogenesis as opposed to angiogenesis, apparently due to the inability of the cancer cells to bind endothelial cells (180). If classic angiogenesis is not the predominant mechanism by which a tumor gets its blood supply, can antiangiogenic therapy be expected to succeed? This question still requires answering at the clinical level, but preclinical models have begun to explore the relative resistance of alternative blood supplies to antiangiogenic agents. It has been demonstrated that cancers characterized by vasculogenic mimicry are resistant to the antiangiogenic agent endostatin (181). In contrast, bone marrow–derived endothelial cells remain sensitive to VEGF-targeting therapy (182).

Preclinical studies of novel agents often fail to anticipate the dynamic nature of interactions between drug, host, and tumor. It is a peculiar failing of modern science that the study of simple-indeed, simplistic model systems at the molecular or cellular level is considered more scientific than the study of complex, whole systems. And yet the failure of many antiangiogenic agents clearly occurs exactly at the higher-order levels of complexity seen in whole systems. These failures represent what might be termed pharmacokinetic resistance: the inability to deliver the right dose of a biologically active agent to the right cells for the right period of time.

25.7 OVERCOMING RESISTANCE TO ANTIANGIOGENIC THERAPY

The reality of human tumors and initial clinical experience with novel antiangiogenic agents confirms that resistance remains an obstacle. Acknowledging the persistent problem of resistance does not imply that antiangiogenic therapy will prove fruitless, nor that resistance will befall all patients in this setting. Rather it is meant to suggest that the justified enthusiasm for a novel therapeutic modality should not blind us to the real challenges ahead. Understanding the potential mechanisms of antiangiogenesis resistance suggests several possible means to ameliorate or bypass such resistance (Table 25.7).

Extensive preclinical data support a combined approach, with multiple agents that have additive or synergistic combinatorial activity (76,77,183,184). The mechanistic rationale for many of these combinations is poorly understood, and not intuitive, as both radiotherapy and chemotherapy depend on an effective blood supply for therapeutic efficacy. A potential explanation may lie in the inherent inefficiency of the tumor vasculature. Antiangiogenic therapy normalizes flow initially resulting in improved tissue oxygenation and decreased interstitial pressure, increasing delivery of cytotoxic agents (185).

As tumor progression is associated with expression of increasing numbers of proangiogenic factors, the use of different antiangiogenic agents to simultaneously attack this

TABLE 25.7
Overcoming Resistance to Antiangiogenic Therapy

- Combine antiangiogenic agents with standard cytotoxic regimens
- Combine multiple antiangiogenic agents
- Combine antiangiogenic agents with other biologically targeted agents
- Use antiangiogenic therapy as adjuvant therapy
- Use antiangiogenic therapy as targeted therapy

multiply redundant process may overcome resistance to individual agents. This approach is, of course, not unique to antiangiogenic therapy, as it has previously been used to limit resistance to cytotoxic, antimicrobial, and antiviral therapies. The combination of antiangiogenic agents has been tested in preclinical models (186). Combined blockade of the VEGF and EGFR (103) pathways as well as the VEGF and HER-2 pathways (97) is actively studied. As our understanding of tumor biology improves, concomitant blockade of other pathways will certainly be explored as well.

Rarely is a treatment more effective against large tumors than small tumors. Tumor progression results in increased resistance to anticancer therapies. One mean of overcoming the development of drug resistance is to treat cancers when they are small. The adjuvant setting is the logical place to accomplish this goal. The use of antiangiogenics as adjuvant therapy has its own potential barriers. The toxicity of chronic antiangiogenic therapy remains largely unexplored, as is the toxicity of combinations of chemotherapy with antiangiogenic therapy. Although intuitively the impact of angiogenesis inhibition is expected to be greatest in patients with micrometastatic disease, proof of this concept requires commitment of substantial human and financial resources to a randomized adjuvant trial.

Antiangiogenic therapy has been applied as a general therapy given on a population basis, rather than as targeted therapy given to the patients with a specific molecular phenotype. It is reasonable to ask whether we can call failure to respond to a therapy resistance if the target at which the therapy is directed is not present in the tumor. If a patient's tumor does not express VEGF and therefore fails to respond to anti-VEGF therapy, is the tumor resistant to the therapy or is the therapy merely misguided? As insensitivity due to lack of therapeutic targets results in resistance at the patient level, proper targeting is a means of overcoming such resistance. Ideal targets are biologically relevant, reproducibly measurable, and definably correlated with clinical benefits.

25.8 CONCLUSION

The results of E2100 have provided the first proof-of-concept evidence for antiangiogenic therapy in breast cancer. The results of this trial will provide a continuing stimulus to investigators in coming years. It is already clear that the next major challenge will involve the transition of antiangiogenic therapy to the adjuvant setting. Other challenges include the discovery and validation of surrogate markers, which could help gauge response to therapy, as well as allow tailoring treatments according to differential expression of markers. Further work is needed to optimize combination of antiangiogenics and chemotherapy, as well combining agents that target different pathways in the biology of angiogenesis to overcome resistance. It seems likely that this field will continue to grow over the next years, and will improve the survival and likelihood of ultimate cure for patients with breast cancer.

ACKNOWLEDGMENT

This work was supported by a grant from the Breast Cancer Research Foundation.

REFERENCES

1. Folkman J. Tumor angiogenesis: Therapeutic implications. *N Eng J Med* 285: 1182–1186, 1971.
2. Brem SS, Gullino PM, Medina D. Angiogenesis: A marker for neoplastic transformation of mammary papillary hyperplasia. *Science* 195: 880–882, 1977.
3. Jensen HM, Chen I, DeVault MR, et al. Angiogenesis induced by "normal" human breast cell tissue: A probable marker for precancer. *Science* 218: 293–295, 1982.
4. Miller K, Sledge GW. Dimming the blood tide: Angiogenesis, antiangiogenic therapy and breast cancer, in Nabholtz JM (ed): *Breast Cancer Management Application of Clinical and Translational Evidence to Patient Care* (2nd edn). Philadelphia, PA, Lippicott Williams & Wilkins, 2003, pp 287–308.
5. Bos R, Zhong H, Hanrahan CF, et al. Levels of hypoxia-inducible factor 1 alpha during breast carcinogenesis. *J Natl Cancer Inst* 93: 309–314, 2001.
6. Chia SK, Wykoff CC, Watson PH, et al. Prognostic significance of a novel hypoxia-regulated marker, a carbonic anhydrase IX, in invasive breast carcinoma. *J Clin Oncol* 19: 3660–3668, 2001.
7. Wykoff CC, Beasley NJ, Watson PH, et al. Hypoxia-inducible expression of tumor associated carbonic anhydrases. *Cancer Res* 60: 7075–7083, 2000.
8. Guinebretiere JM, LeMonique G, Gavoille A, et al. Angiogenesis and risk of breast cancer in women with fibrocystic disease. *J Natl Cancer Inst* 86: 635–636, 1994.
9. Engels K, Fox SB, Whitehouse RM, et al. Distinct angiogenic patterns are associated with high-grade in situ ductal carcinomas of the breast. *J Pathol* 181: 207–212, 1997.
10. Guidi AJ, Fischer L, Harris JR, et al. Microvessel density and distribution in ductal carcinoma in situ (DCIS) of the breast. *J Natl Cancer Inst* 86: 614–619, 1994.
11. Guidi AJ, Schnitt SJ, Fischer L, et al. Vascular permeability factor (vascular endothelial growth factor) expression and angiogenesis in patients with ductal carcinoma in situ (DCIS) of the breast. *Cancer* 80: 1945–1953, 1997.
12. Weidner N, Semple J, Welch W, et al. Tumor angiogenesis and metastasis-correlation in invasive breast cancer. *N Engl J Med* 324: 1–8, 1991.
13. Weidner N, Folkman J, Pozza F, et al. Tumor angiogenesis: A new significant and independent prognostic indicator in early-stage breast carcinoma. *J Natl Cancer Inst* 84: 1875–1887, 1992.
14. Horak E, Leek R, Klenk N. Angiogenesis, assessed by platelet/endothelial cell adhesion molecule antibodies, as an indicator of node metastasis and survival in breast cancer. *Lancet* 340: 1120–1124, 1992.
15. Martin L, Green B, Renshaw C, et al. Examining the technique of angiogenesis assessment in invasive breast cancer. *Br J Cancer* 76: 40–43, 1997.
16. Simpson J, Ahn C, Battifora H, et al. Endothelial area as a prognostic indicator for invasive breast carcinoma. *Cancer* 77: 2077–2085, 1996.
17. Vermeulen PB, Libura M, Libura J, et al. Influence of investigator experience and microscopic field size on microvessel density in node-negative breast carcinoma. *Breast Cancer Treat Res* 42: 165–172, 1997.
18. Weidner N. Current pathologic methods for measuring intratumoral microvessel density within breast carcinoma and other solid tumors. *Breast cancer Treat Res* 36: 169–180, 1995.
19. Fox SB, Leek RD, Weeks MP, et al. Quantitation and prognostic value of breast cancer angiogenesis: Comparison of microvessel density. Chalkley count, and computer image analysis. *J Pathol* 177: 275–283, 1995.
20. de Jong JS, van Diest PJ, Baak JP. Heterogeneity and reproducibility of microvessel counts in breast cancer. *Lab Invest* 73: 922–926, 1995.
21. Uzzan B, Nicolas P, Cucherat M, et al. Microvessel density as a prognostic factor in women with breast cancer: A systematic review of the literature and meta-analysis. *Cancer Res* 64: 2941–2955, 2004.
22. Relf M, LeJeune S, Scott PA, et al. Expression of the angiogenic factors vascular endothelial cell growth factor, acidic and basic fibroblast growth factor, tumor growth factor beta-1, platelet-derived endothelial cell growth factor, placenta growth factor, and pleiotrophin in human primary breast cancer and its relation to angiogenesis. *Cancer Res* 57: 963–969, 1997.

23. Krippl P, Langsenlehner U, Renner W, et al. A common 936 C/T gene polymorphism of vascular endothelial growth factor is associated with decreased breast cancer risk. *Int J Cancer* 106: 468–471, 2003.
24. Langsenlehner U, Samonigg H, Krippl P. Vascular endothelial growth factor gene polymorphism decreases the risk for breast cancer metastasis. *Breast Cancer Res Treat* 88: S72, 2004 (abstr 1090).
25. Gasparini G, Toi M, Gion M, et al. Prognostic significance of vascular endothelial growth factor protein in node-negative breast carcinoma. *J Natl Cancer Inst* 89: 139–147, 1997.
26. Gasparini G, Toi M, Miceli R, et al. Clinical relevance of vascular endothelial growth factor and thymidine phosphorylase in patients with node-positive breast cancer treated with either adjuvant chemotherapy or hormone therapy. *Cancer J Sci Am* 5: 101–111, 1999.
27. Foekens JA, Peters HA, Grebenchtchikov N, et al. High tumor levels of vascular endothelial growth factor predict poor response to systemic therapy in advanced breast cancer. *Cancer Res* 61: 5407–5414, 2001.
28. Ragaz J, Miller K, Badve S, et al. Adverse association of expressed vascular endothelial growth factor (VEGF), Her2, Cox2, uPA and EMSY with long-term outcome on stage I-III breast cancer (BrCa). Results from the British Columbia Tissue Microarray Project. *Proc Am Soc Clin Oncol* 23: 8, 2004 (abstr 524).
29. Dirix L, Vermeulen P, Pawinski A, et al. Elevated levels of the angiogenic cytokines basic fibroblast growth factor and vascular endothelial growth factor in the sera of cancer patients. *Br J Cancer* 76: 238–243, 1997.
30. Salven P, Perhoniemi V, Tykka H, et al. Serum VEGF levels in women with a benign breast tumor or breast cancer. *Breast Cancer Res Treat* 53: 161–166, 1999.
31. Yamamoto Y, Toi M, Kondo S, et al. Concentrations of vascular endothelial growth factor in the sera of normal controls and cancer patients. *Clin Cancer Res* 2: 821–826, 1996.
32. Verheul H, Hoekman K, Luykx-de Bakker S, et al. Platelet: Transporter of vascular endothelial growth factor. *Clin Cancer Res* 3: 2187–2190, 1997.
33. Wynendaele W, Derua R, Hoylaerts M, et al. Vascular endothelial growth factor measured in platelet poor plasma allows optimal separation between cancer patients and volunteers: A key to study and angiogenic marker in vivo? *Ann Oncol* 10: 965–971, 1999.
34. Adams J, Carder P, Downey S, et al. Vascular endothelial growth factor (VEGF) in breast cancer: Comparison of plasma, serum and tissue VEGF and microvessel density and effects of tamoxifen. *Cancer Res* 60: 2898–2905, 2000.
35. Braybrooke JP, O'Byrne KJ, Propper DJ, et al. A phase II study of razoxane, an antiangiogenic topoisomerase II inhibitor, in renal cell cancer with assessment of potential surrogate markers of angiogenesis. *Clin Cancer Res* 6: 4697–4704, 2000.
36. Zon R, Neuberg D, Wood W, et al. Correlation of plasma VEGF with clinical outcomes in patients with metastatic breast cancer. *Proc Am Soc Clin Oncol* 17: 185, 1998.
37. Byrne G, Blann A, Venizelos J, et al. Serum soluble VCAM: A surrogate marker of angiogenesis. *Breast Cancer Treat Res* 50: 330, 1998.
38. Banks RE, Gearing AJ, Hemingway IK, et al. Circulating intercellular adhesion molecule-1 (ICAM-1), E-selectine and vascular cell adhesion molecule-1 (VCAM-1) in human malignancies. *Br J Cancer* 68: 122–124, 1993.
39. Kossakowska AE, Huchcroft SA, Urbanski SJ, et al. Comparative analysis of the expression patterns of metalloproteinases and their inhibitors in breast neoplasia, sporadic colorectal neoplasia, pulmonary carcinomas and malignant non-Hodgkin's lymphomas in humans. *Br J Cancer* 73: 1401–1408, 1996.
40. Colomer R, Aparicio J, Montero S, et al. Low levels of basic fibroblast growth factor (bFGF) are associated with prognosis in breast cancer. *Br J Cancer* 76: 1215–1220, 1997.
41. Nguyen M, Watanabe H, Budson A, et al. Elevated levels of an angiogenic peptide, basic fibroblast growth factor, in the urine of patients with a spectrum of cancers. *J Natl Cancer* Inst 86: 356–361, 1994.
42. Folkman J. Tumor angiogenesis: Diagnostic and therapeutic implications. *Am Assoc Cancer Res* 34: 571–572, 1993.
43. Sliutz G, Tempfer C, Obermair A, et al. Serum evaluation of basic FGF in breast cancer patients. *Anticancer Res* 15: 2675–2677, 1995.

44. Ferrara KW, Merritt CR, Burns PN, et al. Evaluation of tumor angiogenesis with US: Imaging, Doppler and contrast agents. *Acad Radiol* 7: 824–839, 2000.
45. Foltz RM, McLendon RE, Friedman HS, et al. A pial window model for the intracranial study of human glioma microvascular function. *Neurosurgery* 36: 976–984, 1995.
46. Chaudhari MH, Forsberg F, Voodarla A, et al. Breast tumor vascularity identified by contrast enhanced ultrasound and pathology: Initial results. *Ultrasonics* 38: 105–109, 2000.
47. Jansson T, Westlin JE, Ahlstrom H, et al. Positron emission tomography studies in patients with locally advanced and/or metastatic breast cancer: A method for early therapy evaluation? *J Clin Oncol* 13: 1470–1477, 1995.
48. Bassa P, Kim EE, Inoue T, et al. Evaluation of preoperative chemotherapy using PET with fluorine-18-fluorodeoxyglucose in breast cancer. *J Nucl Med* 37: 931–938, 1996.
49. Wahl RL, Zasadny K, Helvie M, et al. Metabolci monitoring of breast cancer chemotherapy using positron emission tomography: Initial evaluation. *J Clin Oncol* 11: 2101–2111, 1993.
50. Buck AK, Schirrmeister H, Mattfeldt T, et al. Biological characterization of breast cancer by means of PET. *Eur J Nucl Med Mol Imaging* 31: S80–S87, 2004.
51. Bos R, van der Hoeven JM, van der Wall E, et al. Biologic correlates of ^{18}Fluorodeoxyglucose uptake in human breast cancer measured by positron emission tomography. *J Clin Oncol* 20: 379–387, 2002.
52. Arslan N, Ozturk E, Ilgan S, et al. 99Tcm-MIBI scintimammography in the evaluation of breast lesions and axillary involvement: A comparison with mammography and histopathological diagnosis. *Nucl Med Commun* 20: 317–325, 1999.
53. Mankhoff D, Dunnwald L, Gralow J, et al. Monitoring the response of patients with locally advanced breast carcinoma to neoadjuvant chemotherapy using technetium 99m-sestambibi scintimammography. *Cancer* 85: 2410–2423, 1999.
54. Yoon JH, Bom HS, et al. Double-phase Tc-99m sestambibi scintimammography to assess angiogenesis with P-glycoprotein expression in patients with untreated breast cancer. *Clin Nucl Med* 24: 314–318, 1999.
55. Wilson CB, Lammertsma AA, Jones T, et al. Measurements of blood flow and exchanging water space in breast tumors using positron emission tomography: A rapid and noninvasive dynamic method. *Cancer Res* 52: 1592–1597, 1992.
56. Martin G, Caldwell J, Graham M, et al. Noninvasive detection of hypoxic myocardium using fluorin-18-fluoromisonidazole and positron emission tomography. *J Nucl Med* 33: 2202–2208, 1992.
57. Ogawa T, Uenemura K, Shishido F, et al. Changes of cerebral blood flow and oxygen and glucose metabolism following radiochemotherapy of gliomas: A PET study. *J Comput Assist Tomog* 12: 290–297, 1988.
58. Blakenberg FG, Eckelman WC, Strauss HW, et al. Role of radionuclide imaging in trials of antiangiogenic therapy. *Acad Radiol* 7: 851–867, 2000.
59. Stejskal E. Use of spin echos in a pulsed magnetic field gradient to study anisotropic, restricted diffusion and flow. *J Chem Phys* 43: 3597–3603, 1965.
60. Yamada I, Aung W, Himeno Y, et al. Diffusion coefficients in abdominal organs and hepatic lesions: Evaluation with intravoxel incoherent motion echo-planar MR imaging. *Radiology* 210: 617–623, 1999.
61. van Ziji P, Eleff S, Ulatowski J, et al. Quantitative assessment of blood flow, blood volume and blood oxygenation effects in functional magnetic resonance imaging. *Nat Med* 4: 159–167, 1998.
62. Abramovitch R, Frenkiel D, Neeman M. Analysis of subcutaneous angiogenesis by gradient echo magnetic resonance imaging. *Mag Reson Med* 39: 813–824, 1998.
63. Kuhl CK, Schild HH. Dynamic image interpretation of MRI of the breast. *J Mag Reson Imaging* 12: 965–974, 2000.
64. Su YM, Cheung YC, Fruehauf JP, et al. Correlation of dynamic contrast enhancement MRI parameters with microvessel density and VEGF for assessment of angiogenesis in breast cancer. *J Magn Reson Imaging* 18: 467–477, 2003.
65. Miller KD, Soule SE, Calley C, et al. Randomized phase II trial of the anti-angiogenic potential of doxorubicin and docetaxel; primary chemotherapy as biomarker discovery laboratory. *Breast Cancer Res Treat* 89: 187–197, 2005.

66. Linderholm B, Lindh B, Beckman L, et al. The prognostic value of vascular endothelial growth factor (VEGF) and basic fibroblast growth factor (bFGF) and associations to first metastases site in 1307 patients with primary breast cancer. *Proc Am Soc Clin Oncol* 20: 4a, 2001.
67. Miller KD, Sweeney CJ, Sledge GW, Jr. Redefining the target: Chemotherapeutics as antiangiogenics. *J Clin Oncol* 19: 1195–1206, 2001.
68. Wiseman H. Tamoxifen: New membrane-mediated mechanisms of action and therapeutic advances. *Trends Pharmacol Sci* 15: 83–89, 1994.
69. Gagliardi AR, Collins DC. Inhibition of angiogenesis by antiestrogens. *Cancer Res* 53: 533–535, 1993.
70. Gagliardi AR, Henning B, Collins DC. Antiestrogens inhibit endothelial cell growth stimulated by angiogenic growth factors. *Anticancer Res* 16: 1101–1106, 1996.
71. Haran EF, Maretzek AF, Goldberg I, et al. Tamoxifen enhances cell death in implanted MCF7 breast cancer by inhibiting endothelium growth. *Cancer Res* 54: 5511–5514, 1994.
72. Lindner DJ, Borden EC. Effects of tamoxifen and interferon-beta or the combination on tumor induced angiogenesis. *Int J Cancer* 71: 456–461, 1997.
73. Silva ID, Salicioni AM, Russo IH, et al. Tamoxifen down-regulates CD36 messenger RNA levels in normal and neoplastic human breast tissues. *Cancer Res* 57: 378–381, 1997.
74. Sweeney CR, Sledge GW, Jr. Chemotherapy agents as antiangiogenic therapy. *Cancer Conf Highlights* 3: 2–4, 1999.
75. Slaton JW, Perrote P, Inoue K, et al. Interferon-alpha mediated down-regulation of angiogenesis related genes and therapy of bladder cancer are dependent on optimization of biological dose and schedule. *Clin Cancer Res* 5: 2726–2734, 1999.
76. Browder T, Butterfield CE, Kraling BM, et al. Antiangiogenic scheduling of chemotherapy improves efficacy against experimental drug-resistant cancer. *Cancer Res* 60: 1878–1886, 2000.
77. Klement G, Baruchel S, Rak J, et al. Continuous low-dose therapy with vinblastine and VEGF receptor-2 antibody induces sustained tumor regression without overt toxicity. *J Clin Invest* 105: R15–R24, 2000.
78. Wild R, Ding R, Subramanian I, et al. Carboplatin selectively induces the VEGF stress response in endothelial cells: Potentiation of antitumor activity by combination treatment with antibody to VEGF. *Int J Cancer* 110: 343–351, 2004.
79. Hanhanan D, Bergers G, Bergsland E. Less is more, regularly: Metronomic dosing of cytotoxic drugs can target tumor angiogenesis in mice. *J Clin Invest* 105: 1045–1047, 2000.
80. Bocci G, Francia G, Man S, et al. Thormbospondin 1, a mediator of the antiangiogenic effects of low-dose metronomic chemotherapy. *Proc Natl Acad Sci USA* 100: 12917–12922, 2003.
81. Colleoni M, Rocca A, Sandri MT, et al. Low-dose oral methotrexate and cyclophosphamide in metastatic breast cancer: Antitumor activity and correlation with vascular endothelial growth factor levels. *Ann Oncol* 13: 73–80, 2002.
82. Hurwitz H, Fehrenbacher L, Novotny W, et al. Bevacizumab plus irinotecan, fluorouracil, and leucovorin for metastatic colorectal cancer. *N Engl J Med* 350: 2335–2342, 2004.
83. Hainsworth J, Sosma J, Spiegal D, et al. Phase II trial of bevacizuma and erlotinib in patients with metastatic renal carcinoma (RCC). *Proc Am Soc Clin Oncol* 23: 381, 2004 (abstr 4502).
84. Yang JC, Haworth L, Sherry RM, et al. A randomized trial of bevacizumab, an anti-vascular endothelial growth factor antibody, for metastatic renal cancer. *N Eng J Med* 349: 427–434, 2003.
85. Sledge GW, Miller KD, Novotny W, et al. A phase II trial of single agent rhumab VEGF (recombinant humanized monoclonal antibody to vascular endothelial growth factor) in patients with relapsed metastatic breast cancer. *Proc Am Soc Clin Oncol* 19: 3a, 2000 (abstr 5c).
86. Gordon MS, Margolin K, Talpaz M, et al. Phase I safety and pharmacokinetic study of recombinant human anti-vascular endothelial growth factor in patients with advanced cancer. *J Clin Oncol* 19: 843–850, 2001.
87. Burnstein H, Parker L, Savoie J, et al. Phase II trial of the anti-VEGF antibody bevacizumab in combination with vinorelbine for refractory advanced breast cancer. *Breast Cancer Res Treat* 76: S115, 2002 (abstr 224).
88. Ramaswamy B, Rhoades C, Kendra K, et al. CTEP-sponsored phase II trial of bevacizumab (Avastin) in combination with docetaxel (Taxotere) in metastatic breast cancer. *Breast Cancer Res Treat* 82: S50, 2003 (abstr 224).

89. Miller K, Rugo H, Cobleigh M, et al. Phase III trial of capecitabine (Xeloda) plus bevacizumab (Avastin) versus capacitabine alone in women with metastatic breast cancer (MBC) previously treated with an anthracycline and a taxane. *Breast Cancer Res Treat* 88: S106, 2004 (abstr 2088).
90. Hillan K, Koeppen H, Tobin P, et al. The role of VEGF expression in response to bevacizumab plus capecitabine in metastatic breast cancer (MBC). *Proc Am Soc Clin Oncol* 22: 191, 2003 (abstr 766).
91. Overmoyer B, Silverman P, Leeming R, et al. Phase II trial of neoadjuvant docetaxel with or without bevacizumab in patients with locally advanced breast cancer. *Breast Cancer Res Treat* 88: S106, 2004 (abstr 2088).
92. Wedam S, Low J, Yang X, et al. A pilot study to evaluate response and angiogenesis after treatment with bevacizumab in patients with inflammatory breast cancer. *Proc Am Soc Clin Oncol* 23: 21, 2004 (abstr 578).
93. Koukourakis MI, Giatromanolaki A, O'Byrne KJ, et al. bcl-2 and c-erbB-2 proteins are involved in the regulation of VEGF and of thymidine phosphorylase angiogenic activity in non-small-cell lung cancer. *Clin Exp Metastasis* 17: 545–554, 1999.
94. Laughner E, Taghavi P, Chiles K, et al. HER2 (neu) signaling increases the rate of hypoxia-inducible factor 1alpha (HIF-1alpha) synthesis: Novel mechanism for HIF-1-mediated vascular endothelial growth factor expression. *Mol Cell Biol* 21: 3995–4004, 2001.
95. Epstein M, Ayala R, Tchekmedyina N, et al. HE2-overexpressing human breast cancer xenografts exhibit increased angiogenic potential mediated by vascular endothelial growth factor (VEGF). *Breast Cancer Res Treat* 76: S143, 2002 (abstr 570).
96. Konecny GE, Meng YG, Untch M, et al. Association between HER-2/neu and vascular endothelial growth factor expression predicts clinical outcome in primary breast cancer patients. *Clin Cancer Res* 10: 1706–1716, 2004.
97. Pegram M, Yeon C, Ku N, et al. Phase I combined biological therapy of breast cancer using two humanized monoclonal antibodies directed against HER2 proto-oncogene and vascular endothelial growth factor (VEGF). *Breast Cancer Res Treat* 88: S124, 2004 (abstr 3039).
98. Maity A, Pore N, Lee J, et al. Epidermal growth factor receptor transcriptionally up-regulates vascular endothelial growth factor expression in human glioblastoma cells via pathway involving phosphatidylinositol 3″-kinase and distinct from that induced by hypoxia. *Cancer Res* 60: 5879–5886, 2000.
99. Clarke K, Smith K, Gullick WJ, et al. Mutant epidermal growth factor receptor enhances induction of vascular endothelial growth factor by hypoxia and insulin-like growth factor-1 via PI3 kinase dependant pathway. *Br J Cancer* 84: 1322–1329, 2001.
100. Bruns CJ, Solorzano CC, Harbison MT, et al. Blockade of the epidermal growth factor receptor signaling by a novel tyrosine kinase inhibitor leads to apoptosis of endothelial cells and therapy of human pancreatic carcinoma. *Cancer Res* 60: 2926–2935, 2000.
101. Petit AM, Rak J, Hung MC, et al. Neutralizing antibodies against epidermal growth factor and ErbB-2/neu receptor tyrosine kinases down-regulate vascular endothelial growth factor production by tumor cells in vitro and in vivo: Angiogenic implications for signal transduction therapy of solid tumors. *Am J Pathol* 151: 1523–1530, 1997.
102. Viloria-Petit A, Crombet T, Jothy S, et al. Acquired resistance to the antitumor effect of epidermal growth factor receptor-blocking antibodies in vivo: A role for altered tumor angiogenesis. *Cancer Res* 61: 5090–5101, 2001.
103. Rugo H, Dickler M, Cott J, et al. Circulating endothelial cell (CEC) and tumor cell (CTC) analysis in patients (pts) receiving bevacizumab and erlotinib for metastatic breast cancer (MBC). *Breast Cancer Res Treat* 88: S142, 2004 (abstr 3088).
104. Zhu Z, Witte L. Inhibition of tumor growth and metastasis by targeting tumor-associated angiogenesis with antagonists to the receptors of vascular endothelial growth factor. *Invest New Drugs* 17: 195–212, 1999.
105. Kozin S, Boucher Y, Hicklin D, et al. Vascular endothelial growth factor receptor-2-blocking antibody potentiates radiation-induced long term control of human tumor xenografts. *Cancer Res* 61: 39–44, 2001.
106. Lin P, Snakar S, Shan J, et al. Inhibition of tumor growth by targeting tumor endothelium using a soluble vascular endothelial growth factor receptor. *Cell Growth Differ* 9: 49–58, 1998.

107. Abrams TJ, Murray LJ, Presenti E, et al. Preclinical evaluation of the tyrosine kinase inhibitor SU11248 as a single agent and in combination with "standard of care" therapeutic agents for the treatment of breast cancer. *Mol Cancer Ther* 2: 1001–1021, 2003.
108. Murray LJ, Abrams TJ, Long KR, et al. SU011248 inhibits tumor growth and CSF-1R-dependent osteolysis in an experimental breast cancer bone metastasis model. *Clin Exp Metastasis* 20: 757–766, 2003.
109. Miller K, Miller M, Mehrotra S, et al. The search for surrogates—physiologic imaging in a breast cancer xenograft model during treatment with SU011248. *Breast Cancer Res Treat* 82: S18, 2003 (abstr 38).
110. Drevs J, Hofmann I, Hugenschmidt H, et al. Effects of PTK787/ZK 222584, a specific inhibitor of vascular endothelial growth factor tyrosine kinases, on primary tumor, metastasis, vessel density, and blood flow in murine renal cell carcinoma model. *Cancer Res* 60: 4819–4824, 2000.
111. Wood JM, Bold G, Buchdunger E, et al. PTK787/ZK 222584, a novel and potent inhibitor of vascular endothelial growth factor receptor tyrosine kinases, impairs vascular endothelial growth factor-induced responses and tumor growth after oral administration. *Cancer Res* 60: 2178–2189, 2000.
112. Heffelfinger SC, Yan M, Gear RB, et al. Inhibition of VEGFR2 prevents DMBA-induced mammary tumor formation. *Lab Invest* 84: 989–998, 2004.
113. Miller K, Trigo J, Stone A, et al. A phase II trial of ZD6474, a vascular endothelial growth factor receptor-2 (VEGFR-2) and epidermal growth factor receptor (EGFR) tyrosine kinase inhibitor, in patients with previously treated metastatic breast cancer (MBC). *Breast Cancer Res Treat* 88: S240, 2004 (abstr 6060).
114. Weng D, Weiss P, Kellackey C, et al. Angiozyme pharmacokinetic and safety results: A phase I/II study in patients with refractory solid tumors. *Proc Am Soc Clin Oncol* 20: 99a, 2001.
115. Sandber S, Parker V, Blanchar K, et al. Pharmacokinetics and tolerability of anti-angiogenic ribozyme (Angiozyme) in healthy volunteers. *Pharmacology* 40: 1462–1469, 2000.
116. Im S, Kim J, Gomez-Manzano C, et al. Inhibition of breast cancer growth in vivo by antiangiogenesis gene therapy with adenovirus mediated antisense-VEGF. *Br J Cancer* 84: 1252–1257, 2001.
117. Sandberg JA, Bouhana KS, Gallegos AM, et al. Pharmacokinetics of an antiangiogenic ribozyme (Angiozyme) in the mouse. *Antisense Nucleic Acid Drug Dev* 9: 271–277, 1999.
118. Sandberg JA, Sproul CD, Blanchard KS, et al. Acute toxicology and pharmacokinetic assessment of a ribozyme (Angiozyme) targeting vascular endothelial growth factor mRNA in the cynomolgus monkey. *Antisense Nucleic Acid Drug Dev* 10: 153–162, 2000.
119. Hortobagyi GN, Weng D, Elias A, et al. Angiozyme treatment of stage IV metastatic breast cancer patients: Assessment of serum markers of angiogenesis. *Breast Cancer Res Treat* 76: S97, 2002.
120. Ferrante K, Winograd B, Canneta R. Promising new developments in cancer chemotherapy. *Cancer Chemother Pharmacol* 43 [suppl 1]: 561–568, 1999.
121. Nemunaitis J, Poole C, Primrose J, et al. Combined analysis of studies of the effects of the matrix metalloproteinase inhibitor marimastat on serum tumor markers in advanced cancer: Selection of biologically active and tolerable dose for long-term studies. *Clin Cancer Res* 4: 1101–1109, 1998.
122. Miller KD, Gradishar W, Schuchter L, et al. A randomized phase II pilot study of adjuvant marimastat in patients with early-stage breast cancer. *Ann Oncol* 13: 1220–1224, 2002.
123. Sparano JA, Bernardo P, Stephenson P, et al. Randomized phase III trial of marimastat versus placebo in patients with metastatic breast cancer who have responding or stable disease after first-line chemotherapy: Eastern Cooperative Oncology Group Trial E2196. *J Clin Oncol* 22: 4631–4638, 2004.
124. Shepperd FA, Giaccone G, Seymour L, et al. Prospective, randomized, double blind, placebo controlled trial of marimastat after response to first-line chemotherapy in patients with small cell lung cancer: A trial of the National Cancer Institute of Canada-Clinical Trials Group and The European Organization for Research and Treatment of Cancer. *J Clin Oncol* 20: 4434–4439, 2002.
125. Bramhall S, Rosemurgy A, Brown PD, et al. Marimastat as first-line therapy for patients with unresectable pancreatic cancer: A randomized trial. *J Clin Oncol* 19: 3447–3455, 2001.
126. Moore M, Hamm J, Eisenberg P, et al. A comparison between gemcitabine (GEM) and the matrix metalloproteinase inhibitor (MMP) BAY12–9566 in patients with advanced pancreatic cancer. *Proc Am Soc Clin Oncol* 19: 240a, 2000 (abstr 930).

127. Phuphanich S, Levin V, Yung W, et al. A multicenter, randomized, double blind, placebo controlled (PB) trial of marimastat (MT) in patients with glioblastoma multiforme or gliosarcoma following completion of conventional first-line treatment. *Proc Am Soc Clin Oncol* 20: 52a, 2001 (abstr 205).
128. Naglich J, Trail P. *Evaluation of the Anti-Metastatic Potential of a Matrix Etalloproteinase Inhibitor, Bms-275291 in a Murine B16bl6 Melanoma Model.* New York: Bristol-Myers Squibb, 1998.
129. Naglich S, Sure-Kunkel M, Gupta E, et al. Inhibition of angiogenesis and metastasis in two murine models by the metalloproteinase inhibitor BMS-275291. *Cancer Res* 61: 8480–8485, 2001.
130. Trail P. *Combination Therapy with Bms-275291 and Carboplatin, Paclitaxel or Cisplatin.* New York: Bristol-Myers Squibb, 1998.
131. Miller KD, Saphner S, Waterhouse D, et al. A randomized phase II feasibility trial of BMS-275291 in patients with early stage breast cancer. *Clin Cancer Res* 10: 1971–1975, 2004.
132. Ingber D, Fujita T, Kishimoto S, et al. Synthetic analogues of fumangillin that inhibit angiogenesis and suppress tumor growth. *Nature* 348: 555–557, 1990.
133. Kusaka M, Sudo K, Fujita T, et al. Potent anti-angiogenic action of AGM-1470: Comparison to the fumangillin parent. *Biochem Biophys Res Commun* 174: 1070–1076, 1991.
134. O'Reilly M, Brem H, Folkmann J. Treatment of murine hemangioendotheliomas with the angiogenesis inhibitor AGM-1470. *J Pediatr Surg* 30: 325–329, 1997.
135. Singh Y, Shikata N, Kiyozuka Y, et al. Inhibition of tumor growth and metastasis by angiogenesis inhibitor TNP-470 on breast cancer cell lines in vitro and in vivo. *Breast Cancer Res Treat* 45: 15–27, 1997.
136. Twadowski P, Gradishar W, et al. Clinical trials of anti-angiogenic agents. *Curr Opin Oncol* 9: 584–589, 1997.
137. Bhargave P, Marshall JL, Rizvi N, et al. A phase I and pharmacokinetic study of TNP-470 administered weekly to patients with advanced cancer. *Clin Cancer Res* 5: 1989–1995, 1999.
138. Sadler WM, Kuzel T, Shapiro C, et al. Multi-institutional study of the angiogenesis inhibitor TNP-470 in metastatic renal carcinoma. *J Clin Oncol* 17: 2541–2545, 1999.
139. Dezube BJ, von Roenn JH, Holden-Wiltse J, et al. Fumangillin analog in the treatment of Kaposi's sarcoma: A phase I AIDS Clinical Trial Group Study. AIDS Clinical Trial Group No. 215 Team. *J Clin Oncol* 16: 1444–1449, 1998.
140. Logothetis CJ, Wu KK, Finn LD, et al. Phase I trial of the angiogenesis inhibitor TNP-470 for progressive androgen-independent prostate cancer. *Clin Cancer Res* 7: 1198–1203, 2001.
141. Brooks PC, Stromblad S, Klamke R, et al. Anti-integrin alpha V beta 3 blocks human breast cancer growth and angiogenesis in human skin. *J Clin Invest* 96: 1815–1822, 1995.
142. Gutheil J, Campbell T, Pierce J, et al. Phase I study of Vitaxin, an anti-angiogenic humanized monoclonal antibody to vascular integrin avb3. *Proc Am Soc Clin Oncol* 17: 215a, 1998.
143. Posey AJ, Khazaeli MB, DelGrosso A, et al. A pilot trial of Vitaxin, a humanized anti-vitronectin receptor (anti-$\alpha V\beta 3$) antibody in patients with metastatic cancer. *Cancer Biother Radiopharm* 16: 125–132, 2001.
144. Burke P, Denardo S, Miers L, et al. Cilengitide targeting of alpha V beta 3 integrin receptor synergizes with radioimmunotherapy to increase efficacy and apoptosis in breast cancer xenografts. *Cancer Res* 62: 4263–4272, 2001.
145. Seon B, Matsumoto F, Haruto Y, et al. Long lasting complete inhibition of human solid tumors in SCID mice by targeting endothelial cells of tumor vasculature with antihuman endoglin immunotoxin. *Clin Cancer Res* 3:1031–1044, 1997.
146. Matsuno F, Haruto Y, Kondo M, et al. Induction of lasting complete regression of preformed distinct solid tumors by targeting the tumor vasculature using two new anti-endoglin monoclonal antibodies. *Clin Cancer Res* 5: 371–382, 1999.
147. DeMoraes E, Fogler W, Grant D, et al. Recombinant human angiostatin: A phase I clinical trial assessing safety, pharmacokinetics and pharmacodynamics. *Proc Am Soc Clin Oncol* 20: 3a, 2001.
148. Gorski DH, Mauceri HJ, Salloum RM, et al. Potentiation of the antitumor effect of ionizing radiation by brief concomitant exposures to angiostatin. *Cancer Res* 58: 5686–5689, 1998.

149. Gorski DH, Mauceri HJ, Salloum RB, et al. Prolonged treatment with angiostatin reduces metastatic burden during radiation therapy. *Cancer Res* 63: 308–331, 2003.
150. Fogler W, Song M, Supko J, et al. Recombinant human endostatin demonstrates consistent and predictable pharmacokinetics following intravenous bolus administration to cancer patients. *Proc Am Soc Clin Oncol* 20: 69a, 2001.
151. Mundhenke C, Thomas J, Neider R, et al. Endothelial cell kinetics in skin wounds and tumors of patients receiving endostatin. *Proc Am Soc Clin Oncol* 20: 70a, 2001.
152. Yue TL, Wang X, Louden CS, et al. 2-Methoxyestradiol, an endogenous estrogen metabolite, induces apoptosis in endothelial cells and inhibits angiogenesis: Possible role for stress-activated protein kinase signaling pathway and Fas expression. *Mol Pharmacol* 51: 951–962, 1997.
153. Klauber N, Parangi S, Flynn E, et al. Inhibition of angiogenesis in breast cancer in mice by the microtubule inhibitors 2-methoxyestradiol and Taxol. *Cancer Res* 57: 81–86, 1997.
154. Sledge G, Miller K, Haney L, et al. A phase I study of 2-methoxyestradiol (2-ME_2) in patients (pts) with refractory metastatic breast cancer (MBC). *Proc Am Soc Clin Oncol* 21:111a, 2002 (abstr 441).
155. Miller K, Murry D, Curry E, et al. A phase I study of 2-methoxyestradiol (2ME_2) plus docetaxel (D) in patients (pts) with metastatic breast cancer (MBC). *Proc Am Soc Clin Oncol* 21: 111a, 2002 (abstr 442).
156. Kerbel RS. A cancer therapy resistant to resistance. *Nature* 390: 335–336, 1997.
157. Boehm T, Folkman J, Browder T, et al. Antiangiogenic therapy of experimental cancer does not induce acquired drug resistance. *Nature* 390: 404–407, 1997.
158. Fidler IJ, Ellis LM. Neoplastic angiogenesis—not all blood vessels are created equal. *N Eng J Med* 351: 215–216, 2004.
159. Streubel B, Chott A, Huber D, et al. Lymphoma-specific genetic aberrations in microvascular endothelial cells in B-cell lymphomas. *N Eng J Med* 351: 250–259, 2004.
160. Evans WE, McLeod HL. Pharmacogenomics-drug disposition, drug targets, and side effects. *N Eng J Med* 348: 538–549, 2003.
161. Ghilardi G, Biondi ML, Cecchini F, et al. Vascular invasion in human breast cancer is correlated to T-786C polymorphism of NOS3 gene. *Nitric Oxide* 9: 118–122, 2003.
162. Cahill DP, Kinzler KW, Vogelstein B, et al. Genetic instability and Darwinian selection in tumors. *Trends Cell Biol* 9: M57-M60, 1999.
163. Kerbel RS, Yu J, Tran J, et al. Possible mechanisms of acquired resistance to anti-angiogenic drugs: Implications for the use of combination therapy approaches. *Cancer Metastasis Rev* 20: 79–86, 2001.
164. Gohongi T, Fukurama D, Boucher Y, et al. Tumor–host interactions in the gallbladder suppress distal angiogenesis and tumor growth: Involvement of transforming growth factor beta1. *Nat Med* 5: 1203–1208, 1999.
165. Liu W, Davis DW, Ramirez K, et al. Endothelial cell apoptosis is inhibited by a soluble factor secreted by the human colon cancer cells. *Int J Cancer* 92: 26–30, 2001.
166. Low JA, Johnson MD, Bone EA, et al. The matrix metalloproteinase inhibitor batimastat (BB-94) retards human breast cancer solid tumor growth but not ascites formation in nude mice. *Clin Cancer Res* 2: 1207–1214, 1996.
167. Pluen A, Boucher Y, Ramanujan S, et al. Role of tumor–host interactions in interstitial diffusion of macromolecules; cranial vs. subcutaneous tumors. *Proc Natl Acad Sci USA* 98: 4628–4633, 2001.
168. Overmoyer B, Robertson K, Persons M. A phase I pharmacokinetic and pharmacodynamic study of SU5416 and adriamycin in inflammatory breast cancer. *Breast Cancer Res Treat* 69: 284, 2001 (abstr 432).
169. Durand RE, LePard NE. Contributions of transient blood flow to tumor hypoxia in mice. *Acta Oncol* 34: 317–323, 1995.
170. Djonov V, Andres AC, Ziemiecki A. Vascular remodeling during the normal and malignant life cycle of the mammary gland. *Microsc Res Tech* 52: 182–189, 2001.
171. Folber R, Hendrix MJ, Maniotis AJ. Vasculogenic mimicry and tumor angiogenesis. *Am J Pathol* 156: 361–381, 2000.
172. Kunkel P, Ulbricht U, Bohlen P, et al. Inhibition of glioma angiogenesis and growth in vivo by systemic treatment with a monoclonal antibody against the vascular endothelial growth factor receptor-2. *Cancer Res* 61: 6624–6628, 2001.

173. Passlidou E, Trivella M, Singh N, et al. Vascular phenotype in angiogenic and non-angiogenic lung non-small cell carcinomas. *Br J Cancer* 86: 244–249, 2002.
174. Patan S, Munn LL, Jain RK. Intussusceptive microvascular growth in human colon adenocarcinoma xenograft: A novel mechanism of tumor angiogenesis. *Microvasc Res* 51: 260–272, 1996.
175. Patan S, Tanda S, Roberge S, et al. Vascular morphogenesis and remodeling in a human tumor xenograft: Blood vessel formation and growth after ovariectomy and tumor implantation. *Circ Res* 89: 732–739, 2001.
176. Stessels F, van den Eynden G, van der Auwera I, et al. Breast adenocarcinoma liver metastases, in contrast to colorectal cancer liver metastasis, display a non-angiogenic growth pattern that preserves the stroma and lacks hypoxia. *Br J Cancer* 90: 1429–1436, 2004.
177. Asahara T, Masuda H, takahashi T, et al. Bone marrow origin of endothelial progenitor cells responsible for postnatal vasculogenesis in physiological and pathological neovascularization. *Circ Res* 85: 221–228, 1999.
178. Bolontrade MF, Zhou RR, Kleinerman ES. Vasculogenesis plays a role in the growth of Ewing's sarcoma in vivo. *Clin Cancer Res* 8: 3622–3627, 2002.
179. Hattori K, Dias S, Heissing B, et al. Vascular endothelial growth factor and angiopoetin-1 stimulate postnatal hematopoesis by recruitment of vasculogenic and hematopoetic stem cells. *J Exp Med* 193: 1005–1014, 2001.
180. Alpaugh M, Barsky S. the molecular basis of inflammatory breast cancer. *Breast Cancer Res Treat* 69: 312, 2001 (abstr 563).
181. van der Schaft D, Seftor E, Hess A, et al. The differential effects of angiogenesis inhibitors on vascular network formation by endothelial cells versus aggressive melanoma tumor cells. *Proc Am Assoc Cancer Res* 44: 696, 2003 (abstr 3046).
182. Lyden D, Hattori K, Dias S, et al. Impaired recruitment of bone-marrow-derived endothelial and hematopoietic precursor cells block tumor angiogenesis and growth. *Nat Med* 7: 1194–1201, 2001.
183. Sweeney CJ, Miller KD, Sisson SE, et al. The antiangiogenic property of docetaxel is synergistic with recombinant humanized monoclonal antibody against vascular endothelial growth factor or 2-methoxyestradiol but antagonized by endothelial growth factors. *Cancer Res* 61: 3369–3372, 2001.
184. Teicher BA, Sotomayor EA, Huang ZD. Antiangiogenic agents potentiate cytotoxic cancer therapies against primary and metastatic disease. *Cancer Res* 52: 6702–6704, 1992.
185. Jain RK. Normalizing tumor vasculature with anti-angiogenic therapy: A new paradigm for combination therapy. *Nat Med* 7: 987–989, 2001.
186. Brem H, Gresser I, Grosfeld J, et al. The combination of antiangiogenic agents to inhibit primary tumor growth and metastasis. *J Pediatr Surg* 28: 1253–1257, 1993.
187. Hall N, Fish D, Hunt N, et al. Is the relationship between angiogenesis and metastasis in breast cancer real? *Surg Oncol* 1: 223–229, 1995.
188. Bosari S, Lee A, DeLellis R, et al. Microvessel quantitation and prognosis in invasive breast carcinoma. *Hum Pathol* 23: 755–761, 1992.
189. van Hoef M, Knox W, Dhesi S, et al. Assessment of tumor vascularity as a prognostic factor in lymph node negative invasive breast cancer. *Eur J Cancer* 29A: 1141–1145, 1993.
190. Obermair A, Kurz C, Czerwenka K, et al. Microvessel density and vessel invasion in lymph-node-negative breast cancer: effect on recurrence-free survival. *Int J Cancer* 62: 126–131, 1995.
191. Inada K, Toi M, Hoshina S, et al. Significance of tumor angiogenesis as an independent prognostic factor in axillary node-negative breast cancer. *Japanese journal of cancer and chemotherapy* 22 Suppl 1: 59–65, 1995.
192. Ogawa Y, Chung YS, Nakata B, et al. Microvessel quantitation in invasive breast cancer by staining for factor VIII-related antigen. *Br J Cancer* 71: 1297–1301, 1995.
193. Axelsson K, Ljung BM, Moore DH, et al. Tumor angiogenesis as a prognostic assay for invasive ductal breast carcinoma. *J Natl Cancer Inst* 87: 997–1008, 1995.
194. Costello P, McCann A, Carney DN, et al. Prognostic significance or microvessel density in lymph node negative breast carcinoma. *Hum Pathol* 26: 1181–1184, 1995.
195. Morphopoulos G, Pearson M, Ryder WD, et al. Tumor angiogenesis as a prognostic marker in infiltrating lobular carcinoma of the breast. *J Pathol* 180: 44–49, 1996.

196. Karaiossifidi H, Kouri E, Arvaniti H, et al. Tumor angiogenesis in node-negative breast cancer: relationship with relapse free survival. *Anticancer Res* 16: 4001–4002, 1996.
197. Ozer E, Canda T, Kurtodlu B. The role of angiogenesis, laminin and CD44 expression in metastatic behavior of early-stage low-grade invasive breast carcinomas. *Cancer Lett* 121: 119–123, 1997.
198. Tynninen O, von Boguslawski K, Aronen HJ, et al. Prognostic value of vascular density and cell proliferation in breast cancer patients. *Pathol Res Pract* 195: 31–37, 1999.
199. Medri L, Nanni O, Volpi A, et al. Tumor microvessel density and prognosis in node-negative breast cancer. *Int J Cancer* 89: 74–80, 2000.
200. Vincent-Salomon A, Carton M, Zafrani B, et al. Long term outcome of small size invasive breast carcinomas independent from angiogenesis in a series of 685 cases. *Cancer* 92: 249–256, 2001.
201. Fox SB, Leek RD, Smith K, et al. Tumor angiogenesis in node negative breast carcinomas-relationship with epidermal growth factor receptor, estrogen receptor, and survival. *Breast Cancer Res Treat* 29: 109–116, 1994.
202. Gasparini G, Weidner N, Bevilacqua P, et al. Tumor microvessel density, p53 expression, tumor size and peritumoral lymphatic vessel invasion are relevant prognostic markers in node-negative breast carcinoma. *J Clin Oncol* 12: 454–466, 1994.
203. Bevilacqua P, Barbareschi M, Verderio P, et al. Prognostic value of intratumoral microvessel density, a measure of tumor angiogenesis, in node negative breast carcinoma-results of a multiparametric study. *Breast Cancer Res Treat* 36: 205–217, 1995.
204. Fox SB, Turner GD, Leek RD, et al. The prognostic value of quantitative angiogenesis in breast cancer and role of adhesion molecule expression in tumor endothelium. *Breast Cancer Res Treat* 36: 219–226, 1995.
205. Gasparini G, Toi M, Verderio P, et al. Prognostic significance of p53, angiogenesis and other conventional features in operable breast cancer: subanalysis in node-positive and node-negative patients. *Int J Oncol* 12: 1117–1125, 1998.
206. Heimann R, Ferguson D, Powers C, et al. Angiogenesis as a predictor of long-term survival for patients with node-negative breast cancer. *J Natl Cancer Inst* 88: 1764–1769, 1996.
207. Kumar S, Ghellal A, Li C, et al. Breast carcinoma: vascular density determined using CD105 antibody correlates with tumor prognosis. *Cancer Res* 59: 856–861, 1999.
208. Hansen S, Grabau DA, Sorensen FB, et al. The prognostic value of angiogenesis by Chalkley counting in a confirmatory study design of 836 breast cancer patients. *Pathol Res Pract* 195: 31–37, 1999.
209. Guidi AJ, Berry DA, Broadwater G, et al. Association of angiogenesis in lymph node metastases with outcome of breast cancer. *J Natl Cancer Inst* 92: 486–492, 2000.
210. Gunel N, Akcali Z, Coskun U, et al. Prognostic importance of tumor angiogenesis in breast carcinoma with adjuvant chemotherapy. *Path, Res & Pract* 198: 7–12, 2002.
211. Fridman V, Humblet C, Bonjean K, et al. Assessment of tumor angiogenesis in invasive breast carcinomas: absence of correlation with prognosis and pathological factors. *Virchows Archiv* 473: 611–617, 2000.
212. Tas F, Yavuz E, Aydiner A, et al. Angiogenesis and p53 protein expression in breast cancer: prognostic roles and interrelationships. *Am J Clin Oncol* 23: 546–553, 2000.
213. Kato T, Kameoka S, Kimura T, et al. Angiogenesis as a predictor of long-term survival for 377 japanese patients with breast cancer. *Breast Cancer Res Treat* 70: 65–74, 2001.
214. Guidi AJ, Berry DA, Broadwater G, et al. Association of angiogenesis and disease outcome in node positive breast cancer patients treated with adjuvant cyclophosphamide, doxorubicin, and fluorouracil: a cancer and leukemia group B correlative science study from protocols 8541/8869. *J Clin Oncol* 20: 732–742, 2002.
215. Buadu L, Murakami J, Murayama S, et al. Breast lesions: correlation of contrast medium enhancement patterns on MR images with histopathologic findings and tumor angiogenesis. *Radiology* 200: 639–649, 1996.
216. Stomper P, Winston J, Klippenstein D, et al. Angiogenesis and dynamic MR imaging gadolinium enhancement of malignant and benign breast lesions. *Breast Cancer Res Treat* 45: 39–46, 1997.
217. Buckley D, Drew P, Mussurakis S, et al. Microvessel density of invasive breast cancer assessed by dynamic Gd-DTPA enhanced MRI. *J Magn Reson Imaging* 7: 461–464, 1997.

218. Hulka C, Edmister W, Smith B, et al. Dynamic echo-planar imaging of the breast: experience in diagnosing breast carcinoma and correlation with tumor angiogenesis. *Radiology* 1997: 837–842, 1997.
219. Ikeda O, Yamashita Y, Takahashi M, et al. Gd-enhanced dynamic magnetic resonance imaging of breast masses. *Top Magn Reson Imaging* 10: 143–151, 1999.
220. Knopp M, Weiss E, Sinn H, et al. Pathophysiologic basis of contrast enhancement in breast tumors. *J Magn Reson Imaging* 10: 260–266, 1999.
221. Matsubayashi R, Matsuo Y, Edakuni G, et al. Breast masses with peripheral rim enhancement on dynamic contrast-enhanced MR images: correlation of MR findings with histologic features and expression of growth factors. *Radiology* 217: 841–848, 2000.

26 Antiangiogenic Therapy for Lung Malignancies

David J. Stewart

CONTENTS

Lung cancer is the leading cause of cancer death in the world (1). Only 15% of lung cancer patients survive 5 years from diagnosis (1). With respect to non-small cell lung cancer (NSCLC) (i.e., squamous cell cancer, adenocarcinoma, and large cell lung cancer), surgery is generally the treatment of choice for stage I–II disease, and a minority of selected patients with stage IIIA or IIIB may also be considered for surgery. Postoperative adjuvant chemotherapy improves long-term survival by up to 15% (2), and preoperative neoadjuvant chemotherapy also improves outcome in patients with operable stage IIIA disease (3). Definitive radiotherapy may lead to long-term control of disease in 5%–15% of patients with inoperable stage III NSCLC, with both local therapy and development of distant metastases limiting prognosis. Addition of chemotherapy to radiotherapy for inoperable stage III disease also increases long-term survival rates, with reduction in both local failure in the radiotherapy field as well as a reduction in rate of development of distant metastatic disease (4). Chemotherapy may improve local control both by directly killing cancer cells as well as by potentiating the effect of radiation, for example by reducing the ability of cancer cells to repair sublethal damage caused by the radiation. Concurrent administration of chemotherapy and radiotherapy is superior to sequential administration (4), suggesting that the radiopotentiating effect of chemotherapy may be particularly important.

In NSCLC patients with incurable disease (distant metastases, malignant effusion, or inoperable recurrence in a radiotherapy field), chemotherapy can palliate symptoms and provide modest prolongation of survival (5,6). Modern chemotherapy yields partial remissions in 20%–40% of patients and improves symptoms in up to 70% of patients. Randomized studies, comparing chemotherapy with best supportive care (e.g., palliative radiotherapy, analgesics, etc.) in NSCLC, have demonstrated that chemotherapy improves quality of life and duration of survival in the average patient. Average life expectancy increases from around 3 months with best supportive care to up to 6–9 months or longer with the addition of chemotherapy. The proportion of patients alive at 1 year increases from 10% with best supportive care to up to 20%–40% with the addition of chemotherapy, and 5%–15% of patients treated with chemotherapy for advanced NSCLC remain alive at 2 years (5,6).

The fact that the proportion of patients with symptomatic improvement substantially exceeds the proportion achieving partial remission indicates that lesser degrees of tumor shrinkage (classified as stable disease) do translate into symptomatic improvement. Similarly, the survival advantage seen with chemotherapy is as great in NSCLC patients classified as having stable disease as in patients classified as having partial responses (7).

For patients with advanced NSCLC progressing after front-line chemotherapy, second-line chemotherapy with docetaxel is superior to best supportive care with respect to both quality of life and life expectancy (8). Pemetrexed appears to be as effective as docetaxel in this setting, and may be somewhat less toxic (9). Erlotinib, a small molecule inhibitor of the epidermal growth factor receptor (EGFR) tyrosine kinase, was also shown in randomized studies to be superior to best supportive care in the second-line and third-line setting in NSCLC (10).

Hence, currently available chemotherapy and targeted therapies are of unequivocal benefit in NSCLC, but the degree of benefit remains small.

Small cell lung cancer (SCLC) is rapidly growing and highly aggressive. At the time of diagnosis, it is amenable to surgery with curative intent in only a small proportion of patients since most patients will have either distant metastases or extensive involvement of the mediastinum. Local therapies such as surgery and radiotherapy are able to cure fewer than 5% of patients. However, SCLC is generally sensitive to chemotherapy. Up to 50% of patients with limited disease (i.e., those without distant metastases) will achieve complete remission with chemotherapy, and up to 20% of patients are curable with chemotherapy alone. Addition of thoracic radiation to chemotherapy increases the probability of cure by a further 5%. For patients achieving complete remission, prophylactic cranial radiation also improves probability of cure by ~5%. Median survival in patients with limited SCLC is ~3 months without chemotherapy and 12–16 months with chemotherapy (11,12).

With extensive SCLC (i.e., with distant metastases), most patients respond to chemotherapy, but it is generally not considered curable. Median survival with extensive SCLC is ~6 weeks without chemotherapy and 10–12 months with chemotherapy (11,12).

Whereas chemonaïve SCLC is highly sensitive to chemotherapy, SCLC that progresses on front-line chemotherapy or recurs within 3 months of completion of front-line chemotherapy tends to be quite resistant. SCLC that recurs >3 months after completion of front-line chemotherapy will often respond to further chemotherapy (including to chemotherapy that had previously been used front line), but will generally not be curable (13).

Various targeted therapies have been tested in lung cancer. Of these, the most experience is with EGFR inhibitors. NSCLC tumor cells often express the EGFR receptor, and EGFR receptor antagonists are active against NSCLC in preclinical systems. Clinically, the EGFR tyrosine kinase inhibitors gefitinib and erlotinib may induce tumor regression in up to 17% of patients as second-line therapy (14). Tumors with an EGFR-activating mutation are particularly likely to respond (14). EGFR mutation and response to EGFR inhibitors is particularly likely in females, Asians, never-smokers, and in adenocarcinomas.

When compared with best supportive care, erlotinib significantly improved survival as second or third-line therapy (10). Neither erlotinib nor gefitinib improved outcome when added to frontline NSCLC chemotherapy (14). Although response rates were improved in the subset of patients with EGFR mutations, patients with a K-ras mutation had a worse survival when an EGFR inhibitor was added than when treated with chemotherapy alone (15).

Studies are underway assessing anti-EGFR monoclonal antibodies (e.g., cetuximab and panitumumab) in NSCLC.

Expression of various antiangiogenic and proangiogenic factors correlates with prognosis in lung cancer, suggesting that targeting tumor vasculature could prove useful therapeutically.

26.1 ANTIANGIOGENIC FACTORS

26.1.1 Endostatin

Endostatin is an angiogenesis inhibitor that is endogenously produced as a proteolytic fragment of type XVIII collagen. Endostatin was detected in both benign and malignant pleural effusions in lung cancer patients, with average levels being higher in malignant than in benign effusions (16). Tumors from 64% of NSCLC patients had positive immunohistochemistry (IHC) staining with anticollagen XVIII polyclonal antibody, with strong staining in 12%, and tumor collagen XVIII expression correlated negatively with survival in both univariate and multivariate analyses (17). Preoperative serum endostatin levels were higher in NSCLC patients than in controls (17,18), and correlated with tumor collagen XVIII expression (17). They were higher in high stage patients than in early stage patients, and high endostatin levels and high platelet counts were associated with poor prognosis (18).

26.1.2 Angiostatin

By IHC, 34 of 143 (24%) primary NSCLCs stained positively for the antiangiogenic factor angiostatin. Patients whose tumors stained positively for angiostatin lived significantly longer than did those whose tumors stained negatively. The difference was particularly pronounced between patients with angiostatin-positive/vascular endothelial growth factor (VEGF)-negative tumors and those with angiostatin-negative/VEGF-positive tumors (19). Hence, though both endostatin and angiostatin are angiogenesis inhibitors, expression of endostatin was associated with relatively poor prognosis, whereas expression of angiostatin was associated with improved prognosis.

26.1.3 METH-2

For the antiangiogenic factor METH-2 (ADAMTS-8), both mRNA and expression by IHC were substantially lower in NSCLC tumor than in surrounding normal lung tissue. Abnormal hypermethylation of the proximal promoter region of the gene was found in 67% of lung adenocarcinomas and 50% of lung squamous cell carcinomas (20).

26.1.4 Thrombospondin

In resected NSCLC tumor samples, expression of thrombospondin (TSP) (an extracellular matrix glycoprotein with antiangiogenic activity) was lower than expression in adjacent normal tissues (21,22). With respect to an association of TSP with tumor vascularization and prognosis, there has been disagreement between studies. A significant inverse association between TSP-1 mRNA expression and fibroblast growth factor (FGF) protein expression was noted in one study, although no association was found between TSP mRNA expression and expression of other angiogenic factors, tumor neovascularization, or p53 mutations (23). In another study, TSP-1 correlated independently with patient survival in multivariate analysis. The 5 year survival rate was 77% in patients whose tumors expressed TSP-1, and only 55% in patients whose tumors had reduced TSP-1 expression. The correlation with prognosis was seen primarily in patients whose tumors had a high microvessel count, suggesting that TSP-1 may have an effect on prognosis, which is independent of its antiangiogenic effects (24). In yet another study, vascularity in the tumor correlated inversely with *TSP-2* gene expression. Patients with adenocarcinoma of the lung whose tumors tested positively for *TSP-2* gene expression had significantly better prognosis than did patients without tumor *TSP-2* gene expression. *TSP-1* gene expression did not correlate with tumor vascularity or with prognosis in this study (22). In still another series, TSP expression did not correlate with microvessel count or with survival (21). Hence, the prognostic significance of TSP expression in NSCLC is uncertain, and may possibly depend on whether it is TSP-1 or TSP-2 that is expressed.

26.2 PROANGIOGENIC FACTORS

26.2.1 Tumor Expression of VEGF in NSCLC

VEGF regulates both angiogenesis and vascular permeability in NSCLC and promotes both tumor progression and development of malignant pleural effusions (25), although evidence from murine NSCLC xenograft models suggests that different isoforms of VEGF may have different biological effects on angiogenesis and on other tumor properties (26). VEGF expression is detectable by IHC in tumor of ~58%–85% of NSCLC patients (27–29). VEGF expression by IHC in NSCLC correlates with CD105 expression (30) and angiogenesis or microvessel density (23,31,32), with tumor vessel invasion (30), and with lymph node involvement (27), although there was no correlation with microvessel density in occasional trials (33). In one study, tumor microvessel density correlated with *stromal* VEGF-A expression in squamous cell carcinomas (34).

In a meta-analysis of 15 studies involving 1549 NSCLC patients, tumor VEGF expression was found to be an unfavorable prognostic factor, with a hazard ratio of 1.48 (95% confidence intervals, 1.27–1.72) (35). However, within individual trials, there has not been universal agreement on the prognostic importance of VEGF expression and related vascular findings in NSCLC. Patients whose tumors had VEGF overexpression, high microvessel density by CD34, or tumor vessel invasion had reduced overall survival in some studies (27,30,31,36,37), but not in others (28,29,33). In one study, impact was stage-dependent: each of tumor vascularity, VEGF-A, VEGF-C, and E-cadherin was significantly associated with survival in stage I patients but not in stage II–III (38). VEGF expression itself did not correlate with stage (31).

VEGF expression may vary with NSCLC tumor type. In one study, levels of VEGF in adenocarcinoma of the lung were significantly higher than in squamous cell carcinoma of the lung (36). In another study, intratumoral expression of VEGF-A was significantly higher in adenocarcinomas than in squamous cell carcinomas, but stromal expression of VEGF-A was higher in squamous cell carcinomas than in adenocarcinomas. Among squamous cell carcinomas, both intratumoral and stromal VEGF-A expressions were higher in tumors that expressed Wnt1 than in tumors that did not express it. In adenocarcinomas, there was no correlation between intratumoral Wnt1 expression and VEGF-A expression. This suggests that in squamous cell carcinomas, intratumoral Wnt1 induces stromal VEGF-A expression, and that this plays a role in tumor angiogenesis (34).

In squamous cell lung cancer tumor samples, tumors from smokers had significantly lower microvessel density and VEGF expression than tumors from nonsmokers, whereas TSP was significantly increased in tumors from smokers (39).

There has been no consistent correlation found between tumor differentiation or necrosis and angiogenesis. In one study, angiogenesis (which correlated with VEGF expression) was more prominent in poorly differentiated tumors than in well-differentiated tumors (32), whereas it did not vary with grade in another study (31). Similarly, in one study, the expression of both VEGF and IL-8 in NSCLC tumors was significantly higher in tumors with central necrosis than in tumors without (40), but no correlation with necrosis was found in another study (31).

Mature blood vessels stain positively with the monoclonal antibody LH39, whereas immature blood vessels do not. The median proportion of blood vessels in resected NSCLCs that stained positively for LH39 was 46% (range, 15%–90%). LH39 staining did not correlate with VEGF expression. However, extent of LH39 staining did correlate inversely with lymph node metastases and also correlated negatively with thymidine phosphorylase (TR) expression, suggesting intense vascular remodeling in tumors with high TR expression (41).

The existence of association between p53 status and VEGF expression in NSCLC remains unclear. In one study of resected NSCLC tumors, an association was found between the presence of p53 mutations and high VEGF protein expression and neovascularization (23). Similarly, in lung cancers that had metastasized to brain, tumors with loss of wild-type p53 exhibited higher levels of VEGF, and in vitro studies demonstrated that, unlike cells with wild-type p53, mutant p53 cells modulate VEGF levels via the MAPK pathway (42). However, in another study, the p53 status did not correlate with angiogenesis, but wild-type p53 expression was inversely associated with VEGF expression (31). In this latter study, there was no significant association of bcl-2 or c-erbB-2 expression or of Ki67 proliferation index with VEGF expression (31).

In an NSCLC cell line, prostaglandin E2 (PGE2) and vasoactive intestinal peptide (VIP) induced increased VEGF expression, and this was inhibited by H-89, a protein kinase A inhibitor (43). In other preclinical studies using a PC14PE6/AS2 adenocarcinoma murine lung cancer model, activation of STAT3 by autocrine expression of IL-6 appeared to be important in the upregulation of VEGF, which in turn led to increased microvessel density, vascular permeability, and malignant pleural effusion formation (42,44).

26.2.2 Tumor Expression of VEGFR in NSCLC

VEGF receptors (VEGFR) may also be expressed directly on NSCLC cells. The significance of this is uncertain. This suggests the possibility that VEGF could directly stimulate tumor cell growth, in addition to supporting angiogenesis (45). VEGFR-1 (Flt-1) can exist in the membrane-bound (mFlt-1) or soluble isoform. mFlt-1 expression was significantly higher in resected lung adenocarcinomas than in lung squamous cell carcinomas. Expression of mFlt-1 correlated positively with expression of VEGF. In this study, there was no significant correlation of intratumoral microvessel density with mFlt-1 or VEGF expression. Prognosis did not correlate with expression of mFlt-1 or VEGF, but patients with a low mFlt-1/VEGF ratio had higher intratumoral microvessel density and worse survival than did patients with a high mFlt-1/VEGF expression ratio. The mFlt-1/VEGF ratio achieved statistical significance as a favorable prognostic factor in multivariate analysis (33). Overall, a meta-analysis concluded that there were insufficient data at this time to determine the prognostic significance of VEGFR-1 or VEGFR-2 expression in NSCLC (35).

26.2.3 Serum VEGF Levels in NSCLC

NSCLC patients had higher serum (18,46) endostatin and basic FGF (bFGF) levels than did healthy controls, whereas the NSCLC patients had lower serum and plasma TSP-1 levels than did controls (18,46). Pretreatment serum VEGF levels correlated inversely with survival in patients with advanced NSCLC treated with chemotherapy (18,46) and also correlated inversely with response to gefitinib (47). In this latter study, treatment of NSCLC patients with gefitinib did not alter serum or plasma levels of VEGF, bFGF, matrix metalloproteinase (MMP-2), MMP-9, TIMP-1, or TIMP-2 (47). In patients with normal platelet counts, serum VEGF levels during and after radiotherapy for locally advanced NSCLC correlated with prognosis (48). Hence, with a variety of modalities of therapy, high pretreatment serum VEGF levels predict a poor outcome.

Serum VEGF-C, VEGF, and plasma MMP-9 concentrations were significantly higher in patients with lymph node metastases from NSCLC than in patients without lymph node metastases (49). Serum activity of semicarbazide-sensitive amine oxidase (SSAO) (an enzyme associated with the vascular system that catalyzes the deamination of primary monoamines) correlated positively with serum VEGF levels but did not correlate with survival (50).

26.2.4 VEGF in SCLC

A meta-analysis concluded that there were insufficient data to determine the prognostic value of VEGF expression in SCLC (35). However, some studies suggest that it may be of value. In patients undergoing surgery followed by adjuvant chemotherapy for SCLC, prognosis was adversely affected by tumor stage, node involvement, microvessel density, and VEGF expression by IHC, but not by p53 alterations. In multivariate analysis, VEGF expression was an independent prognostic factor (51). Microvessel density appears to be higher in SCLC than in NSCLC (32,51). However, in SCLC, tumor vascularization did not correlate strongly with VEGF expression, suggesting that other pathways may be important (32,51).

In addition to playing a role in tumor angiogenesis, VEGF may also have direct stimulatory effects on tumor cells. In SCLC cell lines, VEGF, VEGF-C, VEGFR-2, and VEGFR-3 were detected by Western blotting in all five lines that were examined. Effect of hypoxia was tested in one line, and expression of both VEGF and VEGFR was significantly increased by hypoxia. Addition of VEGF or VEGF-D caused phosphorylation of both VEGFRs and MAPK, and induced tumor cell proliferation and migration (52).

In SCLC, pretreatment serum-soluble VEGF (s-VEGF) levels were significantly lower in patients who subsequently responded to therapy than in nonresponders. Serum s-VEGF levels correlated negatively with survival in both univariate and multivariate analysis (53).

26.2.5 Basic Fibroblast Growth Factor

The angiogenesis stimulator bFGF was overexpressed in 72% of tumors from patients with metastatic NSCLC (54). A positive association was found between high NSCLC tumor levels of FGF and numbers of microvessels (23). Tumor concentration of bFGF was significantly higher in NSCLC patients with metastatic nodal involvement than in those without nodal involvement (36). Expression did not differ between adenocarcinomas and squamous cell carcinomas. Level of expression of bFGF correlated negatively with survival (36,48), particularly in those who also had high VEGF concentrations (36). In multivariate analysis, lymph node involvement, bFGF expression, and VEGF expression were independent prognostic factors in NSCLC patients who had undergone curative resection (36). However, an association of bFGF expression with survival has not been seen in all studies (55).

26.2.6 Angiopoietin

Angiopoietin-1 and angiopoietin-2 are potent angiogenic factors, which potentiate the effect of VEGF. By IHC, 43% of resected NSCLC tumors expressed angiopoietin-1 and 17% expressed angiopoietin-2 (56). There was no significant correlation between expression of angiopoietin-1 and either intratumoral microvessel density (as assessed by CD34 or

CD105 staining) or survival. However, average CD105 intratumoral microvessel density was significantly higher for angiopoietin-2-positive tumors than for negative ones, provided that VEGF expression was also high. Angiopoietin-2 expression also predicted poor survival in patients with high VEGF expression. The 5 year survival rate was 41% with angiopoietin-2 expression and high VEGF expression, 63% with angiopoietin-2 expression and low VEGF expression, and 72% in patients whose tumors did not express angiopoietin-2 and had low VEGF expression (56).

Tumor vascularity and expression of angiopoietin both correlated significantly with IL-10 expression. In IL-10-positive NSCLC tumor samples, 97% demonstrated gene expression for angiopoietin-1 and 94% for angiopoietin-2 (57).

26.2.7 Cyclooxygenase-2

Cyclooxygenase (COX-2) is overexpressed in NSCLC and may play a role in angiogenesis, tumor invasion, resistance to apoptosis, and suppression of antitumor immunity. COX-2 suppression with nonsteroidal anti-inflammatory agents is associated with a reduced incidence of lung cancer (58).

In IHC studies of resected NSCLC, nitric oxide synthase 2 (NOS2) and COX-2 were both expressed in 48% of tumors, and their levels correlated with microvessel density at the stromal–tumor interphase (29). In autopsy studies, other components of the prostaglandin biosynthetic pathway also correlated with microvessel density and with bFGF expression and with the extent of metastatic disease in NSCLC (59). More adenocarcinomas and large cell carcinomas than squamous cell carcinomas overexpressed NOS2 (29), and in autopsy studies, COX-2 expression was higher in adenocarcinomas than in SCLC (59). NOS2 (29) and COX-2 expression (29,59) both correlated with VEGF expression. Neither NOS2, COX-2, nor microvessel density correlated with survival, but microvessel density did correlate with tumor stage (29).

In NSCLC cell lines, IL-20 expression is decreased compared with that of normal cells, and ability of both IL-10 and IL-20 to regulate COX-2 is impaired in NSCLC cell lines (60). IL-6 induces expression of VEGF in NSCLC cells, and COX-2 may activate STAT3 and VEGF expression by inducing IL-6 production in NSCLC (54,61). IL-4 downregulates COX-2 expression in NSCLC cells, and this inhibition appears to be mediated predominantly by STAT6 (62).

26.2.8 Nuclear Factor-κB (NF-κB)

NF-κB is sequestered in the cytoplasm by the inhibitor of kappa B (IKB) and released to the nucleus by the inhibitor of kappa B kinase (IKK). In a Lewis lung carcinoma model, expression of IKKα in endothelial cells was associated with more rapid tumor growth, increased vessel density, fewer tumor hypoxic regions, and decreased tumor cell apoptosis compared with tumors in which the endothelial cells did not express IKKα (63).

26.2.9 Hypoxia-Inducible Factor

The transcription factor, hypoxia-inducible factor 1 (HIF-1) regulates genes involved in metabolism, angiogenesis, proliferation, and apoptosis (64). Hypoxia induced expression of VEGF mRNA to a greater extent in Lewis lung cancer variance with high metastatic potential compared with those with low metastatic potential. HIF-1α expression but not HIF-1β expression was constitutively upregulated in the cells with high metastatic potential, which also had a higher level of HIF-1α protein in response to hypoxia, suggesting that the HIF-1α expression played a role in the increased VEGF expression (65).

In lung cancer tumor samples, HIF-1α expression varied with histology. It was present in a higher proportion of SCLCs than NSCLCs (64), and in a higher proportion of squamous cell carcinomas than adenocarcinomas (66,67). No significant association was seen between HIF-1α expression and microvessel density (66) or VEGF expression (66).

In some studies, there was a positive correlation between HIF-1α expression and clinical stage (64) and presence of metastases (64), but this association with stage was not seen in other studies (66,67). An association between HIF-1α expression and tumor necrosis has also been reported (67).

Most studies have not reported any association between HIF-1α expression and patient survival (55,66,68), although very high levels of HIF-1α expression were found to be associated with poor prognosis in one NSCLC study (67), whereas patients with high expression of HIF-1α and HIF-1β were actually found to have improved survival in another study (69).

In one study, there was an inverse relationship between the expression of HIF-1α and bcl-2 and a positive relationship between the expression of HIF-1α and Bax (64), whereas no association with BCL-2 expression has been seen in other studies (67). A correlation has been reported between HIF-1α and HIF-1β expression and expression of caspase-3, Fas, and Fas ligand (69). There was no relationship observed between HIF-1α and PCNA (64,66) or other markers of proliferation (69). Overall, available data suggest that HIF-1α may correlate with markers of apoptosis but not with markers of proliferation.

HIF-1α immunostaining has been reported to correlate with expression of EGFR, MMP-9, carbonic anhydrase IX (67), and pAKT (70). In a study in which 80% of tumors from NSCLC patients undergoing curative surgery were found to express HIF-1α, erythropoietin expression was found in 50%, and erythropoietin receptor expression in 96% (71). Overall, although HIF-1α expression is frequent in lung cancer, its prognostic significance is uncertain.

26.2.10 Lactate Dehydrogenase

Lactate dehydrogenase-5 (LDH-5) catalyzes the conversion of pyruvate to lactate during anaerobic glycolysis, and expression of LDH-5 is induced by HIF-1. High serum LDH levels are associated with poor prognosis and chemotherapy resistance, and serum LDH levels correlated with LDH-5 expression in tumor cells (72). Tumor LDH-5 overexpression also correlated with HIF-1α and HIF-2α expression, but not with carbonic anhydrase IX expression. Tumor LDH-5 overexpression was associated with poor prognosis, particularly if the tumor also overexpressed HIF-1α or HIF-2α (72).

26.2.11 Matrix Metalloproteinases

MMPs probably play a role in endothelial cell penetration into tissue during angiogenesis. In IHC studies of the expression in NSCLC of MMPs and MMP inhibitors (TIMPs), all 10 MMPs and 4 TIMPs examined were expressed, with frequencies ranging from 41% for MMP-2 to 68% for MMP-13 (73). Stromal staining ranged from 6% for TIMP-4 to 87% for MMP-13. Staining patterns for adenocarcinomas differed from those for squamous cell carcinomas. Staining did not correlate with presence of ras mutations. TIMP-1 overexpression, which was seen in 27% of resected NSCLC tumors (74), correlated with more advanced disease (73) and, in multivariate analyses, with poor survival (74).

Expression of RECK (a membrane-anchored glycoprotein that negatively regulates MMPs) was reduced in resected NSCLC compared with normal lung tissues (75). RECK expression correlated negatively with intratumoral microvessel density, and reduced RECK expression was significantly associated with poor prognosis in univariate and multivariate analysis, particularly in adenocarcinomas, poorly differentiated tumors, and in higher stage disease (75).

High serum MMP-9 and VEGF levels correlated with poor survival in patients with resected stage I/II (but not stage III) NSCLC in both univariate and multivariate analysis (76).

In SCLC, IHC and in situ hybridization were used to evaluate expression of seven MMPs and four TIMPs (77). IHC staining was positive for MMP-1 and MMP-9 in 60%–70% of tumor cells, and for MMP-11, MMP-13, and MMP-14 and TIMP-2 and TIMP-3 in 70%–100% of tumor cells. Stromal staining of TIMP-1 and TIMP-3 was present in <30% of specimens. Several clinical characteristics were tested in multivariate analysis for association with response, and only stage and decreased tumor expression of TIMP-1 were significant. Significant negative factors for survival included tumor stage, weight loss, and high-tumor cell expression of MMP-3, MMP-11 and MMP-14. MMPs and TIMPs did not vary with stage (77).

26.2.12 Microphthalmia-Associated Transcription Factor

Expression of microphthalmia-associated transcription factor (MITF) in tumors from patients with NSCLC correlated negatively with survival and disease-free survival. In NSCLC lung cancer cell lines, MITF regulates genes involving angiogenesis and various other functions (78).

26.2.13 Macrophage Migration Inhibitory Factor

Angiogenic potential (as evidenced by endothelial cell chemotactic activity) was significantly increased in response to conditioned medium generated from cocultures of monocytes and NSCLC cells, compared with conditioned media from either cell type alone (79). This effect appeared to be predominantly due to expression of CXC chemokines rather than VEGF, and macrophages appeared to be the primary source of increased CXC chemokine production. Macrophage migration inhibitory factor (MIF) derived from tumor cells appears to be responsible for the increased expression of angiogenic activity arising from macrophages (79).

Data from murine lung cancer models suggest that MIF produced by tumor cells promotes macrophage infiltration into the tumor, and the infiltrating macrophages produce angiogenic CXC chemokines and VEGF, thereby stimulating angiogenesis in the tumor and promoting tumor growth (80). Resected human NSCLCs express MIF, and levels of MIF expression (80,81) and microvessel density (81) correlated with expression of angiogenic CXC chemokines. In patients with high-tumor MIF expression, vessel density also correlated with MIF level (81). In multivariate analysis, risk of recurrence of NSCLC was significantly associated with high levels of glutamic acid–leucine–arginine amino acid motif CXC chemokines, MIF, and VEGF (81).

26.2.14 Raf-1 Kinase

Nicotine stimulates cell proliferation and angiogenesis, and nicotine and its metabolites induce Raf-1 kinase activity (82). Raf-1 kinase binds and inactivates Rb. In NSCLC cell lines, nicotine induces binding of Raf-1 kinase to Rb, and elevated Rb–Raf-1 interactions have been observed in NSCLC tumor samples. In NSCLC cell lines, this Rb–Raf-1 interaction is abrogated by c-Src inhibitors, and other compounds that disrupt the Rb–Raf-1 interaction also inhibited nicotine-induced cell proliferation and angiogenesis (82).

26.2.15 Blood Flow and Glucose Metabolism

$^{15}O_2$ PET scanning was used to assess tumor blood flow in patients with stage III NSCLC. Blood flow (which showed substantial interpatient variability) did not correlate with glucose metabolism as measured by FDG-PET (83). Similarly, in patients with resected lung metastases,there was no correlation between microvessel density and preoperative FDG uptake on PET scanning. FDG uptake did correlate with tumor size and type (84).

26.3 ANTIANGIOGENIC STRATEGIES IN LUNG CANCER PRECLINICAL MODELS

26.3.1 Endostatin

In vitro, recombinant human endostatin inhibited both human and rodent endothelial cell proliferation, endothelial cell migration, and tube formation of murine endothelial cells, and caused degeneration of preexisting tubes (85). It also delayed growth of subcutaneously xenotransplanted human NSCLC, with no change in microvessel density, but with a significant reduction of proliferating tumor cells and an increase in bFGF and VEGF expression. Recombinant human endostatin also significantly prolonged survival of mice with human NSCLC tumors orthotopically implanted in their lungs (85).

Systemic administration of a recombinant adenovirus that expresses murine endostatin resulted in persistently high serum endostatin levels in mice, and reduced growth rate of Lewis lung carcinoma, with prevention of pulmonary metastases and decreased tumor blood vessel formation (86).

26.3.2 Angiostatin

Normal and tumor cells cleave plasminogen into angiostatin, consisting of 3–4 kringle-containing antiangiogenic fragments (87). The novel 22 kDa fragment p22, consisting of amino acids 78–180, selectively inhibits proliferation of endothelial cells (but not other cell types) in vitro. In vivo, it prevents vascular growth of chick chorioallantoic membrane (87). Murine angiostatin suppressed the growth of Lewis lung carcinoma primary tumors and metastases in mice (87–89), and also reduced tumor microvessel density in this model (88).

When added to chemotherapy in vitro, angiostatin did not enhance cyclophosphamide cytotoxicity versus human umbilical vein endothelial cells (HUVECs), and cyclophosphamide did not enhance angiostatin-mediated inhibition of migration, but tube formation was inhibited more by the combination than by either drug alone (90). In Lewis lung carcinoma implanted in mice, combined treatment with angiostatin and cyclophosphamide was no better than cyclophosphamide alone in inhibiting growth of the primary tumor, but the addition of angiostatin did significantly reduce the number of pulmonary metastases (90).

26.3.3 VEGF/VEGFR Inhibitors

Several small molecules have been developed that inhibit VEGFR tyrosine kinases, and many also inhibit one or more other receptors. The VEGFR tyrosine kinase inhibitor SU5416 also inhibits kit signaling, and blocked kit-mediated growth of SCLC in vitro (91). In vivo, it decreased tumor microvessel density and also decreased growth of both chemotherapy-resistant and chemotherapy-sensitive murine xenograft SCLC models. It was as effective as the chemotherapy agent carboplatin, but the combination of SU5416 with carboplatin was no better than either agent alone. In addition to its ability to block VEGFR, SU5416 also reduced kit-induced VEGF production by tumor cells, thereby reducing angiogenesis both by blocking VEGFR and by reducing production of ligand (91). SU5416 also completely inhibited endogenous COX-2 expression in the H460 human lung cancer cell line, and when combined with radiation in xenograft models, enhanced the tumor growth delay and downregulated COX-2 expression in tumors compared with controls (92).

Though combining SU5416 with carboplatin did not appear to increase efficacy, combining it in vitro with the direct angiogenesis inhibitor endostatin enhanced inhibition of endothelial cell proliferation, survival, migration/invasion, and tube formation compared with either agent alone (93). Similarly, combining SU5416 with endostatin in vivo was more effective than either agent alone at inhibiting tumor growth of human prostate (PC3), lung (A459), and glioma (U87) xenograft models, and reduced functional microvessel density, tumor microcirculation, and blood perfusion. The authors postulated that the direct antiangiogenic agent endostatin might suppress alternative angiogenic pathways upregulated in the tumor in response to the SU5416-induced indirect, specific pathway inhibition (93).

GFA-116 is a synthetic small molecule that potently inhibits binding of VEGF to VEGFR-2 (Flk-1) in Flk-1-overexpressing cell lines, and inhibits VEGF-stimulated Flk-1 tyrosine phosphorylation and subsequent activation of Erk1/2 MAP kinases. GFA-116 inhibits angiogenesis as assessed by inhibition of migration and formation of capillary-like structures by human endothelial cells (94). In A-594 human lung tumors implanted in nude mice, GFA-116 inhibited tumor growth and angiogenesis (94).

Other small molecule VEGFR inhibitors with activity versus human lung cancer xenografts include ZD4190 (95) and KRN951 (96). KRN951 also markedly reduced tumor xenograft vascular permeability (with tumor rim affected more than the center) and numbers of functional tumor vessels (96).

Radiation-induced inhibition of growth of the moderately radiosensitive human SCLC 54A xenograft in mice was enhanced by administration of an anti-VEGFR-2 antibody (97).

Intrapleural administration of adenovirus-mediated gene therapy with a vector expressing an antiangiogenesis soluble, secreted, extracellular portion of the VEGFR-1 (Flt-1) receptor administered into the pleural cavity decreased growth of lung metastases generated by intravenous injection in mice of CT26/CL25 tumor cells and prolonged animal survival (98).

The orally available small molecule PTC299 inhibits the posttranscriptional expression of hypoxia-induced VEGF-A, and reduces tumor growth and VEGF production in xenograft lung cancer models, as well as in xenografts of other tumor types (99).

26.3.4 Agents Inhibiting Both VEGFR and EGFR

The small molecule ZD6474 is a potent inhibitor of VEGFR-2 (KDR), and also inhibits EGFR at high concentrations. The PC-9/ZD lung adenocarcinoma cell line is resistant to the EGFR inhibitor gefitinib, and is cross-resistant to ZD6474 in vitro. ZD6474 decreased EGFR phosphorylation in the gefitinib-sensitive parent PC-9 cell line, but only partially inhibited phosphorylation in the resistant PC-9/ZD cells (100). When these lines were implanted in athymic mice, ZD6474 resulted in tumor regression in the PC-9 tumor, but also led to growth inhibition of PC-9/ZD tumors, suggesting that the therapeutic effect of ZD6474 was mediated at least in part by its effect on VEGFR (100). ZD6474 also inhibited tumor-induced angiogenesis and tumor growth in A549 human lung cancer xenografts implanted intradermally in mice, and induced a significant reduction in microvessel density in nonnecrotic regions in the Calu-6 lung cancer xenograft model (101). In the A549 human lung adenocarcinoma xenograft model, the combination of ZD6474 and a COX-2 inhibitor was superior to either agent alone, with eradication of tumor in 4 of 10 mice (102).

ZD6474 reduced tumor growth and microvessel density in bone metastases of a human lung adenocarcinoma xenograft in nude mice. However, this effect was partially antagonized by the chemotherapy agent paclitaxel (103).

Low-dose ZD6474 enhanced the antiangiogenic, antivascular, and antitumor effects of radiotherapy in a human lung cancer grown orthotopically in mice to a greater extent than did paclitaxel added to radiotherapy (103), and also reduced tumor perfusion and substantially increased tumor regrowth delay after radiotherapy in the Calu-6 NSCLC murine tumor model (104).

Similarly, the small molecule AEE788 inhibits tyrosine kinases of both VEGFR and EGFR. In enzyme systems, it inhibited Her-2/neu (ErbB2), VEGFR-2 (Flk-1/KDR), and VEGFR-1 (Flt-1), and in cells, it inhibited ligand-induced phosphorylation of EGFR and ErbB2 (105). AEE788 inhibited proliferation of EGFR and ErbB2 expressing cell lines and inhibited EGF- and VEGF-stimulated proliferation of HUVECs (105). In vivo, it was effective versus several animal tumor models, including tumors overexpressing EGFR and ErbB2. AEE788 inhibited growth factor–induced phosphorylation of EGFR and ErbB2 in tumors, and this correlated with antitumor activity. AEE788 also inhibited VEGF-induced angiogenesis in murine tumors, and decreased tumor vascular permeability and interstitial leakage space (105).

26.3.5 Agents Inhibiting Both VEGFR and PDGFR

SU11248 (which targets tyrosine kinases of each of VEGFR, PDGFR, KIT, and FLT3 receptor) inhibited stem-cell factor-stimulated KIT phosphotyrosine levels and proliferation of the KIT-expressing NCI-H526 SCLC cell line in vitro (106). In mice with this tumor implanted subcutaneously, SU11248 inhibited tumor growth and expression of phospho-KIT and tumor stromal phospho-PDGFR-β. It also potentiated the effect of cisplatin. In this model, SU11248 was superior to imatinib (which inhibits KIT, PDGFR-β, and Bcr-abl, but not VEGFR) (106). In other murine lung cancer models, SU11248 enhanced radiation-induced endothelial cytotoxicity, with increased tumor vascular destruction and tumor control. SU11248 administration after completion of radiotherapy or chemotherapy resulted in prolongation of tumor control (107).

BAY 57-9352 is an orally active small molecule inhibitor of VEGFR-2 and PDGFR-β tyrosine kinases, and has activity against a broad spectrum of human tumor xenograft models, including lung (108). In a lung cancer xenograft model, the combination of BAY 57-9352 plus paclitaxel was more effective than was either agent alone (109).

The VEGFR/PDGFR tyrosine kinase inhibitor PTK 787 reduced pleural effusion formation in mice bearing the PC14PE6 human lung adenocarcinoma xenograft without reducing the number of lung metastases. It significantly suppressed vascular hyperpermeability induced by the tumor but did not affect the VEGF concentration in the pleural effusion or expression of VEGF protein in the lung tumors. This suggested that the impact of the drug on pleural effusion formation was mediated mainly by its effect on vascular permeability (110).

AG013736, which also inhibits VEGFR and PDGFR tyrosine kinases, inhibited VEGF-mediated endothelial cell survival, receptor phosphorylation, angiogenesis, and tumor growth in xenograft models. In the murine syngeneic Lewis lung carcinoma model, docetaxel enhanced the efficacy of AG013736 though the tumor was resistant to single-agent docetaxel (111).

26.3.6 Other PDGFR Inhibitors

There is also preclinical experience with agents that inhibit PDGFR ± EGFR, but not VEGFR. Aptamers (oligonucleotides with high specificity and affinity) against PDGF-B resulted in reduced pericyte coverage of blood vessels and reduced blood vessel density in a subcutaneous lung carcinoma model (112). The EGFR and PDGFR tyrosine kinase inhibitor ABT-869 decreased VEGF and bFGF-induced angiogenesis, and reduced tumor growth in a human SCLC xenograft model (113).

26.3.7 COX-2 Inhibitors

The human lung adenocarcinoma line A549 has low basal levels of COX-2 protein. In vitro, the COX-2 inhibitor celecoxib inhibited cell growth and caused DNA damage, whereas in vivo, celecoxib inhibited tumor growth, was additive with radiotherapy, marginally reduced both total and perfused blood vessels, and reduced VEGF protein levels (114). In a Lewis lung carcinoma model expressing human IL-1β, administration of a COX-2 inhibitor resulted in significant reduction of tumor growth, angiogenesis, and infiltration of macrophages (115).

26.3.8 Protein Kinase C Inhibitors

The protein kinase C (PKC) inhibitor 317615·2HCl was relatively ineffective against cultured Calu-6 NSCLC cells (116). However, it is a potent inhibitor of VEGF-stimulated HUVEC proliferation, and administration of 317615·2HCL to nude mice bearing Calu-6 human lung cancer xenografts decreased the number of intratumoral blood vessels, and it substantially potentiated the antitumor efficacy of paclitaxel and carboplatin. The drug was also effective versus Lewis lung carcinoma and augmented tumor growth delay when added to each of paclitaxel, gemcitabine, carboplatin, and radiation (116). It also induced a growth delay and decreased tumor microvessel density in nude mice bearing SW2 human SCLC xenografts, and potentiated the effect of paclitaxel plus carboplatin therapy against this xenograft (117).

26.3.9 Heat Shock Protein-90 Antagonists

Treatment of nude mice bearing H358 human NSCLC xenografts with the Hsp90 antagonist 17-allylamino geldanamycin substantially enhanced the efficacy of paclitaxel, with increased tumor cell apoptosis, suppression of tumor growth, prolongation of survival, reduced tumor VEGF expression, and reduced tumor microvasculature (118).

26.3.10 MMP Inhibitors

In mice bearing human MV522 NSCLC xenografts, the metalloproteinase inhibitor AG3340 inhibited angiogenesis and tumor growth, and induced tumor necrosis, and potentiated both paclitaxel and carboplatin (119).

26.3.11 Thalidomide

In human A549 lung adenocarcinoma cells, the immunomodulator thalidomide downregulated the mRNA and protein expression of bFGF. VEGF mRNA and protein expression was increased with low-dose thalidomide, but decreased with higher thalidomide concentrations (120). Similar effects were seen in a cisplatin-resistant variant, A549DDP (121). Thalidomide also inhibited growth of various other NSCLC cell lines, particularly large cell lung cancer (122). It resulted in a dose-dependent increase in peroxisome proliferator-activated receptor-gamma (PPAR-γ) protein and activity, decreased NF-κB activity, increased apoptosis, and decreased production of angiogenic factors (122). In lung cancer xenograft models, intratumoral thalidomide reduced tumor growth, and was associated with increased PPAR-γ expression (122).

26.3.12 Other Immunomodulators

Topical application of the immunomodulator imiquimod reduced angiogenesis in a murine L1 lung sarcoma model. This inhibition of angiogenesis appeared to be mediated by an increased production of IL-18. IL-18 promotes production of interferon-γ, a potent inhibitor of angiogenesis (123). Systemic delivery of interleukin-24 (which has tumor suppressor, antiangiogenic and cytokine properties) in DOTAP:cholesterol nanoparticles resulted in suppression of tumor growth and reduced tumor vascularization in primary and metastatic human lung cancer xenografts (124).

26.3.13 Heparins

In the chorioallantoic membrane tumor implant model, the anticoagulant low molecular weight heparin tinzaparin inhibited growth of various human cancers, including lung cancers, and was a potent antiangiogenic agent (125). Its activity was dependent on the relatively higher molecular weight tinzaparin fragments and was independent of the angiogenic factor tested (125). Its effect was mediated by cellular release of tissue pathway factor inhibitor (TFPI). Recombinant-TFPI also inhibited angiogenesis regardless of proangiogenic factor (125). Hydrophobic heparin nanoparticles also inhibited angiogenesis in the chick chorioallantoic membrane assay and reduced lung tumor growth in a murine model (126). In the H460 lung carcinoma model, heparin oligosaccharides reduced tumor growth, angiogenesis, and microvessel density (127).

26.3.14 Prothrombin Kringles

Kringles are triple-disulfide-loop folding domains found in several blood proteins. Various prothrombin kringles have potent antiangiogenic activity. They inhibited bFGF-stimulated growth of bovine capillary endothelial cells, angiogenesis in the chick embryo chorioallantoic membrane assay, and growth of Lewis lung carcinoma in vivo models (128).

26.3.15 Metastatin

The hyaluronan-binding complex metastatin, isolated from bovine cartilage, blocked formation of tumor nodules in lungs of mice inoculated with Lewis lung carcinoma cells, and also blocked migration and proliferation of cultured endothelial cells and VEGF-induced angiogenesis in chorioallantoic membranes (129).

26.3.16 Copper-Complexing Agents

The copper-complexing agent, tetrathiolmolybdate (which has previously demonstrated antiangiogenic activity), was additive with radiotherapy in a Lewis lung cancer mouse tumor model, despite having no direct effect on Lewis lung carcinoma cells (130).

26.3.17 TNP-470

Exposure in vitro of two NSCLC cell lines to the antiangiogenic agent TNP-470 inhibited growth when used alone, and schedule-dependent synergism was noted when it was used with paclitaxel, the optimal schedule being paclitaxel followed by TNP-470 (131). TNP-470 also potentiated the effect in mice bearing Lewis lung carcinomas of a low-dose chronic administration schedule of cyclophosphamide used as an antiangiogenic strategy (132). This therapy induced endothelial cell apoptosis that preceded tumor cell apoptosis. The effect was particularly pronounced in p53-null mice (132). In other Lewis lung cancer studies, each of the antiangiogenic agents TNP-740, minocycline, suramin, and genistein decreased tumor blood vessels (133). The antiangiogenic agents potentiated chemotherapy more effectively when used in two-drug antiangiogenic combinations than when a single agent was used. TNP-470 plus minocycline was the most effective of the antiangiogenic combinations tested (133).

26.3.18 Histone Deacetylase Inhibitors

In the CL-1 human lung cancer cell line, treatment with the histone deacetylase (HDAC) inhibitor trichostatin-A upregulated expression of a membrane-anchored glycoprotein, which negatively regulates MMPs and inhibits tumor metastasis and angiogenesis (134).

26.3.19 Squalamine

The antiangiogenic aminosterol squalamine reduced growth of MV-522 human NSCLC xenografts in nude mice, and also potentiated the effect of cisplatin and of the paclitaxel/carboplatin combination. In vitro, following exposure of HUVECs to VEGF, squalamine induced disorganization of F-actin stress fibers and reduced membrane VE-cadherin, both of which could play a role in reduction in angiogenesis (135).

26.3.20 Retinoids

The retinoid bexarotene reduced angiogenesis induced by NSCLC cells, decreasing both endothelial cell growth and migration. It decreased tumor cell secretion of angiogenic factors and MMPs, while increasing secretion by tumor of tissue inhibitor of MMPs (136).

26.3.21 Inhibitors of HIF-1α Expression

HIF-1α is expressed in response to hypoxia and plays an important role in angiogenesis. Several agents including farnesyl transferase inhibitors (FTI), flavonoids, investigational cytotoxic agents, and HDAC inhibitors have been shown to decrease HIF-1α expression and to reduce angiogenesis in lung cancer models. The FTI SCH66336 exhibited antiangiogenic activity and decreased VEGF and HIF-1α expression in NSCLC cells in vitro and in vivo. It decreased insulin growth factor 1- or hypoxia-stimulated HIF-1α levels. It reduced the half-life of HIF-1α by blocking the interaction between HIF-1α and Hsp90 (137). Silibinin, an anticancer flavonoid, decreased the ability of a cytokine mixture to increase production of HIF-1α and inducible NOS (iNOS) in an NSCLC cell line, suggesting that it would reduce production of the potent angiogenic factor nitric oxide (138). Another dietary flavonoid, apigenin, inhibited VEGF expression by decreasing HIF-1α, but not HIF-1β expression, and significantly inhibited tumor-induced angiogenesis and growth of human A549 lung cancer xenografts (139).

The hypoxic cytotoxins TX-402, TX-1102, and tirapazamine inhibited angiogenesis in response to the human NSCLC cell line H1299 grown in the chick chorioallantoic membrane assay. TX-402 reduced the inducible expression under hypoxic conditions of VEGF and glucose transporter type 3 (GLUT-3) and suppressed mRNA and protein expression of HIF-1α, suggesting this as a mechanism of angiogenesis suppression by these agents (140).

When human lung adenocarcinoma CL-1–5 cells were exposed to the investigational topoisomerase-II inhibitor GL331, the conditioned medium from the cells had reduced ability to induce in vitro angiogenesis. GL331-treated cells had reduced expression of both HIF-1α and VEGF. The drug decreased binding of tumor cell–derived nuclear components to the promoter of the HIF-1*α* gene (141).

In hypoxic cells, the HDAC inhibitor FK228 inhibited HIF-1α induction and activity, and also decreased VEGF expression. In a Lewis lung carcinoma model, FK228 blocked the angiogenesis induced by hypoxia (142).

26.3.22 Other Cytotoxins

Some other chemotherapy agents also have antiangiogenic effects. Endothelial cells were 10–100-fold more sensitive to taxanes than were tumor cells. Angiogenesis was blocked by inhibition of proliferation and differentiation and by induction of cell death. Docetaxel was a more potent angiogenesis inhibitor than was paclitaxel (143). In nude mice bearing human lung cancer xenografts treated with paclitaxel weekly or q3wk, the two schedules had the same effect on tumor growth and tumor microvasculature (144).

When nude mice bearing A549 human lung cancer subcutaneous xenografts were treated with ST1481, a novel lipophilic camptothecin, prolonged daily low-dose administration schedules inhibited tumor growth and decreased tumor microvessel density to a greater extent than did intermittent high-dose administration (145). Reduction in microvessel density correlated with inhibition of tumor growth. In vitro, ST1481 also had an antimotility effect on endothelial cells, and inhibited vascularization in the Matrigel assay, and downregulated expression of bFGF in A549 cells in association with inhibition of the Akt pathway (145).

Conversely, the new agent arsenic trioxide causes a dose-dependent increase in vessel density in the chick chorioallantoic membrane assay, increased vessel density in the mouse Matrigel assay, and increased melanoma tumor growth and lung metastases in nude mice (146).

26.3.23 Antivascular Agents

In addition to agents that suppress angiogenesis, investigations are also under way of new agents that are directly toxic to tumor blood vessels. ZD6126 (ANG453) selectively disrupts the cytoskeleton of tumor endothelial cells (147). In mice, ZD6126 caused bleeding and necrotic changes in PC14PE6 human lung adenocarcinoma and H226 human lung squamous cell carcinoma pulmonary metastases, and also caused apoptotic changes in CD31-positive endothelial cells in tumor but not in normal lung. Administration of ZD6126 for 14 days led to decreased size of lung metastases without a reduction in number of metastases (147). ZD6126 also decreased the growth of human A549 lung cancer xenografts, and resulted in protracted tumor regression when added to radiation (148). ZD6126 plus gefitinib caused greater tumor regression than either agent alone, and this was further improved by addition of radiation. Both gefitinib and ZD6126 substantially reduced tumor angiogenesis. ZD6126 caused substantial vessel destruction, with loss of endothelial cells and thrombosis, and increased the necrosis caused by radiation. The combination of radiation, ZD6126 and gefitinib induced greater effects on tumor growth and angiogenesis than did any of these approaches alone (148). In a human lung adenocarcinoma PC14PE6 xenograft model, the combination of ZD6126 plus cisplatin increased both tumor cell and endothelial cell apoptosis compared with either treatment alone. Tumor burden was also reduced (149).

The antivascular agent combretastatin had an antiproliferative effect on human NSCLC cells grown in vitro. In vivo, it delayed growth of subcutaneously implanted lung cancer and prolonged survival in a metastasizing lung orthotopic xenotransplant model (150). Similarly, the N-cadherin antagonist ADH-1 inhibits angiogenesis and is directly toxic to vessels, and in the Lewis lung carcinoma model, reduced microvessel density and tumor growth rate (151).

26.4 CLINICAL TRIALS OF ANTIANGIOGENIC AGENTS IN LUNG CANCER

26.4.1 Endostatin

In a phase I study of endostatin, two patients with NSCLC were entered. No responses were seen in the study (152).

26.4.2 Thrombospondin

In a phase IB trial of the TSP-1 mimetic peptide ABT-510, prolonged stable disease was seen in 1 of 2 NSCLC patients entered, and a partial remission was seen in a soft tissue sarcoma patient (153).

26.4.3 Agents Inactivating VEGF

The VEGF/VEGFR pathway has been targeted by using agents that act directly on VEGF as well as by using agents that inhibit VEGFR tyrosine kinases.

Bevacizumab is a recombinant humanized monoclonal antibody that binds and inactivates VEGF. Ninety-nine patients with chemonaïve-advanced NSCLC were entered in a phase II study in which they were randomized to receive carboplatin plus paclitaxel alone, or combined with bevacizumab 7.5 or 15 mg/kg (154). Compared with the chemotherapy alone arm, the group receiving bevacizumab 15 mg/kg with chemotherapy had a higher response rate (31.5% versus 18.8%), longer time to progression (7.4 months versus 4.2 months), and prolongation of median survival time (17.7 months versus 14.9 months). Of 19 patients on the control arm who subsequently received single-agent bevacizumab, five achieved stable disease and the 1 year survival rate was 47%. Six patients experienced major life-threatening bleeding with hemoptysis or hematemesis, and four of these events were fatal. These events occurred in patients with centrally located tumors close to major blood vessels, and five cases had tumor cavitation either before or during treatment. Four of the patients experiencing severe hemorrhage had squamous cell carcinomas of lung (154).

A total of the 878 chemonaïve patients with advanced NSCLC were then randomized to receive paclitaxel plus carboplatin with or without bevacizumab. Compared with the control, bevacizumab resulted in improvement with respect to response rate (27% versus 10%), progression-free survival time (6.4 months versus 4.5 months) and median survival time (12.5 months versus 10.2 months). The incidence of grade $\geq$4 neutropenia, grade 3/4 thrombosis/embolism, and hemorrhage was all higher on the bevacizumab arm. Of 11 treatment-related deaths (two with chemotherapy alone and nine with chemotherapy plus bevacizumab), five on the bevacizumab arm were due to hemoptysis (155).

In a phase I/II study in 40 previously treated patients with NSCLC, the combination of bevacizumab plus erlotinib resulted in a response rate of 20%, and an additional 65% of patients had stable disease. The median overall survival time was 12.6 months for the 34 patients treated at the phase II dose, and the progression-free survival was 6.2 months in this group (156).

Veglin is a VEGF antisense oligonucleotide that targets VEGF-A, VEGF-C, and VEGF-D. A phase I study was undertaken in patients with advanced malignancies that included two patients with lung cancer, neither of whom responded, although responses were seen in some other malignancies (157).

26.4.4 Inhibitors of VEGFR Tyrosine Kinases

Efficacy has also been seen in NSCLC patients treated with small molecules that inhibit tyrosine kinases of VEGFR and other receptors, but as with bevacizumab, tumor cavitation and fatal hemoptysis have been seen in some patients:

ZD6474 inhibits tyrosine kinases of VEGFR and EGFR. In a phase I study of ZD6474 doses conducted in Japan, four of nine patients with refractory NSCLC achieved partial responses (158). In a randomized study comparing ZD6474 with gefitinib in patients with advanced NSCLC, patients receiving ZD6474 had a significant prolongation of progression-free survival compared with those receiving gefitinib (11.9 weeks versus 8.1 weeks) (159).

In a randomized phase II study comparing docetaxel alone with docetaxel plus ZD6474 doses of 100 or 300 mg/day, patients receiving ZD6474 had an increase in median progression-free survival (12.0 weeks for docetaxel alone versus 18.7 weeks for docetaxel plus ZD6474 doses 100 mg versus 17.0 weeks for docetaxel plus ZD6474 doses 300 mg) (159).

In a phase I trial of XL647 (an orally bioavailable small molecule inhibitor of multiple receptor tyrosine kinases including EGFR, erbB2, VEGFR-2, and EphB4), there were no objective responses, but 1 of 5 patients with NSCLC experienced a 22% reduction in tumor size and continued to remain stable at last follow-up at 22 weeks (160).

AG013736 is an orally administered receptor tyrosine kinase inhibitor of VEGFR, PDGFR, and c-kit. In a phase I study that included patients with NSCLC, three NSCLC patients underwent an evaluation by DCE-MRI of changes in tumor blood flow, as assessed by the volume transfer constant (K^{trans}) and the initial area under the curve (IAUC). All three had a reduction of >33% in both K^{trans} and IAUC, of whom two also experienced tumor cavitation (161). Both patients subsequently died of hemoptysis, and one of these deaths was attributed to AG013736 (162).

GW786034 is an orally available tyrosine kinase inhibitor of VEGFR-1, VEGFR-2, and VEGFR-3, PDGFR-α and PDGFR-β, and c-kit. In a phase I study, suggestions of efficacy were seen in a number of tumor types, including stable disease lasting >6 months in a patient with lung cancer (163).

In a phase I study of SU6668 (a small molecule inhibitor of the receptor tyrosine kinases VEGFR-1, PDGFR, and FGFR), three lung cancer patients were included. None of these three had stable disease or response (164).

BAY 43-9006 (sorafenib) is an orally administered inhibitor of Raf kinase and VEGFR tyrosine kinases. In a phase I study that included 10 patients with NSCLC, 1 NSCLC patient had a partial response and 2 had stable disease (165). In a phase II study of sorafenib in 54 patients with advanced previously treated NSCLC, 4 patients (8%) experienced reduction in maximum tumor diameter of ≥30%, and another 11 patients (23%) had tumor shrinkage but <30%. Four patients with stable disease developed tumor cavitation. Median progression-free survival was 11.9 weeks. One patient died from hemoptysis (166).

26.4.5 COX-2 Inhibitors

The COX-2 inhibitor celecoxib has also been tested in combination with chemotherapy and radiotherapy in NSCLC, and has generally been well tolerated. The combination of docetaxel plus celecoxib was tested in patients with recurrent NSCLC previously treated with one or more chemotherapy regimens (167). Serum VEGF levels fell from an average of 661.2 pg/mL prior to celecoxib to 483.6 pg/mL afterward—a 29% decrease. Serum endostatin levels increased after celecoxib, and the serum VEGF:endostatin ratio decreased after celecoxib. High COX-2 expression was detected in 9 of 12 evaluable tumors. Intratumoral PGE2 levels also decreased with celecoxib in this study (167), whereas in another study of preoperative paclitaxel plus carboplatin plus daily celecoxib in stages I–IIIA NSCLC, the response rate was 65%, and celecoxib prevented the usual rise in PGE2 that is seen in NSCLC following chemotherapy (168).

A phase I study was undertaken in patients with inoperable NSCLC in which celecoxib 200–800 mg was administered daily beginning 5 days before radiotherapy and continuing concurrently with the radiotherapy. Patients received either 45 Gy radiation in 15 fractions, 66 Gy in 33 fractions, or 63 Gy in 35 fractions, depending on their clinical situation. No unexpected radiation toxicity was seen. Of 37 evaluable patients, 73% responded, including 38% with a complete response. The local progression-free survival rate was 66% at 1 year and 42% at 2 years (169).

In a phase I trial of celecoxib plus erlotinib in patients with advanced, refractory NSCLC, there were no unexpected toxicities. Of 12 evaluable subjects, 4 achieved partial responses and 3 patients had stable disease. Responses were seen both with and without EGFR-activating mutations. A significant decline in urinary PGE-M was demonstrated (170).

26.4.6 Protein Kinase C Inhibitors

PKC412 (*N*-benzoyl staurosporine) inhibits PKC and also inhibits angiogenesis in preclinical systems. In a phase I trial of PKC412 combined with gemcitabine and cisplatin in NSCLC, nausea, vomiting, and diarrhea were dose limiting. Further trials would be needed to define efficacy (171).

26.4.7 Matrix Metalloproteinase Inhibitors

In a randomized, double-blind, placebo-controlled phase II trial involving 75 patients with NSCLC, the matrix metalloproteinase inhibitor (MMPI) BMS-275291 was added to paclitaxel and carboplatin. The response rate was 22% in the BMS-275291 arm and 36% in the placebo arm (172). In a larger phase III study, 774 NSCLC patients were randomized to receive paclitaxel plus carboplatin with BMS-275291 versus placebo. In the BMS-275291 versus placebo arms, respectively, the overall survival time was 8.6 months versus 9.2 months, progression-free survival was 4.9 months versus 5.3 months, and response rate was 25.8% versus 33.7% (173). Hence, neither the phase II nor the phase III randomized trials suggested efficacy of the MMPI.

Similarly, the MMPI prinomastat added to chemotherapy did not improve outcome in advanced NSCLC compared to chemotherapy alone. 362 patients were randomized to receive gemcitabine plus cisplatin combined with prinomastat versus placebo. There was no statistically significant difference between prinomastat versus placebo, respectively, for response rate (27% versus 26%), median overall survival time (11.5 months versus 10.8 months), 1 year survival rate (43% versus 38%), or progression-free survival time (6.1 months versus 5.5 months) (174). In studies involving 1023 patients with advanced NSCLC treated with chemotherapy plus prinomastat versus placebo, prinomastat approximately doubled the hazard ratio for the development of venous thromboembolism (175).

In a trial involving 279 patients with limited SCLC and 253 with extensive SCLC, patients were randomized to the MMPI marimastat versus placebo after response to front-line chemotherapy. The median time to progression was 4.3 months for marimastat patients and 4.4 months for placebo patients, and median survival times were 9.3 and 9.7 months, respectively. Marimastat toxicity adversely impacted quality of life (176). Hence, MMPIs have not improved therapeutic outcome in either NSCLC or SCLC.

26.4.8 Interferons

Interferons exert a variety of biological effects, including inhibiting angiogenesis. There have been several trials of interferon-α in lung cancer, and a few trials of interferon-β and interferon-γ.

26.4.8.1 Interferon-α in SCLC

Forty patients with extensive SCLC were treated on a phase II trial with a combination of cisplatin plus etoposide plus interferon-α2a, followed by maintenance with interferon and megestrol acetate. The overall response rate was 89%, including three complete remissions. Median survival was 46 weeks. Hence, there was no indication that treatment efficacy was improved by the addition to chemotherapy of interferon plus megestrol acetate (177).

In a phase III trial, patients with SCLC were treated with the combination of cisplatin plus etoposide plus interferon-α, with 78 patients receiving chemotherapy alone, 75 receiving chemotherapy plus natural interferon-α, and 66 receiving chemotherapy plus recombinant interferon-α. There was no difference in survival between the three arms, with median survival times of 10.2, 10.0, and 10.1 months in the three arms, respectively, and 2 year survival rates of 15%, 3%, and 11%, respectively. Hence, the interferon did not improve outcome (178).

However, in another phase III trial in which patients with chemonaïve SCLC were randomly assigned to treatment with chemotherapy alone versus chemotherapy plus interferon-α2a, there was an indication of benefit from the interferon. In this trial, responding patients on each arm received radiotherapy to the primary site and prophylactic cranial radiation, followed by maintenance chemotherapy with cyclophosphamide orally each day. Response rates were 90% versus 86% in the two arms, but the complete remission rate on the interferon arm of 38% was significantly higher than the 28% complete remission rate on the chemotherapy alone arm. Furthermore, patients on the interferon arm had a significantly longer median survival time than did patients on the chemotherapy alone arm in this trial. The benefit of interferon was predominantly in patients with limited disease (179).

In a third randomized trial, patients with SCLC were randomized to receive treatment with chemotherapy alone (three cycles of cyclophosphamide/vincristine/doxorubicin and three cycles of cisplatin/etoposide) or chemotherapy plus interferon-α2c. Complete remission rates (30% versus 15%) and partial remission rates (42% versus 29%) were higher in the interferon arm than in the chemotherapy alone arm. Differences in time to tumor progression (7.6 months versus 5.4 months) did not reach statistical significance but overall survival was significantly longer in the interferon arm. The 2 year survival rate was 14% on the interferon arm versus 0% on the chemotherapy alone arm (180). Hence, two of three of these trials suggest a possible benefit of adding interferon to frontline SCLC chemotherapy.

In some SCLC trials, interferon was given as maintenance therapy after response to chemotherapy ± radiation. Limited stage SCLC patients who had responded to chemotherapy were randomized to receive either maintenance interferon-α2a versus observation. Median time from randomization to progression was 9 months on the interferon arm versus 10 months on the observation arm, and overall median survival was 13 months on the interferon arm and 16 months on the observation arm (181). In another randomized phase III trial, patients with SCLC were treated with four cycles of chemotherapy (cyclophosphamide plus vincristine plus etoposide) plus radiotherapy. Responding patients were then randomly assigned to low-dose interferon-α versus maintenance chemotherapy versus no maintenance therapy. Median survival times were comparable in the three groups (11, 11, and 10 months, respectively), but there appeared to be an improvement in long-term survival in the limited disease group favoring the interferon-α arm. Ten percent of patients in the interferon group survived for 5 years or more, versus 2% of patients in the chemotherapy maintenance group or control group ($p = 0.04$) (182). In still another study, patients with SCLC who achieved complete remission with chemotherapy were randomized to receive maintenance therapy with interferon-α2b versus no maintenance therapy. Twenty-six patients were randomized, including 14 to interferon maintenance and 12 to the control arm. Median time to progression from randomization was 12 months in patients on the interferon arm versus 7 months for patients on the control arm. Median survival was 15 months versus 9 months. These differences did not reach statistical significance, possibly due to low statistical power (183).

In a randomized phase II trial, 60 SCLC patients responding to front-line chemotherapy were randomly assigned to a maintenance arm (for 1 year) consisting of interferon-α2a plus retinoic acid versus trophosphamide versus no maintenance. Median survival was 17.1 months on the interferon plus retinoic acid arm, 12.4 months on the trophosphamide arm, and 13.5 months on the control arm, with the 1 year survival rates of 82%, 56%, and 55%, respectively. Time to progression was 8.6, 8.0, and 6.8 months, respectively. These differences were not statistically significant. Patients on the interferon-α2a plus retinoic acid arm lived longer after the onset of progressive disease than did patients on the other two arms (184). Overall, there is suggestive evidence of benefit of interferon-α in SCLC patients, but the data are inconclusive.

26.4.8.2 Interferon-α Plus Radiation or Chemotherapy in NSCLC

Interferon-α has also been tested extensively in advanced NSCLC. In a small trial involving 20 patients with inoperable NSCLC, patients were randomized to receive hyperfractionated radiation 60 Gy or the same radiotherapy concurrently with intramuscular plus inhaled interferon-α. Toxicity may have been increased by the interferon. Efficacy did not appear to be increased by the addition of interferon (185).

The combination of cisplatin plus interferon-α2a was investigated in a phase II trial of chemonaïve patients with advanced NSCLC. The response rate was 33% among evaluable patients. Median survival time was 6.4 months. Response rate was 45% in stage IIIA disease and 22% in stage IV disease (186). In another small phase II trial involving 10 patients with advanced NSCLC, the combination of cisplatin plus interferon-α resulted in a response rate of 50% and a median survival time of 8 months (187). When the combination of carboplatin plus interferon-α was administered to 44 chemonaïve patients with advanced NSCLC, 3 of 41 evaluable patients (7.3%) achieved partial remissions. The overall median survival time was 6 months. It was concluded that the addition of interferon to carboplatin did not improve efficacy (188).

In a phase I study of gemcitabine plus interferon-α2b, seven patients with NSCLC were included of whom one achieved a partial remission (189). In 18 chemonaïve NSCLC patients treated with interferon-α2b plus 5-fluorouracil plus leucovorin, the response rate was 39% with a median survival time of 10 months (190).

Interferon has also been given with combination chemotherapy. In 34 chemonaïve patients with advanced NSCLC treated with the combination of cyclophosphamide, epidoxorubicin, cisplatin, and recombinant interferon-α2b, the response rate was 19.3%. The median overall survival and progression-free survival were 37 and 20 weeks, respectively (191). In a follow-up randomized trial in which 182 chemonaïve patients with advanced NSCLC were randomized to receive cisplatin plus epidoxorubicin plus cyclophosphamide with or without recombinant interferon-α, the median survival was 6.0 months on the chemotherapy plus interferon arm and 5.5 months on the chemotherapy alone arm. Progression-free survival was similar for the two arms. The response rate was 18.9% on the chemotherapy plus interferon arm versus 7.6% on the chemotherapy alone arm. Hence, the interferon may have increased the response rate but did not improve survival (192).

The combination of mitomycin-C plus vindesine plus cisplatin plus interferon-α was tested in 35 patients with advanced NSCLC. The overall response rate was 51%, with median time to treatment failure of 6 months and a median survival of 9.5 months (193). When 22 patients with advanced NSCLC were randomized to receive ifosfamide alone or ifosfamide followed by thymosin-α-1 plus low-dose interferon-α, the response rates were 10% and 33%, respectively, and time to progression was significantly longer in the group receiving thymosin-α-1 plus interferon compared than in the group receiving chemotherapy alone (194).

Overall, as with SCLC, some studies in NSCLC suggest a modest benefit when interferon-α is added to chemotherapy, but most studies were negative.

26.4.8.3 Interferon-α Plus Retinoids or Interleukin-2 in NSCLC

Interferon-α has also been administered with retinoids in some trials. In one trial, NSCLC patients with advanced disease were treated with a combination of *trans*-retinoic acid plus interferon-α. Objective responses were seen in 4 of 29 patients (195). In another trial, patients with advanced squamous cell carcinomas of the lung and other sites were treated on a phase II study of cisplatin plus all-*trans*-retinoic acid plus interferon-α. Of 38 patients treated, 7 showed objective responses (196). When patients with squamous cell carcinomas of the lung and other sites were treated with a combination of high-dose etretinate and interferon-α, 3 of 24 patients experienced partial remissions and 8 stable disease (197).

In one study in which 10 patients with squamous cell carcinomas of the lung or head and neck were treated with the combination of 13-*cis*-retinoic acid plus interferon-α, no responses were observed (198), whereas in another phase II study of the combination of interferon-α plus 13-*cis*-retinoic acid in patients with advanced squamous cell lung cancer, 1 of 17 evaluable patients achieved a partial remission (199).

A phase I/II trial of interleukin-2 plus interferon-α was undertaken in patients with advanced NSCLC. In the 11 patients treated, no responses were seen, and 9 patients developed progressive disease during the first 5 weeks of treatment, indicating that the treatment was ineffective in this tumor type (200).

Overall, efficacy of the combination of interferon-α plus retinoids or interleukins-2 in NSCLC appears to be limited.

26.4.8.4 Single-Agent Interferon-β in NSCLC

Forty-one patients with advanced NSCLC were treated on a phase II trial of interferon-β. No responses were seen (201).

26.4.8.5 Interferon-β Plus Radiotherapy in NSCLC

A phase I/II study was undertaken in patients with inoperable stage III NSCLC, with varying doses of interferon-β administered with radiotherapy. The response rate was 81%, with 44% achieving complete remission. Median survival was 19.7 months and the 5 year survival rate was 31%. No unexpected toxicity was seen (202). In another trial, 15 patients with stage II–III NSCLC were treated with 60 Gy radiation combined with escalating doses of interferon-β1a. One patient died of pneumonia, sepsis, adult respiratory distress syndrome, and radiation pneumonitis. One patient achieved complete remission, six had partial responses, and six had stable disease (203). Fourteen patients with stage IIIB NSCLC were treated with radiation 59.4 Gy, with intravenous interferon. Of these, four (20.6%) patients had complete responses and seven (50%) had partial responses, for an overall response rate of 70.6%. Median survival was 13 months, with the 3 year survival rate of 37.5% (204).

In a phase III study in stage III NSCLC, patients were treated with radiotherapy 60 Gy over 6 weeks with versus without interferon-β 16 million IU i.v. 3 days/week in weeks 1, 3, and 5 of the radiation. Radiotherapy was completed in 94% of control arm patients and in 82% of interferon patients. In the interferon arm, 76% of patients completed interferon, with toxicity as the major reason that it was not completed. There was no significant difference in median survival (9.5 months versus 10.3 months for arms without versus with interferon) or in percentage of 1 year survival (44% versus 42%) (205).

26.4.8.6 Interferon-β Plus Other Systemic Agents in NSCLC

Patients with advanced NSCLC were randomized between interleukin-2 alone versus interleukin-2 plus interferon-β in a randomized phase II study. The overall response rate on the study was only 4%, but median survival was 33 weeks (206). In another trial, patients with chemonaïve advanced NSCLC were treated with cisplatin, 5-fluorouracil, vindesine, interferon-β, and retinyl palmitate. Responders were maintained with the same doses of interferon and retinyl palmitate. Of 40 evaluable patients, 17 (42%) responded. Median overall survival was 9.1 months (207).

Overall, there is little indication that interferon-β is therapeutically active in NSCLC.

26.4.8.7 Interferon-γ

In advanced NSCLC, 80 patients were randomly assigned to receive chemotherapy alone versus chemotherapy with either interferon-γ or interferon-γ plus interferon-α. Median survival in all arms was 6 to 7 months. The addition of interferon to chemotherapy was not associated with improvement in survival (208).

26.4.9 TNP-470

In a phase I clinical trial combining the antiangiogenesis agent TNP-470 with carboplatin and paclitaxel in patients with NSCLC and other malignancies, hematological toxicity, efficacy, and carboplatin pharmacokinetics were comparable with what would be expected with chemotherapy alone. The TNP-470 was associated with reversible mild to moderate neurocognitive impairment (209).

In a phase I trial of the antiangiogenic agent TNP-470 with paclitaxel in adults with solid tumors, the MTD was 60 mg/m^2 three times per week for TNP-470 with paclitaxel 225 mg/m^2 every 3 weeks. Paclitaxel myelosuppression was not increased by TNP-470, although paclitaxel clearance was decreased slightly. Mild reversible neurocognitive impairment was seen. Median survival was 14.1 months. Partial responses were seen in 8 (25%) of 32 patients overall, including 6 (38%) of 16 patients with NSCLC, 60% of whom had had prior chemotherapy (210).

26.4.10 Squalamine

Squalamine, an agent with antiangiogenic activity in animal models, that acts directly in activated endothelial cells, was combined with carboplatin and paclitaxel in a phase I/IIA trial in patients with advanced NSCLC. Arthralgias and myalgias limited squalamine dose. There was no evidence of pharmacokinetic interactions between the drugs. The response rate of 28%, the median survival time of 10 months, and the 1 year survival rate of 40% were similar to what would be expected with chemotherapy alone (211). In a second phase II study combining weekly squalamine (with a randomization between 100 and 200 mg/m^2) with carboplatin plus paclitaxel in NSCLC, the response rate was 23% (212).

26.4.11 AE-941 (Neovastat)

Neovastat, an antiangiogenic compound derived from shark cartilage, inhibits blood vessel formation and endothelial cell proliferation in the chicken embryo vascularization assay, blocks formation of blood vessels in Matrigel implants containing bFGF, inhibits the activity of the MMP-2, MMP-9, and MMP-12, and inhibits VEGF binding to endothelial cells, VEGF-dependent tyrosine phosphorylation, and VEGF-induced vascular permeability in mice (213). It also decreased the number of lung metastases in a Lewis lung carcinoma model and was additive to cisplatin in this model (213). In a phase I/II study that included 48 patients with advanced NSCLC, there was evidence of a dose–response effect in the NSCLC patients, in that median survival was statistically significantly longer in patients receiving >2.6 mL/kg/day than in patients receiving lower doses (6.1 months versus 4.6 months). Although no objective responses were seen, 26% of the patients in the high-dose group had stable disease versus 14% of the low-dose group (214). Neovastat is currently undergoing clinical trials in a randomized study of chemoradiation with or without Neovastat in stage III NSCLC (213).

26.5 ANTIANGIOGENIC THERAPY AND PROBLEMS UNIQUE TO THE LUNG ENVIRONMENT

As noted earlier, in clinical trials of antiangiogenic therapies in patients with lung cancers, some patients have developed life-threatening or fatal hemoptysis (154,155,162,166). Risk factors for development of life-threatening hemoptysis have included development of cavitation (154,162,166), presence of squamous cell lung carcinoma (154), and central location of the tumor adjacent to large blood vessel (154).

Why this is a greater problem in lung cancer than in other tumor types is unknown, but a variety of issues are worth noting. First, most standard chemotherapy agents and radiation are more effective against well-oxygenated tumor cells than against hypoxic tumor cells (215,216). Blood flow and tumor oxygenation are generally better at the periphery of the tumor than in the tumor center (217–223). Simplistically, this may in part explain why most tumors shrink concentrically when treated with radiation and standard chemotherapy: the most sensitive tumor cells are at the periphery and hence tumor cells at the periphery die first.

On the other hand, tumor cells at the center tend to be hypoxic, with decreased blood flow, as noted earlier. The observation of increased cavitation with exposure to antiangiogenic agents would suggest that these agents are further potentiating the natural tumor necrosis that occurs based on the usual central tumor hypoxia. In tumors in organs other than lung, such a predilection for antiangiogenic therapy to be most effective against the hypoxic center of the tumor might be reflected in decreased central attenuation of tumor on computed tomography (CT) scan, as is seen with treatment of gastrointestinal stromal tumors with imatinib (224). One might speculate that the reason why one sees cavitation in lung tumors, rather than decreased attenuation is due simply to the proximity of lung tumors to airways: the central necrotic debris is cleared through the airway and replaced by air. With respect to why this might have a higher tendency to bleed than do tumors in other areas, one might envision that the necrotic debris filling the center of tumor would tend to tamponade blood vessels, thereby reducing bleeding. If this central necrotic debris is replaced by air, this tamponade effect would be lost.

Various strategies are currently considered to try to reduce the problem of bleeding into lung tumors after antiangiogenic therapy. Strategies include giving radiotherapy before treatment with an antiangiogenic agent for large tumors that are at particularly high risk (e.g., in close proximity to major blood vessels, squamous cell carcinomas). An alternate approach that might be explored would be to monitor patients after therapy with antiangiogenic agents and to resect or focally radiate any tumor that does cavitate.

Another issue is whether antiangiogenic therapy should be given concurrently with or sequentially with standard therapies. Theoretically, giving antiangiogenic therapy concurrently with other standard therapies might have the advantage of decreasing accelerated repopulation between cycles or fractions of these other therapies. Hence, by reducing the regrowth of the cancer between cycles of standard therapy, these antiangiogenic therapies could augment tumor shrinkage even if not causing any direct tumor regression themselves.

26.6 FUTURE RESEARCH STRATEGIES FOR COMBINING ANTIANGIOGENIC THERAPY WITH STANDARD THERAPY IN LUNG CANCER

Consideration is currently given to combining antiangiogenic agents with vascular toxic agents such as Exherin. In addition, there are theoretical advantages to giving agents, which would decrease VEGF or other ligands (or would decrease the interaction of ligand with relevant receptor) with an inhibitor of the receptor tyrosine kinase. In addition, if antiangiogenic agents exert their effect by infarcting central tumor cells, then combining the antiangiogenic agent with an antihypertensive agent to reduce blood pressure (and hence tumor blood flow) might augment cell kill. Conversely, following induction of hypoxic injury, additional therapeutic advantage might be gained by increasing blood pressure. In theory, this increased blood flow and oxygenation in tumor might translate into increased reperfusion injury in tumor.

26.7 CONCLUSION

There is substantial evidence that lung cancers rely on angiogenesis for growth, that increased expression of angiogenic factors correlates with poor prognosis in lung cancer clinically, and that a number of antiangiogenic strategies are effective in some lung cancer preclinical models. Early studies indicate that some of these antiangiogenic strategies are of limited value clinically, but others (e.g., bevacizumab) are of proven benefit or appear promising. Substantial additional investigation is required for us to learn how to optimally use antiangiogenic therapies in combination with standard approaches.

REFERENCES

1. SEER cancer statistics review. http://seer.ca.gov, 1975–2001.
2. Domont, J., Soria, J.C., and Le Chevalier, T. Adjuvant chemotherapy in early-stage non-small cell lung cancer. *Semin Oncol, 32*: 279–283, 2005.
3. Belani, C.P. Adjuvant and neoadjuvant therapy in non-small cell lung cancer. *Semin Oncol, 32* (suppl 2): S9–S15, 2005.
4. Farray, D., Mirkovic, N., and Albain, K.S. Multimodality therapy for stage III non-small-cell lung cancer. *J Clin Oncol, 23*: 3257–3269, 2005.
5. Laskin, J.J. and Sandler, A.B. State of the art in therapy for non-small cell lung cancer. *Cancer Invest, 23*: 427–442, 2005.
6. Raez, L.E. and Lilenbaum, R. Chemotherapy for advanced non-small-cell lung cancer. *Clin Adv Hematol Oncol, 2*: 173–178, 2004.
7. Rapp, E., Pater, J.L., Willan, A., Cormier, Y., Murray, N., Evans, W.K., Hodson, D.I., Clark, D.A., Feld, R., and Arnold, A.M. Chemotherapy can prolong survival in patients with advanced non-small-cell lung cancer—report of a Canadian multicenter randomized trial. *J Clin Oncol, 6*: 633–641, 1988.
8. Shepherd, F.A., Fossella, F.V., Lynch, T., Armand, J.P., Rigas, J.R., and Kris, M.G. Docetaxel (Taxotere) shows survival and quality-of-life benefits in the second-line treatment of non-small cell lung cancer: a review of two phase III trials. *Semin Oncol, 28*: 4–9, 2001.
9. Hanna, N., Shepherd, F.A., Fossella, F.V., Pereira, J.R., De Marinis, F., von Pawel, J., Gatzemeier, U., Tsao, T.C., Pless, M., Muller, T., Lim, H.L., Desch, C., Szondy, K., Gervais, R., Shaharyar, Manegold, C., Paul, S., Paoletti, P., Einhorn, L., and Bunn, P.A., Jr. Randomized phase III trial of pemetrexed versus docetaxel in patients with non-small-cell lung cancer previously treated with chemotherapy. *J Clin Oncol, 22*: 1589–1597, 2004.
10. Shepherd, F.A., Rodrigues Pereira, J., Ciuleanu, T., Tan, E.H., Hirsh, V., Thongprasert, S., Campos, D., Maoleekoonpiroj, S., Smylie, M., Martins, R., van Kooten, M., Dediu, M., Findlay, B., Tu, D., Johnston, D., Bezjak, A., Clark, G., Santabarbara, P., and Seymour, L. Erlotinib in previously treated non-small-cell lung cancer. *N Engl J Med, 353*: 123–132, 2005.
11. Thatcher, N., Faivre-Finn, C., and Lorigan, P. Management of small-cell lung cancer. *Ann Oncol, 16* (suppl 2): ii235–ii239, 2005.
12. Kurup, A. and Hanna, N.H. Treatment of small cell lung cancer. *Crit Rev Oncol Hematol, 52*: 117–126, 2004.
13. Ardizzoni, A. Topotecan in the treatment of recurrent small cell lung cancer: an update. *Oncologist, 9* (suppl 6): 4–13, 2004.
14. Herbst, R.S. and Sandler, A.B. Overview of the current status of human epidermal growth factor receptor inhibitors in lung cancer. *Clin Lung Cancer, 6* (suppl 1): S7–S19, 2004.
15. Eberhard, D.A., Johnson, B.E., Amler, L.C., Goddard, A.D., Heldens, S.L., Herbst, R.S., Ince, W.L., Janne, P.A., Januario, T., Johnson, D.H., Klein, P., Miller, V.A., Ostland, M.A., Ramies, D.A., Sebisanovic, D., Stinson, J.A., Zhang, Y.R., Seshagiri, S., and Hillan, K.J. Mutations in the epidermal growth factor receptor and in KRAS are predictive and prognostic indicators in patients with non-small-cell lung cancer treated with chemotherapy alone and in combination with erlotinib. *J Clin Oncol, 23*: 5900–5909, Epub 2005 July 5925, 2005.

16. Sumi, M., Kagohashi, K., Satoh, H., Ishikawa, H., Funayama, Y., and Sekizawa, K. Endostatin levels in exudative pleural effusions. *Lung, 181*: 329–334, 2003.
17. Iizasa, T., Chang, H., Suzuki, M., Otsuji, M., Yokoi, S., Chiyo, M., Motohashi, S., Yasufuku, K., Sekine, Y., Iyoda, A., Shibuya, K., Hiroshima, K., and Fujisawa, T. Overexpression of collagen XVIII is associated with poor outcome and elevated levels of circulating serum endostatin in non-small cell lung cancer. *Clin Cancer Res, 10*: 5361–5366, 2004.
18. Suzuki, M., Iizasa, T., Ko, E., Baba, M., Saitoh, Y., Shibuya, K., Sekine, Y., Yoshida, S., Hiroshima, K., and Fujisawa, T. Serum endostatin correlates with progression and prognosis of non-small cell lung cancer. *Lung Cancer, 35*: 29–34, 2002.
19. Volm, M., Mattern, J., and Koomagi, R. Angiostatin expression in non-small cell lung cancer. *Clin Cancer Res, 6*: 3236–3240, 2000.
20. Dunn, J.R., Panutsopulos, D., Shaw, M.W., Heighway, J., Dormer, R., Salmo, E.N., Watson, S.G., Field, J.K., and Liloglou, T. METH-2 silencing and promoter hypermethylation in NSCLC. *Br J Cancer, 91*: 1149–1154, 2004.
21. Mascaux, C., Martin, B., Paesmans, M., Verdebout, J.M., Verhest, A., Vermylen, P., Bosschaerts, T., Ninane, V., and Sculier, J.P. Expression of thrombospondin in non-small cell lung cancer. *Anticancer Res, 22*: 1273–1277, 2002.
22. Oshika, Y., Masuda, K., Tokunaga, T., Hatanaka, H., Kamiya, T., Abe, Y., Ozeki, Y., Kijima, H., Yamazaki, H., Tamaoki, N., Ueyama, Y., and Nakamura, M. Thrombospondin 2 gene expression is correlated with decreased vascularity in non-small cell lung cancer. *Clin Cancer Res, 4*: 1785–1788, 1998.
23. Fontanini, G., Boldrini, L., Calcinai, A., Chine, S., Lucchi, M., Mussi, A., Angeletti, C.A., Basolo, F., and Bevilacqua, G. Thrombospondins I and II messenger RNA expression in lung carcinoma: relationship with p53 alterations, angiogenic growth factors, and vascular density. *Clin Cancer Res, 5*: 155–161, 1999.
24. Yamaguchi, M., Sugio, K., Ondo, K., Yano, T., and Sugimachi, K. Reduced expression of thrombospondin-1 correlates with a poor prognosis in patients with non-small cell lung cancer. *Lung Cancer, 36*: 143–150, 2002.
25. Yano, S., Nishioka, Y., Goto, H., and Sone, S. Molecular mechanisms of angiogenesis in non-small cell lung cancer, and therapeutics targeting related molecules. *Cancer Sci, 94*: 479–485, 2003.
26. Huang, H.-T., Yuan, A., Chen, J.-J.W., Shih, F.-U., Kuo, T.-H., Luh, K.-T., Lee, Y.-C., Yu, C.-J., and Yang, P.-C. Overexpression of four VEGF isoforms induces different patterns of neovasculature, increases in vitro cancer cell invasion and tumorigenesis in non-small cell lung cancer. *Proc Amer Assoc Cancer Res, 46*: 914, 2005.
27. Ren, Y., Li, X., and Zhang, J. [Correlation between the expression of vascular endothelial growth factor and prognosis of lung carcinoma]. *Zhonghua Jie He He Hu Xi Za Zhi, 22*: 538–540, 1999.
28. Toomey, D., Smyth, G., Condron, C., Kay, E., Conroy, R., Foley, D., Hong, C., Hogan, B., Toner, S., McCormick, P., Broe, P., Kelly, C., and Bouchier-Hayes, D. Immune function, telomerase, and angiogenesis in patients with primary, operable nonsmall cell lung carcinoma: tumor size and lymph node status remain the most important prognostic features. *Cancer, 92*: 2648–2657, 2001.
29. Marrogi, A.J., Travis, W.D., Welsh, J.A., Khan, M.A., Rahim, H., Tazelaar, H., Pairolero, P., Trastek, V., Jett, J., Caporaso, N.E., Liotta, L.A., and Harris, C.C. Nitric oxide synthase, cyclooxygenase-2, and vascular endothelial growth factor in the angiogenesis of non-small cell lung carcinoma. *Clin Cancer Res, 6*: 4739–4744, 2000.
30. Mineo, T.C., Ambrogi, V., Baldi, A., Rabitti, C., Bollero, P., Vincenzi, B., and Tonini, G. Prognostic impact of VEGF, CD31, CD34, and CD105 expression and tumour vessel invasion after radical surgery for IB-IIA non-small cell lung cancer. *J Clin Pathol, 57*: 591–597, 2004.
31. Giatromanolaki, A., Koukourakis, M.I., Kakolyris, S., Turley, H., O'Byrne, K., Scott, P.A., Pezzella, F., Georgoulias, V., Harris, A.L., and Gatter, K.C. Vascular endothelial growth factor, wild-type p53, and angiogenesis in early operable non-small cell lung cancer. *Clin Cancer Res, 4*: 3017–3024, 1998.
32. Stefanou, D., Batistatou, A., Arkoumani, E., Ntzani, E., and Agnantis, N.J. Expression of vascular endothelial growth factor (VEGF) and association with microvessel density in small-cell and non-small-cell lung carcinomas. *Histol Histopathol, 19*: 37–42, 2004.

33. Takenaka, K., Tanaka, F., Katakura, H., Chen, F., Ogawa, E., Adachi, M., and Wada, H. The ratio of membrane-bound form Flt-1 mRNA to VEGF mRNA was correlated with tumor angiogenesis and prognosis in non-small cell lung cancer. *Proc Amer Soc Clin Oncol, 23*: 857s, 2005.
34. Ishikawa, S., Liu, D., Nakano, J., Nakashima, T., Masuya, K., Kontani, H., and Yokomise, C.H. The tumor–stromal interaction between the intratumoral Wnt1 and the stromal VEGF-A is associated with angiogenesis in squamous cell carcinomas of the lung. *Proc Amer Soc Clin Oncol, 23*: 668s, 2005.
35. Delmotte, P., Martin, B., Paesmans, M., Berghmans, T., Mascaux, C., Meert, A.P., Steels, E., Verdebout, J.M., Lafitte, J.J., and Sculier, J.P. [VEGF and survival of patients with lung cancer: a systematic literature review and meta-analysis]. *Rev Mal Respir, 19*: 577–584, 2002.
36. Iwasaki, A., Kuwahara, M., Yoshinaga, Y., and Shirakusa, T. Basic fibroblast growth factor (bFGF) and vascular endothelial growth factor (VEGF) levels, as prognostic indicators in NSCLC. *Eur J Cardiothorac Surg, 25*: 443–448, 2004.
37. Tonini, G., Ambrogi, V., Rabitti, C., Baldi, A., Santini, D., Vincenzi, B., and Mineo, T.C. Prognostic impact of VEGF expression, CD31, CD34, CD105 and tumor vessel invasion after radical surgery for ib-iia non-small cell lung cancer. *Proc Amer Soc Clin Oncol, 22*: 659, 2003.
38. Huang, C., Liu, D., Masuya, D., Nakashima, T., Kameyama, K., Ishikawa, S., Ueno, M., Haba, R., and Yokomise, H. Clinical application of biological markers for treatments of resectable non-small-cell lung cancers. *Br J Cancer, 92*: 1231–1239, 2005.
39. Volm, M., Koomagi, R., and Mattern, J. Angiogenesis and cigarette smoking in squamous cell lung carcinomas: an immunohistochemical study of 28 cases. *Anticancer Res, 19*: 333–336, 1999.
40. Saji, H., Nakamura, H., Awut, I., Kawasaki, N., Hagiwara, M., Ogata, A., Hosaka, M., Saijo, T., Kato, Y., and Kato, H. Significance of expression of TGF-beta in pulmonary metastasis in non-small cell lung cancer tissues. *Ann Thorac Cardiovasc Surg, 9*: 295–300, 2003.
41. Kakolyris, S., Giatromanolaki, A., Koukourakis, M., Kaklamanis, L., Kouroussis, C.H., Bozionelou, V., Georgoulias, V., Gatter, K.C., and Harris, A.L. Assessment of vascular maturation in lung and breast carcinomas using a novel basement membrane component, LH39. *Anticancer Res, 21*: 4311–4316, 2001.
42. Harter, D., Hillard, V., Gulati, A., Braun, A., Melamed, M., Benzil, D., Murali, R., and Jhanwar-Uniyal, M. p53 gene in cancer metastasis to brain. *Proc Amer Assoc Cancer Res, 46*: 1315, 2005.
43. Casibang, M., Purdom, S., Jakowlew, S., Neckers, L., Zia, F., Ben-Av, P., Hla, T., You, L., Jablons, D.M., and Moody, T.W. Prostaglandin E2 and vasoactive intestinal peptide increase vascular endothelial cell growth factor mRNAs in lung cancer cells. *Lung Cancer, 31*: 203–212, 2001.
44. Yeh, H. Autocrine IL-6 induced Stat3 activation contributes to the pathogenesis of lung adenocarcinoma and malignant pleural effusion. *Proc Amer Assoc Cancer Res, 46*: 36, 2005.
45. Hicklin, D.J. and Ellis, L.M. Role of the vascular endothelial growth factor pathway in tumor growth and angiogenesis. *J Clin Oncol, 23*: 1011–1027, Epub 2004 December 1017, 2005.
46. Dudek, A.Z. and Mahaseth, H. Circulating angiogenic cytokines in patients with advanced non-small cell lung cancer: correlation with treatment response and survival. *Cancer Invest, 23*: 193–200, 2005.
47. Yoshimoto, A., Kasahara, K., Nishio, M., Hourai, T., Sone, T., Kimura, H., Fujimura, M., and Nakao, S. Changes in angiogenic growth factor levels after gefitinib treatment in non-small cell lung cancer. *Jpn J Clin Oncol, 35*: 233–238, Epub 2005 May 2010, 2005.
48. Brattstrom, D., Bergqvist, M., Hesselius, P., Larsson, A., Wagenius, G., and Brodin, O. Serum VEGF and bFGF adds prognostic information in patients with normal platelet counts when sampled before, during and after treatment for locally advanced non-small cell lung cancer. *Lung Cancer, 43*: 55–62, 2004.
49. Tamura, M., Oda, M., Matsumoto, I., Tsunezuka, Y., Kawakami, K., Ohta, Y., and Watanabe, G. The combination assay with circulating vascular endothelial growth factor (VEGF)-C, matrix metalloproteinase-9, and VEGF for diagnosing lymph node metastasis in patients with non-small cell lung cancer. *Ann Surg Oncol, 11*: 928–933, Epub 2004 September 2020, 2004.
50. Garpenstrand, H., Bergqvist, M., Brattstrom, D., Larsson, A., Oreland, L., Hesselius, P., and Wagenius, G. Serum semicarbazide-sensitive amine oxidase (SSAO) activity correlates with VEGF in non-small-cell lung cancer patients. *Med Oncol, 21*: 241–250, 2004.

51. Lucchi, M., Mussi, A., Fontanini, G., Faviana, P., Ribechini, A., and Angeletti, C.A. Small cell lung carcinoma (SCLC): the angiogenic phenomenon. *Eur J Cardiothorac Surg*, *21*: 1105–1110, 2002.
52. Tanno, S., Ohsaki, Y., Nakanishi, K., Toyoshima, E., and Kikuchi, K. Human small cell lung cancer cells express functional VEGF receptors, VEGFR-2 and VEGFR-3. *Lung Cancer*, *46*: 11–19, 2004.
53. Salven, P., Ruotsalainen, T., Mattson, K., and Joensuu, H. High pre-treatment serum level of vascular endothelial growth factor (VEGF) is associated with poor outcome in small-cell lung cancer. *Int J Cancer*, *79*: 144–146, 1998.
54. Arena, F.P., Kurzyna-Solinas, A., DePasquale, V., and Park, S. Screening of metastatic tumors for the production of basic fibroblast growth factor (bFGF). *Proc Amer Soc Clin Oncol*, *21*: 46b, 2002.
55. Volm, M., Koomagi, R., Mattern, J., and Efferth, T. Expression profile of genes in non-small cell lung carcinomas from long-term surviving patients. *Clin Cancer Res*, *8*: 1843–1848, 2002.
56. Tanaka, F., Ishikawa, S., Yanagihara, K., Miyahara, R., Kawano, Y., Li, M., Otake, Y., and Wada, H. Expression of angiopoietins and its clinical significance in non-small cell lung cancer. *Cancer Res*, *62*: 7124–7129, 2002.
57. Hatanaka, H., Abe, Y., Naruke, M., Tokunaga, T., Oshika, Y., Kawakami, T., Osada, H., Nagata, J., Kamochi, J., Tsuchida, T., Kijima, H., Yamazaki, H., Inoue, H., Ueyama, Y., and Nakamura, M. Significant correlation between interleukin 10 expression and vascularization through angiopoietin/TIE2 networks in non-small cell lung cancer. *Clin Cancer Res*, *7*: 1287–1292, 2001.
58. Sandler, A.B. and Dubinett, S.M. COX-2 inhibition and lung cancer. *Semin Oncol*, *31*: 45–52, 2004.
59. Yoshimoto, A., Kasahara, K., Kawashima, A., Fujimura, M., and Nakao, S. Characterization of the prostaglandin biosynthetic pathway in non-small cell lung cancer: a comparison with small cell lung cancer and correlation with angiogenesis, angiogenic factors and metastases. *Oncol Rep*, *13*: 1049–1057, 2005.
60. Heuze-Vourc'h, N., Liu, M.Q., Dalwadi, H., Baratelli, F.E., Zhu, L., Goodglick, L., Solomon, J., WInter, L., Pold, M., Ramirez, R., Shay, J.W., DMinna, J.D., and Dubinett, S.M. Impaired IL-20 expression in lung cancer contributes to high levels of COX-2/PGE_2 and angiogenesis. *Proc Amer Assoc Cancer Res*, *46*: 1096, 2005.
61. Dalwadi, H.N., Krysan, K., Heuze-Vourc'h, N., Pold, M., Dohadwala, M., Elashoff, D., Sharma, S., Cacalano, N., Lichtenstein, A., and Dubinett, S. Cyclooxygenase-2-dependent activation of signal transducer and activator of transcription 3 by interleukin-6 in non-small cell lung cancer. *Proc Amer Assoc for Cancer Res*, *46*: 1237–1238, 2005.
62. Cui, X., Yang, S.C., Dalwadi, H., Sharma, S., Heuze-Vourc'h, N., Calano, N., and Dubinett, S. STAT6 mediates IL-4 inhibition of cyclooxygenase-2 in non-small cell lung cancer. *Proc Amer Assoc Cancer Res*, *46*: 1185, 2005.
63. Debusk, L.M. and Lin, C. IKK alpha induces angiogenesis. *Proc Amer Assoc Cancer Res*, *46*: 1100, 2005.
64. Fan, L.F., Diao, L.M., Chen, D.J., Liu, M.Q., Zhu, L.Q., Li, H.G., Tang, Z.J., Xia, D., Liu, X., and Chen, H.L. [Expression of HIF-1 alpha and its relationship to apoptosis and proliferation in lung cancer]. *Ai Zheng*, *21*: 254–258, 2002.
65. Koshikawa, N., Iyozumi, A., Gassmann, M., and Takenaga, K. Constitutive upregulation of hypoxia-inducible factor-1alpha mRNA occurring in highly metastatic lung carcinoma cells leads to vascular endothelial growth factor overexpression upon hypoxic exposure. *Oncogene*, *22*: 6717–6724, 2003.
66. Lee, C.H., Lee, M.K., Kang, C.D., Kim, Y.D., Park do, Y., Kim, J.Y., Sol, M.Y., and Suh, K.S. Differential expression of hypoxia inducible factor-1 alpha and tumor cell proliferation between squamous cell carcinomas and adenocarcinomas among operable non-small cell lung carcinomas. *J Korean Med Sci*, *18*: 196–203, 2003.
67. Swinson, D.E., Jones, J.L., Cox, G., Richardson, D., Harris, A.L., and O'Byrne, K.J. Hypoxia-inducible factor-1alpha in non small cell lung cancer: relation to growth factor, protease and apoptosis pathways. *Int J Cancer*, *111*: 43–50, 2004.
68. Felip, E., Taron, M., Rosell, R., Mendez, P., Queralt, C., Ronco, M.S., Sanchez, J.J., Sanchez, J.M., Maestre, J., and Majo, J. Clinical significance of hypoxia-inducible factor-1a messenger RNA expression in locally advanced non-small-cell lung cancer after platinum agent and gemcitabine chemotherapy followed by surgery. *Clin Lung Cancer*, *6*: 299–303, 2005.

69. Volm, M. and Koomagi, R. Hypoxia-inducible factor (HIF-1) and its relationship to apoptosis and proliferation in lung cancer. *Anticancer Res*, *20*: 1527–1533, 2000.
70. Hirami, Y., Aoe, M., Tsukuda, K., Hara, F., Otani, Y., Koshimune, R., Hanabata, T., Nagahiro, I., Sano, Y., Date, H., and Shimizu, N. Relation of epidermal growth factor receptor, phosphorylated-Akt, and hypoxia-inducible factor-1alpha in non-small cell lung cancers. *Cancer Lett*, *214*: 157–164, 2004.
71. Dagnon, K., Pacary, E., Commo, F., Antoine, M., Bernaudin, M., Bernaudin, J.F., and Callard, P. Expression of erythropoietin and erythropoietin receptor in non-small cell lung carcinomas. *Clin Cancer Res*, *11*: 993–999, 2005.
72. Koukourakis, M.I., Giatromanolaki, A., Sivridis, E., Bougioukas, G., Didilis, V., Gatter, K.C., and Harris, A.L. Lactate dehydrogenase-5 (LDH-5) overexpression in non-small-cell lung cancer tissues is linked to tumour hypoxia, angiogenic factor production and poor prognosis. *Br J Cancer*, *89*: 877–885, 2003.
73. Thomas, P., Khokha, R., Shepherd, F.A., Feld, R., and Tsao, M.S. Differential expression of matrix metalloproteinases and their inhibitors in non-small cell lung cancer. *J Pathol*, *190*: 150–156, 2000.
74. Aljada, I.S., Ramnath, N., Donohue, K., Harvey, S., Brooks, J.J., Wiseman, S.M., Khoury, T., Loewen, G., Slocum, H.K., Anderson, T.M., Bepler, G., and Tan, D. Upregulation of the tissue inhibitor of metalloproteinase-1 protein is associated with progression of human non-small-cell lung cancer. *J Clin Oncol*, *22*: 3218–3229, Epub 2004 July 3212, 2004.
75. Takenaka, K., Tanaka, F., Ishikawa, S., Kawano, Y., Miyahara, R., Yanagihara, K., Otake, Y., Takahasi, C., Noda, M., and Wada, H. RECK expression and its clinical significance in correlation with angiogenesis in non-small cell lung cancer. *Proc Amer Soc Clin Oncol*, *22*: 656, 2003.
76. Laack, E., Kohler, A., Kugler, C., Dierlamm, T., Knuffmann, C., Vohwinkel, G., Niestroy, A., Dahlmann, N., Peters, A., Berger, J., Fiedler, W., and Hossfeld, D.K. Pretreatment serum levels of matrix metalloproteinase-9 and vascular endothelial growth factor in non-small-cell lung cancer. *Ann Oncol*, *13*: 1550–1557, 2002.
77. Michael, M., Babic, B., Khokha, R., Tsao, M., Ho, J., Pintilie, M., Leco, K., Chamberlain, D., and Shepherd, F.A. Expression and prognostic significance of metalloproteinases and their tissue inhibitors in patients with small-cell lung cancer. *J Clin Oncol*, *17*: 1802–1808, 1999.
78. Yu, S.-L., Chen, J.-W.J., Chiu, S.-C., Chen, H.-Y., Chang, G.-C., and Yang, P.-C. Expression of microphthalmia-associated transcription factor (MITF) suppresses cancer cell invasion and tumorigenesis in lung cancer. *Proc Amer Assoc Cancer Res*, *46*, 2005.
79. White, E.S., Strom, S.R., Wys, N.L., and Arenberg, D.A. Non-small cell lung cancer cells induce monocytes to increase expression of angiogenic activity. *J Immunol*, *166*: 7549–7555, 2001.
80. Arenberg, D., Carskadon, S.L., White, E.S., and McClelland, M. Macrophage migration inhibitory factor (MIF)-dependent induction of angiogenic CXC chemokines in an animal model of NSCLC. *Proc Amer Assoc Cancer Res*, *46*: 1099, 2005.
81. White, E.S., Flaherty, K.R., Carskadon, S., Brant, A., Iannettoni, M.D., Yee, J., Orringer, M.B., and Arenberg, D.A. Macrophage migration inhibitory factor and CXC chemokine expression in non-small cell lung cancer: role in angiogenesis and prognosis. *Clin Cancer Res*, *9*: 853–860, 2003.
82. Dasgupta, P., Rastogi, S., Ordonez, D., Morris, M.M., Haura, E., and Chellappan, S.P. Nicotine-induced cell proliferation and angiogenesis requires Raf-1 mediated inactivation of Rb. *Proc Amer Assoc Cancer Res*, *46*: 1238, 2005.
83. Hoekstra, C.J., Stroobants, S.G., Hoekstra, O.S., Smit, E.F., Vansteenkiste, J.F., and Lammertsma, A.A. Measurement of perfusion in stage IIIA-N2 non-small cell lung cancer using H(2)(15)O and positron emission tomography. *Clin Cancer Res*, *8*: 2109–2115, 2002.
84. Veronesi, G., Landoni, C., Pelosi, G., Picchio, M., Sonzogni, A., Leon, M.E., Solli, P.G., Leo, F., Spaggiari, L., Bellomi, M., Fazio, F., and Pastorino, U. Fluoro-deoxy-glucose uptake and angiogenesis are independent biological features in lung metastases. *Br J Cancer*, *86*: 1391–1395, 2002.
85. Boehle, A.S., Kurdow, R., Schulze, M., Kliche, U., Sipos, B., Soondrum, K., Ebrahimnejad, A., Dohrmann, P., Kalthoff, H., Henne-Bruns, D., and Neumaier, M. Human endostatin inhibits growth of human non-small-cell lung cancer in a murine xenotransplant model. *Int J Cancer*, *94*: 420–428, 2001.

86. Sauter, B.V., Martinet, O., Zhang, W.J., Mandeli, J., and Woo, S.L. Adenovirus-mediated gene transfer of endostatin in vivo results in high level of transgene expression and inhibition of tumor growth and metastases. *Proc Natl Acad Sci USA*, *97*: 4802–4807, 2000.
87. Kwon, M., Yoon, C.S., Fitzpatrick, S., Kassam, G., Graham, K.S., Young, M.K., and Waisman, D.M. p22 is a novel plasminogen fragment with antiangiogenic activity. *Biochemistry*, *40*: 13246–13253, 2001.
88. Schmitz, V., Wang, L., Barajas, M., Peng, D., Prieto, J., and Qian, C. A novel strategy for the generation of angiostatic kringle regions from a precursor derived from plasminogen. *Gene Ther*, *9*: 1600–1606, 2002.
89. Sim, B.K., O'Reilly, M.S., Liang, H., Fortier, A.H., He, W., Madsen, J.W., Lapcevich, R., and Nacy, C.A. A recombinant human angiostatin protein inhibits experimental primary and metastatic cancer. *Cancer Res*, *57*: 1329–1334, 1997.
90. Mauceri, H.J., Seetharam, S., Beckett, M.A., Schumm, L.P., Koons, A., Gupta, V.K., Park, J.O., Manan, A., Lee, J.Y., Montag, A.G., Kufe, D.W., and Weichselbaum, R.R. Angiostatin potentiates cyclophosphamide treatment of metastatic disease. *Cancer Chemother Pharmacol*, *50*: 412–418. Epub 2002 September 2018, 2002.
91. Litz, J., Sakuntala Warshamana-Greene, G., Sulanke, G., Lipson, K.E., and Krystal, G.W. The multi-targeted kinase inhibitor SU5416 inhibits small cell lung cancer growth and angiogenesis, in part by blocking Kit-mediated VEGF expression. *Lung Cancer*, *46*: 283–291, 2004.
92. Saha, D., Choy, H., Cao, Q., Kim, J., Hallahan, D., Beauchamp, D., and Johnson, D.H. Modulation of COX-2 expression and enhanced radio-sensitization in human lung carcinoma cells by tyrosine kinase inhibitor, SU-5416. *Proc Amer Soc Clin Oncol*, *21*: 445a, 2002.
93. Abdollahi, A., Lipson, K.E., Sckell, A., Zieher, H., Klenke, F., Poerschke, D., Roth, A., Han, X., Krix, M., Bischof, M., Hahnfeldt, P., Grone, H.J., Debus, J., Hlatky, L., and Huber, P.E. Combined therapy with direct and indirect angiogenesis inhibition results in enhanced antiangiogenic and antitumor effects. *Cancer Res*, *63*: 8890–8898, 2003.
94. Sun, J., Blaskovich, M.A., Jain, R.K., Delarue, F., Paris, D., Brem, S., Wotoczek-Obadia, M., Lin, Q., Coppola, D., Choi, K., Mullan, M., Hamilton, A.D., and Sebti, S.M. Blocking angiogenesis and tumorigenesis with GFA-116, a synthetic molecule that inhibits binding of vascular endothelial growth factor to its receptor. *Cancer Res*, *64*: 3586–3592, 2004.
95. Wedge, S.R., Ogilvie, D.J., Dukes, M., Kendrew, J., Curwen, J.O., Hennequin, L.F., Thomas, A.P., Stokes, E.S., Curry, B., Richmond, G.H., and Wadsworth, P.F. ZD4190: an orally active inhibitor of vascular endothelial growth factor signaling with broad-spectrum antitumor efficacy. *Cancer Res*, *60*: 970–975, 2000.
96. Nakamura, K., Taguchi, E., Miura, T., Bichat, F., Just, N., Duchamp, O., Walker, P., Guilbaud, N., and Iso, T. The VEGF receptor inhibitor KRN951 decreases vascular permeability in tumors and inhibits tumor growth: an analysis using dynamic contrast-enhanced magnetic resonance imaging. *Proc Amer Assoc Cancer Res*, *46*: 1375, 2005.
97. Kozin, S.V., Boucher, Y., Hicklin, D.J., Bohlen, P., Jain, R.K., and Suit, H.D. Vascular endothelial growth factor receptor-2-blocking antibody potentiates radiation-induced long-term control of human tumor xenografts. *Cancer Res*, *61*: 39–44, 2001.
98. Mae, M. and Crystal, R.G. Gene transfer to the pleural mesothelium as a strategy to deliver proteins to the lung parenchyma. *Hum Gene Ther*, *13*: 1471–1482, 2002.
99. Davis, T.W., Cao, L., Lennox, W., Bombard, J., Preston, E., Hedrick, J., Corson, D., Romfo, C., Risher, N., Sheedy, J., Yeh, S., Ullner, P., Kandel, J., Yamashiro, D., Qi, H., Tamilarasu, N., Choi, S., Weetall, M., Moon, Y.-C., and Colacino, J. Post-transcriptional control as a novel approach to anti-angiogenesis: development of a small molecule that reduces the production of tumor vascular endothelial growth factor A (VEGF). *Proc Amer Assoc Cancer Res*, *46*: 912, 2005.
100. Taguchi, F., Koh, Y., Koizumi, F., Tamura, T., Saijo, N., and Nishio, K. Anticancer effects of ZD6474, a VEGF receptor tyrosine kinase inhibitor, in gefitinib ("Iressa")-sensitive and resistant xenograft models. *Cancer Sci*, *95*: 984–989, 2004.
101. Wedge, S.R., Ogilvie, D.J., Dukes, M., Kendrew, J., Chester, R., Jackson, J.A., Boffey, S.J., Valentine, P.J., Curwen, J.O., Musgrove, H.L., Graham, G.A., Hughes, G.D., Thomas, A.P., Stokes, E.S., Curry, B., Richmond, G.H., Wadsworth, P.F., Bigley, A.L., and Hennequin, L.F.

ZD6474 inhibits vascular endothelial growth factor signaling, angiogenesis, and tumor growth following oral administration. *Cancer Res*, *62*: 4645–4655, 2002.

102. Tuccillo, C., Romano, M., Troiani, T., Martinelli, E., Morgillo, F., De Vita, F., Bianco, R., Fontanini, G., Bianco, R.A., Tortora, G., and Ciardiello, F. Antitumor activity of ZD6474, a vascular endothelial growth factor-2 and epidermal growth factor receptor small molecule tyrosine kinase inhibitor, in combination with SC-236, a cyclooxygenase-2 inhibitor. *Clin Cancer Res*, *11*: 1268–1276, 2005.
103. Shintani, T., Lewis, V.O., Komaki, R., Wu, W., Ryan, A.J., Herbst, R.S., and O'Reilly, M.S. ZD6474 inhibits human lung cancer bone metastases in a murine model by targeting both the tumor and its vasculature. *Proc Amer Assoc Cancer Res*, *46*: 1375, 2005.
104. Williams, K.J., Telfer, B.A., Brave, S., Kendrew, J., Whittaker, L., Stratford, I.J., and Wedge, S.R. ZD6474, a potent inhibitor of vascular endothelial growth factor signaling, combined with radiotherapy: schedule-dependent enhancement of antitumor activity. *Clin Cancer Res*, *10*: 8587–8593, 2004.
105. Traxler, P., Allegrini, P.R., Brandt, R., Brueggen, J., Cozens, R., Fabbro, D., Grosios, K., Lane, H.A., McSheehy, P., Mestan, J., Meyer, T., Tang, C., Wartmann, M., Wood, J., and Caravatti, G. AEE788: a dual family epidermal growth factor receptor/ErbB2 and vascular endothelial growth factor receptor tyrosine kinase inhibitor with antitumor and antiangiogenic activity. *Cancer Res*, *64*: 4931–4941, 2004.
106. Abrams, T.J., Lee, L.B., Murray, L.J., Pryer, N.K., and Cherrington, J.M. SU11248 inhibits KIT and platelet-derived growth factor receptor beta in preclinical models of human small cell lung cancer. *Mol Cancer Ther*, *2*: 471–478, 2003.
107. Hallahan, D., Schueneman, A., Himmelfarb, E., and Geng, L. SU11248 maintenance therapy prevents tumor regrowth following fractionated irradiation of murine tumor models. *Proc Am Soc Clin Oncol*, *22*: 873, 2003.
108. Dixon, J.A., Boyer, S.J., Brini, W., Dumas, J., Ehrgott, F., Elting, J., Hong, Z., Jones, R., Kluender, H., Lee, W., Ma, X., Sibley, R., Turner, T., and Zhang, Y. Identification of novel VEGFR-2 inhibitors. *Proc Amer Assoc Cancer Res*, *46*: 913, 2005.
109. Chang, Y.S., Cortes, C., Polony, B., Brink, C., and Elting, J.J. Preclinical chemotherapy with the VEGFR-2 and PDGFR inhibitor, BAY 57-9352, in combination with capecitabine and paclitaxel. *Proc Amer Assoc Cancer Res*, *46*: 475, 2005.
110. Yano, S., Herbst, R.S., Shinohara, H., Knighton, B., Bucana, C.D., Killion, J.J., Wood, J., and Fidler, I.J. Treatment for malignant pleural effusion of human lung adenocarcinoma by inhibition of vascular endothelial growth factor receptor tyrosine kinase phosphorylation. *Clin Cancer Res*, *6*: 957, 2000.
111. Hu-Lowe, D.D. and Grazzini, M.L. Significant enhancement of anti-tumor efficacy of VEGF PDGF receptor tyrosine kinase inhibitor AG-013736 in combination with docetaxel in chemo-refractory and/or orthotopic xenograft tumor models in mice. *Proc Amer Assoc Cancer Res*, *46*: 475, 2005.
112. Sennino, B., Falcon, B.L., Grate, D., Espstein, D.M., and McDonald, D.M. Aptamers that bind platelet derived growth factor-B reduce pericyte coverage of tumor blood vessels. *Proc Amer Assoc Cancer Res*, *46*: 1098, 2005.
113. Albert, D.H., Tapang, P., Magoc, T.J., Pease, L.J., Reuter, D.R., Li, J., Guo, J., Ghoreishi-Haack, N.S., Brooks, J.V., Burkofzer, G.T., Wang, Y.-C., Stravropoulos, J.J., Hartandi, K., Niquette, A.L., Wei, R.-Q., McCall, J.O., Bouska, J.J., Luo, Y., Donawho, C.K., Dai, Y., Marcotte, P.A., Glaser, K.B., Michaelides, M.R., and Davidsen, S.K. Preclinical activity of ABT-869, a multi-targeted receptor tyrosine kinase inhibitor. *Proc Amer Assoc Cancer Res*, *46*: 159–160, 2005.
114. Liu, W., Chen, Y., Wang, W., Keng, P., Finkelstein, J., Hu, D., Liang, L., Guo, M., Fenton, B., Okunieff, P., and Ding, I. Combination of radiation and celebrex (celecoxib) reduce mammary and lung tumor growth. *Am J Clin Oncol*, *26*: S103–S109, 2003.
115. Ono, M., Nakao, S., Kuwano, T., Ueda, S.-I., Kimura, Y.N., Oie, S., and Kuwano, M. The control of tumor growth and angiogenesis by inflammatory cytokines and infiltration of macrophages in tumor microenvironment. *Proc Amer Assoc Cancer Res*, *46*: 1096, 2005.
116. Teicher, B.A., Menon, K., Alvarez, E., Galbreath, E., Shih, C., and Faul, M.M. Antiangiogenic and antitumor effects of a protein kinase Cbeta inhibitor in murine lewis lung carcinoma

and human Calu-6 non-small-cell lung carcinoma xenografts. *Cancer Chemother Pharmacol, 48*: 473–480, 2001.

117. Teicher, B.A., Alvarez, E., Menon, K., Esterman, M.A., Considine, E., Shih, C., and Faul, M.M. Antiangiogenic effects of a protein kinase Cbeta-selective small molecule. *Cancer Chemother Pharmacol, 49*: 69–77, 2002.
118. Nguyen, D.M., Lorang, D., Chen, G.A., Stewart, J.H., Tabibi, E., and Schrump, D.S. Enhancement of paclitaxel-mediated cytotoxicity in lung cancer cells by 17-allylamino geldanamycin: in vitro and in vivo analysis. *Ann Thorac Surg, 72*: 378–379, 2001.
119. Shalinsky, D.R., Brekken, J., Zou, H., Bloom, L.A., McDermott, C.D., Zook, S., Varki, N.M., and Appelt, K. Marked antiangiogenic and antitumor efficacy of AG3340 in chemoresistant human non-small cell lung cancer tumors: single agent and combination chemotherapy studies. *Clin Cancer Res, 5*: 1905–1917, 1999.
120. Li, X., Liu, X., Wang, J., Wang, Z., Jiang, W., Reed, E., Zhang, Y., Liu, Y., and Li, Q.Q. Effects of thalidomide on the expression of angiogenesis growth factors in human A549 lung adenocarcinoma cells. *Int J Mol Med, 11*: 785–790, 2003.
121. Li, X., Liu, X., Wang, J., Wang, Z., Jiang, W., Reed, E., Zhang, Y., Liu, Y., and Li, Q.Q. Thalidomide down-regulates the expression of VEGF and bFGF in cisplatin-resistant human lung carcinoma cells. *Anticancer Res, 23*: 2481–2487, 2003.
122. DeCicco, K.L., Tanaka, T., Andreola, F., and De Luca, L.M. The effect of thalidomide on non-small cell lung cancer (NSCLC) cell lines: possible involvement in the PPARgamma pathway. *Carcinogenesis, 25*: 1805–1812, Epub 2004 June 1817, 2004.
123. Majewski, S., Marczak, M., Mlynarczyk, B., Benninghoff, B., and Jablonska, S. Imiquimod is a strong inhibitor of tumor cell-induced angiogenesis. *Int J Dermatol, 44*: 14–19, 2005.
124. Ramesh, R., Ito, I., Saito, Y., Wu, Z., Mhashikar, A.M., Wilson, D.R., Branch, C.D., Roth, J.A., and Chada, S. Local and systemic inhibition of lung tumor growth after nanoparticle-mediated mda-7/IL-24 gene delivery. *DNA Cell Biol, 23*: 850–857, 2004.
125. Mousa, S.A. and Mohamed, S. Anti-angiogenic mechanisms and efficacy of the low molecular weight heparin, tinzaparin: anti-cancer efficacy. *Oncol Rep, 12*: 683–688, 2004.
126. Choi, S., Park, K., Park, S., Chang, H., Byun, Y., and Kim, S.Y. The anti-angiogenic and anti-cancer effect of novel hydrophobic heparin nanoparticles. *Proc Amer Assoc Cancer Res, 46*: 66, 2005.
127. Jayson, G.C., Hasan, J., Backen, A., Shnyder, S., Bibby, M., Bicknell, R., Presta, M., Clamp, A., Rapraeger, A., David, G., McGown, A., and Gallagher, J. Development of oligosaccharides as anti-angiogenic agents. *Proc Amer Assoc Cancer Res, 46*: 714, 2005.
128. Kim, T.H., Kim, E., Yoon, D., Kim, J., Rhim, T.Y., and Kim, S.S. Recombinant human prothrombin kringles have potent anti-angiogenic activities and inhibit Lewis lung carcinoma tumor growth and metastases. *Angiogenesis, 5*: 191–201, 2002.
129. Liu, N., Lapcevich, R.K., Underhill, C.B., Han, Z., Gao, F., Swartz, G., Plum, S.M., Zhang, L., and Green, S.J. Metastatin: a hyaluronan-binding complex from cartilage that inhibits tumor growth. *Cancer Res, 61*: 1022–1028, 2001.
130. Khan, M.K., Miller, M.W., Taylor, J., Gill, N.K., Dick, R.D., Van Golen, K., Brewer, G.J., and Merajver, S.D. Radiotherapy and antiangiogenic TM in lung cancer. *Neoplasia, 4*: 164–170, 2002.
131. Satoh, H., Ishikawa, H., Fujimoto, M., Fujiwara, M., Yamashita, Y.T., Yazawa, T., Ohtsuka, M., Hasegawa, S., and Kamma, H. Combined effects of TNP-470 and taxol in human non-small cell lung cancer cell lines. *Anticancer Res, 18*: 1027–1030, 1998.
132. Browder, T., Butterfield, C.E., Kraling, B.M., Shi, B., Marshall, B., O'Reilly, M.S., and Folkman, J. Antiangiogenic scheduling of chemotherapy improves efficacy against experimental drug-resistant cancer. *Cancer Res, 60*: 1878–1886, 2000.
133. Kakeji, Y. and Teicher, B.A. Preclinical studies of the combination of angiogenic inhibitors with cytotoxic agents. *Invest New Drugs, 15*: 39–48, 1997.
134. Liu, L.T., Chang, H.C., Chiang, L.C., and Hung, W.C. Histone deacetylase inhibitor up-regulates RECK to inhibit MMP-2 activation and cancer cell invasion. *Cancer Res, 63*: 3069–3072, 2003.
135. Williams, J.I., Weitman, S., Gonzalez, C.M., Jundt, C.H., Marty, J., Stringer, S.D., Holroyd, K.J., McLane, M.P., Chen, Q., Zasloff, M., and Von Hoff, D.D. Squalamine treatment of human tumors in nu/nu mice enhances platinum-based chemotherapies. *Clin Cancer Res, 7*: 724–733, 2001.

136. Yen, W.-C., Prudente, R.Y., Corpuz, M.R., Cooke, T.A., Negro-Vilar, A., and Lamph, W.W. A selective retinoid X receptor agonist bexarotene (LGD1069, Targretin) inhibits angiogenesis and metastasis in solid tumors. *Proc Amer Assoc Cancer Res*, *46*: 65, 2005.
137. Han, J.Y., Oh, S.H., Morgillo, F., Myers, J.N., Kim, E., Hong, W.K., and Lee, H.Y. Hypoxia-inducible factor 1alpha and antiangiogenic activity of farnesyltransferase inhibitor SCH66336 in human aerodigestive tract cancer. *J Natl Cancer Inst*, *97*: 1272–1286, 2005.
138. Chittezhath, M., Singh, R.P., Agarwal, C., and Agarwal, R. Silibinin inhibits cytokines-induced STATs, MAPKs and NF-κB activation and down regulates HIF-1α and iNOS in human lung epithelial A549 cells: possible role in angioprevention of lung tumorigenesis. *Proc Amer Assoc Cancer Res*, *46*: 1232, 2005.
139. Liu, L.Z., Fang, J., Zhou, Q., Hu, X., Shi, X., and Jiang, B.H. Apigenin inhibits expression of vascular endothelial growth factor and angiogenesis in human lung cancer cells: implication of chemoprevention of lung cancer. *Mol Pharmacol*, *68*: 635–643, Epub 2005 June 2009, 2005.
140. Nagasawa, H., Mikamo, N., Nakajima, Y., Matsumoto, H., Uto, Y., and Hori, H. Antiangiogenic hypoxic cytotoxin TX-402 inhibits hypoxia-inducible factor 1 signaling pathway. *Anticancer Res*, *23*: 4427–4434, 2003.
141. Chang, H., Shyu, K.G., Lee, C.C., Tsai, S.C., Wang, B.W., Hsien Lee, Y., and Lin, S. GL331 inhibits HIF-1alpha expression in a lung cancer model. *Biochem Biophys Res Commun*, *302*: 95–100, 2003.
142. Mie Lee, Y., Kim, S.H., Kim, H.S., Jin Son, M., Nakajima, H., Jeong Kwon, H., and Kim, K.W. Inhibition of hypoxia-induced angiogenesis by FK228, a specific histone deacetylase inhibitor, via suppression of HIF-1alpha activity. *Biochem Biophys Res Commun*, *300*: 241–246, 2003.
143. Grant, D.S., Williams, T.L., Zahaczewsky, M., and Dicker, A.P. Comparison of antiangiogenic activities using paclitaxel (Taxol) and docetaxel (Taxotere). *Int J Cancer*, *104*: 121–129, 2003.
144. Lau, D., Guo, L., Gandara, D., Young, L.J., and Xue, L. Is inhibition of cancer angiogenesis and growth by paclitaxel schedule dependent? *Anticancer Drugs*, *15*: 871–875, 2004.
145. Petrangolini, G., Pratesi, G., De Cesare, M., Supino, R., Pisano, C., Marcellini, M., Giordano, V., Laccabue, D., Lanzi, C., and Zunino, F. Antiangiogenic effects of the novel camptothecin ST1481 (gimatecan) in human tumor xenografts. *Mol Cancer Res*, *1*: 863–870, 2003.
146. Soucy, N.V., Ihnat, M.A., Kamat, C.D., Hess, L., Post, M.J., Klei, L.R., Clark, C., and Barchowsky, A. Arsenic stimulates angiogenesis and tumorigenesis in vivo. *Toxicol Sci*, *76*: 271–279, Epub 2003 September 2011, 2003.
147. Goto, H., Yano, S., Zhang, H., Matsumori, Y., Ogawa, H., Blakey, D.C., and Sone, S. Activity of a new vascular targeting agent, ZD6126, in pulmonary metastases by human lung adenocarcinoma in nude mice. *Cancer Res*, *62*: 3711–3715, 2002.
148. Raben, D., Bianco, C., Damiano, V., Bianco, R., Melisi, D., Mignogna, C., D'Armiento, F.P., Cionini, L., Bianco, A.R., Tortora, G., Ciardiello, F., and Bunn, P. Antitumor activity of ZD6126, a novel vascular-targeting agent, is enhanced when combined with ZD1839, an epidermal growth factor receptor tyrosine kinase inhibitor, and potentiates the effects of radiation in a human non-small cell lung cancer xenograft model. *Mol Cancer Ther*, *3*: 977–983, 2004.
149. Goto, H., Yano, S., Matsumori, Y., Ogawa, H., Blakey, D.C., and Sone, S. Sensitization of tumor-associated endothelial cell apoptosis by the novel vascular-targeting agent ZD6126 in combination with cisplatin. *Clin Cancer Res*, *10*: 7671–7676, 2004.
150. Boehle, A.S., Sipos, B., Kliche, U., Kalthoff, H., and Dohrmann, P. Combretastatin A-4 prodrug inhibits growth of human non-small cell lung cancer in a murine xenotransplant model. *Ann Thorac Surg*, *71*: 1657–1665, 2001.
151. Blaschuk, O., Lavoie, N., Devamy, E., and Lepekhin, E. Further observations concerning the effects of the N-cadherin antagonist Exherin™ (ADH-1) on endothelial cells and tumor blood vessels. *Proc Amer Assoc Cancer Res*, *46*: 268, 2005.
152. Herbst, R.S., Hess, K.R., Tran, H.T., Tseng, J.E., Mullani, N.A., Charnsangavej, C., Madden, T., Davis, D.W., McConkey, D.J., O'Reilly, M.S., Ellis, L.M., Pluda, J., Hong, W.K., and Abbruzzese, J.L. Phase I study of recombinant human endostatin in patients with advanced solid tumors. *J Clin Oncol*, *20*: 3792–3803, 2002.
153. Gordon, M.S., Mendelson, D., Guirguis, M.S., Knight, R.A., Humerickhouse, R.A., Stopeck, A., and Wang, Q. ABT-510, an anti-angiogenic, thrombospondin-1 (TSP-1) mimetic peptide, exhibits

favorable safety profile and early signals of activity in a randomized phase IB trial. *Proc Amer Soc Clin Oncol, 22*: 195, 2003.

154. Johnson, D.H., Fehrenbacher, L., Novotny, W.F., Herbst, R.S., Nemunaitis, J.J., Jablons, D.M., Langer, C.J., DeVore, R.F., 3rd, Gaudreault, J., Damico, L.A., Holmgren, E., and Kabbinavar, F. Randomized phase II trial comparing bevacizumab plus carboplatin and paclitaxel with carboplatin and paclitaxel alone in previously untreated locally advanced or metastatic non-small-cell lung cancer. *J Clin Oncol, 22*: 2184–2191, 2004.
155. Sandler, A.B., Gray, R., Brahmer, J., Dowlati, A., Schiller, J.H., Perry, M.C., and Johnson, D.H. Randomized phase II/III trial of paclitaxel (P) plus carboplatin (C) with or without bevacizumab (NSC ?704865) in patients with advanced non-squamous non-small cell lung cancer (NSCLC). *Proc Amer Soc Clin Oncol, 23*: 1090s, 2005.
156. Herbst, R.S., Johnson, D.H., Mininberg, E., Carbone, D.P., Henderson, T., Kim, E.S., Blumenschein, G., Jr., Lee, J.J., Liu, D.D., Truong, M.T., Hong, W.K., Tran, H., Tsao, A., Xie, D., Ramies, D.A., Mass, R., Seshagiri, S., Eberhard, D.A., Kelley, S.K., and Sandler, A. Phase I/II trial evaluating the anti-vascular endothelial growth factor monoclonal antibody bevacizumab in combination with the HER-1/epidermal growth factor receptor tyrosine kinase inhibitor erlotinib for patients with recurrent non-small-cell lung cancer. *J Clin Oncol, 23*: 2544–2555, Epub 2005 March 2547, 2005.
157. Levine, A.M., Quinn, D.I., Gorospe, G., Lenz, H.J., and Tulpule, A. Phase I trial of anti-sense oligonucleotide vascular endothelial growth factor (VEGF-AS, Veglin) in patients with relapsed and refractory malignancies. *Proc Amer Soc Clin Oncol, 22*: 3008, 2004.
158. Herbst, R.S., Onn, A., and Sandler, A. Angiogenesis and lung cancer: prognostic and therapeutic implications. *J Clin Oncol, 23*: 3243–3256, 2005.
159. AstraZeneca New phase II data adds to evidence of progression free survival advantage of ZD6474 (Zactima (TM)) in lung cancer. 11th World Conference on Lung Cancer-IASLC, July 6, 2005 (News Release).
160. Wakelee, H., Adjei, A.A., Keer, H., Halsey, J., Hanson, L., Reid, J., Hutchinson, S., Piens, J., Lacy, S., and Sikic, B.I. A Phase I dose-escalation and pharmacokinetic (PK) study of a novel multiple-targeted receptor tyrosine kinase (RTK) inhibitor, XL647, in patients with advanced solid malignancies. *Proc Amer Soc Clin Oncol, 23*: 227s, 2005.
161. Liu, G., Rugo, H.S., Wilding, G., McShane, T.M., Evelhoch, J.L., Ng, C., Jackson, E., Kelcz, F., Yeh, B.M., Lee, F.T., Jr., Charnsangavej, C., Park, J.W., Ashton, E.A., Steinfeldt, H.M., Pithavala, Y.K., Reich, S.D., and Herbst, R.S. Dynamic contrast-enhanced magnetic resonance imaging as a pharmacodynamic measure of response after acute dosing of AG-013736, an oral angiogenesis inhibitor, in patients with advanced solid tumors: results from a phase I study. *J Clin Oncol, 23*: 5464–5473, Epub 2005 July 5418, 2005.
162. Rugo, H.S., Herbst, R.S., Liu, G., Park, J.W., Kies, M.S., Steinfeldt, H.M., Pithavala, Y.K., Reich, S.D., Freddo, J.L., and Wilding, G. Phase I trial of the oral antiangiogenesis agent AG-013736 in patients with advanced solid tumors: pharmacokinetic and clinical results. *J Clin Oncol, 23*: 5474–5483, Epub 2005 July 5418, 2005.
163. Hurwitz, H., Dowlati, A., Savage, S., Fernando, N., Lasalvia, S., Whitehead, B., Suttle, B., Collins, D., Ho, P., and Pandite, L. Safety, tolerability and pharmacokinetics of oral administration of GW786034 in pts with solid tumors. *Proc Amer Soc Clin Oncol, 23*: 195, 2005.
164. Brahmer, J.R., Kelsey, S., Scigalla, P., Hill, G., Bello, C., Elza-Brown, K., and Donehower, R. A phase I study of SU6668 in patients with refractory solid tumors. *Proc Amer Soc Clin Oncol, 21*: 84a, 2002.
165. Minami, H., Kawada, K., Ebi, H., Kitagawa, K., Kim, Y., Araki, K., Mukai, H., Tahara, M., Nakajima, H., and Nakajima, K. A phase I study of BAY 43-9006, a dual inhibitor of Raf and VEGFR kinases, in Japanese patients with solid cancers. *Proc Amer Soc Clin Oncol, 23*: 207, 2005.
166. Blumenschein, G.R., Gatzenmeier, U., Fossella, F., Burk, K., and Reck, M. A phase II multicenter uncontrolled trial of single agent sorafenib (BAY 43-9006) in patients with relapsed or refractory advanced non-small cell lung carcinoma. *Proc AACR-NCI-EORTC International Conference*, "Molecular Targets and Cancer Therapeutics": 46, 2005.
167. Johnson, D.H., Csiki, I., Gonzalez, A., Carbone, D.P., Gautam, S., Campbell, N., Morrow, J., and Sandler, A. Cyclooxygenase-2 (COX-2) inhibition in non-small cell lung cancer (NSCLC): preliminary results of a phase II trial. *Proc Amer Soc Clin Oncol, 22*: 640, 2003.

168. Altorki, N.K., Keresztes, R.S., Port, J.L., Libby, D.M., Korst, R.J., Flieder, D.B., Ferrara, C.A., Yankelevitz, D.F., Subbaramaiah, K., Pasmantier, M.W., and Dannenberg, A.J. Celecoxib, a selective cyclo-oxygenase-2 inhibitor, enhances the response to preoperative paclitaxel and carboplatin in early-stage non-small-cell lung cancer. *J Clin Oncol*, *21*: 2645–2650, 2003.
169. Liao, Z., Komaki, R., Milas, L., Yuan, C., Kies, M., Chang, J.Y., Jeter, M., Guerrero, T., Blumenschien, G., Smith, C.M., Fossella, F., Brown, B., and Cox, J.D. A phase I clinical trial of thoracic radiotherapy and concurrent celecoxib for patients with unfavorable performance status inoperable/unresectable non-small cell lung cancer. *Clin Cancer Res*, *11*: 3342–3348, 2005.
170. Reckamp, K.L., Dubinett, S.M., Krysan, K., and Figlin, R.A. A phase I trial of targeted COX-2 and EGFR TK inhibition in advanced NSCLC. *Proc Amer Soc Clin Oncol*, *23*: 648, 2005.
171. Monnerat, C., Henriksson, R., Le Chevalier, T., Novello, S., Berthaud, P., Faivre, S., and Raymond, E. Phase I study of PKC412 (*N*-benzoyl-staurosporine), a novel oral protein kinase C inhibitor, combined with gemcitabine and cisplatin in patients with non-small-cell lung cancer. *Ann Oncol*, *15*: 316–323, 2004.
172. Douillard, J.Y., Peschel, C., Shepherd, F., Paz-Ares, L., Arnold, A., Davis, M., Tonato, M., Smylie, M., Tu, D., Voi, M., Humphrey, J., Ottaway, J., Young, K., Vreckem, A.V., and Seymour, L. Randomized phase II feasibility study of combining the matrix metalloproteinase inhibitor BMS-275291 with paclitaxel plus carboplatin in advanced non-small cell lung cancer. *Lung Cancer*, *46*: 361–368, 2004.
173. Leighl, N.B., Paz-Ares, L., Douillard, J.Y., Peschel, C., Arnold, A., Depierre, A., Santoro, A., Betticher, D.C., Gatzemeier, U., Jassem, J., Crawford, J., Tu, D., Bezjak, A., Humphrey, J.S., Voi, M., Galbraith, S., Hann, K., Seymour, L., and Shepherd, F.A. Randomized phase III study of matrix metalloproteinase inhibitor BMS-275291 in combination with paclitaxel and carboplatin in advanced non-small-cell lung cancer: National Cancer Institute of Canada-Clinical Trials Group Study BR.18. *J Clin Oncol*, *23*: 2831–2839, 2005.
174. Bissett, D., O'Byrne, K.J., von Pawel, J., Gatzemeier, U., Price, A., Nicolson, M., Mercier, R., Mazabel, E., Penning, C., Zhang, M.H., Collier, M.A., and Shepherd, F.A. Phase III study of matrix metalloproteinase inhibitor prinomastat in non-small-cell lung cancer. *J Clin Oncol*, *23*: 842–849, 2005.
175. Behrendt, C.E. and Ruiz, R.B. Venous thromboembolism among patients with advanced lung cancer randomized to prinomastat or placebo, plus chemotherapy. *Thromb Haemost*, *90*: 734–737, 2003.
176. Shepherd, F.A., Giaccone, G., Seymour, L., Debruyne, C., Bezjak, A., Hirsh, V., Smylie, M., Rubin, S., Martins, H., Lamont, A., Krzakowski, M., Sadura, A., and Zee, B. Prospective, randomized, double-blind, placebo-controlled trial of marimastat after response to first-line chemotherapy in patients with small-cell lung cancer: a trial of the National Cancer Institute of Canada-Clinical Trials Group and the European Organization for Research and Treatment of Cancer. *J Clin Oncol*, *20*: 4434–4439, 2002.
177. Khuri, F.R., Fossella, F.V., Lee, J.S., Murphy, W.K., Shin, D.M., Markowitz, A.B., and Glisson, B.S. Phase II trial of recombinant IFN-alpha2a with etoposide/cisplatin induction and interferon/megestrol acetate maintenance in extensive small cell lung cancer. *J Interferon Cytokine Res*, *18*: 241–245, 1998.
178. Ruotsalainen, T.M., Halme, M., Tamminen, K., Szopinski, J., Niiranen, A., Pyrhonen, S., Riska, H., Maasilta, P., Jekunen, A., Mantyla, M., Kajanti, M., Joensuu, H., Sarna, S., Cantell, K., and Mattson, K. Concomitant chemotherapy and IFN-alpha for small cell lung cancer: a randomized multicenter phase III study. *J Interferon Cytokine Res*, *19*: 253–259, 1999.
179. Zarogoulidis, K., Ziogas, E., Papagiannis, A., Charitopoulos, K., Dimitriadis, K., Economides, D., Maglaveras, N., and Vamvalis, C. Interferon alpha-2a and combined chemotherapy as first line treatment in SCLC patients: a randomized trial. *Lung Cancer*, *15*: 197–205, 1996.
180. Prior, C., Oroszy, S., Oberaigner, W., Schenk, E., Kummer, F., Aigner, K., Hausmaninger, H., Peschel, C., and Huber, H. Adjunctive interferon-alpha-2c in stage IIIB/IV small-cell lung cancer: a phase III trial. *Eur Respir J*, *10*: 392–396, 1997.
181. Kelly, K., Crowley, J.J., Bunn, P.A., Jr., Hazuka, M.B., Beasley, K., Upchurch, C., Weiss, G.R., Hicks, W.J., Gandara, D.R., Rivkin, S., et al. Role of recombinant interferon alfa-2a maintenance in patients with limited-stage small-cell lung cancer responding to concurrent chemoradiation: a Southwest Oncology Group study. *J Clin Oncol*, *13*: 2924–2930, 1995.

182. Mattson, K., Niiranen, A., Ruotsalainen, T., Maasilta, P., Halme, M., Pyrhonen, S., Kajanti, M., Mantyla, M., Tamminen, K., Jekunen, A., Sarna, S., and Cantell, K. Interferon maintenance therapy for small cell lung cancer: improvement in long-term survival. *J Interferon Cytokine Res*, *17*: 103–105, 1997.
183. Tummarello, D., Mari, D., Graziano, F., Isidori, P., Cetto, G., Pasini, F., Santo, A., and Cellerino, R. A randomized, controlled phase III study of cyclophosphamide, doxorubicin, and vincristine with etoposide (CAV-E) or teniposide (CAV-T), followed by recombinant interferon-alpha maintenance therapy or observation, in small cell lung carcinoma patients with complete responses. *Cancer*, *80*: 2222–2229, 1997.
184. Ruotsalainen, T., Halme, M., Isokangas, O.P., Pyrhonen, S., Mantyla, M., Pekonen, M., Sarna, S., Joensuu, H., and Mattson, K. Interferon-alpha and 13-*cis*-retinoic acid as maintenance therapy after high-dose combination chemotherapy with growth factor support for small cell lung cancer—a feasibility study. *Anticancer Drugs*, *11*: 101–108, 2000.
185. Maasilta, P., Holsti, L.R., Halme, M., Kivisaari, L., Cantell, K., and Mattson, K. Natural alpha-interferon in combination with hyperfractionated radiotherapy in the treatment of non-small cell lung cancer. *Int J Radiat Oncol Biol Phys*, *23*: 863–868, 1992.
186. Kataja, V. and Yap, A. Combination of cisplatin and interferon-alpha 2a (Roferon-A) in patients with non-small cell lung cancer (NSCLC). An open phase II multicentre study. *Eur J Cancer*, *31A*: 35–40, 1995.
187. Chao, T.Y., Hwang, W.S., Yang, M.J., Chang, J.Y., Wang, C.C., Hseuh, E.J., Huang, S.H., and Chen, W.C. Combination chemoimmunotherapy with interferon-alpha and cisplatin in patients with advanced non-small cell lung cancer. *Zhonghua Yi Xue Za Zhi (Taipei)*, *56*: 232–238, 1995.
188. Mandanas, R., Einhorn, L.H., Wheeler, B., Ansari, R., Lutz, T., and Miller, M.E. Carboplatin (CBDCA) plus alpha interferon in metastatic non-small cell lung cancer. A Hoosier Oncology Group phase II trial. *Am J Clin Oncol*, *16*: 519–521, 1993.
189. Fuxius, S., Mross, K., Mansouri, K., and Unger, C. Gemcitabine and interferon-alpha 2b in solid tumors: a phase I study in patients with advanced or metastatic non-small cell lung, ovarian, pancreatic or renal cancer. *Anticancer Drugs*, *13*: 899–905, 2002.
190. Quan, W.D., Jr., Casal, R., Rosenfeld, M., and Walker, P.R. Alpha interferon-2b, leucovorin, and 5-fluorouracil (ALF) in non-small cell lung cancer. *Cancer Biother Radiopharm*, *11*: 229–234, 1996.
191. Ardizzoni, A., Rosso, R., Salvati, F., Scagliotti, G., Soresi, E., Ferrara, G., Pennucci, C., Baldini, E., Cruciani, A.R., Antilli, A., et al. Combination chemotherapy and interferon alpha 2b in the treatment of advanced non-small-cell lung cancer. The Italian Lung Cancer Task Force (FONICAP). *Am J Clin Oncol*, *14*: 120–123, 1991.
192. Ardizzoni, A., Salvati, F., Rosso, R., Bruzzi, P., Rubagotti, A., Pennucci, M.C., Mariani, G.L., De Marinis, F., Pallotta, G., Antilli, A., et al. Combination of chemotherapy and recombinant alpha-interferon in advanced non-small cell lung cancer. Multicentric Randomized FONICAP Trial Report. The Italian Lung Cancer Task Force. *Cancer*, *72*: 2929–2935, 1993.
193. Silva, R.R., Bascioni, R., Rossini, S., Zuccatosta, L., Mattioli, R., Pilone, A., Delprete, S., Battelli, N., Gasparini, S., and Battelli, T. A phase II study of mitomycin C, vindesine and cisplatin combined with alpha interferon in advanced non-small cell lung cancer. *Tumori*, *82*: 68–71, 1996.
194. Salvati, F., Rasi, G., Portalone, L., Antilli, A., and Garaci, E. Combined treatment with thymosin-alpha1 and low-dose interferon-alpha after ifosfamide in non-small cell lung cancer: a phase-II controlled trial. *Anticancer Res*, *16*: 1001–1004, 1996.
195. Athanasiadis, I., Kies, M.S., Miller, M., Ganzenko, N., Joob, A., Marymont, M., Rademaker, A., and Gradishar, W.J. Phase II study of all-*trans*-retinoic acid and alpha-interferon in patients with advanced non-small cell lung cancer. *Clin Cancer Res*, *1*: 973–979, 1995.
196. Goncalves, A., Camerlo, J., Bun, H., Gravis, G., Genre, D., Bertucci, F., Resbeut, M., Pech-Gourg, F., Durand, A., Maraninchi, D., and Viens, P. Phase II study of a combination of cisplatin, all-*trans*-retinoic acid and interferon-alpha in squamous cell carcinoma: clinical results and pharmacokinetics. *Anticancer Res*, *21*: 1431–1437, 2001.
197. Roth, A.D., Morant, R., and Alberto, P. High dose etretinate and interferon-alpha—a phase I study in squamous cell carcinomas and transitional cell carcinomas. *Acta Oncol*, *38*: 613–617, 1999.

198. Roth, A.D., Abele, R., and Alberto, P. 13-*cis*-retinoic acid plus interferon-alpha: a phase II clinical study in squamous cell carcinoma of the lung and the head and neck. *Oncology*, *51*: 84–86, 1994.
199. Rinaldi, D.A., Lippman, S.M., Burris, H.A., 3rd, Chou, C., Von Hoff, D.D., and Hong, W.K. Phase II study of 13-*cis*-retinoic acid and interferon-alpha 2a in patients with advanced squamous cell lung cancer. *Anticancer Drugs*, *4*: 33–36, 1993.
200. Jansen, R.L., Slingerland, R., Goey, S.H., Franks, C.R., Bolhuis, R.L., and Stoter, G. Interleukin-2 and interferon-alpha in the treatment of patients with advanced non-small-cell lung cancer. *J Immunother*, *12*: 70–73, 1992.
201. Wheeler, R.H., Herndon, J.E., Clamon, G.H., and Green, M.R. A phase II study of recombinant beta-interferon at maximum tolerated dose in patients with advanced non-small cell lung cancer: a cancer and leukemia group B study. *J Immunother Emphasis Tumor Immunol*, *15*: 212–216, 1994.
202. McDonald, S., Chang, A.Y., Rubin, P., Wallenberg, J., Kim, I.S., Sobel, S., Smith, J., Keng, P., and Muhs, A. Combined Betaseron R (recombinant human interferon beta) and radiation for inoperable non-small cell lung cancer. *Int J Radiat Oncol Biol Phys*, *27*: 613–619, 1993.
203. Byhardt, R.W., Vaickus, L., Witt, P.L., Chang, A.Y., McAuliffe, T., Wilson, J.F., Lawton, C.A., Breitmeyer, J., Alger, M.E., and Borden, E.C. Recombinant human interferon-beta (rHuIFN-beta) and radiation therapy for inoperable non-small cell lung cancer. *J Interferon Cytokine Res*, *16*: 891–902, 1996.
204. Bund, J., Eberhardt, K., Hartmann, W., and Habermalz, H.J. [Treatment of stage IIIB loco-regionally advanced non-small-cell bronchial carcinomas with radiation and interferon-beta. Preliminary results of aphase II study]. *Strahlenther Onkol*, *174*: 300–305, 1998.
205. Bradley, J.D., Scott, C.B., Paris, K.J., Demas, W.F., Machtay, M., Komaki, R., Movsas, B., Rubin, P., and Sause, W.T. A phase III comparison of radiation therapy with or without recombinant beta-interferon for poor-risk patients with locally advanced non-small-cell lung cancer (RTOG 93-04). *Int J Radiat Oncol Biol Phys*, *52*: 1173–1179, 2002.
206. Tester, W.J., Kim, K.M., Krigel, R.L., Bonomi, P.D., Glick, J.H., Asbury, R.F., Kirkwood, J.M., Blum, R.H., and Schiller, J.H. A randomized Phase II study of interleukin-2 with and without beta-interferon for patients with advanced non-small cell lung cancer: an Eastern Cooperative Oncology Group study (PZ586). *Lung Cancer*, *25*: 199–206, 1999.
207. Recchia, F., Sica, G., De Filippis, S., Rea, S., and Frati, L. Combined chemotherapy and differentiation therapy in the treatment of advanced non-small-cell lung cancer. *Anticancer Res*, *17*: 3761–3765, 1997.
208. Halme, M., Maasilta, P.K., Pyrhonen, S.O., and Mattson, K.V. Interferons combined with chemotherapy in the treatment of stage III-IV non-small cell lung cancer—a randomised study. *Eur J Cancer*, *30A*: 11–15, 1994.
209. Tran, H.T., Blumenschein, G.R., Jr., Lu, C., Meyers, C.A., Papadimitrakopoulou, V., Fossella, F.V., Zinner, R., Madden, T., Smythe, L.G., Puduvalli, V.K., Munden, R., Truong, M., and Herbst, R.S. Clinical and pharmacokinetic study of TNP-470, an angiogenesis inhibitor, in combination with paclitaxel and carboplatin in patients with solid tumors. *Cancer Chemother Pharmacol*, *54*: 308–314, Epub 2004 June 2004, 2004.
210. Herbst, R.S., Madden, T.L., Tran, H.T., Blumenschein, G.R., Jr., Meyers, C.A., Seabrooke, L.F., Khuri, F.R., Puduvalli, V.K., Allgood, V., Fritsche, H.A., Jr., Hinton, L., Newman, R.A., Crane, E.A., Fossella, F.V., Dordal, M., Goodin, T., and Hong, W.K. Safety and pharmacokinetic effects of TNP-470, an angiogenesis inhibitor, combined with paclitaxel in patients with solid tumors: evidence for activity in non-small-cell lung cancer. *J Clin Oncol*, *20*: 4440–4447, 2002.
211. Herbst, R.S., Hammond, L.A., Carbone, D.P., Tran, H.T., Holroyd, K.J., Desai, A., Williams, J.I., Bekele, B.N., Hait, H., Allgood, V., Solomon, S., and Schiller, J.H. A phase I/IIA trial of continuous five-day infusion of squalamine lactate (MSI-1256F) plus carboplatin and paclitaxel in patients with advanced non-small cell lung cancer. *Clin Cancer Res*, *9*: 4108–4115, 2003.
212. Rose, V., Schiller, J.H., Wood, A., Eskander, E., Holroyd, K.J., Desai, A., Lee, J.T., Ahmed, M., and Kim, B. Randomized phase II trial of weekly squalamine, carboplatin, and paclitaxel as first line therapy for advanced non-small cell lung cancer. *Proc Amer Soc Clin Oncol*, *22*: 643, 2004.
213. Falardeau, P., Champagne, P., Poyet, P., Hariton, C., and Dupont, E. Neovastat, a naturally occurring multifunctional antiangiogenic drug, in phase III clinical trials. *Semin Oncol*, *28*: 620–625, 2001.

214. Latreille, J., Batist, G., Laberge, F., Champagne, P., Croteau, D., Falardeau, P., Levinton, C., Hariton, C., Evans, W.K., and Dupont, E. Phase I/II trial of the safety and efficacy of AE-941 (Neovastat) in the treatment of non-small-cell lung cancer. *Clin Lung Cancer*, *4*: 231–236, 2003.
215. Harrison, L. and Blackwell, K. Hypoxia and anemia: factors in decreased sensitivity to radiation therapy and chemotherapy? *Oncologist*, *9* (suppl 5): 31–40, 2004.
216. Koch, S., Mayer, F., Honecker, F., Schittenhelm, M., and Bokemeyer, C. Efficacy of cytotoxic agents used in the treatment of testicular germ cell tumours under normoxic and hypoxic conditions in vitro. *Br J Cancer*, *89*: 2133–2139, 2003.
217. Baredes, S. and Kim, S. Distribution of blood flow in an experimental palatal carcinoma. *Am J Otolaryngol*, *9*: 155–160, 1988.
218. Dewhirst, M.W., Ong, E.T., Klitzman, B., Secomb, T.W., Vinuya, R.Z., Dodge, R., Brizel, D., and Gross, J.F. Perivascular oxygen tensions in a transplantable mammary tumor growing in a dorsal flap window chamber. *Radiat Res*, *130*: 171–182, 1992.
219. Feldmann, H.J., Molls, M., Hoederath, A., Krumpelmann, S., and Sack, H. Blood flow and steady state temperatures in deep-seated tumors and normal tissues. *Int J Radiat Oncol Biol Phys*, *23*: 1003–1008, 1992.
220. Groothuis, D.R., Pasternak, J.F., Fischer, J.M., Blasberg, R.G., Bigner, D.D., and Vick, N.A. Regional measurements of blood flow in experimental RG-2 rat gliomas. *Cancer Res*, *43*: 3362–3367, 1983.
221. Jain, R.K. and Baxter, L.T. Mechanisms of heterogeneous distribution of monoclonal antibodies and other macromolecules in tumors: significance of elevated interstitial pressure. *Cancer Res*, *48*: 7022–7032, 1988.
222. Nakagawa, T., Tanaka, R., Takeuchi, S., and Takeda, N. Haemodynamic evaluation of cerebral gliomas using XeCT. *Acta Neurochir (Wien)*, *140*: 223–233, 1998.
223. Taniguchi, H., Koyama, H., Masuyama, M., Takada, A., Mugitani, T., Tanaka, H., Hoshima, M., and Takahashi, T. Angiotensin-II-induced hypertension chemotherapy: evaluation of hepatic blood flow with oxygen-15 PET. *J Nucl Med*, *37*: 1522–1523, 1996.
224. Choi, H., Charnsangavej, C., de Castro Faria, S., Tamm, E.P., Benjamin, R.S., Johnson, M.M., Macapinlac, H.A., and Podoloff, D.A. CT evaluation of the response of gastrointestinal stromal tumors after imatinib mesylate treatment: a quantitative analysis correlated with FDG PET findings. *AJR Am J Roentgenol*, *183*: 1619–1628, 2004.

27 Antiangiogenic Therapy for Prostate Cancer

Leslie K. Walker, Glenn Liu, and George Wilding

CONTENTS

Prostate cancer is the most common malignancy in men and accounts for nearly 232,090 new cases and approximately 30,350 deaths in the United States during the year 2005 alone.[1] For patients with localized disease, primary treatment with surgery or radiation therapy can be offered, but despite our best efforts, as many as 40% of these men experienced biochemical failure as evident by a rise in their serum prostate-specific antigen (PSA).[2] For patients with advanced or recurrent disease, standard initial therapy consists of androgen deprivation therapy. Whereas the majority of patients respond to androgen ablation, eventually all patients develop hormone-refractory prostate cancer (HRPC) after a median of months.[3] Chemotherapy is typically reserved for symptom palliation in patients with metastatic HRPC. More recently, chemotherapy with docetaxel has been shown to provide a significant, but modest, improvement in median survival.[4,5] Nevertheless, for patients with metastatic HRPC, the overall median survival remains poor at around 18 months. Clearly, newer therapies for prostate cancer are needed.

27.1 NATURAL HISTORY OF PROSTATE CANCER

The temporal history of prostate cancer tumorigenesis is summarized in Figure 27.1. It depicts the many genetic events that can occur as a prostate cell transforms from normal epithelial cells to early neoplastic cells, localized to metastatic cancer, and finally androgen-sensitive to androgen-refractory disease. Each of these events is characterized by various mutations, each resulting in a cancer phenotype that is more virulent and resistant to existing therapies. While most novel therapies are focused in the metastatic and HRPC population, this period accounts for only a minor fraction of the natural history of this disease process. The implications of this is that there remains a large portion of time in the natural history of prostate cancer that may be successfully altered by either use of existing therapies earlier in the disease process or with novel therapies that interrupt the process of tumorigenesis, invasion, and metastasis. One such approach is targeting angiogenesis.

Angiogenesis is required for the growth and metastasis of all solid tumors. As a result, inhibiting angiogenesis may be a valid approach in prostate cancer treatments. The following chapter addresses the importance of angiogenesis in the development and progression of prostate cancer, the results of agents that have reached clinical trials in patients with advanced prostate cancer and the specific considerations necessary in developing antiangiogenic approaches for the treatment of patients with this disease.

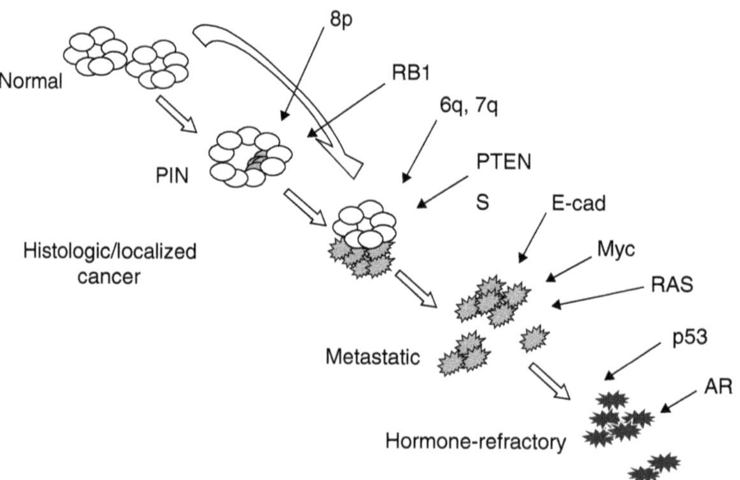

FIGURE 27.1 This cartoon depicts the natural history of prostate cancer development and progression. It emphasizes the fact that numerous genetic events (mutations) must occur during tumorigenesis, each of which results in a tumor that is phenotypically more capable of growth, invasion, and resistance to apoptosis. (Abbreviations: PIN, prostatic intraepithelial neoplasia; 8p, short-arm of chromosome 8; RB1, retinoblastoma 1; 6q, long-arm of chromosome 6; 7q, long-arm of chromosome 7; PTEN, phosphatase and tensin homolog deleted on chromosome 10; E-cad, E-cadherin; Myc, Myc proto-oncogene; RAS, Ras proto-oncogene; AR, mutated-androgen receptor.)

27.2 ANGIOGENESIS AND PROSTATE CANCER

Angiogenesis refers to the complex process of endothelial cell migration toward angiogenic stimuli, followed by endothelial cell proliferation and reorganization into a network of tubules of new blood vessels.[6] This process is regulated by a host of stimulatory and inhibitory factors released by the tumor and its microenvironment.[7] Extensive laboratory studies support that angiogenesis is important in prostate cancer development, growth, and metastasis.

27.2.1 MICROVESSEL DENSITY

Increased microvessel formation is a feature of many cancers including prostate cancer. The quantification of microvessel formation is referred to as microvessel density (MVD) and involves staining tissue with specific antiendothelial antibodies and scoring the vessels under a microscope. Of note, antibodies employed to measure MVD have included anti-PECAM-1, anti-von Willebrand Factor (vWF), anti-CD31, and anti-CD34.[8]

Weidner et al. initially examined the relationship between MVD and prostate cancer. Tumor specimens from 74 patients with invasive prostate carcinoma (29 with metastasis and 45 without metastasis) were evaluated for the study. Tissue was obtained from radical prostatectomy specimens and endothelial cells were stained using a factor VIII-related antigen. MVD was found to be greater in prostate cancer patients with metastatic disease than in those without metastases (76.8 versus 39.2 microvessels per 200 field) ($p < 0.0001$). MVD was also found to increase with increasing Gleason's score ($p < 0.0001$) but this increase was present predominantly in the poorly differentiated tumors.[9]

MVD has been associated, in additional studies, with tumor stage and grade as well as cancer recurrence and cancer-specific survival.[10–15] In a prospective trial by Borre et al., MVD at diagnosis was correlated with disease-specific survival in a noncurative population of prostate cancer patients. MVD was quantified using staining with factor VIII-related antigen in 221 prostate cancer patients. The median length of follow-up was 15 years. The authors found MVD correlated with clinical stage ($p < 0.0001$), histopathological grade ($p < 0.001$), and was shown to be significantly associated with disease-specific survival ($p = 0.0001$).[16]

27.2.2 VEGF

Vascular endothelial growth factor (VEGF) has been demonstrated in preclinical data with prostate cancer cell lines to be important in stimulating cell proliferation as well as angiogenesis and lymphangiogenesis.[17] In experimental prostate cancer models, VEGF expression has also been shown to be upregulated in prostate and prostate cancer tissue by androgens and the levels have been found to fall after castration.[18–20]

VEGF expression has also been examined in human prostatic tissue and found to be highly expressed in most prostate cancers.[21,22] Ferrer et al. initially evaluated VEGF levels in prostate cancer by immunohistochemical staining of archival prostate cancer specimens. The authors demonstrated that prostate cancer cells stained positively for VEGF (20/25 specimens) more often than BPH tissue (2/11) or normal prostate tissue (0/11).[23]

VEGF tissue expression has been correlated with biochemical relapse after prostatectomy.[10] VEGF tissue expression also predicted death from prostate cancer in a cohort that underwent observation for clinically localized disease and two other cohorts with HRPC.[24,25] In the evaluation by George et al., the plasma VEGF levels of 197 patients who had been enrolled in a phase III trial of suramin (CALBG 9480) were examined and in a prospective evaluation, found to be inversely correlated with survival time ($p = 0.002$).[26]

Serum or urine levels of VEGF have also been shown to be elevated in patients with prostate cancer.[27] Duque et al. demonstrated that patients with metastatic prostate cancer had a higher median plasma VEGF level (28.5 pg/mL) than patients with localized disease (7.0 pg/mL) and controls (0.0 pg/mL). Those values were statistically significant both when levels in patients with metastatic disease were compared with those in patients with localized disease and those in normal patients.[28] Elevated serum or urine VEGF levels in patients undergoing radical prostatectomy have been shown to be predicative of earlier disease progression.[29,30] Serum VEGF has also been noted to decrease after prostatectomy.[31]

27.2.2.1 Basic Fibroblast Growth Factor

The angiogenic effect of basic fibroblast growth factor (bFGF) was initially described in a rabbit cornea model.[32] In a subsequent study, the production of bFGF and expression of the bFGF receptor were examined in prostate cell lines including LNCaP, DU145, and PC3. Both the production of bFGF and the expression of its receptor were found to be more significant in cell lines of greater metastatic potential, DU145 and PC3, than in the LNCaP cell line.[33] bFGF levels have also been examined in a nude mouse models. DU145 and PC3 cell lines were injected into mice and bFGF levels were determined by immunoassays. bFGF levels were found to be higher in the PC3 tumors than the DU145 tumors ($p < 0.005$). Given that angiogenesis activity, as determined by factors including vessel count, was found to be greater in the PC3 tumors, the authors suggested that bFGF expression may be associated with an angiogenic phenotype.[34]

Despite those preclinical findings, studies evaluating serum and urinary bFGF measurements in humans have shown no consistent evidence for their use as a prognostic factor. When serum concentrations of bFGF were studied in patients with both localized and metastatic prostate cancer as well as patients with BPH, a difference was found in the serum concentrations of bFGF between patients with local and advanced prostate cancer but no difference was seen between patients with BPH and metastatic disease.[35] Urinary levels of bFGF were prospectively collected from 100 patients undergoing therapy with suramin as part of CALGB 9480. Levels were determined by an ELISA assay. While VEGF levels were found to be predictive of survival in the study, no significant correlations between bFGF levels or change in levels and survival were found.[36]

27.2.2.2 Transforming Growth Factor Beta 1

Transforming growth factor beta 1 (TGF-β_1) is a cytostatic inhibitor of epithelial cell growth and may stimulate the growth of stromal cells such as fibroblasts. Prostate tumors express high levels of TGF-β_1 but seem to acquire resistance to its antiproliferative effects with tumor progression. In addition, TGF-β_1 has been speculated to be associated with tumor-promoting processes in prostate cancer including angiogenesis.

Immunoreactivity for TGF-β_1 and its receptors type I and II were followed in 73 cases of prostate cancer. It was found that patients with tumor overproduction of TGF-β_1 had shorter median cancer-specific survival than patients with normal TGF-β_1 immunoreactivity (5 versus 10 years, $p = 0.006$). Furthermore, increased TGF-β_1 staining was associated with tumor grade, high vascular counts, and metastasis ($p = 0.02, 0.02, 0.01$, respectively). Patients with loss of tumor TGF-β receptor type II expression in addition with TGF-β_1 overproduction showed particularly short survivals (2.6 versus 10 years, $p = 0.0001$), when compared with patients with normal immunoreactivity.[37] Of late, other studies have demonstrated that reduced levels and loss of expression of TGF-β_1 membrane receptors are correlated with tumor grade and prognosis in prostate cancer patients.[38–40] Further data suggesting that TGF-β_1 plays a role in cancer angiogenesis included a study in which TGF-β_1 latency-associated peptide (LAP) or TGF-β_1 neutralizing antibody was used to treat LNCaP cells in a reactive stroma xenograft model. Tumors treated with either inhibitor exhibited a reduction in blood vessels. The MVD of LAP treated tumors was decreased by 3.5-fold when compared with control tumors. The average weight of LAP treated tumors was also reduced by 46%.[41]

27.2.3 Interleukin-8

Interleukin-8 (IL-8) is a macrophage-derived mediator of angiogenesis. It was initially shown to promote angiogenesis in a rat cornea model.[42] While IL-8 expression has been demonstrated in normal prostate epithelial cells, it has also been shown to stain positively more often in adenocarcinoma cells than either normal prostate cells or BPH cells.[43,44]

IL-8 has been detected in a number of prostate cancer cell lines including DU145 and PC-3.[17,44,45] IL-8 mRNA was found to be expressed at higher levels in the highly metastatic human PC-3M-LN4 prostate cell line when compared with PC-3M cells or poorly metastatic cell line.[44] IL-8's role in angiogenesis was further demonstrated in another study involving the injection of low and high IL-8 producing PC-3 clones into nude mice. PC-3 expressing high levels of IL-8 was highly tumorigenic, producing rapidly growing, highly vascularized prostate tumors with 100% incidence of lymph node metastasis whereas the low IL-8 expressing PC-3 cells were less tumorigenic, producing slower growing and less vascularized primary tumors and significantly lower incidence of metastasis. In situ hybridization analysis of the tumors for IL-8 in addition to matrix metalloproteinase (MMP), VEGF, and E-cadherin corresponded with MVD and biological behavior of the prostate cancers in nude mice.[45]

27.3 ANGIOGENESIS INHIBITORS

A number of agents have been developed and tested in patients with prostate cancer. The following represents those agents, which have reached testing in clinical trials.

Angiogenesis inhibitors represent a broad and ill-defined group of therapeutic agents. There is no one accepted classification of these agents and often, while the agents demonstrate antiangiogenic properties, the mechanisms by which they exert those effects are unclear and, thereby, defy easy categorization. The following agents are, therefore, presented, partly, in order of development and testing in patients with prostate cancer, but also more broadly into those agents that have not demonstrated activity in this patient population and those that have produced more promising results.

Suramin was the first angiogenesis inhibitor to be developed and tested in patients with prostate cancer. The MMPs followed including prinomastat. The initial trials with suramin were promising and suramin in addition to prinomastat reached phase III trials before further tests were discontinued due to negative results. Other agents, including TNP-470, CAI, SU5416, also were studied in phase I and II trials in patients with advanced prostate cancer. Due to the negative results of those trials, further testing of those agents in that patient population has come to a halt. Other agents including atrasentan, bevacizumab, thalidomide, and 2-ME have had more promising results in patients with HRPC. Atrasentan has already published the results of a phase III trial of the agent in patients with HRPC. Phase III trials are ongoing for both thalidomide and bevacizumab.

27.3.1 Suramin

Suramin is a polysulfonated naphthylurea, which was first synthesized in 1916 by Bayer AG.[46] The drug has a broad and unusual variety of biologic features and has, in the course of its development, been used as a treatment for disorders including African trypanosomes and onchocerchiasis due to its trypanocidal activity as well as a treatment for acquired immunodeficiency syndrome (AIDS) due to the finding that the agent inhibited reverse transcriptase. The antitumor effects of suramin became apparent in the treatment of AIDS patients with suramin as the agent showed activity in HIV associated malignancies such as Kaposi's Sarcoma and Non-Hodgkins lymphoma. Suramin's antitumor and antiangiogenic effects are thought, in part, to relate to the binding of the agent to a number of growth factors including fibroblast growth factors (FGFs), thereby blocking the action of those factors. However, the degree to which the antitumor effects of suramin are mediated by that mechanism is not clear. Other potential mechanisms include inhibition of DNA topoisomerase II and protein kinase C as well as adrenolytic effects, which has obvious significance in patients with prostate cancer.[47]

Preclinical data has shown in vitro antiproliferative activity against human prostate cancer cell lines (LNCaP, PC-3, and DU145).[48,49] Therefore, suramin was evaluated for antineoplastic properties in patients with prostate cancer.

Of interest, Thalmann et al. reported the effect of suramin on an in vivo model of prostate cancer (C4-2 cells induced in castrated, athymic nude mice) as well as in vitro prostate cancer models (C4-2 and LNCaP cells). While suramin was found to affect the growth of LNCaP cells, it did not inhibit the growth of C4-2 cells either in vivo or in vitro. Suramin significantly decreased PSA mRNA expression in both cell lines in vitro and depressed serum PSA levels in the in vivo model. These results led to the question of whether PSA levels are reliable markers in assessing response to suramin therapy.[50]

Suramin was evaluated in a number of phase I and II trials conducted in patients with HRPC. While the initial results were promising, the trials had often not controlled for confounding variables such as antiandrogen withdrawal responses and the effects of hydrocortisone.[48,51,52] Subsequent trials, which controlled for those variables, reported lower response rates and thus, the use of suramin in patients with advanced prostate cancer remained a question.[53–55]

Suramin was tested in patients with advanced prostate cancer as part of large double-blind, multicenter phase III trial. Nearly 500 patients with symptomatic, HRPC were enrolled in the trial and randomized to receive suramin plus hydrocortisone or a placebo plus hydrocortisone. The primary end point was palliation of pain, which was assessed using pain scores and changes in opioid analgesic use. The results indicated that treatment with suramin resulted in a greater palliation of symptoms in patients treated with suramin than in those who received the placebo as measured by pain scores (43% versus 28%, respectively; $p = 0.001$). Overall mean reductions in combined pain and opioid analgesic intake were also greater for the therapeutic arm than the placebo arm ($p = 0.0001$). Despite those benefits, however, responses in measurable disease were infrequent (<4%) and no survival advantage was seen. While PSA responses were noted more often among the treated patients than the placebo patients, those results were questioned given the previously mentioned findings of Thalmann et al. Quality of life (QoL) measures were quite similar in both arms despite somewhat worse toxicity and more adverse events in the suramin arm. In summary, the trial demonstrated that suramin provided, at best, a modest pain palliation, but with no overall impact on quality of life or survival.[25]

Another phase III trial of suramin in patients with advanced prostate cancer was presented following the trial mentioned earlier. The trial employed a simplified suramin dosing regimen and the authors randomized patients to treatment with three different doses of suramin to assess whether the efficacy and toxicity of suramin was dose-dependent. The results demonstrated objective response rates in <20% of patients on all three arms and no significant differences were found to indicate a dose–response relationship. PSA responses were seen in 24%–34% of patients depending on the treatment arm; however, again no dose–response relationship was demonstrated. Of interest, a significant correlation was found between PSA response (50% decline in PSA at 20 weeks) and survival in a landmark analysis. The authors did find increased toxicities on the higher doses of suramin. Those toxicities reported to be considerable and the authors recommended using the low dose for any subsequent trials; however, any role for suramin plus a glucocorticoid for use as first-line treatment for HRPC effectively ended when the Oncology Advisory Panel of the FDA voted against the approval of suramin for that indication.[56]

27.3.2 MMP Inhibitors

The MMP is a family of zinc- and calcium-dependent endopeptidases that are involved in remodeling the extracellular matrix, a process necessary for functions including cellular migration and angiogenesis. More than 20 members of this family have been identified.[57]

MMPs have been found to be upregulated in most human tumors. A number of studies have examined MMP expression in prostate cancer tissues and found increased expression of the following MMPs in prostate cancer; MMP-9 (gelatinase-B), MMP-2 (gelatinase-A), MMP-7 (matrilysin), and MTI-MMP (MMP-14).[58–68]

Inhibitors of MMP are in clinical development and three have been tested in patients with HRPC: marimastat, prinomastat, and BMS-275291.

27.3.3 Marimastat (BB-2516)

Marimastat (BB-2516) was the first matrix metalloproteinase inhibitor (MMPI) to enter clinical trials. It inhibits all known classes of MMPs but is inactive against other metallo-enzymes. Of note, marimastat represents the chemical modification of batimastat (BB-94), an MMPI developed and tested in the preclinical setting. Further development of batimastat was hampered by poor bioavailability and, thus, marimastat was developed as an agent that retained the activity of batimastat but had improved bioavailability.

Batimastat (BB-94) has been studied in prostate cancer models including DU-145 cells both in vitro and injected into SCID mice as well as MatLyLu cancer cells both in vitro and in a rat orthotopic cancer (R3327 Dunning tumor) model.[69,70] Batimastat was found to inhibit tumor growth in both the in vitro and in vivo models in those studies.

Nemunaitis et al. initially investigated marimastat in patients with advanced prostate cancer. The study recruited 415 patients with advanced cancer including 88 patients with prostate cancer. As marimastat was not expected to result in reduction of tumor volume, the authors used surrogate markers to assess biological response such as PSA in the case of the patients with advanced prostate cancer. After an observation period of 4 weeks during which the PSA was required to rise by $\geq$25%, the patients initiated treatment with marimastat in doses ranging from 2 mg daily to 75 mg twice daily. The most frequently occurring toxicities included arthralgia, myalgia, and tenditis. The authors recommended doses between 10 and 25 mg twice daily for future studies. The study reported evidence for biological activity in patients with advanced malignancy as well as a reduction in the rate of rise in PSA levels to <25% in 50%–70% of the patients who received, at least, 10–50 mg of the MMPI. The median rate of rise over comparable 4 week periods decreased from 53% before to 29% during marimastat treatment ($p = 0.0001$).[71,72]

27.3.4 Prinomastat (AG3340)

Prinomastat (AG3340) is an MMPI with a relative selectivity for MMPs—2, 3, 9, 13, and 14 as opposed to MMP-1. The agent was developed as a more selective inhibitor of MMP with the objective of avoiding the arthritis and arthragias associated thought to be related to inhibition of MMP-1.[73]

Prinomastat has been studied in a number of human tumor models including prostate. Prinomastat was reported to inhibit tumor growth and increased the survival of nude mice bearing PC-3 prostatic tumor. Prinomastat was also shown to decrease angiogenesis in PC-3 tumors.[74]

Hande et al. recently reported the phase I results for prinomastat in 75 patients with advanced cancer (4 with prostate cancer). No documented tumor responses were reported in this trial. The primary toxicities identified were joint- and muscle-related pain. Those toxicities were noted usually 2–3 months after initiation of therapy and reported in >25% of patients treated at doses >25 mg twice daily. Doses of 5–10 mg twice daily were recommended for additional trials.[75]

Prinomastat has been studied in patients with HRPC. The preliminary results of a phase III trial were presented by Ahman et al., in which, 553 patients were randomized to receive either 5 or 10 mg of prinomastat or placebo twice daily in combination with mitoxantrone and prednisone. Toxicities were comparable with musculoskeletal effects more often noted in the prinomastat arms. No differences were observed among treatment arms in PSA response (≥75% PSA reduction), progression-free survival by radiography, PSA (50% increase), or symptoms. No difference was reported in overall survival and 1 year survival.[76]

27.3.5 BMS-275291

BMS-275291 is an MMPI designed to spare a class of closely related metalloproteinases known as sheddases. Apart from a novel zinc-binding group, BMS-275291 is similar in structure to marimastat and inhibits a similar broad spectrum of MMPs including, like marimastat, MMP-1; however, unlike marimastat, BMS-275291 is designed to potently inhibit MMP activity although minimally affecting those of other metalloproteinases (e.g., sheddases) involved in the release of cell-associated molecules such as tumor necrosis factor-α, tumor necrosis factor-α receptor, interleukin-6 receptor, or L-selectin. Inadvertent sheddase inhibition has been hypothesized to play a role in the dose-limiting joint toxicities occurring with hydroxamate-based MMP inhibitors.[77]

BMS-275291 has been studied in a phase I trial in 44 patients with advanced cancer. The toxicities experienced were reported to be mild and not clearly related to therapy. The recommended dosing for subsequent studies was 1200 mg/daily.[78]

Lara et al. presented the results of a phase II study of BMS-275291 in patients with HRPC. The patients were randomized between two doses (1200 or 2400 mg daily). There were no responders; however, the authors reported change in the mean PSA slope pre-/posttreatment, which was not statistically significant.[79]

27.3.6 TNP-470

TNP-470 is an antiangiogenic synthetic analog of fumagillin, an antibiotic produced by the fungus *Aspergillus fumigatus* Fresenius. The antiangiogenic effects of fumagillin and its analog, AGM-1470 (TNP-470), were initially described by Ingber et al.[80] While the mechanism of action of TNP-470 remains unclear, recent studies have identified a protein, methionine aminopeptidase (MetAP-2), which is a cell cycle regulator, as a target of fumigillin and TNP-470.[81,82]

The activity of TNP-470 against prostate cancer has been studied in a number of preclinical trials. Yamaoka et al. showed antitumor activity of TNP-470 in prostate cancer xenografts (PC-3 cells subcutaneously implanted in BALB/c nu/nu mice).[83] In another study, the effect of TNP-470 was examined on both local and metastatic growth in a prostate cancer model (AT6.3 cells in male BALB/c nu/nu mice). TNP-470 inhibited the growth of AT6.3 cells in vitro as well as decreased the number and size of lung metastases from AT6.3 cells inoculated either percutaneously or intravenously.[84]

TNP-470 has been combined with a number of other agents in preclinical trials. In addition to studying the antitumor effects of TNP-470, Yamaoka et al. tested TNP-470 in combination with chemotherapeutic agents and found an additive effect of TNP-470 against the PC-3 cells when combined with cisplatin.[83] In addition, TNP-470 was combined with nicardipine. This combination demonstrated inhibition of in vivo growth of PC-3 cells.[85] TNP-470 has also been combined with gene therapy involving a cell–cell adhesion molecule 1 (C–CAM1). The combination evidently demonstrated a synergistic antitumor effect in DU15 cells both in vitro and in vivo.[86]

Of interest, the effect of TNP-470 on PSA was examined after a phase I study of TNP-470 in patients with HRPC had observed elevated PSA levels in patients treated with TNP-470, which reversed after discontinuation with the drug. While it was found that the highest TNP-470 concentration produced a decrease in cell number, the PSA secretion per cell was induced 1.1- to 1.5-fold following TNP-470 exposure. PSA and androgen receptors were transcriptionally upregulated by 1.4-fold and 1.2-fold respectively, after exposure to TNP-470. It was concluded that PSA was an unreliable marker for measuring activity of TNP-470.[87]

The results of a phase I study of TNP-470 in patients with HRPC have been reported by Logothetis et al. The study accrued 33 patients to the dose escalation trial of intravenous TNP-470. The dose-limiting toxicity was a characteristic neuropsychiatric symptom complex (anesthesia, gait disturbance, and agitation) that resolved on cessation of therapy. No definite antitumor activity of TNP-470 was observed; however, transient stimulation of the serum PSA concentration occurred in some of the patients treated. That issue was addressed by Horti et al. as mentioned earlier. While the authors did recommend a dose for any subsequent studies, they also concluded that the etiology and treatment of the neuropsychiatric effects should be investigated before conducting other studies with the agent.[88]

27.3.7 Carboxyamido-Triazole (CAI)

Carboxyamido-triazole (CAI) is an inhibitor of signal transduction via voltage gated and nonvoltage gated calcium channels. Inhibition of angiogenesis is one potential downstream effect of CAI's ability to modulate calcium-mediated signal transduction. CAI appears to inhibit several components of the angiogenesis process in vitro, including bFGF-induced tyrosine phosphorylation and human umbilical vein endothelial cell (HUVEC) proliferation, adhesion, and motility.[89] In addition, CAI reduces expression of MMP-2, apparently through inhibition of FGF-stimulated MMP2 synthesis.[90]

As prostate tumors develop androgen independence, a variety of stroma- and epithelium-derived autocrine and paracrine growth factors relay signals that influence the behavior of malignant prostate tissue.[91,92] Wasilenko et al. observed that calcium mobilization is a common signaling response in three advanced prostate tumor cell lines (DU-145, PC3, PPC-1) treated with a variety of neuropeptides, including bombesin and endothelin, which are associated with prostate tumor progression.[93] Those findings led to the testing of CAI in prostate cancer models.

Wasilenko et al. published the initial preclinical data in which CAI was tested in prostate cell lines. The investigators examined the effect of CAI on the anchorage-dependent of DU-145, PPC-1, PC3, and LNCaP tumor cells. The 50% inhibitory concentrations ranged from 10 to 30 μM. In a direct cell enumeration assay, the growth-suppressing activity of CAI toward DU-145 cells was reversible, indicating a cytostatic effect on the drug on tumor cells. The drug also inhibited the proliferation of several immortalized human prostatic epithelial cell lines. The proliferation of HaCaT- and RHEK-I-immortalized keratinocyte cell lines was relatively insensitive to CAI. Additionally, invasion by DU-145, PC3, and PPC-1 cells through Martigel in vitro was reduced approximately 60%–70% by 10 μM CAI. Other cellular effects of CAI included an attenuation of the elevation of intracellular free calcium in response to bombesin and carbachol in PC3 cells and a marked dose-dependent inhibition of PSA secretion in LNCa-P cell cultures.[94]

Several formulations have been used in the phase I setting. Early investigation of this compound used either a liquid or a liquid gelcap preparation. A phase I study at the NCI demonstrated a 49% (24/49) rate of disease stabilization using oral formulations.[95] In a concurrent phase I investigation using an oral preparation of CAI at the University of Wisconsin two patients with prostate cancer were observed to have disease stabilization.[96] The oral formulations, however, were not well tolerated due to compliance limiting gastrointestinal toxicity. An encapsulated micronized powder formulation was, therefore, prepared. In a phase I study at the NCI using the micronized formulation, 47% (10/21) experienced disease stabilization, and the formulation was found to be generally well tolerated, but with dose-limiting toxicities of cerebellar ataxia, and confusion.[97] These dose-limiting toxicities occurred at a CAI dose of 350 mg/m^2/day, and thus the maximum tolerated dose was determined to be 300 mg/m^2/day.

Bauer et al. conducted a phase II clinical trial using pharmacokinetic assessment to guide drug dosing. Fifteen patients with stage D2 androgen-independent prostate cancer, with soft tissue metastases, were enrolled. Since CAI previously had been shown to decrease PSA secretion in vitro, this marker was not used to assess disease status. The dose of CAI used in the study was calculated such that plasma steady state maximum concentrations between 2.0 and 5.0 μg/mL would be maintained. Following the initial dosage adjustment, 93% (14 of 15) of patients were within the predicted range. All of the 14 evaluable patients demonstrated progressive disease at approximately 2 months. Twelve patients progressed by CT and/or bone scan at 2 months, while two patients demonstrated clinical progression at 1.5 and 2 months. One patient was removed from the study at 6 weeks due to grade II peripheral neuropathy lasting greater than 1 month. The conclusion was that CAI did not possess clinical activity in patients with androgen-independent prostate cancer and soft tissue metastases. Pharmokinetically guided dosing was not found to be of practical benefit and grade III toxicity was still observed, resulting in drug discontinuation.[98]

27.3.8 SU5416 (Semaxanib)

SU5416 is a novel synthetic compound, a potent and selective inhibitor of the Flk-1/KDR tyrosine kinase receptor.

SU5416 was initially identified and characterized by Fong et al. SU5416 was shown to inhibit the mitogenic response of (HUVECs) stimulated with VEGF. In contrast, SU5416 did not inhibit the in vitro proliferation of a variety of tumor cells in a complete growth medium. The findings confirmed that endothelial cells are the target of SU5416 activity. In in vivo studies, SU5416 was found to be efficacious against a range of tumor types including LNCAP in subcutaneous xenograft models (BALB/c nu/nu female mice).[99,100]

Of interest, Huss et al. examined the effect of SU5416 in a "progression switch" model of prostate cancer (autochthonous TRAMP model). The authors had previously demonstrated that the expression of VEGFR-1 is associated with early and more differentiated disease, whereas the expression of VEGFR-2 is associated with advanced and more poorly differentiated disease. To test the hypothesis that stage-specific inhibition of VEGF signaling could be used as therapy for autochthonous prostate cancer, SU5416 was administered to TRAMP mice earlier in the course of disease (between 10 and 16 weeks of age) and later in the course (between 16 and 22 weeks). In the early intervention trial, administration of SU5416 to TRAMP mice did not influence angiogenesis or tumor progression as opposed to late intervention trial, during which a significant decrease in tumor-associated mean vessel density, increased apoptotic index, and pronounced regions of cell death were observed.[101]

Phase I studies testing SU5416 in advanced cancer have performed.[102,103] Rosen et al. reported the first use of SU5416 in humans. Investigators tested SU5416 in 63 patients with a variety of advanced malignancies. While there were no reported responses among the patients, seven patients were noted to have disease stabilization for a period of more than 6 months.[102] In another phase I study, investigators tested SU5416 in 22 patients with advanced cancer. One patient reportedly had a partial response and three other patients were noted to disease stabilization for a period of, at least, 12 weeks. The toxicities reported in both studies primarily consisted of headaches, nausea, and vomiting.[103] Of note, the two studies examined different methods of drug administration. In the first phase I study, SU5416 was dosed twice weekly, whereas in the second study, the drug was administered weekly after a 5 day loading period. In each study, the dose of 145 mg/m^2 (either twice weekly or once weekly following a loading period) was recommended for further testing.

It is notable that SU5416 has been studied in combination with other cytotoxic chemotherapies. In a phase I trial of SU5416 in combination with cisplatin and gemcitabine, one patient with advanced prostate cancer was reported to have stable disease; however, 8 of 19 patients on that trial developed thromboembolic events. While the authors noted that an increased frequency of thromboembolic events had not been seen in all other combinations of SU5416 and cytotoxic chemotherapies, the authors concluded that further testing of the regimen should be discouraged.[104]

Stadler et al. proceeded to test SU5416 in patients with HRPC. The rationale for using SU5416 in the prostate cancer population was the reported findings that VEGF levels correlated with patient outcomes in patients with localized and advanced prostate cancer.[24,26,28,36] A total of 36 patients were accrued to the study and randomized to treatment with dexamethasone alone or in combination with SU5416 (145 mg/m^2 twice weekly). Of note, the study included a cross-over arm for the patients who progressed on the control arm. The investigators found no effect of SU5416 on either PSA secretion or time to progression (TTP). The toxicity included headaches and fatigue as well as the effects attributable to the dexamethasone and the required central line. Given those findings, the authors stated that further evaluation of SU5416 in prostate cancer patients was unwarranted.[105]

27.3.9 Atrasentan (ABT-627)

Atrasentan is a selective endothelin A receptor antagonist that blocks the proliferative effects of endothelin-1 (ET-1).

The endothelin family of peptides has recently been identified as contributing to the pathophysiology of prostate cancer.[106] ET-1, acting primarily through the endothelin A (ET-A) receptor, is integrally involved in multiple facets of prostate cancer progression, including cell growth, inhibition of apoptosis, angiogenesis, development, and progression of bone metastases, and mediation of the pain responses.[107] In the normal prostate gland, ET-1 is produced by epithelial cells and appears in high concentrations in the seminal fluid.[108,109] In prostate cancer, key components of the ET-1 clearance pathway, the endothelin-B (ET-B) receptors, and the degradative enzyme neutral endopeptidase[110] are diminished, resulting in an increase in local ET-1 concentrations. Expression of the ET-A receptor also increases with tumor stage and grade in prostate cancer.[111]

There are multiple pathways by which the ET-1/ET-A axis may promote prostate cancer progression. ET-1 is a mitogen for prostate cancer cell lines in vitro and acts synergistically with other peptide growth factors.[112] ET-1 is also a mitogen for osteoblasts, the cell type pivotal in the hallmark osteoblastic response of bone to metastatic prostate cancer.[106,113] In addition, ET-1 modulates apoptosis, nociception, and blood flow indicating other potential benefits of ET-A receptor antagonism in prostate cancer.[114–116]

Patients with HRPC have been the subjects of two phase I trials of atrasentan. Zonnenberg et al. investigated atrasentan in 20 patients with HRPC. The agent was well tolerated with no maximum tolerated dose identified. The authors did not comment on whether the drug had any antitumor effect.[117] A second phase I study included 15 patients with advanced prostate cancer. In that study, atrasentan was reported to have antitumor effects with patients experiencing declines in PSA levels. The toxicities were again noted to be mild.[118] The most commonly experienced toxicities reported in both studies included rhinitis, headaches, peripheral edema, mild anorexia, and fatigue.

The results of phase II studies of atrasentan in patients with HRPC have been published. The patients included on one study had disease-related pain, whereas the patients on a second study were asymptomatic. In both studies, patients were randomized to receive either one of two doses of atrasentan (2.5 or 10 mg) or a placebo. In the first study, 131 patients were evaluated for changes in disease-related pain and tumor markers as well as for safety. The authors reported a trend toward improvement in pain and analgesia consumption. The authors also noted an alteration in the rate of rise in PSA in the active treatment arms.[119] The primary objective in the second study was time to disease progression. Two hundred and eighty-eight patients were enrolled and treated on one of the three treatment arms. The authors found a trend in TTP favoring treatment with 10 mg of atrasentan as compared with placebo. Those results became significant when the authors evaluated the TTP in only the evaluable patients.[120] The effects of atrasentan on bone deposition (total alkaline phosphatase and bone alkaline phosphatase) and resorption (N-telopeptides, C-telopeptides and deoxypyridinoline) markers were also examined. In patients receiving the placebo, patients were seen to have significant increases in bone deposition markers as compared with those patients receiving 10 mg of atrasentan. The changes in bone resorption markers were also seen to favor treatment. Consistent with the marker responses, atrasentan demonstrated a trend toward a slower progression of lesions as measured by a bone scan index (a quantification of bone scan results for comparison) than placebo.[121]

The results of a phase III trial of atrasentan in asymptomatic patients with HRPC were recently published. In the study, 809 patients were randomized to receive either a placebo or 10 mg of atrasentan. The findings were similar to the phase II findings in that the time to disease progression trended in favor of treatment for all patients and were significant when examined in the protocol-complaint subset. In addition, the effects of atrasentan on bone markers were also investigated and both the mean increase in bone and total alkaline phosphatase were found to be less than in the placebo group.[122]

Of interest, Carducci et al. presented a meta-analysis of all patients in the phase II and III trials at ASCO in 2004 and again in 2005.[122,123] In the meta-analysis, Carducci reported a significant delay in time to disease progression when the intent to treat populations from both studies was combined. Based on the results of that meta-analysis, atrasentan is to be considered for FDA approval in patients with HRPC.

Given the role of ET-1 in the development and progression of osteoblastic bone metastases and its association with pain responses, it has been speculated that treatment with atrasentan could improve pain control and quality of life in patients with HRPC. In a phase II study of atrasentan in patients with HRPC and disease-related pain, atrasentan resulted in a nonsignificant trend toward improved pain and analgesia use.[119] In two other studies, atrasentan's impact on quality of life was assessed. Singh et al. reported the effect of atrasentan on patients' quality of life, which was assessed using The European Organization for Research and Treatment of Cancer (EORTC) questionnaires (QLQ C-30 and The Functional Assessment of Cancer Therapy—Prostate (FACT-P)). The results of 288 patients with HRPC who had been treated with either placebo or two different doses of atrasentan were presented. The quality-adjusted TTP (defined as the interval from initiation of therapy to the time of documented clinical progression adjusted by the patients Quality of Life (QoL) score) for patients treated with 10 mg atrasentan was significantly longer than the placebo group for most domains of EORTC and FACT and showed strong trends in favor of treatment for other domains. Similar results were observed in comparing subjects treated with 2.5 mg atrasentan to the placebo group.[124] In a second study, Yount et al. assessed the effect of atrasentan on quality of life using the Prostate Cancer Subscale (PCS) and FACT Advanced Prostate Symptom Index (FAPSI) scores. The study included those patients on the phase III study of atrasentan in patients with HRPC.[123] Patients treated with placebo were found to have a greater deterioration in both the PCS and FAPSI than those treated. In a subset analysis, it was found that the difference was greater for those with bone metastases.[125]

27.3.10 Bevacizumab (anti-VEGF rhuMAb)

Bevacizumab is a recombinant humanized anti-VEGF monoclonal antibody composed of human IgG1 framework regions and antigen-binding complementarity-determining regions from a murine monoclonal antibody (muAB VEGF A.4.6.1) that blocks the binding of human VEGF to its receptors. Approximately 93% of the amino acid sequence, including most of the antibody framework, is derived from human IgG1 and approximately 7% of the sequence is derived from the murine antibody.

Borgstrom et al. initially evaluated the anti-VEGF antibody (A4.6.1) in a prostate cell line (DU145). The antibody was evaluated for effects on the growth and angiogenic activity of spheroids of the human prostatic cell line DU145 implanted subcutaneously in nude mice. In the control animals, tumors induced angiogenesis with high vascular density, as opposed to the animals treated with the anti-VEGF antibody in which a complete inhibition of tumor neovascularization was observed.[126]

The effect of VEGF inhibition on primary tumor growth and metastases in an in vivo model of prostate cancer (DU145 subcutaneously injected in a C.B.-17 scid/scid mice) was examined in another study. The study demonstrated suppression of primary tumor growth as well as inhibition of metastatic dissemination to the lung. In addition, the treatment was found to inhibit the progression of primary tumors as well as the progression of metastatic disease even when treatment was delayed until after the tumors were well established.[127] No significant toxicities were reported in either study.

Bevacizumab has been studied in a number of trials both as a single agent and in combination with other cytoxic chemotherapies in a variety of cancers. Reese et al. presented the findings of 15 patients with advanced prostate cancer who had been treated with bevacizumab. In the study, patients were treated with 10 mg/kg of bevacizumab every 14 days for 6 infusions (one cycle) followed by additional treatment for selected patients with a response or stable disease. Of the 14 evaluable patients, seven were reported to have stable disease. No patients were found to have an objective response and none achieved a PSA response (50% decrease in serum PSA). The authors also reported no significant changes on the McGill Present Pain Intensity Index. The toxicities were reportedly mild, with the most frequently observed toxicity being asthenia in 6 of 15 patients.[128]

Bevacizumab was also examined in combination with docetaxel and estramustine in patients with HRPC. The early results of a phase II trial reported that 9/17 patients (53%) were reported to have a PR and 13/20 (65%) had a PSA response (>50% decline in PSA). Toxicities included a high rate of uncomplicated neutropenia with only one reported case of febrile neutropenia. While no episodes of bleeding were reported, one patient had a deep vein thrombosis and another patient died of a mesenteric vein thrombosis.[129] While the results appear promising, whether the findings will be significant is, as of yet, unanswered. Given the favorable appearance of the data and the recent positive results of combined therapy with bevacizumab in patients with metastatic colorectal cancer (ref), a phase III trial is presently ongoing (CALBG 9040) in which patients are randomized between placebo and bevacizumab in combination with docetaxel and prednisone.

27.3.11 Thalidomide

Thalidomide (*N*-phthalidoglutarimide) was originally marketed as a sedative in the 1950s. It gained notoriety for its teratogenic effects and was withdrawn from the market about 30 years ago. The compound was later discovered to be effective in lepromatous leprosy and received USFDA approval in 1998 for the treatment of leprosy.

The antiangiogenic activity of thalidomide was initially described in a rabbit cornea model.[130] Thalidomide was shown to inhibit bFGF-induced angiogenesis. A metabolite of thalidomide was subsequently demonstrated to be responsible for this antiangiogenic activity.[131]

Of interest, the effect of thalidomide was tested on the secretion of PSA in human prostate cell lines (LNCaP). The results demonstrated an increase in PSA/cell levels at all concentrations tested as compared with untested control cells. In the same study, the effect of thalidomide was tested on the growth and viability of prostate cell lines (LNCaP and PC-3). Thalidomide was found to have cytostatic effects on the LNCaP cell line and no effect on the PC-3 cell line.[132]

Thalidomide has been studied as a single agent in patients with HRPC in two clinical trials. In a phase II study, Figg et al. randomized 63 patients to either a low-dose thalidomide arm (200 mg/day) or a dose escalation arm (initial dose of 200 mg/day escalating to 1200 mg/day as tolerated). The investigators reported a PSA response (decline of >50%) in 18% of the patients on the low-dose arm and in none of the patients on the high-dose arm. It was notable that four patients were maintained on treatment for >150 days. The most frequent side effects included constipation, dizziness, edema, fatigue, mood changes, and peripheral neuropathy.

In their conclusions, the authors noted that a total of 27% of all patients had a decline in PSA of >40%, which was often associated with an improvement in clinical symptoms.

The authors stated that the evidence suggested a modest antitumor effect for thalidomide in prostate cancer patients with hormone-refractory disease.[133]

In a second study, Drake et al. also examined thalidomide as a single agent in 20 patients with HRPC. The patients were treated with 100 mg of thalidomide daily. The results were not dissimilar to the results reported by Figg et al. Three men (15%) demonstrated a PSA response (>50% PSA decline). The authors also reported that of the 16 men treated for at least 2 months, 6 (37.5%) showed a fall in absolute PSA by a median of 48%. The side effects encountered included constipation, drowsiness, dizziness, and rash. The authors concluded that thalidomide appeared to have an effect on this patient population.[134]

Thalidomide has been used in combination with cytotoxic chemotherapy in three reported phase II trials. Shevrin et al. reported the results of a combination of mitoxantrone, prednisone, and thalidomide in 15 patients with HRPC. Objective responses (PR) were observed in five patients (33%). A PSA response (>50% PSA reduction) was seen in four patients. The most commonly experienced toxicities included constipation and somnolence. Five patients experienced grade 3 neutropenia, three patients experienced peripheral neuropathy, and two developed venous thromboses. While the study population was small, the author noted that the response rate was similar to that reported when mitoxantrone and prednisone were used alone without thalidomide and the combination only resulted in additional toxicities.[135] Frank et al. reported the results of a combination of 6 months of docetaxel, estramustine, and thalidomide followed by maintenance thalidomide for an additional 6 months (TET regimen). The study included only nine patients with HRPC. Of note, coumadin (2 mg/day) was given as prophylaxis during the course of the study. The authors found that six patients had a PSA response (>50% PSA reduction) and of five patients with measurable disease, two had a partial response. Of the two patients that entered the maintenance phase of treatment, one completed that phase and experienced an additional 9 months of stable disease. The only grade 3 toxicities reported included asthenia in two patients and constipation in one patient. Of note, one patient experienced a deep venous thrombosis despite the addition of prophylactic coumadin during the study.[136]

Dahut et al. studied the combination of docetaxel and thalidomide in 75 patients with HRPC. In the study, patients were randomized to receive either docetaxel as a single agent or docetaxel in combination with thalidomide. The results were favorable for the combined treatment arm when compared with the docetaxel-alone arm; PSA response (53% versus 37%, respectively), median progression-free survival (5.9 versus 3.7 months), and 18 month survival (68.2% versus 42.9%). However, none of those differences were statistically significant. The most concerning toxicities reported were thromboembolic incidences on the combined treatment arm. Twelve of the first 43 patients, all on the combined treatment arm, experienced thromboembolic events (venous thrombosis in nine patients and either TIAs or stroke in three other patients). Due to that finding, subsequent patients enrolled on the combined treatment arm were offered prophylactic anticoagulation with low-molecular-weight heparin. While, the results did not reach statistical significance, the findings were felt to be promising.[137] A phase III trial of thalidomide in combination with docetaxel is currently underway.

Several thalidomide analogs are currently in development. One of those analogs, CC-4047, has been tested in patients with HRPC. The preliminary results of a phase II trial indicated that the agent may have some activity in patients with advanced prostate cancer and has manageable toxicities.[138]

27.3.12 2-Methoxyestradiol (Panzem)

2-Methoxyestradiol (2-ME) is an endogenous metabolite of estrogen hormone, 17 B-estradiol that has been shown to potently inhibit angiogenesis and tumor growth both in vitro and in vivo.[139] The growth inhibitory properties of 2-ME are not completely understood but have been associated with effects on tubulin polymerization as well as an induction of apoptosis potentially through the activation of p53, death receptor 5 or inhibition of HIF-1 alpha, a key angiogenic transcription factor.[140–145] In addition, 2-ME has been found to effect the levels and activity of a number of cofactors of DNA replication and repair such as proliferating cell nuclear antigen (PCNA), cell division cycle kinases and regulators, for example, p34 and cyclin B, and transcription factor modulators, for example, SAPK/JNK.[146–150]

Preclinical data examining 2-ME in prostate cancer models has been reported in three studies. Kumar et al. reported the in vitro inhibition of LNCaP and DU145 cell proliferation by 2-ME. The authors also described the effects of 2-ME including a twofold increase in the G2/M population and induction of apoptosis. Of note, the inhibitory effects of 2-ME were found to be p53 independent.[151] In a second study, 2-ME was investigated both in vitro models (LNCaP, DU145, PC-3, and ALVA-31 prostate cancer cells) as well as in an in vivo mouse model (Gy/T-15 transgenic mouse). The findings demonstrated in vitro inhibition of prostate cancer cells as well inhibition of tumor progression in the mouse model. The authors also noted an accumulation of cells in the G2/M phase of the cell cycle as well as evidenced of apoptosis in vitro.[152] Bu et al. reported the effects of 2-ME in prostate cells (PC-3, LNCaP, and DU-145) as well as a rat prostate tumor model (Dunning R3327-PAP prostate tumors subcutaneously injected in a Copenhagen x Fisher F1 male rat). 2-ME was found to induce apoptosis in the prostate carcinoma cell lines as well inhibit the growth of the rat prostate tumor. In addition, the authors found that 2-ME treatment led to activation of c-Jun

N-terminal kinase (JNK) and phosphorylation of Bcl-2. Those events were noted to precede the induction of apoptosis, which led the authors to conclude that Bcl-2 expression is important for 2-ME induced apoptosis.[153]

2-ME has been studied in two phase I trials. In the first trial, 2-ME was studied exclusively in patients with metastatic breast cancer. The MTD had not been reached at the time the results were presented; however, investigators had noted disease stability and a reduction in bone pain and analgesic use in some of the patients.[154] In a second phase I trial, 2-ME was administered to patients with advanced solid tumors including a patient with prostate cancer. Toxicities were reportedly mild with the exception of one patient with grade 4 angioedema. Of interest, the patient with prostate cancer was noted to have disease stabilization for a period of 10 months.[155]

2-ME has been examined in patients with HRPC as a part of a phase II trial. In the trial, 33 men were randomized between two doses of 2-ME (400 and 1200 mg/day). The most notable toxicity related to therapy was an elevation in hepatic transaminases in three patients (one grade 2 and two grade 3). Other reported toxicities included two patients with deep venous thromboses. While there were no serological or objective responses reported, the PSA was reported to decline between 21% and 40% in seven patients in the 1200 mg group and in one patient in the 400 mg group. The higher dose group also showed significantly decreased PSA velocity ($p = 0.037$) and compared with the 400 mg dose, a longer median time to PSA progression (109 versus 67 days, $p = 0.094$) and a longer time on study (126 versus 61 days, $p = 0.024$). The authors described the change in PSA velocity as a "broken arrow phenomenon" resulting in disease stabilization. The oral bioavailability of the drug was found to be limited, however, and the drug has since been reformulated and is currently evaluated in phase I studies.[156]

27.4 PHASE I

A number of angiogenesis inhibitors are in clinical development. The following agents represent those agents for which phase I testing in men with HRPC has been presented.

27.4.1 PTK787/ZK 222584 (PTK/ZK)

PTK/ZK is a synthetic, low-molecular-weight molecule designed to potently and selectively block the VEGF/VEGF receptor tyrosine kinases. The pharmacologic profile of the inhibitor was initially described by Wood et al. The agent was noted, more specifically, to have activity against KDR, the human VEGF-R2 with slightly weaker inhibition of Flt-1, the human VEGF-R1 in addition to the kinases of other receptors. Of interest, the agent is an orally administered agent. The agent was shown to inhibit tumor growth in several different tumor models including a prostate tumor models (PC-3, DU145, CWR-22). Actually, the best effects were seen against the prostate tumors, particularly the CWR-22 model.[157]

PTK/ZK has been tested in patients with HRPC as part of a phase I study examining the effect of high-fat meals on the pharmacokinetics of the agent. The most common adverse events associated with PTK/ZK included nausea, dizziness, and headaches, all of which were reported in more than 50% of participants. Grade 3 adverse events including hypertension and syncope, were infrequent, consisting of only about 5% of reported events. The food effect was reported not to be clinically significant and three patients were found to have PSA reductions of >40%.[158]

27.4.2 AZD2171

AZD2171 is an indole ether quinazoline derivative and a potent adenosine triphosphate (ATP) competitive inhibitor of recombinant VEGFR-1 and VEGFR-2 tyrosine kinase activity.

It is also an oral agent. The agent was initially described by Hennequin et al.[159] In subsequent studies, AZD2171 was reported to demonstrate inhibition of tumor growth in a number of human tumor xenografts including prostate.[160]

The preliminary results of a phase I trial of AZD2171 in patients with advanced prostate cancer have been reported. The results for 20 patients were included. The drug was found to be well tolerated with adverse events reported as grade 1–2 with the exception of grade 3 toxicities in two patients including fatigue, weakness, and hypertension. No serologic or objective responses had been seen. The study was noted to be ongoing and no recommended dose was included.[161,162]

27.4.3 ZD4054

ZD4054 is an endothelin A receptor antagonist. Curwen et al. initially described the ZD4054 noting that the agent was found to specifically bind to the endothelin A receptor without displacement of endothelin-1 from the endothelin B receptor. The drug was also reported to be orally absorbed, with a favorable toxicity profile.[163]

A phase II trial of ZD4054 in 16 patients with advanced prostate cancer demonstrated that the medication was well tolerated. The most frequent adverse events included headache, peripheral edema, fatigue, and nasal congestion. Dose-limiting toxicities occurred at 22.5 mg and the recommended daily dose was 15 mg.[164] The observed toxicities support specific ET-A antagonism and thus further evaluation of this agent in prostate cancer is planned.

27.5 CONCLUSION

Despite the early optimism regarding potential of angiogenesis inhibitors in the treatment of cancer, the results for patients with HRPC are at best, modest. A number of agents have shown no significant responses in this patient population including the MMPIs, TNP-470, CAI, and SU5416.[71,75,76,88,98,105] While early studies with suramin were encouraging, phase III studies demonstrated only minor palliative benefits and an oncology advisory board for the FDA ultimately recommended against approval of the agent for men with advanced prostate cancer.[56] Other agents including atrasentan, bevacizumab, thalidomide, and 2-ME continue to be evaluated in ongoing clinic trials.[120–123,128,129,133–137,156] In summary, the progress of angiogenesis inhibitors in treating patients with HRPC to date has been disappointing.

Given the success of angiogenesis inhibitors in other cancers, the question then stands; why has not a significant benefit been appreciated in prostate cancer patients?[165–168] With the exception of bevacizumab and thalidomide, none of the agents presented has generated significant results in other cancers. Is it, therefore, reasonable to suggest that the negative results seen in a number of these trials could be attributed to ineffective agents rather than a disease resistant to therapy with angiogenesis inhibitors? Or, are we actually abandoning effective therapies due to our failure to measure the appropriate response end point due to particularities of this disease? These and many other questions need to be addressed before committing further valuable resources into phase II and III trials.

27.5.1 Measurement of Response

The primary goal in assessing novel compounds is to be able to quickly test each agent for early signs of significant drug activity, thus selecting out agents that would be most promising in the context of a larger phase III given the limited patient resources available. While this may sound simple enough, the truth is that the majority of compounds entering phase III testing are eventually found to be ineffective. Why is this so? In most situations, during phase

II testing we are focused on objective responses (i.e., radiographic responses), and thus tend to assume that agents that cause objective tumor regression are the ones most likely to be promising in phase III trials when one is now looking for a survival improvement. This has led to the routine use of objective response criteria that predominate clinical trials to date, as a surrogate marker for clinical benefit.[169] While acceptable, this surrogate marker may not be suitable in all solid tumors and with novel, noncytotoxic agents.

The difficulty in measuring response in HRPC is augmented by the fact that most patients have disease that has primarily metastasized to the bone. It is well known that these bone scintigraphy positive lesions are difficult to evaluate and slow to regress. In addition, only a small subset of patients have soft tissue lesions readily assessable by standard unidimensional or bidimensional measurement criteria, and thus limiting studies to this population only would eliminate up to 80% of patients with HRPC.[170] To make up for this difficulty in assessing response, consensus guidelines have been established supporting the use of PSA as a measurement of outcome.[171] The advantages of using a serum marker for response are severalfold: (1) it is convenient and easy to frequently follow a serum PSA level; (2) it allows assessment of patients with bone only metastatic HRPC; and (3) it is elevated in nearly 95% of patients with metastatic prostate cancer.[170]

Does a PSA response predict clinical benefit? Is a PSA response rate the most appropriate end point for this disease in a phase II trial? If one believes that PSA is reflective of tumor burden, then the answer is clearly yes. In the case of cytotoxic chemotherapy, we would expect to see a drop in PSA in proportion to decreases in tumor burden. However, we are only now beginning to realize that PSA responses are not entirely predictive of overall survival. For example, the TAX 327 study assessed docetaxel chemotherapy (weekly and every 3 week schedule) plus prednisone to mitoxantrone plus prednisone and showed an overall survival using every 3 week docetaxel of 18.9 months, weekly docetaxel of 17.4 months, and mitoxantrone of 16.5 months, with a PSA response rate of 45%, 48%, and 32%, respectively. Despite the similar response rates in the both docetaxel arms, only the every 3 week arm survival was statistically superior to mitoxantrone ($p = 0.009$).[5] Using a regression analysis, PSA response accounted for <50% of the treatment effect (every 3 week docetaxel versus mitoxantrone) on overall survival when PSA response was included in the Cox model implying that PSA response alone cannot replace overall survival in phase III trials in men with HRPC, and that a lack of a PSA response does not mean no benefit to treatment.[172] While the consensus criteria allow more uniformity in reporting PSA response, it is important to remember that some of these novel agents (i.e., TNP-470) may in fact modulate PSA (up- or downregulate) independent of its effects on tumor cell proliferation.[87] Therefore, the use of PSA, a surrogate marker for response, may be leading to the abandonment of many active agents assessed in phase II trials in HRPC, and help explain the lack of activity observed in early HRPC trials with some of these agents to date.

How should the effects of angiogenesis inhibitors best assessed? As angiogenesis inhibitors exert a cytostatic effect on tumor growth, the clinical effect of these drugs would be tumor stabilization rather than shrinkage, which would be the goal of cytotoxic therapy. This has been illustrated by the recent trial of bevacizumab in metastatic renal cell cancer, in which there was a significant increase in the TTP in patients treated with bevacizumab without significant differences in response.[173] Given the cytostatic nature of many of these agents, is TTP a suitable end point to consider in HRPC studies? As prostate cancer is a heterogeneous disease as reflected by a number of variables including variable PSA doubling times, different Gleason scores, and location of metastasis (bone and visceral), there can be marked variation in TTP (both clinical and radiographical). With the variable natural history of prostate cancer, the effects of different therapies can be difficult to discern without large randomized studies, or risk-stratifying patients based on known prognostic variables. Perhaps, alternative end points like changes in PSA doubling time (pretreatment versus treatment) and PSA slope may become more meaningful in the assessment of angiogenesis inhibitors in HRPC. Although improving overall survival is ultimately the end point of choice, this is not feasible in the setting of a phase II trial. What is needed is a better way to assess drug activity, and thus help us narrow down which agents would be worthy of phase III testing in order to answer the survival issue.

All of the above underscores the need to develop suitable biomarkers of antiangiogenic activity. These may include the examination of MVD, the measurement of angiogenic factors in the body fluids (e.g., VEGF), and radiologic imaging of tumor perfusion (e.g., DCE-MRI) and metabolic activity (e.g., positron emission tomography, PET). Likewise, assessments of other markers of angiogenesis including circulating endothelial cells may be promising.[174] Until a suitable biomarker can be identified, then optimization of dose and schedule cannot readily be performed and many agents possibly were abandoned because of presumed inactivity.

As discussed earlier, MVD has been associated with tumor stage and grade as well as cancer recurrence and cancer-specific survival.[10–15] Previous studies have compared vascular densities in radical prostatectomies from patients receiving neoadjuvant hormonal treatment with patients without prior treatment. In one study the vascular density was decreased whereas in the other study it was unaffected.[175,176] Certainly, further studies are needed in order to answer the question whether an estimate of tumor angiogenesis is an indicator of therapeutic effect in patients with prostate cancer. However, the technical difficulty not to mention the inconvenience to patients that would be involved in monitoring tumor progression by this method should be acknowledged.

Available data clearly shows that the levels of VEGF as well as other angiogenic factors are increased in localized or metastatic prostate cancer when compared with healthy controls and the idea of following those levels in addition to other angiogenic levels in the body fluids has been proposed as another method of monitoring tumor angiogenesis.[28,177,178] The results of a number of trials in which angiogenic factors were observed during the course of treatment have been equivocal.[36,98,133,134] While the data is conflicting, it should be emphasized that the angiogenic activity of a tumor is believed to be the sum total of positive and negative regulators. Therefore, it is not expected that quantification of a single factor in tumor tissue of body fluids would provide uniform prognostic information especially since both the quantity and the composition and inhibitors probably change during tumor progression.[179,180,181]

Noninvasive assessment of tumor vascularity by doppler flow signals, dynamic contrast-enhanced magnetic resonance imaging (MRI), or PET has been proposed as another alternative for the assessment of tumor vascularity and blood flow.[182,183] Increased prostate tumor blood flow has been related to tumor stage and grade in localized prostate cancer, the use of dynamic MRI has been shown to discriminate between normal and tumor tissue in the prostate.[184,185] Of great interest are the findings that prostate blood flow decreases earlier than the reduction in prostate volume after castration treatment and that blood vessel permeability decreases in the prostate after hormonal ablation therapy.[186,187] These findings suggest that changes in prostate blood flow and permeability may be indicators of therapeutic effects in prostate cancer patients. Further development of imaging as a surrogate for following the effects of angiogenesis inhibitors is in progress.

What is the future of angiogenesis inhibitors in HRPC? Certainly the success of bevacizumab in combination with irinotecan, fluorouracil, and leucovorin for patients with metastatic colorectal cancer validates the idea that these agents may be effective in combination with cytotoxic chemotherapy.[165] While angiogenesis plays a role in the growth and spread of prostate cancer, it is clear that these cancers use a number of complex disease mechanisms and pathways. Combining angiogenesis inhibitors and cytotoxic therapies, theorically then, allows for the targeting more of those mechanisms and pathways than either therapy would alone. Both bevacizumab and thalidomide have shown activity in patients with advanced prostate cancer in phase II trials in which those agents were combined with cytotoxic therapy: docetaxel plus estramustine and docetaxel, respectively. Phase III trials of those combinations are presently ongoing. As more angiogenesis inhibitors are developed for use in patients with prostate cancer, additional studies combining those agents with cytotoxic chemotherapy are likely to be designed. In addition, combinations with other angiogenesis inhibitors and radiation therapy have shown promise in preclinical studies.[188–196]

To date, studies of angiogenesis inhibitors in patients with prostate cancer have included only patients with advanced, hormone-refractory disease. It has been suggested that angiogenesis inhibitors might be most effective in treating patients with smaller tumor burdens and tumors that have not yet accumulated all the mutations of more advanced tumors. In future trials of angiogenesis inhibitors in patients with advanced cancer, it would be interesting to examine the effects of those agents in either the neoadjuvant or adjuvant setting, rising PSA only situation, as well as in hormone sensitive disease. As stated earlier, the natural history of prostate cancer is quite variable and in designing studies, it may be necessary to stratify patients by prognostic factors, thus allowing the assessment of alternative end points, and define who would most likely benefit from this type of therapy.

Certainly, angiogenesis inhibition represents an innovative approach to tumor control. The novel mechanisms by which these agents act and the generally favorable side effect profile of the agents make angiogenesis inhibitors a promising contribution to the development of better therapies for patients with prostate cancer.

REFERENCES

1. Jemal, A., et al., Cancer Statistics, 2005, *CA Cancer J. Clin.*, 55, 10, 2005.
2. Han, M., et al., Biochemical (prostate specific antigen) recurrence probability following radical prostatectomy for clinically localized prostate cancer, *J. Urol.*, 169(2), 517, 2003.
3. Prostate Cancer Trialists' Collaborative Group, Maximum androgen blockade in advanced prostate cancer: an overview of the randomized trials, *Lancet*, 355, 149, 2000.
4. Petrylak, D.P., et al., Docetaxel and estramustine compared with mitoxantrone and prednisone for advanced refractory prostate cancer, *N. Engl. J. Med.*, 351(15), 1513, 2004.
5. Tannock, I.F., et al., Docetaxel plus prednisone or mitoxantrone plus prednisone for advanced prostate cancer, *N. Engl. J. Med.*, 351(15), 1502, 2004.
6. Folkman, J., et al., Induction of angiogenesis during the transition from hyperplasia to neoplasia, *Nature*, 339, 58, 1989.
7. Liotta, L.A., Steeg, P.A., and Stetler-Stevenson, W.G., Cancer metastasis and angiogenesis: an imbalance of positive and negative regulation, *Cell*, 64, 327, 1991.
8. Nicholson, B. and Theodorescu, D., Angiogenesis and prostate cancer tumor growth, *J. Cell. Biochem.*, 91, 125, 2004.
9. Weidner, N., et al., Tumor angiogenesis correlates with metastasis in invasive prostate carcinoma, *Am. J. Path.*, 143(2), 401, 1993.
10. Strohmeyer, D., et al., Tumor angiogenesis is associated with progression after radical prostatectomy in pT2/pT3 prostate cancer, *Prostate*, 42(1), 26, 2000.
11. Offerson, B.V., Borre, M., and Overgaard, J., Immunohistochemical determination of tumor angiogenesis measured by the maximal microvessel density in human prostate cancer, *APMIS*, 106(4), 463, 1998.
12. Bettencourt, M.C., et al., CD34 immunohistochemical assessment of angiogenesis as a prognostic marker for prostate cancer recurrence after radical prostatectomy, *J. Urol.*, 160(2), 459, 1998.
13. Mydlo, J.H., An analysis of microvessel density, androgen receptor, p53 and HER-2/neu expression and Gleason score in prostate cancer. Preliminary results and therapeutic implications, *Eur. Urol.*, 34(5), 426, 1998.

14. de la Taille, A., et al., Microvessel density as a predicator of PSA recurrence after radical prostatectomy. A comparison of CD34 and CD31, *Am. J. Clin. Pathol.*, 113(4), 555, 2000.
15. Halvorson, O.J., et al., Independent prognostic importance of microvessel density in clinically localized prostate cancer, *Anticancer Res.*, 20(5C), 3791, 2000.
16. Borre, M., et al., Microvessel density predicts survival in prostate cancer patients subjected to watchful waiting, *Br. J. Cancer*, 78(7), 940, 1998.
17. Ferrer, F.A., et al., Angiogenesis and prostate cancer: in vivo and in vitro expression of angiogenesis factors by prostate cancer cells, *Urology*, 51(1), 161, 1998.
18. Joseph, I.B. and Issacs, J.T., Potentiation of the antiangiogenic ability of linomide by androgen ablation involves down-regulation of vascular endothelial growth factor in human androgen-responsive prostatic cancer, *Cancer Res.*, 57, 1054, 1997.
19. Stewart, R.J., Vascular endothelial growth factor cxpression and tumor angiogenesis are regulated by androgens in hormone responsive human prostate carcinoma: evidence for androgen dependent destabilization of vascular endothelial growth factor transcripts, *J. Urol.*, 165(2), 688, 2001.
20. Sordello, S., Bertrand, N., and Plouet, J., Vascular endothelial growth factor is upregulated in vitro and in vivo by androgens, *Biochem. Biophys. Res. Commun.*, 251, 287, 1998.
21. Jackson, M.V., Bentel, J.M., and Tilley, W.D., Vascular endothelial growth factor (VEGF) expression in prostate cancer and benign prostatic hyperplasia, *J. Urol.*, 157(6), 2323, 1997.
22. Ferrer, F.A., et al., Expression of vascular endothelial growth factor receptors in human prostate cancer, *Urology*, 54, 567, 1999.
23. Ferrer, F.A., et al., Vascular endothelial growth factor (VEGF) expression in human prostate cancer: in situ and in vitro expression of VEGF by human prostate cancer cells, *J. Urol.*, 157(6), 2329, 1997.
24. Borre, M., Nerstrom, B., and Overgaard, J., Association between immunohistochemical expression of vascular endothelial growth factor (VEGF), VEGF-expressing neuroendocrine-differentiated tumor cells, and outcome in prostate cancer patients subjected to watchful waiting, *Clin. Cancer Res.*, 6, 1882, 2000.
25. Small, E.J., et al., Suramin therapy for patients with symptomatic hormone refractory prostate cancer: results of a randomized phase III trial comparing suramin plus hydrocortisone to placebo plus hydrocortisone, *J. Clin. Oncol.*, 18, 1440, 2000.
26. George, D.J., et al., Prognostic significance of plasma vascular endothelial growth factor levels in patients with hormone-refractory prostate cancer treated on cancer and leukemia group B9480, *Clin. Cancer Res.*, 7, 1932, 2001.
27. Kohli, M., et al., Prospective study of circulating angiogenic markers in prostate-specific antigen (PSA)-stable and PSA-progressive hormone-sensitive advanced prostate cancer, *Urology*, 61(4), 765, 2003.
28. Duque, J.L., et al., Plasma levels of vascular endothelial growth factor are increased in patients with metastatic prostate cancer, *Urology*, 54, 523, 1999.
29. Shariat, S.F., et al., Association of preoperative plasma levels of vascular endothelial growth factor and soluble vascular cell adhesion molecule-1 with lymph node status and biochemical progression after radical prostatectomy, *J. Clin. Oncol.*, 22(9), 1655, 2004.
30. Chan, L.W., et al., Urinary VEGF and MMP levels as predictive markers of 1-year progression-free survival in cancer patients treated with radiation therapy: a longitudinal study of protein kinetics throughout tumor progression and therapy, *J. Clin. Oncol.*, 22, 499, 2004.
31. George, D.J., et al., Radical prostatectomy lowers plasma vascular endothelial growth factor levels in patients with prostate cancer, *Urology*, 63(2), 327, 2004.
32. Gaudric, A., et al., Quantification of angiogenesis due to basic fibroblast growth factor in a modified rabbit corneal model, *Ophthalmic Res.*, 24(3), 181, 1992.
33. Nakamoto, T., et al., Basic fibroblast growth factor in human prostate cancer cells, *Cancer Res.*, 52(3), 571, 1992.
34. Connolly, J.M. and Rose, D.P., Angiogenesis in two human prostate cancer cell lines with differing metastatic potential when growing as solid tumors in nude mice, *J. Urol.*, 160, 132, 1998.
35. Walsh, K., et al., Angiogenic peptides in prostatic disease, *BJU Int.*, 84(9), 1081, 1999.
36. Bok, R.A., et al., Vascular endothelial growth factor and basic fibroblast growth factor urine levels as predictors of outcome in hormone-refractory prostate cancer patients: a cancer and leukemia group B study, *Cancer Res.*, 61(6), 2533, 2001.

37. Wikstrom, P., et al., Transforming growth factor beta1 is associated with angiogenesis, metastasis, and poor clinical outcome in prostate cancer, *Prostate*, 37(1), 19, 1998.
38. Williams, R.H., et al., Reduced levels of transforming growth factor beta receptor type II in human prostate cancer: an immunohistochemical study, *Clin. Cancer Res.*, 2(4), 635, 1996.
39. Kim, I.Y., et al., Loss of expression of transforming growth factor beta type I and type II receptors correlates with tumor grade in human prostate cancer tissues, *Clin. Cancer Res.*, 2(8), 12555, 1996.
40. Kim, I.Y., et al., Loss of expression of transforming growth factor-beta receptors is associated with poor prognosis in prostate cancer patients, *Clin. Cancer Res.*, 4(7), 1625, 1998.
41. Tuxhorn, J.A., et al., Inhibition of transforming growth factor-beta activity decreases angiogenesis in a human prostate cancer-reactive stroma xenograft model, *Cancer Res.*, 62(21), 6021, 2002.
42. Koch, A.E., et al., Interleukin-8 as a macrophage-derived mediator of angiogenesis, *Science*, 258(5089), 1798, 1992.
43. Campbell, C.L., et al., Expression of multiple angiogenic cytokines in cultured normal human prostate epithelial cells: predominance of vascular endothelial growth factor, *Int. J. Cancer*, 80, 868, 1999.
44. Greene, G.F., et al., Correlation of metastasis-related gene expression with metastatic potential in human prostate carcinoma cells implanted in nude mice using an in situ messenger RNA hybridization technique, *Am. J. Path.*, 150(5), 1571, 1997.
45. Kim, S.J., et al., Expression of interleukin-8 correlates with angiogenesis, tumorigenicity, and metastasis of human prostate cancer cells implanted orthotopically in nude mice, *Neoplasia*, 3(1), 33, 2001.
46. Armand, J.P. and Cvitkovic, E., Suramin: a new therapeutic concept, *Eur. J. Cancer*, 26, 417, 1990.
47. Kaur, M., et al., Suramin's development: what did we learn? *Invest. New Drugs*, 20, 209, 2002.
48. Myers, C., et al., Suramin: a novel growth factor antagonist with activity in hormone-refractory metastatic prostate cancer, *J. Clin. Oncol.*, 10, 881, 1992.
49. Stein, C.A, et al., Suramin—an anticancer drug with a unique mechanism of action, *J. Clin. Oncol.*, 7, 449, 1989.
50. Thalmann, G.N., et al., Suramin-induced decrease in prostate specific antigen expression with no effect on tumor growth in the LNCaP model of human prostate cancer, *J. Natl. Cancer Inst.*, 88(12), 779, 1996.
51. Eisenberger, M.A., et al., Suramin, an active drug for prostate cancer: interim observations in a phase I trial, *J. Natl. Cancer Inst.*, 85(8), 611, 1993.
52. Kobayshi, K., et al., Phase I study of suramin given by intermittent infusion without adaptive control in patients with advanced cancer, *J. Clin. Oncol.*, 13, 2196, 1995.
53. Dawson, N.A., et al., Antitumor activity of suramin in hormone-refractory prostate cancer controlling for hydrocortisone treatment and flutamide withdrawal as potentially confounding variables, *Cancer*, 76, 453, 1995.
54. Rosen, P.J., et al., Suramin in hormone refractory metastatic prostate cancer: a drug with limited efficacy, *J. Clin. Oncol.*, 14, 1626, 1996.
55. Kelly, W.K., et al., Prospective evaluation of hydrocortisone and suramin in patients with androgen-independent prostate cancer, *J. Clin. Oncol.*, 13(9), 2008, 1995.
56. Small, E.J., et al., Randomized study of three different doses of suramin administered with a fixed dosing schedule in patients with advanced prostate cancer: results of intergroup 0159, Cancer and Leukemia Group B 9480, *J. Clin. Oncol.*, 20(16), 3369, 2002.
57. Ramnath, N. and Creaven, P.J., matrix metalloproteinase inhibitors, *Curr. Oncol. Reports*, 6, 96, 2004.
58. Pajouh, M.S., et al., Expression of metalloproteinase genes in human prostate cancer, *J. Cancer Res. Clin. Oncol.*, 117, 144, 1991.
59. Boag, A.H. and Young, I.D., Immunohistochemical analysis of type IV collagenase expression in prostatic hyperplasia and adenocarcinoma, *Mod. Pathol.*, 6(1), 65, 1993.
60. Stearns, M.E. and Wang, M., Type IV collagenase (M(r) 72,000) expression in human prostate: benign and malignant tissue, *Cancer Res.*, 53(4), 878, 1993.
61. Lokeshwar, B.L., et al., Secretion of matrix metalloproteinases and their inhibitors (tissue inhibitor of metalloproteinases) by human prostate in explant cultures: reduced tissue inhibitor of metalloproteinase secretion by malignant tissues, *Cancer Res.*, 53, 4493, 1993.

62. Hamdy, F.C., Matrix metalloproteinase 9 expression in primary human prostatic adenocarcinoma and benign prostatic hyperplasia, *Br. J. Cancer*, 69, 177, 1994.
63. Stearns, M. and Stearns, M.E., Evidence for increased activated metalloproteinase 2 (MMP-2a) expression associated with human prostate cancer progression, *Oncol. Res.*, 8(2), 69, 1996.
64. Stearns, M.E. and Stearns, M., Immunohistochemical studies of activated matrix metalloproteinase-2 (MMP-2a) expression in human prostate cancer, *Oncol. Res.*, 8(2), 63, 1996.
65. Festuccia, C., et al., Increased matrix metalloproteinase-9 secretion in short-term tissue cultures of prostate tumor cells, *Int. J. Cancer*, 69(5), 386, 1996.
66. Wood, M., et al., In situ hybridization studies of metalloproteinases 2 and 9 and TIMP-1 and TIMP-2 expression in human prostate cancer, *Clin. Exp. Metastasis*, 15(3), 246, 1997.
67. Upadhyay, J., et al., Membrane type 1-matrix metalloproteinase (MT1-MMP) and MMP-2 immunolocalization in human prostate: change in cellular localization associated with high grade prostatic intraepithelial neoplasia, *Clin. Cancer Res.*, 5, 4105, 1999.
68. Still, K., et al., Localization and quantification of mRNA for matrix metalloproteinase-2 (MMP-2) and tissue inhibitor of matrix metalloproteinase-2 (TIMP-2) in human benign and malignant prostatic tissue, *Prostate*, 42, 18, 2000.
69. Knox, J.D., et al., Synthetic matrix metalloproteinase inhibitor, BB-94, inhibits the invasion of neoplastic human prostate cells in a mouse model, *Prostate*, 1998, 248, 1998.
70. Lein, M., et al., Synthetic inhibitor of matrix metalloproteinases (Batimastat) reduces prostate cancer growth in an orthotopic rat model, *Prostate*, 43, 77, 2000.
71. Nemunaitis, J., et al., Combined analysis of studies of the effects of the matrix metalloproteinase inhibitor marimastat on serum tumor markers in advanced cancer: selection of a biologically active and tolerable dose for longer-term studies, *Clin. Cancer Res.*, 4, 1101, 1998.
72. Boasberg, P., et al., Marimastat in patients with hormone refractory prostate cancer: a dose-finding study, *Proc. Am. Clin. Oncol.*, Abstract No. 1126, 1997.
73. Vincenti, M.P. and Brinckerhoff, C.E., Transcriptional regulation of collagenase (MMP-1, MMP-13) genes in arthritis: integration of complex signaling pathways for the recruitment of gene-specific transcription factors, *Arthritis Res.*, 4, 157, 2002.
74. Shalinsky, D.R., et al., Broad antitumor and antiangiogenic activities of AG3340, a potent MMP inhibitor undergoing advanced oncology clinical trials, *Ann. NY Acad. Sci.*, 878, 236, 1999.
75. Hande, K.R., et al., Phase I and pharmacokinetic study of prinomastat, a matrix metalloproteinase inhibitor, *Clin. Cancer Res.*, 10, 909, 2004.
76. Ahmann, F.R., et al., Interium results of a phase III study of the matrix metalloprotease inhibitor prinomastat in patients having metastatic, hormone refractory prostate cancer, *Proc. Am. Clin. Oncol.*, Abstract No. 692, 2001.
77. Lombard, M.A., Synthetic matrix metalloproteinase inhibitors and tissue inhibitor of metalloproteinase (TIMP)-2, but not TIMP-1, inhibit shedding of tumor necrosis factor-receptors in a human colon adenocarcinoma (Colo 205) cell line, *Cancer Res.*, 58, 4001, 1998.
78. Rizvi, N.A., et al., A phase I study of oral BMS-275291, a novel nonhydroxamate sheddase-sparing matrix metalloproteinase inhibitor, in patients with advanced or metastatic cancer, *Clin. Cancer Res.*, 10(6), 1963, 2004.
79. Lara, P.N., et al., Angiogenesis inhibition in metastatic hormone refractory prostate cancer (HRPC): a randomized phase II trial of two doses of the matrix metalloproteinase inhibitor (MMPI)—BMS-275291, *Proc. Am. Clin. Oncol.*, Abstract No. 4647, 2004.
80. Ingber, D., et al., Synthetic analogues of fumagillin that inhibit angiogenesis and suppress tumour growth, *Nature*, 348, 555, 1990.
81. Sin, N., et al., The anti-angiogenic agent fumagillin covalently binds and inhibits and methionine aminopeptidase, MetAP-2, *Proc. Natl. Acad. Sci.*, 94(12), 6099, 1997.
82. Turk, B.E., et al., Selective inhibition of amino-terminal methionine processing by TNP-470 and ovalicin in endothelial cells, *Chem. Biol.*, 6, 823, 1999.
83. Yamaoka, M., et al., Angiogenesis inhibitor TNP-470 (AGM-1470) potently inhibits the tumor growth of hormone-independent human breast and prostate carcinoma cell lines, *Cancer Res.*, 53, 5233, 1993.
84. Miki, T., et al., Angiogenesis inhibitor TNP-470 inhibits growth and metastasis of a hormone-independent rat prostatic carcinoma cell line, *J. Urol.*, 160(1), 210, 1998.

85. Arisawa, C., et al., TNP-470 combined with nicardipine suppresses in vivo growth of PC-3, a human prostate cancer cell line, *Urol. Oncol.*, 7, 229, 2002.
86. Pu, Y.S., et al., Enhanced suppression of prostate tumor growth by combining C–CAM1 gene therapy and angiogenesis inhibitor TNP-470, *Anticancer Drugs*, 13, 743, 2002.
87. Horti, J., et al., Increased transcriptional activity of prostate specific antigen in the presence of TNP-470, an angiogenesis inhibitor, *Br. J. Cancer*, 79, 1588, 1999.
88. Logothetis, C.J., et al., Phase I trial of the angiogenesis inhibitor TNP-470 for progressive androgen-independent prostate cancer, *Clin. Cancer Res.*, 7, 1198, 2001.
89. Kohn, E.C., et al., Angiogenesis: role of calcium-mediated signal transduction, *Proc. Natl. Acad. Sci.*, 92, 1307, 1995.
90. Kohn, E.C., et al., Calcium influx modulates expression of matrix metalloproteinase-2 (72 kDa type IV collagenase, gelatinase A), *J. Biol. Chem.*, 269, 21505, 1994.
91. Chung, L.W.K., Fibroblasts are critical determinants in prostatic cancer growth and dissemination, *Cancer Metastasis Rev.*, 10, 263, 1991.
92. Ware, J.L., et al., Growth factors and their receptors as determinants in the proliferation and metastasis of human prostate cancer, *Cancer Metastasis Rev.*, 12, 286, 1993.
93. Waslienko, W.J., et al., Calcium signaling in prostate cancer cells: evidence for multiple receptors and enhanced sensitivity to bombesin/GRP, *Prostate*, 30, 167, 1997.
94. Wasilenko, W.J., et al., Effects of the calcium influx inhibitor carboxyamido-triazole on the proliferation and invasiveness of human prostate cell lines, *Intl. J. Cancer*, 68, 259, 1996.
95. Kohn, E.C., et al., Clinical investigation of a cytostatic calcium influx inhibitor in patients with refractory cancers, *Cancer Res.*, 56(3), 569, 1996.
96. Berlin, J., et al., Phase I clinical and pharmacokinetic study of oral carboxyamido-triazole, a signal transduction inhibitor, *J. Clin. Oncol.*, 15(2), 781, 1997.
97. Kohn, E.C., et al., Phase I trial of micronized formulation carboxyamidotriazole in patients with refractory solid tumors: pharmacokinetics, clinical outcome, and comparison of formulations, *J. Clin. Oncol.*, 15(5), 1985, 1997.
98. Bauer, K.S., et al., A pharmacokinetically guided phase II study of carboxyamidotriazole in AIPC, *Clin. Cancer Res.*, 5, 2324, 1999.
99. Fong, T.A., et al., SU5416 is a potent and selective inhibitor of the vascular endothelial growth factor receptor (Flk-1/KDR) that inhibits tyrosine kinase catalysis, tumor vascularization, and growth of multiple tumor types, *Cancer Res.*, 59, 99, 1999.
100. Mendel, D.B., et al., Development of SU5416, a selective small molecule inhibitor of VEGF receptor tyrosine activity, as an anti-angiogenesis agent, *Anticancer Drug Des.*, 15, 29, 2000.
101. Huss, W.J., et al., SU5416 selectively impairs angiogenesis to induce prostate cancer specific apoptosis, *Mol. Cancer Ther.*, 2, 611, 2003.
102. Rosen, L., et al., Phase I dose-escalating trial of SU5416, a novel angiogenesis inhibitor in patients with advanced malignancies, *Proc. Am. Clin. Oncol.*, Abstract No. 618, 1999.
103. Stopeck, A., et al., Results of a phase I dose-escalating study of the anti-angiogenic agent, SU5416, in patients with advanced malignancies, *Clin. Cancer Res.*, 8, 2798, 2002.
104. Kuenen, B.C., et al., Dose-finding and pharmacokinetic study of cipslatin, gemcitabine and SU5416 in patients with solid tumors, *J. Clin. Oncol.*, 20(6), 1657, 2002.
105. Stadler, W.M., et al., A randomized phase II trial of the antiangiogenic agent SU5416 in hormone-refractory prostate cancer, *Clin. Cancer Res.*, 10, 3365, 2004.
106. Nelson, J.B., et al., Identification of endothelin-1 in the pathophysiology of metastatic adenocarcinoma of the prostate, *Nat. Med.*, 9, 944, 1995.
107. Lassieter, L.K. and Carducci, M.A., Endothelin receptor antagonists in the treatment of prostate cancer, *Semin. in Oncol.*, 30(5), 678, 2003.
108. Langenstroer, P., et al., Endothelin-1 in the human prostate: tissue levels, source of production and isometric tension studies, *J. Urol.*, 150, 495, 1993.
109. Casey, M.L., et al., Massive amounts of immunoreactive endothelin in human seminal fluid, *J. Clin. Endocrinol. Metab.*, 74(1), 223, 1992.
110. Usami, B.A., et al., Methylation of the neutral endopeptidase gene promoter in human prostate cancers, *Clin. Cancer Res.*, 6, 1664, 2000.

111. Gohji, K., et al., Expression of endothelin receptor A associated with prostate cancer progression, *J. Urol.*, 165(3), 1033, 2001.
112. Nelson, J.B., et al., Endothelin-1 production and decreased endothelin B receptor expression in advanced prostate cancer, *Cancer Res.*, 56(4), 663, 1996.
113. Takuwa, Y., Masaki, T., and Yamashita, K., The effects of the endothelin family peptides on cultured osteoblastic cells from rat calvarie, *Biochem. Biophys. Res. Commun.*, 170, 998, 1990.
114. Levin, E.R., Endothelins, *NEJM*, 333(6), 356, 1995.
115. Davar, G., et al., Behavioral signs of acute pain produced by application of endothelin-1 to rat sciatic nerve, *Neuroreport*, 9(10), 2279, 1998.
116. Pflug, B., Udan, M.S., and Nelson, J.B., Endothelin-1 promotes survival in prostate and renal carcinoma cell lines by inhibition of apoptosis involving the AKT pathway, *91st Annual Meeting of the Am. Assoc. Cancer Res.*, Abstract No. 5241, 2000.
117. Zonnenberg, B.A., et al., Phase I dose-escalation study of the safety and pharmacokinetics of atrasentan: an endothelial receptor antagonist for refractory prostate cancer, *Clin. Cancer Res.*, 9, 2965, 2003.
118. Carducci, M.A., et al., ABT-627, an endothelin-receptor antagonist for refractory adenocarcinomas: phase I and pharmacologic evaluation, *Proc. Am. Clin. Oncol.*, Abstract No. 625, 1999.
119. Carducci, M.A., et al., A placebo controlled phase II dose-ranging evaluation of an endothelin-A receptor antagonist for men with HRPC and disease related pain, *Proc. Am. Clin. Oncol.*, Abstract No. 1314, 2000.
120. Carducci, M.A., et al., Effect of endothelin-A receptor blockade with atrasentan on tumor progression in men with HRPC: a randomized, phase II, placebo-controlled trial, *J. Clin. Oncol.*, 21(4), 679, 2003.
121. Nelson, J.B., et al., Suppression of prostate cancer induced bone remodeling by the endothelin receptor A antagonist atrasentan, *J. Urol.*, 169, 1143, 2003.
122. Carducci, M.A., et al., Effects of atrasentan on disease progression and biological markers in men with metastatic HRPC: phase 3 study, *Proc. Am. Clin. Oncol.*, Abstract No. 4508, 2004.
123. Vogelzang, N.J., et al., Meta-analysis of clinical trials of atrasentan 10 mg in metastatic hormone-refractory prostate cancer, *Proc. Am. Clin. Oncol.*, Abstract No. 4563, 2005.
124. Singh, A., Padley, R.J., and Ashraf, T., The selective endothelin-A receptor antagonist atrasentan improves quality of life adjusted time to progression (QATTP) in HRPC, *Proc. Am. Clin. Oncol.*, Abstract No. 1567, 2001.
125. Yount, S., et al., Impact of atrasentan on prostate-specific outcomes with hormone-refractory prostate cancer, *Proc. Am. Clin. Oncol.*, Abstract No. 4582, 2004.
126. Borgstrom, P., et al., Neutralizing anti-vascular endothelial growth factor antibody completely inhibits angiogenesis and growth of human prostate carcinoma micro tumors in vivo, *Prostate*, 35, 1, 1998.
127. Melnyk, O., et al., Neutralizing anti-vascular endothelial growth factor antibody inhibits further growth of established prostate cancer and metastases in a preclinical model, *J. Urol.*, 161, 960, 1999.
128. Reese, D., et al., A phase II trial of humanized monoclonal antivascular endothelial growth factor antibody (rhuMAb VEGF) in HRPC, *Proc. Am. Clin. Oncol.*, Abstrast No. 1355, 1999.
129. Picus, J., et al., The use of bevacizumab with docetaxel and estramustine in hormone refractory prostate cancer: initial results of CALBG 90006, *Proc. Am. Clin. Oncol.*, Abstract No. 1578, 2003.
130. D'Amato, R.J., et al., Thalidomide is an inhibitor of angiogenesis, *Proc. Natl. Acad. Sci.*, 91, 4082, 1994.
131. Bauer, K.S., Inhibition of angiogenesis by thalidomide requires metabolic activation, which is species dependent, *Biochem. Pharmacol.*, 55, 1827, 1998.
132. Dixon, S.C., et al., Thalidomide up-regulates prostate-specific antigen secretion from LNCaP cells, *Cancer Chemother. Pharmacol.*, 43, S78, 1999.
133. Figg, W.D., et al., A randomized phase II trial of thalidomide, an angiogenesis inhibitor, in patients with androgen-independent prostate cancer, *Clin. Cancer Res.*, 7, 1888, 2001.
134. Drake, M.J., et al., An open-label phase II study of low dose thalidomide in androgen-independent prostate cancer, *Br. J. Cancer*, 88, 822, 2003.

135. Shevrin, D.H., et al., Phase II study of thalidomide an mitoxantrone/prednisone in patients with HRPC, *Proc. Am. Clin. Oncol.*, Abstract No. 1787, 2003.
136. Frank, R.C., et al., Low dose docetaxel, estramustine and thalidomide followed by maintenance thalidomide for the treatment of HRPC: a phase II community based trial, *Proc. Am. Clin. Oncol.*, Abstract No. 4681, 2004.
137. Dahut, W.L., et al., Randomized phase II trial of docetaxel plus thalidomide in AIPC, *J. Clin. Oncol.*, 22(13), 2532, 2004.
138. Sison, B., et al., Phase II study of CC-4047 in patients with metastatic HRPC, *J. Clin. Oncol.*, 22(14S), 2701, 2004.
139. Fotsis, T., et al., The endogenous oestrogen metabolite 2-methoxyestradiol inhibits angiogenesis and suppresses tumour growth, *Nature*, 368, 237, 1994.
140. D'Amato, R.J., et al., 2-methoxyestradiol, an endogenous mammalian metabolite, inhibits tubulin polymerization by interacting at the colchicines site, *Proc. Natl. Acad. Sci.*, 91, 3964, 1994.
141. Hamel, E., et al., Interactions of 2-methoxyestradiol, an endogenous mammalian metabolite, with unpolymerized tubulin and with tubulin polymers, *Biochemistry*, 35, 1304, 1996.
142. Mukhopadhyay, T. and Roth, J.A., Induction of apoptosis in human lung cancer cells after wild-type p53 activation by methoxyestradiol, *Oncogene*, 14(3), 379, 1997.
143. Seegers, J.C., et al., The mammalian metabolite, 2-methoxyestradiol, affects p53 levels and apoptosis induction in transformed cells but not in normal cells, *J. Steroid Biochem. Mol. Biol.*, 62(4), 253, 1997.
144. LaVallee, T.M., et al., 2-Methoxyestradiol up-regulates death receptor 5 and induces apoptosis through activation of the extrinsic pathway, *Cancer Res.*, 63(2), 468, 2003.
145. Mooberry, S.L., Mechanism of action of 2-methoxyestradiol: new developments, *Drug Resist. Update*, 6, 355, 2003.
146. Lottering, M.L., et al., 17 beta-Estradiol metabolites affect some regulators of the MCF-7 cell cycle, *Cancer Lett.*, 110, 181, 1996.
147. Attalla, H., et al., 2-Methoxyestradiol arrests in mitosis without depolymerizing tubulin, *Biochem. Biophys. Res. Commun.*, 228, 467, 1996.
148. Zoubine, M.N., et al., 2-Methoxyestradiol-induced growth suppression and lethality in estrogen-responsive MCF-7 cells may be mediated by down regulation of p34cdc2 and cyclin B1 expression, *Int. J. Oncol.*, 15(4), 639, 1999.
149. Yue, T.-L., et al., 2-Methoxyestradiol, an endogenous estrogen metabolite, induces apoptosis in endothelial cells and inhibits angiogenesis: possible role for stress-activated protein kinase signaling pathway and fas expression, *Mol. Pharm.*, 51, 951, 1997.
150. Attalla, H., et al., 2-Methoxyestradiol-induced phosphorylation of Bcl-2: uncoupling from JNK/SAPK activation, *Biochem. Biophys. Res. Commun.*, 247(3), 616, 1998.
151. Kumar, A.P., Garcia, G.E., and Slaga, T.J., 2-Methoxyestradiol blocks cell-cycle progression at G2/M phase and inhibits growth of human prostate cancer cells, *Molecular Carinogenesis*, 31, 111, 2001.
152. Qadan, L.R., et al., 2-Methoxyestradiol induces G2/M arrest and apoptosis in prostate cancer, *Biochem. Biophys. Res. Commun.*, 285, 1259, 2001.
153. Bu, S., et al., Mechanisms for 2-methoxyestradiol-induced apoptosis of prostate cancer cells, *FEBS Letters*, 531, 141, 2002.
154. Miller, K.D., et al., A phase I safety, pharmacokinetic and pharmacodynamic study of 2-methoxyestradiol (2-ME2) in patients with refractory metastatic breast cancer (MBC), *Proc. Am. Clin. Oncol.*, Abstract No. 170, 2001.
155. Dahut, W.L., et al., A phase I study of 2-methoxyestradiol (2-ME2) in patients with solid tumors, *Proc. Am. Soc. Clin. Oncol.*, Abstract No. 833, 2003.
156. Sweeney, C.J., et al., A phase II multicenter, randomized, double-blind, safety trial assessing the pharmacokinetics, pharmacodynamics and efficacy of oral 2-ME in HRPC, *Clin. Cancer Res.*, 11, 6625, 2005.
157. Wood, J.M., et al., PTK787/ZK 2222584, a novel and potent inhibitor of vascular endothelial growth factor receptor kinases, impairs vascular endothelial growth factor induced responses and tumor growth after oral administration, *Cancer Res.*, 60, 2178, 2000.

158. George, D., et al., Phase I study of the novel, oral angiogenesis inhibitor PTK787/ZK 222584 (PTK/ZK): evaluating the pharmacokinetic effect of a high-fat meal in patients with hormone-refractory prostate cancer, *Proc. Am. Clin. Oncol.*, Abstract No. 4689, 2004.
159. Hennequin, L.F., et al., Structure–activity relationship, physicochemical and pharmacokinetic properties of AZD2171: a highly potent inhibitor of VEGF receptor tyrosine kinases, *Proc. Am. Assoc. Cancer Res.*, 45, Abstract No. 2539, 2004.
160. Ogilvie, D.J., et al., AZD2171, a highly potent inhibitor of VEGF receptor signaling in primary human endothelial cells, exhibits broad-spectrum activity in tumor xenograft models, *Proc. Am. Assoc. Cancer Res.*, 45, Abstract No. 4553, 2004.
161. Ryan, C., et al., Safety and tolerability of AZD2171, a highly potent VEGFR inhibitor, in patients with advanced prostate adenocarcinoma, *2005 Prostate Cancer Symposium, ASCO*, Abstract No. 253, 2005.
162. Ryan, C., et al., Safety and tolerability of AZD2171, a highly potent VEGFR inhibitor, in patients with advanced prostate adenocarcinoma, *Proc. Am. Clin. Oncol.*, Abstract No. 3049, 2005.
163. Curwen, J.O. and Wilson, C., ZD4054: A specific endothelin A receptor antagonist with potential utility in prostate cancer and metastatic bone disease, *Eur. J. Cancer*, 38, S102, Abstract No. 340, 2002.
164. Liu, G., et al., Tolerability profile of ZD4054 is consistent with the effects of endothelin A receptor specific antagonism, *Proc. Am. Clin. Oncol.*, Abstract No. 4628, 2005.
165. Hurwitz, H., et al., Bevacizumab plus irinotecan, fluorouracil, and leucovorin for metastatic colorectal cancer, *New Engl. J. Med.*, 350(23), 2335, 2004.
166. Sandler, A.B., et al., Randomized phase II/III trial of paclitaxel plus carboplatin with or without bevacizumab in patients with advanced non-squamous non-small cell lung cancer (NSCLC): an Eastern Cooperative Oncology Group (ECOG) Trial—E4599, *Proc. Am. Clin. Oncol.*, Abstract No. LBA4, 2005.
167. Motzer, R.J., et al., Phase 2 trials of SU11248 show antitumor activity in second-line therapy for patients with metastatic renal cell carcinoma (RCC), *Proc. Am. Clin. Oncol.*, Abstract No. 4508, 2005.
168. Escudier, B., et al., Randomized phase III trial of the Raf kinase and VEGFR inhibitor sorafenib (BAY 43-9006) in patients with advanced renal cell carcinoma (RCC), *Proc. Am. Clin. Oncol.*, Abstract No. LBA4510, 2005.
169. Therasse, P., et al., New Guidelines for assessing the response to treatment in solid tumors, *JNCI*, 92(3), 205, 2000.
170. Figg, W.D., et al., Lack of correlation between prostate-specific antigen and the presence of measurable soft tissue metastasis in hormone-refractory prostate cancer, *Cancer Invest.*, 14, 5137, 1996.
171. Bubley, G., et al., Eligibility and response guidelines for phase II clinical trials in androgen-independent prostate cancer, *J. Clin. Oncol.*, 17(11), 3461, 1999.
172. Roessner, M., et al., Prostate-specific antigen (PSA) response as a surrogate endpoint for overall survival (OS): analysis of the TAX 327 study comparing docetaxel plus prednisone with mitoxantrone plus prednisone in advanced prostate cancer, *Proc. Am. Clin. Oncol.*, Abstract No. 4554, 2005.
173. Yang, J.C., et al., A randomized trial of bevacizumab, an anti-vascular endothelial growth factor antibody, for metastatic renal cancer, *New Engl. J. Med.*, 349(5), 427, 2003.
174. Schuch, G., et al., Endostatin inhibits the vascular endothelial growth factor-induced mobilization of endothelial progenitor cells, *Cancer Res.*, 63(23), 8345, 2003.
175. Benjamin, L.E., et al., Selective ablation of immature blood vessels in established human tumors follows vascular endothelial growth factor withdrawal, *J. Clin. Invest.*, 103(2), 159, 1999.
176. Matsushima, H., et al., Correlation between proliferation, apoptosis, and angiogenesis in prostate carcinoma and their relation to androgen ablation, *Cancer*, 85(8), 1822, 1999.
177. Salven, P., et al., Serum vascular endothelial growth factor is often elevated in disseminated cancer, *Clin. Cancer Res.*, 3(5), 647, 1997.
178. Jones, A., et al., Elevated serum vascular endothelial growth factor in patients with hormone-escaped prostate cancer, *BJU*, 85(3), 276, 2000.

179. Rak, J., Filmus, J., and Kerbel, R.S., Reciprocal paracrine interactions between tumour cells and endothelial cells: the "angiogenesis progression" hypothesis, *Eur. J. Cancer*, 32A, 2438, 1996.
180. Yoshji, H., Harris, S.R., and Thorgeirsson, U.P., Vascular endothelial growth factor is essential for initial but not continued in vivo growth of human breast carcinoma cells, *Cancer Res.*, 5, 3924, 1997.
181. Lissbrant, I.F., et al., Blood vessels are regulators of growth, diagnostic markers and therapeutic targets in prostate cancer, *Scand. J. Urol. Nephrol.*, 35, 437, 2001.
182. Fanelli, M., et al., Assessment of tumor vascularization: immunohistochemical and non-invasive methods, *Int. J. Biol. Markers*, 14(4), 218, 1999.
183. Weidner, N., *Angiogenesis: Models, Modulators, and Clinical Applications*, Plenum Press, New York, 389, 1999.
184. Padhani, A.R., et al., Dynamic contrast enhanced MRI of prostate cancer: correlation with morphology and tumour stage, histological grade and PSA, *Clin. Radiol.*, 55(2), 99, 2000.
185. Ismail, M., et al., Color Doppler imaging in predicting the biologic behavior of prostate cancer: correlation with disease-free survival, *Urology*, 50(6), 906, 1997.
186. Okihara, K., Watanabe, H., and Kojima, M., *Ultrasound Med. Biol.*, 25(1), 89, 1999.
187. Padhani, A.R., Effects of androgen deprivation on prostatic morphology and vascular permeability evaluated with MR imaging, *Radiology*, 218, 365, 2001.
188. Abdollahi, A., et al., Combined therapy with direct and indirect angiogenesis inhibition results in enhanced antiangiogenic and antitumor effects, *Cancer Res.*, 63, 8890, 2003.
189. Mauceri, H.J., et al., Combined effects of angiostatin and ionizing radiation in anti-tumor therapy, *Nature* (London), 394, 287, 1998.
190. Ning, S., et al., The antiangiogenic agents of SU5416 and SU6668 increase the antitumor effects of fractionated irradiation, *Radiat. Res.*, 157, 45, 2002.
191. Abdollahi, A., et al., SU5416 and SU6668 attenuate the angiogenic effects of radiation-induced tumor cell growth production and amplify the direct anti-endothelial action of radiation in vitro, *Cancer Res.*, 63, 3755, 2003.
192. Lee, C.G., et. al., Anti-VEGF treatments augments tumor radiation response under normoxic or hypoxic conditions, *Cancer Res.*, 60, 5565, 2000.
193. Kozin, S.V., et al., VEGFR2-blocking antibody potentiates radiation induced long-term control of human tumor xenografts, *Cancer Res.*, 61, 39, 2001.
194. Geng, L., et al., Inhibition of VEGFR signaling leads to reversal of tumor resistance to radiotherapy, *Cancer Res.*, 61, 2413, 2001.
195. Hess, C., et al., Effect of VEGF receptor inhibitor PTK787/ZK22258 combined with ionizing radiation on endothelial cells and tumor growth, *Br. J. Cancer*, 85, 2010, 2001.
196. Bischof, M., et al., Triple combination of irradiation, chemotherapy (pemetrexed) and VEGFR inhibition (SU5416) in human endothelial and (SU5416) in human endothelial and tumor cells, *Int. J. Rad. Oncol., Bio. Phys.*, 60(4), 1220, 2004.

28 Antiangiogenic Therapy for Hematologic Malignancies

Karen W.L. Yee and Francis J. Giles

CONTENTS

Poor response to therapy and survival in patients with hematologic malignancies has been associated with genetic and molecular abnormalities.[1–9] However, recent evidence highlights the importance of the microenvironment, and potentially, bone marrow angiogenesis (mature endothelial cell–derived generation of new blood vessels) and vasculogenesis (progenitor cell–derived new vessel generation) in the pathophysiology of a number of malignancies, such as acute lymphoblastic leukemia (ALL),[10,11] acute myeloid leukemia (AML),[10,12–20] chronic lymphocytic leukemia (CLL),[10,11,21,22] myelodysplastic syndrome (MDS),[10,13,16,19,20,23] myeloproliferative disorders,[19,24–30] lymphoma,[11,31,32] multiple myeloma (MM),[11,16,33,34] and possibly Waldenstrom's macroglobulinemia (WM).[35,36] Several, but not all, studies have also indicated that increased microvessel density (MVD) and proangiogenic factors (e.g., VEGF and bFGF-2) may be associated with an inferior outcome in patients with ALL,[10,11,16,37] AML,[10,13,17,38] agnogenic myeloid metaplasia (AMM),[27] CLL,[10,11,16,22,38] chronic myelomonocytic leukemia (CMML),[10] chronic myeloid leukemia in blastic phase (CML-BP),[10,39] Hodgkin's lymphoma (HL),[31] MDS,[10,13,19,40] MM,[11,16,33,34,38,41–43] and non-Hodgkin's lymphoma (NHL).[11,16,32,38,44] Therefore, evidence supports a clinically significant role for angiogenesis in the pathophysiology of these diseases.

A balance between proangiogenic and antiangiogenic factors generated by both the tumor and the microenvironment (e.g., endothelial cells, marrow stromal cells) regulates angiogenesis.[45] Proangiogenic factors include vascular endothelial growth factor (VEGF), angiopoietin-1, basic fibroblastic growth factor (bFGF), insulin-like growth factor-I (IGF-I), platelet-derived growth factor (PDGF), tumor necrosis factor-α (TNF-α), transforming growth factor-α (TGF-α), and TGF-β.[45,46] VEGF is a key mediator of angiogenesis.[47] The VEGF isoforms, including the prototype VEGF and its homolog, placental growth factor (PlGF), regulate several endothelial cell functions, including mitogenesis, permeability, and the production of vasoactive molecules involved in vessel budding and tube formation. VEGF is also a survival factor required for the maintenance of new blood vessels, vascular tone, and interstitial fluid pressure.[47] The activity of VEGF is mediated through four receptors: VEGF receptor 1 (VEGFR-1; Flt-1), VEGFR-2 (Flk-1; KDR), VEGFR-3 (Flt-4), and the coreceptor neuropilin-1 (NRP-1).[47] PlGF, which binds to VEGFR-1 but not VEGFR-2, potentiates angiogenic responses to VEGF. Antiangiogenic factors include endostatin (internal C-terminal fragment of XVIII collagen), thrombospondin (TSP), angiostatin (internal fragments of plasminogen), interferon-α (IFN-α), IFN-β, IFN-γ, and tissue inhibitors of matrix metalloproteinases (TIMP).[45]

Therapeutic vascular targeting strategies have been developed to prevent the neovascularization of malignancies or disruption of the established tumor vasculature (Table 28.1). IFN-α was one of the first successful antiangiogenic agents in use for the treatment of a life-threatening pulmonary hemangioma in a child.[48] IFN-α also possesses direct antiproliferative effects and immunomodulatory activity leading to its use in the treatment of a number of hematological malignancies.[49] However, its use is associated with frequent side effects resulting in treatment discontinuation.[49,50] Antiangiogenic agents currently in clinical development include those targeting the VEGF pathway, other growth factors and their receptors (such as PDGF), cell adhesion molecules on endothelial cells (e.g., integrins), matrix metalloproteinases (MMPs), and cyclooxygenase (COX), as well as those which modulate the tumor microenvironment (Table 28.1). Several agents have been shown to have activity against hematological malignancies in vitro (e.g., angiostatin,[51] endostatin,[51–55] GW654652,[56,57]

TABLE 28.1
List of Selected Antiangiogenic Agents in Clinical Trials for Cancer

Agent	Company	Mechanism of Action
Cyclooxygenase-2 inhibitors		
Celecoxib (Celebrex)[a,b]	Pfizer	Induces apoptosis via cyclooxygenase-2-dependent and -independent pathways
Endogenous inhibitors		
Angiostatin	EntreMed	Binds to multiple endothelial sites (including $\alpha_v\beta_3$ integrin)
Endostatin (Endostar; YH-16)	EntreMed	Binds to endothelial cell $\alpha_v\beta_1$ integrin
Interferons[a]	Roche/Schering/GSK	Antiangiogenesis; immunomodulatory
VEGF inhibitors		
Bevacizumab (Avastin)[a,b]	Genetech Biotech	Humanized monoclonal IgG1$_\kappa$ antibody to VEGF-A
HuMV833	Protein Design Labs	Humanized monoclonal IgG4$_\kappa$ antibody to VEGF-A
VEGF-Trap	Regeneron	Soluble receptor for VEGF-A and VEGF-B, and PlGF
VEGF-AS (Veglin)	VasGene	Antisense oligonucleotide
VEGFR inhibitors		
Angiozyme	Ribozyme	Ribozyme cleavage of VEGFR-1 mRNA
AE-941 (Neovastat)	AEterna	Shark cartilage extract inhibitor of VEGF–VEGFR binding, MMP-2, MMP-9, and MMP-12
AG-013736[a,b]	Pfizer	Small molecule RTK inhibitor of VEGFR-1–3, PDGFR-β, and c-kit
AMG 706	Amgen	Small molecule RTK inhibitor of VEGFR-1–2, PDGFR-β, and c-kit
AZD2171	AstraZeneca	Small molecule RTK inhibitor of VEGFR-1–3, PDGFR-β, and c-kit
BAY 43-9006 (sorafenib)[a,b]	Bayer/Onyx	Small molecule RTK inhibitor of VEGFR-2, PDGFR-β, FLT-3, and c-kit
BMS-582664	BMS	Small molecule RTK inhibitor of VEGFR-2 and FGFR
CEP-7055	Cephalon	Small molecule RTK inhibitor of VEGFR-1–3, FGFR, PDGFR, and c-kit
CHIR-258[a,b]	Chiron	Small molecule RTK inhibitor of VEGFR-1–2, FGFR-1, and FGFR-3
CP-547632	Pfizer	Small molecule RTK inhibitor of VEGFR-2
GW654652	GSK	Small molecule RTK inhibitor of VEGFR-1–3
GW786034	GSK	Small molecule RTK inhibitor of VEGFR-1–3, PDGFR-β, and c-kit
IMC-1C11	ImClone	Chimeric monoclonal antibody against VEGFR-2
KRN-951	Kirin Brewery	Small molecule RTK inhibitor of VEGFR-1–3, PDGFR-β, and c-kit
OSI-930	OSI	Small molecule RTK inhibitor of VEGFR and c-kit
PTK787/ZK222584 (vatalanib)[a,b]	Novartis	Small molecule RTK inhibitor of VEGFR-1–3, PDGFR-β, and c-kit
SU5416 (TSU-16; semaxanib)[a,b]	Merck	Small molecule RTK inhibitor of VEGFR-1–2, PDGFR-β, FLT-3, and c-kit
SU11248 (sunitinb; Sutent)[a,b]	Merck	Small molecule RTK inhibitor of VEGFR-2, EGFR, and RET
SU6668 (TSU-68)	Merck	Small molecule RTK inhibitor of VEGFR-2, PDGFR-β, FGFR, and c-kit
XL999	Exelixis	Small molecule RTK inhibitor of VEGFR, PDGFR, FGFR, and FLT-3
ZD6474 (AZD6474; Zactima)[a,b]	AstraZeneca	Small molecule RTK inhibitor of VEGFR-2, EGFR, and RET
ZK-CDK	Schering	Small molecule ATP-competitive kinase inhibitor of CDKs, VEGFRs, and PDGFR
		Monoclonal antibody to VEGFR-2

TABLE 28.1 (continued)
List of Selected Antiangiogenic Agents in Clinical Trials for Cancer

Agent	Company	Mechanism of Action
Vascular damaging agents		
AVE8062A	Aventis	Colchicine analog; C4A analog; causes endothelial tubulin depolymerization
Combretastatin A-4 (CA4) prodrug	OxiGene	Colchicine analog; causes endothelial tubulin depolymerization
DMXAA (AS1404)	Antisoma	Flavone-8-acetic acid analog; microtubule-independent vascular damaging agent; TNF induction
ZD6126	AstraZeneca	Colchicine analog; prodrug of the tubulin-binding agent *N*-acetylcolchinol causes endothelial tubulin depolymerization
Immunomodulatory agents		
CC-4047 (Actimid)[a,b]	Celgene	Unclear; inhibits neoangiogenesis and TNF production; decreases cell adhesion molecules; and has immunomodulatory activity
Lenalidomide (CC-5013; Revlimid)[a,b]	Celgene	
Thalidomide (Thalimid)[a,b]	Celgene	
Integrin antagonists		
Vitaxin (Medi-522)	MedImmune	Humanized monoclonal $IgG1_{\kappa}$ antibody to the $\alpha_v\beta_3$ integrin complex
EMD-121974	Merck	Cyclic peptide inhibitor of α_v integrins
Matrix metalloproteinase inhibitors		
AE-941 (Neovastat)	AEterna	Shark cartilage extract inhibitor of VEGF–VEGFR binding, MMP-2, MMP-9, and MMP-12
Marimastat (BB2516)	British Biotech	Collagen peptidomimetic inhibitor of MMP-1, MMP-2, MMP-3, MMP-7, and MMP-9
Prinomastat (AG-3340)	Agouron	Nonpeptidic inhibitor of MMP-2, MMP-9, and MMP-14
Tanomastat (BMS-275291)	Bayer	Nonpeptidic biphenyl inhibitor of MMP-2, MMP-3, and MMP-9
Thrombospondin analog		
ABT-510[a,b]	Abbott	Nonapeptide mimics antiangiogenic activity of endogenous thrombospondin-1

BMS, Bristol–Myers Squibb; GSK, GlaxoSmithKline.

[a] In clinical trials in hematological malignancies and FDA approved.
[b] Further details in the text.

GW786034,[58] IMC-1C11,[59–61] OSI-930,[62,63] combretastatin A-4 prodrug,[64] DMXAA,[65] marimastat,[66] and Neovastat[67]); however, no information with respect to their clinical activity and toxicity in patients with hematological diseases is available yet. This chapter summarizes the literature on the clinical use of antiangiogenic agents in the treatment of patients with hematological malignancies.

28.1 ANTIANGIOGENIC AGENTS

28.1.1 Immunomodulatory Drugs

The tumor microenvironment and cytokine milieu are important for survival, growth, and resistance of a variety of hematological malignancies.[28,68–74] The malignant cells and marrow stromal cells secrete a number of cytokines and growth factors, such as VEGF and TNF-α, which act in an autocrine or paracrine fashion not only to increase malignant cell growth and survival, but also to render to bone marrow milieu more conducive for cell growth, resistance, and survival.[10,11,21,22,70,75–90]

The precise mechanisms of action of thalidomide and its derivatives are unknown. Several activities of this class of drugs may contribute to its clinical effects. Antiangiogenic (e.g., inhibition of VEGF- and bFGF-induced angiogenesis) and immunomodulatory effects (e.g., T-cell costimulation, innate immune system activation), as well as modification of cell surface adhesion molecules, have been reported.[91–101] The immunomodulatory effects are mediated by modulation of cytokines and their downstream effectors, such as the inhibition of NF-κB activation.[91,102,103] Costimulation of T cells by thalidomide and its analogs also leads to inhibition of TNF-receptor 2 (TNF-R2) and IL-2-dependent upregulation of TNF-α production in cancer patients, in contrast with their TNF inhibitory effects during inflammatory stimuli.[104]

Second generation immunomodulatory drugs (IMiDs) have been synthesized to increase the immunological and anticancer effects of and to decrease the toxicities associated with thalidomide. Lenalidomide (CC-5013; Revlimid) and CC-4047 (Actimid) were generated by the addition of an amino group to position 4 of the phthaloyl ring on the thalidomide backbone.[92] For lenalidomide, a carbonyl group was also removed from the 4-amino-substituted phthaloyl ring.[92] Lenalidomide and CC-4047 have different activity profiles, pharmacokinetics, and plasma drug stability.[92] Lenalidomide is up to 2000 times more potent at stimulating T cells and inhibiting TNF-α than the parent compound.[93] It is also more effective than thalidomide at inhibiting tumor growth, decreasing angiogenesis, and prolonging survival in a human plasmacytoma mouse model,[105] and increasing cytotoxicity of natural killer (NK) cells.[92] CC-4047 is the more potent T-cell costimulator, although, it appears to have similar antiangiogenic activity compared with lenalidomide.[92] CC-4047 is able to enhance protective and long-lasting tumor-specific immunity (Th1-type responses) in vivo[106] and proliferation and cytokine production by natural killer T (NKT) cells.[107]

28.2 THALIDOMIDE (THALIMID)

28.2.1 Acute Myeloid Leukemia

28.2.1.1 Thalidomide Monotherapy

A phase I/II study evaluated the safety and efficacy of thalidomide at daily doses of 200–400 mg in 20 patients with AML (14 relapsed or refractory; 6 untreated poor risk).[108] Median duration of therapy with thalidomide was 7 weeks. Seven patients did not complete 4 weeks of therapy (three due to disease progression and death, two due to patient request, and two due to drug intolerance while on the lowest dose level). No patient tolerated doses of >400 mg daily without adverse events (grade ≥2); especially fatigue (60%), constipation (35%), rash (15%), and neuropathy (15%). One patient required dose reduction for grade 3 constipation not alleviated by laxatives. Of the 13 patients receiving thalidomide for >4 weeks, the overall response rate (ORR) was 31% (partial response [PR] 31%), with one additional patient achieving a hematologic improvement (HI). In parallel, MVD and plasma levels of bFGF were significantly decreased in these five patients after 1 month of therapy with thalidomide ($P < 0.05$ and $P = 0.045$, respectively). Median duration of responses was 3 months. Follow-up reporting indicated that one responding patient with relapsed AML achieved a complete response (CR) after 15 months of therapy, despite a dose reduction at 6 months from 200 to 100 mg daily for fatigue.[109] A further dose reduction to 50 mg daily was required for progressive sensory neuropathy after 18 months of therapy. The patient remains in CR for >12 months.

A second phase II trial evaluated thalidomide at doses of 200–800 mg orally daily in 16 patients with relapsed or refractory AML.[110] Dose escalation was not attempted in three patients because of progressive disease or toxicity. Dose escalation to 800 mg was possible in only 5 of the 13 remaining patients. Median daily dose was 400 mg; median duration of therapy was 27 days. The ORR was 6% (CR 6%). Duration of response was 36 months. There was no correlation between response and markers of angiogenesis (e.g., VEGF). Therefore, thalidomide monotherapy has significant toxicity and modest activity in patients with relapsed or refractory AML.

28.2.1.2 Thalidomide and Chemotherapy

A phase II study was performed to determine whether the administration of thalidomide in conjunction with chemotherapy (either liposomal daunorubicin and cytarabine (DA) or liposomal daunorubicin and topotecan [DT]) could improve the outcome of patients with poor prognosis AML or MDS (i.e., refractory anemia with excess blasts (RAEB) or RAEB in transformation [RAEB-t]).[111] Patients were first randomized to receive either DA or DT and subsequently, within each arm, a second randomization to receive chemotherapy alone (DA or DT) or with thalidomide (DATh or DTTh). Thalidomide was initiated at a dose of 400 mg daily with escalation to 600 mg daily if tolerated. Median dose received by all 41 patients randomized to thalidomide was 61% of the targeted maximum dose. None of the 11 patients treated with DT or DTTh responded; therefore, these two arms were closed. Treatment of 17 out of 37 patients (46%) with DA and 16 out of 36 patients (44%) with DATh achieved a CR ($P = 0.71$). Median response duration and survival were similar in both arms (38 vs. 34 weeks, respectively; $P = 0.57$ and 35 vs. 28 weeks, respectively; $P = 0.15$). There were no significant differences in the changes in VEGF levels or MVD after therapy regardless of whether patients received or did not receive thalidomide. These results indicate that there was no benefit from the addition of thalidomide to chemotherapy in poor risk AML or MDS patients.

28.2.2 Chronic Lymphocytic Leukemia

28.2.2.1 Previously Untreated Patients

28.2.2.1.1 Thalidomide and Chemotherapy

A phase I trial evaluated three dose levels of thalidomide (100, 200, and 300 mg orally daily for 5 months) in combination with standard dose fludarabine (administered on Day 7 every 4 weeks for four or six cycles) as initial therapy in 13 patients with CLL.[112] Low-dose coumadin was administered for prophylaxis against venous thromboembolism (VTE). Dose-limiting toxicity (DLT) was not reached. The most common toxicities were fatigue, constipation, and peripheral sensory neuropathy. Fifteen percent of patients developed VTE. The ORR was 100% (CR 56%; nodular PR [NPR] 44%). At a median follow-up of 15+ months, none of the patients have relapsed and median time to disease progression has not been reached. Responses were noted at all dose levels and appear to be higher than historical responses to single-agent fludarabine.[113] Gene expression profiling indicated that thalidomide induced proapoptotic gene expression and decreased prosurvival responses involving the NF-κB pathway.[114] Therefore, the phase II component of this study is evaluating thalidomide 200 mg orally daily in combination with standard dose fludarabine every 4 weeks.[115]

28.2.2.2 Previously Treated Patients

28.2.2.2.1 Thalidomide Monotherapy

A multicenter trial administered single-agent thalidomide initiated at a dose of 200 mg with escalation to 1000 mg daily to 28 patients with relapsed CLL.[116] Common toxicities included neutropenia, fatigue, thrombocytopenia, and anemia. Tumor flares (i.e., increased lymph nodes, spleen, absolute lymphocyte count [ALC], and decrease in hemoglobin and platelets) were observed in 15 patients. Patients with grade >3 tumor flare reactions held treatment until resolution of symptoms. Patients were then restarted at 50 mg daily with weekly 50 mg increments as tolerated. Only seven patients remain on therapy. The study was closed early due to lack of accrual. The maximum tolerated dose (MTD) was not specified. Eighty-five percent of patients had a decrease in ALC, with the greatest decrease occurring at a median of 2.5 cycles of therapy. The ORR was 4% (PR 4%) with 20 patients having stable disease (SD) for at least one cycle of therapy.

28.2.2.2.2 Thalidomide and Chemotherapy

A phase II trial randomized 16 patients who had received prior fludarabine (including six who were fludarabine-refractory) to either thalidomide alone ($n = 8$) or in combination with fludarabine (FT) ($n = 8$).[117] Thalidomide was started at a dose of 200 mg orally daily and adjusted as tolerated. Patients assigned to combination therapy also received fludarabine 25 mg/m^2/day on Days 1–5 every 4 weeks for a total of six cycles. The median dose of thalidomide was 200 mg daily. Grade 3 toxicities included neuropathy, sedation, thrombosis, neutropenic fever, and infection. The distribution of the toxicities in the two arms was not specified. Tumor flares occurred in five patients receiving thalidomide alone; symptoms resolved with either continued therapy ($n = 3$) or temporary discontinuation of therapy ($n = 2$). Of the four evaluable patients treated with thalidomide alone, there was one PR, one HI, and one minor response. Of the six evaluable patients receiving FT therapy, there was one CR, three PRs, one HI, and one minor response. Accrual is ongoing.

Based on the encouraging results obtained with fludarabine and thalidomide in treatment-naive patients with CLL,[112] the efficacy and toxicity of thalidomide (100 mg daily on Days 1–180) combined with fludarabine (30 mg/m^2/day for three consecutive days) and cyclophosphamide (250 mg/m^2/day for three consecutive days) every 28 days for six cycles was evaluated in five previously treated patients with progressive CLL.[118] All patients had received prior fludarabine and alkylating agents. Median duration of thalidomide administration was 60 days. Four patients stopped therapy because of disease progression and one because of neurological toxicity. Serum TNF-α levels increased in all patients during therapy. Therefore, this treatment regimen is not effective in heavily pretreated patients with CLL.

28.2.3 Hodgkin's Lymphoma

Increased angiogenic activity has been documented in patients with HL. Elevated serum levels of cytokines such as VEGF and TNF-α have been observed in patients with HL and decreased significantly after standard treatment.[119] Furthermore, increased levels of TNF-α were observed with more advanced disease.[119] There has been limited experience with thalidomide monotherapy in patients with relapsed or refractory HL.[120] In an attempt to improve on the activity of single-agent vinblastine in patients with recurrent HL (ORR 59%; CR 12%),[121] thalidomide was combined with vinblastine.[122] Eleven patients with previously treated HL, including prior autologous stem cell transplant (ASCT) ($n = 8$), received thalidomide 200 mg daily. After 2 weeks, vinblastine was administered weekly for six doses on an 8 week cycle. Toxicity was mild. The ORR was 36% (PR 36%). Median duration of response was 9 months.

28.2.4 Multiple Myeloma

28.2.4.1 Previously Untreated Patients

28.2.4.1.1 Induction Therapy

Therapeutic strategies for patients with newly diagnosed MM are typically determined by their candidacy for ASCT. High-dose chemotherapy followed by ASCT is currently the treatment of choice for myeloma patients of <65–70 years who do not have severe comorbidities.[123,124] For patients who are not candidates for ASCT, most patients receive an alklylator-based regimen. An overview of over 6000 patients from 27 randomized trials performed by the Myeloma Trialists' Collaborative Group indicated that more intensive combination chemotherapy did not improve overall survival (OS) compared with oral melphalan and prednisone (MP).[125] Therapy with MP can induce ORRs of 53% with a median survival of 29 months. However, it is important to limit exposure to alkylating agents, such as melphalan, in primary therapy to avoid compromising future stem cell harvesting in patients who may be eligible for ASCT. Therefore, for patients who are eligible for ASCT, high-dose dexamethasone alone or high-dose dexamethasone-based regimens, such as vincristine, doxorubicin, and dexamethasone (VAD), have been used as primary induction therapy because of their high response rates and apparent lack of toxicity to stem cells.[126,127] However, the VAD regimen is associated with several disadvantages: risks of anthracycline-induced cardiomyopathy, vincristine-induced neuropathy, and steroid-induced toxicities, and the need for placement of a central venous catheter with risks of line-associated infections and thrombosis. Therefore, alternative therapeutic regimens have been evaluated.

28.2.4.1.1.1 Thalidomide Monotherapy

Immediate treatment of patients with early stage, smouldering, or asymptomatic MM did not have a survival advantage over deferred therapy at the time of development of symptoms.[128,129] Although thalidomide monotherapy in patients with asymptomatic MM can induce ORRs of ~35%,[130,131] only one study reported a prolongation in the median time to progression (TTP) and median response duration (i.e., not reached after 2 years).[130] At the current time, thalidomide's role in delaying progression to symptomatic disease remains to be defined. The efficacy of single-agent thalidomide in patients with symptomatic MM has not been reported.

28.2.4.1.1.2 Thalidomide and Dexamethasone

Single-agent dexamethasone yields ORRs of ~45% in patients with newly diagnosed MM and has been used as single-agent induction therapy for patients who are eligible for ASCT.[132,133] Thalidomide acts synergistically with dexamethasone against myeloma cells.[134] Three randomized trials comparing thalidomide and dexamethasone (TD) with either dexamethasone ± placebo or MP in previously untreated patients with symptomatic MM have demonstrated either improved response rates or disease control in favor of TD (Table 28.2).[132,135,136] However, none have demonstrated an improvement in OS. Baseline MVD or myeloma cell secretion of cytokines does not appear to predict resistance or response to treatment with TD or dexamethasone alone.[137] Median time to response appears to be reduced in patients treated with TD compared with thalidomide alone.[131] No randomized trials have compared VAD therapy to TD. Despite this, the Food and Drug Administration (US FDA) approved the use of thalidomide in conjunction with dexamethasone for the treatment of patients with newly diagnosed MM on May 26, 2006. Therefore, dexamethasone alone or dexamethasone plus thalidomide are currently two common induction regimens for patients who are eligible for ASCT.

28.2.4.1.1.3 Thalidomide and Chemotherapy

Due to thalidomide's unique mechanism of activity and nonoverlapping toxicities, the addition of thalidomide to melphalan-, anthracycline-, or cyclophosphamide-based chemotherapy has been evaluated in an attempt to improve responses (Table 28.2 and Table 28.3).[138–143] In elderly patients (>65 years) with newly diagnosed MM, thalidomide in combination with MP (MPT) significantly improved ORR and disease control and OS compared with MP alone (Table 28.2).[142,143] ORRs (83%–100% with CR 17%–36%) achieved by combining thalidomide with an anthracycline- or cyclophosphamide-based chemotherapy appear to be

TABLE 28.2
Results of Randomized Trials Using Thalidomide-Based Regimens in Untreated Patients with Multiple Myeloma

	Patient Characteristics					Response		
Investigator (y) (Reference)	Median Age (y)	N	ISS Stage[482] (%)	Median F/U (mos)	Chemotherapy Regimen	Response Rate (%)	Response Duration	Overall Survival
Thalidomide and dexamethasone								
Rajkumar (2006)[132]	65 (38–83)	199	I/II (49), III (12)	NR	TD vs. D	OR 63 vs. 41[a] CR 4 vs. 0	NR	~71 vs. 73 at 20 mos
Rajkumar (2006)[135]	65	470	NR	25	TD vs. placebo + D	NR	TTP 17.4 mos vs. 6.4 mos[b]	Not reached both arms
Ludwig (2005)[136]	72	190 (125)	I/II (37), III (63)	NR	TD vs. MP[c]	OR 52 vs. 37[d] CR 10 vs. 3	NR	NR
Thalidomide and chemotherapy								
Palumbo (2006)[143]	72	331 (255)	Durie-Salmon II (42), III (58)	17.6 vs. 15.2	MPT vs. MP	OR 76 vs. 47.6 CR 15.5 vs. 2.4	$EFS_{2\ y}$ 54% vs. 27%[e]	80% vs. 64% at 3 y[f]
Facon (2005)[142]	NR	436	NR	32.2	MPT vs. MP vs. MEL100	NR	PFS 27.6 mos vs. 17.1 mos vs. 19 mos[g]	Not reached vs. 30.3 mos vs. 38.6 mos[h]

CR, complete response; D, dexamethasone; EFS, event-free survival; F/U, follow-up; IFN, interferon-α2b; ISS, International Staging System; MEL100-VAD chemotherapy (vincristine, adriamycin, and dexamethasone) followed by cyclophosphamide mobilized stem cell collection, intermediate dose melphalan, and stem cell transplant; mos, months; MP, melphalan and prednisone; MPT, melphalan, prednisone, and thalidomide; MR, minor response; NR, not reported; OR, overall response (CR + PR); PFS, progression-free survival; PR, partial response; TD, thalidomide and dexamethasone; T/IFN, thalidomide and interferon-α2b; TTP, time to progression; y, year.

[a] $P = 0.0017$.
[b] $P < 0.000065$.
[c] Patients achieving a response or disease stabilization were randomized to maintenance T/IFN or IFN alone.
[d] $P < 0.05$ for OR (CR + PR + MR) 67% vs. 48%.
[e] $P = 0.0006$.
[f] $P = 0.19$.
[g] $P < 0.0001$ for MPT vs. MP, $P = 0.12$ for MP vs. MEL100, and $P = 0.0001$ for MPT vs. MEL100.
[h] $P = 0.0009$ for MPT vs. MP, $P = 0.38$ for MP vs. MEL100, and $P = 0.022$ for MPT vs. MEL100.

TABLE 28.3
Results of Phase II Trials of Thalidomide and Chemotherapy in Untreated Patients with Multiple Myeloma

Investigator (y) (Reference)	Patient Characteristics: Median Age (y) (Range)	*N* (Evaluable)	Prior Therapy (%)	Median F/U (mos)	Treatment Regimen	Response: Response Rate (%)	Response Duration	Overall Survival
Thalidomide and cyclophosphamide-based regimens								
Williams (2004)[139]	55 (31–73)	61[a]	74	NR	T 100–200 mg/day, D 40 mg/day Days 1–4 and 15–18 + CY 500 mg Days 1, 8, and 15 q 28 days × 2–6 cycles then T maintenance	OR 100[b] CR 20	NR	NR
Thalidomide and anthracycline-based regimens								
Hassoun (2006)[140]	59 (35–82)	45 (42)	0	NR	D 40 mg/day Days 1–4, 9–12, and 17–20 + doxorubicin 9 mg m^2 IV Days 1–4 q 28 days × 2–3 cycles then T 100–200/mg/day+D/40/mg/day Days 1–4, 9–12, and 17–20 q 28 days × 2 cycles	OR 90 CR 17; nCR 21	NR	NR
Offidani (2006)[141]	71.5 (65–78)	50	0	18	T 100 mg/day, D 40 mg/day Days 1–4 and 9–12 + Doxil 40 mg/m^2 IV Day 1 q 28 days × 5–6 cycles	OR 84 CR 34; nCR 14	$TTP_{3\ y}$ 60 $EFS_{3\ y}$ 57%	$OS_{3\ y}$ 74%
Hussein (2006)[138]	58.6 (51–67)	55	0	50	T 50–400 mg/day, D 40 mg/day Days 1–4, VCR 2 mg IV Day 1 + Doxil 40 mg/m^2 IV Day 1 q 28 days × 6 cycles then TP maintenance	OR 83 CR 36	PFS 28.8 mos[c]	Not reached[c]

CR, complete response rate; CY, cyclophosphamide; D, dexamethasone; EFS, event-free survival; F/U, follow-up; mos, months; nCR, near complete response; NR, not reported; OR, overall response (CR + nCR + PR); OS, overall survival; PFS, progression-free survival; PR, partial response; T, thalidomide; TP, thalidomide and prednisone; TTP, time to progression; VCR, vincristine; y, year.

[a] Includes untreated ($n = 15$) and relapsed or refractory ($n = 46$) patients.
[b] OR of the 15 untreated patients only.
[c] Median PFS and OS for newly diagnosed patients ($n = 53$) only.

higher than those observed with TD.[138–141] Although ORRs achieved with the addition of thalidomide to VAD chemotherapy (VAD-t)[144] were superior to those historically achieved with VAD chemotherapy (91% vs. 52%–55%, respectively),[145,146] toxicities were severe and vincristine was suspected to exacerbate toxicity attributed to thalidomide.[144,147] As promising results were obtained with thalidomide, dexamethasone, and liposomal doxorubicin,[148] a multicenter, randomized trial is comparing thalidomide plus dexamethasone to thalidomide, dexamethasone, and liposomal doxorubicin in newly diagnosed patients with MM. Combination therapy with cyclophosphamide, thalidomide, and dexamethasone (CTD) has also been demonstrated to yield high response rates in patients with newly diagnosed MM[139]; therefore, a phase III MRC trial is comparing induction therapy with the CTD regimen with infusional cyclophosphamide, vincristine, adriamycin, and dexamethasone (CVAD) in newly diagnosed younger patients with MM. The exact role of these regimens, and others in various stages of development, in the treatment of patients with newly diagnosed MM remains to be defined.

28.2.4.1.1.4 Thalidomide and Targeted Therapy

Thalidomide is also being evaluated in conjunction with arsenic trioxide in patients with high risk previously untreated MM.[149]

28.2.4.1.1.5 Maintenance Therapy

In an attempt to prolong the duration of response, maintenance therapy with various agents, including steroids[150–152] and IFN,[150,153–159] after stem cell transplant (SCT) or chemotherapy has been evaluated. At the current time, the role of maintenance therapy in the treatment of MM is unclear. Thalidomide has been evaluated as maintenance therapy following ASCT. The Inter-Groupe Francophone du Myeloma (IFM) randomized 597 patients with de novo MM without progressive disease after two ASCTs to either observation only, maintenance therapy with pamidronate only, or maintenance therapy with pamidronate and thalidomide.[160] Patients who received maintenance therapy with thalidomide had a superior 3 year event-free survival (EFS) (36% vs. 37% vs. 52%, respectively; $P < 0.009$) and 4 year OS (77% vs. 74% vs. 87%, respectively; $P < 0.04$). Similarly, the National Cancer Institute of Canada (NCIC)[161] and the Dutch-Belgian Hemato-Oncology Cooperative Group (HOVON)/German-Speaking Myeloma Multicenter Group (GMMG)[162] are currently evaluating maintenance therapy with thalidomide and prednisone vs. no therapy or nonmyeloablative allogeneic SCT (for those with a HLA-identical sibling) vs. IFN-α vs. single-agent thalidomide post-ASCT, respectively.

28.2.4.2 Previously Treated Patients

28.2.4.2.1 Induction Therapy, Thalidomide Monotherapy

Despite the lack of a randomized phase III trial, many physicians consider the use of thalidomide in patients with relapsed or refractory MM as part of standard therapy. A systematic review of phase II trials indicated that thalidomide monotherapy can induce responses in 29.4% (CR 1.6%; PR 27.8%) of patients with relapsed or refractory MM, with a median OS of 14 months.[163] Grade 3 or 4 adverse events included somnolence (11%), constipation (16%), neuropathy (6%), rash (3%), thromboembolism (3%), and cardiac events (2%). However, only 13.5% of patients discontinued thalidomide because of adverse events. Although toxicity was dose-related, there was no clear dose-dependent response. In contrast, others have demonstrated that the cumulative dose of thalidomide predicts for response and survival.[164,165] Therefore, the optimal effective dose or schedule of thalidomide is unknown.

In an attempt to address this issue, a multicenter, randomized trial compared therapy with thalidomide at a dose of 100–400 mg daily in 400 previously treated patients with MM.[166] OS at 1 year was similar in both arms; however, more patients dosed at 100 mg daily failed therapy and required the addition of dexamethasone. As expected, dosing at 100 mg daily was better tolerated than dosing at 400 mg daily, with significantly less somnolence, constipation, and peripheral neuropathy. However, the incidence of VTEs was similar. Other reported predictors of response and survival to thalidomide have been early response,[167] serum levels of lactate dehydrogenase (LDH), albumin, and β2-microglobulin,[168,169] and gene expression profiling.[170]

28.2.4.2.1.1 *Thalidomide and Dexamethasone*

TD combination therapy has been used in patients with refractory MM (Table 28.4).[171–177] The ORRs were 25%–57% (CR 0%–13%). Median time to response appears to be reduced in patients treated with TD compared with thalidomide alone.[175] Activity was observed in patients resistant to both single-agent TD or to ASCT.

Certain macrolides, including clarithromycin, can suppress the synthesis of several cytokines, including interleukin (IL)-6, TNF-α, G-CSF, and GM-CSF.[178–181] Furthermore, clarithromycin may inhibit tumor-induced angiogenesis,[182] and has variable activity against myeloma in the clinical setting.[183–185] A prospective phase II study evaluated the nonmyelosuppressive combination of clarithromycin (500 mg orally twice daily continuously), low-dose thalidomide, and dexamethasone (40 mg orally once per week) (BLT-D) in predominantly previously treated patients with MM.[186–188] Neurotoxicity, although usually mild-to-moderate, was the primary reason for treatment discontinuation. Two patients developed pulmonary emboli and one experienced a cerebrovascular accident. Four patients (8%) died, including three sudden deaths in patients with severe cardiopulmonary disease and one from disease progression. After the occurrence of cardiac and thrombotic complications, all patients received low-dose aspirin with no further cardiac or thromboembolic events observed. Of 40 evaluable patients, the ORR was 93% (CR 13%; near CR [nCR] 40%; PR 40%). Therefore, BLT-D appears to be an effective regimen for treatment of patients with MM, although caution should be exercised for patients with severe cardiopulmonary disease or underlying neuropathy.

28.2.4.2.1.2 *Thalidomide and Chemotherapy*

As thalidomide does not cause myelosuppression, several trials have evaluated TD in combination with cyclophosphamide- or anthracycline-based chemotherapy (Table 28.5 and Table 28.6).[138,139,141,189–197] These trials have demonstrated ORRs of 32%–87% with CR rates of 0%–26%.[138,139,141,189–197] As expected, thalidomide-related toxicities were predominantly somnolence, constipation, and neuropathy. Chemotherapy-related toxicities were myelosuppression and infection. Thalidomide with chemotherapy has been reported to be associated with increased occurrences of VTEs compared with thalidomide monotherapy or thalidomide and steroids,[131,132,145,172,173,175,198–200] but may be preventable with aspirin[201,202] or anticoagulation.[131,200,203] Thalidomide in combination with chemotherapy has activity and is feasible and safe. However, randomized trials are required to compare the outcome of thalidomide and chemotherapy with other treatment options in patients with MM.

28.2.4.2.1.3 *Thalidomide and Targeted Therapy*

Thalidomide is also evaluated in combination with the proteosome inhibitor, bortezomib, as salvage therapy in patients with relapsed or refractory MM (Table 28.7).[191,192,204–207] ORRs of 38%–67% (CR 0%–13%) have been obtained in these heavily pretreated patients. However, response duration and OS have been infrequently reported.[204] Thalidomide is also or has been

TABLE 28.4
Results of Phase II Trials Using Thalidomide Plus Dexamethasone in Previously Treated Patients with Multiple Myeloma

Investigator (y) (Reference)	Patient Characteristics: Median Age (y) (Range)	Patient Characteristics: *N* (Evaluable)	Median F/U (mos)	Treatment Regimen	Response: Response Rate (%)	Response: Response Duration	Response: Overall Survival
Weber (1999)[171]	NR	46 (44)	NR	T 200–800 mg/day × 3 mos (n = 44) then nonresponders (*n* = 26) received T at the MTD continuously + D 20 mg/m^2/day Days 1–4, 9–12, and 17–20 then monthly on Days 1–4	OR 25 CR 0	NR	NR
Dimopoulos (2001)[172]	67 (38–87)	44	NR	T 200–400 mg/day + D 20 mg/m^2/day Days 1–4, 9–12, and 17–20 then monthly on Days 1–4	OR 55 CR 0	TTP 4.2 mos	12.6 mos
Anagnostopoulos (2003)[173]	48 (31–77)	47	NR	T 200–600 mg/day + D 20 mg/m^2/day Days 1–5 q 15 days until response then T 100–150 mg/day + D 20 mg/m^2/day Days 1–5 monthly	OR 47 CR 13	NR	38 mos
Palumbo (2002, 2004)[174,175]	63	120	18	T 100 mg/day + D 40 mg/day Days 1–5 monthly	OR 52	PFS 17 mos	Not reached; $OS_{3\ y}$ 60%
Tosi (2004)[176]	66.5	20	13	T 100–400 mg/day (*n* = 8) or T 200 mg/day + D 40 mg/day Days 1–4 q 28 days (*n* = 12)	OR 45[a] CR 0	7	7 mos
Terpos (2005)[177]	63 (44–79)	35	22	T 200 mg/day + D 40 mg/day Days 1–4 q 15 days until maximal response then monthly on Days 1–4 + zoledronic acid 4 mg q 28 days	OR 57 CR 3	PFS 8 mos	19.5 mos

CR, complete response; D, dexamethasone; F/U, follow-up; mos, months; MTD, maximally tolerated dose; NR, not reported; OR, overall response (CR + PR); OS, overall survival; PFS, progression-free survival; PR, partial response; T, thalidomide; TTP, time to progression; y, year.

[a] In 7 of 12 patients treated with thalidomide + dexamethasone and in 2 of 8 treated with thalidomide only.

TABLE 28.5
Results of Phase II Trials of Thalidomide and Cyclophosphamide-Based Chemotherapy in Previously Treated Patients with Myeloma

	Patient Characteristics					Response		
Investigator (y) (Reference)	Median Age (y) (Range)	N (Evaluable)	Prior Therapy (%)	Median F/U (mos)	Treatment Regimen	Response Rate (%)	Response Duration	Overall Survival
Kropff (2003)[193]	NR	60 (57)	100	NR	T 100–400 mg/day, D 20 mg/m^2/day Days 1–4, 9–12, and 17–20 (during first cycle with optional reduction to once monthly pulses in cycles 2–6) + CY 300 mg/m^2 q 12 h × 6 doses Days 1–3 × 2–6 mos then TD maintenance	OR 72 CR 4	EFS 11 mos	19 mos
Williams (2004)[139]	55 (31–73)	61[a]	74	NR	T 100–200 mg/day, D 40 mg/day Days 1–4 and 15–18 + CY 500 mg Days 1, 8, and 15 q 28 days × 2–6 cycles then T maintenance	OR 78[b] CR not specified	NR	NR
Garcia-Sanz (2004)[194]	NR	71 (66)	100	18	T 200–800 mg/day, D 40 mg/day for 4 days + CY 50 mg/day q 21 days	OR 57 CR 2	$PFS_{2\,y}$ 57%	$OS_{2\,y}$ 66%
Dimopoulos (2004)[195]	64 (36–86)	53	100	NR	T 400 mg/day Days 1–5 and 14–18, D 20 mg/m^2/day Days 1–5 and 14–18 + CY 150 mg/m^2 q 21 h Days 1–5 q 28 days (CTD) × 3 cycles then responders received CTD maintenance on Days 1–5 q monthly	OR 60 CR 5	TTP 8.2 mos	17.5 mos
Kyriakou (2005)[196]	58 (34–75)	52	100	18	T 50–300 mg/day, D 40 mg/day × 4 days + CY 300 mg/m^2 q weekly every month until best response then T maintenance	OR 79 CR 17	$EFS_{2\,y}$ 34%	$OS_{2\,y}$ 73%
Glasmacher (2005)[189]	NR	39	NR	NR	T 400 mg/day, D 320 mg/cycle, CY 800 mg/m^2/cycle + idarubicin 40 mg/m^2/cycle × 3–8 cycles then patients with ≥SD randomized to T alone or T + oral idarubicin maintenance × 1 y	OR 57	NR	NR

(continued)

TABLE 28.5 (continued)
Results of Phase II Trials of Thalidomide and Cyclophosphamide-Based Chemotherapy in Previously Treated Patients with Myeloma

	Patient Characteristics					Response		
Investigator (y) (Reference)	Median Age (y) (Range)	*N* (Evaluable)	Prior Therapy (%)	Median F/U (mos)	Treatment Regimen	Response Rate (%)	Response Duration	Overall Survival
Moehler (2001)[197]	NR	56 (50)	100	14	T 400 mg/day, D 40 mg/day Days 1–4, CY 400 mg/m^2 IV Days 1–4 + etoposide 40 mg/m^2 IV Days 1–4 q 28 days × 3–6 cycles	OR 68 CR 4	PFS 16 mos	$OS_{1\ y}$ 62.6%
Lee (2003)[190]	60 (31–84)	236	100	NR	D 40 mg/day × 4 days, T 400 mg/day, cisplatin 10 mg/m^2/day ci × 4 days, doxorubicin 10 mg/m^2/day ci × 4 days, CY 400 mg/m^2/day ci × 4 days + etoposide 40 mg/m^2/day ci × 4 days	OR 32 CR 7; nCR 9	NR	NR

ci, continuous infusion; CR, complete response; CTD, cyclophosphamide, thalidomide, and dexamethasone; CY, cyclophosphamide; D, dexamethasone; EFS, event-free survival; F/U, follow-up; mos, months; NR, not reported; OR, overall response (CR + PR); OS, overall survival; PFS, progression-free survival; PR, partial response; SD, stable disease; T, thalidomide; TD, thalidomide and dexamethasone; TTP, time to progression; y, years.

[a] Includes untreated ($n = 15$) and relapsed or refractory ($n = 46$) patients.

[b] OR of the 46 previously treated patients only.

TABLE 28.6
Results of Trials of Thalidomide and Anthracycline-Based Chemotherapy in Previously Treated Patients with Multiple Myeloma

Investigator (y) (Reference)	Patient Characteristics: Median Age (y) (Range)	N (Evaluable)	Prior Therapy (%)	Median F/U (mos)	Treatment Regimen	Response: Response Rate (%)	Response Duration	Overall Survival
Glasmacher (2005)[189]	NR	39	NR	NR	T 400 mg/day, D 320 mg/cycle, CY 800 mg/m^2/cycle + idarubicin 40 mg/m^2/cycle × 3–8 cycles then patients with ≥SD randomized to T alone or T + oral idarubicin maintenance × 1 y	OR 57	NR	NR
Offidani (2006)[148]	68.5 (41–82)	50	100	12	T 100 mg/day, D 40 mg/day Days 1–4 and 9–12 + Doxil 40 mg/m^2 IV Day 1 q 28 days × 3–6 cycles	OR 76 CR 26; nCR 6	PFS 22 mos EFS 17 mos	Not reached
Hussein (2006)[138]	65.5 (57–71)	50	48	28[a]	T 50–400 mg/day, D 40 mg/day Days 1–4, VCR 2 mg IV Day 1 + Doxil 40 mg/m^2 IV Day 1 q 28 days × 6 cycles then TP maintenance	OR 76 CR 20	PFS 15.5 mos[b]	39.9 mos[b]
Lee (2003)[190]	60 (31–84)	236	100	NR	D 40 mg/day × 4 days, T 400 mg/day, cisplatin 10 mg/m^2/day ci × 4 days, doxorubicin 10 mg/m^2/day ci × 4 days, CY 400 mg/m^2/day ci × 4 days + etoposide 40 mg/m^2/day ci × 4 days	OR 32 CR 7; nCR 9	NR	NR
Chanan-Khan (2004)[191]	56 (44–80)[b]	18 (13)[b]	100	NR	T 200 mg/day, Doxil 20 mg/m^2 IV Days 1 and 15 + V1.3 mg/m^2 IV Days 1, 4, 15, and 18 q 28 days × 4–6 cycles	OR 38[c] CR 0	NR	NR
Hollmig (2004)[192]	NR	20 (14)	100	NR	T 50 or 100 mg/day Days 1–12, doxorubicin 2.5–10 mg/m^2 IV Days 1–4 and 9–12 + V 1 or 1.3 mg/m^2 IV Days 1, 4, 9, and 11 q 21 days	OR 50 CR 0	NR	NR

ci, continuous infusion; CR, complete response; CY, cyclophosphamide; D, dexamethasone; EFS, event-free survival; F/U, follow-up; M, melphalan; MM, multiple myeloma; mos, months; nCR, near complete response; NR, not reported; OR, overall response (CR + nCR + PR); OS, overall survival; P, prednisone; PFS, progression-free survival; PR, partial response; SCT, stem cell transplant; SD, stable disease; T, thalidomide; TP, thalidomide and prednisone; TTP, time to progression; V, Velcade (bortezomib); VCR, vincristine; WM, Waldenstrom's macroglobulinemia; y, year.

[a] Includes relapsed/refractory ($n = 50$) and newly diagnosed ($n = 55$) patients.
[b] Median PFS and OS for newly diagnosed patients ($n = 49$) only.
[c] Includes patients with MM ($n = 16$) and WM ($n = 2$).

TABLE 28.7
Results of Trials of Thalidomide and Bortezomib-Based Chemotherapy in Previously Treated Patients with Multiple Myeloma

	Patient Characteristics					Response		
Investigator (y) (Reference)	Median Age (y) (Range)	*N* (Evaluable)	Prior Therapy (%)	Median F/U (mos)	Treatment Regimen	Response Rate (%)	Response Duration	Overall Survival
Chanan-Khan (2004)[191]	56 (44–80)[a]	18 (13)[a]	100	NR	T 200 mg/day, Doxil 20 mg/m^2 IV Days 1 and 15 + V1.3 mg/m^2 IV Days 1, 4, 15, and 18 q 28 days × 4–6 cycles	OR 38[a] CR 0	NR	NR
Hollmig (2004)[192]	NR	20 (14)	100	NR	T 50 or 100 mg/day Days 1–12, doxorubicin 2.5–10 mg/m^2 IV Days 1–4 and 9–12 + V1 or 1.3 mg/m^2 IV Days 1, 4, 9, and 11 q 21 days	OR 50 CR 0	NR	NR
Zangari (2005)[204]	NR	85	NR	NR	V1 or 1.3 mg/m^2 IV days 1, 4, 8, and 11 q 21 days + T 50–200 mg/day q 21 days (added for cycle 2 onwards); D 20 mg/day on the day of and the day following each V dose q 21 days with cycle 4 if <PR achieved	OR 55 CR 0; nCR 16	EFS 9 mos	22 mos
Ciolli (2006)[205]	63 (53–76)	18 (17)	100	11	V1 mg/m^2 IV days 1, 4, 8, and 11, D 20 mg/day on the day of and the day following each V dose + T 100 mg/day every 28 days	OR 47 CR 12	NR	NR
Terpos (2005)[206]	66 (45–83)	31 (25)	100	NR	V1 mg/m^2 IV days 1, 4, 8, and 11, M 0.15 mg/kg/day Days 1–4, T 100 mg/day + D 12 mg/m^2/day Days 1–4 and 17–20 q 28 days × 8 cycles	OR 56 CR 8	NR	NR
Palumbo (2005)[207]	65 (38–73)	20 (15)	100	5	V 0.7–1.6 mg/m^2 IV days 1, 4, 15, and 22, M6 mg/m^2/day Days 1–5, T 100 mg/day + P60 mg/m^2/day Days 1–5 q 35 days × 6 cycles	OR 67 CR 13; nCR 6	NR	NR

CR, complete response; CY, cyclophosphamide; D, dexamethasone; EFS, event-free survival; F/U, follow-up; M, melphalan; MM, multiple myeloma; mos, months; nCR, near complete response; NR, not reported; OR, overall response (CR + nCR + PR); OS, overall survival; P, prednisone; PFS, progression-free survival; PR, partial response; T, thalidomide; TP, thalidomide and prednisone; TTP, time to progression; V, Velcade (bortezomib); VCR, vincristine; WM, Waldenstrom's macroglobulinemia; y, year.

[a] Includes patients with MM ($n = 16$) and WM ($n = 2$).

evaluated in conjunction with a variety of other agents, including the monoclonal anti-VEGF inhibitor bevacizumab,[208] the bcl-2 antisense oligonucleotide oblimersen,[209] the COX-2 inhibitor celecoxib,[210] and arsenic trioxide.[149]

28.2.4.2.2 Maintenance Therapy

The role of thalidomide as maintenance therapy following ASCT has been evaluated in previously treated patients with MM.[211,212] The median tolerated dose of thalidomide was 200 mg/day.[211,212] A retrospective comparison suggested that patients who received thalidomide as maintenance or salvage therapy post-ASCT had an improved median OS compared with those who did not (65.5 months vs. 44.5 months, respectively; $P = 0.09$).[212] Further analysis indicated that patients who received thalidomide as maintenance therapy had a superior OS compared with those who received thalidomide as salvage therapy (65 months vs. 54 months, respectively; $P = 0.05$).[212]

28.2.4.3 Toxicities

Thalidomide is associated with nonmyelosuppressive toxicities, which are usually dose-related, leading to intolerance. Teratogenicity is a well-known serious adverse effect of thalidomide; thus, to prevent fetal exposure to thalidomide, a restricted distribution risk management program, the System for Thalidomide Education and Prescribing Safety (STEPS), has been established in the United States.[213,214] Other serious adverse effects associated with its use are thrombotic complications and peripheral neuropathy. Venous thrombotic complications appear to be higher when thalidomide is used in combination with other agents, including dexamethasone, than as monotherapy (0%–35% vs. 4%–5%),[131,132,145,172,173,175,198–200] but may be reduced with aspirin[201,202] or therapeutic anticoagulation with low molecular weight heparin or wafarin.[131,200,203] In addition to VTEs, several cases of arterial thrombosis have been reported with thalidomide therapy; the mechanism for these events, like those for VTEs, is unclear.[215–217] Peripheral neuropathy, due to axonal damage,[218] appears to be dose-dependent and usually occurs after prolonged exposure to thalidomide or in patients with preexisting neuropathy. Neuropathy can occur in up to 80% of patients, whereas severe grade 3 or 4 neuropathy occurs in ~3%–5% of patients.[219] Neuropathy may be irreversible if the drug is not promptly dose reduced or discontinued. Other common side effects are somnolence and fatigue (>75%) and rash (>40%), which are responsive to dose reduction.[220–223] Constipation (80%–90%) is manageable with a high fiber diet and laxatives.[220,221,223] Less common toxicities include bradycardia (25%; severe 1%–3%), peripheral edema (15%), xerostomia (10%), and subclinical hypothyroidism (5%–20%; myxedema rare).[221,223] Infrequent adverse events include elevated serum transaminases and Steven–Johnson syndrome.[223,224]

28.2.5 Myelodysplastic Syndrome

The specific pathogenesis of MDS is unknown, but may involve elevated cytokines, such as TNF-α, which may promote apoptosis of intramedullary hematopoietic cells,[83,225–230] immune-mediated suppression of hematopoiesis,[231–233] or increased bone marrow angiogenesis.[10,13,16,19,20,23]

28.2.5.1 Thalidomide Monotherapy

Thalidomide has been used in patients with MDS to improve hematopoiesis (Table 28.8).[234–240] Responses were observed in 18%–66% of patients (CR 0%) at doses of 100–1000 mg/day, but responses were limited with the development of toxicities. Common toxicities included

TABLE 28.8
Results of Phase II Trials Using Thalidomide Patients with Myelodysplastic Syndrome

	Patient Characteristics						Response	
Investigator (y) (Reference)	Median Age (y) (Range)	*N* (Evaluable)	IPSS Score[324] (%)	Prior Therapy (%)	Median F/U (mos)	Thalidomide Dose (Median)/Regimen	Response Rate (%)	Response Duration
Thalidomide monotherapy								
Raza (2001)[236]	67	83 (51)	Low (25), Int-1 (45), Int-2 (14), high (16)	NR	NR	100–400 mg/day	OR 31 HI 31	10.2 mos[a]
Musto (2002, 2004)[237,238]	66 (48–85)	40 (25)	Low (45), Int-1 (30), Int-2 (18), high (8)	NR[b]	NR	100–300 mg/day	OR 32	11.5+ mos
Strupp (2002)[239]	67 (54–83)	34 (29)	Low (12), Int-1 (41), Int-2 (26), high (21)	32	13	100 to MTD (400 mg/day)	OR 66 PR 31; HI 34	10 mos
Bowen (2005)[235]	64 (51–76)	12 (11)	Int-1 (median)	NR	NR	100–800 mg/day	OR 18 HI 18	NR
Bouscary (2005)[234]	67	47 (39)	Low (17), Int-1 (55), Int-2 (15), high (0), Indeterminate (13)	NR	NR	200–800 mg/day	OR 59	8.7 mos
Moreno-Aspitia (2006)[240]	71–73 (51–89)	72 (68)	Low (19), Int-1 (38), Int-2 (32), high (10)	NR	23	200–1000 mg/day (300 mg/day)	OR 9 PR 2; HI 7	NR

Thalidomide combination therapy								
Steurer (2003)[241c]	66 (62–69)	7	Low and Int (100)	NR	NR	T 100 mg/day + darbepoietin-α 2.25 sc μg/kg/day	N/A	N/A
Musto (2006)[242]	63 (41–81)	30 (27)	Low (33), Int-1 (67)	100	10[d]	T 200 mg/day + r-EPO 40,000 U sc q weekly × 12 wks then T or EPO alone	OR 23 HI-E 23	7+ to 16+ mos
Raza (2004)[243]	65	28 (28)	Low (21), Int-1 (21), Int-2 (11), high (46)	NR	NR	T 100 mg/day + ATO 0.25 mg/kg IV × 5 days weekly for 2 wks followed by 2 wks off	OR 25	4.5 mos
Raza (2006)[244]	68 (52–79)	45 (38)	NR	NR	NR	Topotecan 1.25 mg/m^2 IV Days 1–5 q 21 days × 3–5 cycles then T 100–300 mg/day	OR 24 PR 24	NR
Westervelt (2006)[245]	70	29 (25)[e]	Low (6), Int-1 (41), Int-2 (29), high (24)	NR	NR	T 50–100 mg/day + azacitidine 75 mg/kg sc × 5 days q 28 days	OR 56[e] CR 24; HI 32[e]	NR

AML, acute myeloid leukemia; ATO, arsenic trioxide; CR, complete response; F/U, follow-up; HI, hematologic improvement; HI-E, hematologic improvement-erythroid; IPSS, International Prognostic Scoring System; MDS, myelodysplastic syndrome; mos, months; MTD, maximum tolerated dose; N/A, not applicable; NR, not reported; OR, overall response (CR + PR + HI); PR, partial response; r-EPO, recombinant erythropoietin; sc, subcutaneous; T, thalidomide; wks, weeks; y, year.

[a] For 15 responding patients only.

[b] Includes previously treated patients (n = not specified).

[c] Study closed prematurely because 3 of 7 patients (43%) developed venothrombotic events.

[d] For responding patients only.

[e] Includes patients with MDS ($n = 17$), AML ($n = 10$), and unknown ($n = 2$).

constipation (7%–82%), fatigue/somnolence (24%–79%), orthostatic symptoms/dizziness (25%–40%), rash (3%–31%), peripheral neuropathy (9%–29%), nausea (4%–27%), and VTEs (0%–6%).

28.2.5.2 Thalidomide and Targeted Therapy

Thalidomide has also been evaluated in conjunction with erythropoietin,[241,242] arsenic trioxide,[243] topotecan,[244] and azacitidine[245] (Table 28.8). Increased VTEs have been associated with the administration of thalidomide and darbepoietin,[241] but not shorter acting erythropoietin.[242]

28.2.6 Myelofibrosis with Myeloid Metaplasia

The term myelofibrosis with myeloid metaplasia (MMM) encompasses chronic idiopathic myelofibrosis (i.e., AMM) and secondary MMM (i.e., postthrombocythemic myeloid metaplasia (PTMM) and postpolycythemic myeloid metaplasia [PPMM]). Several cytokines, including bFGF, VEGF, and TNF, are implicated in the pathogenesis of the marrow fibrosis observed in patients with MMM.[10,28,246] TNF has been implicated in the inhibition of normal hematopoiesis, stimulation of fibroblast proliferation, pathogenesis of bone marrow fibrosis, and cancer-related cachexia.[247–250]

28.2.6.1 Thalidomide Monotherapy

Although responses of 0%–67% have been reported (Table 28.9),[251–258] a true estimate of the effect of thalidomide on patients with MMM is hampered by the lack of uniform response criteria between the studies.[259,260] Furthermore, responses refer to HI in anemia, leukopenia or leukocytosis, and thrombocytopenia and clinical improvements consisting of decreased splenomegaly and improvement of constitutional symptoms, and not the traditionally defined CR or PR state. None of the studies reported OS and only two groups indicated duration of response.[251,256] Common toxicities included fatigue/somnolence, constipation, peripheral neuropathy, orthostatic symptoms/dizziness, edema, rash, and abdominal discomfort. VTEs occurred in 3%–33% of patients.[253,257,258] Thrombocytosis and leukocytosis were observed in 7%–31% and 9%–23% of patients, respectively.[252,255,257,258]

In an attempt to overcome these limitations, an individual patient data meta-analysis was performed on 62 patients enrolled onto five different studies.[251–253,255,260,261] The median thalidomide dose was not specified; however, 51.6% of patients received >100 mg/day of thalidomide. Improvements in moderate to severe anemia, moderate to severe thrombocytopenia, and high-grade splenomegaly were observed in 29%, 38%, and 41% of patients, respectively. Whereas only 12.2% of patients had a decrease in the Dupriez prognostic score,[262] 44.9% of patients had a decrease in the investigator-developed "severity" score.[260] Sixty-six percent of patients discontinued the drug before 6 months of treatment due to intolerance. The most common toxicities were constipation (62.9%), fatigue (46.7%), paresthesias (24.2%), and sedation (22.6%). Neutropenia (grade ≤ 3) occurred in 9.7% of patients. One patient developed a VTE.

28.2.6.2 Thalidomide and Prednisone

The combination of thalidomide with prednisone improved tolerability, where only one of 21 patients discontinued the treatment[263] compared with 25% of patients receiving thalidomide at doses of 50 mg/day.[257] ORRs of 62% were observed. It remains to be determined whether improved and durable responses are obtained.

TABLE 28.9
Results of Phase II Trials Using Thalidomide in Patients with Myelofibrosis

	Patient Characteristics							Response	
Investigator (y) (Reference)	Median Age (y) (Range)	*N* (Evaluable)	Dupreiz Score[262] (%)	Median mos from Diagnosis (Range)	Prior Therapy (%)	Median F/U (mos)	Thalidomide Dose (Median)/Regimen	Response Rate (%)	Response Duration
Thalidomide monotherapy									
Canepa (2001)[251]	56 and 58[a] (35–74)	10 (10)	NR	5 and 17[a] (1–54)	70	9.5 and 6.2[a]	200–800 mg/day (400 mg/day)	OR 30	12.8 mos
Barosi (2001)[252]	66.5 (41–81)	21 (13)	Low (19), Int (67), high (14)	60 (9–216)	100[b]	NR	100–400 mg/day	OR 61.5	NR
Piccaluga (2002)[253]	63 (52–77)	12 (11)	Low (58), Int (25), high (17)	132 (60–192)	100	NR	100–600 mg/day	OR 64	NR
Merup (2002)[254]	68 (43–77)	15 (14)	NR	NR	40	NR	100–800 mg/day (400 mg/day)	OR 0	N/A
Elliott (2002)[255]	65 (41–79)	15 (13)	Low (20), Int (53), high (27)	36 (2–180)	NR	NR	200–1000 mg/day, reinitiated at 50 mg/day for at least 1 y	OR 20 + (anemia 20; splenomegaly 8)	NR
Strupp (2004)[256]	59 (52–78)	16 (15)	NR	NR	31	9[c]	100–400 mg/day (300 mg/day then 200 mg/day in responding patients)	OR 67	9 mos
Marchetti (2004)[257]	68 (43–80)	63 (49)	Low (29), Int (52), high (19)	51 (3–263)	30+[d]	NR	50–400 mg/day (100 mg/day)	OR 40 (anemia 22; splenomegaly 19; thrombocytopenia 41)	NR
Thomas (2006)[258]	65 (27–85)	44 (41)	Low (27), Int (48), high (25)	15 (1–122)	64	NR	200–800 mg/day (400 mg/day)	OR 41	NR
Thalidomide and prednisone									
Mesa (2003)[263]	63 (46–74)	21 (20)	Low (24), Int (48), high (28)	NR	NR	NR	T 50 mg/day + P 0.5 mg/kg/day tapered over 3 mos	OR 62	NR

F/U, follow-up; Int, intermediate; mos, months; N/A, not applicable; NR, not reported; OR, overall response; P, prednisone; T, thalidomide; y, year.

[a] Agnogenic and secondary myelofibrosis with myeloid metaplasia, respectively.
[b] Nine patients receiving concurrent hydroxyurea ($n = 8$) or danazol ($n = 1$).
[c] Responding patients only.
[d] Proportion of patients receiving concurrent cytostatics or steroids NR.

28.2.6.3 Thalidomide and Targeted Therapy

Constitutional symptoms (e.g., fatigue, fever, night sweats) and cachexia occur in 20% of patients with MMM.[259] Etanercept, a soluble recombinant form of the extracellular domain of the human p75 TNF-R fused to the Fc fragment of human IgG1, can alleviate constitutional symptoms (60%) and improve cytopenias and splenomegaly (20%) in patients with myelofibrosis.[264] Low-dose thalidomide (50 mg daily) in conjunction with prednisone (0.5 mg/kg/day tapered over 3 months) and etanercept (25 mg subcutaneously twice a week) (PET) was evaluated in 15 patients with symptomatic myelofibrosis.[265] Twelve patients (80%) completed the planned three cycles of therapy; nine (60%) patients with evidence of response continued on an additional 3 months of thalidomide and etanercept alone. Anemia and platelets were improved in 6 of 11 (54%) and 7 of 7 (100%) of patients, respectively. A 50% or more reduction in splenomegaly or hepatomegaly was observed in three (25% of eligible) and one (25% of eligible) patient(s), respectively. Constitutional symptoms resolved or were significantly improved. Grade 3 or 4 toxicities included elevated bilirubin, infection, blurred vision, and anemia. The study confirms the ability of etanercept in alleviating constitutional symptoms, but the PET regimen does not appear to induce more hematologic or organ responses compared with thalidomide and prednisone.

28.2.7 Non-Hodgkin's Lymphoma

28.2.7.1 Thalidomide Monotherapy

Single-agent thalidomide (escalating doses of 200–800 mg daily) was evaluated in 19 patients with previously treated NHL ($n = 17$) and HL ($n = 2$).[120] Three patients did not complete 4 weeks of therapy because of pancytopenia ($n = 1$) or progressive disease ($n = 2$). The median thalidomide dose was 400 mg daily. Only seven patients received the planned 800 mg daily dose of thalidomide. The ORR was 5% (one CR in a patient with mucosa-associated lymphoid tissue (MALT) lymphoma which was associated with decreased levels of VEGF and b-FGF).

A phase II multicenter study evaluated escalating doses of thalidomide (50–800 mg daily) in 25 previously treated patients with follicular lymphoma and small lymphocytic lymphoma.[266] The median daily dose was 400 mg. Grade 3 or 4 toxicities include neutropenia, anemia, dyspnea, fatigue, neuropathy, somnolence, dizziness, depression, and anxiety. The ORR was 8% (CR 4%; PR 4%). Median EFS was 2.6 months with a median OS of 23.3 months. There have also been anecdoctal reports of activity of thalidomide in patients with angioimmunoblastic T-cell lymphoma.[267,268] Despite an acceptable toxicity profile, single-agent thalidomide appears to have minimal activity in previously treated patients with NHL.

28.2.7.2 Thalidomide and Targeted Therapy

A phase II study evaluated a treatment strategy targeting both lymphoma cells (by rituximab) and the microenvironment (by thalidomide) in 16 patients with relapsed or refractory mantle cell lymphoma (MCL).[269] Rituximab was administered at standard dose and schedule for four weekly doses concomitantly with thalidomide (200 mg daily escalated to 400 mg on Day 15); therapy continued until disease progression or relapse. The planned thalidomide dose of 400 mg could only be achieved in six patients because of thalidomide-related toxicities. Maintenance dose (50–200 mg daily) of thalidomide was adapted on an individual basis. Grade 4 neutropenia occurred in one patient and VTEs occurred in two patients. The ORR was 81% (CR 25%; unverified CR [CRu] 6%; PR 50%). Median progression-free survival (PFS) for all patients was 20.4 months, with an estimated 3 year OS of 75%. In patients achieving a CR, PFS after rituximab and thalidomide was longer than that obtained after the preceding chemotherapy. Therefore, further evaluation of a similar rituximab-based regimen is warranted.

28.2.8 Plasma Cell Leukemia

Less than a dozen of patients with plasma cell leukemia (PCL) have been treated with thalidomide either as a single agent or in combination with either dexamethasone or dexamethasone, doxorubicin, cyclophosphamide, and etoposide followed by bortezomib; all administered predominantly in the salvage setting.[270–276] Therefore, its role in the treatment of PCL is unclear.

28.2.9 Systemic Amyloidosis

Amyloidosis is a plasma cell dyscrasia with a small monoclonal population of plasma cells in the bone marrow.[277] Responses have been observed with similar treatment regimens as those employed in patients with MM.[277]

28.2.9.1 Thalidomide Monotherapy

Escalating doses of thalidomide (100–400 mg) have been administered to six patients with previously treated primary amyloidosis and renal involvement.[278] Four patients also received oral corticosteroids. Two patients were also diagnosed with MM ($n = 1$) and WM ($n = 1$). The individual MTDs ranged from 100 to 400 mg. Median duration of therapy was 5 months. Five patients showed improvement, including recovery of macroglossia and speech impairment (2 of 2 patients), improvement of joint symptoms (2 of 2 patients), and resolution of diarrhea (1 of 3 patients). Four patients also experienced improvement in renal symptoms with decreased peripheral edema and decreased or discontinued diuretic use; three patients had a decrease in proteinuria (by >50% in two patients). Quality of life improved in four patients. Two patients died from disease progression. The remaining four patients continued to do well on thalidomide and corticosteroids.

A phase I/II trial evaluated thalidomide in 16 patients with primary AL amyloidosis, most of whom had failed prior therapy with high-dose melphalan and ASCT.[279] Thalidomide was administered at a starting dose of 200 mg daily, and escalated every 2 weeks as tolerated. Fourteen patients had renal involvement, four cardiac involvement, four liver involvement, and two had predominantly lymph node or soft tissue involvement. The median MTD was 300 mg. Fatigue and other central nervous system side effects represented the major DLTs. Toxicities, infrequently reported for other patient populations, included exacerbation of peripheral and pulmonary edema and worsening azotemia. Fifty percent of patients experienced grade 3 or 4 toxicities; 25% discontinued thalidomide due to side effects. No complete hematologic responses were observed. Although 25% of patients had a significant reduction in Bence-Jones proteinuria, none had a significant reduction in their nephrotic range proteinuria.

Results of a phase II study of thalidomide in 12 patients with primary amyloidosis have been presented.[280] Significant toxicities were noted and included progressive edema, cognitive difficulties, and constipation, and dyspnea, dizziness, and rash. Five patients developed progressive renal insufficiency. One patient each developed a VTE and syncope. Median time on study was 2.4 months. All 12 patients have discontinued therapy (six due to toxicities, four disease progression, and two death). No responses were observed possibly due to rapid patient attrition. Thalidomide at doses of ≥200 mg/day is poorly tolerated in patients with primary light-chain amyloidosis.[279–281]

A phase II trial evaluated lower doses of thalidomide monotherapy (50–400 mg/day) in 18 patients with systemic amyloidosis.[282] Thirteen patients had received prior therapy, including ASCT ($n = 3$). The median tolerated dose was 100 mg/day. Median time on study was 5.6 months. The overall organ response rate was 11%. There were no hematologic responses. The most common drug-related adverse events were constipation, edema, sinus bradycardia, dyspnea, and paresthesias. Light-headedness, elevated creatinine, thrombosis, infection, syncope, and rash occurred in 5.6% of patients. Therefore, low-dose thalidomide is tolerable but has limited efficacy in this group of patients.

28.2.9.2 Thalidomide and Dexamethasone

Previous studies have demonstrated that high-dose dexamethasone can achieve a 35% response rate in unselected patients with primary amyloidosis.[283] Therefore, the efficacy of thalidomide (100 mg orally daily escalated to 400 mg daily) combined with dexamethasone (20 mg orally on Days 1–4 every 21 days for up to nine cycles) was evaluated in 31 patients with primary amyloidosis who did not respond to or whose disease relapsed after first-line therapy.[284] Patients did not receive prophylaxis against VTEs. Sixty-one percent of patients had two or more organs involved (71% renal, 38% cardiac, 23% hepatic, 10% skin, and 10% gastrointestinal). Thirty-two percent had received prior ASCT. Only 11 patients (35%) tolerated the target thalidomide dose of 400 mg daily for at least 1 month. Median MTD was 300 mg/day. Sixty percent of patients experienced grade 3 or 4 thalidomide-related toxicity (eight patients with symptomatic bradycardia; four sedation/fatigue; two constipation; two acute dyspnea; and one each with VTE, skin lesions, epilepsy, and renal failure). The overall hematologic response rate was 48% (CR 19%; 26% functional improvement of organs involved). The response rate was higher among patients receiving the 400 mg/day dose (overall hematologic response 73% vs. 35%, respectively; CR 27% vs. 15%, respectively; organ response 36% vs. 20%, respectively). Cardiac amyloidosis did not improve. Median time to response was 3.6 months. Duration of response was not reported. Median follow-up was 32 months for surviving patients. Nine patients died after a median follow-up of 9 months (seven patients due to heart failure; one sudden death; and one renal failure).

28.2.9.3 Thalidomide and Chemotherapy

Thalidomide-based therapy was administered to 99 patients with systemic AL amyloidosis in whom cytotoxic therapy was deemed either ineffective or too toxic to pursue.[285] Thalidomide was administered as monotherapy ($n = 56$), in conjunction with dexamethasone ($n = 12$), cyclophosphamide ($n = 8$), cyclophosphamide and dexamethasone ($n = 13$), melphalan ($n = 5$), melphalan and dexamethasone ($n = 4$), or with maintenance antirejection therapy in a recipient of a solid organ transplant ($n = 1$). Thalidomide was administered for a median of 5 months at a median dose of 100 mg/day. Adverse events occurred in 75 patients (76%), including fatigue/somnolence, neuropathy, significant constipation, mental changes, and edema. Six patients (6%) developed VTEs. Thalidomide was discontinued in 41% of patients due to therapy-related toxicities. No CRs occurred. Partial hematologic responses (defined as either a >50% reduction in free light chains [FLC] or paraproteins) were observed in 34% and 36% of patients, respectively. Organ function improved in 23% of patients and sarum amyloid (SAP) scintigraphy showed regression of amyloid in 18% of patients. Median follow-up was 11 months. Median OS from thalidomide therapy was 26 months. Nineteen patients died from disease progression and one from a fatal pulmonary embolus. Survival was significantly better when dexamethasone was part of the regimen ($P < 0.012$). Thus, although lower doses of thalidomide may be better tolerated, fewer CRs are achieved.

Preliminary results of risk-adapted thalidomide-based combination therapy using cyclophosphamide 500 mg once weekly, thalidomide 100–200 mg/day continuously, and dexamethasone 40 mg Days 1–4 and 9–12 every 21 days (CTD) in 43 patients with AL amyloidosis have been reported.[286] The regimen was attenuated in patients over the age of 70 years, with heart failure, or significant fluid overload to cyclophosphamide 500 mg Days 1, 8, and 15, thalidomide 50–200 mg/day continuously, and dexamethasone 20 mg Days 1–4 and 15–18 every 28 days (CTDa). Thirty-five patients received CTD and eight CTDa. Median number of organs involved was two. Median follow-up was 7 months. Toxicities occurred in 39% of patients (3 receiving CTDa and 14 CTD) necessitating dose reduction (25%), dexamethasone omission (14%), thalidomide omission (2%), and complete regimen discontinuation (9%). Treatment-related toxicities included worsening heart failure, neuropathy, infections, somnolence, neutropenia, fatigue, renal impairment, and constipation. No VTEs or treatment-related mortality was observed. Of the 35 evaluable patients, the overall hematologic response was 74% (complete 34%; partial 40%). Median survival has not been reached at 36 months. This regimen appears to be more effective than thalidomide with or without dexamethasone; however, a significant proportion of patients required dose reduction or discontinuation of therapy.

As melphalan, thalidomide, and dexamethasone all have activity in patients with primary amyloidosis, a phase II study of risk-adapted melphalan followed by TD in patients with newly diagnosed, previously untreated primary systemic amyloidosis is currently underway.

28.2.10 Waldenstrom's Macroglobulinemia

28.2.10.1 Thalidomide Monotherapy

Single-agent thalidomide has been evaluated in 20 patients with WM (10 untreated; 10 primary refractory or relapsed).[287] Thalidomide was administered at a starting dose of 200 mg daily with escalation to a maximum of 600 mg. All previously treated patients had received an alkylating agent–based regimen. The average daily dose of thalidomide was 300 mg. The target dose of 600 mg was administered only to five patients. Adverse events were common but reversible and consisted primarily of grade 1 or 2 constipation, somnolence, fatigue, and mood changes. Grade 3 constipation ($n = 2$) and somnolence ($n = 2$) were observed. In seven patients (35%), thalidomide was discontinued because of intolerance. One patient with chronic atrial fibrillation died of an embolic cerebrovascular accident 5 months after initiation of thalidomide. The ORR was 25% (PR 25%). Responses occurred in 3 of 10 previously untreated and 2 of 10 pretreated patients. None of the patients treated during refractory relapse or with disease duration exceeding 2 years responded. Median time to response was 2.5 months. After response was achieved, responding patients were maintained on low-dose thalidomide (200 mg daily [$n = 3$] or 100 mg daily [$n = 2$]). Median response duration was 11 months. Therefore, thalidomide has activity in WM but only low doses were tolerated in this older population.

28.2.10.2 Thalidomide and Dexamethasone

Two prospective phase II studies have evaluated the nonmyelosuppressive combination of clarithromycin (500 mg orally twice daily continuously), low-dose thalidomide, and dexamethasone (40 mg orally once per week) (BLT-D) in previously treated patients with WM.[186–188] Coleman et al. treated 12 patients with WM with the BLT-D regimen for a minimum of 6 weeks.[186,188] Thalidomide was administered at a dose of 50 mg daily and escalated to

200 mg/day. All patients received enteric-coated aspirin 81 mg orally daily. Grade 3 or 4 toxicities were neurological (42%), endocrinological (17%), and gastrointestinal (9%). Other toxicities, including cardiac disease and edema, were minimal (8%). Eight percent of patients developed grade 1 or 2 thrombotic events; no thrombotic events occurred after the initiation of aspirin. The ORR was 83% (nCR 25%; major response 25%; PR 33%; minor response 17%). Patients with minor responses were unable to tolerate thalidomide doses >50 mg daily due to toxicity. Most responding patients achieved >50% of their maximal response within 2.5 months of therapy. At the time of reporting, only 17% remained on therapy, with a median time on treatment of 7 months. Therapy was discontinued because of neurotoxicity (42%), treatment changes (17%), and therapeutic resistance (25%). These responses appear to be higher than those achieved with single-agent thalidomide (OR 25%; PR 25%)[287]; however, the ORR, quality, rapidity, and duration of responses are lower than that achieved in patients with MM.[188] Similarly, treatment-related toxicity, predominantly neurotoxicity, was also increased in patients with WM compared with those with MM. Increased neurotoxicity may be due to subclinical neuropathy from the macroglobulinemia in patients with WM.[186]

In the second phase II study, thalidomide was administered at a dose of 200 mg orally daily 12 previously treated patients with WM.[187] Four patients had previously been treated with thalidomide without a response. Adverse events occurred in all patients and were usually mild-to-moderate and reversible on dose reduction (constipation, somnolence/fatigue, edema/Cushing's-like syndrome, gastric irritation, tremor, headache, peripheral neuropathy, hyperglycemia, and proximal muscle weakness). However, residual neuropathy persisted despite discontinuation of thalidomide. Thalidomide dose reduction was required in 42% of patients because of toxicities, with two patients discontinuing therapy because of neuropathy. The ORR was 42% (PR 25%; minor response 17%). Two of the three patients who achieved a PR had received prior thalidomide. Median time to PR was 1.5 months. Median response duration was 8+ months.

28.2.10.3 Thalidomide and Targeted Therapy

Rituximab, a chimeric monoclonal anti-CD20 antibody, can induce responses ranging from 27% to 75% in patients with both untreated and previously treated WM.[288–294] Therefore, a phase II study evaluated combination therapy with rituximab and thalidomide in 25 patients with WM who had not received prior therapy with rituximab or thalidomide (20 were previously untreated).[295] Thalidomide was administered continuously for 52 weeks at an initial dose of 200 mg daily then escalated to 400 mg daily and rituximab was administered at standard weekly dose on weeks 2–5 and 13–16. Grade 3 or 4 thalidomide-related toxicities included neuropathy (60%), somnolence or confusion (48%), rash (28%), tremors (8%), bradycardia (8%), which led to its discontinuation in 11 of 19 patients. Six other patients were removed from study due to lack of response ($n = 4$) and death unrelated to treatment regimen ($n = 2$). All evaluable patients received the intended rituximab therapy. Paradoxical IgM spikes following rituximab therapy were observed in 48% of patients, similar to previous reports.[296,297] Of the 23 evaluable patients, the ORR was 65% (CR 4%; PR 52%; minor response 9%). None of the 19 patients with SD or better have progressed with a median follow-up of 10 months. Response to therapy was associated with a higher cumulative dose of thalidomide. In view of these encouraging results and potential for improved toxicity profile, a phase II study evaluating lenalidomide and rituximab in this patient population is currently underway. Furthermore, the combination of thalidomide, liposomal doxorubicin, and bortezomib is evaluated in this patient population.[191]

28.3 LENALIDOMIDE (REVLIMID; CC-5013)

28.3.1 Chronic Lymphocytic Leukemia

28.3.1.1 Lenalidomide and Targeted Therapy

A phase II study is currently evaluating lenalidomide with or without rituximab in patients with refractory or relapsed CLL.[298] Twenty-nine patients have been treated with lenalidomide 25 mg daily for 21 days every 28 days for up to 12 courses. Patients with progressive disease are to receive rituximab 375 mg/m^2 intravenously (IV) on Days 1, 8, and 15 during the first course of treatment and on Days 1 and 15 of all subsequent courses. The most common grade 3 or 4 toxicities consisted of neutropenia and thrombocytopenia. Flare reactions and tumor lysis syndrome were observed in 79% of patients; flare reaction did not correlate with response.[299] Hematologic toxicity was the most common adverse event requiring dose reduction. Of the 19 evaluable patients, the ORR was 68% (CR 16% of whom two achieved a molecular CR; PR 53%). No one has received rituximab. Further follow-up is required to determine the durability of the responses. The clinical activity of lenalidomide does not appear to be mediated by a direct apoptotic effect on the CLL cells, but potentially by modulation of the tumor microenvironment.[300] A slow dose escalation schema, starting at 15 mg, is investigated.

28.3.2 Multiple Myeloma

28.3.2.1 Previously Untreated Patients

28.3.2.1.1 Lenalidomide and Dexamethasone

A phase II study indicated that lenalidomide and dexamethasone (LD) was highly active in newly diagnosed patients with MM with fewer nonhematologic toxicities compared with thalidomide (ORR 91% vs. 52%–63%, respectively; CR 6% vs. 4%–10%, respectively).[132,136,301] Therefore, a Southwest Oncology Group (SWOG) phase III study is comparing front-line therapy with LD to dexamethasone alone in patients with MM.[302] A second phase III study by the Eastern Cooperative Oncology Group (ECOG) is comparing lenalidomide and standard dose dexamethasone with lenalidomide and low-dose dexamethasone in newly diagnosed patients with MM, in an attempt to reduce toxicity while preserving efficacy. Results have not been reported.

In an attempt to improve responses achievable with lenalidomide and steroids, clarithromycin (BiRD)[303] has been combined with this regimen. A phase II trial has evaluated BiRD in 46 untreated patients with MM.[303] The ORR was 95% (CR 25%; nCR 5%; PR 65%). Nineteen patients experienced grade 3 toxicities: anemia, neutropenia, thrombocytopenia, and thrombosis (15%). Four of the seven thrombotic events occurred while off aspirin; two were fatal.

28.3.2.1.2 Lenalidomide and Chemotherapy

Preliminary results of the phase I trial of lenalidomide, melphalan, and prednisone (R-MP) in 50 patients (older than 65 years) with symptomatic MM have been presented.[304] Significant responses were obtained after one cycle of therapy (PR 51%). The ORR after three cycles of therapy was 70% (CR 10%; PR 60%). Toxicities were manageable and consisted on neutropenia, thrombocytopenia, cutaneous eruptions, infections, and neutropenic fever. There were no VTEs.

28.3.2.2 Previously Treated Patients

28.3.2.2.1 Lenalidomide Monotherapy

Two phase I dose escalation trials of lenalidomide in advanced MM reported 25 mg daily as the MTD, with responses occurring in the majority of patients, including those who had received prior thalidomide.[305,306] These studies provided the rationale for the subsequent phase II studies with single-agent lenalidomide (Table 28.10).[307–309] Five different dosing schedules have been evaluated in two separate randomized phase II trials.[307,308] In one study, preliminary results after the initial 38 patients had been treated indicated that lenalidomide 50 mg daily for 10 days every 28 days appeared to be inferior to 25 mg daily for 20 days every 28 days (ORR 21% vs. 42%, $P = 0.162$); therefore, the dosing schedule was modified to compare lenalidomide 50 mg daily every other day for 10 days every 28 days with lenalidomide 25 mg daily for 20 days every 28 days.[307] No significant neurotoxicity or sedation was observed. Myelosuppression, especially thrombocytopenia, was apparently dose-limiting (however, no further details are available).[307] At the time of analysis, the ORR was 25% vs. 9% in favor of the 25 mg daily arm.

A second trial compared lenalidomide 15 mg twice daily for 21 days every 28 days to 30 mg once daily for 21 days every 28 days.[308] Although no difference in responses (OR 14% vs. 18%, respectively; CR 0% vs. 6%, respectively) or toxicities were observed between the two arms, the time to first occurrence of clinically significant grade 3 or 4 myelosuppression was shorter in the twice daily arm (1.8 months vs. 5.5 months, $P = 0.05$).[308] Based on the encouraging response rates and PFS, including in patients who had previously received thalidomide, a larger multicenter phase II trial is evaluating single-agent lenalidomide in patients with relapsed or refractory MM.[309]

28.3.2.2.2 Lenalidomide and Dexamethasone

In the multicenter phase II trial of lenalidomide, 68 of 102 patients with MM who had progressive or stable disease after two cycles of therapy continued lenalidomide treatment and received dexamethasone 40 mg daily for 4 days every 14 days.[308] Fifteen (22%) patients achieved a response (CR 1%; PR 21%). A phase IV expanded access, multicenter study is currently evaluating lenalidomide and high-dose dexamethasone (Days 1–4, 9–12, and 17–20 for the first four 28 day cycles and on Days 1–4 for all subsequent cycles).

Two pivotal phase III randomized, double-blind, placebo-controlled trials (MM-009 and MM-010) comparing LD with dexamethasone alone for patients with relapsed or refractory MM have been performed (Table 28.11).[310,311] Grade 3 or 4 toxicities were increased in the LD arm (16.5% vs. 1.2%, respectively).[310] Although grade 3 or 4 neutropenia was increased in the LD arm (24% vs. 3.5%, respectively),[311] grade 3 or 4 infections were similar between the two arms.[310,311] VTEs were documented in 8.5%–15% of patients receiving LD and in 3.5%–4.5% of patients receiving dexamethasone only.[310,311] ORRs were significantly higher for patients treated with LD than with dexamethasone alone (58%–59% vs. 21%–22%, respectively; $P < 0.001$). In March 2005, the studies were unblinded based on recommendations from an independent Data Safety Monitoring Committee, as they exceeded the prespecified interim efficacy end point ($P < 0.0015$) for disease progression; lenalidomide was then offered to all patients treated on the dexamethasone arm.[312] Furthermore, the MM-009 trial demonstrated an improved OS for patients receiving LD (not reached vs. 24 months, respectively; $P = 0.0125$).[311]

28.3.2.2.3 Lenalidomide and Chemotherapy

Two phase I/II trials are evaluating the safety and efficacy of lenalidomide in combination with doxorubicin-based regimens (i.e., lenalidomide, doxorubicin, and dexamethasone [RAD] and liposomal doxorubicin, vincristine, reduced frequency dexamethasone, and lenalidomide

TABLE 28.10
Results of Trials Using Lenalidomide-Based Regimens in Previously Treated Patients with Multiple Myeloma

Investigator (y) (Reference)	Patient Characteristics: Median Age (y) (Range)	Patient Characteristics: *N* (Evaluable)	Median F/U (mos)	Treatment Regimen	Response: Response Rate (%)	Response: Response Duration	Response: Overall Survival
Lenalidomide monotherapy							
Barlogie (2003)[307]	NR	47	NR	L 25 mg daily × 20 days or 50 mg daily × 10 days[a] or 50 mg q other day × 10 doses q 28 days	OR 25 vs. 9[b]	NR	NR
Richardson (2006)[308]	60 (38–90)	102	31	L 15 mg bid ($n = 35$) or 30 mg daily ($n = 67$) × 21 days q 28 days[c]	OR 14 vs. 18[c] CR 0 vs. 6[c]	23 mos vs. 19 mos[c] PFS 3.5 mos vs. 8.3 mos[c] PFS 3.9 mos vs. 7.7 mos	27 mos vs. 28 mos
Richardson (2005)[309]	NR	222 (212)	NR	L 30 mg daily × 21 days q 28 days	OR 25	TTP 5.6 mos	Not reached
Lenalidomide and chemotherapy							
Hussein (2004)[313]	62 (53–71)	25 (21)	NR	L 5–15 mg/day × 21 days, Doxil 40 mg/m² IV Day 1, VCR 2 mg IV Day 1 + D 40 mg/day × 4 days q 28 days followed by L ± P maintenance	OR 67 CR 14; nCR 19	NR	NR
Gerecke (2005)[314]	NR	6	NR	L 10 or 15 mg/day × 21 days, doxorubicin 4, 6, or 9 mg/m² IV Days 1–4 + D 40 mg/day Days 1–4 and 17–20 q 28 days	NR	NR	NR
Lenalidomide and targeted therapy							
Richardson (2005)[317]	NR	19 (17)	NR	L 5–20 mg/day Days 1–14 + V 1 or 1.3 mg/m² Days 1, 4, 8, and 11 q 21 days	OR 59 CR 6; nCR 6	NR	NR

P = not significant unless otherwise specified.

CR, complete response; D, dexamethasone; F/U, follow-up; L, lenalidomide; mos, months; nCR, near CR; NR, not reported; NS, not significant; OR, overall response (CR + nCR + PR); P, prednisone; PFS, progression-free survival; PR, partial response; TTP, time to progression; V, Velcade (bortezomib); VCR, vincristine; y, year.

[a] After entry of 38 patients, lenalidomide 50 mg daily × 10 days appeared inferior to 25 mg daily × 20 days (OR 21% vs. 42%, $P = 0.162$); therefore, changed dosing to 50 mg every other day × 10 doses.

[b] OR for 25 mg daily vs. 50 mg daily and 50 mg q other day dosing scheduled.

[c] 68 patients ($n = 27$ on twice daily arm and $n = 41$ on once daily arm) also received dexamethasone (responses censored at the time of dexamethasone addition).

TABLE 28.11
Results of Phase III Trials Using Lenalidomide-Based Regimens in Previously Treated Patients with Multiple Myeloma

	Patient Characteristics				Response		
Investigator (y) (Reference)	Median Age (y) (Range)	N (Evaluable)	Median F/U (mos)	Treatment Regimen	Response Rate (%)	Response Duration	Overall Survival
Dimopoulos (2005) MM-010[310]	NR	351	NR	LD vs. placebo + D	OR 58 vs. 22[a]	TTP 13.3 mos vs. 5.1 mos[b]	NR
Weber (2006) MM-009[311]	NR	354	NR	LD vs. placebo + D	OR 59.4 vs. 21.1[a] CR 12.9 vs. 0.6	TTP 11.1 mos vs. 4.7 mos[b]	Not reached vs. 24 mos[c]

CR, complete response; D, dexamethasone; F/U, follow-up; LD, lenalidomide and dexamethasone; NR, not reported; mos, months; OR, overall response; TTP, time to progression; y, year.

[a] $P < 0.001$.

[b] $P < 0.000001$.

[c] $P = 0.0125$.

[DVd-R]) in patients with relapsed or refractory MM (Table 28.10).[313,314] Of the 21 evaluable patients treated with DVd-R, the DLT was non-neutropenic sepsis or septic shock that occurred in two patients treated at dose level 3 (i.e., lenalidomide 15 mg daily for 21 days every 28 days).[313] Therefore, the MTD for lenalidomide was 10 mg daily for the DVd-R regimen. Grade 3 neutropenia and neuropathy occurred in 10% and 5% of patients, respectively. One patient developed a pulmonary embolus. Responses were observed in 67% (CR 14%, nCR 19%, PR 33%) of patients. Therefore, despite the concern that lenalidomide, unlike thalidomide, causes myelosuppression, DVd-R appears to be an effective regimen in patients with refractory MM with minimal toxicity. However, the durability of these responses is unclear.

28.3.2.2.4 Lenalidomide and Targeted Therapy

Preclinical data indicate that the combination of lenalidomide with rapamycin or bortezomib has synergistic or additive effects inhibiting the growth of myeloma cells.[315,316] Therefore, lenalidomide is currently evaluated in combination with bortezomib-based regimens as salvage therapy in patients with relapsed or refractory MM (Table 28.10).[317]

28.3.2.3 Toxicities

Lenalidomide is effective in patients refractory to thalidomide, with myelosuppression as the most common side effect (up to 80%), which is usually reversible.[306] Unlike thalidomide, lenalidomide causes almost no sedation and is associated with less fatigue (31%), constipation (24%), and neurotoxicity (8%).[224,312] However, the addition of dexamethasone or erythropoietin to lenalidomide is associated with a greater risk of thromboembolism than single-agent dexamethasone in patients with MM (11%–75% vs. 4%–7%, respectively).[302,311,318,319] It is unclear whether administration of lower doses of dexamethasone with lenalidomide will decrease the thrombotic incidence.[320] At the current time, prophylactic anticoagulation or low-dose aspirin is indicated and possibly minimizing administration of erythropoietin to patients with MM receiving LD.[202,321,322] Other side effects include rash and pruritus (36%–42%), nausea (23%), and arthralgias (22%).[224]

28.3.3 Myelodysplastic Syndrome

28.3.3.1 Lenalidomide Monotherapy

Forty-three patients with MDS and transfusion-dependent or symptomatic anemia received single-agent lenalidomide at doses of 10 or 25 mg orally daily continuously or 10 mg orally per day for 21 days every 28 days.[323] All patients had either failed therapy with recombinant erythropoietin or had an endogenous serum level of >500 U/L. Eighty-eight percent of patients had a low- or intermediate-1 (Int-1) risk International Prognostic Scoring System (IPSS) score.[324] Neutropenia and thrombocytopenia occurred in 65% and 74% of patients, respectively, and necessitated treatment interruption or dose reductions in 58% of patients. Other common toxicities were pruritus, diarrhea, and urticaria. The ORR was 56% (major HI-erythroid [HI-E] 49%; minor HI-E 7%).[325] Responses were observed at all dose levels; median time to response increased from 9 to 11.5 weeks for patients treated with the 25 mg dose and those treated with 10 mg daily for 21 out of 28 days, respectively. The response rate was highest among those patients with a clonal interstitial deletion involving chromosome 5q31.1 compared with those with a normal karyotype or other abnormal karyotypes (83% vs. 57% vs. 12%, respectively; $P = 0.007$). However, the pharmacological target in the chromosome 5q31 region remains to be defined.[326,327] Disease duration, FAB classification, IPSS risk category, and number of prior therapies did not correlate with response. Of the 20 patients with clonal cytogenetic abnormalities, 10 had a complete cytogenetic remission (CCyR). After a median follow-up of 81 weeks, the median response duration for major HI-E had not been reached (after >48 weeks) and the median hemoglobin level was 132 g/L compared with baseline hemoglobin levels of 80–83 g/L.

A confirmatory phase II multicenter study evaluated the clinical efficacy of lenalidomide in 148 patients with low- or Int-1 risk MDS, clonal interstitial deletion involving chromosome 5q31, and transfusion-dependent anemia.[328] Lenalidomide was administered at a dose/schedule of either 10 mg orally daily for 21 days every 28 days ($n = 45$) or 10 mg orally daily continuously ($n = 103$). Low- or Int-1 risk MDS was confirmed in only 120 patients. The most frequently reported toxicities were thrombocytopenia (58%) and neutropenia (57%). Grade 3 or 4 thrombocytopenia or neutropenia was observed in 54% and 55%, respectively. Other common adverse events were pruritus (32%), rash (28%), diarrhea (24%), and fatigue (12%). Eighty percent of patients required dose reductions. Twenty-four percent of patients were removed before evaluation of response at 24 weeks (due to a lack of benefit in 5%, adverse event 11%, and other 7%). In total, 15 deaths occurred, 2 of which were suspected to be drug related. In an intent-to-treat analysis, the ORR was 75%; transfusion independence was achieved in 66% of patients. A minor HI-E was observed in 9% of patients. Of 115 patients, 51 (44%) achieved a CCyR. Median time to response was 4.4 weeks. Probability of hematologic and cytogenetic response was independent of karyotype complexity and chromosomal deletion break point. Hematologic and cytogenetic responses were significantly higher in patients with preserved thrombopoiesis and lower transfusion requirements. Disease progression may be associated with the inability of lenalidomide to inhibit angiogenesis in the bone marrow ($P = 0.0005$).[329] After a median follow-up of 58 weeks, the median response duration for transfusion independence had not been reached (after >47 weeks). Lenalidomide appears to be a cost-effective treatment option in transfusion-dependent patients with low- or Int-1 risk MDS and an associated deletion 5q31 abnormality.[330]

Based on the earlier findings, on December 27, 2005, the FDA granted Subpart H approval (restricted distribution) to lenalidomide for use in patients with transfusion-dependent anemia due to low- or Int-1 risk MDS associated with a deletion 5q cytogenetic abnormality with or without additional cytogenetic abnormalities. A randomized, double-blinded, European trial is currently evaluating these two doses of lenalidomide compared with placebo in low- or Int-1 risk MDS patients with transfusion-dependent anemia and a deletion 5q31 abnormality (http://www.leukemia-net.org).

A third phase II study is evaluating the effect of lenalidomide in patients with low- or Int-1 risk MDS and transfusion-dependent anemia without $5q^-$ aberrations.[331,332] Lenalidomide was administered at 10 mg orally daily for 21 days every 28 days or 10 mg orally daily continuously. Of 215 patients enrolled in the study, 169 had confirmed low- or Int-1 risk MDS. Grade 3 or 4 neutropenia and thrombocytopenia occurred in 24% and 19% of patients, respectively, and necessitated dose interruption or dose reduction. In an intent-to-treat analysis, ORR was 44% (transfusion independence 27%; minor HI-E 18%). Median duration of transfusion independence was 43 weeks; median increase in hemoglobin was 33 g/L. Of the 169 patients with low- or Int-1 risk, the ORR was 51% (transfusion independence 33%; minor HI-E 18%). Median duration of transfusion independence was 41 weeks; median increase in hemoglobin was 32 g/L. This study confirms the activity of lenalidomide in patients with low- or Int-1 risk MDS.

28.3.4 Myelofibrosis with Myeloid Metaplasia

28.3.4.1 Lenalidomide Monotherapy

Updated results of two phase II studies of lenalidomide at a dose of 10 mg orally daily continuously in patients with myelofibrosis have been presented.[333–335] The Mayo Clinic treated 27 patients with myelofibrosis.[334,335] Median duration of disease was 40 months. Seventy-eight percent of patients were transfusion-dependent. Cytogenetic abnormalities were detected in 13 patients (48%) and JAK2 V617F mutations detected in 56% (homozygous 16%; heterozygous 40%). Median grade of reticulin fibrosis, MVD, and osteosclerosis was 4, 3, and 2, respectively. Thirty percent of patients had received prior thalidomide. Grade 3 or 4 toxicities included neutropenia (30%), fatigue (15%), respiratory distress with hypoxia (15%), anemia (11%), thrombocytopenia (7%), and rash (7%). Three patients developed marked drug-induced thrombocytosis (peak platelet count $727 \times 10^9/L$ to $1598 \times 10^9/L$) and one patient each with PTMM or AMM experienced erythrocytosis or biopsy-proven disseminated extramedullary hematopoiesis, respectively. Thirteen patients (48%) did not complete three cycles of treatment. Six patients (22%) required dose modification. Median follow-up was 19 months. The ORR, in terms of either anemia or splenomegaly, was 37% (major anemia response 15%; minor anemia response 7%; minor spleen response without associated improvement in hemoglobin level 15%). Two of the four patients with a major response displayed a reduction in bone marrow fibrosis and angiogenesis; one of whom had a documented treatment-associated reduction in the percentage of cells bearing the del(5)(q13q33) abnormality from 78% to 0% with an associated reduction in JAK2 V617F mutation burden. This patient remains in remission >3 months after discontinuation of therapy. None of the three other patients with major responses had a del5q abnormality. Duration of major anemia responses for the remaining three patients was 5+, 2, and 6 months after treatment discontinuation. Responses may be associated with antecedent polycythemia vera or essential thrombocythemia ($P = 0.01$) and degree of marrow fibrosis ($P = 0.002$). Neither JAK2 mutational status nor degree of bone marrow angiogenesis were associated with response to lenalidomide. None of the four patients with major responses, but both patients with minor responses, were previously exposed to thalidomide therapy. Eight deaths have occurred, all occurring after discontinuation of therapy and none attributable to the drug.

Forty-one patients were treated at the M.D. Anderson Cancer Center with lenalidomide 10 mg orally daily if platelet count $\geq 100 \times 10^9/L$ at study entry or 5 mg daily if platelet count

$\geq 100 \times 10^9/L$ at study entry.[333,335] Median duration of disease was 6 months. Thirty-two percent of patients were transfusion-dependent. Cytogenetic abnormalities were detected in 17 patients (42%) and JAK2 V617F mutations detected in 18 of 35 patients (51%). Bone marrow angiogenesis was not evaluated. Thirty-two percent of patients had received prior thalidomide. Therapy was well tolerated with grade 3 or 4 toxicities: neutropenia (32%), thrombocytopenia (27%), rash (5%), fatigue (7%), and pruritus (2%). Six patients (3 PTMM, 2 AMM, and 1 PPMM) experienced drug-induced thrombocytosis (peak platelet count $933 \times 10^9/L$ to $1631 \times 10^9/L$). Sixteen patients (39%) did not complete three cycles of treatment. Ten patients (24%) required dose modification. Median follow-up was 5 months. The ORR, in terms of anemia ($n = 2$), splenomegaly ($n = 6$), or both ($n = 2$), was 24%. A major anemia response was observed in 4 of 19 patients (21%); two of whom had an enlarged spleen before therapy and became impalpable in one patient and decreased in size from 10 to 2 cm in a second patient. All four patients had de novo AMM. Only one of four major anemia responders had a baseline cytogenetic abnormality (trisomy 9) that was not affected by therapy. Major anemia response duration was 1.5 to 9 months in three patients while on active treatment. The fourth patient relapsed off therapy. Six of the eight patients with a minor response in splenomegaly did not qualify for anemia response because of either a baseline hemoglobin of >100 g/L or lack of concomitant response in anemia. Two of these six patients have relapsed while on therapy. A platelet response was observed in 6 of 12 patients (50%) with thrombocytopenia, with a median platelet increase of $60 \times 10^9/L$ over pretreatment values. Four of six patients with a JAK2 mutation responded. Only 1 of 68 patients in both studies developed a VTE.[333–335] Five deaths have occurred. One patient died while on therapy due to non-neutropenic sepsis and four after discontinuation of therapy; none were attributable to the drug. Twenty-one patients (51%) remain on study. On the basis of these two studies, continued therapy with lenalidomide may be necessary to sustain lenalidomide-induced remissions in patients with myelofibrosis.

28.3.5 Non-Hodgkin's Lymphoma

28.3.5.1 Lenalidomide Monotherapy

Phase II studies are evaluating lenalidomide monotherapy in patients with previously treated indolent or aggressive NHL[336] and cutaneous T-cell lymphoma.[337] Decreased tumor burden was noted. There were no tumor flares.

28.3.6 Systemic Amyloidosis

28.3.6.1 Lenalidomide and Dexamethasone

A phase II trial evaluated lenalidomide (25 mg daily for 21 days every 28 days) in 23 patients with symptomatic systemic amyloidosis with a measurable plasma cell disorder and adequate organ and marrow reserve.[338] If there was disease progression before three cycles of therapy or no evidence of response after three cycles, dexamethasone 40 mg orally Days 1–4 and 15–18 was added. Median number of organs involved was two; cardiac (61%), renal (70%), hepatic (22%), and nerve (13%). Grade 3 or 4 toxicities included neutropenia (43%), thrombocytopenia (26%), rash (17%), dyspnea (9%), fatigue (9%), and edema (4%). There were five deaths, with four patients having severe cardiac involvement and at least three organs affected by amyloid. An additional seven patients withdrew from study (two due to adverse events and two progressive disease). Median follow-up for the remaining 11 patients was 6.2 months. Of 12 evaluable patients who received at least 3 months of therapy (all but one received dexamethasone), the ORR was 83% (confirmed hematologic PR 33%; unconfirmed hematologic PR 25%; renal response 17%; liver response 8%). Whereas thalidomide has had limited utility in this disease because of its toxicities,[277,279,280] LD appear to be better tolerated and has activity. Accrual is ongoing.

Preliminary results of a second study evaluating lenalidomide (25 mg daily for 21 days every month) in 24 patients with AL amyloidosis have been presented.[339] If there was no evidence of response after three cycles, dexamethasone 40 mg orally for 4 days thrice a month every other month was added. Organ involvement consisted of renal (75%), cardiac (38%), hepatic or gastrointestinal (18%), neuropathy (17%), and soft tissue or lymphadenopathy (13%). The majority had received prior therapy (92%), including SCT (50%). As none of the first six patients tolerated three treatment cycles at 25 mg/day, the starting dose was reduced to 15 mg/day. Five patients were removed from study (two due to rashes, one due to gout and severe fatigue, one patient request, and one death from complications of the disease). Grade 3 toxicities included fatigue (33%), hypoalbuminemia (29%), worsening performance status (29%), neutropenia (21%), rash (21%), dizziness (13%), gout (13%), dyspnea (8%), muscle weakness (8%), and pneumonia (8%). One patient (4%) had a VTE on LD. The ORR was 62% (complete or near complete hematologic CR 23%); 4 of 14 (28%) patients had improvement in the proteinuria.

28.4 CC-4047 (ACTIMID)

28.4.1 Multiple Myeloma

28.4.1.1 CC-4047 Monotherapy

A phase I trial evaluated four dose levels of single-agent CC-4047 (1, 2, 5, and 10 mg orally daily) in 24 patients with relapsed or refractory MM.[340] DLTs consisted of VTE in four (17%) patients (one at the 1 mg dose and who was subsequently found to have a malignant melanoma with lymphadenopathy, two at 2 mg, and one at 5 mg) and grade 4 neutropenia in six patients (two at the 2 mg dose, two at 5 mg, and two at 10 mg). Grade 3 neutropenia developed in a further eight patients (four at the 5 mg dose). Therefore, the MTD was 2 mg orally daily. Common grade 1 or 2 toxicities were skin toxicity, neuropathy, and constipation. All patients showed increased CD45RO expression on $CD4^+$ and $CD8^+$ cells, with a concomitant decrease in $CD4^+/CD45RA^+$ and $CD8^+/CD45RA^+$ cells, suggesting a switch from resting memory or naïve cells to activated effector T cells. Furthermore, CC-4047 therapy was associated with a significant increase in serum interleukin (IL)-2 receptor and IL-12 levels, which was consistent with activation of T cells, monocytes, and macrophages. Pharmacokinetic data showed CC-4047 was well absorbed orally, with peak plasma concentrations occurring 2.5–4 h after dose and a median half-life of 7 h. The ORR was 71% (CR 17%; very good PR 12.5%; PR 25%; minimal response 17%). Responses were noted at all dose levels, and occurred despite dose reductions. Median time to maximum response was 21 weeks. Median duration of therapy was 28 weeks. The median EFS, PFS, and OS were 28, 39, and 90 weeks, respectively.

A second phase I study evaluated alternate day dosing in patients with relapsed or refractory MM.[341] Four dose levels (1, 2, 5, and 10 mg orally every other day) were evaluated in 30 patients. Patients who relapsed following achievement of a response could receive dexamethasone 20 mg orally daily for 4 days every other week. Sixty-five percent of patients had received prior SCT and 85% prior thalidomide (85%). DLTs consisted of grade 4 neutropenia in three patients receiving 10 mg every other day. Therefore, the MTD was defined as 5 mg every other day. Three patients withdrew from study due to recurrence of relapsing/remitting fevers of unknown origin, development of transverse myelitis, and withdrawal of consent. With a median follow-up of 12 months, the ORR was 60% (CR 10%; very good PR 40%; minor response 10%). Two of six patients who received dexamethasone achieved a very good PR, one a minor response, and the remaining three SD.[342] Assessment of serum FLC ratios during therapy may allow for early prediction of patients likely to respond to therapy with CC-4047. Median time to maximum response and TTP were 4 and 10 months, respectively. Five patients died due to disease progression; there were no treatment-related deaths.

28.5 INHIBITORS OF CYCLOOXYGENASE-2

28.5.1 Celecoxib (Celebrex)

COX-2 is overexpressed in a number of hematological malignancies,[343–345] including MM[346–348] and lymphoma,[349–352] and may be associated with a poor outcome.[345,347,348,351] COX-2 metabolizes arachidonic acid to prostaglandins and thromboxanes, which leads to increased angiogenesis and decreased apoptosis.[353] The selective COX-2 inhibitor, celecoxib, induced apoptosis in a variety of leukemia and lymphoma cell lines.[343,354,355] However, celecoxib has been shown to have activity against COX-2-negative hematopoietic and epithelial cell lines mediated via activation of a novel mitochondrial apoptosis signaling pathway.[356,357] Recent studies have raised concerns over an increased risk of cardiovascular events associated with selective COX-2 inhibitors.[358,359] At this time, it is not clear whether all members of this class of drugs are equally culpable.[360–363]

28.5.1.1 Chronic Lymphocytic Leukemia

Thirteen patients with CLL not requiring treatment (eight Rai stage 0, two stage 1, and three stage 2) were treated with celecoxib 400 mg daily for 3 months.[364] No patients were removed from study due to adverse effects. COX-2 expression in the CLL cells was not assessed. No significant changes in the white blood cell count or lymphocyte count were observed.

28.5.1.2 Multiple Myeloma

Sixty-six patients with relapsed or resistant MM following prior systemic combination chemotherapy were treated with thalidomide 200 mg daily escalated to 800 mg daily, as tolerated, and celecoxib 400 mg twice daily.[210] Median individual tolerated dose of thalidomide was 400 mg/day. Median average dose of celecoxib administered was 738 mg/day. Thalidomide-related toxicities were expected and consisted of constipation (50%), fatigue (45%), and sensory (32%) and motor (15%) neuropathy. Three patients (5%) had nonfatal thromboembolism. Seventy-seven percent of patients discontinued thalidomide therapy; 20% due to toxicity. Celecoxib-related toxicities consisted of fluid retention (peripheral edema [30%] and pulmonary congestion [18%]), elevated creatinine (8%), epigrastic/esophageal discomfort (11%), and hematemesis/melena stools (3%); no myocardial or cerebrovascular events occurred. Eighty-percent of patients discontinued celecoxib therapy; 57% due to toxicities. Fifteen (23%) patients remained on therapy at the time of analysis. Median duration of follow-up was 20 months. ORR rate was 42% (CR 3%; PR 39%). Median PFS and OS for all patients were 6.8 and 21.4 months, respectively. One year PFS and OS were 37% and 65%, respectively. Subgroup analysis indicated that ORR (62% vs. 30%, $P = 0.021$), PFS (12.7 months vs. 4.6 months, $P = 0.039$), and OS (29.6+ months vs. 18.9 months, $P = 0.035$) were superior in patients who received a total celecoxib dose of $\geq$40 g during the first 8 weeks of therapy. However, this regimen was associated with significant toxicities.

28.5.1.3 Non-Hodgkin's Lymphoma

Twenty-nine patients with relapsed and refractory aggressive NHL and measurable disease were treated with low-dose oral cyclophosphamide (50 mg daily) and celecoxib 400 mg twice daily.[365]

Eight patients had received a prior ASCT. Median follow-up was 4 months. Grade 3 or 4 toxicities included neutropenia (2), thrombocytopenia (3), hypertension (1), atrial arrhythmia (1), and rash (1). Of the 25 evaluable patients, ORR was 40% (CR or CRu 8%; PR 32%). No patient with preceding chemoresistance responded. Responding patients had a decrease in circulating endothelial cells (CECs) and circulating endothelial precursors (CEPs). Plasma VEGF and TSP-1 levels did not correlate with response. Two patients in PR discontinued treatment because of a cumulative cyclophosphamide exposure exceeding 40 g; both progressed within 2 to 3 months of discontinuation. Four of ten patients remain in PR; one of whom was in PR at 22 months.

28.6 INHIBITORS OF VASCULAR ENDOTHELIAL GROWTH FACTOR

28.6.1 Bevacizumab (Avastin)

Bevacizumab (Avastin) is a recombinant humanized IgG1 monoclonal antibody that blocks the binding of all VEGF-A isoforms to its cognate receptors, but does not bind to other VEGF family members, such as PlGF or VEGF-B.[366] It was approved by the FDA on February 26, 2004 for the treatment of patients with metastatic colon cancer in conjunction with chemotherapy.

28.6.1.1 Acute Myeloid Leukemia

In an attempt to exploit drug-induced changes, bevacizumab was administered sequentially after chemotherapy at the predicted time of peak cell regeneration.[367] It is postulated that multiple cytokines, such as VEGF, may drive the proliferation of hematopoietic precursors and endothelial cells in an attempt to reconstitute the bone marrow after damage by cytotoxic agents. Forty-eight patients with AML (45 relapsed or refractory; 3 untreated poor-risk) received induction therapy consisting of cytarabine 2 g/m^2 as a continuous infusion over 72 h starting of Day 1, mitoxantrone 40 mg/m^2 IV on Day 4, and bevacizumab 10 mg/kg IV on Day 8. The majority of patients had received prior anthracycline- or anthraquinone-based (92%) therapy; 90% had received prior cytarabine-based therapy. There was no prolonged myelosuppression compared with other timed sequential regimens. Toxicities during first cycle of therapy included decreased in ejection fraction (6%; all three patients had received prior anthracyclines or anthraquinones), cerebrovascular bleed (4%), and death (15%). No thrombotic events were observed. VEGFR-1 expression on marrow blasts did not correlate with clinical outcome; VEGFR-2 was not detectable. There was a marked decrease in MVD after bevacizumab therapy. Serum VEGF levels were detectable in 14 of 21 (67%) patients, had increased by Day 8 in 52% of cases, and decreased in 93% (67% undetectable) of cases 2 h after bevacizumab administration. The ORR was 48% (CR 33%; PR 15%). In the 14 primary refractory patients, the ORR was 54% (CR 27%; PR 27%). Eighteen patients (14 CR; 4 PR) received a second cycle of therapy and five patients (three CR; two PR) underwent an allogeneic SCT. Median OS for all patients was 8.4 months, with 35% survival at 1 year and 18% survival at 2 years. Median disease-free survival (DFS) and OS for patients in CR were 7 and 16.2 months, respectively, with 1 year DFS and OS rates of 35% and 64%, respectively.

A randomized multicenter phase II trial is currently evaluating idarubicin and cytarabine administered with or without bevacizumab in patients with newly diagnosed AML. Patients are stratified according to age, Flt-3 (Fms-like tyrosine kinase-3) status, cytogenetics, and de novo vs. secondary or therapy-related AML.

28.6.1.2 Multiple Myeloma

Preliminary results of a randomized phase II study in patients with MM comparing bevacizumab monotherapy (10 mg/kg IV every 2 weeks) in thalidomide-exposed patients with bevacizumab and thalidomide 50 to 400 mg/day in thalidomide-naive patients have been presented.[208] Twelve patients have been treated (6 per arm). Six patients had received a prior ASCT. Grade 3 toxicities in the bevacizumab arm consisted of fatigue and neutropenia ($n = 1$), hypertension ($n = 1$), and hyponatremia ($n = 1$). In the bevacizumab and thalidomide arm, grade 3 lymphopenia was observed during cycle 3 ($n = 1$) and one patient was removed from study due to exacerbation of preexisting (diet pill induced) pulmonary hypertension. Immunohistochemical studies demonstrated VEGF expression on myeloma cells in 7 of 8 bone marrow samples with weak and weak-to-moderate cell surface staining of VEGFR-1 and VEGFR-2, respectively, in five cases. ORRs were not reported. Median TTP in the bevacizumab arm was 2 months. Median PFS for the five evaluable patients on the bevacizumab and thalidomide arm was 8+ months. Two patients in the bevacizumab and thalidomide arm were removed from study (one patient request and one SCT). The study has been closed to further accrual due to slow study enrollment; two patients continue to receive therapy with bevacizumab and thalidomide.

28.6.1.3 Myelodysplastic Syndrome

A phase II study evaluated single-agent bevacizumab at a dose of 10 mg/kg IV every 2 weeks for 4 months in patients with MDS.[368] The dose was escalated to 15 mg/kg for an additional 2 months in patients without a hematologic response. At the time of publication, 15 patients (seven refractory anemia [RA]; three RA with ringed sideroblasts [RARS]; one CMML; and four RA with excess blasts [RAEB]) had been enrolled. Patients had IPSS scores of Int-1 ($n = 11$) or Int-2 ($n = 4$). Three patients were removed from study early because of worsening cytopenias ($n = 2$) or disease progression ($n = 1$). Other toxicities, in addition to the worsening thrombocytopenia or neutropenia ($n = 3$) included epistaxis (grade 3; $n = 2$), hypertension (grades 2 or 3; $n = 2$), and fatigue (grade 2; $n = 3$). Correlative studies are being performed. Of the 10 evaluable patients, 7 have completed at least 4 months of therapy. Of these seven patients, the ORR was 28% (major HI-E 28%). Accrual is ongoing.

28.6.1.4 Non-Hodgkin's Lymphoma

28.6.1.4.1 Previously Untreated Patients

28.6.1.4.1.1 Bevacizumab and Chemotherapy

Rituximab plus cyclophosphamide, doxorubicin, vincristine, and prednisone (R-CHOP) or a CHOP-like regimen improved response rates, and prolonged time to treatment failure and survival in patients with diffuse large B-cell lymphoma (DLBCL) compared with chemotherapy alone.[369,370] In an attempt to improve on these results, especially in patients with high-intermediate and high-risk International Prognostic Index (IPI) scores,[371] bevacizumab was combined with R-CHOP (RA-CHOP) for the treatment of patients with newly diagnosed DLBCL.[372] Bevacizumab was administered at a dose of 15 mg/kg IV on Day 1 of each cycle. R-CHOP was administered on Day 2 of cycle 1 and on Day 1 of subsequent cycles. Twelve patients have been enrolled to date. Grade 3 or 4 toxicities included neutropenia ($n = 7$) associated with neutropenic fever ($n = 6$) (four of whom had a central line infection), central line-associated VTE ($n = 2$), elevated liver enzymes ($n = 1$), herpes simplex virus esophagitis ($n = 1$), and transient hypertension ($n = 2$). Bevacizumab was held in five cycles in three patients due to elevated liver enzymes in a hepatitis C seropositive patient, transient proteinuria, and central line-associated VTE. Pharmacokinetic analyses of bevacizumab and rituximab, and correlative studies evaluating circulating VEGF levels in urine and serum and VEGF expression in tumor samples are performed. Of the 10 evaluable patients, the ORR was 70% (CR 30%; PR 40%). Enrollment is ongoing. RA-CHOP is also evaluated in older patients with previously untreated bulky stage II or advanced stage DLBCL.

28.6.1.4.2 *Previously Treated Patients*

28.6.1.4.2.1 *Bevacizumab Monotherapy*

A phase II trial evaluated single-agent bevacizumab (10 mg/kg IV every 2 weeks) in 46 patients with relapsed advanced stage aggressive NHL.[373] Grade 3 toxicities occurred in 15 patients: cytopenias ($n = 10$); fatigue, malaise, or lethargy ($n = 5$); dyspnea ($n = 3$); hypertension ($n = 2$); hyponatremia ($n = 2$); and dehydration ($n = 2$). Two patients have died on study: one from a pulmonary embolus and one probably from progressive disease although cannot exclude a possible relationship to the drug. Of the 19 patients with adequate samples for immunohistochemical analysis, 14 (74%) expressed VEGF and 13 (68%) expressed one or both VEGFRs.[373,374] Over the course of therapy, CECs ($P = 0.12$) and plasma VEGF ($P < 0.001$) levels decreased whereas plasma VCAM ($P = 0.02$) increased. The ORR was 5% (PR 5%).[373] Median TTP for the eight patients with SD and two with PR was 5 months. The estimated 6 month PFS and OS for all patients were 14% and 58%, respectively. Single-agent bevacizumab may also have activity in relapsed angioimmunoblastic T-cell lymphoma.[375]

28.6.1.4.2.2 *Bevacizumab and Chemotherapy*

Several phase II studies are evaluating bevacizumab in combination with other agents in patients with NHL, including bevacizumab and rituximab for rituximab-refractory aggressive B-cell NHL (http://www.cancer.gov/clinicaltrials). Bevacizumab is also administered with chemotherapy in patients with primary effusion lymphoma and peripheral T-cell or NK cell neoplasms.

28.7 INHIBITORS OF VASCULAR ENDOTHELIAL GROWTH FACTOR RECEPTOR

28.7.1 AG-013736

AG-013736 is an orally available potent small molecule receptor tyrosine kinase (RTK) inhibitor of VEGFR-1 and VEGFR-2 phosphorylation (inhibitory constant [K_i] values of 8.3 and 1.1 nM, respectively; median inhibitory concentration [IC_{50}] values of 1.2 [in the presence of albumin] and 0.25 nM, respectively, in cellular phosphorylation assays).[376] The closely related c-kit, VEGFR-3, and PDGFR-β were also inhibited by AG-013736 to a significant degree (IC_{50} values of 2.0, 0.29, and 2.5 nM, respectively).[376]

28.7.1.1 Acute Myeloid Leukemia and Myelodysplastic Syndrome

Expression of VEGFR-1 is detected in up to 76% of AML patients and 52% of patients with MDS, whereas VEGFR-2 is expressed in up to 20% of patients with AML and 13% of patients with MDS[14,23]; therefore, autocrine and paracrine effects of VEGF on leukemic blasts are possible.[23,376] Bone marrow VEGF and VEGFR-2 levels correlated with MVD in patients with AML; there was a reduction in VEGFR-2 levels to normal values in patients who achieved a CR after induction chemotherapy.[17] VEGFR-3 expression is detected in up to 36% of patients with AML.[377] The c-kit receptor is overexpressed in 60%–70% of patients with AML and 75% of patients with MDS.[378,379] PDGFR-β is expressed in 90% of primary AML blasts[380]; PDGF isoforms are produced by the leukemic blasts.[380]

A phase II study was conducted to determine the clinical efficacy of AG-013736 in patients with poor prognosis AML ($n = 6$) and MDS ($n = 6$).[376] AG-013736 was administered at a total dose of 10 mg administered orally daily in two divided doses on a continuous basis. Grade 3 or 4 drug-related toxicities included hypertension (42%), mucositis (8%), and VTE (8%). Hypertension was reversible with discontinuation of the drug. No CRs or PRs were observed; two patients with MDS (one RAEB and one RAEB-t) had stable disease for 8.3 and 6.2 months, respectively. One patient had a transient decrease in bone marrow and peripheral blood blasts by ≥46%. Overall median survival was 48 weeks. A sustained decrease in soluble VEGFR-2 plasma levels with concomitant elevation in plasma VEGF and PlGF levels was demonstrated with therapy; thus, providing indirect evidence of the biological activity of AG-013736. Single-agent AG-01736 had minimal biologic and clinical activity in this predominantly elderly patient population with poor prognosis AML and MDS. Studies of AG-013736 in combination with other therapeutic agents may be warranted in this population.

28.7.2 CHIR-258

CHIR-258 is an orally active small molecule RTK inhibitor, which exhibits activity against VEGFR-1, VEGFR-2, and VEGFR-3 (IC_{50} 10, 13, and 8 nM, respectively), c-kit (IC_{50} 2 nM), Flt-3 (IC_{50} 1 nM), FGFR-1 and FGFR-3 (IC_{50} 8 and 5–9 nM, respectively), colony stimulating factor receptor (CSF-1R; c-fms; M-CSF-1R; IC_{50} 36 nM), and PDGFR-β (IC_{50} 27 nM).[381,382] Preclinical data have indicated that CHIR-258 is a potent inhibitor of Flt3-ITD (internal tandem duplication), compared with wild-type (Flt3-WT) kinase, with regression and eradication of AML cells from the bone marrows of the xenograft mice.[383] CHIR-258 has activity against FGFR-3-transformed hematopoietic cell lines, human myeloma cell lines expressing either wild-type or mutant FGFR-3, and primary myeloma cells from t(4;14) (i.e., FGFR3/MMSET) patients.[382] Furthermore, CHIR-258 had activity against a xenograft mouse model of FGFR3 MM.[382,384]

28.7.2.1 Acute Myeloid Leukemia

Activating mutations of Flt-3 (Flt3-ITD and Flt3-TKD [tyrosine kinase domain]) are also detected in 20% to 30% and 7% to 10% of patients with AML and 3% each of patients with MDS.[385–389] MDS patients with Flt3-ITD mutations have a higher risk of leukemic transformation and a shorter OS.[385–388] Patients with AML and Flt3-ITD duplications have an increased risk of relapse or shorter OS.[390–396] In contrast, the prognostic significance of Flt3-TKD mutants is unclear.[387,390–392,397,398] Additionally, as both MDS and AML cells express Flt3 ligand, there may be a role for autocrine and paracrine stimulation.[399,400]

A phase I dose escalating trial of CHIR-258 administered orally to patients with relapsed or refractory AML on a 7 day on/7 day off schedule followed by continuous daily dosing for 28 days (i.e., one cycle of therapy).[401] Subsequent 28 day cycles were permitted. As of May 2005, eight patients have been treated at two dose levels (50 and 100 mg). Five patients remain on study with a median of three treatment cycles. There have been no DLTs to date. Of the eight patients treated, one had a Flt3-ITD mutation and seven had Flt3-WT. The patient with the Flt3-ITD mutation, who was treated at the 100 mg dose level, had near complete clearing of the blasts in the bone marrow and peripheral blood. This patient later died from a presumed fungal infection unrelated to treatment with CHIR-258. Dose escalation to 200 mg has been tolerated and accrual is ongoing.

28.7.2.2 Multiple Myeloma

CHIR-258 is administered once daily in a dose-escalating phase I trial to patients with relapsed or refractory MM.[402] Nine patients have been treated to date. All had received prior thalidomide, eight prior bortezomib, and eight prior SCT. Four patients had FGFR3-expressing myeloma cells. CHIR-258 (at doses of 50, 100, and 200 mg/day) was generally well tolerated. One DLT (neutropenia) was observed at the 200 mg/day dose level. Five patients received dexamethasone in conjunction with CHIR-258. No CRs or PRs have been observed.

However, FGFR3-positive patients receiving dexamethasone and CHIR-258 had a greater decline in urine and serum paraprotein compared with CHIR-258 alone. Plasma exposure and C_{max} increased proportionally with increasing doses. Pharmacodynamic evaluations are being performed. Four patients remain on study (three of whom are FGFR3-positive). Enrollment is ongoing.

28.7.3 Semaxanib (SU5416)

Semaxanib (SU5416; Z-3-50 mg-2-indolinone) is a small, lipophilic, highly protein-bound, synthetic RTK competitive inhibitor of VEGFR-1 and VEGFR-2 (K_i of 0.16 μM; median IC_{50} of 0.1 μM in cellular phosphorylation assays),[403–405] c-kit (IC_{50} between 0.1 and 1 μM for ligand-dependent phosphorylation),[406] and Flt3-WT and Flt3-ITD (IC_{50} of 0.1 to 0.25 μM for ligand-independent phosphorylation).[407] Although semaxanib was active against PDGFR-β (K_i of 0.32 μM), it was a 20-fold less potent inhibitor of PDGFR-β (IC_{50} of ~20 μM) in cellular phosphorylation assays.[403,405] Semaxanib is a weak inhibitor of FGF-R, Met, Src, Abl, and Lck (IC_{50} values of at least 10 μM)[403,405] and does not have any activity against cells expressing epidermal growth factor (EGF) or insulin receptors (IC_{50} >100 μM).[405]

RTK inhibitors are likely to require chronic administration. Because of hypersensitivity and injection reactions associated with administration of semaxanib, the inconvenience of a chronic intravenous therapy administered twice weekly, and its short half-life with the potential for inadequate free drug plasma concentrations, further development of semaxanib has been halted.

28.7.3.1 Acute Myeloid Leukemia and Myelodysplastic Syndrome

Based on a report of achievement of a CR and normalization of the elevated bone marrow MVD in a patient with chemorefractory AML following a 12 week course of twice weekly semaxanib therapy,[408] two phase II studies have evaluated semaxanib therapy (at the recommended phase II dose of 145 mg/m^2 twice weekly via a central venous catheter) in patients with relapsed AML and MDS.[409,410] Fiedler et al. treated 42 patients with c-kit-positive AML who were refractory to therapy or elderly and deemed not medically fit for induction chemotherapy.[410] Seven patients had Flt3-ITD mutations. Twenty-five patients completed one cycle (4 weeks) of therapy (13 had progressive disease and 4 died during the first cycle of therapy without reevaluation of their disease status). Treatment was generally well tolerated, with most toxicities of grade 1 or 2 (nausea, bone/musculoskeletal pain, headache, insomnia, vomiting, vertigo, and fatigue/malaise). Seven percent of patients developed grade 3 or 4 bony pain, which was believed to be drug related. No thrombotic events or hypertension was reported. Four patients discontinued therapy due to study drug-related adverse event. Pharmacokinetic analyses were not performed. Of the 25 evaluable patients, the ORR was 32% (CR 4%; PR 28%). On an intent-to-treat analysis, which included all patients who received at least one dose of semaxanib, the ORR was 19%. Mean response duration was 1.6 months, with all patients relapsing despite ongoing therapy with semaxanib. Patients with leukemic blasts expressing high levels of VEGF mRNA had a higher response rate ($P = 0.059$) and reduction in bone marrow MVD ($P = 0.033$) than patients with low VEGF levels. No correlation between Flt3 or VEGFR-1 expression and response was noted. Semaxanib has a short half-life; analysis of Flt3 phosphorylation in limited samples indicated that there was nonsustained weak Flt3 inhibition at 1 h after semaxanib infusion.[400]

In the second trial, semaxanib was administered to 55 patients with relapsed or refractory AML ($n = 33$) or advanced MDS (14 RAEB and 8 RAEB-t) without a baseline c-kit requirement.[409] Eleven patients did not complete 4 weeks of therapy (10 due to progressive disease; 1 due to adverse event). Grade 3 or 4 drug-related toxicities included headaches

(14%), infusion-related reactions (11%), dyspnea (14%), fatigue (7%), thrombotic events (7%) consisting of ischemic strokes and deep vein thrombosis (DVT), bone pain (5%), and gastrointestinal disturbances (4%). No hypertensive episodes were reported. The relative low rate of thrombotic events observed in the current study may be the result of it administered as monotherapy compared with monotherapy combined with chemotherapy (7% vs. 22%–42%, respectively)[411–413] and the use of low-dose prophylactic anticoagulation in some patients (24%). Five percent of patients experienced hypersensitivity reactions associated with drug infusion and 43% had grade 1 or 2 injection site reactions attributable to the formulation of the drug (i.e., high concentration of polyoxyl 35 castor oil [Cremophor] and hyperosmolarity of the dilute drug product). The ORR was 7% (PR 5% including two AML and one MDS; HI 2%). Although Flt3 phosphorylation was inhibited immediately following a 1 h semaxanib infusion in 7 of 17 patient blood samples, especially with plasma drug levels >15 μM, this did not correlate with clinical response.[400] However, the duration of Flt3 inhibition may have been inadequate to translate into biological activity; this was not evaluated in the current trial. Median EFS was 5.2 months. Median OS for patients with AML was 3 months and had not yet been reached for patients with MDS.

28.7.3.2 Myeloproliferative Diseases and Myelodysplastic/Myeloproliferative Diseases

Bone marrow and serum VEGF levels are significantly elevated in patients with AMM,[24,246,414] CMML,[10,415] and CML.[10,24] Furthermore, in patients with CMML and CML-BP, leukemic blast cellular levels of VEGF and VEGFR-2 are elevated.[10,20,23] Increased bone marrow VEGF levels are associated with reduced CR rates, DFS, and OS in patients with CMML.[20] Elevated plasma levels of PDGF and bFGF have been detected in patients with essential thrombocythemia, myelofibrosis, polycythemia rubra vera, AMM, and other myeloproliferative diseases (MPD).[28,416–418] Furthermore, tyrosine kinase fusion genes involving *PDGFR-β* or *FGF-R1* have been detected in patients with chronic MPDs and CMML,[419–422] leading to deregulated hematopoiesis.

A phase II study evaluated semaxanib 145 mg/m^2 twice weekly intravenously in 32 patients with advanced or refractory chronic MPDs (19 *bcr-abl*-negative MPD, 6 CMML, 4 CML-BP, and 3 AMM).[423] Twenty-one patients withdrew from study (nine due to disease progression; two due to inconvenience of treatment schedule; none due to toxicity). Three patients received only a single infusion of semaxanib. Eleven patients received only one cycle of therapy. Grade 3 or 4 toxicities consisted of abdominal pain (13%), infusion-related dyspnea or headache (9% and 6%, respectively), bone pain (9%), diarrhea (6%), fatigue (6%), and catheter site reactions (3%). Thrombosis was observed in two patients (one of which was a central venous catheter-associated thrombosis). Of the 26 evaluable patients, the ORR was 4%. One patient with AMM achieved a PR of >12 months in duration and continued on therapy with semaxanib at the time of publication.

28.7.3.3 Multiple Myeloma

Therapy with semaxanib, at a fixed dose/schedule of 145 mg/m^2 twice weekly intravenously, was evaluated in 25 patients with relapsed or refractory MM.[424] Sixty-seven percent had failed prior therapy with thalidomide and 63% prior ASCT. Two patients withdrew before completing one cycle of therapy (one with worsening headaches deemed to be due to semaxanib). Grade 3 or 4 toxicities consisted of thrombocytopenia (12%) and neutropenia (4%). Thrombotic events occurred in four patients (one each of pulmonary embolus, phlebitis, subclavian vein thrombosis, and cerebrovascular accident). The majority of patients ($n = 19$) received prophylactic anticoagulation while on study; there was no difference in the incidence of thrombotic events in patients receiving anticoagulation compared with those not receiving anticoagulation (11% vs. 13%). Median time on study was 1.7 months. Of the 23 evaluable patients, no objective responses were observed. In patients with progressive disease, median VEGF levels remained unchanged. In the seven patients with SD, median VEGF levels decreased from 117.77 pg/mL on Day 1 to 66.44 pg/mL on Day 25; however, there was no correlation between decreases in VEGF levels and decline in M protein, percent plasmacytosis, or extended cycles of therapy. All patients subsequently developed progressive disease after three to nine cycles of therapy. Median survival was 10.5 months.

28.7.4 Sorafenib Tosylate (Bay 43-9006; Nexavar)

Sorafenib is an orally available cytostatic biaryl urea, which inhibits the serine/threonine kinases c-Raf (Raf-1) and B-Raf (IC_{50} of ~6 and 22 nM, respectively, in biochemical assays), and the RTKs c-kit (IC_{50} of ~68 nM in biochemical assays), VEGFR-2, murine VEGFR-3, PDGFR-β, Flt-3, and RET (IC_{50} values of ~30, 100, 80, 20, and 50 nM, respectively, in cellular phosphorylation assays).[425,426] The US FDA granted approval to sorafenib tosylate on December 20, 2005 for use in patients with advanced renal cell carcinoma.

28.7.4.1 Acute Myeloid Leukemia and Myelodysplastic Syndrome

A randomized phase I trial of sorafenib in patients with AML or MDS has been conducted by the NCIC.[427,428] Preliminary results of 36 patients (32 AML; 4 MDS) have been presented. Patients were randomized to two different schedules of sorafenib: continuous twice daily dosing for 4 weeks or twice daily dosing for 2 weeks followed by a rest period of 2 weeks. Escalating doses of sorafenib were administered (100, 200, 300, and 400 mg). Forty-seven percent of patients had received prior chemotherapy. DLT was seen in 2 of 12 patients receiving sorafenib at a dose of 200 mg twice daily and in 1 of 17 patients at a dose of 400 mg twice daily. However, at the 400 mg twice daily dose, six patients received <14 days of therapy due to grade 1 or 2 toxicity (abdominal pain, nausea, vomiting, rash, stroke, and thrombocytopenia), indicating that the dose was not tolerable. No cases of thyroid dysfunction were documented. Accrual to the 300 mg twice daily dosing for 28 days has been completed, but results have not been reported. Flow cytometry was used to monitor sorafenib activity in patient blast cells and lymphocytes.[427] There was no consistent effect of treatment on the activation of ERK by phorbol myristate acetate (PMA) in either the lymphocyte or blast cell populations. However, activation of ERK in the blast cells by stem cell factor was decreased with sorafenib therapy. Biological effects (i.e., reduction in peripheral blood and bone marrow blasts) have been observed in 4 of 27 evaluable patients: 2 at the 200 mg twice daily dose (1 each in the 4 week continuous and 2 week on/2 week off dosing schedules) and 2 at 400 mg twice daily dose (both on the 4 week continuous dosing schedule). One patient with AML and a Flt3 mutation, who received 400 mg twice daily, achieved a CR without platelet recovery (CRp) of 84 days duration.

Phase I or II studies are currently evaluating single-agent sorafenib in patients with relapsed or refractory AML, ALL, CML-BP, MM, DLBCL, and CLL (http://www.cancer.gov/clinicaltrials). A phase II trial evaluating sorafenib in patients with imatinib mesylate-resistant chronic phase CML was performed; however, results have not been reported.

28.7.5 Sunitinib Malate (SU11248; SU011248; Sutent)

Sunitinib malate is a multitargeted, small molecule RTK inhibitor of VEGFR-1, VEGFR-2, and VEGFR-3, PDGFR-α and PDFGR-β, c-kit, RET, and Flt3 with IC_{50} values in the 4–14 nM range.[429–431] Sunitinib competitively inhibited VEGFR-2 and PDGFR-β with K_i

values of 9 and 8 nM, respectively, and with IC_{50} values of ~10 nM for ligand-dependent autophosphorylation.[431] Sunitinib blocked VEGF- and FGF-dependent proliferation of human umbilical vein endothelial cells (HUVECs) with a potency of 40 and 700 nM, respectively.[430,431] Furthermore, sunitinib inhibited phosphorylation of CSF-1R (IC_{50} of 50–100 nM).[432] Sunitinib was granted approval on January 26, 2006 by the US FDA for the treatment of gastrointestinal stromal tumor (GIST) after disease progression on or intolerance to imatinib mesylate and advanced renal cell carcinoma.

28.7.5.1 Acute Myeloid Leukemia and Myelodysplastic Syndrome

A multicenter, dose-escalating phase I study was conducted to assess the degree and the duration of wild-type and mutant Flt-3 inhibition and the safety, tolerability, and pharmacokinetics of a single dose of sunitinib in 29 patients with AML.[433] Sunitinib was administered at 50 to 350 mg. Study drug-related adverse events occurred in 31% of patients, mainly grade 1 or 2 diarrhea and nausea, at the 250 and 350 mg dose levels. Two patients developed serious toxicities possibly related to sunitinib (one patient receiving a 200 mg dose developed transient grade 3 hypertension and one patient with underlying coronary artery disease receiving 350 mg developed a transient grade 1 ventricular tachycardia requiring hospitalization for observation). Serial electrocardiograms, performed in all patients, did not reveal any clinically relevant or consistent changes from baseline. Thyroid dysfunction was not reported.[434,435] Inhibition of Flt3 phosphorylation occurred in 50% of Flt3-WT patients at doses ≥200 mg and in 100% of Flt3-ITD patients at doses <200 mg. Achieving and maintaining plasma drug concentrations of ≥100 ng/mL were required for strong (>50%) and sustained Flt3-WT inhibition, whereas less is required for inhibition of Flt3-ITD (mean C_{max} 34 ± 17 ng/mL). Downstream effector molecules were also inhibited; STAT5 phosphorylation was reduced primarily in Flt3-ITD patients and at late time points in Flt3-WT patients, whereas ERK1/2 and MEK1/2 phosphorylation was reduced in the majority of patients, independent of Flt3 inhibition. Furthermore, decreased in blast counts were observed in five patients.

Initial phase I trials with sunitinib were limited to either treatment continuously for 4 weeks followed by a rest period of 2 weeks or 2 weeks with a rest period of 2 weeks for safety reasons.[436,437] DLTs appeared to be associated with combined (sunitinib and its major active metabolite SU012662) trough plasma concentrations of ≥100 ng/mL, which was not observed with sunitinib doses of <50 mg continuously and rarely with doses of 50 mg continuously.[436] Therefore, a multicenter phase I study evaluated two dose levels (50 and 75 mg) orally daily for 4 weeks followed by a 2 week rest period in 15 patients with relapsed or refractory AML or not amenable to conventional therapy.[437] This was amended to shorten the rest period when it was noted that the 2 week washout period led to almost complete elimination of sunitinib and SU012662.[437] This was accompanied by increased phosphorylation of Flt3 and an increase in blasts in 4 of 7 responding patients at the end of the treatment-free interval. Reinstitution of therapy led to a drop in blast counts again. Hence, a third cohort of patients received sunitinib 50 mg daily for 4 weeks followed by 1 week of rest.

No DLTs were observed in patients with AML treated with sunitinib at doses of 50 mg.[437] At the 75 mg dose level ($n = 2$), one case each of grade 4 fatigue, hypertension, and cardiac failure was observed. Thyroid dysfunction was not reported.[434,435] Therefore, the MTD was 50 mg dose. The ORR was 40% (CRp 7%; PR 33%). All four patients with Flt3 mutations achieved a CRp or PR compared with 2 of 10 evaluable patients with Flt3-WT. Responses, although longer in patients with mutated Flt3, were of short duration (4 to 16 weeks). Reductions in cellularity and phosphorylated c-kit, VEGFR-2, STAT5, and Akt cells were observed in the bone marrow.

In an attempt to maintain consistent therapeutic drug levels, without increased toxicity, phase II studies in patients with MDS administer sunitinib continuously at a dose of 37.5 mg orally daily.

28.7.6 Vatalanib (PTK787; ZK222584)

Vatalanib (PTK787; ZK 222584; 1-[4-chloroanilino]-4-[4-pyridylmethyl] phthalazine succinate) is a competitive small molecule tyrosine kinase inhibitor of VEGFR-1, VEGFR-2, and VEGFR-3 (IC_{50} of 77, 37, and 660 nM, respectively).[438] It also inhibits PDGFR-β, c-kit, and c-Fms, but at higher concentrations (i.e., 580, 730, and 1.4 μM, respectively). It is not active against EGFR, FGFR-1, c-Met, and Tie-2, c-Src, c-Abl, and protein kinase C-α (PKC-α) (IC_{50} >10 μM).[438]

28.7.6.1 Acute Myeloid Leukemia and Myelodysplastic Syndrome

A phase I multicenter study evaluated escalating doses of vatalanib (500, 750, and 1000 mg orally twice a day) as monotherapy ($n = 46$) or in combination with daunorubicin (45 to 60 $mg/m^2/day$ for 3 days) and cytarabine (100 $mg/m^2/day$ for 7 days) chemotherapy ($n = 17$) in 63 patients with relapsed or refractory AML ($n = 43$) or advanced MDS ($n = 20$).[439] Of the 17 patients who received combination therapy with vatalanib and chemotherapy, 7 patients were treated initially with monotherapy followed by combination therapy with vatalanib and chemotherapy, and 10 were treated with the combination upfront. DLTs were observed at the 1000 mg twice a day dose level and consisted of lethargy, hypertension, nausea, emesis, and anorexia. Therefore, the MTD for vatalanib monotherapy was 750 mg twice daily. In the combination therapy arm, one DLT consisting of hyperbilirubinemia was identified. However, four patients receiving combination therapy were removed from study (one each for drug-related thrombocytopenia and urinary frequency, both of which occurred after the 28 day DLT evaluation period; one patient developed atrial fibrillation complicated by a cerebrovascular accident and myocardial infarction unrelated to study drug; and a fourth patient who received combination therapy after 1 month of vatalanib monotherapy, developed a pulmonary embolus 9 days after the combination was started). Four patients died while on study due to leukemic complications. Pharmacokinetic analyses indicated that steady state was achieved by Day 14. Peak vatalanib levels were well above the IC_{50} for VEGFR-2. Vatalanib did not affect the pharmacokinetics of cytarabine, but it appeared to decrease daunorubicin clearance by ~36%. No responses were observed with vatalanib monotherapy. The ORR for the 17 patients treated with induction chemotherapy and vatalanib was 47% (CR 29%; CRp 12%; PR 6%).

The Cancer and Leukemia Group B (CALGB) have administered vatalanib at a continuous daily dose of 1250 mg orally to 80 patients with MDS in a phase II trial.[440] Therapy was continued until disease progression or unacceptable toxicity. Preliminary toxicity and pharmacokinetic data on 56 patients indicated that the most common grade ≥2 nonhematological toxicities were fatigue, nausea, vomiting, dizziness, and ataxia. There was no evidence indicating that variability in drug exposure was associated with nonhematological toxicities of vatalanib. Responses were not reported.

28.7.7 ZD6474 (AZD6474; Zactima)

ZD6474 (AZD6474; Zactima; *N*-(4-bromo-2-fluorophenyl)-6-methoxy-7-[(1-methylpiperidin-4-yl)methoxy]quinazolin-4-amine), a heteroaromatic-substituted anilinoquinazoline, is an orally bioavailable small molecule reversible RTK inhibitor VEGFR-2 (IC_{50} 40–140 nM and 60 nM in biochemical and cellular assays, respectively), VEGFR-3 (IC_{50} 110 nM), EGFR (IC_{50} 500 and 170 nM in biochemical and cellular assays, respectively), and

RET (IC_{50} 130 nM).[441–443] Much higher doses of ZD6474 were required to inhibit PDGFR-β (IC_{50} 1.1 μM in biochemical assays) and FGFR (IC_{50} 3.6 μM in biochemical assays) activities.[442]

28.7.7.1 Multiple Myeloma

A phase II trial evaluated ZD6474 at a dose of 100 mg orally daily in 17 patients with relapsed MM.[444] The majority (82%) of patients had received prior SCT. Treatment-related nonhematological toxicities included nausea (35%), vomiting (29%), fatigue (29%), rash (29%), prurtis (18%), diarrhea (12%), dizziness (12%), sensory neuropathy (12%), and headache (12%). One patient had grade 3 anemia. There were no drug-related serious adverse events. Pharmacokinetic analysis indicated that trough levels of ZD6474 achieved or exceeded the IC_{50} (190 ng/mL) required for inhibition of VEGFR-2 activity. However, this was not reflected in clinical benefit as no responses, as defined by reduction in M protein, were observed.

28.7.8 Toxicities

Similar to that observed in patients with solid tumors, the most consistent toxicities that have been observed with anti-VEGF agents are hypertension, which may be related in part to the inhibition of the release of the potent vasodilator nitric oxide, and asymptomatic proteinuria.[424,437,445–451] Hypertension is usually manageable with medical therapy and continued therapy and decreased with cessation of therapy. Other serious but rare toxicities include thromboembolic events, such as DVT, ischemic cerebrovascular accidents,[409,423,424] and congestive heart failure.[437] Although life-threatening or fatal hemorrhage has also been reported, it is usually attributed to progressive disease in this thrombocytopenic patient population.[437] As more patients receive these agents for prolonged period of time, other therapy-related toxicities may emerge.

28.8 THROMBOSPONDIN ANALOGS

28.8.1 ABT-510

TSP-1 is a multifunctional extracellular matrix protein involved in regulating extracellular matrix structure and endothelial cell adhesion, migration, and invasion.[452,453] It has also been demonstrated to inhibit angiogenesis; this antiangiogenic activity has been mapped to a properdin region hexapeptide motif in the N-terminal of TSP-1.[454] TSP-1 promotes endothelial cell apoptosis in part by interacting with the CD36 cell surface receptor with resultant upregulation of FAS ligand expression and induction of Fas/Fas ligand-mediated apoptosis.[455–457] Preclinical studies revealed low levels of TSP-1 mRNA transcripts in human leukemia, lymphoma, and myeloma cell lines, as well as primary leukemia samples.[458] Low levels of TSP-1 are secreted by B-CLL cells.[80] Similarly, patients with AML have lower plasma TSP-1 levels compared with normal controls,[459] and abnormal forms of platelet TSP-1 have been described in patients with essential thrombocythemia.[460]

ABT-510 is a nonapeptide analog of the active heptapeptide, which mimics the antiangiogenic activity of TSP-1.[461–463] ABT-510 competes with TSP-1 for binding to endothelial cells, induces Fas ligand expression on endothelial cells, inhibits VEGF- and bFGF-stimulated migration of human microvascular endothelial cells (HMVECs), and suppresses murine CEPs.[461–464] Preclinically, ABT-510 caused tumor regression in dogs with malignancies (including lymphoma) and suppressed tumor growth in xenograft mouse models for a variety of solid tumors and myelomatous SCID-hu (severe combined immunodeficiency [SCID] mice previously given implants of a human fetal bone chip) mice.[461,465]

28.8.1.1 Non-Hodgkin's Lymphoma

Preliminary results of a randomized phase II trial evaluating ABT-510 at doses of 10 or 100 mg subcutaneously twice daily in 67 patients with relapsed or refractory lymphoma have been reported.[466] Doses of ≥20 mg daily exceeded the pharmacokinetic target of 100 ng/mL for ≥3 h. After 15 patients were enrolled, the 10 mg twice daily arm was discontinued to consolidate recruitment to the higher dose level. The most frequent toxicities were asthenia (39%), injection site reactions (39%), diarrhea (24%), and anorexia (20%). Grade 3 to 4 toxicities occurred in 19% of patients; four patients developed anemia, two thrombocytopenia, and one each asthenia, chest pain, hemorrhage, gastrointestinal hemorrhage, leucopenia, and hypercalcemia. Of the 56 evaluable patients (11 HL; 45 NHL), ORR was 5% (PR 5%); all responses occurred in the 100 mg dose arm. Twelve month PFS was 15% (0% in the 10 mg arm and 24% in the 100 mg arm). Six patients remain on active therapy. ABT-510 is well tolerated and combination with other agents is warranted.

28.9 CONCLUSION

The use of antiangiogenic agents is an exciting strategy for the treatment of hematological malignancies. As the vast majority of preclinical studies demonstrate that antiangiogenic drugs inhibit tumor growth rather than cause regression of established tumors,[467] nonconventional end points in clinical trial design are required to determine appropriate dosing,[468,469] i.e., surrogate markers that reliably predict efficacy[470] and therefore, hopefully, the patients most likely to respond to or benefit from therapy. Several other questions remain to be answered, including the most effective treatment strategies (e.g., combination with chemotherapy or other targeted agents, schedule and timing of administration [induction vs. maintenance]), and the long-term consequences of prolonged antiangiogenic therapy on normal tissues and physiological angiogenesis. An emerging issue is the occurrence of resistance to antiangiogenic therapies.[471–473] There is evidence that antiangiogenic drug resistance may be mediated via alternate angiogenic pathways (i.e., redundancy of proangiogenic growth factors) as the disease progresses,[474–478] selection for more mature, stabilized blood vessels that are intrinsically less responsive to antiangiogenic drugs,[479] and selection of hypoxia-resistant subpopulations that are less dependent on blood vessels and oxygen for survival.[480,481] Furthermore, as the antiangiogenic drug may directly inhibit proliferation or induce apoptosis of the malignant hematologic clone (e.g., leukemic cells), which may contain receptors to the proangiogenic factors, the malignant clone itself may acquire resistance to the drug. This suggests that optimal therapy will be combining antiangiogenic agents with standard chemotherapy or other targeted agents, including other antiangiogenesis drugs. Trials are underway that address some of these issues and will help clarify the efficacy and toxicities of antiangiogenic agents as sensitizing agents for chemotherapy or other targeted agents.

REFERENCES

1. Byrd, J.C., Stilgenbauer, S., and Flinn, I.W., Chronic lymphocytic leukemia, *Hematology* (*Am Soc Hematol Educ Program*), 163–183, 2004.
2. Kienle, D.L., Korz, C., Hosch, B., Benner, A., Mertens, D., Habermann, A., Krober, A., Jager, U., Lichter, P., Dohner, H., and Stilgenbauer, S., Evidence for distinct pathomechanisms in genetic subgroups of chronic lymphocytic leukemia revealed by quantitative expression analysis of cell cycle, activation, and apoptosis-associated genes, *J Clin Oncol* 23 (16), 3780–3792, 2005.
3. Hideshima, T., Chauhan, D., Richardson, P., and Anderson, K.C., Identification and validation of novel therapeutic targets for multiple myeloma, *J Clin Oncol* 23 (26), 6345–6350, 2005.

4. Ghobrial, I.M., Gertz, M.A., and Fonseca, R., Waldenstrom macroglobulinaemia, *Lancet Oncol* 4 (11), 679–685, 2003.
5. Lossos, I.S., Molecular pathogenesis of diffuse large B-cell lymphoma, *J Clin Oncol* 23 (26), 6351–6357, 2005.
6. Frohling, S., Scholl, C., Gilliland, D.G., and Levine, R.L., Genetics of myeloid malignancies: pathogenetic and clinical implications, *J Clin Oncol* 23 (26), 6285–6295, 2005.
7. Re, D., Kuppers, R., and Diehl, V., Molecular pathogenesis of Hodgkin's lymphoma, *J Clin Oncol* 23 (26), 6379–6386, 2005.
8. Fernandez, V., Hartmann, E., Ott, G., Campo, E., and Rosenwald, A., Pathogenesis of mantle-cell lymphoma: all oncogenic roads lead to dysregulation of cell cycle and DNA damage response pathways, *J Clin Oncol* 23 (26), 6364–6369, 2005.
9. Marcucci, G., Mrozek, K., and Bloomfield, C.D., Molecular heterogeneity and prognostic biomarkers in adults with acute myeloid leukemia and normal cytogenetics, *Curr Opin Hematol* 12 (1), 68–75, 2005.
10. Aguayo, A., Kantarjian, H., Manshouri, T., Gidel, C., Estey, E., Thomas, D., Koller, C., Estrov, Z., O'Brien, S., Keating, M., Freireich, E., and Albitar, M., Angiogenesis in acute and chronic leukemias and myelodysplastic syndromes, *Blood* 96 (6), 2240–2245, 2000.
11. Ribatti, D., Scavelli, C., Roccaro, A.M., Crivellato, E., Nico, B., and Vacca, A., Hematopoietic cancer and angiogenesis, *Stem Cells Dev* 13 (5), 484–495, 2004.
12. Aguayo, A., Estey, E., Kantarjian, H., Mansouri, T., Gidel, C., Keating, M., Giles, F., Estrov, Z., Barlogie, B., and Albitar, M., Cellular vascular endothelial growth factor is a predictor of outcome in patients with acute myeloid leukemia, *Blood* 94 (11), 3717–3721, 1999.
13. Aguayo, A., Kantarjian, H.M., Estey, E.H., Giles, F.J., Verstovsek, S., Manshouri, T., Gidel, C., O'Brien, S., Keating, M.J., and Albitar, M., Plasma vascular endothelial growth factor levels have prognostic significance in patients with acute myeloid leukemia but not in patients with myelodysplastic syndromes, *Cancer* 95 (9), 1923–1930, 2002.
14. Fiedler, W., Graeven, U., Ergun, S., Verago, S., Kilic, N., Stockschlader, M., and Hossfeld, D.K., Vascular endothelial growth factor, a possible paracrine growth factor in human acute myeloid leukemia, *Blood* 89 (6), 1870–1875, 1997.
15. Hussong, J.W., Rodgers, G.M., and Shami, P.J., Evidence of increased angiogenesis in patients with acute myeloid leukemia, *Blood* 95 (1), 309–313, 2000.
16. Moehler, T.M., Ho, A.D., Goldschmidt, H., and Barlogie, B., Angiogenesis in hematologic malignancies, *Crit Rev Oncol Hematol* 45 (3), 227–244, 2003.
17. Padro, T., Bieker, R., Ruiz, S., Steins, M., Retzlaff, S., Burger, H., Buchner, T., Kessler, T., Herrera, F., Kienast, J., Muller-Tidow, C., Serve, H., Berdel, W.E., and Mesters, R.M., Overexpression of vascular endothelial growth factor (VEGF) and its cellular receptor KDR (VEGFR-2) in the bone marrow of patients with acute myeloid leukemia, *Leukemia* 16 (7), 1302–1310, 2002.
18. Padro, T., Ruiz, S., Bieker, R., Burger, H., Steins, M., Kienast, J., Buchner, T., Berdel, W.E., and Mesters, R.M., Increased angiogenesis in the bone marrow of patients with acute myeloid leukemia, *Blood* 95 (8), 2637–2644, 2000.
19. Pruneri, G., Bertolini, F., Soligo, D., Carboni, N., Cortelezzi, A., Ferrucci, P.F., Buffa, R., Lambertenghi-Deliliers, G., and Pezzella, F., Angiogenesis in myelodysplastic syndromes, *Br J Cancer* 81 (8), 1398–1401, 1999.
20. Verstovsek, S., Estey, E., Manshouri, T., Giles, F.J., Cortes, J., Beran, M., Rogers, A., Keating, M., Kantarjian, H., and Albitar, M., Clinical relevance of vascular endothelial growth factor receptors 1 and 2 in acute myeloid leukaemia and myelodysplastic syndrome, *Br J Haematol* 118 (1), 151–156, 2002.
21. Chen, H., Treweeke, A.T., West, D.C., Till, K.J., Cawley, J.C., Zuzel, M., and Toh, C.H., In vitro and in vivo production of vascular endothelial growth factor by chronic lymphocytic leukemia cells, *Blood* 96 (9), 3181–3187, 2000.
22. Molica, S., Vacca, A., Ribatti, D., Cuneo, A., Cavazzini, F., Levato, D., Vitelli, G., Tucci, L., Roccaro, A.M., and Dammacco, F., Prognostic value of enhanced bone marrow angiogenesis in early B-cell chronic lymphocytic leukemia, *Blood* 100 (9), 3344–3351, 2002.

23. Bellamy, W.T., Richter, L., Sirjani, D., Roxas, C., Glinsmann-Gibson, B., Frutiger, Y., Grogan, T.M., and List, A.F., Vascular endothelial cell growth factor is an autocrine promoter of abnormal localized immature myeloid precursors and leukemia progenitor formation in myelodysplastic syndromes, *Blood* 97 (5), 1427–1434, 2001.
24. Di Raimondo, F., Palumbo, G.A., Molica, S., and Giustolisi, R., Angiogenesis in chronic myeloproliferative diseases, *Acta Haematol* 106 (4), 177–183, 2001.
25. Lundberg, L.G., Lerner, R., Sundelin, P., Rogers, R., Folkman, J., and Palmblad, J., Bone marrow in polycythemia vera, chronic myelocytic leukemia, and myelofibrosis has an increased vascularity, *Am J Pathol* 157 (1), 15–19, 2000.
26. Massa, M., Rosti, V., Ramajoli, I., Campanelli, R., Pecci, A., Viarengo, G., Meli, V., Marchetti, M., Hoffman, R., and Barosi, G., Circulating $CD34^+$, $CD133^+$, and vascular endothelial growth factor receptor 2-positive endothelial progenitor cells in myelofibrosis with myeloid metaplasia, *J Clin Oncol* 23 (24), 5688–5695, 2005.
27. Mesa, R.A., Hanson, C.A., Rajkumar, S.V., Schroeder, G., and Tefferi, A., Evaluation and clinical correlations of bone marrow angiogenesis in myelofibrosis with myeloid metaplasia, *Blood* 96 (10), 3374–3380, 2000.
28. Tefferi, A., Myelofibrosis with myeloid metaplasia, *N Engl J Med* 342 (17), 1255–1265, 2000.
29. Thiele, J., Rompcik, V., Wagner, S., and Fischer, R., Vascular architecture and collagen type IV in primary myelofibrosis and polycythaemia vera: an immunomorphometric study on trephine biopsies of the bone marrow, *Br J Haematol* 80 (2), 227–234, 1992.
30. Verstovsek, S., Kantarjian, H., Manshouri, T., Cortes, J., Giles, F.J., Rogers, A., and Albitar, M., Prognostic significance of cellular vascular endothelial growth factor expression in chronic phase chronic myeloid leukemia, *Blood* 99 (6), 2265–2267, 2002.
31. Korkolopoulou, P., Thymara, I., Kavantzas, N., Vassilakopoulos, T.P., Angelopoulou, M.K., Kokoris, S.I., Dimitriadou, E.M., Siakantaris, M.P., Anargyrou, K., Panayiotidis, P., Tsenga, A., Androulaki, A., Doussis-Anagnostopoulou, I.A., Patsouris, E., and Pangalis, G.A., Angiogenesis in Hodgkin's lymphoma: a morphometric approach in 286 patients with prognostic implications, *Leukemia* 19 (6), 894–900, 2005.
32. Salven, P., Orpana, A., Teerenhovi, L., and Joensuu, H., Simultaneous elevation in the serum concentrations of the angiogenic growth factors VEGF and bFGF is an independent predictor of poor prognosis in non-Hodgkin lymphoma: a single-institution study of 200 patients, *Blood* 96 (12), 3712–3718, 2000.
33. Pruneri, G., Ponzoni, M., Ferreri, A.J., Decarli, N., Tresoldi, M., Raggi, F., Baldessari, C., Freschi, M., Baldini, L., Goldaniga, M., Neri, A., Carboni, N., Bertolini, F., and Viale, G., Microvessel density, a surrogate marker of angiogenesis, is significantly related to survival in multiple myeloma patients, *Br J Haematol* 118 (3), 817–820, 2002.
34. Rajkumar, S.V., Mesa, R.A., Fonseca, R., Schroeder, G., Plevak, M.F., Dispenzieri, A., Lacy, M.Q., Lust, J.A., Witzig, T.E., Gertz, M.A., Kyle, R.A., Russell, S.J., and Greipp, P.R., Bone marrow angiogenesis in 400 patients with monoclonal gammopathy of undetermined significance, multiple myeloma, and primary amyloidosis, *Clin Cancer Res* 8 (7), 2210–2216, 2002.
35. Hayman, S.R., Fonseca, R., Dispenzieri, A., Lacy, M.Q., Wellik, L., Plevak, M., Witzig, T.E., Gertz, M.A., Kyle, R.A., Griepp, P.R., Lust, J.A., and Rajkumar, S.V., Bone marrow angiogenesis in Waldenstroms macroglobulinemia (WM), *Blood* 96 (11), 754a, 2000.
36. Rajkumar, S.V., Hayman, S., and Greipp, P.R., Angiogenesis in Waldenstrom's macroglobulinemia, *Semin Oncol* 30 (2), 262–264, 2003.
37. Faderl, S., Do, K.A., Johnson, M.M., Keating, M., O'Brien, S., Jilani, I., Ferrajoli, A., Ravandi-Kashani, F., Aguilar, C., Dey, A., Thomas, D.A., Giles, F.J., Kantarjian, H.M., and Albitar, M., Angiogenic factors may have a different prognostic role in adult acute lymphoblastic leukemia, *Blood* 106 (13), 4303–4307, 2005.
38. Moehler, T.M., Neben, K., Ho, A.D., and Goldschmidt, H., Angiogenesis in hematologic malignancies, *Ann Hematol* 80 (12), 695–705, 2001.
39. Korkolopoulou, P., Viniou, N., Kavantzas, N., Patsouris, E., Thymara, I., Pavlopoulos, P.M., Terpos, E., Stamatopoulos, K., Plata, E., Anargyrou, K., Androulaki, A., Davaris, P., and Yataganas, X., Clinicopathologic correlations of bone marrow angiogenesis in chronic myeloid leukemia: a morphometric study, *Leukemia* 17 (1), 89–97, 2003.

40. Korkolopoulou, P., Apostolidou, E., Pavlopoulos, P.M., Kavantzas, N., Vyniou, N., Thymara, I., Terpos, E., Patsouris, E., Yataganas, X., and Davaris, P., Prognostic evaluation of the microvascular network in myelodysplastic syndromes, *Leukemia* 15 (9), 1369–1376, 2001.
41. Munshi, N.C. and Wilson, C., Increased bone marrow microvessel density in newly diagnosed multiple myeloma carries a poor prognosis, *Semin Oncol* 28 (6), 565–569, 2001.
42. Rajkumar, S.V., Leong, T., Roche, P.C., Fonseca, R., Dispenzieri, A., Lacy, M.Q., Lust, J.A., Witzig, T.E., Kyle, R.A., Gertz, M.A., and Greipp, P.R., Prognostic value of bone marrow angiogenesis in multiple myeloma, *Clin Cancer Res* 6 (8), 3111–3116, 2000.
43. Vacca, A., Ribatti, D., Presta, M., Minischetti, M., Iurlaro, M., Ria, R., Albini, A., Bussolino, F., and Dammacco, F., Bone marrow neovascularization, plasma cell angiogenic potential, and matrix metalloproteinase-2 secretion parallel progression of human multiple myeloma, *Blood* 93 (9), 3064–3073, 1999.
44. Koster, A., van Krieken, J.H., Mackenzie, M.A., Schraders, M., Borm, G.F., van der Laak, J.A., Leenders, W., Hebeda, K., and Raemaekers, J.M., Increased vascularization predicts favorable outcome in follicular lymphoma, *Clin Cancer Res* 11 (1), 154–161, 2005.
45. Mangi, M.H. and Newland, A.C., Angiogenesis and angiogenic mediators in haematological malignancies, *Br J Haematol* 111 (1), 43–51, 2000.
46. Reinmuth, N., Stoeltzing, O., Liu, W., Ahmad, S.A., Jung, Y.D., Fan, F., Parikh, A., and Ellis, L.M., Endothelial survival factors as targets for antineoplastic therapy, *Cancer J* 7 (Suppl 3), S109–S119, 2001.
47. Ferrara, N., Gerber, H.P., and LeCouter, J., The biology of VEGF and its receptors, *Nat Med* 9 (6), 669–676, 2003.
48. White, C.W., Sondheimer, H.M., Crouch, E.C., Wilson, H., and Fan, L.L., Treatment of pulmonary hemangiomatosis with recombinant interferon alfa-2a, *N Engl J Med* 320 (18), 1197–1200, 1989.
49. Jonasch, E. and Haluska, F.G., Interferon in oncological practice: review of interferon biology, clinical applications, and toxicities, *Oncologist* 6 (1), 34–55, 2001.
50. Al-Zahrani, H., Gupta, V., Minden, M.D., Messner, H.A., and Lipton, J.H., Vascular events associated with alpha interferon therapy, *Leuk Lymphoma* 44 (3), 471–475, 2003.
51. Scappaticci, F.A., Smith, R., Pathak, A., Schloss, D., Lum, B., Cao, Y., Johnson, F., Engleman, E.G., and Nolan, G.P., Combination angiostatin and endostatin gene transfer induces synergistic antiangiogenic activity in vitro and antitumor efficacy in leukemia and solid tumors in mice, *Mol Ther* 3 (2), 186–196, 2001.
52. Schuch, G., Oliveira-Ferrer, L., Loges, S., Laack, E., Bokemeyer, C., Hossfeld, D.K., Fiedler, W., and Ergun, S., Antiangiogenic treatment with endostatin inhibits progression of AML in vivo, *Leukemia* 19 (8), 1312–1317, 2005.
53. Capillo, M., Mancuso, P., Gobbi, A., Monestiroli, S., Pruneri, G., Dell'Agnola, C., Martinelli, G., Shultz, L., and Bertolini, F., Continuous infusion of endostatin inhibits differentiation, mobilization, and clonogenic potential of endothelial cell progenitors, *Clin Cancer Res* 9 (1), 377–382, 2003.
54. Eisterer, W., Jiang, X., Bachelot, T., Pawliuk, R., Abramovich, C., Leboulch, P., Hogge, D., and Eaves, C., Unfulfilled promise of endostatin in a gene therapy–xenotransplant model of human acute lymphocytic leukemia, *Mol Ther* 5 (4), 352–359, 2002.
55. Iversen, P.O., Sorensen, D.R., and Benestad, H.B., Inhibitors of angiogenesis selectively reduce the malignant cell load in rodent models of human myeloid leukemias, *Leukemia* 16 (3), 376–381, 2002.
56. Podar, K., Catley, L.P., Tai, Y.T., Shringarpure, R., Carvalho, P., Hayashi, T., Burger, R., Schlossman, R.L., Richardson, P.G., Pandite, L.N., Kumar, R., Hideshima, T., Chauhan, D., and Anderson, K.C., GW654652, the pan-inhibitor of VEGF receptors, blocks the growth and migration of multiple myeloma cells in the bone marrow microenvironment, *Blood* 103 (9), 3474–3479, 2004.
57. Le Gouill, S., Podar, K., Amiot, M., Hideshima, T., Chauhan, D., Ishitsuka, K., Kumar, S., Raje, N., Richardson, P.G., Harousseau, J.L., and Anderson, K.C., VEGF induces Mcl-1 up-regulation and protects multiple myeloma cells against apoptosis, *Blood* 104 (9), 2886–2892, 2004.
58. Podar, K., Simoncini, M., Le Gouill, S., Tai, Y.T., Kumar, S., Tassone, P., Catley, L.P., Pandite, L.N., Kumar, R., Hideshima, T., Chauhan, D., and Anderson, K.C., Effects of the indazolylpyrimidine GW786034 on angiogenesis and MM cell growth: therapeutic implications, *Blood* 104 (11), 673a, 2004.

59. Wang, E.S., Teruya-Feldstein, J., Wu, Y., Zhu, Z., Hicklin, D.J., and Moore, M.A., Targeting autocrine and paracrine VEGF receptor pathways inhibits human lymphoma xenografts in vivo, *Blood* 104 (9), 2893–2902, 2004.
60. Zhu, Z., Hattori, K., Zhang, H., Jimenez, X., Ludwig, D.L., Dias, S., Kussie, P., Koo, H., Kim, H.J., Lu, D., Liu, M., Tejada, R., Friedrich, M., Bohlen, P., Witte, L., and Rafii, S., Inhibition of human leukemia in an animal model with human antibodies directed against vascular endothelial growth factor receptor 2. Correlation between antibody affinity and biological activity, *Leukemia* 17 (3), 604–611, 2003.
61. Dias, S., Hattori, K., Zhu, Z., Heissig, B., Choy, M., Lane, W., Wu, Y., Chadburn, A., Hyjek, E., Gill, M., Hicklin, D.J., Witte, L., Moore, M.A., and Rafii, S., Autocrine stimulation of VEGFR-2 activates human leukemic cell growth and migration, *J Clin Invest* 106 (4), 511–521, 2000.
62. Garton, A.J., Crew, A.P., Franklin, M., Cooke, A.R., Wynne, G.M., Castaldo, L., Kahler, J., Winski, S.L., Franks, A., Brown, E.N., Bittner, M.A., Keily, J.F., Briner, P., Hidden, C., Srebernak, M.C., Pirrit, C., O'Connor, M., Chan, A., Vulevic, B., Henninger, D., Hart, K., Sennello, R., Li, A.H., Zhang, T., Richardson, F., Emerson, D.L., Castelhano, A.L., Arnold, L.D., and Gibson, N.W., OSI-930: a novel selective inhibitor of Kit and kinase insert domain receptor tyrosine kinases with antitumor activity in mouse xenograft models, *Cancer Res* 66 (2), 1015–1024, 2006.
63. Petti, F., Thelemann, A., Kahler, J., McCormack, S., Castaldo, L., Hunt, T., Nuwaysir, L., Zeiske, L., Haack, H., Sullivan, L., Garton, A., and Haley, J.D., Temporal quantitation of mutant Kit tyrosine kinase signaling attenuated by a novel thiophene kinase inhibitor OSI-930, *Mol Cancer Ther* 4 (8), 1186–1197, 2005.
64. Nabha, S.M., Wall, N.R., Mohammad, R.M., Pettit, G.R., and Al-Katib, A.M., Effects of combretastatin A-4 prodrug against a panel of malignant human B-lymphoid cell lines, *Anticancer Drugs* 11 (5), 385–392, 2000.
65. Woon, S.T., Reddy, C.B., Drummond, C.J., Schooltink, M.A., Baguley, B.C., Kieda, C., and Ching, L.M., A comparison of the ability of DMXAA and xanthenone analogues to activate NF-κB in murine and human cell lines, *Oncol Res* 15 (7–8), 351–364, 2005.
66. Hansen, H.P., Matthey, B., Barth, S., Kisseleva, T., Mokros, T., Davies, S.J., Beckett, R.P., Foelster-Holst, R., Lange, H.H., Engert, A., and Lemke, H., Inhibition of metalloproteinases enhances the internalization of anti-CD30 antibody Ki-3 and the cytotoxic activity of Ki-3 immunotoxin, *Int J Cancer* 98 (2), 210–215, 2002.
67. Gingras, D., Boivin, D., Deckers, C., Gendron, S., Barthomeuf, C., and Beliveau, R., Neovastat—a novel antiangiogenic drug for cancer therapy, *Anticancer Drugs* 14 (2), 91–96, 2003.
68. Caligaris-Cappio, F. and Hamblin, T.J., B-cell chronic lymphocytic leukemia: a bird of a different feather, *J Clin Oncol* 17 (1), 399–408, 1999.
69. Kay, N.E., Hamblin, T.J., Jelinek, D.F., Dewald, G.W., Byrd, J.C., Farag, S., Lucas, M., and Lin, T., Chronic lymphocytic leukemia, *Hematology* (*Am Soc Hematol Educ Program*), 193–213, 2002.
70. Yasui, H., Hideshima, T., Richardson, P.G., and Anderson, K.C., Novel therapeutic strategies targeting growth factor signalling cascades in multiple myeloma, *Br J Haematol* 132 (4), 385–397, 2006.
71. Liesveld, J.L., Rosell, K.E., Lu, C., Bechelli, J., Phillips, G., Lancet, J.E., and Abboud, C.N., Acute myelogenous leukemia—microenvironment interactions: role of endothelial cells and proteasome inhibition, *Hematology* 10 (6), 483–494, 2005.
72. Milojkovic, D., Devereux, S., Westwood, N.B., Mufti, G.J., Thomas, N.S., and Buggins, A.G., Antiapoptotic microenvironment of acute myeloid leukemia, *J Immunol* 173 (11), 6745–6752, 2004.
73. de Jong, D., Molecular pathogenesis of follicular lymphoma: a cross talk of genetic and immunologic factors, *J Clin Oncol* 23 (26), 6358–6363, 2005.
74. Flores-Figueroa, E., Gutierrez-Espindola, G., Montesinos, J.J., Arana-Trejo, R.M., and Mayani, H., In vitro characterization of hematopoietic microenvironment cells from patients with myelodysplastic syndrome, *Leuk Res* 26 (7), 677–686, 2002.
75. Foa, R., Massaia, M., Cardona, S., Tos, A.G., Bianchi, A., Attisano, C., Guarini, A., di Celle, P.F., and Fierro, M.T., Production of tumor necrosis factor-alpha by B-cell chronic lymphocytic leukemia cells: a possible regulatory role of TNF in the progression of the disease, *Blood* 76 (2), 393–400, 1990.

76. van Kooten, C., Rensink, I., Aarden, L., and van Oers, R., Cytokines and intracellular signals involved in the regulation of B-CLL proliferation, *Leuk Lymphoma* 12 (1–2), 27–33, 1993.
77. Ferrajoli, A., Keating, M.J., Manshouri, T., Giles, F.J., Dey, A., Estrov, Z., Koller, C.A., Kurzrock, R., Thomas, D.A., Faderl, S., Lerner, S., O'Brien, S., and Albitar, M., The clinical significance of tumor necrosis factor-alpha plasma level in patients having chronic lymphocytic leukemia, *Blood* 100 (4), 1215–1219, 2002.
78. Ferrajoli, A., Manshouri, T., Estrov, Z., Keating, M.J., O'Brien, S., Lerner, S., Beran, M., Kantarjian, H.M., Freireich, E.J., and Albitar, M., High levels of vascular endothelial growth factor receptor-2 correlate with shortened survival in chronic lymphocytic leukemia, *Clin Cancer Res* 7 (4), 795–799, 2001.
79. Kay, N.E., The angiogenic status of B-CLL B cells: role of the VEGF receptors, *Leuk Res* 28 (3), 221–222, 2004.
80. Kay, N.E., Bone, N.D., Tschumper, R.C., Howell, K.H., Geyer, S.M., Dewald, G.W., Hanson, C.A., and Jelinek, D.F., B-CLL cells are capable of synthesis and secretion of both pro- and anti-angiogenic molecules, *Leukemia* 16 (5), 911–919, 2002.
81. Bairey, O., Boycov, O., Kaganovsky, E., Zimra, Y., Shaklai, M., and Rabizadeh, E., All three receptors for vascular endothelial growth factor (VEGF) are expressed on B-chronic lymphocytic leukemia (CLL) cells, *Leuk Res* 28 (3), 243–248, 2004.
82. Waage, A. and Espevik, T., TNF receptors in chronic lymphocytic leukemia, *Leuk Lymphoma* 13 (1–2), 41–46, 1994.
83. Sawanobori, M., Yamaguchi, S., Hasegawa, M., Inoue, M., Suzuki, K., Kamiyama, R., Hirokawa, K., and Kitagawa, M., Expression of TNF receptors and related signaling molecules in the bone marrow from patients with myelodysplastic syndromes, *Leuk Res* 27 (7), 583–591, 2003.
84. Tsimberidou, A.M. and Giles, F.J., TNF-alpha targeted therapeutic approaches in patients with hematologic malignancies, *Expert Rev Anticancer Ther* 2 (3), 277–286, 2002.
85. He, B., Chadburn, A., Jou, E., Schattner, E.J., Knowles, D.M., and Cerutti, A., Lymphoma B cells evade apoptosis through the TNF family members BAFF/BLyS and APRIL, *J Immunol* 172 (5), 3268–3279, 2004.
86. Tassone, P., Neri, P., Kutok, J.L., Tournilhac, O., Santos, D.D., Hatjiharissi, E., Munshi, V., Venuta, S., Anderson, K.C., Treon, S.P., and Munshi, N.C., A SCID-hu in vivo model of human Waldenstrom macroglobulinemia, *Blood* 106 (4), 1341–1345, 2005.
87. Gupta, D., Treon, S.P., Shima, Y., Hideshima, T., Podar, K., Tai, Y.T., Lin, B., Lentzsch, S., Davies, F.E., Chauhan, D., Schlossman, R.L., Richardson, P., Ralph, P., Wu, L., Payvandi, F., Muller, G., Stirling, D.I., and Anderson, K.C., Adherence of multiple myeloma cells to bone marrow stromal cells upregulates vascular endothelial growth factor secretion: therapeutic applications, *Leukemia* 15 (12), 1950–1961, 2001.
88. Kumar, S., Witzig, T.E., Timm, M., Haug, J., Wellik, L., Fonseca, R., Greipp, P.R., and Rajkumar, S.V., Expression of VEGF and its receptors by myeloma cells, *Leukemia* 17 (10), 2025–2031, 2003.
89. Podar, K., Tai, Y.T., Davies, F.E., Lentzsch, S., Sattler, M., Hideshima, T., Lin, B.K., Gupta, D., Shima, Y., Chauhan, D., Mitsiades, C., Raje, N., Richardson, P., and Anderson, K.C., Vascular endothelial growth factor triggers signaling cascades mediating multiple myeloma cell growth and migration, *Blood* 98 (2), 428–435, 2001.
90. Podar, K., Tai, Y.T., Lin, B.K., Narsimhan, R.P., Sattler, M., Kijima, T., Salgia, R., Gupta, D., Chauhan, D., and Anderson, K.C., Vascular endothelial growth factor-induced migration of multiple myeloma cells is associated with beta 1 integrin- and phosphatidylinositol 3-kinase-dependent PKC alpha activation, *J Biol Chem* 277 (10), 7875–7881, 2002.
91. Teo, S.K., Stirling, D.I., and Zeldis, J.B., Thalidomide as a novel therapeutic agent: new uses for an old product, *Drug Discov Today* 10 (2), 107–114, 2005.
92. Bartlett, J.B., Dredge, K., and Dalgleish, A.G., The evolution of thalidomide and its IMiD derivatives as anticancer agents, *Nat Rev Cancer* 4 (4), 314–322, 2004.
93. Muller, G.W., Chen, R., Huang, S.Y., Corral, L.G., Wong, L.M., Patterson, R.T., Chen, Y., Kaplan, G., and Stirling, D.I., Amino-substituted thalidomide analogs: potent inhibitors of TNF-alpha production, *Bioorg Med Chem Lett* 9 (11), 1625–1630, 1999.

94. Dredge, K., Marriott, J.B., Macdonald, C.D., Man, H.W., Chen, R., Muller, G.W., Stirling, D., and Dalgleish, A.G., Novel thalidomide analogues display anti-angiogenic activity independently of immunomodulatory effects, *Br J Cancer* 87 (10), 1166–1172, 2002.
95. Lentzsch, S., Rogers, M.S., LeBlanc, R., Birsner, A.E., Shah, J.H., Treston, A.M., Anderson, K.C., and D'Amato, R.J., *S*-3-amino-phthalimido-glutarimide inhibits angiogenesis and growth of B-cell neoplasias in mice, *Cancer Res* 62 (8), 2300–2305, 2002.
96. Haslett, P.A., Corral, L.G., Albert, M., and Kaplan, G., Thalidomide costimulates primary human T lymphocytes, preferentially inducing proliferation, cytokine production, and cytotoxic responses in the $CD8^+$ subset, *J Exp Med* 187 (11), 1885–1892, 1998.
97. Davies, F.E., Raje, N., Hideshima, T., Lentzsch, S., Young, G., Tai, Y.T., Lin, B., Podar, K., Gupta, D., Chauhan, D., Treon, S.P., Richardson, P.G., Schlossman, R.L., Morgan, G.J., Muller, G.W., Stirling, D.I., and Anderson, K.C., Thalidomide and immunomodulatory derivatives augment natural killer cell cytotoxicity in multiple myeloma, *Blood* 98 (1), 210–216, 2001.
98. LeBlanc, R., Hideshima, T., Catley, L.P., Shringarpure, R., Burger, R., Mitsiades, N., Mitsiades, C., Cheema, P., Chauhan, D., Richardson, P.G., Anderson, K.C., and Munshi, N.C., Immunomodulatory drug costimulates T cells via the B7-CD28 pathway, *Blood* 103 (5), 1787–1790, 2004.
99. Dredge, K., Horsfall, R., Robinson, S.P., Zhang, L.H., Lu, L., Tang, Y., Shirley, M.A., Muller, G., Schafer, P., Stirling, D., Dalgleish, A.G., and Bartlett, J.B., Orally administered lenalidomide (CC-5013) is anti-angiogenic in vivo and inhibits endothelial cell migration and Akt phosphorylation in vitro, *Microvasc Res* 69 (1–2), 56–63, 2005.
100. D'Amato, R.J., Loughnan, M.S., Flynn, E., and Folkman, J., Thalidomide is an inhibitor of angiogenesis, *Proc Natl Acad Sci USA* 91 (9), 4082–4085, 1994.
101. Kenyon, B.M., Browne, F., and D'Amato, R.J., Effects of thalidomide and related metabolites in a mouse corneal model of neovascularization, *Exp Eye Res* 64 (6), 971–978, 1997.
102. Keifer, J.A., Guttridge, D.C., Ashburner, B.P., and Baldwin, A.S., Jr., Inhibition of NF-kappa B activity by thalidomide through suppression of IkappaB kinase activity, *J Biol Chem* 276 (25), 22382–22387, 2001.
103. Majumdar, S., Lamothe, B., and Aggarwal, B.B., Thalidomide suppresses NF-kappa B activation induced by TNF and H_2O_2, but not that activated by ceramide, lipopolysaccharides, or phorbol ester, *J Immunol* 168 (6), 2644–2651, 2002.
104. Marriott, J.B., Clarke, I.A., Dredge, K., Muller, G., Stirling, D., and Dalgleish, A.G., Thalidomide and its analogues have distinct and opposing effects on TNF-alpha and TNFR2 during co-stimulation of both CD4(+) and CD8(+) T cells, *Clin Exp Immunol* 130 (1), 75–84, 2002.
105. Lentzsch, S., LeBlanc, R., Podar, K., Davies, F., Lin, B., Hideshima, T., Catley, L., Stirling, D.I., and Anderson, K.C., Immunomodulatory analogs of thalidomide inhibit growth of Hs Sultan cells and angiogenesis in vivo, *Leukemia* 17 (1), 41–44, 2003.
106. Dredge, K., Marriott, J.B., Todryk, S.M., Muller, G.W., Chen, R., Stirling, D.I., and Dalgleish, A.G., Protective antitumor immunity induced by a costimulatory thalidomide analog in conjunction with whole tumor cell vaccination is mediated by increased Th1-type immunity, *J Immunol* 168 (10), 4914–4919, 2002.
107. Chang, D.H., Liu, N., Klimek, V., Hassoun, H., Mazumder, A., Nimer, S.D., Jagannath, S., and Dhodapkar, M.V., Enhancement of ligand-dependent activation of human natural killer T cells by lenalidomide: therapeutic implications, *Blood* 108 (2), 618–621, 2006.
108. Steins, M.B., Padro, T., Bieker, R., Ruiz, S., Kropff, M., Kienast, J., Kessler, T., Buechner, T., Berdel, W.E., and Mesters, R.M., Efficacy and safety of thalidomide in patients with acute myeloid leukemia, *Blood* 99 (3), 834–839, 2002.
109. Steins, M.B., Bieker, R., Padro, T., Kessler, T., Kienast, J., Berdel, W.E., and Mesters, R.M., Thalidomide for the treatment of acute myeloid leukemia, *Leuk Lymphoma* 44 (9), 1489–1493, 2003.
110. Thomas, D.A., Estey, E., Giles, F.J., Faderl, S., Cortes, J., Keating, M., O'Brien, S., Albitar, M., and Kantarjian, H., Single agent thalidomide in patients with relapsed or refractory acute myeloid leukaemia, *Br J Haematol* 123 (3), 436–441, 2003.
111. Cortes, J., Kantarjian, H., Albitar, M., Thomas, D., Faderl, S., Koller, C., Garcia-Manero, G., Giles, F., Andreeff, M., O'Brien, S., Keating, M., and Estey, E., A randomized trial of liposomal daunorubicin and cytarabine versus liposomal daunorubicin and topotecan with or without

thalidomide as initial therapy for patients with poor prognosis acute myelogenous leukemia or myelodysplastic syndrome, *Cancer* 97 (5), 1234–1241, 2003.

112. Chanan-Khan, A., Miller, K.C., Takeshita, K., Koryzna, A., Donohue, K., Bernstein, Z.P., Mohr, A., Klippenstein, D., Wallace, P., Zeldis, J.B., Berger, C., and Czuczman, M.S., Results of a phase 1 clinical trial of thalidomide in combination with fludarabine as initial therapy for patients with treatment-requiring chronic lymphocytic leukemia (CLL), *Blood* 106 (10), 3348–3352, 2005.
113. Yee, K.W., O'Brien, S.M., and Giles, F.J., An update on the management of chronic lymphocytic leukaemia, *Expert Opin Pharmacother* 5 (7), 1535–1554, 2004.
114. Chanan-Khan, A.A., Padmanabhan, S., Stein, L., Panzarella, J., Miller, K.C., and Hawthorne, L., Validating molecular targets of thalidomide in CLL: net effect of increased apoptosis through the intrinsic pathway and down regulation of NF-κB signaling—validation using gene expression profile from the phase I/II clinical trial of thalidomide and fludarabine, *Blood* 106 (11), 342b, 2005.
115. Chanan-Khan, A.A., Miller, K.C., Marshall, P., Padmanabhan, S., Brady, W., Bernstein, Z.P., Wallace, P., and Czuczman, M.S., Thalidomide (T) in combination with fludarabine (F) as initial therapy for patients (pts) with treatment naive chronic lymphocytic leukemia (CLL): preliminary results of a phase I/II clinical trial, *Blood* 106 (11), 834a, 2005.
116. Kay, N., Geyer, S., Yaqoob, I., Phyliky, R., Kutteh, L., and Li, C.Y., Thalidomide (Td) treatment in chronic lymphocytic leukemia (CLL): a North Central Cancer Treatment Group (NCCTG) study, *Blood* 102 (11), 359b, 2003.
117. Furman, R.R., Leonard, J.P., Allen, S.L., Coleman, M., Rosenthal, T., and Gabrilove, J.L., Thalidomide alone or in combination with fludarabine are effective treatments for patients with fludarabine-relapsed and refractory CLL, *J Clin Oncol* 23, 595s, 2005.
118. Laurenti, L., Piccioni, P., Tarnani, M., De Padua, L., Garzia, M., Efremov, D.G., Piccirillo, N., Chiusolo, P., Sica, S., and Leone, G., Low-dose thalidomide in combination with oral fludarabine and cyclophosphamide is ineffective in heavily pre-treated patients with chronic lymphocytic leukemia, *Leuk Res* 31 (2), 253–256, 2007.
119. Passam, F.H., Alexandrakis, M.G., Moschandrea, J., Sfiridaki, A., Roussou, P.A., and Siafakas, N.M., Angiogenic molecules in Hodgkin's disease: results from sequential serum analysis, *Int J Immunopathol Pharmacol* 19 (1), 161–170, 2006.
120. Pro, B., Younes, A., Albitar, M., Dang, N.H., Samaniego, F., Romaguera, J., McLaughlin, P., Hagemeister, F.B., Rodriguez, M.A., Clemons, M., and Cabanillas, F., Thalidomide for patients with recurrent lymphoma, *Cancer* 100 (6), 1186–1189, 2004.
121. Little, R., Wittes, R.E., Longo, D.L., and Wilson, W.H., Vinblastine for recurrent Hodgkin's disease following autologous bone marrow transplant, *J Clin Oncol* 16 (2), 584–588, 1998.
122. Kuruvilla, J., Song, K., Mollee, P., Panzarella, T., McCrae, J., Nagy, T., Crump, M., and Keating, A., A phase II study of thalidomide and vinblastine for palliative patients with Hodgkin's lymphoma, *Hematology* 11 (1), 25–29, 2006.
123. Attal, M., Harousseau, J.L., Stoppa, A.M., Sotto, J.J., Fuzibet, J.G., Rossi, J.F., Casassus, P., Maisonneuve, H., Facon, T., Ifrah, N., Payen, C., and Bataille, R., A prospective, randomized trial of autologous bone marrow transplantation and chemotherapy in multiple myeloma. Intergroupe Francais du Myelome, *N Engl J Med* 335 (2), 91–97, 1996.
124. Child, J.A., Morgan, G.J., Davies, F.E., Owen, R.G., Bell, S.E., Hawkins, K., Brown, J., Drayson, M.T., and Selby, P.J., High-dose chemotherapy with hematopoietic stem-cell rescue for multiple myeloma, *N Engl J Med* 348 (19), 1875–1883, 2003.
125. Myeloma Trialists' Collaborative Group, Combination chemotherapy versus melphalan plus prednisone as treatment for multiple myeloma: an overview of 6,633 patients from 27 randomized trials, *J Clin Oncol* 16 (12), 3832–3842, 1998.
126. Alexanian, R., Barlogie, B., and Tucker, S., VAD-based regimens as primary treatment for multiple myeloma, *Am J Hematol* 33 (2), 86–89, 1990.
127. Samson, D., Gaminara, E., Newland, A., Van de Pette, J., Kearney, J., McCarthy, D., Joyner, M., Aston, L., Mitchell, T., Hamon, M., et al., Infusion of vincristine and doxorubicin with oral dexamethasone as first-line therapy for multiple myeloma, *Lancet* 2 (8668), 882–885, 1989.
128. Hjorth, M., Hellquist, L., Holmberg, E., Magnusson, B., Rodjer, S., and Westin, J., Initial versus deferred melphalan–prednisone therapy for asymptomatic multiple myeloma stage I—a randomized study, Myeloma Group of Western Sweden, *Eur J Haematol* 50 (2), 95–102, 1993.

129. Grignani, G., Gobbi, P.G., Formisano, R., Pieresca, C., Ucci, G., Brugnatelli, S., Riccardi, A., and Ascari, E., A prognostic index for multiple myeloma, *Br J Cancer* 73 (9), 1101–1107, 1996.
130. Rajkumar, S.V., Gertz, M.A., Lacy, M.Q., Dispenzieri, A., Fonseca, R., Geyer, S.M., Iturria, N., Kumar, S., Lust, J.A., Kyle, R.A., Greipp, P.R., and Witzig, T.E., Thalidomide as initial therapy for early-stage myeloma, *Leukemia* 17 (4), 775–779, 2003.
131. Weber, D., Rankin, K., Gavino, M., Delasalle, K., and Alexanian, R., Thalidomide alone or with dexamethasone for previously untreated multiple myeloma, *J Clin Oncol* 21 (1), 16–19, 2003.
132. Rajkumar, S.V., Blood, E., Vesole, D., Fonseca, R., and Greipp, P.R., Phase III clinical trial of thalidomide plus dexamethasone compared with dexamethasone alone in newly diagnosed multiple myeloma: a clinical trial coordinated by the Eastern Cooperative Oncology Group, *J Clin Oncol* 24 (3), 431–436, 2006.
133. Alexanian, R., Dimopoulos, M.A., Delasalle, K., and Barlogie, B., Primary dexamethasone treatment of multiple myeloma, *Blood* 80 (4), 887–890, 1992.
134. Hideshima, T., Chauhan, D., Shima, Y., Raje, N., Davies, F.E., Tai, Y.T., Treon, S.P., Lin, B., Schlossman, R.L., Richardson, P., Muller, G., Stirling, D.I., and Anderson, K.C., Thalidomide and its analogs overcome drug resistance of human multiple myeloma cells to conventional therapy, *Blood* 96 (9), 2943–2950, 2000.
135. Rajkumar, S.V., Hussein, M., Catalano, J., Jedrzejcak, W., Sirkovich, S., Olesnyckyj, M., Yu, Z., Knight, R., Zeldis, J.B., and Blade, J., A multicenter, randomized, double-blind, placebo-controlled trial of thalidomide plus dexamethasone versus dexamethasone alone as initial therapy for newly diagnosed multiple myeloma, *J Clin Oncol* 24 (18S), 426s, 2006.
136. Ludwig, H., Drach, J., Tothova, E., Gisslinger, H., Linkesch, W., Jaksic, B., Fridik, M., Thaler, J., Lang, A., Hajek, R., Zojer, N., Greil, R., Kuhn, I., Hinke, A., and Labar, B., Thalidomide–dexamethasone versus melphalan–prednisolone as first line treatment in elderly patients with multiple myeloma: an interim analysis, *Blood* 106 (11), 231a, 2005.
137. Kumar, S., Greipp, P.R., Haug, J.L., Gertz, M.A., Blood, E., and Rajkumar, S.V., Correlation of bone marrow angiogenesis and response to thalidomide dexamethasone in multiple myeloma, *J Clin Oncol* 24 (18S), 451s, 2006.
138. Hussein, M.A., Baz, R., Srkalovic, G., Agrawal, N., Suppiah, R., Hsi, E., Andresen, S., Karam, M.A., Reed, J., Faiman, B., Kelly, M., and Walker, E., Phase 2 study of pegylated liposomal doxorubicin, vincristine, decreased-frequency dexamethasone, and thalidomide in newly diagnosed and relapsed-refractory multiple myeloma, *Mayo Clin Proc* 81 (7), 889–895, 2006.
139. Williams, C.D., Byrne, J.L., Sidra, G., Zaman, S., and Russell, N.H., Combination chemotherapy with cyclophosphamide, thalidomide and dexamethasone achieves a high response rate in patients with newly diagnosed, VAD-refractory and relapsed myeloma, *Blood* 104 (11), 419a, 2004.
140. Hassoun, H., Reich, L., Klimek, V.M., Dhodapkar, M., Cohen, A., Kewalramani, T., Zimman, R., Drake, L., Riedel, E.R., Hedvat, C.V., Teruya-Feldstein, J., Filippa, D.A., Fleisher, M., Nimer, S.D., and Comenzo, R.L., Doxorubicin and dexamethasone followed by thalidomide and dexamethasone is an effective well tolerated initial therapy for multiple myeloma, *Br J Haematol* 132 (2), 155–161, 2006.
141. Offidani, M., Corvatta, L., Piersantelli, M.N., Visani, G., Alesiani, F., Brunori, M., Galieni, P., Catarini, M., Burattini, M., Centurioni, R., Ferranti, M., Rupoli, S., Scortechini, A.R., Giuliodori, L., Candela, M., Capelli, D., Montanari, M., Olivieri, A., Poloni, A., Polloni, C., Marconi, M., and Leoni, P., Thalidomide, dexamethasone and pegylated liposomal doxorubicin (ThaDD) for newly diagnosed multiple myeloma patients over 65 years, *Blood* 108 (7), 2159–2164, 2006.
142. Facon, T., Mary, J.Y., Hulin, C., Benboubker, L., Attal, M., Renaud, M., Harrousseau, J.L., Pegourie, B., Guillerm, G., Chaleteix, C., Dib, M., Voillat, L., Maisonneuve, H., Troncy, J., Dorvaux, V., Monconduit, M., Martin, C., Casassus, P., Jaubert, J., Jardel, H., Kolb, B., and Bauters, F., Major superiority of melphalan–prednisone (MP) + thalidomide (THAL) over MP and autologous stem cell transplantation in the treatment of newly diagnosed elderly patients with multiple myeloma, *Blood* 106 (11), 230a, 2005.
143. Palumbo, A., Bringhen, S., Caravita, T., Merla, E., Capparella, V., Callea, V., Cangialosi, C., Grasso, M., Rossini, F., Galli, M., Catalano, L., Zamagni, E., Petrucci, M.T., De Stefano, V., Ceccarelli, M., Ambrosini, M.T., Avonto, I., Falco, P., Ciccone, G., Liberati, A.M., Musto, P., and Boccadoro, M., Oral melphalan and prednisone chemotherapy plus thalidomide compared

with melphalan and prednisone alone in elderly patients with multiple myeloma: randomised controlled trial, *Lancet* 367 (9513), 825–831, 2006.

144. Chanan-Khan, A.A., Miller, K.C., McCarthy, P., Koryzna, A., Kouides, P., Donohue, K., Mohr, A., Bernstein, Z.P., Alm, A., and Czuczman, M.S., VAD-t (vincristine, adriamycin, dexamethasone and low-dose thalidomide) is an effective initial therapy with high response rates for patients with treatment naive multiple myeloma (MM), *Blood* 104 (11), 943a–944a, 2004.
145. Cavo, M., Zamagni, E., Tosi, P., Tacchetti, P., Cellini, C., Cangini, D., de Vivo, A., Testoni, N., Nicci, C., Terragna, C., Grafone, T., Perrone, G., Ceccolini, M., Tura, S., and Baccarani, M., Superiority of thalidomide and dexamethasone over vincristine–doxorubicindexamethasone (VAD) as primary therapy in preparation for autologous transplantation for multiple myeloma, *Blood* 106 (1), 35–39, 2005.
146. Jimenez-Zepeda, V.H. and Dominguez-Martinez, V.J., Vincristine, doxorubicin, and dexamethasone or thalidomide plus dexamethasone for newly diagnosed patients with multiple myeloma? *Eur J Haematol* 77 (3), 239–244, 2006.
147. Greipp, P., Treatment paradigms for the newly diagnosed patient with multiple myeloma, *Semin Hematol* 42 (4 Suppl 4), S16–S21, 2005.
148. Offidani, M., Corvatta, L., Marconi, M., Visani, G., Alesiani, F., Brunori, M., Galieni, P., Catarini, M., Burattini, M., Centurioni, R., Rupoli, S., Scortechini, A.R., Giuliodori, L., Candela, M., Capelli, D., Montanari, M., Olivieri, A., Piersantelli, M.N., and Leoni, P., Low-dose thalidomide with pegylated liposomal doxorubicin and high-dose dexamethasone for relapsed/refractory multiple myeloma: a prospective, multicenter, phase II study, *Haematologica* 91 (1), 133–136, 2006.
149. Baz, R.C., Kelly, M., Reed, J., Karam, M., Faiman, B., Andresen, S., and Hussein, M.A., Phase II study of dexamethasone, ascorbic acid, thalidomide and arsenic trioxide (DATA) in high risk previously untreated (PU) and relapsed/refractory (RR) multiple myeloma (MM), *J Clin Oncol* 24 (18S), 682s, 2006.
150. Alexanian, R., Weber, D., Dimopoulos, M., Delasalle, K., and Smith, T.L., Randomized trial of alpha-interferon or dexamethasone as maintenance treatment for multiple myeloma, *Am J Hematol* 65 (3), 204–209, 2000.
151. Berenson, J.R., Crowley, J.J., Grogan, T.M., Zangmeister, J., Briggs, A.D., Mills, G.M., Barlogie, B., and Salmon, S.E., Maintenance therapy with alternate-day prednisone improves survival in multiple myeloma patients, *Blood* 99 (9), 3163–3168, 2002.
152. Shustik, C., Belch, A., Robinson, S.P., Rubin, S., Dolan, S., Kovacs, M.J., Djurfeldt, M., Shepherd, L., Ding, K., and Meyer, R.M., Dexamethasone (dex) maintenance versus observation (obs) in patients with previously untreated multiple myeloma: a National Cancer Institute of Canada Clinical Trials Group study: MY. 7, *J Clin Oncol* 22 (14S), 560s, 2004.
153. Barlogie, B., Kyle, R.A., Anderson, K.C., Greipp, P.R., Lazarus, H.M., Hurd, D.D., McCoy, J., Dakhil, S.R., Lanier, K.S., Chapman, R.A., Cromer, J.N., Salmon, S.E., Durie, B., and Crowley, J.C., Standard chemotherapy compared with high-dose chemoradiotherapy for multiple myeloma: final results of phase III US Intergroup Trial S9321, *J Clin Oncol* 24 (6), 929–936, 2006.
154. Fritz, E. and Ludwig, H., Interferon-alpha treatment in multiple myeloma: meta-analysis of 30 randomised trials among 3948 patients, *Ann Oncol* 11 (11), 1427–1436, 2000.
155. Mandelli, F., Avvisati, G., Amadori, S., Boccadoro, M., Gernone, A., Lauta, V.M., Marmont, F., Petrucci, M.T., Tribalto, M., Vegna, M.L., et al., Maintenance treatment with recombinant interferon alfa-2b in patients with multiple myeloma responding to conventional induction chemotherapy, *N Engl J Med* 322 (20), 1430–1434, 1990.
156. Schaar, C.G., Kluin-Nelemans, H.C., Te Marvelde, C., le Cessie, S., Breed, W.P., Fibbe, W.E., van Deijk, W.A., Fickers, M.M., Roozendaal, K.J., and Wijermans, P.W., Interferon-alpha as maintenance therapy in patients with multiple myeloma, *Ann Oncol* 16 (4), 634–639, 2005.
157. Barlogie, B., Tricot, G., Rasmussen, E., Anaissie, E., van Rhee, F., Zangari, M., Fassas, A., Hollmig, K., Pineda-Roman, M., Shaughnessy, J., Epstein, J., and Crowley, J., Total therapy 2 without thalidomide in comparison with total therapy 1: role of intensified induction and posttransplantation consolidation therapies, *Blood* 107 (7), 2633–2638, 2006.
158. Segeren, C.M., Sonneveld, P., van der Holt, B., Vellenga, E., Croockewit, A.J., Verhoef, G.E., Cornelissen, J.J., Schaafsma, M.R., van Oers, M.H., Wijermans, P.W., Fibbe, W.E., Wittebol, S., Schouten, H.C., van Marwijk Kooy, M., Biesma, D.H., Baars, J.W., Slater, R., Steijaert, M.M.,

Buijt, I., and Lokhorst, H.M., Overall and event-free survival are not improved by the use of myeloablative therapy following intensified chemotherapy in previously untreated patients with multiple myeloma: a prospective randomized phase 3 study, *Blood* 101 (6), 2144–2151, 2003.

159. Bjorkstrand, B., Svensson, H., Goldschmidt, H., Ljungman, P., Apperley, J., Mandelli, F., Marcus, R., Boogaerts, M., Alegre, A., Remes, K., Cornelissen, J.J., Blade, J., Lenhoff, S., Iriondo, A., Carlson, K., Volin, L., Littlewood, T., Goldstone, A.H., San Miguel, J., Schattenberg, A., and Gahrton, G., Alpha-interferon maintenance treatment is associated with improved survival after high-dose treatment and autologous stem cell transplantation in patients with multiple myeloma: a retrospective registry study from the European Group for Blood and Marrow Transplantation (EBMT), *Bone Marrow Transplant* 27 (5), 511–515, 2001.
160. Attal, M., Harousseau, J.L., Leyvraz, S., Doyen, C., Hulin, C., Benboubker, L., Yakoub Agha, I., Bourhis, J.H., Garderet, L., Pegourie, B., Dumontet, C., Renaud, M., Voillat, L., Berthou, C., Marit, G., Monconduit, M., Caillot, D., Grobois, B., Avet-Loiseau, H., Moreau, P., and Facon, T., Maintenance therapy with thalidomide improves survival in multiple myeloma patients, *Blood* 108 (10), 3289–3294, 2006.
161. Stewart, A.K., Chen, C.I., Howson-Jan, K., White, D., Roy, J., Kovacs, M.J., Shustik, C., Sadura, A., Shepherd, L., Ding, K., Meyer, R.M., and Belch, A.R., Results of a multicenter randomized phase II trial of thalidomide and prednisone maintenance therapy for multiple myeloma after autologous stem cell transplant, *Clin Cancer Res* 10 (24), 8170–8176, 2004.
162. Goldschmidt, H., Sonneveld, P., Cremer, F.W., van der Holt, B., Westveer, P., Breitkreutz, I., Benner, A., Glasmacher, A., Schmidt-Wolf, I.G., Martin, H., Hoelzer, D., Ho, A.D., and Lokhorst, H.M., Joint HOVON-50/GMMG-HD3 randomized trial on the effect of thalidomide as part of a high-dose therapy regimen and as maintenance treatment for newly diagnosed myeloma patients, *Ann Hematol* 82 (10), 654–659, 2003.
163. Glasmacher, A., Hahn, C., Hoffmann, F., Naumann, R., Goldschmidt, H., von Lilienfeld-Toal, M., Orlopp, K., Schmidt-Wolf, I., and Gorschluter, M., A systematic review of phase-II trials of thalidomide monotherapy in patients with relapsed or refractory multiple myeloma, *Br J Haematol* 132 (5), 584–593, 2006.
164. Neben, K., Moehler, T., Benner, A., Kraemer, A., Egerer, G., Ho, A.D., and Goldschmidt, H., Dose-dependent effect of thalidomide on overall survival in relapsed multiple myeloma, *Clin Cancer Res* 8 (11), 3377–3382, 2002.
165. Barlogie, B., Desikan, R., Eddlemon, P., Spencer, T., Zeldis, J., Munshi, N., Badros, A., Zangari, M., Anaissie, E., Epstein, J., Shaughnessy, J., Ayers, D., Spoon, D., and Tricot, G., Extended survival in advanced and refractory multiple myeloma after single-agent thalidomide: identification of prognostic factors in a phase 2 study of 169 patients, *Blood* 98 (2), 492–494, 2001.
166. Yakoub-Agha, I., Doyen, C., Hulin, C., Marit, G., Voillat, L., Grosbois, B., Harousseau, J.L., Duguet, C., Zerbib, R., Facon, T., and Mary, J., A multicenter prosepctive randomized study testing non-inferiority of thalidomide 100 mg/day as compared with 400 mg/day in patients with refractory/relapsed multiple myeloma: results of the final analysis of the IFM 01-02 study, *J Clin Oncol* 24 (18S), 427s, 2006.
167. Waage, A., Gimsing, P., Juliusson, G., Turesson, I., Gulbrandsen, N., Eriksson, T., Hjorth, M., Nielsen, J.L., Lenhoff, S., Westin, J., and Wisloff, F., Early response predicts thalidomide efficiency in patients with advanced multiple myeloma, *Br J Haematol* 125 (2), 149–155, 2004.
168. Hus, I., Dmoszynska, A., Manko, J., Hus, M., Jawniak, D., Soroka-Wojtaszko, M., Hellmann, A., Ciepluch, H., Skotnicki, A., Wolska-Smolen, T., Sulek, K., Robak, T., Konopka, L., and Kloczko, J., An evaluation of factors predicting long-term response to thalidomide in 234 patients with relapsed or resistant multiple myeloma, *Br J Cancer* 91 (11), 1873–1879, 2004.
169. Anagnostopoulos, A., Gika, D., Hamilos, G., Zervas, K., Zomas, A., Pouli, A., Zorzou, M., Kastritis, E., Anagnostopoulos, N., Tassidou, A., Anagnostou, D., and Dimopoulos, M.A., Treatment of relapsed/refractory multiple myeloma with thalidomide-based regimens: identification of prognostic factors, *Leuk Lymphoma* 45 (11), 2275–2279, 2004.
170. Kumar, S., Greipp, P.R., Haug, J., Kline, M., Chng, W.J., Blood, E., Bergsagel, L., Lust, J.A., Gertz, M.A., Fonseca, R., and Rajkumar, S.V., Gene expression profiling of myeloma cells at diagnosis can predict response to therapy with thalidomide and dexamethasone combination, *Blood* 106 (11), 152a, 2005.

171. Weber, D.M., Gavino, M., Delasalle, K., Rankin, K., Giralt, S., and Alexanian, R., Thalidomide alone or with dexamethasone for multiple myeloma, *Blood* 94 (Suppl 1), 604a, 1999.
172. Dimopoulos, M.A., Zervas, K., Kouvatseas, G., Galani, E., Grigoraki, V., Kiamouris, C., Vervessou, E., Samantas, E., Papadimitriou, C., Economou, O., Gika, D., Panayiotidis, P., Christakis, I., and Anagnostopoulos, N., Thalidomide and dexamethasone combination for refractory multiple myeloma, *Ann Oncol* 12 (7), 991–995, 2001.
173. Anagnostopoulos, A., Weber, D., Rankin, K., Delasalle, K., and Alexanian, R., Thalidomide and dexamethasone for resistant multiple myeloma, *Br J Haematol* 121 (5), 768–771, 2003.
174. Palumbo, A., Bertola, A., Cavallo, F., Falco, P., Bringhen, S., Giaccone, L., Musto, P., Pregno, P., and Boccadoro, M., Low-dose thalidomide and dexamethasone improves survival in advanced multiple myeloma, *Blood* 100 (11), 211a, 2002.
175. Palumbo, A., Bertola, A., Falco, P., Rosato, R., Cavallo, F., Giaccone, L., Bringhen, S., Musto, P., Pregno, P., Caravita, T., Ciccone, G., and Boccadoro, M., Efficacy of low-dose thalidomide and dexamethasone as first salvage regimen in multiple myeloma, *Hematol J* 5 (4), 318–324, 2004.
176. Tosi, P., Zamagni, E., Cellini, C., Cangini, D., Tacchetti, P., Tura, S., Baccarani, M., and Cavo, M., Thalidomide alone or in combination with dexamethasone in patients with advanced, relapsed or refractory multiple myeloma and renal failure, *Eur J Haematol* 73 (2), 98–103, 2004.
177. Terpos, E., Mihou, D., Szydlo, R., Tsimirika, K., Karkantaris, C., Politou, M., Voskaridou, E., Rahemtulla, A., Dimopoulos, M.A., and Zervas, K., The combination of intermediate doses of thalidomide with dexamethasone is an effective treatment for patients with refractory/relapsed multiple myeloma and normalizes abnormal bone remodeling, through the reduction of sRANKL/osteoprotegerin ratio, *Leukemia* 19 (11), 1969–1976, 2005.
178. Morikawa, K., Watabe, H., Araake, M., and Morikawa, S., Modulatory effect of antibiotics on cytokine production by human monocytes in vitro, *Antimicrob Agents Chemother* 40 (6), 1366–1370, 1996.
179. Morikawa, K., Oseko, F., Morikawa, S., and Iwamoto, K., Immunomodulatory effects of three macrolides, midecamycin acetate, josamycin, and clarithromycin, on human T-lymphocyte function in vitro, *Antimicrob Agents Chemother* 38 (11), 2643–2647, 1994.
180. Matsuoka, N., Eguchi, K., Kawakami, A., Tsuboi, M., Kawabe, Y., Aoyagi, T., and Nagataki, S., Inhibitory effect of clarithromycin on costimulatory molecule expression and cytokine production by synovial fibroblast-like cells, *Clin Exp Immunol* 104 (3), 501–508, 1996.
181. Adachi, T., Motojima, S., Hirata, A., Fukuda, T., Kihara, N., Kosaku, A., Ohtake, H., and Makino, S., Eosinophil apoptosis caused by theophylline, glucocorticoids, and macrolides after stimulation with IL-5, *J Allergy Clin Immunol* 98 (6 Pt 2), S207–S215, 1996.
182. Yatsunami, J., Turuta, N., Wakamatsu, K., Hara, N., and Hayashi, S., Clarithromycin is a potent inhibitor of tumor-induced angiogenesis, *Res Exp Med (Berl)* 197 (4), 189–197, 1997.
183. Durie, B.G.M., Urnovitz, H.B., and Martin, D.S., Biaxin (Clarithromycin) as treatment for multiple myeloma, *Cancer Investig* 17, 54, 1999.
184. Shannon, K.J., Dave, H.P.G., and Schechter, G.P., Clarithromycin therapy ineffective in multiple myeloma (MM), *Blood* 94 (10 Suppl 1), 314b, 1999.
185. Stewart, A.K., Trudel, S., Al-Berouti, B.M., Sutton, D.M., and Meharchand, J., Lack of response to short-term use of clarithromycin (BIAXIN) in multiple myeloma, *Blood* 93 (12), 4441, 1999.
186. Coleman, M., Leonard, J., Lyons, L., Szelenyi, H., and Niesvizky, R., Treatment of Waldenstrom's macroglobulinemia with clarithromycin, low-dose thalidomide, and dexamethasone, *Semin Oncol* 30 (2), 270–274, 2003.
187. Dimopoulos, M.A., Tsatalas, C., Zomas, A., Hamilos, G., Panayiotidis, P., Margaritis, D., Matsouka, C., Economopoulos, T., and Anagnostopoulos, N., Treatment of Waldenstrom's macroglobulinemia with single-agent thalidomide or with the combination of clarithromycin, thalidomide and dexamethasone, *Semin Oncol* 30 (2), 265–269, 2003.
188. Coleman, M., Leonard, J., Lyons, L., Pekle, K., Nahum, K., Pearse, R., Niesvizky, R., and Michaeli, J., BLT-D (clarithromycin [Biaxin], low-dose thalidomide, and dexamethasone) for the treatment of myeloma and Waldenstrom's macroglobulinemia, *Leuk Lymphoma* 43 (9), 1777–1782, 2002.
189. Glasmacher, A., Hahn, C., Hoffmann, F., Furkert, K., Von Lilienfeld-Toal, M., Orlopp, K., Naumann, R., Goldschmidt, H., Schmidt-Wolf, I.G.H., and Gorschluter, M., Thalidomide in

relapsed or refractory patients with multiple myeloma: monotherapy or combination therapy? A report from systematic reviews, *Blood* 106 (11), 364b, 2005.

190. Lee, C.K., Barlogie, B., Munshi, N., Zangari, M., Fassas, A., Jacobson, J., van Rhee, F., Cottler-Fox, M., Muwalla, F., and Tricot, G., DTPACE: an effective, novel combination chemotherapy with thalidomide for previously treated patients with myeloma, *J Clin Oncol* 21 (14), 2732–2739, 2003.
191. Chanan-Khan, A.A., Miller, K.C., McCarthy, P., DiMiceli, L.A., Yu, J., Bernstein, Z.P., and Czuczman, M.S., A phase II study of Velcade (V), Doxil (D) in combination with low-dose thalidomide (T) as salvage therapy for patients (pts) with relapsed (rel) or refractory (ref) multiple myeloma (MM) and Waldenstrom's macroglobulinemia (WM): encouraging preliminary results, *Blood* 104 (11), 665a–666a, 2004.
192. Hollmig, K., Stover, J., Talamo, G., Fassas, A., Lee, C.-K., Anaissie, E., Tricot, G., and Barlogie, B., Bortezomib (Velcade™) + Adriamycin™ + thalidomide + dexamethasone (VATD) as an effective regimen in patients with refractory or relapsed multiple myeloma (MM), *Blood* 104 (11), 659a, 2004.
193. Kropff, M.H., Lang, N., Bisping, G., Domine, N., Innig, G., Hentrich, M., Mitterer, M., Sudhoff, T., Fenk, R., Straka, C., Heinecke, A., Koch, O.M., Ostermann, H., Berdel, W.E., and Kienast, J., Hyperfractionated cyclophosphamide in combination with pulsed dexamethasone and thalidomide (HyperCDT) in primary refractory or relapsed multiple myeloma, *Br J Haematol* 122 (4), 607–616, 2003.
194. Garcia-Sanz, R., Gonzalez-Porras, J.R., Hernandez, J.M., Polo-Zarzuela, M., Sureda, A., Barrenetxea, C., Palomera, L., Lopez, R., Grande-Garcia, C., Alegre, A., Vargas-Pabon, M., Gutierrez, O.N., Rodriguez, J.A., and San Miguel, J.F., The oral combination of thalidomide, cyclophosphamide and dexamethasone (ThaCyDex) is effective in relapsed/refractory multiple myeloma, *Leukemia* 18 (4), 856–863, 2004.
195. Dimopoulos, M.A., Hamilos, G., Zomas, A., Gika, D., Efstathiou, E., Grigoraki, V., Poziopoulos, C., Xilouri, I., Zorzou, M.P., Anagnostopoulos, N., and Anagnostopoulos, A., Pulsed cyclophosphamide, thalidomide and dexamethasone: an oral regimen for previously treated patients with multiple myeloma, *Hematol J* 5 (2), 112–117, 2004.
196. Kyriakou, C., Thomson, K., D'Sa, S., Flory, A., Hanslip, J., Goldstone, A.H., and Yong, K.L., Low-dose thalidomide in combination with oral weekly cyclophosphamide and pulsed dexamethasone is a well tolerated and effective regimen in patients with relapsed and refractory multiple myeloma, *Br J Haematol* 129 (6), 763–770, 2005.
197. Moehler, T.M., Neben, K., Benner, A., Egerer, G., Krasniqi, F., Ho, A.D., and Goldschmidt, H., Salvage therapy for multiple myeloma with thalidomide and CED chemotherapy, *Blood* 98 (13), 3846–3848, 2001.
198. Zangari, M., Siegel, E., Barlogie, B., Anaissie, E., Saghafifar, F., Fassas, A., Morris, C., Fink, L., and Tricot, G., Thrombogenic activity of doxorubicin in myeloma patients receiving thalidomide: implications for therapy, *Blood* 100 (4), 1168–1171, 2002.
199. Zangari, M., Anaissie, E., Barlogie, B., Badros, A., Desikan, R., Gopal, A.V., Morris, C., Toor, A., Siegel, E., Fink, L., and Tricot, G., Increased risk of deep-vein thrombosis in patients with multiple myeloma receiving thalidomide and chemotherapy, *Blood* 98 (5), 1614–1615, 2001.
200. Zangari, M., Barlogie, B., Anaissie, E., Saghafifar, F., Eddlemon, P., Jacobson, J., Lee, C.K., Thertulien, R., Talamo, G., Thomas, T., Van Rhee, F., Fassas, A., Fink, L., and Tricot, G., Deep vein thrombosis in patients with multiple myeloma treated with thalidomide and chemotherapy: effects of prophylactic and therapeutic anticoagulation, *Br J Haematol* 126 (5), 715–721, 2004.
201. Baz, R., Li, L., Kottke-Marchant, K., Srkalovic, G., McGowan, B., Yiannaki, E., Karam, M.A., Faiman, B., Jawde, R.A., Andresen, S., Zeldis, J., and Hussein, M.A., The role of aspirin in the prevention of thrombotic complications of thalidomide and anthracycline-based chemotherapy for multiple myeloma, *Mayo Clin Proc* 80 (12), 1568–1574, 2005.
202. Niesvizky, R., Martinez-Banos, D.M., Gelbshtein, U.Y., Cho, H.J., Pearse, R.N., Zafar, F., Pekle, K., Furman, R., Leonard, J.P., and Coleman, M., Prophylactic low-dose aspirin is effective as antithrombotic therapy in patients receiving combination thalidomide or lenalidomide, *Blood* 106 (11), 964a, 2005.

203. Minnema, M.C., Breitkreutz, I., Auwerda, J.J., van der Holt, B., Cremer, F.W., van Marion, A.M., Westveer, P.H., Sonneveld, P., Goldschmidt, H., and Lokhorst, H.M., Prevention of venous thromboembolism with low molecular-weight heparin in patients with multiple myeloma treated with thalidomide and chemotherapy, *Leukemia* 18 (12), 2044–2046, 2004.
204. Zangari, M., Barlogie, B., Burns, M.J., Bolejack, V., Hollmig, K.A., van Rhee, F., Pineda-Roman, M., Elice, F., and Tricot, G.J., Velcade (V)–thalidomide (T)–dexamethasone (D) for advanced and refractory multiple myeloma (MM): long-term follow-up of phase I–II trial UARK 2001-37: superior outcome in patients with normal cytogenetics and no prior T, *Blood* 106 (11), 717a, 2005.
205. Ciolli, S., Leoni, F., Gigli, F., Rigacci, L., and Bosi, A., Low dose Velcade, thalidomide and dexamethasone (LD-VTD): an effective regimen for relapsed and refractory multiple myeloma patients, *Leuk Lymphoma* 47 (1), 171–173, 2006.
206. Terpos, E., Anagnostopoulos, A., Kastritis, E., Zomas, A., Poziopoulos, C., Anagnostopoulos, N., Tsionos, K., and Dimopoulos, M.A., The combination of bortezomib, melphalan, dexamethasone and intermittent thalidomide (VMDT) is an effective treatment for relapsed/refractory myeloma: results of a phase II clinical trial, *Blood* 106 (11), 110a, 2005.
207. Palumbo, A., Ambrosini, M.T., Pregno, P., Pescosta, N., Callea, V., Cangialosi, C., Caravita, T., Morabito, F., Omede, P., Gay, F., Avonto, I., Falco, P., Bringhen, S., and Boccadoro, M., Velcade® plus melphalan, prednisone, and thalidomide (V-MPT) for advanced multiple myeloma, *Blood* 106 (11), 717a, 2005.
208. Somlo, G., Bellamy, W., Zimmerman, T.M., Frankel, P., Tuscano, J., O'Donnell, M., Mohrbacher, A., Forman, S., Chen, H., Doroshow, J., and Gandara, D., Phase II randomized trial of bevacizumab versus bevacizumab and thalidomide for relapsed/refractory multiple myeloma, *Blood* 106 (11), 723a, 2005.
209. Badros, A.Z., Goloubeva, O., Rapoport, A.P., Ratterree, B., Gahres, N., Meisenberg, B., Takebe, N., Heyman, M., Zwiebel, J., Streicher, H., Gocke, C.D., Tomic, D., Flaws, J.A., Zhang, B., and Fenton, R.G., Phase II study of G3139, a Bcl-2 antisense oligonucleotide, in combination with dexamethasone and thalidomide in relapsed multiple myeloma patients, *J Clin Oncol* 23 (18), 4089–4099, 2005.
210. Prince, H.M., Mileshkin, L., Roberts, A., Ganju, V., Underhill, C., Catalano, J., Bell, R., Seymour, J.F., Westerman, D., Simmons, P.J., Lillie, K., Milner, A.D., Iulio, J.D., Zeldis, J.B., and Ramsay, R., A multicenter phase II trial of thalidomide and celecoxib for patients with relapsed and refractory multiple myeloma, *Clin Cancer Res* 11 (15), 5504–5514, 2005.
211. Sahebi, F., Spielberger, R., Kogut, N.M., Fung, H., Falk, P.M., Parker, P., Krishnan, A., Rodriguez, R., Nakamura, R., Nademanee, A., Popplewell, L., Frankel, P., Ruel, C., Tin, R., Ilieva, P., Forman, S.J., and Somlo, G., Maintenance thalidomide following single cycle autologous peripheral blood stem cell transplant in patients with multiple myeloma, *Bone Marrow Transplant* 37 (9), 825–829, 2006.
212. Brinker, B.T., Waller, E.K., Leong, T., Heffner, L.T., Jr., Redei, I., Langston, A.A., and Lonial, S., Maintenance therapy with thalidomide improves overall survival after autologous hematopoietic progenitor cell transplantation for multiple myeloma, *Cancer* 106 (10), 2171–2180, 2006.
213. Uhl, K., Cox, E., Rogan, R., Zeldis, J.B., Hixon, D., Furlong, L.A., Singer, S., Holliman, T., Beyer, J., and Woolever, W., Thalidomide use in the US: experience with pregnancy testing in the S.T.E.P.S. programme, *Drug Saf* 29 (4), 321–329, 2006.
214. Zeldis, J.B., Williams, B.A., Thomas, S.D., and Elsayed, M.E., S.T.E.P.S.: a comprehensive program for controlling and monitoring access to thalidomide, *Clin Ther* 21 (2), 319–330, 1999.
215. Scarpace, S.L., Hahn, T., Roy, H., Brown, K., Paplham, P., Chanan-Khan, A., van Besien, K., and McCarthy, P.L., Jr., Arterial thrombosis in four patients treated with thalidomide, *Leuk Lymphoma* 46 (2), 239–242, 2005.
216. Bowcock, S.J., Rassam, S.M., Ward, S.M., Turner, J.T., and Laffan, M., Thromboembolism in patients on thalidomide for myeloma, *Hematology* 7 (1), 51–53, 2002.
217. Jego, C., Barbou, F., Laurent, P., Gisserot, O., Cellarier, G., Bonal, J., Bouchiat, C., Landais, C., de Jaureguiberry, J.P., and Dussarat, G.V., [Left atrial thrombus in multiple myeloma treated with thalidomide], *Arch Mal Coeur Vaiss* 96 (10), 1006–1010, 2003.
218. Fullerton, P.M. and O'Sullivan, D.J., Thalidomide neuropathy: a clinical electrophysiological, and histological follow-up study, *J Neurol Neurosurg Psychiatry* 31 (6), 543–551, 1968.

219. Tosi, P., Zamagni, E., Cellini, C., Plasmati, R., Cangini, D., Tacchetti, P., Perrone, G., Pastorelli, F., Tura, S., Baccarani, M., and Cavo, M., Neurological toxicity of long-term (>1 yr) thalidomide therapy in patients with multiple myeloma, *Eur J Haematol* 74 (3), 212–216, 2005.
220. Grover, J.K., Uppal, G., and Raina, V., The adverse effects of thalidomide in relapsed and refractory patients of multiple myeloma, *Ann Oncol* 13 (10), 1636–1640, 2002.
221. Dimopoulos, M.A. and Eleutherakis-Papaiakovou, V., Adverse effects of thalidomide administration in patients with neoplastic diseases, *Am J Med* 117 (7), 508–515, 2004.
222. Hall, V.C., El-Azhary, R.A., Bouwhuis, S., and Rajkumar, S.V., Dermatologic side effects of thalidomide in patients with multiple myeloma, *J Am Acad Dermatol* 48 (4), 548–552, 2003.
223. Ghobrial, I.M. and Rajkumar, S.V., Management of thalidomide toxicity, *J Support Oncol* 1 (3), 194–205, 2003.
224. Kumar, S. and Rajkumar, S.V., Thalidomide and lenalidomide in the treatment of multiple myeloma, *Eur J Cancer* 42 (11), 1612–1622, 2006.
225. Verhoef, G.E., De Schouwer, P., Ceuppens, J.L., Van Damme, J., Goossens, W., and Boogaerts, M.A., Measurement of serum cytokine levels in patients with myelodysplastic syndromes, *Leukemia* 6 (12), 1268–1272, 1992.
226. Koike, M., Ishiyama, T., Tomoyasu, S., and Tsuruoka, N., Spontaneous cytokine overproduction by peripheral blood mononuclear cells from patients with myelodysplastic syndromes and aplastic anemia, *Leuk Res* 19 (9), 639–644, 1995.
227. Shetty, V., Mundle, S., Alvi, S., Showel, M., Broady-Robinson, L., Dar, S., Borok, R., Showel, J., Gregory, S., Rifkin, S., Gezer, S., Parcharidou, A., Venugopal, P., Shah, R., Hernandez, B., Klein, M., Alston, D., Robin, E., Dominquez, C., and Raza, A., Measurement of apoptosis, proliferation and three cytokines in 46 patients with myelodysplastic syndromes, *Leuk Res* 20 (11–12), 891–900, 1996.
228. Kitagawa, M., Saito, I., Kuwata, T., Yoshida, S., Yamaguchi, S., Takahashi, M., Tanizawa, T., Kamiyama, R., and Hirokawa, K., Overexpression of tumor necrosis factor (TNF)-alpha and interferon (IFN)-gamma by bone marrow cells from patients with myelodysplastic syndromes, *Leukemia* 11 (12), 2049–2054, 1997.
229. Mundle, S.D., Ali, A., Cartlidge, J.D., Reza, S., Alvi, S., Showel, M.M., Mativi, B.Y., Shetty, V.T., Venugopal, P., Gregory, S.A., and Raza, A., Evidence for involvement of tumor necrosis factor-alpha in apoptotic death of bone marrow cells in myelodysplastic syndromes, *Am J Hematol* 60 (1), 36–47, 1999.
230. Parcharidou, A., Raza, A., Economopoulos, T., Papageorgiou, E., Anagnostou, D., Papadaki, T., and Raptis, S., Extensive apoptosis of bone marrow cells as evaluated by the in situ end-labelling (ISEL) technique may be the basis for ineffective haematopoiesis in patients with myelodysplastic syndromes, *Eur J Haematol* 62 (1), 19–26, 1999.
231. Smith, M.A. and Smith, J.G., The occurrence subtype and significance of haemopoietic inhibitory T cells (HIT cells) in myelodysplasia: an in vitro study, *Leuk Res* 15 (7), 597–601, 1991.
232. Sugawara, T., Endo, K., Shishido, T., Sato, A., Kameoka, J., Fukuhara, O., Yoshinaga, K., and Miura, A., T cell-mediated inhibition of erythropoiesis in myelodysplastic syndromes, *Am J Hematol* 41 (4), 304–305, 1992.
233. Molldrem, J.J., Jiang, Y.Z., Stetler-Stevenson, M., Mavroudis, D., Hensel, N., and Barrett, A.J., Haematological response of patients with myelodysplastic syndrome to antithymocyte globulin is associated with a loss of lymphocyte-mediated inhibition of CFU-GM and alterations in T-cell receptor Vbeta profiles, *Br J Haematol* 102 (5), 1314–1322, 1998.
234. Bouscary, D., Legros, L., Tulliez, M., Dubois, S., Mahe, B., Beyne-Rauzy, O., Quarre, M.C., Vassilief, D., Varet, B., Aouba, A., Gardembas, M., Giraudier, S., Guerci, A., Rousselot, P., Gaillard, F., Moreau, A., Rousselet, M.C., Ifrah, N., Fenaux, P., and Dreyfus, F., A non-randomised dose-escalating phase II study of thalidomide for the treatment of patients with low-risk myelodysplastic syndromes: the Thal-SMD-2000 trial of the Groupe Francais des Myelodysplasies, *Br J Haematol* 131 (5), 609–618, 2005.
235. Bowen, D., MacIlwaine, L., Cavanagh, J., Killick, S., Culligan, D., Thomson, E., and Mufti, G., Thalidomide therapy for low-risk myelodysplasia, *Leuk Res* 29 (2), 235–236, 2005.
236. Raza, A., Meyer, P., Dutt, D., Zorat, F., Lisak, L., Nascimben, F., du Randt, M., Kaspar, C., Goldberg, C., Loew, J., Dar, S., Gezer, S., Venugopal, P., and Zeldis, J., Thalidomide produces

transfusion independence in long-standing refractory anemias of patients with myelodysplastic syndromes, *Blood* 98 (4), 958–965, 2001.

237. Musto, P., Falcone, A., Sanpaolo, G., Bisceglia, M., Matera, R., and Carella, A.M., Thalidomide abolishes transfusion-dependence in selected patients with myelodysplastic syndromes, *Haematologica* 87 (8), 884–886, 2002.
238. Musto, P., Thalidomide therapy for myelodysplastic syndromes: current status and future perspectives, *Leuk Res* 28 (4), 325–332, 2004.
239. Strupp, C., Germing, U., Aivado, M., Misgeld, E., Haas, R., and Gattermann, N., Thalidomide for the treatment of patients with myelodysplastic syndromes, *Leukemia* 16 (1), 1–6, 2002.
240. Moreno-Aspitia, A., Colon-Otero, G., Hoering, A., Tefferi, A., Niedringhaus, R.D., Vukov, A., Li, C.Y., Menke, D.M., Geyer, S.M., and Alberts, S.R., Thalidomide therapy in adult patients with myelodysplastic syndrome: a North Central Cancer Treatment Group phase II trial, *Cancer* 107 (4), 767–772, 2006.
241. Steurer, M., Sudmeier, I., Stauder, R., and Gastl, G., Thromboembolic events in patients with myelodysplastic syndrome receiving thalidomide in combination with darbepoietin-alpha, *Br J Haematol* 121 (1), 101–103, 2003.
242. Musto, P., Falcone, A., Sanpaolo, G., and Bodenizza, C., Combination of erythropoietin and thalidomide for the treatment of anemia in patients with myelodysplastic syndromes, *Leuk Res* 30 (4), 385–388, 2006.
243. Raza, A., Buonamici, S., Lisak, L., Tahir, S., Li, D., Imran, M., Chaudary, N.I., Pervaiz, H., Gallegos, J.A., Alvi, M.I., Mumtaz, M., Gezer, S., Venugopal, P., Reddy, P., Galili, N., Candoni, A., Singer, J., and Nucifora, G., Arsenic trioxide and thalidomide combination produces multi-lineage hematological responses in myelodysplastic syndromes patients, particularly in those with high pre-therapy EVI1 expression, *Leuk Res* 28 (8), 791–803, 2004.
244. Raza, A., Lisak, L., Billmeier, J., Pervaiz, H., Mumtaz, M., Gohar, S., Wahid, K., and Galili, N., Phase II study of topotecan and thalidomide in patients with high-risk myelodysplastic syndromes, *Leuk Lymphoma* 47 (3), 433–440, 2006.
245. Westervelt, P., Amirifeli, S., Mehdi, M., Mumtaz, M., Alhomsi, S., Wang, S., Miron, P., Lata, C., Galili, N., and Raza, A., Low dose Vidaza and thalidomide is an effective combination for patients with myelodysplastic syndromes (MDS) and acute myeloid leukemia (AML), *J Clin Oncol* 24 (18S), 354s, 2006.
246. Di Raimondo, F., Azzaro, M.P., Palumbo, G.A., Bagnato, S., Stagno, F., Giustolisi, G.M., Cacciola, E., Sortino, G., Guglielmo, P., and Giustolisi, R., Elevated vascular endothelial growth factor (VEGF) serum levels in idiopathic myelofibrosis, *Leukemia* 15 (6), 976–980, 2001.
247. Rusten, L.S. and Jacobsen, S.E., Tumor necrosis factor (TNF)-alpha directly inhibits human erythropoiesis in vitro: role of p55 and p75 TNF receptors, *Blood* 85 (4), 989–996, 1995.
248. Battegay, E.J., Raines, E.W., Colbert, T., and Ross, R., TNF-alpha stimulation of fibroblast proliferation. Dependence on platelet-derived growth factor (PDGF) secretion and alteration of PDGF receptor expression, *J Immunol* 154 (11), 6040–6047, 1995.
249. Dinarello, C.A., Cannon, J.G., Wolff, S.M., Bernheim, H.A., Beutler, B., Cerami, A., Figari, I.S., Palladino, M.A., Jr., and O'Connor, J.V., Tumor necrosis factor (cachectin) is an endogenous pyrogen and induces production of interleukin 1, *J Exp Med* 163 (6), 1433–1450, 1986.
250. Selleri, C., Sato, T., Anderson, S., Young, N.S., and Maciejewski, J.P., Interferon-gamma and tumor necrosis factor-alpha suppress both early and late stages of hematopoiesis and induce programmed cell death, *J Cell Physiol* 165 (3), 538–546, 1995.
251. Canepa, L., Ballerini, F., Varaldo, R., Quintino, S., Reni, L., Clavio, M., Miglino, M., Pierri, I., and Gobbi, M., Thalidomide in agnogenic and secondary myelofibrosis, *Br J Haematol* 115 (2), 313–315, 2001.
252. Barosi, G., Grossi, A., Comotti, B., Musto, P., Gamba, G., and Marchetti, M., Safety and efficacy of thalidomide in patients with myelofibrosis with myeloid metaplasia, *Br J Haematol* 114 (1), 78–83, 2001.
253. Piccaluga, P.P., Visani, G., Pileri, S.A., Ascani, S., Grafone, T., Isidori, A., Malagola, M., Finelli, C., Martinelli, G., Ricci, P., Baccarani, M., and Tura, S., Clinical efficacy and antiangiogenic activity of thalidomide in myelofibrosis with myeloid metaplasia. A pilot study, *Leukemia* 16 (9), 1609–1614, 2002.

254. Merup, M., Kutti, J., Birgergard, G., Mauritzson, N., Bjorkholm, M., Markevarn, B., Maim, C., Westin, J., and Palmblad, J., Negligible clinical effects of thalidomide in patients with myelofibrosis with myeloid metaplasia, *Med Oncol* 19 (2), 79–86, 2002.
255. Elliott, M.A., Mesa, R.A., Li, C.Y., Hook, C.C., Ansell, S.M., Levitt, R.M., Geyer, S.M., and Tefferi, A., Thalidomide treatment in myelofibrosis with myeloid metaplasia, *Br J Haematol* 117 (2), 288–296, 2002.
256. Strupp, C., Germing, U., Scherer, A., Kundgen, A., Modder, U., Gattermann, N., and Haas, R., Thalidomide for the treatment of idiopathic myelofibrosis, *Eur J Haematol* 72 (1), 52–57, 2004.
257. Marchetti, M., Barosi, G., Balestri, F., Viarengo, G., Gentili, S., Barulli, S., Demory, J.L., Ilariucci, F., Volpe, A., Bordessoule, D., Grossi, A., Le Bousse-Kerdiles, M.C., Caenazzo, A., Pecci, A., Falcone, A., Broccia, G., Bendotti, C., Bauduer, F., Buccisano, F., and Dupriez, B., Low-dose thalidomide ameliorates cytopenias and splenomegaly in myelofibrosis with myeloid metaplasia: a phase II trial, *J Clin Oncol* 22 (3), 424–431, 2004.
258. Thomas, D.A., Giles, F.J., Albitar, M., Cortes, J.E., Verstovsek, S., Faderl, S., O'Brien, S.M., Garcia-Manero, G., Keating, M.J., Pierce, S., Zeldis, J., and Kantarjian, H.M., Thalidomide therapy for myelofibrosis with myeloid metaplasia, *Cancer* 106 (9), 1974–1984, 2006.
259. Tefferi, A., Barosi, G., Mesa, R.A., Cervantes, F., Deeg, H.J., Reilly, J.T., Verstovsek, S., Dupriez, B., Silver, R.T., Odenike, O., Cortes, J., Wadleigh, M., Solberg, L.A., Jr., Camoriano, J.K., Gisslinger, H., Noel, P., Thiele, J., Vardiman, J.W., Hoffman, R., Cross, N.C., Gilliland, D.G., and Kantarjian, H., International Working Group (IWG) consensus criteria for treatment response in myelofibrosis with myeloid metaplasia: on behalf of the IWG for myelofibrosis research and treatment (IWG-MRT), *Blood* 108 (5), 1497–1503, 2006.
260. Giovanni, B., Michelle, E., Letizia, C., Filippo, B., Paolo, P.P., Giuseppe, V., Monia, M., Gabriele, P., Francesca, Z., and Ayalew, T., Thalidomide in myelofibrosis with myeloid metaplasia: a pooled-analysis of individual patient data from five studies, *Leuk Lymphoma* 43 (12), 2301–2307, 2002.
261. Pozzato, G., Zorat, F., Nascimben, F., Comar, C., Kikic, F., and Festini, G., Thalidomide therapy in compensated and decompensated myelofibrosis with myeloid metaplasia, *Haematologica* 86 (7), 772–773, 2001.
262. Dupriez, B., Morel, P., Demory, J.L., Lai, J.L., Simon, M., Plantier, I., and Bauters, F., Prognostic factors in agnogenic myeloid metaplasia: a report on 195 cases with a new scoring system, *Blood* 88 (3), 1013–1018, 1996.
263. Mesa, R.A., Steensma, D.P., Pardanani, A., Li, C.Y., Elliott, M., Kaufmann, S.H., Wiseman, G., Gray, L.A., Schroeder, G., Reeder, T., Zeldis, J.B., and Tefferi, A., A phase 2 trial of combination low-dose thalidomide and prednisone for the treatment of myelofibrosis with myeloid metaplasia, *Blood* 101 (7), 2534–2541, 2003.
264. Steensma, D.P., Mesa, R.A., Li, C.Y., Gray, L., and Tefferi, A., Etanercept, a soluble tumor necrosis factor receptor, palliates constitutional symptoms in patients with myelofibrosis with myeloid metaplasia: results of a pilot study, *Blood* 99 (6), 2252–2254, 2002.
265. Mesa, R.A., Stensma, D.P., Li, C.Y., Hoering, A., Allred, J.B., Powell, H.L., and Tefferi, A., Phase II study of the combination of low-dose thalidomide, prednisone, and etanercept (PET regimen) in the treatment of anemia, splenomegaly, and constitutional symptoms associated with myelofibrosis with myeloid metaplasia (MMM), *Blood* 106 (11), 724a, 2005.
266. Grinblatt, D.L., Johnson, J., Niedzwicki, D., Rizzieri, D.A., Bartlett, N., and Cheson, B.D., Phase II study of thalidomide in escalating doses for follicular (F-NHL) and small lymphocytic lymphoma (SLL): CALGB study 50002, *Blood* 104 (11), 897a–898a, 2004.
267. Strupp, C., Aivado, M., Germing, U., Gattermann, N., and Haas, R., Angioimmunoblastic lymphadenopathy (AILD) may respond to thalidomide treatment: two case reports, *Leuk Lymphoma* 43 (1), 133–137, 2002.
268. Dogan, A., Ngu, L.S., Ng, S.H., and Cervi, P.L., Pathology and clinical features of angioimmunoblastic T-cell lymphoma after successful treatment with thalidomide, *Leukemia* 19 (5), 873–875, 2005.
269. Kaufmann, H., Raderer, M., Wohrer, S., Puspok, A., Bankier, A., Zielinski, C., Chott, A., and Drach, J., Antitumor activity of rituximab plus thalidomide in patients with relapsed/refractory mantle cell lymphoma, *Blood* 104 (8), 2269–2271, 2004.

270. Johnston, R.E. and Abdalla, S.H., Thalidomide in low doses is effective for the treatment of resistant or relapsed multiple myeloma and for plasma cell leukaemia, *Leuk Lymphoma* 43 (2), 351–354, 2002.
271. Bauduer, F., Efficacy of thalidomide in the treatment of VAD-refractory plasma cell leukaemia appearing after autologous stem cell transplantation for multiple myeloma, *Br J Haematol* 117 (4), 996–997, 2002.
272. Tsiara, S., Chaidos, A., Kapsali, H., Tzouvara, E., and Bourantas, K.L., Thalidomide administration for the treatment of resistant plasma cell leukemia, *Acta Haematol* 109 (3), 153–155, 2003.
273. Rodriguez, C., Pont, J.C., Gouin-Thibault, I., Andrieu, A.G., Molina, T., Le Tourneau, A., Le Garff-Tavernier, M., Siguret, V., and Chaibi, P., [Plasma cell leukaemia], *Ann Biol Clin (Paris)* 63 (5), 535–539, 2005.
274. Fowler, R. and Imrie, K., Thalidomide-associated hepatitis: a case report, *Am J Hematol* 66 (4), 300–302, 2001.
275. Wohrer, S., Ackermann, J., Baldia, C., Seidl, S., Raderer, M., Simonitsch, I., and Drach, J., Effective treatment of primary plasma cell leukemia with thalidomide and dexamethasone—a case report, *Hematol J* 5 (4), 361–363, 2004.
276. Caldera, H.J., Fernandez, G.L., and Leon, B., Treatment of plasma cell leukemia (PCL) with bortezomib and thalidomide: a case report and literature review, *Blood* 106 (11), 362b–363b, 2005.
277. Gertz, M.A., Merlini, G., and Treon, S.P., Amyloidosis and Waldenstrom's macroglobulinemia, *Hematology (Am Soc Hematol Educ Program)*, 257–282, 2004.
278. Vescio, R.A. and Berenson, J.R., Thalidomide is an effective agent for patients with primary amyloidosis, *Blood* 96 (11 part 2), 296b, 2000.
279. Seldin, D.C., Choufani, E.B., Dember, L.M., Wiesman, J.F., Berk, J.L., Falk, R.H., O'Hara, C., Fennessey, S., Finn, K.T., Wright, D.G., Skinner, M., and Sanchorawala, V., Tolerability and efficacy of thalidomide for the treatment of patients with light chain-associated (AL) amyloidosis, *Clin Lymphoma* 3 (4), 241–246, 2003.
280. Dispenzieri, A., Lacy, M.Q., Rajkumar, S.V., Geyer, S.M., Witzig, T.E., Fonseca, R., Lust, J.A., Greipp, P.R., Kyle, R.A., and Gertz, M.A., Poor tolerance to high doses of thalidomide in patients with primary systemic amyloidosis, *Amyloid* 10 (4), 257–261, 2003.
281. Gertz, M.A., Lacy, M.Q., and Dispenzieri, A., Therapy for immunoglobulin light chain amyloidosis: the new and the old, *Blood Rev* 18 (1), 17–37, 2004.
282. Dispenzieri, A., Lacy, M.Q., Geyer, S.M., Griepp, P.R., Witzig, T.E., Lust, J.A., Zeldenrust, S.R., Rajkumar, S.V., Kyle, R.A., Fonseca, R., and Gertz, M.A., Low dose single agent thalidomide is tolerated in patients with primary systemic amyloidosis, but responses are limited, *Blood* 104 (11), 312b, 2004.
283. Palladini, G., Anesi, E., Perfetti, V., Obici, L., Invernizzi, R., Balduini, C., Ascari, E., and Merlini, G., A modified high-dose dexamethasone regimen for primary systemic (AL) amyloidosis, *Br J Haematol* 113 (4), 1044–1046, 2001.
284. Palladini, G., Perfetti, V., Perlini, S., Obici, L., Lavatelli, F., Caccialanza, R., Invernizzi, R., Comotti, B., and Merlini, G., The combination of thalidomide and intermediate-dose dexamethasone is an effective but toxic treatment for patients with primary amyloidosis (AL), *Blood* 105 (7), 2949–2951, 2005.
285. Goodman, H.J.B., Lachmann, H.J., Gallimore, R., Bradwell, A.R., and Hawkins, P.N., Thalidomide treatment in 99 patients with AL amyloidosis: tolerability, clonal disease response and clinical outcome, *Haematologica* 90 (Suppl 1), 202a, 2005.
286. Wechalekar, A.D., Goodman, H.J.B., Gillmore, J.D., Lachmann, H.J., Offer, M., Bradwell, A.R., and Hawkins, P.N., Efficacy of risk adapted cyclophosphamide, thalidomide and dexamethasone in systemic AL amyloidosis, *Blood* 106 (11), 976a, 2005.
287. Dimopoulos, M.A., Zomas, A., Viniou, N.A., Grigoraki, V., Galani, E., Matsouka, C., Economou, O., Anagnostopoulos, N., and Panayiotidis, P., Treatment of Waldenstrom's macroglobulinemia with thalidomide, *J Clin Oncol* 19 (16), 3596–3601, 2001.
288. Byrd, J.C., White, C.A., Link, B., Lucas, M.S., Velasquez, W.S., Rosenberg, J., and Grillo-Lopez, A.J., Rituximab therapy in Waldenstrom's macroglobulinemia: preliminary evidence of clinical activity, *Ann Oncol* 10 (12), 1525–1527, 1999.

289. Treon, S.P., Agus, T.B., Link, B., Rodrigues, G., Molina, A., Lacy, M.Q., Fisher, D.C., Emmanouilides, C., Richards, A.I., Clark, B., Lucas, M.S., Schlossman, R., Schenkein, D., Lin, B., Kimby, E., Anderson, K.C., and Byrd, J.C., CD20-directed antibody-mediated immunotherapy induces responses and facilitates hematologic recovery in patients with Waldenstrom's macroglobulinemia, *J Immunother* 24 (3), 272–279, 2001.
290. Weber, D.M., Gavino, M., Huh, Y., and Alexanian, R., Phenotypic and clinical evidence supports rituximab for Waldenstrom's macroglobulinemia, *Blood* 94 (10 Suppl 1), 125a, 1999.
291. Foran, J.M., Rohatiner, A.Z., Cunningham, D., Popescu, R.A., Solal-Celigny, P., Ghielmini, M., Coiffier, B., Johnson, P.W., Gisselbrecht, C., Reyes, F., Radford, J.A., Bessell, E.M., Souleau, B., Benzohra, A., and Lister, T.A., European phase II study of rituximab (chimeric anti-CD20 monoclonal antibody) for patients with newly diagnosed mantle-cell lymphoma and previously treated mantle-cell lymphoma, immunocytoma, and small B-cell lymphocytic lymphoma, *J Clin Oncol* 18 (2), 317–324, 2000.
292. Gertz, M.A., Rue, M., Blood, E., Kaminer, L.S., Vesole, D.H., and Griepp, P.R., Rituximab for Waldenstrom's macroglobulinemia (WM) (E3A98): an ECOG phase II pilot study for untreated or previously treated patients, *Blood* 102 (11), 148a, 2003.
293. Dimopoulos, M.A., Zervas, C., Zomas, A., Kiamouris, C., Viniou, N.A., Grigoraki, V., Karkantaris, C., Mitsouli, C., Gika, D., Christakis, J., and Anagnostopoulos, N., Treatment of Waldenstrom's macroglobulinemia with rituximab, *J Clin Oncol* 20 (9), 2327–2333, 2002.
294. Treon, S.P., Emmanouilides, C., Kimby, E., Kelliher, A., Preffer, F., Branagan, A.R., Anderson, K.C., and Frankel, S.R., Extended rituximab therapy in Waldenstrom's macroglobulinemia, *Ann Oncol* 16 (1), 132–138, 2005.
295. Branagan, A., Hunter, Z., Santos, D., Moran, J., and Treon, S.P., Thalidomide and rituximab in Waldenstrom's macroglobulinemia, *Blood* 104 (11), 415a, 2004.
296. Ghobrial, I.M., Fonseca, R., Greipp, P.R., Blood, E., Rue, M., Vesole, D., and Gertz, M., The initial "flare" of IgM level after rituximab therapy in patients diagnosed with Waldenstrom macroglobulinemia (WM): an Eastern Cooperative Oncology Group (ECOG) study, *Blood* 102 (11), 1637a, 2003.
297. Treon, S.P., Branagan, A.R., and Anderson, K.C., Paradoxical increases in serum IgM levels and serum viscosity following rituximab therapy in patients with Waldenstrom's macroglobulinemia (WM), *Blood* 102 (11), 690a, 2003.
298. Miller, K., Czuczman, M.S., DiMiceli, L., Pandmanabhan, S., Lawrence, D., Bernstein, Z., Takeshita, K., Spaner, D., Byrne, C., Crystal, C., and Chanan-Khan, A.A., Lenalidomide (L) induces high response rates with molecular remission in patients (pts) with relapsed (rel) or refractory (ref) chronic lymphocytic leukemia (CLL), *J Clin Oncol* 24 (18S), 341s, 2006.
299. DiMiceli, L., Miller, K.C., Rickert, M., Wallace, P., Marshall, P., DePaolo, D., Padmanabhan, S., Landrigan, B., and Chanan-Khan, A.A., Characterization of IMiDs (immunomodulating agents) induced "flare reaction" in patients with chronic lymphocytic leukemia (CLL) and correlation with changes in serum cytokine levels, *Blood* 106 (11), 344b, 2005.
300. Chanan-Khan, A.A., Padmanabhan, S., Miller, K.C., Pera, P., DiMiceli, L., Ersing, N., Wallace, P., Rickert, M., and Porter, C., In vivo evaluation of immunomodulating effects of lenalidomide (L) on tumor cell microenvironment as a possible underlying mechanism of the antitumor effects observed in patients (pts) with chronic lymphocytic leukemia (CLL), *Blood* 106 (11), 834a, 2005.
301. Rajkumar, S.V., Hayman, S.R., Lacy, M.Q., Dispenzieri, A., Geyer, S.M., Kabat, B., Zeldenrust, S.R., Kumar, S., Greipp, P.R., Fonseca, R., Lust, J.A., Russell, S.J., Kyle, R.A., Witzig, T.E., and Gertz, M.A., Combination therapy with lenalidomide plus dexamethasone (Rev/Dex) for newly diagnosed myeloma, *Blood* 106 (13), 4050–4053, 2005.
302. Zonder, J.A., Durie, B.G.M., McCoy, J., Crowley, J., Zeldis, J.B., Ghannam, L., and Barlogie, B., High incidence of thrombotic events observed in patients receiving lenalidomide (L) + dexamethasone (D) (LD) as first-line therapy for multiple myeloma (MM) without aspirin (ASA) prophylaxis, *Blood* 106 (11), 964a, 2005.
303. Niesvizky, R., Jayabalan, D.S., Furst, J.R., Cho, H.J., Pearse, R.N., Zafar, F., Lent, R.W., Tepler, J., Schuster, M.W., Leonard, J.P., and Coleman, M., Clarithromycin, lenalidomide and dexamethasone combination therapy as primary treatment of multiple myeloma, *J Clin Oncol* 24 (18S), 433s, 2006.

304. Palumbo, A., Falco, P., Benevolo, G., Canepa, L., D'Ardia, S., Gozzetti, A., Nozza, A., Zeldis, J., Boccadoro, M., and Petrucci, M.T., Oral lenalidomide plus melphalan and prednisone (R-MP) for newly diagnosed multiple myeloma, *J Clin Oncol* 24 (18S), 426s, 2006.
305. Zangari, M., Tricot, G., Zeldis, J., Eddlemon, P., Saghafifar, F., and Barlogie, B., Results of a phase I study of CC-5013 for the treatment of multiple myeloma (MM) patients who relapse after high dose chemotherapy (HDCT), *Blood* 98, 775a, 2001.
306. Richardson, P.G., Schlossman, R.L., Weller, E., Hideshima, T., Mitsiades, C., Davies, F., LeBlanc, R., Catley, L.P., Doss, D., Kelly, K., McKenney, M., Mechlowicz, J., Freeman, A., Deocampo, R., Rich, R., Ryoo, J.J., Chauhan, D., Balinski, K., Zeldis, J., and Anderson, K.C., Immunomodulatory drug CC-5013 overcomes drug resistance and is well tolerated in patients with relapsed multiple myeloma, *Blood* 100 (9), 3063–3067, 2002.
307. Barlogie, B., Jacobson, J., Lee, C.K., Zangari, M., Badros, A., Shaughnessy, J., and Tricot, G., Thalidomide (thal), Revlimid (Rev) and Velcade (Vel) in advanced multiple myeloma., *Hematol J* 4 (Suppl 1), S5–S7, 2003.
308. Richardson, P.G., Blood, E., Mitsiades, C.S., Jagannath, S., Zeldenrust, S.R., Alsina, M., Schlossman, R.L., Rajkumar, S.V., Desikan, K.R., Hideshima, T., Munshi, N.C., Kelly-Colson, K., Doss, D., McKenney, M.L., Gorelik, S., Warren, D., Freeman, A., Rich, R., Wu, A., Olesnyckyj, M., Wride, K., Dalton, W.S., Zeldis, J., Knight, R., Weller, E., and Anderson, K.C., A randomized phase 2 study of lenalidomide therapy for patients with relapsed or relapsed and refractory multiple myeloma, *Blood* 108, 3458–3464, 2006.
309. Richardson, P., Jagannath, S., Hussein, M., Berenson, J., Singhal, S., Irwin, D., Williams, S.F., Bensinger, W., Badros, A.Z., Vescio, R., Kenvin, L., Yu, Z., Olesnyckyj, M., Faleck, H., Zeldis, J., Knight, R., and Anderson, K.C., A multicenter, single-arm, open-label study to evaluate the efficacy and safety of single-agent lenalidomide in patients with relapsed and refractory multiple myeloma; preliminary results, *Blood* 106 (11), 449a, 2005.
310. Dimopoulos, M.A., Spencer, A., Attal, M., Prince, M., Harousseau, J.-L., Dmoszynska, A., Yu, Z., Olesnyckyj, M., Zeldis, J., and Knight, R., Study of lenalidomide plus dexamethasone versus dexamethasone alone in relapsed or refractory multiple myeloma (MM): results of a phase 3 study (MM-010), *Blood* 106 (11), 6a–7a, 2005.
311. Weber, D.M., Chen, C., Niesvizky, R., Wang, M., Belch, A., Stadtmauer, E., Yu, Z., Olesnyckyj, M., Zeldis, J., and Knight, R., Lenalidomide plus high-dose dexamethasone provides improved overall survival compared to high-dose dexamethasone alone for relapsed or refractory multiple myeloma (MM): results of a North American phase III study (MM-009), *J Clin Oncol* 24 (18S), 427s, 2006.
312. Richardson, P., Management of the relapsed/refractory myeloma patient: strategies incorporating lenalidomide, *Semin Hematol* 42 (4 Suppl 4), S9–S15, 2005.
313. Hussein, M.A., Karam, M.A., Brand, C., Pearce, G.L., Reed, J., Bruening, K., Sartori, P., Srkalovic, G., Olesnyckyj, M., Knight, R., Balinski, K., and Zeldis, J., Doxil (D), vincristine (V), reduced frequency dexamethasone (d) and Revlimid (R) (DVd-R) a phase I/II trial in advanced relapsed/refractory multiple myeloma (Rmm) patients, *Blood* 104 (11), 63a–64a, 2004.
314. Gerecke, C., Knop, S., Topp, M.S., Kotkiewitz, S., Gollasch, H., Liebisch, P., Hess, G., Watters, K., Einsele, H., and Bargou, R.C., A multicenter phase I/II trial evaluating the safety and efficacy of lenalidomide [Revlimidä (R), CC-5013] in combination with doxorubicin and dexamethasone (RAD) in patients with relapsed or refractory multiple myeloma, *Blood* 106 (11), 367b, 2005.
315. Raje, N., Kumar, S., Hideshima, T., Ishitsuka, K., Chauhan, D., Mitsiades, C., Podar, K., Le Gouill, S., Richardson, P., Munshi, N.C., Stirling, D.I., Antin, J.H., and Anderson, K.C., Combination of the mTOR inhibitor rapamycin and CC-5013 has synergistic activity in multiple myeloma, *Blood* 104 (13), 4188–4193, 2004.
316. Mitsiades, N., Mitsiades, C.S., Poulaki, V., Chauhan, D., Richardson, P.G., Hideshima, T., Munshi, N.C., Treon, S.P., and Anderson, K.C., Apoptotic signaling induced by immunomodulatory thalidomide analogs in human multiple myeloma cells: therapeutic implications, *Blood* 99 (12), 4525–4230, 2002.
317. Richardson, P., Schlossman, R., Munshi, N., Avigan, D., Jagannath, S., Alsina, M., Doss, D., McKenney, M., Hande, K., Farrell, M., Gorelik, S., Colson, K., Warren, D., Lunde, L., Michelle, R., Cole, G., Mitsages, C., Hideshima, T., Myers, T., Knight, R., and Anderson, K., A phase 1 trial of

lenalidomide (Revlimid®) with bortezomib (Velcade™) in relapsed and refractory multiple myeloma, *Blood* 106 (11), 110a–111a, 2005.

318. Niesvizky, R., Spencer, A., Wang, M., Weber, D., Chen, C., Dimopoulos, M.A., Yu, Z., Delap, R., Zeldis, J., and Knight, R.D., Increased risk of thrombosis with lenalidomide in combination with dexamethasone and erythropoietin, *J Clin Oncol* 24 (18S), 423s, 2006.
319. Angelotta, C., Lurie, A.J., Lurie, A.J., Yarnold, P.R., Singhal, S., Mehta, J., Lyons, E.A., and Bennett, C.L., Black box warning on lenalidomide-associated venous thromboembolism (VTE) in off-label setting: a preemptive and unusual safety initiative, *J Clin Oncol* 24 (18S), 319s, 2006.
320. Rajkumar, S.V. and Blood, E., Lenalidomide and venous thrombosis in multiple myeloma, *N Engl J Med* 354 (19), 2079–2080, 2006.
321. Zonder, J.A., Barlogie, B., Durie, B.G., McCoy, J., Crowley, J., and Hussein, M.A., Thrombotic complications in patients with newly diagnosed multiple myeloma treated with lenalidomide and dexamethasone: benefit of aspirin prophylaxis, *Blood* 108 (1), 403; author reply 404, 2006.
322. Rajkumar, S.V. and Gertz, M.A., Lenalidomide therapy and deep-vein thrombosis in multiple myeloma., *Blood* 108 (1), 404, 2006.
323. List, A., Kurtin, S., Roe, D.J., Buresh, A., Mahadevan, D., Fuchs, D., Rimsza, L., Heaton, R., Knight, R., and Zeldis, J.B., Efficacy of lenalidomide in myelodysplastic syndromes, *N Engl J Med* 352 (6), 549–557, 2005.
324. Greenberg, P., Cox, C., LeBeau, M.M., Fenaux, P., Morel, P., Sanz, G., Sanz, M., Vallespi, T., Hamblin, T., Oscier, D., Ohyashiki, K., Toyama, K., Aul, C., Mufti, G., and Bennett, J., International scoring system for evaluating prognosis in myelodysplastic syndromes, *Blood* 89 (6), 2079–2088, 1997.
325. Cheson, B.D., Bennett, J.M., Kantarjian, H., Pinto, A., Schiffer, C.A., Nimer, S.D., Lowenberg, B., Beran, M., de Witte, T.M., Stone, R.M., Mittelman, M., Sanz, G.F., Wijermans, P.W., Gore, S., and Greenberg, P.L., Report of an international working group to standardize response criteria for myelodysplastic syndromes, *Blood* 96 (12), 3671–3674, 2000.
326. Giagounidis, A.A., Germing, U., and Aul, C., Biological and prognostic significance of chromosome 5q deletions in myeloid malignancies, *Clin Cancer Res* 12 (1), 5–10, 2006.
327. Gandhi, A.K., Naziruddin, S., Verhelle, D., Brady, H., Schafer, P., and Stirling, D., Antiproliferative activity of CC-5013 in 5q-myelodysplastic syndrome (MDS) and acute lymphocytic leukemia (ALL) cell lines, *J Clin Oncol* 22 (14S), 587s, 2004.
328. List, A.F., Dewald, G., Bennett, J., Giagounadis, A., Raza, A., Feldman, E., Powell, B., Greenberg, P., Zeldis, J., and Knight, R., Hematologic and cytogenetic (CTG) response to lenalidomide (CC-5013) in patients with transfusion-dependent (TD) myelodysplastic syndrome (MDS) and chromosome 5q31.1 deletion: results of the multicenter MDS-003 study, *J Clin Oncol* 23 (16S), 2s, 2005.
329. Buesche, G., Dieck, S., Giagounidis, A., Bock, O., Wilkens, L., Schlegelberger, B., Knight, R., Bennett, J., Aul, C., and Kriepe, H.H., Anti-angiogenic in vivo effect of lenalidomide (CC-5013) in myelodysplastic syndrome with del(5q) chromosome abnormality and its relation to the course of the disease, *Blood* 106 (11), 113a, 2005.
330. Goss, T.F., Szende, A., Schaefer, C., Totten, P.J., Wang, M.Y., and List, A.F., Cost-effectiveness of lenalidomide in treating patients with transfusion-dependent myelodysplastic syndromes (MDS) in the United States, *J Clin Oncol* 24 (18S), 332s, 2006.
331. List, A.F., Gewald, G., Bennett, J., Giagounadis, A., Raza, A., Feldman, E., Powell, B., Greenberg, P., Faleck, H., Zeldis, J., and Knight, R., Results of the MDS-002 and -003 international phase II studies evaluating lenalidomide (CC-5013; Revlimida) in the treatment of transfusion-dependent (TD) patients with myelodysplastic syndrome (MDS), *Haematologica* 90 (S2), 307–308, 2005.
332. Bartlett, J.B., Tozer, A., Stirling, D., and Zeldis, J.B., Recent clinical studies of the immunomodulatory drug (IMiD) lenalidomide, *Br J Cancer* 93 (6), 613–619, 2005.
333. Cortes, J., Thomas, D., Verstovsek, S., Giles, F., Beran, M., Koller, C., and Kantarjian, H., Phase II study of lenalidomide (CC-5013, Revlimid®) for patients (pts) with myelofibrosis (MF), *Blood* 106 (11), 114a, 2005.
334. Tefferi, A., Mesa, R.A., Hogan, W.J., Shaw, T.A., Reyes, G.E., Allred, J.B., Ma, C.X., Dy, G.K., Wolanskyj, A.P., Litzow, M.L., Steensma, D.P., Call, T.G., and McClure, R.F., Lenalidomide

(CC-5013) treatment for anemia associated with myelofibrosis with myeloid metaplasia, *Blood* 106 (11), 726a, 2005.
335. Tefferi, A., Cortes, J., Verstovsek, S., Mesa, R.A., Thomas, D., Lasho, T.L., Hogan, W.J., Litzow, M.R., Allred, J.B., Jones, D., Byrne, C., Zeldis, J.B., Ketterling, R.P., McClure, R.F., Giles, F., and Kantarjian, H.M., Lenalidomide therapy in myelofibrosis with myeloid metaplasia, *Blood* 108, 1158–1164, 2006.
336. Wiernik, P.H., Lossos, I.S., Justice, G., Zeldis, J.B., Takeshita, K., Piertronigro, D., Habermann, T.M., and Witzig, T.E., Preliminary results from two phase II studies of lenalidomide monotherapy in relapsed/refractory non-Hodgkin's lymphoma, *J Clin Oncol* 24 (18S), 685s, 2006.
337. Querfeld, C., Kuzel, T.M., Guitart, J., and Rosen, S.T., Preliminary results of a phase II study of CC-5013 (lenalidomide, Revlimid®) in patients with cutaneous T-cell lymphoma, *Blood* 106 (11), 936a–937a, 2005.
338. Dispenzieri, A., Lacy, M.Q., Zeldenrust, S.R., geyer, S.M., Lust, J.A., Hayman, S.R., Kumar, S.K., Allred, J.B., Rajkumar, S.V., Kabat, B., Witzig, T.E., Greipp, P.R., Russell, S.R., Ghobrial, I., and Gertz, M.A., Lenalidomide has activity in a phase II trial in patients with primary systemic amyloidosis, *Blood* 106 (11), 77a, 2005.
339. Sanchorawala, V., Wright, D.G., Rosenzweig, M., Finn, K.T., Skinner, M., and Seldin, D.C., A phase II trial of lenalidomide for patients with AL amyloidosis, *J Clin Oncol* 24 (18S), 428s, 2006.
340. Schey, S.A., Fields, P., Bartlett, J.B., Clarke, I.A., Ashan, G., Knight, R.D., Streetly, M., and Dalgleish, A.G., Phase I study of an immunomodulatory thalidomide analog, CC-4047, in relapsed or refractory multiple myeloma, *J Clin Oncol* 22 (16), 3269–3276, 2004.
341. Streetly, M., Gyertson, K., Kazmi, M., Zeldis, J., and Schey, S.A., Alternate day Actimid™ (CC-4047) is well tolerated and is active when used to treat relapsed/refractory myeloma, *Blood* 104 (11), 98a, 2004.
342. Patten, P.E., Ahsan, G., Kazmi, M., Fields, P.A., Chick, G.W., Jones, R.R., Bradwell, A.R., and Schey, S.A., The early use of the serum free light chain assay in patients with relapsed refractory myeloma receiving treatment with a thalidomide analogue (CC-4047), *Blood* 102 (11), 449a, 2003.
343. Lilly, M.B., Drapiza, L., Sheth, M., Zemskova, M., Bashkirova, S., and Morris, J., Expression of cyclooxygenase-2 (COX-2) in human leukemias and hematopoietic cells, *Blood* 104 (11), 168b, 2004.
344. Ryan, E.P., Pollock, S.J., Kaur, K., Felgar, R.E., Bernstein, S.H., Chiorazzi, N., and Phipps, R.P., Constitutive and activation-inducible cyclooxygenase-2 expression enhances survival of chronic lymphocytic leukemia B cells, *Clin Immunol* 120 (1), 76–90, 2006.
345. Giles, F.J., Kantarjian, H.M., Bekele, B.N., Cortes, J.E., Faderl, S., Thomas, D.A., Manshouri, T., Rogers, A., Keating, M.J., Talpaz, M., O'Brien, S., and Albitar, M., Bone marrow cyclooxygenase-2 levels are elevated in chronic-phase chronic myeloid leukaemia and are associated with reduced survival, *Br J Haematol* 119 (1), 38–45, 2002.
346. Hoang, B., Zhu, L., Shi, Y., Frost, P., Yan, H., Sharma, S., Sharma, S., Goodglick, L., Dubinett, S., and Lichtenstein, A., Oncogenic RAS mutations in myeloma cells selectively induce cox-2 expression, which participates in enhanced adhesion to fibronectin and chemoresistance, *Blood* 107 (11), 4484–4490, 2006.
347. Cetin, M., Buyukberber, S., Demir, M., Sari, I., Sari, I., Deniz, K., Eser, B., Altuntas, F., Camci, C., Ozturk, A., Turgut, B., Vural, O., and Unal, A., Overexpression of cyclooxygenase-2 in multiple myeloma: association with reduced survival, *Am J Hematol* 80 (3), 169–173, 2005.
348. Ladetto, M., Vallet, S., Trojan, A., Dell'Aquila, M., Monitillo, L., Rosato, R., Santo, L., Drandi, D., Bertola, A., Falco, P., Cavallo, F., Ricca, I., De Marco, F., Mantoan, B., Bode-Lesniewska, B., Pagliano, G., Francese, R., Rocci, A., Astolfi, M., Compagno, M., Mariani, S., Godio, L., Marino, L., Ruggeri, M., Omede, P., Palumbo, A., and Boccadoro, M., Cyclooxygenase-2 (COX-2) is frequently expressed in multiple myeloma and is an independent predictor of poor outcome, *Blood* 105 (12), 4784–4791, 2005.
349. Wun, T., McKnight, H., and Tuscano, J.M., Increased cyclooxygenase-2 (COX-2): a potential role in the pathogenesis of lymphoma, *Leuk Res* 28 (2), 179–190, 2004.
350. Tzankov, A., Heiss, S., Ebner, S., Sterlacci, W., Schaefer, G., Augustin, F., Fiegl, M., and Dirnhofer, S., Angiogenesis in nodal B-cell lymphomas: a high throughput study, *J Clin Pathol*, 2006, June 21. [Epub ahead of print].

351. Hazar, B., Ergin, M., Seyrek, E., Erdogan, S., Tuncer, I., and Hakverdi, S., Cyclooxygenase-2 (Cox-2) expression in lymphomas, *Leuk Lymphoma* 45 (7), 1395–1399, 2004.
352. Li, H.L., Sun, B.Z., and Ma, F.C., Expression of COX-2, iNOS, p53 and Ki-67 in gastric mucosa-associated lymphoid tissue lymphoma, *World J Gastroenterol* 10 (13), 1862–1866, 2004.
353. Gately, S. and Li, W.W., Multiple roles of COX-2 in tumor angiogenesis: a target for antiangiogenic therapy, *Semin Oncol* 31 (2 Suppl 7), 2–11, 2004.
354. Subhashini, J., Mahipal, S.V., and Reddanna, P., Anti-proliferative and apoptotic effects of celecoxib on human chronic myeloid leukemia in vitro, *Cancer Lett* 224 (1), 31–43, 2005.
355. Zhang, G.S., Liu, D.S., Dai, C.W., and Li, R.J., Antitumor effects of celecoxib on K562 leukemia cells are mediated by cell-cycle arrest, caspase-3 activation, and downregulation of Cox-2 expression and are synergistic with hydroxyurea or imatinib, *Am J Hematol* 81 (4), 242–255, 2006.
356. Waskewich, C., Blumenthal, R.D., Li, H., Stein, R., Goldenberg, D.M., and Burton, J., Celecoxib exhibits the greatest potency amongst cyclooxygenase (COX) inhibitors for growth inhibition of COX-2-negative hematopoietic and epithelial cell lines, *Cancer Res* 62 (7), 2029–2033, 2002.
357. Jendrossek, V., Handrick, R., and Belka, C., Celecoxib activates a novel mitochondrial apoptosis signaling pathway, *FASEB J* 17 (11), 1547–1549, 2003.
358. Bombardier, C., Laine, L., Reicin, A., Shapiro, D., Burgos-Vargas, R., Davis, B., Day, R., Ferraz, M.B., Hawkey, C.J., Hochberg, M.C., Kvien, T.K., and Schnitzer, T.J., Comparison of upper gastrointestinal toxicity of rofecoxib and naproxen in patients with rheumatoid arthritis. VIGOR Study Group, *N Engl J Med* 343 (21), 1520–1528, 2 p following 1528, 2000.
359. Mukherjee, D., Nissen, S.E., and Topol, E.J., Risk of cardiovascular events associated with selective COX-2 inhibitors, *JAMA* 286 (8), 954–959, 2001.
360. Silverstein, F.E., Faich, G., Goldstein, J.L., Simon, L.S., Pincus, T., Whelton, A., Makuch, R., Eisen, G., Agrawal, N.M., Stenson, W.F., Burr, A.M., Zhao, W.W., Kent, J.D., Lefkowith, J.B., Verburg, K.M., and Geis, G.S., Gastrointestinal toxicity with celecoxib vs nonsteroidal anti-inflammatory drugs for osteoarthritis and rheumatoid arthritis: the CLASS study: a randomized controlled trial. Celecoxib Long-term Arthritis Safety Study, *JAMA* 284 (10), 1247–1255, 2000.
361. Pitt, B., Pepine, C., and Willerson, J.T., Cyclooxygenase-2 inhibition and cardiovascular events, *Circulation* 106 (2), 167–169, 2002.
362. Chenevard, R., Hurlimann, D., Bechir, M., Enseleit, F., Spieker, L., Hermann, M., Riesen, W., Gay, S., Gay, R.E., Neidhart, M., Michel, B., Luscher, T.F., Noll, G., and Ruschitzka, F., Selective COX-2 inhibition improves endothelial function in coronary artery disease, *Circulation* 107 (3), 405–409, 2003.
363. Celik, T., Iyisoy, A., Kursaklioglu, H., and Yuksel, U.C., Cyclooxygenase-2 inhibitors dilemma in cardiovascular medicine: are all family members bad players? *Int J Cardiol* 116 (2), 259–260, 2007.
364. Kara, I.O. and Sahin, B., COX-2 inhibitory treatment in chronic lymphocytic leukemia: a preliminary clinical study, *Leuk Lymphoma* 45 (7), 1495–1496, 2004.
365. Buckstein, R., Crump, M., Shaked, Y., Spaner, D., Piliotis, E., Imrie, K., Nayar, R., Foden, C., Turner, R., Taylor, D., Man, S., Baruchel, S., Stempak, D., Quirt, I., Sturgeon, J., Bertolini, F., and Kerbel, R.S., High dose celecoxib and metronomic 'low-dose' cyclophosphamide is effective therapy in patients with relapsed and refractory aggressive histology NHL, *Blood* 104 (11), 239b, 2004.
366. Ferrara, N., Hillan, K.J., and Novotny, W., Bevacizumab (Avastin), a humanized anti-VEGF monoclonal antibody for cancer therapy, *Biochem Biophys Res Commun* 333 (2), 328–335, 2005.
367. Karp, J.E., Gojo, I., Pili, R., Gocke, C.D., Greer, J., Guo, C., Qian, D., Morris, L., Tidwell, M., Chen, H., and Zwiebel, J., Targeting vascular endothelial growth factor for relapsed and refractory adult acute myelogenous leukemias: therapy with sequential 1-beta-D-arabinofuranosylcytosine, mitoxantrone, and bevacizumab, *Clin Cancer Res* 10 (11), 3577–3585, 2004.
368. Gotlib, J., Jamieson, C.H.M., List, A., Cortes, J., Albitar, M., Sridhar, K., Dugan, K., Quesada, S., Diaz, G., Pate, O., Novotny, W., Chen, H., and Greenberg, P.L., Phase II study of bevacizumab (anti-VEGF humanized monoclonal antibody) in patients with myelodysplastic syndrome (MDS): preliminary results, *Blood* 102 (11), 425a, 2003.
369. Feugier, P., Van Hoof, A., Sebban, C., Solal-Celigny, P., Bouabdallah, R., Ferme, C., Christian, B., Lepage, E., Tilly, H., Morschhauser, F., Gaulard, P., Salles, G., Bosly, A., Gisselbrecht, C., Reyes, F., and Coiffier, B., Long-term results of the R-CHOP study in the treatment of elderly

patients with diffuse large B-cell lymphoma: a study by the Groupe d'Etude des Lymphomes de l'Adulte, *J Clin Oncol* 23 (18), 4117–4126, 2005.

370. Pfreundschuh, M., Trumper, L., Osterborg, A., Pettengell, R., Trneny, M., Imrie, K., Ma, D., Gill, D., Walewski, J., Zinzani, P.L., Stahel, R., Kvaloy, S., Shpilberg, O., Jaeger, U., Hansen, M., Lehtinen, T., Lopez-Guillermo, A., Corrado, C., Scheliga, A., Milpied, N., Mendila, M., Rashford, M., Kuhnt, E., and Loeffler, M., CHOP-like chemotherapy plus rituximab versus CHOP-like chemotherapy alone in young patients with good-prognosis diffuse large-B-cell lymphoma: a randomised controlled trial by the MabThera International Trial (MInT) Group, *Lancet Oncol* 7 (5), 379–391, 2006.
371. The International Non-Hodgkin's Lymphoma Prognostic Factors Project, a predictive model for aggressive non-Hodgkin's lymphoma, *N Engl J Med* 329 (14), 987–994, 1993.
372. Ganjoo, K.N., Gordon, L., Robertson, M.J., and Horning, S.J., Rituximab, bevacizumab (Avastin) and CHOP (RA-CHOP) for patients with diffuse large B-cell lymphoma (DLBCL), *Blood* 104 (11), 389a, 2004.
373. Stopeck, A.T., Bellamy, W., Unger, J., Rimsza, L., Iannone, M., Fisher, R.I., and Miller, T.P., Phase II trial of single agent bevacizumab (Avastin) in patients with relapsed, aggressive non-Hodgkin's lymphoma (NHL): Southwest Oncology Group study S0108, *J Clin Oncol* 23 (16S), 583s, 2005.
374. Stopeck, A., Iannone, M., Rimsza, L., Miller, T., Fisher, R., and Bellamy, W., Expression of VEGF, VEGF receptors, and other angiogenic markers in relapsed aggressive non-Hodgkin's lymphoma: correlative studies for the SWOG S0108 trial, *Blood* 104 (11), 629a, 2004.
375. Bruns, I., Fox, F., Reinecke, P., Kobbe, G., Kronenwett, R., Jung, G., and Haas, R., Complete remission in a patient with relapsed angioimmunoblastic T-cell lymphoma following treatment with bevacizumab, *Leukemia* 19 (11), 1993–1995, 2005.
376. Giles, F.J., Bellamy, W.T., Estrov, Z., O'Brien, S.M., Verstovsek, S., Ravandi, F., Beran, M., Bycott, P., Pithavala, Y., Steinfeldt, H., Reich, S.D., List, A.F., and Yee, K.W., The anti-angiogenesis agent, AG-013736, has minimal activity in elderly patients with poor prognosis acute myeloid leukemia (AML) or myelodysplastic syndrome (MDS), *Leuk Res* 30 (7), 801–811, 2006.
377. Fielder, W., Graeven, U., Ergun, S., Verago, S., Kilic, N., Stockschlader, M., and Hossfeld, D.K., Expression of FLT4 and its ligand VEGF-C in acute myeloid leukemia, *Leukemia* 11 (8), 1234–1237, 1997.
378. Wang, C., Curtis, J.E., Geissler, E.N., McCulloch, E.A., and Minden, M.D., The expression of the proto-oncogene C-kit in the blast cells of acute myeloblastic leukemia, *Leukemia* 3 (10), 699–702, 1989.
379. Wells, S.J., Bray, R.A., Stempora, L.L., and Farhi, D.C., CD117/CD34 expression in leukemic blasts, *Am J Clin Pathol* 106 (2), 192–195, 1996.
380. Foss, B., Ulvestad, E., and Bruserud, O., Platelet-derived growth factor (PDGF) in human acute myelogenous leukemia: PDGF receptor expression, endogenous PDGF release and responsiveness to exogenous PDGF isoforms by in vitro cultured acute myelogenous leukemia blasts, *Eur J Haematol* 67 (4), 267–278, 2001.
381. Lee, S.H., Lopes de Menezes, D., Vora, J., Harris, A., Ye, H., Nordahl, L., Garrett, E., Samara, E., Aukerman, S.L., Gelb, A.B., and Heise, C., In vivo target modulation and biological activity of CHIR-258, a multitargeted growth factor receptor kinase inhibitor, in colon cancer models, *Clin Cancer Res* 11 (10), 3633–3641, 2005.
382. Trudel, S., Li, Z.H., Wei, E., Wiesmann, M., Chang, H., Chen, C., Reece, D., Heise, C., and Stewart, A.K., CHIR-258, a novel, multitargeted tyrosine kinase inhibitor for the potential treatment of t(4;14) multiple myeloma, *Blood* 105 (7), 2941–2948, 2005.
383. Lopes de Menezes, D.E., Peng, J., Garrett, E.N., Louie, S.G., Lee, S.H., Wiesmann, M., Tang, Y., Shephard, L., Goldbeck, C., Oei, Y., Ye, H., Aukerman, S.L., and Heise, C., CHIR-258: a potent inhibitor of FLT3 kinase in experimental tumor xenograft models of human acute myelogenous leukemia, *Clin Cancer Res* 11 (14), 5281–5291, 2005.
384. Xin, X., Tang, Y., Oei, Y., Ye, H., Menezes, D., Rendahl, K., Hollenbach, P., Pryer, N.K., Trudel, S., Mendel, D., Jallal, B., and Heise, C., CHIR-258 efficacy in a newly developed preclinical bone marrow model of t(4;14) multiple myeloma, *Blood* 106 (11), 963a, 2005.

385. Shih, L.Y., Huang, C.F., Wang, P.N., Wu, J.H., Lin, T.L., Dunn, P., and Kuo, M.C., Acquisition of FLT3 or N-ras mutations is frequently associated with progression of myelodysplastic syndrome to acute myeloid leukemia, *Leukemia* 18 (3), 466–475, 2004.
386. Shih, L.Y., Lin, T.L., Wang, P.N., Wu, J.H., Dunn, P., Kuo, M.C., and Huang, C.F., Internal tandem duplication of fms-like tyrosine kinase 3 is associated with poor outcome in patients with myelodysplastic syndrome, *Cancer* 101 (5), 989–998, 2004.
387. Yamamoto, Y., Kiyoi, H., Nakano, Y., Suzuki, R., Kodera, Y., Miyawaki, S., Asou, N., Kuriyama, K., Yagasaki, F., Shimazaki, C., Akiyama, H., Saito, K., Nishimura, M., Motoji, T., Shinagawa, K., Takeshita, A., Saito, H., Ueda, R., Ohno, R., and Naoe, T., Activating mutation of D835 within the activation loop of FLT3 in human hematologic malignancies, *Blood* 97 (8), 2434–2439, 2001.
388. Horiike, S., Yokota, S., Nakao, M., Iwai, T., Sasai, Y., Kaneko, H., Taniwaki, M., Kashima, K., Fujii, H., Abe, T., and Misawa, S., Tandem duplications of the FLT3 receptor gene are associated with leukemic transformation of myelodysplasia, *Leukemia* 11 (9), 1442–1446, 1997.
389. Reindl, C., Bagrintseva, K., Vempati, S., Schnittger, S., Ellwart, J.W., Wenig, K., Hopfner, K.P., Hiddemann, W., and Spiekermann, K., Point mutations found in the juxtamembrane domain of FLT3 define a new class of activating mutations in AML, *Blood* 107 (9), 3700–3707, 2006.
390. Yanada, M., Matsuo, K., Suzuki, T., Kiyoi, H., and Naoe, T., Prognostic significance of FLT3 internal tandem duplication and tyrosine kinase domain mutations for acute myeloid leukemia: a meta-analysis, *Leukemia* 19 (8), 1345–1349, 2005.
391. Frohling, S., Schlenk, R.F., Breitruck, J., Benner, A., Kreitmeier, S., Tobis, K., Dohner, H., and Dohner, K., Prognostic significance of activating FLT3 mutations in younger adults (16 to 60 years) with acute myeloid leukemia and normal cytogenetics: a study of the AML Study Group Ulm, *Blood* 100 (13), 4372–4380, 2002.
392. Thiede, C., Steudel, C., Mohr, B., Schaich, M., Schakel, U., Platzbecker, U., Wermke, M., Bornhauser, M., Ritter, M., Neubauer, A., Ehninger, G., and Illmer, T., Analysis of FLT3-activating mutations in 979 patients with acute myelogenous leukemia: association with FAB subtypes and identification of subgroups with poor prognosis, *Blood* 99 (12), 4326–4335, 2002.
393. Whitman, S.P., Archer, K.J., Feng, L., Baldus, C., Becknell, B., Carlson, B.D., Carroll, A.J., Mrozek, K., Vardiman, J.W., George, S.L., Kolitz, J.E., Larson, R.A., Bloomfield, C.D., and Caligiuri, M.A., Absence of the wild-type allele predicts poor prognosis in adult de novo acute myeloid leukemia with normal cytogenetics and the internal tandem duplication of FLT3: a cancer and leukemia group B study, *Cancer Res* 61 (19), 7233–7239, 2001.
394. Verhaak, R.G., Goudswaard, C.S., van Putten, W., Bijl, M.A., Sanders, M.A., Hugens, W., Uitterlinden, A.G., Erpelinck, C.A., Delwel, R., Lowenberg, B., and Valk, P.J., Mutations in nucleophosmin (NPM1) in acute myeloid leukemia (AML): association with other gene abnormalities and previously established gene expression signatures and their favorable prognostic significance, *Blood* 106 (12), 3747–3754, 2005.
395. Dohner, K., Schlenk, R.F., Habdank, M., Scholl, C., Rucker, F.G., Corbacioglu, A., Bullinger, L., Frohling, S., and Dohner, H., Mutant nucleophosmin (NPM1) predicts favorable prognosis in younger adults with acute myeloid leukemia and normal cytogenetics: interaction with other gene mutations, *Blood* 106 (12), 3740–3746, 2005.
396. Stirewalt, D.L., Kopecky, K.J., Meshinchi, S., Engel, J.H., Pogosova-Agadjanyan, E.L., Linsley, J., Slovak, M.L., Willman, C.L., and Radich, J.P., Size of FLT3 internal tandem duplication has prognostic significance in patients with acute myeloid leukemia, *Blood* 107 (9), 3724–3726, 2006.
397. Schnittger, S., Schoch, C., Kern, W., Mecucci, C., Tschulik, C., Martelli, M.F., Haferlach, T., Hiddemann, W., and Falini, B., Nucleophosmin gene mutations are predictors of favorable prognosis in acute myelogenous leukemia with a normal karyotype, *Blood* 106 (12), 3733–3739, 2005.
398. Abu-Duhier, F.M., Goodeve, A.C., Wilson, G.A., Care, R.S., Peake, I.R., and Reilly, J.T., Identification of novel FLT-3 Asp835 mutations in adult acute myeloid leukaemia, *Br J Haematol* 113 (4), 983–988, 2001.
399. Drexler, H.G., Expression of FLT3 receptor and response to FLT3 ligand by leukemic cells, *Leukemia* 10 (4), 588–599, 1996.

400. O'Farrell, A.M., Yuen, H.A., Smolich, B., Hannah, A.L., Louie, S.G., Hong, W., Stopeck, A.T., Silverman, L.R., Lancet, J.E., Karp, J.E., Albitar, M., Cherrington, J.M., and Giles, F.J., Effects of SU5416, a small molecule tyrosine kinase receptor inhibitor, on FLT3 expression and phosphorylation in patients with refractory acute myeloid leukemia, *Leuk Res* 28 (7), 679–689, 2004.
401. Morgan, G.J., Parker, A., Cavet, J., Cavenaugh, J., Heise, C., and Garzon, F., A phase I trial of CHIR-258, a multitargeted RTK inhibitor, in acute myeloid leukemia (AML), *Blood* 106 (11), 783a–784a, 2005.
402. Lonial, S., Alsina, M., Anderson, K., Richardson, P., Stewart, K., Fonseca, R., Heise, C., Fox, J., Allen, A., and Michelson, G., Phase I trial of chir-258 in multiple myeloma, *J Clin Oncol* 24 (18S), 680s, 2006.
403. Mendel, D.B., Schreck, R.E., West, D.C., Li, G., Strawn, L.M., Tanciongco, S.S., Vasile, S., Shawver, L.K., and Cherrington, J.M., The angiogenesis inhibitor SU5416 has long-lasting effects on vascular endothelial growth factor receptor phosphorylation and function, *Clin Cancer Res* 6 (12), 4848–4858, 2000.
404. Sukbuntherng, J., Cropp, G., Hannah, A., Wagner, G.S., Shawver, L.K., and Antonian, L., Pharmacokinetics and interspecies scaling of a novel VEGF receptor inhibitor, SU5416, *J Pharm Pharmacol* 53 (12), 1629–1636, 2001.
405. Fong, T.A., Shawver, L.K., Sun, L., Tang, C., App, H., Powell, T.J., Kim, Y.H., Schreck, R., Wang, X., Risau, W., Ullrich, A., Hirth, K.P., and McMahon, G., SU5416 is a potent and selective inhibitor of the vascular endothelial growth factor receptor (Flk-1/KDR) that inhibits tyrosine kinase catalysis, tumor vascularization, and growth of multiple tumor types, *Cancer Res* 59 (1), 99–106, 1999.
406. Smolich, B.D., Yuen, H.A., West, K.A., Giles, F.J., Albitar, M., and Cherrington, J.M., The antiangiogenic protein kinase inhibitors SU5416 and SU6668 inhibit the SCF receptor (c-kit) in a human myeloid leukemia cell line and in acute myeloid leukemia blasts, *Blood* 97 (5), 1413–1421, 2001.
407. Yee, K.W., O'Farrell, A.M., Smolich, B.D., Cherrington, J.M., McMahon, G., Wait, C.L., McGreevey, L.S., Griffith, D.J., and Heinrich, M.C., SU5416 and SU5614 inhibit kinase activity of wild-type and mutant FLT3 receptor tyrosine kinase, *Blood* 100 (8), 2941–2949, 2002.
408. Mesters, R.M., Padro, T., Bieker, R., Steins, M., Kreuter, M., Goner, M., Kelsey, S., Scigalla, P., Fiedler, W., Buchner, T., and Berdel, W.E., Stable remission after administration of the receptor tyrosine kinase inhibitor SU5416 in a patient with refractory acute myeloid leukemia, *Blood* 98 (1), 241–243, 2001.
409. Giles, F.J., Stopeck, A.T., Silverman, L.R., Lancet, J.E., Cooper, M.A., Hannah, A.L., Cherrington, J.M., O'Farrell, A.M., Yuen, H.A., Louie, S.G., Hong, W., Cortes, J.E., Verstovsek, S., Albitar, M., O'Brien, S.M., Kantarjian, H.M., and Karp, J.E., SU5416, a small molecule tyrosine kinase receptor inhibitor, has biologic activity in patients with refractory acute myeloid leukemia or myelodysplastic syndromes, *Blood* 102 (3), 795–801, 2003.
410. Fiedler, W., Mesters, R., Tinnefeld, H., Loges, S., Staib, P., Duhrsen, U., Flasshove, M., Ottmann, O.G., Jung, W., Cavalli, F., Kuse, R., Thomalla, J., Serve, H., O'Farrell, A.M., Jacobs, M., Brega, N.M., Scigalla, P., Hossfeld, D.K., and Berdel, W.E., A phase 2 clinical study of SU5416 in patients with refractory acute myeloid leukemia, *Blood* 102 (8), 2763–2767, 2003.
411. Kuenen, B.C., Rosen, L., Smit, E.F., Parson, M.R., Levi, M., Ruijter, R., Huisman, H., Kedde, M.A., Noordhuis, P., van der Vijgh, W.J., Peters, G.J., Cropp, G.F., Scigalla, P., Hoekman, K., Pinedo, H.M., and Giaccone, G., Dose-finding and pharmacokinetic study of cisplatin, gemcitabine, and SU5416 in patients with solid tumors, *J Clin Oncol* 20 (6), 1657–1667, 2002.
412. Rosen, P.J., Kabbinavar, F., Figlin, R.A., Parson, M., Laxa, B., Hernandez, L., Mayers, A., Cropp, G.F., Hannah, A.L., and Rosen, L.S., A phase I/II trial and pharmacokinetic (PK) study of SU5416 in combination with paclitaxel/carboplatin, *Proc Am Soc Clin Oncol* 20, 98a, 2001.
413. Rosen, P., Amado, R., Hecht, J., Chang, D., Mulay, M., Parson, M., Laxa, B., Brown, J., Cropp, G., Hannah, A., and Rosen, L., A phase I/II study of SU5416 in combination with 5-FU/leucovorin in patients with metastatic colorectal cancer, *Proc Am Soc Clin Oncol* 19, 3a, 2000.
414. Musolino, C., Calabro, L., Bellomo, G., Martello, F., Loteta, B., Pezzano, C., Rizzo, V., and Alonci, A., Soluble angiogenic factors: implications for chronic myeloproliferative disorders, *Am J Hematol* 69 (3), 159–163, 2002.

415. Brunner, B., Gunsilius, E., Schumacher, P., Zwierzina, H., Gastl, G., and Stauder, R., Blood levels of angiogenin and vascular endothelial growth factor are elevated in myelodysplastic syndromes and in acute myeloid leukemia, *J Hematother Stem Cell Res* 11 (1), 119–125, 2002.
416. Gersuk, G.M., Carmel, R., and Pattengale, P.K., Platelet-derived growth factor concentrations in platelet-poor plasma and urine from patients with myeloproliferative disorders, *Blood* 74 (7), 2330–2334, 1989.
417. Rueda, F., Pinol, G., Marti, F., and Pujol-Moix, N., Abnormal levels of platelet-specific proteins and mitogenic activity in myeloproliferative disease, *Acta Haematol* 85 (1), 12–15, 1991.
418. Dalley, A., Smith, J.M., Reilly, J.T., and Neil, S.M., Investigation of calmodulin and basic fibroblast growth factor (bFGF) in idiopathic myelofibrosis: evidence for a role of extracellular calmodulin in fibroblast proliferation, *Br J Haematol* 93 (4), 856–862, 1996.
419. Cross, N.C. and Reiter, A., Tyrosine kinase fusion genes in chronic myeloproliferative diseases, *Leukemia* 16 (7), 1207–1212, 2002.
420. Apperley, J.F., Gardembas, M., Melo, J.V., Russell-Jones, R., Bain, B.J., Baxter, E.J., Chase, A., Chessells, J.M., Colombat, M., Dearden, C.E., Dimitrijevic, S., Mahon, F.X., Marin, D., Nikolova, Z., Olavarria, E., Silberman, S., Schultheis, B., Cross, N.C., and Goldman, J.M., Response to imatinib mesylate in patients with chronic myeloproliferative diseases with rearrangements of the platelet-derived growth factor receptor beta, *N Engl J Med* 347 (7), 481–487, 2002.
421. Levine, R.L., Wadleigh, M., Sternberg, D.W., Wlodarska, I., Galinsky, I., Stone, R.M., DeAngelo, D.J., Gilliland, D.G., and Cools, J., KIAA1509 is a novel PDGFRB fusion partner in imatinib-responsive myeloproliferative disease associated with a t(5;14)(q33;q32), *Leukemia* 19 (1), 27–30, 2005.
422. Magnusson, M.K., Meade, K.E., Nakamura, R., Barrett, J., and Dunbar, C.E., Activity of STI571 in chronic myelomonocytic leukemia with a platelet-derived growth factor beta receptor fusion oncogene, *Blood* 100 (3), 1088–1091, 2002.
423. Giles, F.J., Cooper, M.A., Silverman, L., Karp, J.E., Lancet, J.E., Zangari, M., Shami, P.J., Khan, K.D., Hannah, A.L., Cherrington, J.M., Thomas, D.A., Garcia-Manero, G., Albitar, M., Kantarjian, H.M., and Stopeck, A.T., Phase II study of SU5416—a small-molecule, vascular endothelial growth factor tyrosine-kinase receptor inhibitor—in patients with refractory myeloproliferative diseases, *Cancer* 97 (8), 1920–1928, 2003.
424. Zangari, M., Anaissie, E., Stopeck, A., Morimoto, A., Tan, N., Lancet, J., Cooper, M., Hannah, A., Garcia-Manero, G., Faderl, S., Kantarjian, H., Cherrington, J., Albitar, M., and Giles, F.J., Phase II study of SU5416, a small molecule vascular endothelial growth factor tyrosine kinase receptor inhibitor, in patients with refractory multiple myeloma, *Clin Cancer Res* 10 (1 Pt 1), 88–95, 2004.
425. Wilhelm, S.M., Carter, C., Tang, L., Wilkie, D., McNabola, A., Rong, H., Chen, C., Zhang, X., Vincent, P., McHugh, M., Cao, Y., Shujath, J., Gawlak, S., Eveleigh, D., Rowley, B., Liu, L., Adnane, L., Lynch, M., Auclair, D., Taylor, I., Gedrich, R., Voznesensky, A., Riedl, B., Post, L.E., Bollag, G., and Trail, P.A., BAY 43-9006 exhibits broad spectrum oral antitumor activity and targets the RAF/MEK/ERK pathway and receptor tyrosine kinases involved in tumor progression and angiogenesis, *Cancer Res* 64 (19), 7099–7109, 2004.
426. Carlomagno, F., Anaganti, S., Guida, T., Salvatore, G., Troncone, G., Wilhelm, S.M., and Santoro, M., BAY 43-9006 inhibition of oncogenic RET mutants, *J Natl Cancer Inst* 98 (5), 326–334, 2006.
427. Tong, F.K., Chow, S., and Hedley, D., Pharmacodynamic monitoring of BAY 43-9006 (Sorafenib) in phase I clinical trials involving solid tumor and AML/MDS patients, using flow cytometry to monitor activation of the ERK pathway in peripheral blood cells, *Cytometry B Clin Cytom* 70B (3), 107–114, 2006.
428. Crump, M., Leber, B., Kassis, J., Hedley, D., Minden, M., Buckstein, R., McIntosh, L., Eisenhauer, E., and Seymour, L., A randomized phase I clinical and biologic study of two schedules of BAY 43-9006 in patients with myelodysplastic syndrome (MDS) or acute myeloid leukemia (AML): a National Cancer Institute of [Canada] Cancer Clinical Trials Group study, *J Clin Oncol* 22 (14S), 585s, 2004.
429. Abrams, T.J., Lee, L.B., Murray, L.J., Pryer, N.K., and Cherrington, J.M., SU11248 inhibits KIT and platelet-derived growth factor receptor beta in preclinical models of human small cell lung cancer, *Mol Cancer Ther* 2 (5), 471–478, 2003.

430. Mendel, D.B., Laird, A.D., Xin, X., Louie, S.G., Christensen, J.G., Li, G., Schreck, R.E., Abrams, T.J., Ngai, T.J., Lee, L.B., Murray, L.J., Carver, J., Chan, E., Moss, K.G., Haznedar, J.O., Sukbuntherng, J., Blake, R.A., Sun, L., Tang, C., Miller, T., Shirazian, S., McMahon, G., and Cherrington, J.M., In vivo antitumor activity of SU11248, a novel tyrosine kinase inhibitor targeting vascular endothelial growth factor and platelet-derived growth factor receptors: determination of a pharmacokinetic/pharmacodynamic relationship, *Clin Cancer Res* 9 (1), 327–337, 2003.
431. Sakamoto, K.M., Su-11248 Sugen, *Curr Opin Investig Drugs* 5 (12), 1329–1339, 2004.
432. Murray, L.J., Abrams, T.J., Long, K.R., Ngai, T.J., Olson, L.M., Hong, W., Keast, P.K., Brassard, J.A., O'Farrell, A.M., Cherrington, J.M., and Pryer, N.K., SU11248 inhibits tumor growth and CSF-1R-dependent osteolysis in an experimental breast cancer bone metastasis model, *Clin Exp Metastasis* 20 (8), 757–766, 2003.
433. O'Farrell, A.M., Foran, J.M., Fiedler, W., Serve, H., Paquette, R.L., Cooper, M.A., Yuen, H.A., Louie, S.G., Kim, H., Nicholas, S., Heinrich, M.C., Berdel, W.E., Bello, C., Jacobs, M., Scigalla, P., Manning, W.C., Kelsey, S., and Cherrington, J.M., An innovative phase I clinical study demonstrates inhibition of FLT3 phosphorylation by SU11248 in acute myeloid leukemia patients, *Clin Cancer Res* 9 (15), 5465–5476, 2003.
434. Schoeffski, P., Wolter, P., Himpe, U., Dychter, S.S., Baum, C.M., Prenen, H., Wildiers, H., Bex, M., and Dumez, H., Sunitinib-related thyroid dysfunction: a single-center retrospective and prospective evaluation, *J Clin Oncol* 24 (18S), 143s, 2006.
435. Shaheen, P.E., Tamaskar, I.R., Salas, R.N., Rini, B.I., Garcia, J., Wood, L., Dreicer, R., and Bukowski, R.M., Thyroid function tests (TFTs) abnormalities in patients (pts) with metastatic renal cell carcinoma (mRCC) treated with sunitinib, *J Clin Oncol* 24 (18S), 242s, 2006.
436. Faivre, S., Delbaldo, C., Vera, K., Robert, C., Lozahic, S., Lassau, N., Bello, C., Deprimo, S., Brega, N., Massimini, G., Armand, J.P., Scigalla, P., and Raymond, E., Safety, pharmacokinetic, and antitumor activity of SU11248, a novel oral multitarget tyrosine kinase inhibitor, in patients with cancer, *J Clin Oncol* 24 (1), 25–35, 2006.
437. Fiedler, W., Serve, H., Dohner, H., Schwittay, M., Ottmann, O.G., O'Farrell, A.M., Bello, C.L., Allred, R., Manning, W.C., Cherrington, J.M., Louie, S.G., Hong, W., Brega, N.M., Massimini, G., Scigalla, P., Berdel, W.E., and Hossfeld, D.K., A phase 1 study of SU11248 in the treatment of patients with refractory or resistant acute myeloid leukemia (AML) or not amenable to conventional therapy for the disease, *Blood* 105 (3), 986–993, 2005.
438. Wood, J.M., Bold, G., Buchdunger, E., Cozens, R., Ferrari, S., Frei, J., Hofmann, F., Mestan, J., Mett, H., O'Reilly, T., Persohn, E., Rosel, J., Schnell, C., Stover, D., Theuer, A., Towbin, H., Wenger, F., Woods-Cook, K., Menrad, A., Siemeister, G., Schirner, M., Thierauch, K.H., Schneider, M.R., Drevs, J., Martiny-Baron, G., and Totzke, F., PTK787/ZK 222584, a novel and potent inhibitor of vascular endothelial growth factor receptor tyrosine kinases, impairs vascular endothelial growth factor-induced responses and tumor growth after oral administration, *Cancer Res* 60 (8), 2178–2189, 2000.
439. Roboz, G.J., Giles, F.J., List, A.F., Cortes, J.E., Carlin, R., Kowalski, M., Bilic, S., Masson, E., Rosamilia, M., Schuster, M.W., Laurent, D., and Feldman, E.J., Phase 1 study of PTK787/ZK 222584, a small molecule tyrosine kinase receptor inhibitor, for the treatment of acute myeloid leukemia and myelodysplastic syndrome, *Leukemia* 20, 952–957, 2006.
440. Gupta, P., Miller, A.A., Owzar, K., Murry, D.J., Sanford, B.L., Vij, R., Yu, D., Hasserjian, R.P., Larson, R.A., and Ratain, M.J., Pharamcokinetics of an oral VEGF receptor tyrosine kinase inhibitor (PTK787/ZK222584) in patients with myelodysplastic syndrome (MDS): Cancer and Leukemia Group B study 10105, *J Clin Oncol* 24 (18S), 355s, 2006.
441. Ryan, A.J. and Wedge, S.R., ZD6474—a novel inhibitor of VEGFR and EGFR tyrosine kinase activity, *Br J Cancer* 92 (Suppl 1), S6–S13, 2005.
442. Wedge, S.R., Ogilvie, D.J., Dukes, M., Kendrew, J., Chester, R., Jackson, J.A., Boffey, S.J., Valentine, P.J., Curwen, J.O., Musgrove, H.L., Graham, G.A., Hughes, G.D., Thomas, A.P., Stokes, E.S., Curry, B., Richmond, G.H., Wadsworth, P.F., Bigley, A.L., and Hennequin, L.F., ZD6474 inhibits vascular endothelial growth factor signaling, angiogenesis, and tumor growth following oral administration, *Cancer Res* 62 (16), 4645–4655, 2002.
443. Amino, N., Ideyama, Y., Yamano, M., Kuromitsu, S., Tajinda, K., Samizu, K., Hisamichi, H., Matsuhisa, A., Shirasuna, K., Kudoh, M., and Shibasaki, M., YM-359445, an orally bioavailable

vascular endothelial growth factor receptor-2 tyrosine kinase inhibitor, has highly potent antitumor activity against established tumors, *Clin Cancer Res* 12 (5), 1630–1638, 2006.

444. Kovacs, M.J., Reece, D.E., Marcellus, D., Meyer, R.M., Mathews, S., Dong, R.P., and Eisenhauer, E., A phase II study of ZD6474 (Zactima®), a selective inhibitor of VEGFR and EGFR tyrosine kinase in patients with relapsed multiple myeloma-NCIC CTG IND. 145, *Invest New Drugs* 24 (6), 529–535, 2006.
445. Giles, F.J., Bellamy, W.T., Estrov, Z., O'Brien S.M., Verstovsek, S., Ravandi, F., Beran, M., Bycott, P., Pithavala, Y., Steinfeldt, H., Reich, S.D., List, A.F., and Yee, K.W., The anti-angiogenesis agent, AG-013736, has minimal activity in elderly patients with poor prognosis acute myeloid leukemia (AML) or myelodysplastic syndrome (MDS), *Leuk Res* 30 (7), 801–811, 2006.
446. Cobleigh, M.A., Langmuir, V.K., Sledge, G.W., Miller, K.D., Haney, L., Novotny, W.F., Reimann, J.D., and Vassel, A., A phase I/II dose-escalation trial of bevacizumab in previously treated metastatic breast cancer, *Semin Oncol* 30 (5 Suppl 16), 117–124, 2003.
447. Miller, K.D., Chap, L.I., Holmes, F.A., Cobleigh, M.A., Marcom, P.K., Fehrenbacher, L., Dickler, M., Overmoyer, B.A., Reimann, J.D., Sing, A.P., Langmuir, V., and Rugo, H.S., Randomized phase III trial of capecitabine compared with bevacizumab plus capecitabine in patients with previously treated metastatic breast cancer, *J Clin Oncol* 23 (4), 792–799, 2005.
448. Herbst, R.S., Johnson, D.H., Mininberg, E., Carbone, D.P., Henderson, T., Kim, E.S., Blumenschein, G., Jr., Lee, J.J., Liu, D.D., Truong, M.T., Hong, W.K., Tran, H., Tsao, A., Xie, D., Ramies, D.A., Mass, R., Seshagiri, S., Eberhard, D.A., Kelley, S.K., and Sandler, A., Phase I/II trial evaluating the anti-vascular endothelial growth factor monoclonal antibody bevacizumab in combination with the HER-1/epidermal growth factor receptor tyrosine kinase inhibitor erlotinib for patients with recurrent non-small-cell lung cancer, *J Clin Oncol* 23 (11), 2544–2555, 2005.
449. Johnson, D.H., Fehrenbacher, L., Novotny, W.F., Herbst, R.S., Nemunaitis, J.J., Jablons, D.M., Langer, C.J., DeVore, R.F., 3rd, Gaudreault, J., Damico, L.A., Holmgren, E., and Kabbinavar, F., Randomized phase II trial comparing bevacizumab plus carboplatin and paclitaxel with carboplatin and paclitaxel alone in previously untreated locally advanced or metastatic non-small-cell lung cancer, *J Clin Oncol* 22 (11), 2184–2191, 2004.
450. Langmuir, V.K., Cobleigh, M.A., Herbst, R.S., Holmgren, E., Hurwitz, H., Kabbinavar, F., Miller, K., and Novotny, W., Successful long-term therapy with bevacizumab (Avastin) in solid tumors, *Proc Am Soc Clin Oncol* 21, 9a, 2002.
451. Veronese, M.L., Mosenkis, A., Flaherty, K.T., Gallagher, M., Stevenson, J.P., Townsend, R.R., and O'Dwyer, P.J., Mechanisms of hypertension associated with BAY 43-9006, *J Clin Oncol* 24 (9), 1363–1369, 2006.
452. Lawler, J., Thrombospondin-1 as an endogenous inhibitor of angiogenesis and tumor growth, *J Cell Mol Med* 6 (1), 1–12, 2002.
453. Ren, B., Yee, K.O., Lawler, J., and Khosravi-Far, R., Regulation of tumor angiogenesis by thrombospondin-1, *Biochim Biophys Acta* 1765 (2), 178–188, 2006.
454. Dawson, D.W., Volpert, O.V., Pearce, S.F., Schneider, A.J., Silverstein, R.L., Henkin, J., and Bouck, N.P., Three distinct D-amino acid substitutions confer potent antiangiogenic activity on an inactive peptide derived from a thrombospondin-1 type 1 repeat, *Mol Pharmacol* 55 (2), 332–338, 1999.
455. Jimenez, B., Volpert, O.V., Crawford, S.E., Febbraio, M., Silverstein, R.L., and Bouck, N., Signals leading to apoptosis-dependent inhibition of neovascularization by thrombospondin-1, *Nat Med* 6 (1), 41–48, 2000.
456. Crawford, S.E., Stellmach, V., Murphy-Ullrich, J.E., Ribeiro, S.M., Lawler, J., Hynes, R.O., Boivin, G.P., and Bouck, N., Thrombospondin-1 is a major activator of TGF-beta1 in vivo, *Cell* 93 (7), 1159–1170, 1998.
457. Volpert, O.V., Zaichuk, T., Zhou, W., Reiher, F., Ferguson, T.A., Stuart, P.M., Amin, M., and Bouck, N.P., Inducer-stimulated Fas targets activated endothelium for destruction by anti-angiogenic thrombospondin-1 and pigment epithelium-derived factor, *Nat Med* 8 (4), 349–357, 2002.
458. Cikojevic, A. and Wang, E.S., Preclinical anti-tumor efficacy of ABT-510 (thrombospondin-1 mimetic peptide), a novel anti-angiogenic agent, in hematologic malignancies, *Blood* 106 (11), 681a, 2005.

459. Ozatli, D., Kocoglu, H., Haznedaroglu, I.C., Kosar, A., Buyukasik, Y., Ozcebe, O., Kirazli, S., and Dundar, S.V., Circulating thrombomodulin, thrombospondin, and fibronectin in acute myeloblastic leukemias, *Haematologia (Budap)* 29 (4), 277–283, 1999.
460. Lawler, J., Cohen, A.M., Chao, F.C., and Moriarty, D.J., Thrombospondin in essential thrombocythemia, *Blood* 67 (2), 555–558, 1986.
461. Haviv, F., Bradley, M.F., Kalvin, D.M., Schneider, A.J., Davidson, D.J., Majest, S.M., McKay, L.M., Haskell, C.J., Bell, R.L., Nguyen, B., Marsh, K.C., Surber, B.W., Uchic, J.T., Ferrero, J., Wang, Y.C., Leal, J., Record, R.D., Hodde, J., Badylak, S.F., Lesniewski, R.R., and Henkin, J., Thrombospondin-1 mimetic peptide inhibitors of angiogenesis and tumor growth: design, synthesis, and optimization of pharmacokinetics and biological activities, *J Med Chem* 48 (8), 2838–2846, 2005.
462. Reiher, F.K., Volpert, O.V., Jimenez, B., Crawford, S.E., Dinney, C.P., Henkin, J., Haviv, F., Bouck, N.P., and Campbell, S.C., Inhibition of tumor growth by systemic treatment with thrombospondin-1 peptide mimetics, *Int J Cancer* 98 (5), 682–689, 2002.
463. Hoekstra, R., de Vos, F.Y., Eskens, F.A., Gietema, J.A., van der Gaast, A., Groen, H.J., Knight, R.A., Carr, R.A., Humerickhouse, R.A., Verweij, J., and de Vries, E.G., Phase I safety, pharmacokinetic, and pharmacodynamic study of the thrombospondin-1-mimetic angiogenesis inhibitor ABT-510 in patients with advanced cancer, *J Clin Oncol* 23 (22), 5188–5197, 2005.
464. Shaked, Y., Bertolini, F., Man, S., Rogers, M.S., Cervi, D., Foutz, T., Rawn, K., Voskas, D., Dumont, D.J., Ben-David, Y., Lawler, J., Henkin, J., Huber, J., Hicklin, D.J., D'Amato, R.J., and Kerbel, R.S., Genetic heterogeneity of the vasculogenic phenotype parallels angiogenesis; implications for cellular surrogate marker analysis of antiangiogenesis, *Cancer Cell* 7 (1), 101–111, 2005.
465. Yaccoby, S., Ling, W., Saha, R., Yata, K., and Tricot, G., Systemic treatment with the antiangiogenic agent ABT-510 inhibited myeloma-induced microvessels and tumor growth in myelomatous SCID-hu mice, *Blood* 106 (11), 963a, 2005.
466. Levine, A.M., Schwartzberg, L., Smith, S., Belt, R., Knight, R., Carlson, D., Hippensteel, R., and Kahl, B., A phase 2 study of the thrombospondin-mimetic peptide ABT-510 in patients with refractory lymphoma, *Blood* 106 (11), 430a, 2005.
467. Shaheen, R.M., Davis, D.W., Liu, W., Zebrowski, B.K., Wilson, M.R., Bucana, C.D., McConkey, D.J., McMahon, G., and Ellis, L.M., Antiangiogenic therapy targeting the tyrosine kinase receptor for vascular endothelial growth factor receptor inhibits the growth of colon cancer liver metastasis and induces tumor and endothelial cell apoptosis, *Cancer Res* 59 (21), 5412–5416, 1999.
468. Korn, E.L., Arbuck, S.G., Pluda, J.M., Simon, R., Kaplan, R.S., and Christian, M.C., Clinical trial designs for cytostatic agents: are new approaches needed? *J Clin Oncol* 19 (1), 265–272, 2001.
469. Ratain, M.J. and Stadler, W.M., Clinical trial designs for cytostatic agents, *J Clin Oncol* 19 (12), 3154–3155, 2001.
470. Jain, R.K., Duda, D.G., Clark, J.W., and Loeffler, J.S., Lessons from phase III clinical trials on anti-VEGF therapy for cancer, *Nat Clin Pract Oncol* 3 (1), 24–40, 2006.
471. Ferrara, N. and Kerbel, R.S., Angiogenesis as a therapeutic target, *Nature* 438 (7070), 967–974, 2005.
472. Kerbel, R.S., Yu, J., Tran, J., Man, S., Viloria-Petit, A., Klement, G., Coomber, B.L., and Rak, J., Possible mechanisms of acquired resistance to anti-angiogenic drugs: implications for the use of combination therapy approaches, *Cancer Metastasis Rev* 20 (1–2), 79–86, 2001.
473. Sweeney, C.J., Miller, K.D., and Sledge, G.W., Jr., Resistance in the anti-angiogenic era: nay-saying or a word of caution? *Trends Mol Med* 9 (1), 24–29, 2003.
474. Casanovas, O., Hicklin, D.J., Bergers, G., and Hanahan, D., Drug resistance by evasion of antiangiogenic targeting of VEGF signaling in late-stage pancreatic islet tumors, *Cancer Cell* 8 (4), 299–309, 2005.
475. Kerbel, R. and Folkman, J., Clinical translation of angiogenesis inhibitors, *Nat Rev Cancer* 2 (10), 727–739, 2002.
476. Tam, B.Y., Wei, K., Rudge, J.S., Hoffman, J., Holash, J., Park, S.K., Yuan, J., Hefner, C., Chartier, C., Lee, J.S., Jiang, S., Niyak, N.R., Kuypers, F.A., Ma, L., Sundram, U., Wu, G., Garcia, J.A., Schrier, S.L., Maher, J.J., Johnson, R.S., Yancopoulos, G.D., Mulligan, R.C., and Kuo, C.J., VEGF modulates erythropoiesis through regulation of adult hepatic erythropoietin synthesis, *Nat Med* 12 (7), 793–800, 2006.

477. Heeschen, C., Aicher, A., Lehmann, R., Fichtlscherer, S., Vasa, M., Urbich, C., Mildner-Rihm, C., Martin, H., Zeiher, A.M., and Dimmeler, S., Erythropoietin is a potent physiologic stimulus for endothelial progenitor cell mobilization, *Blood* 102 (4), 1340–1346, 2003.
478. Kertesz, N., Wu, J., Chen, T.H., Sucov, H.M., and Wu, H., The role of erythropoietin in regulating angiogenesis, *Dev Biol* 276 (1), 101–110, 2004.
479. Glade Bender, J., Cooney, E.M., Kandel, J.J., and Yamashiro, D.J., Vascular remodeling and clinical resistance to antiangiogenic cancer therapy, *Drug Resist Updat* 7 (4–5), 289–300, 2004.
480. Yu, J.L., Rak, J.W., Coomber, B.L., Hicklin, D.J., and Kerbel, R.S., Effect of p53 status on tumor response to antiangiogenic therapy, *Science* 295 (5559), 1526–1528, 2002.
481. Rak, J., Yu, J.L., Kerbel, R.S., and Coomber, B.L., What do oncogenic mutations have to do with angiogenesis/vascular dependence of tumors?, *Cancer Res* 62 (7), 1931–1934, 2002.
482. Greipp, P.R., San Miguel, J., Durie, B.G., Crowley, J.J., Barlogie, B., Blade, J., Boccadoro, M., Child, J.A., Avet-Loiseau, H., Kyle, R.A., Lahuerta, J.J., Ludwig, H., Morgan, G., Powles, R., Shimizu, K., Shustik, C., Sonneveld, P., Tosi, P., Turesson, I., and Westin, J., International staging system for multiple myeloma, *J Clin Oncol* 23 (15), 3412–3420, 2005.

29 Antiangiogenic Therapy for Gliomas

Heinrich Elinzano and Howard A. Fine

CONTENTS

Gliomas are primary tumors of the central nervous system (CNS) that have predominantly glial histopathological and genetic features. The broad category of gliomas can be subdivided into different histological subtypes, which includes the astrocytic tumors, oligodendroglial tumors, and mixed oligoastrocytomas. The 2000 World Health Organization (WHO) classification of tumors of the nervous system further grades these tumors into varying degrees of malignancy based on the presence or absence of morphological features such as pleomorphism (or nuclear atypia), high mitotic activity, microvascular proliferation (MVP), and necrosis. Although the WHO grading system is currently the accepted gold standard criteria for grading gliomas, it is simply an estimate of malignancy for most tumors of the nervous system including some gliomas. For the diffuse glial tumors, however, it is a true grading system since a spectrum of progression from low- (WHO Grade II) to high-grade (WHO Grades III and IV) actually exists.[1] Glial tumors designated as WHO Grade II have only pleomorphism present, whereas WHO Grade III gliomas have both pleomorphism and high mitotic activity present. Finally, WHO Grade IV gliomas have any three of the four morphological features present. WHO Grade I is usually reserved for circumscribed gliomas, most often found in children, such as pilocytic astrocytoma.

Gliomas are a significant proportion of primary brain tumors in adults. The 2004–2005 Central Brain Tumor Registry of the United States (CBRTUS) statistical report on primary brain tumors using data collected from 1997 to 2001 and adjusted to the 2000 US standard million population estimates that gliomas account for about 42% of all adult primary brain and CNS tumors and of these, glioblastomas (GBMs), the most aggressive of all gliomas, make up about 50.5%.[2] There is a large variation in survival estimates between histological subtypes such that the 5 year survival rates exceed 90% for pilocytic astrocytomas (WHO Grade I) but are less than 4% for GBMs (WHO Grade IV) and survival generally decreases at older age of diagnosis.[3] Analysis of the 1973–1997 data obtained from the population-based Surveillance, Epidemiology, and End Results Program (SEER) has shown that although the relative survival rate of individuals with primary malignant brain tumor of various histological subtypes including some gliomas increased over the past three decades, there was no change in survival rate over time for individuals with GBM.[4] Despite advances in diagnostic and therapeutic modalities, the overall survival rate for high-grade gliomas remains poor and this is heavily weighted on GBMs, the most common primary malignant brain tumor in adults. GBMs are also considered as molecularly distinct tumors that appear histologically the same[5] and so management of these tumors can be particularly challenging. On the other hand, the so-called low-grade gliomas may undergo malignant transformation to high-grade gliomas over time in part by acquiring genetic mutations that confer tumor cell survival advantage. Furthermore, gliomas can cause significant mortality and morbidity regardless of the histological subtype or grade because of the involvement of the nervous system in the disease process. Still, brain and other nervous system tumors continue to be one of the top 15 primary cancer sites in the United States.[6] There is therefore a pressing need to develop new therapeutic approaches in the management of nervous system tumors particularly of malignant gliomas.

Angiogenesis, defined elsewhere in this book as the formation of new blood vessels from existing ones, does have an important role in the brain during embryogenesis but there is hardly any new blood vessel formation in the normal adult brain. Angiogenesis in the adult brain is evident in certain pathological reactive states to injury in the brain such as in the AIDS-dementia complex[7] and in experimental models of ischemia where vascular endothelial growth factor (VEGF) and its receptors are upregulated.[8] Angiopoeitin-2 (Ang-2) is also overexpressed in the first few hours of ischemia following middle cerebral artery (MCA) occlusion in rats with accompanying increase in vessel density in the periinfarct area and coinciding with endothelial proliferation.[9] Angiogenesis has long been established as an important factor for tumor growth and metastasis.[10] It has been recognized that gliomas are highly angiogenic and increasing vascularity has been correlated with increasing grade in astrocytoma.[11] The amount of neovascularization in high-grade gliomas is also closely correlated with the prognosis of the patients.[12] GBMs are among the most highly vascularized tumors with extremely elevated levels of proangiogenic factors in the nervous system.[13] MVP is a diagnostic hallmark of GBMs and is typically florid.[14] Glomeruloid bodies are the most exaggerated form of MVP and are most characteristic in GBM.[13] The degree of neovascularization in high-grade gliomas makes them an attractive candidate for antiangiogenic therapy.

Inhibition of angiogenesis therefore certainly has important therapeutic implication. Since the clinical results of angiogenic inhibitor drugs such as endostatin and angiostatin have initially fallen short of expectations, the anticancer potential of angiogenesis inhibitors in some systemic solid tumors has previously been highly controversial.[15] The dynamic science of tumor angiogenesis and the exciting developments in the field of molecular biology and genetics of gliomas, however, have ushered in a renewed interest in antiangiogenic therapy for gliomas. Drugs with presumed antiangiogenic properties continue to be investigated both in translational research and in clinical trials. In this chapter, specific antiangiogenic therapies for gliomas and GBMs in particular are reviewed and the potential of and obstacles to various antiangiogenic strategies are discussed.

29.1 ANGIOGENESIS IN GLIOMAS

Low-grade infiltrating astrocytomas have a vessel density that is comparable to or slightly greater than adjacent normal brain and they first acquire their blood supply by co-opting existing normal brain vasculature without initiating angiogenesis.[16] A switch to the angiogenic phenotype in high-grade gliomas is believed to occur when proangiogenic factors outweigh antiangiogenic factors in the tumor microenvironment and this angiogenic switch has been reported to occur in several steps: (1) the breakdown of the endothelial basement membrane and extracellular matrix (ECM), (2) the migration and proliferation of endothelial cells, (3) the differentiation of cells to allow for adherence and formation of new basement membrane, (4) the formation of new lumen, (5) the recruitment of supporting cells, including pericytes and smooth muscle cells, and (6) the synthesis of new ECM components.[17] As with tumor angiogenesis in other solid organs, the process of angiogenesis in gliomas is regulated by the balance of proangiogenic and antiangiogenic signals. In this section, the molecular mechanisms of angiogenesis relevant to gliomas are reviewed.

29.1.1 Proangiogenic Signals

The vascular permeability factor VEGF is the most studied and probably the most important angiogenic growth factor secreted by human glioma cells.[18] Although embryonic astrocytes synthesize VEGF during embryonic and early postnatal development,[19] nonmalignant astrocytes of adults including glial cells surrounding the tumor do not produce VEGF.[20] VEGF expression also correlates with glioma vascularity and grade such that high-grade gliomas produce more VEGF than low-grade astrocytomas.[21] VEGF is a potent mitogen for endothelial cells through signaling receptors VEGFR-1 (also termed *fms*-like tyrosine kinase Flt-1) and VEGFR-2 (also termed kinase insert domain receptor KDR or—as murine homologue—fetal liver kinase-1 Flk-1) coexpressed in endothelial cells. VEGF also induces the migration of endothelial cells and the expression of several genes involved in ECM degradation including matrix metalloproteases and serine proteases (urokinase- and tissue-type plasminogen activators).[20] VEGF therefore induces proliferation, sprouting, migration, and tube formation of endothelial cells[22] and it has also been shown to be a potent survival factor for endothelial cells by inducing the expression of antiapoptotic proteins in the endothelial cells during physiological and tumor angiogenesis.[23] Some glioma cells have been reported to express low levels of VEGFR-1 and VEGFR-2 but their role on proliferation is negligible.[24]

Although in vitro studies have not conclusively established whether placental growth factor (PIGF), a member of the VEGF family, transmits angiogenic signals, it has been shown to either stimulate endothelial cell growth and migration or to have minimal effects on proliferation, migration, and permeability of endothelial cells.[25] Several hypervascular primary brain tumors including some gliomas were found to have PIGF mRNA expression whereas some hypovascular primary nervous system tumors such schwannomas or germ cell tumors do not.[26]

A number of other tumor-secreted growth factors and their corresponding receptors can directly and indirectly regulate or modify glioma angiogenesis. These include fibroblast growth factor (FGF-1 [acidic FGF], -2 [basic FGF], and possibly -4 [hst-1/Kaposi FGF]), platelet-derived growth factor (PDGF), epidermal growth factor (EGF), and transforming growth factors-*a* and -*b* (TGF-*a* and TGF-*b*). In vitro and in vivo immunohistochemical mRNA expression studies have shown that increased levels of these growth factors and their receptors in malignant gliomas mediate their angiogenesis effects by (1) direct endothelial action and tube formation, (2) indirectly stimulating endothelial cell proliferation through increasing VEGF and other growth factor expression from tumor or endothelial cells or both, and (3) mediating the upregulation of key proteases on endothelial cells to remodel surrounding ECM and permitting endothelial cell migration.[27] Pleiotrophin (PTN) is a heparin-binding peptide growth factor produced by glioma cells, but not by regular adult glial cells, that also acts as mitogen and chemoattractant for endothelial cells.[20] Hepatocyte growth factor/scatter factor (HGF/SF) has been found through immunohistochemical and in situ hybridization studies to be strongly expressed in high-grade gliomas as well as endothelial cells in human GBMs.[28] It has been shown to induce chemotaxis and proliferation of human cerebral microvascular endothelial cells and may stimulate VEGF secretion and contribute to vessel maturation by facilitating the recruitment of perivascular cells.[28] HGF/SF was originally identified as a liver regeneration factor and mitogen for hepatocytes[29] and was independently described as a potent mitogen that dissociates or scatters colonies of epithelial cells.[30] Insulin-like growth factor-1 (IGF-1) and its receptor IGF-1R mRNA expression have also been found to correlate with both histopathologic grade and Ki-67 labeling indices in diffusely infiltrating astrocytomas. In GBMs, the number of IGF-1 immunoreactive cells was frequently more pronounced in perivascular tumor cells surrounding microvascular hyperplasia and palisading necrosis, in reactive astrocytes at the margins of tumor infiltration, and in microvascular cells exhibiting endothelial/pericytic hyperplasia suggesting the association of the IGF-1 signaling pathway with the development of the malignant phenotype.[31] The increased mRNA and subsequent protein secretion of most of these growth factors have been frequently shown to correlate with the microvessel density and grade of gliomas but not significantly with other primary nervous system tumors.

The angiopoietins Ang-1 and Ang-2 are angiogenesis factors that act mainly on endothelial cells and unlike VEGF, they appear to act in later stages of vascular development during vascular remodeling and maturation, probably by playing a role in the interaction of endothelial cells with other vascular elements such as the smooth muscles and the pericytes.[32] Their tyrosine kinase receptor Tie2/Tek is selectively expressed in endothelial cells. It was previously reported that Ang-1 is expressed by glioma cells and endothelial cells whereas Ang-2 is expressed by endothelial cells only[19] but a more recent study has shown that Ang-1 expression was detected in a few astrocytes of low-grade astrocytomas and it remained constant during progression from low- to high-grade astrocytomas while Ang-2 expression was absent in low-grade astrocytomas but present in anaplastic astrocytomas and GBMs.[33] The intensity of Ang-1 expression and the number of positive vessels increased during progression from low- to high-grade astrocytomas. Ang-2 expression involved a higher number of astrocytes grouped in areas with high vessel density.[33]

There are also other important molecules that are increasingly regarded as proangiogenic in gliomas. A potent activator of angiogenesis in gliomas is the hypoxia-inducible factor 1 (HIF-1). Activation of the HIF-1 pathway can be mediated by hypoxia, by growth factor-induced activation of receptor tyrosine kinases (RTKs) through PI3K/AKT/FRAP/mTOR pathway, by genetic alteration in gliomas such as epidermal growth factor receptor (EGFR) gene amplification and overexpression, by PTEN loss of function, and by p53 mutation.[34] This results in the activation of VEGF, VEGFR, matrix metalloproteinases (MMPs), plasminogen activator inhibitor type 1 (PAI-1), endothelin-1 (ET-1), inducible nitric oxide synthase (NOS), adrenomedullin (ADM), erythropoietin (Epo), and TGF-*a* and -*b,* which all affect glioma angiogenesis.[34] Another mediator of angiogenesis in gliomas is interleukin-8 (IL-8), which is a member of the chemokine family defined by their ability to cause directed migration of leukocytes. Through its receptors CXCR1 and CXCR2 on the endothelial cells, IL-8 enhances endothelial cell proliferation, chemotaxis, and survival (by inhibiting endothelial cell apoptosis), and promotes protease activation (by increased endothelial cell mRNA expression of MMPs-2 and -9).[35] IL-8 mRNA and protein are expressed in grades II, III and IV astrocytomas and anaplastic oligodendrogliomas can be detected in the cyst fluid of primary astrocytic tumor, and is highest in the perivascular region[36] and in pseudopalisading cells surrounding necrosis suggesting that IL-8 expression is also induced by hypoxia.[37] Both HIF-1 and IL-8 therefore may act synergistically with or alternatively to VEGF in promoting glioma angiogenesis.

29.1.2 Antiangiogenic Signals

There are a number of endogenous inhibitors of angiogenesis that are expressed in malignant gliomas and several of these mediate their antiangiogenic effect either through multiple protein–protein interactions that inhibit the action of proangiogenic molecules rather than through specific receptor-mediated signaling events [e.g., angiostatin, endostatin, PEX, tissue inhibitors of MMPs (TIMPs), platelet factor-4 (PF-4), and pigment epithelial-derived factor (PEDF)] or partly through a specific receptor [e.g., CD36 for thrombospondin (TSP)-1 and -2].[38]

A number of proteolytic fragments of proteins act as inhibitors of angiogenesis and these include angiostatin, endostatin, and PEX. Angiostatin is an internal fragment of plasminogen[39] generated by a series of proteolytic cleavage of plasminogen by urinary-type plasminogen activator or tissue-type plasminogen activator, phosphoglycerate kinase, serine protease, and MMP. It inhibits angiogenesis in gliomas through three potential receptors (and resulting inhibitory mechanisms): (1) integrin *avB*3 (inhibiting matrix ligands, cell attachment and adhesion, and cell migration), (2) ATP synthase (inhibiting energy for angiogenic process), and (3) angiomotin (inhibiting promigratory protein and cell migration).[38] Endostatin is a 20 kDa carboxyl-terminal fragment of collagen type XVIII[40] generated by a two-step proteolytic cleavage process. It inhibits angiogenesis via interactions with [and resulting inhibitory mechanisms]: (1) heparin sulfate proteoglycans (HSPGs) [inhibiting *b*FGF and angiogenesis], (2) proangiogenic molecules (MMP-2 and Flk-1) [inhibiting angiogenesis], (3) *a*5*B*1 [inhibiting matrix ligands, FAK/ERK, and cell migration and tube formation], and (4) tropomyosin [inhibiting microfilament integrity and cell migration].[38] PEX is a 210 amino acid fragment of MMP-2[41] and it competitively binds with integrin *avB*3 and preventing the proangiogenic interaction of this integrin with MMP-2.[38] Integrin *avB*3 expression is upregulated in tumor cells of grades III and IV malignant astrocytoma biopsies[42] and PEX has been detected in several malignant glioma cell lines.[43]

TIMPs are reported to inhibit angiogenesis directly by binding MMPs and inhibiting their activity, by blocking cell proliferation, and by downregulating VEGF expression.[44] TIMP-1 and -4 expression increased with increasing tumor grade in gliomas.[45] PF-4 protein and its carboxyl fragment promote antiangiogenesis by competing with proangiogenic growth factors for binding to HSPG (inhibiting *b*FGF and VEGF and angiogenesis).[46] Prolonged animal survival, slower-growing tumors, and decreased tumor blood vessel were found by the injection of human glioma cells virally transduced with full-length PF-4 in an intracranial mouse model of malignant glioma.[47]

PEDF is a 50 kDa glycoprotein[48] that might mediate its antiangiogenic effect by modulating the expression of proangiogenic molecules (e.g., *b*FGF, VEGF, and MMP-9) or other antiangiogenic molecules (e.g., TSP-1) as seen in overexpression of exogenous PEDF in U251 MG human glioblastoma cell lines in vitro[49]. It has also been demonstrated that PEDF can induce apoptosis of endothelial cells via activation of the c-jun N-terminal protein kinase (JNK) and the Fas ligand pathways but a specific PEDF receptor promoting a direct antiangiogenic effect of PEDF has not been identified. PEDF mRNA and protein expression were found to be inversely correlated with tumor histologic grade.[50]

Both TSP-1 and -2 contain three type-1 repeat domains[51] and both can exert their antiangiogenic effect, at least in part, by binding to CD36. Binding the CD36 receptor can induce endothelial cell apoptosis potentially leading to inhibition of angiogenesis through an interaction with pro-MMP-2 and -9 with MMP-2 and -9 or by inducing endothelial cell cycle arrest.[38] TSP-1 and -2 expression has been reported to be localized in vascular mesenchymal cells with minimal tumor cell expression. Whereas TSP-1 expression was not consistently correlated with tumor grade, TSP-2 mRNA expression appeared to inversely correlate with a higher histologic grade.[52]

The *ING4* gene has been recently reported to be involved in regulating brain tumor growth and angiogenesis by the interaction of ING4 with p65 (RelA) subunit of nuclear factor NF-k*B* and through transcriptional repression of NF-k*B*-responsive genes.[53] Xenografts of human glioblastoma U87MG cells overexpressing ING4 in mice were found to be more vascularized and had a higher vessel density and a higher frequency of hemorrhage than control tumors. The ING4 expression was significantly reduced in gliomas compared with normal brain and the loss of expression correlated with tumor grades.

29.2 SECONDARY EFFECTS OF GLIOMA ANGIOGENESIS IN THE BRAIN

The vascular changes that accompany malignant progression of gliomas are consistent with the observation that angiogenesis is an important factor for tumor growth[27] but additional consequences of glioma angiogenesis in the brain have equally important clinical implications. The brain is a highly vascularized organ and it is much dependent on blood circulation for precise physiologic functioning. The cerebral microvasculature is anatomically unique in that the endothelial cells in these blood vessels form extensive tight junction, have absent fenestrations, have few endocytic vesicles, and possess pericytes that wrap around the endothelial cells and astrocytic processes that form end-feet lining the cerebral blood vessels and underlying basement membrane.[54,55,56] This unit collectively referred to as the blood–brain barrier allows selective permeability to certain molecules thereby controlling movement of molecules to and from the brain tissue. The endothelial cells, astrocyte end-feet, and pericytes are the cellular components of the blood–brain barrier but it is the tight junctions present between the endothelial cells that form a diffusion barrier.[57] The endothelial tight junction is maintained by transmembrane proteins occludin and claudin that bind across adjacent endothelial cell membranes, effectively gluing them together, and these structures are stabilized by the cytoplasmic proteins ZO-1, ZO-2, and ZO-3 that bind the transmembrane proteins and the actin cytoskeleton.[58] The cytoplasmic/intracellular proteins of the endothelial tight junction are found to be members of the membrane activated guanylate kinase (MAGUK) family and may respond to second messengers causing conformational changes in the transmembrane proteins that open or close tight junctions.[58]

Glioma microvessels appear to have three principal origins: (1) neovascularization by sprouting from preexisting vessels or angiogenesis, (2) tumor cell takeover of host vessels or cooption, and (3) partition of the vessel lumen by insertion of interstitial tissue columns or intussusception.[59] A fourth possible mechanism of glioma microvessel formation is by the recruitment of circulating endothelial progenitor cells or vasculogenesis as reported by studies on glioma animal models but this may play a supportive rather a constitutive role in glioma neovascularization.[60] The glioma microvessels are structurally, molecularly, and functionally different from the normal cerebral microvasculature. Glioma microvessel endothelium ultrastructure studies have shown defects in tight junctions, an increase in the number of pinocytotic vesicles, and the presence of fenestrations and these become more pronounced with increasing glioma malignancy.[61,62] Evidence also suggests that the component proteins of tight junctions are either not expressed (such as claudin-1 or -5) or abnormally regulated (such as the downregulated expression of ZO-1) in human GBMs or express a nonfunctional form of the protein (such as claudin) in malignant astrocytomas.[63,64] VEGF has also been found to increase vascular permeability by formation of fenestrations, vesicular organelles, and transcellular gaps within endothelial cells[65] causing further endothelial barrier leakage.

The continuous maintenance of the blood–brain barrier by the brain tissue microenvironment is therefore interrupted in glioma growth[66] and its associated neovascularization leading to the destruction of the blood–brain barrier resulting in increased capillary permeability and vasogenic peritumoral edema.[67] This is of great concern as the brain, enclosed in a rigid vault, is very sensitive to increased intracranial pressure and so most patients with gliomas manifest with signs and symptoms of increased intracranial pressure at some stage during the course of their disease.[68] Increased intracranial pressure can manifest with global neurological dysfunction such as headaches, vomiting, and altered consciousness or with focal neurological dysfunction such as a rapidly progressive focal neurological deficit. Neurological signs and symptoms are primarily due to the disruption of neural function secondary to the mass effect exerted by the tumor and the peritumoral vasogenic edema but neurological dysfunction can also be due to obstructive hydrocephalus, ischemia from compression of surrounding normal blood vessels, and toxic inhibition of local neuronal activity secondary to metabolic abnormalities.[69] Additional consequences of the recruitment of the compromised vessels to the brain tumor bed include ischemic steal from hyperemia and increased likelihood of spontaneous hemorrhage is also of great concern in an organ of exquisite oxygen dependence.[70,71] Altered expression of antithrombotic molecules such as reduced expression of antithrombin III and the overexpression of thrombomodulin can cause the increased tendency of glioma microvessels to develop intratumoral hemorrhage and intravascular thrombosis.[72] The alterations seen in molecules involved in the coagulation cascade may play an important role in the high frequency of thromboembolic events seen in patients with malignant gliomas. Angiogenesis therefore does not only have an important role in the growth and malignant progression of gliomas but it is also a significant cause of the clinical morbidity in patients with brain tumors.

29.3 SPECIFIC ANTIANGIOGENIC THERAPIES FOR GLIOMAS

There have been numerous preclinical studies and clinical trials that evaluated various agents with antiangiogenic potential and many more are underway. There are various ways to target the vasculature for anticancer treatment and these could involve other strategies of delivering the antiangiogenic agents to the brain tumor or combinations of antiangiogenic agents. Several antiangiogenic therapies can also affect multiple and overlapping steps in the angiogenic process. The following is a review of antiangiogenic agents relevant to gliomas.

29.3.1 Inhibitors of the Endothelial Basement Membrane and Extracellular Matrix Breakdown

The breakdown of the endothelial basement membrane and ECM is an important step in angiogenesis and although endogenous inhibitors of MMPs exist such as TIMP-1 and -2, they are not absorbed orally and have limited tissue penetration,[73] hence are not practical for clinical use. Marimastat is a low molecular-weight hydroxamic acid derivative based on the structure of the natural MMP substrate collagen, orally bioavailable, and is a broad spectrum, competitive, and reversible MMP inhibitor.[74] Marimastat however has failed to demonstrate any benefit by the primary end points of progression-free survival or overall survival in a recently completed Phase III trial of patients with newly diagnosed glioblastoma treated with or without marimastat following standard radiation therapy.[75] Prinomastat (AG3340) is another synthetic MMP inhibitor but unlike marimastat it is a more selective inhibitor of four MMPs. It has demonstrated activity against gliomas in vivo in a SCID mouse model and is currently in a Phase I trial for glioblastoma.[73] COL-3 (6-demethyl-6-deoxy-4-dedimethylaminotetracycline), a high lipophilic, chemically modified tetratcycline without antimicrobial properties but directly inhibits several MMPs and is directly toxic to a number of cell lines, is currently in Phase II trial for brain tumors.[74]

29.3.2 Inhibitors of Angiogenic Growth Factors

Endothelial migration and proliferation is primarily achieved by proangiogenic growth factors. Inhibition of VEGF and its receptor has both antiangiogenic and antitumor growth effects and this inhibition has been achieved in various ways. Inhibition of tumor formation by antisense VEGF constructs has been shown in animal studies using malignant glioma cell lines[76] and C6 rat glioma cells.[77] Inhibition of tumor growth, decreased vascularization, and increased survival were also seen with dominant negative inhibition of VEGF-receptor signaling in a mouse model utilizing a retroviral construct expressing a mutant VEGF receptor Flk-1.[78] The tumorigenicity of human glioblastoma cell lines can also be significantly suppressed by monoclonal anti-VEGF-A antibodies.[79] Bevacizumab (Avastin), an anti-VEGF monoclonal antibody and the first angiogenesis inhibitor that is FDA-approved for first-line treatment for patients with metastatic colorectal cancer, is currently in various phases of clinical development for various advanced cancers and is currently being studied in malignant gliomas. Bleeding identified as toxicity related to bevacizumab in the previous Phase II colorectal cancer raises some concern for the likelihood of this toxicity occurring in the brain, which can cause further neurological damage in a patient already with brain tumor.

Most antiangiogenic agents that inhibit growth factors are small-molecular protein tyrosine kinase (PTK) inhibitors which specifically occupy ATP-binding cassette pocket of a particular PTK thereby blocking access of ATP, and consequently preventing phosphorylation of cellular substrates (adenine mimicry).[80] PTK inhibitors are low molecular-weight compounds and thus non-immunogenic but systemic administration of these angiogenic inhibitors has inherent limitation because of the short half-life of these compounds necessitating repeated administration, the need for treatment to be continuous to sustain tumor response or stable disease, risk of toxicity, and high cost of treatment.[16] Several compounds were developed to inhibit VEGF-receptor function by inhibiting its phosphorylation status. SU5416 is a small peptide tyrosine kinase inhibitor of VEGF receptor type 2 (Flk-1/KDR) that blocks VEGF signaling.[74] Although a 70% decrease in tumor volume was obtained in heterotopically implanted C6 rat glioma xenografts in SU5416-treated animals, SU5416 appeared to target neovascularization directly since it inhibited endothelial cell proliferation but not six glioma cells in vitro.[81] The derivative SU6668 was designed as a broad spectrum receptor tyrosine kinase inhibitor of VEGFR-2, PDGFR, and FGFR, which in animal models show greater efficacy but greater toxicity than SU5416.[82] A Phase I/II trial was conducted using SU5416 for patients who had recurrent high-grade gliomas and several had prolonged stabilization of disease but final results from that trial are pending.[83] SU6668 has currently entered clinical trials and its safety profiles are expected to be published shortly.[16] Another similar drug SU11248 has been found to be active against glioma cell lines in animal models[84] but has not been evaluated in clinical trials for patients with malignant gliomas.

ZD1839 (Iressa) is a PTK inhibitor of EGFR-tyrosine kinase, which has demonstrated antiangiogenic activity[85] in a series of human tumor xenografts. However, the treatment of xenografts expressing EGFRvIII, the mutated variant of EGFR commonly found in gliomas, resulted in a partial decrease in EGFR autophorylation and to an overall increase in EGFRvIII expression.[86] ZD6474 is a potent inhibitor of the tyrosine kinase activity of kinase insert domain-containing receptor (KDR), an endothelial cell receptor for VEGF, that also has activity against the EGFR. ZD6474 has been evaluated in preclinical studies and was reported to significantly reduce the growth of glioma cell line BT4C and the transformed rat brain endothelial cell line RBE4 in animal models.[87] It is currently undergoing evaluation in Phase I/II trials in patients with malignant gliomas.

Imatinib mesylate or STI571 (Gleevec) specifically targets the RTKs of the BRC-Abl, PDGF, and c-kit. It was originally developed as an agent for chronic myelogenous leukemia but has been shown to have some antiangiogenic activity and could affect glioma cells directly but does not penetrate the blood–brain barrier effectively.[83] A multi-institutional Phase I/II trial for patients with recurrent malignant gliomas, unfortunately demonstrated no activity in GBMs and resulted in a 10% incidence of intracranial/intratumoral hemorrhage. Whether the bleeding episodes were attributable to STI571 or merely a function of the propensity of malignant gliomas to spontaneously hemorrhage remains unclear at this time. Nevertheless, destabilization of tumor vasculature with resultant intratumoral bleeding has been a theoretical concern for the entire class of antiangiogenic agents from their inception.

Suramin is a polysulfonated naphthylurea antiparasitic agent that inhibits the function of growth factors and growth factor receptors implicated in glioma progression, angiogenesis, and radioresistance including VEGF, *b*FGF, and PDGF.[88,89] Other mechanisms of action had also been advocated such as its adrenolytic effects, inhibition of protein kinase C (PKC), and topoisomerase II.[73] It has been shown to slow glioma cell growth and glioma angiogenesis. A Phase I trial in recurrent high-grade glioma showed safety and possible efficacy[90] but a Phase II study administering suramin using an intermittent fixed-dosing regimen during cranial RT in newly diagnosed glioblastoma, although generally well tolerated, showed that overall survival is not significantly improved.[91]

A trial of directly instilling PF-4, an endogenous antiangiogenic protein synthesized by platelets (see above) into recurrent high-grade gliomas with an Ommaya reservoir was initiated but discontinued because of drug availability issues.[75]

29.3.3 Inhibitors of Endothelial Cell Adherence and the Formation of New Basement Membrane

There are four classes of adhesion molecules required by proliferating endothelial cells to establish contact with the ECM and they include: integrins, selectins, cadherins, and the immunoglobulin family of molecules.[73] Endothelial cells use this contact with the ECM to generate the traction required for changes in morphology and physical cell movement and physical movement through the ECM in response to humoral stimuli.[92,93] The adhesion molecule, *avB*3 integrin, has been demonstrated to be overexpressed in the regions of vascular proliferation in GBMs.[94] EMD121974 (Cilengitide), a pentapeptide *avB*3 and *avB*5 integrin antagonist, has been reported to markedly suppress human glioma growth in an orthotopic but not in a subcutaneous model in nude mice.[95] It is currently in Phase I/II trial for glioblastoma.[96]

29.3.4 Thalidomide

The mechanism of action of thalidomide or *a*-(*N*-phthalimido)glutaride is complex and may include TNF-*a* inhibition, inhibition of angiogenesis, and a variety of effects on the immune system and cell surface receptors.[74] Thalidomide has been shown to inhibit *b*FGF-induced angiogenesis using a rabbit cornea micropocket assay.[97] In gliomas, results of previously published Phase II trial of thalidomide and carmustine (BCNU) for patients with recurrent high-grade gliomas showed an objective radiographic response rate of 24%, compared favorably with data from historical controls than of either agent alone.[98] In an earlier published Phase II trial of thalidomide in 36 patients with recurrent high-grade gliomas, 6% had objective radiographic partial response, 6% had minor response, and 33% had stable disease.[99] A Phase II clinical trial of thalidomide and procarbazine is currently ongoing for patients with recurrent or progressive malignant glioma. Immunomodulatory drugs (IMiDs) are a series of compounds that were developed by using the first-generation IMiD thalidomide as the lead compound in a drug discovery compound[100] and recent results have confirmed that the IMiDs, in particular the clinical lead compounds CC-5013 (Revimid) and CC-4047 are antiangiogenic.[101] A Phase I clinical trial of CC-5013 for recurrent high-grade glioma is currently ongoing.[102]

29.3.5 Cytokines

Although initially implicated in tumor necrosis and cachexia (hence the name, cachectin), TNF-*a* acts through its receptors TNFR1 and TNFR2 on endothelial cells as a proangiogenic molecule that can induce angiogenesis both in vitro and in vivo.[73] It also has a complex role in endothelial cell survival and migration and induces the production of other proangiogenic molecules, including VEGF by endothelial cells.[103] CDC-501, a TNF-*a* inhibitor manufactured by Celgene Corporation, has been used in a Phase II trial for patients with gliomas but results are not yet available.[96]

Interferon-*a* and -*B* have been shown to have antiangiogenic properties by inhibiting angiogenic factors (lowering *b*FGF and possibly VEGF production),[104] MMP-9 expression,[83] and vascular smooth muscle proliferation.[105] Interferon-*B* was found to significantly inhibit glioma growth in an orthotopic xenograft model of human in nude mice particularly when begun early in the tumor course.[106] Interferons have entered clinical trials against malignancies including gliomas but mostly in combination with cytotoxic agents such as interferon-*a* with 5-fluorouracil (5-FU) or with temozolomide against recurrent malignant gliomas or interferon-*B* as adjuvant therapy in patients who respond favorably to radiation therapy.[73,107] PEG-intron or pegylated interferon-*a* 2b with or without thalidomide is currently being evaluated in Phase I and II trials in patients with recurrent high-grade gliomas. Other cytokines such as gamma, interferon-inducible protein-10 (IP-10), IL-12, and IL-18 have been found to have potent antiangiogenic activity and are in early clinical trials for gliomas.[75]

29.3.6 LY317615

PKC is a gene family consisting of at least 12 isoforms and it is likely that PKC is an important pathway component for both the intracellular signal transduction pathways for VEGF and *b*FGF in endothelial cells.[108] A Phase II trial of LY317615 (Enzastaurin), an orally bioavailable PKC*B* selective inhibitor, in patients with recurrent high-grade gliomas has recently closed and preliminary data demonstrates clear antiglioma activity but final results are not yet published.[109] A modified Phase I trial using a different dosing schedule in recurrent high-grade gliomas is currently ongoing and a multinational Phase III randomized trial of Enzastaurin versus CCNU in patients with recurrent GBM is planned.

29.3.7 2-Methoxyestradiol

2-Methoxyestradiol (2ME) is a metabolite of estradiol that inhibits proliferation of various tumor cells including human glioblastoma cell lines in vitro and was found to have a strong antiproliferative effect on human glioblastoma. Glioma cells exposed to 2ME were blocked in the G2/M phase of the cell cycle and underwent apoptosis, possibly through the upregulation of wild type p53.[110] 2ME is a natural compound with inhibitory activity against HIF-1alpha, a transcription factor that promotes expression of angiogenesis factors and resistance to programmed and therapy-induced cell death (see above). Phase I and II clinical trials for advanced solid tumors, multiple myeloma,[111] and malignant gliomas are ongoing.[16]

29.3.8 Carboxyamido-Triazole

Carboxyamido-triazole (CAI) is an inhibitor of nonvoltage-gated calcium channels and has inhibitory effects on tumor cell invasion and motility as well as inhibitory effects on human endothelial cell proliferation, migration, and adhesion to the basement membrane. Biological assays in glioma have shown CAI to be active in inhibiting invasion and angiogenesis but the exact mechanism of action is not clearly understood. It has been suggested that CAI works through inhibition of calcium influx in several signal transduction pathways that inhibit cell cycle progression. CAI-induced apoptosis in bovine aortic endothelial cells and a human glioma cell line (U251N) was found to be both dose and time dependent in micromolar concentrations achievable in brain tissue in vivo.[112] A Phase I/II trial of CAI for recurrent high-grade gliomas is currently ongoing.[75]

29.3.9 Antivascular Agents

Drugs that target existing tumor vasculature rather than prevent new blood vessel growth have also been explored at least in the animal glioma model. Combretastatin A4 disodium phosphate (CA-4) is a tubulin-binding drug, similar to cholchicine, that inhibit proliferating endothelial cells.[113] CA-4 has been reported to induce a gradual reduction in tumor blood flow, which can be exploited for hyperthermia sensitization in the subcutaneous BT4An rat glioma model[114] but it is unclear if CA-4 has yet been entered into clinical trials. The reported dose-limiting toxicities of cerebellar ataxia and confusion may confound the clinical assessment of response to treatment or worsen the neurological deficits already present in patients with gliomas.

29.3.10 Cyclooxygenase-2 Inhibitors

Celecoxib is a nonsteroidal anti-inflammatory drug that selectively inhibits cyclooxygenase-2 (COX-2) and has been reported to have antitumor activity through pro-apoptotic and antiangiogenic mechanisms. COX-2 produces PGE-2, which induces the proangiogenic molecules VEGF, FGF, TGF-*b*, PDGF, and ET-1.[115] COX-2 has been reported to be upregulated in the majority of high-grade gliomas and that there is a potential role of COX-2 inhibitors as an adjuvant therapy for brain tumors.[116] Along with its potential antiangiogenic activities, celecoxib exerts COX-2-independent antiproliferative effects on glioblastoma cell growth. These effects appear to be more potent than those of other selective COX-2 inhibitors or traditional NSAIDs, which are thought to be mediated via the transcriptional inhibition of cyclin A and cyclin B.[117] A recently published study reported that CPT-11 plus celecoxib can be safely administered concurrently at full dose levels and that this regimen has encouraging activity among heavily pretreated patients with recurrent malignant gliomas.[118] A Phase I/II clinical trial using celecoxib concurrently with twice daily dose of temozolomide in patients with documented histological diagnosis of relapsed or refractory GBM with promising response rates was published as an abstract but there is no published report to date.[119] Given the recently appreciated cardiovascular effects of Vioxx and other COX-2 inhibitors, the clinical future for this class of compounds remains uncertain.[120]

29.3.11 Copper Chelating Agents

The stimulation of vessel growth in the rabbit corneal assay by CuSO4 and ceruplasmin, the high-affinity binding of copper with angiogenic growth factors VEGF and FGF suggesting that the activity of some angiogenic activators is copper dependent, the higher levels of copper in tumor tissue than in nontumor tissue, and the selective inhibitory activity of polyamine donor chelating agents toward the proliferation of human umbilical vein endothelial cells (HUVEC) compared with fibroblast and human glioma cells support the role of copper in angiogenesis.[121] Copper depletion and chelation have been shown to suppress tumor growth and tumor angiogenesis in experimental 9L gliosarcoma model.[122,123] In addition to copper chelation, penicillamine blocks endothelial cell migration and proliferation and is also an inhibitor of urokinase plasminogen activator.[74] Penicillamine is currently in a Phase II study of adults who have newly diagnosed GBM.[83]

29.3.12 Endothelin Receptor Antagonist

The endothelins (ETs), 21-aa peptides ET-1, ET-2, and ET-3, are potent vasoconstricting peptides involved in the pathophysiology of different malignancies including CNS tumors.[124] The production of biologically active ET-1 formed by intracellular and membrane-bound endothelin-converting enzyme isoforms is stimulated by cytokines and growth factors, including IL-1*B*, TNF-*a*, TGF-*b*, PDGF, vasopressin, hypoxia, and shear stress.[125] ETs are mitogens for endothelial cells, vascular smooth muscle, fibroblasts, and pericytes and are also angiogenic factors. Additionally, ET-1 modulates various stages of neovascularization and promotes VEGF production through HIF-1*a*.[125] ET_A receptors are expressed in and affect cell proliferation of human glioblastoma cell lines[126] and there is enhanced expression of ET_A receptors in capillaries of human glioblastoma.[127] ABT-627 (Atrasentan) is an orally bioavailable selective ET_A receptor antagonist that is currently in Phase II clinical trial for patients with glioblastoma.[74]

29.3.13 Inhibition of Angiogenesis through Gene Delivery

Gene therapy has been applied to glioma models and includes approaches such as transduction of oncolytic viruses,[128,129] prodrug activation,[130,131] and immunomodulation.[132] Antiangiogenic gene therapy has the potential advantages of sustained high-level expression after single application, inhibition of multiple pathways simultaneously with the delivery of more than one transgene by using bi- or tricitronic vectors, and potential reduction in the cost of treatment.[133] The adenoviral vectors have been the most widely used for gene delivery; however, one major disadvantage of adenoviral vectors for gene delivery is the host-immune response to the vector limiting long-term delivery of antiangiogenic proteins.[16] Compared with systemic administration, repeated and more prolonged transgene expression is nevertheless achieved when adenoviral vector is delivered into the brain due to the immune sanctuary status of the CNS.[134] Some of the genes delivered in this fashion have included those for angiostatin, endostatin, PF-4, TIMP-2, a urokinase antagonist, IP-10, proliferin-related protein (PRP).[75] Adeno-associated virus (AAV) carrying the angiostatin gene have been found to delay tumor growth.[135] AAV has the advantages of low pathogenecity, the ability to infect both dividing and nondividing cells, long-term expression of antiangiogenic proteins, and the ability to penetrate solid tumors including glioma spheroids.[136,137] Retargeted adenoviral vectors have also been obtained with adenoviral vectors lacking coxsackievirus–adenovirus receptor and enhanced targeting capacity toward the *a*v-integrin and EGFR with promising results in glioma models.[138] These retargeted adenoviral vectors may circumvent the native viral tropism and increase selective binding to protein overexpressed in glioma cells such as the αv-integrin and EGFR reducing transduction in normal brain tissue and increasing therapeutic index. Despite some promising preclinical data, antiangiogenic gene therapy has not yet been translated to clinical trials.

29.3.14 Local Delivery of Inhibitors by Encapsulated Producer Cells

Since maintenance of plasma levels of that are sufficient to inhibit tumor growth is limited by the short half-life of endostatins (2–10 h), it appears at least in glioma xenografts that the local delivery of angiogenesis inhibitors such as endostatin is most effective when a constant level of the inhibitor is maintained in the tumor environment.[139] Local implantation of transfected cells engineered to express endostatin encapsulated in a sodium alginate polymer derived from seaweed has been developed to enhance local deliver of angiogenic inhibitors.[140] A reduction in tumor growth of over 70% was found in orthotopic glioma tumor models[141,142] and reduction of vessel density, perfusion of the tumors, and mean vessel diameter C6 rat glioma model when endostatin-producer beads were implanted together with glioma spheroids.[140] Local delivery of angiogenesis inhibitors by encapsulated producer cells has not been applied in clinical trials at this time.

29.3.15 Combined Strategies in Antiangiogenic Treatment

To overcome the complex regulation of tumor-associated angiogenesis, it seems logical that to increase the effectiveness of antiangiogenic treatment, multiple pathways involved in angiogenesis have to be targeted simultaneously and a multimodal approach may be necessary to optimize clinical efficacy.[16] Studies using glioma xenograft model[143,144] have shown that there is enhanced inhibition of tumor growth when radiotherapy and intratumoral injections of a replication-defective adenovirus carrying the angiostatin gene are combined compared to either treatments alone. The mechanism underlying this additive effect appears to be that ionizing radiation elicits a stress response in malignant cells and induces upregulation of VEGF-A expression via the mitogen-activated protein kinase (MAPK) dependent pathway.[145,146] Radiation combined with treatment with an anti-VEGF antibody resulted in greater than additive delay in tumor growth of U87 glioma xenografts than with either treatment alone.[147] Combination of radiotherapy with SU5416 (see above) was associated with enhanced endothelial cell apoptosis and reversal of radioresistance in a GL261 GBM tumor model in one study[148] and that tumor xenografts grown in apoptosis-resistant mice showed marked reduce response to radiotherapy in another study.[149] These observations seem to suggest that the radioresponsiveness of angiogenic tumors is in part regulated by microvascular damage and through expression of pro-endothelial cell survival cytokines such as VEGF and PDGF. The tumor endothelium therefore is a susceptible target for combined therapeutic strategies since these cytokines are overexpressed in the tumor vasculature. Combining antiangiogenic treatment with cytotoxic drugs can also potentiate the effects of chemotherapy by normalization of the tumor vasculature to improve delivery of therapeutic compounds and oxygen.[150] The hyperpermeable state of the atypical leaky tumor vessels alters the vascular and interstitial pressure gradient resulting in interstitial hypertension and subsequently impairing the flow of macromolecules and oxygen.[151] The interstitial pressure in malignant tumors has been shown to be elevated[152] and it is lowered by antiangiogenic therapy.[147] Treatment with SU5416 of an orthotopically implanted glioma xenograft overexpressing VEGF-A was shown to significantly increase intratumoral concentrations of temozolomide compared with controls in a murine glioma model.[153] Based on the results reported in these animal studies, patients with malignant glioma are currently enrolled in several clinical trials evaluating the efficacy of similar therapeutic approaches combining antiangiogenic agent with other treatment modalities including ZD1839 [Iressa] and temozolomide (EGFR PTK inhibitor and chemotherapy), LMW-heparin [dalteparin] and radiotherapy (VEGF-binding by heparin and enhancement of radiation efficacy), R115777 and temozolomide (farnesyltransferase inhibitor shown to have antiangiogenic properties by decreasing VEGF expression and chemotherapy to enhance cytotoxic effects).[16] Additionally, arsenic trioxide, found to cause acute vascular shutdown and induce rapid tumor necrosis in animal studies,[154,155] is currently in clinical trial for newly diagnosed glioblastoma in combination with radiation therapy.[160] The encouraging results of combined strategies of antiangiogenic agent with other treatment modalities in animal models, however, do not always translate to similar results in clinical trials. In a Phase II study of temozolomide and thalidomide combined with radiation therapy and then both temozolomide and thalidomide continued for 1 year or until disease progression or unacceptable toxicity in patients with newly diagnosed GBM, the survival outcome was similar to those who have not received thalidomide. Although the combination of thalidomide with radiation and temozolomide was relatively well tolerated, the added advantage of thalidomide was unclear.[157]

29.4 ASSESSING RESPONSE TO ANTIANGIOGENIC THERAPIES FOR GLIOMAS

There have been significant discrepancies between results obtained in preclinical studies and clinical trials and this is in part due to the differences in the vasculature, immunological interactions between tumor and host, and molecular and morphological heterogeneity between the human tumors and the experimentally induced tumors in animals.[16,158] Since most antiangiogenic agents are cytostatic rather than cytotoxic in nature, stabilization of disease rather than tumor shrinkage is to be expected more often.[159] Thus, assessment of tumor response by classical means such as clinical examination and measurement of tumor size using conventional contrast-enhanced magnetic resonance imaging (MRI), although still mandatory, may not always be reflective of the true status of the disease or tumor burden in patients with gliomas. Neurological decline in a patient with glioma may not always be due to disease progression and failure of current therapy but due to several causes including ongoing seizures, concurrent neurological or systemic infection, neurotoxicity of antineoplastic drugs, radiation effects, other concurrent neurological illness (stroke, seizures, migraines, neuropathy, myopathy), and concurrent medical illnesses (renal failure, hepatic failure, hypertension).[160] Measuring the largest perpendicular diameters of the contrast-enhancing lesion is the standard approach currently used in most clinical trials.[161] Tumor response is therefore categorized as complete or partial response and stable or progressive disease. This standard approach, however, is subject to interobserver variability and tumor margins may not be well delineated due to the infiltrative nature of these tumors. It is also often difficult to evaluate glioma progression on conventional contrast-enhanced MRI alone because abnormal enhancement, although representing breakdown of the blood–brain barrier, is nonspecific and cannot differentiate tumor progression from treatment effects such as cerebral radionecrosis. Using a more valid method or methods of monitoring efficacy of antiangiogenic agents therefore has important implications in research and treatment planning. It may also be an important tool in selecting potential responders to a particular therapy thus increasing therapeutic index and minimizing unnecessary toxicity. However, identification of the most appropriate method or methods of assessing the efficacy of antiangiogenic agents continues to be one of the major obstacles in angiogenesis research.

Surrogate end points for evaluating antiangiogenic treatment efficacy in the clinic are explored in tumors of all types including gliomas. The quantification of pro- and antiangiogenic factors has been evaluated as a surrogate end point. Urinary VEGF and MMP levels were previously reported to correlate with 1 year progression-free survival in a prospective study of cancer patients with nonmetastatic disease, including a few brain tumor patients, undergoing radiation therapy.[162] In a study of glioma xenografts expressing EGFRvIII, however, only partial decrease of EGFR autophosphorylation and an overall increase in EGFRvIII expression were seen following treatment with ZD1839 (Iressa), an RTK inhibitor of EGFR-tyrosine kinase (see earlier).[85] In the previously published Phase II trial of thalidomide and carmustine (BCNU) for patients with recurrent high-grade gliomas, authors found no correlation peak serum levels of *b*FGF and radiographic response to therapy, progression-free survival, or overall survival.[98] The quantification of angiogenic factors therefore may not always be a valid biologic end point in assessing glioma response to antiangiogenic therapy. Other biologic end points have also been explored. The quantification of microvessel density, although it has been shown to be an independent prognostic indictor of survival in glioma patients,[12] is probably not very useful or practical as an indicator of antiangiogenic therapeutic efficacy.[163] CD133-expressing circulating endothelial progenitor cells (CEPC) are explored as relevant biologic end points in antiangiogenic therapy but have not been validated in glioma patients.[156]

Imaging end points can be alternatives to biologic end points. The three main physiological differences of tumor microcirculation from normal brain i.e. flow characteristics and blood volume of the microvasculature, microvascular permeability, and increased fractional volume of extravascular (EES) and extracellular space or interstitial volume fraction (IVF), lead to blood flow that is spatially and temporally more heterogeneous than normal vasculature.[156] Since the marked alteration in the relative volumes of major tissue components: vascular (VVF), intracellular (ICF), and EES/IVF affects the trapping and clearance of agents such as contrast material in tumors,[156] MRI techniques that measure the uptake or clearance of contrast material can provide physiologic data on the state of the tumor vasculature. Physiologic imaging using recent advances in MR technology may provide more relevant information in assessing tumor angiogenesis. The relative regional cerebral blood volumes (rCBV) and regional cerebral flow (rCBF) in solid tumors can be measured by dynamic contrast-enhanced MRI from the first pass of a bolus of contrast agent through the microcirculation and permeability can be estimated by using a slower, steady-state dynamic imaging method.[164] Both rCBV and permeability have been shown to correlate with tumor grading and histologic microvascular density.[165] This MR imaging technique has been strongly correlated with efficacy of antiangiogenic response at least in animal models.[166] It has also been reported to be feasible in a study examining patients with low-grade astrocytoma.[167] In a study of patients with recurrent gliomas treated with thalidomide and carboplatin, rCBV values decreased significantly in all patients between the start of therapy and the first follow-up.[168] The difference in rCBV values in the patients who are clinically stable and those with progressive disease was also statistically significant with the progressive group having higher values. Dynamic contrast-enhanced MR imaging, although yet to be definitively validated, may be a potential surrogate measure of antiangiogenic response in gliomas.

Molecular imaging using positron-emission tomography (PET) scan has also been developed to image targeted angiogenic factor such as VEGF in animal models[169] and ongoing research offers hope of applying this modality to gliomas as well. At present several radiotracers such as 18-FDG, H20–15, and 11-CO are currently in clinical trials of angiogenesis imaging.[16,75,83]

29.5 CONCLUSION

Gliomas account for approximately half of all adult primary brain and CNS tumors and GBM, the most aggressive of all gliomas, constitute the majority of these. Despite advances in diagnostic and therapeutic modalities, there has been no change in survival rate over time for individuals with GBM over the past three decades with the 5 year survival rates less than 4%.

Gliomas are highly angiogenic and increasing vascularity has been correlated with increasing grade. Although the vascular changes that accompany malignant progression of gliomas are consistent with the observation that angiogenesis is an important factor for tumor growth, glioma angiogenesis in the brain has other consequences such as vasogenic edema, increased intracranial pressure, cerebral ischemia, and intratumoral hemorrhage that can cause clinically significant neurologic sequelae. The degree of neovascularization in high-grade gliomas makes them an attractive candidate for antiangiogenic therapy. Various ways to target the vasculature for anticancer treatment involve antiangiogenic agents or combinations of antiangiogenic agents and different strategies for delivering the antiangiogenic agents to the brain tumor. Numerous promising preclinical studies evaluating various antiangiogenic agents have been conducted but translation of these studies into the clinical setting has yet to be proven definitively successful. Several clinical trials using antiangiogenic agents for gliomas are currently underway; however, effective glioma antiangiogenic therapy with a favorable side effect profile has yet to be realized. An important roadblock to antiangiogenic therapy for gliomas is identifying the appropriate measure of assessing glioma response to antiangiogenic therapy. Surrogate end points for assessing glioma response to antiangiogenic therapy, such as biological or imaging end points, may be necessary in order to identify the most biologically and clinically active therapies.

REFERENCES

1. Kleihues, P. et al., The WHO classification of tumors of the nervous system, *J. Neuropathol. Exp. Neurol.*, 61, 215, 2002.
2. CBTRUS (2004–2005), Statistical report: primary brain tumors in the United States, 1997–2001, published by the Central Brain Tumor Registry of the United States.
3. Estimated by CBTRUS using Surveillance, Epidemiology, and End Results (SEER) Program public use CD-ROM (1973–2001). National Cancer Institute, DCPC, Surveillance Program, Cancer Statistics Branch, issued April 2004, based on the November 2003 submission.
4. Barnholtz-Sloan, J.S., Sloan, A.E., and Schwartz, A.G., Relative survival rates and patterns of diagnosis analyzed by time period for individuals with primary malignant brain tumor, 1973–1997, *J. Neurosurg.*, 99, 458, 2003.
5. Senger, D., Cairncross, G., and Forsythe, P., Long-term survivors of glioblastoma: statistical aberration or important unrecognized subtype, *Cancer J.*, 9, 14, 2003.
6. Jemal, A. et al., Annual report to the nation on the status of cancer, 1975–2001, with a special feature regarding survival, *Cancer*, 101, 3, 2004.
7. Gerhardt, H. and Betsholtz, C., Endothelial–pericyte interaction in angiogenesis, *Cell Tissue Res.*, 314, 15, 2003.
8. Plate, K.H., Mechanisms of angiogenesis in brain, *J. Neuropathol. Exp. Neurol.*, 58, 313, 1999.
9. Beck, H. et al., Expression of angiopoeitin-1, angiopoeitin-2, and tie receptors after middle cerebral artery occlusion in the rat, *Am. J. Pathol.*, 157, 1473, 2000.
10. Folkman, J., Tumor angiogenesis: therapeutic implications, *New Engl. J. Med.*, 285, 1182, 1971.
11. Wesseling, P. et al., Quantitative analysis of microvascular changes in diffuse astrocytic neoplasms with increasing grade of malignancy, *Hum. Pathol.*, 29, 352, 1998.
12. Leon, S.P., Folkert, R.D., and Black, P.M., Microvessel density is a prognostic indicator of patients with astroglial tumors, *Cancer*, 77, 362, 1996.
13. Kaur, B. et al., Genetic and hypoxic regulation of angiogenesis in gliomas, *J. Neurooncol.*, 70, 229, 2004.
14. Eberhard, A. et al., Heterogeneity of angiogenesis and blood vessel maturation in human tumors: implications for antiangiogenic tumor therapies, *Cancer Res.*, 60, 1388, 2000.
15. Novak, K., Angiogenesis inhibitors revised and reviewed at AACR, *Nat. Med.*, 8, 427, 2002.
16. Jansen, M. et al., Current perspectives on antiangiogenesis strategies in the treatment of malignant gliomas, *Brain Res. Brain Res. Rev.*, 45, 143, 2004.

17. Fisher, M.J. and Adamson, P.C., Anti-angiogenic agents for the treatment of brain tumors, *Neuroimag. J. N. Am.*, 12, 477, 2002.
18. Libermann, T.A. et al., An angiogenic growth factor is expressed in human glioma cells, *EMBO*, 6, 1627, 1987.
19. Rosenstein, J.M. et al., Patterns of brain angiogenesis after vascular endothelial growth factor administration and in vitro and in vivo, *Proc. Natl. Acad. Sci. U.S.A.*, 95, 7086, 1998.
20. Mentlein, R. and Held-Feindt, J., Angiogenesis factors in gliomas: a new key to tumour therapy, *Naturwissenschaften*, 90, 385, 2003.
21. Chaudry, I.H. et al., Vascular endothelial growth factor expression correlates with tumour grade and vascularity in gliomas, *Histopathology*, 39, 409, 2001.
22. Ferrara, N., Gerber, H.P., and LeCouter, J., The biology of VEGF and its receptors, *Nat. Med.*, 9, 669, 2003.
23. Gerber, H.P., Dixit, V., and Ferrara, N., Vascular endothelial growth factor induces expression of the antiapoptotic proteins Bcl-2 and A1 in vascular endothelial cells, *J. Biol. Chem.*, 273, 13313–6.
24. Herold-Mende, C. et al., Expression and functional significance of vascular endothelial growth factor receptors in human tumor cells, *Lab. Invest.*, 79, 1573, 1999.
25. Lutton, A. et al., Genetic dissection of tumor angiogenesis: are PIGF and VEGFR-1 novel anti-cancer targets, *Biochim. Biophys. Acta*, 1654, 79, 2004.
26. Nomura, M. et al., Placenta growth factor (PIGF) mRNA expression in brain tumors, *J. Neurooncol.*, 40, 123, 1998.
27. Dunn, I.F., Heese, O., and Black, P.McL., Growth factors in glioma angiogenesis: FGFs, PDGF, EGF, and TGF, *J. Neurooncol.*, 50, 121, 2000.
28. Lamszus, K., Heese, O., and Westphal, M., Angiogenesis-related growth factors in brain tumors, *Cancer Treat. Res.*, 117, 169, 2004.
29. Nakanura, T., Teramoto, H., and Ichihara, A., Purification and characterization of a growth factor from rat platelets for mature parenchymal hepatocytes in primary cultures, *Proc. Natl. Acad. Sci. U.S.A.*, 83, 6489, 1986.
30. Stoker, M. et al., Scatter factor is a fibroblast-derived modulator of epithelial cell mobility, *Nature*, 327, 239, 1987.
31. Hirano, H. et al., Insulin-like growth factor-1 content and pattern of expression correlates with histopathologic grade in diffusely-infiltrating astrocytomas, *Neurooncol.*, 1, 109, 1999.
32. Lopes, M.B., Angiogenesis in brain tumors, *Microrsc. Res. Tech.*, 60, 225, 2003.
33. Audero, E. et al., Expression of Angiopoeitin-1 in human glioblastomas regulates tumor-induced angiogenesis in vivo and in vitro studies, *Arterioscler. Thromb. Vasc. Biol.*, 21, 513, 2001.
34. Kaur, B. et al., Hypoxia and the hypoxia-inducible factor pathway in glioma growth and angiogenesis, *Neurooncol.*, 7, 134, 2005.
35. Brat, D.J., Bellail, A.C., and Van Meir, E.G., The role of interleukin-8 and its receptors in gliomagenesis and tumoral angiogenesis, *Neurooncol.*, 7, 122, 2005.
36. Desbaillets, I. et al., Upregulation of interleukin 8 by oxygen-deprived cells in glioblastoma suggests a role in leukocyte activation, chemotaxis, and angiogenesis, *J. Exp. Med.*, 186, 1201, 1997.
37. Desbaillets, I. et al., Regulation of interleukin-8 expression by reduced oxygen pressure in human glioblastoma, *Oncogene*, 18, 1447, 1999.
38. Rege, T.A., Fears, C.Y., and Gladson, C.L., Endogenous inhibitors of angiogenesis in malignant gliomas: nature's antiangiogenic therapy, *Neurooncol.*, 7, 106, 2005.
39. O'Reilly, M.S. et al., Angiostatin: a novel angiogenesis inhibitor that mediates the suppression of metastases by Lewis lung carcinoma, *Cell*, 79, 315, 1994.
40. O'Reilly, M.S. et al., Endostatin: an endogenous inhibitor of angiogenesis and tumor growth, *Cell*, 88, 277, 1997.
41. Brooks, P.C. et al., Disruption of angiogenesis by PEX, a noncatalytic metalloproteinase fragment with integrin binding activity, *Cell*, 92, 391, 1998.
42. Gingras, M.C. et al., Comparison of cell adhesion molecule expression between glioblastoma multiforme and autologous normal brain tissue, *J. Neuroimmunol.*, 57, 143, 1995.
43. Bello, L. et al., Simultaneous inhibition of glioma angiogenesis, cell proliferation, and invasion by naturally occurring fragment of human metalloproteinase 2, *Cancer Res.*, 61, 8730, 2001.

44. Jiang, T., Goldberg, I.D., and Shi, Y.E., Complex roles of tissue inhibitors of metalloproteinases in cancer, *Oncogene*, 21, 2245, 2002.
45. Groft, L.L. et al., Differential expression and localization of TIMP-1 and TIMP-4 in human gliomas, *Br. J. Cancer*, 85, 55, 2001.
46. Bikfalvi, A., Recent developments in the inhibition of angiogenesis: examples from studies on platelet factor-4 and the VEGF/VEGFR system, *Biochem. Pharmachol.*, 68, 1017, 2004.
47. Tanaka, T. et al., Viral vector-mediated transduction of a modified platelet factor 4 cDNA inhibits angiogenesis and tumor growth, *Nat. Med.*, 3, 437, 1997.
48. Bouck, N., PEDF: anti-angiogenic guardian of ocular function, *Trends Mol. Med.*, 8, 330, 2002.
49. Guan, M. et al., Inhibition of glioma invasion by overexpression of pigment epithelium-derived factor, *Cancer Gene Ther.*, 11, 325, 2004.
50. Guan, M. et al., Loss of pigment epithelium derived factor expression in glioma progression, *J. Clin. Pathol.*, 56, 277, 2003.
51. Adams, J.C. and Lawler, J., The thrombospondins, *Int. J. Biochem. Cell Biol.*, 36, 961, 2004.
52. Kazuno, M. et al., Thrombospondin-2 (TSP-2) expression is inversely correlated with vascularity in glioma, *Eur. J. Cancer*, 35, 502, 1999.
53. Garkavtsev, I. et al., The candidate tumor suppressor protein ING4 regulates brain tumor growth and angiogenesis, *Nature*, 428, 328, 2004.
54. Dermietzel, R. and Krause, D., Molecular anatomy of the blood brain barrier as defined by immunohistochemistry, *Int. Rev. Cytol.*, 127, 57, 1999.
55. Janzer, R.C. and Raff, R.C., Astrocytes induce blood–barrier properties in endothelial cells, *Nature*, 325, 253, 1987.
56. Risau, W. and Wolburg, H., Development of blood–brain barrier, *Trends Neurosci.*, 13, 174, 1990.
57. Ballabh, P., Braun, A., and Nedergaard, M., The blood–brain barrier: an overview structure, regulation, and clinical implications, *Neurobiol. Dis.*, 16, 1, 2004.
58. Papadopoulos, M.C. et al., Emerging molecular brain tumor mechanisms of brain tumor edema, *Br. J. Neurosurg.*, 15, 101, 2001.
59. Vajkoczy, P. and Menger, M.D., Vascular microenvironment in gliomas, *J. Neurooncol.*, 50, 99, 2000.
60. Ferrari, N. et al., Bone marrow–derived, endothelial progenitor-like cells as angiogenesis-selective gene-targeting vectors, *Gene Ther.*, 10, 647, 2003.
61. Grieg, N.H., Brain tumors and the blood–brain barrier, in *Implications of the Blood–Brain Barrier and Its Manipulation*, Neuwelt, E.A., ed., Plenum Press, New York, 1989, 77–106.
62. Shibata, T., Ultrastructure of capillary walls in human brain tumors, *Acta Neuropathol. (Berl.)*, 78, 561, 1989.
63. Liebner, S. et al., Caludin-1 and claudin-5 expression and tight junction morphology are altered in blood vessels of human glioblastoma multiforme, *Acta Neuropathol. (Berl.)*, 100, 323, 2000.
64. Sawada, T., Kobayashi, M., and Takekekawa, Y., Immunohistochemical study of tight junction-related protein in neovasculature in astrocytic tumor, *Brain Tumor Pathol*, 17, 1, 2000.
65. Bates, D.O. et al., Regulation of microvascular permeability by vascular endothelial growth factors, *J. Anat.*, 200, 581, 2002.
66. Rosenstein, J.M. et al., Patterns of brain angiogenesis after vascular endothelial growth factor administration and in vitro and in vivo, *Proc. Natl. Acad. Sci. U.S.A.*, 95, 7086, 1998.
67. Reijneveld, J.C., Voest, E.E., and Taphoorn, M.J.B., Angiogenesis in malignant primary and metastatic brain tumors, *J. Neurol.*, 247, 597, 2000.
68. Mentlein, R. and Held-Feindt, J., Angiogenesis factors in gliomas: a new key to tumour therapy, *Naturwissenschaften*, 90, 385, 2003.
69. Chaudry, I.H. et al., Vascular endothelial growth factor expression correlates with tumour grade and vascularity in gliomas, *Histopathology*, 39, 409, 2001.
70. Ferrara, N., Gerber, H.P., and LeCouter, J., The biology of VEGF and its receptors, *Nat. Med.*, 9, 669, 2003.
71. Gerber, H.P., Dixit, V., and Ferrara, N., Vascular endothelial growth factor induces expression of the antiapoptotic proteins Bcl-2 and A1 in vascular endothelial cells, *J. Biol. Chem.*, 273, 13313, 1998.
72. Isaka, T. et al., Altered expression of antithrombotic molecules in human glioma vessels, *Acta Neuropathol. (Berl.)*, 87, 81, 1994.

73. Puduvalli, V.K. and Sawaya, R., Antiangiogenesis-therapeutic strategies and clinical implications for brain tumors, *J. Neurooncol.*, 50, 189, 2000.
74. Deplangue, G. and Harris, A.L., Anti-angiogenic agents: clinical trial design and therapies in development, *Eur. J. Cancer*, 36, 1713, 2000.
75. Purow, B. and Fine, H.A., Progress report on the potential of angiogenesis inhibitors for neuro-oncology, *Cancer Investig.*, 22, 577, 2004.
76. Cheng, S.Y. et al., Suppression of glioblastoma angiogenicity and tumorigenicity by inhibition of endogenous expression of vascular endothelial growth factor, *Proc. Natl. Acad. Sci. U.S.A.*, 93, 8502, 1996.
77. Saleh, M., Stacker, S.A., and Wilks, A.F., Inhibition of growth of C6 glioma cells in vivo by expression of antisense vascular endothelial growth factor sequence, *Cancer Res.*, 56, 393, 1996.
78. Millauer, B. et al., Glioblastoma growth inhibited in vivo by a dominant negative FLK-1 mutant, *Nature*, 367, 576, 1994.
79. Kim, K.J. et al., Inhibition of vascular endothelial growth factor-induced angiogenesis suppresses tumour growth in vivo, *Nature*, 362, 841, 1993.
80. Fabbro, D., Parkinson, D., and Matter, A., Protein tyrosine kinase inhibitors: new treatment modalities, *Curr. Opin. Pharmacol.*, 2, 374, 2002.
81. Fong, T.A. et al., SU5416 is a potent and selective inhibitor of the vascular endothelial growth factor receptor (FLK-1/KDR) that inhibits tyrosine kinase catalysis, tumor vascularization, and growth of multiple tumor types, *Cancer Res.*, 59, 99, 1999.
82. Laird, A.D. et al., SU6668 is a potent antiangiogenic and antitumor agent that induces regression of established tumors, *Cancer Res.*, 60, 4152, 2000.
83. Purow, B. and Fine, H.A., Antiangiogenic therapy for primary and metastatic brain tumors, *Hematol. Oncol. Clin. N. Am.*, 18, 1161, 2004.
84. Mendel, D.B. et al., In vivo antitumor activity of SU11248, a novel tyrosine kinase inhibitor targeting vascular endothelial growth factor and platelet-derived growth factor receptors: determination of a pharmacokinetic/pharmacodynamic relationship, *Clin. Cancer Res.*, 9, 327, 2003.
85. Hirata, A. et al., ZD1839 (Iressa) induces antiangiogenic effects through inhibition of epidermal growth factor receptor tyrosine kinase, *Cancer Res.*, 62, 2554, 2002.
86. Heimberger, A.B. et al., Brain tumors in mice are susceptible to blockade of epidermal growth factor receptor (EGFR) with the oral, specific, EGFR-tyrosine kinase inhibitor ZD1839 (Iressa), *Clin. Cancer Res.*, 8, 3496, 2002.
87. Sandstrom, M. et al., The tyrosine kinase inhibitor ZD6474 inhibits tumour growth in an intracerebral rat glioma model, *Br. J. Cancer*, 91, 1174, 2004.
88. Takano, S. et al., Suramin inhibits glioma cell proliferation in vitro and in the brain, *J. Neurooncol.*, 21, 189, 1994.
89. Coomber, B.L., Suramin inhibits C6 glioma-induced angiogenesis in vitro, *J. Cell Biochem.*, 58, 199, 1995.
90. Grossman, S.A. et al., Toxicity, efficacy, and pharmacology of suramin in adults with recurrent high-grade gliomas, *J. Clin. Oncol.*, 19, 3260, 2001.
91. Laterra, J.J. et al., Suramin and radiotherapy in newly-diagnosed: phase 2 NABTT CNS consortium study, *Neurooncol.*, 6, 15, 2004.
92. Ingber, D.E. et al., Cell shape, cytoskeletal mechanics, and cell cycle control in angiogenesis, *J. Biomech.*, 28, 1471, 1995.
93. Mizejewski, G.J., Role of integrins in cancer: survey of expression patterns. *Proc. Soc. Exp. Biol. Med.*, 222, 124, 1999.
94. Gladson, C.L., Expression of integrin *avB*3 in small blood vessels of glioblastoma tumors, *J. Neuropathol. Exp. Neurol.*, 55, 1143, 1996.
95. MacDonald, T.J. et al., Preferential susceptibility of brain tumors to the angiogenic effects of an alpha(v) integrin antagonist, *Neurosurgery*, 48, 151, 2001.
96. Scappaticci, F.A., The therapeutic potential of novel antiangiogenic therapies, *Expert Opin. Investig. Drugs*, 12, 923, 2003.
97. D'Amato, R.J. et al., Thalidomide is an inhibitor of angiogenesis, *Proc. Natl. Acad. Sci. U.S.A.*, 91, 4082, 1994.

98. Fine, H.A. et al., Phase II trial of thalidomide and carmustine for patients with recurrent high-grade gliomas, *J. Clin. Oncol.*, 21, 2299, 2003.
99. Fine, H.A. et al., Phase II trial of antiangiogenic agent thalidomide in patients with recurrent high-grade gliomas, *J. Clin. Oncol.*, 18, 708, 2000.
100. Bartlett, J.B., Dredge, K., and Dalgleish, A.G., The evolution of thalidomide and its IMiD derivatives as anticancer agents, *Nature. Rev. Cancer*, 4, 314, 2004.
101. Dredge, K. et al., Novel thalidomide analogues display anti-angiogenic activity independently of immunomodulatory effects, *Br. J. Cancer*, 87, 1166, 2002.
102. Fine, H.A. et al., A phase I trial of CC-5103, a potent thalidomide analog, in patients with recurrent high-grade gliomas and other refractory CNS malignancies, presented at the 39th ASCO Annual Meeting, Chicago, IL, May 31–June 3, 2003.
103. Ryuto, M. et al., Induction of vascular endothelial growth factor by tumor necrosis factor alpha in human glioma cells. Possible roles of SP-1, *J. Biol. Chem.*, 271, 28220, 1996.
104. Singh, R.K. et al., Interferons alpha and beta down-regulate the expression of basic fibroblast growth factor in human carcinomas, *Proc. Natl. Acad. Sci. U.S.A.*, 92, 4562, 1995.
105. Heyns, A.D. et al., The antiproliferative effect of interferon and the mitogenic activity of growth factors are independent cell cycle events. Studies with vascular smooth muscle cells and endothelial cells, *Exp. Cell. Res.*, 161, 297, 1985.
106. Hong, Y.K. et al., Efficient inhibition of in vivo human malignant glioma growth and angiogenesis by interferon-beta treatment at early stage of tumor development, *Clin. Cancer Res.*, 6, 3354, 2000.
107. Fine, H.A. et al., A phase I trial of a new recombinant human beta-interferon (BG9015) for the treatment of patients with recurrent gliomas, *Clin. Cancer Res.*, 3, 381, 1997.
108. Teicher, B.A. et al., Antiangiogenic and antitumor effects of a protein kinase c*B* inhibitor in human T98g glioblastoma multiforme xenografts, *Clin. Cancer Res.*, 7, 634, 2001.
109. Fine, H.A. et al., Results from phase II trial of Enzastaurin (LY317615) in patients with recurrent high-grade gliomas, presented at the 2005 ASCO Annual Meeting Proceedings, 2005.
110. Lis, A. et al., 2-Methoxyestradiol inhibits proliferation of normal and neoplastic glial cells, and induces cell death, in vitro, *Cancer Lett.*, 213, 57, 2004.
111. Ricker, J.L. et al., 2-methoxyestradiol inhibits hypoxia-inducible factor 1alpha, tumor growth, and angiogenesis and augments paclitaxel efficacy in head and neck squamous cell carcinoma, *Clin. Cancer Res.*, 10, 8665, 2004.
112. Ge, S. et al., Carboxyamido-triazole induces apoptosis in bovine aortic endothelial and human glioma cells, *Clin. Cancer Res.*, 6, 1248, 2000.
113. Tozer, G.M. et al., Mechanisms associated with tumor vascular shut-down induced by combretastatin A-4 phosphate: intravital microscopy and measurement of vascular permeability, *Cancer Res.*, 61, 6413, 2001.
114. Eikesdal, H.P. et al., The new tubulin-inhibitor combretastatin A-4 enhances thermal damage in the BT4An rat glioma, *Int. J. Radiat. Oncol. Biol. Phys.*, 46, 645, 2000.
115. Yamada, M. et al., The effect of selective cyclooxygenase-2 inhibitor on corneal angiogenesis in the rat, *Curr. Eye Res.*, 19, 300, 1999.
116. Joki, T. et al., Expression of cyclooxygenase 2 (COX-2) in human glioma and in vitro inhibition by a specific COX-2 inhibitor, NS-398, *Cancer Res.*, 60, 4926, 2000.
117. Kardosh, A. et al., Differential effects of selective COX-2 inhibitors on cell cycle regulation and proliferation of glioblastoma cell lines, *Cancer Biol. Ther.*, 3, 55, 2004.
118. Reardon, D.A. et al., Phase II trial of irinotecan plus celecoxib in adults with recurrent malignant glioma, *Cancer*, 103, 329, 2005.
119. Pannullo, S. et al., Temozolomide plus celecoxib for treatment of malignant gliomas, *Proc. Am. Soc. Clin. Oncol.*, 22, 114, 2003 (abstract 455).
120. Brophy, J.M., Celecoxib and cardiovascular risks, *Expert Opin. Drug Saf.*, 4, 1005, 2005.
121. Camphausen, K. et al., Evaluation of chelating agents as anti-angiogenic therapy through copper chelation, *Bioorg. Med. Chem.*, 12, 5133, 2004.
122. Yoshida, D., Ikeda, Y., and Nakazawa, S., Suppression of tumor growth of experimental 9L gliosarcoma model by copper depletion, *Neurol. Med. Chir. (Tokyo)*, 35, 133, 1995.
123. Yoshida, D., Ikeda, Y., and Nakazawa, S., Copper chelation inhibits tumor angiogenesis in the experimental 9L gliosarcoma model by copper depletion, *Neurosurgery*, 37, 287, 1995.

124. Nelson, J. et al., The endothelin axis: emerging role in cancer, *Nat. Rev. Cancer*, 3, 110, 2003.
125. Bagnato, A. and Natali, P.G., Endothelin receptors as novel targets in tumor therapy, *J. Transl. Med.*, 2, 1, 2004.
126. Takahashi, K. et al., Three vasoactive peptides, endothelin-I, adrenomedullin and urotensin-II, in human tumour cell lines of different origin: expression and effects on proliferation, *Clin. Sci.*, 103, 35S, 2000.
127. Tsutsumi, K. et al., Enhanced expression of an endothelin ET_A receptor in capillaries of human glioblastoma: a quantitative receptor autoradiographic analysis using a radioluminographic imaging plate system, *J. Neurochem.*, 63, 2240, 1994.
128. Ring, C.J., Cytolytic viruses as potential anti-cancer agents, *J. Gen. Virol.*, 83, 491, 2002.
129. Shinoura, N. et al., Highly augmented cytopathic effect of a fiber-mutant E1B-defective adenovirus for gene therapy of gliomas, *Cancer Res.*, 59, 3411, 1999.
130. Miller, C.R. et al., Intratumoral 5-fluorouracil produced by cytosine deaminase/5-fluorocytosine gene therapy is effective for experimental human glioblastomas, *Cancer Res.*, 62, 773, 2002.
131. Nanda, D. et al., Treatment of malignant gliomas with a replicating adenoviral vector expressing herpes simplex virus-thymidine kinase, *Cancer Res.*, 61, 8743, 2001.
132. Iwadate, Y. et al., Induction of immunity in peripheral tissues combined with intracerebral transplantation of interleukin-2-producing cells eliminates established brain tumors, *Cancer Res.*, 61, 8769, 2001.
133. Lam, P.Y. and Breakefield, X.O., Potential of gene therapy for brain tumors, *Hum. Mol. Genet.*, 10, 777, 2001.
134. Parr, M.J. et al., Immune parameters affecting adenoviral vector gene therapy in the brain, *J. Neurovirol.*, 4, 194, 1998.
135. Ma, H.I. et al., Suppression of intracranial glioma growth after intramuscular administration of an adeno-associated viral vector expressing angiostatin, *Cancer Res.*, 62, 756, 2002.
136. Ponnazhagan, S. et al., Adeno-associated virus for cancer gene therapy, *Cancer Res.*, 61, 6313, 2001.
137. Enger, P.O. et al., Adeno-associated viral vectors penetrate human solid tumor tissue in vivo more effectively than adenoviral vectors, *Hum. Gene Ther.*, 13, 1115, 2002.
138. Kerbel, R.S., Clinical trials of antiangiogenic drugs: opportunities, problems, and assessment of initial results, *J. Clin. Oncol.*, 19, 45S, 2001.
139. Giussani, C. et al., Local intracerebral delivery of endogenous inhibitors by osmotic minipumps effectively suppresses glioma growth in vivo, *Cancer Res.*, 63, 2499, 2003.
140. Read, T.A. et al., Intravital microscopy reveals novel antivascular and antitumor effects of endostatin delivered locally by alginate-encapsulated cells, *Cancer Res.*, 61, 6830, 2001.
141. Joki, T. et al., Continuous release of endostatin from microencapsulated engineered cells for tumor therapy, *Nat. Biotechnol.*, 19, 35, 2001.
142. Read, T.A. et al., Local endostatin treatment of gliomas administered by microencapsulated producers cells, *Nat. Biotechnol.*, 19, 29, 2001.
143. Mauceri, H.J. et al., Combined effects of angiostatin and ionizing radiation in antitumor therapy, *Nature*, 394, 287, 1998.
144. Griscelli, F. et al., Combined effects of radiotherapy and angiostatin gene therapy in glioma tumor model, *Proc. Natl. Acad. Sci. U.S.A.*, 97, 6698, 2000.
145. Gorski, D.H. et al., Blockage of the vascular endothelial growth factor stress response increases the antitumor effects of ionizing radiation, *Cancer Res.*, 59, 3374, 1999.
146. Park, J.S. et al., Ionizing radiation modulates vascular endothelial growth factor (VEGF) expression through multiple mitogen activated protein kinase dependent pathways, *Oncogene*, 20, 3266, 2001.
147. Lee, C.G. et al., Anti-vascular endothelial growth factor treatment augments tumor radiation response under normoxic or hypoxic conditions, *Cancer Res.*, 60, 5565, 2000.
148. Geng, L. et al., Inhibition of vascular endothelial growth factor receptor signaling leads to reversal of tumor resistance to radiotherapy, *Cancer Res.*, 61, 2413, 2001.
149. Garcia-Barros, M. et al., Tumor response to radiotherapy regulated by endothelial apoptosis, *Science*, 300, 1155, 2003.
150. Jain, R.K., Normalizing tumor vasculature with antiangiogenic therapy: a new paradigm for combination therapy, *Nat. Med.*, 7, 987, 2001.

151. Jain, R.K., Munn, L.L., and Fukumura, D., Dissecting tumor pathophysiology using intravital microscopy, *Nat. Rev. Cancer*, 2, 295, 2002.
152. Stohrer, M. et al., Oncotic pressure in solid tumors is elevated, *Cancer Res.*, 60, 4251, 2000.
153. Ma, J. et al., Pharmacodynamic-mediated effects of the angiogenesis inhibitor SU5416 on the tumor deposition of temozolomide in subcutaneous and intracerebral glioma xenograft models, *J. Pharmacol. Ther.*, 305, 833, 2003.
154. Lew, Y.S. et al., Synergistic interaction with arsenic trioxide and fractionated radiation in locally advanced murine tumor, *Cancer Res.*, 62, 4202, 2002.
155. Lew, Y.S. et al., Arsenic trioxide causes selective necrosis in solid murine tumors by vascular shutdown, *Cancer Res.*, 59, 6033, 1999.
156. Bogler, O. and Mikkelsen, T., Angiogenesis in glioma: molecular mechanisms and roadblocks to translation, *Cancer J.*, 9, 205, 2003.
157. Chang, S. et al., Phase II study of temozolomide and thalidomide with radiation therapy for newly diagnosed glioblastoma multiforme, *Int. J. Radiat. Oncol. Biol. Phys.*, 60, 353, 2004.
158. Holand, E.C. et al., Combined activation of Ras and Akt in neural progenitors induces glioblastoma formation in mice, *Nat. Genet.*, 25, 55, 2000.
159. Burke, P.A. and De Nardo, S.J., Antiangiogenic agents and their promising potential in combined therapy, *Crit. Rev. Oncol./Hematol.*, 39, 155, 2001.
160. Plotkin, S.R. and Wen, P.Y., Neurologic complications of cancer therapy, *Neurol. Clin. N. Am.*, 21, 279, 2003.
161. Macdonald, D.R. et al., Response criteria for phase II studies of supratentorial malignant glioma. *J. Clin. Oncol.*, 8, 1277, 1990.
162. Chan, L.W. et al., Urinary VEGF and MMP levels as predictive markers of 1-year progression-free survival in cancer patients treated with radiation therapy: a longitudinal study of protein kinetics throughout tumor progression and therapy, *J. Clin. Oncol.*, 22, 499, 2004.
163. Hlatky, L., Hahnfeldt, P., and Folkman, J., Clinical application of antiangiogenic therapy: microvessel density, what it does and doesn't tell us, *J. Natl. Cancer Inst.*, 94, 883, 2002.
164. Nelson, S.J. and Cha, S., Imaging glioblastoma multiforme, *Cancer J.*, 9, 134, 2003.
165. Law, M. et al., Glioma grading: sensitivity, specificity, and predictive values of perfusion MR imaging and proton MR spectroscopic imaging compared with conventional MR imaging, *AJNR Am. J. Neuroradiol.*, 24, 1989, 2003.
166. Gossmann, A. et al., Dynamic contrast-enhanced magnetic resonance imaging as a surrogate marker of tumor response to anti-angiogenic therapy in a xenograft model of glioblastoma, *J. Magn. Reson. Imaging*, 15, 233, 2002.
167. Fuss, M. et al., Tumor angiogenesis of low-grade astrocytomas measured by dynamic susceptibility contrast-enhanced MRI (DSC-MRI) is predictive of local tumor control after radiation therapy, *Int. J. Radiat. Oncol. Biol. Phys.*, 51, 478, 2001.
168. Cha, S. et al., Dynamic contrast-enhanced T2*-weighted MR imaging of recurrent malignant gliomas treated with thalidomide and carboplatin, *ANJR Am. J. Neuroradiol.*, 21, 881, 2000.
169. Collingridge, D.R. et al., The development of [(124)I]iodinated-VG76e: a novel tracer for imaging vascular endothelial growth factor in vivo using positron emission tomography, *Cancer Res.*, 62, 5912, 2002.

30 Antiangiogenic Therapy for Kaposi's Sarcoma

Henry B. Koon, Liron Pantanowitz, and Bruce J. Dezube

CONTENTS

Kaposi's sarcoma (KS) is a multifocal angioproliferative disease that occurs in several clinical–epidemiological settings. It is a common tumor in human immunodeficiency virus (HIV)-infected patients and is considered an acquired immune deficiency syndrome (AIDS)-defining illness. The pathogenesis of KS is multifactorial and involves Kaposi's sarcoma herpesvirus/human herpesvirus-8 (KSHV/HHV8), altered expression and response to cytokines, genetic reprogramming of lesional cells, and in cases of AIDS-related KS, stimulation of KS growth by the HIV-1-transactivating protein Tat. The creation of an angiogenic-inflammatory state by HHV8 is a critical step in the development of KS. This chapter provides a detailed overview of how KS lesions are formed, discusses the lymphatic phenotype of KS tumor cells, and provides an in-depth description of the many angiogenic factors that participate in KS tumorigenesis. Advances in the understanding of the pathogenesis of KS have uncovered many potential targets for KS therapies. Pathogenesis-driven therapies, which have been and currently are in clinical development, include COL-3, imatinib, thalidomide, interferon-α, halofuginone, veglin, and antiretroviral agents. Many of these noncytotoxic therapeutic approaches have resulted in marked clinical benefit to patients coping with the physical and psychological consequences of this vascular tumor. Many in vitro and animal models are available to study KS and for in vitro testing of potential therapies. Moreover, because of the relative ease with which KS tissue may be obtained, clinical KS trials provide a perfect opportunity for those studying angiogenesis to assess the biologic and therapeutic effects of agents on these targets in vivo.

30.1 EPIDEMIOLOGY

Four clinical–epidemiological forms of KS are described.

30.1.1 Classic (Sporadic) KS

Afflicted patients, particularly males of Mediterranean or Jewish Ashkenazic origin, present primarily with KS lesions on their distal extremities. The male-to-female ratio is approximately 10:1. Rare familial cases have been reported.

30.1.2 African (Endemic) KS

Before the appearance of HIV, KS in equatorial Africa was largely an endemic disease. However, KS in sub-Saharan Africa has now reached epidemic proportions because of the explosive spread of AIDS. As a result, women and children there are more frequently affected, and patients have a very high KS tumor burden with rapid disease progression, which results in limited life expectancy.[1,2]

30.1.3 AIDS-Associated (Epidemic) KS

The AIDS epidemic was heralded by an increased incidence of KS in 1981.[3] In 1982, the U.S. Centers for Disease Control and Prevention proposed that KS be an AIDS-defining malignancy. KS is a common neoplasm in HIV-infected patients and is observed in patients of varying ages. In Western AIDS patients, most cases of KS occur in the homosexual risk group. Since the introduction of highly active antiretroviral therapy (HAART), the rate of KS in the HIV-positive population has dropped significantly.[4] Nevertheless, KS continues to be a substantial cause of morbidity in these individuals.[5]

30.1.4 Immunosuppression-Associated (Iatrogenic) KS

This is a rare disease presenting in patients receiving immunosuppressive therapy, especially in relation to transplantation.[6] In addition, postirradiation primary KS has been described.[7]

30.2 CLINICAL FEATURES

Classic KS presents mainly as indolent tumors on the lower limbs, with rare progression to systemic disease. Reports of classic KS confined solely to visceral organs are uncommon.[8] Endemic African KS may present (i) in young children with generalized lymphadenopathy, or (ii) in middle-aged men with similar clinical features to classic KS. The clinical presentation of AIDS- and immunosuppression-associated KS varies from minimal to fulminant disease, with frequent multifocal, mucocutaneous, and visceral involvement.

30.2.1 Cutaneous Disease

Skin lesions appear mainly on the lower extremities (Figure 30.1), face, and genitalia. They are characterized by multifocal blue-brown-violet colored macules, plaques, or nodules. Several morphologic variants have been described such as telangiectatic, ecchymotic, and keloidal KS.[9] Exophytic and fungating lesions can occur, with breakdown of the overlying skin. There may be deep local destruction, including involvement of underlying bone. Associated lymphedema, particularly in lymphostatic sites such as the face, genitalia, and lower extremities, may be extensive. Lymphedema may result from local destruction of lymphatics, KS involvement of regional lymph nodes, and liberation of cytokines.[10] Some authors believe that chronic lymphedema may predispose to the development of early KS.[11–13]

30.2.2 Extracutaneous Disease

Extracutaneous spread of KS is most typical in HIV-infected individuals.[14,15] AIDS-associated KS in the oral cavity is particularly common, as is gastrointestinal (Figure 30.2) and pulmonary

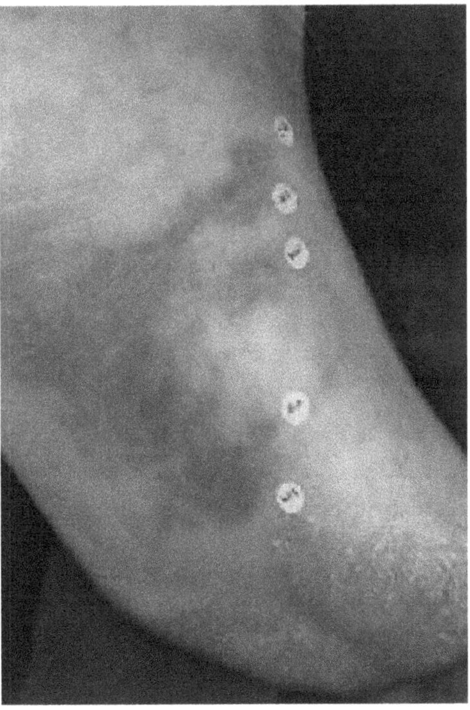

FIGURE 30.1 (See color insert following page 558.) Cutaneous Kaposi's sarcoma. Erythematous irregular plaques of the foot. Numbers are used to monitor the response of individual lesions to therapy. (Reproduced from Dezube, B.J., Pantanowitz, L., and Aboulafia, D.M., *AIDS Read* 14(5), 237, 2004. With permission.)

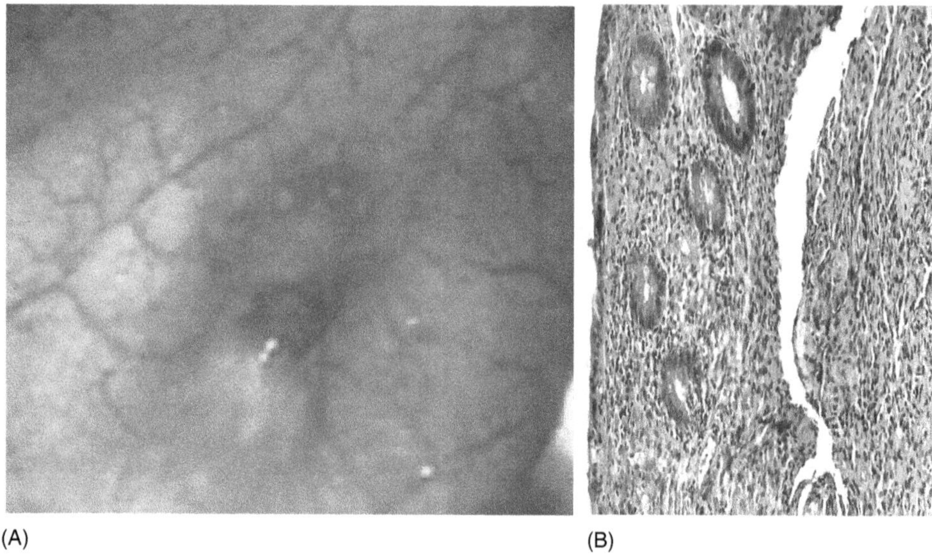

(A) (B)

FIGURE 30.2 (See color insert following page 558.) Colonic Kaposi's sarcoma seen (A) on endoscopy. (Reproduced from Dezube, B.J., Pantanowitz, L., and Aboulafia, D.M., *AIDS Read* 14(5), 237, 2004. With permission.) (B) On mucosal biopsy.

involvement. Visceral KS in a subset of patients may occur without evidence of mucocutaneous disease. Autopsy series in AIDS patients have described KS in lymph nodes, liver, pancreas, heart, testes, and bone marrow.[15]

30.2.3 Exacerbation

There are several clinical circumstances that may result in the exacerbation of KS including corticosteroid therapy and opportunistic infections.[14,16,17] In addition, AIDS-related KS may dramatically flare following the initiation of effective HAART, and may represent a protean manifestation of the immune reconstitution syndrome.[18]

30.2.4 Regression

Regression has been observed with transplant-related KS following the withdrawal of immunosuppressive agents, spontaneously in 2%–10% of classic KS cases and only rarely in AIDS-associated KS.[19] More commonly, KS regression occurs in HIV-positive individuals following effective antiretroviral use or chemotherapy (Figure 30.3). Interestingly, KS lesions in transplant recipients often reappear at the same sites as previously healed lesions when immunosuppressive therapy is reintroduced.[20] Regression is characterized histologically (Figure 30.4) by a complete or partial loss of spindle cells, increased perivascular lymphocytes, and dermal siderophage deposition. Occasionally, regression may be associated with dense fibrosis.[21]

30.2.5 Staging

Clinical staging of KS disease is predictive of survival and frequently performed to facilitate therapeutic decisions. A proposed staging system for classic KS is comprised of four stages:

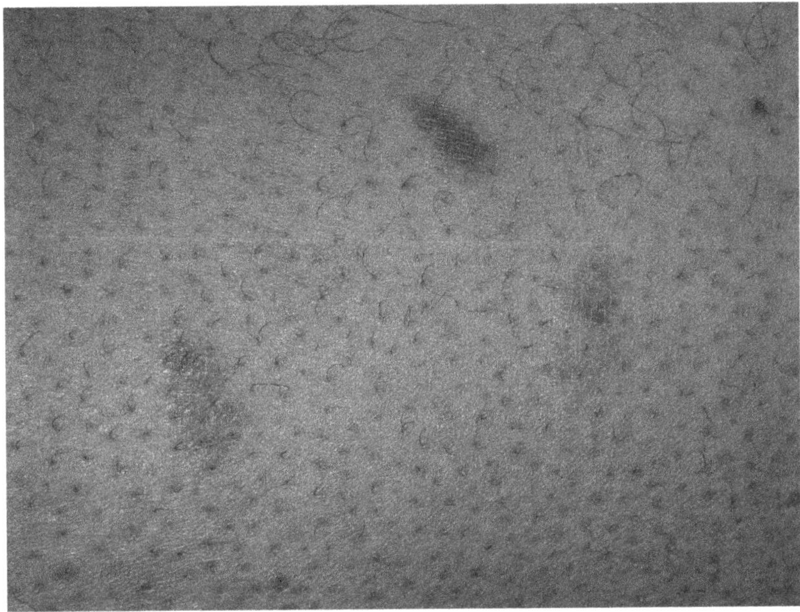

FIGURE 30.3 (See color insert following page 558.) Regressed pigmented flat Kaposi's sarcoma lesions following systemic therapy.

(i) maculonodular stage with lesions localized to the lower extremity; (ii) infiltrative stage where KS involves wide areas on the lower limbs; (iii) florid stage for exuberant or ulcerated lesions involving one or more limbs; and (iv) disseminated stage where KS extends to cutaneous sites in addition to the limbs.[22] The most utilized staging system for AIDS-related KS is the AIDS Clinical Trials Group (ACTG) classification system, which divides patients into good and poor risk groups based on their tumor burden (T), immune function (I) as measured by $CD4^+$ T-lymphocyte count, and the presence of systemic illness (S).[23] A more recent prospective evaluation of this staging system conducted in the HAART era showed that only the combination of poor tumor state and poor systemic disease state adequately identified patients with an unfavorable prognosis.[24]

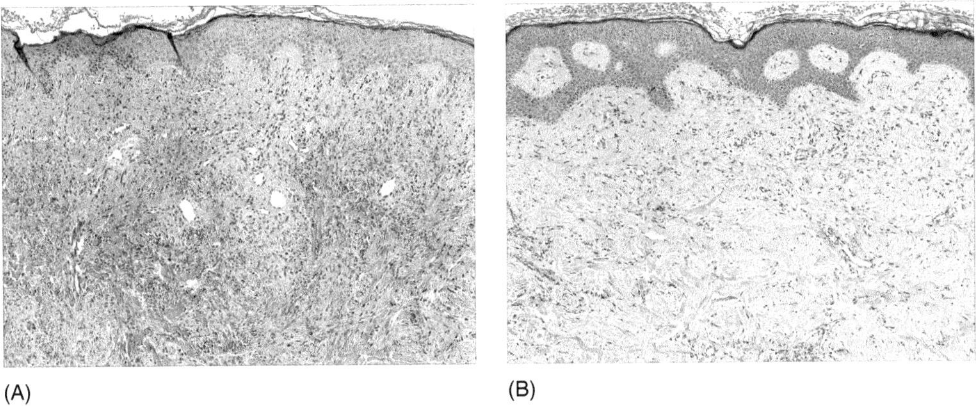

FIGURE 30.4 (See color insert following page 558.) (A) Plaque stage cutaneous lesion (B) shown to have undergone near complete regression following effective chemotherapy.

30.3 HISTOGENESIS

30.3.1 Histologic Appearance

All cases of KS show similar histological features regardless of the clinical subgroup. Gene expression profiling has shown KS lesions to consist of a mixture of aberrant endothelial cells and an inflammatory infiltrate.[25] Early lesions (patch stage) evolve into more advanced lesions (plaque stage) that may eventually become tumors (nodular stage).[26] In some cases, a mixed pattern may be seen. Distinct stages are best characterized in the skin.[27,28] In the *patch stage* (Figure 30.5), small irregular anastomosing lymphatic-like vascular channels lined by a single layer of endothelial cells (Figure 30.6) proliferate in the reticular dermis (forming angiomatoid and glomeruloid structures), as well as around preexisting blood vessels and adnexal structures (so-called promontory sign). Only during this early stage is there a discontinuous basement membrane lining vascular channels,[29,30] similar to normal lymphatic capillaries. There is often associated red blood cell extravasation and a variable lymphoplasmacytic infiltrate. Very early in situ KS resembles early patch stage lesions.[31] In the *plaque stage* (Figure 30.7), lesions expand and there is a more pronounced spindle cell component. Erythrophagocytosis is a characteristic finding, and hemosiderin deposition becomes more prominent. In the *nodular stage* (Figure 30.8), monomorphic spindle cells form tumors comprised of intersecting fascicles. KS lesional cells only show mild cytologic atypia. Slit-like (if sectioned tangentially, see Figure 30.8B) or sieve-like (if viewed in cross section, see Figure 30.8C) vascular spaces are present between spindle cells and contain abundant erythrocytes. Unlike in vivo KS lesions, vasoformative KS cells in culture may vary from spindle to epithelioid morphology.[32] Antigen-presenting dendrocytes (Factor XIIIa-positive), which belong to the reticuloendothelial system, are increased within and around KS lesions and likely contribute to neoangiogenesis.[33] Dendrocyte density is decreased in cases of immunosuppression-associated KS.[34] Unusual histologic variants include lymphangiomatous, keloidal, pleomorphic, and anaplastic (controversial) KS.[35]

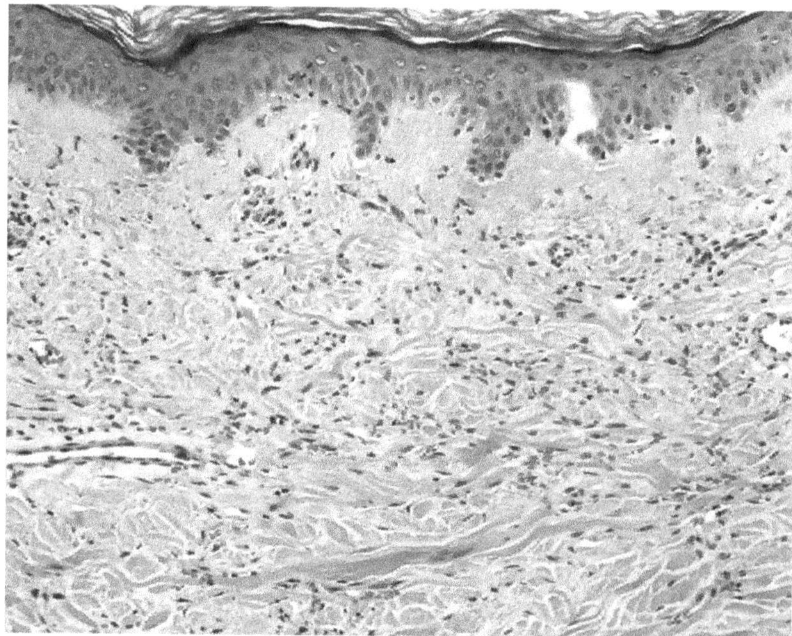

FIGURE 30.5 (See color insert following page 558.) Patch stage cutaneous Kaposi's sarcoma lesion showing dissecting vascular spaces throughout the superficial dermis.

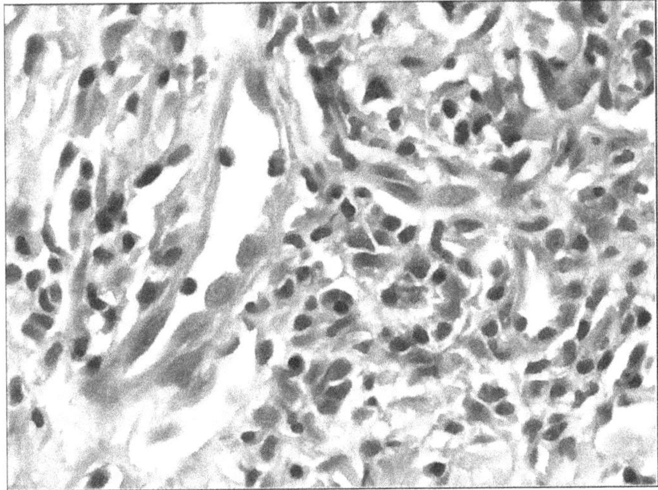

FIGURE 30.6 (See color insert following page 558.) Neoangiogenesis in an early KS lesion. Abnormal vessels are lined by plump endothelial-like cells.

30.3.2 Cellular Origin

For many years, the histogenesis of KS lesional cells was disputed. Data based on ultrastructural studies, enzyme histochemistry, immunohistochemistry, and more recently molecular experiments, implied that KS tumor cells may be derived from endothelium, dendritic cells, neural cells, macrophages, fibroblasts, or smooth muscle cells. Controversy in clarifying the origin of KS cells may be attributed to the various biomarkers, tissue processing and techniques employed, as well as the cellular heterogeneity of KS lesions.[36,37] Such heterogeneity may explain why, unlike their phenotype in vivo, culture-derived KS cells preferentially give rise to mesenchymal cells that lack vascular markers.[38,39] In vivo, KS tumor cells have been shown to express a plethora of vascular markers (Figure 30.9) including CD31 (also known as platelet–endothelial cell adhesion molecule-1 [PECAM-1]), the human progenitor cell antigen CD34, Factor VIII-related antigen (von Willebrand factor), the actin-bundling motility-associated protein fascin, and less frequently anti-HLA-DR/Ia, VE-cadherin, *Ulex*

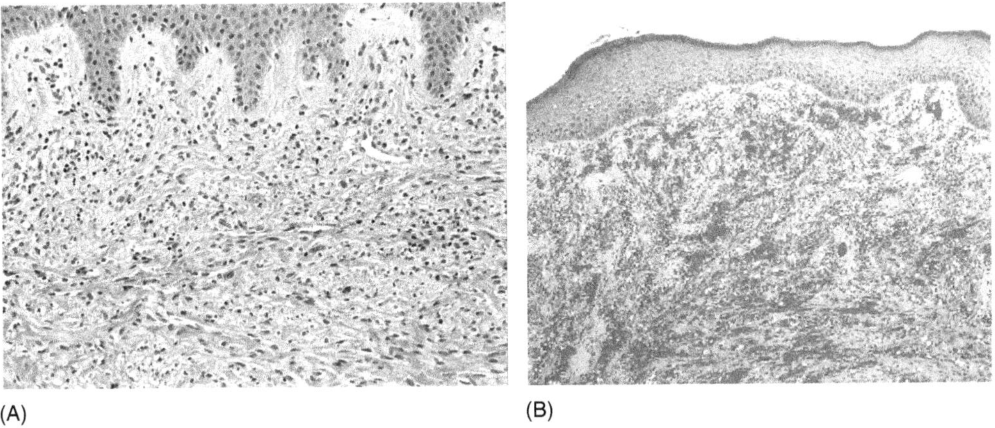

FIGURE 30.7 (See color insert following page 558.) Plaque stage (A) cutaneous and (B) gingival Kaposi's sarcoma lesions showing infiltrating abnormal vessels and a conspicuous spindle cell component.

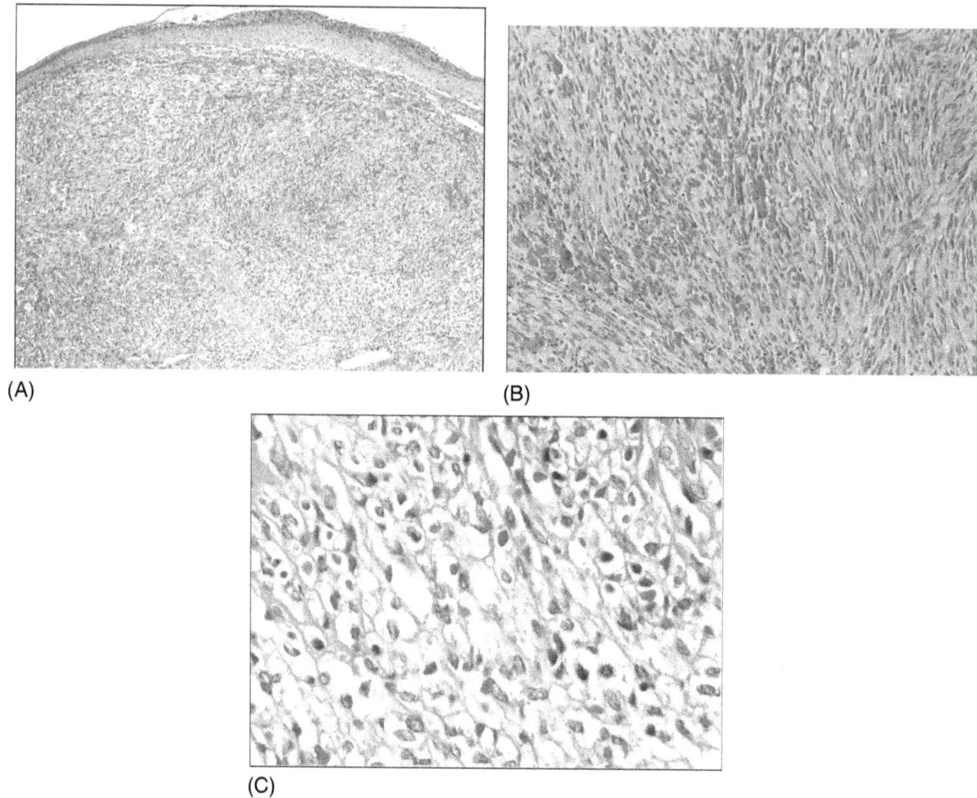

FIGURE 30.8 (See color insert following page 558.) (A) Nodular cutaneous Kaposi's sarcoma lesion (low-power magnification), (B) at high-power magnification shown to be characterized by intersecting fascicles of spindle cells with intervening blood-filled spaces. (C) Cross section demonstrates spindle cells with a sieve-like appearance.

europaeus agglutinin I (UEA-I) lectin, B721, E431, and EN-4.[32,40–47] KS cells may also coexpress macrophage markers (e.g., CD68, PAM-1, CD14, and mannose receptor), similar to the sinus-lining macrophages found in the spleen and lymph nodes.[47,48]

KS lesional cells have a lymphatic phenotype.[49] This conclusion is supported by the unique distribution of KS lesions in anatomic sites rich in lymphatics (skin, gastrointestinal tract, and lymph nodes). KS involvement of the thoracic duct with subsequent chylothorax formation has been documented.[31,50] Interestingly, most transgenic mice that express the HHV8 latent-cycle gene k-cyclin (*kCYC*) and vascular endothelial growth factor receptor (VEGF-R3) die within 6 months as a consequence of the progressive accumulation of chylous pleural fluid.[51] In addition, cutaneous lesions often follow the course of draining lymphatics,[52] and pulmonary KS tends to develop along lymphatics. Ultrastructurally, KS spindle cells show some morphological traits resembling lymphatics (e.g., scant basal lamina and absent Weibel-Palade bodies).[53,54] Observations utilizing enzyme histochemistry provided some of the initial evidence favoring the derivation of KS spindle cells from lymphatic endothelium.[55,56] More recently, the lymphatic origin of KS was confirmed by immunohistochemistry using the monoclonal antibody D2-40 (Figure 30.10), which is directed against a fixation-resistant sialoglycoprotein epitope present on lymphatic endothelium.[57–59] In cutaneous KS, D2-40 positively stains lesional cells at all stages. The expression of other lymphatic endothelial markers including podoplanin, as well as VEGFC and its receptors VEGF-R2 and VEGF-R3, in KS cells provides further lineage evidence.[60–63] Moreover, gene expression profiling demonstrates a resemblance between KS and lymphatic endothelial

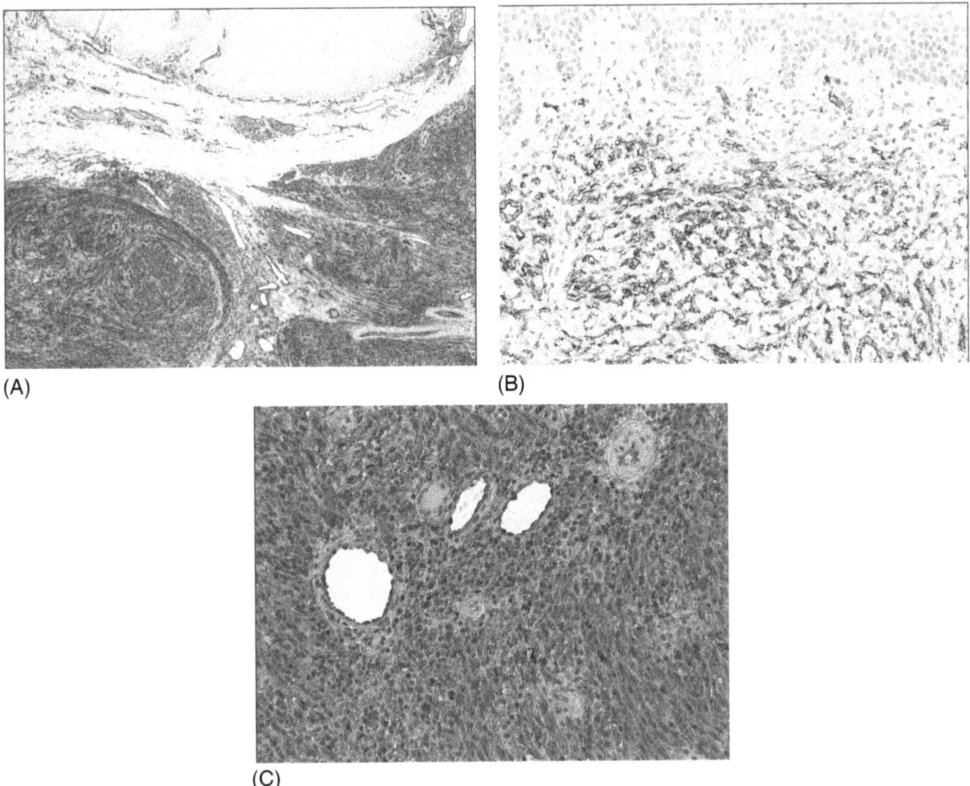

FIGURE 30.9 (See color insert following page 558.) Kaposi's sarcoma tumor cells demonstrate strong immunoreactivity for the vascular markers (A) CD31, (B) CD34, and (C) fascin.

cells.[64] Following HHV8 infection (see Section 30.4) of endothelial cells, the expression of several lymphatic lineage–specific genes, including LYVE-1, PROX-1, reelin, follistatin, desmoplakin, and leptin receptor is significantly increased.[65–67] LYVE-1, a member of

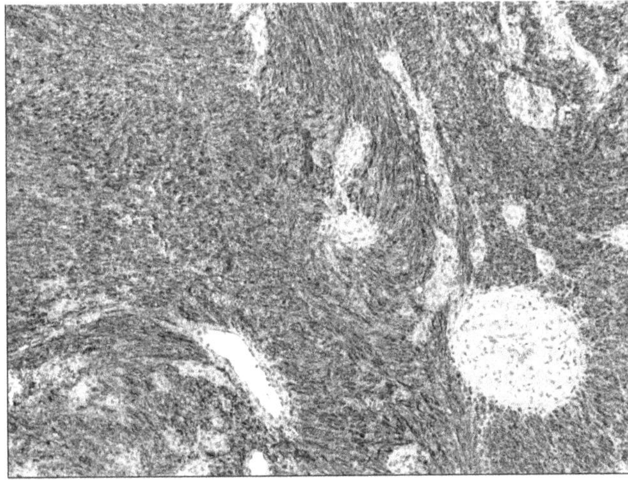

FIGURE 30.10 (See color insert following page 558.) Kaposi's sarcoma lesional cells are strongly immunoreactive for the lymphatic endothelial marker D2-40. Low-power magnification showing lack of immunostaining in endothelial blood vessels.

the Link protein family and a homolog of the CD44 glycoprotein, is the lymphatic receptor for the extracellular matrix mucopolysaccharide hyaluronan.[68] Within KS lesions, LYVE-1 stains areas that fail to stain with CD31, CD34, and Factor VIII-related antigen.[67] PROX-1 is a homeobox transcription factor that regulates lymphatic vessel development and differentiation. Overexpression of PROX-1 in differentiated blood vascular endothelial cells is capable of not only inducing lymphatic endothelial cell–specific gene transcription, but also the simultaneous downregulation of many blood vascular endothelial cell–specific genes.[66,69] VEGF-R3, LYVE-1, and PROX-1 are essential for the embryonic development of the mammalian lymphatic system.[70,71] Therefore, it is plausible that the expression of these genes in KS tumor cells may represent reiteration of an embryonic expression pattern.

30.4 ETIOLOGY

The search for a transmissible infectious agent as the cause of KS led to the discovery in 1994 of the etiologic agent KSHV, also known as HHV8.[72] We now know that HHV8, a γ-herpesvirus, is associated with all forms of KS,[73] and that HHV8 superinfection is possible.[74] Although necessary for the development of KS, the presence of HHV8 is not sufficient to cause KS. HHV8 viremia has been shown to serve as an early marker of KS, and the risk of developing KS increases with increasing HHV8 serum antibody titers.[75,76] A study from southern Africa showed that the level of HHV8 viremia was significantly lower in patients with true endemic KS, than in those with AIDS who had similar tumor burdens.[1] This supports the premise that HIV infection augments HHV8 replication.

The oncogenic role of HHV8 in KS development is associated with the expression of both latent (e.g., latent nuclear antigen or LANA) and lytic viral genes (e.g., viral G-protein-coupled receptor [vGPCR]), several of which are homologous to human interleukins (e.g., IL-6), as well as chemokines of the macrophage inflammatory family, cell cycle regulators of the cyclin family, and antiapoptosis markers of the bcl-2 family.[77] In KS lesions, HHV8 gene products are expressed in spindle-shaped lesional cells, monocytes, and possibly in some infiltrating lymphocytes. The spindle cells primarily harbor HHV8 as circular episomes and synthesize latency-associated gene products,[78] whereas the smaller monocyte subpopulation is infected with HHV8 during the lytic or replicative period of the viral life cycle.[79] For its infectious entry into cells, HHV8 uses clathrin-mediated endocytosis and a low-pH intracellular environment,[80] and for initial binding to target cells, it uses ubiquitous heparan sulfate molecules via its envelope-associated glycoproteins.[81]

Immunohistochemistry is often used to localize HHV8 in KS lesions. The monoclonal antibody most often used for immunohistochemical work is directed against the latency-associated nuclear antigen (LANA-1) encoded by ORF-73. Given the strong association between HHV8 and KS, positive immunoreactivity for LANA-1 (Figure 30.11) has proved to be a useful diagnostic tool to help differentiate KS from its mimics.[82,83] However, the presence of HHV8 appears not to be fully restricted to KS, as HHV8 has rarely been detected in other vascular lesions (e.g., angiosarcoma) and in several non-KS lesions within HIV-positive patients.[84–87] In the latter cases, the detection of HHV8 may simply reflect blood-borne viremia. Although HHV8 is capable of infecting both lymphatic and blood vascular endothelial cells in vitro; the former cells are more competent hosts.[64]

In addition to HHV8 infection, the development of KS lesions is associated with activation of a Th-1 response of CD8 T cells leading to production of inflammatory cytokines (ICs). These ICs include γ-IFN, interleukin-1 (IL-1), interleukin-6 (IL-6), and tumor necrosis factor-α (TNF-α), which recruit lymphocytes and macrophages that produce more of these ICs.[88,89] This may help explain why in early stage KS lesions, a leukocyte infiltrate precedes the appearance of spindle-shaped cells and the formation of prominent microvasculature. Early stage KS lesions appear to be polyclonal and, typically in

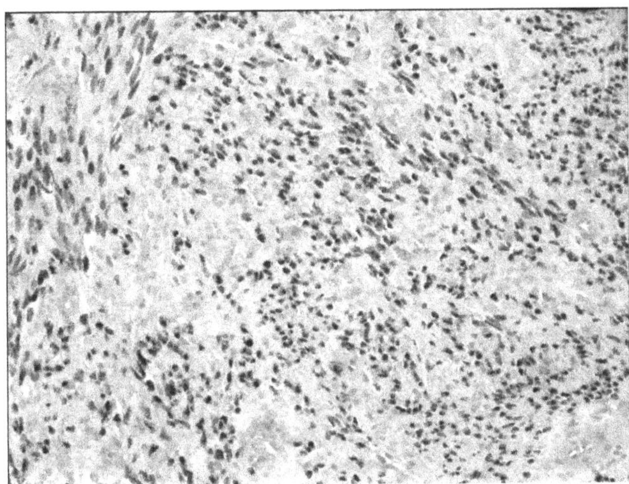

FIGURE 30.11 (See color insert following page 558.) Kaposi's sarcoma lesional cells demonstrate strong immunoreactivity for LANA-1.

non-AIDS-associated KS, may regress without any therapy. In these early KS lesions, only 10% of lesional spindle cells express LANA.[90] The ICs activate latently infected HHV8 spindle cells, as well as the infiltrating inflammatory cells, to produce a number of autocrine and paracrine growth factors that lead to the further development of the KS lesion. It is unclear if HHV8-infected cells precipitate the initial IC burst or are just beneficiaries. In late stage KS, the KS spindle cells appear to be clonal and >90 of these cells express LANA.[90] In these advanced lesions, the spindle cells have evidence of microsatellite instability.[91] Additionally, the KS spindle cells from these lesions have been shown to express mutated p53 and ras.[92–94] These observations suggest a model of KS progression that begins as a cytokine-driven reactive process, and then as mutations are acquired, culminates in the autonomous growth of the lesion. These data support the view that KS is indeed a true sarcoma.

30.5 CYTOKINES/CHEMOKINES IN KS-INDUCED ANGIOGENESIS

Cytokines and chemokines play a critical role in inducing the angiogenic milieu of the KS lesion. Endogenous cytokines/chemokines as well as HHV8 homologs of cytokines/chemokines are angiogenic themselves or stimulate angiogenic growth factors.

30.5.1 Interleukin-6/Viral Interleukin-6/Glycoprotein-130 Receptor

IL-6 is an IC which plays an important role in immunity (reviewed in Refs. 95,96). IL-6 is induced in HHV8-infected cells by the early lytic genes Rta and the vGPCR.[97–99] Rta is an HHV8 transcriptional activator that regulates lytic gene expression, and vGPCR is a homolog of the IL-8 receptor (discussed later).[100] The K2 open reading frame (ORF) is a lytic *HHV8* gene that produces a viral homolog of interleukin-6 (vIL-6). Both IL-6 and vIL-6 signal through the glycoprotein-130 receptor (GP130).[101–103]

GP130 is a shared subunit of the receptor complex for IL-6, leukemia inhibitory factor, ciliary neurotrophic factor, cardiotrophin-1, cardiotrophin-like cytokine, oncostatin-M, interleukin-11, and interleukin-27 (reviewed in Ref. 104). The IL-6 receptor is composed of an α-chain (CD126) in addition to GP130.[104] The α-chain does not contain a cytoplasmic portion. However, both subunits are required for IL-6 activation of downstream-signaling events. Jak1, Jak2, and Tyk2 are members of the JAK family tyrosine kinases and constitutively associate with GP130.[105,106] Binding of IL-6 to its receptor results in the activation of JAK kinases and phosphorylation of GP130.[105,106] Activation of GP130 in turn results in the phosphorylation and subsequent activation of STAT1 and STAT3, which are members of the signal transducer and activator of transcription (STAT) family. STATs play a critical role in cytokine signaling as well as in transformation (reviewed in Ref. 107). GP130 activation of STAT3 induces cell cycle progression via induction of cyclins D2, D3, A, and cdc25 with concurrent downregulation of the cell cycle inhibitors p27 and p21.[108] GP130 activates the phosphatidylinositol 3-kinase (PI3K)/AKT pathway as well as members of the mitogen-activated protein kinase (MAPK) family, which includes p38, ERK, and JNK. The PI3K and MAPK pathways promote growth and antiapoptotic pathways.[109–112] GP130 also regulates cell growth through the mammalian target of rapamycin (mTor) pathway.[113] The PI3K/AKT/mTor pathway also appears to regulate angiogenesis through regulation of the transcription factor known as hypoxia inducible factor-1α (HIF1α).[114,115] Like IL-6, vIL-6 activates the JAK/STAT, MAPK, and PI3K/AKT/mTor pathways through GP130. However, vIL-6 is unique in that it can activate GP130 without interacting with any other receptor subunits.[103] Since GP130 has a broad expression pattern, vIL-6 affects many more tissue types than IL-6.

IL-6 and vIL-6 have been shown to induce VEGF expression through MAPK and JAK/STAT dependent pathways.[116–118] Human umbilical vein endothelial cells (HUVECs) in a matrigel assay have been shown to proliferate in response to conditioned media from NIH3T3 cells stably expressing vIL-6, and this proliferation was blocked by anti-VEGF antibody.[119] When these stably transfected cells were implanted in athymic mice, the vIL-6-expressing tumors grew more rapidly and were more vascular than control tumors. The vIL-6-expressing tumors were also found to have higher expression of VEGF than the control tumors.[119] These data support a critical role for vIL-6 and IL-6 in HHV8-induced angiogenesis, growth, and proliferation.

30.5.2 Interleukin-1

Interleukin-1 (IL-1) is an IC, which is not only important in HHV8 activation but also in an autocrine growth factor for KS spindle cells. IL-1β is upregulated in KS spindle cells compared with those cell lining vessels, and when IL-1β binding is blocked, the growth of KS spindle cells is inhibited.[120] IL-1β induces fibroblast growth factor 2 (FGF-2 or basic FGF), a potent angiogenesis growth factor (discussed later).[121] Additionally, in conjunction with platelet-derived growth factor B (PDGFB), IL-1β increases VEGF expression in KS spindle cells.[122] Thus, IL-1β is important in KS spindle cell mitogenesis and KS-induced angiogenesis.

30.5.3 Viral Chemokines (vCCL1, vCCL2, vCCL3)

Chemokines are structurally related glycoproteins that induce leukocyte activation and chemotaxis. The four families of chemokines (C-C, C-X-C, C-X3-C, and C) are segregated by the amino acid sequence around conserved N-terminal cystines (reviewed in Ref. 123). HHV8 infection also produces viral homologs of macrophage inhibitory factors (MIPs), a member of the C-C chemokine family. The viral chemokines vCCL1 (vMIP-I or ORF K6), vCCL2 (vMIP-II or ORF K4), and vCCL3 (vMIP-III or ORF K4.1) are lytic genes, which are 25%–40% homologous to MIP and bind to endogenous chemokine receptors.[124–126] The evidence suggests that vCCLs not only modulate host immunity, but also play an important role in HHV8-induced angiogenesis.

Chemokines exert their effects through GPCRs. GPCRs are also known as serpentine receptors and associate with a G-protein complex that consists of an α, β, and γ subunit.

When activated, the GAP-bound α-subunit and the βγ-subunit activate other downstream effectors such as phospholipase C, MAPKs, and PI3K (reviewed in Ref. 127). vCCL1 appears to be a specific agonist of the CCR8 chemokine receptor.[128] By activating CCR8, vCCL1 promotes migration of monocytes, endothelial cells, and Th-2 lymphocytes.[128] CCR4 is activated by vCCL3 and also recruits Th-2 T cells. By recruiting Th-2 cells, vCCL2 and vCCL3 blunt the immune response to HHV8 infection. Although vCCL2 exhibits binding to a broad spectrum of chemokine receptors including CCR8, vCCL2 functions as an antagonist to a number of chemokines as measured by its ability to block the induction of Ca^{2+} flux by their endogenous ligands of these receptors.[129] It has been proposed that vCCL2 restricts the immune response to HHV8 infection by blocking this broad spectrum of receptors.

In addition to their effects on the immune system, all three vCCLs have been shown to be angiogenic in the chick chorioallantoic membrane assay.[125,130] VEGFA and VEGFB are induced by vCCL1.[118] The mechanism by which vCCL2 and vCCL3 induce angiogenesis remains to be clarified. Additionally, both vCCL1 and vCCL2 have been shown to have antiapoptotic effects on HHV8-infected lymphocytes.[118] Thus, the vCCLs acting in concert manipulate immune and angiogenic pathways to facilitate a microenvironment that allows HHV8-infected cells to survive.

30.5.4 Viral G-Protein-Coupled Receptor

In addition to chemokine homologs, ORF74 of HHV8 codes for a lytic gene product that is a GCPR, which is homologous to CXCR1 and CXCR2.[131,132] The vGPCR has a mutation (Asp142Val) that renders it constitutively active.[133] It has been shown to bind to multiple chemokines that modulate its activity with some functioning as agonists and others functioning as antagonists.[134,135] In a retroviral infection model, vGPCR was the only HHV8 ORF that was able to induce HHV8 lesions in mice.[136] The vGPCR has been shown to activate the PI3K/AKT/mTor and MAPK pathways which would provide a survival advantage to cells.[137] The vGPCR also activates HIF1α, a critical transcriptional regulator of angiogenesis.[138]

HIF1α is one of three HIFα isoforms. HIFα is a hypoxia-inducible transcription factor that binds to HIFβ which is constitutively expressed.[139,140] HIF1α expression is regulated posttranscriptionally by oxygen-dependent proteasomal degradation.[141–149] Additionally, HIF1α activity is also modulated by the MAPK and PI3K/AKT/mTOR pathways.[114,115,150] The HIF heterodimer binds to DNA to induce transcription of a broad spectrum of gene products, which are critical for angiogenesis, cell survival, and glycolysis (reviewed in Ref. 151).

The activation of these signaling pathways by vGPCR also leads to the expression of a number of angiogenic factors. In primary effusion lymphoma cells, vGPCR induces the angiogenic factors VEGF and vIL-6.[152] In endothelial cells, vGPCR induces expression of IL-6, IL-8, Groα, VEGFA, and VEGF-R2.[153,154] It is unknown if vGPCR mediates the expression of any of the other known HIF targets that are expressed in KS such as PDGFB, TGFα, Ang2, and VEGF-R1. One model for the role of vGPCR and the vCCLs is that they induce an angiogenic and immunosuppressed microenvironment through autocrine and paracrine mechanisms that are permissive for other HHV8-infected cells to survive and enter the lytic phase.

30.6 ANGIOGENIC GROWTH FACTORS IN KS DEVELOPMENT

30.6.1 Vascular Endothelial Growth Factor

A number of the effects of the viral cytokines and vGPCR are mediated through VEGF or VEGFA. VEGF is a member of a gene family that includes placental growth factor (PLGF),

VEGFB, VEGFC, and VEGFD. HIF1α is a key regulator of VEGF expression. VEGF-R1 (Flt1), VEGF-R2 (KDR), and VEGF-R3 (Flt4) are type III receptor tyrosine kinases that are activated differentially by the VEGF family members. VEGF-R1 can be activated by VEGFA, VEGFB, or PLGF; VEGF-R2 can be activated by VEGFA, VEGFC, or VEGFD; and VEGF-R3 binds to VEGFC and VEGFD. VEGF-R1 and VEGF-R2 are expressed on vascular endothelial cells and hematopoietic cells. VEGF-R2 is the major mediator of angiogenic and mitotic signaling by VEGF, whereas VEGF-R1 appears to be important in mediating monocyte chemotaxis and VEGF-R3 is restricted to lymphatic endothelial cells in adults but is unregulated in neoangiogenesis in tumors as well as KS lesions (reviewed in Ref. 155).

As previously noted, the angiogenic effects of some viral cytokines and the vGPCR are, either in part or in whole, mediated through the HIF1α-mediated expression of VEGFA. VEGFC and its receptors VEGF-R2 and VEGF-R3 are expressed in KS lesions.[60,63,156] However, which pathways lead to their expression is unknown. Recent genomic array studies have confirmed this finding and also demonstrated that the PLGF is upregulated in KS lesions.[64] The evidence supports a central role for members of the VEGF/VEGF-R family in the development of KS.

30.6.2 Fibroblast Growth Factors

The FGF family is composed of 22 members. FGFs regulate a myriad of effects in endothelial cells including mitogenesis, chemotaxis, and angiogenesis (reviewed in Ref. 157). The FGFs mediate their effects by activating one or more of the FGF receptors. Binding of FGFs to their receptors is modulated by heparin proteoglycans.[158] The four FGF receptors (FGF-R1, FGF-R2, FGF-R3, FGF-R4) are tyrosine kinases. The effects of these receptors are mediated by multiple downstream-signaling pathways including the MAPK and PI3K/AKT/mTor pathways.[159,160] FGFs implicated in angiogenesis appear to act through multiple mechanisms. FGF-1 (acidic FGF) and FGF-2 (basic FGF) act directly on the endothelial cells to induce migration, proliferation, and tube formation, whereas other family members such as FGF-4 (kaposi-FGF) can induce the expression of VEGF.

FGF-2 was one of the first autocrine growth factors described for HHV8.[161] FGF-2 expression is upregulated in KS spindle cells compared with vascular endothelial cells.[161] In KS spindle cell cultures, blocking the binding of FGF-2 results in decreased cellular proliferation.[161] FGF-2 synergizes with VEGF to induce angiogenesis in KS lesions.[162] Thus, FGF-2 is important in KS spindle cell mitogenesis and KS-induced angiogenesis. FGF-4 was originally isolated from KS cells and has been shown to induce expression of VEGF. HIV-1-transactivating protein Tat induces FGF-1 and FGF-2.[163,164]

30.6.3 Platelet-Derived Growth Factor

The platelet-derived growth factors exist as either homo- or heterodimers of the A, B, C, and D isoforms. These dimers bind to the PDGF receptor (PDGF-R), which belong to the type III receptor tyrosine kinase family. PDGF-R exists as α- and β-isoforms, which can form homo- or heterodimers (reviewed in Ref. 165). Activated PDGF-R simulates multiple signaling pathways through downstream effectors including the PI3K/AKT/mTOR and MAPK pathways.[166–169] Activation of the PDGF-R by autocrine/paracrine mechanisms, activating mutations, or by chromosomal translocation has been implicated in the pathogenesis of multiple malignancies.

The PDGF pathway also plays an important role in angiogenesis. Pericytes, which stabilize normal and tumor vasculature, have been shown to require PDGFB and PDGF-Rβ. Recent data suggest that pericytes also stabilize tumor vasculature. PDGF-Rβ, staining tumor capillary pericytes, has been reported in lung, colon, breast, and prostate carcinomas. Murine models have demonstrated decreased pericyte recruitment to tumor vasculature in mice with a PDGF-Rβ mutant that cannot activate the PI3K/AKT pathway when compared with the wild-type receptor.

These data would suggest that PDGF is important for tumor angiogenesis as well as normal angiogenesis.

Several studies have addressed the role of PDGF-R in KS. When KS spindle cells are cultured, they express both α- and β-PDGF receptors.[170] Cultured KS spindle cells undergo growth arrest when placed in PDGF-depleted media, and this arrest can be mitigated by the addition of recombinant PDGF.[171] Whether cultured cells reflect KS spindle cell behavior in vivo is unclear. Moreover, the source of the PDGF in vivo remains unanswered. Indeed, only PDGF-Rβ appears to be expressed in KS specimens as assayed by in situ hybridization and immunohistochemistry.[172] These data have been confirmed by recent genomic profiling studies, which demonstrated that PDGFB and PDGF-Rβ are upregulated in HHV8-infected cells and KS lesions. Additionally, PDGFB in conjunction with IL1-β induces the expression of VEGF by cultured KS spindle cells.[122] PDGF-R is involved in two critical pathways in KS development: the induction of (i) growth of KS spindle cells and (ii) angiogenesis.

30.6.4 Angiopoietins

The angiopoietin family of vascular growth factors includes four members, angiopoietin-1 through angiopoietin-4.[173–176] Tie-2 is a receptor tyrosine kinase that binds all four angiopoietins.[173–176] Tie-1 is an orphan receptor for which the ligand is unknown. The expression of both Tie receptors is restricted to vascular endothelial cells. Mice deficient in either receptor demonstrate abnormalities of vascular development. Tie-2 was the first receptor to be shown to have a natural antagonist, angiopoietin-2 (Ang2), as well as an agonist, angiopoietin-1 (Ang1).[175,176] The Tie-2/Ang1 pathway acts in conjunction with VEGF to promote vascular stabilization.[177] Ang1 is widely expressed in human endothelial cells suggesting that it has a role in maintaining vascular endothelial cell stability. Ang2 appears to be upregulated at the sites of vascular remodeling. In the presence of high VEGF, Ang2 antagonism of Tie-2 leads to vessel sprouting; but in the absence of VEGF, Ang2 expression leads to vascular regression.[178,179] These data would suggest that Ang2 disruption of Ang1-mediated endothelial cell stability might play a role in neoangiogenesis seen in tumors.

KS has been shown to express Tie-1, Tie-2, Ang1, Ang2, and Ang4.[180] Ang2 was expressed to a greater level than Ang1 in KS lesions.[180] The role of angiopoietins in the development of KS and how their expression is mediated in KS lesions are unclear. However, if one extrapolates from other systems, Ang2 likely facilitates VEGF-mediated vascular remodeling, and Ang2 expression is likely mediated by HIF1α. These hypotheses need to be confirmed in KS.

30.6.5 Matrix Metalloproteinases

Although not direct angiogenic growth factors, the matrix metalloproteinases (MMPs) play a potential role in KS-related angiogenesis.[181] The MMPs are a family of endopeptidases involved in the destruction of extracellular matrix proteins. By destroying the underlying extracellular matrix, MMPs facilitate angiogenesis. Naturally occurring tissue inhibitors of MMPs block tumor cell invasion and angiogenesis.[182–184] The supernatant containing MMPs obtained from AIDS-related KS have been demonstrated in vitro to induce normal endothelial cells to invade through the reconstituted basement membrane matrigel, which correlates with the angiogenic potential of KS cells in vivo.[185] KS cells in vitro and in vivo have been shown to constitutively overexpress MMP-1, MMP-2, MMP-3, and MMP-9 (Figure 30.12).[186–189] Increased levels of MMP-2 have been found in plasma from patients with AIDS-KS, compared

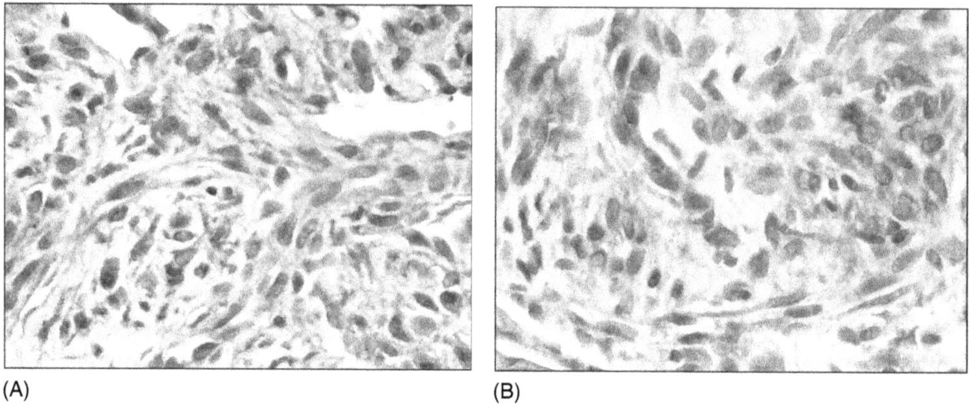

FIGURE 30.12 (See color insert following page 558.) Kaposi's sarcoma tumor cells demonstrate (A) MMP-1 and (B) MMP-2 immunoreactivity.

with HIV-uninfected individuals with classic KS.[190] MMP-2 has also been shown in vivo to be strongly expressed in KS lesions, but not in uninvolved skin from the same individuals. KS spindle-shaped tumor cells as well as admixed chronic inflammatory cells are capable of producing MMPs. Therefore, it is not surprising that some authors have noted significant MMP expression in infiltrating leukocytes and macrophages in KS lesions.[191]

30.7 ANTIANGIOGENIC CANCER THERAPY IN ACTION: CLINICAL TRAILS

Given the highly significant role that angiogenesis plays in the pathogenesis of KS, it is not surprising that many of the angiogenesis inhibitors in development have been or are currently tested in patients with AIDS-related KS. Examples of some of the potential targets evaluated in pathogenesis-based trials are MMPs, PDGF-R, VEGF, FGF, and in cases of AIDS-related KS, the HIV-1 transactivating protein Tat.

30.7.1 ANTIRETROVIRALS

Most, if not all, AIDS-related KS patients should receive HAART that will maximally decrease HIV-1 viral load.[14] The dramatic impact of HAART on KS is underscored by a large Swiss cohort study of HIV-infected patients.[4] The relative risk (hazard risk) of KS development between 1997 and 1998 (after the introduction of HAART) compared with 1992 and 1994 (before the introduction of HAART) was 0.08, representing a dramatic reduction in KS. Although HAART may not be considered classic antiangiogenic therapy, HAART indirectly inhibits angiogenesis by inhibiting the HIV-1 Tat protein, which synergizes with FGF-1 to induce VEGF expression, and by ameliorating the immune response against HHV8, which induces lymphatic endothelial cell–specific gene transcription. In addition, direct antiangiogenic activity has been demonstrated by protease inhibitors.[192,193]

30.7.2 MATRIX METALLOPROTEINASE INHIBITORS

COL-3 (6-demethyl-6-deoxy-4-dedimethylaminotetracycline, Collagenex Pharmaceuticals, Newtown, PA) is a chemically modified tetracycline,[194] which inhibits the activity of activated neutrophil gelatinase and the expression of MMPs in human cancer cell lines, the invasion of such lines into matrigel, and the invasion of human melanoma cell lines through basement membrane matrix.

In phase I and II trials of COL-3 in patients with AIDS-related KS conducted by the AIDS Malignancy Consortium (AMC) of the National Cancer Institute (NCI), overall response rates (complete and partial responses) of 44% were noted with a median response duration of >25 weeks.[195,196] The most common adverse event was photosensitivity. Of note, there was a significant difference between responders and nonresponders with respect to change in MMP-2 plasma levels from baseline to minimum value on treatment.

Halofuginone (Collgard Pharmaceuticals, Israel) is a small molecule that inhibits several essential stages of angiogenesis—endothelial cell proliferation, MMP-2 expression, basement membrane invasion, collagen type I synthesis during angiogenic sprouting, and FGF-2-induced neovascularization.[197] Halofuginone is in an early exploratory trial in AIDS-related KS patients, which is conducted by the AMC/NCI and is translational in nature—biologic end points will be correlated with tumor response.

30.7.3 Tyrosine Kinase Inhibitors

PDGF-R is involved in two critical pathways in KS development: (i) induction of KS spindle cell growth and (ii) induction of angiogenesis through VEGF. Imatinib mesylate (Gleevec, Novartis Pharmaceuticals, East Hanover, NJ) selectively inhibits the tyrosine kinases c-kit, PDGF-R, and Abl. In an exploratory study, imatinib was administered to patients with AIDS-related KS.[198] Clinical and histologic regression of cutaneous KS lesions occurred within 4 weeks of therapy. The most common adverse event noted was diarrhea. Tumor biopsies, obtained at baseline and following 4 weeks of therapy, demonstrated a decrease of phosphorylation (activation) of the PDGF and MAPK pathways. Only minimal or no activation of AKT was noted in pretreatment tumor specimens. This study suggests that the MAPK pathway, in contrast to the PI3K/AKT pathway, may be critical for proliferation of KS lesions. This is in contrast to other cell systems in which signaling from PDGF-R is often transmitted through AKT.[199]

30.7.4 Anti-VEGF Therapies

The remarkable finding that vIL-6, vCCL, and vGPCR induce expression of VEGF provides a missing link to the chain of events by which HHV8 creates an inflammatory–angiogenic environment.[119] Veglin (VasGene Therapeutics, Los Angeles, CA) is an antisense oligonucleotide to the *VEGF* gene and is studied in early exploratory trials in AIDS-related KS patients, which are conducted by the AMC/NCI. Although, to our knowledge, the FDA-approved humanized monoclonal antibody to VEGF, bevacizumab (Avastin, Genentech, South San Francisco, CA), has not been reported in AIDS-related KS, we would expect it to be efficacious in this setting.

30.7.5 Interferon-α and Thalidomide

Both thalidomide (Celgene, Warren, NJ) and interferon-α (Roche, Nutley, NJ; Schering, Kenilworth, NJ) are approved agents with a broad spectrum of biologic and antiangiogenic activities. Both agents have demonstrated efficacy in AIDS-related KS. In an ACTG AIDS-related KS study, low dose (1 million units daily) and moderate dose (10 million units daily) interferon-α produced similar response rates when each was combined with didanosine.[200] However, interferon therapy is often associated with significant systemic toxicity including fever, chills, neutropenia, nephrotoxicity, and cognitive impairment. Poor tumor response and drug-related toxicity are particularly notable in advanced AIDS patients (those with a $CD4^+$ T-lymphocyte count below 200 cells/μL). In an NCI AIDS-related KS trial, thalidomide, given at doses ranging from 200 to 1000 mg/day, produced tumor regression.[201]

Thalidomide activity, however, was offset in part by its side effects (drowsiness, depression, and neuropathy). Although circulating cytokines and growth factors, including VEGF, IL-6, FGF-2 (b-FGF), and TNF-α, were assessed, these did not prove to be useful biologic correlates of clinical activity.

30.7.6 Other Potential Agents and Targets

As our knowledge of the signaling pathways pirated by HHV8 has increased, so have the opportunities to design therapeutic trials that target these pathways. Many of the investigational signal transduction inhibitors currently in clinical trials, such as sorafenib (Bay 43-9006) or Sutent (SU11248), inhibit multiple tyrosine kinases including VEGF-R2 and PDGF-R. These agents would be expected to demonstrate activity in KS. Additionally, pharmacologic inhibitors of specific MMPs or of the FGF pathway would similarly be expected to be active in this disease.

Since a number of the angiogenic pathways, which appear to be activated in KS, use common intracellular signaling, these pathways may also represent therapeutic targets. The PI3K/AKT pathway has been shown to regulate mTOR activity through the tuberous sclerosis (*TSC 1/2*) gene complex.[111] This complex appears to regulate VEGF expression through mTOR-dependent and -independent mechanisms.[114,115] Whether this pathway is important for KS-mediated angiogenesis is unknown; however, the recent report of transplant-related KS responding to rapamycin suggests that mTOR may play a critical role in KS development.[202] If these results were reproducible in other forms of KS, it would suggest that inhibition of other parts of the PI3K/AKT/mTor pathway is a valid therapeutic target. Alternatively, inhibition of the MAPK and JAK/STAT pathways may be an additional target given their roles in KS cytokine signaling.

REFERENCES

1. Campbell, T.B., Borok, M., White, I.E., Gudza, I., Ndemera, B., Taziwa, A., Weinberg, A., and Gwanzura, L., Relationship of Kaposi sarcoma (KS)-associated herpesvirus viremia and KS disease in Zimbabwe, *Clin Infect Dis* 36 (9), 1144–1151, 2003.
2. Dourmishev, L.A., Dourmishev, A.L., Palmeri, D., Schwartz, R.A., and Lukac, D.M., Molecular genetics of Kaposi's sarcoma-associated herpesvirus (human herpesvirus-8) epidemiology and pathogenesis, *Microbiol Mol Biol Rev* 67 (2), 175–212, table of contents, 2003.
3. Hymes, K.B., Cheung, T., Greene, J.B., Prose, N.S., Marcus, A., Ballard, H., William, D.C., and Laubenstein, L.J., Kaposi's sarcoma in homosexual men—a report of eight cases, *Lancet* 2 (8247), 598–600, 1981.
4. Ledergerber, B., Telenti, A., and Egger, M., Risk of HIV related Kaposi's sarcoma and non-Hodgkin's lymphoma with potent antiretroviral therapy: prospective cohort study. Swiss HIV Cohort Study, *BMJ* 319 (7201), 23–24, 1999.
5. Von Roenn, J.H., Clinical presentations and standard therapy of AIDS-associated Kaposi's sarcoma, *Hematol Oncol Clin North Am* 17 (3), 747–762, 2003.
6. Marcelin, A.G., Roque-Afonso, A.M., Hurtova, M., Dupin, N., Tulliez, M., Sebagh, M., Arkoub, Z.A., Guettier, C., Samuel, D., Calvez, V., and Dussaix, E., Fatal disseminated Kaposi's sarcoma following human herpesvirus 8 primary infections in liver-transplant recipients, *Liver Transpl* 10 (2), 295–300, 2004.
7. De Pasquale, R., Nasca, M.R., and Micali, G., Postirradiation primary Kaposi's sarcoma of the head and neck, *J Am Acad Dermatol* 40 (2 Pt 2), 312–314, 1999.
8. Pedulla, F., Sisteron, O., Chevallier, P., Piche, T., Saint-Paul, M.C., and Bruneton, J.N., Kaposi's sarcoma confined to the colorectum: a case report, *Clin Imaging* 28 (1), 33–35, 2004.
9. Schwartz, R.A., Kaposi's sarcoma: an update, *J Surg Oncol* 87 (3), 146–151, 2004.

10. Witte, M.H., Fiala, M., McNeill, G.C., Witte, C.L., Williams, W.H., and Szabo, J., Lymphangioscintigraphy in AIDS-associated Kaposi sarcoma, *AJR Am J Roentgenol* 155 (2), 311–315, 1990.
11. Ruocco, V., Schwartz, R.A., and Ruocco, E., Lymphedema: an immunologically vulnerable site for development of neoplasms, *J Am Acad Dermatol* 47 (1), 124–127, 2002.
12. Schwartz, R.A., Cohen, J.B., Watson, R.A., Gascon, P., Ahkami, R.N., Ruszczak, Z., Halpern, J., and Lambert, W.C., Penile Kaposi's sarcoma preceded by chronic penile lymphoedema, *Br J Dermatol* 142 (1), 153–156, 2000.
13. Simonart, T., Dobbeleer, G.D., Peny, M., Fayt, I., Parent, D., Vooren, J., and Noel, J., Pre-Kaposi's sarcoma: an expansion of the spectrum of Kaposi's sarcoma lesions, *Eur J Dermatol* 9 (6), 480–482, 1999.
14. Dezube, B.J., Pantanowitz, L., and Aboulafia, D.M., Management of AIDS-related Kaposi sarcoma: advances in target discovery and treatment, *AIDS Read* 14 (5), 236–238, 243–244, 251–253, 2004.
15. Ioachim, H.L., Adsay, V., Giancotti, F.R., Dorsett, B., and Melamed, J., Kaposi's sarcoma of internal organs. A multiparameter study of 86 cases, *Cancer* 75 (6), 1376–1385, 1995.
16. Gill, P.S., Loureiro, C., Bernstein-Singer, M., Rarick, M.U., Sattler, F., and Levine, A.M., Clinical effect of glucocorticoids on Kaposi sarcoma related to the acquired immunodeficiency syndrome (AIDS), *Ann Intern Med* 110 (11), 937–940, 1989.
17. Trattner, A., Hodak, E., David, M., and Sandbank, M., The appearance of Kaposi sarcoma during corticosteroid therapy, *Cancer* 72 (5), 1779–1783, 1993.
18. Aboulafia, D.M., Kaposi sarcoma flares during effective antiretroviral treatment, *AIDS Read* 15 (4), 190–191, 2005.
19. Pantanowitz, L., Dezube, B.J., Pinkus, G.S., and Tahan, S.R., Histological characterization of regression in acquired immunodeficiency syndrome-related Kaposi's sarcoma, *J Cutan Pathol* 31 (1), 26–34, 2004.
20. al-Sulaiman, M.H., Mousa, D.H., Dhar, J.M., and al-Khader, A.A., Does regressed posttransplantation Kaposi's sarcoma recur following reintroduction of immunosuppression? *Am J Nephrol* 12 (5), 384–386, 1992.
21. Eng, W. and Cockerell, C.J., Histological features of Kaposi sarcoma in a patient receiving highly active antiviral therapy, *Am J Dermatopathol* 26 (2), 127–132, 2004.
22. Brambilla, L., Boneschi, V., Taglioni, M., and Ferrucci, S., Staging of classic Kaposi's sarcoma: a useful tool for therapeutic choices, *Eur J Dermatol* 13 (1), 83–86, 2003.
23. Krown, S.E., Metroka, C., and Wernz, J.C., Kaposi's sarcoma in the acquired immune deficiency syndrome: a proposal for uniform evaluation, response, and staging criteria. AIDS Clinical Trials Group Oncology Committee, *J Clin Oncol* 7 (9), 1201–1207, 1989.
24. Nasti, G., Talamini, R., Antinori, A., Martellotta, F., Jacchetti, G., Chiodo, F., Ballardini, G., Stoppini, L., Di Perri, G., Mena, M., Tavio, M., Vaccher, E., D'Arminio Monforte, A., and Tirelli, U., AIDS-related Kaposi's sarcoma: evaluation of potential new prognostic factors and assessment of the AIDS Clinical Trial Group Staging System in the Haart Era—the Italian Cooperative Group on AIDS and Tumors and the Italian Cohort of Patients Naive from Antiretrovirals, *J Clin Oncol* 21 (15), 2876–2882, 2003.
25. Cornelissen, M., van der Kuyl, A.C., van den Burg, R., Zorgdrager, F., van Noesel, C.J., and Goudsmit, J., Gene expression profile of AIDS-related Kaposi's sarcoma, *BMC Cancer* 3 (1), 7, 2003.
26. Facchetti, F., Lucini, L., Gavazzoni, R., and Callea, F., Immunomorphological analysis of the role of blood vessel endothelium in the morphogenesis of cutaneous Kaposi's sarcoma: a study of 57 cases, *Histopathology* 12 (6), 581–593, 1988.
27. Francis, N.D., Parkin, J.M., Weber, J., and Boylston, A.W., Kaposi's sarcoma in acquired immune deficiency syndrome (AIDS), *J Clin Pathol* 39 (5), 469–474, 1986.
28. Gottlieb, G.J. and Ackerman, A.B., Kaposi's sarcoma: an extensively disseminated form in young homosexual men, *Hum Pathol* 13 (10), 882–892, 1982.
29. Dictor, M. and Andersson, C., Lymphaticovenous differentiation in Kaposi's sarcoma. Cellular phenotypes by stage, *Am J Pathol* 130 (2), 411–417, 1988.
30. Green, T.L., Meyer, J.R., Daniels, T.E., Greenspan, J.S., de Souza, Y., and Kramer, R.H., Kaposi's sarcoma in AIDS: basement membrane and endothelial cell markers in late-stage lesions, *J Oral Pathol* 17 (6), 266–272, 1988.

31. Pantanowitz, L., Tahan, S.R., Pinkus, G.S., and Dezube, B.J., Morphologic and immunophenotypic evidence of in-situ Kaposi's sarcoma (abstract), *Arch Pathol Lab Med* 129, 556, 2005.
32. Roth, W.K., Brandstetter, H., and Sturzl, M., Cellular and molecular features of HIV-associated Kaposi's sarcoma, *AIDS* 6 (9), 895–913, 1992.
33. Nickoloff, B.J. and Griffiths, C.E., The spindle-shaped cells in cutaneous Kaposi's sarcoma. Histologic simulators include factor XIIIa dermal dendrocytes, *Am J Pathol* 135 (5), 793–800, 1989.
34. Kanitakis, J. and Roca-Miralles, M., Factor-XIIIa-expressing dermal dendrocytes in Kaposi's sarcoma. A comparison between classical and immunosuppression-associated types, *Virchows Arch A Pathol Anat Histopathol* 420 (3), 227–231, 1992.
35. Chor, P.J. and Santa Cruz, D.J., Kaposi's sarcoma. A clinicopathologic review and differential diagnosis, *J Cutan Pathol* 19 (1), 6–20, 1992.
36. Kaaya, E.E., Parravicini, C., Ordonez, C., Gendelman, R., Berti, E., Gallo, R.C., and Biberfeld, P., Heterogeneity of spindle cells in Kaposi's sarcoma: comparison of cells in lesions and in culture, *J Acquir Immune Defic Syndr Hum Retrovirol* 10 (3), 295–305, 1995.
37. Regezi, J.A., MacPhail, L.A., Daniels, T.E., DeSouza, Y.G., Greenspan, J.S., and Greenspan, D., Human immunodeficiency virus-associated oral Kaposi's sarcoma. A heterogeneous cell population dominated by spindle-shaped endothelial cells, *Am J Pathol* 143 (1), 240–249, 1993.
38. Huang, Y.Q., Friedman-Kien, A.E., Li, J.J., and Nickoloff, B.J., Cultured Kaposi's sarcoma cell lines express factor XIIIa, CD14, and VCAM-1, but not factor VIII or ELAM-1, *Arch Dermatol* 129 (10), 1291–1296, 1993.
39. Simonart, T., Degraef, C., Heenen, M., Hermans, P., Van Vooren, J.P., and Noel, J.C., Expression of the fibroblast/macrophage marker 1B10 by spindle cells in Kaposi's sarcoma lesions and by Kaposi's sarcoma-derived tumor cells, *J Cutan Pathol* 29 (2), 72–78, 2002.
40. Pantanowitz, L., Dezube, B.J., and Pinkus, G.S., Dendritic cells in the pathogenesis of Kaposi's sarcoma (abstract), *Blood* 102, 281–282a, 2003.
41. Corbeil, J., Evans, L.A., Vasak, E., Cooper, D.A., and Penny, R., Culture and properties of cells derived from Kaposi sarcoma, *J Immunol* 146 (9), 2972–2976, 1991.
42. Nadimi, H., Saatee, S., Armin, A., and Toto, P.D., Expression of endothelial cell markers PAL-E and EN-4 and Ia-antigens in Kaposi's sarcoma, *J Oral Pathol* 17 (8), 416–420, 1988.
43. Nickoloff, B.J., The human progenitor cell antigen (CD34) is localized on endothelial cells, dermal dendritic cells, and perifollicular cells in formalin-fixed normal skin, and on proliferating endothelial cells and stromal spindle-shaped cells in Kaposi's sarcoma, *Arch Dermatol* 127 (4), 523–529, 1991.
44. Nickoloff, B.J., PECAM-1 (CD31) is expressed on proliferating endothelial cells, stromal spindle-shaped cells, and dermal dendrocytes in Kaposi's sarcoma, *Arch Dermatol* 129 (2), 250–251, 1993.
45. Rutgers, J.L., Wieczorek, R., Bonetti, F., Kaplan, K.L., Posnett, D.N., Friedman-Kien, A.E., and Knowles, D.M., 2nd, The expression of endothelial cell surface antigens by AIDS-associated Kaposi's sarcoma. Evidence for a vascular endothelial cell origin, *Am J Pathol* 122 (3), 493–499, 1986.
46. Scully, P.A., Steinman, H.K., Kennedy, C., Trueblood, K., Frisman, D.M., and Voland, J.R., AIDS-related Kaposi's sarcoma displays differential expression of endothelial surface antigens, *Am J Pathol* 130 (2), 244–251, 1988.
47. Uccini, S., Ruco, L.P., Monardo, F., Stoppacciaro, A., Dejana, E., La Parola, I.L., Cerimele, D., and Baroni, C.D., Co-expression of endothelial cell and macrophage antigens in Kaposi's sarcoma cells, *J Pathol* 173 (1), 23–31, 1994.
48. Uccini, S., Sirianni, M.C., Vincenzi, L., Topino, S., Stoppacciaro, A., Lesnoni La Parola, I., Capuano, M., Masini, C., Cerimele, D., Cella, M., Lanzavecchia, A., Allavena, P., Mantovani, A., Baroni, C.D., and Ruco, L.P., Kaposi's sarcoma cells express the macrophage-associated antigen mannose receptor and develop in peripheral blood cultures of Kaposi's sarcoma patients, *Am J Pathol* 150 (3), 929–938, 1997.
49. Cheung, L. and Rockson, S.G., The lymphatic biology of Kaposi's sarcoma, *Lymphat Res Biol* 3 (1), 25–35, 2005.
50. Marais, B.J., Pienaar, J., and Gie, R.P., Kaposi sarcoma with upper airway obstruction and bilateral chylothoraces, *Pediatr Infect Dis J* 22 (10), 926–928, 2003.
51. Sugaya, M., Watanabe, T., Yang, A., Starost, M.F., Kobayashi, H., Atkins, A.M., Borris, D.L., Hanan, E.A., Schimel, D., Bryant, M.A., Roberts, N., Skobe, M., Staskus, K.A., Kaldis, P., and

Blauvelt, A., Lymphatic dysfunction in transgenic mice expressing KSHV k-cyclin under the control of the VEGFR-3 promoter, *Blood* 105 (6), 2356–2363, 2005.

52. Witte, M.H., Stuntz, M., and Witte, C.L., Kaposi's sarcoma. A lymphologic perspective, *Int J Dermatol* 28 (9), 561–570, 1989.
53. Dictor, M., Carlen, B., Bendsoe, N., and Flamholc, L., Ultrastructural development of Kaposi's sarcoma in relation to the dermal microvasculature, *Virchows Arch A Pathol Anat Histopathol* 419 (1), 35–43, 1991.
54. McNutt, N.S., Fletcher, V., and Conant, M.A., Early lesions of Kaposi's sarcoma in homosexual men. An ultrastructural comparison with other vascular proliferations in skin, *Am J Pathol* 111 (1), 62–77, 1983.
55. Beckstead, J.H., Wood, G.S., and Fletcher, V., Evidence for the origin of Kaposi's sarcoma from lymphatic endothelium, *Am J Pathol* 119 (2), 294–300, 1985.
56. Dorfman, R.F., Kaposi's sarcoma. The contribution of enzyme histochemistry to the identification of cell types, *Acta Unio Int Contra Cancrum* 18, 464–476, 1962.
57. Pantanowitz, L., Tahan, S.R., Dezube, B.J., and Pinkus, G.S., Dendritic, endothelial and lymphatic cell markers in Kaposi's sarcoma (abstract), *Mod Pathol* 17, 97A, 2004.
58. Fukunaga, M., Expression of D2-40 in lymphatic endothelium of normal tissues and in vascular tumours, *Histopathology* 46 (4), 396–402, 2005.
59. Kahn, H.J., Bailey, D., and Marks, A., Monoclonal antibody D2-40, a new marker of lymphatic endothelium, reacts with Kaposi's sarcoma and a subset of angiosarcomas, *Mod Pathol* 15 (4), 434–440, 2002.
60. Folpe, A.L., Veikkola, T., Valtola, R., and Weiss, S.W., Vascular endothelial growth factor receptor-3 (VEGFR-3): a marker of vascular tumors with presumed lymphatic differentiation, including Kaposi's sarcoma, kaposiform and Dabska-type hemangioendotheliomas, and a subset of angiosarcomas, *Mod Pathol* 13 (2), 180–185, 2000.
61. Jussila, L., Valtola, R., Partanen, T.A., Salven, P., Heikkila, P., Matikainen, M.T., Renkonen, R., Kaipainen, A., Detmar, M., Tschachler, E., Alitalo, R., and Alitalo, K., Lymphatic endothelium and Kaposi's sarcoma spindle cells detected by antibodies against the vascular endothelial growth factor receptor-3, *Cancer Res* 58 (8), 1599–1604, 1998.
62. Weninger, W., Partanen, T.A., Breiteneder-Geleff, S., Mayer, C., Kowalski, H., Mildner, M., Pammer, J., Sturzl, M., Kerjaschki, D., Alitalo, K., and Tschachler, E., Expression of vascular endothelial growth factor receptor-3 and podoplanin suggests a lymphatic endothelial cell origin of Kaposi's sarcoma tumor cells, *Lab Invest* 79 (2), 243–251, 1999.
63. Skobe, M., Brown, L.F., Tognazzi, K., Ganju, R.K., Dezube, B.J., Alitalo, K., and Detmar, M., Vascular endothelial growth factor-C (VEGF-C) and its receptors KDR and flt-4 are expressed in AIDS-associated Kaposi's sarcoma, *J Invest Dermatol* 113 (6), 1047–1053, 1999.
64. Wang, H.W., Trotter, M.W., Lagos, D., Bourboulia, D., Henderson, S., Makinen, T., Elliman, S., Flanagan, A.M., Alitalo, K., and Boshoff, C., Kaposi sarcoma herpesvirus-induced cellular reprogramming contributes to the lymphatic endothelial gene expression in Kaposi sarcoma, *Nat Genet* 36 (7), 687–693, 2004.
65. Carroll, P.A., Brazeau, E., and Lagunoff, M., Kaposi's sarcoma-associated herpesvirus infection of blood endothelial cells induces lymphatic differentiation, *Virology* 328 (1), 7–18, 2004.
66. Hong, Y.K., Foreman, K., Shin, J.W., Hirakawa, S., Curry, C.L., Sage, D.R., Libermann, T., Dezube, B.J., Fingeroth, J.D., and Detmar, M., Lymphatic reprogramming of blood vascular endothelium by Kaposi sarcoma-associated herpesvirus, *Nat Genet* 36 (7), 683–685, 2004.
67. Xu, H., Edwards, J.R., Espinosa, O., Banerji, S., Jackson, D.G., and Athanasou, N.A., Expression of a lymphatic endothelial cell marker in benign and malignant vascular tumors, *Hum Pathol* 35 (7), 857–861, 2004.
68. Banerji, S., Ni, J., Wang, S.X., Clasper, S., Su, J., Tammi, R., Jones, M., and Jackson, D.G., LYVE-1, a new homologue of the CD44 glycoprotein, is a lymph-specific receptor for hyaluronan, *J Cell Biol* 144 (4), 789–801, 1999.
69. Petrova, T.V., Makinen, T., Makela, T.P., Saarela, J., Virtanen, I., Ferrell, R.E., Finegold, D.N., Kerjaschki, D., Yla-Herttuala, S., and Alitalo, K., Lymphatic endothelial reprogramming of vascular endothelial cells by the Prox-1 homeobox transcription factor, *EMBO J* 21 (17), 4593–4599, 2002.

70. Oliver, G. and Detmar, M., The rediscovery of the lymphatic system: old and new insights into the development and biological function of the lymphatic vasculature, *Genes Dev* 16 (7), 773–783, 2002.
71. Wigle, J.T., Harvey, N., Detmar, M., Lagutina, I., Grosveld, G., Gunn, M.D., Jackson, D.G., and Oliver, G., An essential role for Prox1 in the induction of the lymphatic endothelial cell phenotype, *EMBO J* 21 (7), 1505–1513, 2002.
72. Chang, Y., Cesarman, E., Pessin, M.S., Lee, F., Culpepper, J., Knowles, D.M., and Moore, P.S., Identification of herpesvirus-like DNA sequences in AIDS-associated Kaposi's sarcoma, *Science* 266 (5192), 1865–1869, 1994.
73. Schwartz, E.J., Dorfman, R.F., and Kohler, S., Human herpesvirus-8 latent nuclear antigen-1 expression in endemic Kaposi sarcoma: an immunohistochemical study of 16 cases, *Am J Surg Pathol* 27 (12), 1546–1550, 2003.
74. Beyari, M.M., Hodgson, T.A., Cook, R.D., Kondowe, W., Molyneux, E.M., Scully, C.M., Teo, C.G., and Porter, S.R., Multiple human herpesvirus-8 infection, *J Infect Dis* 188 (5), 678–689, 2003.
75. Engels, E.A., Biggar, R.J., Marshall, V.A., Walters, M.A., Gamache, C.J., Whitby, D., and Goedert, J.J., Detection and quantification of Kaposi's sarcoma-associated herpesvirus to predict AIDS-associated Kaposi's sarcoma, *AIDS* 17 (12), 1847–1851, 2003.
76. Newton, R., Ziegler, J., Bourboulia, D., Casabonne, D., Beral, V., Mbidde, E., Carpenter, L., Reeves, G., Parkin, D.M., Wabinga, H., Mbulaiteye, S., Jaffe, H., Weiss, R., and Boshoff, C., The sero-epidemiology of Kaposi's sarcoma-associated herpesvirus (KSHV/HHV-8) in adults with cancer in Uganda, *Int J Cancer* 103 (2), 226–232, 2003.
77. Verma, S.C. and Robertson, E.S., Molecular biology and pathogenesis of Kaposi sarcoma-associated herpesvirus, *FEMS Microbiol Lett* 222 (2), 155–163, 2003.
78. Dezube, B.J., Zambela, M., Sage, D.R., Wang, J.F., and Fingeroth, J.D., Characterization of Kaposi sarcoma-associated herpesvirus/human herpesvirus-8 infection of human vascular endothelial cells: early events, *Blood* 100 (3), 888–896, 2002.
79. Blasig, C., Zietz, C., Haar, B., Neipel, F., Esser, S., Brockmeyer, N.H., Tschachler, E., Colombini, S., Ensoli, B., and Sturzl, M., Monocytes in Kaposi's sarcoma lesions are productively infected by human herpesvirus 8, *J Virol* 71 (10), 7963–7968, 1997.
80. Akula, S.M., Naranatt, P.P., Walia, N.S., Wang, F.Z., Fegley, B., and Chandran, B., Kaposi's sarcoma-associated herpesvirus (human herpesvirus 8) infection of human fibroblast cells occurs through endocytosis, *J Virol* 77 (14), 7978–7990, 2003.
81. Akula, S.M., Pramod, N.P., Wang, F.Z., and Chandran, B., Human herpesvirus 8 envelope-associated glycoprotein B interacts with heparan sulfate-like moieties, *Virology* 284 (2), 235–249, 2001.
82. Audard, V., Lok, C., Trabattoni, M., Wechsler, J., Brousse, N., and Fraitag, S., [Misleading Kaposi's sarcoma: usefulness of anti HHV-8 immunostaining], *Ann Pathol* 23 (4), 345–348, 2003.
83. Hong, A., Davies, S., and Lee, C.S., Immunohistochemical detection of the human herpesvirus 8 (HHV8) latent nuclear antigen-1 in Kaposi's sarcoma, *Pathology* 35 (5), 448–450, 2003.
84. Pantanowitz, L., Pinkus, G.S., Dezube, B.J., and Tahan, S.R., HHV8 is not restricted to Kaposi's sarcoma, *Mod Pathol* 18 (8), 1148, 2005.
85. Hammock, L., Reisenauer, A., Wang, W., Cohen, C., Birdsong, G., and Folpe, A.L., Latency-associated nuclear antigen expression and human herpesvirus-8 polymerase chain reaction in the evaluation of Kaposi sarcoma and other vascular tumors in HIV-positive patients, *Mod Pathol* 18 (4), 463–468, 2005.
86. Kazakov, D.V., Schmid, M., Adams, V., Cathomas, G., Muller, B., Burg, G., and Kempf, W., HHV-8 DNA sequences in the peripheral blood and skin lesions of an HIV-negative patient with multiple eruptive dermatofibromas: implications for the detection of HHV-8 as a diagnostic marker for Kaposi's sarcoma, *Dermatology* 206 (3), 217–221, 2003.
87. McDonagh, D.P., Liu, J., Gaffey, M.J., Layfield, L.J., Azumi, N., and Traweek, S.T., Detection of Kaposi's sarcoma-associated herpesvirus-like DNA sequence in angiosarcoma, *Am J Pathol* 149 (4), 1363–1368, 1996.
88. Oxholm, A., Epidermal expression of interleukin-6 and tumour necrosis factor-alpha in normal and immunoinflammatory skin states in humans, *APMIS Suppl* 24, 1–32, 1992.
89. Fiorelli, V., Gendelman, R., Sirianni, M.C., Chang, H.K., Colombini, S., Markham, P.D., Monini, P., Sonnabend, J., Pintus, A., Gallo, R.C., and Ensoli, B., gamma-Interferon produced by $CD8^+$ T cells infiltrating Kaposi's sarcoma induces spindle cells with angiogenic phenotype and synergy

with human immunodeficiency virus-1 Tat protein: an immune response to human herpesvirus-8 infection? *Blood* 91 (3), 956–967, 1998.
90. Dupin, N., Fisher, C., Kellam, P., Ariad, S., Tulliez, M., Franck, N., van Marck, E., Salmon, D., Gorin, I., Escande, J.P., Weiss, R.A., Alitalo, K., and Boshoff, C., Distribution of human herpesvirus-8 latently infected cells in Kaposi's sarcoma, multicentric Castleman's disease, and primary effusion lymphoma, *Proc Natl Acad Sci USA* 96 (8), 4546–4551, 1999.
91. Bedi, G.C., Westra, W.H., Farzadegan, H., Pitha, P.M., and Sidransky, D., Microsatellite instability in primary neoplasms from HIV+ patients, *Nat Med* 1 (1), 65–68, 1995.
92. Noel, J.C., De Thier, F., Simonart, T., Andre, J., Hermans, P., Van Vooren, J.P., and Heenen, M., p53 Protein overexpression is a common but late event in the pathogenesis of iatrogenic and AIDS-related Kaposi's sarcoma, *Arch Dermatol Res* 289 (11), 660–661, 1997.
93. Scinicariello, F., Dolan, M.J., Nedelcu, I., Tyring, S.K., and Hilliard, J.K., Occurrence of human papillomavirus and p53 gene mutations in Kaposi's sarcoma, *Virology* 203 (1), 153–157, 1994.
94. Nicolaides, A., Huang, Y.Q., Li, J.J., Zhang, W.G., and Friedman-Kien, A.E., Gene amplification and multiple mutations of the K-ras oncogene in Kaposi's sarcoma, *Anticancer Res* 14 (3A), 921–926, 1994.
95. Ishihara, K. and Hirano, T., IL-6 in autoimmune disease and chronic inflammatory proliferative disease, *Cytokine Growth Factor Rev* 13 (4–5), 357–368, 2002.
96. Yoshizaki, K., Nishimoto, N., Matsumoto, K., Tagoh, H., Taga, T., Deguchi, Y., Kuritani, T., Hirano, T., Hashimoto, K., Okada, N., et al., Interleukin 6 and expression of its receptor on epidermal keratinocytes, *Cytokine* 2 (5), 381–387, 1990.
97. Pati, S., Cavrois, M., Guo, H.G., Foulke, J.S., Jr., Kim, J., Feldman, R.A., and Reitz, M., Activation of NF-kappaB by the human herpesvirus 8 chemokine receptor ORF74: evidence for a paracrine model of Kaposi's sarcoma pathogenesis, *J Virol* 75 (18), 8660–8673, 2001.
98. Schwarz, M. and Murphy, P.M., Kaposi's sarcoma-associated herpesvirus G protein-coupled receptor constitutively activates NF-kappa B and induces proinflammatory cytokine and chemokine production via a C-terminal signaling determinant, *J Immunol* 167 (1), 505–513, 2001.
99. Deng, H., Song, M.J., Chu, J.T., and Sun, R., Transcriptional regulation of the interleukin-6 gene of human herpesvirus 8 (Kaposi's sarcoma-associated herpesvirus), *J Virol* 76 (16), 8252–8264, 2002.
100. West, J.T. and Wood, C., The role of Kaposi's sarcoma-associated herpesvirus/human herpesvirus-8 regulator of transcription activation (RTA) in control of gene expression, *Oncogene* 22 (33), 5150–5163, 2003.
101. Nicholas, J., Ruvolo, V.R., Burns, W.H., Sandford, G., Wan, X., Ciufo, D., Hendrickson, S.B., Guo, H.G., Hayward, G.S., and Reitz, M.S., Kaposi's sarcoma-associated human herpesvirus-8 encodes homologues of macrophage inflammatory protein-1 and interleukin-6, *Nat Med* 3 (3), 287–292, 1997.
102. Neipel, F., Albrecht, J.C., Ensser, A., Huang, Y.Q., Li, J.J., Friedman-Kien, A.E., and Fleckenstein, B., Human herpesvirus 8 encodes a homolog of interleukin-6, *J Virol* 71 (1), 839–842, 1997.
103. Mullberg, J., Geib, T., Jostock, T., Hoischen, S.H., Vollmer, P., Voltz, N., Heinz, D., Galle, P.R., Klouche, M., and Rose-John, S., IL-6 receptor independent stimulation of human gp130 by viral IL-6, *J Immunol* 164 (9), 4672–4677, 2000.
104. Muller-Newen, G., The cytokine receptor gp130: faithfully promiscuous, *Sci STKE* 2003 (201), PE40, 2003.
105. Stahl, N., Boulton, T.G., Farruggella, T., Ip, N.Y., Davis, S., Witthuhn, B.A., Quelle, F.W., Silvennoinen, O., Barbieri, G., Pellegrini, S., et al., Association and activation of Jak-Tyk kinases by CNTF-LIF-OSM-IL-6 beta receptor components, *Science* 263 (5143), 92–95, 1994.
106. Lutticken, C., Wegenka, U.M., Yuan, J., Buschmann, J., Schindler, C., Ziemiecki, A., Harpur, A.G., Wilks, A.F., Yasukawa, K., Taga, T., et al., Association of transcription factor APRF and protein kinase Jak1 with the interleukin-6 signal transducer gp130, *Science* 263 (5143), 89–92, 1994.
107. Bromberg, J. and Darnell, J.E., Jr., The role of STATs in transcriptional control and their impact on cellular function, *Oncogene* 19 (21), 2468–2473, 2000.
108. Fukada, T., Ohtani, T., Yoshida, Y., Shirogane, T., Nishida, K., Nakajima, K., Hibi, M., and Hirano, T., STAT3 orchestrates contradictory signals in cytokine-induced G1 to S cell-cycle transition, *EMBO J* 17 (22), 6670–6677, 1998.

109. Tee, A.R., Fingar, D.C., Manning, B.D., Kwiatkowski, D.J., Cantley, L.C., and Blenis, J., Tuberous sclerosis complex-1 and -2 gene products function together to inhibit mammalian target of rapamycin (mTOR)-mediated downstream signaling, *Proc Natl Acad Sci USA* 99 (21), 13571–13576, 2002.
110. Yao, R. and Cooper, G.M., Requirement for phosphatidylinositol-3 kinase in the prevention of apoptosis by nerve growth factor, *Science* 267 (5206), 2003–2006, 1995.
111. Manning, B.D., Tee, A.R., Logsdon, M.N., Blenis, J., and Cantley, L.C., Identification of the tuberous sclerosis complex-2 tumor suppressor gene product tuberin as a target of the phosphoinositide 3-kinase/akt pathway, *Mol Cell* 10 (1), 151–162, 2002.
112. Fukada, T., Hibi, M., Yamanaka, Y., Takahashi-Tezuka, M., Fujitani, Y., Yamaguchi, T., Nakajima, K., and Hirano, T., Two signals are necessary for cell proliferation induced by a cytokine receptor gp130: involvement of STAT3 in anti-apoptosis, *Immunity* 5 (5), 449–460, 1996.
113. Oh, H., Fujio, Y., Kunisada, K., Hirota, H., Matsui, H., Kishimoto, T., and Yamauchi-Takihara, K., Activation of phosphatidylinositol 3-kinase through glycoprotein 130 induces protein kinase B and p70 S6 kinase phosphorylation in cardiac myocytes, *J Biol Chem* 273 (16), 9703–9710, 1998.
114. Brugarolas, J.B., Vazquez, F., Reddy, A., Sellers, W.R., and Kaelin, W.G., Jr., TSC2 regulates VEGF through mTOR-dependent and -independent pathways, *Cancer Cell* 4 (2), 147–158, 2003.
115. Brugarolas, J., Lei, K., Hurley, R.L., Manning, B.D., Reiling, J.H., Hafen, E., Witters, L.A., Ellisen, L.W., and Kaelin, W.G., Jr., Regulation of mTOR function in response to hypoxia by REDD1 and the TSC1/TSC2 tumor suppressor complex, *Genes Dev* 18 (23), 2893–2904, 2004.
116. Funamoto, M., Fujio, Y., Kunisada, K., Negoro, S., Tone, E., Osugi, T., Hirota, H., Izumi, M., Yoshizaki, K., Walsh, K., Kishimoto, T., and Yamauchi-Takihara, K., Signal transducer and activator of transcription 3 is required for glycoprotein 130-mediated induction of vascular endothelial growth factor in cardiac myocytes, *J Biol Chem* 275 (14), 10561–10566, 2000.
117. Naruishi, K., Nishimura, F., Yamada-Naruishi, H., Omori, K., Yamaguchi, M., and Takashiba, S., C-jun N-terminal kinase (JNK) inhibitor, SP600125, blocks interleukin (IL)-6-induced vascular endothelial growth factor (VEGF) production: cyclosporine A partially mimics this inhibitory effect, *Transplantation* 76 (9), 1380–1382, 2003.
118. Liu, C., Okruzhnov, Y., Li, H., and Nicholas, J., Human herpesvirus 8 (HHV-8)-encoded cytokines induce expression of and autocrine signaling by vascular endothelial growth factor (VEGF) in HHV-8-infected primary-effusion lymphoma cell lines and mediate VEGF-independent anti-apoptotic effects, *J Virol* 75 (22), 10933–10940, 2001.
119. Aoki, Y., Jaffe, E.S., Chang, Y., Jones, K., Teruya-Feldstein, J., Moore, P.S., and Tosato, G., Angiogenesis and hematopoiesis induced by Kaposi's sarcoma-associated herpesvirus-encoded interleukin-6, *Blood* 93 (12), 4034–4043, 1999.
120. Ensoli, B., Nakamura, S., Salahuddin, S.Z., Biberfeld, P., Larsson, L., Beaver, B., Wong-Staal, F., and Gallo, R.C., AIDS-Kaposi's sarcoma-derived cells express cytokines with autocrine and paracrine growth effects, *Science* 243 (4888), 223–226, 1989.
121. Louie, S., Cai, J., Law, R., Lin, G., Lunardi-Iskandar, Y., Jung, B., Masood, R., and Gill, P., Effects of interleukin-1 and interleukin-1 receptor antagonist in AIDS-Kaposi's sarcoma, *J Acquir Immune Defic Syndr Hum Retrovirol* 8 (5), 455–460, 1995.
122. Cornali, E., Zietz, C., Benelli, R., Weninger, W., Masiello, L., Breier, G., Tschachler, E., Albini, A., and Sturzl, M., Vascular endothelial growth factor regulates angiogenesis and vascular permeability in Kaposi's sarcoma, *Am J Pathol* 149 (6), 1851–1869, 1996.
123. Homey, B., Muller, A., and Zlotnik, A., Chemokines: agents for the immunotherapy of cancer? *Nat Rev Immunol* 2 (3), 175–184, 2002.
124. Moore, P.S., Boshoff, C., Weiss, R.A., and Chang, Y., Molecular mimicry of human cytokine and cytokine response pathway genes by KSHV, *Science* 274 (5293), 1739–1744, 1996.
125. Stine, J.T., Wood, C., Hill, M., Epp, A., Raport, C.J., Schweickart, V.L., Endo, Y., Sasaki, T., Simmons, G., Boshoff, C., Clapham, P., Chang, Y., Moore, P., Gray, P.W., and Chantry, D., KSHV-encoded CC chemokine vMIP-III is a CCR4 agonist, stimulates angiogenesis, and selectively chemoattracts TH2 cells, *Blood* 95 (4), 1151–1157, 2000.
126. Sun, R., Lin, S.F., Staskus, K., Gradoville, L., Grogan, E., Haase, A., and Miller, G., Kinetics of Kaposi's sarcoma-associated herpesvirus gene expression, *J Virol* 73 (3), 2232–2242, 1999.

127. Sodhi, A., Montaner, S., and Gutkind, J.S., Viral hijacking of G-protein-coupled-receptor signalling networks, *Nat Rev Mol Cell Biol* 5 (12), 998–1012, 2004.
128. Dairaghi, D.J., Fan, R.A., McMaster, B.E., Hanley, M.R., and Schall, T.J., HHV8-encoded vMIP-I selectively engages chemokine receptor CCR8. Agonist and antagonist profiles of viral chemokines, *J Biol Chem* 274 (31), 21569–21574, 1999.
129. Kledal, T.N., Rosenkilde, M.M., Coulin, F., Simmons, G., Johnsen, A.H., Alouani, S., Power, C.A., Luttichau, H.R., Gerstoft, J., Clapham, P.R., Clark-Lewis, I., Wells, T.N., and Schwartz, T.W., A broad-spectrum chemokine antagonist encoded by Kaposi's sarcoma-associated herpesvirus, *Science* 277 (5332), 1656–1659, 1997.
130. Boshoff, C., Endo, Y., Collins, P.D., Takeuchi, Y., Reeves, J.D., Schweickart, V.L., Siani, M.A., Sasaki, T., Williams, T.J., Gray, P.W., Moore, P.S., Chang, Y., and Weiss, R.A., Angiogenic and HIV-inhibitory functions of KSHV-encoded chemokines, *Science* 278 (5336), 290–294, 1997.
131. Bais, C., Santomasso, B., Coso, O., Arvanitakis, L., Raaka, E.G., Gutkind, J.S., Asch, A.S., Cesarman, E., Gershengorn, M.C., and Mesri, E.A., G-protein-coupled receptor of Kaposi's sarcoma-associated herpesvirus is a viral oncogene and angiogenesis activator, *Nature* 391 (6662), 86–89, 1998.
132. Arvanitakis, L., Geras-Raaka, E., Varma, A., Gershengorn, M.C., and Cesarman, E., Human herpesvirus KSHV encodes a constitutively active G-protein-coupled receptor linked to cell proliferation, *Nature* 385 (6614), 347–350, 1997.
133. Rosenkilde, M.M., Kledal, T.N., Holst, P.J., and Schwartz, T.W., Selective elimination of high constitutive activity or chemokine binding in the human herpesvirus 8 encoded seven transmembrane oncogene ORF74, *J Biol Chem* 275 (34), 26309–26315, 2000.
134. Gershengorn, M.C., Geras-Raaka, E., Varma, A., and Clark-Lewis, I., Chemokines activate Kaposi's sarcoma-associated herpesvirus G protein-coupled receptor in mammalian cells in culture, *J Clin Invest* 102 (8), 1469–1472, 1998.
135. Rosenkilde, M.M., Kledal, T.N., Brauner-Osborne, H., and Schwartz, T.W., Agonists and inverse agonists for the herpesvirus 8-encoded constitutively active seven-transmembrane oncogene product, ORF-74, *J Biol Chem* 274 (2), 956–961, 1999.
136. Montaner, S., Sodhi, A., Molinolo, A., Bugge, T.H., Sawai, E.T., He, Y., Li, Y., Ray, P.E., and Gutkind, J.S., Endothelial infection with KSHV genes in vivo reveals that vGPCR initiates Kaposi's sarcomagenesis and can promote the tumorigenic potential of viral latent genes, *Cancer Cell* 3 (1), 23–36, 2003.
137. Montaner, S., Sodhi, A., Pece, S., Mesri, E.A., and Gutkind, J.S., The Kaposi's sarcoma-associated herpesvirus G protein-coupled receptor promotes endothelial cell survival through the activation of Akt/protein kinase B, *Cancer Res* 61 (6), 2641–2648, 2001.
138. Sodhi, A., Montaner, S., Patel, V., Zohar, M., Bais, C., Mesri, E.A., and Gutkind, J.S., The Kaposi's sarcoma-associated herpesvirus G protein-coupled receptor up-regulates vascular endothelial growth factor expression and secretion through mitogen-activated protein kinase and p38 pathways acting on hypoxia-inducible factor 1alpha, *Cancer Res* 60 (17), 4873–4880, 2000.
139. Hoffman, E.C., Reyes, H., Chu, F.F., Sander, F., Conley, L.H., Brooks, B.A., and Hankinson, O., Cloning of a factor required for activity of the Ah (dioxin) receptor, *Science* 252 (5008), 954–958, 1991.
140. Wang, G.L. and Semenza, G.L., General involvement of hypoxia-inducible factor 1 in transcriptional response to hypoxia, *Proc Natl Acad Sci USA* 90 (9), 4304–4308, 1993.
141. Huang, L.E., Gu, J., Schau, M., and Bunn, H.F., Regulation of hypoxia-inducible factor 1alpha is mediated by an O_2-dependent degradation domain via the ubiquitin-proteasome pathway, *Proc Natl Acad Sci USA* 95 (14), 7987–7992, 1998.
142. Bruick, R.K. and McKnight, S.L., A conserved family of prolyl-4-hydroxylases that modify HIF, *Science* 294 (5545), 1337–1340, 2001.
143. Ivan, M., Kondo, K., Yang, H., Kim, W., Valiando, J., Ohh, M., Salic, A., Asara, J.M., Lane, W.S., and Kaelin, W.G., Jr., HIFalpha targeted for VHL-mediated destruction by proline hydroxylation: implications for O_2 sensing, *Science* 292 (5516), 464–468, 2001.
144. Jaakkola, P., Mole, D.R., Tian, Y.M., Wilson, M.I., Gielbert, J., Gaskell, S.J., Kriegsheim, A., Hebestreit, H.F., Mukherji, M., Schofield, C.J., Maxwell, P.H., Pugh, C.W., and Ratcliffe, P.J., Targeting of HIF-alpha to the von Hippel-Lindau ubiquitylation complex by O_2-regulated prolyl hydroxylation, *Science* 292 (5516), 468–472, 2001.

145. Maxwell, P.H., Wiesener, M.S., Chang, G.W., Clifford, S.C., Vaux, E.C., Cockman, M.E., Wykoff, C.C., Pugh, C.W., Maher, E.R., and Ratcliffe, P.J., The tumour suppressor protein VHL targets hypoxia-inducible factors for oxygen-dependent proteolysis, *Nature* 399 (6733), 271–275, 1999.
146. Ohh, M., Park, C.W., Ivan, M., Hoffman, M.A., Kim, T.Y., Huang, L.E., Pavletich, N., Chau, V., and Kaelin, W.G., Ubiquitination of hypoxia-inducible factor requires direct binding to the beta-domain of the von Hippel-Lindau protein, *Nat Cell Biol* 2 (7), 423–427, 2000.
147. Cockman, M.E., Masson, N., Mole, D.R., Jaakkola, P., Chang, G.W., Clifford, S.C., Maher, E.R., Pugh, C.W., Ratcliffe, P.J., and Maxwell, P.H., Hypoxia inducible factor-alpha binding and ubiquitylation by the von Hippel-Lindau tumor suppressor protein, *J Biol Chem* 275 (33), 25733–25741, 2000.
148. Tanimoto, K., Makino, Y., Pereira, T., and Poellinger, L., Mechanism of regulation of the hypoxia-inducible factor-1 alpha by the von Hippel-Lindau tumor suppressor protein, *EMBO J* 19 (16), 4298–4309, 2000.
149. Kamura, T., Sato, S., Iwai, K., Czyzyk-Krzeska, M., Conaway, R.C., and Conaway, J.W., Activation of HIF1alpha ubiquitination by a reconstituted von Hippel-Lindau (VHL) tumor suppressor complex, *Proc Natl Acad Sci USA* 97 (19), 10430–10435, 2000.
150. Minet, E., Arnould, T., Michel, G., Roland, I., Mottet, D., Raes, M., Remacle, J., and Michiels, C., ERK activation upon hypoxia: involvement in HIF-1 activation, *FEBS Lett* 468 (1), 53–58, 2000.
151. Acker, T. and Plate, K.H., A role for hypoxia and hypoxia-inducible transcription factors in tumor physiology, *J Mol Med* 80 (9), 562–575, 2002.
152. Cannon, M., Philpott, N.J., and Cesarman, E., The Kaposi's sarcoma-associated herpesvirus G protein-coupled receptor has broad signaling effects in primary effusion lymphoma cells, *J Virol* 77 (1), 57–67, 2003.
153. Montaner, S., Sodhi, A., Servitja, J.M., Ramsdell, A.K., Barac, A., Sawai, E.T., and Gutkind, J.S., The small GTPase Rac1 links the Kaposi sarcoma-associated herpesvirus vGPCR to cytokine secretion and paracrine neoplasia, *Blood* 104 (9), 2903–2911, 2004.
154. Bais, C., Van Geelen, A., Eroles, P., Mutlu, A., Chiozzini, C., Dias, S., Silverstein, R.L., Rafii, S., and Mesri, E.A., Kaposi's sarcoma associated herpesvirus G protein-coupled receptor immortalizes human endothelial cells by activation of the VEGF receptor-2/ KDR, *Cancer Cell* 3 (2), 131–143, 2003.
155. Ferrara, N., Gerber, H.P., and LeCouter, J., The biology of VEGF and its receptors, *Nat Med* 9 (6), 669–676, 2003.
156. Masood, R., Cesarman, E., Smith, D.L., Gill, P.S., and Flore, O., Human herpesvirus-8-transformed endothelial cells have functionally activated vascular endothelial growth factor/vascular endothelial growth factor receptor, *Am J Pathol* 160 (1), 23–29, 2002.
157. Eswarakumar, V.P., Lax, I., and Schlessinger, J., Cellular signaling by fibroblast growth factor receptors, *Cytokine Growth Factor Rev* 16 (2), 139–149, 2005.
158. Neufeld, G., Gospodarowicz, D., Dodge, L., and Fujii, D.K., Heparin modulation of the neurotropic effects of acidic and basic fibroblast growth factors and nerve growth factor on PC12 cells, *J Cell Physiol* 131 (1), 131–140, 1987.
159. Creuzet, C., Loeb, J., and Barbin, G., Fibroblast growth factors stimulate protein tyrosine phosphorylation and mitogen-activated protein kinase activity in primary cultures of hippocampal neurons, *J Neurochem* 64 (4), 1541–1547, 1995.
160. Kanda, S., Hodgkin, M.N., Woodfield, R.J., Wakelam, M.J., Thomas, G., and Claesson-Welsh, L., Phosphatidylinositol 3′-kinase-independent p70 S6 kinase activation by fibroblast growth factor receptor-1 is important for proliferation but not differentiation of endothelial cells, *J Biol Chem* 272 (37), 23347–23353, 1997.
161. Sinkovics, J.G., Kaposi's sarcoma: its 'oncogenes' and growth factors, *Crit Rev Oncol Hematol* 11 (2), 87–107, 1991.
162. Samaniego, F., Markham, P.D., Gendelman, R., Watanabe, Y., Kao, V., Kowalski, K., Sonnabend, J.A., Pintus, A., Gallo, R.C., and Ensoli, B., Vascular endothelial growth factor and basic fibroblast growth factor present in Kaposi's sarcoma (KS) are induced by inflammatory cytokines and synergize to promote vascular permeability and KS lesion development, *Am J Pathol* 152 (6), 1433–1443, 1998.

163. Barillari, G., Sgadari, C., Palladino, C., Gendelman, R., Caputo, A., Morris, C.B., Nair, B.C., Markham, P., Nel, A., Sturzl, M., and Ensoli, B., Inflammatory cytokines synergize with the HIV-1 Tat protein to promote angiogenesis and Kaposi's sarcoma via induction of basic fibroblast growth factor and the alpha v beta 3 integrin, *J Immunol* 163 (4), 1929–1935, 1999.
164. Opalenik, S.R., Shin, J.T., Wehby, J.N., Mahesh, V.K., and Thompson, J.A., The HIV-1 TAT protein induces the expression and extracellular appearance of acidic fibroblast growth factor, *J Biol Chem* 270 (29), 17457–17467, 1995.
165. Betsholtz, C., Karlsson, L., and Lindahl, P., Developmental roles of platelet-derived growth factors, *Bioessays* 23 (6), 494–507, 2001.
166. Brennan, P., Babbage, J.W., Burgering, B.M., Groner, B., Reif, K., and Cantrell, D.A., Phosphatidylinositol 3-kinase couples the interleukin-2 receptor to the cell cycle regulator E2F, *Immunity* 7 (5), 679–689, 1997.
167. Chaudhary, L.R. and Avioli, L.V., Extracellular-signal regulated kinase signaling pathway mediates downregulation of type I procollagen gene expression by FGF-2, PDGF-BB, and okadaic acid in osteoblastic cells, *J Cell Biochem* 76 (3), 354–359, 2000.
168. Franke, T.F., Yang, S.I., Chan, T.O., Datta, K., Kazlauskas, A., Morrison, D.K., Kaplan, D.R., and Tsichlis, P.N., The protein kinase encoded by the Akt proto-oncogene is a target of the PDGF-activated phosphatidylinositol 3-kinase, *Cell* 81 (5), 727–736, 1995.
169. Pearson, M.A., O'Farrell, A.M., Dexter, T.M., Whetton, A.D., Owen-Lynch, P.J., and Heyworth, C.M., Investigation of the molecular mechanisms underlying growth factor synergy: the role of ERK 2 activation in synergy, *Growth Factors* 15 (4), 293–306, 1998.
170. Werner, S., Hofschneider, P.H., Heldin, C.H., Ostman, A., and Roth, W.K., Cultured Kaposi's sarcoma-derived cells express functional PDGF A-type and B-type receptors, *Exp Cell Res* 187 (1), 98–103, 1990.
171. Roth, W.K., Werner, S., Schirren, C.G., and Hofschneider, P.H., Depletion of PDGF from serum inhibits growth of AIDS-related and sporadic Kaposi's sarcoma cells in culture, *Oncogene* 4 (4), 483–487, 1989.
172. Sturzl, M., Roth, W.K., Brockmeyer, N.H., Zietz, C., Speiser, B., and Hofschneider, P.H., Expression of platelet-derived growth factor and its receptor in AIDS-related Kaposi sarcoma in vivo suggests paracrine and autocrine mechanisms of tumor maintenance, *Proc Natl Acad Sci USA* 89 (15), 7046–7050, 1992.
173. Lee, H.J., Cho, C.H., Hwang, S.J., Choi, H.H., Kim, K.T., Ahn, S.Y., Kim, J.H., Oh, J.L., Lee, G.M., and Koh, G.Y., Biological characterization of angiopoietin-3 and angiopoietin-4, *FASEB J* 18 (11), 1200–1208, 2004.
174. Kim, I., Kwak, H.J., Ahn, J.E., So, J.N., Liu, M., Koh, K.N., and Koh, G.Y., Molecular cloning and characterization of a novel angiopoietin family protein, angiopoietin-3, *FEBS Lett* 443 (3), 353–356, 1999.
175. Maisonpierre, P.C., Suri, C., Jones, P.F., Bartunkova, S., Wiegand, S.J., Radziejewski, C., Compton, D., McClain, J., Aldrich, T.H., Papadopoulos, N., Daly, T.J., Davis, S., Sato, T.N., and Yancopoulos, G.D., Angiopoietin-2, a natural antagonist for Tie2 that disrupts in vivo angiogenesis, *Science* 277 (5322), 55–60, 1997.
176. Davis, S., Aldrich, T.H., Jones, P.F., Acheson, A., Compton, D.L., Jain, V., Ryan, T.E., Bruno, J., Radziejewski, C., Maisonpierre, P.C., and Yancopoulos, G.D., Isolation of angiopoietin-1, a ligand for the TIE2 receptor, by secretion-trap expression cloning, *Cell* 87 (7), 1161–1169, 1996.
177. Suri, C., Jones, P.F., Patan, S., Bartunkova, S., Maisonpierre, P.C., Davis, S., Sato, T.N., and Yancopoulos, G.D., Requisite role of angiopoietin-1, a ligand for the TIE2 receptor, during embryonic angiogenesis, *Cell* 87 (7), 1171–1180, 1996.
178. Oh, H., Takagi, H., Suzuma, K., Otani, A., Matsumura, M., and Honda, Y., Hypoxia and vascular endothelial growth factor selectively up-regulate angiopoietin-2 in bovine microvascular endothelial cells, *J Biol Chem* 274 (22), 15732–15739, 1999.
179. Lobov, I.B., Brooks, P.C., and Lang, R.A., Angiopoietin-2 displays VEGF-dependent modulation of capillary structure and endothelial cell survival in vivo, *Proc Natl Acad Sci USA* 99 (17), 11205–11210, 2002.

180. Brown, L.F., Dezube, B.J., Tognazzi, K., Dvorak, H.F., and Yancopoulos, G.D., Expression of Tie1, Tie2, and angiopoietins 1, 2, and 4 in Kaposi's sarcoma and cutaneous angiosarcoma, *Am J Pathol* 156 (6), 2179–2183, 2000.
181. Pantanowitz, L. and Dezube, B.J., Advances in the pathobiology and treatment of Kaposi sarcoma, *Curr Opin Oncol* 16 (5), 443–449, 2004.
182. Alvarez, O.A., Carmichael, D.F., and DeClerck, Y.A., Inhibition of collagenolytic activity and metastasis of tumor cells by a recombinant human tissue inhibitor of metalloproteinases, *J Natl Cancer Inst* 82 (7), 589–595, 1990.
183. Johnson, M.D., Kim, H.R., Chesler, L., Tsao-Wu, G., Bouck, N., and Polverini, P.J., Inhibition of angiogenesis by tissue inhibitor of metalloproteinase, *J Cell Physiol* 160 (1), 194–202, 1994.
184. Kruger, A., Fata, J.E., and Khokha, R., Altered tumor growth and metastasis of a T-cell lymphoma in Timp-1 transgenic mice, *Blood* 90 (5), 1993–2000, 1997.
185. Benelli, R., Adatia, R., Ensoli, B., Stetler-Stevenson, W.G., Santi, L., and Albini, A., Inhibition of AIDS-Kaposi's sarcoma cell induced endothelial cell invasion by TIMP-2 and a synthetic peptide from the metalloproteinase propeptide: implications for an anti-angiogenic therapy, *Oncol Res* 6 (6), 251–257, 1994.
186. Bernhard, E.J., Gruber, S.B., and Muschel, R.J., Direct evidence linking expression of matrix metalloproteinase 9 (92-kDa gelatinase/collagenase) to the metastatic phenotype in transformed rat embryo cells, *Proc Natl Acad Sci USA* 91 (10), 4293–4297, 1994.
187. Kawamata, H., Kameyama, S., Kawai, K., Tanaka, Y., Nan, L., Barch, D.H., Stetler-Stevenson, W.G., and Oyasu, R., Marked acceleration of the metastatic phenotype of a rat bladder carcinoma cell line by the expression of human gelatinase A, *Int J Cancer* 63 (4), 568–575, 1995.
188. Meade-Tollin, L.C., Way, D., and Witte, M.H., Expression of multiple matrix metalloproteinases and urokinase type plasminogen activator in cultured Kaposi sarcoma cells, *Acta Histochem* 101 (3), 305–316, 1999.
189. Ray, J.M. and Stetler-Stevenson, W.G., Gelatinase A activity directly modulates melanoma cell adhesion and spreading, *EMBO J* 14 (5), 908–917, 1995.
190. Toschi, E., Barillari, G., Sgadari, C., Bacigalupo, I., Cereseto, A., Carlei, D., Palladino, C., Zietz, C., Leone, P., Sturzl, M., Butto, S., Cafaro, A., Monini, P., and Ensoli, B., Activation of matrix-metalloproteinase-2 and membrane-type-1-matrix-metalloproteinase in endothelial cells and induction of vascular permeability in vivo by human immunodeficiency virus-1 Tat protein and basic fibroblast growth factor, *Mol Biol Cell* 12 (10), 2934–2946, 2001.
191. Impola, U., Cuccuru, M.A., Masala, M.V., Jeskanen, L., Cottoni, F., and Saarialho-Kere, U., Preliminary communication: matrix metalloproteinases in Kaposi's sarcoma, *Br J Dermatol* 149 (4), 905–907, 2003.
192. Pati, S., Pelser, C.B., Dufraine, J., Bryant, J.L., Reitz, M.S., Jr., and Weichold, F.F., Antitumorigenic effects of HIV protease inhibitor ritonavir: inhibition of Kaposi sarcoma, *Blood* 99 (10), 3771–3779, 2002.
193. Sgadari, C., Barillari, G., Toschi, E., Carlei, D., Bacigalupo, I., Baccarini, S., Palladino, C., Leone, P., Bugarini, R., Malavasi, L., Cafaro, A., Falchi, M., Valdembri, D., Rezza, G., Bussolino, F., Monini, P., and Ensoli, B., HIV protease inhibitors are potent anti-angiogenic molecules and promote regression of Kaposi sarcoma, *Nat Med* 8 (3), 225–232, 2002.
194. Golub, L.M., Ramamurthy, N.S., McNamara, T.F., Greenwald, R.A., and Rifkin, B.R., Tetracyclines inhibit connective tissue breakdown: new therapeutic implications for an old family of drugs, *Crit Rev Oral Biol Med* 2 (3), 297–321, 1991.
195. Cianfrocca, M., Cooley, T.P., Lee, J.Y., Rudek, M.A., Scadden, D.T., Ratner, L., Pluda, J.M., Figg, W.D., Krown, S.E., and Dezube, B.J., Matrix metalloproteinase inhibitor COL-3 in the treatment of AIDS-related Kaposi's sarcoma: a phase I AIDS malignancy consortium study, *J Clin Oncol* 20 (1), 153–159, 2002.
196. Dezube, B.J., Krown, S.E., Lee, J.Y., Bauer, K.S., and Aboulafia, D.M., Randomized phase II trial of matrix metalloproteinase inhibitor COL-3 in AIDS-related Kaposi's sarcoma: an AIDS Malignancy Consortium Study, *J Clin Oncol* 24 (9), 1389–1394, 2006.
197. Elkin, M., Miao, H.Q., Nagler, A., Aingorn, E., Reich, R., Hemo, I., Dou, H.L., Pines, M., and Vlodavsky, I., Halofuginone: a potent inhibitor of critical steps in angiogenesis progression, *FASEB J* 14 (15), 2477–2485, 2000.

198. Koon, H.B., Bubley, G.J., Pantanowitz, L., Masiello, D., Smith, B., Crosby, K., Proper, J., Weeden, W., Miller, T.E., Chatis, P., Egorin, M.J., Tahan, S.R., and Dezube, B.J., Imatinib-induced regression of AIDS-related Kaposi's sarcoma, *J Clin Oncol* 23 (5), 982–989, 2005.
199. Chaudhary, L.R. and Hruska, K.A., The cell survival signal Akt is differentially activated by PDGF-BB, EGF, and FGF-2 in osteoblastic cells, *J Cell Biochem* 81 (2), 304–311, 2001.
200. Krown, S.E., Li, P., Von Roenn, J.H., Paredes, J., Huang, J., and Testa, M.A., Efficacy of low-dose interferon with antiretroviral therapy in Kaposi's sarcoma: a randomized phase II AIDS clinical trials group study, *J Interferon Cytokine Res* 22 (3), 295–303, 2002.
201. Little, R.F., Wyvill, K.M., Pluda, J.M., Welles, L., Marshall, V., Figg, W.D., Newcomb, F.M., Tosato, G., Feigal, E., Steinberg, S.M., Whitby, D., Goedert, J.J., and Yarchoan, R., Activity of thalidomide in AIDS-related Kaposi's sarcoma, *J Clin Oncol* 18 (13), 2593–2602, 2000.
202. Stallone, G., Schena, A., Infante, B., Di Paolo, S., Loverre, A., Maggio, G., Ranieri, E., Gesualdo, L., Schena, F.P., and Grandaliano, G., Sirolimus for Kaposi's sarcoma in renal-transplant recipients, *N Engl J Med* 352 (13), 1317–1323, 2005.

31 Antiangiogenic Therapy for Melanoma

Keith Dredge and Angus G. Dalgleish

CONTENTS

New blood vessel formation and secretion of associated growth factors are essential for the development of metastatic melanoma. Various mediators of angiogenesis may provide correlations between tumor angiogenesis and clinical outcome, while angiogenesis inhibitors have been demonstrated to block tumor growth and metastases. However, clinical data using angiogenesis inhibitors in melanoma trials have been disappointing. Nonetheless, new strategies and agents have ensured their ongoing clinical development, such as their use in conjunction with other modalities. Therefore, despite initial setbacks, the role of antiangiogenic therapy in the treatment of melanoma remains a real possibility.

31.1 MELANOMA

Melanoma is the leading cause of death from skin tumors worldwide, with an annual increase in incidence over the past decade [1]. There are approximately 133,000 new cases of melanoma worldwide each year, of which 80% occur in the Caucasian population of North America, Europe, Australia, and New Zealand [2]. About 54,000 people in the United States each year are diagnosed with melanoma and about 7,600 people die of the disease [3]. Melanoma grows from pigment cells called melanocytes in the outer layer of the skin and can take more than a decade to develop [3]. Sun exposure, particularly recreational or intermittent sun exposure, is the major known etiologic factor for melanoma, and it may interact with other important risk factors such as nevi and tanning ability [4].

In particular, one significant risk factor for the development of melanoma is the number of pigmented nevi (moles). The indirect effect of sunlight on melanoma development is to stimulate nevogenesis. One of the risk-modifying genes is the gene coding for melanocortin-1-receptor (MC1R). The presence of some gene variants has been found to lead to changes in melanin synthesis and is associated with a higher risk of melanoma. This situation can lead to hypermutability and genetic instability. [5]. The clinical features of melanoma are asymmetry, a coastline border, or multiple colors, that is, signs reflecting the heterogeneity of the melanoma cells with one tumor [6]. However, the best prognostic indicator of clinical outcome in primary melanoma is essentially determined by vertical tumor thickness and the presence or absence of ulceration [7]. Invasive melanomas are clinically aggressive and prone to metastasis, leading to disseminated disease [8,9].

Metastases from tumors such as melanoma may remain inactive for many years before relapse [10]. This is not due to slower growth of metastatic cells but to a state of dormancy from which they ultimately escape to become clinically evident [11,12]. Dormancy can occur for years and represents an equilibrium between tumor-cell proliferation and tumor-cell apoptosis [13]. Micrometastatic deposits may escape from dormancy by removal of a circulating inhibitor of angiogenesis or by a switch to the angiogenic phenotype [14]. Therefore, mediators of angiogenesis seem to play a role in the development of metastatic disease encountered in patients with malignant melanoma.

31.2 MEDIATORS OF TUMOR ANGIOGENESIS

Angiogenesis was first recognized as a therapeutically interesting process by Judah Folkman in 1971, when he suggested that tumor growth and metastases are angiogenesis-dependent and that some tumors are highly vascularized. This suggests that blocking angiogenesis could be a strategy to arrest tumor growth [15]. In normal circumstances when the angiogenic switch is "off" proangiogenic and endogenous antiangiogenic molecules are balanced but in most cancers including melanoma, proangiogenic molecules are switched "on" to drive the angiogenic process [16,17]. Table 31.1 indicates some of the pro- and antiangiogenic molecules that have the ability to switch angiogenesis on or off, respectively.

The molecules referred to in Table 31.1 begin to illustrate the complexity of the angiogenic process. In neoplastic conditions hypoxia is a strong stimulus for angiogenesis activated by hypoxia-inducible transcription factors (HIFs), which induce expression of several proangiogenic markers including vascular endothelial growth factor (VEGF), nitric oxide synthase (NOS), platelet-derived growth factor (PDGF), and others [18]. Angiogenesis depends mainly on proper activation, proliferation, adhesion, migration, and maturation of endothelial cells (ECs) [18]. Therefore, once angiogenic activators have been upregulated the process of angiogenesis commences. Vasodilatation, which involves NO and vascular permeability, also increases in response to VEGF thereby allowing extravasation of plasma proteins that form a matrix to aid migration of ECs [19].

Permeability is mediated by the formation of fenestrations and the redistribution of platelet EC adhesion molecule (PECAM-1) and vascular endothelial (VE)-cadherin, and involves Src kinases [20]. Angiopoietin-1 (Ang-1), a ligand of the endothelial Tie-2 receptor, is a natural inhibitor of permeability, and prevents excessive vessel leakage that may otherwise

TABLE 31.1
Regulators Involved in the Angiogenic Process

Proangiogenic Factors	Antiangiogenic Factors
Vascular endothelial growth factor (VEGF)	Thrombospondin-1 (TSP-1)
Basic fibroblast growth factor (bFGF or FGF-2)	Tissue inhibitor of MMPs (TIMPs)
Acidic fibroblast growth factor (aFGF or FGF-1)	Angiostatin
Hypoxia-inducible factor (HIF)	Endostatin
Integrin $\alpha v\beta_3$ and $\alpha v\beta_5$	Angiopoietin-1 and -2 (Ang-1 and -2)
Transforming growth factor-β (TGF-β)	Prolactin
Thymidine phosphorylase (TP)	Interferon-α (IFN-α)
Tumor necrosis factor (TNF-α)	Interferon-β (IFN-β)
Platelet-derived growth factor (PDGF)	Interferon-γ (IFN-γ)
Matrix metalloproteinases (MMPs)	IFN-γ-inducible protein (IP-10)
Interleukin-8 (IL-8)	Monokine induced by γ-IFN (MIG)
Prostaglandin E_1, E_2 (PGE_1, PGE_2)	Interleukin-12 (IL-12)
Cyclooxygenase-2 (COX-2)	Interleukin-1 (IL-1)
Platelet-activating factor	Interleukin-4 (IL-4)
Substance P	Plasminogen activator inhibitors
Hepatocyte growth factor/scatter factor (HGF/SF)	Antiangiogenic antithrombin (aaAT)
Nitric oxide (NO)	Angiopoietin-2[a] (Ang-2)
Placental growth factor (PlGF)	Insulin-like growth factor (IGF)
Plasminogen activator (PA)	Metallospondins (METH-1 and -2)
Urokinase plasminogen activator (uPA)	Pigment epithelium-derived factor (PEDF)
Estrogen	Retinoic acid
Ephrins	Vasostatin/calreticulin
AC133[b]	Platelet factor-4
Id1/Id3[c]	Neuropilin-1 (NRP-1)
Granulocyte colony stimulating factor	Osteopontin fragment
Granulocyte macrophage colony stimulating factor	Maspin

Source: Modified from Dredge, K., Dalgleish, A.G., and Marriott, J.B., *Expert Opin. Biol Ther.*, 2(8), 953, 2002. With permission.

[a] Ang-2 can have opposite effects in some contexts by acting as an Ang-1 antagonist.
[b] Novel 5-transmembrane antigen present on a CD34 (bright) subset of human hematopoietic stem cells.
[c] *Id1* and *Id3* are two developmental genes that, if deleted in mice, result in the death of the embryos due to pulmonary hemorrhages. However, if one allele (pair) of *Id1* is deleted and both alleles of *Id3* are deleted, the mice survive and cannot initiate a normal angiogenic response if implanted with tumor cells. As a result tumors do not grow, or metastasize.

result in vascular collapse [19]. On the other hand Ang-2, an inhibitor of Tie-2 signaling, has been proposed to be involved in loosening the matrix that allows EC migration [21]. More recently Ang-2 has been implicated in the regulation of tumor angiogenesis [22] and one study found melanoma tissue overexpressing Ang-2 mRNA compared with spleen, liver, and bone marrow of normal mice, suggesting its role during melanoma progression [23]. Any subsequent migration of ECs or invasion of tumor cells occurs following degradation of the basement membranes by matrix metalloproteinases (MMPs), tissue serine proteinases such as tissue plasminogen, factor, urokinase, thrombin and plasmin, adamalysin-related membrane proteinases, and the bone morphogenetic protein-1-type metalloproteinases [24]. The process is outlined in Figure 31.1.

Recent interest has focussed on the urokinase plasminogen activator (uPA) system, found to initiate multiple cascades associated with angiogenesis and metastases [13,16,25]. The receptor (uPAR) has also been suggested to be a useful target for putative cancer therapeutics [26]. For example, the human melanoma cell line IGR-1 was found to have

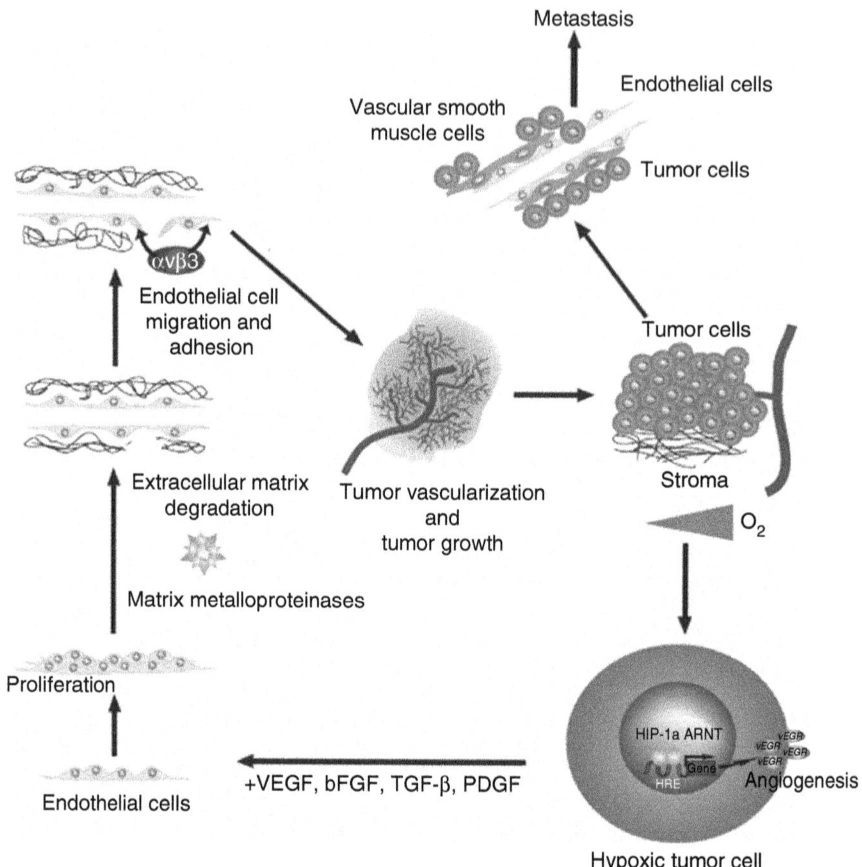

FIGURE 31.1 (See color insert following page 558.) The induction of hypoxia-inducible factor (HIF-1) by tumor cells can lead to the production of growth factors such as VEGF, bFGF, TGF-β, and PDGF. Such factors induce proliferation of endothelial cells and tumor cells and lead to the induction of enzymes like matrix metalloproteases, which degrade the ECM. Subsequent migration and cell adhesion lead to angiogenesis and tumor vascularization. This further enhances tumor growth and tumor cells along endothelial vessels that may metastasize to distant regions leading to metastatic disease.

increased expression of uPA and uPAR on the cell surface [27]. The integrin vitronectin receptor alpha (v) beta (3) (αvβ3) is a mediator of cellular migration and invasion and has been identified as a marker of progression in malignant melanoma. This receptor is also coordinately expressed with uPAR and can enhance invasion by regulating the uPAR/uPA/plasmin system of proteolysis, probably with PKC beta as an intermediate in the activation pathway [28]. The determination of uPA and PAI-1 can provide significant additional prognostic information for melanoma patients [29]. Therefore, inhibitors of uPA are currently being investigated but have not yet entered clinical trials.

From a preclinical perspective, inhibition of B16BL6 melanoma invasion by tyrosine and phenylalanine deprivation is associated with decreased secretion of plasminogen activators and increased plasminogen activator inhibitors [30]. Moreover, inhibition of establishment of primary and micrometastatic tumors by a urokinase plasminogen activator receptor antagonist has been reported, further supporting a role for uPA as an angiogenic factor in melanoma [31]. A novel antiangiogenic agent Cambodian Phellinus linteus (CPL) is a fungus recently found to downregulate the expression of uPA, and dose-dependently inhibit the pulmonary metastatic colonies in mice injected with B16BL6 melanoma cells. As CPL did not significantly affect the expression of MMP-2, tissue inhibitor of metalloproteinase 2 (TIMP-2) or exhibit cytotoxicity against the tumor cells, inhibition metastasis could be at least partly via regulation of uPA [32].

The migration and proliferation of ECs are controlled by three classes of angiogenesis inducers [33]. The first class consists of VEGF and the angiopoietins that act specifically on ECs. The second class includes direct-acting molecules such as cytokines basic fibroblast growth factor (bFGF), chemokines [34], and angiogenic enzymes [35,36] that activate many other cell types. Finally, there are indirect-acting molecules such as TNF-α and TGF-β, which are produced by macrophages, endothelial or tumor cells and increase proangiogenic factors such as VEGF, IL-8, or bFGF which in turn induce angiogenesis [37,38]. As tumors develop, they begin to produce a wider array of angiogenic molecules so blocking one growth factor (e.g., VEGF) may result in a tumor switching to another molecules (e.g., bFGF or IL-8), suggesting that in the case of tumor angiogenesis a cocktail of antiangiogenic therapies is required for successful treatment [39]. With this in mind, two important considerations come into play. First, for antiangiogenic therapy to be effective the regulators/growth factors of each disease/each tumor at each stage must be elucidated. Secondly, although antiangiogenic therapies are thought to be of low toxicity, the long-term effects on normal physiological processes, for example, embryonic development or wound healing are unknown [40].

31.3 ANGIOGENESIS AND MELANOMA

Tumor progression in melanoma is a multistep process involving proliferation, neovascularization, lymphangiogenesis, invasion, circulation, embolism, and extravasation [41]. Although the spread of tumor cells to regional lymph node is known as an indicator of tumor aggressiveness and unfavorable prognosis [42], progression toward the metastasizing phenotype was also shown to be accompanied by prominent angiogenesis [43]. VEGF is known to stimulate angiogenesis [44,45] and in particular VEGF-C was shown to correlate with the incidence of lymph node metastases [46]. It has also been found overexpressed in human melanoma cells transplanted on nude mice, enlarged lymphatic vessels in the tumor periphery, and enhanced tumor angiogenesis, indicating a coordinated regulation of lymphangiogenesis and angiogenesis [42]. Overproduction of VEGF concomitantly expressed with its receptors favors cell growth and survival of melanoma cells through MAPK and PI3K signaling pathways [47]. Figure 31.2 highlights these pathways.

Human melanoma progresses through different steps: nevocellular nevi, dysplastic nevi, in situ melanoma, radial growth phase melanoma (Breslow index $\leq$0.75 mm), vertical growth phase melanoma (Breslow index $>$0.75 mm), and metastatic melanoma. The acquisition of a vascular supply during this progression occurs in concert with an increasing proportion of tumor cells expressing the laminin receptor, which enables their adhesion to the vascular wall. Hence, both phenomena favor tumor cell extravasation and metastases. Melanocytic cells produce and release bFGF, mainly in the steps of dysplastic nevus and melanoma in vertical growth phase. Melanoma cells also secrete VEGF, in parallel with the switch from the radial to the vertical growth phase and the metastatic phase [48]. Others have also suggested that VEGF expression may be a later event in the progression of melanoma than bFGF expression while pointing out that both angiogenic factors are important for the formation of vessels in tumors [49]. Therefore, it is becoming evident that antiangiogenic therapy interferes with melanoma progression.

Proteolytic and migratory activity, the expression of adhesion molecules, and the deposition of extracellular matrix (ECM) by stromal and tumor cells also induce metastatic spread. Adhesion molecules of the cadherin, integrin, and immunoglobulin families are important for growth and metastasis of cutaneous melanoma [50]. For instance, a positive correlation has been observed between the endothelial expression of vascular cell adhesion molecule-1 and

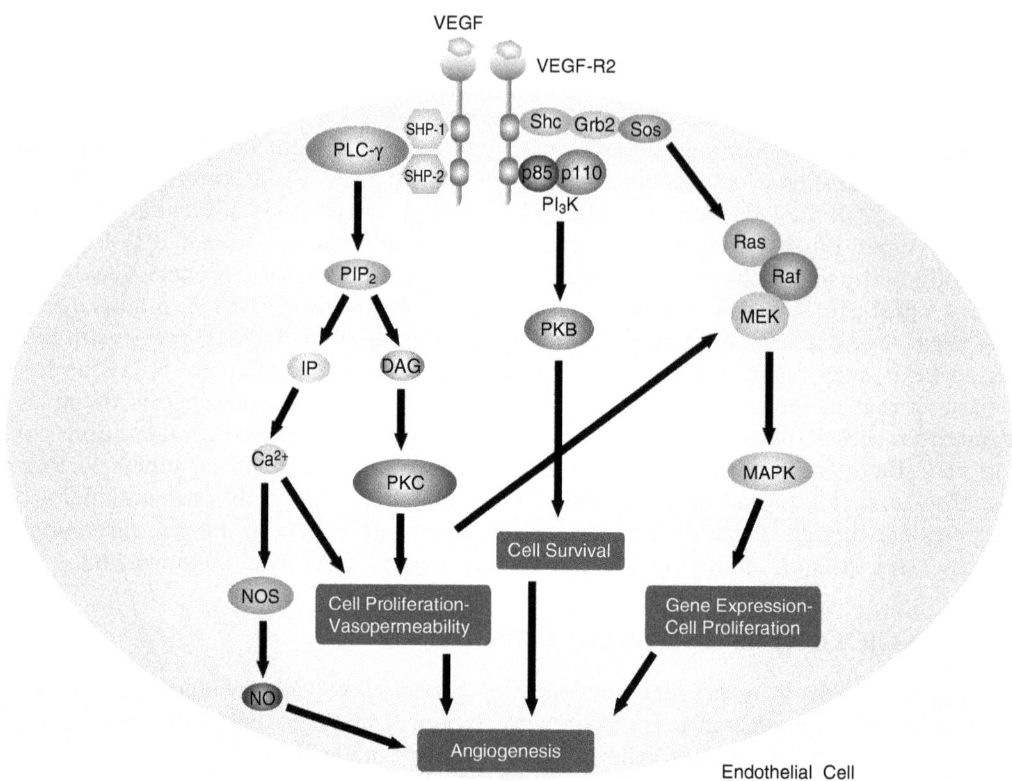

FIGURE 31.2 (See color insert following page 558.) Activation of VEGF receptors (and other growth factor receptors) leads to overexpression of distinct multiple intracellular signaling molecules. Proangiogenic signals can initiate several pathways in endothelial cells by direct binding to the cytoplasmic region of the VEGFR-2. Phospholipase C catalyses the hydrolysis of PIP2 to the second messengers IP3 and DAG. These help to regulate intracellular Ca^{2+} levels. PKC induces cell proliferation and permeability and subsequent production of NO, all of which contribute to angiogenesis. The p85 subunit of IP3-kinase recruits the catalytic p110 subunit leading to further activation of second messengers that enhance cell survival. Protein tyrosine phosphorylation of Shr, Grb2, and Sos activate the Ras/Raf/MEK kinases, part of the MAPK pathway also involved in angiogenesis. Some pathways such as Ras/Raf/MEK have also been noted in melanoma cells as potential targets for therapeutic intervention.

tumor burden in cutaneous melanoma [51], whereas the integrin $\alpha v\beta 3$ is associated with increased tumor growth in vitro and may promote tumor invasion [52].

Overexpression and mutation of hepatocyte growth factor (HGF) pathway members have been implicated in the origin or progression of many cancer types, including melanoma [53]. HGF binds to the c-met proto-oncogene tyrosine kinase receptor. HGF signals invasive patterning in normal tissues such as axons and the mammary gland. However, a recent study showing that hypoxia can stimulate HGF-driven tumor cell invasion has raised the possibility that induction of hypoxia by antiangiogenic strategies may have unintended consequences such as the stimulation of invasion. The authors proposed that an "invasive switch" in response to hypoxia permits tumor cells to escape and survive. [54]. Therefore, the development of combination therapies to simultaneously halt angiogenesis and tumor cell invasion may represent an attractive potential solution [55].

At the genetic level it is interesting to note that while certain mutations in genes have been linked to progressive disease [56–58], the melanomas ascribed to specific genetic defects are the exception rather than the rule. However, a mutation in the BRAF gene, localized on chromosome 7, has recently been reported in melanoma [59,60]. The role of Raf proteins in proliferation control is well established and is the first element of the classical cytoplasmic cascade (Raf/MEK/Erk), which transmits growth signals from activated Ras at the plasma membrane to transcription factors in the nucleus [61]. Activating mutations of BRAF are found in melanomas, with highest mutational rates of up to 68% [59,60]. A recent study found that constitutive activation of the Ras/Raf signaling pathway in melanoma metastases is significantly associated with a poorer prognosis [61]. The fact that BRAF mutations are more frequent in advanced melanoma (as opposed to primary melanoma) suggests that activation of this pathway favors the metastatic process [62]. Inhibitors of Raf kinase such as Sorafenib (BAY 43-9006), also regarded as an antiangiogenic agent, are discussed later in the chapter. However, a recent study concluded BRAF mutations are detectable in conjunctival but not uveal melanomas [63], suggesting that such treatments are not suited for all melanoma types.

31.4 ANTIANGIOGENIC THERAPIES FOR MELANOMA

There are several antiangiogenic therapies currently under preclinical and clinical investigation. These range from heparanase inhibitors, Raf kinase inhibitors like BAY 43-9006, VEGF receptor tyrosine kinase (RTK) inhibitors, and "natural" and "endogenous" angiogenesis inhibitors to small molecular weight compounds with ill-defined modes of action. Table 31.2 shows some inhibitors currently in clinical trials for advanced cancer/melanoma. The next section highlights recent findings concerning these inhibitors of angiogenesis.

31.4.1 Heparanase Inhibitors

Heparanase (HPSE-1) is an endo-beta-D-glucuronidase that cleaves heparan sulfate proteoglycans (HSPG), and its expression has been associated with increased growth, metastasis, and angiogenesis of tumors. A recent study found a 29-fold upregulation of HPSE-1 expression in metastatic melanoma compared with normal lung tissue in experimental animals and selective HPSE-1 staining in human metastatic melanoma when compared with primary melanoma tumors from the same patient. Therefore, HPSE-1 probably plays important roles in regulating the in vivo growth and progression of melanoma. These results further emphasize the importance that therapies designed to block HPSE-1 activity may aid in controlling this type of cancer [64].

PI-88 is a heparanase inhibitor possessing antiangiogenic and antimetastatic properties. It has received orphan drug status by the FDA for melanoma and has a good safety and tolerability profile in addition to evidence of efficacy in Phase I and II trials. Results of a recent phase I trial found PI-88 exhibited antitumor activity as evidenced by one patient with metastatic melanoma who had an objective response maintained >50 months and five patients with stable disease for 7 to >38 months [65]. A Phase II trial commenced in January 2004 in stage II, III, and IV melanoma using PI-88 as a monotherapy. The mechanism of PI-88 is a heparan sulfate mimicry and this is manifested in three important ways [66–68]:

1. Heparan sulfate is an important structural component of the ECM. Heparanase, expressed on the surface of tumor cells, degrades the ECM to allow metastasis. This degradation also releases angiogenic growth factors held in the ECM, creating locally increased concentrations of these angiogenic stimuli. Therefore, PI-88 inhibits metastasis and growth factors release (thus angiogenesis) by inhibiting heparanase activity.
2. PI-88, due to heparan sulfate mimicry, competitively binds to angiogenic growth factors (which are heparin-binding proteins) in a selective fashion.

TABLE 31.2
Some Examples of Antiangiogenic Compounds in Use as Monotherapies or Part of Multiple Modalities in Current Trial for Melanoma

Compound	Stage of Disease	Phase	Mechanism
Thalidomide (Celgene)	Advanced melanoma	I/II	Mechanism unknown. Evidence for inhibition of bFGF and VEGF
Thalidomide and Docetaxel	Advanced cancers	I	
Thalidomide and Temozolomide	Brain metastases secondary to melanoma	II	
Revlimid/Lenalidomide/CC-5013 (Celgene)	Refractory metastatic disease	I	Mechanism unknown. Evidence for inhibition of bFGF
SU6668 (Sugen)	Advanced tumors	I	Blocks VEGF, FGF, and PDGF signaling
SU11248 (Sugen)	Advanced cancer	I	Blocks VEGF, FGF, and PDGF signaling
IFN-α (Amarillo Biosciences)	Melanoma (not specified)	II/III	Inhibition of bFGF/VEGF production
IFN-α and interleukin-12	Malignant melanoma	I	
PEG-interferon α-2b	Melanoma IV	II	
Bevacuzimab/Avastin (Genetech)	Melanoma I, II, III		Anti-VEGF receptor antibody
Bevacizumab and Imatinib/Gleevac	Melanoma III or recurrent	I/II	
Bevacizumab plus IFN-α	Metastatic malignant melanoma	II	
PI-88 (Progen)	Melanoma II, III	II/III	Heparanase and growth factor inhibition
Sorafenib/BAY 43-9006 (Bayer)	Metastatic melanoma	II	PDGFR and VEGR-2 inhibitor
Sorafenib and carboplatin plus paclitaxel	Metastatic melanoma	II	
Cilengitide (Merck)	Melanoma III or IV	II	Selective inhibitor of integrins
17-N-allylamino-17-demethoxygeldanamycin [17-AAG] (Kosan Biosciences)	Melanoma II/III or recurrent	II	Inhibits heat shock protein 90, a recently discovered target for antiangiogenesis treatment
AG-013736 (Pfizer)	Metastatic malignant melanoma (stage IV)	II	Inhibits VEGF and PDGF receptor protein tyrosine kinases

Source: www.clinicaltrials.gov and www.cancer.gov/clinical trials.

3. The mimicking of heparan sulfate also acts to increase the release of tissue factor pathway inhibitor (TFPI), an important antiangiogenic and antimetastatic agent.

31.4.2 BAY 43-9006 (SORAFENIB)

The Ras/Raf signaling pathway is an important mediator of tumor-cell proliferation and angiogenesis and activation of Raf kinases have been associated with poor prognosis in melanoma [61]. The novel bi-aryl urea BAY 43-9006 is a potent inhibitor of Raf-1, a member of the Raf/MEK/Erk signaling pathway [69]. However, BAY 43-9006 has also demonstrated significant activity against several receptor tyrosine kinases involved in neovascularization and tumor progression, including VEGFR-2, VEGFR-3, PDGFR, Flt-3, and c-KIT [70]. Therefore, BAY 43-9006 is a novel dual action Raf kinase and VEGFR inhibitor that targets tumor-cell proliferation and tumor angiogenesis. Clinical trials have found that BAY 43-9006 was well tolerated and appeared to provide some clinical benefits in patients with advanced refractory solid tumors [71]. Phase II trials in melanoma for BAY 43-9006 in combination with carboplatin plus paclitaxel are ongoing and Phase III trials were in the planning stages [72].

31.4.3 Tyrosine Kinase Inhibitors

Growth factors such as VEGF and bFGF drive the angiogenic process and thus are important targets for therapeutic intervention in angiogenesis-dependent disorders [73]. The main approach (other than anti-VEGF antibodies) is the blockade of tyrosine kinase receptors, which neutralize the effects of these growth factors. Recent studies have found that ZD6474, a KDR/VEGFR-2 inhibitor, could reduce VEGF signaling, angiogenesis, and established tumor growth following oral administration to athymic mice [74]. Furthermore, this compound has been found to block oncogenic RET kinases (rearranged during transfection; RET is a cell surface tyrosine kinase receptor), prevalent in papillary carcinomas of the thyroid, which suggests its use for carcinomas sustaining oncogenic activation of RET [75].

SU5416 and SU6668 (Pharmacia/SUGEN) are specific inhibitors of VEGFR-2 signaling. However, antiangiogenic activity by SU5416 is also partly attributable to inhibition of Flt-1 (VEGFR-1) receptor signaling [76]. Both SU5416 and SU6668 inhibited colon cancer liver metastases in a mouse model and also microvessel density and cell proliferation while inducing EC apoptosis [77]. SU6668 has also been found to destroy vessels and established tumors in mice [78]. A Phase I trial of SU5416 in patients with advanced malignancies was well tolerated and showed signs of biological activity as four patients showing clinical benefit had reduced levels of urinary VEGF [79]. However, Phase III clinical trials for SU5416 in advanced colorectal cancer were halted despite the fact that some individual cancer patients did experience dramatic tumor responses [80]. A Phase II study of SU5416, a preferential inhibitor of one of the VEGF RTKs called Flk-1, found it to be relatively well tolerated although only modest clinical activity and limited effects on tumor vascularity question the ongoing clinical development of VEGF inhibition in melanoma [81]. SU6668 is currently evaluated in other Phase I trials (see Table 31.2). SU11248 is a recently described selective inhibitor with selectivity for split kinase domain RTK. SU11248 also has potent activity against the receptor Flt-3 and induces apoptosis in vitro while regressing subcutaneous tumors in vivo [82].

Compounds found in green tea have been discovered to possess antiangiogenic activity by inhibiting tyrosine kinase receptors. Catechins, novel inhibitors of VEGFR-2, demonstrate similar activity to the VEGFR-2 antagonist SU5416, which correlated with suppression of in vitro angiogenesis [83]. Of the four types of catechins investigated, epigallocatechin gallate (EGCG) had the most potent inhibitory effects on angiogenesis [84]. Animal models of angiogenesis have also shown that green tea can inhibit angiogenesis in vivo [85].

AG013736 inhibits VEGF signaling and treatment with this inhibitor was observed to cause robust and early changes in ECs, pericytes, and basement membrane of vessels in various tumor models. Vascular basement membrane persisted after ECs degenerated, providing a ghost-like record of pretreatment vessel number and location and a potential scaffold for vessel re-growth. The potent antivascular action observed is evidence that AG013736 does more than just inhibit angiogenesis [86]. It is currently in trials for stage IV malignant metastatic melanoma (see Table 31.2).

31.4.4 MMP Inhibitors

The tetracycline analogs minocycline and doxycycline are inhibitors of MMPs and have been shown to inhibit angiogenesis in vivo and in an in vitro model of angiogenesis, aortic sprouting in fibrin gels. Doxycycline was found to inhibit collagenase, gelatinase A, and stromelysin with IC50s of 452, 56, and 32 μM, respectively. Minocycline was found to inhibit only stromelysin in the micromolar range with an IC50 of 290 μM. However, these compounds did not inhibit in vitro angiogenesis by an MMP-dependent mechanism, which could have implications for the mechanism of action of tetracycline analogs, particularly where they are being considered for the treatment of disorders of ECM degradation including tumor angiogenesis [87].

Preclinical data have revealed that one MMP inhibitor, Neovastat (AE-941), induces apoptosis in ECs (but not other cell types) by increasing caspase-3 activity [88]. However, the lack of success in the clinic has been attributed to the fact that MMP inhibitors seem effective only at early or intermediate stages of disease, or potentially in combination with other modalities [89].

31.4.5 Monoclonal Antibody Therapy

The best example is probably, Bevacizumab, a humanized monoclonal antibody generated by engineering the VEGF-binding residues of a murine neutralizing antibody into the framework of a normal human immunoglobulin G (IgG). It has been found to possess antiangiogenic and preclinical antitumor activities and is scheduled for development in breast and colorectal cancer [90]. CNTO 95 is a fully human antibody that recognizes the alpha (v) family of integrins, which in vitro inhibited human melanoma cell adhesion, migration, and invasion [91]. In a rat aortic ring sprouting assay, CNTO completely inhibited sprouting and inhibited growth of human melanoma tumors in nude mice by approximately 80%. Based on these preclinical data, a dose-escalating Phase I clinical trial in cancer patients has been initiated [91].

31.4.6 Cytokines

Interferon-alpha (IFN-α) is a cytokine associated with antiangiogenic properties [92] and plays a critical role in cancers such as melanoma [93]. A beneficial effect for IFN-α in high-risk operable malignant melanoma [93] has led to further Eastern Cooperative Oncology Group (ECOG) studies to evaluate the role of IFN-α in patients with intermediate risk disease [94].

Human interferon-inducible protein 10 (IP-10) is a CXC chemokine that contains binding domains for both the chemokine receptor CXCR3 and glycosaminoglycans. IP-10 has been recently demonstrated to be a potent angiostatic protein in vivo [95]. It is also interesting to note that CXCR3 receptor binding, but not glycosaminoglycan binding, is essential for the tumor angiostatic activity of IP-10 [96].

31.4.7 Natural Inhibitors

Many natural products have antiangiogenic properties. For example, bryoanthrathiophene is a new antiangiogenic constituent from the bryozoan Watersipora subtorquata [97], whereas flaxseed inhibits metastasis and decreases VEGF in a human xenograft model [98]. Extracts of wild berries have been found to inhibit hydrogen peroxide (H_2O_2) and tumor necrosis factor-α-induced VEGF expression by human keratinocytes and reduced angiogenesis in a matrigel assay using human dermal microvascular ECs [99]. Kallistatin, a serine proteinase inhibitor (serpin) and a heparin-binding protein, inhibits angiogenesis in vitro and in vivo [100].

Curcumin, the active component of turmeric, has been found to inhibit tumor cell accumulation in vivo and angiogenesis in vitro [101]. Moreover, a novel synthetic component of curcumin (hydrazinocurcumin) also inhibited EC proliferation suggesting that this may be the active antiangiogenic component [102]. Other examples of natural inhibitors include bioactive soybean components, which inhibited orthotopic growth and metastasis of androgen-sensitive human prostate tumors in mice [103], ginkgo biloba, which inhibited VEGF in rat ECs [104], and green tea (see tyrosine kinase inhibitors).

31.4.8 "Endogenous" Inhibitors

A number of inhibitors of angiogenesis found naturally in the body have been developed as antiangiogenic agents. Endostatin is a 20 kDa C-terminal fragment of collagen XVIII found to inhibit tumor growth in mouse model by specifically inhibiting endothelial proliferation and hence angiogenesis [105]. Endostatin blocks mitogen-activated protein kinase (MAPK) activation in ECs [106]. The efficacy of endostatin was assessed in the mouse xenograft tumor models BxPC-3 and HT1080 by comparing subcutaneous or intraperitoneal (i.p.) routes with administration via an i.p. implanted mini-osmotic pump [107]. Continuous administration of endostatin resulted in highly effective tumor suppression, suggesting that continuous administration may enhance the efficacy of antiangiogenic therapies. Importantly, tolerability or resistance following long-term treatment does not seem to be a problem [108].

Angiostatin is an internal fragment of plasminogen and has antiangiogenic and antimetastatic properties. It has been found to induce EC apoptosis and inhibit EC migration in mouse models [109]. By 2001, both endostatin and angiostatin were in Phase I clinical trials [110] and a Phase II endostatin clinical trial commenced in China in 2003 [111].

Thrombospondin-1 (TSP-1), a multifunctional ECM protein, was one of the first endogenous inhibitors of angiogenesis to be discovered. TSP-1 and TSP-2 can inhibit EC proliferation, migration, and capillary formation. The overexpression of TSP-1 or TSP-2 on tumor cells reduces tumor vascularization and growth [112]. TSP-1 has been found to block the early stages of tumor initiation and development affecting both angiogenesis and tumor-cell apoptosis in an intestinal carcinogenesis model [113]. However, it failed to inhibit tumor-associated lymphangiogenesis or lymphatic tumor spread to regional lymph nodes [114].

31.4.9 Gene Therapy

Gene therapy may offer a new tool for the treatment of a number of cancers. Endostatin gene-transfected therapy, using cationic vector-mediated intravenous gene transfer, might be feasible for organ-targeted prevention and regulation of possible disseminated cancers [115]. A retroviral-mediated gene delivery system to affect sustained autocrine expression of the endogenous angiogenesis inhibitor TIMP-3 showed that its overexpression could inhibit tumor angiogenesis in melanoma tumor models [116].

An apoptosis modulator, bcl-2, has recently been found to play a crucial role in melanoma angiogenesis as downregulation of bcl-2 by antisense treatment inhibits angiogenesis [117]. The antiangiogenic activity of bcl-2 antisense oligonucleotides supports the clinical development of agents such as G319, which is currently in Phase III trials in melanoma patients [118].

Normal melanocytes require FGFs such as bFGF for growth in culture [119] and express FGFRs [120], but do not express bFGF [121]. However, bFGF is expressed by melanoma cells [121,122] as are other FGFs and their receptors, both in vitro and in vivo [123–125]. The biological significance of this autocrine loop has been demonstrated using antisense oligonucleotides that targeted bFGF, resulting in inhibition of melanoma cell proliferation both in vitro [126] and in vivo [127]. In addition, neutralizing anti-bFGF antibodies introduced into the cytoplasm of melanoma cells led to inhibition of proliferation [128]. Finally, increased expression of bFGF has been shown to affect both the transformed phenotype and tumorigenesis of melanocytic cells in a variety of systems [129–132]. Together, these data indicate that autocrine production of FGFs may promote melanoma tumor progression and may be an attractive target for gene therapy. Using an approach to disrupt FGF signaling via adenoviral-mediated expression of a dominant negative FGF receptor, Ozen et al. showed inhibition of FGF signaling decreases the proliferation of melanoma cells by a novel mechanism involving a decrease in cdc2 kinase activity, an increase in p21 expression, inhibition of

G2 progression, inhibition of cytokinesis, and increased cell death. In contrast, disruption of FGF signaling had only slight effects on normal melanocytes. Therefore, disruption of FGF signaling via the expression of the dominant negative FGF receptor is a potential therapeutic approach in human melanoma [133].

31.4.10 Thalidomide and Its Analogs

Thalidomide is antiangiogenic in vivo [134,135] although a metabolite is most likely responsible for this effect [136]. Thalidomide significantly inhibits EC proliferation in vitro [137], although the required concentration is 10-fold higher than that required to inhibit angiogenesis in the rat aorta assay [138]. This indicates that inhibition of angiogenesis by thalidomide is unlikely to be due to inhibition of ECs. In a recent Phase II trial, thalidomide (as a monotherapy) showed poor activity, but acceptable toxicity, in patients with metastatic melanoma. Nonetheless, the authors suggest exploration of this agent in combination with other biological agents or cytotoxic agents, such as temozolomide [139].

Potent structural analogs with reduced toxicity form two distinct classes of compound: Selective Cytokine Inhibitory Drugs (SelCIDs) and Immunomodulatory Drugs (IMiDs) [140]. IMiDs are forwarded into Phase II/III studies in patients with advanced cancer [141]. A recent Phase I study in patients with metastatic malignant melanoma found an IMiD (REVLIMID™) to be safe, well tolerated, and signs of some clinical efficacy [142]. A pivotal program for REVLIMID for refractory metastatic melanoma comprised two studies in 274 patients whose disease has progressed on treatment with dacarbazine, IL-2, IFN-α, or IFN-β. The primary end point was overall survival. The first trial to be conducted in the United States compared two different doses of the drug, whereas the second was to be conducted internationally to compare 25 mg/day of REVUMID versus placebo. By January 2004, enrollment in the melanoma trials has been competed [143].

31.4.11 Miscellaneous Compounds

NAMI-A (an acronym for new antitumor and metastasis inhibitor) is a ruthenium-based compound with selective antimetastasis activity in experimental models of solid tumors. It inhibits EC proliferation, chemotaxis, secretion of MMP-2, and displays dose-dependent antiangiogenic activity in the chorioallantoic membrane model [144]. Inhibitors of thymidine phosphorylase (an enzyme with angiogenic properties) are antiangiogenic in vitro suggesting that they may be clinically useful [145,146]. In addition, oral administration of the novel tubulin-binding drug BTO-956 (SRI International) to rats has been shown to inhibit mammary carcinoma growth and angiogenesis [147].

Cilengitide (EMD 121974) is a cyclic peptide intended to inhibit the integrins αvβ3 and αvβ5. In preclinical models it blocks ligand binding to αvβ3 integrins at nanomolar concentrations, thus altering the interaction of ECs with the ECM. This potentially results in the induction of apoptosis in activated ECs and causes the tumors to starve (courtesy of EMD pharmaceuticals—www.emdpharmaceutical.com). The compound is currently in Phase II trials for melanoma (see Table 31.2).

17-allylamino-17-demethoxygeldanamycin (17-AAG) targets heat shock protein 90 (Hsp90) and inhibit angiogenesis. Daily oral administration of 17-AAG affected the angiogenic response in matrigel in a dose-dependent manner. The hemoglobin content in the matrigel implants was significantly inhibited, and the histological analysis confirmed a decrease of CD31 positive ECs and of structures organized in cord and erythrocyte-containing vessels. In vitro, the compound inhibited dose-dependently the migration and the ECM invasiveness of HUVEC and their capacity to form capillary-like structures in matrigel [148].

17-AAG is currently in Phase I trials as a single agent and in combination with chemotherapy for advanced cancer [149].

31.5 CYTOTOXIC AGENTS IN CONJUNCTION WITH ANGIOGENESIS INHIBITORS

Some cytotoxic agents alone possess antiangiogenic activities, such as the inhibition of EC migration in vitro and microvessel formation in vivo [150]. The strategy using cytotoxic agents administered at low concentrations and over long but constant intervals is known as "metronomic" dosing. This approach has shown that cytotoxic drugs possess antiangiogenic activity and that the benefits of metronomic dosing include decreased side effects and obligatory rest periods [151,152]. Frequent administration of low doses of certain chemotherapeutic drugs may provide a safe and stable way to circumvent multidrug resistance in established orthotopically growing tumors when used in combination with antiangiogenic therapy.

Preclinical and clinical studies are assessing the use of angiogenesis inhibitors in conjunction with established anticancer agents [153]. For example, the antiangiogenic agent TNP-470 has been used in combination with cytotoxic agents such as mitomycin C, adriamycin, cisplatin, and 5-fluorouracil using the B16BL6 melanoma model. The effect of combination therapy was additive and dose-dependent, and the earlier fractionated dosing schedule exerted more enhanced antitumor effects. TNP-470 reduced the body weight by approximately 10% of control at maximum, but this toxicity was reversible and was not affected by addition of the cytotoxic agents. Therefore, the combination of angiogenesis inhibitors such as TNP and standard cytotoxic agents may be a beneficial addition to the treatment of tumors such as melanoma [154].

Another study found evidence of genistein possessing a higher antiangiogenic (rather than cytotoxic effect) in the B16 melanoma mouse tumor model when used in combination with cyclophosphamide [155]. The antiangiogenic effects of TNP-470 and IL-12 have been enhanced using a suboptimal concentration of cisplatin in a melanoma mouse model, indicating that the antitumor activity is probably due to concomitant effects on both endothelial and tumor cell compartments [156]. Other studies also found increased rates of survival in animals treated with a combination of antiangiogenic and cytotoxic therapies [157].

31.6 LIMITATIONS OF ANTIANGIOGENIC THERAPY

Despite recent advances in antiangiogenic therapy, other studies have revealed the need for caution when considering antiangiogenic strategies, particularly in the case of cancer. Genetic alterations can decrease the vascular dependence of tumor cells as evidenced by a study that demonstrated that mice-bearing tumors derived from $p53^{-/-}$ HCT116 human colorectal cancer cells were less responsive to antiangiogenic combination therapy than mice with $p53^{+/+}$ tumors [158]. Although the target of antiangiogenic therapy, the EC is genetically stable in comparison with tumor cells, this study exemplifies the importance of other factors such as genetics of the tumor cell. However, it has been suggested that increasing the dose or combining two or more angiogenesis inhibitors (that have shown virtually no toxicity in animals and humans, for example, endostatin) should obviate the problem of $p53^{-/-}$ tumor cells [158]. Other recent observations that some non-small cell lung carcinomas do not undergo any neo-angiogenesis suggest that this type of cancer may be resistant to some antiangiogenic therapy [159].

It has become evident in recent years why effective antiangiogenic agents in animal models that moved rapidly into the clinic demonstrated little or no efficacy. As mentioned earlier, the trials differed in many ways from preclinical studies, particularly in terms of drug schedules and stage of tumor progression [89]. Failures have also been linked to the need to characterize tumors before patients are included in trials of molecularly targeted therapeutics [160]. Moreover, clinical protocols used to determine efficacy of cytotoxic agents do not always apply to antiangiogenic therapy [161]. For example, while "stable disease" following a cytotoxic drug may be considered a failure because the tumor will eventually become resistant, this is less of an issue, especially with direct antiangiogenic therapies. Furthermore, a cytotoxic regimen usually destroys tumor tissue rapidly while antiangiogenic therapy is slow, requiring up to a year before any tumor regression is observed. Recent evidence showing that slower-growing tumors are as responsive as fast-growing tumors to antiangiogenic therapy indicates its potential in treating slow-growing indolent tumors [161].

Clinical end points designed to assess conventional chemotherapy (which is less effective against slow-growing tumors) still requires reevaluation. One of the major factors in evaluating a response to antiangiogenic therapy is the discovery of a reliable surrogate marker. In the case of VEGF, serum levels have been suggested to be a prognostic marker in melanoma [162] and hypoxia-induced VEGF expression in tumors has also been suggested as a useful surrogate for the effectiveness of angiogenesis inhibitors [163]. Most recently, a prospective cohort study using 324 patients with cutaneous melanoma at different clinical stages was investigated over 2 years measuring VEGF as a marker for tumor angiogenesis. A significant increase in plasma VEGF levels was found in patients with melanoma compared with healthy controls, with statistically significant differences between patients in stages I, II, and III versus those in stage IV, but not between patients in stages I, II, and III. The study confirmed that blood VEGF levels are significantly increased in patients with melanoma and, more interestingly, that the absence of plasma VEGF level increase during follow-up appears to be associated with remission [164]. Preliminary data indicate that circulating endothelial precursors might be useful as a surrogate marker of tumor angiogenesis in animals and patients receiving endostatin [160]. However, it still remains unlikely that the quantification of circulating factors alone will serve as useful surrogate markers. Ultimately, knowledge gained from earlier clinical trials and recent laboratory data should now enable investigators to set up improved clinical protocols, which should reveal the efficacy of these compounds, particularly in certain cancers.

It is now widely believed that the use of angiogenesis inhibitors in the treatment of cancer will necessitate chronic, life-long therapy to maintain disease control. Gene therapy is one avenue gaining momentum as a possible future success in the delivery of endogenous inhibitors of angiogenesis such as endostatin. Interfering with specific targeted genes may inhibit angiogenesis, but it should be borne in mind that the wide array of angiogenic factors, which become expressed during tumor progression, suggests that combination therapy will be essential for effective treatment. Furthermore, as such factors may not be disease or tumor-specific, the use of specific gene therapy as a single modality may also have its limitations.

Detailed evaluation of current angiogenesis inhibitors in terms of dosage, administration schedules, and combination strategies is required to ensure maximum efficacy of these agents. Angiogenesis inhibitors, despite exhibiting some dose-related adverse effects, will improve "quality of life" for many cancer patients due to reduced use of chemotherapy. However, if as expected, long-term administration of antiangiogenic agents is required, the side effects of chronic exposure remain unknown. In such a scenario, the interactions between angiogenesis inhibitors and chemotherapeutic agents and common medications would also require further investigation [153].

31.7 CONCLUSION

Despite a disappointing introduction, angiogenesis inhibitors are emerging as effective anticancer agents. Novel inhibitors of angiogenesis have been discovered from food constituents, existing cytotoxic agents, and newly synthesized compounds. However, detailed preclinical and clinical evaluation is required to ensure maximum efficacy of these agents. The vast range of inhibitors now identified, together with improved knowledge in terms of administration schedules, combination strategies, and gene therapy, suggest these compounds could make a profound impact on cancer therapy. In particular, if life-long administration of these agents is required, the side effects of chronic exposure must be determined. The interactions between angiogenesis inhibitors and chemotherapeutic agents and common medications would also require further investigation. Therefore, while it is prudent to retain a cautious approach using these agents, their introduction as a first-line treatment for melanoma is twofold. First, they should improve quality of life for many cancer patients by reducing chemotherapy and secondly reduce tumor burden and metastases by blocking angiogenesis [165].

REFERENCES

1. Stahl, S. et al., Genetics in melanoma. *Isr. Med. Assoc. J.*, 6(12), 774, 2004.
2. Ferlay, J., Parkin, F., and Pisani, P., Cancer Incidence and Mortality Worldwide. In: *IARC Cancer Bases*. Volume 5. IARC Press, 2001.
3. Christensten, D., Data still cloudy on association between sunscreen use and melanoma risk. *J. Natl Cancer Inst.*, 95(13), 932–933, 2003.
4. Armstrong, B.K. and Kricker, A., How much melanoma is caused by sun exposure? *Melanoma Res.*, 3, 395, 1993.
5. Pavel, S. and Smit, N.P., Risk factors for skin melanoma: genetic factors probably more important than exposure to sunlight. *Ned. Tijdschr Geneeskd.*, 148(46), 2267, 2004.
6. Roland, H. et al., Constitutive activation of the Ras-Raf signaling pathway in metastatic melanoma is associated with poor prognosis. *J. Carcinogenesis*, 3, 6, 2004.
7. Balch, C.M. et al., Prognostic factors analysis of 17,600 melanoma patients: validation of the American Joint Committee on Cancer melanoma staging system. *J. Clin. Oncol.*, 19(16), 3622, 2001.
8. Hofmann, U.B., Matrix metalloproteinases in human melanoma. *J. Invest. Dermatol.*, 115, 337, 2000.
9. Nikkola, J. et al., High expression levels of collagenase-1 and stromelysin-1 correlate with shorter disease-free survival in human metastatic melanoma. *Int. J. Cancer*, 97, 432, 2002.
10. McDonnell, C.O. et al., Tumour micrometastases: the influence of angiogenesis. *Eur. J. Sur. Oncol.*, 26, 105, 2000.
11. Demicheli, R. et al., Local recurrences following masectomy: support for the concept of tumour dormancy. *J. Natl. Cancer Inst.*, 85, 45, 1994.
12. Crowley, N.N.J. and Seigler, H.F., Relationship between disease free interval and survival in patients with recurrent melanoma. *Arch. Surg.*, 127, 1303, 1992.
13. Holmgren, L., O'Reilly, M.S., and Folkman, J., Dormancy for micrometastases: balanced proliferation and apoptosis in the presence of angiogenesis suppression. *Nature Med.*, 1, 149, 1995.
14. O'Reilly, M.S. et al., Angiostatin induces and sustains dormancy of human primary tumours in mice. *Nature Med.*, 2, 689, 1996.
15. Folkman, J., Tumor Angiogenesis: therapeutic implications. *N. Engl. J. Med.*, 181, 182, 1971.
16. Hanahan, D. and Weinberg, R.A., The hallmarks of cancer. *Cell*, 100, 57, 2000.
17. Bouck, N., Stellmach, V., and Hsu, S.C., How tumors become angiogenic. *Adv. Cancer Res.*, 69, 135, 1996.
18. Semenza, G.L., Hypoxia-inducible factor 1: master regulation of O_2 homeostasis. *Curr. Opin. Genet. Dev.*, 8, 588, 1998.
19. Carmeliet, P., Mechanisms of angiogenesis and arteriogenesis. *Nat. Med.*, 6, 389, 2000.
20. Eliceiri, B.P. et al., Selective requirement for Src kinases during VEGF-induced angiogenesis and vascular permeability. *Mol. Cell*, 4, 915, 1999.
21. Maisonpierre, P.C. et al., Angiopoietin-2, a natural antagonist for Tie2 that disrupts in vivo angiogenesis. *Science*, 277, 55, 1997.
22. Yu, Q. and Stamenkovic, I., Angiopoietin-2 is implicated in the regulation of tumor angiogenesis. *Am. J. Pathol.*, 158(2), 563, 2001.

23. Pomyje, J. et al., Angiopoietin-1, angiopoietin-2 and Tie-2 in tumour and non-tumour tissues during growth of experimental melanoma. *Melanoma Res.*, 11(6), 639, 2001.
24. Liotta, L.A. and Kohn, E.C., The microenvironment of the tumor-host interface. *Nature*, 411, 375, 2001.
25. Rabbani, S.A. and Mazar, A.P., The role of the plasminogen activation system in angiogenesis and metastasis. *Surg. Oncol. N. Am.*, 2, 393, 2001.
26. Mazar, A.P., The urokinase plasminogen activator receptor (uPAR) as a target for the diagnosis and therapy of cancer. *Anticancer Drugs*, 5, 387, 2001.
27. Laube, F., Co-localization of CD44 and urokinase-type plasminogen activator on the surface of human melanoma cells. *Anticancer Res.*, 20(6D), 5045, 2000.
28. Khatib, A.M. et al., Regulation of urokinase plasminogen activator/plasmin-mediated invasion of melanoma cells by the integrin vitronectin receptor alphaV beta3. *Int. J. Cancer*, 91(3), 300, 2001.
29. Stabuc, B. et al., Urokinase-type plasminogen activator and plasminogen activator inhibitor type 1 and type 2 in stage I malignant melanoma. *Oncol. Rep.*, 10(3), 635, 2003.
30. Pelayo, B.A., Fu, Y.M., and Meadows, G.G., Inhibition of B16BL6 melanoma invasion by tyrosine and phenylalanine deprivation is associated with decreased secretion of plasminogen activators and increased plasminogen activator inhibitors. *Clin. Exp. Metastasis*, 17(10), 841, 1999.
31. Ignar, D.M. et al., Inhibition of establishment of primary and micrometastatic tumors by a urokinase plasminogen activator receptor antagonist. *Clin. Exp. Metastasis*, 16(1), 9, 1998.
32. Lee, H.J. et al., Cambodian Phellinus linteus inhibits experimental metastasis of melanoma cells in mice via regulation of urokinase type plasminogen activator. *Biol. Pharm. Bull.*, 28(1), 27, 2005.
33. Klagsbrun, M. and Moses, M.A. Molecular angiogenesis. *Chem. Biol.*, 6, 217, 1999.
34. Moore, B.B. et al., CXC chemokines modulation of angiogenesis: the importance of balance between angiogenic and angiostatic members of the family. *J. Investig. Med.*, 46, 113, 1998.
35. Brown, N.S. and Bicknell, R., Thymidine phosphorylase, 2-deoxy-d-ribose and angiogenesis. *Biochem. J.*, 334, 1, 1998.
36. Chiarugi, V., Magnelli, L., and Gallo, O. Cox-2, iNOS and p53 as playmakers of tumor angiogenesis. *Int. J. Mol. Med.*, 2, 715, 1998.
37. Pinatavorn, P. and Ballerman, B.J. TGF-β and the endothelium during immune injury. *Kidney Int.*, 51, 1401, 1997.
38. Yoshina, S. et al., Involvement of interleukin-8, vascular endothelial growth factor, and basic fibroblast growth factor in tumor necrosis alpha-dependent angiogenesis. *Mol. Cell. Biol.*, 17, 4015, 1997.
39. Carmeliet, P. and Jain, R.K., Angiogenesis in cancer and other diseases. *Nature*, 407, 249, 2000.
40. Griffioen, A.W. and Molema, G., Angiogenesis: potentials for pharmacologic intervention in the treatment of cancer, cardiovascular diseases, and chronic inflammation. *Pharmacol. Rev.*, 52, 237, 2000.
41. Fidler, I.J., The biology of melanoma metastasis. *J. Dermatol. Surg. Oncol.*, 14, 875, 1998.
42. Skobe, M. et al., Concurrent induction of lymphangiogenesis, angiogenesis, and macrophage recruitment by vascular endothelial growth factor-C in melanoma. *Am. J. Pathol.*, 159, 893, 2001.
43. Kashani-Sabet, M. et al., Tumor vascularity in the prognostic assessment of primary cutaneous melanoma. *J. Clin. Oncol.*, 20, 1826, 2002.
44. Joukov, V. et al., A novel vascular endothelial growth factor, VEGF-C, is a ligand for the Flt4 (VEGFR-3) and KDR (VEGFR-2) receptor tyrosine kinases. *EMBO J.*, 15, 1751, 1996.
45. Lee, J. et al., Vascular endothelial growth factor-related protein: a ligand and specific activator of the tyrosine kinase receptor Flt4. *Proc. Natl. Acad. Sci. U.S.A.*, 93, 1988, 1996.
46. Padera, T.P. et al., Lymphatic metastasis in the absence of functional intratumor lymphatics. *Science*, 296, 1883, 2002.
47. Graells, J. et al., Overproduction of VEGF concomitantly expressed with its receptors promotes growth and survival of melanoma cells through MAPK and PI3K signaling. *J. Invest. Dermatol.*, 123(6), 1151, 2004.
48. Vacca, A., Angiogenesis and tumor progression in melanoma. *Recenti. Prog. Med.*, 91(11), 581, 2000.
49. Birck, A. et al., Expression of basic fibroblast growth factor and vascular endothelial growth factor in primary and metastatic melanoma from the same patients. *Melanoma Res.*, 9(4), 375, 1999.
50. Leiter, U. et al., The natural course of cutaneous melanoma. *J. Surg. Oncol.*, 86(4), 172, 2004.

51. Langley, R.R. et al., Endothelial expression of vascular cell adhesion molecule-1 correlates with metastatic pattern in spontaneous melanoma. *Microcirculation*, 8, 335, 2001.
52. McGary, E.C., Lev, D.C., and Bar-Eli, M., Cellular adhesion pathways and metastatic potential of human melanoma. *Cancer Biol. Ther.*, 1, 459, 2002.
53. Trusolino, L. and Comoglio, P., Scatter-factor and semaphorin receptors: cell signalling for invasive growth. *Nature Med.*, 2, 289, 2002.
54. Pennacchietti, S. et al., Hypoxia promotes invasive growth by transcriptional activation of the met protooncogene. *Cancer Cell*, 3, 347, 2003.
55. Steeg, P.S., Angiogenesis inhibitors: motivators of metastasis? *Nature Med.*, 9, 822, 2003.
56. Schaffer, J.V. and Bolognia, J.L., The melanocortin-1 receptor: red hair and beyond. *Arch. Dermatol.*, 137(11),1477, 2001.
57. Halachmi, S. and Gilchrest, B.A., Update on genetic events in the pathogenesis of melanoma. *Curr. Opin. Oncol.*, 13(2), 129, 2001.
58. Pollock, P.M. and Trent, J.M., The genetics of cutaneous melanoma. *Clin. Lab. Med.*, 20(4), 667, 2000.
59. Davies, H. et al., Mutations of the BRAF gene in human cancer. *Nature*, 417, 949, 2002.
60. Pollock, P.M. et al., High frequency of BRAF mutations in nevi. *Nature Genet.*, 33, 19, 2003.
61. Houben, R. et al., Constitutive activation of the Ras–Raf signaling pathway in metastatic melanoma is associated with poor prognosis. *J. Carcinogen.*, 3, 6, 2004. Mutations in nevi.
62. Dong, J., BRAF oncogenic mutations correlate with progression rather than initiation of human melanoma. *Cancer Res.*, 63(14), 3883, 2003.
63. Spendlove, H.E. et al., BRAF mutations are detectable in conjunctival but not uveal melanomas. *Melanoma Res.*, 14(6), 449, 2004.
64. Murry, B.P., Selective heparanase localization in malignant melanoma. *Int. J. Oncol.*, 26(2), 345, 2005.
65. Basche, M. et al., A phase I biological and pharmacologic study of the heparanase inhibitor PI-88 in patients with advanced solid tumors. *Clin. Cancer Res.*, 12, 5471, 2002.
66. Parish, C.R. et al., Identification of Sulfated Oligosaccharide-based Inhibitors of Tumor Growth and Metastasis Using Novel in Vitro Assays for Angiogenesis and Heparanase Activity. *Cancer Res.*, 59, 3433, 1999.
67. Cochran, S. et al., Probing the interactions of phosphosulfomannans with angiogenic growth factor by surface plasmon resonance. *J. Med. Chem.*, 46, 4601, 2003.
68. Demir, M. et al., Anticoagulant and antiprotease profiles of a novel natural heparinomimetic mannopentaose phosphate sulfate (PI-88). *Clin. Appl. Thromb. Hemost.*, 7, 131, 2001.
69. Lee, J.T. and McCubrey, J.A., Targeting the Raf kinase cascade in cancer therapy–novel molecular targets and therapeutic strategies. *Expert Opin. Ther. Targets*, 6(6), 659, 2002.
70. Wilhelm, S.M. et al., BAY 43-9006 exhibits broad spectrum oral antitumor activity and targets the RAF/MEK/ERK pathway and receptor tyrosine kinases involved in tumor progression and angiogenesis. *Cancer Res.*, 64(19), 7099, 2004.
71. Strumberg, D. et al., Phase I clinical and pharmacokinetic study of the novel Raf kinase and vascular endothelial growth factor receptor inhibitor BAY 43-9006 in patients with advanced refractory solid tumors. *J. Clin. Oncol.*, 23, 965, 2005.
72. Lee, J.T. and McCubrey, J.A. BAY-43-9006 Bayer/Onyx. *Curr. Opin. Investig. Drugs*, 4(6), 757, 2003.
73. Rak, J. and Kerbal, R.S., bFGF and tumor angiogenesis—back in the limelight? *Nature Med.*, 3(10), 1083, 1997.
74. Wedge, S.R., ZD6474 inhibits vascular endothelial growth factor signaling, angiogenesis, and tumor growth following oral administration. *Cancer Res.*, 62, 4645, 2002.
75. Carlomagno, F. et al., ZD6474, an orally available inhibitor of KDR tyrosine kinase activity, efficiently blocks oncogenic RET kinases. *Cancer Res.*, 62, 7284, 2002.
76. Itokawa, T., Antiangiogenic effect of SU5416 is partly attributable to inhibition of Flt-1 receptor signaling. *Mol. Cancer Ther.*, 1, 295, 2002.
77. Shaheen, R.M. et al., Antiangiogenic therapy targeting the tyrosine kinase receptor for vascular endothelial growth factor receptor inhibits the growth of colon cancer liver metastasis and induces tumor and endothelial cell apoptosis. *Cancer Res.*, 59(21), 5412, 1999.

78. Laird, A.D. et al., SU6668 inhibits Flk-1/KDR and PDGFb in vivo, resulting in rapid apoptosis of tumor vasculature and tumor regression in mice. *FASEB*, 16(7), 681, 2002.
79. Stopeck, A. et al., Results of a phase I dose-escalating study of the antiangiogenic agent, SU5416, in patients with advanced malignancies. *Clin. Cancer Res.*, 8, 2798, 2002.
80. Fogarty, M., Learning from angiogenesis trial failures. *The Scientist*, 18, 31, 2002.
81. Peterson, A.C. et al., Phase II study of the Flk-1 tyrosine kinase inhibitor SU5416 in advanced melanoma. *Clin. Cancer Res.*, 10, 4048, 2004.
82. O'Farrell, A.M. et al., SU11248 is a novel FLT3 tyrosine kinase inhibitor with potent activity in vitro and in vivo. *Blood*, 101(9), 3597, 2003.
83. Lamy, S., Gingras, D., and Beliveau, R., Green tea catechins inhibit vascular endothelial growth factor receptor phosphorylation. *Cancer Res.*, 62(2), 381, 2002.
84. Kondo, T. et al., Tea catechins inhibit angiogenesis in vitro, measured by human endothelial cell growth, migration and tube formation, through inhibition of VEGF receptor binding. *Cancer Lett.*, 180(2), 139, 2002.
85. Cao, Y. and Cao, R., Angiogenesis inhibited by drinking tea. *Nature*, 398(6726), 381, 1999.
86. Inai, T. et al., Inhibition of vascular endothelial growth factor (VEGF) signaling in cancer causes loss of endothelial fenestrations, regression of tumor vessels, and appearance of basement membrane ghosts. *Am. J. Pathol.*, 165(1), 35, 2004.
87. Gilbertson-Beadling, S. et al., The tetracycline analogs minocycline and doxycycline inhibit angiogenesis in vitro by a non-metalloproteinase-dependent mechanism. *Cancer Chemother. Pharmacol.*, 37(1–2), 194, 1995.
88. Boivin, D. et al., The antiangiogenic agent Neovastat (AE-941) induces endothelial cell apoptosis. *Mol. Cancer Ther.*, 1, 795, 2002.
89. Coussens, L.M., Fingleton, B., and Matrisian, L.M., Matrix metalloproteinase inhibitors and cancer: trials and tribulations. *Science*, 295(5564), 2387, 2002.
90. Sorbera, L.A., Leeson, P.A., and Bayes, M., Bevacizumab-Oncolytic. *Drugs Future*, 27, 625, 2002.
91. Trikha, M. et al., CNTO 95, a fully human monoclonal antibody that inhibits alphav integrins, has antitumor and antiangiogenic activity in vivo. *Int. J. Cancer*, 110(3), 326, 2004.
92. Linder, D.J., Interferons as antiangiogenic agents. *Curr. Oncol. Rep.*, 4, 510, 2002.
93. Kirkwood, J.M. et al., High- and low-dose interferon a-2b in high-risk melanoma: first analysis of intergroup trial E1690/S9111/C9190. *J. Clin Oncol.*, 18, 2444, 2000.
94. Sparano, J.A. et al., Evaluating antiangiogenesis agents in the clinic: The Eastern Cooperative Oncology Group Portfolio of Clinical Trials. *Clin. Cancer Res.*, 10, 1206, 2004.
95. Strieter, R.M. et al., Interferon gamma-inducible protein 10 (IP-10), a member of the C-X-C chemokine family, is an inhibitor of angiogenesis. *Biochem. Biophys. Res. Commun.*, 210(1), 51, 1995.
96. Yang, J. and Richmond, A., The angiostatic activity of interferon-inducible protein-10/CXCL10 in human melanoma depends on binding to CXCR3 but not to glycosaminoglycan. *Mol. Ther.*, 9(6), 846, 2004.
97. Jeong, S.J. et al., Bryoanthrathiophene, a new antiangiogenic constituent from the bryozoan Watersipora subtorquata (d'Orbigny, 1852). *J. Natural Products*, 65, 1344, 2002.
98. Dabrosin, C. et al., Flaxseed inhibits metastasis and decreases vascular endothelial growth factor in human breast cancer xenografts. *Cancer Lett.*, 185, 31, 2002.
99. Roy, S. et al., Anti-angiogenic property of edible berries. *Free Radical Res.*, 36, 1023, 2002.
100. Miao, R.Q. et al., Kallistatin is a new inhibitor of angiogenesis and tumor growth. *Blood*, 100, 3245, 2002.
101. Gururaj, A.E. et al., Molecular mechanisms of anti-angiogenic effect of curcumin. *Biochem. Biophys. Res. Comm.*, 297, 934, 2002.
102. Shim, J.S. et al., Hydrazinocurcumin, a novel synthetic curcumin derivative, is a potent inhibitor of endothelial cell proliferation. *Bio. Med. Chem.*, 10, 2987, 2002.
103. Zhou, J.R. et al., Inhibition of orthotopic growth and metastasis of androgen-sensitive human prostate tumors in mice by bioactive soybean components. *Prostate*, 53, 143, 2002.
104. Zhang, L. et al., Inhibitory effects of Ginkgo biloba extract on vascular endothelial growth factor in rat aortic endothelial cells. *Acta. Pharma. Sinica*, 23, 919, 2002.

105. O'Reilly, M.S. et al., Angiostatin: a novel angiogenesis inhibitor of metastases by a Lewis lung carcinoma. *Cell*, 79(2), 185, 1994.
106. Sim, B.K.L., Angiostatin and endostatin: endogenous inhibitors of tumor growth. *Cancer Metastasis Rev.*, 19, 181, 2000.
107. Kisker, O. et al., Continuous administration of endostatin by intraperitoneally implanted osmotic pump improves the efficacy and potency of therapy in a mouse xenograft tumor model. *Cancer Res.*, 61, 7669, 2001.
108. Matter, A., Tumor angiogenesis as a therapeutic target. *Drug Discov. Today*, 6(19), 1005, 2001.
109. O'Reilly, M.S. et al., Endostatin: an endogenous inhibitor of angiogenesis and tumor growth. *Cell*, 88(2), 277, 1997.
110. Li, W.W. et al., Global development of angiogenesis inhibitors for cancer: principles, progress and new paradigms. *Jpn. J. Cancer Chemother.*, 29(1), 50, 2002.
111. Phase II endostatin clinical trial to begin in people's republic of China: Clinical updates-oncology (January 2002), Angiogenesis Foundation. http://www.angio.org
112. Vailhe, B. and Feige, J.J., Thrombospondins as anti-angiogenic therapeutic agents. *Curr. Pharm. Design*, 9, 583, 2003.
113. Gutierrez, L.S. et al., Thrombospondin 1—a regulator of adenoma growth and carcinoma progression in the APC(min/+) mouse model. *Carcinogenesis*, 24, 199, 2003.
114. Hawighorst, T. et al., Thrombospondin-1 selectively inhibits early stage carcinogenesis and angiogenesis but not tumor lymphangiogenesis and lymphatic metastasis in transgenic mice. *Oncogene*, 21, 7945, 2002.
115. Nakashima, Y. et al., Endostatin gene therapy on murine lung metastases model utilizing cationic vector mediated intravenous gene delivery. *Gene Ther.*, 10(2), 123, 2003.
116. Spurbeck, W.W. et al., Enforced expression of tissue inhibitor of matrix metalloproteinase-3 affects functional capillary morphogenesis and inhibits tumor growth in a murine tumor model. *Blood*, 100(9), 3361, 2002.
117. Del Bufalo, D. et al., Treatment of melanoma cells with a bcl-2/bcl-xL antisense oligonucleotide induces antiangiogenic activity. *Oncogene*, 22(52), 8441, 2003.
118. Jansen, B. and Zangemeister-Wittke, U. Antisense therapy for cancer—the time of truth. *Lancet Oncol.*, 3(11), 672, 2002.
119. Halaban, R. et al., Basic fibroblast growth factor from human keratinocytes is a natural mitogen for melanocytes. *J. Cell Biol.*, 107, 1611, 1998.
120. Lee, S.T., Strunk, K.M., and Spritz, R.A., A survey of protein tyrosine kinase mRNAs expressed in normal human melanocytes. *Oncogene*, 8, 3403, 1993.
121. Rodeck, U. et al., Constitutive expression of multiple growth factor genes by melanoma cells but not normal melanocytes. *J. Invest. Dermatol.*, 97, 20, 1991.
122. Birck, A. et al., Expression of basic fibroblast growth factor and vascular endothelial growth factor in primary and metastatic melanoma from the same patients. *Melanoma Res.*, 9, 375, 1999.
123. Albino, A.P., Davis, B.M., and Nanus, D.M., Induction of growth factor RNA expression in human malignant melanoma: markers of transformation. *Cancer Res.*, 51, 4815, 1991.
124. Ahmed, N.U., Expression of fibroblast growth factor receptors in naevus–cell naevus and malignant melanoma. *Melanoma Res.*, 7, 299, 1997.
125. Easty, D.J., Herlyn, M., and Bennett, D.C., Abnormal protein tyrosine kinase gene expression during melanoma progression and metastasis. *Int. J. Cancer*, 60, 129, 1995.
126. Becker, D., Meier, C.B., and Herlyn, M., Proliferation of human malignant melanomas is inhibited by antisense oligodeoxynucleotides targeted against basic fibroblast growth factor. *EMBO J.*, 8, 3685, 1989.
127. Wang, Y. and Becker, D., Antisense targeting of basic fibroblast growth factor and fibroblast growth factor receptor-1 in human melanomas blocks intratumoral angiogenesis and tumor growth. *Nature Med.*, 3, 887, 1997.
128. Halaban, R. et al., bFGF as an autocrine growth factor for human melanomas. *Oncogene Res.*, 3, 177, 1998.
129. Dotto, G.P. et al., Transformation of murine melanocytes by basic fibroblast growth factor cDNA and oncogenes and selective suppression of the transformed phenotype in a reconstituted cutaneous environment. *J. Cell Biol.*, 109, 3115, 1989.

130. Nesbit, M. et al., Basic fibroblast growth factor induces a transformed phenotype in normal human melanocytes. *Oncogene*, 18, 6469, 1999.
131. Berking, C. et al., Basic fibroblast growth factor and ultraviolet B transform melanocytes in human skin. *Am. J. Pathol.*, 158, 943, 2001.
132. Meier, F. et al., Human melanoma progression in skin reconstructs: biological significance of bFGF. *Am. J. Pathol.*, 156, 193, 2000.
133. Ozen, M., Medrano, E.E., and Ittmann, M., Inhibition of proliferation and survival of melanoma cells by adenoviral-mediated expression of dominant. *Melanoma Res.*, 14, 13, 2004.
134. D'Amato, J.D. et al., Thalidomide is an inhibitor of angiogenesis. *Proc. Natl. Acad. Sci.*, 91, 4082, 1994.
135. Kenyon, B.M., Browne, F., and D'Amato, R.J., Effects of thalidomide and related metabolites in a mouse corneal model of neovascularization. *Exp. Eye Res.*, 64, 971, 1997.
136. Bauer, K.S., Dixon, S.C., and Figg, W.D., Inhibition of angiogenesis by thalidomide requires metabolic activation, which is species-dependent. *Biochem. Pharmacol.*, 55, 1827, 1998.
137. Moriera, A.L. et al., Thalidomide and a thalidomide analogue inhibit endothelial cell proliferation in vitro. *J. Neuro-Oncol.*, 43, 109, 1999.
138. Dredge, K. et al., Novel thalidomide analogues display anti-angiogenic activity independently of immunomodulatory effects. *Br. J. Cancer*, 87(10), 1166, 2002.
139. Reiriz, A.B. et al., Phase II study of thalidomide in patients with metastatic malignant melanoma. *Melanoma Res.*, 14(6), 527, 2004.
140. Corral, L.G. et al., Differential cytokine modulation and T cell activation by two distinct classes of Thd analogs that are potent inhibitors of TNF-a. *J. Immunol.*, 163, 380, 1999.
141. Marriott, J.B. et al., Immunotherapeutic and antitumor potential of thalidomide analogs. *Expert Opin. Biol. Ther.*, 1(4), 1, 2001.
142. Bartlett, J.B. et al., A Phase I study to determine the safety, tolerability and immunostimulatory activity of thalidomide analogue CC-5013 in patients with metastatic malignant melanoma and other advanced cancers. *Br. J. Cancer*, 90(5), 955, 2004.
143. Mitsiades, C.S. and Mitsiades, N., CC-5013 (Celgene). *Curr. Opin. Investig. Drugs*, 5(6), 635, 2004.
144. Vacca, A. et al., Inhibition of endothelial cell functions and of angiogenesis by the metastasis inhibitor NAMI-A. *Br. J. Cancer*, 86(6), 993, 2002.
145. Liekens, S. et al., Anti-angiogenic activity of a novel multisubstrate analogue inhibitor of thymidine phosphorylase. *FEBS Lett.*, 510(1–2), 83, 2002.
146. Klein, R.S. et al., Novel 6-substituted uracil analogs as inhibitors of the angiogenic actions of thymidine phosphroylase. *Biochem. Pharmacol.*, 62(9), 1257, 2001.
147. Shan, S. et al., The novel tubulin binding drug BTO-956 inhibits R3230AC mammary carcinoma growth and angiogenesis in Fischer 344 rats. *Clin. Cancer Res.*, 7(8), 2590, 2001.
148. Kaur, G. et al., Antiangiogenic properties of 17-(dimethylaminoethylamino)-17-demethoxygeldanamycin: an orally bioavailable heat shock protein 90 modulator. *Clin. Cancer Res.*, 10(14), 4813, 2004.
149. Goetz, M.P. et al., Phase I trial of 17-allylamino-17-demethoxygeldanamycin in patients with advanced cancer. *J. Clin. Oncol.*, 23(6), 1078, 2005.
150. Hotchkiss, K.A. et al., Inhibition of endothelial cell function in vitro and angiogenesis in vivo by docetaxel (Taxotere): association with impaired repositioning of the microtubule organizing center. *Mol. Cancer Ther.*, 1, 1191, 2002.
151. Browder, T. et al., Antiangiogenic scheduling of chemotherapy improves efficacy against experimental drug-resistant cancer. *Cancer Res.*, 60(7), 1878, 2000.
152. Hanahan, D., Bergers, G., and Bergland, E., Less is more, regularly: metronomic dosing of cytotoxic drugs can target tumor angiogenesis in mice. *J. Clin. Invest.*, 105(8), 1045, 2000.
153. Dredge, K., Dalgleish, A.G., and Marriott, J.B., Recent development in antiangiogenic therapy. *Expert Opin. Biol Ther.*, 2(8), 953, 2002.
154. Kato, T., Sato, K., Kakinuma, H., and Matsuda, Y., Enhanced suppression of tumor growth by combination of angiogenesis inhibitor O-(chloroacetyl-carbamoyl)fumagillol (TNP-470) and cytotoxic agents in mice. *Cancer Res.*, 54(19), 5143, 1994.
155. Wietrzyk, J. et al., Antiangiogenic and antitumor effects in vivo of Genistein applied alone or combined with cyclophosphamide. *Anticancer Res.*, 21, 3893, 2001.

156. Dabrowska-Iwanicka, A. et al., Augmented antitumor effects of combination therapy with TNP-470 and chemoimmunotherapy in mice. *J. Cancer Res. Clin. Oncol.*, 128, 433, 2002.
157. Kakeji, Y. and Teicher, B.A., Preclinical studies of the combination of angiogenic inhibitors with cytotoxic agents. *Invest New Drugs*, 15(1), 39, 1997.
158. Yu, J.L. et al., Effect of p53 status on tumour response to anti-angiogenic therapy. *Science*, 295(5559), 1526, 2002.
159. Folkman, J., Looking for a good endothelial address. *Cancer Cell*, 1(2), 113, 2002.
160. Kerbel, R. and Folkman, J., Clinical translation of angiogenesis inhibitors. *Nature Reviews Cancer*, 2, 727, 2002.
161. Beecken, W.D. et al., Effect of anti-angiogenic therapy on slowly growing, poorly vascularised tumours in mice. *J. Natl. Cancer Inst.*, 93(5), 382, 2001.
162. Ascierto, P.A. et al., Prognostic value of serum VEGF in melanoma patients: a pilot study. *Anticancer Res.*, 24(6), 4255, 2004.
163. Kim, E.S. et al., Potent VEGF blockade causes regression of coopted vessels in a model of neuroblastoma. *Proc. Natl. Acad. Sci. USA*, 99(17), 11399, 2002.
164. Pelletier, F. et al., Circulating vascular endothelial growth factor in cutaneous malignant melanoma. *Br. J. Dermatol.*, 152(4), 685, 2005.
165. Dredge, K., Dalgleish, A.G., and Marriott, J.B., Angiogenesis inhibitors in cancer therapy. *Current Opin. Invest. Drugs*, 4(6), 667, 2003.

32 Sunitinib and Renal Cell Carcinoma

Robert J. Motzer and Sakina Hoosen

CONTENTS

32.1 INTRODUCTION

The majority of renal tumor malignancies are classified as renal cell carcinoma (RCC), which ranks 10th as the leading cause of cancer deaths, and constitutes 3% of all solid malignancies [1]. In the United States, approximately 31,000 new cases of RCC were predicted for 2005, leading to an estimated 12,000 deaths [2] and in Europe approximately 40,000 patients are newly diagnosed each year, leading to an estimated 20,000 deaths [3].

If detected early, RCC can be treated by radical nephrectomy or nephron-sparing surgery, although disease recurrence is thought to occur in up to 40% of patients [4]. Furthermore, RCC is often asymptomatic or associated with nonspecific symptoms such as fatigue, weight loss, malaise, fever, and night sweats [5]; consequently, up to one-third of patients are initially diagnosed with locally invasive or stage IV disease [4,6]. Approximately half of all cases are

TABLE 32.1
Historical Experience with Randomized Trials of Cytokine Therapy

Regimen	*N*	ORR (%)	Survival Benefit	Reference
IFN-α	117	14	Yes	[54]
MPA	97	2		
IFN-α	169	8	No	[55]
IFN-α + thalidomide	175	3		
IL-2	138	6.5		
IFN-α	147	7.5	No	[56]
IL-2 + IFN-α	140	19		
High dose IL-2	96	21		
Low dose IL-2 (IV)	92	11	No	[57]
Low dose IL-2 (SC)	93	10		

ORR: Objective response rate; MPA: medroxyprogesterone acetate; IV: intravenous; SC: subcutaneous.

thought to be detected incidentally as a result of the widespread use of abdominal imaging (e.g., ultrasound, CT, and MRI) for other medical conditions [7,8].

The progression of advanced RCC is notoriously unpredictable [9] and metastases are highly resistant to conventional chemotherapy, radiotherapy, and hormonal therapy [10,11]. Until 2006, standard treatment comprised interferon (IFN)-α and/or interleukin (IL)-2. However, only a limited subset of patients benefited [12]. Table 32.1 highlights the historical experience with cytokine therapy. Furthermore, no effective therapy was available to patients whose disease did not respond, progressed after an initial response, or who were unable to tolerate the significant toxicity associated with cytokine-based treatment [3]. One study reported that among patients randomized to IFN-α ($n = 122$), IL-2 ($n = 125$), or IFN-α plus IL-2 ($n = 122$), there were 78, 143, and 209 grade 3/4 toxic events, respectively [13]. In general, 90% of patients treated with IFN-α experience flu-like symptoms, 85% experience nausea and vomiting, and 30%–65% experience neurological toxicity [3]. Until recently, high-dose IL-2 was the only treatment approved by the United States Food and Drug Administration (US FDA) for the treatment of metastatic RCC (mRCC) and is favored by some due to the modest proportion of patients who obtain a complete response (CR) and long-term survival [14]. However, the 5 year survival rate remains 7%–9% [15].

32.2 TUMOR PATHOBIOLOGY

Considering the high rate of disease recurrence and the limited number of patients who obtain benefit from cytokine treatment, developing a new and effective therapy has been a high priority in recent years, and research into the molecular foundation of RCC paved the way for the development of targeted therapy. Four subtypes of renal neoplasms have been reported: clear-cell, papillary, chromophobe, and oncocytoma [12]. Each arises from distinct regions of the renal epithelia and is caused by a distinct set of gene mutations. The majority of kidney tumors (up to 80%) demonstrate clear-cell carcinoma histology [4,9].

The minority of RCC incidence is due to hereditary clear-cell carcinoma [16–18], which is associated with inactivation of the von Hippel-Lindau (VHL) tumor suppressor gene (through deletion, mutation, or methylation). Defects in the VHL gene also appear to be responsible for as many as 80% of sporadic clear-cell RCC cases [16,19]. Loss of VHL gene function leads to overexpression of a number of hypoxia-responsive proteins, such as vascular endothelial growth factor (VEGF) and platelet-derived growth factor (PDGF), which act as agonists of their respective receptor tyrosine kinases (RTKs), VEGFR and PDGFR. Overexpression of RTKs is implicated in tumor growth and angiogenesis [20] and may contribute to the vascular nature of the disease. As such, the VEGF-VEGFR and PDGF-PDGFR signaling cascades are rational targets for anticancer therapy, and particularly in clear-cell RCC.

32.3 RTKs AND MOLECULARLY TARGETED THERAPIES

RTKs are cell surface receptors comprising an extracellular ligand-binding domain, a transmembrane domain, and a cytoplasmic catalytic domain. Once activated, phosphorylation of tyrosine residues located in the intracellular domain initiates downstream signaling cascades resulting in altered gene expression and cellular function [21], including DNA synthesis, cell division, differentiation, growth, progression, migration, and death. Aberrant RTK signaling has been implicated in a variety of cancers [20] and targeting multiple molecules that drive tumor proliferation and survival represents a logical therapeutic strategy.

Molecularly targeted therapies that inhibit a single or limited number of RTKs, such as the stem cell factor receptor (KIT) by imatinib mesylate, the epidermal growth factor receptor (EGFR) by erlotinib, the human epidermal growth factor receptor-2 (HER2/neu) by trastuzumab, and the VEGFR ligand (VEGF) by bevacizumab, have demonstrated clinical activity thus validating this approach. However, tumor growth, proliferation, and angiogenesis rarely result from one single, well-defined genetic mutation and the majority involve multiple signaling mutations [22]. Therefore, inhibition of multiple RTKs may result in increased clinical activity.

32.4 TARGETING VEGFR AND PDGFR PATHWAYS IN RCC

A number of studies have indicated an association between overexpression of VEGF and increased microvascular density, cancer recurrence, and ultimately decreased survival [23]. VEGF is required for endothelial cell proliferation promoting the vascularization required for primary tumor growth and for establishing new metastatic foci [23,24]. An early study by Kim et al. reported that inhibition of VEGF-induced angiogenesis suppressed tumor growth in a variety of tumor models [25]. A more recent phase II study of the anti-VEGF antibody bevacizumab (high dose; 10 mg/kg every other week) in mRCC also reported moderate clinical activity [26]. However, the investigators recommended that future treatments should inhibit VEGF in combination with other targets.

Evidence gained to date indicates that PDGFR-α and -β contribute to angiogenesis, thus making these appropriate additional targets. PDGFR-β is expressed on pericytes/smooth muscle cells, which provide mechanical support to vasculature; receptor inhibition elicits detachment of pericytes and disruption of tumor vascularity [27,28]. PDGFR-α and -β signals also induce expression of growth factors and proangiogenesis signals (including VEGF) in endothelial cells [29]. VEGFR-1 and -2 contribute to the angiogenic process; thus, inhibition of multiple VEGFRs may ensure more effective signal blockage. In preclinical experiments, selective and simultaneous inhibition of both VEGFR and PDGFR has been shown to halt tumor cell proliferation and induce tumor cell apoptosis and tumor necrosis, resulting in a far greater antitumor and antiangiogenic response than inhibition of either receptor type alone [29,30].

32.5 SUNITINIB MALATE (SU11248, SUTENT®)

Sunitinib is a rationally designed, small molecule, RTK inhibitor with antiangiogenic and antitumor activity. Based on clinical activity demonstrated in two phase II studies (discussed in more detail later) [31,32], sunitinib was approved in 2006 by the US FDA for the treatment of advanced RCC and by the European Medicines Agency (EMEA) for use in advanced and/or metastatic RCC. (The EMEA review included the interim phase III first-line data for sunitinib [33].) Sunitinib was also approved by the US FDA for the treatment of gastrointestinal stromal tumor (GIST) after disease progression on or intolerance to imatinib mesylate treatment and by the EMEA for use in unresectable and/or metastatic malignant GIST after failure of imatinib treatment due to resistance or intolerance.

32.6 PRECLINICAL ACTIVITY OF SUNITINIB

Sunitinib (Figure 32.1) emerged from a research program investigating a chemical library of indolinone compounds, which were evaluated against >80 kinases. In biochemical and cellular assays, sunitinib was identified as a potent inhibitor of a number of the class III/V split kinase domain family of RTKs including VEGFR, PDGFR, KIT, colony stimulating factor receptor-1 (CSF-1R), Fms-like tyrosine kinase-3 receptor (FLT3), and the glial cell-line derived neurotrophic factor receptor (rearranged during transfection; RET), and inhibition of function has been demonstrated in cell proliferation assays [Pfizer Inc., data on file; 34–37]. Pharmacokinetic studies revealed that sunitinib is metabolized by cytochrome P450 (CYP) 3A4 to an active primary metabolite (SU12662), which exhibits similar potency to sunitinib and similar binding to human plasma protein (90% and 95%, respectively) [38].

In vivo, sunitinib demonstrated broad and potent antitumor activity, causing tumor regression in murine models of renal (786-O), colon (Colo205 and HT-29), breast (MDA-MB-435), lung (NCI-H226 and H460), and prostate (PC3-3M-luc) cancers, while suppressing tumor growth in others, such as human glioma xenografts (SF763T) and melanoma lung metastases (B16) [39]. In addition, a 30% reduction in tumor microvessel density confirmed inhibition of tumor-related angiogenesis [40]. In mouse xenograft models, sunitinib time- and dose-dependently inhibited the phosphorylation of PDGFR-β, VEGFR-2, and KIT at plasma concentrations ≥50 ng/mL, and daily administration was sufficient to inhibit receptor phosphorylation for 12 h. Figure 32.2 shows the mechanism of action of sunitinib in RCC.

32.7 SAFETY AND CLINICAL ACTIVITY OF SUNITINIB: PHASE I STUDIES

The pharmacokinetics and safety of sunitinib were assessed in phase I dose-escalation studies [41–43]. Sunitinib was administered once daily at doses ranging from 25 to 150 mg in three different dosing regimens or cycles: 2 weeks on treatment followed by 1 week off (2/1 schedule); 2 weeks on treatment followed by 2 weeks off (2/2 schedule); or 4 weeks on treatment followed by 2 weeks off (4/2 schedule). The off-treatment period was incorporated into the regimens based on preliminary preclinical toxicology and pharmacokinetic data. Adverse events (AEs)

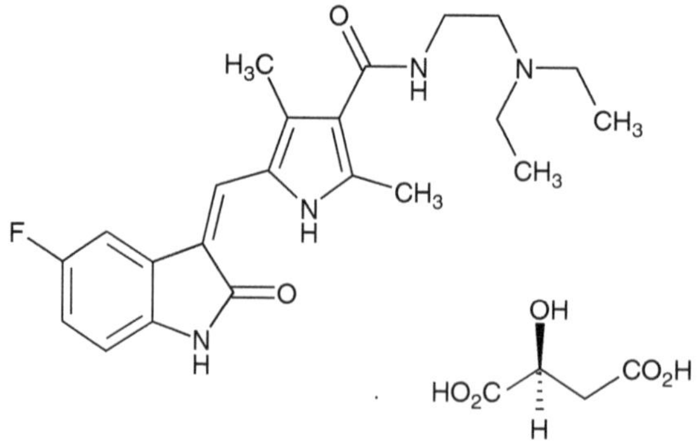

FIGURE 32.1 Chemical structure of sunitinib (SUTENT [sunitinib malate] prescribing information; 2007).

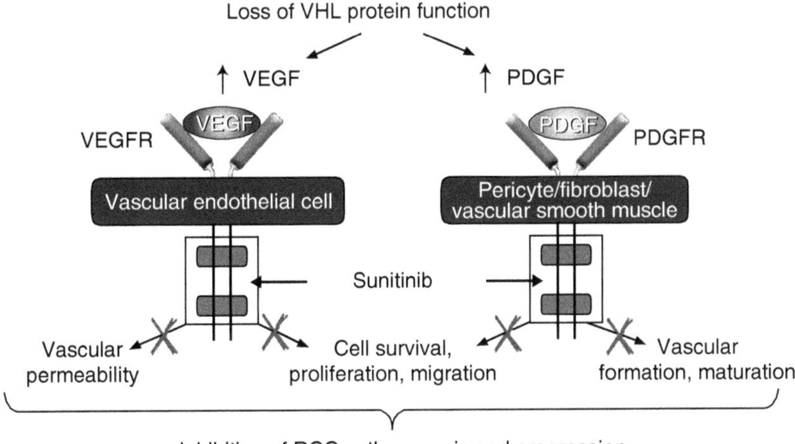

FIGURE 32.2 (See color insert following page 558.) Sunitinib mechanism of action in RCC.

were monitored throughout the phase I studies and were graded according to the National Cancer Institute (NCI) Common Toxicity Criteria (CTC), Version 2.0. The most common AEs experienced by patients included fatigue, nausea, vomiting, diarrhea, neutropenia, thrombocytopenia, and hair/skin discoloration. The severity of AEs was generally dose-dependent and fatigue/asthenia tended to increase and decrease in association with on- and off-treatment periods, respectively. The maximum tolerated dose was defined as 50 mg/day. Patients receiving this dose achieved trough plasma concentrations >50 ng/mL (sunitinib plus SU12662); in preclinical studies, this concentration was found to be sufficient to inhibit receptor phosphorylation, thereby bringing about tumor regression. In addition, based on preliminary tumor response data in phase I studies [43], the recommended dosing schedule for phase II and III studies was the 4/2 schedule, which provided optimal plasma concentration levels and acceptable toxicity. (Additional schedules are being investigated, such as continuous dosing at 37.5 mg/day.)

Pharmacokinetic data obtained from phase I and subsequent clinical trials show the terminal half-lives of sunitinib and SU12662 are 40–60 and 80–110 h, respectively, with maximum plasma concentrations reached between 6 and 12 h, postadministration. Sunitinib and SU12662 pharmacokinetics are not significantly altered by repeated daily dosing or repeated cycles, and steady-state plasma concentration levels are achieved after 10–14 days [Pfizer Inc., data on file; 38].

Preclinical observations suggesting substantial antitumor and antiangiogenic activity in a broad range of solid tumors were confirmed by clinically relevant responses achieved by patients receiving single-agent sunitinib in these phase I studies. A total of 117 patients with various advanced solid tumors (including RCC, GIST, neuroendocrine tumors, colorectal cancer, and prostate cancer) were enrolled in phase I trials. Based on Response Evaluation Criteria in Solid Tumors (RECIST) [44], 25 patients (21%) achieved either a partial response (PR; $n = 17$) or stable disease (SD; $n = 8$). Four patients with mRCC achieved a PR, lending further clinical support to this therapeutic strategy in mRCC [Pfizer Inc., data on file; 42,43].

32.8 EFFICACY AND SAFETY OF SUNITINIB IN CYTOKINE-REFRACTORY mRCC: PHASE II STUDIES

Two consecutively conducted, independent, open-label, multicenter phase II studies evaluated the efficacy and tolerability of single-agent sunitinib in patients with cytokine-refractory mRCC [31,32]. The similar design of both trials permitted a subsequent independent analysis of the pooled data. In both trials, patients were required to have measurable disease and failure of one prior cytokine therapy. In addition, patients in the first study were permitted to have any RCC histology, while in study 2, clear-cell histology was a requirement. Other eligibility criteria specific to study 2 included radiographic documentation of progression and prior nephrectomy. Patients received sunitinib 50 mg once daily on the 4/2 schedule. Treatment continued until disease progression, unacceptable toxicity, or withdrawal of consent. Based on individual patient tolerability, dose reduction to 37.5 mg/day and then 25 mg/day was permitted. The primary end point in both trials was objective response rate (ORR) by RECIST (assessed every 1–2 cycles), defined as the proportion of patients with a confirmed CR or PR. Secondary end points included time to tumor progression (TTP) in the first study, progression-free survival (PFS) in the second study, overall survival (OS), and safety.

32.8.1 Efficacy

In trial 1 [31], 63 patients with mRCC (87% with clear-cell histology) received a median of 9 months of therapy (range <1–24+ months). The ORR was 40% (25 patients with PR) and a further 17 patients (27%) had SD $\geq$ 3 months (Table 32.2). Median TTP was 8.7 months (95% CI: 5.5–10.7) and median OS was 16.4 months (95% CI: 10.8–not available). In addition, the majority of patients in trial 1 had a reduction in measurable disease following sunitinib treatment. Computed tomography scan images from a patient who achieved a PR in trial 1 are shown in Figure 32.3.

In trial 2 (ongoing) [32], 106 patients received a median of 5 cycles of therapy (range 0–11). Of 105 patients evaluable for efficacy (one patient with a diagnosis of clear-cell RCC was withdrawn after a repeat biopsy showed a diagnosis of a different cancer type), the ORR was 44%; 1 patient (1%) achieved a CR and 45 patients (43%) a PR. A further 23 patients (22%) had SD $\geq$ 3 months (Table 32.2). Median PFS based on third-party independent assessment was 8.3 months (95% CI: 7.8–14.5) and at the time of analysis, median OS had not been reached; the 6-month survival was 79% (95% CI: 70%–86%).

In a pooled analysis [32], which combined the demographics and efficacy data from the two phase II trials ($N = 168$), the ORR was 42% (Table 32.2) and the median PFS was 8.2 months (95% CI: 7.8–10.4) (Figure 32.4). Patients who responded to treatment demonstrated a longer period without disease progression. Median PFS for responders was 14.8 months compared with 7.9 months for patients with SD $\geq$ 3 months and 2.1 months for patients with SD < 3 months or progressive disease (PD). The pooled analysis also evaluated various

TABLE 32.2
Responses Achieved with Sunitinib in Phase II Studies in Patients with Cytokine-Refractory mRCC

Response, *n* (%)	Trial 1 (N=63)	Trial 2[a] (N=105)	Pooled Analysis (N=168)
Overall response	25 (40)	46 (44)	71 (42)
CR	0	1 (1)	1 (1)
PR	25 (40)	45 (43)	70 (42)
SD $\geq$ 3 months	17 (27)	23 (22)	40 (24)
PD, SD < 3 months or not evaluable	21 (33)	36 (34)	57 (34)

Source: Reproduced from Motzer, R.J., Rini, B.I., Bukowski, R.M., Curti, B.D., George, D.J., Hudes, G.R., Redman, B.G., Margolin, K.A., Merchan, J.R., Wilding, G., Ginsberg, M.S., Bacik, J., Kim, S.T., Baum, C.M., and Michaelson, M.D., *JAMA*, 295(21), 2516, 2006. With permission.

[a] Study ongoing.

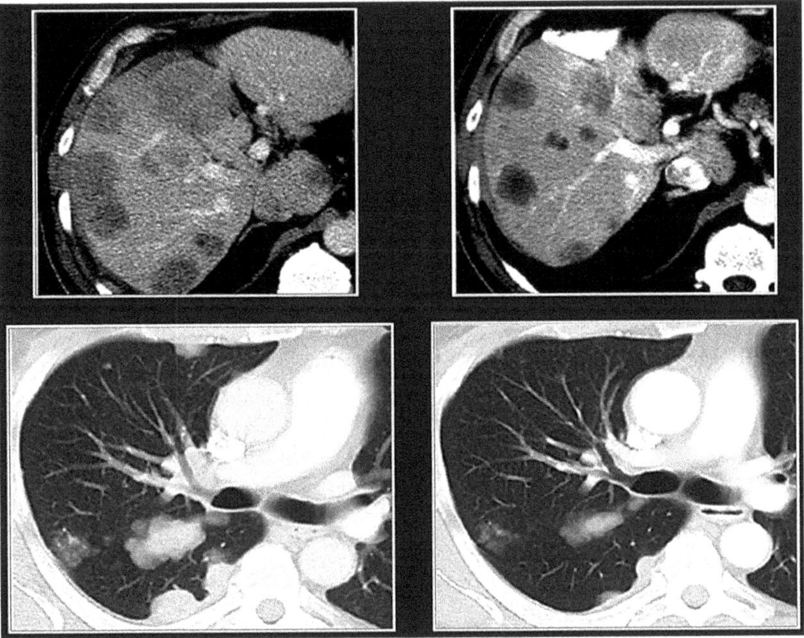

FIGURE 32.3 Tumor responses in a patient with multiple hepatic, lung, and pleural metastases following sunitinib treatment in a phase II trial of mRCC (trial 1). (Reproduced from Motzer, R.J., Michaelson, M.D., Redman, B.G., Hudes, G.R., Wilding, G., Figlin, R.A., Ginsberg, M.S., Kim, S.T., Baum, C.M., DePrimo, S.E., Li, J.Z., Bello, C.L., Theuer, C.P., George, D.J., and Rini, B.I., *J. Clin. Oncol.*, 24(1), 16, 2006. With permission.)

pretreatment clinical features as potential indicators of response. Overall, 78% of responders versus 43% of nonresponders ($P < 0.001$, 2-tailed Fisher exact test) had normal baseline hemoglobin levels rather than levels below the lower limit of normal, and 66% of responders versus 45% of nonresponders ($P < 0.008$, Fisher exact test) had an Eastern Cooperative Oncology Group performance status (ECOG PS) of 0 rather than 1. A univariate analysis

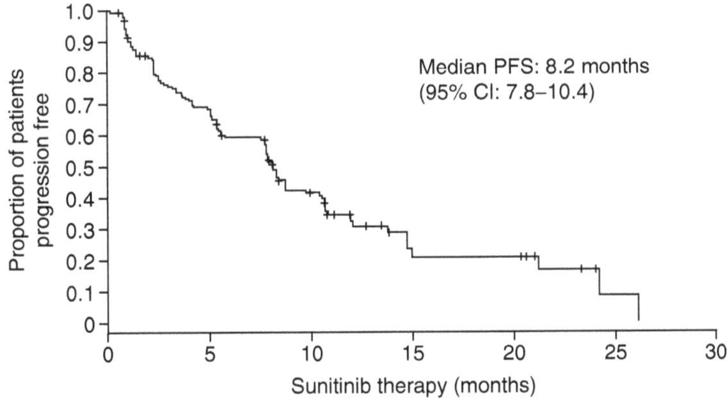

FIGURE 32.4 Kaplan–Meier plot of PFS in the pooled analysis of two phase II studies in mRCC ($N = 168$). (Reproduced from Motzer, R.J., Rini, B.I., Bukowski, R.M., Curti, B.D., George, D.J., Hudes, G.R., Redman, B.G., Margolin, K.A., Merchan, J.R., Wilding, G., Ginsberg, M.S., Bacik, J., Kim, S.T., Baum, C.M., and Michaelson, M.D., *JAMA*, 295(21), 2516, 2006. With permission.)

also demonstrated these clinical characteristics (normal serum hemoglobin and favorable ECOG PS) were associated with longer PFS. Poor performance in addition to anemia has been previously reported to be associated with reduced survival in second-line therapy for mRCC [45].

The unprecedented responses and durability of response achieved by sunitinib treatment in these phase II trials compared with historical observations of conventional second-line therapies led to accelerated US FDA and EMEA approval for use in advanced RCC.

32.8.2 Safety

In trials 1 and 2, sunitinib demonstrated a similar and favorable tolerability profile; the majority of AEs reported were manageable with temporary delays, dose reduction, and/or standard medical interventions. In trial 1, the most commonly reported NCI CTC grade 3, nonhematological treatment-related AEs were fatigue (11%), diarrhea (3%), nausea (3%), and vomiting (3%); no nonhematological AEs were experienced at grade 4 severity. The most commonly reported grade 3/4 laboratory abnormalities were lymphopenia (32%), elevated lipase (21%; not associated with symptoms of pancreatitis), neutropenia (13%), and anemia (10%). Twenty-two patients had their sunitinib dose reduced to 37.5 mg/day, and two of these patients underwent further dose reduction to 25 mg/day. The reasons for dose reductions were asymptomatic elevated lipase or amylase and fatigue. Five patients had their sunitinib dose increased to 62.5 mg/day and one patient to 75 mg/day; however, no evidence of improved response was observed at these dosing levels.

In trial 2, the most commonly reported grade 3 (NCI Common Terminology Criteria for Adverse Events Version 3.0 [CTCAE]), nonhematological treatment-related AEs were fatigue (11%), hand–foot syndrome (7%), hypertension (6%), stomatitis (5%), and diarrhea (3%); no nonhematological AEs were experienced at grade 4 severity. The most commonly reported grade 3/4 laboratory abnormalities were elevated lipase (17%; not associated with symptoms of pancreatitis), neutropenia (16%; not associated with fever or sepsis), thrombocytopenia (6%), and anemia (6%). Sunitinib was discontinued due to AEs in 12 patients (11%).

32.8.3 Biomarkers of Response to Sunitinib

The advent of molecularly targeted therapies for mRCC has sparked a similar level of research into molecular tumor markers, the consequences of which are likely to allow an increased variety of treatment strategies based on individual patient profiles. In trial 1, the plasma levels of serially collected VEGF and VEGFR-2 were analyzed for correlation with drug exposure and objective responses. At the end of the first cycle, VEGF levels increased greater than threefold (relative to baseline) in 24 of 54 cases (44%) and levels of soluble VEGFR-2 (sVEGFR-2) decreased by at least 30% in 50 of 55 cases (91%) and by at least 20% in all cases. In addition, of 25 patients achieving a PR, significantly larger changes in both VEGF and sVEGFR-2 were observed compared with 32 patients who had SD or PD (only patients with a baseline and one posttreatment time point were included) [46].

32.8.4 Patient Reported Outcomes

In trial 1, patient reported fatigue was assessed weekly using the Functional Assessment of Chronic Illness Therapy-Fatigue (FACIT-Fatigue) scale during the first four cycles of treatment (24 weeks). The FACIT-Fatigue scale is a 13-item, validated questionnaire that assesses the impact of patient reported fatigue on daily activities and function. The possible score range is 0–52, with 0 representing the most fatigue and 52 the least. As a reference, the US general population has a mean score of 43.6, nonanemic cancer patients 40.0, and anemic

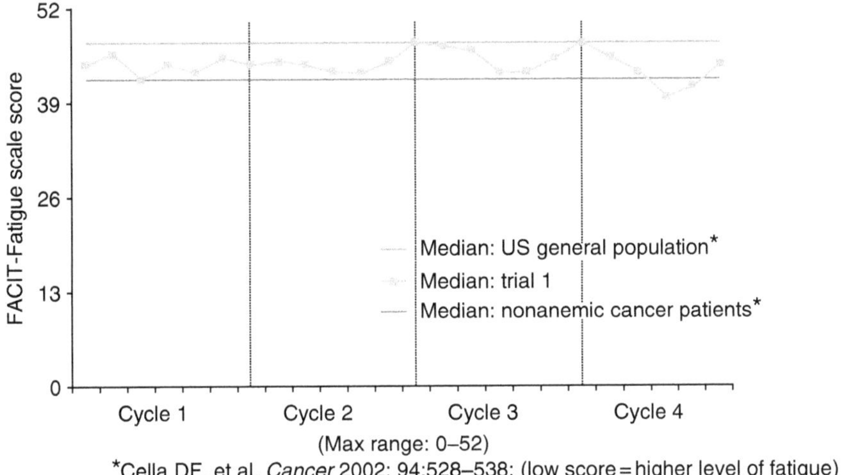

FIGURE 32.5 (See color insert following page 558.) FACIT-Fatigue scores obtained by patients receiving sunitinib in trial 1 (phase II study). (Reproduced from Motzer, R.J., Michaelson, M.D., Redman, B.G., Hudes, G.R., Wilding, G., Figlin, R.A., Ginsberg, M.S., Kim, S.T., Baum, C.M., DePrimo, S.E., Li, J.Z., Bello, C.L., Theuer, C.P., George, D.J., and Rini, B.I., *J. Clin. Oncol.*, 24(1), 16, 2006. With permission.)

cancer patients 23.9. For the FACIT-Fatigue scale, the established minimally important difference is a 3–4 point change [47]. The mean baseline FACIT-Fatigue score was 40.4, representing more fatigue than the US general population. Baseline FACIT-Fatigue scores correlated with baseline ECOG PS ($P < 0.001$); patients with a better ECOG PS reported less fatigue [48]. During the first four cycles of therapy, a mean change from baseline ($\pm$ standard deviation) of -2.2 ± 7.8 was reported, which is less than the established minimally important difference [47]. In general, within treatment cycles, mean scores decreased (more fatigue) and increased (less fatigue) during on- and off-treatment periods, respectively, but not to a level that was considered significant (Figure 32.5).

32.9 EFFICACY AND SAFETY OF SUNITINIB AS FIRST-LINE THERAPY IN mRCC: PIVOTAL PHASE III RANDOMIZED TRIAL

As a result of the exceptional activity demonstrated by sunitinib in the second-line setting for mRCC, which exceeded the response rates reported for cytokines as first-line treatment, sunitinib versus IFN-α is currently being investigated as a first-line treatment in an international, randomized, phase III trial [33]. Patients with clear-cell mRCC receive oral sunitinib at a dose of 50 mg/day on the 4/2 schedule or subcutaneous injection of IFN-α at a dose of 9 MU three times per week (3 MU in the first week, 6 MU in the second week, and 9 MU in all subsequent weeks). The primary end point is PFS, and secondary end points included ORR (by RECIST), OS, and safety. Data from a planned interim analysis were recently published [33].

A total of 750 patients were randomized to treatment, 375 to sunitinib and 375 to IFN-α, and baseline characteristics were well balanced between the two groups; the median age was 60 years and 90% of patients had prior nephrectomy. At the time of analysis, 248 patients (66%) were continuing treatment with sunitinib versus 126 patients (34%) with IFN-α. Sunitinib treatment resulted in a statistically significant improvement in both PFS and ORR. Median PFS by independent central review was 11 months for sunitinib compared with 5 months for IFN-α (hazard ratio [HR] 0.42; 95% CI: 0.32–0.54; $P < 0.001$). The ORR by independent central review was 31% for sunitinib versus 6% for IFN-α ($P < 0.001$), and by investigator assessment was 37% for sunitinib versus 9% for IFN-α ($P < 0.001$). Eight percent of patients in the sunitinib group withdrew from the study due to an AE versus 13% in the IFN-α group ($P = 0.05$).

These data complement the findings of the second-line phase II trials and establish sunitinib as a new reference standard for first-line treatment of clear-cell mRCC.

32.10 OTHER SUNITINIB STUDIES IN mRCC

In order to strive for continued improvement in mRCC therapy by providing patients with tailored treatment and thus potentially optimal therapy, a number of ongoing phase II clinical trials with sunitinib have been initiated including sunitinib administered at a different dosing schedule, in different patient populations, and in combination with other targeted agents.

32.10.1 Sunitinib Administered in a Continuous Dosing Regimen

In a continuous dosing study [49], 107 patients with measurable mRCC, an ECOG PS of 0 or 1, and failure of one prior cytokine-based regimen were randomized to receive sunitinib either in the morning ($n = 54$) or in the evening ($n = 53$), given once daily at a starting dose of 37.5 mg. Dose reduction to 25 mg/day was permitted based on individual patient tolerability. The primary end point was RECIST-defined ORR.

Sixty-six patients received continuous sunitinib treatment at 37.5 mg/day for up to 28 weeks, and 23 patients (21.5%) discontinued treatment (14 due to disease progression, 7 due to AEs, and 2 due to consent withdrawal). Eighteen (16.8%) patients required dose reduction to 25 mg/day due to grade 2 or 3 AEs, which included mucositis, thrombocytopenia, nausea/diarrhea, neutropenia, hand–foot syndrome, and asthenia. These preliminary results were comparable between patients who received sunitinib in the morning versus evening.

32.10.2 Sunitinib in Patients with Bevacizumab-Refractory mRCC

In a phase II study [50], patients with measurable, clear-cell mRCC and disease progression (as defined by RECIST) within 3 months of bevacizumab-based therapy received sunitinib (50 mg/day on the 4/2 schedule). Of 61 patients for whom interim data were available, some degree of tumor shrinkage was observed in 48 patients (87%), including 6 patients (11%; 95% CI: 4%–22%) with PR. The most commonly reported AEs included fatigue, diarrhea, dysgeusia, and nausea, and serious treatment-related AEs included fatigue, diarrhea, nausea, and one fatal cerebral hemorrhage. Three patients (5%) discontinued therapy due to AEs. Preliminary efficacy data suggest that sunitinib may inhibit signaling pathways involved in bevacizumab resistance, although additional studies are needed to determine the precise mechanism of response to sunitinib.

32.10.3 Combination Study of Sunitinib plus Gefitinib

While the above studies have investigated sunitinib as monotherapy, a phase I study has also explored sunitinib plus gefitinib (an epidermal growth factor receptor [EGFR] inhibitor) as combination therapy in cytokine-refractory mRCC [51]. The primary objective was to determine the maximum tolerated dose and overall safety of sunitinib plus gefitinib. Eleven patients with measurable, clear-cell mRCC were given sunitinib on the 4/2 schedule at one of two doses: 37.5 mg/day ($n = 4$) or 50 mg/day ($n = 7$). Gefitinib was coadministered daily at a dose of 250 mg, starting on day 10 of the first sunitinib cycle. The maximum tolerated dose of sunitinib when given in combination with gefitinib was established as 37.5 mg/day. Grade 3 AEs included diarrhea ($n = 2$) and hand–foot syndrome ($n = 2$). Grade 3 laboratory abnormalities included neutropenia ($n = 2$; one patient experienced grade 4) and thrombocytopenia ($n = 2$).

Overall, 5 of 11 patients had demonstrated PR. The preliminary data support preclinical observations that the combination of VEGFR and EGFR inhibition may provide a broader spectrum of tyrosine kinase inhibition and act synergistically to enhance antitumor activity. Patient accrual to a phase II portion of the study with this combination is now under way.

32.10.4 Sunitinib Expanded-Access Study

The aim of this ongoing expanded-access study was to make sunitinib available to patients who were ineligible for participation in any ongoing phase I, II, or III clinical trials but who it was felt would derive clinical benefit from sunitinib, as assessed by the investigating physician [52]. Sunitinib was made available to these patients on a compassionate-use basis, to expand access to the drug and also allow patients to be treated in countries where regulatory approval was yet to be granted. In addition, this study further assesses the safety and efficacy (end points: ORR, TTP, PFS, and OS) of sunitinib. Sunitinib 50 mg/day is administered on the 4/2 schedule with dose reduction permitted for toxicity. It is hoped that the results of this study will provide data from a large pool of patients and may give a better insight into dosing and other patient populations who may derive benefit from sunitinib.

32.11 DISCUSSION AND FUTURE DIRECTIONS

The preclinical and phase I, II and III studies with single-agent sunitinib conducted in the second- and first-line setting demonstrate that the targets of sunitinib are integral to the signaling cascades that directly and indirectly regulate tumor growth, survival, and angiogenesis, and that inhibition of these multiple RTKs results in clinical activity [Pfizer Inc., data on file; 31–37,43]. The efficacy results reported here are among the most encouraging presented, to date, for second-line and first-line treatment of mRCC. Activity in bevacizumab-refractory disease [50] and in combination with gefitinib [51] further confirms that sunitinib has substantial antitumor activity and is well tolerated in patients with mRCC.

The safety profile of sunitinib has been generally consistent across a variety of solid tumor types, with similar AEs reported in a number of phase II trials in mRCC, a phase III trial in mRCC and a placebo-controlled phase III study of sunitinib in patients with GIST [53]. Sunitinib has a favorable tolerability profile; the majority of AEs experienced by patients are mild to moderate in severity, and are generally manageable with standard medical intervention, dose interruptions, or dose reductions. The safety profile of sunitinib in a dosing schedule of 37.5 mg/day administered continuously is comparable with that observed with the current recommended intermittent dosing regimen of 50 mg/day at the 4/2 schedule. In addition, administration of sunitinib orally once a day with no regard to meals offers greater convenience for patients.

Recent phase III first-line data confirm the results of the second-line phase II trials and show clinical and statistically significant (more than twofold) improvement in PFS for sunitinib compared with conventional IFN-α treatment. These data demonstrate that sunitinib is now the new reference standard for first-line therapy of mRCC.

Other ongoing and future approaches with sunitinib in RCC include further combination studies (with chemotherapy, biologic therapy, and with other targeted agents), studies in the adjuvant and neoadjuvant setting, and further biomarker research to identify patients who are most likely to respond to sunitinib therapy.

The past 5 years have seen rapid advances in our understanding of the molecular mechanisms associated with RCC and the future of anticancer therapeutics may largely focus on drugs such as sunitinib, which target the molecules or mechanisms that drive tumor proliferation and survival. Data obtained to date have validated mechanism-directed RCC therapy based on tumor-specific molecular features.

REFERENCES

1. Cohen HT and McGovern FJ. Renal-cell carcinoma. *N Engl J Med* 2005;353(23):2477–2490.
2. Jemal A, Murray T, Ward E, Samuels A, Tiwari RC, Ghafoor A, Feuer EJ, and Thun MJ. Cancer statistics, 2005. *CA Cancer J Clin* 2005;55(1):10–30.
3. Schoffski P, Dumez H, Clement P, Hoeben A, Prenen H, Wolter P, Joniau S, Roskams T, and Van Poppel H. Emerging role of tyrosine kinase inhibitors in the treatment of advanced renal cell cancer: a review. *Ann Oncol* 2006;17(8):1185–1196.
4. Lam JS, Leppert JT, Belldegrun AS, and Figlin RA. Novel approaches in the therapy of metastatic renal cell carcinoma. *World J Urol* 2005;23(3):202–212.
5. Godley PA and Taylor M. Renal cell carcinoma. *Curr Opin Oncol* 2001;13(3):199–203.
6. Janzen NK, Kim HL, Figlin RA, and Belldegrun AS. Surveillance after radical or partial nephrectomy for localized renal cell carcinoma and management of recurrent disease. *Urol Clin North Am* 2003;30(4):843–852.
7. Porena M, Vespasiani G, Rosi P, Costantini E, Virgili G, Mearini E, and Micali F. Incidentally detected renal cell carcinoma: role of ultrasonography. *J Clin Ultrasound* 1992;20(6):395–400.
8. Jayson M, Sanders H. Increased incidence of serendipitously discovered renal cell carcinoma. *Urology* 1998;51(2):203–205.
9. Pantuck AJ, Zeng G, Belldegrun AS, and Figlin RA. Pathobiology, prognosis, and targeted therapy for renal cell carcinoma: exploiting the hypoxia-induced pathway. *Clin Cancer Res* 2003;9(13):4641–4652.
10. Lilleby W and Fossa SD. Chemotherapy in metastatic renal cell cancer. *World J Urol* 2005;23(3):175–179.
11. Rohrmann K, Staehler M, Haseke N, Bachmann A, Stief CG, and Siebels M. Immunotherapy in metastatic renal cell carcinoma. *World J Urol* 2005;23(3):196–201.
12. Motzer RJ, Bander NH, and Nanus DM. Renal-cell carcinoma. *N Engl J Med* 1996;335(12):865–875.
13. Negrier S, Perol D, Ravaud A, Chevreau C, Bay JO, Delva R, Sevin E, Caty A, Tubiana-Mathieu N, and Escudier B and the French Immunotherapy Intergroup. Do cytokines improve survival in patients with metastatic renal cell carcinoma (mRCC) of intermediate prognosis? Results of the prospective randomized PERCY Quattro trial. *Proc Am Soc Clin Oncol* 2005;23(16s) (Abstract 4511).
14. Motzer RJ, Mazumdar M, Bacik J, Russo P, Berg WJ, and Metz EM. Effect of cytokine therapy on survival for patients with advanced renal cell carcinoma. *J Clin Oncol* 2000;18(9):1928–1935.
15. Chow WH, Devesa SS, Warren JL, and Fraumeni JF, Jr. Rising incidence of renal cell cancer in the United States. *JAMA* 1999;281(17):1628–1631.
16. Gnarra JR, Tory K, Weng Y, Schmidt L, Wei MH, Li H, Latif F, Liu S, Chen F, Duh FM, et al. Mutations of the VHL tumour suppressor gene in renal carcinoma. *Nat Genet* 1994;7(1):85–90.
17. Gnarra JR, Zhou S, Merrill MJ, Wagner JR, Krumm A, Papavassiliou E, Oldfield EH, Klausner RD, and Linehan WM. Post-transcriptional regulation of vascular endothelial growth factor mRNA by the product of the VHL tumor suppressor gene. *Proc Natl Acad Sci USA* 1996;93(20): 10589–10594.
18. George DJ and Kaelin WG, Jr. The von Hippel-Lindau protein, vascular endothelial growth factor, and kidney cancer. *N Engl J Med* 2003;349(5):419–421.
19. Kim WY and Kaelin WG. Role of VHL gene mutation in human cancer. *J Clin Oncol* 2004;22(24):4991–5004.
20. Krause DS and Van Etten RA. Tyrosine kinases as targets for cancer therapy. *N Engl J Med* 2005;353(2):172–187.
21. Schlessinger J. Cell signaling by receptor tyrosine kinases. *Cell* 2000;103(2):211–225.
22. Hanahan D and Weinberg RA. The hallmarks of cancer. *Cell* 2000;100(1):57–70.
23. Parikh AA and Ellis LM. The vascular endothelial growth factor family and its receptors. *Hematol Oncol Clin North Am* 2004;18(5):951–971, vii.
24. Ferrara N, Gerber HP, and LeCouter J. The biology of VEGF and its receptors. *Nat Med* 2003;9(6):669–676.
25. Kim KJ, Li B, Winer J, Armanini M, Gillett N, Phillips HS, and Ferrara N. Inhibition of vascular endothelial growth factor-induced angiogenesis suppresses tumour growth in vivo. *Nature* 1993; 362(6423):841–844.

26. Yang JC, Haworth L, Sherry RM, Hwu P, Schwartzentruber DJ, Topalian SL, Steinberg SM, Chen HX, and Rosenberg SA. A randomized trial of bevacizumab, an anti-vascular endothelial growth factor antibody, for metastatic renal cancer. *N Engl J Med* 2003;349(5):427–434.
27. Plate KH, Breier G, Weich HA, and Risau W. Vascular endothelial growth factor is a potential tumour angiogenesis factor in human gliomas in vivo. *Nature* 1992;359(6398):845–848.
28. Sundberg C, Ljungstrom M, Lindmark G, Gerdin B, and Rubin K. Microvascular pericytes express platelet-derived growth factor-beta receptors in human healing wounds and colorectal adenocarcinoma. *Am J Pathol* 1993;143(5):1377–1388.
29. Bergers G, Song S, Meyer-Morse N, Bergsland E, and Hanahan D. Benefits of targeting both pericytes and endothelial cells in the tumor vasculature with kinase inhibitors. *J Clin Invest* 2003;111(9):1287–1295.
30. Erber R, Thurnher A, Katsen AD, Groth G, Kerger H, Hammes HP, Menger MD, Ullrich A, and Vajkoczy P. Combined inhibition of VEGF and PDGF signaling enforces tumor vessel regression by interfering with pericyte-mediated endothelial cell survival mechanisms. *Faseb J* 2004;18(2): 338–340.
31. Motzer RJ, Michaelson MD, Redman BG, Hudes GR, Wilding G, Figlin RA, Ginsberg MS, Kim ST, Baum CM, DePrimo SE, Li JZ, Bello CL, Theuer CP, George DJ, and Rini BI. Activity of SU11248, a multitargeted inhibitor of vascular endothelial growth factor receptor and platelet-derived growth factor receptor, in patients with metastatic renal cell carcinoma. *J Clin Oncol* 2006;24(1):16–24.
32. Motzer RJ, Rini BI, Bukowski RM, Curti BD, George DJ, Hudes GR, Redman BG, Margolin KA, Merchan JR, Wilding G, Ginsberg MS, Bacik J, Kim ST, Baum CM, and Michaelson MD. Sunitinib in patients with metastatic renal cell carcinoma. *JAMA* 2006;295(21):2516–2524.
33. Motzer RJ, Hutson TE, Tomczak P, Michaelson MD, Bukowski RM, Rixe O, Oudard S, Negrier S, Szczylik C, Kim ST, Chen I, Bycott PW, Baum CM, and Figlin RA. Sunitinib versus interferon alfa in metastatic renal-cell carcinoma. *N Engl J Med* 2007;356(2):115–124.
34. Abrams TJ, Lee LB, Murray LJ, Pryer NK, and Cherrington JM. SU11248 inhibits KIT and platelet-derived growth factor receptor beta in preclinical models of human small cell lung cancer. *Mol Cancer Ther* 2003;2(5):471–478.
35. Mendel DB, Laird AD, Xin X, Louie SG, Christensen JG, Li G, Schreck RE, Abrams TJ, Ngai TJ, Lee LB, Murray LJ, Carver J, Chan E, Moss KG, Haznedar JO, Sukbuntherng J, Blake RA, Sun L, Tang C, Miller T, Shirazian S, McMahon G, and Cherrington JM. In vivo antitumor activity of SU11248, a novel tyrosine kinase inhibitor targeting vascular endothelial growth factor and platelet-derived growth factor receptors: determination of a pharmacokinetic/pharmacodynamic relationship. *Clin Cancer Res* 2003;9(1):327–337.
36. Murray LJ, Abrams TJ, Long KR, Ngai TJ, Olson LM, Hong W, Keast PK, Brassard JA, O'Farrell AM, Cherrington JM, and Pryer NK. SU11248 inhibits tumor growth and CSF-1R-dependent osteolysis in an experimental breast cancer bone metastasis model. *Clin Exp Metastasis* 2003; 20(8):757–766.
37. O'Farrell AM, Abrams TJ, Yuen HA, Ngai TJ, Louie SG, Yee KW, Wong LM, Hong W, Lee LB, Town A, Smolich BD, Manning WC, Murray LJ, Heinrich MC, and Cherrington JM. SU11248 is a novel FLT3 tyrosine kinase inhibitor with potent activity in vitro and in vivo. *Blood* 2003;101(9):3597–3605.
38. SUTENT (sunitinib malate) prescribing information. New York, NY: Pfizer Inc., February 2007. Available at www.pfizer.com. 2006.
39. Sakamoto JM. SU-11248 SUGEN. *Curr Opin Investig Drugs* 2004;5(12):1329–1339.
40. Marzola P, Degrassi A, Calderan L, Farace P, Nicolato E, Crescimanno C, Sandri M, Giusti A, Pesenti E, Terron A, Sbarbati A, and Osculati F. Early antiangiogenic activity of SU11248 evaluated in vivo by dynamic contrast-enhanced magnetic resonance imaging in an experimental model of colon carcinoma. *Clin Cancer Res* 2005;11(16):5827–5832.
41. Demetri DG, George S, Heinrich MC, et al. Clinical activity and tolerability of the multi-targeted tyrosine kinase inhibitor SU11248 in patients with metastatic gastrointestinal stromal tumor refractory to imatinib mesylate. *Proc Am Soc Clin Oncol* 2003;22:814 (Abstract 3273).

42. Rosen L, Mulay M, Long J, Wittner J, Brown J, Martino A-M, Bello CL, Walter S, Scigalla P, and Zhu J. Phase I trial of SU11248, a novel tyrosine kinase inhibitor in advanced solid tumors. *Proc Am Soc Clin Oncol* 2003;22:191 (Abstract 765).
43. Faivre S, Delbaldo C, Vera K, Robert C, Lozahic S, Lassau N, Bello C, Deprimo S, Brega N, Massimini G, Armand JP, Scigalla P, and Raymond E. Safety, pharmacokinetic, and antitumor activity of SU11248, a novel oral multitarget tyrosine kinase inhibitor, in patients with cancer. *J Clin Oncol* 2006;24(1):25–35.
44. Therasse P, Arbuck SG, Eisenhauer EA, Wanders J, Kaplan RS, Rubinstein L, Verweij J, Van Glabbeke M, van Oosterom AT, Christian MC, and Gwyther SG. New guidelines to evaluate the response to treatment in solid tumors. European Organization for Research and Treatment of Cancer, National Cancer Institute of the United States, National Cancer Institute of Canada. *J Natl Cancer Inst* 2000;92(3):205–216.
45. Motzer RJ, Bacik J, Schwartz LH, Reuter V, Russo P, Marion S, and Mazumdar M. Prognostic factors for survival in previously treated patients with metastatic renal cell carcinoma. *J Clin Oncol* 2004;22(3):454–463.
46. Deprimo SE, Bello CL, Smeraglia J, et al. Soluble protein biomarkers of pharmacodynamic activity of the multitargeted tyrosine kinase inhibitor SU11248 in patients with metastatic renal cell carcinoma [abstract]. 96th Annual Meeting of the American Association for Cancer Research, 16–20 April 2005, Anaheim, California.
47. Cella D, Lai JS, Chang CH, Peterman A, and Slavin M. Fatigue in cancer patients compared with fatigue in the general United States population. *Cancer* 2002;94(2):528–538.
48. Beaumont JL, Cella D, Li JZ, Huang X, Bycott P, Baum C, Kulke MH, Demetri DG, and Motzer RJ, Sunitinib Malate (SU11248): Efficacy and Tolerability in Solid Tumors. Presented at the Cancer and Kinases Meeting, Santa Fe, New Mexico, USA, February 14–19, 2006.
49. Escudier B, Roigas J, Gillessen S, Srinivas S, Pisa P, Vogelzang N, Fountzilas G, Peschel C, Baum C, and De Mulder P. Continuous daily administration of sunitinib malate (SU11248)—a phase II study in patients (pts) with cytokine-refractory metastatic renal cell carcinoma (mRCC). Presented at the 31st European Society for Medical Oncology Congress, Istanbul, Turkey, 29 September–3 October, 2006.
50. Rini BI, George DJ, Michaelson MD, Rosenberg JE, Bukowski RM, Sosman JA, Stadler WM, Margolin K, Hutson TE, and Baum C. Phase II study of sunitinib malate (SU11248) in bevacizumab-refractory metastatic renal cell carcinoma (mRCC). Presented at the 31st European Society for Medical Oncology Congress, Istanbul, Turkey, 29 September–3 October, 2006.
51. Ronnen EA, Kondagunta GV, Lau C, Fischer P, Ginsberg MS, Baum M, Kim ST, Chen I, Baum CM, and Motzer RJ. A phase I study of sunitinib malate (SU11248) in combination with gefitinib in patients with metastatic renal cell carcinoma (mRCC). (Oral presentation) *Proc Am Soc Clin Oncol* 2006;24:225s (Abstract 4537).
52. Bukowski RM, Szyzylik C, Porta C, Bodrogi I, Eisen T, Oudard S, Bjarnason G, Hawkins R, Wilner K, and Chen I. Preliminary results of an expanded-access trial of sunitinib malate (SU11248) for the treatment of patients with refractory metastatic renal cell carcinoma. Presented at the 31st European Society for Medical Oncology Congress, Istanbul, Turkey, 29 September–3 October, 2006.
53. Demetri DG, van Oosterom AT, Garrett CR, Blackstein ME, Shah MH, Verweij J, McArthur G, Judson IR, Heinrich MC, Morgan JA, Desai J, Fletcher CD, George S, Bello CL, Huang X, Baum CM, and Casali PG. Efficacy and safety of sunitinib in patients with advanced gastrointestinal stromal tumour after failure of imatinib: a randomized controlled trial. *Lancet* 2006;368:1329–1338.
54. MRCRCC. Interferon-alpha and survival in metastatic renal carcinoma: early results of a randomised controlled trial. Medical Research Council Renal Cancer Collaborators. *Lancet* 1999;353(9146):14–17.
55. Gordon MS, Manola J, Fairclough D, Cella D, Richardson R, Sosman J, Kasimis B, Dutcher JP, and Wilding G. Low dose interferon-α2b (IFN) + thalidomide (T) in patients (pts) with previously untreated renal cell cancer (RCC). Improvement in progression-free survival (PFS) but not quality of life (QoL) or overall survival (OS). A phase III study of the Eastern Cooperative Oncology Group (E2898). *Proc Am Soc Clin Oncol* 2004;22:384s (Abstract 4516).
56. Negrier S, Escudier B, Lasset C, Douillard JY, Savary J, Chevreau C, Ravaud A, Mercatello A, Peny J, Mousseau M, Philip T, and Tursz T. Recombinant human interleukin-2, recombinant

human interferon alfa-2a, or both in metastatic renal-cell carcinoma. Groupe Francais d'Immunotherapie. *N Engl J Med* 1998;338(18):1272–1278.

57. Yang JC, Sherry RM, Steinberg SM, Topalian SL, Schwartzentruber DJ, Hwu P, Seipp CA, Rogers-Freezer L, Morton KE, White DE, Liewehr DJ, Merino MJ, and Rosenberg SA. Randomized study of high-dose and low-dose interleukin-2 in patients with metastatic renal cancer. *J Clin Oncol* 2003;21(16):3127–3132.

Index

A

B

C

D

E

F

G

H

I

J

K

L

M

P

Q

R

S

T

W

X

Z

Milton Keynes UK
Ingram Content Group UK Ltd.
UKHW050458071024
449327UK00015B/433